List of the Elements with Their Symbols and Atomic Masses*

Element	Symbol	Atomic Number	Atomic Mass†	Element	Symbol	Atomic Number	Atomic Mass†
Actinium	Ac	89	(227)	Mendelevium	Md	101	(256)
Aluminum	Al	13	26.98	Mercury	Hg	80	200.6
Americium	Am	95	(243)	Molybdenum	Mo	42	95.94
Antimony	Sb	51	121.8	Neodymium	Nd	60	144.2
Argon	Ar	18	39.95	Neon	Ne	10	20.18
Arsenic	As	33	74.92	Neptunium	Np	93	(237)
Astatine	At	85	(210)	Nickel	Ni	28	58.69
Barium	Ba	56	137.3	Niobium	Nb	41	92.91
Berkelium	Bk	97	(247)	Nitrogen	N	7	14.01
Beryllium	Be	4	9.012	Nobelium	No	102	(253)
Bismuth	Bi	83	209.0	Osmium	Os	76	190.2
Bohrium	Bh	107	(262)	Oxygen	O	8	16.00
Boron	B	5	10.81	Palladium	Pd	46	106.4
Bromine	Br	35	79.90	Phosphorus	P	15	30.97
Cadmium	Cd	48	112.4	Platinum	Pt	78	195.1
Calcium	Ca	20	40.08	Plutonium	Pu	94	(242)
Californium	Cf	98	(249)	Polonium	Po	84	(210)
Carbon	C	6	12.01	Potassium	K	19	39.10
Cerium	Ce	58	140.1	Praseodymium	Pr	59	140.9
Cesium	Cs	55	132.9	Promethium	Pm	61	(147)
Chlorine	Cl	17	35.45	Protactinium	Pa	91	(231)
Chromium	Cr	24	52.00	Radium	Ra	88	(226)
Cobalt	Co	27	58.93	Radon	Rn	86	(222)
Copper	Cu	29	63.55	Rhenium	Re	75	186.2
Curium	Cm	96	(247)	Rhodium	Rh	45	102.9
Darmstadtium	Ds	110	(269)	Roentgenium	Rg	111	(272)
Dubnium	Db	105	(260)	Rubidium	Rb	37	85.47
Dysprosium	Dy	66	162.5	Ruthenium	Ru	44	101.1
Einsteinium	Es	99	(254)	Rutherfordium	Rf	104	(257)
Erbium	Er	68	167.3	Samarium	Sm	62	150.4
Europium	Eu	63	152.0	Scandium	Sc	21	44.96
Fermium	Fm	100	(253)	Seaborgium	Sg	106	(263)
Fluorine	F	9	19.00	Selenium	Se	34	78.96
Francium	Fr	87	(223)	Silicon	Si	14	28.09
Gadolinium	Gd	64	157.3	Silver	Ag	47	107.9
Gallium	Ga	31	69.72	Sodium	Na	11	22.99
Germanium	Ge	32	72.59	Strontium	Sr	38	87.62
Gold	Au	79	197.0	Sulfur	S	16	32.07
Hafnium	Hf	72	178.5	Tantalum	Ta	73	180.9
Hassium	Hs	108	(265)	Technetium	Tc	43	(99)
Helium	He	2	4.003	Tellurium	Te	52	127.6
Holmium	Ho	67	164.9	Terbium	Tb	65	158.9
Hydrogen	H	1	1.008	Thallium	Tl	81	204.4
Indium	In	49	114.8	Thorium	Th	90	232.0
Iodine	I	53	126.9	Thulium	Tm	69	168.9
Iridium	Ir	77	192.2	Tin	Sn	50	118.7
Iron	Fe	26	55.85	Titanium	Ti	22	47.88
Krypton	Kr	36	83.80	Tungsten	W	74	183.9
Lanthanum	La	57	138.9	Uranium	U	92	238.0
Lawrencium	Lr	103	(257)	Vanadium	V	23	50.94
Lead	Pb	82	207.2	Xenon	Xe	54	131.3
Lithium	Li	3	6.941	Ytterbium	Yb	70	173.0
Lutetium	Lu	71	175.0	Yttrium	Y	39	88.91
Magnesium	Mg	12	24.31	Zinc	Zn	30	65.39
Manganese	Mn	25	54.94	Zirconium	Zr	40	91.22
Meitnerium	Mt	109	(266)				

*All atomic masses have four significant figures. These values are recommended by the Committee on Teaching of Chemistry, International Union of Pure and Applied Chemistry.

†Approximate values of atomic masses for radioactive elements are given in parentheses.

General Chemistry

About the Cover

The cover shows a diatomic molecule being irradiated with laser light of appropriate frequency. As a result, the molecule is promoted to a highly excited vibrational energy level, which subsequently leads to dissociation into atomic species.

General Chemistry

The Essential Concepts

Sixth Edition

Raymond Chang
Williams College

Jason Overby
The College of Charleston

GENERAL CHEMISTRY: THE ESSENTIAL CONCEPTS, SIXTH EDITION

Published by McGraw-Hill, a business unit of The McGraw-Hill Companies, Inc., 1221 Avenue of the Americas, New York, NY 10020.

This book is printed on acid-free paper.

1 2 3 4 5 6 7 8 9 0 DOW/DOW 1 0 9 8 7 6 5 4 3 2 1 0

ISBN 978–0–07–337563–2
MHID 0–07–337563–2

Publisher: *Ryan Blankenship*
Senior Sponsoring Editor: *Tamara L. Hodge*
Director of Development: *Kristine Tibbetts*
Senior Developmental Editor: *Shirley R. Oberbroeckling*
Senior Marketing Manager: *Todd L. Turner*
Senior Project Manager: *Gloria G. Schiesl*
Senior Production Supervisor: *Kara Kudronowicz*
Lead Media Project Manager: *Judi David*
Senior Designer: *Laurie B. Janssen*
Cover Illustration: *Precision Graphics*
Senior Photo Research Coordinator: *John C. Leland*
Photo Research: *Toni Michaels/PhotoFind, LLC*
Supplement Producer: *Mary Jane Lampe*
Compositor: *Aptara, Inc.*
Typeface: *10/12 Times Roman*
Printer: *R. R. Donnelley*

Library of Congress Cataloging-in-Publication Data

Chang, Raymond.
General chemistry : the essential concepts / Raymond Chang. — 6th ed. / Jason Overby.
p. cm.
Includes index.
ISBN 978–0–07–337563–2 — ISBN 0–07–337563–2 (hard copy : alk. paper) 1. Chemistry—Textbooks. I. Overby, Jason Scott, 1970- II. Title.
QD33.2.C48 2011
540—dc22

2009034749

www.mhhe.com

About the Authors

Raymond Chang was born in Hong Kong and grew up in Shanghai and Hong Kong. He received his B.Sc. degree in chemistry from London University, England, and his Ph.D. in chemistry from Yale University. After doing postdoctoral research at Washington University and teaching for a year at Hunter College of the City University of New York, he joined the chemistry department at Williams College, where he has taught since 1968.

Professor Chang has served on the American Chemical Society Examination Committee, the National Chemistry Olympiad Examination Committee, and the Graduate Record Examinations (GRE) Committee. He is an editor of *The Chemical Educator*. Professor Chang has written books on physical chemistry, industrial chemistry, and physical science. He has also coauthored books on the Chinese language, children's picture books, and a novel for young readers.

For relaxation, Professor Chang maintains a forest garden; plays tennis, Ping-Pong, and the harmonica; and practices the violin.

Jason Overby was born in Bowling Green, Kentucky, and grew up in Clarksville, Tennessee. He received his B.S. in chemistry and political science from the University of Tennessee at Martin and his Ph.D. in inorganic chemistry from Vanderbilt University. After postdoctoral research at Dartmouth College, he began his academic career at the College of Charleston in 1999.

Professor Overby maintains research interests in synthetic and computational inorganic and organometallic chemistry. His educational pursuits include inorganic chemistry laboratory pedagogy and the use of digital technology, including online homework, as tools in the classroom.

In his spare time, Professor Overby enjoys cooking, computers, and spending time with his family.

Brief Contents

Contents

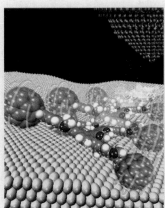

CHAPTER 7

The Electronic Structure of Atoms 211

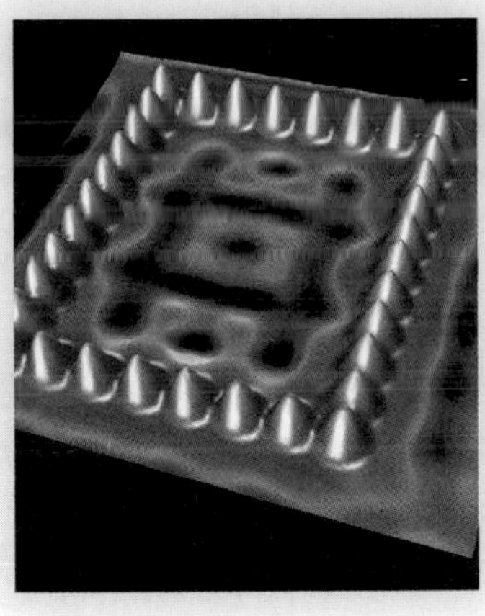

CHAPTER 8

The Periodic Table 251

CHAPTER 9

Chemical Bonding I: The Covalent Bond 285

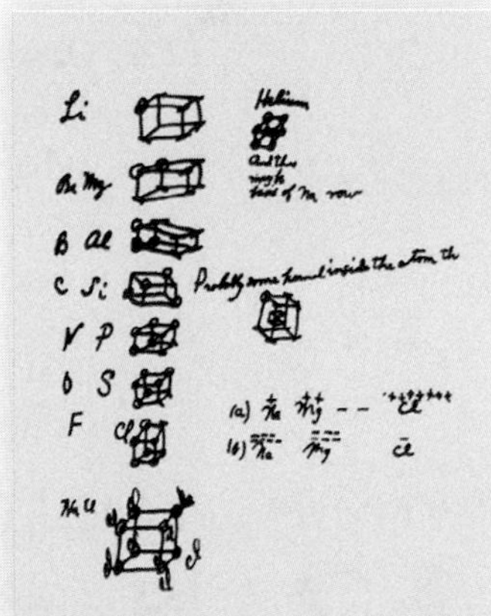

HO
H
N
O
CH_3
O

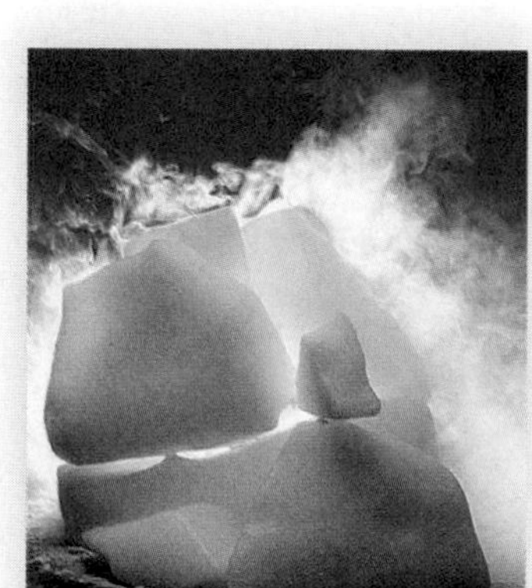

CHAPTER 17 Acid-Base Equilibria and Solubility Equilibria 590

CHAPTER 18 Thermodynamics 628

CHAPTER 19 Redox Reactions and Electrochemistry 661

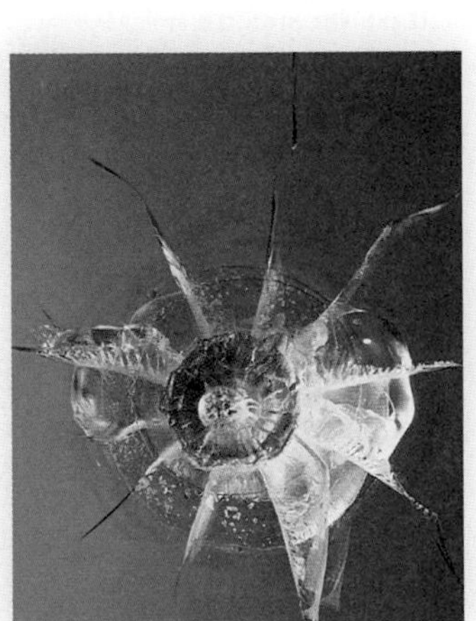

List of Animations

The animations listed below are correlated to ***General Chemistry*** within each chapter in two ways. The first is the Student Interactive Activities found in the opening pages of every chapter. Then within the chapter are icons letting the student and the instructor know that an animation is available for a specific topic and where to find the animation for viewing on our Chang *General Chemistry* ARIS website.

Chang Animations

Absorption of color (20.4)
Acid-base titrations (17.3)
Acid ionization (16.5)
Activation energy (14.4)
Alpha, beta, and gamma rays (2.2)
Alpha-particle scattering (2.2)
Atomic and ionic radius (8.3)
Base ionization (16.6)
Buffer solutions (17.2)
Catalysis (14.6)
Cathode ray tube (2.2)
Chemical equilibrium (15.1)
Chirality (11.5)
Collecting a gas over water (5.5)
Diffusion of gases (5.6)
Dissolution of an ionic and a covalent compound (13.2)
Electron configurations (7.8)
Emission spectra (7.3)
Equilibrium vapor pressure (12.6)
Formal charge calculations (9.7)
Galvanic cells (19.2)
Gas laws (5.3)
Heat flow (6.4)
Hybridization (10.4)
Hydration (4.1)
Ionic versus covalent bonding (9.4)
Le Châtelier's principle (15.4)
Limiting reagent (3.9)
Making a solution (4.5)
Millikan oil drop (2.2)
Neutralization reactions (4.3)
Nuclear fission (21.5)
Orientation of collision (14.4)
Osmosis (13.6)
Oxidation-reduction reactions (4.4)
Packing spheres (12.4)
Polarity of molecules (10.2)
Precipitation reactions (4.2)
Preparing a solution by dilution (4.5)
Radioactive decay (21.3)
Resonance (9.8)
Sigma and pi bonds (10.5)
Strong electrolytes, weak electrolytes, and nonelectrolytes (4.1)
VSEPR (10.1)

McGraw-Hill Animations

Atomic line spectra (7.3)
Charles' law (5.3)
Cubic unit cells and their origins (12.4)
Dissociation of strong and weak acids (16.5)
Dissolving table salt (4.1)
Electronegativity (9.3)
Equilibrium (15.1)
Exothermic and endothermic reactions (6.2)
Formal charge calculations (9.5)
Formation of an ionic compound (9.3)
Formation of the covalent bond in H_2 (10.4)
Half-life (14.3)
Influence of shape on polarity (10.2)
Law of conservation of mass (2.1)
Molecular shape and orbital hybridization (10.4)
Nuclear medicine (21.7)
Operation of voltaic cell (19.2)
Oxidation-reduction reaction (4.4 & 19.1)
Phase diagrams and the states of matter (12.7)
Reaction rate and the nature of collisions (14.4)
Three states of matter (1.3)
Using a buffer (17.2)
VSEPR theory and the shapes of molecules (10.1)

Simulations

Stoichiometry (Chapter 3)
Ideal gas law (Chapter 5)
Kinetics (Chapter 14)
Equilibrium (Chapter 15)
Titration (Chapter 17)
Electrochemistry (Chapter 19)
Nuclear (Chapter 21)

PREFACE

The sixth edition of *General Chemistry: The Essential Concepts*, continues the tradition of presenting only the material that is essential to a one-year general chemistry course. As with previous editions, it includes all the core topics that are necessary for a solid foundation in general chemistry without sacrificing depth, clarity, or rigor.

General Chemistry covers these topics in the same depth and at the same level as 1100-page texts. All essential topics are in the text with the exception of descriptive chemistry. Therefore, this book is not a condensed version of a big text. Our hope is that this concise-but-thorough approach will appeal to efficiency-minded instructors and will please value-conscious students. The positive feedback from users over the years shows that there is a strong need for such a text. So we have written a text containing all of the core concepts necessary for a solid foundation in general chemistry.

What's New in This Edition?

- The most obvious change is the addition of a coauthor, Jason Overby, who brings new pedagogical insights to the text.
- **NEW** to the chapters is the Review of Concepts feature. This is a quick knowledge test for the student to gauge his or her understanding of the concept just presented. The answers to the Review of Concepts are available in the Student Solutions Manual and on the companion ARIS (Assessment, Review, and Instruction System) website.

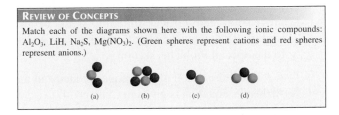

REVIEW OF CONCEPTS

Match each of the diagrams shown here with the following ionic compounds: Al_2O_3, LiH, Na_2S, $Mg(NO_3)_2$. (Green spheres represent cations and red spheres represent anions.)

(a) (b) (c) (d)

- **NEW** are powerful connections to electronic homework. All of the practice exercises for the Worked Examples in all chapters are now found within the McGraw-Hill ARIS (Assessment, Review, and Instruction System) electronic homework system. Each end-of-chapter problem in ARIS is noted in the Electronic Homework Problem section by an icon.

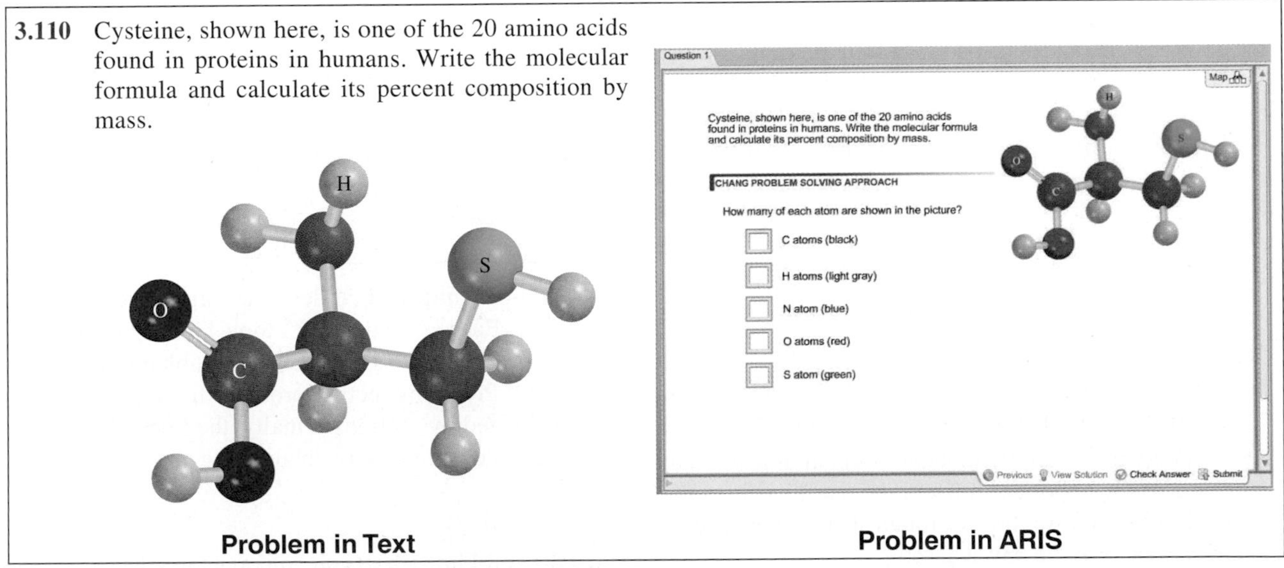
3.110 Cysteine, shown here, is one of the 20 amino acids found in proteins in humans. Write the molecular formula and calculate its percent composition by mass.
H
S
O
C
Question 1
Map
Cysteine, shown here, is one of the 20 amino acids found in proteins in humans. Write the molecular formula and calculate its percent composition by mass.
CHANG PROBLEM SOLVING APPROACH
How many of each atom are shown in the picture?
C atoms (black)
H atoms (light gray)
N atom (blue)
O atoms (red)
S atom (green)
Previous
View Solution
Check Answer
Submit
Problem in Text

Problem in ARIS

- Many sections have been revised and updated based on the comments from reviewers and users. Some examples are

  —A revised treatment of amounts of reactants and products is given in Chapter 3.

  —A revised explanation of thermochemical equations is presented in Chapter 6.

  —Expanded coverage of effective nuclear charge appears in Chapter 8.

  —**New** computer-generated molecular orbital diagrams are presented in Chapter 10.

  —Many **new** end-of-chapter problems with molecular art have been added to test the conceptual comprehension and critical thinking skills of the student. The more challenging problems are added to the Special Problems section.

  —A revised discussion of the frequency factor in the Arrhenius equation is given in Chapter 14.

  —The ARIS electronic homework system is available for the sixth edition. ARIS will enhance the student learning experience, administer assignments, track student progress, and administer an instructor's course. The students can locate the animations and interactives noted in the text margins in ARIS. Quizzing and homework assigned by the instructor is available in the ARIS electronic homework program.

Problem Solving

The development of problem-solving skills has always been a major objective of this text. The two major categories of learning are the worked examples and end-of-chapter problems. Many of them present extra tidbits of knowledge and enable the student to solve a problem that a chemist would solve. The examples and problems show students the real world of chemistry and applications to everyday life situations.

Worked examples follow a proven step-by-step strategy and solution.

- **Problem** statement is the reporting of the facts needed to solve the problem based on the question posed.
- **Strategy** is a carefully thought-out plan or method to serve as an important function of learning. In some cases, students are shown a rough sketch, which helps them visualize the physical setup.
- **Solution** is the process of solving a problem given in a stepwise manner.
- **Check** enables the student to compare and verify with the source information to make sure that the answer is reasonable.
- **Practice Exercise** provides the opportunity to solve a similar problem in order to become proficient in this problem type. The Practice Exercises are available in the ARIS electronic homework system. The marginal note lists additional similar problems to work in the end-of-chapter problem section.

EXAMPLE 3.13

The food we eat is degraded, or broken down, in our bodies to provide energy for growth and function. A general overall equation for this very complex process represents the degradation of glucose ($C_6H_{12}O_6$) to carbon dioxide (CO_2) and water (H_2O):

$$C_6H_{12}O_6 + 6O_2 \longrightarrow 6CO_2 + 6H_2O$$

If 968 g of $C_6H_{12}O_6$ is consumed by a person over a certain period, what is the mass of CO_2 produced?

Strategy Looking at the balanced equation, how do we compare the amount of $C_6H_{12}O_6$ and CO_2? We can compare them based on the *mole ratio* from the balanced equation. Starting with grams of $C_6H_{12}O_6$, how do we convert to moles of $C_6H_{12}O_6$? Once moles of CO_2 are determined using the mole ratio from the balanced equation, how do we convert to grams of CO_2?

Solution We follow the preceding steps and Figure 3.8.

Step 1: The balanced equation is given in the problem.

Step 2: To convert grams of $C_6H_{12}O_6$ to moles of $C_6H_{12}O_6$, we write

$$968 \text{ g } \cancel{C_6H_{12}O_6} \times \frac{1 \text{ mol } C_6H_{12}O_6}{180.2 \text{ g } \cancel{C_6H_{12}O_6}} = 5.372 \text{ mol } C_6H_{12}O_6$$

Step 3: From the mole ratio, we see that 1 mol $C_6H_{12}O_6 \simeq 6$ mol CO_2. Therefore, the number of moles of CO_2 formed is

$$5.372 \text{ } \cancel{\text{mol } C_6H_{12}O_6} \times \frac{6 \text{ mol } CO_2}{1 \text{ } \cancel{\text{mol } C_6H_{12}O_6}} = 32.23 \text{ mol } CO_2$$

Step 4: Finally, the number of grams of CO_2 formed is given by

$$32.23 \text{ } \cancel{\text{mol } CO_2} \times \frac{44.01 \text{ g } CO_2}{1 \text{ } \cancel{\text{mol } CO_2}} = 1.42 \times 10^3 \text{ g } CO_2$$

After some practice, we can combine the conversion steps

$$\text{grams of } C_6H_{12}O_6 \longrightarrow \text{moles of } C_6H_{12}O_6 \longrightarrow \text{moles of } CO_2 \longrightarrow \text{grams of } CO_2$$

into one equation:

$$\text{mass of } CO_2 = 968 \text{ g } \cancel{C_6H_{12}O_6} \times \frac{1 \text{ } \cancel{\text{mol } C_6H_{12}O_6}}{180.2 \text{ g } \cancel{C_6H_{12}O_6}} \times \frac{6 \text{ } \cancel{\text{mol } CO_2}}{1 \text{ } \cancel{\text{mol } C_6H_{12}O_6}} \times \frac{44.01 \text{ g } CO_2}{1 \text{ } \cancel{\text{mol } CO_2}}$$
$$= 1.42 \times 10^3 \text{ g } CO_2$$

Check Does the answer seem reasonable? Should the mass of CO_2 produced be larger than the mass of $C_6H_{12}O_6$ reacted, even though the molar mass of CO_2 is considerably less than the molar mass of $C_6H_{12}O_6$? What is the mole ratio between CO_2 and $C_6H_{12}O_6$?

Practice Exercise Methanol (CH_3OH) burns in air according to the equation

$$2CH_3OH + 3O_2 \longrightarrow 2CO_2 + 4H_2O$$

If 209 g of methanol are used up in a combustion process, what is the mass of H_2O produced?

- **End-of-Chapter Problems** are organized in various ways. Each section under a topic heading begins with Review Questions followed by Problems. The Additional Problems section provides more problems not organized by sections. Finally, the Special Problems section contains more challenging problems.

Visualization

Graphs and Flow Charts are important in science. In *General Chemistry*, flow charts show the thought process of a concept and graphs present data to comprehend the concept.

Molecular Art appears in various formats to serve different needs. You will find molecular art incorporated in all facets of the text and homework. Molecular models help students to visualize the three dimensional arrangement of atoms in a molecule. Electrostatic potential maps illustrate the electron density distribution in molecules. Finally, there is the macroscopic-to-microscopic art helping students understand processes at the molecular level.

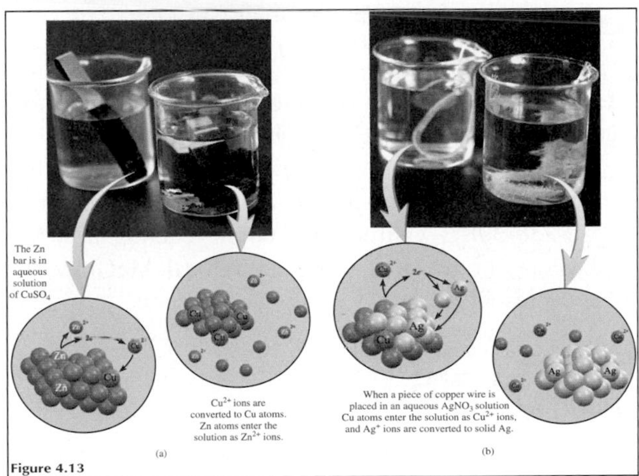

The Zn bar is in aqueous solution of $CuSO_4$
Zn
Cu
Cu
Ag
Ag
Cu^{2+} ions are converted to Cu atoms. Zn atoms enter the solution as Zn^{2+} ions.
When a piece of copper wire is placed in an aqueous $AgNO_3$ solution Cu atoms enter the solution as Cu^{2+} ions, and Ag^+ ions are converted to solid Ag.
(a)
(b)

Figure 4.13

Photos are used to help students become familiar with chemicals and understand how chemical reactions appear in reality.

Figures of Apparatus enable the student to visualize the practical arrangement in a chemistry laboratory.

Study Aids

Setting the Stage

On the chapter opening page for each chapter the Chapter Outline, Student Interactive Activities, and Essential Concepts appear.

- **Chapter Outline** enables the student to see at a glance the big picture and focus on the main ideas of the chapter.
- **Student Interactive Activities** show where the electronic media are used in the chapter. A list of the animations and questions in McGraw-Hill ARIS homework is given. Within the chapter, icons are used to refer to the items shown in the Student Interactive Activities list.
- **Essential Concepts** summarizes the main topics to be presented in the chapter.

Tools to Use for Studying

Useful aids for studying are plentiful in *General Chemistry* and should be used constantly to reinforce the comprehension of chemical concepts.

- **Marginal Notes** are used to provide hints and feedback to enhance the knowledge base for the student.
- **Worked Examples** along with the accompanying Practice Exercises are very important tools for learning and mastering chemistry. The problem solving steps guide the student through the critical thinking necessary for succeeding in chemistry. Using sketches helps student understand the inner workings of a problem. A marginal note lists similar problems in the end-of-chapter problems section, enabling the student to apply the new skill to other problems of the same type. Answers to the Practice Exercises are listed at the end of the chapter problems.
- **Review of Concepts** enable students to quickly evaluate whether they understand the concept presented in the section. Answers to the Review of Concepts can be found in the Student Solution Manual and online in the accompanying ARIS companion website.
- **Key Equations** are highlighted within the chapter, drawing the student's eye to material that needs to be understood and retained. The key equations are also presented in the chapter summary materials for easy access in review and study.
- **Summary of Facts and Concepts** provides a quick review of concepts presented and discussed in detail within the chapter.
- **Key Words** are a list of all the important terms to help the student understand the language of chemistry.

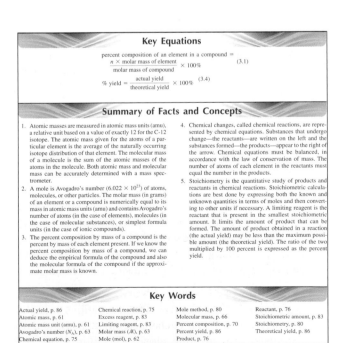

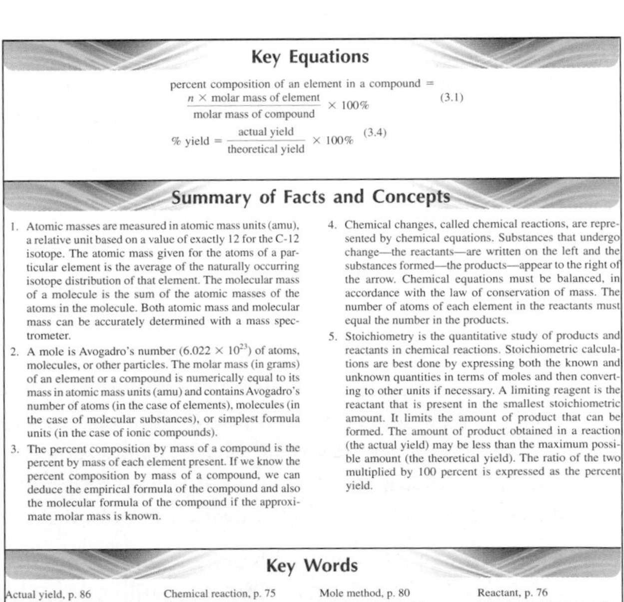

Key Equations

percent composition of an element in a compound =

$$\frac{n \times \text{molar mass of element}}{\text{molar mass of compound}} \times 100\% \qquad (3.1)$$

$$\% \text{ yield} = \frac{\text{actual yield}}{\text{theoretical yield}} \times 100\% \qquad (3.4)$$

Summary of Facts and Concepts

1. Atomic masses are measured in atomic mass units (amu), a relative unit based on a value of exactly 12 for the C-12 isotope. The atomic mass given for the atoms of a particular element is the average of the naturally occurring isotope distribution of that element. The molecular mass of a molecule is the sum of the atomic masses of the atoms in the molecule. Both atomic mass and molecular mass can be accurately determined with a mass spectrometer.
2. A mole is Avogadro's number (6.022×10^{23}) of atoms, molecules, or other particles. The molar mass (in grams) of an element or a compound is numerically equal to its mass in atomic mass units (amu) and contains Avogadro's number of atoms (in the case of elements), molecules (in the case of molecular substances), or simplest formula units (in the case of ionic compounds).
3. The percent composition by mass of a compound is the percent by mass of each element present. If we know the percent composition by mass of a compound, we can deduce the empirical formula of the compound and also the molecular formula of the compound if the approximate molar mass is known.
4. Chemical changes, called chemical reactions, are represented by chemical equations. Substances that undergo change—the reactants—are written on the left and the substances formed—the products—appear to the right of the arrow. Chemical equations must be balanced, in accordance with the law of conservation of mass. The number of atoms of each element in the reactants must equal the number in the products.
5. Stoichiometry is the quantitative study of products and reactants in chemical reactions. Stoichiometric calculations are best done by expressing both the known and unknown quantities in terms of moles and then converting to other units if necessary. A limiting reagent is the reactant that is present in the smallest stoichiometric amount. It limits the amount of product that can be formed. The amount of product obtained in a reaction (the actual yield) may be less than the maximum possible amount (the theoretical yield). The ratio of the two multiplied by 100 percent is expressed as the percent yield.

Key Words

Actual yield, p. 86
Atomic mass, p. 61
Atomic mass unit (amu), p. 61
Avogadro's number (N_A), p. 63
Chemical equation, p. 75
Chemical reaction, p. 75
Excess reagent, p. 83
Limiting reagent, p. 83
Molar mass ($\mathcal{M}$), p. 63
Mole (mol), p. 62
Mole method, p. 80
Molecular mass, p. 66
Percent composition, p. 70
Percent yield, p. 86
Product, p. 76
Reactant, p. 76
Stoichiometric amount, p. 83
Stoichiometry, p. 80
Theoretical yield, p. 86

Testing Your Knowledge

Review of Concepts lets students pause and check to see if they understand the concept presented and discussed in the section. Answers to the **Review of Concepts** can be found in the Student Solution Manual and online in the accompanying ARIS companion website.

End-of-Chapter Problems enable the student to practice critical thinking and problem-solving skills. The problems are broken into various types:

- By chapter section. Starting with Review Questions to test basic conceptual understanding, followed by Problems to test the student's skill in solving problems for that particular section of the chapter.
- Additional Problems use knowledge gained from the various sections and/or previous chapters to solve the problem.
- The Special Problems section contains more challenging problems that are suitable for group projects.

16.114 The diagrams here show three weak acids HA (A = X, Y, or Z) in solution. (a) Arrange the acids in order of increasing K_a. (b) Arrange the conjugate bases in increasing order of K_b. (c) Calculate the percent ionization of each acid. (d) Which of the 0.1 *M* sodium salt solutions (NaX, NaY, or NaZ) has the lowest pH? (The hydrated proton is shown as a hydronium ion. Water molecules are omitted for clarity.)

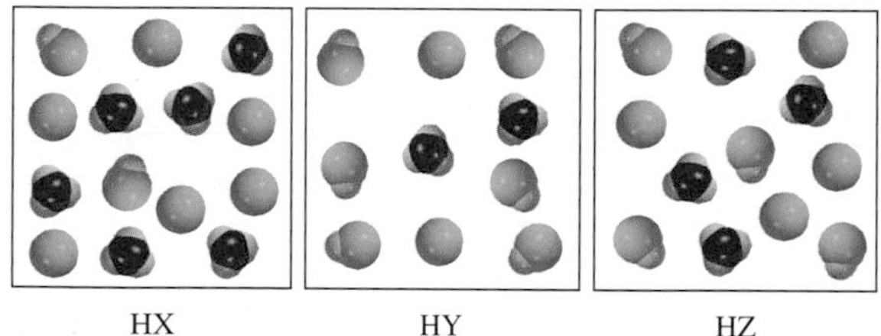

HX HY HZ

Media

The Student Interactive Activities on the chapter opening page enables the student and instructor to see at a glance the media that can be incorporated into the learning process. Within the text, an icon shows the student where the concept in the animation or interactive is introduced. The icon directs the student to the ARIS website for viewing. For the instructor, there are also directions for finding the animation or interactive in the instructor materials.

- **Animations**—We have a library of animations that support the sixth edition. The animations visually bring to life the areas in chemistry that are difficult to understand by reading alone. The animations are marked by an icon and located within ARIS for student use.
- **Electronic Homework (ARIS)**—The Practice Exercises from the Worked Examples and many end-of-chapter problems are in the electronic homework system ARIS. Each exercise and end-of-chapter problem contained in ARIS is marked by

Instructor Resources

McGraw-Hill offers various tools and technology products to support the *General Chemistry, Sixth Edition.*

Instructors can obtain teaching aides by calling the McGraw-Hill Customer Service Department at 1-800-338-3987, visiting our online catalog at www.mhhe.com, or by contacting their local McGraw-Hill sales representative.

The Assessment, Review, and Instruction System, also known as McGraw-Hill ARIS, is an electronic homework and course management system designed for greater flexibility, power, and ease of use than any other system. Whether you are looking for a preplanned course or one you can customize to fit your course needs, ARIS is your solution.

In addition to having access to all student digital learning objects, ARIS enables instructors to:

Build Assignments

- Choose from prebuilt assignments or create your own custom content by importing your own content or editing an existing assignment from the prebuilt assignment.
- Assignments can include quiz questions, animations, and videos—anything found on the website.
- Create announcements and utilize full course or individual student communcation tools.
- Assign questions developed following the problem-solving strategy used within the textual material, enabling students to continue the learning process from the text into their homework assignments in a structured manner.
- Instructors can choose the assignment setting for an individual student to help manage missed assignments, special needs students, and any specific situations that arise during the semester.
- Assign algorithmic questions, providing students with multiple chances to practice and gain skill at problem solving on the same concept.

Track Student Progress

- Assignments are automatically graded.
- Gradebook functionality enables full course management, including:
  - —Dropping the lowest grades
  - —Weighting grades/manually adjusting grades
  - —Exporting your gradebook to Excel, WebCT, or BlackBoard
  - —Manipulating data, enabling you to track student progress through multiple reports
  - —Providing a visual representation of key grade book reports
  - —Offering the opportunity to select an assignment and view detailed statistics on student performance for each question

Offer More Flexibility

- **Sharing Course Materials with Colleagues**—Instructors can create and share course materials and assignments with colleagues with a few clicks of the mouse, allowing for multiple section courses with many instructors (and TAs) to continually be in synch if desired.
- **Integration with BlackBoard or WebCT**—Once a student is registered in the course, all student activity within McGraw-Hill ARIS is automatically recorded and available to the instructor through a fully integrated grade book that can be downloaded to Excel, WebCT, or Blackboard.

Presentation Center

The **Presentation Center** is a complete set of electronic book images and assets for instructors. You can build instructional materials wherever, whenever, and however you want! Accessed from your textbook's ARIS website, the Presentation Center is an online digital library containing photos, artwork, animations, and other media types that can be used to create customized lectures, visually enhanced tests and quizzes, compelling course websites, or attractive printed support materials. All assets are copyrighted by McGraw-Hill Higher Education, but can be used by instructors for classroom purposes. The visual resources in this collection include:

- **Art** Full-color digital files of all illustrations in the book can be readily incorporated into lecture presentations, exams, or custom-made classroom materials. In addition, all files are preinserted into PowerPoint® slides for ease of lecture preparation.
- **Photos** The photo collection contains digital files of photographs from the text, which can be reproduced for multiple classroom uses.
- **Tables** Every table that appears in the text has been saved in electronic form for use in classroom presentations and/or quizzes.
- **Animations** Numerous full-color animations illustrating important processes are also provided. Harness the visual impact of concepts in motion by importing these files into classroom presentations or online course materials.

Also residing on your textbook's ARIS website are:

- **PowerPoint Lecture Outlines** Ready-made presentations that combine art, and lecture notes are provided for each chapter of the text.
- **PowerPoint Slides** For instructors who prefer to create their lectures from scratch, all illustrations, photos, and tables are preinserted by chapter into blank PowerPoint slides.
- **Instructor Solution Manual** Solutions are provided for all end-of-chapter problems in the text.

Access to your book, access to all books!

The Presentation Center library includes thousands of assets from many McGraw-Hill titles. This ever-growing resource gives instructors the power to utilize assets specific to an adopted textbook as well as content from all other books in the library.

Nothing could be easier!

Accessed from the instructor side of your textbook's ARIS website, the Presentation Center's dynamic search engine enables you to explore by discipline, course, textbook chapter, asset type, or keyword. Simply browse, select, and download the files you need to build engaging course materials. All assets are copyrighted by McGraw-Hill Higher Education but can be used by instructors for classroom purposes. Instructors: To access ARIS, request registration information from your McGraw-Hill sales representative.

Computerized Test Bank Online

A comprehensive bank of test questions by Ken Goldsby (Florida State University) and Jason Overby (College of Charleston) is provided within a computerized test bank, enabling you to create paper and online tests or quizzes in this easy-to-use program. Imagine being able to create and access your test or quiz anywhere, at any time.

Instructors can create or edit questions and drag-and-drop questions to create tests quickly and easily. The test can be published automatically online to your course and course management system, or you can print them for paper-based tests.

The test bank contains multiple-choice, true/false, and short answer questions. The questions, which are graded in difficulty, are comparable to the problems in the text.

Student Response System

Wireless technology brings interactivity into the classroom or lecture hall. Instructors and students receive immediate feedback through wireless response pads that are easy to use and engage students. This system can be used by instructors to

- Take attendance
- Administer quizzes and tests
- Create a lecture with intermittent questions
- Manage lectures and student comprehension through the use of the grade book
- Integrate interactivity into their PowerPoint presentations

Content Delivery Flexibility

General Chemistry by Raymond Chang and Jason Overby is available in many formats in addition to the traditional textbook to give instructors and students more choices when deciding on the format of their chemistry text. Choices include:

Color Custom by Chapter

For even more flexibility, we offer the Chang/Overby ***General Chemistry*** text in a full-color, custom version that enables instructors to pick the chapters they want to include. Students pay for only what the instructor chooses.

eBook

If you or your students are ready for an alternative version of the traditional textbook, McGraw-Hill brings you innovative and inexpensive electronic textbooks. By purchasing ebooks from McGraw-Hill, students can save as much as 50% on selected titles delivered on the most advanced ebook platform.

eBooks from McGraw-Hill are smart, interactive, searchable, and portable with a powerful suite of built-in tools that enable detailed searching, highlighting, note taking, and student-to-student or instructor-to-student note sharing. In addition, the media-rich ebook for ***General Chemistry*** integrates relevant animations and videos into the textbook content for a true multimedia learning experience. ebooks from McGraw-Hill will help students study smarter and quickly find the information they need. And they will save money. Contact your McGraw-Hill sales representative to discuss ebook packaging options.

Mc Graw Hill tegrity campus

McGraw-Hill Tegrity Campus is a service that makes class time available all the time by automatically capturing every lecture in a searchable format for students to review when they study and complete assignments. With a simple one-click start and stop process, you capture all computer screens and corresponding audio. Students replay any part of any class with easy-to-use browser-based viewing on a PC or Mac.

Educators know that the more students can see, hear, and experience class resources, the better they learn. With Tegrity Campus, students quickly recall key moments by using Tegrity Campus's unique search feature. This search helps students efficiently find what they need, when they need it across an entire semester of class recordings. Help turn all your students' study time into learning moments immediately supported by your lecture.

To learn more about Tegrity watch a 2 minute Flash demo at tegritycampus.mhhe.com.

Cooperative Chemistry Laboratory Manual

By Melanie Cooper (Clemson University). This innovative guide features open-ended problems designed to simulate experience in a research lab. Working in groups, students investigate one problem over a period of several weeks, so that they might complete three or four projects during the semester, rather than one preprogrammed experiment per class. The emphasis here is on experimental design, analysis problem solving, and communication.

Student Resources

McGraw-Hill offers various tools and technology products to support the *General Chemistry*, Sixth Edition.

Students can order supplemental study materials by contacting their campus bookstore, calling 1-800-262-4729, or online at www.shopmcgraw-hill.com.

Problem-Solving Workbook with Solutions

By Brandon J. Cruickshank (Northern Arizona University) and Raymond Chang, this workbook is a success guide written for use with ***General Chemistry***. It aims to help students hone their analytical and problem-solving skills by presenting detailed approaches to solving chemical problems. Solutions for all of the text's even-numbered problems are included.

McGraw-Hill ARIS (Assessment, Review, and Instruction System) is an electronic study system that offers students a digital portal of knowledge.

Students can readily access a variety of **digital learning objects,** which include:

- Chapter-level quizzing
- Animations
- Interactives
- MP3 and MP4 downloads of selected content

Acknowledgments

Reviewers

We would like to thank the following individuals who reviewed or participated in various McGraw-Hill symposia on general chemistry. Their insight into the needs of students and instructors were invaluable to us in preparing this revision.

DeeDee A. Allen *Wake Technical Community College*
Vladimir Benin *University of Dayton*
Elizabeth D. Blue *Wake Technical Community College*
R. D. Braun *University of Louisiana at Lafayette*
William Broderick *Montana State University*
Christopher M. Burba *Northeastern State University*
Charles Carraher *Florida Atlantic University*
John P. DiVincenzo *Middle Tennessee State University*
Ajit S. Dixit *Wake Technical Community College*
Michael A. Hauser *St. Louis Community College–Meramec*
Andy Holland *Idaho State University*
Daniel King *Drexel University*
Kathleen Knierim *University of Louisiana at Lafayette*
Andrew Langrehr *St. Louis Community College–Meramec*
Terrence A. Lee *Middle Tennessee State University*
Jessica D. Martin *Northeastern State University*
Gordon J. Miller *Iowa State University*
Spence Pilcher *Northeastern State University*
Susanne Raynor *Rutgers University*
John T. Reilly *Coastal Carolina University*
Shirish Shah *Towson University*
Thomas E. Sorensen *University of Wisconsin–Milwaukee*
Zhiqiang (George) Yang *Macomb Community College*

We would also like to thank Dr. Enrique Peacock-Lopez and Desire Gijima of Williams College for the computer-generated molecular orbital diagrams in Chapters 10 and 11.

As always, we have benefited much from discussions with our colleagues at Williams College and the College of Charleston and correspondence with many instructors here and abroad.

It is a pleasure to acknowledge the support given to us by the following members of McGraw-Hill's College Division: Doug Dinardo, Tammy Ben, Thomas Timp, Marty Lange, Kent Peterson, Chad Grall, and Kurt Strand. In particular, we would like to mention Gloria Schiesl for supervising the production, Laurie Janssen for the book design, Daryl Bruflodt and Judi David for the media, and Todd Turner, the marketing manager, for his suggestions and encouragement. Our publisher Ryan Blankenship and our editor Tami Hodge provided advice and support whenever we needed them. Finally, our special thanks go to Shirley Oberbroeckling, the developmental editor, for her care and enthusiasm for the project, and supervision at every stage of the writing of this edition.

Raymond Chang

Jason Overby

A Note to the Student

General chemistry is commonly perceived to be more difficult than most other subjects. There is some justification for this perception. For one thing, chemistry has a very specialized vocabulary. At first, studying chemistry is like learning a new language. Furthermore, some of the concepts are abstract. Nevertheless, with diligence you can complete this course successfully, and you might even enjoy it. Here are some suggestions to help you form good study habits and master the material in this text.

- Attend classes regularly and take careful notes.
- If possible, always review the topics discussed in class the same day they are covered in class. Use this book to supplement your notes.
- Think critically. Ask yourself if you really understand the meaning of a term or the use of an equation. A good way to test your understanding is to explain a concept to a classmate or some other person.
- Do not hesitate to ask your instructor or your teaching assistant for help.

The sixth edition tools for *General Chemistry* are designed to enable you to do well in your general chemistry course. The following guide explains how to take full advantage of the text, technology, and other tools.

- Before delving into the chapter, read the chapter *outline* and the chapter *introduction* to get a sense of the important topics. Use the outline to organize your note taking in class.
- Use the *Student Interactive Activities* icon as a guide to review challenging concepts in motion. The animations are valuable in presenting a concept and enabling the student to manipulate or choose steps so full understanding can happen.
- At the end of each chapter, you will find a summary of facts and concepts, key equations, and a list of key words, all of which will help you review for exams.
- Definitions of the key words can be studied in context on the pages cited in the end-of-chapter list or in the glossary at the back of the book.
- ARIS houses an extraordinary amount of resources. You can explore chapter quizzes, animations, interactivities, simulations, and more.
- Careful study of the worked-out examples in the body of each chapter will improve your ability to analyze problems and correctly carry out the calculations needed to solve them. Also take the time to work through the practice exercise that follows each example to be sure you understand how to solve the type of problem illustrated in the example. The answers to the practice exercises appear at the end of the chapter, following the homework problems. For additional practice, you can turn to similar homework problems referred to in the margin next to the example.
- The questions and problems at the end of the chapter are organized by section.
- The back inside cover shows a list of important figures and tables with page references. This index makes it convenient to quickly look up information when you are solving problems or studying related subjects in different chapters.

If you follow these suggestions and stay up-to-date with your assignments, you should find that chemistry is challenging, but less difficult and much more interesting than you expected.

Raymond Chang

Jason Overby

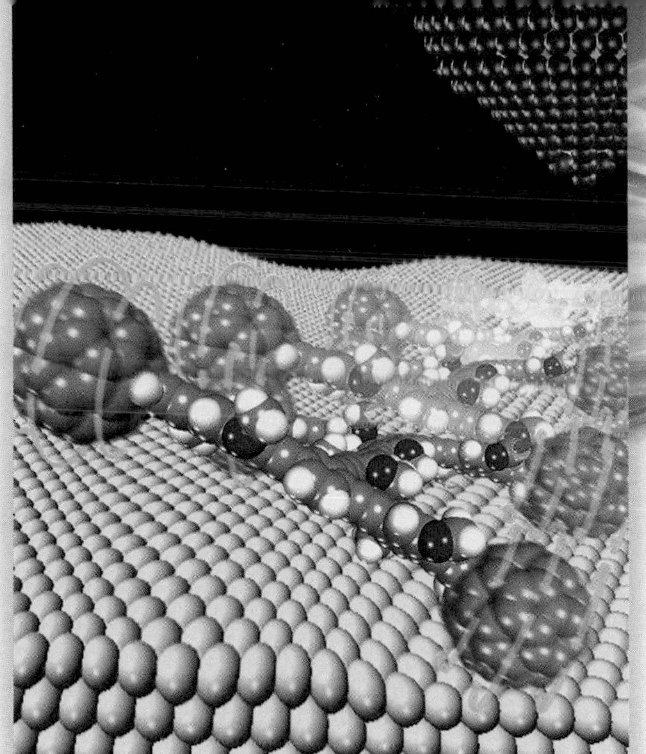

A "nanocar" rolls on a surface of gold atoms as detected by a scanning tunneling microscope. The atomic scale vehicle is assembled using buckminsterfullerene, or "buckyballs," a molecule with 60 carbon atoms in a sphere, in a series of well-defined chemical reactions. The entire nanocar is 20,000 times smaller than a human hair.

CHAPTER 1

Introduction

Chapter Outline

Student Interactive Activities

 **Elecronic Homework**
Example Practice Problems
End-of-Chapter Problems

Essential Concepts

The Study of Chemistry Chemistry is the study of the properties of matter and the changes it undergoes. Elements and compounds are substances that take part in chemical transformation.

Physical and Chemical Properties To characterize a substance, we need to know its physical properties, which can be observed without changing its identity, and chemical properties, which can be demonstrated only by chemical changes.

Measurements and Units Chemistry is a quantitative science and requires measurements. The measured quantities (for example, mass, volume, density, and temperature) usually have units associated with them. The units used in chemistry are based on the international system (SI) of units.

Handling Numbers Scientific notation is used to express large and small numbers, and each number in a measurement must indicate the meaningful digits, called significant figures.

Doing Chemical Calculations A simple and effective way to perform chemical calculations is dimensional analysis. In this procedure, an equation is set up in such a way that all the units cancel except the ones for the final answer.

1.1 The Study of Chemistry

Whether or not this is your first course in chemistry, you undoubtedly have some preconceived ideas about the nature of this science and about what chemists do. Most likely, you think chemistry is practiced in a laboratory by someone in a white coat who studies things in test tubes. This description is fine, up to a point. Chemistry is largely an experimental science, and a great deal of knowledge comes from laboratory research. In addition, however, today's chemists may use a computer to study the microscopic structure and chemical properties of substances or employ sophisticated electronic equipment to analyze pollutants from auto emissions or toxic substances in the soil. Many frontiers in biology and medicine are currently being explored at the level of atoms and molecules—the structural units on which the study of chemistry is based. Chemists participate in the development of new drugs and in agricultural research. What's more, they are seeking solutions to the problem of environmental pollution along with replacements for energy sources. And most industries, whatever their products, have a basis in chemistry. For example, chemists developed the polymers (very large molecules) that manufacturers use to make a wide variety of goods, including clothing, cooking utensils, artificial organs, and toys. Indeed, because of its diverse applications, chemistry is often called the "central science."

How to Study Chemistry

Compared with other subjects, chemistry is commonly perceived to be more difficult, at least at the introductory level. There is some justification for this perception. For one thing, chemistry has a very specialized vocabulary. At first, studying chemistry is like learning a new language. Furthermore, some of the concepts are abstract. Nevertheless, with diligence you can complete this course successfully—and perhaps even pleasurably. Listed here are some suggestions to help you form good study habits and master the material:

- Attend classes regularly and take careful notes.
- If possible, always review the topics you learned in class the *same* day the topics are covered in class. Use this book to supplement your notes.
- Think critically. Ask yourself if you really understand the meaning of a term or the use of an equation. A good way to test your understanding is for you to explain a concept to a classmate or some other person.
- Do not hesitate to ask your instructor or your teaching assistant for help.

You will find that chemistry is much more than numbers, formulas, and abstract theories. It is a logical discipline brimming with interesting ideas and applications.

1.2 The Scientific Method

All sciences, including the social sciences, employ variations of what is called the ***scientific method***—*a systematic approach to research.* For example, a psychologist who wants to know how noise affects people's ability to learn chemistry and a chemist interested in measuring the heat given off when hydrogen gas burns in air follow roughly the same procedure in carrying out their investigations. The first step is carefully defining the problem. The next step includes performing experiments, making careful observations, and recording information, or *data,* about the system—the part of the universe that is under investigation. (In these examples, the systems are the group of people the psychologist will study and a mixture of hydrogen and air.)

The data obtained in a research study may be both ***qualitative,*** *consisting of general observations about the system,* and ***quantitative,*** *comprising numbers obtained by various measurements of the system.* Chemists generally use standardized symbols and equations in recording their measurements and observations. This form of representation not only simplifies the process of keeping records, but also provides a common basis for communications with other chemists. Figure 1.1 summarizes the main steps of the research process.

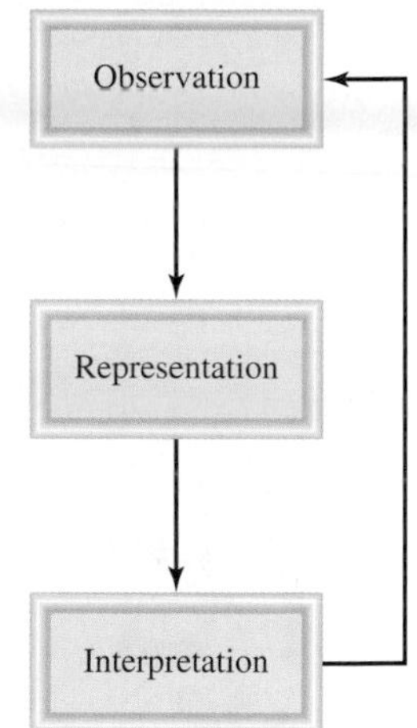

Figure 1.1
The three levels of studying chemistry and their relationships. Observation deals with events in the macroscopic world; atoms and molecules constitute the microscopic world. Representation is a scientific shorthand for describing an experiment in symbols and chemical equations. Chemists use their knowledge of atoms and molecules to explain an observed phenomenon.

When the experiments have been completed and the data have been recorded, the next step in the scientific method is interpretation, meaning that the scientist attempts to explain the observed phenomenon. Based on the data that were gathered, the researcher formulates a ***hypothesis,*** or *tentative explanation for a set of observations.* Further experiments are devised to test the validity of the hypothesis in as many ways as possible, and the process begins anew.

After a large amount of data has been collected, it is often desirable to summarize the information in a concise way, as a law. In science, a ***law*** is *a concise verbal or mathematical statement of a relationship between phenomena that is always the same under the same conditions.* For example, Sir Isaac Newton's second law of motion, which you may remember from high school science, says that force equals mass times acceleration ($F = ma$). What this law means is that an increase in the mass or in the acceleration of an object always increases the object's force proportionally, and a decrease in mass or acceleration always decreases the force.

Hypotheses that survive many experimental tests of their validity may evolve into theories. A ***theory*** is *a unifying principle that explains a body of facts and/or those laws that are based on them.* Theories, too, are constantly being tested. If a theory is disproved by experiment, then it must be discarded or modified so that it becomes consistent with experimental observations. Proving or disproving a theory can take years, even centuries, in part because the necessary technology is not available. Atomic theory, which we will study in Chapter 2, is a case in point. It took more than 2000 years to work out this fundamental principle of chemistry proposed by Democritus, an ancient Greek philosopher.

Scientific progress is seldom, if ever, made in a rigid, step-by-step fashion. Sometimes a law precedes a theory; sometimes it is the other way around. Two scientists may start working on a project with exactly the same objective, but may take drastically different approaches. They may be led in vastly different directions. Scientists are, after all, human beings, and their modes of thinking and working are very much influenced by their backgrounds, training, and personalities.

The development of science has been irregular and sometimes even illogical. Great discoveries are usually the result of the cumulative contributions and experience of many workers, even though the credit for formulating a theory or a law is usually given to only one individual. There is, of course, an element of luck involved in scientific discoveries, but it has been said that "chance favors the prepared mind." It takes an alert and well-trained person to recognize the significance of an accidental discovery and to take full advantage of it. More often than not, the public learns only of spectacular scientific breakthroughs. For every success story, however, there are hundreds of cases in which scientists spent years working on projects that ultimately led to a dead end. Many positive achievements came only after many wrong turns and at such a slow pace that they went unheralded. Yet even the dead ends contribute something to the continually growing body of knowledge about the physical universe. It is the love of the search that keeps many scientists in the laboratory.

Review of Concepts

Which of the following statements is true?

(a) A hypothesis always leads to the formation of a law.

(b) The scientific method is a rigid sequence of steps in solving problems.

(c) A law summarizes a series of experimental observations; a theory provides an explanation for the observations.

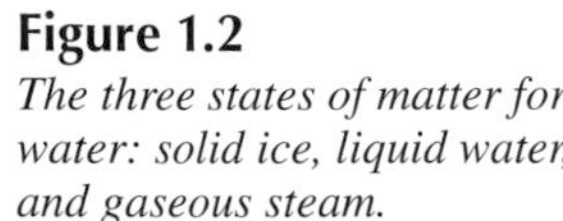

The Chinese characters for chemistry mean "The study of change."

1.3 Classifications of Matter

Matter is *anything that occupies space and has mass,* and ***chemistry*** is *the study of matter and the changes it undergoes.* All matter, at least in principle, can exist in three states: solid, liquid, and gas. Solids are rigid objects with definite shapes. Liquids are less rigid than solids and are fluid—they are able to flow and assume the shape of their containers. Like liquids, gases are fluid, but unlike liquids, they can expand indefinitely.

The three states of matter can be interconverted without changing the composition of the substance. Upon heating, a solid (for example, ice) will melt to form a liquid (water). (The temperature at which this transition occurs is called the *melting point.*) Further heating will convert the liquid into a gas. (This conversion takes place at the *boiling point* of the liquid.) On the other hand, cooling a gas will cause it to condense into a liquid. When the liquid is cooled further, it will freeze into the solid form. Figure 1.2 shows the three states of water. Note that the properties of water are unique

Figure 1.2
The three states of matter for water: solid ice, liquid water, and gaseous steam.

(a)

(b)

Figure 1.3
(a) The mixture contains iron filings and sand. (b) A magnet separates the iron filings from the mixture. The same technique is used on a larger scale to separate iron and steel from nonmagnetic objects such as aluminum, glass, and plastics.

among common substances in that the molecules in the liquid state are more closely packed than those in the solid state.

Substances and Mixtures

A ***substance*** is *matter that has a definite or constant composition and distinct properties.* Examples are water, silver, ethanol, table salt (sodium chloride), and carbon dioxide. Substances differ from one another in composition and can be identified by their appearance, smell, taste, and other properties. At present, over 20 million substances are known, and the list is growing rapidly.

A ***mixture*** is *a combination of two or more substances in which the substances retain their distinct identities.* Some examples are air, soft drinks, milk, and cement. Mixtures do not have constant composition. Therefore, samples of air collected in different cities would probably differ in composition because of differences in altitude, pollution, and so on.

Mixtures are either homogeneous or heterogeneous. When a spoonful of sugar dissolves in water, *the composition of the mixture,* after sufficient stirring, *is the same throughout the solution.* This solution is a ***homogeneous mixture.*** If sand is mixed with iron filings, however, the sand grains and the iron filings remain visible and separate (Figure 1.3). This type of mixture, in which *the composition is not uniform,* is called a ***heterogeneous mixture.*** Adding oil to water creates another heterogeneous mixture because the liquid does not have a constant composition.

Any mixture, whether homogeneous or heterogeneous, can be created and then separated by physical means into pure components without changing the identities of the components. Thus, sugar can be recovered from a water solution by heating the solution and evaporating it to dryness. Condensing the water vapor will give us back the water component. To separate the iron-sand mixture, we can use a magnet to remove the iron filings from the sand, because sand is not attracted to the magnet (see Figure 1.3b). After separation, the components of the mixture will have the same composition and properties as they did to start with.

Elements and Compounds

A substance can be either an element or a compound. An ***element*** is *a substance that cannot be separated into simpler substances by chemical means.* At present, 117 elements have been positively identified. (See the list inside the front cover of this book.)

Table 1.1 Some Common Elements and Their Symbols

Name	Symbol	Name	Symbol	Name	Symbol
Aluminum	Al	Fluorine	F	Oxygen	O
Arsenic	As	Gold	Au	Phosphorus	P
Barium	Ba	Hydrogen	H	Platinum	Pt
Bromine	Br	Iodine	I	Potassium	K
Calcium	Ca	Iron	Fe	Silicon	Si
Carbon	C	Lead	Pb	Silver	Ag
Chlorine	Cl	Magnesium	Mg	Sodium	Na
Chromium	Cr	Mercury	Hg	Sulfur	S
Cobalt	Co	Nickel	Ni	Tin	Sn
Copper	Cu	Nitrogen	N	Zinc	Zn

Chemists use alphabetical symbols to represent the names of the elements. The first letter of the symbol for an element is *always* capitalized, but the second letter is *never* capitalized. For example, Co is the symbol for the element cobalt, whereas CO is the formula for carbon monoxide, which is made up of the elements carbon and oxygen. Table 1.1 shows some of the more common elements. The symbols for some elements are derived from their Latin names—for example, Au from *aurum* (gold), Fe from *ferrum* (iron), and Na from *natrium* (sodium)—although most of them are abbreviated forms of their English names.

Figure 1.4 shows the most abundant elements in Earth's crust and in the human body. As you can see, only five elements (oxygen, silicon, aluminum, iron, and calcium) comprise over 90 percent of Earth's crust. Of these five elements, only oxygen is among the most abundant elements in living systems.

Most elements can interact with one or more other elements to form compounds. We define a ***compound*** as *a substance composed of two or more elements chemically united in fixed proportions.* Hydrogen gas, for example, burns in oxygen gas to form water, a compound whose properties are distinctly different from those of the starting materials. Water is made up of two parts of hydrogen and one part of oxygen. This composition does not change, regardless of whether the water comes from a faucet in the United States, the Yangtze River in China, or the ice caps on Mars. Unlike mixtures, compounds can be separated only by chemical means into their pure components.

The relationships among elements, compounds, and other categories of matter are summarized in Figure 1.5.

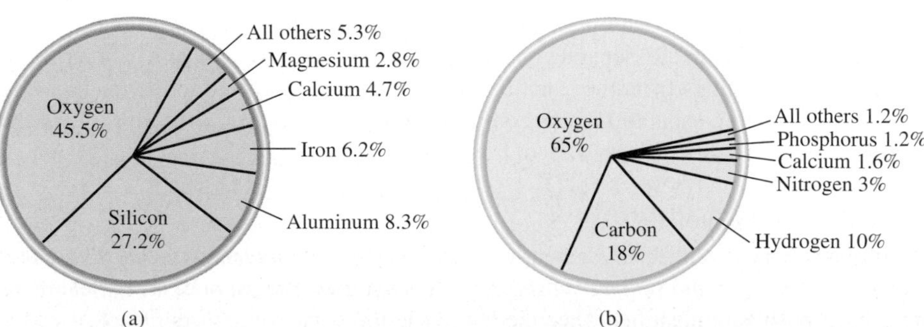

Figure 1.4
(a) Natural abundance of the elements in percent by mass. For example, oxygen's abundance is 45.5 percent. This means that in a 100-g sample of Earth's crust there are, on the average, 45.5 g of the element oxygen.
(b) Abundance of elements in the human body in percent by mass.

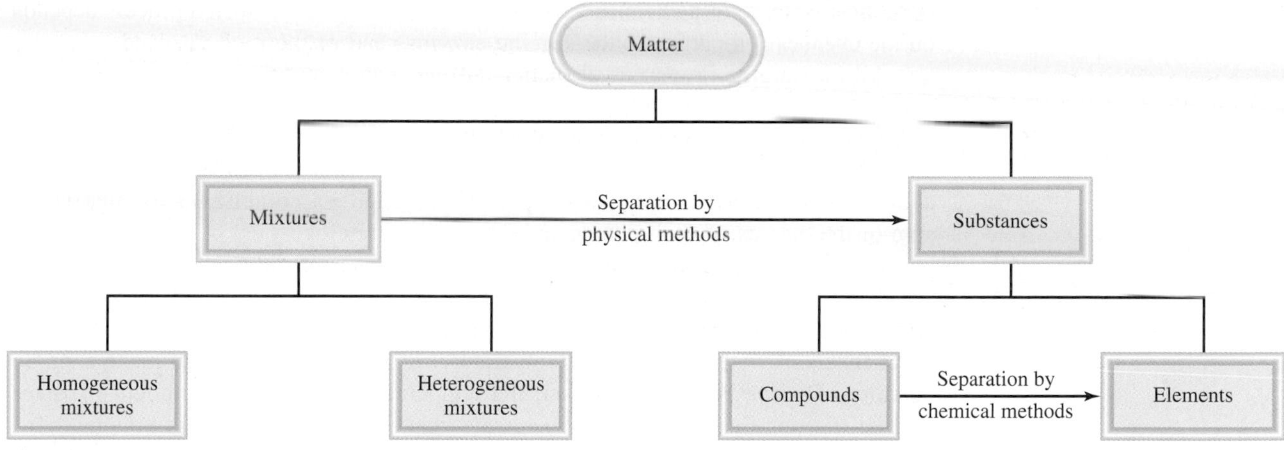

Figure 1.5
Classification of matter.

REVIEW OF CONCEPTS

Which of the following diagrams represent elements and which represent compounds? Each color sphere (or truncated sphere) represents an atom.

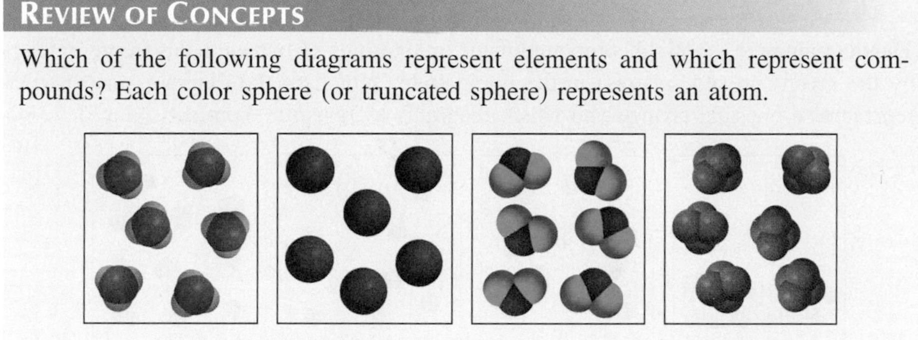

1.4 Physical and Chemical Properties of Matter

Substances are identified by their properties as well as by their composition. Color, melting point, boiling point, and density are physical properties. A ***physical property*** *can be measured and observed without changing the composition or identity of a substance.* For example, we can measure the melting point of ice by heating a block of ice and recording the temperature at which the ice is converted to water. Water differs from ice only in appearance and not in composition, so this is a physical change; we can freeze the water to recover the original ice. Therefore, the melting point of a substance is a physical property. Similarly, when we say that helium gas is lighter than air, we are referring to a physical property.

On the other hand, the statement "Hydrogen gas burns in oxygen gas to form water" describes a ***chemical property*** of hydrogen because *to observe this property we must carry out a chemical change,* in this case burning. After the change, the original substances, hydrogen and oxygen gas, will have vanished and a chemically different substance—water—will have taken their place. We *cannot* recover hydrogen and oxygen from water by a physical change such as boiling or freezing.

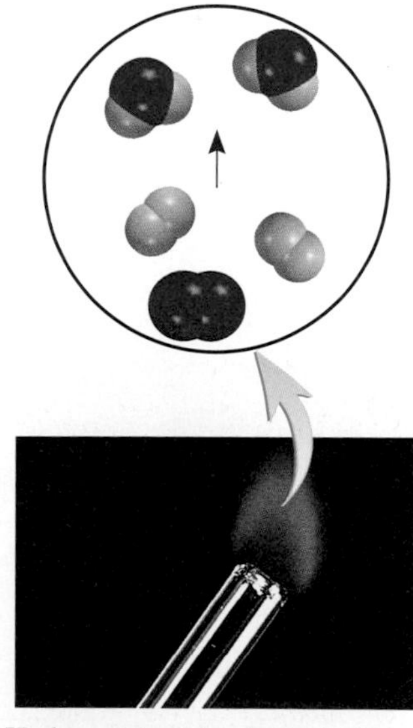
Hydrogen burning in air to form water.

Every time we hard-boil an egg, we bring about a chemical change. When subjected to a temperature of about 100°C, the yolk and the egg white undergo reactions that alter not only their physical appearance but their chemical makeup as well. When eaten, the egg is changed again, by substances in the body called *enzymes.* This digestive action is another example of a chemical change. What happens during such a process depends on the chemical properties of the specific enzymes and of the food involved.

All measurable properties of matter fall into two categories: extensive properties and intensive properties. The measured value of an ***extensive property*** *depends on how much matter is being considered.* Mass, length, and volume are extensive properties. More matter means more mass. Values of the same extensive property can be added together. For example, two copper pennies have a combined mass that is the sum of the masses of each penny, and the total volume occupied by the water in two beakers is the sum of the volumes of the water in each of the beakers.

The measured value of an ***intensive property*** *does not depend on the amount of matter being considered.* Temperature is an intensive property. Suppose that we have two beakers of water at the same temperature. If we combine them to make a single quantity of water in a larger beaker, the temperature of the larger amount of water will be the same as it was in two separate beakers. Unlike mass and volume, temperature and other intensive properties such as melting point, boiling point, and density are not additive.

REVIEW OF CONCEPTS

The diagram in (a) shows a compound made up of atoms of two elements (represented by the green and red spheres) in the liquid state. Which of the diagrams in (b)–(d) represents a physical change and which diagrams represent a chemical change?

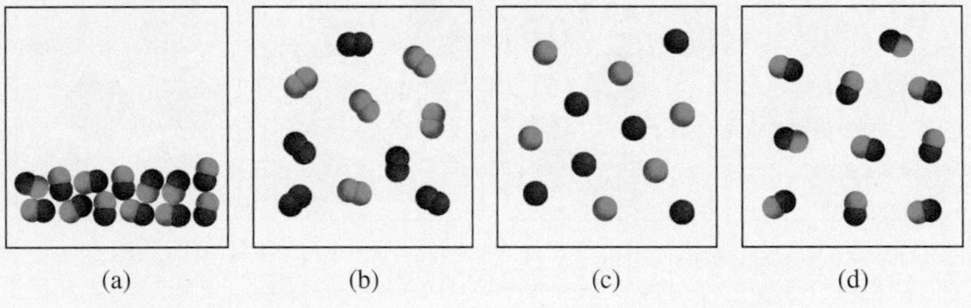

1.5 Measurement

The study of chemistry depends heavily on measurement. For instance, chemists use measurements to compare the properties of different substances and to assess changes resulting from an experiment. A number of common devices enable us to make simple measurements of a substance's properties: The meterstick measures length; the buret, the pipet, the graduated cylinder, and the volumetric flask measure volume (Figure 1.6); the balance measures mass; the thermometer measures temperature. These instruments provide measurements of ***macroscopic properties,*** which *can be determined directly.* ***Microscopic properties,*** *on the atomic or molecular scale, must be determined by an indirect method,* as we will see in Chapter 2.

A measured quantity is usually written as a number with an appropriate unit. To say that the distance between New York and San Francisco by car along a certain route is 5166 is meaningless. We must specify that the distance is 5166 kilometers. In science, units are essential to stating measurements correctly.

SI Units

For many years scientists recorded measurements in *metric units,* which are related decimally, that is, by powers of 10. In 1960, however, the General Conference of Weights and Measures, the international authority on units, proposed a revised metric

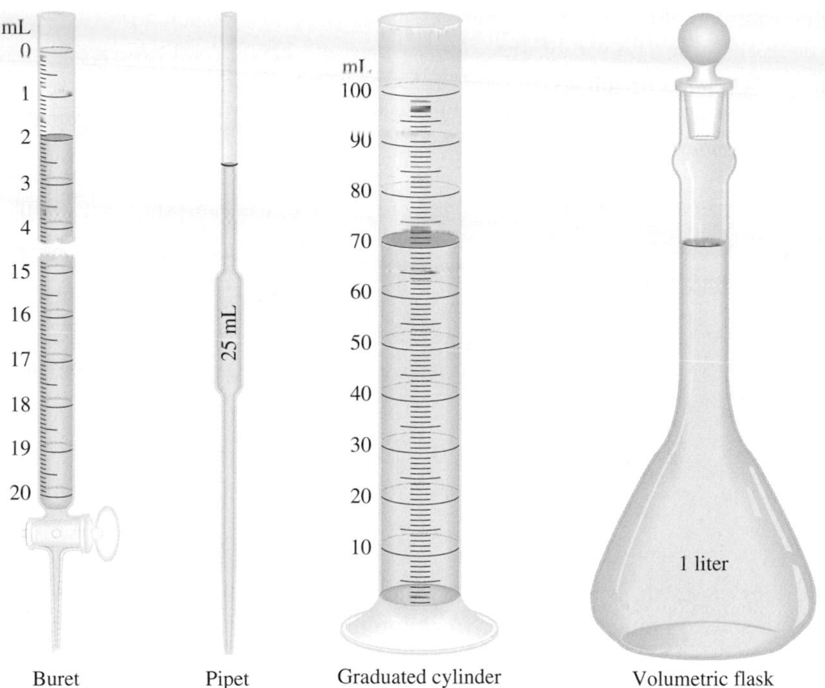

Figure 1.6
Some common measuring devices found in a chemistry laboratory. These devices are not drawn to scale relative to one another. We will discuss the use of these measuring devices in Chapter 4.

system called the ***International System of Units*** (abbreviated *SI,* from the French *S*ystem *I*nternational d'Unites). Table 1.2 shows the seven SI base units. All other SI units of measurement can be derived from these base units. Like metric units, SI units are modified in decimal fashion by a series of prefixes, as shown in Table 1.3. We use both metric and SI units in this book.

Measurements that we will utilize frequently in our study of chemistry include time, mass, volume, density, and temperature.

Mass and Weight

Mass is *a measure of the quantity of matter in an object.* The terms "mass" and "weight" are often used interchangeably, although, strictly speaking, they refer to different quantities. In scientific terms, ***weight*** is *the force that gravity exerts on an object.* An apple that falls from a tree is pulled downward by Earth's gravity. The mass of the apple is constant and does not depend on its location, but its weight does. For example, on the surface of the moon the apple would weigh only one-sixth

Table 1.2 SI Base Units

Base Quantity	Name of Unit	Symbol
Length	meter	m
Mass	kilogram	kg
Time	second	s
Electrical current	ampere	A
Temperature	kelvin	K
Amount of substance	mole	mol
Luminous intensity	candela	cd

Note that a metric prefix simply represents a number:

$1\ \text{mm} = 1 \times 10^{-3}\ \text{m}$

Table 1.3 Prefixes Used with SI Units

Prefix	Symbol	Meaning	Example
tera-	T	1,000,000,000,000, or 10^{12}	1 terameter (Tm) $= 1 \times 10^{12}$ m
giga-	G	1,000,000,000, or 10^{9}	1 gigameter (Gm) $= 1 \times 10^{9}$ m
mega-	M	1,000,000, or 10^{6}	1 megameter (Mm) $= 1 \times 10^{6}$ m
kilo-	k	1,000, or 10^{3}	1 kilometer (km) $= 1 \times 10^{3}$ m
deci-	d	1/10, or 10^{-1}	1 decimeter (dm) = 0.1 m
centi-	c	1/100, or 10^{-2}	1 centimeter (cm) = 0.01 m
milli-	m	1/1,000, or 10^{-3}	1 millimeter (mm) = 0.001 m
micro-	μ	1/1,000,000, or 10^{-6}	1 micrometer (μm) $= 1 \times 10^{-6}$ m
nano-	n	1/1,000,000,000, or 10^{-9}	1 nanometer (nm) $= 1 \times 10^{-9}$ m
pico-	p	1/1,000,000,000,000, or 10^{-12}	1 picometer (pm) $= 1 \times 10^{-12}$ m

An astronaut jumping on the surface of the moon.

what it does on Earth, because of the smaller mass of the moon. This is why astronauts were able to jump about rather freely on the moon's surface despite their bulky suits and equipment. The mass of an object can be determined readily with a balance, and this process, oddly, is called weighing.

The SI base unit of mass is the *kilogram* (kg), but in chemistry the smaller gram (g) is more convenient:

$$1\ \text{kg} = 1000\ \text{g} = 1 \times 10^{3}\ \text{g}$$

Volume

Volume is *length (m) cubed,* so its SI-derived unit is the cubic meter (m^3). Generally, however, chemists work with much smaller volumes, such as the cubic centimeter (cm^3) and the cubic decimeter (dm^3):

$$1\ \text{cm}^3 = (1 \times 10^{-2}\ \text{m})^3 = 1 \times 10^{-6}\ \text{m}^3$$
$$1\ \text{dm}^3 = (1 \times 10^{-1}\ \text{m})^3 = 1 \times 10^{-3}\ \text{m}^3$$

Another common, non-SI unit of volume is the liter (L). A ***liter*** is *the volume occupied by one cubic decimeter.* Chemists generally use L and mL for liquid volume. One liter is equal to 1000 milliliters (mL) or 1000 cubic centimeters:

$$\begin{aligned} 1\ \text{L} &= 1000\ \text{mL} \\ &= 1000\ \text{cm}^3 \\ &= 1\ \text{dm}^3 \end{aligned}$$

and one milliliter is equal to one cubic centimeter:

$$1\ \text{mL} = 1\ \text{cm}^3$$

Figure 1.7 compares the relative sizes of two volumes.

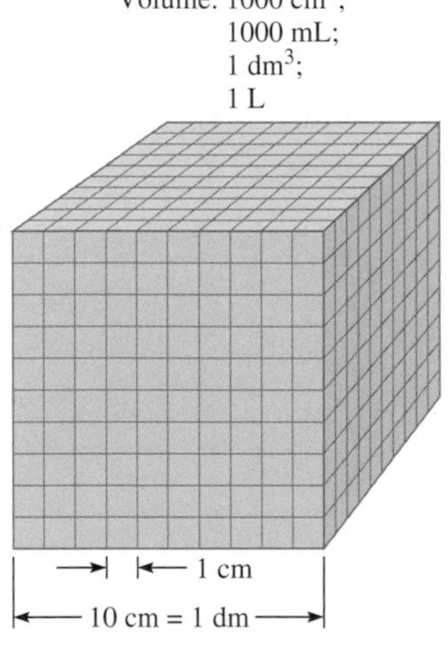

Figure 1.7
Comparison of two volumes, 1 mL and 1000 mL.

Density

Density is *the mass of an object divided by its volume:*

$$\text{density} = \frac{\text{mass}}{\text{volume}}$$

or

$$d = \frac{m}{V} \tag{1.1}$$

where d, m, and V denote density, mass, and volume, respectively. Note that density is an intensive property that does not depend on the quantity of mass present. The reason is that V increases as m does, so the ratio of the two quantities always remains the same for a given material.

The SI-derived unit for density is the kilogram per cubic meter (kg/m^3). This unit is awkwardly large for most chemical applications. Therefore, grams per cubic centimeter (g/cm^3) and its equivalent, grams per milliliter (g/mL), are more commonly used for solid and liquid densities. Table 1.4 lists the densities of several substances.

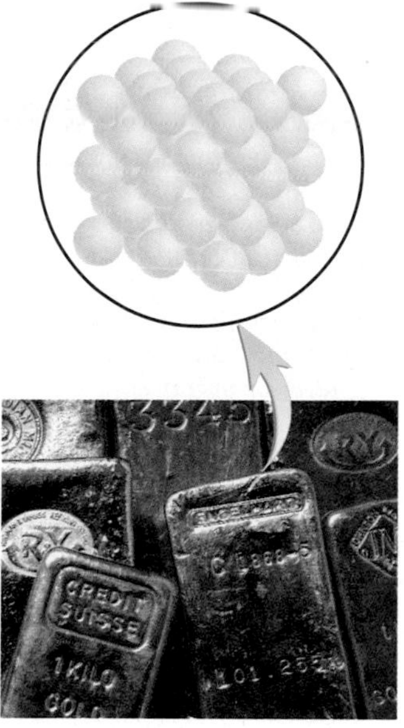

Gold bars and the solid-state arrangement of the gold atoms.

EXAMPLE 1.1

Gold is a precious metal that is chemically unreactive. It is used mainly in jewelry, dentistry, and electronic devices. A piece of gold ingot with a mass of 257 g has a volume of 13.3 cm^3. Calculate the density of gold.

Solution We are given the mass and volume and asked to calculate the density. Therefore, from Equation (1.1), we write

$$\begin{aligned} d &= \frac{m}{V} \\ &= \frac{257\ \text{g}}{13.3\ \text{cm}^3} \\ &= 19.3\ \text{g/cm}^3 \end{aligned}$$

Practice Exercise A piece of platinum metal with a density of 21.5 g/cm^3 has a volume of 4.49 cm^3. What is its mass?

Similar problems: 1.17, 1.18.

Temperature Scales

Three temperature scales are currently in use. Their units are °F (degrees Fahrenheit), °C (degrees Celsius), and K (kelvin). The Fahrenheit scale, which is the most commonly used scale in the United States outside the laboratory, defines the normal freezing and boiling points of water to be exactly 32°F and 212°F, respectively. The Celsius scale divides the range between the freezing point (0°C) and boiling point (100°C) of water into 100 degrees. As Table 1.2 shows, the *kelvin* is the SI base unit of temperature; it is the *absolute* temperature scale. By absolute we mean that the zero on the Kelvin scale, denoted by 0 K, is the lowest temperature that can be attained theoretically. On the other hand, 0°F and 0°C are based on the behavior of an arbitrarily chosen substance, water. Figure 1.8 compares the three temperature scales.

Note that the Kelvin scale does not have the degree sign. Also, temperatures expressed in kelvins can never be negative.

The size of a degree on the Fahrenheit scale is only 100/180, or 5/9, of a degree on the Celsius scale. To convert degrees Fahrenheit to degrees Celsius, we write

$$?°\text{C} = (°\text{F} - 32°\text{F}) \times \frac{5°\text{C}}{9°\text{F}} \tag{1.2}$$

The following equation is used to convert degrees Celsius to degrees Fahrenheit:

$$?°\text{F} = \frac{9°\text{F}}{5°\text{C}} \times (°\text{C}) + 32°\text{F} \tag{1.3}$$

Table 1.4

Densities of Some Substances at 25°C

Substance	Density (g/cm^3)
Air*	0.001
Ethanol	0.79
Water	1.00
Mercury	13.6
Table salt	2.2
Iron	7.9
Gold	19.3
Osmium†	22.6

* Measured at 1 atmosphere.
† Osmium (Os) is the densest element known.

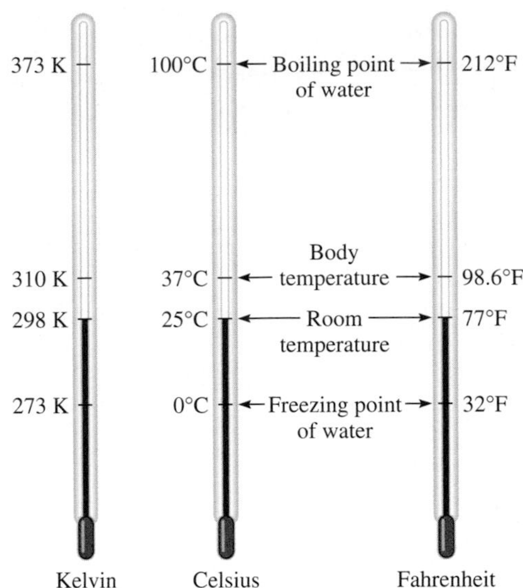

Figure 1.8
Comparison of the three temperature scales: Celsius, Fahrenheit, and the absolute (Kelvin) scales. Note that there are 100 divisions, or 100 degrees, between the freezing point and the boiling point of water on the Celsius scale, and there are 180 divisions, or 180 degrees, between the same two temperature limits on the Fahrenheit scale. The Celsius scale was formerly called the centigrade scale. Note that the Kelvin scale does not have the degree sign. Also, temperature expressed in kelvins can never be negative.

Both the Celsius and the Kelvin scales have units of equal magnitude; that is, one degree Celsius is equivalent to one kelvin. Experimental studies have shown that absolute zero on the Kelvin scale is equivalent to $-273.15°C$ on the Celsius scale. Thus, we can use the following equation to convert degrees Celsius to kelvin:

$$?\ \text{K} = (°\text{C} + 273.15°\text{C})\,\frac{1\ \text{K}}{1°\text{C}} \tag{1.4}$$

Solder is used extensively in the construction of electronic circuits.

EXAMPLE 1.2

(a) Solder is an alloy made of tin and lead that is used in electronic circuits. A certain solder has a melting point of 224°C. What is its melting point in degrees Fahrenheit? (b) Helium has the lowest boiling point of all the elements at −452°F. Convert this temperature to degrees Celsius. (c) Mercury, the only metal that exists as a liquid at room temperature, melts at −38.9°C. Convert its melting point to kelvins.

Solution These three parts require that we carry out temperature conversions, so we need Equations (1.2), (1.3), and (1.4). Keep in mind that the lowest temperature on the Kelvin scale is zero (0 K); therefore, it can never be negative.

(a) This conversion is carried out by writing

$$\frac{9°\text{F}}{5°\text{C}} \times (224°\text{C}) + 32°\text{F} = 435°\text{F}$$

(b) Here we have

$$(-452°\text{F} - 32°\text{F}) \times \frac{5°\text{C}}{9°\text{F}} = -269°\text{C}$$

(c) The melting point of mercury in kelvins is given by

$$(-38.9°\text{C} + 273.15°\text{C}) \times \frac{1\ \text{K}}{1°\text{C}} = 234.3\ \text{K}$$

(Continued)

Similar problems: 1.19, 1.20.

Practice Exercise Convert (a) 327.5°C (the melting point of lead) to degrees Fahrenheit; (b) 172.9°F (the boiling point of ethanol) to degrees Celsius; and (c) 77 K, the boiling point of liquid nitrogen, to degrees Celsius.

Review of Concepts

The density of copper is 8.94 g/cm^3 at 20°C and 8.91 g/cm^3 at 60°C. The decrease in density is the result of which of the following?

(a) The metal expands increasing the volume.
(b) The metal contracts decreasing the volume.
(c) The mass of the metal increases.
(d) The mass of the metal decreases.

1.6 Handling Numbers

Having surveyed some of the units used in chemistry, we now turn to techniques for handling numbers associated with measurements: scientific notation and significant figures.

Scientific Notation

Chemists often deal with numbers that are either extremely large or extremely small. For example, in 1 g of the element hydrogen there are roughly

$$602,200,000,000,000,000,000,000$$

hydrogen atoms. Each hydrogen atom has a mass of only

$$0.00000000000000000000000166 \text{ g}$$

These numbers are cumbersome to handle, and it is easy to make mistakes when using them in arithmetic computations. Consider the following multiplication:

$$0.0000000056 \times 0.00000000048 = 0.000000000000000002688$$

It would be easy for us to miss one zero or add one more zero after the decimal point. Consequently, when working with very large and very small numbers, we use a system called *scientific notation.* Regardless of their magnitude, all numbers can be expressed in the form

$$N \times 10^n$$

where N is a number between 1 and 10 and n, the exponent, is a positive or negative integer (whole number). Any number expressed in this way is said to be written in scientific notation.

Suppose that we are given a certain number and asked to express it in scientific notation. Basically, this assignment calls for us to find n. We count the number of places that the decimal point must be moved to give the number N (which is between

1 and 10). If the decimal point has to be moved to the left, then n is a positive integer; if it has to be moved to the right, n is a negative integer. The following examples illustrate the use of scientific notation:

(1) Express 568.762 in scientific notation:

$$568.762 = 5.68762 \times 10^2$$

Note that the decimal point is moved to the left by two places and $n = 2$.

(2) Express 0.00000772 in scientific notation:

$$0.00000772 = 7.72 \times 10^{-6}$$

Here the decimal point is moved to the right by six places and $n = -6$.

Any number raised to the power zero is equal to one.

Keep in mind the following two points. First, $n = 0$ is used for numbers that are not expressed in scientific notation. For example, 74.6×10^0 ($n = 0$) is equivalent to 74.6. Second, the usual practice is to omit the superscript when $n = 1$. Thus, the scientific notation for 74.6 is 7.46×10 and not 7.46×10^1.

Next, we consider how scientific notation is handled in arithmetic operations.

Addition and Subtraction

To add or subtract using scientific notation, we first write each quantity—say N_1 and N_2—with the same exponent n. Then we combine N_1 and N_2; the exponents remain the same. Consider the following examples:

$$\begin{aligned}
(7.4 \times 10^3) + (2.1 \times 10^3) &= 9.5 \times 10^3 \\
(4.31 \times 10^4) + (3.9 \times 10^3) &= (4.31 \times 10^4) + (0.39 \times 10^4) \\
&= 4.70 \times 10^4 \\
(2.22 \times 10^{-2}) - (4.10 \times 10^{-3}) &= (2.22 \times 10^{-2}) - (0.41 \times 10^{-2}) \\
&= 1.81 \times 10^{-2}
\end{aligned}$$

Multiplication and Division

To multiply numbers expressed in scientific notation, we multiply N_1 and N_2 in the usual way, but *add* the exponents together. To divide using scientific notation, we divide N_1 and N_2 as usual and subtract the exponents. The following examples show how these operations are performed:

$$\begin{aligned}
(8.0 \times 10^4) \times (5.0 \times 10^2) &= (8.0 \times 5.0)(10^{4+2}) \\
&= 40 \times 10^6 \\
&= 4.0 \times 10^7 \\
(4.0 \times 10^{-5}) \times (7.0 \times 10^3) &= (4.0 \times 7.0)(10^{-5+3}) \\
&= 28 \times 10^{-2} \\
&= 2.8 \times 10^{-1} \\
\frac{6.9 \times 10^7}{3.0 \times 10^{-5}} &= \frac{6.9}{3.0} \times 10^{7-(-5)} \\
&= 2.3 \times 10^{12} \\
\frac{8.5 \times 10^4}{5.0 \times 10^9} &= \frac{8.5}{5.0} \times 10^{4-9} \\
&= 1.7 \times 10^{-5}
\end{aligned}$$

Significant Figures

Except when all the numbers involved are integers (for example, in counting the number of students in a class), obtaining the exact value of the quantity under investigation is often impossible. For this reason, it is important to indicate the margin of error in a measurement by clearly indicating the number of ***significant figures,*** which are *the meaningful digits in a measured or calculated quantity.* When significant figures are used, the last digit is understood to be uncertain. For example, we might measure the volume of a given amount of liquid using a graduated cylinder with a scale that gives an uncertainty of 1 mL in the measurement. If the volume is found to be 6 mL, then the actual volume is in the range of 5 mL to 7 mL. We represent the volume of the liquid as (6 ± 1) mL. In this case, there is only one significant figure (the digit 6) that is uncertain by either plus or minus 1 mL. For greater accuracy, we might use a graduated cylinder that has finer divisions, so that the volume we measure is now uncertain by only 0.1 mL. If the volume of the liquid is now found to be 6.0 mL, we may express the quantity as (6.0 ± 0.1) mL, and the actual value is somewhere between 5.9 mL and 6.1 mL. We can further improve the measuring device and obtain more significant figures, but in every case, the last digit is always uncertain; the amount of this uncertainty depends on the particular measuring device we use.

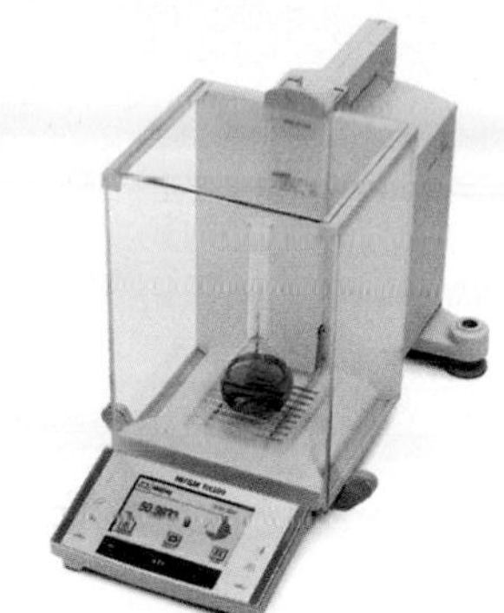

Figure 1.9
A single-pan balance.

Figure 1.9 shows a modern balance. Balances such as this one are available in many general chemistry laboratories; they readily measure the mass of objects to four decimal places. Therefore, the measured mass typically will have four significant figures (for example, 0.8642 g) or more (for example, 3.9745 g). Keeping track of the number of significant figures in a measurement such as mass ensures that calculations involving the data will reflect the precision of the measurement.

Guidelines for Using Significant Figures

We must always be careful in scientific work to write the proper number of significant figures. In general, it is fairly easy to determine how many significant figures a number has by following these rules:

1. Any digit that is not zero is significant. Thus, 845 cm has three significant figures, 1.234 kg has four significant figures, and so on.
2. Zeros between nonzero digits are significant. Thus, 606 m contains three significant figures, 40,501 kg contains five significant figures, and so on.
3. Zeros to the left of the first nonzero digit are not significant. Their purpose is to indicate the placement of the decimal point. For example, 0.08 L contains one significant figure, 0.0000349 g contains three significant figures, and so on.
4. If a number is greater than 1, then all the zeros written to the right of the decimal point count as significant figures. Thus, 2.0 mg has two significant figures, 40.062 mL has five significant figures, and 3.040 dm has four significant figures. If a number is less than 1, then only the zeros that are at the end of the number and the zeros that are between nonzero digits are significant. This means that 0.090 kg has two significant figures, 0.3005 L has four significant figures, 0.00420 min has three significant figures, and so on.
5. For numbers that do not contain decimal points, the trailing zeros (that is, zeros after the last nonzero digit) may or may not be significant. Thus, 400 cm may have one significant figure (the digit 4), two significant figures (40), or three significant figures (400). We cannot know which is correct without more information. By using scientific notation, however, we avoid this ambiguity. In this particular case, we can express the number 400 as 4×10^2 for one significant figure, 4.0×10^2 for two significant figures, or 4.00×10^2 for three significant figures.

EXAMPLE 1.3

Determine the number of significant figures in the following measurements: (a) 394 cm, (b) 5.03 g (c) 0.714 m, (d) 0.052 kg, (e) 2.720×10^{22} atoms, (f) 3000 mL.

Solution (a) Three, because each digit is a nonzero digit. (b) Three, because zeros between nonzero digits are significant. (c) Three, because zeros to the left of the first nonzero digit do not count as significant figures. (d) Two. Same reason as in (c). (e) Four, because the number is greater than one, all the zeros written to the right of the decimal point count as significant figures. (f) This is an ambiguous case. The number of significant figures may be four (3.000×10^3), three (3.00×10^3), two (3.0×10^3), or one (3×10^3). This example illustrates why scientific notation must be used to show the proper number of significant figures.

Similar problems: 1.27, 1.28.

Practice Exercise Determine the number of significant figures in each of the following measurements: (a) 35 mL, (b) 2008 g, (c) 0.0580 m^3, (d) 7.2×10^4 molecules, (e) 830 kg.

A second set of rules specifies how to handle significant figures in calculations.

1. In addition and subtraction, the answer cannot have more digits to the right of the decimal point than either of the original numbers. Consider these examples:

$$\begin{array}{r l} 89.332 & \\ +\ 1.1\ \ \ \ & \longleftarrow \text{one digit after the decimal point} \\ \hline 90.432 & \longleftarrow \text{round off to } 90.4 \end{array}$$

$$\begin{array}{r l} 2.097 & \\ -0.12\ \ & \longleftarrow \text{two digits after the decimal point} \\ \hline 1.977 & \longleftarrow \text{round off to } 1.98 \end{array}$$

The rounding-off procedure is as follows. To round off a number at a certain point we simply drop the digits that follow if the first of them is less than 5. Thus, 8.724 rounds off to 8.72 if we want only two digits after the decimal point. If the first digit following the point of rounding off is equal to or greater than 5, we add 1 to the preceding digit. Thus, 8.727 rounds off to 8.73, and 0.425 rounds off to 0.43.

2. In multiplication and division, the number of significant figures in the final product or quotient is determined by the original number that has the *smallest* number of significant figures. The following examples illustrate this rule:

$$2.8 \times 4.5039 = 12.61092 \longleftarrow \text{round off to } 13$$

$$\frac{6.85}{112.04} = 0.0611388789 \longleftarrow \text{round off to } 0.0611$$

3. Keep in mind that *exact numbers* obtained from definitions (such as 1 ft = 12 in, where 12 is an exact number) or by counting numbers of objects can be considered to have an infinite number of significant figures.

EXAMPLE 1.4

Carry out the following arithmetic operations to the correct number of significant figures: (a) 12,343.2 g + 0.1893 g, (b) 55.67 L − 2.386 L, (c) 7.52 m × 6.9232, (d) 0.0239 kg ÷ 46.5 mL, (e) 5.21×10^3 cm + 2.92×10^2 cm.

(Continued)

Solution In addition and subtraction, the number of decimal places in the answer is determined by the number having the lowest number of decimal places. In multiplication and division, the significant number of the answer is determined by the number having the smallest number of significant figures.

(a) 12,343.2 g
 + 0.1893 g
 12,343.3893 g ⟵ round off to **12,343.4 g**

(b) 55.67 L
 − 2.386 L
 53.284 L ⟵ round off to **53.28 L**

(c) 7.52 m × 6.9232 = 52.06246 m ⟵ round off to **52.1 m**

(d) $\frac{0.0239 \text{ kg}}{46.5 \text{ mL}} = 0.0005139784946 \text{ kg/mL}$ ⟵ round off to **0.000514 kg/mL** or **5.14×10^{-4} kg/mL**

(e) First we change 2.92×10^2 cm to 0.292×10^3 cm and then carry out the addition $(5.21 \text{ cm} + 0.292 \text{ cm}) \times 10^3$. Following the procedure in (a), we find the answer is **5.50×10^3 cm.**

Similar problems: 1.29, 1.30.

Practice Exercise Carry out the following arithmetic operations and round off the answers to the appropriate number of significant figures: (a) 26.5862 L + 0.17 L, (b) 9.1 g − 4.682 g, (c) 7.1×10^4 dm × 2.2654×10^2 dm, (d) 6.54 g ÷ 86.5542 mL, (e) $(7.55 \times 10^4 \text{ m}) - (8.62 \times 10^3 \text{ m})$.

The preceding rounding-off procedure applies to one-step calculations. In *chain calculations,* that is, calculations involving more than one step, we can get a different answer depending on how we round off. Consider the following two-step calculations:

First step: A × B = C
Second step: C × D = E

Let's suppose that A = 3.66, B = 8.45, and D = 2.11. Depending on whether we round off C to three (Method 1) or four (Method 2) significant figures, we obtain a different number for E:

Method 1	Method 2
3.66 × 8.45 = 30.9	3.66 × 8.45 = 30.93
30.9 × 2.11 = 65.2	30.93 × 2.11 = 65.3

However, if we had carried out the calculation as 3.66 × 8.45 × 2.11 on a calculator without rounding off the intermediate answer, we would have obtained 65.3 as the answer for E. Although retaining an additional digit past the number of significant figures for intermediate steps helps to eliminate errors from rounding, this procedure is not necessary for most calculations because the difference between the answers is usually quite small. Therefore, for most examples and end-of-chapter problems where intermediate answers are reported, all answers, intermediate and final, will be rounded.

Accuracy and Precision

In discussing measurements and significant figures it is useful to distinguish between *accuracy* and *precision.* ***Accuracy*** tells us *how close a measurement is to the true value of the quantity that was measured.* To a scientist there is a distinction between accuracy and precision. ***Precision*** *refers to how closely two or more measurements of the same quantity agree with one another* (Figure 1.10).

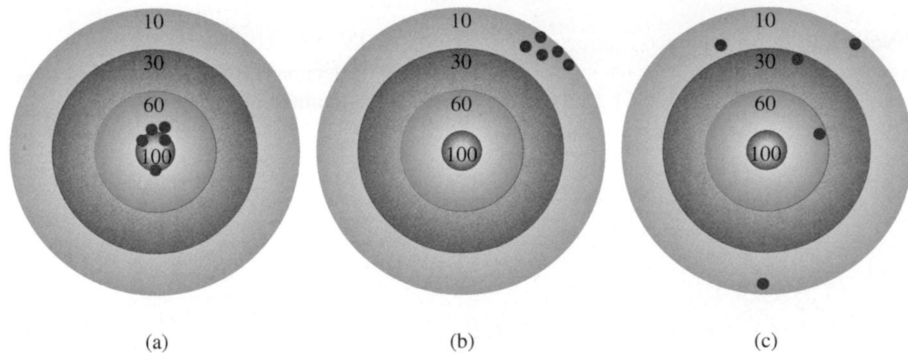

Figure 1.10
The distribution of darts on a dartboard shows the difference between precise and accurate. (a) Good accuracy and good precision. (b) Poor accuracy and good precision. (c) Poor accuracy and poor precision. The blue dots show the positions of the darts.

The difference between accuracy and precision is a subtle but important one. Suppose, for example, that three students are asked to determine the mass of a piece of copper wire. The results of two successive weighings by each student are

	Student A	Student B	Student C
	1.964 g	1.972 g	2.000 g
	1.978 g	1.968 g	2.002 g
Average value	1.971 g	1.970 g	2.001 g

The true mass of the wire is 2.000 g. Therefore, Student B's results are more *precise* than those of Student A (1.972 g and 1.968 g deviate less from 1.970 g than 1.964 g and 1.978 g from 1.971 g), but neither set of results is very *accurate.* Student C's results are not only the most *precise,* but also the most *accurate,* because the average value is closest to the true value. Highly accurate measurements are usually precise too. On the other hand, highly precise measurements do not necessarily guarantee accurate results. For example, an improperly calibrated meterstick or a faulty balance may give precise readings that are in error.

REVIEW OF CONCEPTS

Consider the following measured values: (a) 2.54 g, (b) 0.0034 L, (c) 1.408 × 10^{23} atoms, (d) 80036 m. Which of these quantities has the most significant figures? Which has the least number of significant figures?

1.7 Dimensional Analysis in Solving Problems

Dimensional analysis might also have led Einstein to his famous mass-energy equation $E = mc^2$.

Careful measurements and the proper use of significant figures, along with correct calculations, will yield accurate numerical results. But to be meaningful, the answers also must be expressed in the desired units. The procedure we use to convert between units in solving chemistry problems is called *dimensional analysis* (also called the *factor-label method*). A simple technique requiring little memorization, dimensional analysis is based on the relationship between different units that express the same physical quantity. For example, by definition 1 in = 2.54 cm (exactly). This equivalence enables us to write a conversion factor as follows:

$$\frac{1 \text{ in}}{2.54 \text{ cm}}$$

Because both the numerator and the denominator express the same length, this fraction is equal to 1. Similarly, we can write the conversion factor as

$$\frac{2.54\ \text{cm}}{1\ \text{in}}$$

which is also equal to 1. Conversion factors are useful for changing units. Thus, if we wish to convert a length expressed in inches to centimeters, we multiple the length by the appropriate conversion factor

$$12.00\ \cancel{\text{in}} \times \frac{2.54\ \text{cm}}{1\ \cancel{\text{in}}} = 30.48\ \text{cm}$$

We choose the conversion factor that cancels the unit inches and produces the desired unit, centimeters. Note that the result is expressed in four significant figures because 2.54 is an exact number.

Next, let us consider the conversion of 57.8 meters to centimeters. This problem can be expressed as

$$?\ \text{cm} = 57.8\ \text{m}$$

By definition,

$$1\ \text{cm} = 1 \times 10^{-2}\ \text{m}$$

Because we are converting "m" to "cm," we choose the conversion factor that has meters in the denominator:

$$\frac{1\ \text{cm}}{1 \times 10^{-2}\ \text{m}}$$

and write the conversion as

$$\begin{aligned} ?\ \text{cm} &= 57.8\ \cancel{\text{m}} \times \frac{1\ \text{cm}}{1 \times 10^{-2}\ \cancel{\text{m}}} \\ &= 5780\ \text{cm} \\ &= 5.78 \times 10^{3}\ \text{cm} \end{aligned}$$

Note that scientific notation is used to indicate that the answer has three significant figures. Again, the conversion factor $1\ \text{cm}/1 \times 10^{-2}\ \text{m}$ contains exact numbers; therefore, it does not affect the number of significant figures.

In general, to apply dimensional analysis we use the relationship

$$\text{given quantity} \times \text{conversion factor} = \text{desired quantity}$$

and the units cancel as follows:

$$\cancel{\text{given unit}} \times \frac{\text{desired unit}}{\cancel{\text{given unit}}} = \text{desired unit}$$

In dimensional analysis, the units are carried through the entire sequence of calculations. Therefore, if the equation is set up correctly, then all the units will cancel except the desired one. If this is not the case, then an error must have been made somewhere, and it can usually be spotted by reviewing the solution.

A Note on Problem Solving

At this point you have been introduced to scientific notation, significant figures, and dimensional analysis, which will help you in solving numerical problems. Chemistry is an experimental science and many of the problems are quantitative in nature. The key to success in problem solving is practice. Just as a marathon runner cannot prepare for a race by simply reading books on running and a violinist cannot give a successful concert by only memorizing the musical score, you cannot be sure of your understanding of chemistry without solving problems. The following steps will help to improve your skill at solving numerical problems:

1. Read the question carefully. Understand the information that is given and what you are asked to solve. Frequently it is helpful to make a sketch that will help you to visualize the situation.
2. Find the appropriate equation that relates the given information and the unknown quantity. Sometimes solving a problem will involve more than one step, and you may be expected to look up quantities in tables that are not provided in the problem. Dimensional analysis is often needed to carry out conversions.
3. Check your answer for the correct sign, units, and significant figures.
4. A very important part of problem solving is being able to judge whether the answer is reasonable. It is relatively easy to spot a wrong sign or incorrect units. But if a number (say 8) is incorrectly placed in the denominator instead of in the numerator, the answer would be too small even if the sign and units of the calculated quantity were correct.
5. One way to quickly check the answer is to make a "ball-park" estimate. The idea here is to round off the numbers in the calculation in such a way that we simplify the arithmetic. This approach is sometimes called the "back-of-the-envelope calculation" because it can be done easily without using a calculator. The answer you get will not be exact, but it will be close to the correct one.

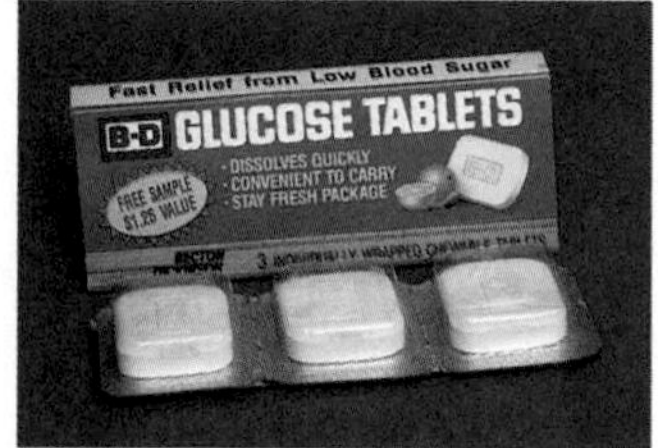

Glucose tablets can provide diabetics with a quick method for raising their blood sugar levels.

Conversion factors for some of the English system units commonly used in the United States for nonscientific measurements (for example, pounds and inches) are provided inside the back cover of this book.

EXAMPLE 1.5

A person's average daily intake of glucose (a form of sugar) is 0.0833 pound (lb). What is this mass in milligrams (mg)? (1 lb = 453.6 g.)

Strategy The problem can be stated as

$$? \text{ mg} = 0.0833 \text{ lb}$$

The relationship between pounds and grams is given in the problem. This relationship will enable conversion from pounds to grams. A metric conversion is then needed to convert grams to milligrams ($1 \text{ mg} = 1 \times 10^{-3} \text{ g}$). Arrange the appropriate conversion factors so that pounds and grams cancel and the unit milligrams is obtained in your answer.

Solution The sequence of conversion is

$$\text{pounds} \longrightarrow \text{grams} \longrightarrow \text{milligrams}$$

Using the following conversion factors:

$$\frac{453.6 \text{ g}}{1 \text{ lb}} \quad \text{and} \quad \frac{1 \text{ mg}}{1 \times 10^{-3} \text{ g}}$$

(Continued)

we obtain the answer in one step:

$$? \text{ mg} = 0.0833 \cancel{\text{lb}} \times \frac{453.6 \cancel{\text{g}}}{1 \cancel{\text{lb}}} \times \frac{1 \text{ mg}}{1 \times 10^{-3} \cancel{\text{g}}} = 3.78 \times 10^4 \text{ mg}$$

Check As an estimate, we note that 1 lb is roughly 500 g and that 1 g = 1000 mg. Therefore, 1 lb is roughly 5 × 10^5 mg. Rounding off 0.0833 lb to 0.1 lb, we get 5 × 10^4 mg, which is close to the preceding quantity.

Similar problem: 1.37(a).

Practice Exercise A roll of aluminum foil has a mass of 1.07 kg. What is its mass in pounds?

As Examples 1.6 and 1.7 illustrate, conversion factors can be squared or cubed in dimensional analysis.

EXAMPLE 1.6

A liquid helium storage tank has a volume of 275 L. What is the volume in m^3?

Strategy The problem can be stated as

$$? \text{ m}^3 = 275 \text{ L}$$

How many conversion factors are needed for this problem? Recall that 1 L = 1000 cm^3 and 1 cm = 1 × 10^{-2} m.

Solution We need two conversion factors here: one to convert liters to cm^3 and one to convert centimeters to meters:

$$\frac{1000 \text{ cm}^3}{1 \text{ L}} \quad \text{and} \quad \frac{1 \times 10^{-2} \text{ m}}{1 \text{ cm}}$$

Because the second conversion deals with length (cm and m) and we want volume here, it must therefore be cubed to give

$$\frac{1 \times 10^{-2} \text{ m}}{1 \text{ cm}} \times \frac{1 \times 10^{-2} \text{ m}}{1 \text{ cm}} \times \frac{1 \times 10^{-2} \text{ m}}{1 \text{ cm}} = \left(\frac{1 \times 10^{-2} \text{ m}}{1 \text{ cm}}\right)^3$$

This means that 1 cm^3 = 1 × 10^{-6} m^3. Now we can write

$$? \text{ m}^3 = 275 \cancel{\text{L}} \times \frac{1000 \text{ cm}^3}{1 \cancel{\text{L}}} \times \left(\frac{1 \times 10^{-2} \text{ m}}{1 \text{ cm}}\right)^3 = 0.275 \text{ m}^3$$

Check From the preceding conversion factors you can show that 1 L = 1 × 10^{-3} m^3. Therefore, a 275-L storage tank would be equal to 275 × 10^{-3} m^3 or 0.275 m^3, which is the answer.

A cryogenic storage tank for liquid helium.

Similar problem: 1.38(g).

Practice Exercise The volume of a room is 1.08 × 10^8 dm^3. What is the volume in m^3?

EXAMPLE 1.7

Liquid nitrogen is obtained from liquefied air and is used to prepare frozen goods and in low-temperature research. The density of the liquid at its boiling point (−196°C or 77 K) is 0.808 g/cm^3. Convert the density to units of kg/m^3.

(Continued)

Liquid nitrogen.

Strategy The problem can be stated as

$$? \text{ kg/m}^3 = 0.808 \text{ g/cm}^3$$

Two separate conversions are required for this problem: g $\longrightarrow$ kg and $cm^3 \longrightarrow m^3$. Recall that 1 kg = 1000 g and 1 cm = 1×10^{-2} m.

Solution In Example 1.6 we saw that 1 cm^3 = 1×10^{-6} m^3. The conversion factors are

$$\frac{1 \text{ kg}}{1000 \text{ g}} \quad \text{and} \quad \frac{1 \text{ cm}^3}{1 \times 10^{-6} \text{ m}^3}$$

Finally,

$$? \text{ kg/m}^3 = \frac{0.808 \text{ g}}{1 \text{ cm}^3} \times \frac{1 \text{ kg}}{1000 \text{ g}} \times \frac{1 \text{ cm}^3}{1 \times 10^{-6} \text{ m}^3} = \boxed{808 \text{ kg/m}^3}$$

Check Because 1 m^3 = 1×10^6 cm^3, we would expect much more mass in 1 m^3 than in 1 cm^3. Therefore, the answer is reasonable.

Similar problem: 1.39.

Practice Exercise The density of the lightest metal, lithium (Li), is 5.34×10^2 kg/m^3. Convert the density to g/cm^3.

Review of Concepts

The Food and Drug Administration recommends no more than 65 g of daily intake of fat. What is this mass in pounds? (1 lb = 453.6 g.)

Key Equations

$d = \frac{m}{V}$ (1.1)	Equation for density
$? ^\circ\text{C} = (^\circ\text{F} - 32^\circ\text{F}) \times \frac{5^\circ\text{C}}{9^\circ\text{F}}$ (1.2)	Converting °F to °C
$? ^\circ\text{F} = \frac{9^\circ\text{F}}{5^\circ\text{C}} \times (^\circ\text{C}) + 32^\circ\text{F}$ (1.3)	Converting °C to °F
$? \text{K} = (^\circ\text{C} + 273.15^\circ\text{C})\frac{1 \text{ K}}{1^\circ\text{C}}$ (1.4)	Converting °C to K

Summary of Facts and Concepts

1. The scientific method is a systematic approach to research that begins with the gathering of information through observation and measurements. In the process, hypotheses, laws, and theories are devised and tested.
2. Chemists study matter and the substances of which it is composed. All substances, in principle, can exist in three states: solid, liquid, and gas. The interconversion between these states can be effected by a change in temperature.
3. The simplest substances in chemistry are elements. Compounds are formed by the combination of atoms of different elements. Substances have both unique physical properties that can be observed without changing the identity of the substances and unique chemical properties that, when they are demonstrated, do change the identity of the substances.
4. SI units are used to express physical quantities in all sciences, including chemistry. Numbers expressed in

scientific notation have the form $N \times 10^n$, where N is between 1 and 10 and n is a positive or negative integer. Scientific notation helps us handle very large and very small quantities. Most measured quantities are inexact to some extent. The number of significant figures indicates the exactness of the measurement.

5. In the dimensional analysis method of solving problems the units are multiplied together, divided into each other, or canceled like algebraic quantities. Obtaining the correct units for the final answer ensures that the calculation has been carried out properly.

Key Words

Questions and Problems

Basic Definitions

Review Questions

1.1 Define these terms: (a) matter, (b) mass, (c) weight, (d) substance, (e) mixture.

1.2 Which of these statements is scientifically correct?
"The mass of the student is 56 kg."
"The weight of the student is 56 kg."

1.3 Give an example of a homogeneous mixture and an example of a heterogeneous mixture.

1.4 What is the difference between a physical property and a chemical property?

1.5 Give an example of an intensive property and an example of an extensive property.

1.6 Define these terms: (a) element, (b) compound.

Problems

1.7 Do these statements describe chemical or physical properties? (a) Oxygen gas supports combustion. (b) Fertilizers help to increase agricultural production. (c) Water boils below 100°C on top of a mountain. (d) Lead is denser than aluminum. (e) Uranium is a radioactive element.

1.8 Does each of these describe a physical change or a chemical change? (a) The helium gas inside a balloon tends to leak out after a few hours. (b) A flashlight beam slowly gets dimmer and finally goes out. (c) Frozen orange juice is reconstituted by adding water to it. (d) The growth of plants depends on the sun's energy in a process called photosynthesis. (e) A spoonful of table salt dissolves in a bowl of soup.

1.9 Which of these properties are intensive and which are extensive? (a) length, (b) volume, (c) temperature, (d) mass.

1.10 Which of these properties are intensive and which are extensive? (a) area, (b) color, (c) density.

1.11 Classify each of these substances as an element or a compound: (a) hydrogen, (b) water, (c) gold, (d) sugar.

1.12 Classify each of these as an element or a compound: (a) sodium chloride (table salt), (b) helium, (c) alcohol, (d) platinum.

Units

Review Questions

1.13 Give the SI units for expressing these: (a) length, (b) area, (c) volume, (d) mass, (e) time, (f) force, (g) energy, (h) temperature.

1.14 Write the numbers for these prefixes: (a) mega-, (b) kilo-, (c) deci-, (d) centi-, (e) milli-, (f) micro-, (g) nano-, (h) pico-.

1.15 Define density. What units do chemists normally use for density? Is density an intensive or extensive property?

1.16 Write the equations for converting degrees Celsius to degrees Fahrenheit and degrees Fahrenheit to degrees Celsius.

Problems

1.17 A lead sphere has a mass of 1.20×10^4 g, and its volume is 1.05×10^3 cm^3. Calculate the density of lead.

1.18 Mercury is the only metal that is a liquid at room temperature. Its density is 13.6 g/mL. How many grams of mercury will occupy a volume of 95.8 mL?

1.19 (a) Normally the human body can endure a temperature of 105°F for only short periods of time without permanent damage to the brain and other vital organs. What is this temperature in degrees Celsius? (b) Ethylene glycol is a liquid organic compound that is used as an antifreeze in car radiators. It freezes at −11.5°C. Calculate its freezing temperature in degrees Fahrenheit. (c) The temperature on the surface of the sun is about 6300°C. What is this temperature in degrees Fahrenheit? (d) The ignition temperature of paper is 451°F. What is the temperature in degrees Celsius?

1.20 (a) Convert the following temperatures to kelvin: (i) 113°C, the melting point of sulfur, (ii) 37°C, the normal body temperature, (iii) 357°C, the boiling point of mercury. (b) Convert the following temperatures to degrees Celsius: (i) 77 K, the boiling point of liquid nitrogen, (ii) 4.2 K, the boiling point of liquid helium, (iii) 601 K, the melting point of lead.

Scientific Notation

Problems

1.21 Express these numbers in scientific notation: (a) 0.000000027, (b) 356, (c) 0.096.

1.22 Express these numbers in scientific notation: (a) 0.749, (b) 802.6, (c) 0.000000621.

1.23 Convert these to nonscientific notation: (a) 1.52×10^4, (b) 7.78×10^{-8}.

1.24 Convert these to nonscientific notation: (a) 3.256×10^{-5}, (b) 6.03×10^6.

1.25 Express the answers to these in scientific notation:

(a) $145.75 + (2.3 \times 10^{-1})$

(b) $79{,}500 \div (2.5 \times 10^2)$

(c) $(7.0 \times 10^{-3}) - (8.0 \times 10^{-4})$

(d) $(1.0 \times 10^4) \times (9.9 \times 10^6)$

1.26 Express the answers to these in scientific notation:

(a) $0.0095 + (8.5 \times 10^{-3})$

(b) $653 \div (5.75 \times 10^{-8})$

(c) $850{,}000 - (9.0 \times 10^5)$

(d) $(3.6 \times 10^{-4}) \times (3.6 \times 10^6)$

Significant Figures

Problems

1.27 What is the number of significant figures in each of these measured quantities? (a) 4867 miles, (b) 56 mL, (c) 60,104 tons, (d) 2900 g.

1.28 What is the number of significant figures in each of these measured quantities? (a) 40.2 g/cm^3, (b) 0.0000003 cm, (c) 70 min, (d) 4.6×10^{19} atoms.

1.29 Carry out these operations as if they were calculations of experimental results, and express each answer in the correct units and with the correct number of significant figures:

(a) 5.6792 m + 0.6 m + 4.33 m

(b) 3.70 g − 2.9133 g

(c) 4.51 cm × 3.6666 cm

(d) $(3 \times 10^4 \text{ g} + 6.827 \text{ g})/(0.043 \text{ cm}^3 - 0.021 \text{ cm}^3)$

1.30 Carry out these operations as if they were calculations of experimental results, and express each answer in the correct units and with the correct number of significant figures:

(a) 7.310 km ÷ 5.70 km

(b) $(3.26 \times 10^{-3} \text{ mg}) - (7.88 \times 10^{-5} \text{ mg})$

(c) $(4.02 \times 10^6 \text{ dm}) + (7.74 \times 10^7 \text{ dm})$

(d) (7.8 m − 0.34 m)/(1.15 s + 0.82 s)

Dimensional Analysis

Problems

1.31 Carry out these conversions: (a) 22.6 m to decimeters, (b) 25.4 mg to kilograms.

1.32 Carry out these conversions: (a) 242 lb to milligrams, (b) 68.3 cm^3 to cubic meters.

1.33 The price of gold on a certain day in 2009 was $932 per troy ounce. How much did 1.00 g of gold cost that day? (1 troy ounce = 31.03 g.)

1.34 Three students (A, B, and C) are asked to determine the volume of a sample of methanol. Each student measures the volume three times with a graduated cylinder. The results in milliliters are A (47.2, 48.2, 47.6); B (46.9, 47.1, 47.2); C (47.8, 47.8, 47.9). The true volume of methanol is 47.0 mL. Which student is the most accurate? Which student is the most precise?

1.35 Three students (X, Y, and Z) are assigned the task of determining the mass of a sample of iron. Each student makes three determinations with a balance. The results in grams are X (61.5, 61.6, 61.4); Y (62.8, 62.2, 62.7); Z (61.9, 62.2, 62.1). The actual mass of the iron is 62.0 g. Which student is the least precise? Which student is the most accurate?

1.36 A slow jogger runs a mile in 13 min. Calculate the speed in (a) in/s, (b) m/min, (c) km/h. (1 mi = 1609 m; 1 in = 2.54 cm.)

1.37 Carry out these conversions: (a) A 6.0-ft person weighs 168 lb. Express this person's height in meters and weight in kilograms. (1 lb = 453.6 g; 1 m = 3.28 ft.) (b) The current speed limit in some states in the United States is 55 miles per hour. What is the speed limit in kilometers per hour? (c) The speed of light is 3.0×10^{10} cm/s. How many miles does light travel in 1 hour? (d) Lead is a toxic substance. The "normal" lead content in human blood is about 0.40 part per million (that is, 0.40 g of lead per million grams of blood). A value of 0.80 part per million (ppm) is considered to be dangerous. How many grams of lead are contained in 6.0×10^3 g of blood (the amount in an average adult) if the lead content is 0.62 ppm?

1.38 Carry out these conversions: (a) 1.42 light-years to miles (a light-year is an astronomical measure of distance—the distance traveled by light in a year, or 365 days), (b) 32.4 yd to centimeters, (c) 3.0×10^{10} cm/s to ft/s, (d) 47.4°F to degrees Celsius, (e) −273.15°C (the lowest temperature) to degrees Fahrenheit, (f) 71.2 cm^3 to m^3, (g) 7.2 m^3 to liters.

1.39 Aluminum is a lightweight metal (density = 2.70 g/cm^3) used in aircraft construction, high-voltage transmission lines, and foils. What is its density in kg/m^3?

1.40 The density of ammonia gas under certain conditions is 0.625 g/L. Calculate its density in g/cm^3.

Additional Problems

1.41 Which of these describe physical and which describe chemical properties? (a) Iron has a tendency to rust. (b) Rainwater in industrialized regions tends to be acidic. (c) Hemoglobin molecules have a red color. (d) When a glass of water is left out in the sun, the water gradually disappears. (e) Carbon dioxide in air is converted to more complex molecules by plants during photosynthesis.

1.42 In 2004 about 87.0 billion pounds of sulfuric acid were produced in the United States. Convert this quantity to tons.

1.43 Suppose that a new temperature scale has been devised on which the melting point of ethanol (−117.3°C) and the boiling point of ethanol (78.3°C) are taken as 0°S and 100°S, respectively, where S is the symbol for the new temperature scale. Derive an equation relating a reading on this scale to a reading on the Celsius scale. What would this thermometer read at 25°C?

1.44 In the determination of the density of a rectangular metal bar, a student made the following measurements: length, 8.53 cm; width, 2.4 cm; height, 1.0 cm; mass, 52.7064 g. Calculate the density of the metal to the correct number of significant figures.

1.45 Calculate the mass of each of these: (a) a sphere of gold of radius 10.0 cm [the volume of a sphere of radius r is $V = (\frac{4}{3})\pi r^3$; the density of gold = 19.3 g/cm^3], (b) a cube of platinum of edge length 0.040 mm (the density of platinum = 21.4 g/cm^3), (c) 50.0 mL of ethanol (the density of ethanol = 0.798 g/mL).

1.46 A cylindrical glass tube 12.7 cm in length is filled with mercury. The mass of mercury needed to fill the tube is found to be 105.5 g. Calculate the inner diameter of the tube. (The density of mercury = 13.6 g/mL.)

1.47 This procedure was carried out to determine the volume of a flask. The flask was weighed dry and then filled with water. If the masses of the empty flask and the filled flask were 56.12 g and 87.39 g, respectively, and the density of water is 0.9976 g/cm^3, calculate the volume of the flask in cubic centimeters.

1.48 A silver (Ag) object weighing 194.3 g is placed in a graduated cylinder containing 242.0 mL of water. The volume of water now reads 260.5 mL. From these data calculate the density of silver.

1.49 The experiment described in Problem 1.48 is a crude but convenient way to determine the density of some solids. Describe a similar experiment that would enable you to measure the density of ice. Specifically, what would be the requirements for the liquid used in your experiment?

1.50 The speed of sound in air at room temperature is about 343 m/s. Calculate this speed in miles per hour (mph).

1.51 The medicinal thermometer commonly used in homes can be read to ±0.1°F, whereas those in the doctor's office may be accurate to ±0.1°C. In degrees Celsius, express the percent error expected from each of these thermometers in measuring a person's body temperature of 38.9°C.

1.52 A thermometer gives a reading of 24.2°C ± 0.1°C. Calculate the temperature in degrees Fahrenheit. What is the uncertainty?

1.53 Vanillin (used to flavor vanilla ice cream and other foods) is the substance whose aroma the human nose detects in the smallest amount. The threshold limit is 2.0×10^{-11} g per liter of air. If the current price of 50 g of vanillin is $112, determine the cost to supply enough vanillin so that the aroma could be detectable in a large aircraft hangar of volume 5.0×10^7 ft^3.

1.54 A resting adult requires about 240 mL of pure oxygen/min and breathes about 12 times every minute. If inhaled air contains 20 percent oxygen by volume and exhaled air 16 percent, what is the volume of air per breath? (Assume that the volume of inhaled air is equal to that of exhaled air.)

1.55 The total volume of seawater is 1.5×10^{21} L. Assume that seawater contains 3.1 percent sodium chloride by mass and that its density is 1.03 g/mL. Calculate the total mass of sodium chloride in kilograms and in tons. (1 ton = 2000 lb; 1 lb = 453.6 g.)

1.56 Magnesium (Mg) is a valuable metal used in alloys, in batteries, and in chemical synthesis. It is obtained mostly from seawater, which contains about 1.3 g of Mg for every kilogram of seawater. Calculate the volume of seawater (in liters) needed to extract 8.0×10^4 tons of Mg, which is roughly the annual production in the United States. (Density of seawater = 1.03 g/mL.)

1.57 A student is given a crucible and asked to prove whether it is made of pure platinum. She first weighs the crucible in air and then weighs it suspended in water (density = 0.9986 g/cm^3). The readings are 860.2 g and 820.2 g, respectively. Given that the density of platinum is 21.45 g/cm^3, what should her conclusion be based on these measurements? (*Hint:* An object suspended in a fluid is buoyed up by the mass of the fluid displaced by the object. Neglect the buoyancy of air.)

1.58 At what temperature does the numerical reading on a Celsius thermometer equal that on a Fahrenheit thermometer?

1.59 The surface area and average depth of the Pacific Ocean are 1.8×10^8 km^2 and 3.9×10^3 m, respectively. Calculate the volume of water in the ocean in liters.

1.60 Percent error is often expressed as the absolute value of the difference between the true value and the experimental value, divided by the true value:

$$\text{Percent error} = \frac{|\text{true value} - \text{experimental value}|}{|\text{true value}|} \times 100\%$$

where the vertical lines indicate absolute value. Calculate the percent error for these measurements: (a) The density of alcohol (ethanol) is found to be 0.802 g/mL. (True value: 0.798 g/mL.) (b) The mass of gold in an earring is analyzed to be 0.837 g. (True value: 0.864 g.)

1.61 Osmium (Os) is the densest element known (density = 22.57 g/cm^3). Calculate the mass in pounds and kilograms of an Os sphere 15 cm in diameter (about the size of a grapefruit). See Problem 1.45 for volume of a sphere.

1.62 A 1.0-mL volume of seawater contains about 4.0×10^{-12} g of gold. The total volume of ocean water is 1.5×10^{21} L. Calculate the total amount of gold in grams that is present in seawater and its worth in dollars, assuming that the price of gold is $930 an ounce. With so much gold out there, why hasn't someone become rich by mining gold from the ocean?

1.63 The thin outer layer of Earth, called the crust, contains only 0.50 percent of Earth's total mass and yet is the source of almost all the elements (the atmosphere provides elements such as oxygen, nitrogen, and a few other gases). Silicon (Si) is the second most abundant element in Earth's crust (27.2 percent by mass). Calculate the mass of silicon in kilograms in Earth's crust. (The mass of Earth is 5.9×10^{21} tons. 1 ton = 2000 lb; 1 lb = 453.6 g.)

1.64 The diameter of a copper (Cu) atom is roughly 1.3×10^{-10} m. How many times can you divide evenly a piece of 10-cm copper wire until it is reduced to two separate copper atoms? (Assume there are appropriate tools for this procedure and that copper atoms are lined up in a straight line, in contact with each other.) Round off your answer to an integer.

1.65 One gallon of gasoline burned in an automobile's engine produces on the average 9.5 kg of carbon dioxide, which is a greenhouse gas, that is, it promotes the warming of Earth's atmosphere. Calculate the annual production of carbon dioxide in kilograms if there are 40 million cars in the United States, and each car covers a distance of 5000 mi at a consumption rate of 20 mi per gallon.

1.66 A sheet of aluminum (Al) foil has a total area of 1.000 ft^2 and a mass of 3.636 g. What is the thickness of the foil in millimeters? (Density of Al = 2.699 g/cm^3.)

1.67 Chlorine is used to disinfect swimming pools. The accepted concentration for this purpose is 1 ppm chlorine or 1 g of chlorine per million g of water. Calculate the volume of a chlorine solution (in milliliters) a homeowner should add to her swimming pool if the solution contains 6.0 percent chlorine by mass and there are 2×10^4 gallons of water in the pool. (1 gallon = 3.79 L; density of liquids = 1.0 g/mL.)

1.68 Fluoridation is the process of adding fluorine compounds to drinking water to help fight tooth decay. A concentration of 1 ppm of fluorine is sufficient for the purpose. (1 ppm means 1 g of fluorine per 1 million g of water.) The compound normally chosen for fluoridation is sodium fluoride, which is also added to some toothpastes. Calculate the quantity of sodium fluoride in kilograms needed per year for a city of 50,000 people if the daily consumption of water per person is 150 gallons. What percent of the sodium fluoride is "wasted" if each person uses only 6.0 L of water a day for drinking and cooking? (Sodium fluoride is 45.0 percent fluorine by mass. 1 gallon = 3.79 L; 1 year = 365 days; density of water = 1.0 g/mL.)

1.69 In water conservation, chemists spread a thin film of certain inert material over the surface of water to cut down the rate of evaporation of water in reservoirs. This technique was pioneered by Benjamin Franklin three centuries ago. Franklin found that 0.10 mL of oil could spread over the surface of water of about 40 m^2 in area. Assuming that the oil forms a *monolayer,* that is, a layer that is only one molecule thick, estimate the length of each oil molecule in nanometers. (1 nm = 1×10^{-9} m.)

1.70 Pheromones are compounds secreted by females of many insect species to attract mates. Typically, 1.0×10^{-8} g of a pheromone is sufficient to reach all targeted males within a radius of 0.50 mi. Calculate the density of the pheromone (in grams per liter) in a cylindrical air space having a radius of 0.50 mi and a height of 40 ft.

1.71 Three different 25.0 g samples of solid pellets are added to 20.0 mL of water in three different cylinders. The results are illustrated here. Given the densities of the three materials used, identify each sample of solid pellets: solid A (2.9 g/cm^3), solid B (8.3 g/cm^3), and solid C (3.3 g/cm^3).

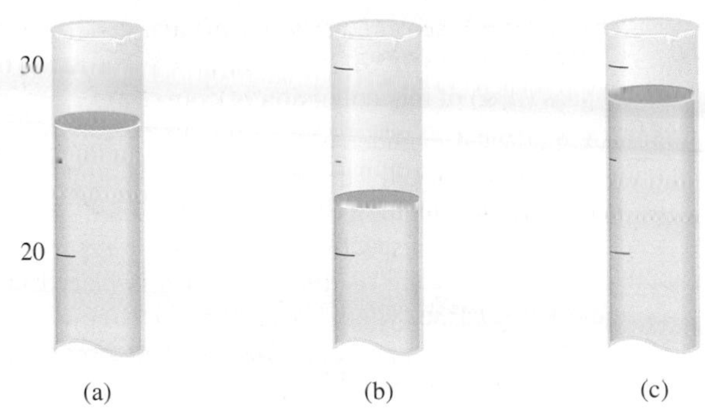

Special Problems

1.72 Dinosaurs dominated life on Earth for millions of years and then disappeared very suddenly. In the experimentation and data-collecting stage, paleontologists studied fossils and skeletons found in rocks in various layers of Earth's crust. Their findings enabled them to map out which species existed on Earth during specific geologic periods. They also revealed no dinosaur skeletons in rocks formed immediately after the Cretaceous period, which dates back some 65 million years. It is therefore assumed that the dinosaurs became extinct about 65 million years ago.

Among the many hypotheses put forward to account for their disappearance were disruptions of the food chain and a dramatic change in climate caused by violent volcanic eruptions. However, there was no convincing evidence for any one hypothesis until 1977. It was then that a group of paleontologists working in Italy obtained some very puzzling data at a site near Gubbio. The chemical analysis of a layer of clay deposited above sediments formed during the Cretaceous period (and therefore a layer that records events occurring *after* the Cretaceous period) showed a surprisingly high content of the element iridium. Iridium is very rare in Earth's crust but is comparatively abundant in asteroids.

This investigation led to the hypothesis that the extinction of dinosaurs occurred as follows. To account for the quantity of iridium found, scientists suggested that a large asteroid several miles in diameter hit Earth about the time the dinosaurs disappeared. The impact of the asteroid on Earth's surface must have been so tremendous that it literally vaporized a large quantity of surrounding rocks, soils, and other objects. The resulting dust and debris floated through the air and blocked the sunlight for months or perhaps years. Without ample sunlight most plants could not grow, and the fossil record confirms that many types of plants did indeed die out at this time. Consequently, of course, many plant-eating animals gradually perished, and then, in turn, meat-eating animals began to starve. Limitation of food sources obviously affects large animals needing great amounts of food more quickly and more severely than small animals. Therefore, the huge dinosaurs vanished because of lack of food.

(a) How does the study of dinosaur extinction illustrate the scientific method?

(b) Suggest two ways to test the hypothesis.

(c) In your opinion, is it justifiable to refer to the asteroid explanation as the theory of dinosaur extinction?

(d) Available evidence suggests that about 20 percent of the asteroid's mass turned to dust and spread uniformly over Earth after eventually settling out of the upper atmosphere. This dust amounted to about 0.02 g/cm^2 of Earth's surface. The asteroid very likely had a density of about 2 g/cm^3. Calculate the mass (in kilograms and tons) of the asteroid and its radius in meters, assuming that it was a sphere. (The area of Earth is 5.1×10^{14} m^2; 1 lb = 453.6 g.) (Source: *Consider a Spherical Cow—A Course in Environmental Problem Solving* by J. Harte, University Science Books, Mill Valley, CA, 1988. Used with permission.)

1.73 You are given a liquid. Briefly describe steps you would take to show whether it is a pure substance or a homogeneous mixture.

1.74 A bank teller is asked to assemble "one-dollar" sets of coins for his clients. Each set is made of three quarters, one nickel, and two dimes. The masses of the coins are: quarter: 5.645 g; nickel: 4.967 g; dime: 2.316 g. What is the maximum number of sets that can be assembled from 33.871 kg of quarters, 10.432 kg of nickels, and 7.990 kg of dimes? What is the total mass (in g) of this collection of coins?

1.75 A graduated cylinder is filled to the 40.00-mL mark with a mineral oil. The masses of the cylinder before and after the addition of the mineral oil are 124.966 g and 159.446 g, respectively. In a separate experiment, a metal ball bearing of mass 18.713 g is placed in the cylinder and the cylinder is again filled to the 40.00-mL mark with the mineral oil. The combined mass of the ball bearing and mineral oil is 50.952 g. Calculate the density and radius of the ball bearing. [The volume of a sphere of radius r is $(4/3)\pi r^3$.]

1.76 Bronze is an alloy made of copper (Cu) and tin (Sn). Calculate the mass of a bronze cylinder of radius 6.44 cm and length 44.37 cm. The composition of the bronze is 79.42 percent Cu and 20.58 percent Sn and the densities of Cu and Sn are 8.94 g/cm^3 and 7.31 g/cm^3, respectively. What assumption should you make in this calculation?

1.77 A chemist in the nineteenth century prepared an unknown substance. In general, do you think it would be more difficult to prove that it is an element or a compound? Explain.

1.78 Tums is a popular remedy for acid indigestion. A typical Tums tablet contains calcium carbonate plus some inert substances. When ingested, it reacts with the gastric juice (hydrochloric acid) in the stomach to give off carbon dioxide gas. When a 1.328-g tablet reacted with 40.00 mL of hydrochloric acid (density: 1.140 g/mL), carbon dioxide gas was given off and the resulting solution weighed 46.699 g. Calculate the number of liters of carbon dioxide gas released if its density is 1.81 g/L.

1.79 A 250-mL glass bottle was filled with 242 mL of water at 20°C and tightly capped. It was then left outdoors overnight, where the average temperature was −5°C. Predict what would happen. The density of water at 20°C is 0.998 g/cm^3 and that of ice at −5°C is 0.916 g/cm^3.

Answers to Practice Exercises

1.1 96.5 g. **1.2** (a) 621.5°F, (b) 78.3°C, (c) −196°C. **1.3** (a) Two, (b) four, (c) three, (d) two. (e) three or two. **1.4** (a) 26.76 L, (b) 4.4 g, (c) 1.6×10^7 dm^2, (d) 0.0756 g/mL, (e) 6.69×10^4 m. **1.5** 2.36 lb. **1.6** 1.08×10^5 m^3. **1.7** 0.534 g/cm^3.

CHAPTER

2

Atoms, Molecules, and Ions

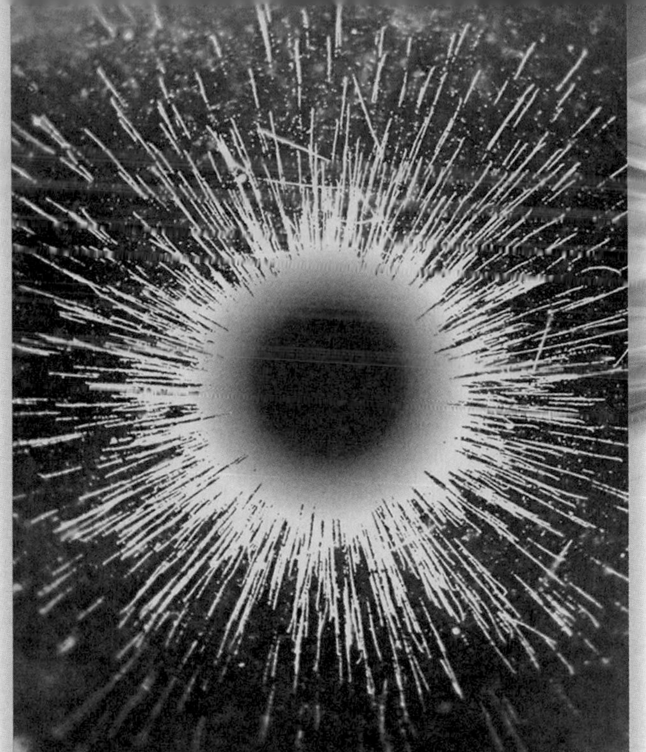

Colored images of the radioactive emission of radium (Ra). Study of radioactivity helped to advance scientists' knowledge about atomic structure.

Chapter Outline

Student Interactive Activities

Animations
Cathode Ray Tube (2.2)
Millikan Oil Drop (2.2)
Alpha, Beta, and Gamma Rays (2.2)
α-Particle Scattering (2.2)

Electronic Homework
Example Practice Problems
End of Chapter Problems

Essential Concepts

Development of the Atomic Theory The search for the fundamental units of matter began in ancient times. The modern version of atomic theory was laid out by John Dalton, who postulated that elements are made of extremely small particles, called atoms, and that all atoms of a given element are identical, but they are different from atoms of all other elements.

The Structure of the Atom An atom is composed of three elementary particles: proton, electron, and neutron. The proton has a positive charge, the electron has a negative charge, and the neutron has no charge. Protons and neutrons are located in a small region at the center of the atom, called the nucleus, and electrons are spread out about the nucleus at some distance from it.

Ways to Identify Atoms Atomic number is the number of protons in a nucleus; atoms of different elements have different atomic numbers. Isotopes are atoms of the same element having different numbers of neutrons. Mass number is the sum of the number of protons and neutrons in an atom.

The Periodic Table Elements can be grouped together according to their chemical and physical properties in a chart called the periodic table. The periodic table enables us to classify elements (as metals, metalloids, and nonmetals) and correlate their properties in a systematic way.

From Atoms to Molecules and Ions Atoms of most elements interact to form compounds, which are classified as molecules or ionic compounds made of positive (cations) and negative (anions) ions. Chemical formulas tell us the type and number of atoms present in a molecule or compound.

Naming Compounds The names of many inorganic compounds can be deduced from a set of simple rules.

Organic Compounds The simplest type of organic compounds is the hydrocarbons.

2.1 The Atomic Theory

In the fifth century B.C., the Greek philosopher Democritus expressed the belief that all matter consists of very small, indivisible particles, which he named *atomos* (meaning uncuttable or indivisible). Although Democritus' idea was not accepted by many of his contemporaries (notably Plato and Aristotle), somehow it endured. Experimental evidence from early scientific investigations provided support for the notion of "atomism" and gradually gave rise to the modern definitions of elements and compounds. It was in 1808 that an English scientist and schoolteacher, John Dalton, formulated a precise definition of the indivisible building blocks of matter that we call atoms.

Dalton's work marked the beginning of the modern era of chemistry. The hypotheses about the nature of matter on which Dalton's atomic theory is based can be summarized as

1. Elements are composed of extremely small particles, called atoms.
2. All atoms of a given element are identical, having the same size, mass, and chemical properties. The atoms of one element are different from the atoms of all other elements.
3. Compounds are composed of atoms of more than one element. In any compound, the ratio of the numbers of atoms of any two of the elements present is either an integer or a simple fraction.
4. A chemical reaction involves only the separation, combination, or rearrangement of atoms; it does not result in their creation or destruction.

Figure 2.1 is a schematic representation of hypotheses 2 and 3.

Dalton's concept of an atom was far more detailed and specific than Democritus'. The second hypothesis states that atoms of one element are different from atoms of all other elements. Dalton made no attempt to describe the structure or composition of atoms—he had no idea what an atom is really like. But he did realize that the different properties shown by elements such as hydrogen and oxygen can be explained by assuming that hydrogen atoms are not the same as oxygen atoms.

The third hypothesis suggests that, to form a certain compound, we need not only atoms of the right kinds of elements, but the specific numbers of these atoms as well. This idea is an extension of a law published in 1799 by Joseph Proust, a French chemist. Proust's ***law of definite proportions*** states that *different samples of the same compound always contain its constituent elements in the same proportion by mass.* Thus, if we were to analyze samples of carbon dioxide gas obtained from different sources, we would find in each sample the same ratio by mass of carbon to oxygen. It stands to reason, then, that if the ratio of the masses of different elements in a given

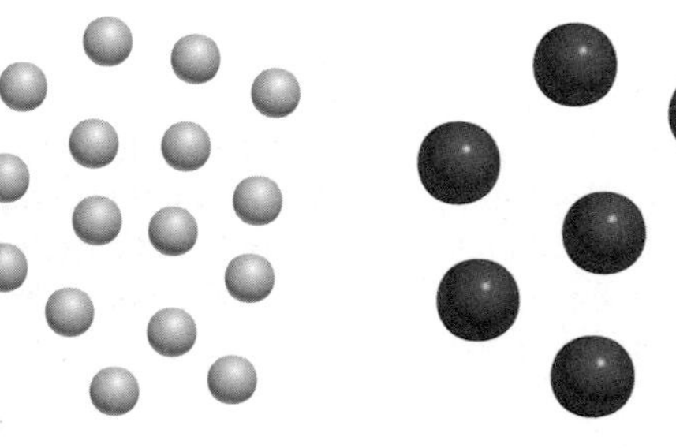

Atoms of element X — Atoms of element Y

(a)

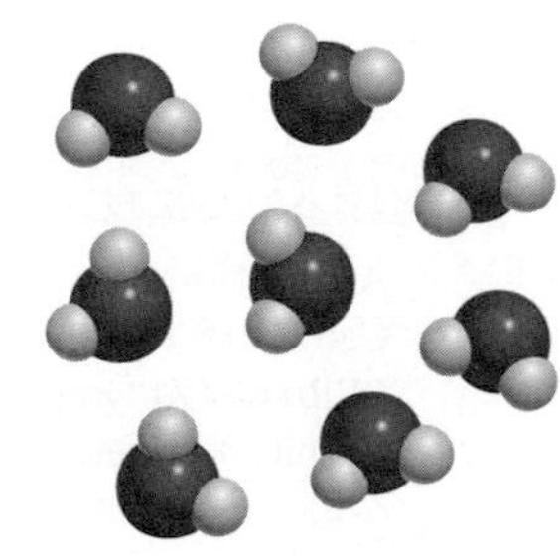

Compound of elements X and Y

(b)

Figure 2.1
(a) According to Dalton's atomic theory, atoms of the same element are identical, but atoms of one element are different from atoms of other elements. (b) Compounds formed from atoms of elements X and Y. In this case, the ratio of the atoms of element X to the atoms of element Y is 2:1.

compound is fixed, the ratio of the atoms of these elements in the compound also must be constant.

Dalton's third hypothesis also supports another important law, the ***law of multiple proportions.*** According to this law, *if two elements can combine to form more than one compound, the masses of one element that combine with a fixed mass of the other element are in ratios of small whole numbers.* Dalton's theory explains the law of multiple proportions quite simply: The compounds differ in the number of atoms of each kind that combine. For example, carbon forms two stable compounds with oxygen, namely, carbon monoxide and carbon dioxide. Modern measurement techniques indicate that one atom of carbon combines with one atom of oxygen in carbon monoxide and that one atom of carbon combines with two oxygen atoms in carbon dioxide. Thus, the ratio of oxygen in carbon monoxide to oxygen in carbon dioxide is 1:2. This result is consistent with the law of multiple proportions because the mass of an element in a compound is proportional to the number of atoms of the element present (Figure 2.2).

Dalton's fourth hypothesis is another way of stating the ***law of conservation of mass,*** which is that *matter can be neither created nor destroyed.*† Because matter is made of atoms that are unchanged in a chemical reaction, it follows that mass must be conserved as well. Dalton's brilliant insight into the nature of matter was the main stimulus for the rapid progress of chemistry during the nineteenth century.

Carbon monoxide

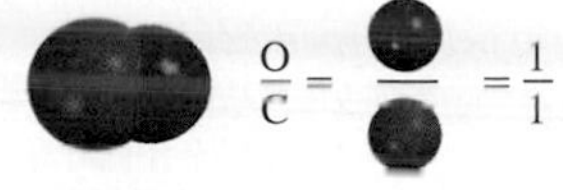

Carbon dioxide

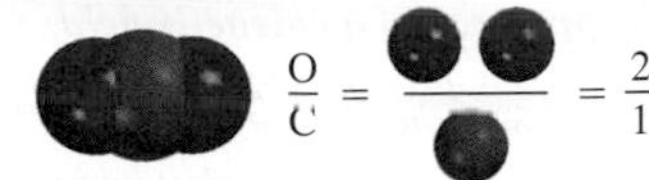

Ratio of oxygen in carbon monoxide to oxygen in carbon dioxide: 1:2

Figure 2.2
An illustration of the law of multiple proportions.

Review of Concepts

The atoms of elements A (blue) and B (yellow) form two compounds shown here. Do these compounds obey the law of multiple proportions?

2.2 The Structure of the Atom

On the basis of Dalton's atomic theory, we can define an ***atom*** as *the basic unit of an element that can enter into chemical combination.* Dalton imagined an atom that was both extremely small and indivisible. However, a series of investigations that began in the 1850s and extended into the twentieth century clearly demonstrated that atoms actually possess internal structure; that is, they are made up of even smaller particles, which are called *subatomic particles.* This research led to the discovery of three such particles—electrons, protons, and neutrons.

The Electron

In the 1890s many scientists became caught up in the study of ***radiation,*** *the emission and transmission of energy through space in the form of waves.* Information gained from this research contributed greatly to our understanding of atomic structure. One device used to investigate this phenomenon was a cathode ray tube, the forerunner of the television tube (Figure 2.3). It is a glass tube from which most of the air has been evacuated. When the two metal plates are connected to a high-voltage source, the

†According to Albert Einstein, mass and energy are alternate aspects of a single entity called *mass-energy.* Chemical reactions usually involve a gain or loss of heat and other forms of energy. Thus, when energy is lost in a reaction, for example, mass is also lost. Except for nuclear reactions (see Chapter 21), however, changes of mass in chemical reactions are too small to detect. Therefore, for all practical purposes mass is conserved.

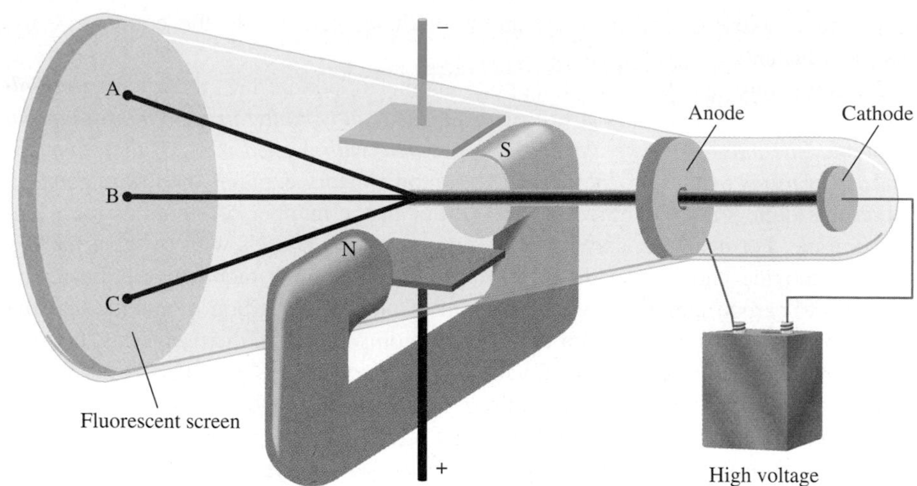

Figure 2.3
A cathode ray tube with an electric field perpendicular to the direction of the cathode rays and an external magnetic field. The symbols N and S denote the north and south poles of the magnet. The cathode rays will strike the end of the tube at A in the presence of a magnetic field, at C in the presence of an electric field, and at B when there are no external fields present or when the effects of the electric field and magnetic field cancel each other.

negatively charged plate, called the *cathode,* emits an invisible ray. The cathode ray is drawn to the positively charged plate, called the *anode,* where it passes through a hole and continues traveling to the other end of the tube. When the ray strikes the specially coated surface, it produces a strong fluorescence, or bright light.

In some experiments, two electrically charged plates and a magnet were added to the *outside* of the cathode ray tube (see Figure 2.3). When the magnetic field is on and the electric field is off, the cathode ray strikes point A. When only the electric field is on, the ray strikes point C. When both the magnetic and the electric fields are off or when they are both on but balanced so that they cancel each other's influence, the ray strikes point B. According to electromagnetic theory, a moving charged body behaves like a magnet and can interact with electric and magnetic fields through which it passes. Because the cathode ray is attracted by the plate bearing positive charges and repelled by the plate bearing negative charges, it must consist of negatively charged particles. We know these *negatively charged particles* as ***electrons.*** Figure 2.4 shows the effect of a bar magnet on the cathode ray.

Electrons are normally associated with atoms. However, they can be studied individually.

An English physicist, J. J. Thomson, used a cathode ray tube and his knowledge of electromagnetic theory to determine the ratio of electric charge to the mass of an individual electron. The number he came up with is -1.76×10^8 C/g, where C stands

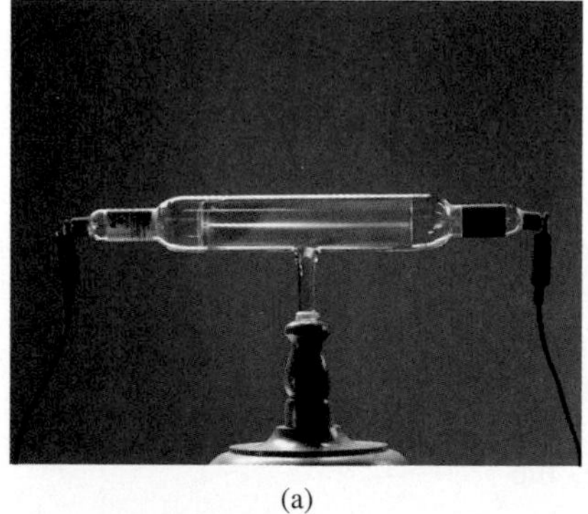
(a)

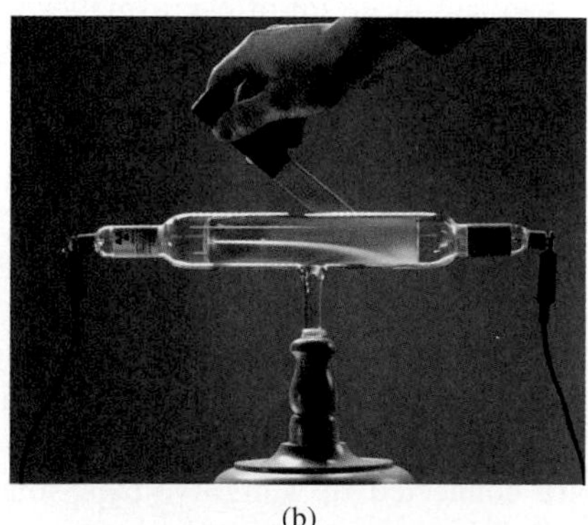
(b)

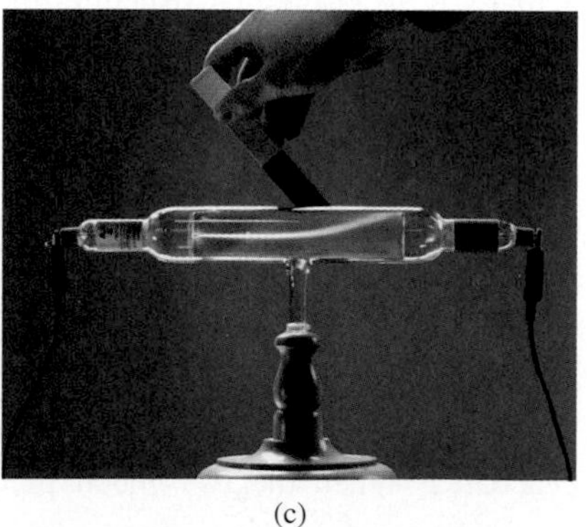
(c)

Figure 2.4
(a) A cathode ray produced in a discharge tube traveling from the cathode (left) to the anode (right). The ray itself is invisible, but the fluorescence of a zinc sulfide coating on the glass causes it to appear green. (b) The cathode ray is bent downward when a bar magnet is brought toward it. (c) When the polarity of the magnet is reversed, the ray bends in the opposite direction.

for *coulomb,* which is the unit of electric charge. Thereafter, in a series of experiments carried out between 1908 and 1917, R. A. Millikan, an American physicist, found the charge of an electron to be -1.6022×10^{-19} C. From these data he calculated the mass of an electron:

$$\begin{aligned}\text{mass of an electron} &= \frac{\text{charge}}{\text{charge/mass}} \\ &= \frac{-1.6022 \times 10^{-19}\ \text{C}}{-1.76 \times 10^{8}\ \text{C/g}} \\ &= 9.10 \times 10^{-28}\ \text{g}\end{aligned}$$

which is an exceedingly small mass.

Radioactivity

In 1895, the German physicist Wilhelm Röntgen noticed that cathode rays caused glass and metals to emit very unusual rays. This highly energetic radiation penetrated matter, darkened covered photographic plates, and caused a variety of substances to fluoresce. Because these rays could not be deflected by a magnet, they could not contain charged particles as cathode rays do. Röntgen called them X rays.

Not long after Röntgen's discovery, Antoine Becquerel, a professor of physics in Paris, began to study fluorescent properties of substances. Purely by accident, he found that exposing thickly wrapped photographic plates to a certain uranium compound caused them to darken, even without the stimulation of cathode rays. Like X rays, the rays from the uranium compound were highly energetic and could not be deflected by a magnet, but they differed from X rays because they were generated spontaneously. One of Becquerel's students, Marie Curie, suggested the name ***radioactivity*** to describe this *spontaneous emission of particles and/or radiation.* Consequently, any element that spontaneously emits radiation is said to be *radioactive.*

Further investigation revealed that three types of rays are produced by the *decay,* or breakdown, of radioactive substances such as uranium. Two of the three kinds are deflected by oppositely charged metal plates (Figure 2.5). ***Alpha*** **(α)** ***rays*** consist of *positively charged particles,* called **α *particles,*** and therefore are deflected by the positively charged plate. ***Beta*** **(β)** ***rays,*** or **β *particles,*** are *electrons* and are deflected

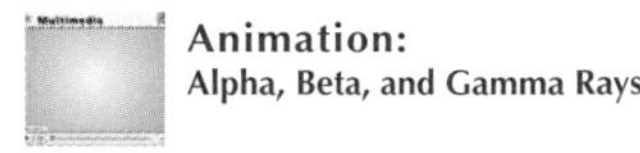

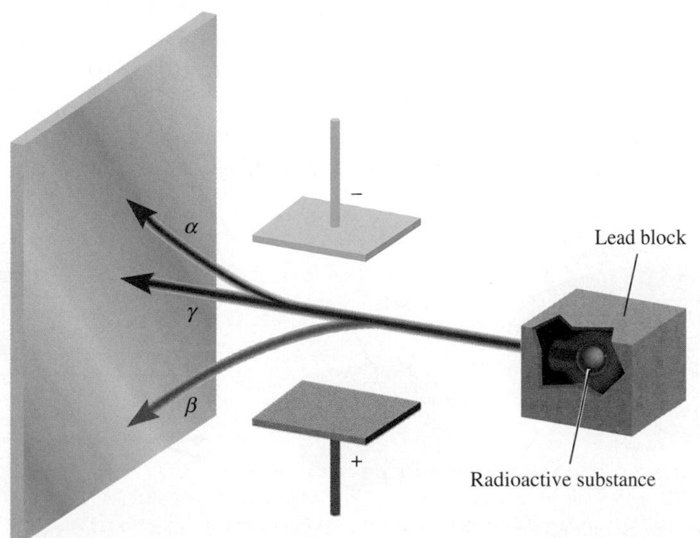

Figure 2.5
Three types of rays emitted by radioactive elements. β rays consist of negatively charged particles (electrons) and are therefore attracted by the positively charged plate. The opposite holds true for α rays—they are positively charged and are drawn to the negatively charged plate. Because γ rays have no charges, their path is unaffected by an external electric field.

Figure 2.6
Thomson's model of the atom, sometimes described as the "plum-pudding" model, after a traditional English dessert containing raisins. The electrons are embedded in a uniform, positively charged sphere.

Animation: α-Particle Scattering

by the negatively charged plate. The third type of radioactive radiation consists of *high–energy rays* called ***gamma (γ) rays.*** Like X rays, γ rays have no charge and are not affected by an external electric or magnetic field.

The Proton and the Nucleus

By the early 1900s, two features of atoms had become clear: They contain electrons, and they are electrically neutral. To maintain electrical neutrality, an atom must contain an equal number of positive and negative charges. On the basis of this information, Thomson proposed that an atom could be thought of as a uniform, positive sphere of matter in which electrons are embedded (Figure 2.6). Thomson's so-called "plum pudding" model was the accepted theory for a number of years.

In 1910 the New Zealand physicist Ernest Rutherford, who had earlier studied with Thomson at Cambridge University, decided to use α particles to probe the structure of atoms. Together with his associate Hans Geiger and an undergraduate named Ernest Marsden, Rutherford carried out a series of experiments using very thin foils of gold and other metals as targets for α particles from a radioactive source (Figure 2.7). They observed that the majority of particles penetrated the foil either undeflected or with only a slight deflection. They also noticed that every now and then an α particle was scattered (or deflected) at a large angle. In some instances, an α particle actually bounced back in the direction from which it had come! This was a most surprising finding, for in Thomson's model the positive charge of the atom was so diffuse (spread out) that the positive α particles were expected to pass through with very little deflection. To quote Rutherford's initial reaction when told of this discovery: "It was as incredible as if you had fired a 15-inch shell at a piece of tissue paper and it came back and hit you."

To explain the results of the α-scattering experiment, Rutherford devised a new model of atomic structure, suggesting that most of the atom must be empty space. This structure would allow most of the α particles to pass through the gold foil with little or no deflection. The atom's positive charges, Rutherford proposed, are all concentrated in the ***nucleus,*** *a dense central core within the atom.* Whenever an α particle came close to a nucleus in the scattering experiment, it experienced a large repulsive force and therefore a large deflection. Moreover, an α particle traveling directly toward a nucleus would experience an enormous repulsion that could completely reverse the direction of the moving particle.

The positively charged particles in the nucleus are called ***protons.*** In separate experiments, it was found that the charge of each proton has the same *magnitude* as that of an electron and that the mass of the proton is 1.67262×10^{-24} g—about 1840 times the mass of the oppositely charged electron.

At this stage of investigation, scientists perceived the atom as follows. The mass of a nucleus constitutes most of the mass of the entire atom, but the nucleus

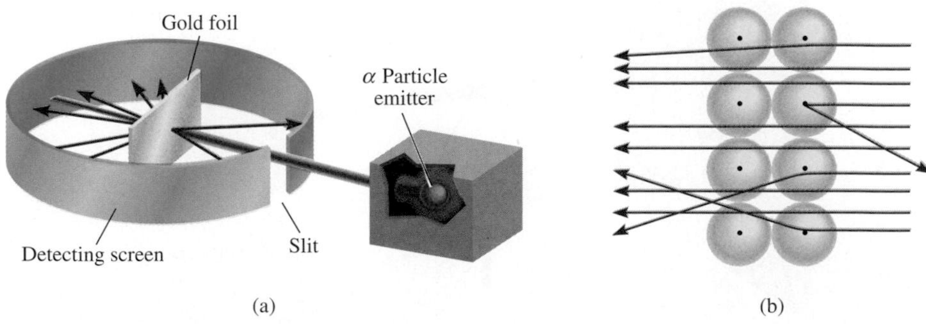

Figure 2.7
(a) Rutherford's experimental design for measuring the scattering of α particles by a piece of gold foil. Most of the α particles passed through the gold foil with little or no deflection. A few were deflected at wide angles. Occasionally an α particle was turned back. (b) Magnified view of α particles passing through and being deflected by nuclei.

occupies only about $1/10^{13}$ of the volume of the atom. We express atomic (and molecular) dimensions in terms of the SI unit called the *picometer (pm)*, and

$$1 \text{ pm} = 1 \times 10^{-12} \text{ m}$$

A common non-SI unit for atomic length is the angstrom (Å; 1 Å = 100 pm).

A typical atomic radius is about 100 pm, whereas the radius of an atomic nucleus is only about 5×10^{-3} pm. You can appreciate the relative sizes of an atom and its nucleus by imagining that if an atom were the size of a sports stadium, the volume of its nucleus would be comparable to that of a small marble. Although the protons are confined to the nucleus of the atom, the electrons are conceived of as being spread out about the nucleus at some distance from it.

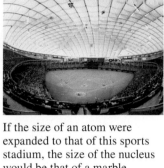

If the size of an atom were expanded to that of this sports stadium, the size of the nucleus would be that of a marble.

The Neutron

Rutherford's model of atomic structure left one major problem unsolved. It was known that hydrogen, the simplest atom, contains only one proton and that the helium atom contains two protons. Therefore, the ratio of the mass of a helium atom to that of a hydrogen atom should be 2:1. (Because electrons are much lighter than protons, their contribution can be ignored.) In reality, however, the ratio is 4:1.

Rutherford and others postulated that there must be another type of subatomic particle in the atomic nucleus; the proof was provided by another English physicist, James Chadwick, in 1932. When Chadwick bombarded a thin sheet of beryllium with α particles, a very high energy radiation similar to γ rays was emitted by the metal. Later experiments showed that the rays actually consisted of *electrically neutral particles having a mass slightly greater than that of protons.* Chadwick named these particles ***neutrons.***

The mystery of the mass ratio could now be explained. In the helium nucleus there are two protons and two neutrons, but in the hydrogen nucleus there is only one proton and no neutrons; therefore, the ratio is 4:1.

Figure 2.8 shows the location of the elementary particles (protons, neutrons, and electrons) in an atom. There are other subatomic particles, but the electron, the proton, and the neutron are the three fundamental components of the atom that are important in chemistry. Table 2.1 shows the masses and charges of these three elementary particles.

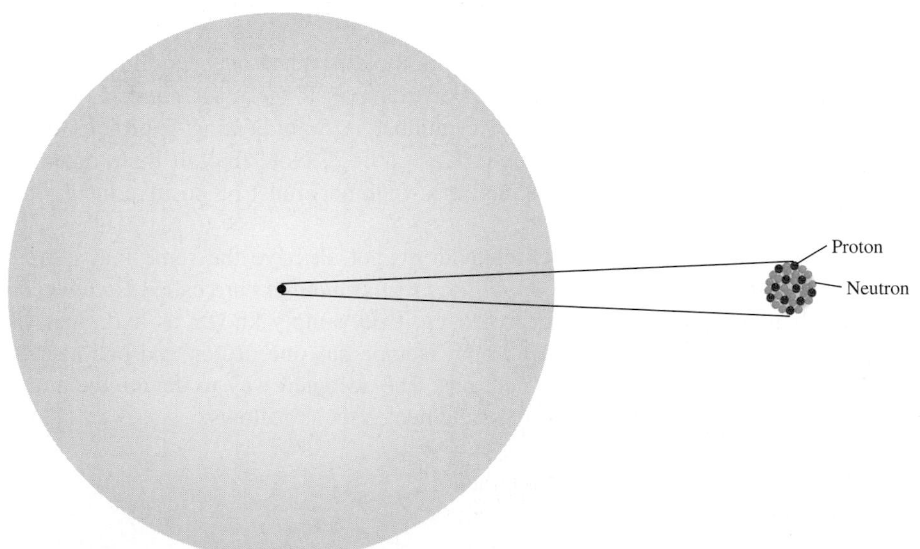

Figure 2.8
The protons and neutrons of an atom are packed in an extremely small nucleus. Electrons are shown as "clouds" around the nucleus.

Table 2.1 Mass and Charge of Subatomic Particles

		Charge	
Particle	**Mass (g)**	**Coulomb**	**Charge Unit**
Electron*	9.10938×10^{-28}	-1.6022×10^{-19}	-1
Proton	1.67262×10^{-24}	$+1.6022 \times 10^{-19}$	$+1$
Neutron	1.67493×10^{-24}	0	0

*More refined experiments have given us a more accurate value of an electron's mass than Millikan's.

Review of Concepts

Which two of the three fundamental particles of the atom have approximately the same mass?

2.3 Atomic Number, Mass Number, and Isotopes

All atoms can be identified by the number of protons and neutrons they contain. *The number of protons in the nucleus of each atom of an element* is called the ***atomic number (Z).*** In a neutral atom the number of protons is equal to the number of electrons, so the atomic number also indicates the number of electrons present in the atom. The chemical identity of an atom can be determined solely by its atomic number. For example, the atomic number of nitrogen is 7; this means that each neutral nitrogen atom has 7 protons and 7 electrons. Or viewed another way, every atom in the universe that contains 7 protons is correctly named "nitrogen."

The ***mass number (A)*** is *the total number of neutrons and protons present in the nucleus of an atom of an element.* Except for the most common form of hydrogen, which has one proton and no neutrons, all atomic nuclei contain both protons and neutrons. In general, the mass number is given by

$$\begin{aligned}\text{mass number} &= \text{number of protons} + \text{number of neutrons}\\ &= \text{atomic number} + \text{number of neutrons}\end{aligned}$$

The number of neutrons in an atom is equal to the difference between the mass number and the atomic number, or $(A - Z)$. For example, if the mass number of a particular boron atom is 12 and the atomic number is 5 (indicating 5 protons in the nucleus), then the number of neutrons is $12 - 5 = 7$. Note that all three quantities (atomic number, number of neutrons, and mass number) must be positive integers, or whole numbers.

In most cases atoms of a given element do not all have the same mass. *Atoms that have the same atomic number but different mass numbers* are called ***isotopes.*** For example, there are three isotopes of hydrogen. One, simply known as hydrogen, has one proton and no neutrons. The deuterium isotope has one proton and one neutron, and tritium has one proton and two neutrons. The accepted way to denote the atomic number and mass number of an atom of element X is as follows:

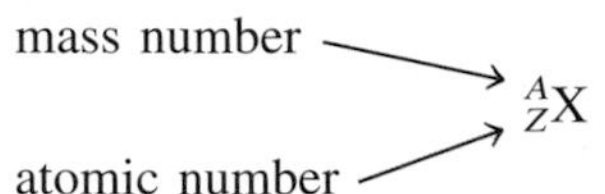

Thus, for the isotopes of hydrogen, we write

$$\underset{\text{hydrogen}}{^1_1H} \qquad \underset{\text{deuterium}}{^2_1H} \qquad \underset{\text{tritium}}{^3_1H}$$

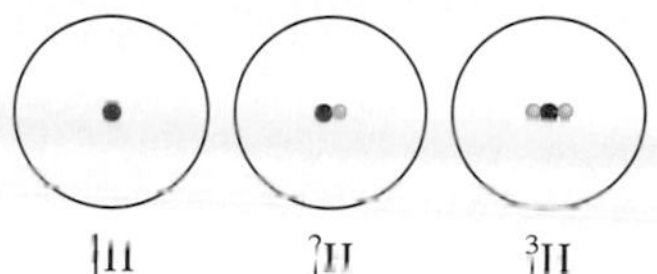

As another example, consider two common isotopes of uranium with mass numbers of 235 and 238, respectively:

$$^{235}_{92}U \qquad ^{238}_{92}U$$

The first isotope is used in nuclear reactors and atomic bombs, whereas the second isotope lacks the properties necessary for these applications. With the exception of hydrogen, isotopes of elements are identified by their mass numbers. Thus, these two isotopes are called uranium-235 (pronounced "uranium two thirty-five") and uranium-238 (pronounced "uranium two thirty-eight").

The chemical properties of an element are determined primarily by the protons and electrons in its atoms; neutrons do not take part in chemical changes under normal conditions. Therefore, isotopes of the same element have similar chemistries, forming the same types of compounds and displaying similar reactivities.

EXAMPLE 2.1

Give the number of protons, neutrons, and electrons in each of the following species: (a) $^{195}_{79}Au$, (b) $^{197}_{79}Au$, (c) ^{18}F, (d) carbon-13.

Strategy Recall that the superscript denotes mass number (A) and the subscript denotes atomic number (Z). Mass number is always greater than atomic number. (The only exception is ^1_1H, where the mass number is equal to the atomic number.) In the case where no subscript is shown, as in parts (c) and (d), the atomic number can be deduced from the element symbol or name. To determine the number of electrons, remember that because atoms are electrically neutral, the number of electrons is equal to the number of protons.

Solution

(a) The atomic number of Au (gold) is 79, so there are 79 protons. The mass number is 195, so the number of neutrons is $195 - 79 = 116$. The number of electrons is the same as the number of protons; that is, 79.

(b) Here the number of protons is the same as in (a), or 79. The mass number is 197, so the number of neutrons is $197 - 79 = 118$. The number of electrons is also the same as in (a), 79. The species in (a) and (b) are chemically similar isotopes of gold.

(c) The atomic number of F (fluorine) is 9, so there are 9 protons. The mass number is 18, so the number of neutrons is $18 - 9 = 9$. The number of electrons is the same as the number of protons; that is, 9.

(d) Carbon-13 can also be represented as ^{13}C. The atomic number of carbon is 6, so there are $13 - 6 = 7$ neutrons. The number of electrons is 6.

Similar problems: 2.13, 2.14.

Practice Exercise How many protons, neutrons, and electrons are in the following isotope of copper: ^{63}Cu?

REVIEW OF CONCEPTS

(a) What is the atomic number of an element containing 12 neutrons and having a mass number of 24?

(b) What is the mass number of a silicon atom with 16 neutrons in its nucleus?

2.4 The Periodic Table

More than half of the elements known today were discovered between 1800 and 1900. During this period, chemists noted that many elements show very strong similarities to one another. Recognition of periodic regularities in physical and chemical behavior and the need to organize the large volume of available information about the structure and properties of elemental substances led to the development of the ***periodic table***—*a chart in which elements having similar chemical and physical properties are grouped together.* Figure 2.9 shows the modern periodic table, in which the elements are arranged by atomic number (shown above the element symbol) in *horizontal rows* called ***periods*** and in *vertical columns* known as ***groups*** or ***families,*** according to similarities in their chemical properties. Note that elements 112, 114, 116, and 118 have recently been synthesized, although they have not yet been named.

The elements can be divided into three categories—metals, nonmetals, and metalloids. A ***metal*** is *a good conductor of heat and electricity,* whereas a ***nonmetal*** is usually *a poor conductor of heat and electricity.* A **metalloid** *has properties that are intermediate between those of metals and nonmetals.* Figure 2.9 shows that the majority of known elements are metals; only seventeen elements are nonmetals, and eight

1 1A	2 2A	3 3B	4 4B	5 5B	6 6B	7 7B	8	9 8B	10	11 1B	12 2B	13 3A	14 4A	15 5A	16 6A	17 7A	18 8A
1 **H**																	2 **He**
3 **Li**	4 **Be**											5 **B**	6 **C**	7 **N**	8 **O**	9 **F**	10 **Ne**
11 **Na**	12 **Mg**											13 **Al**	14 **Si**	15 **P**	16 **S**	17 **Cl**	18 **Ar**
19 **K**	20 **Ca**	21 **Sc**	22 **Ti**	23 **V**	24 **Cr**	25 **Mn**	26 **Fe**	27 **Co**	28 **Ni**	29 **Cu**	30 **Zn**	31 **Ga**	32 **Ge**	33 **As**	34 **Se**	35 **Br**	36 **Kr**
37 **Rb**	38 **Sr**	39 **Y**	40 **Zr**	41 **Nb**	42 **Mo**	43 **Tc**	44 **Ru**	45 **Rh**	46 **Pd**	47 **Ag**	48 **Cd**	49 **In**	50 **Sn**	51 **Sb**	52 **Te**	53 **I**	54 **Xe**
55 **Cs**	56 **Ba**	57 **La**	72 **Hf**	73 **Ta**	74 **W**	75 **Re**	76 **Os**	77 **Ir**	78 **Pt**	79 **Au**	80 **Hg**	81 **Tl**	82 **Pb**	83 **Bi**	84 **Po**	85 **At**	86 **Rn**
87 **Fr**	88 **Ra**	89 **Ac**	104 **Rf**	105 **Db**	106 **Sg**	107 **Bh**	108 **Hs**	109 **Mt**	110 **Ds**	111 **Rg**	112	113	114	115	116	(117)	118

58 **Ce**	59 **Pr**	60 **Nd**	61 **Pm**	62 **Sm**	63 **Eu**	64 **Gd**	65 **Tb**	66 **Dy**	67 **Ho**	68 **Er**	69 **Tm**	70 **Yb**	71 **Lu**
90 **Th**	91 **Pa**	92 **U**	93 **Np**	94 **Pu**	95 **Am**	96 **Cm**	97 **Bk**	98 **Cf**	99 **Es**	100 **Fm**	101 **Md**	102 **No**	103 **Lr**

Metals

Metalloids

Nonmetals

Figure 2.9

The modern periodic table. The elements are arranged according to the atomic numbers above their symbols. With the exception of hydrogen (H), nonmetals appear at the far right of the table. The two rows of metals beneath the main body of the table are conventionally set apart to keep the table from being too wide. Actually, cerium (Ce) should follow lanthanum (La), and thorium (Th) should come right after actinium (Ac). The 1–18 group designation has been recommended by the International Union of Pure and Applied Chemistry (IUPAC), but is not yet in wide use. In this text, we use the standard U.S. notation for group numbers (1A–8A and 1B–8B). No names have been assigned to elements 112–116, and 118. Element 117 has not yet been synthesized.

elements are metalloids. From left to right across any period, the physical and chemical properties of the elements change gradually from metallic to nonmetallic. The periodic table is a handy tool that correlates the properties of the elements in a systematic way and helps us to make predictions about chemical behavior. We will take a closer look at this keystone of chemistry in Chapter 8.

Elements are often referred to collectively by their periodic table group number (Group 1A, Group 2A, and so on). However, for convenience, some element groups have special names. *The Group 1A elements (Li, Na, K, Rb, Cs, and Fr)* are called ***alkali metals,*** and *the Group 2A elements (Be, Mg, Ca, Sr, Ba, and Ra)* are called ***alkaline earth metals.*** *Elements in Group 7A (F, Cl, Br, I, and At)* are known as ***halogens,*** and *those in Group 8A (He, Ne, Ar, Kr, Xe, and Rn)* are called ***noble gases*** (or ***rare gases***). The names of other groups or families will be introduced later.

Review of Concepts

In viewing the periodic table, do chemical properties change more markedly across a period or down a group?

2.5 Molecules and Ions

Of all the elements, only the six noble gases in Group 8A of the periodic table (He, Ne, Ar, Kr, Xe, and Rn) exist in nature as single atoms. For this reason, they are called *monatomic* (meaning a single atom) gases. Most matter is composed of molecules or ions formed by atoms.

Molecules

A ***molecule*** is an *aggregate of at least two atoms in a definite arrangement held together by chemical forces* (also called *chemical bonds*). A molecule may contain atoms of the same element or atoms of two or more elements joined in a fixed ratio, in accordance with the law of definite proportions stated in Section 2.1. Thus, a molecule is not necessarily a compound, which, by definition, is made up of two or more elements. Hydrogen gas, for example, is a pure element, but it consists of molecules made up of two H atoms each. Water, on the other hand, is a molecular compound that contains hydrogen and oxygen in a ratio of two H atoms and one O atom. Like atoms, molecules are electrically neutral.

We will discuss the nature of chemical bonds in Chapters 9 and 10.

The hydrogen molecule, symbolized as H_2, is called a ***diatomic molecule*** because it *contains only two atoms.* Other elements that normally exist as diatomic molecules are nitrogen (N_2) and oxygen (O_2), as well as the Group 7A elements—fluorine (F_2), chlorine (Cl_2), bromine (Br_2), and iodine (I_2). Of course, a diatomic molecule can contain atoms of different elements. Examples are hydrogen chloride (HCl) and carbon monoxide (CO).

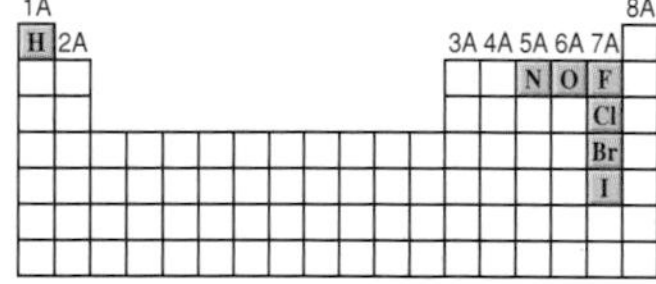

Elements that exist as diatomic molecules.

The vast majority of molecules contain more than two atoms. They can be atoms of the same element, as in ozone (O_3), which is made up of three atoms of oxygen, or they can be combinations of two or more different elements. *Molecules containing more than two atoms* are called ***polyatomic molecules.*** Like ozone, water (H_2O) and ammonia (NH_3) are polyatomic molecules.

Ions

An ***ion*** is *an atom or a group of atoms that has a net positive or negative charge.* The number of positively charged protons in the nucleus of an atom remains the same during ordinary chemical changes (called chemical reactions), but negatively charged electrons

In Chapter 8, we will see why atoms of different elements gain (or lose) a specific number of electrons.

may be lost or gained. The loss of one or more electrons from a neutral atom results in a ***cation,*** *an ion with a net positive charge.* For example, a sodium atom (Na) can readily lose an electron to become a sodium cation, which is represented by Na^+:

Na Atom	Na^+ Ion
11 protons	11 protons
11 electrons	10 electrons

On the other hand, an ***anion*** is *an ion whose net charge is negative* due to an increase in the number of electrons. A chlorine atom (Cl), for instance, can gain an electron to become the chloride ion Cl^-:

Cl Atom	Cl^- Ion
17 protons	17 protons
17 electrons	18 electrons

Sodium chloride (NaCl), ordinary table salt, is called an ***ionic compound*** because it is *formed from cations and anions.*

An atom can lose or gain more than one electron. Examples of ions formed by the loss or gain of more than one electron are Mg^{2+}, Fe^{3+}, S^{2-}, and N^{3-}. These ions, as well as Na^+ and Cl^-, are called ***monatomic ions*** because they *contain only one atom.* Figure 2.10 shows the charges of a number of monatomic ions. With very few exceptions, metals tend to form cations and nonmetals form anions.

In addition, two or more atoms can combine to form an ion that has a net positive or net negative charge. ***Polyatomic ions*** such as OH^- (hydroxide ion), CN^- (cyanide ion), and NH_4^+ (ammonium ion) are *ions containing more than one atom.*

Review of Concepts

Determine the number of protons and electrons for the following ions: (a) Se^{2-} and (b) Cr^{3+}.

1 1A	2 2A	3 3B	4 4B	5 5B	6 6B	7 7B	8 8B	9 8B	10 8B	11 1B	12 2B	13 3A	14 4A	15 5A	16 6A	17 7A	18 8A
Li^+													C^{4-}	N^{3-}	O^{2-}	F^-	
Na^+	Mg^{2+}											Al^{3+}		P^{3-}	S^{2-}	Cl^-	
K^+	Ca^{2+}				Cr^{2+} Cr^{3+}	Mn^{2+} Mn^{3+}	Fe^{2+} Fe^{3+}	Co^{2+} Co^{3+}	Ni^{2+} Ni^{3+}	Cu^+ Cu^{2+}	Zn^{2+}				Se^{2-}	Br^-	
Rb^+	Sr^{2+}									Ag^+	Cd^{2+}		Sn^{2+} Sn^{4+}		Te^{2-}	I^-	
Cs^+	Ba^{2+}									Au^+ Au^{3+}	Hg_2^{2+} Hg^{2+}		Pb^{2+} Pb^{4+}				

Figure 2.10
Common monatomic ions arranged according to their positions in the periodic table. Note that the Hg_2^{2+} ion contains two atoms.

2.6 Chemical Formulas

Chemists use ***chemical formulas*** to *express the composition of molecules and ionic compounds in terms of chemical symbols.* By composition we mean not only the elements present but also the ratios in which the atoms are combined. Here we are mainly concerned with two types of formulas: molecular formulas and empirical formulas.

Molecular Formulas

A ***molecular formula*** *shows the exact number of atoms of each element in the smallest unit of a substance.* In our discussion of molecules, each example was given with its molecular formula in parentheses. Thus, H_2 is the molecular formula for hydrogen, O_2 is oxygen, O_3 is ozone, and H_2O is water. The subscript numeral indicates the number of atoms of an element present. There is no subscript for O in H_2O because there is only one atom of oxygen in a molecule of water, and so the number "one" is omitted from the formula. Note that oxygen (O_2) and ozone (O_3) are allotropes of oxygen. An ***allotrope*** is *one of two or more distinct forms of an element.* Two allotropic forms of the element carbon—diamond and graphite—are dramatically different not only in properties but also in their relative cost.

Molecular Models

Molecules are too small for us to observe directly. An effective means of visualizing them is by the use of molecular models. Two standard types of molecular models are currently in use: *ball-and-stick* models and *space-filling* models (Figure 2.11). In ball-and-stick model kits, the atoms are wooden or plastic balls with holes in them. Sticks or springs are used to represent chemical bonds. The angles they form between atoms approximate the bond angles in actual molecules. With the exception of the H atom, the balls are all the same size and each type of atom is represented by a specific color.

See back end paper for color codes for atoms.

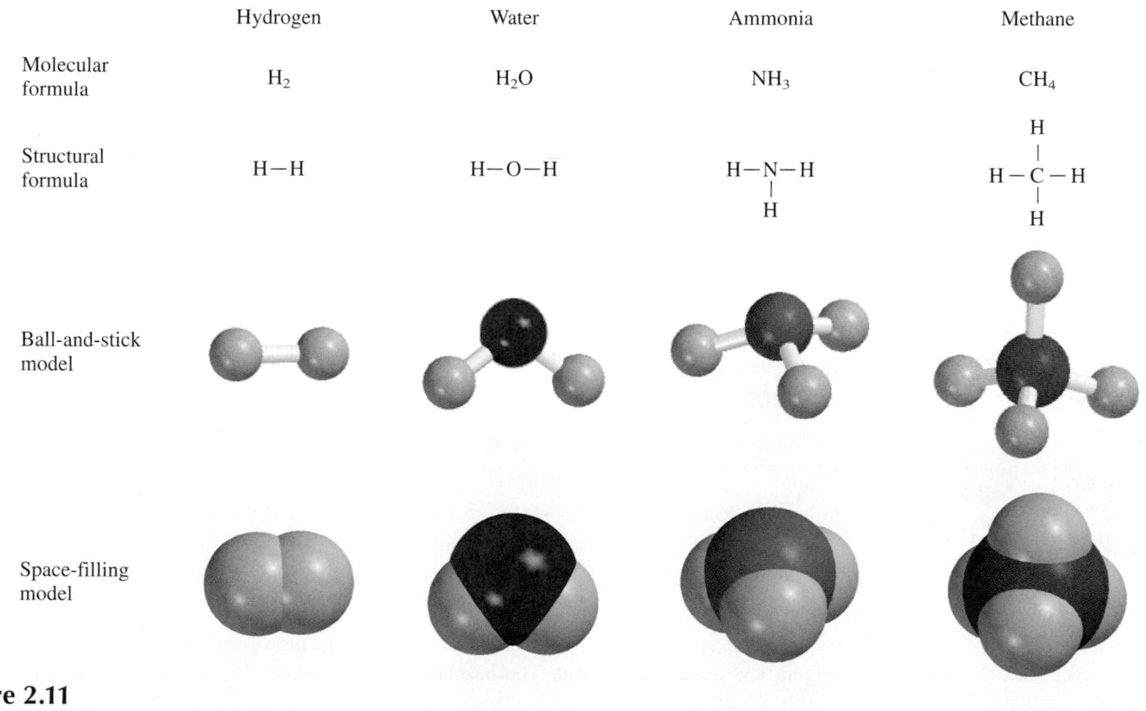

	Hydrogen	Water	Ammonia	Methane
Molecular formula	H_2	H_2O	NH_3	CH_4
Structural formula	H−H	H−O−H	H−N−H (with H below N)	H−C−H (with H above and below C)
Ball-and-stick model				
Space-filling model				

Figure 2.11
Molecular and structural formulas and molecular models of four common molecules.

In space-filling models, atoms are represented by truncated balls held together by snap fasteners, so that the bonds are not visible. The balls are proportional in size to atoms. The first step toward building a molecular model is writing the ***structural formula,*** which *shows how atoms are bonded to one another in a molecule.* For example, it is known that each of the two H atoms is bonded to an O atom in the water molecule. Therefore, the structural formula of water is H—O—H. A line connecting the two atomic symbols represents a chemical bond.

Ball-and-stick models show the three-dimensional arrangement of atoms clearly, and they are fairly easy to construct. However, the balls are not proportional to the size of atoms. Furthermore, the sticks greatly exaggerate the space between atoms in a molecule. Space-filling models are more accurate because they show the variation in atomic size. Their drawbacks are that they are time-consuming to put together and they do not show the three-dimensional positions of atoms very well. We will use both models extensively in this text.

Empirical Formulas

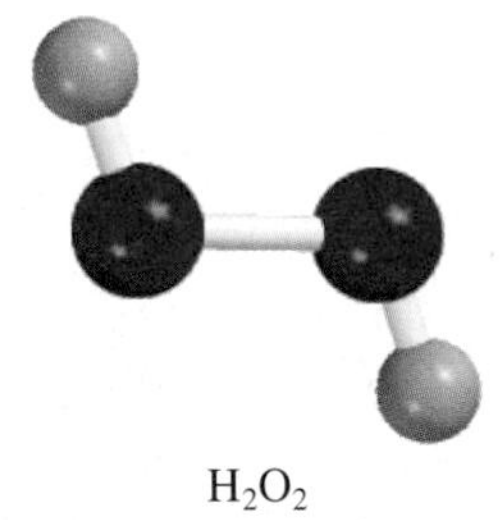

H_2O_2

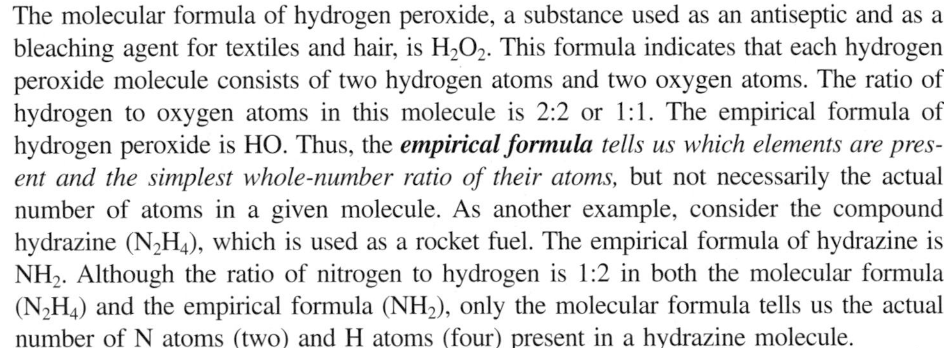

The molecular formula of hydrogen peroxide, a substance used as an antiseptic and as a bleaching agent for textiles and hair, is H_2O_2. This formula indicates that each hydrogen peroxide molecule consists of two hydrogen atoms and two oxygen atoms. The ratio of hydrogen to oxygen atoms in this molecule is 2:2 or 1:1. The empirical formula of hydrogen peroxide is HO. Thus, the ***empirical formula*** *tells us which elements are present and the simplest whole-number ratio of their atoms,* but not necessarily the actual number of atoms in a given molecule. As another example, consider the compound hydrazine (N_2H_4), which is used as a rocket fuel. The empirical formula of hydrazine is NH_2. Although the ratio of nitrogen to hydrogen is 1:2 in both the molecular formula (N_2H_4) and the empirical formula (NH_2), only the molecular formula tells us the actual number of N atoms (two) and H atoms (four) present in a hydrazine molecule.

The word "empirical" means "derived from experiment." As we will see in Chapter 3, empirical formulas are determined experimentally.

Empirical formulas are the *simplest* chemical formulas; they are written by reducing the subscripts in the molecular formulas to the smallest possible whole numbers. Molecular formulas are the *true* formulas of molecules. If we know the molecular formula, we also know the empirical formula, but the reverse is not true. Why, then, do chemists bother with empirical formulas? As we will see in Chapter 3, when chemists analyze an unknown compound, the first step is usually the determination of the compound's empirical formula. With additional information, it is possible to deduce the molecular formula.

For many molecules, the molecular formula and empirical formula are one and the same. Some examples are water (H_2O), ammonia (NH_3), carbon dioxide (CO_2), and methane (CH_4).

Methanol

EXAMPLE 2.2

Write the molecular formula of methanol, an organic solvent and antifreeze, from its ball-and-stick model, shown in the margin.

Solution Refer to the labels (also see back end paper). There are four H atoms, one C atom, and one O atom. Therefore, the molecular formula is CH_4O. However, the standard way of writing the molecular formula for methanol is CH_3OH because it shows how the atoms are joined in the molecule.

Similar problems: 2.43, 2.44.

Practice Exercise Write the molecular formula of chloroform, which is used as a solvent and a cleansing agent. The ball-and-stick model of chloroform is shown in the margin on p. 43.

EXAMPLE 2.3

Write the empirical formulas for the following molecules: (a) acetylene (C_2H_2), which is used in welding torches; (b) glucose ($C_6H_{12}O_6$), a substance known as blood sugar; and (c) nitrous oxide (N_2O), a gas that is used as an anesthetic gas ("laughing gas") and as an aerosol propellant for whipped creams.

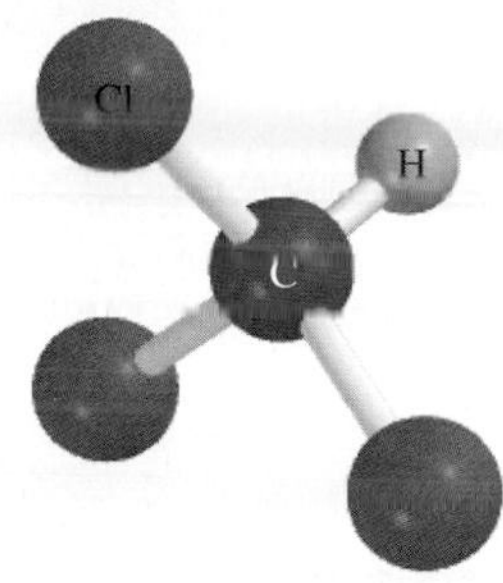

Chloroform

Strategy Recall that to write the empirical formula, the subscripts in the molecular formula must be converted to the smallest possible whole numbers.

Solution

(a) There are two carbon atoms and two hydrogen atoms in acetylene. Dividing the subscripts by 2, we obtain the empirical formula CH.

(b) In glucose there are 6 carbon atoms, 12 hydrogen atoms, and 6 oxygen atoms. Dividing the subscripts by 6, we obtain the empirical formula CH_2O. Note that if we had divided the subscripts by 3, we would have obtained the formula $C_2H_4O_2$. Although the ratio of carbon to hydrogen to oxygen atoms in $C_2H_4O_2$ is the same as that in $C_6H_{12}O_6$ (1:2:1), $C_2H_4O_2$ is not the simplest formula because its subscripts are not in the smallest whole-number ratio.

(c) Because the subscripts in N_2O are already the smallest possible whole numbers, the empirical formula for nitrous oxide is the same as its molecular formula.

Similar problems: 2.41, 2.42.

Practice Exercise Write the empirical formula for caffeine ($C_8H_{10}N_4O_2$), a stimulant found in tea and coffee.

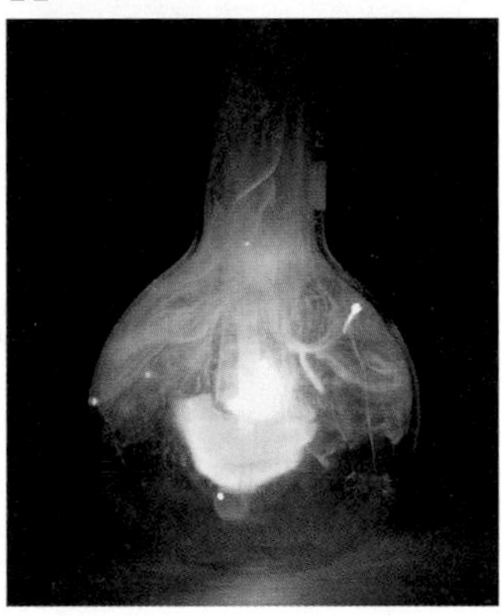
Sodium metal reacting with chlorine gas to form sodium chloride.

Formula of Ionic Compounds

The formulas of ionic compounds are usually the same as their empirical formulas because ionic compounds do not consist of discrete molecular units. For example, a solid sample of sodium chloride (NaCl) consists of equal numbers of Na^+ and Cl^- ions arranged in a three-dimensional network (Figure 2.12). In such a compound, there is a 1:1 ratio of cations to anions so that the compound is electrically neutral. As you can see in Figure 2.12, no Na^+ ion in NaCl is associated with just one particular Cl^- ion. In fact, each Na^+ ion is equally held by six surrounding Cl^- ions and vice versa. Thus, NaCl is the empirical formula for sodium chloride. In other ionic compounds, the actual structure may be different, but the arrangement of cations and anions is

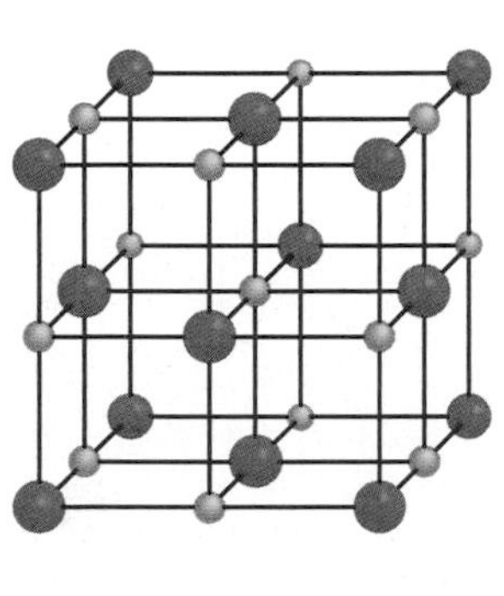
(a)

(b)

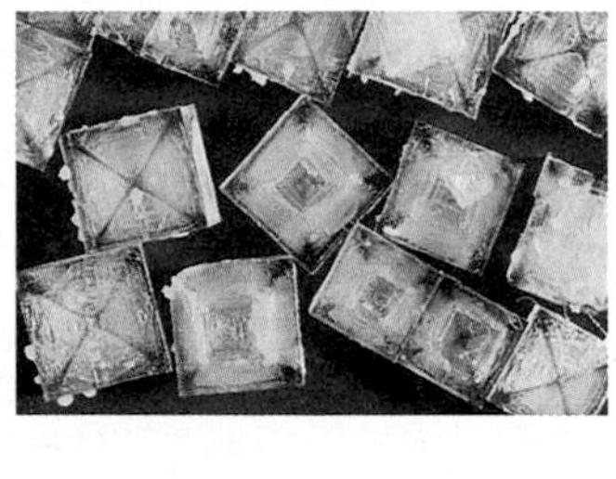
(c)

Figure 2.12
(a) Structure of solid NaCl. (b) In reality, the cations are in contact with the anions. In both (a) and (b), the smaller spheres represent Na^+ ions and the larger spheres, Cl^- ions. (c) Crystals of NaCl.

such that the compounds are all electrically neutral. Note that the charges on the cation and anion are not shown in the formula for an ionic compound.

In order for ionic compounds to be electrically neutral, the sum of the charges on the cation and anion in each formula unit must be zero. If the charges on the cation and anion are numerically different, we apply the following rule to make the formula electrically neutral: *The subscript of the cation is numerically equal to the charge on the anion, and the subscript of the anion is numerically equal to the charge on the cation.* If the charges are numerically equal, then no subscripts are necessary. This rule follows from the fact that because the formulas of most ionic compounds are empirical formulas, the subscripts must always be reduced to the smallest ratios. Let us consider some examples.

Refer to Figure 2.10 for charges of cations and anions.

- **Potassium Bromide.** The potassium cation K^+ and the bromine anion Br^- combine to form the ionic compound potassium bromide. The sum of the charges is $+1 + (-1) = 0$, so no subscripts are necessary. The formula is KBr.
- **Zinc Iodide.** The zinc cation Zn^{2+} and the iodine anion I^- combine to form zinc iodide. The sum of the charges of one Zn^{2+} ion and one I^- ion is $+2 + (-1) = +1$. To make the charges add up to zero we multiply the -1 charge of the anion by 2 and add the subscript "2" to the symbol for iodine. Therefore, the formula for zinc iodide is ZnI_2.
- **Aluminum Oxide.** The cation is Al^{3+} and the oxygen anion is O^{2-}. The following diagram helps us determine the subscripts for the compound formed by the cation and the anion:

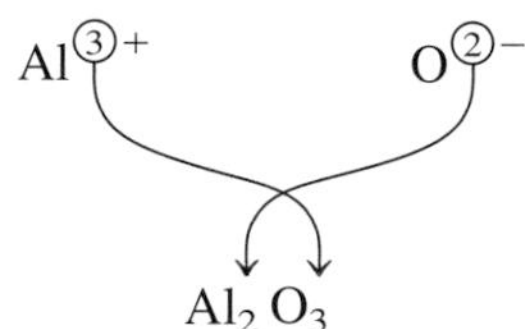

The sum of the charges is $2(+3) + 3(-2) = 0$. Thus, the formula for aluminum oxide is Al_2O_3.

Review of Concepts

Match each of the diagrams shown here with the following ionic compounds: Al_2O_3, LiH, Na_2S, $Mg(NO_3)_2$. (Green spheres represent cations and red spheres represent anions.)

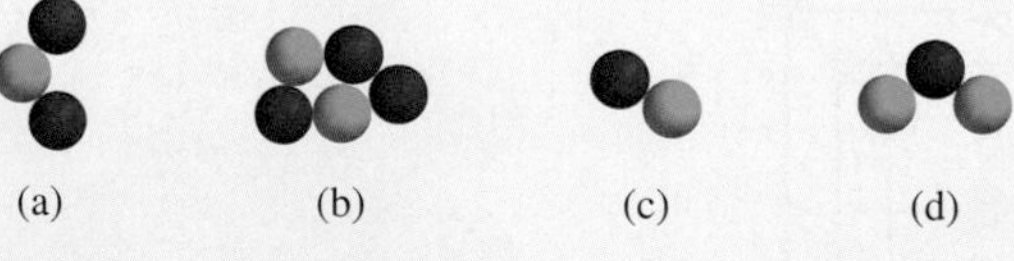

2.7 Naming Compounds

In addition to using formulas to show the composition of molecules and compounds, chemists have developed a system for naming substances on the basis of their composition. First, we divide them into three categories: ionic compounds, molecular compounds, and acids and bases. Then we apply certain rules to derive the scientific name for a given substance.

Table 2.2 The "-ide" Nomenclature of Some Common Monatomic Anions According to Their Positions in the Periodic Table

Group 4A	Group 5A	Group 6A	Group 7A
C Carbide (C^{4-})*	N Nitride (N^{3-})	O Oxide (O^{2-})	F Fluoride (F^-)
Si Silicide (Si^{4-})	P Phosphide (P^{3-})	S Sulfide (S^{2-})	Cl Chloride (Cl^-)
		Se Selenide (Se^{2-})	Br Bromide (Br^-)
		Te Telluride (Te^{2-})	I Iodide (I^-)

*The word "carbide" is also used for the anion C_2^{2-}.

Ionic Compounds

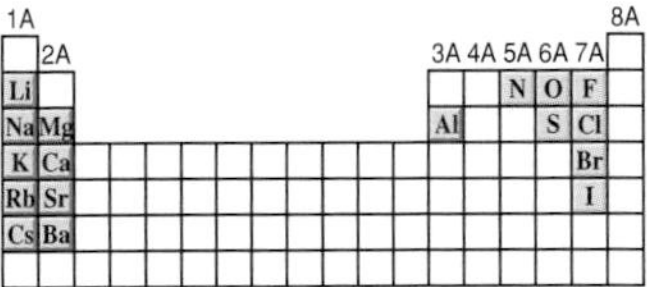

The most reactive metals (green) and the most reactive nonmetals (blue) combine to form ionic compounds.

In Section 2.5 we learned that ionic compounds are made up of cations (positive ions) and anions (negative ions). With the important exception of the ammonium ion, NH_4^+, all cations of interest to us are derived from metal atoms. Metal cations take their names from the elements. For example,

Element		Name of Cation	
Na	sodium	Na^+	sodium ion (or sodium cation)
K	potassium	K^+	potassium ion (or potassium cation)
Mg	magnesium	Mg^{2+}	magnesium ion (or magnesium cation)
Al	aluminum	Al^{3+}	aluminum ion (or aluminum cation)

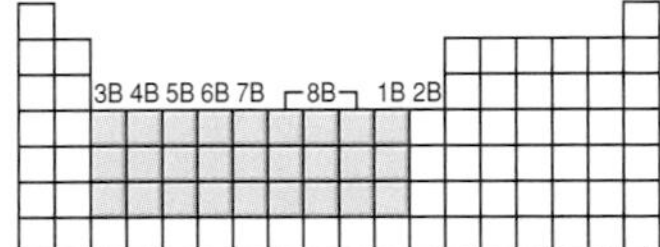

The transition metals are the elements in Groups 1B and 3B–8B (see Figure 2.9).

Many ionic compounds are ***binary compounds,*** or *compounds formed from just two elements.* For binary ionic compounds the first element named is the metal cation, followed by the nonmetallic anion. Thus, NaCl is sodium chloride. The anion is named by taking the first part of the element name (chlorine) and adding "-ide." Potassium bromide (KBr), zinc iodide (ZnI_2), and aluminum oxide (Al_2O_3) are also binary compounds. Table 2.2 shows the "-ide" nomenclature of some common monatomic anions according to their positions in the periodic table.

The "-ide" ending is also used for certain anion groups containing different elements, such as hydroxide (OH^-) and cyanide (CN^-). Thus, the compounds LiOH and KCN are named lithium hydroxide and potassium cyanide. These and a number of other such ionic substances are called ***ternary compounds,*** meaning *compounds consisting of three elements.* Table 2.3 lists alphabetically the names of a number of common cations and anions.

$FeCl_2$ (left) and $FeCl_3$ (right).

Certain metals, especially the *transition metals,* can form more than one type of cation. Take iron as an example. Iron can form two cations: Fe^{2+} and Fe^{3+}. The accepted procedure for designating different cations of the *same* element is to use Roman numerals. The Roman numeral I is used for one positive charge, II for two positive charges, and so on. This is called the *Stock system.* In this system, the Fe^{2+} and Fe^{3+} ions are called iron(II) and iron(III), and the compounds $FeCl_2$ (containing the Fe^{2+} ion) and $FeCl_3$ (containing the Fe^{3+} ion) are called iron-two chloride and iron-three chloride, respectively. As another example, manganese (Mn) atoms can assume several different positive charges:

Mn^{2+}:	MnO	manganese(II) oxide
Mn^{3+}:	Mn_2O_3	manganese(III) oxide
Mn^{4+}:	MnO_2	manganese(IV) oxide

These compound names are pronounced "manganese-two oxide," "manganese-three oxide," and "manganese-four oxide."

Table 2.3 Names and Formulas of Some Common Inorganic Cations and Anions

Cation	Anion
aluminum (Al^{3+})	bromide (Br^-)
ammonium (NH_4^+)	carbonate (CO_3^{2-})
barium (Ba^{2+})	chlorate (ClO_3^-)
cadmium (Cd^{2+})	chloride (Cl^-)
calcium (Ca^{2+})	chromate (CrO_4^{2-})
cesium (Cs^+)	cyanide (CN^-)
chromium(III) or chromic (Cr^{3+})	dichromate ($Cr_2O_7^{2-}$)
cobalt(II) or cobaltous (Co^{2+})	dihydrogen phosphate ($H_2PO_4^-$)
copper(I) or cuprous (Cu^+)	fluoride (F^-)
copper(II) or cupric (Cu^{2+})	hydride (H^-)
hydrogen (H^+)	hydrogen carbonate or bicarbonate (HCO_3^-)
iron(II) or ferrous (Fe^{2+})	hydrogen phosphate (HPO_4^{2-})
iron(III) or ferric (Fe^{3+})	hydrogen sulfate or bisulfate (HSO_4^-)
lead(II) or plumbous (Pb^{2+})	hydroxide (OH^-)
lithium (Li^+)	iodide (I^-)
magnesium (Mg^{2+})	nitrate (NO_3^-)
manganese(II) or manganous (Mn^{2+})	nitride (N^{3-})
mercury(I) or mercurous (Hg_2^{2+})*	nitrite (NO_2^-)
mercury(II) or mercuric (Hg^{2+})	oxide (O^{2-})
potassium (K^+)	permanganate (MnO_4^-)
rubidium (Rb^+)	peroxide (O_2^{2-})
silver (Ag^+)	phosphate (PO_4^{3-})
sodium (Na^+)	sulfate (SO_4^{2-})
strontium (Sr^{2+})	sulfide (S^{2-})
tin(II) or stannous (Sn^{2+})	sulfite (SO_3^{2-})
zinc (Zn^{2+})	thiocyanate (SCN^-)

*Mercury(I) exists as a pair as shown.

EXAMPLE 2.4

Name the following compounds: (a) $Fe(NO_3)_2$, (b) Na_2HPO_4, and (c) $(NH_4)_2SO_3$.

Strategy Our reference for the names of cations and anions is Table 2.3. Keep in mind that if a metal can form cations of different charges (see Figure 2.10), we need to use the Stock system.

Solution

(a) The nitrate ion (NO_3^-) bears one negative charge, so the iron ion must have two positive charges. Because iron forms both Fe^{2+} and Fe^{3+} ions, we need to use the Stock system and call the compound iron(II) nitrate.

(Continued)

(b) The cation is Na^+ and the anion is HPO_4^{2-} (hydrogen phosphate). Because sodium only forms one type of ion (Na^+), there is no need to use sodium(I) in the name. The compound is sodium hydrogen phosphate.

(c) The cation is NH_4^+ (ammonium ion) and the anion is SO_3^{2-} (sulfite ion). The compound is ammonium sulfite.

Similar problems: 2.47(a), (b), (e).

Practice Exercises Name the following compounds: (a) PbO and (b) $LiClO_3$.

EXAMPLE 2.5

Write chemical formulas for the following compounds: (a) mercury(I) nitrate, (b) cesium oxide, and (c) strontium nitride.

Strategy We refer to Table 2.3 for the formulas of cations and anions. Recall that the Roman numerals in the Stock system provide useful information about the charges of the cation.

Solution

(a) The Roman numeral shows that the mercury ion bears a +1 charge. According to Table 2.3, however, the mercury(I) ion is diatomic (that is, Hg_2^{2+}) and the nitrate ion is NO_3^-. Therefore, the formula is $Hg_2(NO_3)_2$.

Note that the subscripts of this ionic compound are not reduced to the smallest ratio because the Hg(I) ion exists as a pair or dimer.

(b) Each oxide ion bears two negative charges, and each cesium ion bears one positive charge (cesium is in Group 1A, as is sodium). Therefore, the formula is Cs_2O.

(c) Each strontium ion (Sr^{2+}) bears two positive charges, and each nitride ion (N^{3-}) bears three negative charges. To make the sum of the charges equal zero, we must adjust the numbers of cations and anions:

$$3(+2) + 2(-3) = 0$$

Thus, the formula is Sr_3N_2.

Similar problems: 2.49(a), (b), (h).

Practice Exercises Write formulas for the following ionic compounds: (a) rubidium sulfate and (b) barium hydride.

Molecular Compounds

Unlike ionic compounds, molecular compounds contain discrete molecular units. They are usually composed of nonmetallic elements (see Figure 2.9). Many molecular compounds are binary compounds. Naming binary molecular compounds is similar to naming binary ionic compounds. We place the name of the first element in the formula first, and the second element is named by adding "-ide" to the root of the element name. Some examples are

HCl	Hydrogen chloride	SiC	Silicon carbide
HBr	Hydrogen bromide		

It is quite common for one pair of elements to form several different compounds. In these cases, confusion in naming the compounds is avoided by the use of Greek prefixes to denote the number of atoms of each element present (Table 2.4). Consider these examples:

CO	Carbon monoxide	SO_3	Sulfur trioxide
CO_2	Carbon dioxide	NO_2	Nitrogen dioxide
SO_2	Sulfur dioxide	N_2O_4	Dinitrogen tetroxide

Table 2.4 Greek Prefixes Used in Naming Molecular Compounds

Prefix	Meaning
mono-	1
di-	2
tri-	3
tetra-	4
penta-	5
hexa-	6
hepta-	7
octa-	8
nona-	9
deca-	10

These guidelines are helpful when you are naming compounds with prefixes:

- The prefix "mono-" may be omitted for the first element. For example, PCl_3 is named phosphorus trichloride, not monophosphorus trichloride. Thus, the absence of a prefix for the first element usually means that only one atom of that element is present in the molecule.
- For oxides, the ending "a" in the prefix is sometimes omitted. For example, N_2O_4 may be called dinitrogen tetroxide rather than dinitrogen tetraoxide.

Exceptions to the use of Greek prefixes are molecular compounds containing hydrogen. Traditionally, many of these compounds are called either by their common, nonsystematic names or by names that do not specifically indicate the number of H atoms present:

B_2H_6	Diborane	PH_3	Phosphine
CH_4	Methane	H_2O	Water
SiH_4	Silane	H_2S	Hydrogen sulfide
NH_3	Ammonia		

Note that even the order of writing the elements in the formulas is irregular. These examples show that H is written first in water and hydrogen sulfide, whereas H is written last in the other compounds.

Writing formulas for molecular compounds is usually straightforward. Thus, the name arsenic trifluoride means that there are one As atom and three F atoms in each molecule and the molecular formula is AsF_3. Note that the order of elements in the formula is the same as that in its name.

Figure 2.13 summarizes the steps for naming ionic and molecular compounds.

EXAMPLE 2.6

Name the following molecular compounds: (a) PBr_5 and (b) As_2O_5.

Strategy We refer to Table 2.4 for the prefixes used in naming molecular compounds.

Solution

(a) Because there are five bromine atoms present, the compound is phosphorus pentabromide.

(b) There are two arsenic atoms and five oxygen atoms present, so the compound is diarsenic pentoxide. Note that the "a" is omitted in "penta."

Similar problems: 2.47(c), (h), (j).

Practice Exercises Name the following molecular compounds: (a) NF_3 and (b) Cl_2O_7.

EXAMPLE 2.7

Write chemical formulas for the following molecular compounds: (a) bromine trifluoride and (b) diboron trioxide.

Strategy We refer to Table 2.4 for the prefixes used in naming molecular compounds.

Solution

(a) Because there are three fluorine atoms and one bromine atom present, the formula is BrF_3.

(b) There are two boron atoms and three oxygen atoms present, so the formula is B_2O_3.

Similar problems: 2.49(f), (g).

Practice Exercises Write chemical formulas for the following molecular compounds: (a) sulfur tetrafluoride and (b) dinitrogen pentoxide.

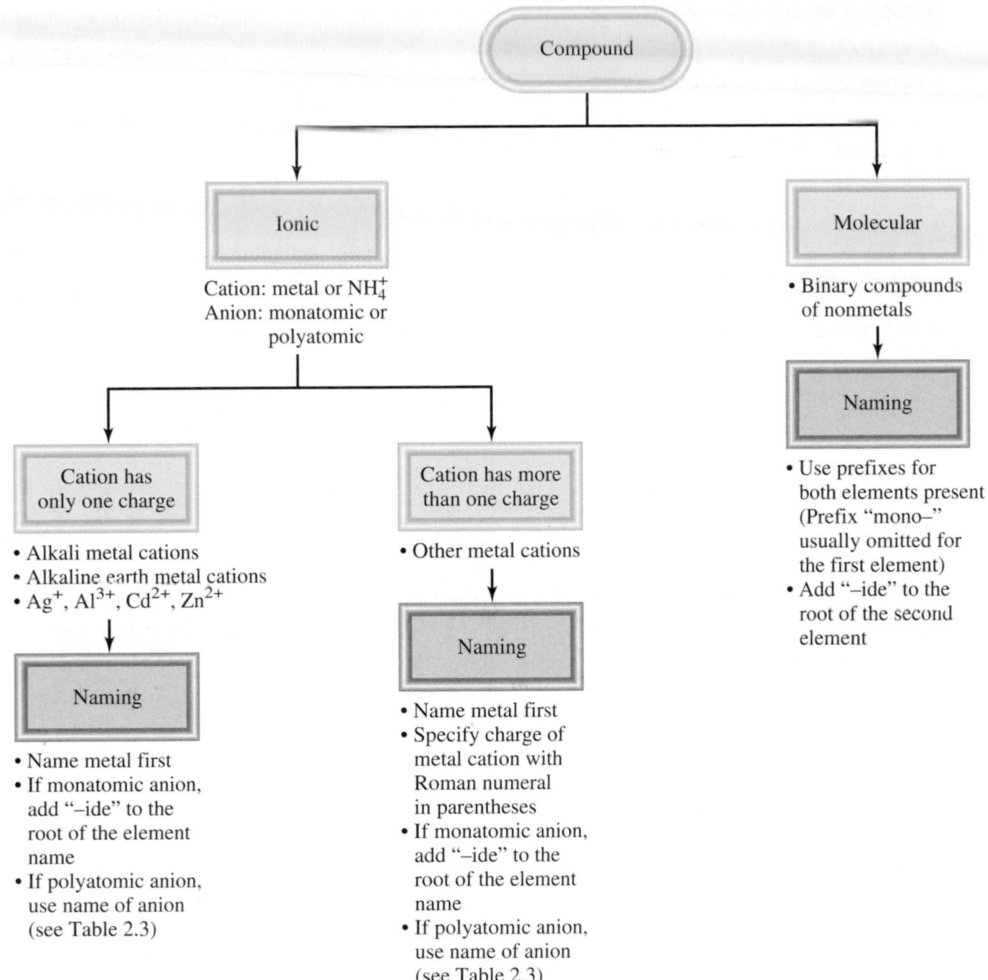

Figure 2.13
Steps for naming ionic and molecular compounds.

Acids and Bases

Naming Acids

An ***acid*** can be described as *a substance that yields hydrogen ions (H^+) when dissolved in water.* (H^+ is equivalent to one proton, and is often referred to that way.) Formulas for acids contain one or more hydrogen atoms as well as an anionic group. Anions whose names end in "-ide" have associated acids with a "hydro-" prefix and an "-ic" ending, as shown in Table 2.5. In some cases two different names are assigned to the same chemical formula. For instance, HCl is known as both hydrogen chloride and hydrochloric acid. The name used for this compound depends on its physical state. In the gaseous or pure liquid state, HCl is a molecular compound called hydrogen chloride. When it is dissolved in water, the molecules break up into H^+ and Cl^- ions; in this condition, the substance is called hydrochloric acid.

Oxoacids are acids that *contain hydrogen, oxygen, and another element (the central element).* The formulas of oxoacids are usually written with the H first, followed

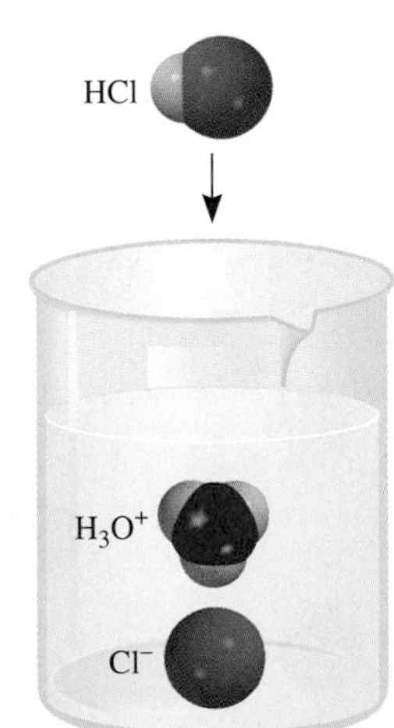

When dissolved in water, the HCl molecule is converted to the H^+ and Cl^- ions. The H^+ ion is associated with one or more water molecules, and is usually represented as H_3O^+.

Note that these acids all exist as molecular compounds in the gas phase.

Table 2.5 Some Simple Acids

Anion	Corresponding Acid
F^- (fluoride)	HF (hydrofluoric acid)
Cl^- (chloride)	HCl (hydrochloric acid)
Br^- (bromide)	HBr (hydrobromic acid)
I^- (iodide)	HI (hydroiodic acid)
CN^- (cyanide)	HCN (hydrocyanic acid)
S^{2-} (sulfide)	H_2S (hydrosulfuric acid)

by the central element and then O. We use the following five common acids as our references in naming oxoacids:

H_2CO_3	carbonic acid	H_3PO_4	phosphoric acid
$HClO_3$	chloric acid	H_2SO_4	sulfuric acid
HNO_3	nitric acid		

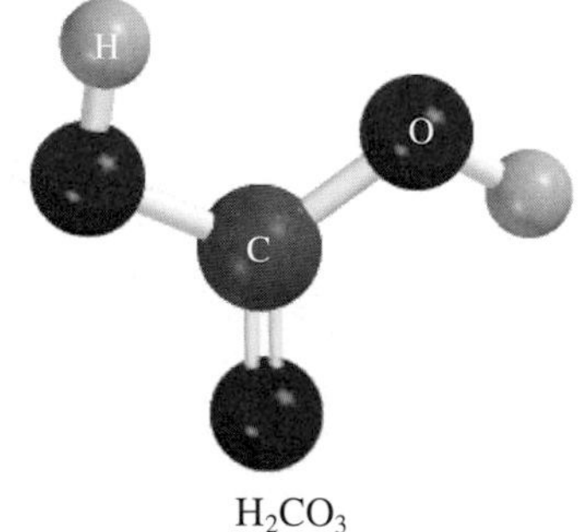

H_2CO_3

Often two or more oxoacids have the same central atom but a different number of O atoms. Starting with our reference oxoacids, whose names all end with "-ic," we use the following rules to name these compounds.

1. Addition of one O atom to the "-ic" acid: The acid is called "per . . . -ic" acid. Thus, adding an O atom to $HClO_3$ changes chloric acid to perchloric acid, $HClO_4$.
2. Removal of one O atom from the "-ic" acid: The acid is called "-ous" acid. Thus, nitric acid, HNO_3, becomes nitrous acid, HNO_2.
3. Removal of two O atoms from the "-ic" acid: The acid is called "hypo . . . -ous" acid. Thus, when $HBrO_3$ is converted to HBrO, the acid is called hypobromous acid.

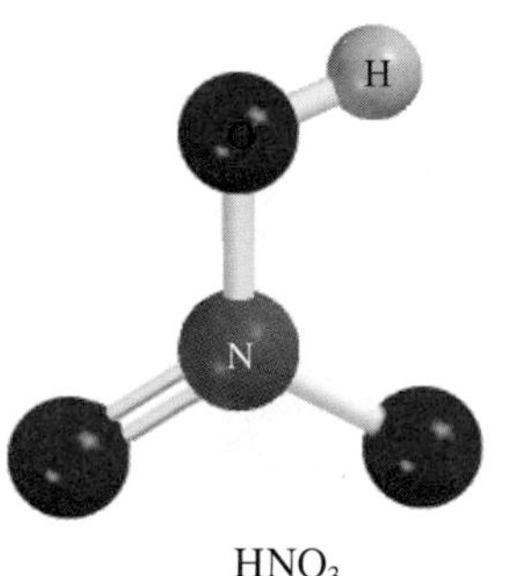

HNO_3

The rules for naming *anions of oxoacids,* called ***oxoanions,*** are

1. When all the H ions are removed from the "-ic" acid, the anion's name ends with "-ate." For example, the anion CO_3^{2-} derived from H_2CO_3 is called carbonate.
2. When all the H ions are removed from the "-ous" acid, the anion's name ends with "-ite." Thus, the anion ClO_2^- derived from $HClO_2$ is called chlorite.
3. The names of anions in which one or more but not all of the hydrogen ions have been removed must indicate the number of H ions present. For example, consider the anions derived from phosphoric acid:

H_3PO_4	Phosphoric acid	HPO_4^{2-}	Hydrogen phosphate
$H_2PO_4^-$	Dihydrogen phosphate	PO_4^{3-}	Phosphate

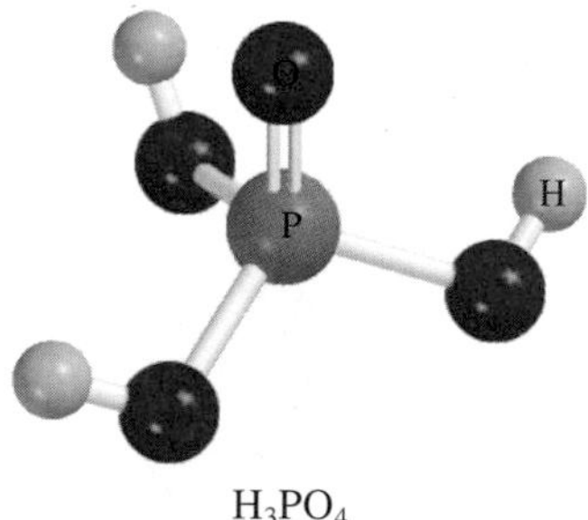

H_3PO_4

Note that we usually omit the prefix "mono-" when there is only one H in the anion. Table 2.6 gives the names of the oxoacids and oxoanions that contain

Table 2.6 Names of Oxoacids and Oxoanions That Contain Chlorine

Acid	Anion
$HClO_4$ (perchloric acid)	ClO_4^- (perchlorate)
$HClO_3$ (chloric acid)	ClO_3^- (chlorate)
$HClO_2$ (chlorous acid)	ClO_2^- (chlorite)
HClO (hypochlorous acid)	ClO^- (hypochlorite)

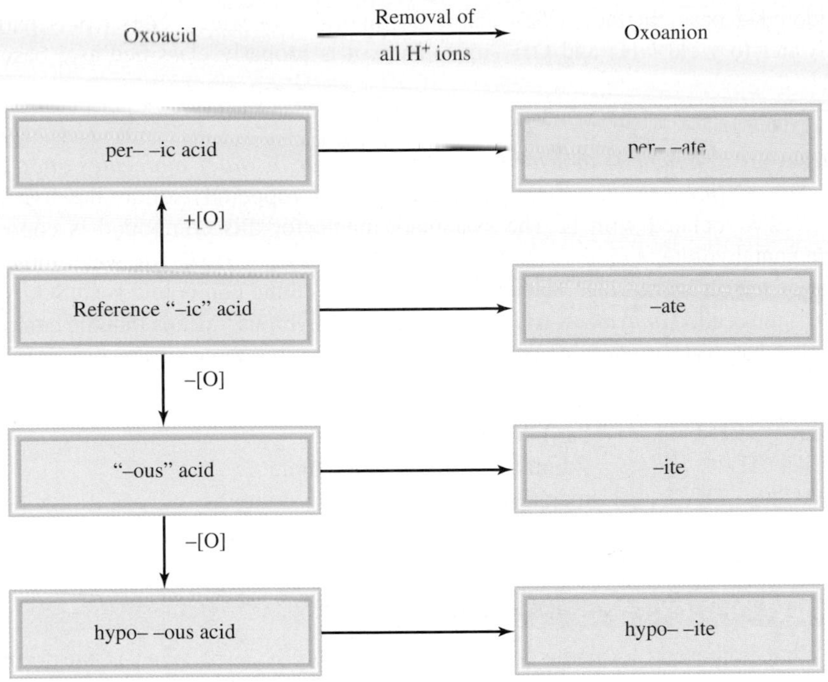

Oxoacid	Removal of all H^+ ions →	Oxoanion
per– –ic acid	→	per– –ate
↑ +[O]		
Reference "–ic" acid	→	–ate
↓ –[O]		
"–ous" acid	→	–ite
↓ –[O]		
hypo– –ous acid	→	hypo– –ite

Figure 2.14 *Naming oxoacids and oxoanions.*

chlorine, and Figure 2.14 summarizes the nomenclature for the oxoacids and oxoanions.

EXAMPLE 2.8

Name the following oxoacid and oxoanion: (a) H_3PO_3 and (b) IO_4^-.

Strategy We refer to Figure 2.14 and Table 2.6 for the conventions used in naming oxoacids and oxoanions.

Solution

(a) We start with our reference acid, phosphoric acid (H_3PO_4). Because H_3PO_3 has one fewer O atom, it is called phosphorous acid.

(b) The parent acid is HIO_4. Because the acid has one more O atom than our reference iodic acid (HIO_3), it is called periodic acid. Therefore, the anion derived from HIO_4 is called periodate.

Practice Exercises Name the following oxoacid and oxoanion: (a) HBrO and (b) HSO_4^-.

Similar problems: 2.48(f), 2.49(c).

Naming Bases

A ***base*** can be described as *a substance that yields hydroxide ions (OH^-) when dissolved in water.* Some examples are

NaOH	Sodium hydroxide	$Ba(OH)_2$	Barium hydroxide
KOH	Potassium hydroxide		

Ammonia (NH_3), a molecular compound in the gaseous or pure liquid state, is also classified as a common base. At first glance this may seem to be an exception to the definition of a base. But note that as long as a substance *yields* hydroxide ions when dissolved in water, it need not contain hydroxide ions in its structure to be considered a base. In fact, when ammonia dissolves in water, NH_3 reacts partially with water to yield NH_4^+ and OH^- ions. Thus, it is properly classified as a base.

Hydrates

$CuSO_4 \cdot 5H_2O$ (left) is blue; $CuSO_4$ (right) is white.

Hydrates are *compounds that have a specific number of water molecules attached to them.* For example, in its normal state, each unit of copper(II) sulfate has five water molecules associated with it. The systematic name for this compound is copper(II) sulfate pentahydrate, and its formula is written as $CuSO_4 \cdot 5H_2O$. The water molecules can be driven off by heating. When this occurs, the resulting compound is $CuSO_4$, which is sometimes called *anhydrous* copper(II) sulfate; "anhydrous" means that the compound no longer has water molecules associated with it (see the photo in the margin). Some other hydrates are

$BaCl_2 \cdot 2H_2O$	barium chloride dihydrate
$LiCl \cdot H_2O$	lithium chloride monohydrate
$MgSO_4 \cdot 7H_2O$	magnesium sulfate heptahydrate

REVIEW OF CONCEPTS

(a) What is the correct name for the compound Mg_3N_2? (b) What is the formula of iodine chloride?

2.8 Introduction to Organic Compounds

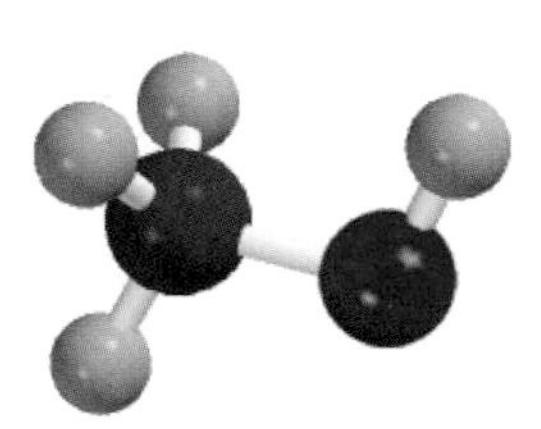

CH_3OH

CH_3NH_2

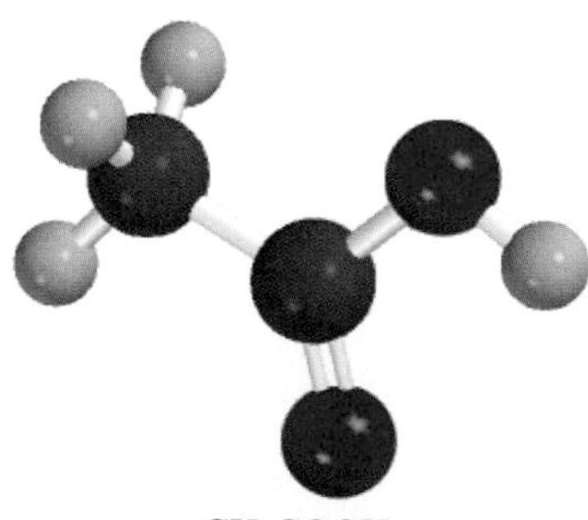

CH_3COOH

The simplest type of organic compounds is the *hydrocarbons,* which contain only carbon and hydrogen atoms. The hydrocarbons are used as fuels for domestic and industrial heating, for generating electricity and powering internal combustion engines, and as starting materials for the chemical industry. One class of hydrocarbons is called the *alkanes.* Table 2.7 shows the names, formulas, and molecular models of the first ten *straight-chain* alkanes, in which the carbon chains have no branches. Note that all the names end with *-ane*. Starting with C_5H_{12}, we use the Greek prefixes in Table 2.4 to indicate the number of carbon atoms present.

The chemistry of organic compounds is largely determined by the *functional groups,* which consist of one or a few atoms bonded in a specific way. For example, when an H atom in methane is replaced by a hydroxyl group (—OH), an amino group (—NH_2), and a carboxyl group (—COOH), the following molecules are generated:

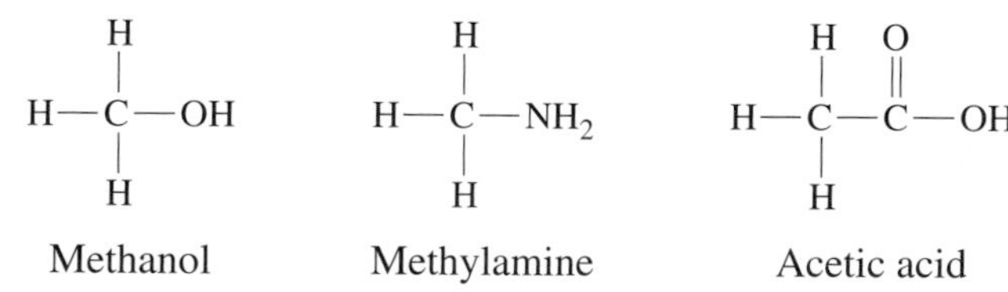

The chemical properties of these molecules can be predicted based on the reactivity of the functional groups. We will frequently use organic compounds as examples to illustrate chemical bonding, acid-base reactions, and other properties throughout the book.

Table 2.7 The First Ten Straight-Chain Alkanes

Name	Formula	Molecular Model
Methane	CH_4	
Ethane	C_2H_6	
Propane	C_3H_8	
Butane	C_4H_{10}	
Pentane	C_5H_{12}	
Hexane	C_6H_{14}	
Heptane	C_7H_{16}	
Octane	C_8H_{18}	
Nonane	C_9H_{20}	
Decane	$C_{10}H_{22}$	

Summary of Facts and Concepts

1. Modern chemistry began with Dalton's atomic theory, which states that all matter is composed of tiny, indivisible particles called atoms; that all atoms of the same element are identical; that compounds contain atoms of different elements combined in whole-number ratios; and that atoms are neither created nor destroyed in chemical reactions (the law of conservation of mass).
2. Atoms of constituent elements in a particular compound are always combined in the same proportions by mass

(law of definite proportions). When two elements can combine to form more than one type of compound, the masses of one element that combine with a fixed mass of the other element are in a ratio of small whole numbers (law of multiple proportions).

3. An atom consists of a very dense central nucleus made up of protons and neutrons, plus electrons that move about the nucleus at a relatively large distance from it. Protons are positively charged, neutrons have no charge, and electrons are negatively charged. Protons and neutrons have roughly the same mass, which is about 1840 times greater than the mass of an electron.
4. The atomic number of an element is the number of protons in the nucleus of an atom of the element; it determines the identity of an element. The mass number is the sum of the number of protons and the number of neutrons in the nucleus. Isotopes are atoms of the same element that have the same number of protons but different numbers of neutrons.
5. Chemical formulas combine the symbols for the constituent elements with whole-number subscripts to show the type and number of atoms contained in the smallest unit of a compound. The molecular formula conveys the specific number and types of atoms combined in each molecule of a compound. The empirical formula shows the simplest ratios of the atoms in a molecule.
6. Chemical compounds are either molecular compounds (in which the smallest units are discrete, individual molecules) or ionic compounds (in which positive and negative ions are held together by mutual attraction). Ionic compounds are made up of cations and anions, formed when atoms lose electrons and gain electrons, respectively.
7. The names of many inorganic compounds can be deduced from a set of simple rules. The formulas can be written from the names of the compounds.
8. The simplest type of organic compounds is the hydrocarbons.

Key Words

Acid, p. 49
Alkali metals, p. 39
Alkaline earth metals, p. 39
Allotrope, p. 41
Alpha (α) particles, p. 33
Alpha (α) rays, p. 33
Anion, p. 40
Atom, p. 31
Atomic number (Z), p. 36
Base, p. 51
Beta (β) particles, p. 33
Beta (β) rays, p. 33
Binary compound, p. 45
Cation, p. 40
Chemical formula, p. 41
Diatomic molecule, p. 39
Electron, p. 32
Empirical formula, p. 42
Families, p. 38
Gamma (γ) rays, p. 34
Groups, p. 38
Halogens, p. 39
Hydrate, p. 52
Ion, p. 39
Ionic compound, p. 40
Isotope, p. 36
Law of conservation of mass, p. 31
Law of definite proportions, p. 30
Law of multiple proportions, p. 31
Mass number (A), p. 36
Metal, p. 38
Metalloid, p. 38
Molecular formula, p. 41
Molecule, p. 39
Monatomic ion, p. 40
Neutron, p. 35
Noble gases, p. 39
Nonmetal, p. 38
Nucleus, p. 34
Oxoacid, p. 49
Oxoanion, p. 50
Periodic Table, p. 38
Periods, p. 38
Polyatomic Ion, p. 40
Polyatomic Molecule, p. 39
Proton, p. 34
Radiation, p. 31
Radioactivity, p. 33
Rare Gases, p. 39
Structural Formula, p. 42
Ternary Compound, p. 45

Questions and Problems

Structure of the Atom

Review Questions

2.1 Define these terms: (a) α particle, (b) β particle, (c) γ ray, (d) X ray.

2.2 List the types of radiation that are known to be emitted by radioactive elements.

2.3 Compare the properties of: α particles, cathode rays, protons, neutrons, and electrons. What is meant by the term "fundamental particle"?

2.4 Describe the contributions of these scientists to our knowledge of atomic structure: J. J. Thomson, R. A. Millikan, Ernest Rutherford, James Chadwick.

2.5 A sample of a radioactive element is found to be losing mass gradually. Explain what is happening to the sample.

2.6 Describe the experimental basis for believing that the nucleus occupies a very small fraction of the volume of the atom.

Problems

2.7 The diameter of a neutral helium atom is about 1×10^2 pm. Suppose that we could line up helium atoms side by side in contact with one another. Approximately how many atoms would it take to make the distance from end to end 1 cm?

2.8 Roughly speaking, the radius of an atom is about 10,000 times greater than that of its nucleus. If an atom were magnified so that the radius of its nucleus became 10 cm, what would be the radius of the atom in miles? (1 mi = 1609 m.)

Atomic Number, Mass Number, and Isotopes

Review Questions

2.9 Define these terms: (a) atomic number, (b) mass number. Why does a knowledge of atomic number enable us to deduce the number of electrons present in an atom?

2.10 Why do all atoms of an element have the same atomic number, although they may have different mass numbers? What do we call atoms of the same element with different mass numbers? Explain the meaning of each term in the symbol $^{A}_{Z}X$.

Problems

2.11 What is the mass number of an iron atom that has 28 neutrons?

2.12 Calculate the number of neutrons of ^{239}Pu.

2.13 For each of these species, determine the number of protons and the number of neutrons in the nucleus:

$$^{3}_{2}He,\ ^{4}_{2}He,\ ^{24}_{12}Mg,\ ^{25}_{12}Mg,\ ^{48}_{22}Ti,\ ^{79}_{35}Br,\ ^{195}_{78}Pt$$

2.14 Indicate the number of protons, neutrons, and electrons in each of these species:

$$^{15}_{7}N,\ ^{33}_{16}S,\ ^{63}_{29}Cu,\ ^{84}_{38}Sr,\ ^{130}_{56}Ba,\ ^{186}_{74}W,\ ^{202}_{80}Hg$$

2.15 Write the appropriate symbol for each of these isotopes: (a) $Z = 11, A = 23$; (b) $Z = 28, A = 64$.

2.16 Write the appropriate symbol for each of these isotopes: (a) $Z = 74, A = 186$; (b) $Z = 80, A = 201$.

The Periodic Table

Review Questions

2.17 What is the periodic table, and what is its significance in the study of chemistry? What are groups and periods in the periodic table?

2.18 Give two differences between a metal and a nonmetal.

2.19 Write the names and symbols for four elements in each of these categories: (a) nonmetal, (b) metal, (c) metalloid.

2.20 Without consulting a periodic table, name each of the lettered groups in the following table. Provide two examples from each group.

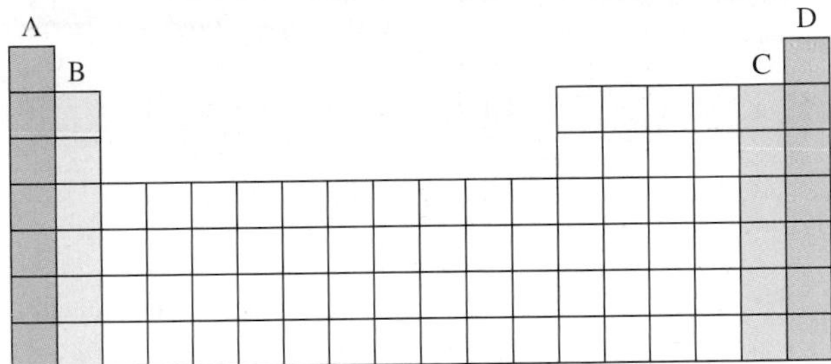

Problems

2.21 Elements whose names end with "-ium" are usually metals; sodium is one example. Identify a nonmetal whose name also ends with "-ium."

2.22 Describe the changes in properties (from metals to nonmetals or from nonmetals to metals) as we move (a) down a periodic group and (b) across the periodic table.

2.23 Consult a handbook of chemical and physical data (ask your instructor where you can locate a copy of the handbook) to find (a) two metals less dense than water, (b) two metals more dense than mercury, (c) the densest known solid metallic element, (d) the densest known solid nonmetallic element.

2.24 Group these elements in pairs that you would expect to show similar chemical properties: K, F, P, Na, Cl, and N.

Molecules and Ions

Review Questions

2.25 What is the difference between an atom and a molecule?

2.26 What are allotropes? Give an example. How are allotropes different from isotopes?

2.27 Describe the two commonly used molecular models.

2.28 Give an example of each of the following: (a) a monatomic cation, (b) a monatomic anion, (c) a polyatomic cation, (d) a polyatomic anion.

Problems

2.29 Which of the following diagrams represent diatomic molecules, polyatomic molecules, molecules that are not compounds, molecules that are compounds, or an elemental form of the substance?

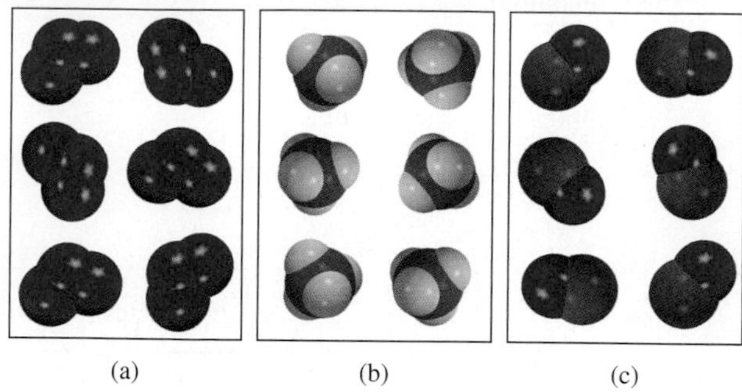

2.30 Which of the following diagrams represent diatomic molecules, polyatomic molecules, molecules that are not compounds, molecules that are compounds, or an elemental form of the substance?

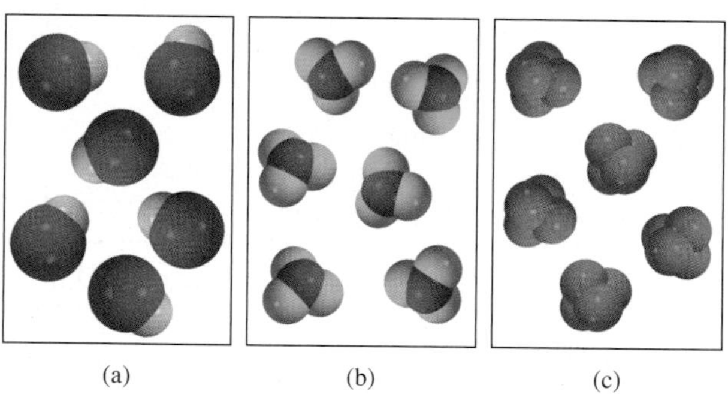

2.31 Identify the following as elements or compounds: NH_3, N_2, S_8, NO, CO, CO_2, H_2, SO_2.

2.32 Give two examples of each of the following: (a) a diatomic molecule containing atoms of the same element, (b) a diatomic molecule containing atoms of different elements, (c) a polyatomic molecule containing atoms of the same element, (d) a polyatomic molecule containing atoms of different elements.

2.33 Give the number of protons and electrons in each of the following common ions: Na^+, Ca^{2+}, Al^{3+}, Fe^{2+}, I^-, F^-, S^{2-}, O^{2-}, N^{3-}.

2.34 Give the number of protons and electrons in each of the following common ions: K^+, Mg^{2+}, Fe^{3+}, Br^-, Mn^{2+}, C^{4-}, Cu^{2+}.

Chemical Formulas

Review Questions

2.35 What does a chemical formula represent? What is the ratio of the atoms in the following molecular formulas? (a) NO, (b) NCl_3, (c) N_2O_4, (d) P_4O_6

2.36 Define molecular formula and empirical formula. What are the similarities and differences between the empirical formula and molecular formula of a compound?

2.37 Give an example of a case in which two molecules have different molecular formulas but the same empirical formula.

2.38 What does P_4 signify? How does it differ from 4P?

2.39 What is an ionic compound? How is electrical neutrality maintained in an ionic compound?

2.40 Explain why the chemical formulas of ionic compounds are usually the same as their empirical formulas.

Problems

2.41 What are the empirical formulas of the following compounds? (a) C_2N_2, (b) C_6H_6, (c) C_9H_{20}, (d) P_4O_{10}, (e) B_2H_6

2.42 What are the empirical formulas of the following compounds? (a) Al_2Br_6, (b) $Na_2S_2O_4$, (c) N_2O_5, (d) $K_2Cr_2O_7$

2.43 Write the molecular formula of glycine, an amino acid present in proteins. The color codes are: black (carbon), blue (nitrogen), red (oxygen), and gray (hydrogen).

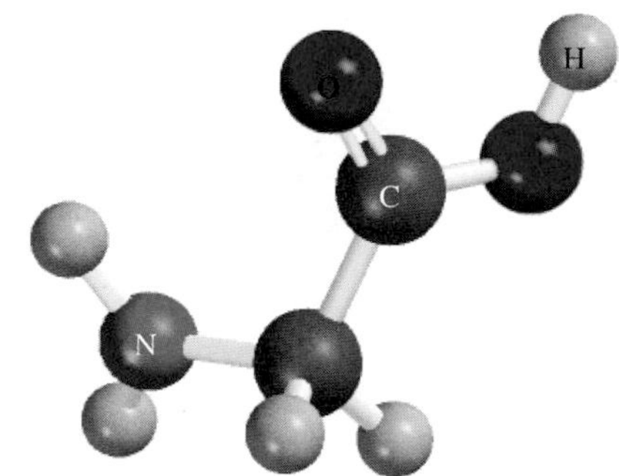

2.44 Write the molecular formula of ethanol. The color codes are: black (carbon), red (oxygen), and gray (hydrogen).

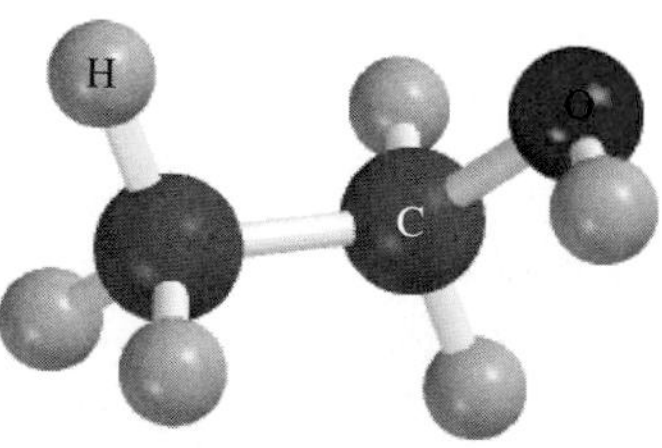

2.45 Which of the following compounds are likely to be ionic? Which are likely to be molecular? $SiCl_4$, LiF, $BaCl_2$, B_2H_6, KCl, C_2H_4

2.46 Which of the following compounds are likely to be ionic? Which are likely to be molecular? CH_4, NaBr, BaF_2, CCl_4, ICl, CsCl, NF_3

Naming Compounds

Problems

2.47 Name these compounds: (a) Na_2CrO_4, (b) K_2HPO_4, (c) HBr (gas), (d) HBr (in water), (e) Li_2CO_3, (f) $K_2Cr_2O_7$, (g) NH_4NO_2, (h) PF_3, (i) PF_5, (j) P_4O_6, (k) CdI_2, (l) $SrSO_4$, (m) $Al(OH)_3$, (n) $Na_2CO_3 \cdot 10H_2O$.

2.48 Name these compounds: (a) KClO, (b) Ag_2CO_3, (c) $FeCl_2$, (d) $KMnO_4$, (e) $CsClO_3$, (f) HIO, (g) FeO, (h) Fe_2O_3, (i) $TiCl_4$, (j) NaH, (k) Li_3N, (l) Na_2O, (m) Na_2O_2, (n) $FeCl_3 \cdot 6H_2O$.

2.49 Write the formulas for these compounds: (a) rubidium nitrite, (b) potassium sulfide, (c) perbromic acid, (d) magnesium phosphate, (e) calcium hydrogen phosphate, (f) boron trichloride, (g) iodine heptafluoride, (h) ammonium sulfate, (i) silver perchlorate, (j) iron(III) chromate, (k) calcium sulfate dihydrate.

2.50 Write the formulas for these compounds: (a) copper(I) cyanide, (b) strontium chlorite, (c) perchloric acid, (d) hydroiodic acid, (e) disodium ammonium phosphate, (f) lead(II) carbonate, (g) tin(II) fluoride, (h) tetraphosphorus decasulfide, (i) mercury(II) oxide, (j) mercury(I) iodide, (k) cobalt(II) chloride hexahydrate.

Additional Problems

2.51 One isotope of a metallic element has a mass number of 65 and has 35 neutrons in the nucleus. The cation derived from the isotope has 28 electrons. Write the symbol for this cation.

2.52 In which one of these pairs do the two species resemble each other most closely in chemical properties? (a) ${}^1_1H$ and ${}^1_1H^+$, (b) ${}^{14}_7N$ and ${}^{14}_7N^{3-}$, (c) ${}^{12}_6C$ and ${}^{13}_6C$.

2.53 This table gives numbers of electrons, protons, and neutrons in atoms or ions of a number of elements. (a) Which of the species are neutral? (b) Which are negatively charged? (c) Which are positively charged? (d) What are the conventional symbols for all the species?

Atom or Ion of Element	A	B	C	D	E	F	G
Number of electrons	5	10	18	28	36	5	9
Number of protons	5	7	19	30	35	5	9
Number of neutrons	5	7	20	36	46	6	10

2.54 What is wrong or ambiguous about these descriptions? (a) 1 g of hydrogen, (b) four molecules of NaCl.

2.55 These phosphorus sulfides are known: P_4S_3, P_4S_7, and P_4S_{10}. Do these compounds obey the law of multiple proportions?

2.56 Which of these are elements, which are molecules but not compounds, which are compounds but not molecules, and which are both compounds and molecules? (a) SO_2, (b) S_8, (c) Cs, (d) N_2O_5, (e) O, (f) O_2, (g) O_3, (h) CH_4, (i) KBr, (j) S, (k) P_4, (l) LiF.

2.57 Determine the molecular and empirical formulas of the compounds shown here. (Black spheres are carbon and gray spheres are hydrogen).

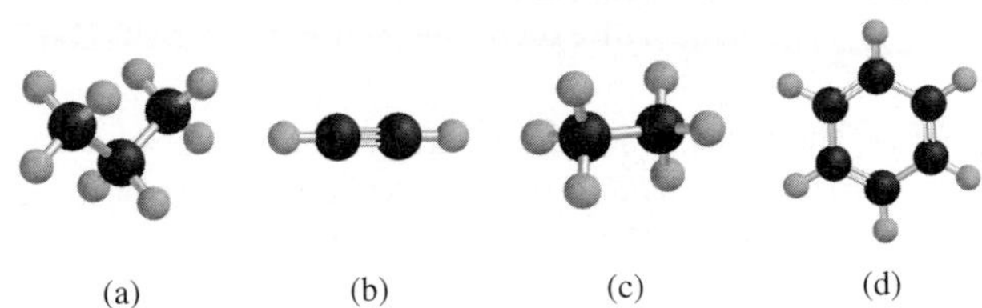

(a) (b) (c) (d)

2.58 Some compounds are better known by their common names than by their systematic chemical names. Consult a handbook, a dictionary, or your instructor for the chemical formulas of these substances: (a) dry ice, (b) table salt, (c) laughing gas, (d) marble (chalk, limestone), (e) quicklime, (f) slaked lime, (g) baking soda, (h) milk of magnesia.

2.59 Fill in the blanks in this table:

Symbol		${}^{54}_{26}Fe^{2+}$			
Protons	5			79	86
Neutrons	6		16	117	136
Electrons	5		18	79	
Net charge			−3		0

2.60 (a) Which elements are most likely to form ionic compounds? (b) Which metallic elements are most likely to form cations with different charges?

2.61 Many ionic compounds contain either aluminum (a Group 3A metal) or a metal from Group 1A or Group 2A and a nonmetal—oxygen, nitrogen, or a halogen (Group 7A). Write the chemical formulas and names of all the binary compounds that can result from such combinations.

2.62 Which of these symbols provides more information about the atom: ${}^{23}Na$ or ${}_{11}Na$? Explain.

2.63 Write the chemical formulas and names of acids that contain Group 7A elements. Do the same for elements in Groups 3A, 4A, 5A, and 6A.

2.64 While most isotopes of light elements such as oxygen and phosphorus contain relatively equal amounts of protons and neutrons in the nucleus, recent results indicate that a new class of isotopes called neutron-rich isotopes can be prepared. These neutron-rich isotopes push the limits of nuclear stability as the large numbers of neutrons approach the "neutron drip line." These neutron-rich isotopes may play a critical role in the nuclear reactions of stars. Determine the number of neutrons in the following neutron-rich isotopes: (a) ^{40}Mg, (b) ^{44}Si, (c) ^{48}Ca, (d) ^{43}Al.

2.65 Group the following elements in pairs that you would expect to show similar chemical properties: K, F, P, Na, Cl, and N.

2.66 List the elements that exist as gases at room temperature. (*Hint:* All except one element can be found in Groups 5A, 6A, 7A, and 8A.)

2.67 The Group 1B metals, Cu, Ag, and Au, are called coinage metals. What chemical properties make them specially suitable for making coins and jewels?

2.68 The elements in Group 8A of the periodic table are called noble gases. Can you guess the meaning of "noble" in this context?

2.69 The formula for calcium oxide is CaO. What are the formulas for magnesium oxide and cesium oxide?

2.70 A common mineral of barium is barytes, or barium sulfate ($BaSO_4$). Because elements in the same periodic group have similar chemical properties, we might expect to find some radium sulfate ($RaSO_4$) mixed with barytes because radium is the last member of Group 2A. However, the only source of radium compounds in nature is in uranium minerals. Why?

2.71 Fluorine reacts with hydrogen (H) and with deuterium (D) to form hydrogen fluoride (HF) and deuterium fluoride (DF) [deuterium (^2_1H) is an isotope of hydrogen]. Would a given amount of fluorine react with different masses of the two hydrogen isotopes? Does this violate the law of definite proportions? Explain.

2.72 Predict the formula and name of a binary compound formed from these elements: (a) Na and H, (b) B and O, (c) Na and S, (d) Al and F, (e) F and O, (f) Sr and Cl.

2.73 Fill the blanks in the following table.

Cation	Anion	Formula	Name
			Magnesium bicarbonate
		$SrCl_2$	
Fe^{3+}	NO_2^-		
			Manganese(II) chlorate
		$SnBr_4$	
Co^{2+}	PO_4^{3-}		
Hg_2^{2+}	I^-		
		Cu_2CO_3	
			Lithium nitride
Al^{3+}	S^{2-}		

2.74 Identify each of the following elements: (a) a halogen whose anion contains 36 electrons, (b) a radioactive noble gas with 86 protons, (c) a Group 6A element whose anion contains 36 electrons, (d) an alkali metal cation that contains 36 electrons, (e) a Group 4A cation that contains 80 electrons.

2.75 Write the molecular formulas for and names of the following compounds.

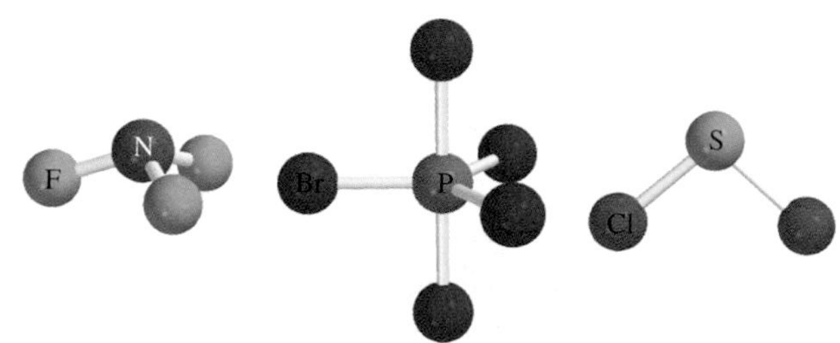

Special Problems

2.76 On p. 31 it was pointed out that mass and energy are alternate aspects of a single entity called *mass-energy.* The relationship between these two physical quantities is Einstein's famous equation, $E = mc^2$, where E is energy, m is mass, and c is the speed of light. In a combustion experiment, it was found that 12.096 g of hydrogen molecules combined with 96.000 g of oxygen molecules to form water and released 1.715×10^3 kJ of heat. Calculate the corresponding mass change in this process and comment on whether the law of conservation of mass holds for ordinary chemical processes. (*Hint:* The Einstein equation can be used to calculate the change in mass as a result of the change in energy. $1 \text{ J} = 1 \text{ kg m}^2/\text{s}^2$ and $c = 3.00 \times 10^8$ m/s.)

2.77 (a) Describe Rutherford's experiment and how it led to the structure of the atom. How was he able to estimate the number of protons in a nucleus from the scattering of the α particles? (b) Consider the ^{23}Na atom. Given that the radius and mass of the nucleus are 3.04×10^{-15} m and 3.82×10^{-23} g, respectively, calculate the density of the nucleus in g/cm^3. The radius of a ^{23}Na atom is 186 pm. Calculate the density of the space occupied by the electrons in the sodium atom. Do your results support Rutherford's model of an atom? [The volume of a sphere is $(4/3)\pi r^3$, where r is the radius.]

2.78 Ethane and acetylene are two gaseous hydrocarbons. Chemical analyses show that in one sample of ethane, 2.65 g of carbon are combined with 0.665 g of hydrogen, and in one sample of acetylene, 4.56 g of carbon are combined with 0.383 g of hydrogen. (a) Are these results consistent with the law of multiple proportions? (b) Write reasonable molecular formulas for these compounds.

2.79 Draw two different structural formulas based on the molecular formula C_2H_6O. Is the fact that you can have more than one compound with the same molecular formula consistent with Dalton's atomic theory?

2.80 A monatomic ion has a charge of +2. The nucleus of the parent atom has a mass number of 55. If the number of neutrons in the nucleus is 1.2 times that of the number of protons, what is the name and symbol of the element?

2.81 Name the following acids:

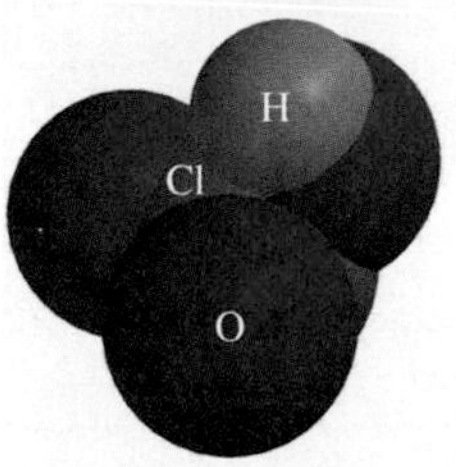

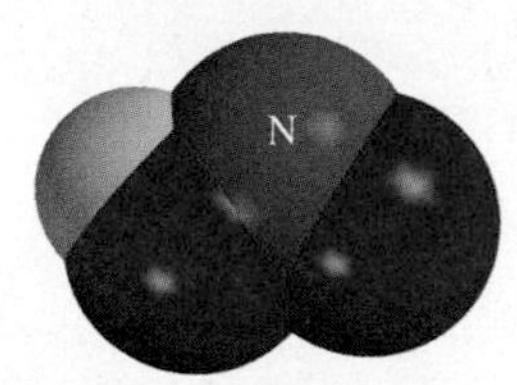

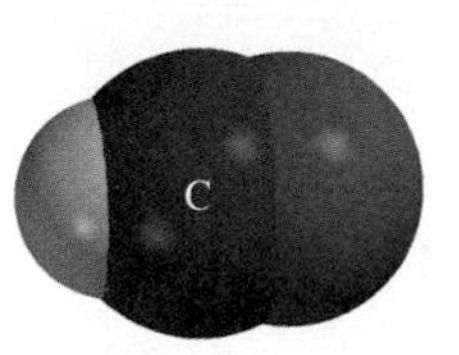

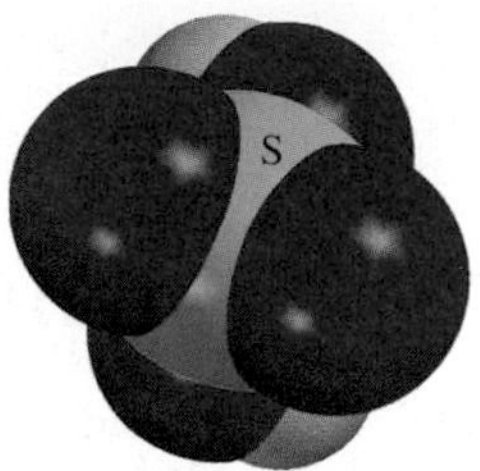

Answers to Practice Exercises

2.1 29 protons, 34 neutrons, and 29 electrons. **2.2** $CHCl_3$. **2.3** $C_4H_5N_2O$. **2.4** (a) Lead(II) oxide, (b) lithium chlorate. **2.5** (a) Rb_2SO_4, (b) BaH_2. **2.6** (a) Nitrogen trifluoride, (b) dichlorine heptoxide. **2.7** (a) SF_4, (b) N_2O_5. **2.8** (a) Hypobromous acid, (b) hydrogen sulfate ion.

CHAPTER

3

Stoichiometry

Sulfur burning in oxygen to form sulfur dioxide. About 50 million tons of SO_2 are released to the atmosphere every year.

Chapter Outline

Student Interactive Activities

Animations
Limiting Reagent (3.9)

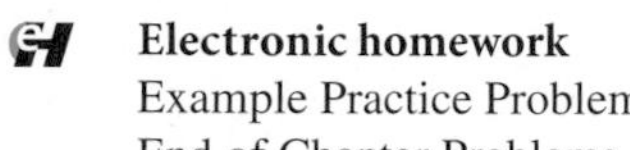

Electronic homework
Example Practice Problems
End of Chapter Problems

Essential Concepts

Atomic Mass and Molar Mass The mass of an atom, which is extremely small, is based on the carbon-12 isotope scale. An atom of the carbon-12 isotope is assigned a mass of exactly 12 atomic mass units (amu). To work with the more convenient scale of grams, chemists use the molar mass. The molar mass of carbon-12 is exactly 12 g and contains an Avogadro's number (6.022×10^{23}) of atoms. The molar masses of other elements are also expressed in grams and contain the same number of atoms. The molar mass of a molecule is the sum of the molar masses of its constituent atoms.

Percent Composition of a Compound The makeup of a compound is most conveniently expressed in terms of its percent composition, which is the percent by mass of each element the compound contains. A knowledge of its chemical formula enables us to calculate the percent composition. Experimental determination of percent composition and the molar mass of a compound enables us to determine its chemical formula.

Writing Chemical Equations An effective way to represent the outcome of a chemical reaction is to write a chemical equation, which uses chemical formulas to describe what happens. A chemical equation must be balanced so that we have the same number and type of atoms for the reactants, the starting materials, and the products, the substances formed at the end of the reaction.

Mass Relationships of a Chemical Reaction A chemical equation enables us to predict the amount of product(s) formed, called the yield, knowing how much reactant(s) was (were) used. This information is of great importance for reactions run on the laboratory or industrial scale. In practice, the actual yield is almost always less than that predicted from the equation because of various complications.

3.1 Atomic Mass

In this chapter, we will use what we have learned about chemical structure and formulas in studying the mass relationships of atoms and molecules. These relationships in turn will help us to explain the composition of compounds and the ways in which composition changes.

The mass of an atom depends on the number of electrons, protons, and neutrons it contains. Knowledge of an atom's mass is important in laboratory work. But atoms are extremely small particles—even the smallest speck of dust that our unaided eyes can detect contains as many as 1×10^{16} atoms! Clearly we cannot weigh a single atom, but it is possible to determine the mass of one atom *relative* to another experimentally. The first step is to assign a value to the mass of one atom of a given element so that it can be used as a standard.

Section 3.4 describes a method for determining atomic mass.

By international agreement, ***atomic mass*** (sometimes called *atomic weight*) is *the mass of the atom in atomic mass units (amu).* One ***atomic mass unit*** is defined as *a mass exactly equal to one-twelfth the mass of one carbon-12 atom.* Carbon-12 is the carbon isotope that has six protons and six neutrons. Setting the atomic mass of carbon-12 at 12 amu provides the standard for measuring the atomic mass of the other elements. For example, experiments have shown that, on average, a hydrogen atom is only 8.400 percent as massive as the carbon-12 atom. Thus, if the mass of one carbon-12 atom is exactly 12 amu, the atomic mass of hydrogen must be 0.084×12.00 amu or 1.008 amu. Similar calculations show that the atomic mass of oxygen is 16.00 amu and that of iron is 55.85 amu. Thus, although we do not know just how much an average iron atom's mass is, we know that it is approximately 56 times as massive as a hydrogen atom.

One atomic mass unit is also called one dalton.

Average Atomic Mass

When you look up the atomic mass of carbon in a table such as the one on the inside front cover of this book, you will find that its value is not 12.00 amu but 12.01 amu. The reason for the difference is that most naturally occurring elements (including carbon) have more than one isotope. This means that when we measure the atomic mass of an element, we must generally settle for the *average* mass of the naturally occurring mixture of isotopes. For example, the natural abundances of carbon-12 and carbon-13 are 98.90 percent and 1.10 percent, respectively. The atomic mass of carbon-13 has been determined to be 13.00335 amu. Thus, the average atomic mass of carbon can be calculated as follows:

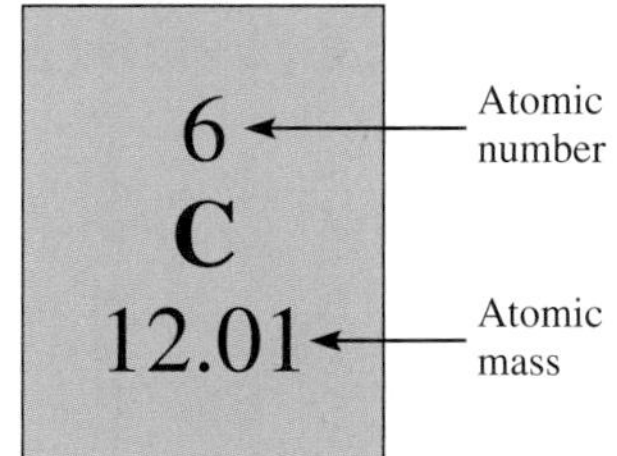

$$\begin{aligned}\text{average atomic mass}\\ \text{of natural carbon} &= (0.9890)(12.00000\text{ amu}) + (0.0110)(13.00335\text{ amu})\\ &= 12.01\text{ amu}\end{aligned}$$

Note that in calculations involving percentages, we need to convert percentages to fractions. For example, 98.90 percent becomes 98.90/100, or 0.9890. Because there are many more carbon-12 atoms than carbon-13 atoms in naturally occurring carbon, the average atomic mass is much closer to 12 amu than to 13 amu.

It is important to understand that when we say that the atomic mass of carbon is 12.01 amu, we are referring to the *average* value. If carbon atoms could be examined individually, we would find either an atom of atomic mass 12.00000 amu or one of 13.00335 amu, but never one of 12.01 amu.

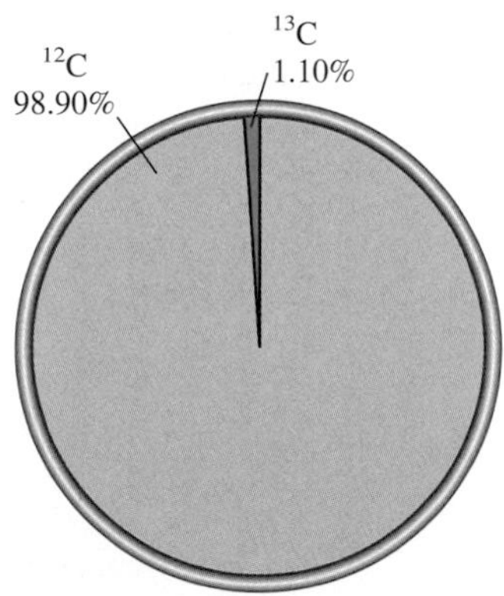

Natural abundances of C-12 and C-13 isotopes.

EXAMPLE 3.1

Copper, a metal known since ancient times, is used in electrical cables and pennies, among other things. The atomic masses of its two stable isotopes, $^{63}_{29}Cu$ (69.09 percent) and $^{65}_{29}Cu$ (30.91 percent), are 62.93 amu and 64.9278 amu, respectively. Calculate the average atomic mass of copper. The relative abundances are given in parentheses.

Strategy Each isotope contributes to the average atomic mass based on its relative abundance. Multiplying the mass of an isotope by its fractional abundance (not percent) will give the contribution to the average atomic mass of that particular isotope.

Solution First the percents are converted to fractions: 69.09 percent to 69.09/100 or 0.6909 and 30.91 percent to 30.91/100 or 0.3091. We find the contribution to the average atomic mass for each isotope, then add the contributions together to obtain the average atomic mass.

$$(0.6909)(62.93 \text{ amu}) + (0.3091)(64.9278 \text{ amu}) = \boxed{63.55 \text{ amu}}$$

Check The average atomic mass should be between the two isotopic masses; therefore, the answer is reasonable. Note that because there are more $^{63}_{29}Cu$ than $^{65}_{29}Cu$ isotopes, the average atomic mass is closer to 62.93 amu than to 64.9278 amu.

Practice Exercise The atomic masses of the two stable isotopes of boron, $^{10}_{5}B$ (19.78 percent) and $^{11}_{5}B$ (80.22 percent), are 10.0129 amu and 11.0093 amu, respectively. Calculate the average atomic mass of boron.

Copper metal and the solid-state structure of copper.

Similar problems: 3.5, 3.6.

The atomic masses of many elements have been accurately determined to five or six significant figures. However, for our purposes we will normally use atomic masses accurate only to four significant figures (see table of atomic masses inside the front cover). For simplicity, we will omit the word "average" when we discuss the atomic masses of the elements.

REVIEW OF CONCEPTS

The atomic mass of helium (He) as reported on the periodic table is 4.003 amu. Given that there are two stable isotopes of He ($^{3}_{2}He$ and $^{4}_{2}He$), what is the probability that a single, randomly selected atom of helium would have a mass of 4.003 amu?

3.2 Avogadro's Number and the Molar Mass of an Element

Atomic mass units provide a relative scale for the masses of the elements. But because atoms have such small masses, no usable scale can be devised to weigh them in calibrated units of atomic mass units. In any real situation, we deal with macroscopic samples containing enormous numbers of atoms. Therefore, it is convenient to have a special unit to describe a very large number of atoms. The idea of a unit to denote a particular number of objects is not new. For example, the pair (2 items), the dozen (12 items), and the gross (144 items) are all familiar units. Chemists measure atoms and molecules in moles.

The adjective formed from the noun "mole" is "molar."

In the SI system the ***mole (mol)*** is *the amount of a substance that contains as many elementary entities (atoms, molecules, or other particles) as there are atoms in*

Figure 3.1
One mole each of several common elements. Carbon (black charcoal powder), sulfur (yellow powder), iron (as nails), copper (wires), and mercury (shiny liquid metal).

exactly 12 g (or 0.012 kg) of the carbon-12 isotope. The actual number of atoms in 12 g of carbon-12 is determined experimentally. This number is called ***Avogadro's number*** ***(N_A)***, in honor of the Italian scientist Amedeo Avogadro. The currently accepted value is

$$N_A = 6.0221415 \times 10^{23}$$

Generally, we round Avogadro's number to 6.022×10^{23}. Thus, just as one dozen oranges contains 12 oranges, 1 mole of hydrogen atoms contains 6.022×10^{23} H atoms. Figure 3.1 shows samples containing 1 mole each of several common elements.

The enormity of Avogadro's number is difficult to imagine. For example, spreading 6.022×10^{23} oranges over the entire surface of Earth would produce a layer 9 mi into space! Because atoms (and molecules) are so tiny, we need a huge number to study them in manageable quantities.

We have seen that 1 mole of carbon-12 atoms has a mass of exactly 12 g and contains 6.022×10^{23} atoms. This mass of carbon-12 is its ***molar mass*** ***($\mathcal{M}$),*** defined as *the mass (in grams or kilograms) of 1 mole of units* (such as atoms or molecules) *of a substance*. Note that the molar mass of carbon-12 (in grams) is numerically equal to its atomic mass in amu. Likewise, the atomic mass of sodium (Na) is 22.99 amu and its molar mass is 22.99 g; the atomic mass of phosphorus is 30.97 amu and its molar mass is 30.97 g; and so on. If we know the atomic mass of an element, we also know its molar mass.

In calculations, the units of molar mass are g/mol or kg/mol.

Knowing the molar mass and Avogadro's number, we can calculate the mass of a single atom in grams. For example, we know the molar mass of carbon-12 is 12.00 g and there are 6.022×10^{23} carbon-12 atoms in 1 mole of the substance; therefore, the mass of one carbon-12 atom is given by

$$\frac{12.00 \text{ g carbon-12 atoms}}{6.022 \times 10^{23} \text{ carbon-12 atoms}} = 1.993 \times 10^{-23} \text{ g}$$

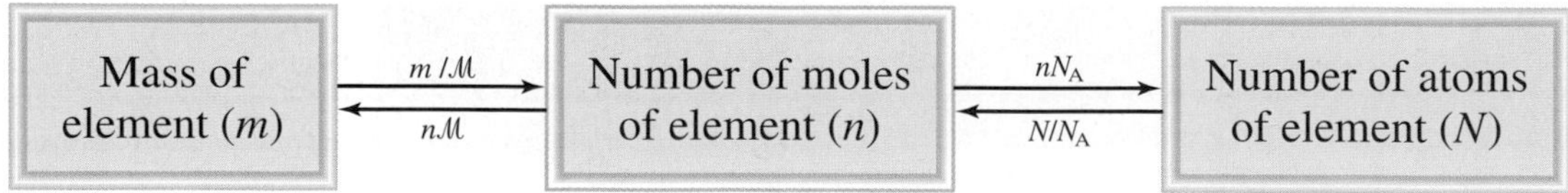

Figure 3.2
The relationships between mass (m in grams) of an element and number of moles of an element (n) and between number of moles of an element and number of atoms (N) of an element. ℳ is the molar mass (g/mol) of the element and N_A is Avogadro's number.

We can use the preceding result to determine the relationship between atomic mass units and grams. Because the mass of every carbon-12 atom is exactly 12 amu, the number of atomic mass units equivalent to 1 gram is

$$\frac{\text{amu}}{\text{gram}} = \frac{12 \text{ amu}}{1 \cancel{\text{carbon-12 atom}}} \times \frac{1 \cancel{\text{carbon-12 atom}}}{1.993 \times 10^{-23} \text{ g}}$$
$$= 6.022 \times 10^{23} \text{ amu/g}$$

Thus,

$$1 \text{ g} = 6.022 \times 10^{23} \text{ amu}$$

and

$$1 \text{ amu} = 1.661 \times 10^{-24} \text{ g}$$

This example shows that Avogadro's number can be used to convert from the atomic mass units to mass in grams and vice versa.

The notions of Avogadro's number and molar mass enable us to carry out conversions between mass and moles of atoms and between moles and number of atoms (Figure 3.2). We will employ the following conversion factors in the calculations:

$$\frac{1 \text{ mol X}}{\text{molar mass of X}} \quad \text{and} \quad \frac{1 \text{ mol X}}{6.022 \times 10^{23} \text{ X atoms}}$$

where X represents the symbol of an element. Using the proper conversion factors we can convert one quantity to another, as Examples 3.2–3.4 show.

Zinc.

EXAMPLE 3.2

Zinc (Zn) is a silvery metal that is used in making brass (with copper) and in plating iron to prevent corrosion. How many moles of Zn are there in 45.9 g of Zn?

Strategy We are trying to solve for moles of Zn. What conversion factor do we need to convert between grams and moles? Arrange the appropriate conversion factor so that grams cancel and the unit mol is obtained for your answer.

Solution The conversion factor needed to convert between grams and moles is the molar mass. In the periodic table (see inside front cover) we see that the molar mass of Zn is 65.39 g. This can be expressed as

$$1 \text{ mol Zn} = 65.39 \text{ g Zn}$$

From this equality, we can write the two conversion factors

$$\frac{1 \text{ mol Zn}}{65.39 \text{ g Zn}} \quad \text{and} \quad \frac{65.39 \text{ g Zn}}{1 \text{ mol Zn}}$$

(Continued)

The conversion factor on the left is the correct one. Grams will cancel, leaving the unit of mol for the answer. The number of moles of Zn is

$$45.9 \text{ g Zn} \times \frac{1 \text{ mol Zn}}{65.39 \text{ g Zn}} = 0.702 \text{ mol Zn}$$

Thus, there is 0.702 mole of Zn in 45.9 g of Zn.

Check Because 45.9 g is less than the molar mass of Zn, we expect the result to be less than 1 mole.

Similar problem: 3.15.

Practice Exercise Calculate the number of grams of lead (Pb) in 12.4 moles of lead.

EXAMPLE 3.3

Sulfur (S) is a nonmetallic element that is present in coal. When coal is burned, sulfur is converted to sulfur dioxide and eventually to sulfuric acid, which gives rise to the acid rain phenomenon. How many atoms are in 25.1 g of S?

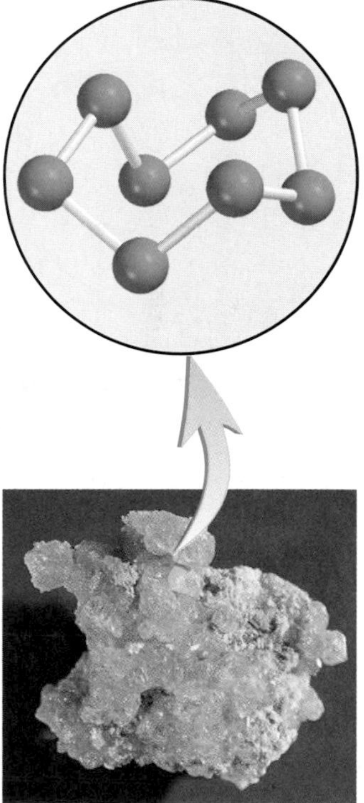

Elemental sulfur (S_8) consists of eight S atoms joined in a ring.

Strategy The question asks for atoms of sulfur. We cannot convert directly from grams to atoms of sulfur. What unit do we need to convert grams of sulfur to in order to convert to atoms? What does Avogadro's number represent?

Solution We need two conversions: first from grams to moles and then from moles to number of particles (atoms). The first step is similar to Example 3.2. Because

$$1 \text{ mol S} = 32.07 \text{ g S}$$

the conversion factor is

$$\frac{1 \text{ mol S}}{32.07 \text{ g S}}$$

Avogadro's number is the key to the second step. We have

$$1 \text{ mol} = 6.022 \times 10^{23} \text{ particles (atoms)}$$

and the conversion factors are

$$\frac{6.022 \times 10^{23} \text{ S atoms}}{1 \text{ mol S}} \quad \text{and} \quad \frac{1 \text{ mol S}}{6.022 \times 10^{23} \text{ S atoms}}$$

The conversion factor on the left is the one we need because it has the number of S atoms in the numerator. We can solve the problem by first calculating the number of moles contained in 25.1 g of S, and then calculating the number of S atoms from the number of moles of S:

$$\text{grams of S} \longrightarrow \text{moles of S} \longrightarrow \text{number of S atoms}$$

We can combine these conversions in one step as follows:

$$25.1 \text{ g S} \times \frac{1 \text{ mol S}}{32.07 \text{ g S}} \times \frac{6.022 \times 10^{23} \text{ S atoms}}{1 \text{ mol S}} = 4.71 \times 10^{23} \text{ S atoms}$$

Thus, there are 4.71×10^{23} atoms of S in 25.1 g of S.

Check Should 25.1 g S contain fewer than Avogadro's number of atoms? What mass of S would contain Avogadro's number of atoms?

Similar problems: 3.20, 3.21.

Practice Exercise Calculate the number of atoms in 0.551 g of potassium (K).

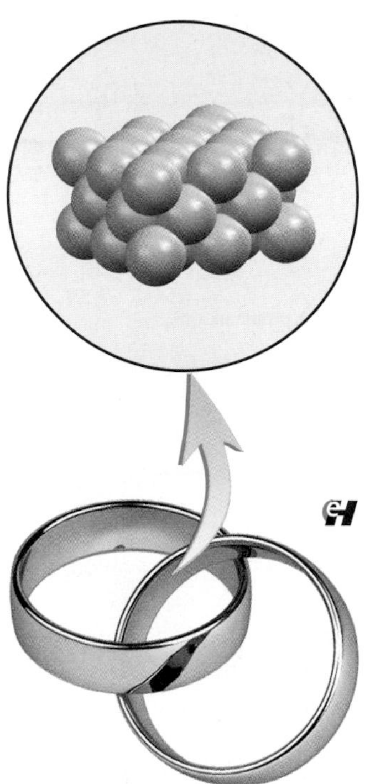

Silver rings and the solid-state structure of silver.

EXAMPLE 3.4

Silver (Ag) is a precious metal used mainly in jewelry. What is the mass (in grams) of one Ag atom?

Strategy The question asks for the mass of one Ag atom. How many Ag atoms are in 1 mole of Ag and what is the molar mass of Ag?

Solution Because 1 mole of Ag atom contains 6.022×10^{23} Ag atoms and has a mass of 107.9 g, we can calculate the mass of one Ag atom as follows:

$$1 \text{ Ag atom} \times \frac{1 \text{ mol Ag}}{6.022 \times 10^{23} \text{ Ag atoms}} \times \frac{107.9 \text{ g}}{1 \text{ mol Ag}} = 1.792 \times 10^{-22} \text{ g}$$

Check Because 6.022×10^{23} atoms of Ag have a mass 107.9 g, one atom of Ag should have a significantly smaller mass.

Practice Exercise What is the mass (in grams) of one iodine (I) atom?

Similar problem: 3.17.

REVIEW OF CONCEPTS

Referring only to the periodic table in the inside front cover and Figure 3.2, determine which of the following contains the largest number of atoms: (a) 2 g of He, (b) 110 g of Fe, and (c) 250 g of Hg.

3.3 Molecular Mass

If we know the atomic masses of the component atoms, we can calculate the mass of a molecule. The ***molecular mass*** (sometimes called *molecular weight*) is *the sum of the atomic masses (in amu) in the molecule.* For example, the molecular mass of H_2O is

$$2(\text{atomic mass of H}) + \text{atomic mass of O}$$

or

$$2(1.008 \text{ amu}) + 16.00 \text{ amu} = 18.02 \text{ amu}$$

In general, we need to multiply the atomic mass of each element by the number of atoms of that element present in the molecule and sum over all the elements.

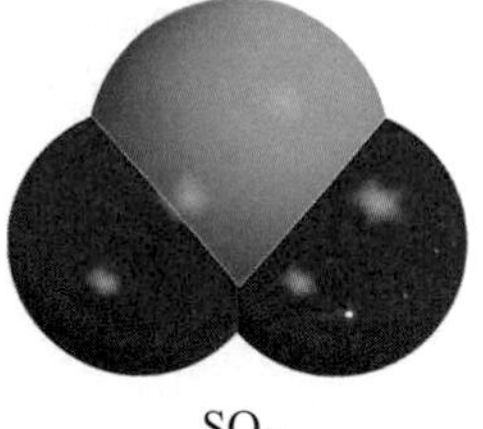

SO_2

EXAMPLE 3.5

Calculate the molecular masses (in amu) of the following compounds: (a) sulfur dioxide (SO_2) and (b) caffeine ($C_8H_{10}N_4O_2$).

Strategy How do atomic masses of different elements combine to give the molecular mass of a compound?

Solution To calculate molecular mass, we need to sum all the atomic masses in the molecule. For each element, we multiply the atomic mass of the element by the number of atoms of that element in the molecule. We find atomic masses in the periodic table (inside front cover).

(a) There are two O atoms and one S atom in SO_2, so that

$$\text{molecular mass of } SO_2 = 32.07 \text{ amu} + 2(16.00 \text{ amu}) = 64.07 \text{ amu}$$

(Continued)

(b) There are eight C atoms, ten H atoms, four N atoms, and two O atoms in caffeine, so the molecular mass of $C_8H_{10}N_4O_2$ is given by

$$8(12.01 \text{ amu}) + 10(1.008 \text{ amu}) + 4(14.01 \text{ amu}) + 2(16.00 \text{ amu}) = 194.20 \text{ amu}$$

Similar problems: 3.23, 3.24.

Practice Exercise What is the molecular mass of methanol (CH_4O)?

From the molecular mass we can determine the molar mass of a molecule or compound. The molar mass of a compound (in grams) is numerically equal to its molecular mass (in amu). For example, the molecular mass of water is 18.02 amu, so its molar mass is 18.02 g. Note that 1 mole of water weighs 18.02 g and contains 6.022×10^{23} H_2O *molecules,* just as 1 mole of elemental carbon contains 6.022×10^{23} carbon *atoms.*

As Examples 3.6 and 3.7 show, a knowledge of the molar mass enables us to calculate the numbers of moles and individual atoms in a given quantity of a compound.

EXAMPLE 3.6

Methane (CH_4) is the principal component of natural gas. How many moles of CH_4 are present in 4.83 g of CH_4?

CH_4

Strategy We are given grams of CH_4 and asked to solve for moles of CH_4. What conversion factor do we need to convert between grams and moles? Arrange the appropriate conversion factor so that grams cancel and the unit moles are obtained for your answer.

Solution The conversion factor needed to convert between grams and moles is the molar mass. First we need to calculate the molar mass of CH_4, following the procedure in Example 3.5:

$$\begin{aligned}\text{molar mass of } CH_4 &= 12.01 \text{ g} + 4(1.008 \text{ g}) \\ &= 16.04 \text{ g}\end{aligned}$$

Methane gas burning on a cooking range.

Because

$$1 \text{ mol } CH_4 = 16.04 \text{ g } CH_4$$

the conversion factor we need should have grams in the denominator so that the unit g will cancel, leaving the unit mol in the numerator:

$$\frac{1 \text{ mol } CH_4}{16.04 \text{ g } CH_4}$$

We now write

$$4.83 \text{ g } CH_4 \times \frac{1 \text{ mol } CH_4}{16.04 \text{ g } CH_4} = 0.301 \text{ mol } CH_4$$

Thus, there is 0.301 mole of CH_4 in 4.83 g of CH_4.

Check Should 4.83 g of CH_4 equal less than 1 mole of CH_4? What is the mass of 1 mole of CH_4?

Similar problem: 3.26.

Practice Exercise Calculate the number of moles of chloroform ($CHCl_3$) in 198 g of chloroform.

Urea.

EXAMPLE 3.7

How many hydrogen atoms are present in 43.8 g of urea [$(NH_2)_2CO$], which is used as a fertilizer, in animal feed, and in the manufacture of polymers? The molar mass of urea is 60.06 g.

Strategy We are asked to solve for atoms of hydrogen in 43.8 g of urea. We cannot convert directly from grams of urea to atoms of hydrogen. How should molar mass and Avogadro's number be used in this calculation? How many moles of H are in 1 mole of urea?

Solution To calculate the number of H atoms, we first must convert grams of urea to moles of urea using the molar mass of urea. This part is similar to Example 3.2. The molecular formula of urea shows there are four moles of H atoms in one mole of urea molecule, so the mole ratio is 4:1. Finally, knowing the number of moles of H atoms, we can calculate the number of H atoms using Avogadro's number. We need two conversion factors: molar mass and Avogadro's number. We can combine these conversions

$$\text{grams of urea} \longrightarrow \text{moles of urea} \longrightarrow \text{moles of H} \longrightarrow \text{atoms of H}$$

into one step:

$$43.8\ \cancel{\text{g}\ (NH_2)_2CO} \times \frac{1\ \cancel{\text{mol}\ (NH_2)_2CO}}{60.06\ \cancel{\text{g}\ (NH_2)_2CO}} \times \frac{4\ \cancel{\text{mol H}}}{1\ \cancel{\text{mol}\ (NH_2)_2CO}} \times \frac{6.022 \times 10^{23}\ \text{H atoms}}{1\ \cancel{\text{mol H}}}$$
$$= 1.76 \times 10^{24}\ \text{H atoms}$$

Check Does the answer look reasonable? How many atoms of H would 60.06 g of urea contain?

Similar problems: 3.27, 3.28.

Practice Exercise How many H atoms are in 72.5 g of isopropanol (rubbing alcohol), C_3H_8O?

For molecules, formula mass and molecular mass refer to the same quantity.

Finally, note that for ionic compounds like NaCl and MgO that do not contain discrete molecular units, we use the term *formula mass* instead. The formula unit of NaCl consists of one Na^+ ion and one Cl^- ion. Thus, the formula mass of NaCl is the mass of one formula unit:

$$\begin{aligned}\text{formula mass of NaCl} &= 22.99\ \text{amu} + 35.45\ \text{amu}\\ &= 58.44\ \text{amu}\end{aligned}$$

and its molar mass is 58.44 g.

REVIEW OF CONCEPTS

Determine the molecular mass and the molar mass of citric acid, $H_3C_6H_5O_7$.

3.4 The Mass Spectrometer

The most direct and most accurate method for determining atomic and molecular masses is mass spectrometry, which is depicted in Figure 3.3. In a *mass spectrometer,* a gaseous sample is bombarded by a stream of high-energy electrons. Collisions between the electrons and the gaseous atoms (or molecules) produce positive ions by dislodging an electron from each atom or molecule. These positive ions (of mass m

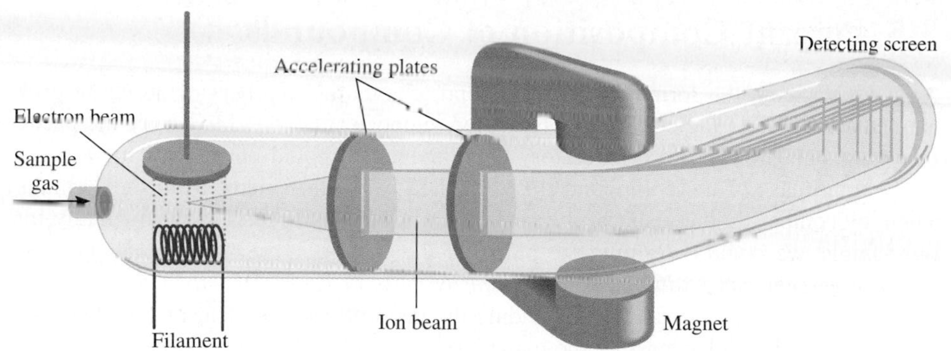

Figure 3.3
Schematic diagram of one type of mass spectrometer.

and charge e) are accelerated by two oppositely charged plates as they pass through the plates. The emerging ions are deflected into a circular path by a magnet. The radius of the path depends on the charge-to-mass ratio (that is, e/m). Ions of smaller e/m ratio trace a wider curve than those having a larger e/m ratio, so that ions with equal charges but different masses are separated from one another. The mass of each ion (and hence its parent atom or molecule) is determined from the magnitude of its deflection. Eventually the ions arrive at the detector, which registers a current for each type of ion. The amount of current generated is directly proportional to the number of ions, so it enables us to determine the relative abundance of isotopes.

The first mass spectrometer, developed in the 1920s by the English physicist F. W. Aston, was crude by today's standards. Nevertheless, it provided indisputable evidence of the existence of isotopes—neon-20 (atomic mass 19.9924 amu and natural abundance 90.92 percent) and neon-22 (atomic mass 21.9914 amu and natural abundance 8.82 percent). When more sophisticated and sensitive mass spectrometers became available, scientists were surprised to discover that neon has a third stable isotope with an atomic mass of 20.9940 amu and natural abundance 0.257 percent (Figure 3.4). This example illustrates how very important experimental accuracy is to a quantitative science like chemistry. Early experiments failed to detect neon-21 because its natural abundance is just 0.257 percent. In other words, only 26 in 10,000 Ne atoms are neon-21. The masses of molecules can be determined in a similar manner by the mass spectrometer.

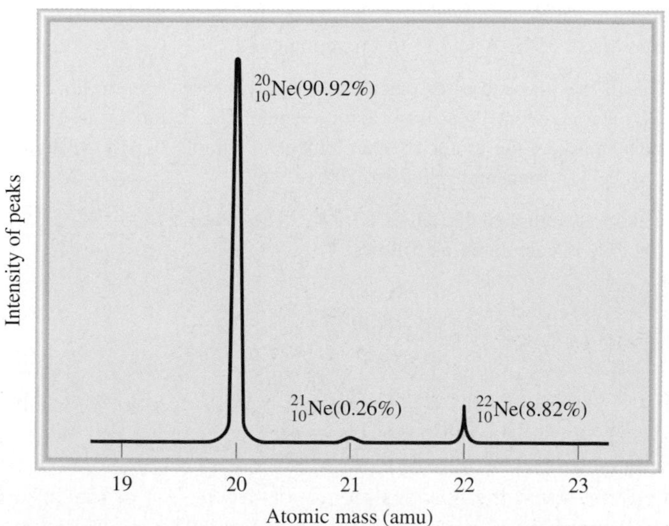

Figure 3.4
The mass spectrum of the three isotopes of neon.

3.5 Percent Composition of Compounds

As we have seen, the formula of a compound tells us the numbers of atoms of each element in a unit of the compound. However, suppose we needed to verify the purity of a compound for use in a laboratory experiment. We could calculate what percent of the total mass of the compound is contributed by each element from the formula. Then, by comparing the result to the percent composition obtained experimentally for our sample, we could determine the purity of the sample.

The ***percent composition*** is the *percent by mass of each element in a compound.* Percent composition is obtained by dividing the mass of each element in 1 mole of the compound by the molar mass of the compound and multiplying by 100 percent. Mathematically, the percent composition of an element in a compound is expressed as

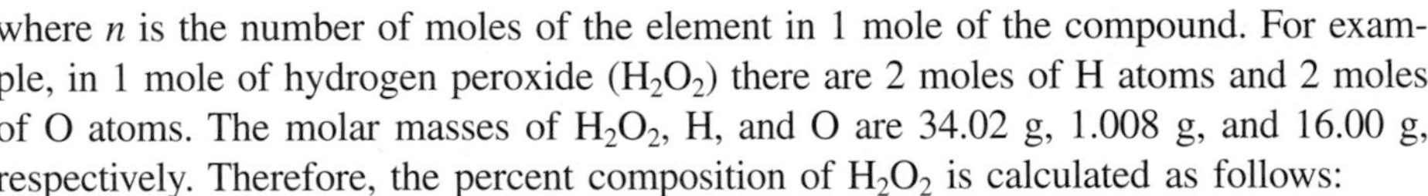

$$\text{percent composition of an element} = \frac{n \times \text{molar mass of element}}{\text{molar mass of compound}} \times 100\% \qquad (3.1)$$

where n is the number of moles of the element in 1 mole of the compound. For example, in 1 mole of hydrogen peroxide (H_2O_2) there are 2 moles of H atoms and 2 moles of O atoms. The molar masses of H_2O_2, H, and O are 34.02 g, 1.008 g, and 16.00 g, respectively. Therefore, the percent composition of H_2O_2 is calculated as follows:

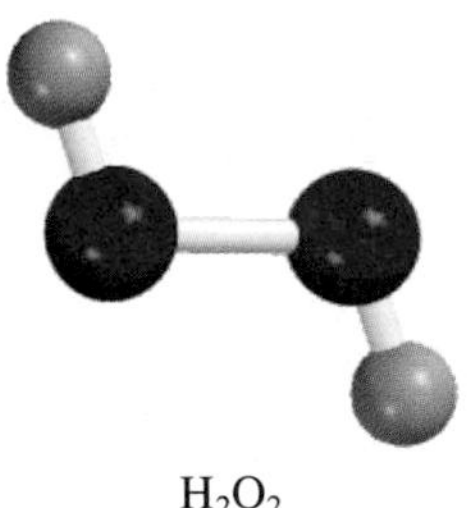

H_2O_2

$$\%\text{H} = \frac{2 \times 1.008 \text{ g}}{34.02 \text{ g}} \times 100\% = 5.926\%$$

$$\%\text{O} = \frac{2 \times 16.00 \text{ g}}{34.02 \text{ g}} \times 100\% = 94.06\%$$

The sum of the percentages is 5.926 percent + 94.06 percent = 99.99 percent. The small discrepancy from 100 percent is due to the way we rounded off the molar masses of the elements. If we had used the empirical formula HO for the calculation, we would have obtained the same percentages. This is so because both the molecular formula and empirical formula tell us the percent composition by mass of the compound.

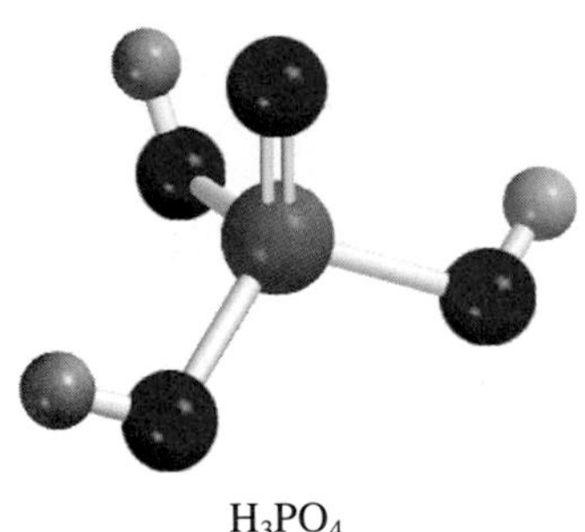

H_3PO_4

EXAMPLE 3.8

Phosphoric acid (H_3PO_4) is a colorless, syrupy liquid used in detergents, fertilizers, toothpastes, and in carbonated beverages for a "tangy" flavor. Calculate the percent composition by mass of H, P, and O in this compound.

Strategy Recall the procedure for calculating a percentage. Assume that we have 1 mole of H_3PO_4. The percent by mass of each element (H, P, and O) is given by the combined molar mass of the atoms of the element in 1 mole of H_3PO_4 divided by the molar mass of H_3PO_4, then multiplied by 100 percent.

Solution The molar mass of H_3PO_4 is 97.99 g. The percent by mass of each of the elements in H_3PO_4 is calculated as follows:

$$\%\text{H} = \frac{3(1.008 \text{ g}) \text{ H}}{97.99 \text{ g } H_3PO_4} \times 100\% = 3.086\%$$

$$\%\text{P} = \frac{30.97 \text{ g P}}{97.99 \text{ g } H_3PO_4} \times 100\% = 31.61\%$$

$$\%\text{O} = \frac{4(16.00 \text{ g}) \text{ O}}{97.99 \text{ g } H_3PO_4} \times 100\% = 65.31\%$$

(Continued)

Check Do the percentages add to 100 percent? The sum of the percentages is (3.086% + 31.61% + 65.31%) = 100.01%. The small discrepancy from 100 percent is due to the way we rounded off.

Practice Exercise Calculate the percent composition by mass of each of the elements in sulfuric acid (H_2SO_4).

Similar problem: 3.40.

The procedure used in Example 3.8 can be reversed if necessary. Given the percent composition by mass of a compound, we can determine the empirical formula of the compound (Figure 3.5). Because we are dealing with percentages and the sum of all the percentages is 100 percent, it is convenient to assume that we started with 100 g of a compound, as Example 3.9 shows.

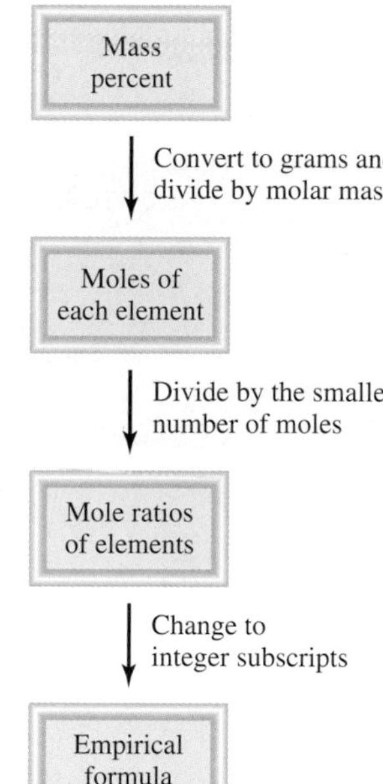

Figure 3.5
Procedure for calculating the empirical formula of a compound from its percent compositions.

EXAMPLE 3.9

Ascorbic acid (vitamin C) cures scurvy. It is composed of 40.92 percent carbon (C), 4.58 percent hydrogen (H), and 54.50 percent oxygen (O) by mass. Determine its empirical formula.

Strategy In a chemical formula, the subscripts represent the ratio of the number of moles of each element that combine to form one mole of the compound. How can we convert from mass percent to moles? If we assume an exactly 100-g sample of the compound, do we know the mass of each element in the compound? How do we then convert from grams to moles?

Solution If we have 100 g of ascorbic acid, then each percentage can be converted directly to grams. In this sample, there will be 40.92 g of C, 4.58 g of H, and 54.50 g of O. Because the subscripts in the formula represent a mole ratio, we need to convert the grams of each element to moles. The conversion factor needed is the molar mass of each element. Let n represent the number of moles of each element so that

$$n_C = 40.92 \text{ g C} \times \frac{1 \text{ mol C}}{12.01 \text{ g C}} = 3.407 \text{ mol C}$$

$$n_H = 4.58 \text{ g H} \times \frac{1 \text{ mol H}}{1.008 \text{ g H}} = 4.54 \text{ mol H}$$

$$n_O = 54.50 \text{ g O} \times \frac{1 \text{ mol O}}{16.00 \text{ g O}} = 3.406 \text{ mol O}$$

Thus, we arrive at the formula $C_{3.407}H_{4.54}O_{3.406}$, which gives the identity and the mole ratios of atoms present. However, chemical formulas are written with whole numbers. Try to convert to whole numbers by dividing all the subscripts by the smallest subscript (3.406):

$$\text{C:} \frac{3.407}{3.406} \approx 1 \quad \text{H:} \frac{4.54}{3.406} = 1.33 \quad \text{O:} \frac{3.406}{3.406} = 1$$

where the $\approx$ sign means "approximately equal to." This gives $CH_{1.33}O$ as the formula for ascorbic acid. Next, we need to convert 1.33, the subscript for H, into an integer. This can be done by a trial-and-error procedure:

$$1.33 \times 1 = 1.33$$
$$1.33 \times 2 = 2.66$$
$$1.33 \times 3 = 3.99 \approx 4$$

Because 1.33 × 3 gives us an integer (4), we multiply all the subscripts by 3 and obtain $C_3H_4O_3$ as the empirical formula for ascorbic acid.

(Continued)

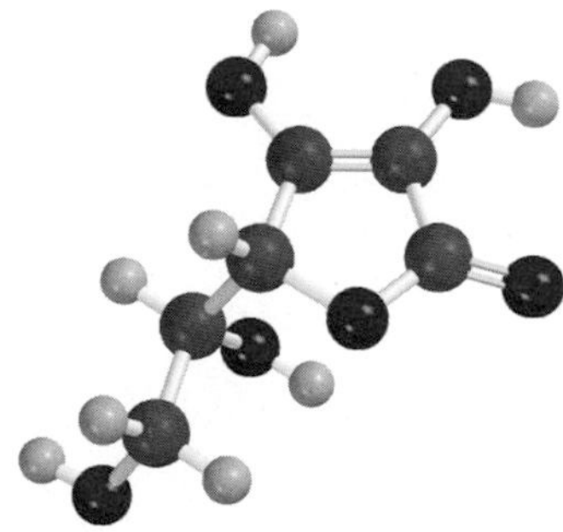

The molecular formula of ascorbic acid is $C_6H_8O_6$.

Similar problems: 3.49, 3.50.

Check Are the subscripts in $C_3H_4O_3$ reduced to the smallest whole numbers?

Practice Exercise Determine the empirical formula of a compound having the following percent composition by mass: K: 24.75 percent; Mn: 34.77 percent; O: 40.51 percent.

Chemists often want to know the actual mass of an element in a certain mass of a compound. For example, in the mining industry, this information will tell the scientists about the quality of the ore. Because the percent composition by mass of the elements in the substance can be readily calculated, such a problem can be solved in a rather direct way.

Chalcopyrite.

EXAMPLE 3.10

Chalcopyrite ($CuFeS_2$) is a principal mineral of copper. Calculate the number of kilograms of Cu in 5.93×10^3 kg of chalcopyrite.

Strategy Chalcopyrite is composed of Cu, Fe, and S. The mass due to Cu is based on its percentage by mass in the compound. How do we calculate mass percent of an element?

Solution The molar mass of Cu and $CuFeS_2$ are 63.55 g and 183.5 g, respectively. The mass percent of Cu is therefore

$$\%\text{Cu} = \frac{\text{molar mass of Cu}}{\text{molar mass of CuFeS}_2} \times 100\%$$

$$= \frac{63.55\text{ g}}{183.5\text{ g}} \times 100\% = 34.63\%$$

To calculate the mass of Cu in a 5.93×10^3 kg sample of $CuFeS_2$, we need to convert the percentage to a fraction (that is, convert 34.63 percent to 34.63/100, or 0.3463) and write

$$\text{mass of Cu in CuFeS}_2 = 0.3463 \times (5.93 \times 10^3\text{ kg}) = 2.05 \times 10^3\text{ kg}$$

We can also solve the problem by reading the formula as the ratio of moles of chalcopyrite to moles of copper using the following conversions:

grams of chalcopyrite $\longrightarrow$ moles of chalcopyrite $\longrightarrow$ moles of Cu $\longrightarrow$ grams of Cu

Try it.

Check As a ballpark estimate, note that the mass percent of Cu is roughly 33 percent, so that a third of the mass should be Cu; that is, $\frac{1}{3} \times 5.93 \times 10^3\text{ kg} \approx 1.98 \times 10^3\text{ kg}$. This quantity is quite close to the answer.

Similar problem: 3.45.

Practice Exercise Calculate the number of grams of Al in 371 g of Al_2O_3.

Review of Concepts

Without doing detailed calculations, estimate whether the percent composition by mass of Hg is greater than or smaller than O in mercury(II) nitrate [$Hg(NO_3)_2$].

3.6 Experimental Determination of Empirical Formulas

The fact that we can determine the empirical formula of a compound if we know the percent composition enables us to identify compounds experimentally. The procedure is as follows. First, chemical analysis tells us the number of grams of each element present in a given amount of a compound. Then, we convert the quantities in grams

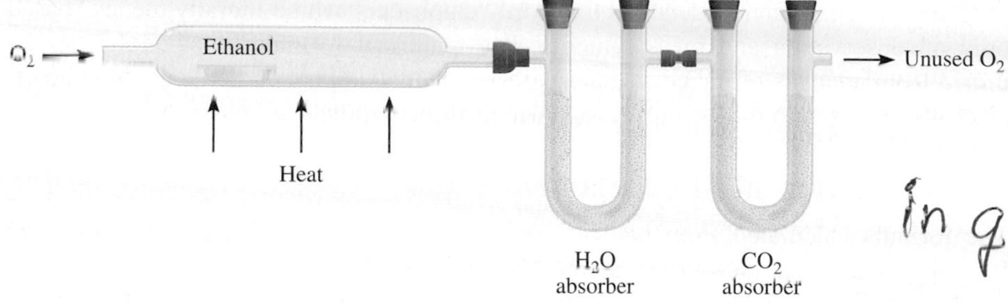

Figure 3.6
Apparatus for determining the empirical formula of ethanol. The absorbers are substances that can retain water and carbon dioxide, respectively.

to number of moles of each element. Finally, using the method given in Example 3.9, we find the empirical formula of the compound.

As a specific example, let us consider the compound ethanol. When ethanol is burned in an apparatus such as that shown in Figure 3.6, carbon dioxide (CO_2) and water (H_2O) are given off. Because neither carbon nor hydrogen was in the inlet gas, we can conclude that both carbon (C) and hydrogen (H) were present in ethanol and that oxygen (O) may also be present. (Molecular oxygen was added in the combustion process, but some of the oxygen may also have come from the original ethanol sample.)

The masses of CO_2 and of H_2O produced can be determined by measuring the increase in mass of the CO_2 and H_2O absorbers, respectively. Suppose that in one experiment the combustion of 11.5 g of ethanol produced 22.0 g of CO_2 and 13.5 g of H_2O. We can calculate the mass of carbon and hydrogen in the original 11.5-g sample of ethanol as follows:

$$\begin{aligned}\text{mass of C} &= 22.0\ \text{g CO}_2 \times \frac{1\ \text{mol CO}_2}{44.01\ \text{g CO}_2} \times \frac{1\ \text{mol C}}{1\ \text{mol CO}_2} \times \frac{12.01\ \text{g C}}{1\ \text{mol C}}\\ &= 6.00\ \text{g C}\end{aligned}$$

$$\begin{aligned}\text{mass of H} &= 13.5\ \text{g H}_2\text{O} \times \frac{1\ \text{mol H}_2\text{O}}{18.02\ \text{g H}_2\text{O}} \times \frac{2\ \text{mol H}}{1\ \text{mol H}_2\text{O}} \times \frac{1.008\ \text{g H}}{1\ \text{mol H}}\\ &= 1.51\ \text{g H}\end{aligned}$$

Thus, 11.5 g of ethanol contains 6.00 g of carbon and 1.51 g of hydrogen. The remainder must be oxygen, whose mass is

$$\begin{aligned}\text{mass of O} &= \text{mass of sample} - (\text{mass of C} + \text{mass of H})\\ &= 11.5\ \text{g} - (6.00\ \text{g} + 1.51\ \text{g})\\ &= 4.0\ \text{g}\end{aligned}$$

The number of moles of each element present in 11.5 g of ethanol is

$$\text{moles of C} = 6.00\ \text{g C} \times \frac{1\ \text{mol C}}{12.01\ \text{g C}} = 0.500\ \text{mol C}$$

$$\text{moles of H} = 1.51\ \text{g H} \times \frac{1\ \text{mol H}}{1.008\ \text{g H}} = 1.50\ \text{mol H}$$

$$\text{moles of O} = 4.0\ \text{g O} \times \frac{1\ \text{mol O}}{16.00\ \text{g O}} = 0.25\ \text{mol O}$$

It happens that the molecular formula of ethanol is the same as its empirical formula.

The formula of ethanol is therefore $C_{0.50}H_{1.5}O_{0.25}$ (we round off the number of moles to two significant figures). Because the number of atoms must be an integer, we divide the subscripts by 0.25, the smallest subscript, and obtain for the empirical formula C_2H_6O.

Now we can better understand the word "empirical," which literally means "based only on observation and measurement." The empirical formula of ethanol is determined from analysis of the compound in terms of its component elements. No knowledge of how the atoms are linked together in the compound is required.

Determination of Molecular Formulas

Note that the molar mass of a compound can be determined experimentally even if we do not know its molecular formula.

The formula calculated from percent composition by mass is always the empirical formula because the subscripts in the formula are always reduced to the smallest whole numbers. To calculate the actual, molecular formula we must know the *approximate* molar mass of the compound in addition to its empirical formula. Knowing that the molar mass of a compound must be an integral multiple of the molar mass of its empirical formula, we can use the molar mass to find the molecular formula, as Example 3.11 demonstrates.

EXAMPLE 3.11

A sample of a compound contains 1.52 g of nitrogen (N) and 3.47 g of oxygen (O). The molar mass of this compound is between 90 g and 95 g. Determine the molecular formula and the accurate molar mass of the compound.

Strategy To determine the molecular formula, we first need to determine the empirical formula. How do we convert between grams and moles? Comparing the empirical molar mass to the experimentally determined molar mass will reveal the relationship between the empirical formula and molecular formula.

Solution We are given grams of N and O. Use molar mass as a conversion factor to convert grams to moles of each element. Let *n* represent the number of moles of each element. We write

$$n_N = 1.52 \text{ g N} \times \frac{1 \text{ mol N}}{14.01 \text{ g N}} = 0.108 \text{ mol N}$$

$$n_O = 3.47 \text{ g O} \times \frac{1 \text{ mol O}}{16.00 \text{ g O}} = 0.217 \text{ mol O}$$

Thus, we arrive at the formula $N_{0.108}O_{0.217}$, which gives the identity and the ratios of atoms present. However, chemical formulas are written with whole numbers. Try to convert to whole numbers by dividing the subscripts by the smaller subscript (0.108). After rounding off, we obtain NO_2 as the empirical formula.

The molecular formula might be the same as the empirical formula or some integral multiple of it (for example, two, three, four, or more times the empirical formula). Comparing the ratio of the molar mass to the molar mass of the empirical formula will show the integral relationship between the empirical and molecular formulas. The molar mass of the empirical formula NO_2 is

$$\text{empirical molar mass} = 14.01 \text{ g} + 2(16.00 \text{ g}) = 46.01 \text{ g}$$

Next, we determine the ratio between the molar mass and the empirical molar mass

$$\frac{\text{molar mass}}{\text{empirical molar mass}} = \frac{90 \text{ g}}{46.01 \text{ g}} \approx 2$$

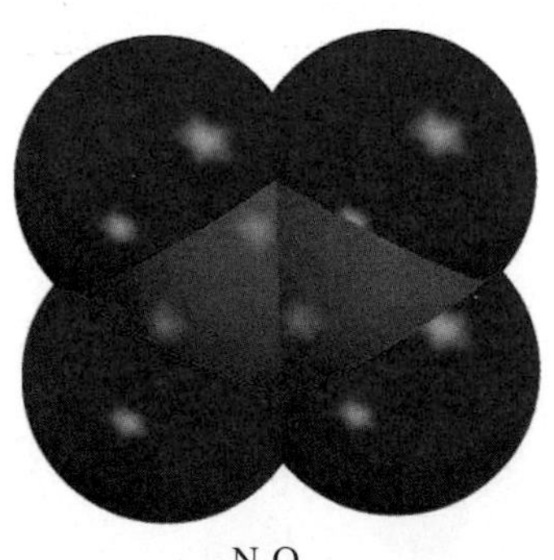

N_2O_4

The molar mass is twice the empirical molar mass. This means that there are two NO_2 units in each molecule of the compound, and the molecular formula is $(NO_2)_2$ or N_2O_4.

The actual molar mass of the compound is two times the empirical molar mass, that is, 2(46.01 g) or 92.02 g, which is between 90 g and 95 g.

(Continued)

Check Note that in determining the molecular formula from the empirical formula, we need only know the *approximate* molar mass of the compound. The reason is that the true molar mass is an integral multiple (1×, 2×, 3×, . . .) of the empirical molar mass. Therefore, the ratio (molar mass/empirical molar mass) will always be close to an integer.

Similar problems: 3.52, 3.53, 3.54.

Practice Exercise A sample of a compound containing boron (B) and hydrogen (H) contains 6.444 g of B and 1.803 g of H. The molar mass of the compound is about 30 g. What is its molecular formula?

Review of Concepts

What is the molecular formula of a compound containing only carbon and hydrogen if combustion of 1.05 g of the compound produces 3.30 g CO_2 and 1.35 g H_2O and its molar mass is about 70 g?

3.7 Chemical Reactions and Chemical Equations

Having discussed the masses of atoms and molecules, we turn next to what happens to atoms and molecules in a ***chemical reaction,*** *a process in which a substance (or substances) is changed into one or more new substances.* To communicate with one another about chemical reactions, chemists have devised a standard way to represent them using chemical equations. A ***chemical equation*** *uses chemical symbols to show what happens during a chemical reaction.* In this section we will learn how to write chemical equations and balance them.

Writing Chemical Equations

Consider what happens when hydrogen gas (H_2) burns in air (which contains oxygen, O_2) to form water (H_2O). This reaction can be represented by the chemical equation

$$H_2 + O_2 \longrightarrow H_2O \tag{3.2}$$

where the "plus" sign means "reacts with" and the arrow means "to yield." Thus, this symbolic expression can be read: "Molecular hydrogen reacts with molecular oxygen to yield water." The reaction is assumed to proceed from left to right as the arrow indicates.

Equation (3.2) is not complete, however, because there are twice as many oxygen atoms on the left side of the arrow (two) as on the right side (one). To conform with the law of conservation of mass, there must be the same number of each type of atom on both sides of the arrow; that is, we must have as many atoms after the reaction ends as we did before it started. We can *balance* Equation (3.2) by placing the appropriate coefficient (2 in this case) in front of H_2 and H_2O:

$$2H_2 + O_2 \longrightarrow 2H_2O$$

When the coefficient is 1, as in the case of O_2, it is not shown.

This *balanced chemical equation* shows that "two hydrogen molecules can combine or react with one oxygen molecule to form two water molecules" (Figure 3.7). Because the ratio of the number of molecules is equal to the ratio of the number of moles, the equation can also be read as "2 moles of hydrogen molecules react with 1 mole of oxygen molecules to produce 2 moles of water molecules." We know the mass of a mole of each of these substances, so we can also interpret the equation as "4.04 g of

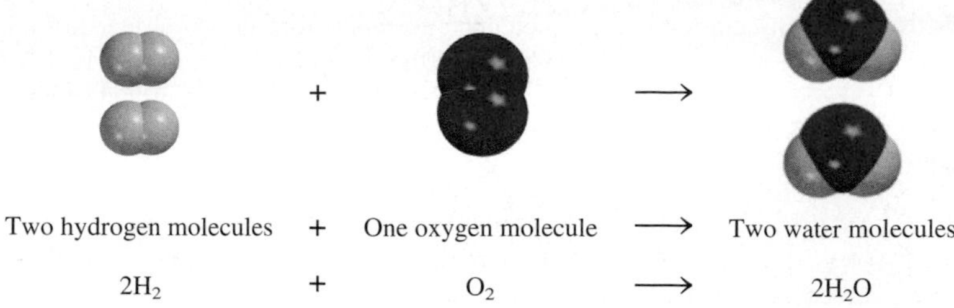

Figure 3.7
Three ways of representing the combustion of hydrogen. In accordance with the law of conservation of mass, the number of each type of atom must be the same on both sides of the equation.

H_2 react with 32.00 g of O_2 to give 36.04 g of H_2O." These three ways of reading the equation are summarized in Table 3.1.

We refer to H_2 and O_2 in Equation (3.2) as ***reactants,*** which are *the starting materials in a chemical reaction.* Water is the ***product,*** which is *the substance formed as a result of a chemical reaction.* A chemical equation, then, is just the chemist's shorthand description of a reaction. In a chemical equation the reactants are conventionally written on the left and the products on the right of the arrow:

$$\text{reactants} \longrightarrow \text{products}$$

To provide additional information, chemists often indicate the physical states of the reactants and products by using the letters *g*, *l*, and *s* to denote gas, liquid, and solid, respectively. For example,

The procedure for balancing chemical equations is shown on p. 77.

$$2CO(g) + O_2(g) \longrightarrow 2CO_2(g)$$
$$2HgO(s) \longrightarrow 2Hg(l) + O_2(g)$$

To represent what happens when sodium chloride (NaCl) is added to water, we write

$$NaCl(s) \xrightarrow{H_2O} NaCl(aq)$$

where *aq* denotes the aqueous (that is, water) environment. Writing H_2O above the arrow symbolizes the physical process of dissolving a substance in water, although it is sometimes left out for simplicity.

Balancing Chemical Equations

Suppose we want to write an equation to describe a chemical reaction that we have just carried out in the laboratory. How should we go about doing it? Because we know the identities of the reactants, we can write their chemical formulas. The identities of products are more difficult to establish. For simple reactions, it is often possible to guess the product(s). For more complicated reactions involving three or more products, chemists may need to perform further tests to establish the presence of specific compounds.

Table 3.1 Interpretation of a Chemical Equation

$2H_2$	$+ O_2$	$\longrightarrow 2H_2O$
Two molecules	+ one molecule	⟶ two molecules
2 moles	+ 1 mole	⟶ 2 moles
2(2.02 g) = 4.04 g	+ 32.00 g	⟶ 2(18.02 g) = 36.04 g
36.04 g reactants		36.04 g product

Once we have identified all the reactants and products and have written the correct formulas for them, we assemble them in the conventional sequence—reactants on the left separated by an arrow from products on the right. The equation written at this point is likely to be *unbalanced;* that is, the number of each type of atom on one side of the arrow differs from the number on the other side. In general, we can balance a chemical equation by the following steps:

1. Identify all reactants and products and write their correct formulas on the left side and right side of the equation, respectively.
2. Begin balancing the equation by trying different coefficients to make the number of atoms of each element the same on both sides of the equation. We can change the coefficients (the numbers preceding the formulas) but not the subscripts (the numbers within formulas). Changing the subscripts would change the identity of the substance. For example, $2NO_2$ means "two molecules of nitrogen dioxide," but if we double the subscripts, we have N_2O_4, which is the formula of dinitrogen tetroxide, a completely different compound.
3. First, look for elements that appear only once on each side of the equation with the same number of atoms on each side. The formulas containing these elements must have the same coefficient. Therefore, there is no need to adjust the coefficients of these elements at this point. Next, look for elements that appear only once on each side of the equation but in unequal numbers of atoms. Balance these elements. Finally, balance elements that appear in two or more formulas on the same side of the equation.
4. Check your balanced equation to be sure that you have the same total number of each type of atoms on both sides of the equation arrow.

Let's consider a specific example. In the laboratory, small amounts of oxygen gas can be prepared by heating potassium chlorate ($KClO_3$). The products are oxygen gas (O_2) and potassium chloride (KCl). From this information, we write

$$KClO_3 \longrightarrow KCl + O_2$$

(For simplicity, we omit the physical states of reactants and products.) All three elements (K, Cl, and O) appear only once on each side of the equation, but only for K and Cl do we have equal numbers of atoms on both sides. Thus, $KClO_3$ and KCl must have the same coefficient. The next step is to make the number of O atoms the same on both sides of the equation. Because there are three O atoms on the left and two O atoms on the right of the equation, we can balance the O atoms by placing a 2 in front of $KClO_3$ and a 3 in front of O_2.

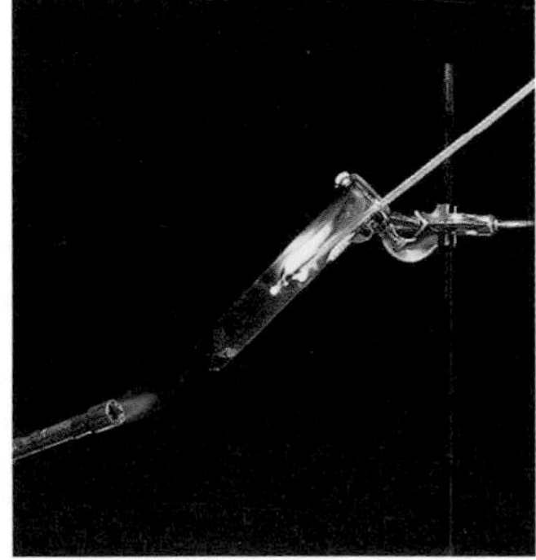

Heating potassium chlorate produces oxygen, which supports the combustion of wood splint.

$$2KClO_3 \longrightarrow KCl + 3O_2$$

Finally, we balance the K and Cl atoms by placing a 2 in front of KCl:

$$2KClO_3 \longrightarrow 2KCl + 3O_2 \qquad (3.3)$$

As a final check, we can draw up a balance sheet for the reactants and products where the number in parentheses indicates the number of atoms of each element:

Reactants	**Products**
K (2)	K (2)
Cl (2)	Cl (2)
O (6)	O (6)

Note that this equation could also be balanced with coefficients that are multiples of 2 (for $KClO_3$), 2 (for KCl), and 3 (for O_2); for example,

$$4KClO_3 \longrightarrow 4KCl + 6O_2$$

However, it is common practice to use the *simplest* possible set of whole-number coefficients to balance the equation. Equation (3.3) conforms to this convention.

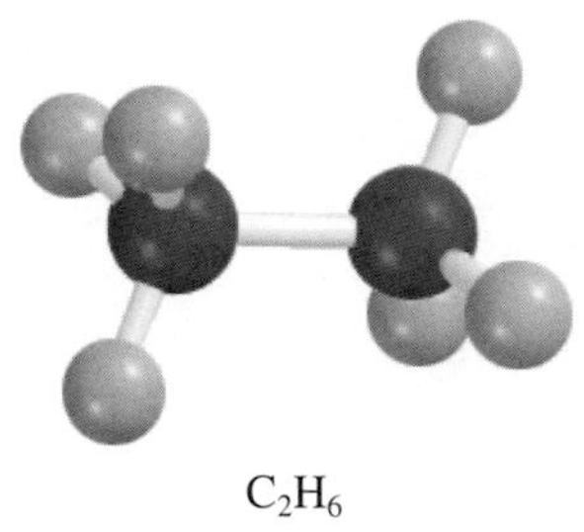

C_2H_6

Now let us consider the combustion (that is, burning) of the natural gas component ethane (C_2H_6) in oxygen or air, which yields carbon dioxide (CO_2) and water. The unbalanced equation is

$$C_2H_6 + O_2 \longrightarrow CO_2 + H_2O$$

We see that the number of atoms is not the same on both sides of the equation for any of the elements (C, H, and O). In addition, C and H appear only once on each side of the equation; O appears in two compounds on the right side (CO_2 and H_2O). To balance the C atoms, we place a 2 in front of CO_2:

$$C_2H_6 + O_2 \longrightarrow 2CO_2 + H_2O$$

To balance the H atoms, we place a 3 in front of H_2O:

$$C_2H_6 + O_2 \longrightarrow 2CO_2 + 3H_2O$$

At this stage, the C and H atoms are balanced, but the O atoms are not because there are seven O atoms on the right-hand side and only two O atoms on the left-hand side of the equation. This inequality of O atoms can be eliminated by writing $\frac{7}{2}$ in front of the O_2 on the left-hand side:

$$C_2H_6 + \tfrac{7}{2}O_2 \longrightarrow 2CO_2 + 3H_2O$$

The "logic" for using $\frac{7}{2}$ as a coefficient is that there were seven oxygen atoms on the right-hand side of the equation, but only a pair of oxygen atoms (O_2) on the left. To balance them we ask how many *pairs* of oxygen atoms are needed to equal seven oxygen atoms. Just as 3.5 pairs of shoes equal seven shoes, $\frac{7}{2}O_2$ molecules equal seven O atoms. As the following tally shows, the equation is now balanced:

Reactants	Products
C (2)	C (2)
H (6)	H (6)
O (7)	O (7)

However, we normally prefer to express the coefficients as whole numbers rather than as fractions. Therefore, we multiply the entire equation by 2 to convert $\frac{7}{2}$ to 7:

$$2C_2H_6 + 7O_2 \longrightarrow 4CO_2 + 6H_2O$$

The final tally is

Reactants	Products
C (4)	C (4)
H (12)	H (12)
O (14)	O (14)

Note that the coefficients used in balancing the last equation are the smallest possible set of whole numbers.

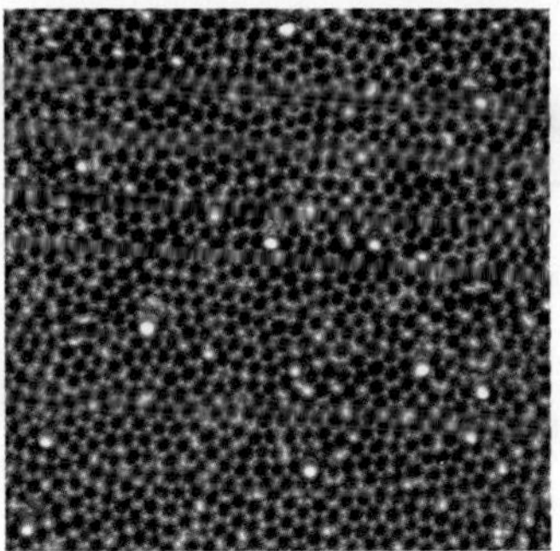

An atomic scale image of aluminum oxide.

EXAMPLE 3.12

When aluminum metal is exposed to air, a protective layer of aluminum oxide (Al_2O_3) forms on its surface. This layer prevents further reaction between aluminum and oxygen, and it is the reason that aluminum beverage cans do not corrode. [In the case of iron, the rust, or iron(III) oxide, that forms is too porous to protect the iron metal underneath, so rusting continues.] Write a balanced equation for the formation of Al_2O_3.

Strategy Remember that the formula of an element or compound cannot be changed when balancing a chemical equation. The equation is balanced by placing the appropriate coefficients in front of the formulas. Follow the procedure described on p. 77.

Solution The unbalanced equation is

$$Al + O_2 \longrightarrow Al_2O_3$$

In a balanced equation, the number and types of atoms on each side of the equation must be the same. We see that there is one Al atom on the reactants side and there are two Al atoms on the product side. We can balance the Al atoms by placing a coefficient of 2 in front of Al on the reactants side.

$$2Al + O_2 \longrightarrow Al_2O_3$$

There are two O atoms on the reactants side, and three O atoms on the product side of the equation. We can balance the O atoms by placing a coefficient of $\frac{3}{2}$ in front of O_2 on the reactants side.

$$2Al + \tfrac{3}{2}O_2 \longrightarrow Al_2O_3$$

This is a balanced equation. However, equations are normally balanced with the smallest set of *whole* number coefficients. Multiplying both sides of the equation by 2 gives whole number coefficients.

$$2(2Al + \tfrac{3}{2}O_2 \longrightarrow Al_2O_3)$$
$$4Al + 3O_2 \longrightarrow 2Al_2O_3$$

Check For an equation to be balanced, the number and types of atoms on each side of the equation must be the same. The final tally is

Reactants	Products
Al (4)	Al (4)
O (6)	O (6)

The equation is balanced.

Similar problems: 3.59, 3.60.

Practice Exercise Balance the equation representing the reaction between iron(III) oxide, Fe_2O_3, and carbon monoxide (CO) to yield iron (Fe) and carbon dioxide (CO_2).

Review of Concepts

Which parts shown here are essential for a balanced equation and which parts are helpful if we want to carry out the reaction in the laboratory?

$$CaH_2(s) + 2H_2O(l) \longrightarrow Ca(OH)_2(aq) + 2H_2(g)$$

3.8 Amounts of Reactants and Products

A basic question raised in the chemical laboratory is "How much product will be formed from specific amounts of starting materials (reactants)?" Or in some cases, we might ask the reverse question: "How much starting material must be used to obtain a specific amount of product?" To interpret a reaction quantitatively, we need to apply our knowledge of molar masses and the mole concept. ***Stoichiometry*** is *the quantitative study of reactants and products in a chemical reaction.*

Whether the units given for reactants (or products) are moles, grams, liters (for gases), or some other units, we use moles to calculate the amount of product formed in a reaction. This approach is called the ***mole method,*** which means simply that *the stoichiometric coefficients in a chemical equation can be interpreted as the number of moles of each substance.* For example, industrially ammonia is synthesized from hydrogen and nitrogen as follows:

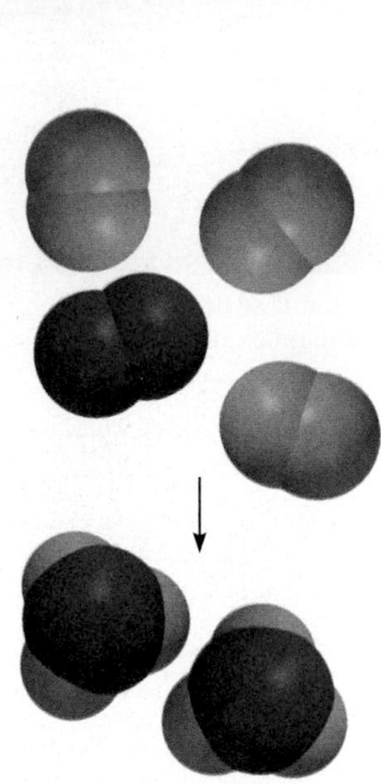

The synthesis of NH_3 from H_2 and N_2.

$$N_2(g) + 3H_2(g) \longrightarrow 2NH_3(g)$$

The stoichiometric coefficients show that one molecule of N_2 reacts with three molecules of H_2 to form two molecules of NH_3. It follows that the relative numbers of moles are the same as the relative number of molecules:

$N_2(g)$	+	$3H_2(g)$	$\longrightarrow$	$2NH_3(g)$
1 molecule		3 molecules		2 molecules
6.022×10^{23} molecules		$3(6.022 \times 10^{23}$ molecules)		$2(6.022 \times 10^{23}$ molecules)
1 mol		3 mol		2 mol

Thus, this equation can also be read as "1 mole of N_2 gas combines with 3 moles of H_2 gas to form 2 moles of NH_3 gas." In stoichiometric calculations, we say that 3 moles of H_2 are equivalent to 2 moles of NH_3, that is,

$$3 \text{ mol } H_2 \simeq 2 \text{ mol } NH_3$$

where the symbol $\simeq$ means "stoichiometrically equivalent to" or simply "equivalent to." This relationship enables us to write the conversion factors

$$\frac{3 \text{ mol } H_2}{2 \text{ mol } NH_3} \quad \text{and} \quad \frac{2 \text{ mol } NH_3}{3 \text{ mol } H_2}$$

Similarly, we have 1 mol $N_2 \simeq$ 2 mol NH_3 and 1 mol $N_2 \simeq$ 3 mol H_2.

Let's consider a simple example in which 6.0 moles of H_2 react completely with N_2 to form NH_3. To calculate the amount of NH_3 produced in moles, we use the conversion factor that has H_2 in the denominator and write

$$\begin{aligned}\text{moles of } NH_3 \text{ produced} &= 6.0 \cancel{\text{mol } H_2} \times \frac{2 \text{ mol } NH_3}{3 \cancel{\text{mol } H_2}} \\ &= 4.0 \text{ mol } NH_3\end{aligned}$$

Now suppose 16.0 g of H_2 react completely with N_2 to form NH_3. How many grams of NH_3 will be formed? To do this calculation, we note that the link between H_2 and NH_3 is the mole ratio from the balanced equation. So we need to first convert grams of H_2 to moles of H_2, then to moles of NH_3, and finally to grams of NH_3. The conversion steps are

$$\text{grams of } H_2 \longrightarrow \text{moles of } H_2 \longrightarrow \text{moles of } NH_3 \longrightarrow \text{grams of } NH_3$$

First, we convert 16.0 g of H_2 to number of moles of H_2, using the molar mass of H_2 as the conversion factor:

$$\begin{aligned}\text{moles of } H_2 &= 16.0 \cancel{\text{g } H_2} \times \frac{1 \text{ mol } H_2}{2.016 \cancel{\text{g } H_2}} \\ &= 7.94 \text{ mol } H_2\end{aligned}$$

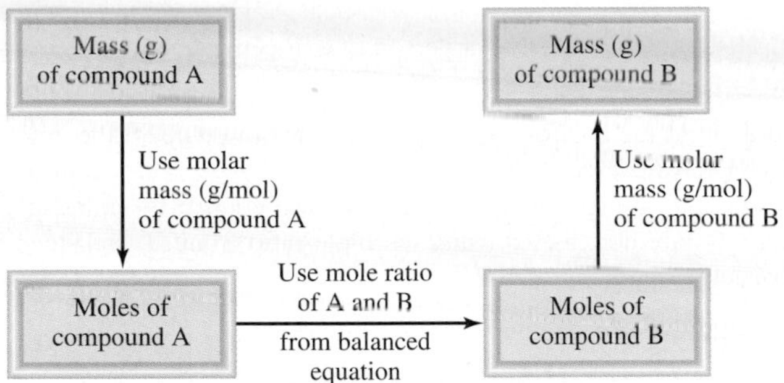

Figure 3.8 *The mole method. First convert the quantity of reactant A (in grams or other units) to number of moles. Next, use the mole ratio in the balanced equation to calculate the number of moles of product B formed. Finally, convert moles of product to grams of product.*

Next, we calculate the number of moles of NH_3 produced:

$$\text{moles of } NH_3 = 7.94\ \cancel{\text{mol } H_2} \times \frac{2 \text{ mol } NH_3}{3\ \cancel{\text{mol } H_2}}$$
$$= 5.29 \text{ mol } NH_3$$

Finally, we calculate the mass of NH_3 produced in grams using the molar mass of NH_3 as the conversion factor:

$$\text{grams of } NH_3 = 5.29\ \cancel{\text{mol } NH_3} \times \frac{17.03 \text{ g } NH_3}{1\ \cancel{\text{mol } NH_3}}$$
$$= 90.1 \text{ g } NH_3$$

These three separate calculations can be combined in a single step as follows:

$$\text{grams of } NH_3 = 16.0\ \cancel{\text{g } H_2} \times \frac{1\ \cancel{\text{mol } H_2}}{2.016\ \cancel{\text{g } H_2}} \times \frac{2\ \cancel{\text{mol } NH_3}}{3\ \cancel{\text{mol } H_2}} \times \frac{17.03 \text{ g } NH_3}{1\ \cancel{\text{mol } NH_3}}$$
$$= 90.1 \text{ g } NH_3$$

Similarly, we can calculate the mass in grams of N_2 consumed in this reaction. The conversion steps are

$$\text{grams of } H_2 \longrightarrow \text{moles of } H_2 \longrightarrow \text{moles of } N_2 \longrightarrow \text{grams of } N_2$$

By using the relationship 1 mol $N_2 \simeq$ 3 mol H_2, we write

$$\text{grams of } N_2 = 16.0\ \cancel{\text{g } H_2} \times \frac{1\ \cancel{\text{mol } H_2}}{2.016\ \cancel{\text{g } H_2}} \times \frac{1\ \cancel{\text{mol } N_2}}{3\ \cancel{\text{mol } H_2}} \times \frac{28.02 \text{ g } N_2}{1\ \cancel{\text{mol } N_2}}$$
$$= 74.1 \text{ g } N_2$$

Figure 3.8 shows the steps involved in stoichiometric calculations using the mole method.

EXAMPLE 3.13

The food we eat is degraded, or broken down, in our bodies to provide energy for growth and function. A general overall equation for this very complex process represents the degradation of glucose ($C_6H_{12}O_6$) to carbon dioxide (CO_2) and water (H_2O):

$$C_6H_{12}O_6 + 6O_2 \longrightarrow 6CO_2 + 6H_2O$$

(Continued)

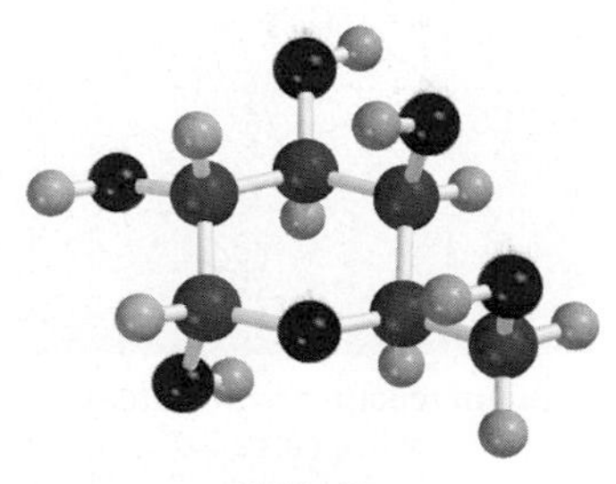

$C_6H_{12}O_6$

If 968 g of $C_6H_{12}O_6$ is consumed by a person over a certain period, what is the mass of CO_2 produced?

Strategy Looking at the balanced equation, how do we compare the amount of $C_6H_{12}O_6$ and CO_2? We can compare them based on the *mole ratio* from the balanced equation. Starting with grams of $C_6H_{12}O_6$, how do we convert to moles of $C_6H_{12}O_6$? Once moles of CO_2 are determined using the mole ratio from the balanced equation, how do we convert to grams of CO_2?

Solution We follow the preceding steps and Figure 3.8.

Step 1: The balanced equation is given in the problem.

Step 2: To convert grams of $C_6H_{12}O_6$ to moles of $C_6H_{12}O_6$, we write

$$968 \text{ g } \cancel{C_6H_{12}O_6} \times \frac{1 \text{ mol } C_6H_{12}O_6}{180.2 \text{ g } \cancel{C_6H_{12}O_6}} = 5.372 \text{ mol } C_6H_{12}O_6$$

Step 3: From the mole ratio, we see that 1 mol $C_6H_{12}O_6 \simeq$ 6 mol CO_2. Therefore, the number of moles of CO_2 formed is

$$5.372 \text{ } \cancel{\text{mol } C_6H_{12}O_6} \times \frac{6 \text{ mol } CO_2}{1 \text{ } \cancel{\text{mol } C_6H_{12}O_6}} = 32.23 \text{ mol } CO_2$$

Step 4: Finally, the number of grams of CO_2 formed is given by

$$32.23 \text{ } \cancel{\text{mol } CO_2} \times \frac{44.01 \text{ g } CO_2}{1 \text{ } \cancel{\text{mol } CO_2}} = 1.42 \times 10^3 \text{ g } CO_2$$

After some practice, we can combine the conversion steps

$$\text{grams of } C_6H_{12}O_6 \longrightarrow \text{moles of } C_6H_{12}O_6 \longrightarrow \text{moles of } CO_2 \longrightarrow \text{grams of } CO_2$$

into one equation:

$$\text{mass of } CO_2 = 968 \text{ g } \cancel{C_6H_{12}O_6} \times \frac{1 \text{ } \cancel{\text{mol } C_6H_{12}O_6}}{180.2 \text{ g } \cancel{C_6H_{12}O_6}} \times \frac{6 \text{ } \cancel{\text{mol } CO_2}}{1 \text{ } \cancel{\text{mol } C_6H_{12}O_6}} \times \frac{44.01 \text{ g } CO_2}{1 \text{ } \cancel{\text{mol } CO_2}}$$

$$= 1.42 \times 10^3 \text{ g } CO_2$$

Check Does the answer seem reasonable? Should the mass of CO_2 produced be larger than the mass of $C_6H_{12}O_6$ reacted, even though the molar mass of CO_2 is considerably less than the molar mass of $C_6H_{12}O_6$? What is the mole ratio between CO_2 and $C_6H_{12}O_6$?

Similar problem: 3.72.

Practice Exercise Methanol (CH_3OH) burns in air according to the equation

$$2CH_3OH + 3O_2 \longrightarrow 2CO_2 + 4H_2O$$

If 209 g of methanol are used up in a combustion process, what is the mass of H_2O produced?

Lithium reacting with water to produce hydrogen gas.

EXAMPLE 3.14

All alkali metals react with water to produce hydrogen gas and the corresponding alkali metal hydroxide. A typical reaction is that between lithium and water:

$$2Li(s) + 2H_2O(l) \longrightarrow 2LiOH(aq) + H_2(g)$$

How many grams of Li are needed to produce 7.79 g of H_2?

(Continued)

Strategy The question asks for number of grams of reactant (Li) to form a specific amount of product (H_2). Therefore, we need to reverse the steps shown in Figure 3.8. From the equation we see that 2 mol Li ≏ 1 mol H_2.

Solution The conversion steps are

$$\text{grams of } H_2 \longrightarrow \text{moles of } H_2 \longrightarrow \text{moles of Li} \longrightarrow \text{grams of Li}$$

Combining these steps into one equation, we write

$$7.79 \text{ g } H_2 \times \frac{1 \text{ mol } H_2}{2.016 \text{ g } H_2} \times \frac{2 \text{ mol Li}}{1 \text{ mol } H_2} \times \frac{6.941 \text{ g Li}}{1 \text{ mol Li}} = 53.6 \text{ g Li}$$

Check There are roughly 4 moles of H_2 in 7.79 g H_2, so we need 8 moles of Li. From the approximate molar mass of Li (7 g), does the answer seem reasonable?

Similar problems: 3.65, 3.71.

Practice Exercise The reaction between nitric oxide (NO) and oxygen to form nitrogen dioxide (NO_2) is a key step in photochemical smog formation:

$$2NO(g) + O_2(g) \longrightarrow 2NO_2(g)$$

How many grams of O_2 are needed to produce 2.21 g of NO_2?

REVIEW OF CONCEPTS

Which of the following statements is correct for the equation shown here?

$$4NH_3(g) + 5O_2(g) \longrightarrow 4NO(g) + 6H_2O(g)$$

(a) One mole of NO is produced per mole of NH_3 reacted.
(b) 6 g of H_2O are produced for every 4 g of NH_3 reacted.
(c) 2 moles of NO are produced for every 3 moles of O_2 reacted.

3.9 Limiting Reagents

Animation:
Limiting Reagent

When a chemist carries out a reaction, the reactants are usually not present in exact ***stoichiometric amounts,*** that is, *in the proportions indicated by the balanced equation.* Because the goal of a reaction is to produce the maximum quantity of a useful compound from the starting materials, frequently a large excess of one reactant is supplied to ensure that the more expensive reactant is completely converted to the desired product. Consequently, some reactant will be left over at the end of the reaction. *The reactant used up first in a reaction* is called the ***limiting reagent,*** because the maximum amount of product formed depends on how much of this reactant was originally present. When this reactant is used up, no more product can be formed. ***Excess reagents*** are the *reactants present in quantities greater than necessary to react with the quantity of the limiting reagent.*

The concept of the limiting reagent is analogous to the relationship between men and women in a dance contest at a club. If there are 14 men and only 9 women, then only 9 female/male pairs can compete. Five men will be left without partners. The number of women thus *limits* the number of men that can dance in the contest, and there is an *excess* of men.

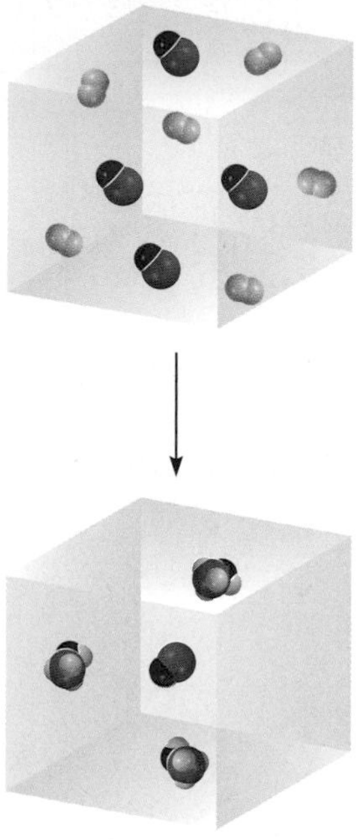

Figure 3.9
At the start of the reaction, there were six H_2 molecules and four CO molecules. At the end, all the H_2 molecules are gone and only one CO molecule is left. Therefore, the H_2 molecule is the limiting reagent and CO is the excess reagent. Each molecule can also be treated as one mole of the substance in this reaction.

Consider the industrial synthesis of methanol (CH_3OH) from carbon monoxide and hydrogen at high temperatures:

$$CO(g) + 2H_2(g) \longrightarrow CH_3OH(g)$$

Suppose initially we have 4 moles of CO and 6 moles of H_2 (Figure 3.9). One way to determine which of the two reactants is the limiting reagent is to calculate the number of moles of CH_3OH obtained based on the initial quantities of CO and H_2. From the preceding definition, we see that only the limiting reagent will yield the *smaller* amount of the product. Starting with 4 moles of CO, we find the number of moles of CH_3OH produced is

$$4 \cancel{\text{mol CO}} \times \frac{1 \text{ mol } CH_3OH}{1 \cancel{\text{mol CO}}} = 4 \text{ mol } CH_3OH$$

and starting with 6 moles of H_2, the number of moles of CH_3OH formed is

$$6 \cancel{\text{mol } H_2} \times \frac{1 \text{ mol } CH_3OH}{2 \cancel{\text{mol } H_2}} = 3 \text{ mol } CH_3OH$$

Because H_2 results in a smaller amount of CH_3OH, it must be the limiting reagent. Therefore, CO is the excess reagent.

In stoichiometric calculations involving limiting reagents, the first step is to decide which reactant is the limiting reagent. After the limiting reagent has been identified, the rest of the problem can be solved as outlined in Section 3.8. Example 3.15 illustrates this approach.

EXAMPLE 3.15

Urea [$(NH_2)_2CO$] is prepared by reacting ammonia with carbon dioxide:

$$2NH_3(g) + CO_2(g) \longrightarrow (NH_2)_2CO(aq) + H_2O(l)$$

In one process, 849.2 g of NH_3 are treated with 1223 g of CO_2. (a) Which of the two reactants is the limiting reagent? (b) Calculate the mass of $(NH_2)_2CO$ formed. (c) How much excess reagent (in grams) is left at the end of the reaction?

(a) Strategy The reactant that forms fewer moles of product is the limiting reagent because it limits the amount of product that can be formed. How do we convert from the amount of reactant to the amount of product? Perform this calculation for each reactant, then compare the moles of product, $(NH_2)_2CO$, formed by the given amounts of NH_3 and CO_2 to determine which reactant is the limiting reagent.

Solution We carry out two separate calculations. First, starting with 849.2 g of NH_3, we calculate the number of moles of $(NH_2)_2CO$ that could be produced if all the NH_3 reacted according to the following conversions:

$$\text{grams of } NH_3 \longrightarrow \text{moles of } NH_3 \longrightarrow \text{moles of } (NH_2)_2CO$$

Combining these steps into one step, we write

$$\text{moles of } (NH_2)_2CO = 849.2 \cancel{\text{g } NH_3} \times \frac{1 \cancel{\text{mol } NH_3}}{17.03 \cancel{\text{g } NH_3}} \times \frac{1 \text{ mol } (NH_2)_2CO}{2 \cancel{\text{mol } NH_3}}$$
$$= 24.93 \text{ mol } (NH_2)_2CO$$

Second, for 1223 g of CO_2, the conversions are

$$\text{grams of } CO_2 \longrightarrow \text{moles of } CO_2 \longrightarrow \text{moles of } (NH_2)_2CO$$

(Continued)

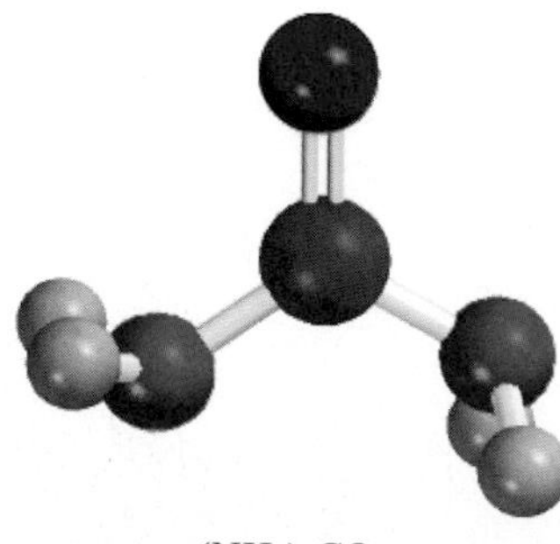
$(NH_2)_2CO$

The number of moles of $(NH_2)_2CO$ that could be produced if all the CO_2 reacted is

$$\text{moles of } (NH_2)_2CO = 1223 \cancel{\text{g } CO_2} \times \frac{1 \cancel{\text{mol } CO_2}}{44.01 \cancel{\text{g } CO_2}} \times \frac{1 \text{ mol } (NH_2)_2CO}{1 \cancel{\text{mol } CO_2}}$$
$$= 27.79 \text{ mol } (NH_2)_2CO$$

It follows, therefore, that NH_3 must be the limiting reagent because it produces a smaller amount of $(NH_2)_2CO$.

(b) Strategy We determined the moles of $(NH_2)_2CO$ produced in part (a), using NH_3 as the limiting reagent. How do we convert from moles to grams?

Solution The molar mass of $(NH_2)_2CO$ is 60.06 g. We use this as a conversion factor to convert from moles of $(NH_2)_2CO$ to grams of $(NH_2)_2CO$:

$$\text{mass of } (NH_2)_2CO = 24.93 \cancel{\text{mol } (NH_2)_2CO} \times \frac{60.06 \text{ g } (NH_2)_2CO}{1 \cancel{\text{mol } (NH_2)_2CO}}$$
$$= 1497 \text{ g } (NH_2)_2CO$$

Check Does your answer seem reasonable? 24.93 moles of product are formed. What is the mass of 1 mole of $(NH_2)_2CO$?

(c) Strategy Working backward, we can determine the amount of CO_2 that reacted to produce 24.93 moles of $(NH_2)_2CO$. The amount of CO_2 left over is the difference between the initial amount and the amount reacted.

Solution Starting with 24.93 moles of $(NH_2)_2CO$, we can determine the mass of CO_2 that reacted using the mole ratio from the balanced equation and the molar mass of CO_2. The conversion steps are

$$\text{moles } (NH_2)_2CO \longrightarrow \text{moles of } CO_2 \longrightarrow \text{mass of } CO_2$$

so that

$$\text{mass of } CO_2 \text{ reacted} = 24.93 \cancel{\text{mol } (NH_2)_2CO} \times \frac{1 \cancel{\text{mol } CO_2}}{1 \cancel{\text{mol } (NH_2)_2CO}} \times \frac{44.01 \text{ g } CO_2}{1 \cancel{\text{mol } CO_2}}$$
$$= 1097 \text{ g } CO_2$$

The amount of CO_2 remaining (in excess) is the difference between the initial amount (1223 g) and the amount reacted (1097 g):

$$\text{mass of } CO_2 \text{ remaining} = 1223 \text{ g} - 1097 \text{ g} = 126 \text{ g}$$

Check Is this answer reasonable? Can the mass of CO_2 remaining be greater than the initial mass of CO_2 used in the reaction?

Similar problem: 3.86.

Practice Exercise The reaction between aluminum and iron(III) oxide can generate temperatures approaching 3000°C and is used in welding metals:

$$2Al + Fe_2O_3 \longrightarrow Al_2O_3 + 2Fe$$

In one process, 124 g of Al are reacted with 601 g of Fe_2O_3. (a) Calculate the mass (in grams) of Al_2O_3 formed. (b) How much of the excess reagent is left at the end of the reaction?

Example 3.15 brings out an important point. In practice, chemists usually choose the more expensive chemical as the limiting reagent so that all or most of it will be consumed in the reaction. In the synthesis of urea, NH_3 is invariably the limiting reagent because it is much more expensive than CO_2.

REVIEW OF CONCEPTS

Consider the following reaction

$$2NO(g) + O_2(g) \longrightarrow 2NO_2(g)$$

Starting with the reactants shown in (a), which of the diagrams shown in (b)–(d) best represents the situation in which the limiting reagent has completely reacted?

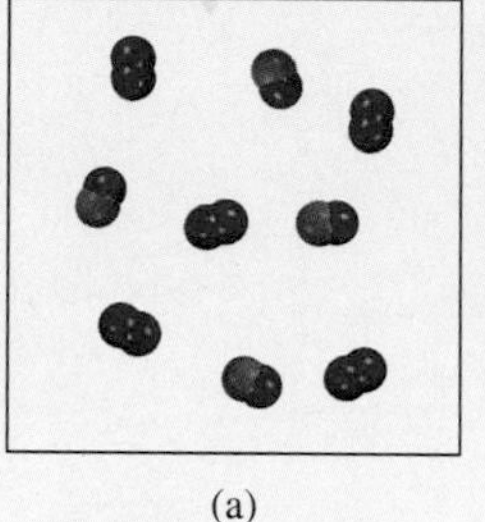

(a)

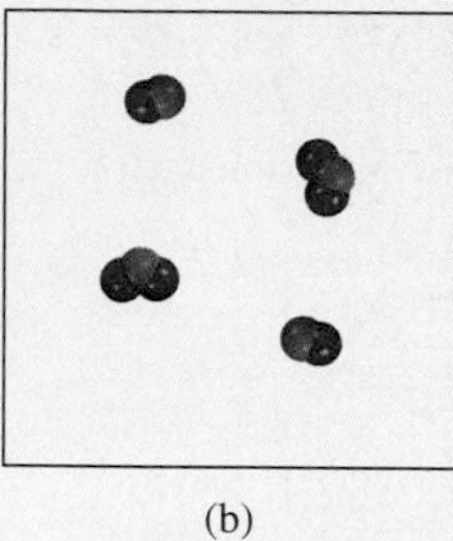

(b)

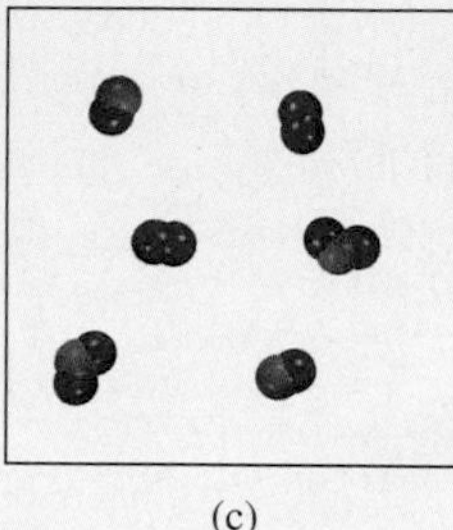

(c)

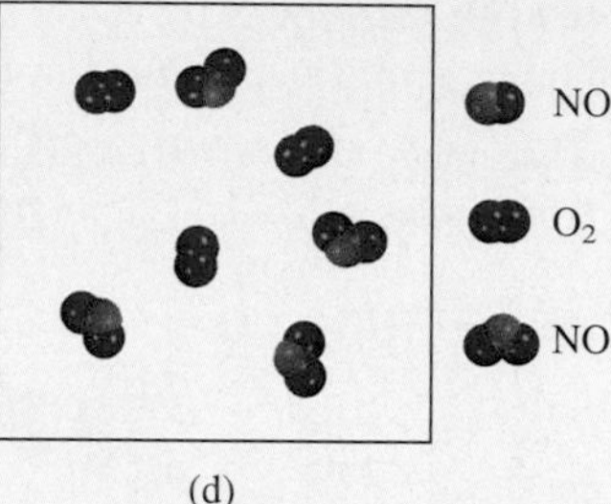

(d)

3.10 Reaction Yield

The amount of limiting reagent present at the start of a reaction determines the ***theoretical yield*** of the reaction, that is, *the amount of product that would result if all the limiting reagent reacted.* The theoretical yield, then, is the *maximum* obtainable yield, predicted by the balanced equation. In practice, the ***actual yield,*** or *the amount of product actually obtained from a reaction,* is almost always less than the theoretical yield. There are many reasons for the difference between actual and theoretical yields. For instance, many reactions are reversible, and so they do not proceed 100 percent from left to right. Even when a reaction is 100 percent complete, it may be difficult to recover all of the product from the reaction medium (say, from an aqueous solution). Some reactions are complex in the sense that the products formed may react further among themselves or with the reactants to form still other products. These additional reactions will reduce the yield of the first reaction.

To determine how efficient a given reaction is, chemists often figure the ***percent yield,*** which describes *the proportion of the actual yield to the theoretical yield.* It is calculated as follows:

$$\%\ \text{yield} = \frac{\text{actual yield}}{\text{theoretical yield}} \times 100\% \tag{3.4}$$

Percent yields may range from a fraction of 1 percent to 100 percent. Chemists strive to maximize the percent yield in a reaction. Factors that can affect the percent yield include temperature and pressure. We will study these effects later.

In Example 3.16, we will calculate the yield of an industrial process.

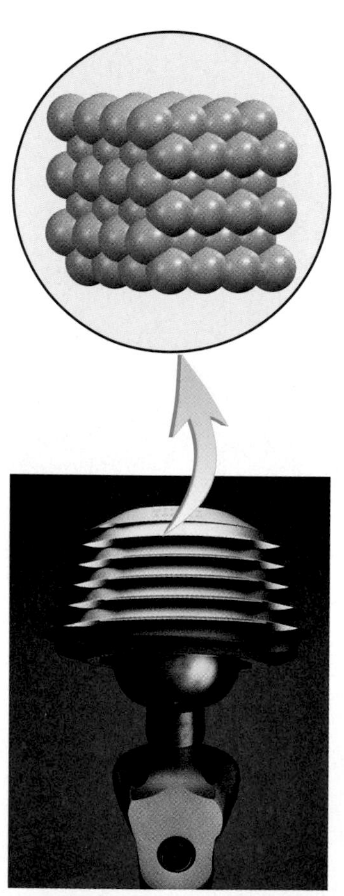

An artificial hip joint made of titanium and the structure of solid titanium.

EXAMPLE 3.16

Titanium is a strong, lightweight, corrosion-resistant metal that is used in aircraft bodies, jet engines, bicycle frames, and artificial joints. It is prepared by the reaction of titantium(IV) chloride with molten magnesium between 950°C and 1150°C:

$$TiCl_4(g) + 2Mg(l) \longrightarrow Ti(s) + 2MgCl_2(l)$$

(Continued)

In a certain industrial operation, 2.84×10^7 g of $TiCl_4$ are reacted with 1.09×10^7 g of Mg. (a) Calculate the theoretical yield of Ti in grams. (b) Calculate the percent yield if 5.97×10^6 g of Ti are actually obtained.

(a) Strategy Because there are two reactants, this is likely to be a limiting reagent problem. The reactant that produces fewer moles of product is the limiting reagent. How do we convert from amount of reactant to amount of product? Perform this calculation for each reactant, then compare the moles of product, Ti, formed.

Solution Carry out two separate calculations to see which of the two reactants is the limiting reagent. First, starting with 2.84×10^7 g of $TiCl_4$, calculate the number of moles of Ti that could be produced if all the $TiCl_4$ reacted. The conversions are

$$\text{grams of } TiCl_4 \longrightarrow \text{moles of } TiCl_4 \longrightarrow \text{moles of Ti}$$

so that

$$\text{moles of Ti} = 2.84 \times 10^7 \cancel{\text{g TiCl}_4} \times \frac{1 \cancel{\text{mol TiCl}_4}}{189.68 \cancel{\text{g TiCl}_4}} \times \frac{1 \text{ mol Ti}}{1 \cancel{\text{mol TiCl}_4}}$$
$$= 1.50 \times 10^5 \text{ mol Ti}$$

Next, we calculate the number of moles of Ti formed from 1.09×10^7 g of Mg. The conversion steps are

$$\text{grams of Mg} \longrightarrow \text{moles of Mg} \longrightarrow \text{moles of Ti}$$

and we write

$$\text{moles of Ti} = 1.09 \times 10^7 \cancel{\text{g Mg}} \times \frac{1 \cancel{\text{mol Mg}}}{24.31 \cancel{\text{g Mg}}} \times \frac{1 \text{ mol Ti}}{2 \cancel{\text{mol Mg}}}$$
$$= 2.24 \times 10^5 \text{ mol Ti}$$

Therefore, $TiCl_4$ is the limiting reagent because it produces a smaller amount of Ti. The mass of Ti formed is

$$1.50 \times 10^5 \cancel{\text{mol Ti}} \times \frac{47.88 \text{ g Ti}}{1 \cancel{\text{mol Ti}}} = \boxed{7.18 \times 10^6 \text{ g Ti}}$$

(b) Strategy The mass of Ti determined in part (a) is the theoretical yield. The amount given in part (b) is the actual yield of the reaction.

Solution The percent yield is given by

$$\% \text{ yield} = \frac{\text{actual yield}}{\text{theoretical yield}} \times 100\%$$
$$= \frac{5.97 \times 10^6 \text{ g Ti}}{7.18 \times 10^6 \text{ g Ti}} \times 100\%$$
$$= \boxed{83.1\%}$$

Check Should the percent yield be less than 100 percent?

Similar problems: 3.89, 3.90.

Practice Exercise Industrially, vanadium metal, which is used in steel alloys, can be obtained by reacting vanadium(V) oxide with calcium at high temperatures:

$$5Ca + V_2O_5 \longrightarrow 5CaO + 2V$$

In one process, 1.54×10^3 g of V_2O_5 react with 1.96×10^3 g of Ca. (a) Calculate the theoretical yield of V. (b) Calculate the percent yield if 803 g of V are obtained.

Review of Concepts

Why can the percent yield for a reaction not be greater than 100%?

Key Equations

percent composition of an element in a compound =

$$\frac{n \times \text{molar mass of element}}{\text{molar mass of compound}} \times 100\% \quad (3.1)$$

$$\% \text{ yield} = \frac{\text{actual yield}}{\text{theoretical yield}} \times 100\% \quad (3.4)$$

Summary of Facts and Concepts

1. Atomic masses are measured in atomic mass units (amu), a relative unit based on a value of exactly 12 for the C-12 isotope. The atomic mass given for the atoms of a particular element is the average of the naturally occurring isotope distribution of that element. The molecular mass of a molecule is the sum of the atomic masses of the atoms in the molecule. Both atomic mass and molecular mass can be accurately determined with a mass spectrometer.
2. A mole is Avogadro's number (6.022×10^{23}) of atoms, molecules, or other particles. The molar mass (in grams) of an element or a compound is numerically equal to its mass in atomic mass units (amu) and contains Avogadro's number of atoms (in the case of elements), molecules (in the case of molecular substances), or simplest formula units (in the case of ionic compounds).
3. The percent composition by mass of a compound is the percent by mass of each element present. If we know the percent composition by mass of a compound, we can deduce the empirical formula of the compound and also the molecular formula of the compound if the approximate molar mass is known.
4. Chemical changes, called chemical reactions, are represented by chemical equations. Substances that undergo change—the reactants—are written on the left and the substances formed—the products—appear to the right of the arrow. Chemical equations must be balanced, in accordance with the law of conservation of mass. The number of atoms of each element in the reactants must equal the number in the products.
5. Stoichiometry is the quantitative study of products and reactants in chemical reactions. Stoichiometric calculations are best done by expressing both the known and unknown quantities in terms of moles and then converting to other units if necessary. A limiting reagent is the reactant that is present in the smallest stoichiometric amount. It limits the amount of product that can be formed. The amount of product obtained in a reaction (the actual yield) may be less than the maximum possible amount (the theoretical yield). The ratio of the two multiplied by 100 percent is expressed as the percent yield.

Key Words

Actual yield, p. 86
Atomic mass, p. 61
Atomic mass unit (amu), p. 61
Avogadro's number (N_A), p. 63
Chemical equation, p. 75
Chemical reaction, p. 75
Excess reagent, p. 83
Limiting reagent, p. 83
Molar mass ($\mathcal{M}$), p. 63
Mole (mol), p. 62
Mole method, p. 80
Molecular mass, p. 66
Percent composition, p. 70
Percent yield, p. 86
Product, p. 76
Reactant, p. 76
Stoichiometric amount, p. 83
Stoichiometry, p. 80
Theoretical yield, p. 86

Questions and Problems

Atomic Mass

Review Questions

3.1 What is an atomic mass unit? Why is it necessary to introduce such a unit?

3.2 What is the mass (in amu) of a carbon-12 atom? Why is the atomic mass of carbon listed as 12.01 amu in the table on the inside front cover of this book?

3.3 Explain clearly what is meant by the statement "The atomic mass of gold is 197.0 amu."

3.4 What information would you need to calculate the average atomic mass of an element?

Problems

3.5 The atomic masses of $^{35}_{17}Cl$ (75.53 percent) and $^{37}_{17}Cl$ (24.47 percent) are 34.968 amu and 36.956 amu, respectively. Calculate the average atomic mass of chlorine. The percentages in parentheses denote the relative abundances.

3.6 The atomic masses of $^{6}_{3}Li$ and $^{7}_{3}Li$ are 6.0151 amu and 7.0160 amu, respectively. Calculate the natural abundances of these two isotopes. The average atomic mass of Li is 6.941 amu.

3.7 What is the mass in grams of 13.2 amu?

3.8 How many amu are there in 8.4 g?

Avogadro's Number and Molar Mass

Review Questions

3.9 Define the term "mole." What is the unit for mole in calculations? What does the mole have in common with the pair, the dozen, and the gross? What does Avogadro's number represent?

3.10 What is the molar mass of an atom? What are the commonly used units for molar mass?

Problems

3.11 Earth's population is about 6.5 billion. Suppose that every person on Earth participates in a process of counting identical particles at the rate of two particles per second. How many years would it take to count 6.0×10^{23} particles? Assume that there are 365 days in a year.

3.12 The thickness of a piece of paper is 0.0036 in. Suppose a certain book has an Avogadro's number of pages; calculate the thickness of the book in light-years. (*Hint:* See Problem 1.38 for the definition of light-year.)

3.13 How many atoms are there in 5.10 moles of sulfur (S)?

3.14 How many moles of cobalt (Co) atoms are there in 6.00×10^9 (6 billion) Co atoms?

3.15 How many moles of calcium (Ca) atoms are in 77.4 g of Ca?

3.16 How many grams of gold (Au) are there in 15.3 moles of Au?

3.17 What is the mass in grams of a single atom of each of the following elements? (a) Hg, (b) Ne.

3.18 What is the mass in grams of a single atom of each of the following elements? (a) As, (b) Ni.

3.19 What is the mass in grams of 1.00×10^{12} lead (Pb) atoms?

3.20 How many atoms are present in 3.14 g of copper (Cu)?

3.21 Which of the following has more atoms: 1.10 g of hydrogen atoms or 14.7 g of chromium atoms?

3.22 Which of the following has a greater mass: 2 atoms of lead or 5.1×10^{-23} mole of helium?

Molecular Mass

Problems

3.23 Calculate the molecular mass or formula mass (in amu) of each of the following substances: (a) CH_4, (b) NO_2, (c) SO_3, (d) C_6H_6, (e) NaI, (f) K_2SO_4, (g) $Ca_3(PO_4)_2$.

3.24 Calculate the molar mass of the following substances: (a) Li_2CO_3, (b) CS_2, (c) $CHCl_3$ (chloroform), (d) $C_6H_8O_6$ (ascorbic acid, or vitamin C), (e) KNO_3, (f) Mg_3N_2.

3.25 Calculate the molar mass of a compound if 0.372 mole of it has a mass of 152 g.

3.26 How many molecules of acetone, shown here, are present in 0.435 g of acetone?

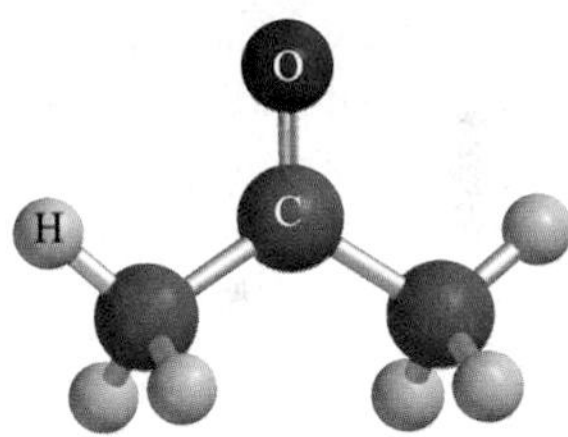

3.27 Calculate the number of C, H, and O atoms in 1.75 g of squaric acid, shown here.

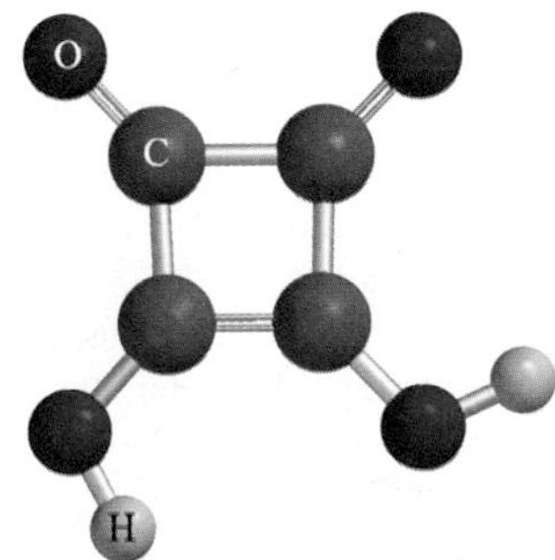

3.28 Urea $[(NH_2)_2CO]$ is used for fertilizer and many other things. Calculate the number of N, C, O, and H atoms in 1.68×10^4 g of urea.

3.29 Pheromones are a special type of compound secreted by the females of many insect species to attract the males for mating. One pheromone has the molecular formula $C_{19}H_{38}O$. Normally, the amount of this pheromone secreted by a female insect is about 1.0×10^{-12} g. How many molecules are there in this quantity?

3.30 The density of water is 1.00 g/mL at 4°C. How many water molecules are present in 2.56 mL of water at this temperature?

Mass Spectrometry

Review Questions

3.31 Describe the operation of a mass spectrometer.

3.32 Describe how you would determine the isotopic abundance of an element from its mass spectrum.

Problems

3.33 Carbon has two stable isotopes, ${}^{12}_{6}C$ and ${}^{13}_{6}C$, and fluorine has only one stable isotope, ${}^{19}_{9}F$. How many peaks would you observe in the mass spectrum of the positive ion of CF_4^+? Assume that the ion does not break up into smaller fragments.

3.34 Hydrogen has two stable isotopes, ${}^{1}_{1}H$ and ${}^{2}_{1}H$, and sulfur has four stable isotopes, ${}^{32}_{16}S$, ${}^{33}_{16}S$, ${}^{34}_{16}S$, and ${}^{36}_{16}S$. How many peaks would you observe in the mass spectrum of the positive ion of hydrogen sulfide, H_2S^+? Assume no decomposition of the ion into smaller fragments.

Percent Composition and Chemical Formulas

Review Questions

3.35 Use ammonia (NH_3) to explain what is meant by the percent composition by mass of a compound.

3.36 Describe how the knowledge of the percent composition by mass of an unknown compound can help us identify the compound.

3.37 What does the word "empirical" in empirical formula mean?

3.38 If we know the empirical formula of a compound, what additional information do we need to determine its molecular formula?

Problems

3.39 Tin (Sn) exists in Earth's crust as SnO_2. Calculate the percent composition by mass of Sn and O in SnO_2.

3.40 For many years chloroform ($CHCl_3$) was used as an inhalation anesthetic in spite of the fact that it is also a toxic substance that may cause severe liver, kidney, and heart damage. Calculate the percent composition by mass of this compound.

3.41 Cinnamic alcohol is used mainly in perfumery, particularly in soaps and cosmetics. Its molecular formula is $C_9H_{10}O$. (a) Calculate the percent composition by mass of C, H, and O in cinnamic alcohol. (b) How many molecules of cinnamic alcohol are contained in a sample of mass 0.469 g?

3.42 All of the substances listed below are fertilizers that contribute nitrogen to the soil. Which of these is the richest source of nitrogen on a mass percentage basis?

(a) Urea, $(NH_2)_2CO$

(b) Ammonium nitrate, NH_4NO_3

(c) Guanidine, $HNC(NH_2)_2$

(d) Ammonia, NH_3

3.43 Allicin is the compound responsible for the characteristic smell of garlic. An analysis of the compound gives the following percent composition by mass: C: 44.4 percent; H: 6.21 percent; S: 39.5 percent; O: 9.86 percent. Calculate its empirical formula. What is its molecular formula given that its molar mass is about 162 g?

3.44 Peroxyacylnitrate (PAN) is one of the components of smog. It is a compound of C, H, N, and O. Determine the percent composition of oxygen and the empirical formula from the following percent composition by mass: 19.8 percent C, 2.50 percent H, 11.6 percent N. What is its molecular formula given that its molar mass is about 120 g?

3.45 The formula for rust can be represented by Fe_2O_3. How many moles of Fe are present in 24.6 g of the compound?

3.46 How many grams of sulfur (S) are needed to react completely with 246 g of mercury (Hg) to form HgS?

3.47 Calculate the mass in grams of iodine (I_2) that will react completely with 20.4 g of aluminum (Al) to form aluminum iodide (AlI_3).

3.48 Tin(II) fluoride (SnF_2) is often added to toothpaste as an ingredient to prevent tooth decay. What is the mass of F in grams in 24.6 g of the compound?

3.49 What are the empirical formulas of the compounds with the following compositions? (a) 2.1 percent H, 65.3 percent O, 32.6 percent S, (b) 20.2 percent Al, 79.8 percent Cl.

3.50 What are the empirical formulas of the compounds with the following compositions? (a) 40.1 percent C, 6.6 percent H, 53.3 percent O, (b) 18.4 percent C, 21.5 percent N, 60.1 percent K.

3.51 The anticaking agent added to Morton salt is calcium silicate, $CaSiO_3$. This compound can absorb up to 2.5 times its mass of water and still remains a free-flowing powder. Calculate the percent composition of $CaSiO_3$.

3.52 The empirical formula of a compound is CH. If the molar mass of this compound is about 78 g, what is its molecular formula?

3.53 The molar mass of caffeine is 194.19 g. Is the molecular formula of caffeine $C_4H_5N_2O$ or $C_8H_{10}N_4O_2$?

3.54 Monosodium glutamate (MSG), a food-flavor enhancer, has been blamed for "Chinese restaurant syndrome," the symptoms of which are headaches and chest pains. MSG has the following composition by mass: 35.51 percent C, 4.77 percent H, 37.85 percent O, 8.29 percent N, and 13.60 percent Na. What is its molecular formula if its molar mass is about 169 g?

Chemical Reactions and Chemical Equations

Review Questions

3.55 Use the formation of water from hydrogen and oxygen to explain the following terms: chemical reaction, reactant, product.

3.56 What is the difference between a chemical reaction and a chemical equation?

3.57 Why must a chemical equation be balanced? What law is obeyed by a balanced chemical equation?

3.58 Write the symbols used to represent gas, liquid, solid, and the aqueous phase in chemical equations.

Problems

3.59 Balance the following equations using the method outlined in Section 3.7:

(a) $C + O_2 \longrightarrow CO$
(b) $CO + O_2 \longrightarrow CO_2$
(c) $H_2 + Br_2 \longrightarrow HBr$
(d) $K + H_2O \longrightarrow KOH + H_2$
(e) $Mg + O_2 \longrightarrow MgO$
(f) $O_3 \longrightarrow O_2$
(g) $H_2O_2 \longrightarrow H_2O + O_2$
(h) $N_2 + H_2 \longrightarrow NH_3$
(i) $Zn + AgCl \longrightarrow ZnCl_2 + Ag$
(j) $S_8 + O_2 \longrightarrow SO_2$
(k) $NaOH + H_2SO_4 \longrightarrow Na_2SO_4 + H_2O$
(l) $Cl_2 + NaI \longrightarrow NaCl + I_2$
(m) $KOH + H_3PO_4 \longrightarrow K_3PO_4 + H_2O$
(n) $CH_4 + Br_2 \longrightarrow CBr_4 + HBr$

3.60 Balance the following equations using the method outlined in Section 3.7:

(a) $N_2O_5 \longrightarrow N_2O_4 + O_2$
(b) $KNO_3 \longrightarrow KNO_2 + O_2$
(c) $NH_4NO_3 \longrightarrow N_2O + H_2O$
(d) $NH_4NO_2 \longrightarrow N_2 + H_2O$
(e) $NaHCO_3 \longrightarrow Na_2CO_3 + H_2O + CO_2$
(f) $P_4O_{10} + H_2O \longrightarrow H_3PO_4$
(g) $HCl + CaCO_3 \longrightarrow CaCl_2 + H_2O + CO_2$
(h) $Al + H_2SO_4 \longrightarrow Al_2(SO_4)_3 + H_2$
(i) $CO_2 + KOH \longrightarrow K_2CO_3 + H_2O$
(j) $CH_4 + O_2 \longrightarrow CO_2 + H_2O$
(k) $Be_2C + H_2O \longrightarrow Be(OH)_2 + CH_4$
(l) $Cu + HNO_3 \longrightarrow Cu(NO_3)_2 + NO + H_2O$
(m) $S + HNO_3 \longrightarrow H_2SO_4 + NO_2 + H_2O$
(n) $NH_3 + CuO \longrightarrow Cu + N_2 + H_2O$

Amounts of Reactants and Products

Review Questions

3.61 On what law is stoichiometry based? Why is it essential to use balanced equations in solving stoichiometric problems?

3.62 Describe the steps involved in the mole method.

Problems

3.63 Which of the following equations best represents the reaction shown in the diagram?

(a) $8A + 4B \longrightarrow C + D$
(b) $4A + 8B \longrightarrow 4C + 4D$
(c) $2A + B \longrightarrow C + D$
(d) $4A + 2B \longrightarrow 4C + 4D$
(e) $2A + 4B \longrightarrow C + D$

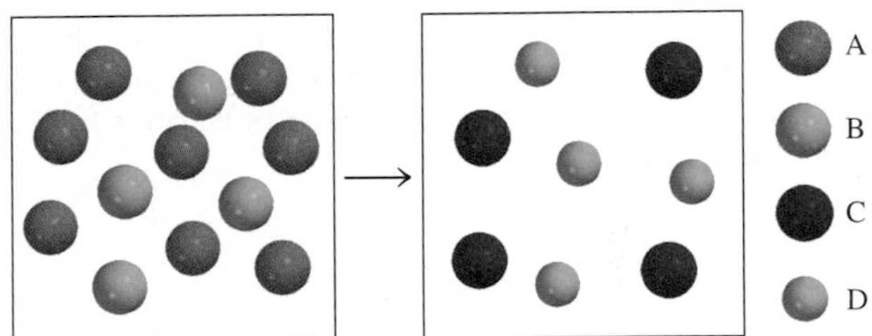

3.64 Which of the following equations best represents the reaction shown in the diagram?

(a) $A + B \longrightarrow C + D$
(b) $6A + 4B \longrightarrow C + D$
(c) $A + 2B \longrightarrow 2C + D$
(d) $3A + 2B \longrightarrow 2C + D$
(e) $3A + 2B \longrightarrow 4C + 2D$

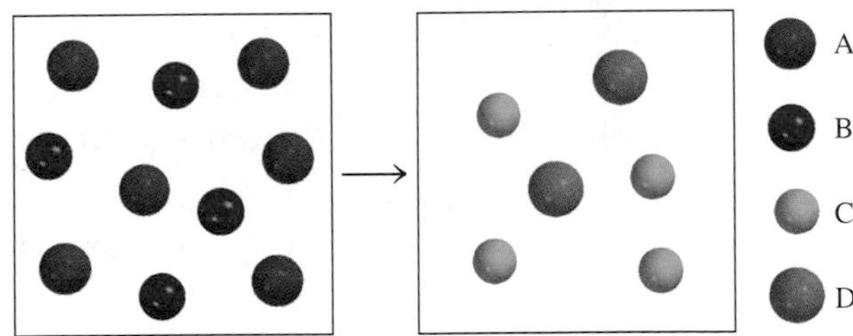

3.65 Consider the combustion of carbon monoxide (CO) in oxygen gas

$$2CO(g) + O_2(g) \longrightarrow 2CO_2(g)$$

Starting with 3.60 moles of CO, calculate the number of moles of CO_2 produced if there is enough oxygen gas to react with all of the CO.

3.66 Silicon tetrachloride ($SiCl_4$) can be prepared by heating Si in chlorine gas:

$$Si(s) + 2Cl_2(g) \longrightarrow SiCl_4(l)$$

In one reaction, 0.507 mole of $SiCl_4$ is produced. How many moles of molecular chlorine were used in the reaction?

3.67 Ammonia is a principal nitrogen fertilizer. It is prepared by the reaction between hydrogen and nitrogen.

$$3H_2(g) + N_2(g) \longrightarrow 2NH_3(g)$$

In a particular reaction, 6.0 moles of NH_3 were produced. How many moles of H_2 and how many moles of N_2 were reacted to produce this amount of NH_3?

3.68 Consider the combustion of butane (C_4H_{10}):

$$2C_4H_{10}(g) + 13O_2(g) \longrightarrow 8CO_2(g) + 10H_2O(l)$$

In a particular reaction, 5.0 moles of C_4H_{10} are reacted with an excess of O_2. Calculate the number of moles of CO_2 formed.

3.69 The annual production of sulfur dioxide from burning coal and fossil fuels, auto exhaust, and other sources is about 26 million tons. The equation for the reaction is

$$S(s) + O_2(g) \longrightarrow SO_2(g)$$

How much sulfur (in tons), present in the original materials, would result in that quantity of SO_2?

3.70 When baking soda (sodium bicarbonate or sodium hydrogen carbonate, $NaHCO_3$) is heated, it releases carbon dioxide gas, which is responsible for the rising of cookies, donuts, and bread. (a) Write a balanced equation for the decomposition of the compound (one of the products is Na_2CO_3). (b) Calculate the mass of $NaHCO_3$ required to produce 20.5 g of CO_2.

3.71 When potassium cyanide (KCN) reacts with acids, a deadly poisonous gas, hydrogen cyanide (HCN), is given off. Here is the equation:

$$KCN(aq) + HCl(aq) \longrightarrow KCl(aq) + HCN(g)$$

If a sample of 0.140 g of KCN is treated with an excess of HCl, calculate the amount of HCN formed, in grams.

3.72 Fermentation is a complex chemical process of wine making in which glucose is converted into ethanol and carbon dioxide:

$$\underset{\text{glucose}}{C_6H_{12}O_6} \longrightarrow \underset{\text{ethanol}}{2C_2H_5OH} + 2CO_2$$

Starting with 500.4 g of glucose, what is the maximum amount of ethanol in grams and in liters that can be obtained by this process? (Density of ethanol = 0.789 g/mL.)

3.73 Each copper(II) sulfate unit is associated with five water molecules in crystalline copper(II) sulfate pentahydrate ($CuSO_4 \cdot 5H_2O$). When this compound is heated in air above 100°C, it loses the water molecules and also its blue color:

$$CuSO_4 \cdot 5H_2O \longrightarrow CuSO_4 + 5H_2O$$

If 9.60 g of $CuSO_4$ are left after heating 15.01 g of the blue compound, calculate the number of moles of H_2O originally present in the compound.

3.74 For many years the recovery of gold—that is, the separation of gold from other materials—involved the use of potassium cyanide:

$$4Au + 8KCN + O_2 + 2H_2O \longrightarrow 4KAu(CN)_2 + 4KOH$$

What is the minimum amount of KCN in moles needed to extract 29.0 g (about an ounce) of gold?

3.75 Limestone ($CaCO_3$) is decomposed by heating to quicklime (CaO) and carbon dioxide. Calculate how many grams of quicklime can be produced from 1.0 kg of limestone.

3.76 Nitrous oxide (N_2O) is also called "laughing gas." It can be prepared by the thermal decomposition of ammonium nitrate (NH_4NO_3). The other product is H_2O. (a) Write a balanced equation for this reaction. (b) How many grams of N_2O are formed if 0.46 mole of NH_4NO_3 is used in the reaction?

3.77 The fertilizer ammonium sulfate [$(NH_4)_2SO_4$] is prepared by the reaction between ammonia (NH_3) and sulfuric acid:

$$2NH_3(g) + H_2SO_4(aq) \longrightarrow (NH_4)_2SO_4(aq)$$

How many kilograms of NH_3 are needed to produce 1.00×10^5 kg of $(NH_4)_2SO_4$?

3.78 A common laboratory preparation of oxygen gas is the thermal decomposition of potassium chlorate ($KClO_3$). Assuming complete decomposition, calculate the number of grams of O_2 gas that can be obtained from 46.0 g of $KClO_3$. (The products are KCl and O_2.)

Limiting Reagents

Review Questions

3.79 Define limiting reagent and excess reagent. What is the significance of the limiting reagent in predicting the amount of the product obtained in a reaction? Can there be a limiting reagent if only one reactant is present?

3.80 Give an everyday example that illustrates the limiting reagent concept.

Problems

3.81 Consider the reaction

$$2A + B \longrightarrow C$$

(a) In the diagram here that represents the reaction, which reactant, A or B, is the limiting reagent? (b) Assuming complete reaction, draw a molecular-model representation of the amounts of reactants and products left after the reaction. The atomic arrangement in C is ABA.

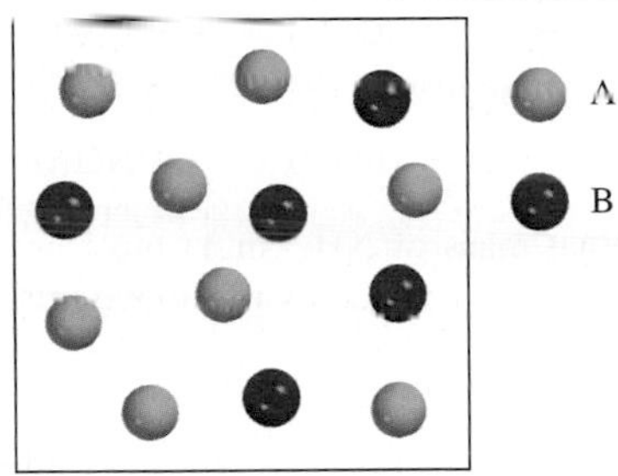

3.82 Consider the reaction

$$N_2 + 3H_2 \longrightarrow 2NH_3$$

Assuming each model represents one mole of the substance, show the number of moles of the product and the excess reagent left after the complete reaction.

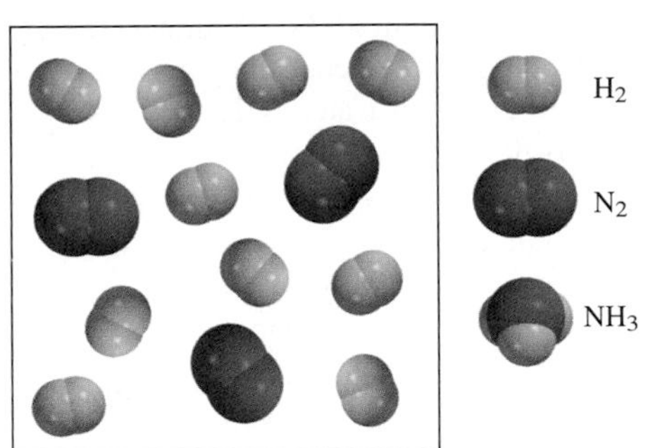

3.83 Nitric oxide (NO) reacts with oxygen gas to form nitrogen dioxide (NO_2), a dark-brown gas:

$$2NO(g) + O_2(g) \longrightarrow 2NO_2(g)$$

In one experiment 0.886 mole of NO is mixed with 0.503 mole of O_2. Calculate which of the two reactants is the limiting reagent. Calculate also the number of moles of NO_2 produced.

3.84 The depletion of ozone (O_3) in the stratosphere has been a matter of great concern among scientists in recent years. It is believed that ozone can react with nitric oxide (NO) that is discharged from the high-altitude jet plane, the SST. The reaction is

$$O_3 + NO \longrightarrow O_2 + NO_2$$

If 0.740 g of O_3 reacts with 0.670 g of NO, how many grams of NO_2 will be produced? Which compound is the limiting reagent? Calculate the number of moles of the excess reagent remaining at the end of the reaction.

3.85 Propane (C_3H_8) is a component of natural gas and is used in domestic cooking and heating. (a) Balance the following equation representing the combustion of propane in air:

$$C_3H_8 + O_2 \longrightarrow CO_2 + H_2O$$

(b) How many grams of carbon dioxide can be produced by burning 3.65 moles of propane? Assume that oxygen is the excess reagent in this reaction.

3.86 Consider the reaction

$$MnO_2 + 4HCl \longrightarrow MnCl_2 + Cl_2 + 2H_2O$$

If 0.86 mole of MnO_2 and 48.2 g of HCl react, which reagent will be used up first? How many grams of Cl_2 will be produced?

Reaction Yield

Review Questions

3.87 Why is the theoretical yield of a reaction determined only by the amount of the limiting reagent?

3.88 Why is the actual yield of a reaction almost always smaller than the theoretical yield?

Problems

3.89 Hydrogen fluoride is used in the manufacture of Freons (which destroy ozone in the stratosphere) and in the production of aluminum metal. It is prepared by the reaction

$$CaF_2 + H_2SO_4 \longrightarrow CaSO_4 + 2HF$$

In one process 6.00 kg of CaF_2 are treated with an excess of H_2SO_4 and yield 2.86 kg of HF. Calculate the percent yield of HF.

3.90 Nitroglycerin ($C_3H_5N_3O_9$) is a powerful explosive. Its decomposition can be represented by

$$4C_3H_5N_3O_9 \longrightarrow 6N_2 + 12CO_2 + 10H_2O + O_2$$

This reaction generates a large amount of heat and many gaseous products. It is the sudden formation of these gases, together with their rapid expansion, that produces the explosion. (a) What is the maximum amount of O_2 in grams that can be obtained from 2.00×10^2 g of nitroglycerin? (b) Calculate the percent yield in this reaction if the amount of O_2 generated is found to be 6.55 g.

3.91 Titanium(IV) oxide (TiO_2) is a white substance produced by the action of sulfuric acid on the mineral ilmenite ($FeTiO_3$):

$$FeTiO_3 + H_2SO_4 \longrightarrow TiO_2 + FeSO_4 + H_2O$$

Its opaque and nontoxic properties make it suitable as a pigment in plastics and paints. In one process 8.00×10^3 kg of $FeTiO_3$ yielded 3.67×10^3 kg of TiO_2. What is the percent yield of the reaction?

3.92 When heated, lithium reacts with nitrogen to form lithium nitride:

$$6Li(s) + N_2(g) \longrightarrow 2Li_3N(s)$$

What is the theoretical yield of Li_3N in grams when 12.3 g of Li are heated with 33.6 g of N_2? If the actual yield of Li_3N is 5.89 g, what is the percent yield of the reaction?

Additional Problems

3.93 The following diagram represents the products (CO_2 and H_2O) formed after the combustion of a hydrocarbon (a compound containing only C and H atoms). Write an equation for the reaction.

(*Hint:* The molar mass of the hydrocarbon is about 30 g.)

3.94 Consider the reaction of hydrogen gas with oxygen gas:

$$2H_2(g) + O_2(g) \longrightarrow 2H_2O(g)$$

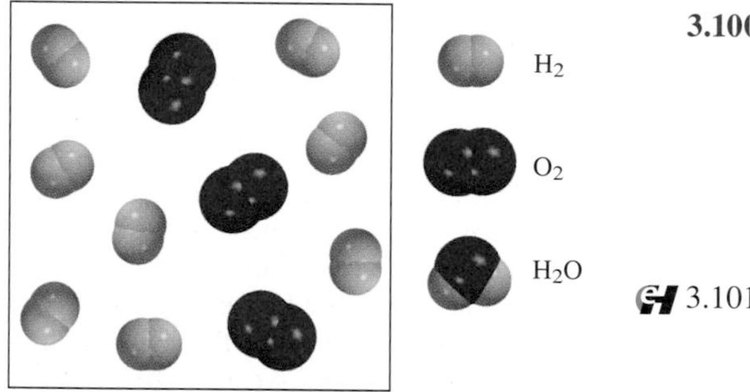

Assuming complete reaction, which of the diagrams shown below represents the amounts of reactants and products left after the reaction?

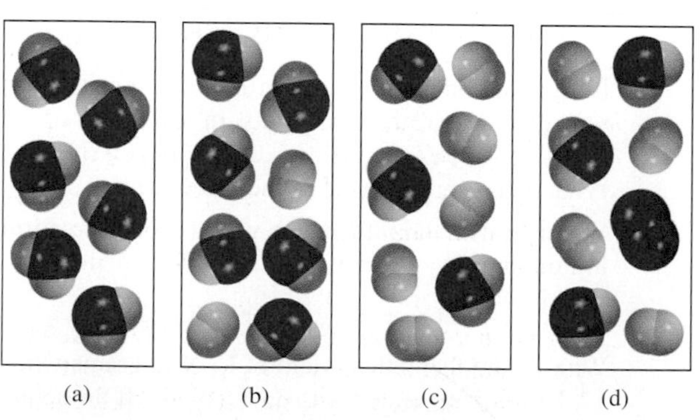

3.95 Industrially, nitric acid is produced by the Ostwald process represented by the following equations:

$$4NH_3(g) + 5O_2(g) \longrightarrow 4NO(g) + 6H_2O(l)$$
$$2NO(g) + O_2(g) \longrightarrow 2NO_2(g)$$
$$2NO_2(g) + H_2O(l) \longrightarrow HNO_3(aq) + HNO_2(aq)$$

What mass of NH_3 (in g) must be used to produce 1.00 ton of HNO_3 by the above procedure, assuming an 80 percent yield in each step? (1 ton = 2000 lb; 1 lb = 453.6 g.)

3.96 A sample of a compound of Cl and O reacts with an excess of H_2 to give 0.233 g of HCl and 0.403 g of H_2O. Determine the empirical formula of the compound.

3.97 The atomic mass of element X is 33.42 amu. A 27.22-g sample of X combines with 84.10 g of another element Y to form a compound XY. Calculate the atomic mass of Y.

3.98 The aluminum sulfate hydrate $[Al_2(SO_4)_3 \cdot xH_2O]$ contains 8.20 percent Al by mass. Calculate x, that is, the number of water molecules associated with each $Al_2(SO_4)_3$ unit.

3.99 An iron bar weighed 664 g. After the bar had been standing in moist air for a month, exactly one-eighth of the iron turned to rust (Fe_2O_3). Calculate the final mass of the iron bar and rust.

3.100 A certain metal oxide has the formula MO, where M denotes the metal. A 39.46-g sample of the compound is strongly heated in an atmosphere of hydrogen to remove oxygen as water molecules. At the end, 31.70 g of the metal is left over. If O has an atomic mass of 16.00 amu, calculate the atomic mass of M and identify the element.

3.101 An impure sample of zinc (Zn) is treated with an excess of sulfuric acid (H_2SO_4) to form zinc sulfate ($ZnSO_4$) and molecular hydrogen (H_2). (a) Write a balanced equation for the reaction. (b) If 0.0764 g of H_2 is obtained from 3.86 g of the sample, calculate the percent purity of the sample. (c) What assumptions must you make in (b)?

3.102 One of the reactions that occurs in a blast furnace, where iron ore is converted to cast iron, is

$$Fe_2O_3 + 3CO \longrightarrow 2Fe + 3CO_2$$

Suppose that 1.64×10^3 kg of Fe are obtained from a 2.62×10^3-kg sample of Fe_2O_3. Assuming that the reaction goes to completion, what is the percent purity of Fe_2O_3 in the original sample?

3.103 Carbon dioxide (CO_2) is the gas that is mainly responsible for global warming (the greenhouse effect). The burning of fossil fuels is a major cause of the increased concentration of CO_2 in the atmosphere. Carbon dioxide is also the end product of metabolism (see Example 3.13). Using glucose as an example of food, calculate the annual human production of CO_2 in grams, assuming that each person consumes 5.0×10^2 g of glucose per day. The world's population is 6.5 billion, and there are 365 days in a year.

3.104 Carbohydrates are compounds containing carbon, hydrogen, and oxygen in which the hydrogen to oxygen ratio is 2:1. A certain carbohydrate contains 40.0 percent carbon by mass. Calculate the empirical and molecular formulas of the compound if the approximate molar mass is 178 g.

3.105 Heating 2.40 g of the oxide of metal X (molar mass of X = 55.9 g/mol) in carbon monoxide (CO) yields the pure metal and carbon dioxide. The mass of the metal product is 1.68 g. From the data given, show that the simplest formula of the oxide is X_2O_3 and write a balanced equation for the reaction.

3.106 A compound X contains 63.3 percent manganese (Mn) and 36.7 percent O by mass. When X is heated, oxygen gas is evolved and a new compound Y containing 72.0 percent Mn and 28.0 percent O is formed. (a) Determine the empirical formulas of X and Y. (b) Write a balanced equation for the conversion of X to Y.

3.107 A sample containing NaCl, Na_2SO_4, and $NaNO_3$ gives the following elemental analysis: Na: 32.08 percent; O: 36.01 percent; Cl: 19.51 percent. Calculate the mass percent of each compound in the sample.

3.108 When 0.273 g of Mg is heated strongly in a nitrogen (N_2) atmosphere, a chemical reaction occurs. The product of the reaction weighs 0.378 g. Calculate the empirical formula of the compound containing Mg and N. Name the compound.

3.109 A mixture of methane (CH_4) and ethane (C_2H_6) of mass 13.43 g is completely burned in oxygen. If the total mass of CO_2 and H_2O produced is 64.84 g, calculate the fraction of CH_4 in the mixture.

3.110 Cysteine, shown here, is one of the 20 amino acids found in proteins in humans. Write the molecular formula and calculate its percent composition by mass.

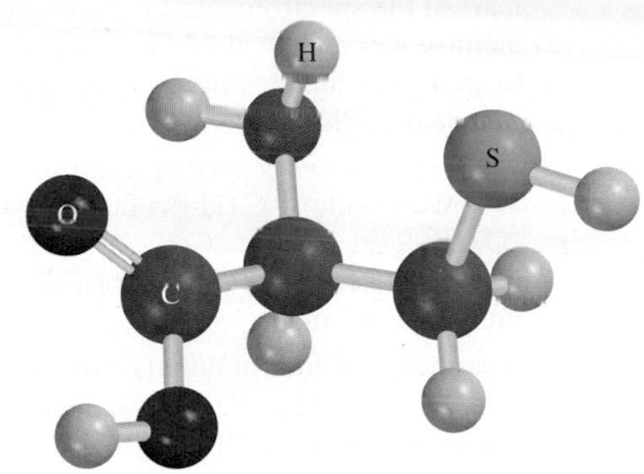

3.111 Isoflurane, shown here, is a common inhalation anesthetic. Write its molecular formula and calculate its percent composition by mass.

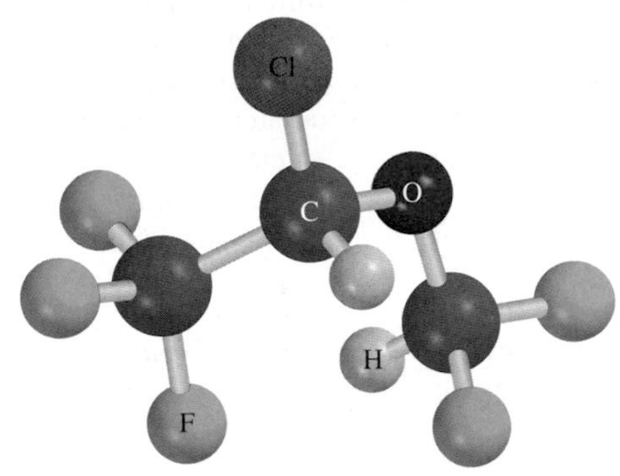

3.112 A reaction having a 90 percent yield may be considered a successful experiment. However, in the synthesis of complex molecules such as chlorophyll and many anticancer drugs, a chemist often has to carry out multiple-step synthesis. What is the overall percent yield for such a synthesis, assuming it is a 30-step reaction with a 90 percent yield at each step?

3.113 A mixture of $CuSO_4 \cdot 5H_2O$ and $MgSO_4 \cdot 7H_2O$ is heated until all the water is lost. If 5.020 g of the mixture gives 2.988 g of the anhydrous salts, what is the percent by mass of $CuSO_4 \cdot 5H_2O$ in the mixture?

Special Problems

3.114 (a) A research chemist used a mass spectrometer to study the two isotopes of an element. Over time, she recorded a number of mass spectra of these isotopes. On analysis, she noticed that the ratio of the taller peak (the more abundant isotope) to the shorter peak (the less abundant isotope) gradually increased with time. Assuming that the mass spectrometer was functioning normally, what do you think was causing this change?

(b) Mass spectrometry can be used to identify the formulas of molecules having small molecular masses. To illustrate this point, identify the molecule which most likely accounts for the observation of a peak in a mass spectrum at: 16 amu, 17 amu, 18 amu, and 64 amu.

(c) Note that there are (among others) two likely molecules that would give rise to a peak at 44 amu, namely, C_3H_8 and CO_2. In such cases, a chemist might try to look for other peaks generated when some of the molecules break apart in the spectrometer. For example, if a chemist sees a peak at 44 amu and also one at 15 amu, which molecule is producing the 44-amu peak? Why?

(d) Using the following precise atomic masses: ^{1}H (1.00797 amu), ^{12}C (12.00000 amu), and ^{16}O (15.99491 amu), how precisely must the masses of C_3H_8 and CO_2 be measured to distinguish between them?

(e) Every year millions of dollars' worth of gold is stolen. In most cases the gold is melted down and shipped abroad. This way the gold retains its value while losing all means of identification. Gold is a highly unreactive metal that exists in nature in the uncombined form. During the mineralization of gold, that is, the formation of gold nuggets from microscopic gold particles, various elements such as cadmium (Cd), lead (Pb), and zinc (Zn) are incorporated into the nuggets. The amounts and types of the impurities or trace elements in gold vary according to the location where it was mined. Based on this knowledge, describe how you would identify the source of a piece of gold suspected of being stolen from Fort Knox, the federal gold depository.

3.115 Potash is any potassium mineral that is used for its potassium content. Most of the potash produced in the United States goes into fertilizer. The major sources of potash are potassium chloride (KCl) and potassium sulfate (K_2SO_4). Potash production is often reported as the potassium oxide (K_2O) equivalent or the amount of K_2O that could be made from a given mineral. (a) If KCl costs \$0.055 per kg, for what price (dollar per kg) must K_2SO_4 be sold in order to supply the same amount of potassium on a per dollar basis? (b) What mass (in kg) of K_2O contains the same number of moles of K atoms as 1.00 kg of KCl?

3.116 A sample of iron weighing 15.0 g was heated with potassium chlorate ($KClO_3$) in an evacuated container. The oxygen generated from the decomposition of $KClO_3$ converted some of the Fe to Fe_2O_3. If the combined mass of Fe and Fe_2O_3 was 17.9 g, calculate the mass of Fe_2O_3 formed and the mass of $KClO_3$ decomposed.

3.117 A certain metal M forms a bromide containing 53.79 percent Br by mass. What is the chemical formula of the compound?

Answers to Practice Exercises

3.1 10.81 amu. **3.2** 2.57×10^3 g. **3.3** 8.49×10^{21} K atoms. **3.4** 2.107×10^{-22} g. **3.5** 32.04 amu. **3.6** 1.66 moles. **3.7** 5.81×10^{24} H atoms. **3.8** H: 2.055%; S: 32.69%; O: 65.25%. **3.9** $KMnO_4$ (potassium permanganate). **3.10** 196 g. **3.11** B_2H_6. **3.12** $Fe_2O_3 + 3CO \longrightarrow 2Fe + 3CO_2$. **3.13** 235 g. **3.14** 0.769 g. **3.15** (a) 234 g, (b) 234 g. **3.16** (a) 863 g, (b) 93.0%.

"Black smoker," insoluble metal sulfides formed on the ocean floor through the lava on a mid-ocean-ridge volcano.

CHAPTER 4

Reactions in Aqueous Solutions

Chapter Outline

Essential Concepts

Reactions in Aqueous Solution Many chemical and almost all biological reactions occur in the aqueous medium. Substances (solutes) that dissolve in water (solvent) can be divided into two categories: electrolytes and nonelectrolytes, depending on their ability to conduct electricity.

Three Major Types of Reactions In a precipitation reaction, the product, an insoluble substance, separates from solution. Acid-base reactions involve the transfer of a proton (H^+) from an acid to a base. In an oxidation-reduction reaction, or redox reaction, electrons are transferred from a reducing agent to an oxidizing agent. These three types of reactions represent the majority of reactions in chemical and biological systems.

Solution Stoichiometry Quantitative studies of reactions in solution require that we know the concentration of the solution, which is usually represented by the molarity unit. These studies include gravimetric analysis, which involves the measurement of mass, and titrations in which the unknown concentration of a solution is determined by reaction with a solution of known concentration.

Student Interactive Activities

Animations
Strong Electrolytes, Weak Electrolytes and Nonelectrolytes (4.1)
Hydration (4.1)
Precipitation Reactions (4.2)
Neutralization Reactions (4.3)
Oxidation-Reduction Reactions (4.4)
Making a Solution (4.5)
Preparing a Solution by Dilution (4.5)

Electronic homework
Example Practice Problems
End of Chapter Problems

4.1 General Properties of Aqueous Solutions

Many chemical reactions and virtually all biological processes take place in an aqueous environment. Therefore, it is important to understand the properties of different substances in solution with water. To start with, what exactly is a solution? A ***solution*** is *a homogeneous mixture of two or more substances. The substance present in a smaller amount* is called the ***solute,*** whereas *the substance present in a larger amount* is called the ***solvent.*** A solution may be gaseous (such as air), solid (such as an alloy), or liquid (seawater, for example). In this section we will discuss only ***aqueous solutions,*** in which *the solute initially is a liquid or a solid and the solvent is water.*

Electrolytes versus Nonelectrolytes

All solutes that dissolve in water fit into one of two categories: electrolytes and nonelectrolytes. An ***electrolyte*** is *a substance that, when dissolved in water, results in a solution that can conduct electricity.* A ***nonelectrolyte*** *does not conduct electricity when dissolved in water.* Figure 4.1 shows an easy and straightforward method of distinguishing between electrolytes and nonelectrolytes. A pair of platinum electrodes is immersed in a beaker of water. To light the bulb, electric current must flow from one electrode to the other, thus completing the circuit. Pure water is a very poor conductor of electricity. However, if we add a small amount of sodium chloride (NaCl), the bulb will glow as soon as the salt dissolves in the water. Solid NaCl, an ionic compound, breaks up into Na^+ and Cl^- ions when it dissolves in water. The Na^+ ions are attracted to the negative electrode and the Cl^- ions to the positive electrode. This movement sets up an electrical current that is equivalent to the flow of electrons along a metal wire. Because the NaCl solution conducts electricity, we say that NaCl is an electrolyte. Pure water contains very few ions, so it cannot conduct electricity.

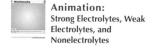

Animation: Strong Electrolytes, Weak Electrolytes, and Nonelectrolytes

Comparing the lightbulb's brightness for the *same molar amounts* of dissolved substances helps us distinguish between strong and weak electrolytes. A characteristic of strong electrolytes is that the solute is assumed to be 100 percent dissociated into ions in solution. (By *dissociation* we mean the breaking up of the compound into cations and anions.) Thus, we can represent sodium chloride dissolving in water as

$$\mathrm{NaCl}(s) \xrightarrow{\mathrm{H_2O}} \mathrm{Na^+}(aq) + \mathrm{Cl^-}(aq)$$

What this equation says is that all the sodium chloride that enters the aqueous solution ends up as Na^+ and Cl^- ions; there are no undissociated NaCl units in solution.

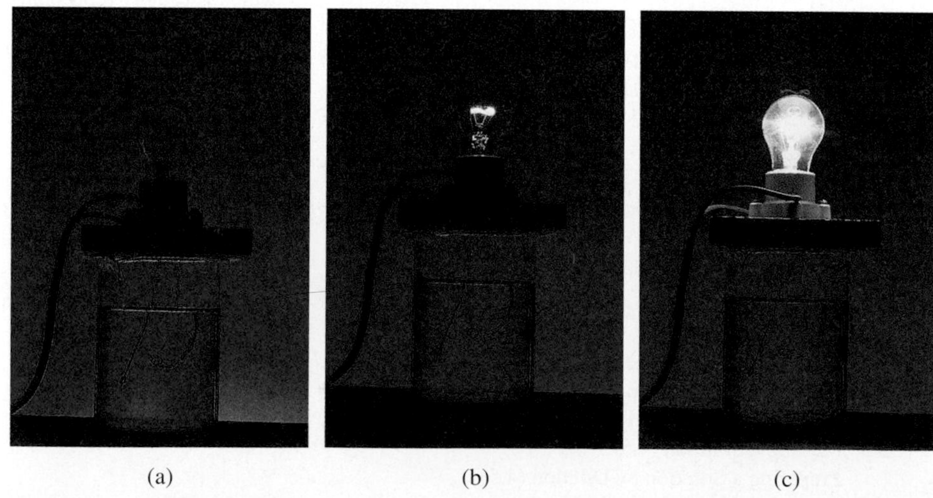

(a) (b) (c)

Figure 4.1 *An arrangement for distinguishing between electrolytes and nonelectrolytes. A solution's ability to conduct electricity depends on the number of ions it contains. (a) A nonelectrolyte solution does not contain ions, and the lightbulb is not lit. (b) A weak electrolyte solution contains a small number of ions, and the lightbulb is dimly lit. (c) A strong electrolyte solution contains a large number of ions, and the lightbulb is brightly lit. The molar amounts of the dissolved solutes are equal in all three cases.*

Table 4.1 Classification of Solutes in Aqueous Solution

Strong Electrolyte	Weak Electrolyte	Nonelectrolyte
HCl	CH_3COOH	$(NH_2)_2CO$ (urea)
HNO_3	HF	CH_3OH (methanol)
$HClO_4$	HNO_2	C_2H_5OH (ethanol)
H_2SO_4*	NH_3	$C_6H_{12}O_6$ (glucose)
NaOH	H_2O†	$C_{12}H_{22}O_{11}$ (sucrose)
$Ba(OH)_2$		
Ionic compounds		

*H_2SO_4 has two ionizable H^+ ions.
†Pure water is an extremely weak electrolyte.

Table 4.1 lists examples of strong electrolytes, weak electrolytes, and nonelectrolytes. Ionic compounds, such as sodium chloride, potassium iodide (KI), and calcium nitrate [$Ca(NO_3)_2$], are strong electrolytes. It is interesting to note that human body fluids contain many strong and weak electrolytes.

Water is a very effective solvent for ionic compounds. Although water is an electrically neutral molecule, it has a positive end (the H atoms) and a negative end (the O atom), or positive and negative "poles"; for this reason, it is often referred to as a *polar* solvent. When an ionic compound such as sodium chloride dissolves in water, the three-dimensional network of the ions in the solid is destroyed, and the Na^+ and Cl^- ions are separated from each other. In solution, each Na^+ ion is surrounded by a number of water molecules orienting their negative ends toward the cation. Similarly, each Cl^- ion is surrounded by water molecules with their positive ends oriented toward the anion (Figure 4.2). *The process in which an ion is surrounded by water molecules arranged in a specific manner* is called ***hydration.*** Hydration helps to stabilize ions in solution and prevents cations from combining with anions.

Acids and bases are also electrolytes. Some acids, including hydrochloric acid (HCl) and nitric acid (HNO_3), are strong electrolytes. These acids ionize completely in water; for example, when hydrogen chloride gas dissolves in water, it forms hydrated H^+ and Cl^- ions:

$$HCl(g) \xrightarrow{H_2O} H^+(aq) + Cl^-(aq)$$

In other words, *all* the dissolved HCl molecules separate into hydrated H^+ and Cl^- ions in solution. Thus, when we write HCl(*aq*), it is understood that it is a solution of only $H^+(aq)$ and $Cl^-(aq)$ ions and there are no hydrated HCl molecules present. On the other hand, certain acids, such as acetic acid (CH_3COOH), which is found in vinegar, ionize to a much lesser extent. We represent the ionization of acetic acid as

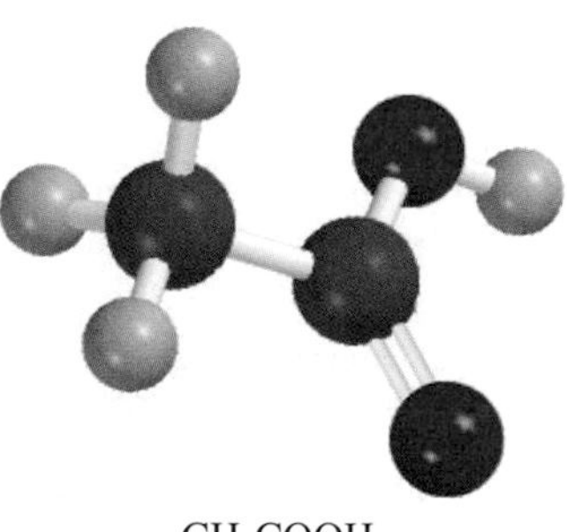
CH_3COOH

$$CH_3COOH(aq) \rightleftharpoons CH_3COO^-(aq) + H^+(aq)$$

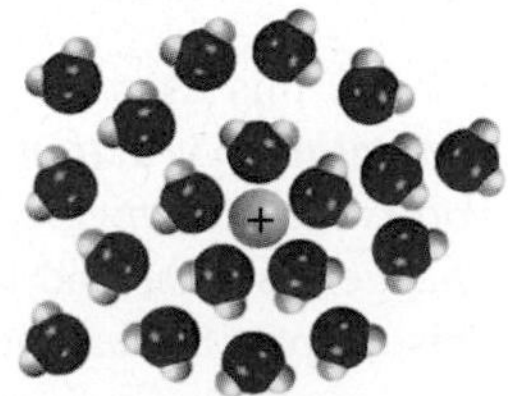

Figure 4.2
Hydration of Na^+ and Cl^- ions.

in which CH_3COO^- is called the acetate ion. (In this book we will use the term *dissociation* for ionic compounds and *ionization* for acids and bases.) By writing the formula of acetic acid as CH_3COOH we indicate that the ionizable proton is in the COOH group.

The double arrow $\rightleftharpoons$ in an equation means that the reaction is ***reversible;*** that is, the *reaction can occur in both directions.* Initially, a number of CH_3COOH molecules break up to yield CH_3COO^- and H^+ ions. As time goes on, some of the CH_3COO^- and H^+ ions recombine to form CH_3COOH molecules. Eventually, a state is reached in which the acid molecules break up as fast as the ions recombine. Such *a chemical state, in which no net change can be observed* (although continuous activity is taking place on the molecular level), is called ***chemical equilibrium.*** Acetic acid, then, is a weak electrolyte because its ionization in water is incomplete. By contrast, in a hydrochloric acid solution, the H^+ and Cl^- ions have no tendency to recombine to form molecular HCl. We use the single arrow to represent complete ionizations.

There are different types of chemical equilibrium. We will return to this very important topic in Chapter 15.

In Sections 4.2–4.4 we will study three types of reactions in the aqueous medium (precipitation, acid-base, and oxidation-reduction) that are of great importance to industrial, environmental, and biological processes. They also play a role in our daily experience.

Review of Concepts

The diagrams show three compounds (a) AB_2, (b) AC_2, and (c) AD_2 dissolved in water. Which is the strongest electrolyte and which is the weakest? (For simplicity, water molecules are not shown.)

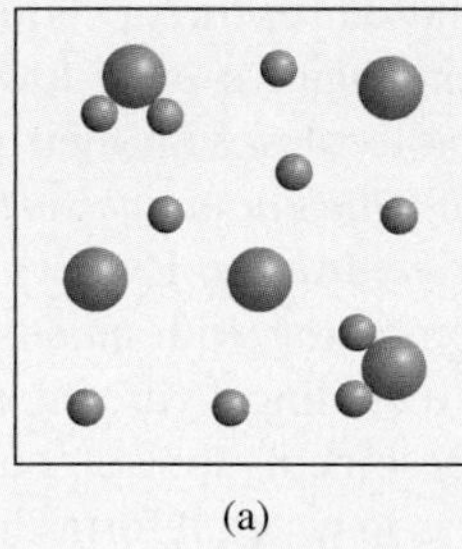
(a)

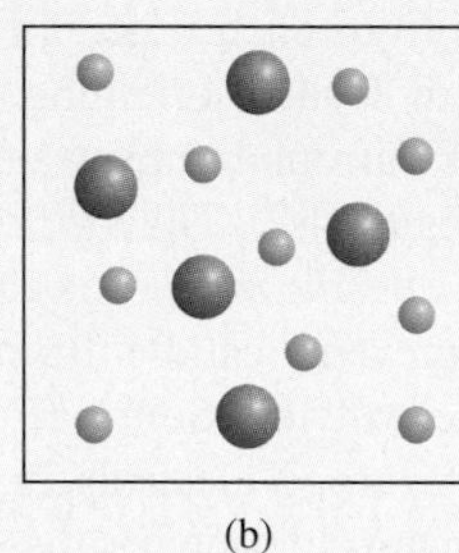
(b)

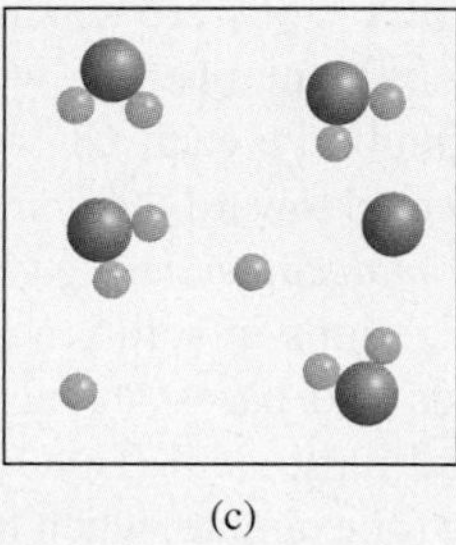
(c)

4.2 Precipitation Reactions

Animation: Precipitation Reactions

One common type of reaction that occurs in aqueous solution is the ***precipitation reaction,*** which *results in the formation of an insoluble product, or precipitate.* A ***precipitate*** is *an insoluble solid that separates from the solution.* Precipitation reactions usually involve ionic compounds. For example, when an aqueous solution of lead(II) nitrate [$Pb(NO_3)_2$] is added to an aqueous solution of potassium iodide (KI), a yellow precipitate of lead iodide (PbI_2) is formed:

$$Pb(NO_3)_2(aq) + 2KI(aq) \longrightarrow PbI_2(s) + 2KNO_3(aq)$$

Potassium nitrate remains in solution. Figure 4.3 shows this reaction in progress.

The preceding reaction is an example of a ***metathesis reaction*** (also called a double displacement reaction), *a reaction that involves the exchange of parts between two compounds.* (In this case, the compounds exchange the NO_3^- and I^- ions.) As we will see, the precipitation reactions discussed in this chapter are examples of metathesis reactions.

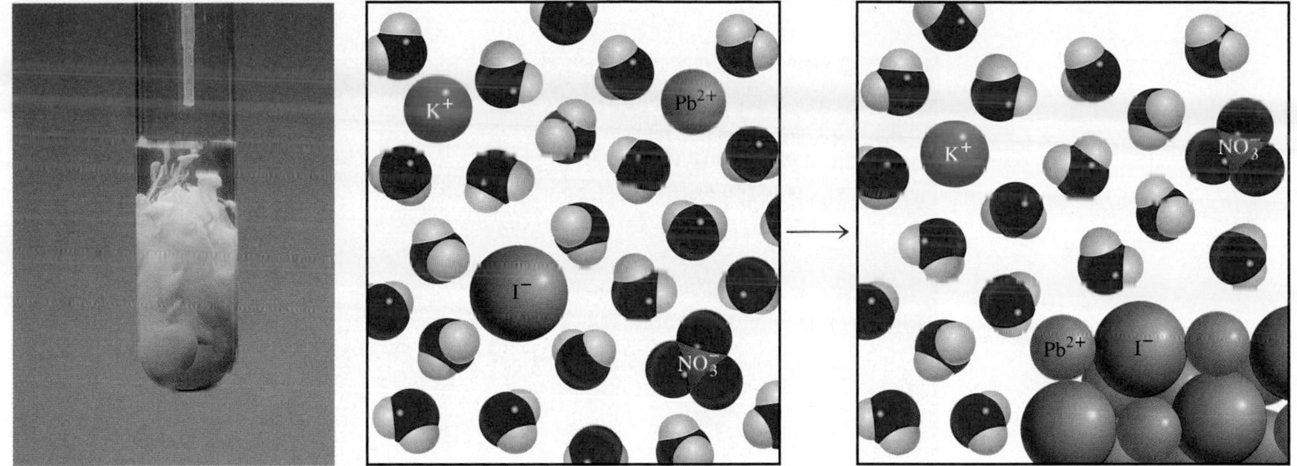

Figure 4.3
Formation of yellow PbI_2 precipitate as a solution of $Pb(NO_3)_2$ is added to a solution of KI.

Solubility

How can we predict whether a precipitate will form when a compound is added to a solution or when two solutions are mixed? It depends on the ***solubility*** of the solute, which is defined as *the maximum amount of solute that will dissolve in a given quantity of solvent at a specific temperature.* Chemists refer to substances as soluble, slightly soluble, or insoluble in a qualitative sense. A substance is said to be soluble if a fair amount of it visibly dissolves when added to water. If not, the substance is described as slightly soluble or insoluble. All ionic compounds are strong electrolytes, but they are not equally soluble.

Table 4.2 classifies a number of common ionic compounds as soluble or insoluble. Keep in mind, however, that even insoluble compounds dissolve to a certain extent. Figure 4.4 shows several precipitates.

Table 4.2 Solubility Rules for Common Ionic Compounds in Water at 25°C

Soluble Compounds	Insoluble Exceptions
Compounds containing alkali metal ions (Li^+, Na^+, K^+, Rb^+, Cs^+) and the ammonium ion (NH_4^+)	
Nitrates (NO_3^-), bicarbonates (HCO_3^-), and chlorates (ClO_3^-)	
Halides (Cl^-, Br^-, I^-)	Halides of Ag^+, Hg_2^{2+}, and Pb^{2+}
Sulfates (SO_4^{2-})	Sulfates of Ag^+, Ca^{2+}, Sr^{2+}, Ba^{2+}, Hg_2^{2+}, and Pb^{2+}
Insoluble Compounds	**Soluble Exceptions**
Carbonates (CO_3^{2-}), phosphates (PO_4^{3-}), chromates (CrO_4^{2-}), and sulfides (S^{2-})	Compounds containing alkali metal ions and the ammonium ion
Hydroxides (OH^-)	Compounds containing alkali metal ions and the Ba^{2+} ion

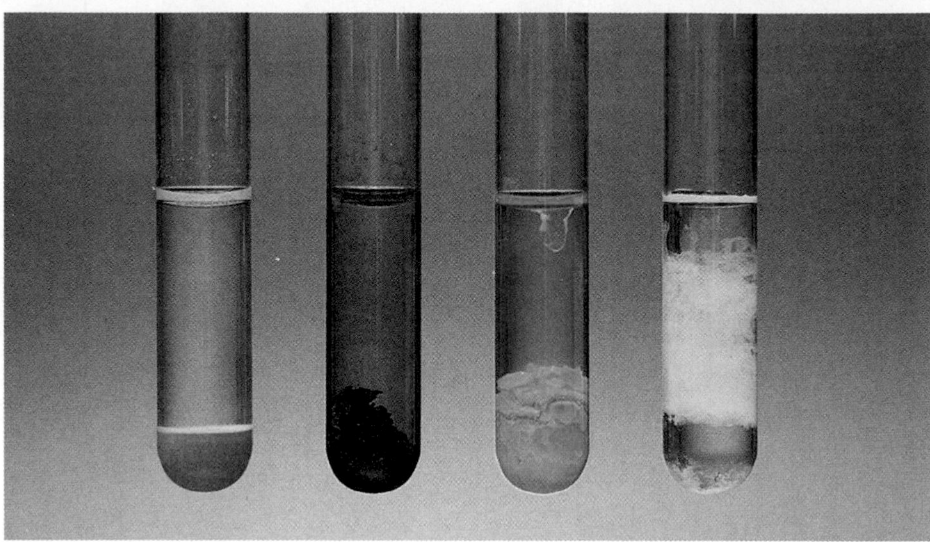

Figure 4.4
Appearance of several precipitates. From left to right: CdS, PbS, $Ni(OH)_2$, $Al(OH)_3$.

EXAMPLE 4.1

Classify the following ionic compounds as soluble or insoluble: (a) lead sulfate ($PbSO_4$), (b) barium carbonate ($BaCO_3$), (c) lithium phosphate (Li_3PO_4).

Strategy Although it is not necessary to memorize the solubilities of compounds, you should keep in mind the following useful rules: all ionic compounds containing alkali metal cations; the ammonium ion; and the nitrate, bicarbonate, and chlorate ions are soluble. For other compounds we need to refer to Table 4.2.

Solution

(a) According to Table 4.2, $PbSO_4$ is insoluble.

(b) This is a carbonate and Ba is a Group 2A metal. Therefore, $BaCO_3$ is insoluble.

(c) Lithium is an alkali metal (Group 1A) so Li_3PO_4 is soluble.

Similar problems: 4.19, 4.20.

Practice Exercise Classify the following ionic compounds as soluble or insoluble: (a) FeS, (b) $Ca(OH)_2$, (c) $Co(NO_3)_3$.

Molecular Equations, Ionic Equations, and Net Ionic Equations

The equation describing the precipitation of lead iodide on page 100 is called a ***molecular equation*** because *the formulas of the compounds are written as though all species existed as molecules or whole units.* A molecular equation is useful because it identifies the reagents (that is, lead nitrate and potassium iodide). If we wanted to bring about this reaction in the laboratory, we would use the molecular equation. However, a molecular equation does not describe in detail what actually is happening in solution.

As pointed out earlier, when ionic compounds dissolve in water, they break apart into their component cations and anions. To be more realistic, the equations should show the dissociation of dissolved ionic compounds into ions. Therefore, returning to the reaction between potassium iodide and lead nitrate, we would write

$$Pb^{2+}(aq) + 2NO_3^-(aq) + 2K^+(aq) + 2I^-(aq) \longrightarrow PbI_2(s) + 2K^+(aq) + 2NO_3^-(aq)$$

The preceding equation is an example of an ***ionic equation,*** which *shows dissolved species as free ions.* To see whether a precipitate might form from this solution, we first combine the cation and anion from different compounds; that is, PbI_2 and KNO_3. Referring to Table 4.2, we see that PbI_2 is an insoluble compound and KNO_3 is soluble. Therefore, the dissolved KNO_3 remains in solution as separate K^+ and NO_3^- ions, which are called ***spectator ions,*** or *ions that are not involved in the overall reaction.* Because spectator ions appear on both sides of an equation, they can be eliminated from the ionic equation

$$Pb^{2+}(aq) + \cancel{2NO_3^-(aq)} + \cancel{2K^+(aq)} + 2I^-(aq) \longrightarrow PbI_2(s) + \cancel{2K^+(aq)} + \cancel{2NO_3^-(aq)}$$

Finally, we end up with the ***net ionic equation,*** which *shows only the species that actually take part in the reaction:*

$$Pb^{2+}(aq) + 2I^-(aq) \longrightarrow PbI_2(s)$$

Looking at another example, we find that when an aqueous solution of barium chloride ($BaCl_2$) is added to an aqueous solution of sodium sulfate (Na_2SO_4), a white precipitate is formed (Figure 4.5). Treating this as a metathesis reaction, the products are $BaSO_4$ and NaCl. From Table 4.2 we see that only $BaSO_4$ is insoluble. Therefore, we write the molecular equation as

$$BaCl_2(aq) + Na_2SO_4(aq) \longrightarrow BaSO_4(s) + 2NaCl(aq)$$

The ionic equation for the reaction is

$$Ba^{2+}(aq) + 2Cl^-(aq) + 2Na^+(aq) + SO_4^{2-}(aq) \longrightarrow BaSO_4(s) + 2Na^+(aq) + 2Cl^-(aq)$$

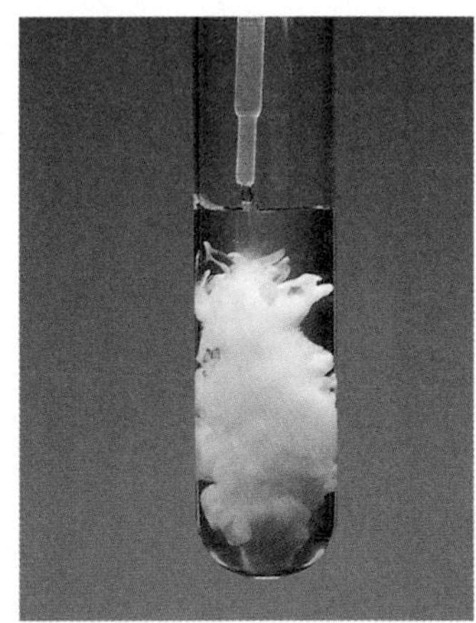

Figure 4.5
Formation of $BaSO_4$ precipitate.

Canceling the spectator ions (Na^+ and Cl^-) on both sides of the equation gives us the net ionic equation

$$Ba^{2+}(aq) + SO_4^{2-}(aq) \longrightarrow BaSO_4(s)$$

The following four steps summarize the procedure for writing ionic and net ionic equations:

1. Write a balanced molecular equation for the reaction, using the correct formulas for the reactant and product ionic compounds. Refer to Table 4.2 to decide which of the products is insoluble and therefore will appear as a precipitate.
2. Write the ionic equation for the reaction. The compound that does not appear as the precipitate should be shown as free ions.
3. Identify and cancel the spectator ions on both sides of the equation. Write the net ionic equation for the reaction.
4. Check that the charges and number of atoms balance in the net ionic equation.

Precipitate formed by the reaction between $K_3PO_4(aq)$ and $Ca(NO_3)_2(aq)$.

EXAMPLE 4.2

Predict what happens when a potassium phosphate (K_3PO_4) solution is mixed with a calcium nitrate [$Ca(NO_3)_2$] solution. Write a net ionic equation for the reaction.

Strategy From the given information, it is useful to first write the unbalanced equation

$$K_3PO_4(aq) + Ca(NO_3)_2(aq) \longrightarrow ?$$

What happens when ionic compounds dissolve in water? What ions are formed from the dissociation of K_3PO_4 and $Ca(NO_3)_2$? What happens when the cations encounter the anions in solution?

Solution In solution, K_3PO_4 dissociates into K^+ and PO_4^{3-} ions and $Ca(NO_3)_2$ dissociates into Ca^{2+} and NO_3^- ions. According to Table 4.2, calcium ions (Ca^{2+}) and phosphate ions (PO_4^{3-}) will form an insoluble compound, calcium phosphate [$Ca_3(PO_4)_2$], while the other product, KNO_3, is soluble and remains in solution. Therefore, this is a precipitation reaction. We follow the stepwise procedure just outlined.

Step 1: The balanced molecular equation for this reaction is

$$2K_3PO_4(aq) + 3Ca(NO_3)_2(aq) \longrightarrow Ca_3(PO_4)_2(s) + 6KNO_3(aq)$$

Step 2: To write the ionic equation, the soluble compounds are shown as dissociated ions:

$$6K^+(aq) + 2PO_4^{3-}(aq) + 3Ca^{2+}(aq) + 6NO_3^-(aq) \longrightarrow 6K^+(aq) + 6NO_3^-(aq) + Ca_3(PO_4)_2(s)$$

Step 3: Canceling the spectator ions (K^+ and NO_3^-) on each side of the equation, we obtain the net ionic equation:

$$3Ca^{2+}(aq) + 2PO_4^{3-}(aq) \longrightarrow Ca_3(PO_4)_2(s)$$

Step 4: Note that because we balanced the molecular equation first, the net ionic equation is balanced as to the number of atoms on each side and the number of positive (+6) and negative (−6) charges on the left-hand side is the same.

Similar problems: 4.21, 4.22.

Practice Exercise Predict the precipitate produced by mixing an $Al(NO_3)_3$ solution with a NaOH solution. Write the net ionic equation for the reaction.

Review of Concepts

Which of the diagrams here accurately describes the reaction between $Ca(NO_3)_2(aq)$ and $Na_2CO_3(aq)$? For simplicity, only the Ca^{2+} (yellow) and CO_3^{2-} (blue) ions are shown.

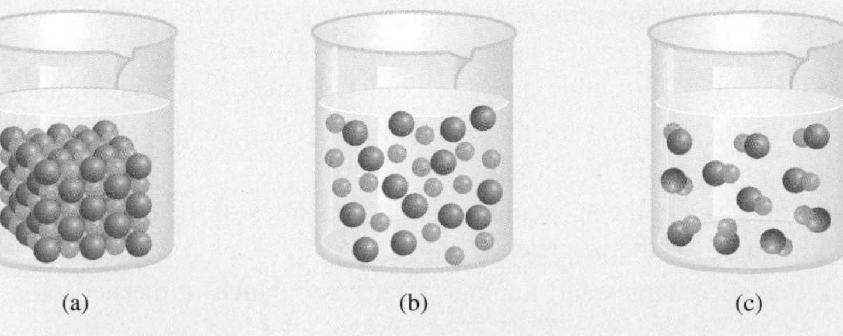

4.3 Acid-Base Reactions

Acids and bases are as familiar as aspirin and milk of magnesia although many people do not know their chemical names—acetylsalicylic acid (aspirin) and magnesium hydroxide (milk of magnesia). In addition to being the basis of many medicinal and household products, acid-base chemistry is important in industrial processes and essential in sustaining biological systems. Before we can discuss acid-base reactions, we need to know more about acids and bases themselves.

General Properties of Acids and Bases

In Section 2.7 we defined acids as substances that ionize in water to produce H^+ ions and bases as substances that ionize in water to produce OH^- ions. These definitions were formulated in the late nineteenth century by the Swedish chemist Svante Arrhenius to classify substances whose properties in aqueous solutions were well known.

Acids

- Acids have a sour taste; for example, vinegar owes its sourness to acetic acid, and lemons and other citrus fruits contain citric acid.
- Acids cause color changes in plant dyes; for example, they change the color of litmus from blue to red.
- Acids react with certain metals, such as zinc, magnesium, and iron, to produce hydrogen gas. A typical reaction is that between hydrochloric acid and magnesium:

$$2HCl(aq) + Mg(s) \longrightarrow MgCl_2(aq) + H_2(g)$$

- Acids react with carbonates and bicarbonates, such as Na_2CO_3, $CaCO_3$, and $NaHCO_3$, to produce carbon dioxide gas (Figure 4.6). For example,

$$2HCl(aq) + CaCO_3(s) \longrightarrow CaCl_2(aq) + H_2O(l) + CO_2(g)$$
$$HCl(aq) + NaHCO_3(s) \longrightarrow NaCl(aq) + H_2O(l) + CO_2(g)$$

- Aqueous acid solutions conduct electricity.

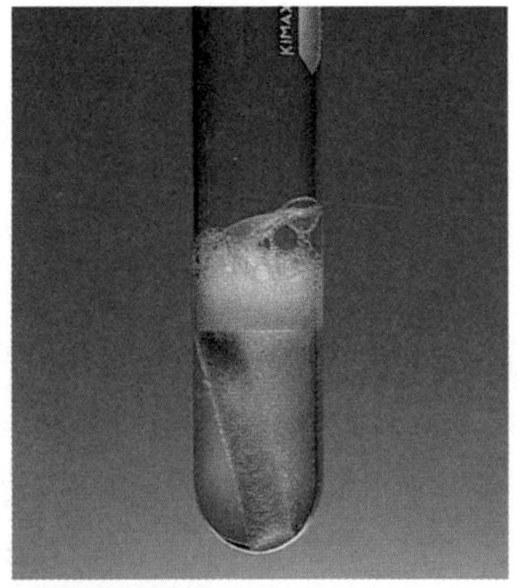

Figure 4.6
A piece of blackboard chalk, which is mostly $CaCO_3$, reacts with hydrochloric acid to produce carbon dioxide gas.

Bases

- Bases have a bitter taste.
- Bases feel slippery; for example, soaps, which contain bases, exhibit this property.
- Bases cause color changes in plant dyes; for example, they change the color of litmus from red to blue.
- Aqueous base solutions conduct electricity.

Brønsted Acids and Bases

Arrhenius's definitions of acids and bases are limited in that they apply only to aqueous solutions. Broader definitions were proposed by the Danish chemist Johannes Brønsted in 1932; a ***Brønsted acid*** is *a proton donor*, and a ***Brønsted base*** is *a proton acceptor*. Note that Brønsted's definitions do not require acids and bases to be in aqueous solution.

Hydrochloric acid is a Brønsted acid because it donates a proton in water:

$$HCl(aq) \longrightarrow H^+(aq) + Cl^-(aq)$$

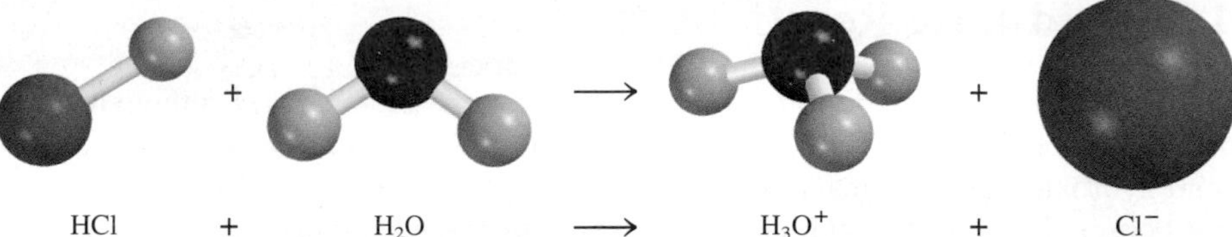

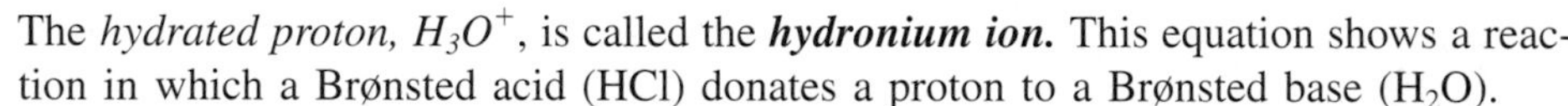

Figure 4.7
Ionization of HCl in water to form the hydronium ion and the chloride ion.

Note that the H^+ ion is a hydrogen atom that has lost its electron; that is, it is just a bare proton. The size of a proton is about 10^{-15} m, compared to a diameter of 10^{-10} m for an average atom or ion. Such an exceedingly small charged particle cannot exist as a separate entity in aqueous solution owing to its strong attraction for the negative pole (the O atom) in H_2O. Consequently, the proton exists in the hydrated form, as shown in Figure 4.7. Therefore, the ionization of hydrochloric acid should be written as

$$HCl(aq) + H_2O(l) \longrightarrow H_3O^+(aq) + Cl^-(aq)$$

The *hydrated proton, H_3O^+*, is called the ***hydronium ion.*** This equation shows a reaction in which a Brønsted acid (HCl) donates a proton to a Brønsted base (H_2O).

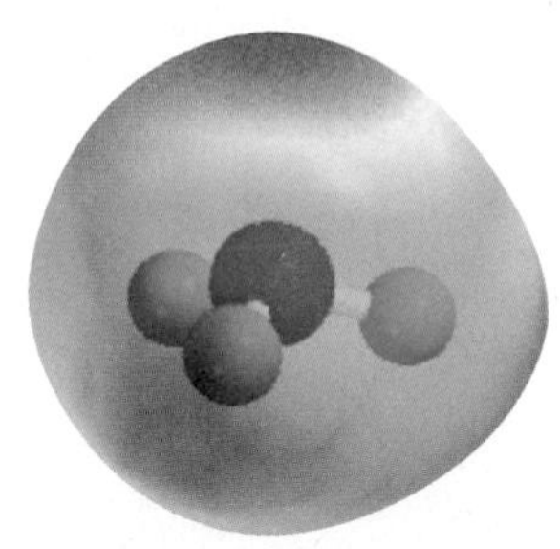

Electrostatic potential map of the H_3O^+ ion. In the rainbow color spectrum representation, the most electron-rich region is red and the most electron-poor region is blue.

Experiments show that the hydronium ion is further hydrated so that the proton may have several water molecules associated with it. Because the acidic properties of the proton are unaffected by the degree of hydration, in this text we will generally use $H^+(aq)$ to represent the hydrated proton. This notation is for convenience, but H_3O^+ is closer to reality. Keep in mind that both notations represent the same species in aqueous solution.

Acids commonly used in the laboratory include hydrochloric acid (HCl), nitric acid (HNO_3), acetic acid (CH_3COOH), sulfuric acid (H_2SO_4), and phosphoric acid (H_3PO_4). The first three are ***monoprotic acids;*** that is, *each unit of the acid yields one hydrogen ion upon ionization:*

$$HCl(aq) \longrightarrow H^+(aq) + Cl^-(aq)$$
$$HNO_3(aq) \longrightarrow H^+(aq) + NO_3^-(aq)$$
$$CH_3COOH(aq) \rightleftharpoons CH_3COO^-(aq) + H^+(aq)$$

As mentioned earlier, because the ionization of acetic acid is incomplete (note the double arrows), it is a weak electrolyte. For this reason it is called a weak acid (see Table 4.1). On the other hand, HCl and HNO_3 are strong acids because they are strong electrolytes, so they are completely ionized in solution (note the use of single arrows).

Sulfuric acid (H_2SO_4) is a ***diprotic acid*** because *each unit of the acid gives up two H^+ ions,* in two separate steps:

$$H_2SO_4(aq) \longrightarrow H^+(aq) + HSO_4^-(aq)$$
$$HSO_4^-(aq) \rightleftharpoons H^+(aq) + SO_4^{2-}(aq)$$

H_2SO_4 is a strong electrolyte or strong acid (the first step of ionization is complete), but HSO_4^- is a weak acid or weak electrolyte, and we need a double arrow to represent its incomplete ionization.

Triprotic acids, which *yield three H^+ ions,* are relatively few in number. The best known triprotic acid is phosphoric acid, whose ionizations are

$$H_3PO_4(aq) \rightleftharpoons H^+(aq) + H_2PO_4^-(aq)$$
$$H_2PO_4^-(aq) \rightleftharpoons H^+(aq) + HPO_4^{2-}(aq)$$
$$HPO_4^{2-}(aq) \rightleftharpoons H^+(aq) + PO_4^{3-}(aq)$$

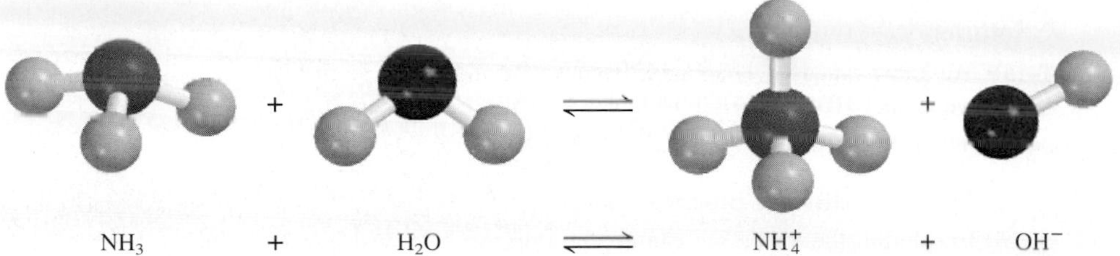

Figure 4.8
Ionization of ammonia in water to form the ammonium ion and the hydroxide ion.

All three species (H_3PO_4, $H_2PO_4^-$, and HPO_4^{2-}) in this case are weak acids, and we use the double arrows to represent each ionization step. Anions such as $H_2PO_4^-$ and HPO_4^{2-} are found in aqueous solutions of phosphates such as NaH_2PO_4 and Na_2HPO_4.

Table 4.1 shows that sodium hydroxide (NaOH) and barium hydroxide [$Ba(OH)_2$] are strong electrolytes. This means that they are completely ionized in solution:

$$NaOH(s) \xrightarrow{H_2O} Na^+(aq) + OH^-(aq)$$
$$Ba(OH)_2(s) \xrightarrow{H_2O} Ba^{2+}(aq) + 2OH^-(aq)$$

The OH^- ion can accept a proton as follows:

$$H^+(aq) + OH^-(aq) \longrightarrow H_2O(l)$$

Thus, OH^- is a Brønsted base.

Ammonia (NH_3) is classified as a Brønsted base because it can accept a H^+ ion (Figure 4.8):

$$NH_3(aq) + H_2O(l) \rightleftharpoons NH_4^+(aq) + OH^-(aq)$$

Ammonia is a weak electrolyte (and therefore a weak base) because only a small fraction of dissolved NH_3 molecules react with water to form NH_4^+ and OH^- ions.

The most commonly used strong base in the laboratory is sodium hydroxide. It is cheap and soluble. (In fact, all of the alkali metal hydroxides are soluble.) The most commonly used weak base is aqueous ammonia solution, which is sometimes erroneously called ammonium hydroxide; there is no evidence that the species NH_4OH actually exists other than the NH_4^+ and OH^- ions in solution. All of the Group 2A elements form hydroxides of the type $M(OH)_2$, where M denotes an alkaline earth metal. Of these hydroxides, only $Ba(OH)_2$ is soluble. Magnesium and calcium hydroxides are used in medicine and industry. Hydroxides of other metals, such as $Al(OH)_3$ and $Zn(OH)_2$ are insoluble and are not used as bases.

Example 4.3 classifies substances as Brønsted acids or Brønsted bases.

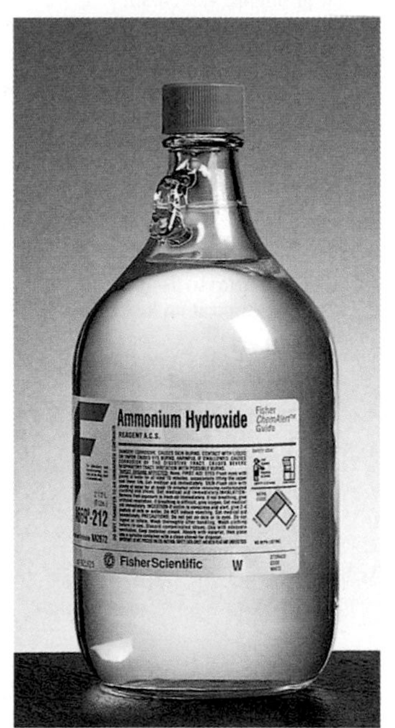

A bottle of aqueous ammonia, which is sometimes erroneously called ammonium hydroxide.

EXAMPLE 4.3

Classify each of the following species in aqueous solution as a Brønsted acid or base: (a) HBr, (b) NO_2^-, (c) HCO_3^-.

Strategy What are the characteristics of a Brønsted acid? Does it contain at least an H atom? With the exception of ammonia, most Brønsted bases that you will encounter at this stage are anions.

(Continued)

Solution

(a) We know that HCl is an acid. Because Br and Cl are both halogens (Group 7A), we expect HBr, like HCl, to ionize in water as follows:

$$HBr(aq) \longrightarrow H^+(aq) + Br^-(aq)$$

Therefore, HBr is a Brønsted acid.

(b) In solution the nitrite ion can accept a proton from water to form nitrous acid:

$$NO_2^-(aq) + H^+(aq) \longrightarrow HNO_2(aq)$$

This property makes NO_2^- a Brønsted base.

(c) The bicarbonate ion is a Brønsted acid because it ionizes in solution as follows:

$$HCO_3^-(aq) \rightleftharpoons H^+(aq) + CO_3^{2-}(aq)$$

It is also a Brønsted base because it can accept a proton to form carbonic acid:

$$HCO_3^-(aq) + H^+(aq) \rightleftharpoons H_2CO_3(aq)$$

Comment The HCO_3^- species is said to be *amphoteric* because it possesses both acidic and basic properties. The double arrows show that this is a reversible reaction.

Similar problems: 4.31, 4.32.

Practice Exercise Classify each of the following species as a Brønsted acid or base: (a) SO_4^{2-}, (b) HI, (c) $H_2PO_4^-$.

Acid-Base Neutralization

Animation: Neutralization Reactions

A ***neutralization reaction*** is *a reaction between an acid and a base.* Generally, aqueous acid-base reactions produce water and a ***salt,*** which is *an ionic compound made up of a cation other than* H^+ *and an anion other than* OH^- *or* O^{2-}:

$$\text{acid} + \text{base} \longrightarrow \text{salt} + \text{water}$$

For example, when a HCl solution is mixed with a NaOH solution, the following reaction occurs:

Acid-base reactions generally go to completion.

$$HCl(aq) + NaOH(aq) \longrightarrow NaCl(aq) + H_2O(l)$$

However, because both the acid and the base are strong electrolytes, they are completely ionized in solution. The ionic equation is

$$H^+(aq) + Cl^-(aq) + Na^+(aq) + OH^-(aq) \longrightarrow Na^+(aq) + Cl^-(aq) + H_2O(l)$$

Therefore, the reaction can be represented by the net ionic equation

$$H^+(aq) + OH^-(aq) \longrightarrow H_2O(l)$$

Both Na^+ and Cl^- are spectator ions.

Now consider the reaction between NaOH with hydrocyanic acid (HCN), which is a weak acid:

$$HCN(aq) + NaOH(aq) \longrightarrow NaCN(aq) + H_2O(l)$$

In this case, the ionic equation is

$$HCN(aq) + Na^+(aq) + OH^-(aq) \longrightarrow Na^+(aq) + CN^-(aq) + H_2O(l)$$

and the net ionic equation is

$$HCN(aq) + OH^-(aq) \longrightarrow CN^-(aq) + H_2O(l)$$

The following are also examples of acid-base neutralization reactions, represented by molecular equations:

$$HF(aq) + KOH(aq) \longrightarrow KF(aq) + H_2O(l)$$
$$H_2SO_4(aq) + 2NaOH(aq) \longrightarrow Na_2SO_4(aq) + 2H_2O(l)$$
$$Ba(OH)_2(aq) + 2HNO_3(aq) \longrightarrow Ba(NO_3)_2(aq) + 2H_2O(l)$$

Acid-Base Reactions Leading to Gas Formation

Certain salts like carbonates (containing the CO_3^{2-} ion), bicarbonates (containing the HCO_3^- ion), sulfites (containing the SO_3^{2-} ion), and sulfides (containing the S^{2-} ion) react with acids to form gaseous products. For example, the molecular equation for the reaction between sodium carbonate (Na_2CO_3) and $HCl(aq)$ is

$$Na_2CO_3(aq) + 2HCl(aq) \longrightarrow 2NaCl(aq) + H_2CO_3(aq)$$

See Figure 4.6 on p. 105.

Carbonic acid is unstable and if present in solution in sufficient concentrations decomposes as follows:

$$H_2CO_3(aq) \longrightarrow H_2O(l) + CO_2(g)$$

Similar reactions involving other mentioned salts are

$$NaHCO_3(aq) + HCl(aq) \longrightarrow NaCl(aq) + H_2O(l) + CO_2(g)$$
$$Na_2SO_3(aq) + 2HCl(aq) \longrightarrow 2NaCl(aq) + H_2O(l) + SO_2(g)$$
$$K_2S(aq) + 2HCl(aq) \longrightarrow 2KCl(aq) + H_2S(g)$$

Review of Concepts

Which of the following diagrams best represents a strong acid? Which represents a weak acid? Which represents a very weak acid? The proton exists in water as the hydronium ion. All of the acids are monoprotic. (For simplicity, water molecules are not shown.)

(a)

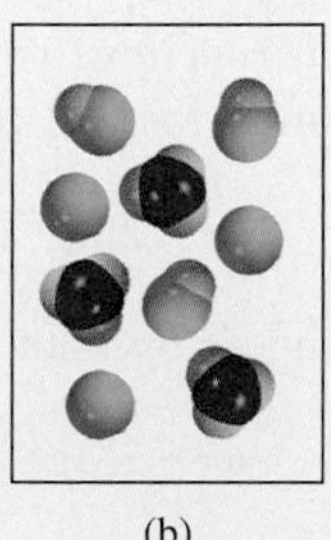
(b)

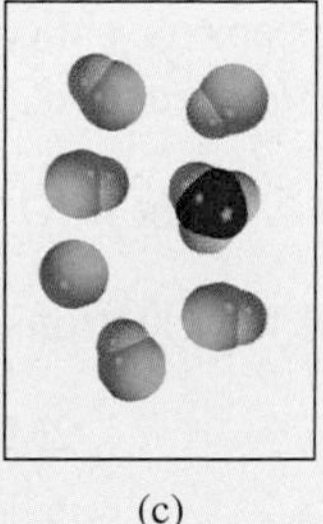
(c)

4.4 Oxidation-Reduction Reactions

Whereas acid-base reactions can be characterized as proton-transfer processes, the class of reactions called ***oxidation-reduction,*** or ***redox, reactions*** are considered *electron-transfer reactions.* Oxidation-reduction reactions are very much a part of the world

Animation:
Oxidation-Reduction Reactions

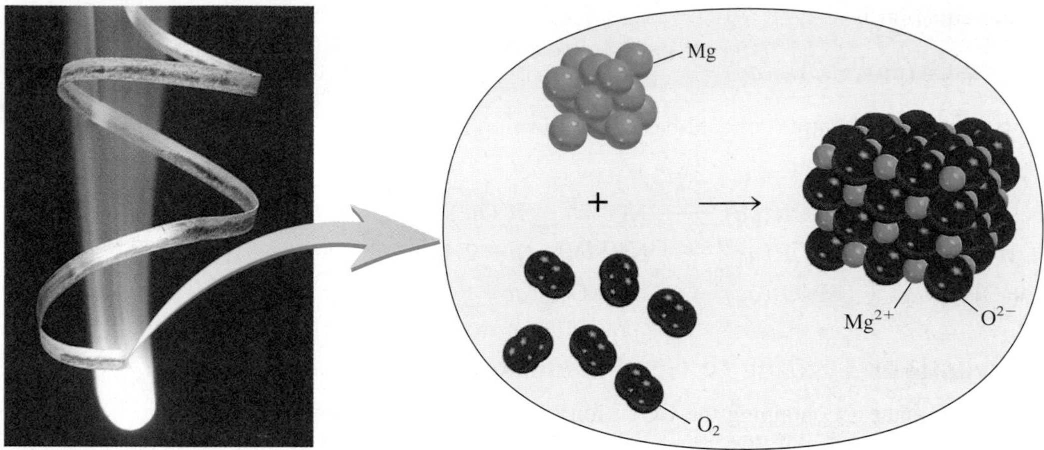

Figure 4.9
Magnesium burns in oxygen to form magnesium oxide.

around us. They range from the burning of fossil fuels to the action of household bleach. Additionally, most metallic and nonmetallic elements are obtained from their ores by the process of oxidation or reduction.

Many important redox reactions take place in water, but not all redox reactions occur in aqueous solution. Nonaqueous redox reactions are less cumbersome to deal with, so we will begin our discussion with a reaction in which two elements combine to form a compound. Consider the formation of magnesium oxide (MgO) from magnesium and oxygen (Figure 4.9):

$$2Mg(s) + O_2(g) \longrightarrow 2MgO(s)$$

Magnesium oxide (MgO) is an ionic compound made up of Mg^{2+} and O^{2-} ions. In this reaction, two Mg atoms give up or transfer four electrons to two O atoms (in O_2). For convenience, we can think of this process as two separate steps, one involving the loss of four electrons by the two Mg atoms and the other being the gain of four electrons by an O_2 molecule:

In oxidation half-reaction, electrons appear as the product; in reduction half-reaction, electrons appear as the reactant.

$$2Mg \longrightarrow 2Mg^{2+} + 4e^-$$
$$O_2 + 4e^- \longrightarrow 2O^{2-}$$

Each of these steps is called a ***half-reaction,*** which *explicitly shows the electrons involved in a redox reaction.* The sum of the half-reactions gives the overall reaction:

$$2Mg + O_2 + 4e^- \longrightarrow 2Mg^{2+} + 2O^{2-} + 4e^-$$

or, if we cancel the electrons that appear on both sides of the equation,

$$2Mg + O_2 \longrightarrow 2Mg^{2+} + 2O^{2-}$$

Finally, the Mg^{2+} and O^{2-} ions combine to form MgO:

$$2Mg^{2+} + 2O^{2-} \longrightarrow 2MgO$$

A useful mnemonic for redox is OILRIG: Oxidation Is Loss (of electrons) and Reduction Is Gain (of electrons).

The term ***oxidation reaction*** refers to the *half-reaction that involves loss of electrons.* Chemists originally used "oxidation" to denote the combination of elements with oxygen. However, it now has a broader meaning that includes reactions not involving oxygen. A ***reduction reaction*** is a *half-reaction that involves gain of electrons.* In the formation of magnesium oxide, magnesium is oxidized. It is said to act as a ***reducing agent*** because it *donates electrons* to oxygen and causes oxygen to be reduced. Oxygen is reduced and acts as an ***oxidizing agent*** because it *accepts electrons* from magnesium, causing magnesium to be oxidized. Note that the extent of oxidation in a redox reaction must be equal to the extent of reduction, that is, the number of electrons lost by a reducing agent must be equal to the number of electrons gained by an oxidizing agent.

Oxidizing agents are always reduced, and reducing agents are always oxidized. This statement may be somewhat confusing, but it is simply a consequence of the definitions of the two processes.

Oxidation Number

The definitions of oxidation and reduction in terms of loss and gain of electrons apply to the formation of ionic compounds such as MgO. However, these definitions do not accurately characterize the formation of hydrogen chloride (HCl) and sulfur dioxide (SO_2):

$$H_2(g) + Cl_2(g) \longrightarrow 2HCl(g)$$
$$S(s) + O_2(g) \longrightarrow SO_2(g)$$

Because HCl and SO_2 are not ionic but molecular compounds, no electrons are actually transferred in the formation of these compounds, as they are in the case of MgO. Nevertheless, chemists find it convenient to treat these reactions as redox reactions because experimental measurements show that there is a partial transfer of electrons (from H to Cl in HCl and from S to O in SO_2).

To keep track of electrons in redox reactions, it is useful to assign oxidation numbers to the reactants and products. An atom's ***oxidation number,*** also called ***oxidation state,*** signifies the *number of charges the atom would have in a molecule (or an ionic compound) if electrons were transferred completely.* For example, we can rewrite the preceding equations for the formation of HCl and SO_2 as follows:

$$\overset{0}{H_2}(g) + \overset{0}{Cl_2}(g) \longrightarrow 2\overset{+1}{H}\overset{-1}{Cl}(g)$$

$$\overset{0}{S}(s) + \overset{0}{O_2}(g) \longrightarrow \overset{+4}{S}\overset{-2}{O_2}(g)$$

The numbers above the element symbols are the oxidation numbers. In both of the reactions shown, there is no charge on the atoms in the reactant molecules. Thus, their oxidation number is zero. For the product molecules, however, it is assumed that complete electron transfer has taken place and that atoms have gained or lost electrons. The oxidation numbers reflect the number of electrons "transferred."

Oxidation numbers enable us to identify elements that are oxidized and reduced at a glance. The elements that show an increase in oxidation number—hydrogen and sulfur in the preceding examples—are oxidized. Chlorine and oxygen are reduced, so their oxidation numbers show a decrease from their initial values. Note that the sum of the oxidation numbers of H and Cl in HCl (+1 and −1) is zero. Likewise, if we add the charges on S (+4) and two atoms of O [$2 \times (-2)$], the total is zero. The reason is that the HCl and SO_2 molecules are neutral, so the charges must cancel.

We use the following rules to assign oxidation numbers:

1. In free elements (that is, in the uncombined state), each atom has an oxidation number of zero. Thus, each atom in H_2, Br_2, Na, Be, K, O_2, and P_4 has the same oxidation number: zero.
2. For ions composed of only one atom (that is, monatomic ions), the oxidation number is equal to the charge on the ion. Thus, Li^+ ion has an oxidation number of +1; Ba^{2+} ion, +2; Fe^{3+} ion, +3; I^- ion, −1; O^{2-} ion, −2; and so on. All

alkali metals have an oxidation number of +1 and all alkaline earth metals have an oxidation number of +2 in their compounds. Aluminum has an oxidation number of +3 in all its compounds.

3. The oxidation number of oxygen in most compounds (for example, MgO and H_2O) is −2, but in hydrogen peroxide (H_2O_2) and peroxide ion (O_2^{2-}), it is −1.
4. The oxidation number of hydrogen is +1, except when it is bonded to metals in binary compounds. In these cases (for example, LiH, NaH, CaH_2), its oxidation number is −1.
5. Fluorine has an oxidation number of −1 in *all* its compounds. Other halogens (Cl, Br, and I) have negative oxidation numbers when they occur as halide ions in their compounds. When combined with oxygen—for example in oxoacids and oxoanions (see Section 2.7)—they have positive oxidation numbers.
6. In a neutral molecule, the sum of the oxidation numbers of all the atoms must be zero. In a polyatomic ion, the sum of oxidation numbers of all the elements in the ion must be equal to the net charge of the ion. For example, in the ammonium ion, NH_4^+, the oxidation number of N is −3 and that of H is +1. Thus, the sum of the oxidation numbers is $-3 + 4(+1) = +1$, which is equal to the net charge of the ion.
7. Oxidation numbers do not have to be integers. For example, the oxidation number of O in the superoxide ion, O_2^-, is $-\frac{1}{2}$.

EXAMPLE 4.4

Assign oxidation numbers to all the elements in the following compounds and ion: (a) Na_2O, (b) HNO_2, (c) $Cr_2O_7^{2-}$.

Strategy In general, we follow the rules just listed for assigning oxidation numbers. Remember that all alkali metals have an oxidation number of +1, and in most cases, hydrogen has an oxidation number of +1 and oxygen has an oxidation number of −2 in their compounds.

Solution

(a) By rule 2, we see that sodium has an oxidation number of +1 (Na^+) and oxygen's oxidation number is −2 (O^{2-}).

(b) This is the formula for nitrous acid, which yields a H^+ ion and a NO_2^- ion in solution. From rule 4, we see that H has an oxidation number of +1. Thus, the other group (the nitrite ion) must have a net oxidation number of −1. Oxygen has an oxidation number of −2, and if we use x to represent the oxidation number of nitrogen, then the nitrite ion can be written as

$$[N^{(x)}O_2^{(2-)}]^-$$

so that $$x + 2(-2) = -1$$

or $$x = +3$$

(c) From rule 6, we see that the sum of the oxidation numbers in the dichromate ion $Cr_2O_7^{2-}$ must be −2. We know that the oxdiation number of O is −2, so all that remains is to determine the oxidation number of Cr, which we call y. The dichromate ion can be written as

$$[Cr_2^{(y)}O_7^{(2-)}]^{2-}$$

so that $$2(y) + 7(-2) = -2$$

or $$y = +6$$

(Continued)

Check In each case, does the sum of the oxidation numbers of all the atoms equal the net charge on the species?

Practice Exercise Assign oxidation numbers to all the elements in the following compound and ion: (a) PF_3, (b) MnO_4^-.

Similar problems: 4.43, 4.45.

Figure 4.10 shows the known oxidation numbers of the familiar elements, arranged according to their positions in the periodic table. We can summarize the content of this figure as follows:

- Metallic elements have only positive oxidation numbers, whereas nonmetallic elements may have either positive or negative oxidation numbers.
- The highest oxidation number an element in Groups 1A–7A can have is its group number. For example, the halogens are in Group 7A, so their highest possible oxidation number is +7.
- The transition metals (Groups 1B, 3B–8B) usually have several possible oxidation numbers.

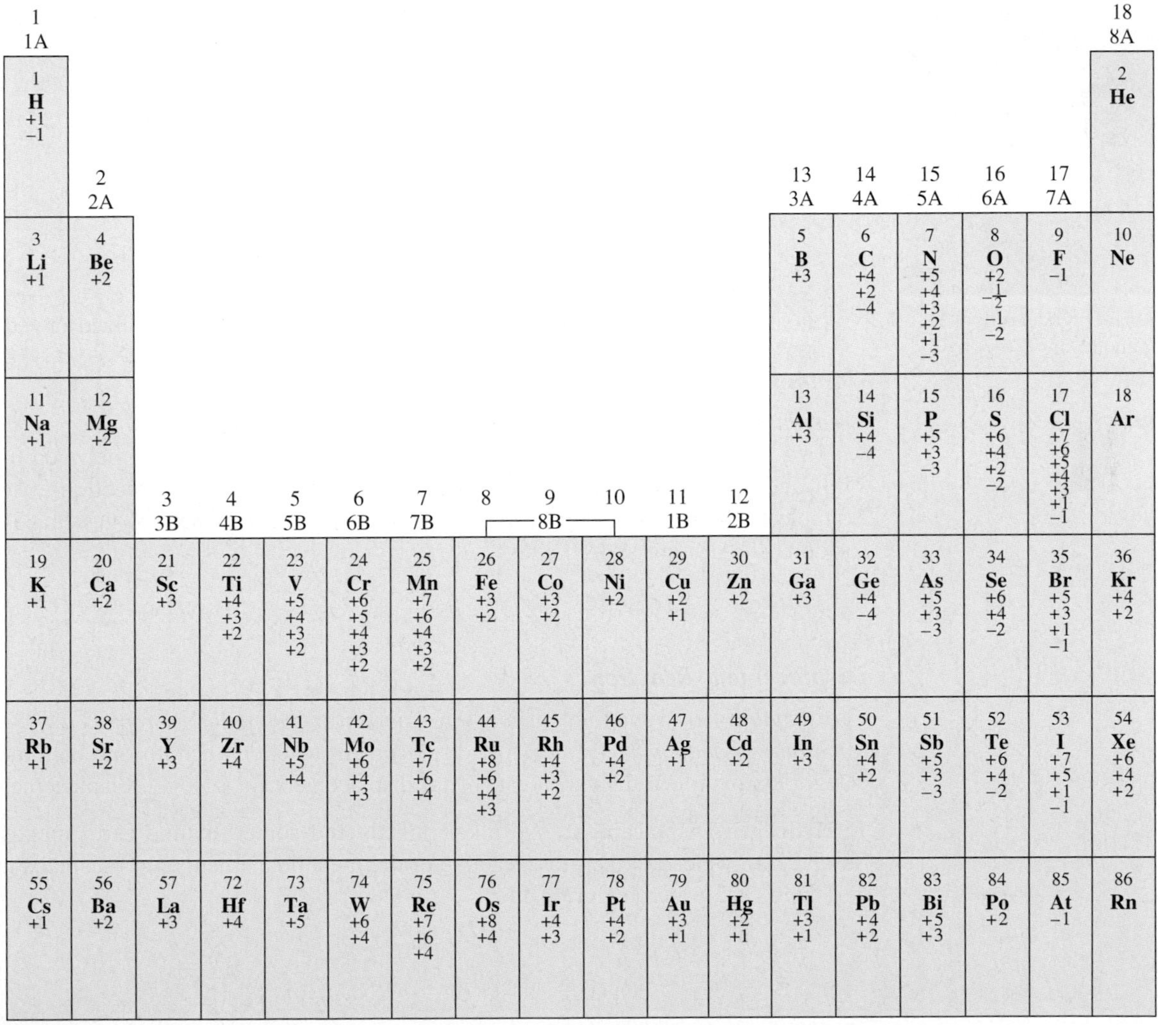

1 1A	2 2A	3 3B	4 4B	5 5B	6 6B	7 7B	8 8B	9 8B	10 8B	11 1B	12 2B	13 3A	14 4A	15 5A	16 6A	17 7A	18 8A
1 **H** +1 −1																	2 **He**
3 **Li** +1	4 **Be** +2											5 **B** +3	6 **C** +4 +2 −4	7 **N** +5 +4 +3 +2 +1 −3	8 **O** +2 $-\frac{1}{2}$ −1 −2	9 **F** −1	10 **Ne**
11 **Na** +1	12 **Mg** +2											13 **Al** +3	14 **Si** +4 −4	15 **P** +5 +3 −3	16 **S** +6 +4 +2 −2	17 **Cl** +7 +6 +5 +4 +3 +1 −1	18 **Ar**
19 **K** +1	20 **Ca** +2	21 **Sc** +3	22 **Ti** +4 +3 +2	23 **V** +5 +4 +3 +2	24 **Cr** +6 +5 +4 +3 +2	25 **Mn** +7 +6 +4 +3 +2	26 **Fe** +3 +2	27 **Co** +3 +2	28 **Ni** +2	29 **Cu** +2 +1	30 **Zn** +2	31 **Ga** +3	32 **Ge** +4 −4	33 **As** +5 +3 −3	34 **Se** +6 +4 −2	35 **Br** +5 +3 +1 −1	36 **Kr** +4 +2
37 **Rb** +1	38 **Sr** +2	39 **Y** +3	40 **Zr** +4	41 **Nb** +5 +4	42 **Mo** +6 +4 +3	43 **Tc** +7 +6 +4	44 **Ru** +8 +6 +4 +3	45 **Rh** +4 +3 +2	46 **Pd** +4 +2	47 **Ag** +1	48 **Cd** +2	49 **In** +3	50 **Sn** +4 +2	51 **Sb** +5 +3 −3	52 **Te** +6 +4 −2	53 **I** +7 +5 +1 −1	54 **Xe** +6 +4 +2
55 **Cs** +1	56 **Ba** +2	57 **La** +3	72 **Hf** +4	73 **Ta** +5	74 **W** +6 +4	75 **Re** +7 +6 +4	76 **Os** +8 +4	77 **Ir** +4 +3	78 **Pt** +4 +2	79 **Au** +3 +1	80 **Hg** +2 +1	81 **Tl** +3 +1	82 **Pb** +4 +2	83 **Bi** +5 +3	84 **Po** +2	85 **At** −1	86 **Rn**

Figure 4.10
The oxidation numbers of elements in their compounds. The more common oxidation numbers are in color.

Sulfur burning in air to form sulfur dioxide.

Some Common Oxidation-Reduction Reactions

Among the most common oxidation-reduction reactions are combination, decomposition, combustion, and displacement reactions.

Combination Reactions

A ***combination reaction*** is *a reaction in which two or more substances combine to form a single product.* For example,

$$\overset{0}{\mathrm{S}}(s) + \overset{0}{\mathrm{O}}_2(g) \longrightarrow \overset{+4}{\mathrm{S}}\overset{-2}{\mathrm{O}}_2(g)$$

$$3\overset{0}{\mathrm{Mg}}(s) + \overset{0}{\mathrm{N}}_2(g) \longrightarrow \overset{+2}{\mathrm{Mg}}_3\overset{-3}{\mathrm{N}}_2(s)$$

Decomposition Reactions

Decomposition reactions are the opposite of combination reactions. Specifically, a ***decomposition reaction*** is *the breakdown of a compound into two or more components.* For example,

On heating, HgO decomposes to give Hg and O_2.

$$2\overset{+2}{\mathrm{Hg}}\overset{-2}{\mathrm{O}}(s) \longrightarrow 2\overset{0}{\mathrm{Hg}}(l) + \overset{0}{\mathrm{O}}_2(g)$$

$$2\mathrm{K}\overset{+5}{\mathrm{Cl}}\overset{-2}{\mathrm{O}}_3(s) \longrightarrow 2\mathrm{K}\overset{-1}{\mathrm{Cl}}(s) + 3\overset{0}{\mathrm{O}}_2(g)$$

$$2\overset{+1}{\mathrm{Na}}\overset{-1}{\mathrm{H}}(s) \longrightarrow 2\overset{0}{\mathrm{Na}}(s) + \overset{0}{\mathrm{H}}_2(g)$$

Note that we show oxidation numbers only for elements that are oxidized or reduced.

Combustion Reactions

All combustions are redox reactions.

A ***combustion reaction*** is *a reaction in which a substance reacts with oxygen, usually with the release of heat and light to produce a flame.* The reactions between magnesium and sulfur with oxygen described earlier are combustion reactions. Another example is the burning of propane (C_3H_8), a component of natural gas that is used for domestic heating and cooking:

$$\mathrm{C_3H_8}(g) + 5\mathrm{O_2}(g) \longrightarrow 3\mathrm{CO_2}(g) + 4\mathrm{H_2O}(l)$$

Displacement Reactions

In a ***displacement reaction,*** *an ion (or atom) in a compound is replaced by an ion (or atom) of another element:* Most displacement reactions fit into one of three subcategories: hydrogen displacement, metal displacement, or halogen displacement.

1. Hydrogen Displacement. All alkali metals and some alkaline earth metals (Ca, Sr, and Ba), which are the most reactive of the metallic elements, will displace hydrogen from cold water (Figure 4.11):

$$2\overset{0}{\mathrm{Na}}(s) + 2\overset{+1}{\mathrm{H}}_2\mathrm{O}(l) \longrightarrow 2\overset{+1}{\mathrm{Na}}\mathrm{O}\overset{+1}{\mathrm{H}}(aq) + \overset{0}{\mathrm{H}}_2(g)$$

$$\overset{0}{\mathrm{Ca}}(s) + 2\overset{+1}{\mathrm{H}}_2\mathrm{O}(l) \longrightarrow \overset{+2}{\mathrm{Ca}}(\mathrm{O}\overset{+1}{\mathrm{H}})_2(s) + \overset{0}{\mathrm{H}}_2(g)$$

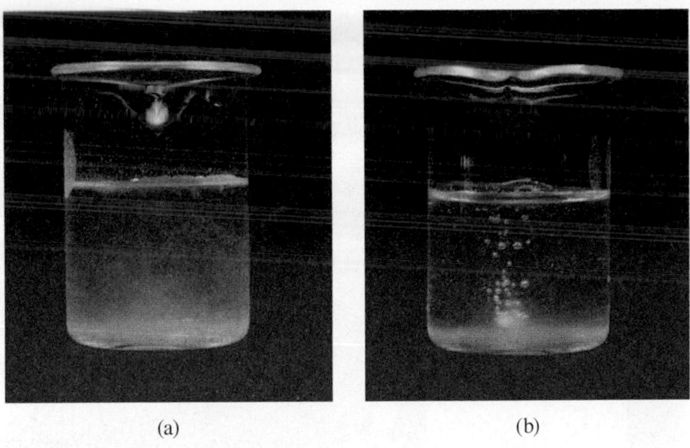

(a) (b)

Figure 4.11
Reactions of (a) sodium (Na) and (b) calcium (Ca) with cold water. Note that the reaction is more vigorous with Na than with Ca.

Many metals, including those that do not react with water, are capable of displacing hydrogen from acids. For example, zinc (Zn) and magnesium (Mg) do not react with cold water but do react with hydrochloric acid, as follows:

$$\overset{0}{\text{Zn}}(s) + 2\overset{+1}{\text{H}}\text{Cl}(aq) \longrightarrow \overset{+2}{\text{Zn}}\text{Cl}_2(aq) + \overset{0}{\text{H}_2}(g)$$

$$\overset{0}{\text{Mg}}(s) + 2\overset{+1}{\text{H}}\text{Cl}(aq) \longrightarrow \overset{+2}{\text{Mg}}\text{Cl}_2(aq) + \overset{0}{\text{H}_2}(g)$$

Figure 4.12 shows the reactions between hydrochloric acid (HCl) and iron (Fe), zinc (Zn), and magnesium (Mg). These reactions are used to prepare hydrogen gas in the laboratory.

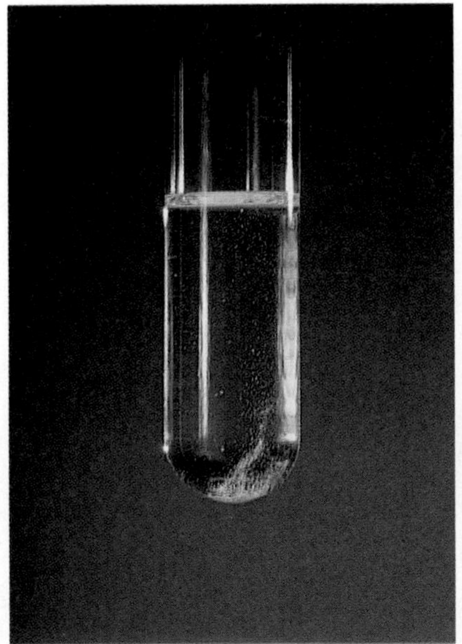

Figure 4.12
Left to right: Reactions of iron (Fe), zinc (Zn), and magnesium (Mg) with hydrochloric acid to form hydrogen gas and the metal chlorides ($FeCl_2$, $ZnCl_2$, $MgCl_2$). The reactivity of these metals is reflected in the rate of hydrogen gas evolution, which is slowest for the least reactive metal, Fe, and the fastest for the most reactive metal, Mg.

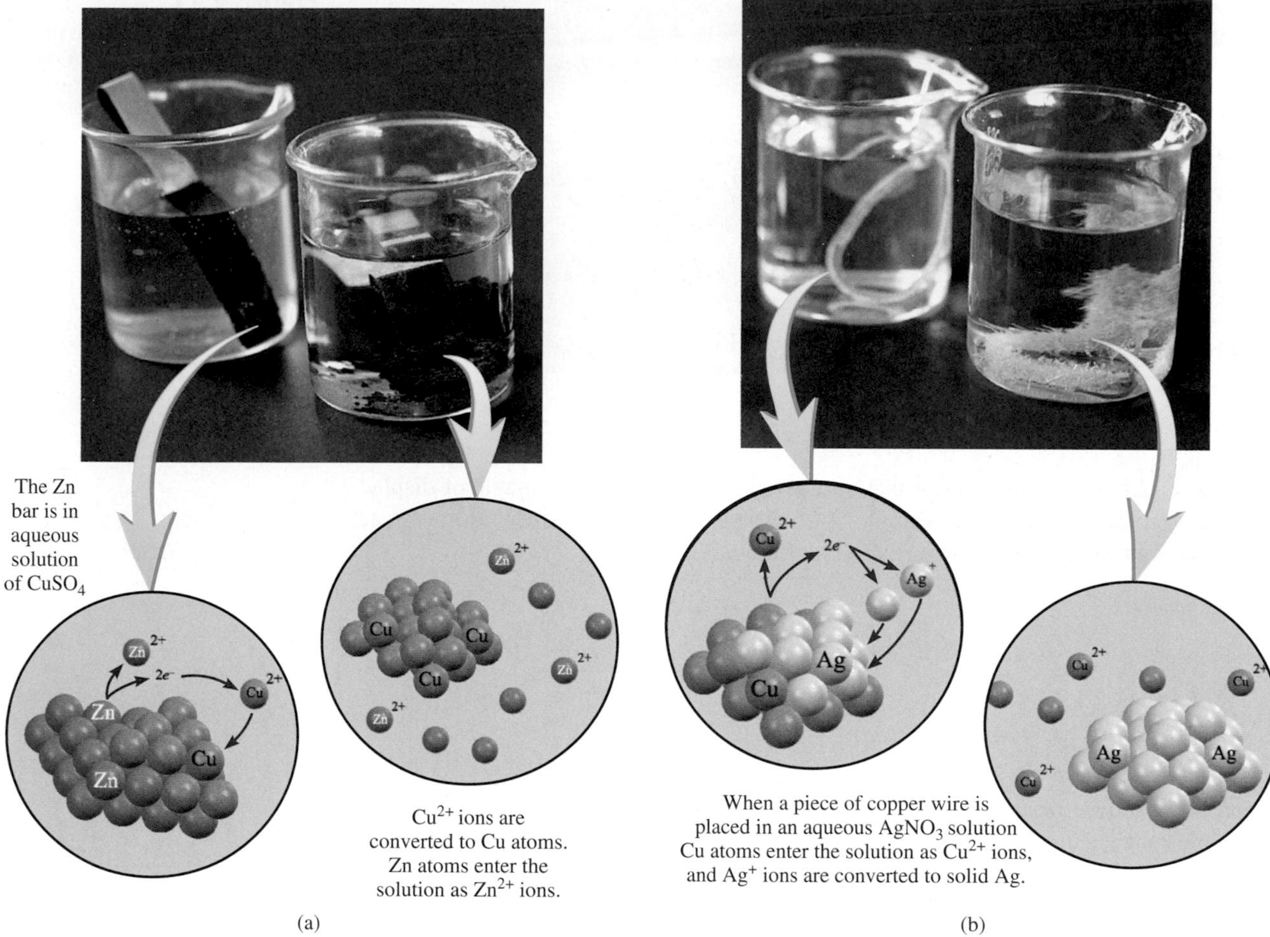
The Zn bar is in aqueous solution of CuSO4
Zn 2+
2e⁻
Cu 2+
Zn
Cu
Zn
Cu
Cu
Cu
Zn 2+
Zn 2+
Cu2+ ions are converted to Cu atoms. Zn atoms enter the solution as Zn2+ ions.
(a)
Cu 2+
2e⁻
Ag +
Ag
Cu
Cu 2+
Cu 2+
Cu 2+
Ag
Ag
When a piece of copper wire is placed in an aqueous AgNO3 solution Cu atoms enter the solution as Cu2+ ions, and Ag+ ions are converted to solid Ag.
(b)

Figure 4.13
Metal displacement reactions in solution.

2. Metal Displacement. A metal in a compound can be displaced by another metal in the uncombined state. For example, when metallic zinc is added to a solution containing copper sulfate ($CuSO_4$), it displaces Cu^{2+} ions from the solution (Figure 4.13):

$$\overset{0}{\mathrm{Zn}}(s) + \overset{+2}{\mathrm{Cu}}\mathrm{SO_4}(aq) \longrightarrow \overset{+2}{\mathrm{Zn}}\mathrm{SO_4}(aq) + \overset{0}{\mathrm{Cu}}(s)$$

The net ionic equation is

$$\overset{0}{\mathrm{Zn}}(s) + \overset{+2}{\mathrm{Cu}^{2+}}(aq) \longrightarrow \overset{+2}{\mathrm{Zn}^{2+}}(aq) + \overset{0}{\mathrm{Cu}}(s)$$

Similarly, metallic copper displaces silver ions from a solution containing silver nitrate ($AgNO_3$) (also shown in Figure 4.13):

$$\overset{0}{\mathrm{Cu}}(s) + 2\overset{+1}{\mathrm{Ag}}\mathrm{NO_3}(aq) \longrightarrow \overset{+2}{\mathrm{Cu}}\mathrm{(NO_3)_2}(aq) + 2\overset{0}{\mathrm{Ag}}(s)$$

The net ionic equation is

$$\overset{0}{\mathrm{Cu}}(s) + 2\overset{+1}{\mathrm{Ag}^{+}}(aq) \longrightarrow \overset{+2}{\mathrm{Cu}^{2+}}(aq) + 2\overset{0}{\mathrm{Ag}}(s)$$

Reducing strength increases ↑

Reaction	
$Li \rightarrow Li^+ + e^-$	React with cold water to produce H_2
$K \rightarrow K^+ + e^-$	
$Ba \rightarrow Ba^{2+} + 2e^-$	
$Ca \rightarrow Ca^{2+} + 2e^-$	
$Na \rightarrow Na^+ + e^-$	
$Mg \rightarrow Mg^{2+} + 2e^-$	React with steam to produce H_2
$Al \rightarrow Al^{3+} + 3e^-$	
$Zn \rightarrow Zn^{2+} + 2e^-$	
$Cr \rightarrow Cr^{3+} + 3e^-$	
$Fe \rightarrow Fe^{2+} + 2e^-$	
$Cd \rightarrow Cd^{2+} + 2e^-$	
$Co \rightarrow Co^{2+} + 2e^-$	React with acids to produce H_2
$Ni \rightarrow Ni^{2+} + 2e^-$	
$Sn \rightarrow Sn^{2+} + 2e^-$	
$Pb \rightarrow Pb^{2+} + 2e^-$	
$H_2 \rightarrow 2H^+ + 2e^-$	
$Cu \rightarrow Cu^{2+} + 2e^-$	Do not react with water or acids to produce H_2
$Ag \rightarrow Ag^+ + e^-$	
$Hg \rightarrow Hg^{2+} + 2e^-$	
$Pt \rightarrow Pt^{2+} + 2e^-$	
$Au \rightarrow Au^{3+} + 3e^-$	

Figure 4.14
The activity series for metals. The metals are arranged according to their ability to displace hydrogen from an acid or water. Li (lithium) is the most reactive metal, and Au (gold) is the least reactive.

Reversing the roles of the metals would result in no reaction. In other words, copper metal will not displace zinc ions from zinc sulfate, and silver metal will not displace copper ions from copper nitrate.

An easy way to predict whether a metal or hydrogen displacement reaction will actually occur is to refer to an ***activity series*** (sometimes called the *electrochemical series*), shown in Figure 4.14. Basically, an activity series is *a convenient summary of the results of many possible displacement reactions* similar to the ones already discussed. According to this series, any metal above hydrogen will displace it from water or from an acid, but metals below hydrogen will not react with either water or an acid. In fact, any metal listed in the series will react with any metal (in a compound) below it. For example, Zn is above Cu, so zinc metal will displace copper ions from copper sulfate.

3. Halogen Displacement. Another activity series summarizes the halogens' behavior in halogen displacement reactions:

$$F_2 > Cl_2 > Br_2 > I_2$$

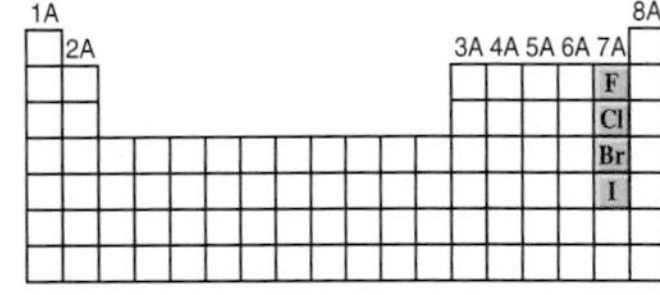

The halogens.

The power of these elements as oxidizing agents decreases as we move down Group 7A from fluorine to iodine, so molecular fluorine can replace chloride, bromide, and iodide ions in solution. In fact, molecular fluorine is so reactive that it also attacks water; thus, these reactions cannot be carried out in aqueous solutions. On the other hand, molecular chlorine can displace bromide and iodide ions in aqueous solution. The displacement equations are

$$\overset{0}{Cl_2}(g) + 2K\overset{-1}{Br}(aq) \longrightarrow 2K\overset{-1}{Cl}(aq) + \overset{0}{Br_2}(l)$$

$$\overset{0}{Cl_2}(g) + 2Na\overset{-1}{I}(aq) \longrightarrow 2Na\overset{-1}{Cl}(aq) + \overset{0}{I_2}(s)$$

Industrially, bromine (a fuming red liquid) is prepared by the action of chlorine on seawater, which is a rich source of Br^- ions.

The ionic equations are

$$\overset{0}{Cl_2}(g) + 2\overset{-1}{Br^-}(aq) \longrightarrow 2\overset{-1}{Cl^-}(aq) + \overset{0}{Br_2}(l)$$

$$\overset{0}{Cl_2}(g) + 2\overset{-1}{I^-}(aq) \longrightarrow 2\overset{-1}{Cl^-}(aq) + \overset{0}{I_2}(s)$$

Molecular bromine, in turn, can displace iodide ion in solution:

$$\overset{0}{Br_2}(l) + 2\overset{-1}{I^-}(aq) \longrightarrow 2\overset{-1}{Br^-}(aq) + \overset{0}{I_2}(s)$$

Reversing the roles of the halogens produces no reaction. Thus, bromine cannot displace chloride ions, and iodine cannot displace bromide and chloride ions.

REVIEW OF CONCEPTS

Which of the following combination reactions is not a redox reaction?

(a) $2Mg(s) + O_2(g) \longrightarrow 2MgO(s)$
(b) $H_2(g) + Cl_2(g) \longrightarrow 2HCl(g)$
(c) $NH_3(g) + HCl(g) \longrightarrow NH_4Cl(s)$
(d) $2Na(s) + S(s) \longrightarrow Na_2S(s)$

4.5 Concentration of Solutions

To study solution stoichiometry, we must know how much of the reactants are present in a solution and also how to control the amounts of reactants used to bring about a reaction in aqueous solution.

The ***concentration of a solution*** is *the amount of solute present in a given amount of solvent, or a given amount of solution.* (For this discussion, we will assume the solute is a liquid or a solid and the solvent is a liquid.) The concentration of a solution can be expressed in many different ways, as we will see in Chapter 13. Here we will consider one of the most commonly used units in chemistry, ***molarity (M)***, or ***molar concentration,*** which is *the number of moles of solute per liter of solution.* Molarity is defined as

$$\text{molarity} = \frac{\text{moles of solute}}{\text{liters of soln}} \tag{4.1}$$

where "soln" denotes "solution." Equation (4.1) can also be expressed algebraically as

$$M = \frac{n}{V} \tag{4.2}$$

where n denotes the number of moles of solute and V is the volume of the solution in liters. Thus, a 1.46 molar glucose ($C_6H_{12}O_6$) solution, written 1.46 M $C_6H_{12}O_6$, contains 1.46 moles of the solute ($C_6H_{12}O_6$) in 1 L of the solution; a 0.52 molar urea [$(NH_2)_2CO$] solution, written 0.52 M $(NH_2)_2CO$, contains 0.52 mole of $(NH_2)_2CO$ (the solute) in 1 L of solution; and so on.

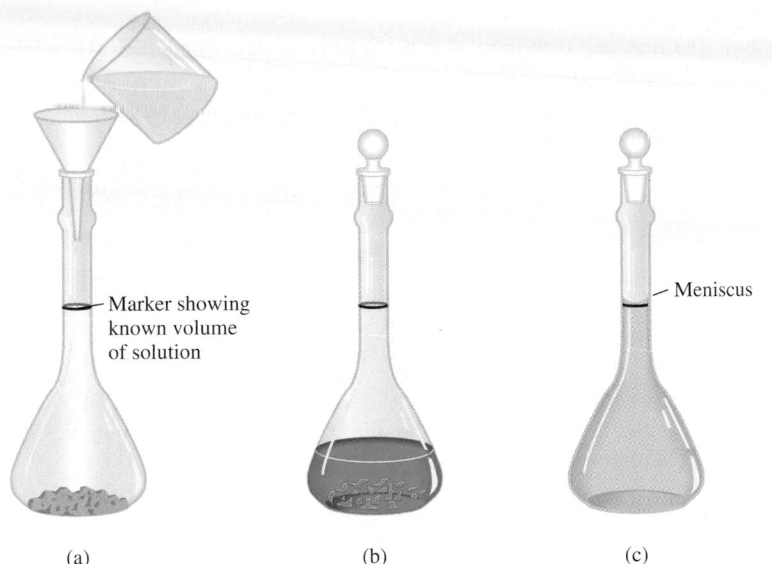

Figure 4.15
Preparing a solution of known molarity. (a) A known amount of a solid solute is transferred into the volumetric flask; then water is added through a funnel. (b) The solid is slowly dissolved by gently swirling the flask. (c) After the solid has completely dissolved, more water is added to bring the level of solution to the mark. Knowing the volume of solution and the amount of solute dissolved in it, we can calculate the molarity of the prepared solution.

Of course, we do not always work with solution volumes of exactly 1 L. This is not a problem as long as we remember to convert the volume of the solution to liters. Thus, a 500-mL solution containing 0.730 mole of $C_6H_{12}O_6$ also has a concentration of 1.46 *M*:

$$M = \text{molarity} = \frac{0.730\ \text{mol}}{0.500\ \text{L}}$$
$$= 1.46\ \text{mol/L} = 1.46\ M$$

As you can see, the unit of molarity is moles per liter, so a 500-mL solution containing 0.730 mole of $C_6H_{12}O_6$ is equivalent to 1.46 mol/L or 1.46 *M*. Note that concentration, like density, is an intensive property, so its value does not depend on how much of the solution is present.

The procedure for preparing a solution of known molarity is as follows. First, the solute is accurately weighed and transferred to a volumetric flask through a funnel (Figure 4.15). Next, water is added to the flask, which is carefully swirled to dissolve the solid. After *all* the solid has dissolved, more water is added slowly to bring the level of solution exactly to the volume mark. Knowing the volume of the solution in the flask and the quantity of compound (the number of moles) dissolved, we can calculate the molarity of the solution using Equation (4.1). Note that this procedure does not require knowing the amount of water added, as long as the volume of the final solution is known.

Animation:
Making a Solution

A $K_2Cr_2O_7$ solution.

EXAMPLE 4.5

How many grams of potassium dichromate ($K_2Cr_2O_7$) are required to prepare a 125-mL solution whose concentration is 1.83 *M*?

Strategy How many moles of $K_2Cr_2O_7$ does a 1-L (or 1000-mL) 1.83 *M* $K_2Cr_2O_7$ solution contain? A 125-mL solution? How would you convert moles to grams?

(Continued)

Solution The first step is to determine the number of moles of $K_2Cr_2O_7$ in 125 mL or 0.125 L of a 1.83 *M* solution:

$$\text{moles of } K_2Cr_2O_7 = 0.125 \cancel{\text{L soln}} \times \frac{1.83 \text{ mol } K_2Cr_2O_7}{1 \cancel{\text{L soln}}}$$
$$= 0.229 \text{ mol } K_2Cr_2O_7$$

The molar mass of $K_2Cr_2O_7$ is 294.2 g, so we write

$$\text{grams of } K_2Cr_2O_7 \text{ needed} = 0.229 \cancel{\text{mol } K_2Cr_2O_7} \times \frac{294.2 \text{ g } K_2Cr_2O_7}{1 \cancel{\text{mol } K_2Cr_2O_7}}$$
$$= 67.4 \text{ g } K_2Cr_2O_7$$

Check As a ball-park estimate, the mass should be given by [molarity (mol/L) × volume (L) × molar mass (g/mol)] or [2 mol/L × 0.125 L × 300 g/mol] = 75 g. So the answer is reasonable.

Similar problems: 4.56, 4.57.

Practice Exercise What is the molarity of a 85.0-mL ethanol (C_2H_5OH) solution containing 1.77 g of ethanol?

EXAMPLE 4.6

In a biochemical assay, a chemist needs to add 4.07 g of glucose to a reaction mixture. Calculate the volume in milliliters of a 3.16 *M* glucose solution she should use for the addition.

Strategy We must first determine the number of moles contained in 4.07 g of glucose and then use Equation (4.2) to calculate the volume.

Solution From the molar mass of glucose, we write

$$4.07 \cancel{\text{g } C_6H_{12}O_6} \times \frac{1 \text{ mol } C_6H_{12}O_6}{180.2 \cancel{\text{g } C_6H_{12}O_6}} = 2.259 \times 10^{-2} \text{ mol } C_6H_{12}O_6$$

Next, we calculate the volume of the solution that contains 2.259×10^{-2} mol of the solute. Rearranging Equation (4.2) gives

$$V = \frac{n}{M}$$
$$= \frac{2.259 \times 10^{-2} \text{ mol } C_6H_{12}O_6}{3.16 \text{ mol } C_6H_{12}O_6/\text{ L soln}} \times \frac{1000 \text{ mL soln}}{1 \text{ L soln}}$$
$$= 7.15 \text{ mL soln}$$

Check One liter of the solution contains 3.16 moles of $C_6H_{12}O_6$. Therefore, the number of moles in 7.15 mL or 7.15×10^{-3} L is ($3.16 \text{ mol} \times 7.15 \times 10^{-3}$) or 2.26×10^{-2} mol. The small difference is due to the different ways of rounding off.

Similar problem: 4.59.

Practice Exercise What volume (in milliliters) of a 0.315 *M* NaOH solution contains 6.22 g of NaOH?

Dilution of Solutions

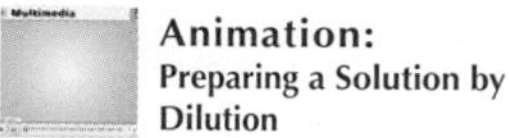

Animation: Preparing a Solution by Dilution

Concentrated solutions are often stored in the laboratory stockroom for use as needed. Frequently we dilute these "stock" solutions before working with them. ***Dilution*** is *the procedure for preparing a less concentrated solution from a more concentrated one.*

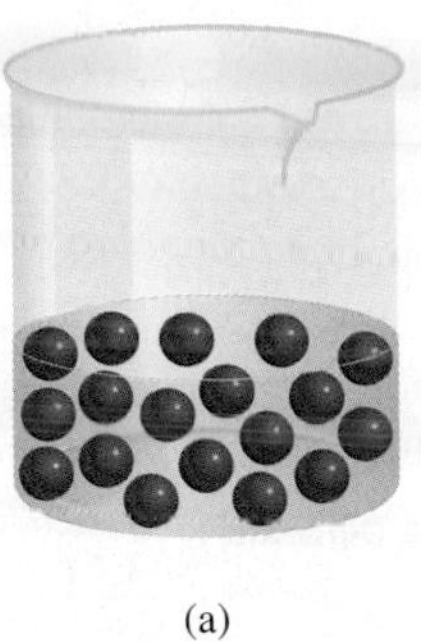
(a)

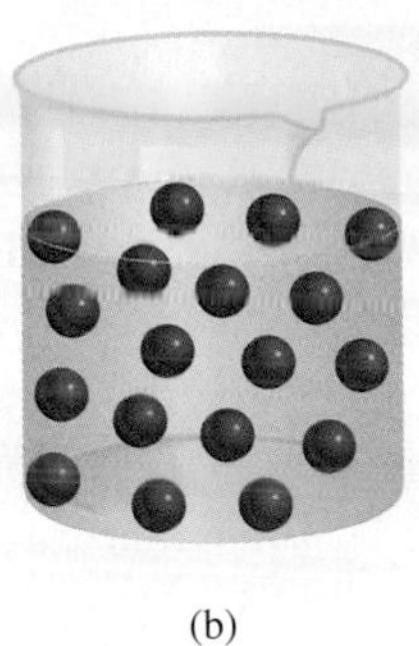
(b)

Figure 4.16
The dilution of a more concentrated solution (a) to a less concentrated one (b) does not change the total number of solute particles (18).

Suppose that we want to prepare 1 L of a 0.400 *M* $KMnO_4$ solution from a solution of 1.00 *M* $KMnO_4$. For this purpose, we need 0.400 mole of $KMnO_4$. Because there is 1.00 mole of $KMnO_4$ in 1 L of a 1.00 *M* $KMnO_4$ solution, there is 0.400 mole of $KMnO_4$ in 0.400 L of the same solution:

$$\frac{1.00\text{ mol}}{1\text{ L soln}} = \frac{0.400\text{ mol}}{0.400\text{ L soln}}$$

Two $KMnO_4$ solutions of different concentrations.

Therefore, we must withdraw 400 mL from the 1.00 *M* $KMnO_4$ solution and dilute it to 1000 mL by adding water (in a 1-L volumetric flask). This method gives us 1 L of the desired solution of 0.400 *M* $KMnO_4$.

In carrying out a dilution process, it is useful to remember that adding more solvent to a given amount of the stock solution changes (decreases) the concentration of the solution without changing the number of moles of solute present in the solution (Figure 4.16). In other words,

$$\text{moles of solute before dilution} = \text{moles of solute after dilution}$$

Because molarity is defined as moles of solute in one liter of solution, we see that the number of moles of solute is given by

$$\underbrace{\frac{\text{moles of solute}}{\text{liters of soln}}}_{M} \times \underbrace{\text{volume of soln (in liters)}}_{V} = \text{moles of solute}$$

or

$$MV = \text{moles of solute}$$

Because all the solute comes from the original stock solution, we can conclude that

$$\underset{\substack{\text{moles of solute}\\ \text{before dilution}}}{M_iV_i} = \underset{\substack{\text{moles of solute}\\ \text{after dilution}}}{M_fV_f} \tag{4.3}$$

where M_i and M_f are the initial and final concentrations of the solution in molarity and V_i and V_f are the initial and final volumes of the solution, respectively. Of course, the units of V_i and V_f must be the same (mL or L) for the calculation to work. To check the reasonableness of your results, be sure that $M_i > M_f$ and $V_f > V_i$.

EXAMPLE 4.7

Describe how you would prepare 2.50×10^2 mL of a 2.25 *M* H_2SO_4 solution, starting with a 7.41 *M* stock solution of H_2SO_4.

Strategy Because the concentration of the final solution is less than that of the original one, this is a dilution process. Keep in mind that in dilution, the concentration of the solution decreases but the number of moles of the solute remains the same.

Solution We prepare for the calculation by tabulating our data:

$$M_i = 7.41\ M \qquad M_f = 2.25\ M$$
$$V_i = ? \qquad V_f = 2.50 \times 10^2\ \text{mL}$$

Substituting in Equation (4.3),

$$(7.41\ M)(V_i) = (2.25\ M)(2.50 \times 10^2\ \text{mL})$$
$$V_i = \frac{(2.25\ M)(2.50 \times 10^2\ \text{mL})}{7.41\ M}$$
$$V_i = 75.9\ \text{mL}$$

Thus, we must dilute 75.9 mL of the 7.41 *M* H_2SO_4 solution with sufficient water to give a final volume of 2.50×10^2 mL in a 250-mL volumetric flask to obtain the desired concentration.

Similar problems: 4.65, 4.66.

Check The initial volume is less than the final volume, so the answer is reasonable.

Practice Exercise How would you prepare 2.00×10^2 mL of a 0.866 *M* KOH solution, starting with a 5.07 *M* stock solution?

REVIEW OF CONCEPTS

What is the final concentration of a 0.6 *M* NaCl solution if its volume is doubled and the number of moles of solute is tripled?

4.6 Solution Stoichiometry

In Chapter 3 we studied stoichiometric calculations in terms of the mole method, which treats the coefficients in a balanced equation as the number of moles of reactants and products. In working with solutions of known molarity, we have to use the relationship MV = moles of solute. We will examine two types of common solution stoichiometry here: gravimetric analysis and acid-base titration.

Gravimetric Analysis

Gravimetric analysis is *an analytical technique based on the measurement of mass.* One type of gravimetric analysis experiment involves the formation, isolation, and mass determination of a precipitate. Generally, this procedure is applied to ionic compounds. A sample substance of unknown composition is dissolved in water and allowed to react with another substance to form a precipitate. The precipitate is filtered off, dried, and weighed. Knowing the mass and chemical formula of the precipitate formed, we can calculate the mass of a particular chemical component (that is, the anion or cation) of the original sample. From the mass of the component and the mass of the original sample, we can determine the percent composition by mass of the component in the original compound.

A reaction that is often studied in gravimetric analysis, because the reactants can be obtained in pure form, is

$$AgNO_3(aq) + NaCl(aq) \longrightarrow NaNO_3(aq) + AgCl(s)$$

The net ionic equation is

$$Ag^+(aq) + Cl^-(aq) \longrightarrow AgCl(s)$$

The precipitate is AgCl (see Table 4.2). As an example, let's say that we are interested in knowing the purity of a sample of NaCl obtained from seawater. To do so we need to determine *experimentally* the percent by mass of Cl in NaCl. First, we would accurately weigh out a sample of the NaCl and dissolve it in water. Next, we would add enough $AgNO_3$ solution to the NaCl solution to cause the precipitation of all the Cl^- ions present in solution as AgCl. In this procedure, NaCl is the limiting reagent and $AgNO_3$ the excess reagent. The AgCl precipitate is separated from the solution by filtration, dried, and weighed. From the measured mass of AgCl, we can calculate the mass of Cl using the percent by mass of Cl in AgCl. Because this same amount of Cl was present in the original NaCl sample, we can calculate the percent by mass of Cl in NaCl and hence deduce its purity. Figure 4.17 shows how this procedure is performed.

Gravimetric analysis is a highly accurate technique, because the mass of a sample can be measured accurately. However, this procedure is applicable only to reactions that go to completion, or have nearly 100 percent yield. Thus, if AgCl were slightly soluble instead of being insoluble, it would not be possible to remove all the Cl^- ions from the NaCl solution and the subsequent calculation would be in error.

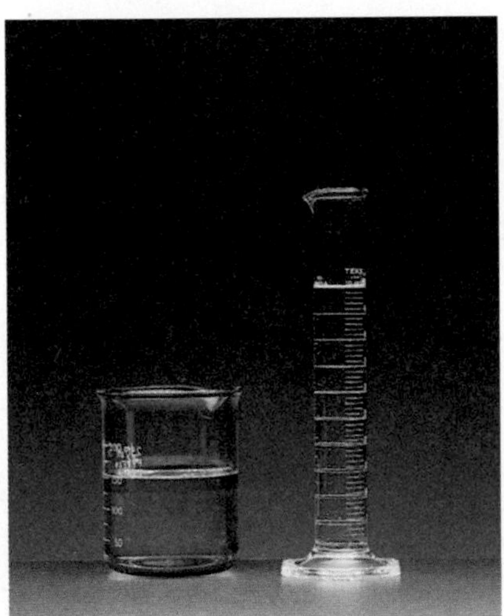

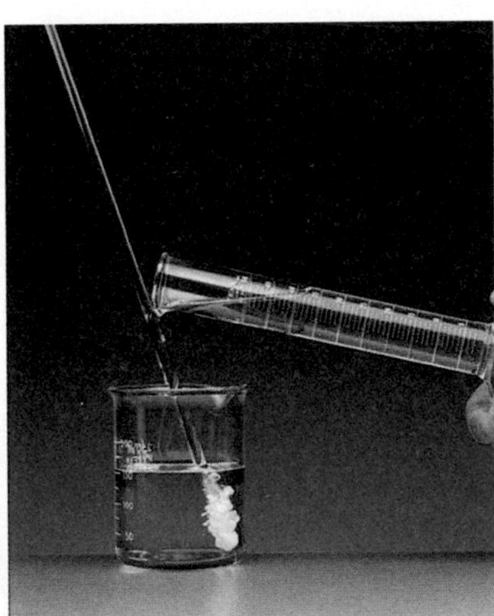

Figure 4.17

Basic steps for gravimetric analysis. (a) A solution containing a known amount of NaCl in a beaker. (b) The precipitation of AgCl upon the addition of $AgNO_3$ solution from a measuring cylinder. In this reaction, $AgNO_3$ is the excess reagent and NaCl is the limiting reagent. (c) The solution containing the AgCl precipitate is filtered through a preweighed sintered-disk crucible, which allows the liquid (but not the precipitate) to pass through. The crucible is then removed from the apparatus, dried in an oven, and weighed again. The difference between this mass and that of the empty crucible gives the mass of the AgCl precipitate.

EXAMPLE 4.8

A 0.7077-g sample of an ionic compound containing chloride ions and an unknown metal is dissolved in water and treated with an excess of $AgNO_3$. If 1.3602 g of AgCl precipitate forms, what is the percent by mass of Cl in the original compound?

Strategy We are asked to calculate the percent by mass of Cl in the unknown sample, which is

$$\%\text{Cl} = \frac{\text{mass of Cl}}{0.7077 \text{ g sample}} \times 100\%$$

The only source of Cl^- ions is the original compound. These chloride ions eventually end up in the AgCl precipitate. Can we calculate the mass of the Cl^- ions if we know the percent by mass of Cl in AgCl?

In general, gravimetric analysis does not establish the identity of the unknown, but it does narrow the possibilities.

Solution The molar massses of Cl and AgCl are 35.45 g and 143.4 g, respectively. Therefore, the percent by mass of Cl in AgCl is given by

$$\begin{aligned}\%\text{Cl} &= \frac{35.45 \text{ g Cl}}{143.4 \text{ g AgCl}} \times 100\% \\ &= 24.72\%\end{aligned}$$

Next, we calculate the mass of Cl in 1.3602 g of AgCl. To do so we convert 24.72 percent to 0.2472 and write

$$\begin{aligned}\text{mass of Cl} &= 0.2472 \times 1.3602 \text{ g} \\ &= 0.3362 \text{ g}\end{aligned}$$

Because the original compound also contained this amount of Cl^- ions, the percent by mass of Cl in the compound is

$$\begin{aligned}\%\text{Cl} &= \frac{0.3362 \text{ g}}{0.7077 \text{ g}} \times 100\% \\ &= 47.51\%\end{aligned}$$

Check Remember that the percent by mass of an element in a compound cannot be greater than 100%. Thus, this answer seems reasonable.

Similar problem: 4.72.

Practice Exercise A sample of 0.3220 g of an ionic compound containing the bromide ion (Br^-) is dissolved in water and treated with an excess of $AgNO_3$. If the mass of the AgBr precipitate that forms is 0.6964 g, what is the percent by mass of Br in the original compound?

Acid-Base Titrations

Quantitative studies of acid-base neutralization reactions are most conveniently carried out using a procedure known as ***titration.*** In a titration experiment, *a solution of accurately known concentration,* called a ***standard solution,*** *is added gradually to another solution of unknown concentration, until the chemical reaction between the two solutions is complete.* If we know the volumes of the standard and unknown solutions used in the titration, along with the concentration of the standard solution, we can calculate the concentration of the unknown solution.

Sodium hydroxide is one of the bases commonly used in the laboratory. However, because it is difficult to obtain solid sodium hydroxide in a pure form, a solution of sodium hydroxide must be *standardized* before it can be used in accurate analytical work. We can standardize the sodium hydroxide solution by titrating it against an acid solution of accurately known concentration. The acid often chosen for this task is a monoprotic acid called potassium hydrogen phthalate (KHP), for which the molecular formula is $KHC_8H_4O_4$. KHP is a white, soluble solid that is commercially available in highly pure form. The reaction between KHP and sodium hydroxide is

Potassium hydrogen phthalate.

$$KHC_8H_4O_4(aq) + NaOH(aq) \longrightarrow KNaC_8H_4O_4(aq) + H_2O(l)$$

The net ionic equation is

$$HC_8H_4O_4^-(aq) + OH^-(aq) \longrightarrow C_8H_4O_4^{2-}(aq) + H_2O(l)$$

The procedure for the titration is shown in Figure 4.18. First, a known amount of KHP is transferred to an Erlenmeyer flask and some distilled water is added to make up a solution. Next, NaOH solution is carefully added to the KHP solution from a buret until we reach the ***equivalence point,*** that is, *the point at which the acid has completely reacted with or been neutralized by the base.* The equivalence point is usually signaled by a sharp change in the color of an indicator in the acid solution. In acid-base titrations, ***indicators*** are *substances that have distinctly different colors in acidic and basic media.* One commonly used indicator is phenolphthalein, which is colorless in acidic and neutral solutions but reddish pink in basic solutions. At the equivalence point, all the KHP present has been neutralized by the added NaOH and the solution is still colorless. However, if we add just one more drop of NaOH solution from the buret, the solution will immediately turn pink because the solution is now basic.

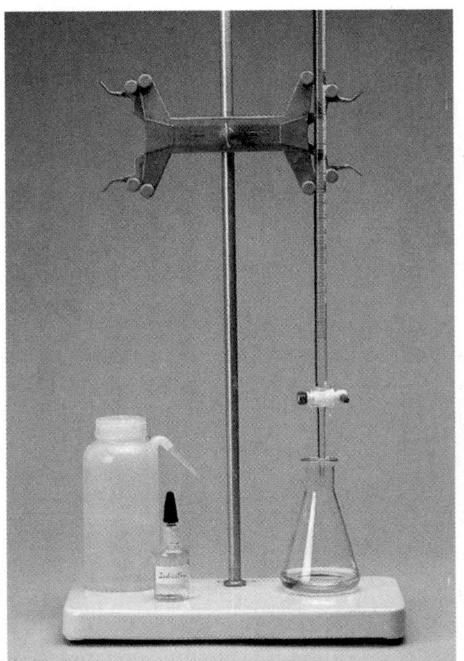

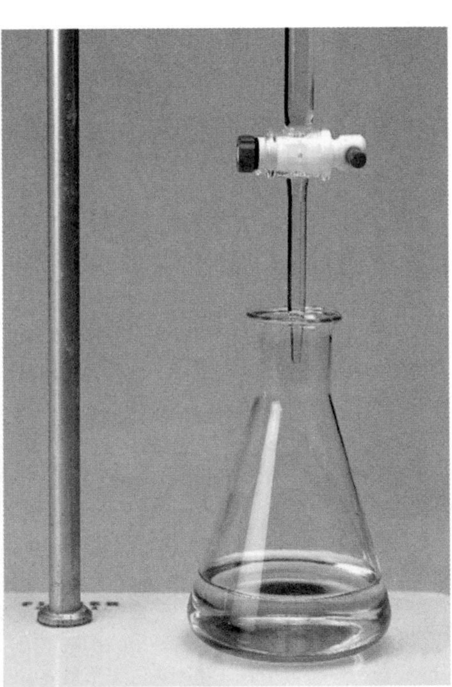

Figure 4.18
(a) Apparatus for acid-base titration. A NaOH solution is added from the burret to a KHP solution in an Erlenmeyer flask. (b) A reddish-pink color appears when the equivalence point is reached. The color here has been intensified for visual display.

EXAMPLE 4.9

In a titration experiment, a student finds that 25.46 mL of a NaOH solution are needed to neutralize 0.6092 g of KHP. What is the concentration (in molarity) of the NaOH solution?

Strategy We want to determine the molarity of the NaOH solution. What is the definition of molarity?

$$\underset{\text{want to calculate}}{\text{molarity of NaOH}} = \frac{\text{mol NaOH} \leftarrow \text{need to find}}{\text{L soln} \leftarrow \text{given}}$$

The volume of NaOH is given in the problem. Therefore, we need to find the number of moles of NaOH to solve for molarity. From the preceding equation for the reaction between KHP and NaOH shown in the text, we see that 1 mole of KHP neutralizes 1 mole of NaOH. How many moles of KHP are contained in 0.6092 g of KHP?

Solution First, we calculate the number of moles of KHP consumed in the titration:

Recall that KHP is $KHC_8H_4O_4$.

$$\text{moles of KHP} = 0.6092 \, \cancel{\text{g KHP}} \times \frac{1 \text{ mol KHP}}{204.2 \, \cancel{\text{g KHP}}}$$
$$= 2.983 \times 10^{-3} \text{ mol KHP}$$

Because 1 mol KHP ≏ 1 mol NaOH, there must be 2.983×10^{-3} mole of NaOH in 25.46 mL of NaOH solution. Finally, we calculate the number of moles of NaOH in 1 L of the solution or the molarity as follows:

$$\text{molarity of NaOH soln} = \frac{2.983 \times 10^{-3} \text{ mol NaOH}}{25.46 \, \cancel{\text{mL soln}}} \times \frac{1000 \, \cancel{\text{mL soln}}}{\text{L soln}}$$
$$= 0.1172 \text{ mol NaOH/L soln} = 0.1172 \, M$$

Similar problems: 4.77, 4.78.

Practice Exercise How many grams of KHP are needed to neutralize 18.64 mL of a 0.1004 *M* NaOH solution?

The neutralization reaction between NaOH and KHP is one of the simplest types of acid-base neutralization known. Suppose, though, that instead of KHP, we wanted to use a diprotic acid such as H_2SO_4 for the titration. The reaction is represented by

$$2\text{NaOH}(aq) + \text{H}_2\text{SO}_4(aq) \longrightarrow \text{Na}_2\text{SO}_4(aq) + 2\text{H}_2\text{O}(l)$$

Because 2 mol NaOH ≏ 1 mol H_2SO_4, we need twice as much NaOH to react completely with a H_2SO_4 solution of the *same* molar concentration and volume as a monoprotic acid like HCl. On the other hand, we would need twice the amount of HCl to neutralize a $Ba(OH)_2$ solution compared to a NaOH solution having the same concentration and volume because 1 mole of $Ba(OH)_2$ yields 2 moles of OH^- ions:

$$2\text{HCl}(aq) + \text{Ba(OH)}_2(aq) \longrightarrow \text{BaCl}_2(aq) + 2\text{H}_2\text{O}(l)$$

In calculations involving acid-base titrations, regardless of the acid or base that takes place in the reaction, keep in mind that the total number of moles of H^+ ions that have reacted at the equivalence point must be equal to the total number of moles of OH^- ions that have reacted.

EXAMPLE 4.10

How many milliliters (mL) of a 0.836 M NaOH solution are needed to neutralize 25.0 mL of a 0.355 M H_2SO_4 solution?

Strategy We want to calculate the volume of the NaOH solution. From the definition of molarity [see Equation (4.1)], we write

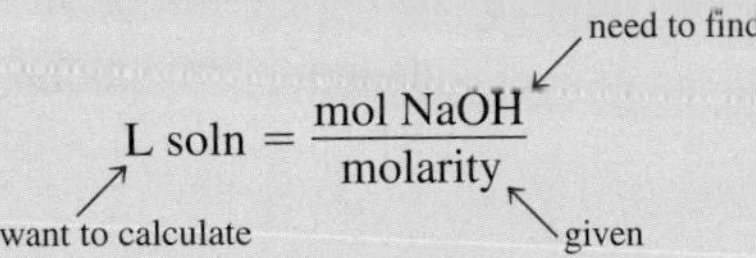

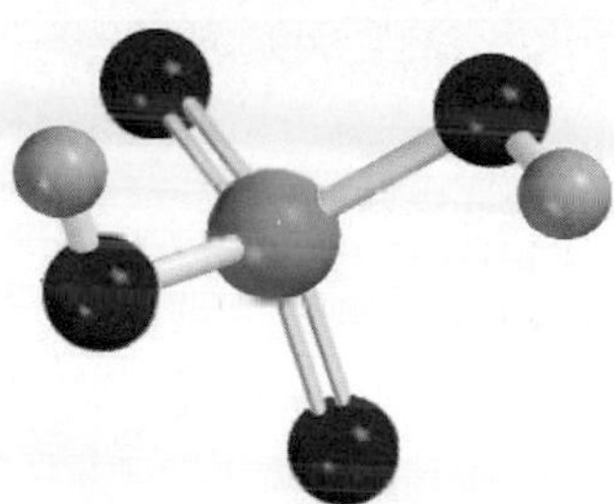

H_2SO_4 has two ionizable protons.

From the equation for the neutralization reaction just shown, we see that 1 mole of H_2SO_4 neutralizes 2 moles of NaOH. How many moles of H_2SO_4 are contained in 25.0 mL of a 0.355 M H_2SO_4 solution? How many moles of NaOH would this quantity of H_2SO_4 neutralize?

Solution First, we calculate the number of moles of H_2SO_4 in a 25.0 mL solution:

$$\text{moles of } H_2SO_4 = \frac{0.355 \text{ mol } H_2SO_4}{1000 \text{ mL soln}} \times 25.0 \text{ mL soln}$$

$$= 8.88 \times 10^{-3} \text{ mol } H_2SO_4$$

From the stoichiometry we see that 1 mol $H_2SO_4 \simeq 2$ mol NaOH. Therefore, the number of moles of NaOH reacted must be $2 \times 8.88 \times 10^{-3}$ mole, or 1.78×10^{-2} mole. From the definition of molarity [see Equation (4.1)], we have

$$\text{liters of soln} = \frac{\text{moles of solute}}{\text{molarity}}$$

or

$$\text{volume of NaOH} = \frac{1.78 \times 10^{-2} \text{ mol NaOH}}{0.836 \text{ mol/L soln}}$$

$$= 0.0213 \text{ L or } 21.3 \text{ mL}$$

Similar problem: 4.79(b), (c).

Practice Exercise How many milliliters of a 1.28 M H_2SO_4 solution are needed to neutralize 60.2 mL of a 0.427 M KOH solution?

Review of Concepts

A NaOH solution is initially mixed with an acid solution shown in (a). Which of the diagrams shown in (b)–(d) corresponds to one of the following acids: HCl, H_2SO_4, H_3PO_4? Color codes: Blue spheres (OH^- ions); red spheres (acid molecules); green spheres (anions of the acid). Assume all the acid-base neutralization reactions go to completion.

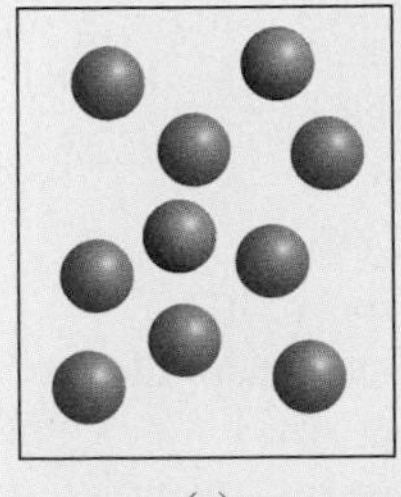

(a)

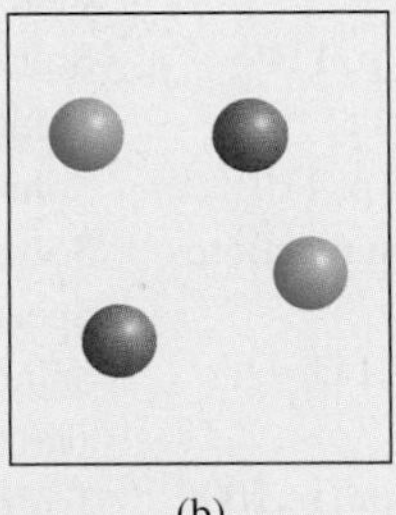

(b)

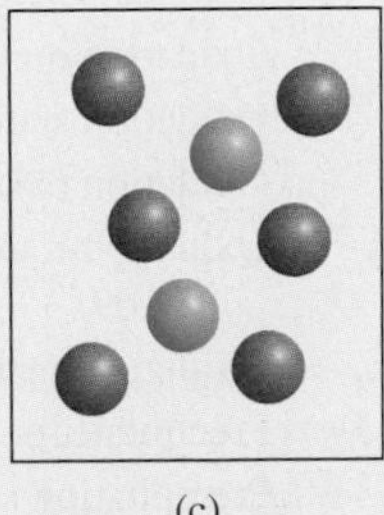

(c)

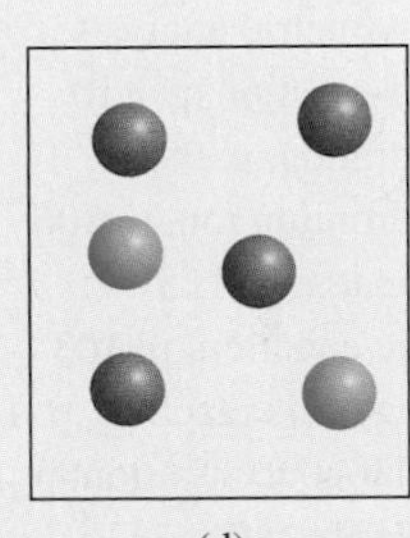

(d)

Key Equation

$$\text{molarity } (M) = \frac{\text{moles of solute}}{\text{liters of soln}} \quad (4.1)$$ Definition of molarity.

$$M = \frac{n}{V} \quad (4.2)$$ Definition of molarity.

$$M_iV_i = M_fV_f \quad (4.3)$$ Dilution of solution.

Summary of Facts and Concepts

1. Aqueous solutions are electrically conducting if the solutes are electrolytes. If the solutes are nonelectrolytes, the solutions do not conduct electricity.
2. Three major categories of chemical reactions that take place in aqueous solution are precipitation reactions, acid-base reactions, and oxidation-reduction reactions.
3. From general rules about solubilities of ionic compounds, we can predict whether a precipitate will form in a reaction.
4. Arrhenius acids ionize in water to give H^+ ions, and Arrhenius bases ionize in water to give OH^- ions. Brønsted acids donate protons, and Brønsted bases accept protons. The reaction of an acid and a base is called neutralization.
5. In redox reactions, oxidation and reduction always occur simultaneously. Oxidation is characterized by the loss of electrons, reduction by the gain of electrons. Oxidation numbers help us keep track of charge distribution and are assigned to all atoms in a compound or ion according to specific rules. Oxidation can be defined as an increase in oxidation number; reduction can be defined as a decrease in oxidation number.
6. The concentration of a solution is the amount of solute present in a given amount of solution. Molarity expresses concentration as the number of moles of solute in 1 L of solution. Adding a solvent to a solution, a process known as dilution, decreases the concentration (molarity) of the solution without changing the total number of moles of solute present in the solution.
7. Gravimetric analysis is a technique for determining the identity of a compound and/or the concentration of a solution by measuring mass. Gravimetric experiments often involve precipitation reactions.
8. In acid-base titration, a solution of known concentration (say, a base) is added gradually to a solution of unknown concentration (say, an acid) with the goal of determining the unknown concentration. The point at which the reaction in the titration is complete is called the equivalence point.

Key Words

Questions and Problems

Properties of Aqueous Solutions

Review Questions

4.1 Define solute, solvent, and solution by describing the process of dissolving a solid in a liquid.

4.2 What is the difference between a nonelectrolyte and an electrolyte? Between a weak electrolyte and a strong electrolyte?

4.3 Describe hydration. What properties of water enable its molecules to interact with ions in solution?

4.4 What is the difference between the following symbols in chemical equations: $\longrightarrow$ and $\rightleftharpoons$?

4.5 Water is an extremely weak electrolyte and therefore cannot conduct electricity. Why are we often cautioned not to operate electrical appliances when our hands are wet?

4.6 Lithium fluoride (LiF) is a strong electrolyte. What species are present in LiF(*aq*)?

Problems

4.7 The aqueous solutions of three compounds are shown in the diagram. Identify each compound as a nonelectrolyte, a weak electrolyte, and a strong electrolyte.

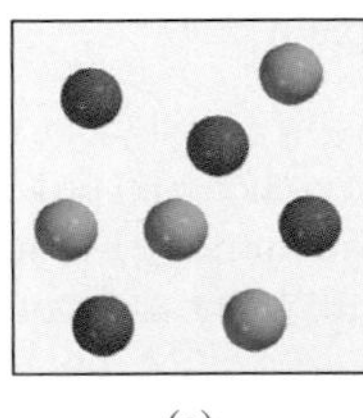
(a)

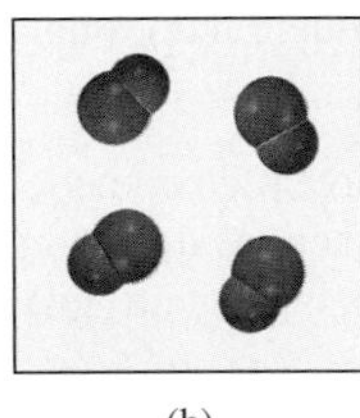
(b)

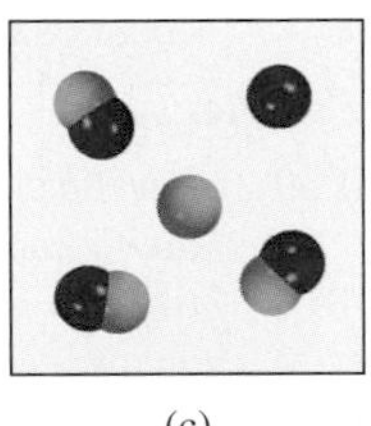
(c)

4.8 Which of the following diagrams best represents the hydration of NaCl when dissolved in water? The Cl^- ion is larger in size than the Na^+ ion.

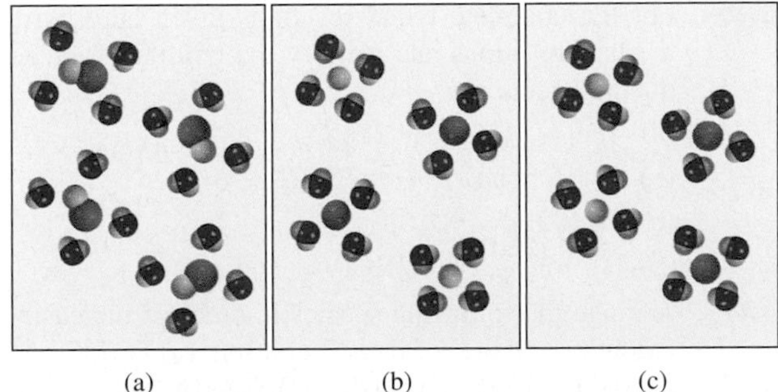
(a) (b) (c)

4.9 Identify each of the following substances as a strong electrolyte, weak electrolyte, or nonelectrolyte: (a) H_2O, (b) KCl, (c) HNO_3, (d) CH_3COOH, (e) $C_{12}H_{22}O_{11}$.

4.10 Identify each of the following substances as a strong electrolyte, weak electrolyte, or nonelectrolyte: (a) $Ba(NO_3)_2$, (b) Ne, (c) NH_3, (d) NaOH.

4.11 The passage of electricity through an electrolyte solution is caused by the movement of (a) electrons only, (b) cations only, (c) anions only, (d) both cations and anions.

4.12 Predict and explain which of the following systems are electrically conducting: (a) solid NaCl, (b) molten NaCl, (c) an aqueous solution of NaCl.

4.13 You are given a water-soluble compound X. Describe how you would determine whether it is an electrolyte or a nonelectrolyte. If it is an electrolyte, how would you determine whether it is strong or weak?

4.14 Explain why a solution of HCl in benzene does not conduct electricity but in water it does.

Precipitation Reactions

Review Questions

4.15 What is the difference between an ionic equation and a molecular equation?

4.16 What is the advantage of writing net ionic equations?

Problems

4.17 Two aqueous solutions of $AgNO_3$ and NaCl are mixed. Which of the following diagrams best represents the mixture? (Ag^+ = gray; Cl^- = orange; Na^+ = green; NO_3^- = blue) (For simplicity, water molecules are not shown.)

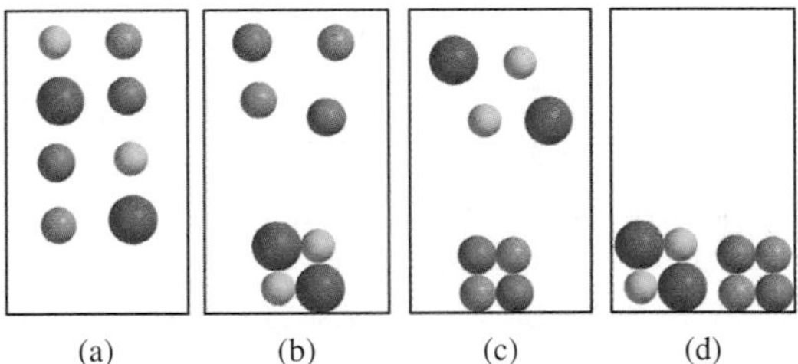
(a) (b) (c) (d)

4.18 Two aqueous solutions of KOH and $MgCl_2$ are mixed. Which of the following diagrams best represents the mixture? (K^+ = purple; OH^- = red; Mg^{2+} = blue; Cl^- = orange) (For simplicity, water molecules are not shown.)

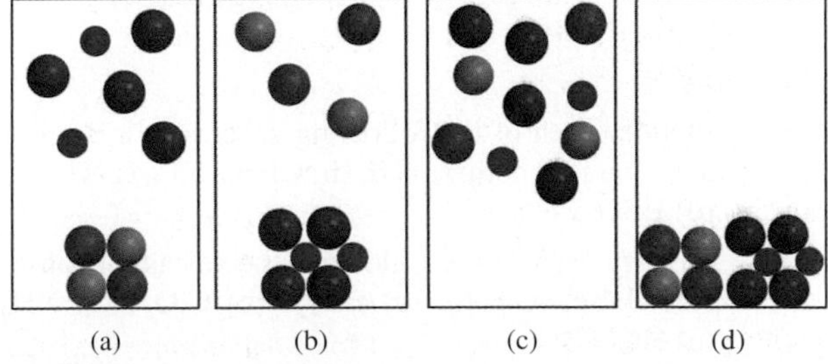
(a) (b) (c) (d)

4.19 Characterize the following compounds as soluble or insoluble in water: (a) $Ca_3(PO_4)_2$, (b) $Mn(OH)_2$, (c) $AgClO_3$, (d) K_2S.

4.20 Characterize the following compounds as soluble or insoluble in water: (a) $CaCO_3$, (b) $ZnSO_4$, (c) $Hg(NO_3)_2$, (d) $HgSO_4$, (e) NH_4ClO_4.

4.21 Write ionic and net ionic equations for the following reactions:

(a) $AgNO_3(aq) + Na_2SO_4(aq) \longrightarrow$

(b) $BaCl_2(aq) + ZnSO_4(aq) \longrightarrow$

(c) $(NH_4)_2CO_3(aq) + CaCl_2(aq) \longrightarrow$

4.22 Write ionic and net ionic equations for the following reactions:

(a) $Na_2S(aq) + ZnCl_2(aq) \longrightarrow$

(b) $K_3PO_4(aq) + 3Sr(NO_3)_2(aq) \longrightarrow$

(c) $Mg(NO_3)_2(aq) + 2NaOH(aq) \longrightarrow$

4.23 Which of the following processes will likely result in a precipitation reaction? (a) Mixing a $NaNO_3$ solution with a $CuSO_4$ solution. (b) Mixing a $BaCl_2$ solution with a K_2SO_4 solution. Write a net ionic equation for the precipitation reaction.

4.24 With reference to Table 4.2, suggest one method by which you might separate (a) K^+ from Ag^+, (b) Ba^{2+} from Pb^{2+}, (c) NH_4^+ from Ca^{2+}, (d) Ba^{2+} from Cu^{2+}. All cations are assumed to be in aqueous solution, and the common anion is the nitrate ion.

Acid-Base Reactions

Review Questions

4.25 List the general properties of acids and bases.

4.26 Give Arrhenius's and Brønsted's definitions of an acid and a base. Why are Brønsted's definitions more useful in describing acid-base properties?

4.27 Give an example of a monoprotic acid, a diprotic acid, and a triprotic acid.

4.28 What are the characteristics of an acid-base neutralization reaction?

4.29 What factors qualify a compound as a salt? Specify which of the following compounds are salts: CH_4, NaF, NaOH, CaO, $BaSO_4$, HNO_3, NH_3, KBr?

4.30 Identify the following as a weak or strong acid or base: (a) NH_3, (b) H_3PO_4, (c) LiOH, (d) HCOOH (formic acid), (e) H_2SO_4, (f) HF, (g) $Ba(OH)_2$.

Problems

4.31 Identify each of the following species as a Brønsted acid, base, or both: (a) HI, (b) CH_3COO^-, (c) $H_2PO_4^-$, (d) HSO_4^-.

4.32 Identify each of the following species as a Brønsted acid, base, or both: (a) PO_4^{3-}, (b) ClO_2^-, (c) NH_4^+, (d) HCO_3^-.

4.33 Balance the following equations and write the corresponding ionic and net ionic equations (if appropriate):

(a) $HBr(aq) + NH_3(aq) \longrightarrow$

(b) $Ba(OH)_2(aq) + H_3PO_4(aq) \longrightarrow$

(c) $HClO_4(aq) + Mg(OH)_2(s) \longrightarrow$

4.34 Balance the following equations and write the corresponding ionic and net ionic equations (if appropriate):

(a) $CH_3COOH(aq) + KOH(aq) \longrightarrow$

(b) $H_2CO_3(aq) + NaOH(aq) \longrightarrow$

(c) $HNO_3(aq) + Ba(OH)_2(aq) \longrightarrow$

Oxidation-Reduction Reactions

Review Questions

4.35 Define the following terms: half-reaction, oxidation reaction, reduction reaction, reducing agent, oxidizing agent, redox reaction.

4.36 What is an oxidation number? How is it used to identify redox reactions? Explain why, except for ionic compounds, oxidation number does not have any physical significance.

4.37 (a) Without referring to Figure 4.10, give the oxidation numbers of the alkali and alkaline earth metals in their compounds. (b) Give the highest oxidation numbers that the Groups 3A–7A elements can have.

4.38 Is it possible to have a reaction in which oxidation occurs and reduction does not? Explain.

Problems

4.39 For the complete redox reactions given here, (i) break down each reaction into its half-reactions; (ii) identify the oxidizing agent; (iii) identify the reducing agent.

(a) $2Sr + O_2 \longrightarrow 2SrO$

(b) $2Li + H_2 \longrightarrow 2LiH$

(c) $2Cs + Br_2 \longrightarrow 2CsBr$

(d) $3Mg + N_2 \longrightarrow Mg_3N_2$

4.40 For the complete redox reactions given here, write the half-reactions and identify the oxidizing and reducing agents:

(a) $4Fe + 3O_2 \longrightarrow 2Fe_2O_3$

(b) $Cl_2 + 2NaBr \longrightarrow 2NaCl + Br_2$

(c) $Si + 2F_2 \longrightarrow SiF_4$

(d) $H_2 + Cl_2 \longrightarrow 2HCl$

4.41 Arrange the following species in order of increasing oxidation number of the sulfur atom: (a) H_2S, (b) S_8, (c) H_2SO_4, (d) S^{2-}, (e) HS^-, (f) SO_2, (g) SO_3.

4.42 Phosphorus forms many oxoacids. Indicate the oxidation number of phosphorus in each of the following acids: (a) HPO_3, (b) H_3PO_2, (c) H_3PO_3, (d) H_3PO_4, (e) $H_4P_2O_7$, (f) $H_5P_3O_{10}$.

4.43 Give the oxidation number of the underlined atoms in the following molecules and ions: (a) $\underline{Cl}F$, (b) $\underline{I}F_7$,

(c) $\underline{C}H_4$, (d) $\underline{C}_2H_2$, (e) $\underline{C}_2H_4$, (f) $K_2\underline{Cr}O_4$, (g) $K_2\underline{Cr}_2O_7$, (h) $K\underline{Mn}O_4$, (i) $NaH\underline{C}O_3$, (j) $\underline{Li}_2$, (k) $Na\underline{I}O_3$, (l) $K\underline{O}_2$, (m) $\underline{P}F_6^-$, (n) $K\underline{Au}Cl_4$.

4.44 Give the oxidation number for the following species: H_2, Se_8, P_4, O, U, As_4, B_{12}.

4.45 Give oxidation numbers for the underlined atoms in the following molecules and ions: (a) $\underline{Cs}_2O$, (b) $Ca\underline{I}_2$, (c) $\underline{Al}_2O_3$, (d) $H_3\underline{As}O_3$, (e) $\underline{Ti}O_2$, (f) $\underline{Mo}O_4^{2-}$, (g) $\underline{Pt}Cl_4^{2-}$, (h) $\underline{Pt}Cl_6^{2-}$, (i) $\underline{Sn}F_2$, (j) $\underline{Cl}F_3$, (k) $\underline{Sb}F_6^-$.

4.46 Give the oxidation numbers of the underlined atoms in the following molecules and ions: (a) $Mg_3\underline{N}_2$, (b) $Cs\underline{O}_2$, (c) $Ca\underline{C}_2$, (d) $\underline{C}O_3^{2-}$, (e) $\underline{C}_2O_4^{2-}$, (f) $Zn\underline{O}_2^{2-}$, (g) $Na\underline{B}H_4$, (h) $\underline{W}O_4^{2-}$.

4.47 Nitric acid is a strong oxidizing agent. State which of the following species is *least* likely to be produced when nitric acid reacts with a strong reducing agent such as zinc metal, and explain why: N_2O, NO, NO_2, N_2O_4, N_2O_5, NH_4^+.

4.48 Which of the following metals can react with water? (a) Au, (b) Li, (c) Hg, (d) Ca, (e) Pt.

4.49 On the basis of oxidation number considerations, one of the following oxides would not react with molecular oxygen: NO, N_2O, SO_2, SO_3, P_4O_6. Which one is it? Why?

4.50 Predict the outcome of the reactions represented by the following equations by using the activity series, and balance the equations.

(a) $Cu(s) + HCl(aq) \longrightarrow$

(b) $I_2(s) + NaBr(aq) \longrightarrow$

(c) $Mg(s) + CuSO_4(aq) \longrightarrow$

(d) $Cl_2(g) + KBr(aq) \longrightarrow$

Concentration of Solutions

Review Questions

4.51 Write the equation for calculating molarity. Why is molarity a convenient concentration unit in chemistry?

4.52 Describe the steps involved in preparing a solution of known molar concentration using a volumetric flask.

Problems

4.53 Calculate the mass of KI in grams required to prepare 5.00×10^2 mL of a 2.80 *M* solution.

4.54 Describe how you would prepare 250 mL of a 0.707 *M* $NaNO_3$ solution.

4.55 How many moles of $MgCl_2$ are present in 60.0 mL of 0.100 *M* $MgCl_2$ solution?

4.56 How many grams of KOH are present in 35.0 mL of a 5.50 *M* solution?

4.57 Calculate the molarity of each of the following solutions: (a) 29.0 g of ethanol (C_2H_5OH) in 545 mL of solution, (b) 15.4 g of sucrose ($C_{12}H_{22}O_{11}$) in 74.0 mL of solution, (c) 9.00 g of sodium chloride (NaCl) in 86.4 mL of solution.

4.58 Calculate the molarity of each of the following solutions: (a) 6.57 g of methanol (CH_3OH) in 1.50×10^2 mL of solution, (b) 10.4 g of calcium chloride ($CaCl_2$) in 2.20×10^2 mL of solution, (c) 7.82 g of naphthalene ($C_{10}H_8$) in 85.2 mL of benzene solution.

4.59 Calculate the volume in mL of a solution required to provide the following: (a) 2.14 g of sodium chloride from a 0.270 *M* solution, (b) 4.30 g of ethanol from a 1.50 *M* solution, (c) 0.85 g of acetic acid (CH_3COOH) from a 0.30 *M* solution.

4.60 Determine how many grams of each of the following solutes would be needed to make 2.50×10^2 mL of a 0.100 *M* solution: (a) cesium iodide (CsI), (b) sulfuric acid (H_2SO_4), (c) sodium carbonate (Na_2CO_3), (d) potassium dichromate ($K_2Cr_2O_7$), (e) potassium permanganate ($KMnO_4$).

Dilution of Solutions

Review Questions

4.61 Describe the basic steps involved in diluting a solution of known concentration.

4.62 Write the equation that enables us to calculate the concentration of a diluted solution. Give units for all the terms.

Problems

4.63 Describe how to prepare 1.00 L of 0.646 *M* HCl solution, starting with a 2.00 *M* HCl solution.

4.64 Water is added to 25.0 mL of a 0.866 *M* KNO_3 solution until the volume of the solution is exactly 500 mL. What is the concentration of the final solution?

4.65 How would you prepare 60.0 mL of 0.200 *M* HNO_3 from a stock solution of 4.00 *M* HNO_3?

4.66 You have 505 mL of a 0.125 *M* HCl solution and you want to dilute it to exactly 0.100 *M*. How much water should you add? Assume volumes are additive.

4.67 A 35.2-mL, 1.66 *M* $KMnO_4$ solution is mixed with 16.7 mL of 0.892 *M* $KMnO_4$ solution. Calculate the concentration of the final solution.

4.68 A 46.2-mL, 0.568 *M* calcium nitrate [$Ca(NO_3)_2$] solution is mixed with 80.5 mL of 1.396 *M* calcium nitrate solution. Calculate the concentration of the final solution.

Gravimetric Analysis

Review Questions

4.69 Describe the basic steps involved in gravimetric analysis. How does this procedure help us determine the identity of a compound or the purity of a compound if its formula is known?

4.70 Distilled water must be used in the gravimetric analysis of chlorides. Why?

Problems

4.71 If 30.0 mL of 0.150 *M* $CaCl_2$ is added to 15.0 mL of 0.100 *M* $AgNO_3$, what is the mass in grams of AgCl precipitate?

4.72 A sample of 0.6760 g of an unknown compound containing barium ions (Ba^{2+}) is dissolved in water and treated with an excess of Na_2SO_4. If the mass of the $BaSO_4$ precipitate formed is 0.4105 g, what is the percent by mass of Ba in the original unknown compound?

4.73 How many grams of NaCl are required to precipitate most of the Ag^+ ions from 2.50×10^2 mL of 0.0113 *M* $AgNO_3$ solution? Write the net ionic equation for the reaction.

4.74 The concentration of Cu^{2+} ions in the water (which also contains sulfate ions) discharged from a certain industrial plant is determined by adding excess sodium sulfide (Na_2S) solution to 0.800 L of the water. The molecular equation is

$$Na_2S(aq) + CuSO_4(aq) \longrightarrow Na_2SO_4(aq) + CuS(s)$$

Write the net ionic equation and calculate the molar concentration of Cu^{2+} in the water sample if 0.0177 g of solid CuS is formed.

Acid-Base Titrations

Review Questions

4.75 Describe the basic steps involved in an acid-base titration. Why is this technique of great practical value?

4.76 How does an acid-base indicator work?

Problems

4.77 A quantity of 18.68 mL of a KOH solution is needed to neutralize 0.4218 g of KHP. What is the concentration (in molarity) of the KOH solution?

4.78 Calculate the concentration (in molarity) of a NaOH solution if 25.0 mL of the solution are needed to neutralize 17.4 mL of a 0.312 *M* HCl solution.

4.79 Calculate the volume in mL of a 1.420 *M* NaOH solution required to titrate the following solutions:
(a) 25.00 mL of a 2.430 *M* HCl solution
(b) 25.00 mL of a 4.500 *M* H_2SO_4 solution
(c) 25.00 mL of a 1.500 *M* H_3PO_4 solution

4.80 What volume of a 0.500 *M* HCl solution is needed to neutralize each of the following:
(a) 10.0 mL of a 0.300 *M* NaOH solution
(b) 10.0 mL of a 0.200 *M* $Ba(OH)_2$ solution

Additional Problems

4.81 Classify these reactions according to the types discussed in the chapter:
(a) $Cl_2 + 2OH^- \longrightarrow Cl^- + ClO^- + H_2O$
(b) $Ca^{2+} + CO_3^{2-} \longrightarrow CaCO_3$
(c) $NH_3 + H^+ \longrightarrow NH_4^+$
(d) $2CCl_4 + CrO_4^{2-} \longrightarrow 2COCl_2 + CrO_2Cl_2 + 2Cl^-$
(e) $Ca + F_2 \longrightarrow CaF_2$
(f) $2Li + H_2 \longrightarrow 2LiH$
(g) $Ba(NO_3)_2 + Na_2SO_4 \longrightarrow 2NaNO_3 + BaSO_4$
(h) $CuO + H_2 \longrightarrow Cu + H_2O$
(i) $Zn + 2HCl \longrightarrow ZnCl_2 + H_2$
(j) $2FeCl_2 + Cl_2 \longrightarrow 2FeCl_3$

4.82 Using the apparatus shown in Figure 4.1, a student found that the lightbulb was brightly lit when the electrodes were immersed in a sulfuric acid solution. However, after the addition of a certain amount of barium hydroxide [$Ba(OH)_2$] solution, the light began to dim even though $Ba(OH)_2$ is also a strong electrolyte. Explain.

4.83 Someone gave you a colorless liquid. Describe three chemical tests you would perform on the liquid to show that it is water.

4.84 You are given two colorless solutions, one containing NaCl and the other sucrose ($C_{12}H_{22}O_{11}$). Suggest a chemical and a physical test that would distinguish between these two solutions.

4.85 Chlorine (Cl_2) is used to purify drinking water. Too much chlorine is harmful to humans. The excess chlorine is often removed by treatment with sulfur dioxide (SO_2). Balance the following equation that represents this procedure:

$$Cl_2 + SO_2 + H_2O \longrightarrow Cl^- + SO_4^{2-} + H^+$$

4.86 Before aluminum was obtained by electrolytic reduction from its ore (Al_2O_3), the metal was produced by chemical reduction of $AlCl_3$. Which metals would you use to reduce Al^{3+} to Al?

4.87 Oxygen (O_2) and carbon dioxide (CO_2) are colorless and odorless gases. Suggest two chemical tests that would enable you to distinguish between them.

4.88 Based on oxidation number, explain why carbon monoxide (CO) is flammable but carbon dioxide (CO_2) is not.

4.89 Which of these aqueous solutions would you expect to be the best conductor of electricity at 25°C? Explain your answer.
(a) 0.20 *M* NaCl
(b) 0.60 *M* CH_3COOH
(c) 0.25 *M* HCl
(d) 0.20 *M* $Mg(NO_3)_2$

4.90 A 5.00×10^2-mL sample of 2.00 *M* HCl solution is treated with 4.47 g of magnesium. Calculate the concentration of the acid solution after all the metal has reacted. Assume that the volume remains unchanged.

4.91 Calculate the volume (in liters) of a 0.156 *M* $CuSO_4$ solution that would react with 7.89 g of zinc.

4.92 Sodium carbonate (Na_2CO_3) can be obtained in very pure form and can be used to standardize acid solutions. What is the molarity of an HCl solution if 28.3 mL of the solution is required to react with 0.256 g of Na_2CO_3?

4.93 A 3.664-g sample of a monoprotic acid was dissolved in water and required 20.27 mL of a 0.1578 *M* NaOH solution for neutralization. Calculate the molar mass of the acid.

4.94 Acetic acid (CH_3COOH) is an important ingredient of vinegar. A sample of 50.0 mL of a commercial vinegar is titrated against a 1.00 *M* NaOH solution. What is the concentration (in *M*) of acetic acid present in the vinegar if 5.75 mL of the base were required for the titration?

4.95 Calculate the mass of precipitate formed when 2.27 L of 0.0820 *M* $Ba(OH)_2$ are mixed with 3.06 L of 0.0664 *M* Na_2SO_4.

4.96 Milk of magnesia is an aqueous suspension of magnesium hydroxide [$Mg(OH)_2$] used to treat acid indigestion. Calculate the volume of a 0.035 *M* HCl solution (a typical acid concentration in an upset stomach) needed to react with two spoonfuls of milk of magnesia [approximately 10.0 mL at 0.080 g $Mg(OH)_2$/mL].

4.97 A 1.00-g sample of a metal X (that is known to form X^{2+} ions) was added to a 0.100 L of 0.500 *M* H_2SO_4. After all the metal had reacted, the remaining acid required 0.0334 L of 0.500 *M* NaOH solution for neutralization. Calculate the molar mass of the metal and identify the element.

4.98 A 60.0-mL 0.513 *M* glucose ($C_6H_{12}O_6$) solution is mixed with 120.0 mL of 2.33 *M* glucose solution. What is the concentration of the final solution? Assume the volumes are additive.

4.99 You are given a soluble compound of unknown molecular formula. (a) Describe three tests that would show that the compound is an acid. (b) Once you have established that the compound is an acid, describe how you would determine its molar mass using an NaOH solution of known concentration. (Assume the acid is monoprotic.) (c) How would you find out whether the acid is weak or strong? You are provided with a sample of NaCl and an apparatus like that shown in Figure 4.1 for comparison.

4.100 Someone spilled concentrated sulfuric acid on the floor of a chemistry laboratory. To neutralize the acid, would it be preferable to pour concentrated sodium hydroxide solution or spray solid sodium bicarbonate over the acid? Explain your choice and the chemical basis for the action.

4.101 These are common household compounds: table salt (NaCl), table sugar (sucrose), vinegar (contains acetic acid), baking soda ($NaHCO_3$), washing soda ($Na_2CO_3 \cdot 10H_2O$), boric acid (H_3BO_3, used in eyewash), epsom salt ($MgSO_4 \cdot 7H_2O$), sodium hydroxide (used in drain openers), ammonia, milk of magnesia [$Mg(OH)_2$], and calcium carbonate. Based on what you have learned in this chapter, describe test(s) that would enable you to identify each of these compounds.

4.102 A 0.8870-g sample of a mixture of NaCl and KCl is dissolved in water, and the solution is then treated with an excess of $AgNO_3$ to yield 1.913 g of AgCl. Calculate the percent by mass of each compound in the mixture.

4.103 Phosphoric acid (H_3PO_4) is an important industrial chemical used in fertilizers, detergents, and in the food industry. It is produced by two different methods. In the *electric furnace method* elemental phosphorus (P_4) is burned in air to form P_4O_{10}, which is then reacted with water to give H_3PO_4. In the *wet process* the mineral phosphate rock [$Ca_5(PO_4)_3F$] is reacted with sulfuric acid to give H_3PO_4 (and HF and $CaSO_4$). Write equations for these processes and classify each step as precipitation, acid-base, or redox reaction.

4.104 Give a chemical explanation for each of these: (a) When calcium metal is added to a sulfuric acid solution, hydrogen gas is generated. After a few minutes, the reaction slows down and eventually stops even though none of the reactants is used up. Explain. (b) In the activity series aluminum is above hydrogen, yet the metal appears to be unreactive toward steam and hydrochloric acid. Why? (c) Sodium and potassium lie above copper in the activity series. Explain why Cu^{2+} ions in a $CuSO_4$ solution are not converted to metallic copper upon the addition of these metals. (d) A metal M reacts slowly with steam. There is no visible change when it is placed in a pale green iron(II) sulfate solution. Where should we place M in the activity series?

4.105 A number of metals are involved in redox reactions in biological systems in which the oxidation state of the metals changes. Which of these metals are most likely to take part in such reactions: Na, K, Mg, Ca, Mn, Fe, Co, Cu, Zn? Explain.

4.106 The recommended procedure for preparing a very dilute solution is not to weigh out a very small mass or measure a very small volume of a stock solution. Instead, it is done by a series of dilutions. A sample of 0.8214 g of $KMnO_4$ was dissolved in water and made up to the volume in a 500-mL volumetric flask. A 2.000-mL sample of this solution was transferred to a 1000-mL volumetric flask and diluted to the mark with water. Next, 10.00 mL of the diluted solution were transferred to a 250-mL flask and diluted to the mark with water. (a) Calculate the concentration (in molarity) of the final solution. (b) Calculate the mass of $KMnO_4$ needed to directly prepare the final solution.

4.107 A 325-mL sample of solution contains 25.3 g of $CaCl_2$. (a) Calculate the molar concentration of Cl^-

in this solution. (b) How many grams of Cl^- are in 0.100 L of this solution?

4.108 Acetylsalicylic acid ($C_9H_8O_4$) is a monoprotic acid commonly known as "aspirin." A typical aspirin tablet, however, contains only a small amount of the acid. In an experiment to determine its composition, an aspirin tablet was crushed and dissolved in water. It took 12.25 mL of 0.1466 *M* NaOH to neutralize the solution. Calculate the number of grains of aspirin in the tablet. (One grain = 0.0648 g.)

4.109 This "cycle of copper" experiment is performed in some general chemistry laboratories. The series of reactions starts with copper and ends with metallic copper. The steps are: (1) A piece of copper wire of known mass is allowed to react with concentrated nitric acid [the products are copper(II) nitrate, nitrogen dioxide, and water]. (2) The copper(II) nitrate is treated with a sodium hydroxide solution to form copper(II) hydroxide precipitate. (3) On heating, copper(II) hydroxide decomposes to yield copper(II) oxide. (4) The copper(II) oxide is reacted with concentrated sulfuric acid to yield copper(II) sulfate. (5) Copper(II) sulfate is treated with an excess of zinc metal to form metallic copper. (6) The remaining zinc metal is removed by treatment with hydrochloric acid, and metallic copper is filtered, dried, and weighed. (a) Write a balanced equation for each step and classify the reactions. (b) Assuming that a student started with 65.6 g of copper, calculate the theoretical yield at each step. (c) Considering the nature of the steps, comment on why it is possible to recover most of the copper used at the start.

4.110 Ammonium nitrate (NH_4NO_3) is one of the most important nitrogen-containing fertilizers. Its purity can be analyzed by titrating a solution of NH_4NO_3 with a standard NaOH solution. In one experiment a 0.2041-g sample of industrially prepared NH_4NO_3 required 24.42 mL of 0.1023 *M* NaOH for neutralization. (a) Write a net ionic equation for the reaction. (b) What is the percent purity of the sample?

4.111 Hydrogen halides (HF, HCl, HBr, HI) are highly reactive compounds that have many industrial and laboratory uses. (a) In the laboratory, HF and HCl can be generated by reacting CaF_2 and NaCl with concentrated sulfuric acid. Write appropriate equations for the reactions. (*Hint:* These are not redox reactions.) (b) Why is it that HBr and HI cannot be prepared similarly, that is, by reacting NaBr and NaI with concentrated sulfuric acid? (*Hint:* H_2SO_4 is a stronger oxidizing agent than both Br_2 and I_2.) (c) HBr can be prepared by reacting phosphorus tribromide (PBr_3) with water. Write an equation for this reaction.

4.112 Referring to Figure 4.14, explain why one must first dissolve the solid completely before making up the solution to the correct volume.

Special Problems

4.113 Magnesium is a valuable, lightweight metal. It is used as a structural metal and in alloys, in batteries, and in chemical synthesis. Although magnesium is plentiful in Earth's crust, it is cheaper to "mine" the metal from seawater. Magnesium forms the second most abundant cation in the sea (after sodium); there are about 1.3 g of magnesium in 1 kg of seawater. The method of obtaining magnesium from seawater employs all three types of reactions discussed in this chapter: precipitation, acid-base, and redox reactions. In the first stage in the recovery of magnesium, limestone ($CaCO_3$) is heated at high temperatures to produce quicklime, or calcium oxide (CaO):

$$CaCO_3(s) \longrightarrow CaO(s) + CO_2(g)$$

When calcium oxide is treated with seawater, it forms calcium hydroxide [$Ca(OH)_2$], which is slightly soluble and ionizes to give Ca^{2+} and OH^- ions:

$$CaO(s) + H_2O(l) \longrightarrow Ca^{2+}(aq) + 2OH^-(aq)$$

The surplus hydroxide ions cause the much less soluble magnesium hydroxide to precipitate:

$$Mg^{2+}(aq) + 2OH^-(aq) \longrightarrow Mg(OH)_2(s)$$

The solid magnesium hydroxide is filtered and reacted with hydrochloric acid to form magnesium chloride ($MgCl_2$):

$$Mg(OH)_2(s) + 2HCl(aq) \longrightarrow MgCl_2(aq) + 2H_2O(l)$$

After the water is evaporated, the solid magnesium chloride is melted in a steel cell. The molten magnesium chloride contains both Mg^{2+} and Cl^- ions. In a process called electrolysis, an electric current is passed through the cell to reduce the Mg^{2+} ions and oxidize the Cl^- ions. The half-reactions are

$$Mg^{2+} + 2e^- \longrightarrow Mg$$
$$2Cl^- \longrightarrow Cl_2 + 2e^-$$

The overall reaction is

$$MgCl_2(l) \longrightarrow Mg(s) + Cl_2(g)$$

This is how magnesium metal is produced. The chlorine gas generated can be converted to hydrochloric acid and recycled through the process.

(a) Identify the precipitation, acid-base, and redox processes.

(b) Instead of calcium oxide, why don't we simply add sodium hydroxide to precipitate magnesium hydroxide?

(c) Sometimes a mineral called dolomite (a combination of $CaCO_3$ and $MgCO_3$) is substituted for limestone ($CaCO_3$) to bring about the precipitation of magnesium hydroxide. What is the advantage of using dolomite?

(d) What are the advantages of mining magnesium from the ocean rather than from Earth's crust?

Magnesium hydroxide was precipitated from processed seawater in settling ponds The Dow Chemical Company once operated.

4.114 A 5.012-g sample of an iron chloride hydrate was dried in an oven. The mass of the anhydrous compound was 3.195 g. The compound was dissolved in water and reacted with an excess of $AgNO_3$. The precipitate of AgCl formed weighed 7.225 g. What is the formula of the original compound?

4.115 A 22.02-mL solution containing 1.615 g $Mg(NO_3)_2$ is mixed with a 28.64-mL solution containing 1.073 g NaOH. Calculate the concentrations of the ions remaining in solution after the reaction is complete. Assume volumes are additive.

4.116 Because acid-base and precipitation reactions discussed in this chapter all involve ionic species, their progress can be monitored by measuring the electrical conductance of the solution. Match the following reactions with the diagrams shown here. The electrical conductance is shown in arbitrary units.

(1) A 1.0 *M* KOH solution is added to 1.0 L of 1.0 *M* CH_3COOH.

(2) A 1.0 *M* NaOH solution is added to 1.0 L of 1.0 *M* HCl.

(3) A 1.0 *M* $BaCl_2$ solution is added to 1.0 L of 1.0 *M* K_2SO_4.

(4) A 1.0 *M* NaCl solution is added to 1.0 L of 1.0 *M* $AgNO_3$.

(5) A 1.0 *M* CH_3COOH solution is added to 1.0 L of 1.0 *M* NH_3.

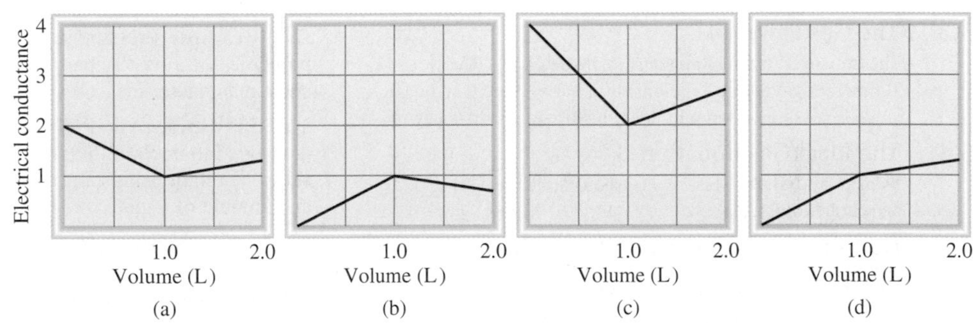

Answers to Practice Exercises

4.1 (a) Insoluble, (b) insoluble, (c) soluble.
4.2 $Al^{3+}(aq) + 3OH^-(aq) \longrightarrow Al(OH)_3(s)$.
4.3 (a) Brønsted base, (b) Brønsted acid, (c) Brønsted acid and Brønsted base.
4.4 (a) P: +3, F: −1; (b) Mn: +7, O: −2. **4.5** 0.452 *M*.
4.6 494 mL. **4.7** Dilute 34.2 mL of the stock solution to 200 mL. **4.8** 92.02%. **4.9** 0.3822 g.
4.10 10.1 mL.

CHAPTER

5

Gases

A hurricane or tropical cyclone is a body of air circulating around a calm center and is characterized by low atmospheric pressure.

Chapter Outline

Student Interactive Activities

Animations
Gas Laws (5.3)
Collecting a Gas over Water (5.5)
Diffusion of Gases (5.6)

Electronic homework
Example Practice Problems
End of Chapter Problems

Essential Concepts

Properties of Gases Gases assume the volume and shape of their containers; they are easily compressible; they mix evenly and completely; and they have much lower densities than liquids and solids.

Gas Pressures Pressure is one of the most readily measurable properties of a gas. A barometer measures atmospheric pressure and a manometer measures the pressure of a gas in the laboratory.

The Gas Laws Over the years, a number of laws have been developed to explain the physical behavior of gases. These laws show the relationships among the pressure, temperature, volume, and amount of a gas.

The Ideal Gas Equation The molecules of an ideal gas possess no volume and exert no forces on one another. At low pressures and high temperatures, most gases can be assumed to behave ideally; their physical behavior is described by the ideal gas equation.

Kinetic Molecular Theory of Gases Macroscopic properties like pressure and temperature of a gas can be related to the kinetic motion of molecules. The kinetic molecular theory of gases assumes that the molecules are ideal, the number of molecules is very large, and that their motions are totally random. Both gas diffusion and gas effusion demonstrate random molecular motion and are governed by the same mathematical laws.

Nonideal Behavior of Gases To account for the behavior of real gases, the ideal gas equation is modified to include the finite volume of molecules and the attractive forces among them.

5.1 Substances That Exist as Gases

We live at the bottom of an ocean of air whose composition by volume is roughly 78 percent N_2, 21 percent O_2, and 1 percent other gases, including CO_2. In the 1990s, the chemistry of this vital mixture of gases became a source of great interest because of the detrimental effects of environmental pollution. Here we will focus generally on the behavior of substances that exist as gases under normal atmospheric conditions, which are defined as 25°C and 1 atmosphere (atm) pressure (see Section 5.2).

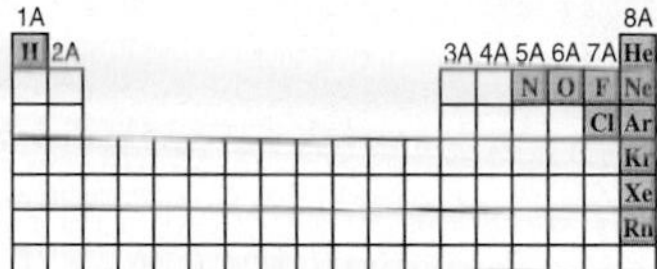

Elements that exist as gases at 25°C and 1 atm. The noble gases (the Group 8A elements) are monatomic species; the other elements exist as diatomic molecules. Ozone (O_3) is also a gas.

Only 11 elements are gases under normal atmospheric conditions. Table 5.1 lists these, along with a number of gaseous compounds. Note that the elements hydrogen, nitrogen, oxygen, fluorine, and chlorine exist as gaseous diatomic molecules. Another form of oxygen, ozone (O_3), is also a gas at room temperature. All the elements in Group 8A, the noble gases, are monatomic gases: He, Ne, Ar, Kr, Xe, and Rn.

Of the gases listed in Table 5.1, only O_2 is essential for our survival. Hydrogen cyanide (HCN) is a deadly poison. Carbon monoxide (CO), hydrogen sulfide (H_2S), nitrogen dioxide (NO_2), O_3, and sulfur dioxide (SO_2) are somewhat less toxic. The gases He and Ne are chemically inert; that is, they do not react with any other substance. Most gases are colorless. Exceptions are F_2, Cl_2, and NO_2. The dark-brown color of NO_2 is sometimes visible in polluted air. All gases have the following physical characteristics:

- Gases assume the volume and shape of their containers.
- Gases are the most compressible of the states of matter.
- Gases will mix evenly and completely when confined to the same container.
- Gases have much lower densities than liquids and solids.

NO_2 gas.

Table 5.1 Some Substances Found as Gases at 1 Atm and 25°C

Elements	Compounds
H_2 (molecular hydrogen)	HF (hydrogen fluoride)
N_2 (molecular nitrogen)	HCl (hydrogen chloride)
O_2 (molecular oxygen)	HBr (hydrogen bromide)
O_3 (ozone)	HI (hydrogen iodide)
F_2 (molecular fluorine)	CO (carbon monoxide)
Cl_2 (molecular chlorine)	CO_2 (carbon dioxide)
He (helium)	NH_3 (ammonia)
Ne (neon)	NO (nitric oxide)
Ar (argon)	NO_2 (nitrogen dioxide)
Kr (krypton)	N_2O (nitrous oxide)
Xe (xenon)	SO_2 (sulfur dioxide)
Rn (radon)	H_2S (hydrogen sulfide)
	HCN (hydrogen cyanide)*

* The boiling point of HCN is 26°C, but it is close enough to qualify as a gas at ordinary atmospheric conditions.

A gas is a substance that is normally in the gaseous state at ordinary temperatures and pressures; a vapor is the gaseous form of any substance that is a liquid or a solid at normal temperatures and pressures. Thus, at 25°C and 1 atm pressure, we speak of water vapor and oxygen gas.

5.2 Pressure of a Gas

Gases exert pressure on any surface with which they come in contact, because gas molecules are constantly in motion. We humans have adapted so well physiologically to the pressure of the air around us that we are usually unaware of it, perhaps as fish are not conscious of the water's pressure on them.

It is easy to demonstrate atmospheric pressure. One everyday example is the ability to drink a liquid through a straw. Sucking air out of the straw reduces the pressure inside the straw. The greater atmospheric pressure on the liquid pushes it up into the straw to replace the air that has been sucked out.

SI Units of Pressure

Pressure is one of the most readily measurable properties of a gas. To understand how we measure the pressure of a gas, it is helpful to know how the units of measurement are derived. We begin with velocity and acceleration.

Velocity is defined as the change in distance with elapsed time; that is,

$$\text{velocity} = \frac{\text{distance moved}}{\text{elapsed time}}$$

The SI unit for velocity is m/s, although we also use cm/s.

Acceleration is the change in velocity with time, or

$$\text{acceleration} = \frac{\text{change in velocity}}{\text{elapsed time}}$$

Acceleration is measured in m/s^2 (or cm/s^2).

The second law of motion, formulated by Sir Isaac Newton in the late seventeenth century, defines another term, from which the units of pressure are derived, namely, *force*. According to this law,

$$\text{force} = \text{mass} \times \text{acceleration}$$

1 N is roughly equivalent to the force exerted by Earth's gravity on an apple.

In this context, the *SI unit of force* is the ***newton (N),*** where

$$1\ \text{N} = 1\ \text{kg m/s}^2$$

Finally, we define ***pressure*** as *force applied per unit area:*

$$\text{pressure} = \frac{\text{force}}{\text{area}}$$

The SI unit of pressure is the ***pascal (Pa),*** defined as *one newton per square meter:*

$$1\ \text{Pa} = 1\ \text{N/m}^2$$

Atmospheric Pressure

The atoms and molecules of the gases in the atmosphere, like those of all other matter, are subject to Earth's gravitational pull. As a consequence, the atmosphere is much denser near the surface of Earth than at high altitudes. (The air outside the pressurized cabin of an airplane at 9 km is too thin to breathe.) In fact, the density of air decreases very rapidly with increasing distance from Earth. Measurements show that about 50 percent of the atmosphere lies within 6.4 km of Earth's surface,

90 percent within 16 km, and 99 percent within 32 km. Not surprisingly, the denser the air is, the greater the pressure it exerts. The force experienced by any area exposed to Earth's atmosphere is equal to *the weight of the column of air above it.* ***Atmospheric pressure*** is *the pressure exerted by Earth's atmosphere* (Figure 5.1). The actual value of atmospheric pressure depends on location, temperature, and weather conditions.

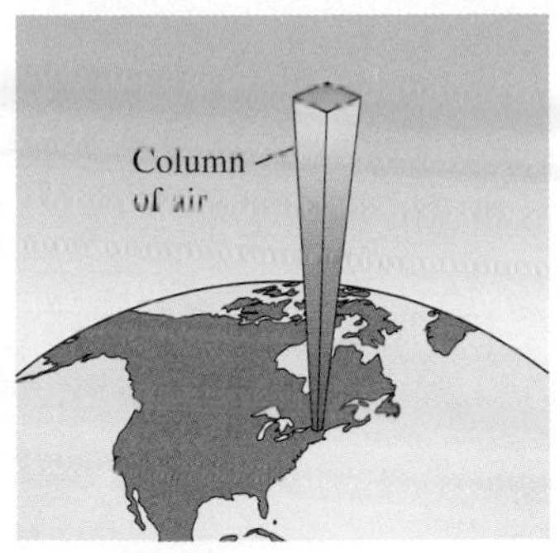

Figure 5.1
A column of air extending from sea level to the upper atmosphere.

Does atmospheric pressure only act downward, as you might infer from its definition? Imagine what would happen, then, if you were to hold a piece of paper tight with both hands above your head. You might expect the paper to bend due to the pressure of air acting on it, but this does not happen. The reason is that air, like water, is a fluid. The pressure exerted on an object in a fluid comes from all directions—downward and upward, as well as from the left and from the right. At the molecular level, air pressure results from collisions between the air molecules and any surface with which they come in contact. The magnitude of pressure depends on how often and how strongly the molecules impact the surface. It turns out that there are just as many molecules hitting the paper from the top as there are from underneath, so the paper stays flat.

How is atmospheric pressure measured? The ***barometer*** is probably the most familiar *instrument for measuring atmospheric pressure.* A simple barometer consists of a long glass tube, closed at one end and filled with mercury. If the tube is carefully inverted in a dish of mercury so that no air enters the tube, some mercury will flow out of the tube into the dish, creating a vacuum at the top (Figure 5.2). The weight of the mercury remaining in the tube is supported by atmospheric pressure acting on the surface of the mercury in the dish. ***Standard atmospheric pressure (1 atm)*** is equal to *the pressure that supports a column of mercury exactly 760 mm (or 76 cm) high at 0°C at sea level.* In other words, the standard atmosphere equals a pressure of 760 mmHg, where mmHg represents the pressure exerted by a column of mercury 1 mm high. The mmHg unit is also called the *torr,* after the Italian scientist Evangelista Torricelli, who invented the barometer. Thus,

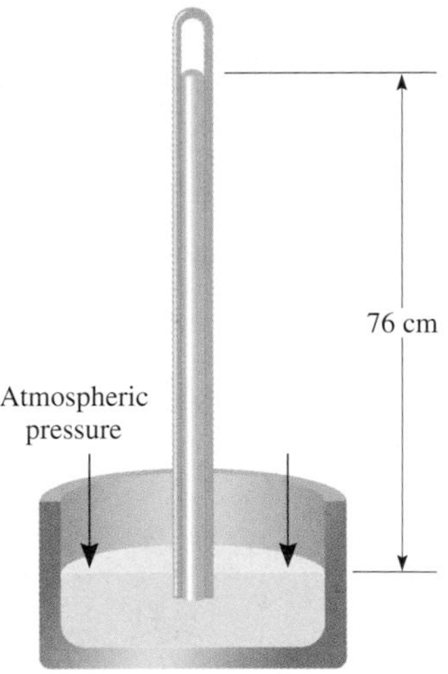

Figure 5.2
A barometer for measuring atmospheric pressure. Above the mercury in the tube is a vacuum. (The space actually contains a very small amount of mercury vapor.) The column of mercury is supported by the atmospheric pressure.

$$1 \text{ torr} = 1 \text{ mmHg}$$

and

$$\begin{aligned} 1 \text{ atm} &= 760 \text{ mmHg} \quad \text{(exactly)} \\ &= 760 \text{ torr} \end{aligned}$$

The relation between atmospheres and pascals (see Appendix 1) is

$$\begin{aligned} 1 \text{ atm} &= 101{,}325 \text{ Pa} \\ &= 1.01325 \times 10^5 \text{ Pa} \end{aligned}$$

and because 1000 Pa = 1 kPa (kilopascal)

$$1 \text{ atm} = 1.01325 \times 10^2 \text{ kPa}$$

EXAMPLE 5.1

The pressure outside a jet plane flying at high altitude falls considerably below standard atmospheric pressure. Therefore, the air inside the cabin must be pressurized to protect the passengers. What is the pressure in atmospheres in the cabin if the barometer reading is 672 mmHg?

(Continued)

Strategy Because 1 atm = 760 mmHg, the following conversion factor is needed to obtain the pressure in atmospheres

$$\frac{1 \text{ atm}}{760 \text{ mmHg}}$$

Solution The pressure in the cabin is given by

$$\text{pressure} = 672 \cancel{\text{mmHg}} \times \frac{1 \text{ atm}}{760 \cancel{\text{mmHg}}}$$
$$= \boxed{0.884 \text{ atm}}$$

Check Because the barometer reading is less than 760 mmHg, the pressure should be less than 1 atm.

Similar problem: 5.14.

Practice Exercise Convert 749 mmHg to atmospheres.

A ***manometer*** is *a device used to measure the pressure of gases other than the atmosphere.* The principle of operation of a manometer is similar to that of a barometer. There are two types of manometers, shown in Figure 5.3. *The closed-tube manometer* is normally used to measure pressures below atmospheric pressure [Figure 5.3(a)], whereas the *open-tube manometer* is better suited for measuring pressures equal to or greater than atmospheric pressure [Figure 5.3(b)].

Nearly all barometers and most manometers use mercury as the working fluid, despite the fact that it is a toxic substance with a harmful vapor. The reason is that mercury has a very high density (13.6 g/mL) compared with most other liquids. Because the height of the liquid in a column is inversely proportional to the liquid's density, this property enables the construction of manageably small barometers and manometers.

Review of Concepts

Rank the following pressure measurements from lowest to highest: (a) 735 mmHg, (b) 1.06×10^5 Pa, (c) 678 torr, (d) 0.926 atm.

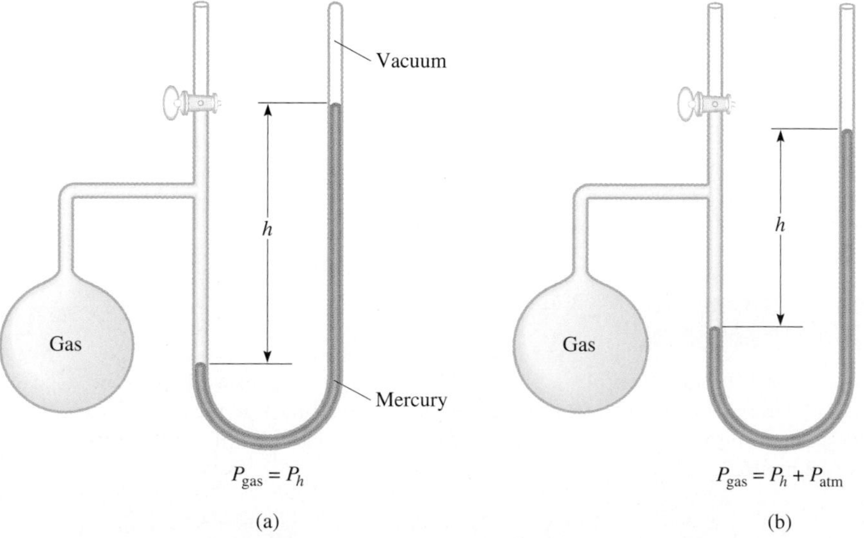

Figure 5.3
Two types of manometers used to measure gas pressures. (a) Gas pressure is less than atmospheric pressure. (b) Gas pressure is greater than atmospheric pressure.

5.3 The Gas Laws

The gas laws we will study in this chapter are the product of countless experiments on the physical properties of gases that were carried out over several centuries. Each of these generalizations regarding the macroscopic behavior of gaseous substances represents a milestone in the history of science. Together they have played a major role in the development of many ideas in chemistry.

Animation: The Gas Laws

The Pressure-Volume Relationship: Boyle's Law

In the seventeenth century, the British chemist Robert Boyle studied the behavior of gases systematically and quantitatively. In one series of studies, Boyle investigated the pressure-volume relationship of a gas sample using an apparatus like that shown in Figure 5.4. In Figure 5.4(a) the pressure exerted on the gas by the mercury added to the tube is equal to atmospheric pressure. In Figure 5.4(b) an increase in pressure due to the addition of more mercury results in a decrease in the volume of the gas and in unequal levels of mercury in the tube. Boyle noticed that when temperature is held constant, the volume (V) of a given amount of a gas decreases as the total applied pressure (P)—atmospheric pressure plus the pressure due to the added mercury—is increased. This relationship between pressure and volume is evident in Figure 5.4. Conversely, if the applied pressure is decreased, the gas volume becomes larger.

The pressure applied to a gas is equal to the gas pressure.

The mathematical expression showing an inverse relationship between pressure and volume is

$$P \propto \frac{1}{V}$$

where the symbol $\propto$ means *proportional to*. To change $\propto$ to an equals sign, we must write

$$P = k_1 \times \frac{1}{V} \tag{5.1a}$$

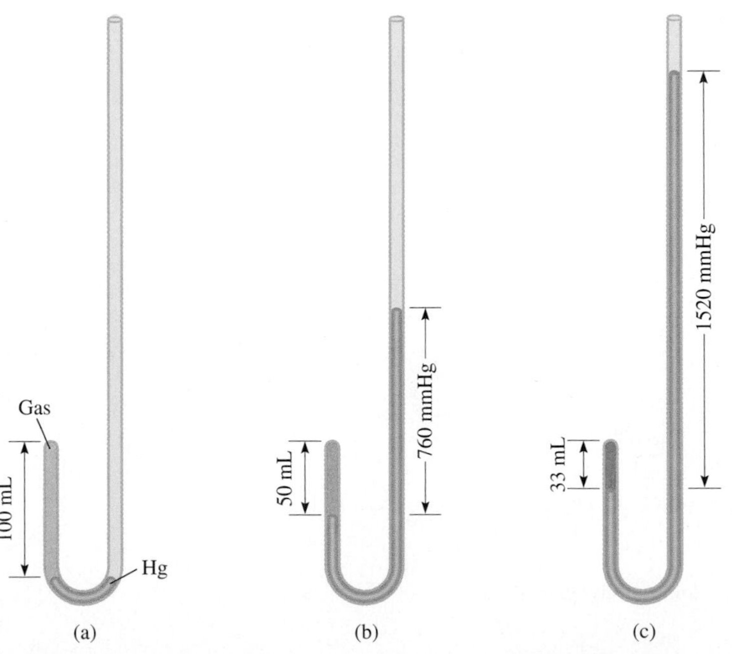

Figure 5.4 *Apparatus for studying the relationship between pressure and volume of a gas. (a) The levels of mercury are equal and the pressure of the gas is equal to the atmospheric pressure (760 mmHg). The gas volume is 100 mL. (b) Doubling the pressure by adding more mercury reduces the gas volume to 50 mL. (c) Tripling the gas pressure decreases the gas volume to one-third of the original volume. The temperature and amount of gas are kept constant.*

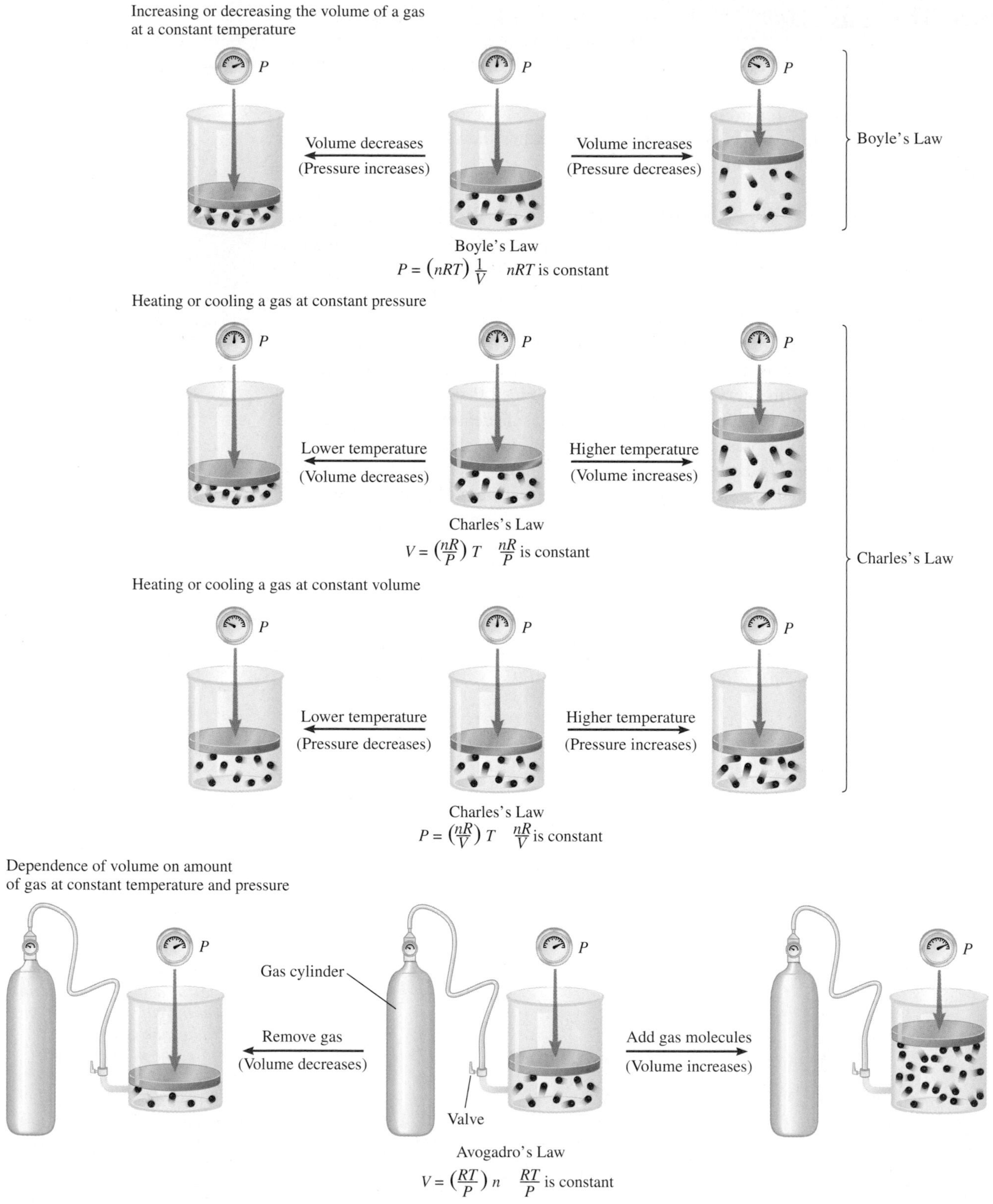

Increasing or decreasing the volume of a gas at a constant temperature

P P P

Volume decreases (Pressure increases)

Volume increases (Pressure decreases)

Boyle's Law

Boyle's Law

$P = (nRT)\frac{1}{V}$ nRT is constant

Heating or cooling a gas at constant pressure

P P P

Lower temperature (Volume decreases)

Higher temperature (Volume increases)

Charles's Law

$V = (\frac{nR}{P})\,T$ $\frac{nR}{P}$ is constant

Charles's Law

Heating or cooling a gas at constant volume

P P P

Lower temperature (Pressure decreases)

Higher temperature (Pressure increases)

Charles's Law

$P = (\frac{nR}{V})\,T$ $\frac{nR}{V}$ is constant

Dependence of volume on amount of gas at constant temperature and pressure

P P P

Gas cylinder

Remove gas (Volume decreases)

Add gas molecules (Volume increases)

Valve

Avogadro's Law

$V = (\frac{RT}{P})\,n$ $\frac{RT}{P}$ is constant

Figure 5.5

Schematic illustrations of Boyle's law, Charles's law, and Avogadro's law.

where k_1 is a constant called the *proportionality constant.* Equation (5.1a) is an expression of ***Boyle's law,*** which states that *the pressure of a fixed amount of gas maintained at constant temperature is inversely proportional to the volume of the gas.* We can rearrange Equation (5.1a) and obtain

$$PV = k_1 \tag{5.1b}$$

This form of Boyle's law says that the product of the pressure and volume of a gas at constant temperature and amount of gas is a constant. Figure 5.5 is a schematic representation of Boyle's law. The quantity n is the number of moles of the gas and R is a constant to be defined in Section 5.4. Thus, the proportionality constant k_1 in Equations (5.1) is equal to nRT.

The concept of one quantity being proportional to another and the use of a proportionality constant can be clarified through the following analogy. The daily income of a movie theater depends on both the price of the tickets (in dollars per ticket) and the number of tickets sold. Assuming that the theater charges one price for all tickets, we write

$$\text{income} = (\text{dollar/ticket}) \times \text{number of tickets sold}$$

Because the number of tickets sold varies from day to day, the income on a given day is said to be proportional to the number of tickets sold:

$$\begin{aligned}\text{income} &\propto \text{number of tickets sold}\\ &= C \times \text{number of tickets sold}\end{aligned}$$

where C, the proportionality constant, is the price per ticket.

Figure 5.6 shows two conventional ways of expressing Boyle's findings graphically. Figure 5.6(a) is a graph of the equation $PV = k_1$; Figure 5.6(b) is a graph of the equivalent equation $P = k_1 \times 1/V$. Note that the latter is a linear equation of the form $y = mx + b$, where $m = k_1$ and $b = 0$.

Although the individual values of pressure and volume can vary greatly for a given sample of gas, as long as the temperature is held constant and the amount of the gas does not change, P times V is always equal to the same constant. Therefore, for a given sample of gas under two different sets of conditions at constant temperature, we have

$$P_1V_1 = k_1 = P_2V_2$$

or

$$P_1V_1 = P_2V_2 \tag{5.2}$$

where V_1 and V_2 are the volumes at pressures P_1 and P_2, respectively.

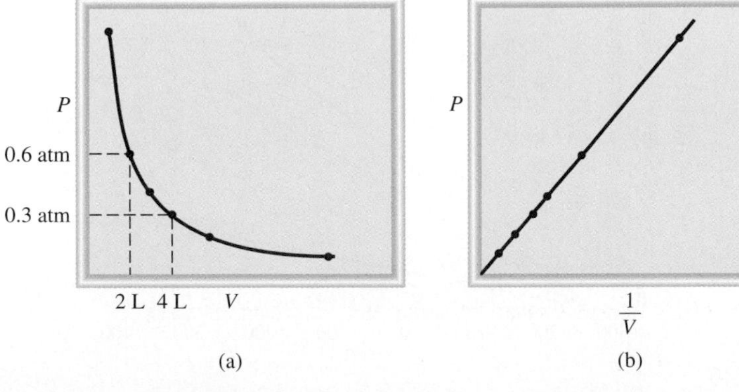

Figure 5.6
Graphs showing variation of the volume of a gas with the pressure exerted on the gas, at constant temperature. (a) P versus V. Note that the volume of the gas doubles as the pressure is halved. (b) P versus 1/V.

The Temperature-Volume Relationship: Charles's and Gay-Lussac's Law

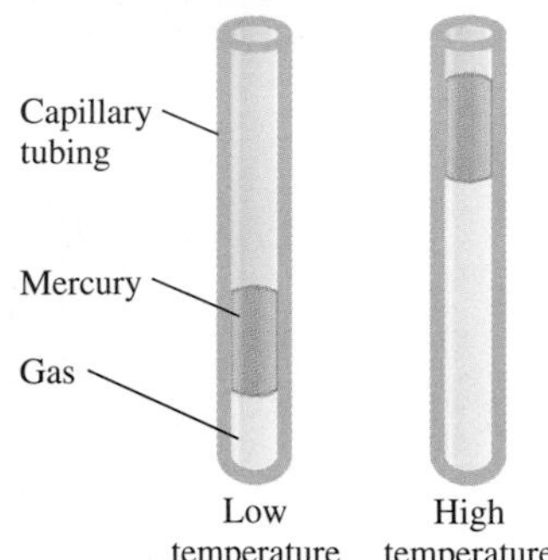

Figure 5.7
Variation of the volume of a gas sample with temperature, at constant pressure. The pressure exerted on the gas is the sum of the atmospheric pressure and the pressure due to the weight of the mercury.

Boyle's law depends on the temperature of the system remaining constant. But suppose the temperature changes: How does a change in temperature affect the volume and pressure of a gas? Let us first look at the effect of temperature on the volume of a gas. The earliest investigators of this relationship were French scientists, Jacques Charles and Joseph Gay-Lussac. Their studies showed that, at constant pressure, the volume of a gas sample expands when heated and contracts when cooled (Figure 5.7). The quantitative relations involved in changes in gas temperature and volume turn out to be remarkably consistent. For example, we observe an interesting phenomenon when we study the temperature-volume relationship at various pressures. At any given pressure, the plot of volume versus temperature yields a straight line. By extending the line to zero volume, we find the intercept on the temperature axis to be −273.15°C. At any other pressure, we obtain a different straight line for the volume-temperature plot, but we get the *same* zero-volume temperature intercept at −273.15°C. (Figure 5.8). (In practice, we can measure the volume of a gas over only a limited temperature range, because all gases condense at low temperatures to form liquids.)

In 1848 the Scottish physicist Lord Kelvin realized the significance of this phenomenon. He identified −273.15°C as ***absolute zero,*** *theoretically the lowest attainable temperature.* Then he set up an ***absolute temperature scale,*** now called the ***Kelvin temperature scale,*** with *absolute zero as the starting point.* On the Kelvin scale, one kelvin (K) is equal *in magnitude* to one degree Celsius. The only difference between the absolute temperature scale and the Celsius scale is that the zero position is shifted. Important points on the two scales match up as follows:

Under special experimental conditions, scientists have succeeded in approaching absolute zero to within a small fraction of a kelvin.

	Kelvin Scale	Celsius Scale
Absolute zero	0 K	−273.15°C
Freezing point of water	273.15 K	0°C
Boiling point of water	373.15 K	100°C

The conversion between °C and K is given in Section 1.5:

$$?\,\text{K} = (^\circ\text{C} + 273.15^\circ\text{C})\,\frac{1\,\text{K}}{1^\circ\text{C}}$$

In most calculations we will use 273 instead of 273.15 as the term relating K and °C. By convention, we use T to denote absolute (kelvin) temperature and t to indicate temperature on the Celsius scale.

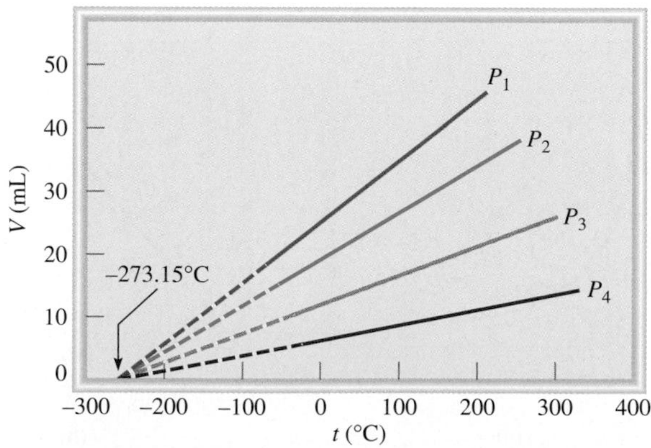

Figure 5.8
Variation of the volume of a gas sample with temperature, at constant pressure. Each line represents the variation at a certain pressure. The pressures increase from P_1 to P_4. All gases ultimately condense (become liquids) if they are cooled to sufficiently low temperatures; the solid portions of the lines represent the temperature region above the condensation point. When these lines are extrapolated, or extended (the dashed portions), they all intersect at the point representing zero volume and a temperature of −273.15°C.

The dependence of the volume of a gas on temperature is given by

$$V \propto T$$

$$V = k_2T$$

or

$$\frac{V}{T} = k_2 \tag{5.3}$$

where k_2 is the proportionality constant. Equation (5.3) is known as ***Charles's and Gay-Lussac's law,*** or simply ***Charles's law,*** which states that *the volume of a fixed amount of gas maintained at constant pressure is directly proportional to the absolute temperature of the gas.* Charles's law is also illustrated in Figure 5.5. We see that the proportionality constant k_2 in Equation (5.3) is equal to nR/P.

Just as we did for pressure-volume relationships at constant temperature, we can compare two sets of volume-temperature conditions for a given sample of gas at constant pressure. From Equation (5.3) we can write

$$\frac{V_1}{T_1} = k_2 = \frac{V_2}{T_2}$$

or

$$\frac{V_1}{T_1} = \frac{V_2}{T_2} \tag{5.4}$$

Temperature must be in kelvins in gas law calculations.

where V_1 and V_2 are the volumes of the gas at temperatures T_1 and T_2 (both in kelvins), respectively.

Another form of Charles's law shows that at constant amount of gas and volume, the pressure of a gas is proportional to temperature

$$P \propto T$$

$$P = k_3T$$

or

$$\frac{P}{T} = k_3 \tag{5.5}$$

From Figure 5.5 we see that $k_3 = nR/V$. Starting with Equation (5.5), we have

$$\frac{P_1}{T_1} = k_3 = \frac{P_2}{T_2}$$

or

$$\frac{P_1}{T_1} = \frac{P_2}{T_2} \tag{5.6}$$

where P_1 and P_2 are the pressures of the gas at temperatures T_1 and T_2, respectively.

The Volume-Amount Relationship: Avogadro's Law

The work of the Italian scientist Amedeo Avogadro complemented the studies of Boyle, Charles, and Gay-Lussac. In 1811 he published a hypothesis stating that at the same temperature and pressure, equal volumes of different gases contain the same number of molecules (or atoms if the gas is monatomic). It follows that the volume of any given gas must be proportional to the number of moles of molecules present; that is,

Avogadro's name first appeared in Section 3.2.

$$V \propto n$$

$$V = k_4n \tag{5.7}$$

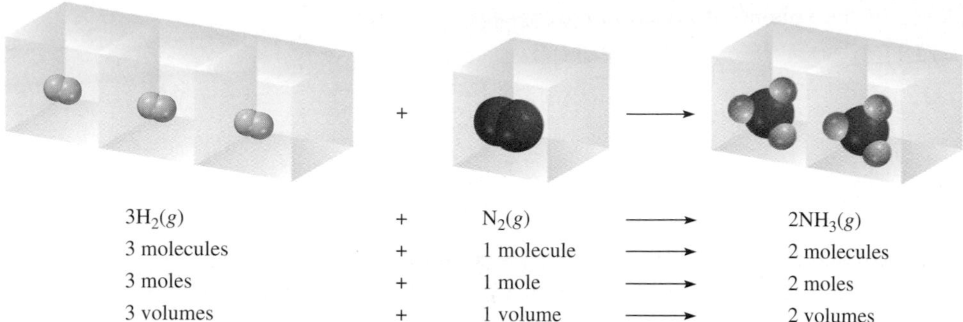

Figure 5.9
Volume relationship of gases in a chemical reaction. The ratio of the volumes of molecular hydrogen to molecular nitrogen is 3:1, and that of ammonia (the product) to molecular hydrogen and molecular nitrogen combined (the reactants) is 2:4, or 1:2.

where n represents the number of moles and k_4 is the proportionality constant. Equation (5.7) is the mathematical expression of ***Avogadro's law,*** which states that *at constant pressure and temperature, the volume of a gas is directly proportional to the number of moles of the gas present.* From Figure 5.5 we see that $k_4 = RT/P$.

According to Avogadro's law we see that when two gases react with each other, their reacting volumes have a simple ratio to each other. If the product is a gas, its volume is related to the volume of the reactants by a simple ratio (a fact demonstrated earlier by Gay-Lussac). For example, consider the synthesis of ammonia from molecular hydrogen and molecular nitrogen:

$$\underset{3\text{ mol}}{3H_2(g)} + \underset{1\text{ mol}}{N_2(g)} \longrightarrow \underset{2\text{ mol}}{2NH_3(g)}$$

Because, at the same temperature and pressure, the volumes of gases are directly proportional to the number of moles of the gases present, we can now write

$$\underset{3\text{ volumes}}{3H_2(g)} + \underset{1\text{ volume}}{N_2(g)} \longrightarrow \underset{2\text{ volumes}}{2NH_3(g)}$$

The volume ratio of molecular hydrogen to molecular nitrogen is 3:1, and that of ammonia (the product) to molecular hydrogen and molecular nitrogen combined (the reactants) is 2:4, or 1:2 (Figure 5.9).

Worked examples illustrating the gas laws are presented in Section 5.4.

Review of Concepts

If the absolute temperature of a gas sample is increased while the volume is held constant, will the pressure of the gas sample increase, decrease, or remain the same?

5.4 The Ideal Gas Equation

Let us summarize the gas laws we have discussed so far:

$$\text{Boyle's law:} \quad V \propto \frac{1}{P} \quad (\text{at constant } n \text{ and } T)$$

$$\text{Charles's law:} \quad V \propto T \quad (\text{at constant } n \text{ and } P)$$

$$\text{Avogadro's law:} \quad V \propto n \quad (\text{at constant } P \text{ and } T)$$

We can combine all three expressions to form a single master equation for the behavior of gases:

$$V \propto \frac{nT}{P}$$

$$V = R\frac{nT}{P}$$

or

$$PV = nRT \qquad (5.8)$$

where ***R,*** *the proportionality constant,* is called the ***gas constant.*** Equation (5.8), which is called the ***ideal gas equation,*** *describes the relationship among the four variables P, V, T, and n.* An ***ideal gas*** is *a hypothetical gas whose pressure-volume-temperature behavior can be completely accounted for by the ideal gas equation.* The molecules of an ideal gas do not attract or repel one another, and their volume is negligible compared with the volume of the container. Although there is no such thing in nature as an ideal gas, discrepancies in the behavior of real gases over reasonable temperature and pressure ranges do not significantly affect calculations. Thus, we can safely use the ideal gas equation to solve many gas problems.

Keep in mind that the ideal gas equation, unlike the gas laws discussed in Section 5.3, applies to systems that do not undergo changes in pressure, volume, temperature, and amount of a gas.

Before we can apply the ideal gas equation to a real system, we must evaluate the gas constant *R*. At 0°C (273.15 K) and 1 atm pressure, many real gases behave like an ideal gas. Experiments show that under these conditions, 1 mole of an ideal gas occupies 22.414 L, which is somewhat greater than the volume of a basketball, as shown in Figure 5.10. *The conditions 0°C and 1 atm* are called ***standard temperature and pressure,*** often abbreviated ***STP.*** From Equation (5.8) we can write

$$\begin{aligned} R &= \frac{PV}{nT} \\ &= \frac{(1\text{ atm})(22.414\text{ L})}{(1\text{ mol})(273.15\text{ K})} \\ &= 0.082057\,\frac{\text{L}\cdot\text{atm}}{\text{K}\cdot\text{atm}} \\ &= 0.082057\text{ L}\cdot\text{atm/K}\cdot\text{mol} \end{aligned}$$

The gas constant can be expressed in different units (see Appendix 2).

Figure 5.10
A comparison of the molar volume at STP (which is approximately 22.4 L) with a basketball.

The dots between L and atm and between K and mol remind us that both L and atm are in the numerator and both K and mol are in the denominator. For most calculations, we will round off the value of R to three significant figures (0.0821 L · atm/K · mol) and use 22.4 L for the molar volume of a gas at STP.

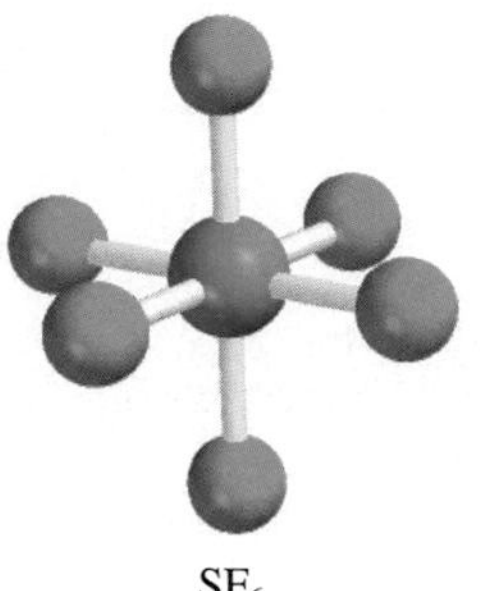

SF_6

EXAMPLE 5.2

Sulfur hexafluoride (SF_6) is a colorless, odorless, very unreactive gas. Calculate the pressure (in atm) exerted by 1.39 moles of the gas in a steel vessel of volume 6.09 L at 55°C.

Strategy The problem gives the amount of the gas and its volume and temperature. Is the gas undergoing a change in any of its properties? What equation should we use to solve for the pressure? What temperature unit should we use?

Solution Because no changes in gas properties occur, we can use the ideal gas equation to calculate the pressure. Rearranging Equation (5.8), we write

$$\begin{aligned} P &= \frac{nRT}{V} \\ &= \frac{(1.39\ \text{mol})(0.0821\ \text{L} \cdot \text{atm/K} \cdot \text{mol})(55 + 273)\ \text{K}}{6.09\ \text{L}} \\ &= \boxed{6.15\ \text{atm}} \end{aligned}$$

Similar problems: 5.32, 5.33.

Practice Exercise Calculate the volume (in liters) occupied by 2.12 moles of nitric oxide (NO) at 6.54 atm and 76°C.

NH_3

EXAMPLE 5.3

Calculate the volume (in liters) occupied by 5.58 g of NH_3 at STP.

Strategy What is the volume of one mole of an ideal gas at STP? How many moles are there in 5.58 g of NH_3?

Solution Recognizing that 1 mole of an ideal gas occupies 22.4 L at STP and using the molar mass of NH_3 (17.03 g), we write the sequence of conversions as

$$\text{grams of } NH_3 \longrightarrow \text{moles of } NH_3 \longrightarrow \text{liters of } NH_3 \text{ at STP}$$

so the volume of NH_3 is given by

$$\begin{aligned} V &= 5.58\ \cancel{\text{g NH}_3} \times \frac{1\ \cancel{\text{mol NH}_3}}{17.03\ \cancel{\text{g NH}_3}} \times \frac{22.4\ \text{L}}{1\ \cancel{\text{mol NH}_3}} \\ &= \boxed{7.34\ \text{L}} \end{aligned}$$

It is often true in chemistry, particularly in gas-law calculations, that a problem can be solved in more than one way. Here the problem can also be solved by first converting 5.58 g of NH_3 to the number of moles of NH_3, and then applying the ideal gas equation ($V = nRT/P$). Try it.

Check Because 5.58 g of NH_3 is smaller than its molar mass, its volume at STP should be smaller than 22.4 L. Therefore, the answer is reasonable.

Similar problems: 5.41, 5.43.

Practice Exercise What is the volume (in liters) occupied by 49.8 g of HCl at STP?

The ideal gas equation is useful for problems that do not involve changes in P, V, T, and n for a gas sample. At times, however, we need to deal with changes in pressure, volume, and temperature, or even in the amount of a gas. When conditions change, we must employ a modified form of the ideal gas equation that takes into account the initial and final conditions. We derive the modified equation as follows. From Equation (5.8),

$$R = \frac{P_1V_1}{n_1T_1} \text{ (before change)} \quad \text{and} \quad R = \frac{P_2V_2}{n_2T_2} \text{ (after change)}$$

Therefore,

$$\frac{P_1V_1}{n_1T_1} = R = \frac{P_2V_2}{n_2T_2} \tag{5.9}$$

The subscripts 1 and 2 denote the initial and final states of the gas. All of the gas laws discussed so far can be derived from Equation (5.9).

If $n_1 = n_2$, as is usually the case because the amount of gas normally does not change, the equation then becomes

$$\frac{P_1V_1}{T_1} = \frac{P_2V_2}{T_2} \tag{5.10}$$

EXAMPLE 5.4

A small bubble rises from the bottom of a lake, where the temperature and pressure are 8°C and 6.4 atm, to the water's surface, where the temperature is 25°C and the pressure is 1.0 atm. Calculate the final volume (in mL) of the bubble if its initial volume was 2.1 mL.

Strategy In solving this kind of problem, where a lot of information is given, it is sometimes helpful to make a sketch of the situation, as shown here:

What temperature unit should be used in the calculation?

Solution According to Equation (5.9)

$$\frac{P_1V_1}{n_1T_1} = \frac{P_2V_2}{n_2T_2}$$

We assume that the amount of air in the bubble remains constant, that is, $n_1 = n_2$ so that

$$\frac{P_1V_1}{T_1} = \frac{P_2V_2}{T_2}$$

(Continued)

which is Equation (5.10). The given information is summarized:

We can use any appropriate units for volume (or pressure) as long as we use the same units on both sides of the equation.

Initial Conditions	Final Conditions
$P_1 = 6.4$ atm	$P_2 = 1.0$ atm
$V_1 = 2.1$ mL	$V_2 = ?$
$T_1 = (8 + 273)$ K $= 281$ K	$T_2 = (25 + 273)$ K $= 298$ K

Rearranging Equation (5.10) gives

$$V_2 = V_1 \times \frac{P_1}{P_2} \times \frac{T_2}{T_1}$$

$$= 2.1\text{ mL} \times \frac{6.4\text{ atm}}{1.0\text{ atm}} \times \frac{298\text{ K}}{281\text{ K}}$$

$$= 14\text{ mL}$$

Check We see that the final volume involves multiplying the initial volume by a ratio of pressures (P_1/P_2) and a ratio of temperatures (T_2/T_1). Recall that volume is inversely proportional to pressure, and volume is directly proportional to temperature. Because the pressure decreases and temperature increases as the bubble rises, we expect the bubble's volume to increase. In fact, here the change in pressure plays a greater role in the volume change.

Similar problems: 5.35, 5.39.

Practice Exercise A gas initially at 4.0 L, 1.2 atm, and 66°C undergoes a change so that its final volume and temperature are 1.7 L and 42°C. What is its final pressure? Assume the number of moles remains unchanged.

Density and Molar Mass of a Gaseous Substance

The ideal gas equation can be applied to determine the density or molar mass of a gaseous substance. Rearranging Equation (5.8), we write

$$\frac{n}{V} = \frac{P}{RT}$$

The number of moles of the gas, n, is given by

$$n = \frac{m}{\mathcal{M}}$$

in which m is the mass of the gas in grams and $\mathcal{M}$ is its molar mass. Therefore,

$$\frac{m}{\mathcal{M}V} = \frac{P}{RT}$$

Because density, d, is mass per unit volume, we can write

$$d = \frac{m}{V} = \frac{P\mathcal{M}}{RT} \tag{5.11}$$

Equation (5.11) enables us to calculate the density of a gas (given in units of grams per liter). More often, the density of a gas can be measured, so this equation can be rearranged for us to calculate the molar mass of a gaseous substance:

$$\mathcal{M} = \frac{dRT}{P} \tag{5.12}$$

In a typical experiment, a bulb of known volume is filled with the gaseous substance under study. The temperature and pressure of the gas sample are recorded, and the total mass of the bulb plus gas sample is determined (Figure 5.11). The bulb is then evacuated (emptied) and weighed again. The difference in mass is the mass of the gas. The density of the gas is equal to its mass divided by the volume of the bulb. Then we can calculate the molar mass of the substance using Equation (5.12).

Figure 5.11
An apparatus for measuring the density of a gas. A bulb of known volume is filled with the gas under study at a certain temperature and pressure. First, the bulb is weighed, and then it is emptied (evacuated) and weighed again. The difference in masses gives the mass of the gas. Knowing the volume of the bulb, we can calculate the density of the gas.

Note that we can determine the molar mass of a compound without knowing its molecular formula.

EXAMPLE 5.5

A chemist has synthesized a greenish-yellow gaseous compound of chlorine and oxygen and finds that its density is 8.14 g/L at 47°C and 3.15 atm. Calculate the molar mass of the compound and determine its molecular formula.

Strategy Because Equations (5.11) and (5.12) are rearrangements of each other, we can calculate the molar mass of a gas if we know its density, temperature, and pressure. The molecular formula of the compound must be consistent with its molar mass. What temperature unit should we use?

Solution From Equation (5.12)

$$\begin{aligned}\mathcal{M} &= \frac{dRT}{P}\\ &= \frac{(8.14\ \text{g/L})(0.0821\ \text{L}\cdot\text{atm/K}\cdot\text{mol})(47 + 273)\ \text{K}}{3.15\ \text{atm}}\\ &= 67.9\ \text{g/mol}\end{aligned}$$

We can determine the molecular formula of the compound by trial and error, using only the knowledge of the molar masses of chlorine (35.45 g) and oxygen (16.00 g). We know that a compound containing one Cl atom and one O atom would have a molar mass of 51.45 g, which is too low, while the molar mass of a compound made up of two Cl atoms and one O atom is 86.90 g, which is too high. Thus, the compound must contain one Cl atom and two O atoms and have the formula ClO_2, which has a molar mass of 67.45 g.

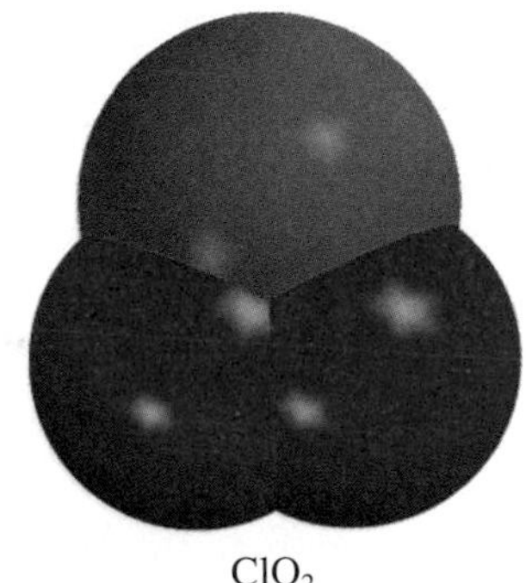

ClO_2

Similar problems: 5.47, 5.49.

Practice Exercise The density of a gaseous organic compound is 3.38 g/L at 40°C and 1.97 atm. What is its molar mass?

Gas Stoichiometry

In Chapter 3 we used relationships between amounts (in moles) and masses (in grams) of reactants and products to solve stoichiometry problems. When the reactants and/or products are gases, we can also use the relationships between amounts (moles, n) and volume (V) to solve such problems (Figure 5.12).

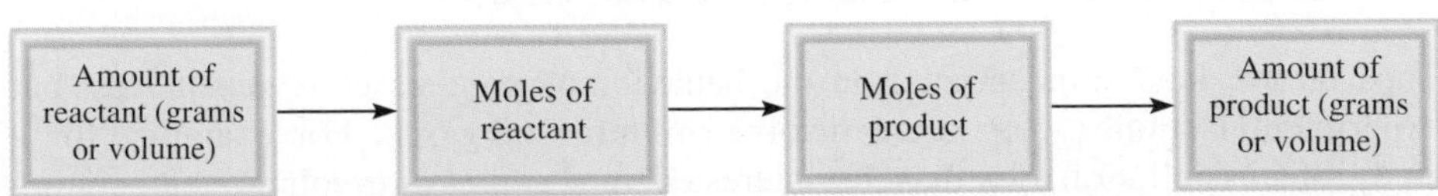

Figure 5.12
Stoichiometric calculations involving gases.

An air bag can protect the driver in an automobile collision.

EXAMPLE 5.6

Sodium azide (NaN_3) is used in some automobile air bags. The impact of a collision triggers the decomposition of NaN_3 as follows:

$$2NaN_3(s) \longrightarrow 2Na(s) + 3N_2(g)$$

The nitrogen gas produced quickly inflates the bag between the driver and the windshield and dashboard. Calculate the volume of N_2 generated at 85°C and 812 mmHg by the decomposition of 50.0 g of NaN_3.

Strategy From the balanced equation we see that 2 mol $NaN_3 \simeq$ 3 mol N_2 so the conversion factor between NaN_3 and N_2 is

$$\frac{3 \text{ mol } N_2}{2 \text{ mol } NaN_3}$$

Because the mass of NaN_3 is given, we can calculate the number of moles of NaN_3 and hence the number of moles of N_2 produced. Finally, we can calculate the volume of N_2 using the ideal gas equation.

Solution The sequence of conversions is as follows:

$$\text{grams of } NaN_3 \longrightarrow \text{moles of } NaN_3 \longrightarrow \text{moles of } N_2 \longrightarrow \text{volume of } N_2$$

First, we calculate the number of moles of N_2 produced by 50.0 g of NaN_3:

$$\text{moles of } N_2 = 50.0 \cancel{\text{g } NaN_3} \times \frac{1 \cancel{\text{mol } NaN_3}}{65.02 \cancel{\text{g } NaN_3}} \times \frac{3 \text{ mol } N_2}{2 \cancel{\text{mol } NaN_3}}$$

$$= 1.15 \text{ mol } N_2$$

The volume of 1.15 mol of N_2 can be obtained by using the ideal gas equation:

$$V = \frac{nRT}{P} = \frac{(1.15 \text{ mol})(0.0821 \text{ L} \cdot \text{atm/K} \cdot \text{mol})(85 + 273) \text{ K}}{(812/760) \text{ atm}}$$

$$= 31.6 \text{ L}$$

Similar problems: 5.51, 5.52.

Practice Exercise The equation for the metabolic breakdown of glucose ($C_6H_{12}O_6$) is the same as the equation for the combustion of glucose in air:

$$C_6H_{12}O_6(s) + 6O_2(g) \longrightarrow 6CO_2(g) + 6H_2O(l)$$

Calculate the volume of CO_2 produced at 37°C and 1.00 atm when 5.60 g of glucose is used up in the reaction.

Review of Concepts

Assuming ideal behavior, which of the following samples of gases will have the greatest volume at STP? Which of these gases will have the greatest density at STP? (a) 0.82 mole of He. (b) 24 g of N_2. (c) 5.0×10^{23} molecules of Cl_2.

5.5 Dalton's Law of Partial Pressures

Thus far we have concentrated on the behavior of pure gaseous substances, but experimental studies very often involve mixtures of gases. For example, for a study of air pollution, we may be interested in the pressure-volume-temperature

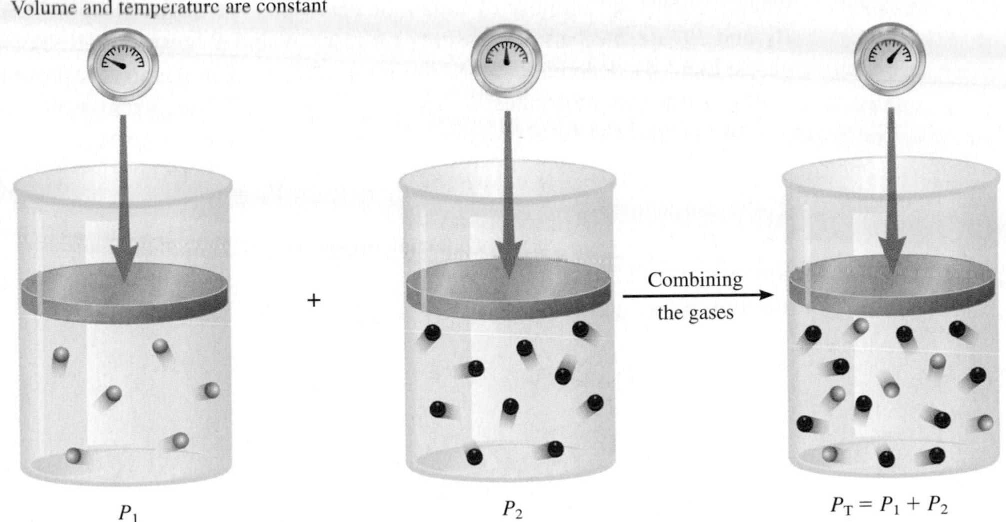

Figure 5.13
Schematic illustration of Dalton's law of partial pressures.

relationship of a sample of air, which contains several gases. In this case, and all cases involving mixtures of gases, the total gas pressure is related to ***partial pressures,*** that is, *the pressures of individual gas components in the mixture.* In 1801 Dalton formulated a law, now known as ***Dalton's law of partial pressures,*** which states that *the total pressure of a mixture of gases is just the sum of the pressures that each gas would exert if it were present alone.* Figure 5.13 illustrates Dalton's law.

As mentioned earlier, gas pressure results from the impact of gas molecules against the walls of the container.

Consider a case in which two gases, A and B, are in a container of volume *V*. The pressure exerted by gas A, according to the ideal gas equation, is

$$P_A = \frac{n_A RT}{V}$$

where n_A is the number of moles of A present. Similarly, the pressure exerted by gas B is

$$P_B = \frac{n_B RT}{V}$$

In a mixture of gases A and B, the total pressure P_T is the result of the collisions of both types of molecules, A and B, with the walls of the container. Thus, according to Dalton's law,

$$\begin{aligned} P_T &= P_A + P_B \\ &= \frac{n_A RT}{V} + \frac{n_B RT}{V} \\ &= \frac{RT}{V}(n_A + n_B) \\ &= \frac{nRT}{V} \end{aligned}$$

where n, the total number of moles of gases present, is given by $n = n_A + n_B$, and P_A and P_B are the partial pressures of gases A and B, respectively. For a mixture of gases, then, P_T depends only on the total number of moles of gas present, not on the nature of the gas molecules.

In general, the total pressure of a mixture of gases is given by

$$P_T = P_1 + P_2 + P_3 + \cdots$$

where P_1, P_2, P_3, . . . are the partial pressures of components 1, 2, 3, ... To see how each partial pressure is related to the total pressure, consider again the case of a mixture of two gases A and B. Dividing P_A by P_T, we obtain

$$\frac{P_A}{P_T} = \frac{n_A RT/V}{(n_A + n_B)RT/V}$$

$$= \frac{n_A}{n_A + n_B}$$

$$= X_A$$

where X_A is called the mole fraction of A. The ***mole fraction*** is *a dimensionless quantity that expresses the ratio of the number of moles of one component to the number of moles of all components present.* In general, the mole fraction of component i in a mixture is given by

$$X_i = \frac{n_i}{n_T} \tag{5.13}$$

where n_i and n_T are the number of moles of component i and the total number of moles present, respectively. The mole fraction is always smaller than 1. We can now express the partial pressure of A as

$$P_A = X_A P_T$$

Similarly,

$$P_B = X_B P_T$$

Note that the sum of the mole fractions for a mixture of gases must be unity. If only two components are present, then

$$X_A + X_B = \frac{n_A}{n_A + n_B} + \frac{n_B}{n_A + n_B} = 1$$

If a system contains more than two gases, then the partial pressure of the ith component is related to the total pressure by

$$P_i = X_i P_T \tag{5.14}$$

How are partial pressures determined? A manometer can measure only the total pressure of a gaseous mixture. To obtain the partial pressures, we need to know the mole fractions of the components, which would involve elaborate chemical analyses. The most direct method of measuring partial pressures is using a mass spectrometer. The relative intensities of the peaks in a mass spectrum are directly proportional to the amounts, and hence to the mole fractions, of the gases present.

EXAMPLE 5.7

A mixture of gases contains 3.85 moles of neon (Ne), 0.92 mole of argon (Ar), and 2.59 moles of xenon (Xe). Calculate the partial pressures of the gases if the total pressure is 2.50 atm at a certain temperature.

Strategy What is the relationship between the partial pressure of a gas and the total gas pressure? How do we calculate the mole fraction of a gas?

Solution According to Equation (5.14), the partial pressure of Ne (P_{Ne}) is equal to the product of its mole fraction (X_{Ne}) and the total pressure (P_T)

$$\underset{\text{want to calculate}}{P_{Ne}} = \overset{\text{need to find}}{X_{Ne}}\underset{\text{given}}{P_T}$$

Using Equation (5.13), we calculate the mole fraction of Ne as follows:

$$X_{Ne} = \frac{n_{Ne}}{n_{Ne} + n_{Ar} + n_{Xe}} = \frac{3.85 \text{ mol}}{3.85 \text{ mol} + 0.92 \text{ mol} + 2.59 \text{ mol}} = 0.523$$

Therefore,

$$\begin{aligned} P_{Ne} &= X_{Ne}P_T \\ &= 0.523 \times 2.50 \text{ atm} \\ &= 1.31 \text{ atm} \end{aligned}$$

Similarly, we can calculate the mole fraction of argon and its partial pressure:

$$\begin{aligned} P_{Ar} &= X_{Ar}P_T \\ &= 0.125 \times 2.50 \text{ atm} \\ &= 0.313 \text{ atm} \end{aligned}$$

Finally, we calculate the mole fraction of xenon and its partial pressure:

$$\begin{aligned} P_{Xe} &= X_{Xe}P_T \\ &= 0.352 \times 2.50 \text{ atm} \\ &= 0.880 \text{ atm} \end{aligned}$$

Check The individual partial pressures must be less than the total pressure and make sure that the sum of the partial pressures is equal to the total pressure; that is, (1.31 + 0.313 + 0.880) atm = 2.50 atm.

Similar problems: 5.57, 5.58.

Practice Exercise A sample of natural gas contains 8.24 moles of methane (CH_4), 0.421 mole of ethane (C_2H_6), and 0.116 mole of propane (C_3H_8). If the total pressure of the gases is 1.37 atm, what are the partial pressures of the gases?

Dalton's law of partial pressures is useful for calculating volumes of gases collected over water. For example, when potassium chlorate ($KClO_3$) is heated, it decomposes to KCl and O_2:

$$2KClO_3(s) \longrightarrow 2KCl(s) + 3O_2(g)$$

The oxygen gas can be collected over water, as shown in Figure 5.14. Initially, the inverted bottle is completely filled with water. As oxygen gas is generated, the gas bubbles rise to the top and displace water from the bottle. This method of collecting a gas is based on the assumptions that the gas does not react with water and that it

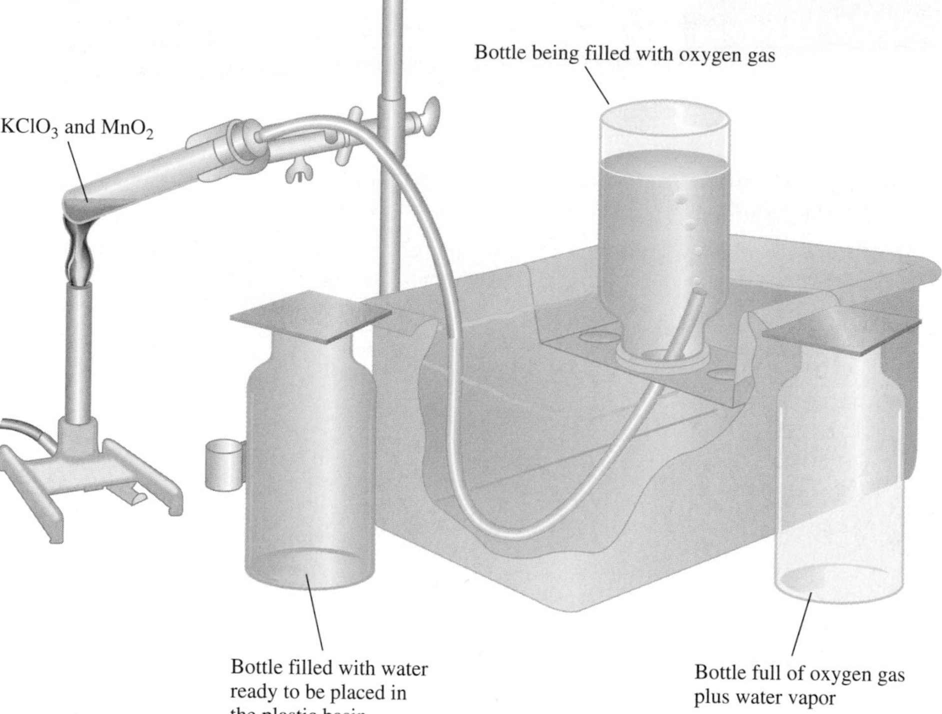

Figure 5.14
An apparatus for collecting gas over water. The oxygen generated by heating potassium chlorate ($KClO_3$) in the presence of a small amount of manganese dioxide (MnO_2), which speeds up the reaction, is bubbled through water and collected in a bottle as shown. Water originally present in the bottle is pushed into the trough by the oxygen gas.

is not appreciably soluble in it. These assumptions are valid for oxygen gas, but not for gases such as NH_3, which dissolves readily in water. The oxygen gas collected in this way is not pure, however, because water vapor is also present in the bottle. The total gas pressure is equal to the sum of the pressures exerted by the oxygen gas and the water vapor:

$$P_T = P_{O_2} + P_{H_2O}$$

Consequently, we must allow for the pressure caused by the presence of water vapor when we calculate the amount of O_2 generated. Table 5.2 shows the pressure of water vapor at various temperatures.

Table 5.2

Pressure of Water Vapor at Various Temperatures

Temperature (°C)	Water Vapor Pressure (mmHg)
0	4.58
5	6.54
10	9.21
15	12.79
20	17.54
25	23.76
30	31.82
35	42.18
40	55.32
45	71.88
50	92.51
55	118.04
60	149.38
65	187.54
70	233.7
75	289.1
80	355.1
85	433.6
90	525.76
95	633.90
100	760.00

EXAMPLE 5.8

Oxygen gas generated by the decomposition of potassium chlorate is collected as shown in Figure 5.14. The volume of oxygen collected at 26°C and atmospheric pressure of 771 mmHg is 141 mL. Calculate the mass (in grams) of oxygen gas obtained. The pressure of the water vapor at 26°C is 25.2 mmHg.

Strategy To solve for the mass of O_2 generated, we must first calculate the partial pressure of O_2 in the mixture. What gas law do we need? How do we convert pressure of O_2 gas to mass of O_2 in grams?

Solution From Dalton's law of partial pressures we know that

$$P_T = P_{O_2} + P_{H_2O}$$

(Continued)

Therefore,

$$\begin{aligned} P_{O_2} &= P_T - P_{H_2O} \\ &= 771 \text{ mmHg} - 25.2 \text{ mmHg} \\ &= 746 \text{ mmHg} \end{aligned}$$

From the ideal gas equation we write

$$PV = nRT = \frac{m}{\mathcal{M}} RT$$

where m and $\mathcal{M}$ are the mass of O_2 collected and the molar mass of O_2, respectively. Rearranging the equation we obtain

$$\begin{aligned} m = \frac{PV\mathcal{M}}{RT} &= \frac{(746/760) \text{ atm } (0.141 \text{ L})(32.00 \text{ g/mol})}{(0.0821 \text{ L} \cdot \text{atm/K} \cdot \text{mol})(273 + 26) \text{ K}} \\ &= 0.180 \text{ g} \end{aligned}$$

Similar problems: 5.63, 5.64.

Practice Exercise Hydrogen gas generated when calcium metal reacts with water is collected as shown in Figure 5.14. The volume of gas collected at 30°C and pressure of 988 mmHg is 641 mL. What is the mass (in grams) of the hydrogen gas obtained? The pressure of water vapor at 30°C is 31.82 mmHg.

REVIEW OF CONCEPTS

Each of the colored spheres represents a different gas molecule. Calculate the partial pressures of the gases if the total pressure is 2.4 atm.

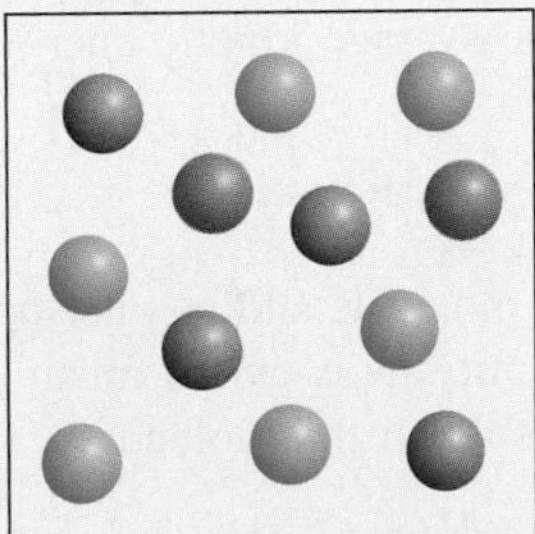

5.6 The Kinetic Molecular Theory of Gases

The gas laws help us to predict the behavior of gases, but they do not explain what happens at the molecular level to cause the changes we observe in the macroscopic world. For example, why does a gas expand on heating?

In the nineteenth century, a number of physicists, notably the Austrian physicist Ludwig Boltzmann and the Scottish physicist James Clerk Maxwell, found that the physical properties of gases can be explained in terms of the motion of individual molecules. This molecular movement is a form of *energy,* which we define as the capacity to do work or to produce change. In mechanics, *work* is defined as force times distance. Because energy can be measured as work, we can write

$$\begin{aligned} \text{energy} &= \text{work done} \\ &= \text{force} \times \text{distance} \end{aligned}$$

The ***joule (J)*** is *the SI unit of energy*

$$1\ \text{J} = 1\ \text{kg m}^2/\text{s}^2$$
$$= 1\ \text{N m}$$

Alternatively, energy can be expressed in kilojoules (kJ):

$$1\ \text{kJ} = 1000\ \text{J}$$

As we will see in Chapter 6, there are many different kinds of energy. ***Kinetic energy (KE)*** is the type of energy expended by a moving object, or *energy of motion.*

The findings of Maxwell, Boltzmann, and others resulted in *a number of generalizations about gas behavior* that have since been known as the ***kinetic molecular theory of gases,*** or simply the *kinetic theory of gases.* Central to the kinetic theory are these assumptions:

1. A gas is composed of molecules that are separated from each other by distances far greater than their own dimensions. The molecules can be considered to be "points"; that is, they possess mass but have negligible volume.
2. Gas molecules are in constant motion in random directions, and they frequently collide with one another. Collisions among molecules are perfectly elastic. In other words, energy can be transferred from one molecule to another as a result of a collision. Nevertheless, the total energy of all the molecules in a system remains the same.
3. Gas molecules exert neither attractive nor repulsive forces on one another.
4. The average kinetic energy of the molecules is proportional to the temperature of the gas in kelvins. Any two gases at the same temperature will have the same average kinetic energy. The average kinetic energy of a molecule is given by

$$\overline{\text{KE}} = \tfrac{1}{2}m\overline{u^2}$$

where m is the mass of the molecule and u is its speed. The horizontal bar denotes an average value. The quantity $\overline{u^2}$ is called mean square speed; it is the average of the square of the speeds of all the molecules:

$$\overline{u^2} = \frac{u_1^2 + u_2^2 + \cdots + u_N^2}{N}$$

where N is the number of molecules.

Assumption 4 enables us to write

$$\overline{\text{KE}} \propto T$$
$$\tfrac{1}{2}m\overline{u^2} \propto T$$

Hence,

$$\overline{\text{KE}} = \tfrac{1}{2}m\overline{u^2} = CT \quad (5.15)$$

where C is the proportionality constant and T is the absolute temperature.

According to the kinetic molecular theory, gas pressure is the result of collisions between molecules and the walls of their container. It depends on the frequency of collision per unit area and on how "hard" the molecules strike the wall. The theory also provides a molecular interpretation of temperature. According to Equation (5.15), the absolute temperature of a gas is a measure of the average kinetic energy of the molecules. In other words, the absolute temperature is a measure of the random motion of the molecules—the higher the temperature, the more energetic the molecules. Because it is related to the temperature of the gas sample, random molecular motion is sometimes referred to as *thermal motion.*

Application to the Gas Laws

Although the kinetic theory of gases is based on a rather simple model, the mathematical details involved are very complex. However, on a qualitative basis, it is possible to use the theory to account for the general properties of substances in the gaseous state. The following examples illustrate the range of its utility:

- **Compressibility of Gases.** Because molecules in the gas phase are separated by large distances (assumption 1), gases can be compressed easily to occupy less volume.
- **Boyle's Law.** The pressure exerted by a gas results from the impact of its molecules on the walls of the container. The collision rate, or the number of molecular collisions with the walls per second, is proportional to the number density (that is, number of molecules per unit volume) of the gas. Decreasing the volume of a given amount of gas increases its number density and hence its collision rate. For this reason, the pressure of a gas is inversely proportional to the volume it occupies; as volume decreases, pressure increases and vice versa.
- **Charles's Law.** Because the average kinetic energy of gas molecules is proportional to the sample's absolute temperature (assumption 4), raising the temperature increases the average kinetic energy. Consequently, molecules will collide with the walls of the container more frequently and with greater impact if the gas is heated, and thus the pressure increases. The volume of gas will expand until the gas pressure is balanced by the constant external pressure (see Figure 5.7).
- **Avogadro's Law.** We have shown that the pressure of a gas is directly proportional to both the density and the temperature of the gas. Because the mass of the gas is directly proportional to the number of moles (n) of the gas, we can represent density by n/V. Therefore,

  $$P \propto \frac{n}{V}T$$

  For two gases, 1 and 2, we write

  $$P_1 \propto \frac{n_1T_1}{V_1} = C\frac{n_1T_1}{V_1}$$

  $$P_2 \propto \frac{n_2T_2}{V_2} = C\frac{n_2T_2}{V_2}$$

  where C is the proportionality constant. Thus, for two gases under the same conditions of pressure, volume, and temperature (that is, when $P_1 = P_2$, $T_1 = T_2$, and $V_1 = V_2$), it follows that $n_1 = n_2$, which is a mathematical expression of Avogadro's law.
- **Dalton's Law of Partial Pressures.** If molecules do not attract or repel one another (assumption 3), then the pressure exerted by one type of molecule is unaffected by the presence of another gas. Consequently, the total pressure is given by the sum of individual gas pressures.

Another way of stating Avogadro's law is that at the same pressure and temperature, equal volumes of gases, whether they are the same or different gases, contain equal numbers of molecules.

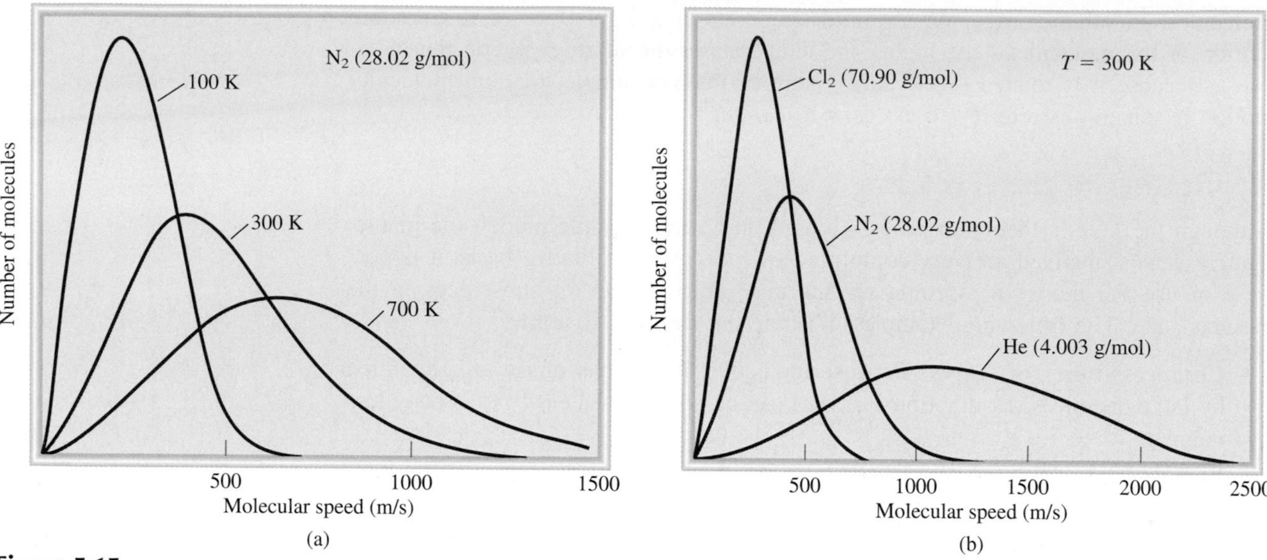

Figure 5.15

(a) The distribution of speeds for nitrogen gas at three different temperatures. At the higher temperatures, more molecules are moving at faster speeds. (b) The distribution of speeds for three gases at 300 K. At a given temperature, the lighter molecules are moving faster, on the average.

Distribution of Molecular Speeds

The kinetic theory of gases enables us to investigate molecular motion in more detail. Suppose we have a large number of gas molecules, say, 1 mole, in a container. As long as we hold the temperature constant, the average kinetic energy and the mean-square speed will remain unchanged as time passes. As you might expect, the motion of the molecules is totally random and unpredictable. At a given instant, how many molecules are moving at a particular speed? To answer this question Maxwell analyzed the behavior of gas molecules at different temperatures.

Figure 5.15(a) shows typical *Maxwell speed distribution curves* for nitrogen gas at three different temperatures. At a given temperature, the distribution curve tells us the number of molecules moving at a certain speed. The peak of each curve represents the *most probable speed,* that is, the speed of the largest number of molecules. Note that the most probable speed increases as temperature increases (the peak shifts toward the right). Furthermore, the curve also begins to flatten out with increasing temperature, indicating that larger numbers of molecules are moving at greater speed. Figure 5.15(b) shows the speed distributions of three gases at the *same* temperature. The difference in the curves can be explained by noting that lighter molecules move faster, on average, than heavier ones.

Root-Mean-Square Speed

How fast does a molecule move, on the average, at any temperature T? One way to estimate molecular speed is to calculate the ***root-mean-square (rms) speed (u_{rms})***, which is *an average molecular speed.* One of the results of the kinetic theory of gases is that the total kinetic energy of a mole of any gas equals $\frac{3}{2}RT$. Earlier we saw that the average kinetic energy of one molecule is $\frac{1}{2}m\overline{u^2}$ and so we can write

$$N_A(\tfrac{1}{2}m\overline{u^2}) = \tfrac{3}{2}RT$$

where N_A is Avogadro's number and m is the mass of a single molecule. Because $N_A m = \mathcal{M}$, where $\mathcal{M}$ is the molar mass, this equation can be rearranged to give

$$\overline{u^2} = \frac{3RT}{\mathcal{M}}$$

Taking the square root of both sides gives

$$\sqrt{\overline{u^2}} = u_{rms} = \sqrt{\frac{3RT}{\mathcal{M}}} \tag{5.16}$$

Equation (5.16) shows that the root-mean-square speed of a gas increases with the square root of its temperature (in kelvins). Because $\mathcal{M}$ appears in the denominator, it follows that the heavier the gas, the more slowly its molecules move. If we substitute 8.314 J/K · mol for R (see Appendix 1) and convert the molar mass to kg/mol, then u_{rms} will be calculated in meters per second (m/s).

EXAMPLE 5.9

Calculate the root-mean-square speeds of helium atoms and nitrogen molecules in m/s at 25°C.

Strategy To calculate the root-mean-square speed we need Equation (5.16). What units should we use for R and $\mathcal{M}$ so that u_{rms} will be expressed in m/s?

Solution To calculate u_{rms}, the units of R should be 8.314 J/K · mol and, because 1 J = 1 kg m²/s², the molar mass must be in kg/mol. The molar mass of He is 4.003 g/mol, or 4.003×10^{-3} kg/mol. From Equation (5.16),

$$\begin{aligned} u_{rms} &= \sqrt{\frac{3RT}{\mathcal{M}}} \\ &= \sqrt{\frac{3(8.314 \text{ J/K} \cdot \text{mol})(298 \text{ K})}{4.003 \times 10^{-3} \text{ kg/mol}}} \\ &= \sqrt{1.86 \times 10^6 \text{ J/kg}} \end{aligned}$$

Using the conversion factor 1 J = 1 kg m²/s² we get

$$\begin{aligned} u_{rms} &= \sqrt{1.86 \times 10^6 \text{ kg m}^2/\text{kg} \cdot \text{s}^2} \\ &= \sqrt{1.86 \times 10^6 \text{ m}^2/\text{s}^2} \\ &= 1.36 \times 10^3 \text{ m/s} \end{aligned}$$

The procedure is the same for N_2, the molar mass of which is 28.02 g/mol, or 2.802×10^{-2} kg/mol so that we write

$$\begin{aligned} u_{rms} &= \sqrt{\frac{3(8.314 \text{ J/K} \cdot \text{mol})(298 \text{ K})}{2.802 \times 10^{-2} \text{ kg/mol}}} \\ &= \sqrt{2.65 \times 10^5 \text{ m}^2/\text{s}^2} \\ &= 515 \text{ m/s} \end{aligned}$$

Check Because He is a lighter gas, we expect it to move faster, on average, than N_2. A quick way to check the answers is to note that the ratio of the two u_{rms} values ($1.36 \times 10^3/515 \approx 2.6$) should be equal to the square root of the ratios of the molar masses of N_2 to He, that is, $\sqrt{28/4} \approx 2.6$.

Similar problems: 5.73, 5.74.

Practice Exercise Calculate the root-mean-square speed of molecular chlorine in m/s at 20°C.

Jupiter. The interior of this massive planet consists mainly of hydrogen.

The calculation in Example 5.9 has an interesting relationship to the composition of Earth's atmosphere. Unlike Jupiter, Earth does not have appreciable amounts of hydrogen or helium in its atmosphere. Why is this the case? A smaller planet than Jupiter, Earth has a weaker gravitational attraction for these lighter molecules. A fairly straightforward calculation shows that to escape Earth's gravitational field, a molecule must possess an escape velocity equal to or greater than 1.1×10^4 m/s. Because the average speed of helium is considerably greater than that of molecular nitrogen or molecular oxygen, more helium atoms escape from Earth's atmosphere into outer space. Consequently, only a trace amount of helium is present in our atmosphere. On the other hand, Jupiter, with a mass about 320 times greater than that of Earth, retains both heavy and light gases in its atmosphere.

Gas Diffusion and Effusion

Gas Diffusion

Diffusion always proceeds from a region of higher concentration to one where the concentration is lower.

Animation: Diffusion of Gases

A direct demonstration of random motion is provided by ***diffusion,*** *the gradual mixing of molecules of one gas with molecules of another by virtue of their kinetic properties.* Despite the fact that molecular speeds are very great, the diffusion process takes a relatively long time to complete. For example, when a bottle of concentrated ammonia solution is opened at one end of a lab bench, it takes some time before a person at the other end of the bench can smell it. The reason is that a molecule experiences numerous collisions while moving from one end of the bench to the other, as shown in Figure 5.16. Thus, diffusion of gases always happens gradually, and not instantly as molecular speeds seem to suggest. Furthermore, because the root-mean-square speed of a light gas is greater than that of a heavier gas (see Example 5.9), a lighter gas will diffuse through a certain space more quickly than will a heavier gas. Figure 5.17 illustrates gaseous diffusion.

Figure 5.16
The path traveled by a single gas molecule. Each change in direction represents a collision with another molecule.

In 1832 the Scottish chemist Thomas Graham found that *under the same conditions of temperature and pressure, rates of diffusion for gases are inversely proportional to the square roots of their molar masses.* This statement, now known as ***Graham's law of diffusion,*** is expressed mathematically as

$$\frac{r_1}{r_2} = \sqrt{\frac{\mathcal{M}_2}{\mathcal{M}_1}} \tag{5.17}$$

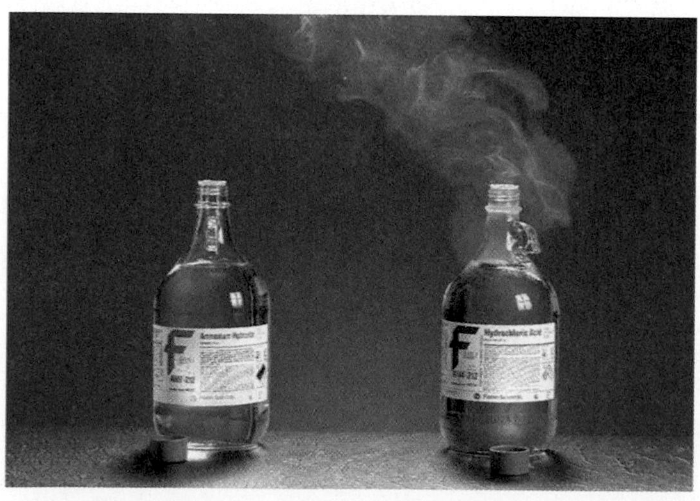

Figure 5.17
A demonstration of gas diffusion. NH_3 gas (from a bottle containing aqueous ammonia) combines with HCl gas (from a bottle containing hydrochloric acid) to form solid NH_4Cl. Because NH_3 is lighter and therefore diffuses faster, solid NH_4Cl first appears nearer the HCl bottle (on the right).

where r_1 and r_2 are the diffusion rates of gases 1 and 2, and $\mathcal{M}_1$ and $\mathcal{M}_2$ are their molar masses, respectively.

Gas Effusion

Whereas diffusion is a process by which one gas gradually mixes with another, ***effusion*** is *the process by which a gas under pressure escapes from one compartment of a container to another by passing through a small opening.* Figure 5.18 shows the effusion of a gas into a vacuum. Although effusion differs from diffusion in nature, the rate of effusion of a gas has the same form as Graham's law of diffusion [see Equation (5.17)]. A helium-filled rubber balloon deflates faster than an air-filled one because the rate of effusion through the pores of the rubber is faster for the lighter helium atoms than for the air molecules. Industrially, gas effusion is used to separate uranium isotopes in the forms of gaseous $^{235}UF_6$ and $^{238}UF_6$. By subjecting the gases to many stages of effusion, scientists were able to obtain highly enriched ^{235}U isotope, which was used in the construction of atomic bombs during World War II.

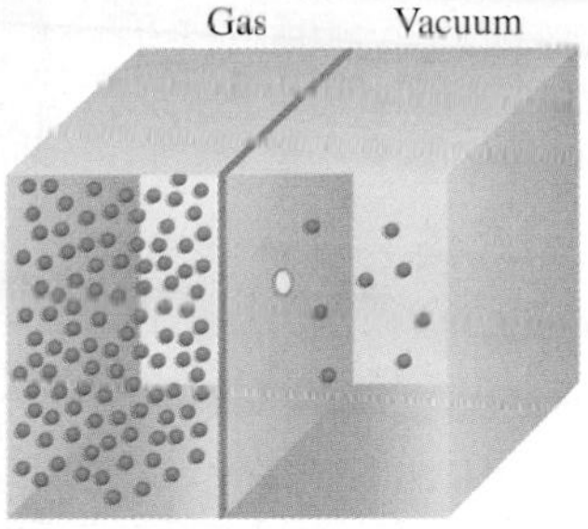

Figure 5.18
Gas effusion. Gas molecules move from a high-pressure region (left) to a low-pressure one through a pinhole.

EXAMPLE 5.10

A flammable gas made up only of carbon and hydrogen is found to effuse through a porous barrier in 3.50 min. Under the same conditions of temperature and pressure, it takes an equal volume of chlorine gas 7.34 min to effuse through the same barrier. Calculate the molar mass of the unknown gas, and suggest what this gas might be.

Strategy The rate of diffusion is the number of molecules passing through a porous barrier in a given time. The longer the time it takes, the slower is the rate. Therefore, the rate is *inversely* proportional to the time required for diffusion. Equation (5.17) can now be written as $r_1/r_2 = t_2/t_1 = \sqrt{\mathcal{M}_2/\mathcal{M}_1}$, where t_1 and t_2 are the times for effusion for gases 1 and 2, respectively.

Solution From the molar mass of Cl_2, we write

$$\frac{3.50 \text{ min}}{7.34 \text{ min}} = \sqrt{\frac{\mathcal{M}}{70.90 \text{ g/mol}}}$$

where $\mathcal{M}$ is the molar mass of the unknown gas. Solving for $\mathcal{M}$, we obtain

$$\mathcal{M} = \left(\frac{3.50 \text{ min}}{7.34 \text{ min}}\right)^2 \times 70.90 \text{ g/mol}$$
$$= 16.1 \text{ g/mol}$$

Because the molar mass of carbon is 12.01 g and that of hydrogen is 1.008 g, the gas is methane (CH_4).

Check Because lighter gases effuse faster than heavier gases, the molar mass of the unknown gas must be smaller than that of chlorine gas. Indeed, the molar mass of methane (16.04 g) is less than the molar mass of chlorine gas (70.90 g).

Similar problems: 5.110, 5.111.

Practice Exercise It takes 192 s for an unknown gas to effuse through a porous wall and 84 s for the same volume of N_2 gas to effuse at the same temperature and pressure. What is the molar mass of the unknown gas?

Review of Concepts

If 0.50 mol $H_2(g)$ and 1.0 mol $He(g)$ are compared at standard temperature and pressure, which of the following quantities will be equal to each other? (a) effusion rates, (b) average molecular speeds, (c) average kinetic energies, (d) volumes.

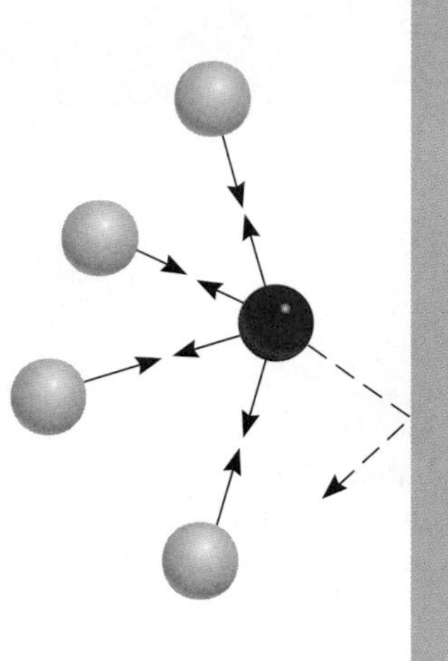

Figure 5.20

Effect of intermolecular forces on the pressure exerted by a gas. The speed of a molecule that is moving toward the container wall (red sphere) is reduced by the attractive forces exerted by its neighbors (gray spheres). Consequently, the impact this molecule makes with the wall is not as great as it would be if no intermolecular forces were present. In general, the measured gas pressure is lower than the pressure the gas would exert if it behaved ideally.

5.7 Deviation from Ideal Behavior

The gas laws and the kinetic molecular theory assume that molecules in the gaseous state do not exert any force, either attractive or repulsive, on one another. The other assumption is that the volume of the molecules is negligibly small compared with that of the container. A gas that satisfies these two conditions is said to exhibit *ideal behavior.*

Although we can assume that real gases behave like an ideal gas, we cannot expect them to do so under all conditions. For example, without intermolecular forces, gases could not condense to form liquids. The important question is: Under what conditions will gases most likely exhibit nonideal behavior?

Figure 5.19 shows PV/RT plotted against P for three real gases and an ideal gas at a given temperature. This graph provides a test of ideal gas behavior. According to the ideal gas equation (for 1 mole of gas), PV/RT equals 1, regardless of the actual gas pressure. (When $n = 1$, $PV = nRT$ becomes $PV = RT$, or $PV/RT = 1$.) For real gases, this is true only at moderately low pressures (≤ 5 atm); significant deviations occur as pressure increases. Attractive forces operate among molecules at relatively short distances. At atmospheric pressure, the molecules in a gas are far apart and the attractive forces are negligible. At high pressures, the density of the gas increases; the molecules are much closer to one another. Intermolecular forces can then be significant enough to affect the motion of the molecules, and the gas will not behave ideally.

Another way to observe the nonideal behavior of gases is to lower the temperature. Cooling a gas decreases the molecules' average kinetic energy, which in a sense deprives molecules of the drive they need to break from their mutual attraction.

To study real gases accurately, then, we need to modify the ideal gas equation, taking into account intermolecular forces and finite molecular volumes. Such an analysis was first made by the Dutch physicist J. D. van der Waals in 1873. Besides being mathematically simple, van der Waals's treatment provides us with an interpretation of real gas behavior at the molecular level.

Consider the approach of a particular molecule toward the wall of a container (Figure 5.20). The intermolecular attractions exerted by its neighbors tend to soften the impact made by this molecule against the wall. The overall effect is a lower gas pressure than we would expect for an ideal gas. Van der Waals suggested that the

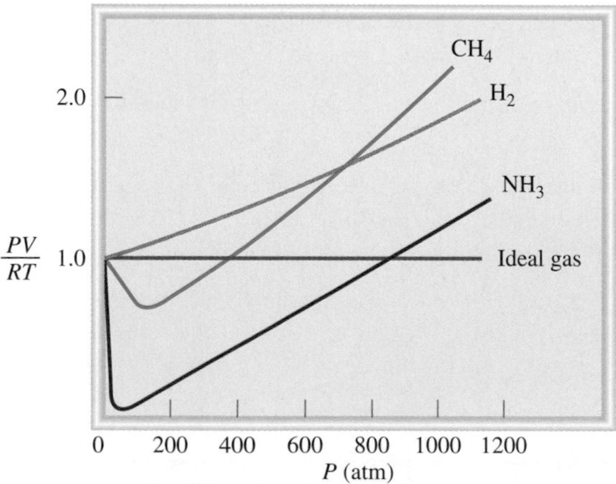

Figure 5.19

Plot of PV/RT versus P of 1 mole of a gas at 0°C. For 1 mole of an ideal gas, PV/RT is equal to 1, no matter what the pressure of the gas is. For real gases, we observe various deviations from ideality at high pressures. At very low pressures, all gases exhibit ideal behavior; that is, their PV/RT values all converge to 1 as P approaches zero.

pressure exerted by an ideal gas, P_{ideal}, is related to the experimentally measured; that is, observed pressure, P_{obs}, by the equation

$$P_{ideal} = \underset{\substack{\uparrow \\ \text{observed} \\ \text{pressure}}}{P_{real}} + \underset{\substack{\uparrow \\ \text{correction} \\ \text{term}}}{\frac{an^2}{V^2}}$$

where a is a constant and n and V are the number of moles and volume of the gas, respectively. The correction term for pressure (an^2/V^2) can be understood as follows. The intermolecular interaction that gives rise to nonideal behavior depends on how frequently any two molecules approach each other closely. The number of such "encounters" increases with the square of the number of molecules per unit volume, $(n/V)^2$, because the presence of each of the two molecules in a particular region is proportional to n/V and so a is just a proportionality constant. The quantity P_{ideal} is the pressure we would measure if there were no intermolecular attractions.

Another correction concerns the volume occupied by the gas molecules. In the ideal gas equation, V represents the volume of the container. However, each molecule does occupy a finite, although small, intrinsic volume, so the effective volume of the gas becomes ($V - nb$), where n is the number of moles of the gas and b is a constant. The term nb represents the volume occupied by n moles of the gas.

Having taken into account the corrections for pressure and volume, we can rewrite the ideal gas equation as follows:

Keep in mind that in Equation (5.18), P is the experimentally measured gas pressure and V is the volume of the gas container.

$$\underbrace{\left(P + \frac{an^2}{V^2}\right)}_{\text{corrected pressure}} \underbrace{(V - nb)}_{\text{corrected volume}} = nRT \tag{5.18}$$

Equation (5.18), *relating P, V, T, and n for a nonideal gas,* is known as the ***van der Waals equation.*** The van der Waals constants a and b are selected to give the best possible agreement between Equation (5.18) and observed behavior of a particular gas.

Table 5.3 lists the values of a and b for a number of gases. The value of a indicates how strongly molecules of a given type of gas attract one another. We see that helium atoms have the weakest attraction for one another, because helium has the smallest a value. There is also a rough correlation between molecular size and b. Generally, the larger the molecule (or atom), the greater b is, but the relationship between b and molecular (or atomic) size is not a simple one.

Table 5.3

van der Waals Constants of Some Common Gases

Gas	a $\left(\frac{\text{atm} \cdot \text{L}^2}{\text{mol}^2}\right)$	b $\left(\frac{\text{L}}{\text{mol}}\right)$
He	0.034	0.0237
Ne	0.211	0.0171
Ar	1.34	0.0322
Kr	2.32	0.0398
Xe	4.19	0.0266
H_2	0.244	0.0266
N_2	1.39	0.0391
O_2	1.36	0.0318
Cl_2	6.49	0.0562
CO_2	3.59	0.0427
CH_4	2.25	0.0428
CCl_4	20.4	0.138
NH_3	4.17	0.0371
H_2O	5.46	0.0305

EXAMPLE 5.11

Given that 2.75 moles of CO_2 occupy 4.70 L at 53°C, calculate the pressure of the gas (in atm) using (a) the ideal gas equation and (b) the van der Waals equation.

Strategy To calculate the pressure of CO_2 using the ideal gas equation, we proceed as in Example 5.2. What corrections are made to the pressure and volume terms in the van der Waals equation?

(Continued)

Solution

(a) We have the following data:

$$V = 4.70 \text{ L}$$
$$T = (53 + 273) \text{ K} = 326 \text{ K}$$
$$n = 2.75 \text{ mol}$$
$$R = 0.0821 \text{ L} \cdot \text{atm/K} \cdot \text{mol}$$

Substituting these values in the ideal gas equation, we write

$$P = \frac{nRT}{V}$$
$$= \frac{(2.75 \text{ mol})(0.0821 \text{ L} \cdot \text{atm/K} \cdot \text{mol})(326 \text{ K})}{4.70 \text{ L}}$$
$$= 15.7 \text{ atm}$$

(b) We need Equation (5.18). It is convenient to first calculate the correction terms in Equation (5.18) separately. From Table 5.3, we have

$$a = 3.59 \text{ atm} \cdot \text{L}^2/\text{mol}^2$$
$$b = 0.0427 \text{ L/mol}$$

so that the correction terms for pressure and volume are

$$\frac{an^2}{V^2} = \frac{(3.59 \text{ atm} \cdot \text{L}^2/\text{mol}^2)(2.75 \text{ mol})^2}{(4.70 \text{ L})^2} = 1.23 \text{ atm}$$
$$nb = (2.75 \text{ mol})(0.0427 \text{ L/mol}) = 0.117 \text{ L}$$

Finally, substituting these values in the van der Waals equation, we have

$$(P + 1.23 \text{ atm})(4.70 \text{ L} - 0.117 \text{ L}) = (2.75 \text{ mol})(0.0821 \text{ L} \cdot \text{atm/K} \cdot \text{mol})(326 \text{ K})$$
$$P = 14.8 \text{ atm}$$

Check Based on your understanding of nonideal gas behavior, is it reasonable that the pressure calculated using the van der Waals equation should be smaller than that using the ideal gas equation? Why?

Similar problems: 5.81, 5.82.

Practice Exercise Using the data shown in Table 5.3, calculate the pressure exerted by 4.37 moles of chlorine gas confined in a volume of 2.45 L at 38°C. Compare the pressure with that calculated using the ideal gas equation.

REVIEW OF CONCEPTS

What pressure and temperature conditions cause the most deviation from ideal behavior?

Key Equations

$P_1V_1 = P_2V_2$	(5.2)	Boyle's law. For calculating pressure or volume changes.
$\frac{V_1}{T_1} = \frac{V_2}{T_2}$	(5.4)	Charles's law. For calculating temperature or volume changes.
$\frac{P_1}{T_1} = \frac{P_2}{T_2}$	(5.6)	Charles's law. For calculating temperature or pressure changes.

$V = k_4 n$	(5.7)	Avogadro's law. Constant P and T.
$PV = nRT$	(5.8)	Ideal gas equation.
$\frac{P_1V_1}{n_1T_1} = \frac{P_2V_2}{n_2T_2}$	(5.9)	Combined ideal gas equations for initial and final states.
$\frac{P_1V_1}{T_1} = \frac{P_2V_2}{T_2}$	(5.10)	For calculating changes in pressure, temperature, or volume when n is constant.
$d = \frac{P\mathcal{M}}{RT}$	(5.11)	For calculating density or molar mass.
$X_i = \frac{n_i}{n_T}$	(5.13)	Definition of mole fraction.
$P_i = X_iP_T$	(5.14)	Dalton's law of partial pressures. For calculating partial pressures.
$u_{rms} = \sqrt{\frac{3RT}{\mathcal{M}}}$	(5.16)	For calculating the root-mean-square speed of gas molecules.
$\frac{r_1}{r_2} = \sqrt{\frac{\mathcal{M}_2}{\mathcal{M}_1}}$	(5.17)	Graham's law of diffusion and effusion.
$\left(P + \frac{an^2}{V^2}\right)(V - nb) = nRT$	(5.18)	van der Waals equation. For calculating the pressure of a nonideal gas.

Summary of Facts and Concepts

1. Under atmospheric conditions, a number of elemental substances are gases: H_2, N_2, O_2, O_3, F_2, Cl_2, and the Group 8A elements (the noble gases).
2. Gases exert pressure because their molecules move freely and collide with any surface in their paths. Gas pressure units include millimeters of mercury (mmHg), torr, pascals, and atmospheres. One atmosphere equals 760 mmHg, or 760 torr.
3. The pressure-volume relationships of ideal gases are governed by Boyle's law: Volume is inversely proportional to pressure (at constant T and n). The temperature-volume relationships of ideal gases are described by Charles's and Gay-Lussac's law: Volume is directly proportional to temperature (at constant P and n). Absolute zero (−273.15°C) is the lowest theoretically attainable temperature. On the Kelvin temperature scale, 0 K is absolute zero. In all gas law calculations, temperature must be expressed in kelvins. The amount-volume relationships of ideal gases are described by Avogadro's law: Equal volumes of gases contain equal numbers of molecules (at the same T and P).
4. The ideal gas equation, $PV = nRT$, combines the laws of Boyle, Charles, and Avogadro. This equation describes the behavior of an ideal gas.
5. Dalton's law of partial pressures states that in a mixture of gases each gas exerts the same pressure as it would if it were alone and occupied the same volume.
6. The kinetic molecular theory, a mathematical way of describing the behavior of gas molecules, is based on the following assumptions: Gas molecules are separated by distances far greater than their own dimensions, they possess mass but have negligible volume, they are in constant motion, and they frequently collide with one another. The molecules neither attract nor repel one another. A Maxwell speed distribution curve shows how many gas molecules are moving at various speeds at a given temperature. As temperature increases, more molecules move at greater speeds.
7. In diffusion, two gases gradually mix with each other. In effusion, gas molecules move through a small opening under pressure. Both processes are governed by the same mathematical law.
8. The van der Waals equation is a modification of the ideal gas equation that takes into account the nonideal behavior of real gases. It corrects for two facts: Real gas molecules do exert forces on each other and they do have volume. The van der Waals constants are determined experimentally for each gas.

Key Words

Absolute temperature scale, p. 144
Absolute zero, p. 144
Atmospheric pressure, p. 139
Avogadro's law, p. 146
Barometer, p. 139
Boyle's law, p. 143
Charles's and Gay-Lussac's law, p. 145
Charles's law, p. 145
Dalton's law of partial pressures, p. 153
Diffusion, p. 162
Effusion, p. 163
Gas constant (R), p. 147
Graham's law of diffusion, p. 162
Ideal gas, p. 147
Ideal gas equation, p. 147
Joule (J), p. 158
Kelvin temperature scale, p. 144
Kinetic energy (KE), p. 158
Kinetic molecular theory of gases, p. 158
Manometer, p. 140
Mole fraction, p. 154
Newton (N), p. 138
Partial pressure, p. 153
Pascal (Pa), p. 138
Pressure, p. 138
Root-mean-square (rms) speed (u_{rms}), p. 160
Standard atmospheric pressure (1 atm), p. 139
Standard temperature and pressure (STP), p. 147
van der Waals equation, p. 165

Questions and Problems

Substances That Exist as Gases

Review Questions

5.1 Name five elements and five compounds that exist as gases at room temperature.

5.2 List the physical characteristics of gases.

Pressure of a Gas

Review Questions

5.3 Define pressure and give the common units for pressure.

5.4 Describe how a barometer and a manometer are used to measure gas pressure.

5.5 Why is mercury a more suitable substance to use in a barometer than water?

5.6 Explain why the height of mercury in a barometer is independent of the cross-sectional area of the tube. Would the barometer still work if the tubing were tilted at an angle, say 15° (see Figure 5.2)?

5.7 Would it be easier to drink water with a straw on top of Mt. Everest or at the foot? Explain.

5.8 Is the atmospheric pressure in a mine that is 500 m below sea level greater or less than 1 atm?

5.9 What is the difference between a gas and a vapor? At 25°C, which of the following substances in the gas phase should be properly called a gas and which should be called a vapor: molecular nitrogen (N_2), mercury?

5.10 If the maximum distance that water may be brought up a well by a suction pump is 34 ft (10.3 m), how is it possible to obtain water and oil from hundreds of feet below the surface of Earth?

5.11 Why is it that if the barometer reading falls in one part of the world, it must rise somewhere else?

5.12 Why do astronauts have to wear protective suits when they are on the surface of the moon?

Problems

5.13 Convert 634 mmHg to atm, torr, and kPa.

5.14 The atmospheric pressure at the summit of Mt. McKinley is 606 mmHg on a certain day. What is the pressure in atm and in kPa?

The Gas Laws

Review Questions

5.15 State the following gas laws in words and also in the form of an equation: Boyle's law, Charles's law, Avogadro's law. In each case, indicate the conditions under which the law is applicable, and give the units for each quantity in the equation.

5.16 Explain why a helium weather balloon expands as it rises in the air. Assume that the temperature remains constant.

Problems

5.17 A gaseous sample of a substance is cooled at constant pressure. Which of the following diagrams best represents the situation if the final temperature is (a) above the boiling point of the substance and (b) below the boiling point but above the freezing point of the substance?

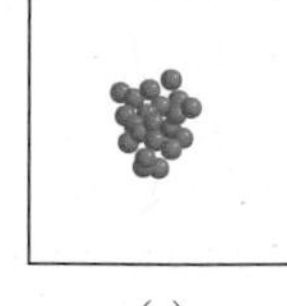
(a)

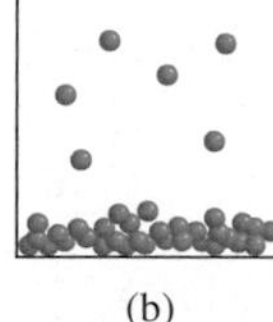
(b)

(c)

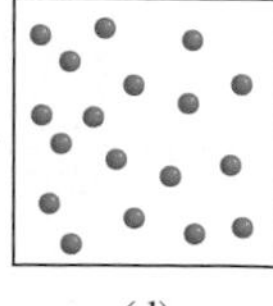
(d)

5.18 Consider the following gaseous sample in a cylinder fitted with a movable piston. Initially there are *n* moles of the gas at temperature *T*, pressure *P*, and volume *V*.

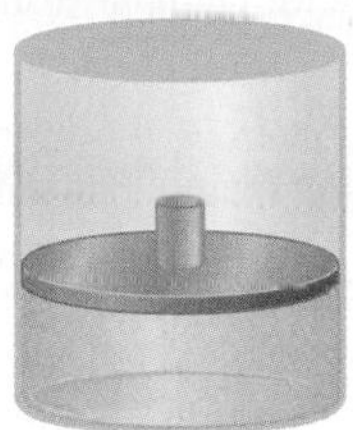

Choose the cylinder shown next that correctly represents the gas after each of the following changes. (1) The pressure on the piston is tripled at constant *n* and *T*. (2) The temperature is doubled at constant *n* and *P*. (3) *n* moles of another gas are added at constant *T* and *P*. (4) *T* is halved and pressure on the piston is reduced to a quarter of its original value.

(a)

(b)

(c)

5.19 A gas occupying a volume of 725 mL at a pressure of 0.970 atm is allowed to expand at constant temperature until its pressure reaches 0.541 atm. What is its final volume?

5.20 At 46°C a sample of ammonia gas exerts a pressure of 5.3 atm. What is the pressure when the volume of the gas is reduced to one-tenth (0.10) of the original value at the same temperature?

5.21 The volume of a gas is 5.80 L, measured at 1.00 atm. What is the pressure of the gas in mmHg if the volume is changed to 9.65 L? (The temperature remains constant.)

5.22 A sample of air occupies 3.8 L when the pressure is 1.2 atm. (a) What volume does it occupy at 6.6 atm? (b) What pressure is required in order to compress it to 0.075 L? (The temperature is kept constant.)

5.23 A 36.4-L volume of methane gas is heated from 25°C to 88°C at constant pressure. What is the final volume of the gas?

5.24 Under constant-pressure conditions a sample of hydrogen gas initially at 88°C and 9.6 L is cooled until its final volume is 3.4 L. What is its final temperature?

5.25 Ammonia burns in oxygen gas to form nitric oxide (NO) and water vapor. How many volumes of NO are obtained from one volume of ammonia at the same temperature and pressure?

5.26 Molecular chlorine and molecular fluorine combine to form a gaseous product. Under the same conditions of temperature and pressure it is found that one volume of Cl_2 reacts with three volumes of F_2 to yield two volumes of the product. What is the formula of the product?

The Ideal Gas Equation

Review Questions

5.27 List the characteristics of an ideal gas.

5.28 Write the ideal gas equation and also state it in words. Give the units for each term in the equation.

5.29 What are standard temperature and pressure (STP)? What is the significance of STP in relation to the volume of 1 mole of an ideal gas?

5.30 Why is the density of a gas much lower than that of a liquid or solid under atmospheric conditions? What units are normally used to express the density of gases?

Problems

5.31 A sample of nitrogen gas kept in a container of volume 2.3 L and at a temperature of 32°C exerts a pressure of 4.7 atm. Calculate the number of moles of gas present.

5.32 Given that 6.9 moles of carbon monoxide gas are present in a container of volume 30.4 L, what is the pressure of the gas (in atm) if the temperature is 62°C?

5.33 What volume will 5.6 moles of sulfur hexafluoride (SF_6) gas occupy if the temperature and pressure of the gas are 128°C and 9.4 atm?

5.34 A certain amount of gas at 25°C and at a pressure of 0.800 atm is contained in a glass vessel. Suppose that the vessel can withstand a pressure of 2.00 atm. How high can you raise the temperature of the gas without bursting the vessel?

5.35 A gas-filled balloon having a volume of 2.50 L at 1.2 atm and 25°C is allowed to rise to the stratosphere (about 30 km above the surface of Earth), where the temperature and pressure are −23°C and 3.00×10^{-3} atm, respectively. Calculate the final volume of the balloon.

5.36 The temperature of 2.5 L of a gas initially at STP is increased to 250°C at constant volume. Calculate the final pressure of the gas in atm.

5.37 The pressure of 6.0 L of an ideal gas in a flexible container is decreased to one-third of its original pressure, and its absolute temperature is decreased by one-half. What is the final volume of the gas?

5.38 A gas evolved during the fermentation of glucose (wine making) has a volume of 0.78 L when measured at 20.1°C and 1.00 atm. What was the volume of this gas at the fermentation temperature of 36.5°C and 1.00 atm pressure?

5.39 An ideal gas originally at 0.85 atm and 66°C was allowed to expand until its final volume, pressure, and temperature were 94 mL, 0.60 atm, and 45°C, respectively. What was its initial volume?

5.40 The volume of a gas at STP is 488 mL. Calculate its volume at 22.5 atm and 150°C.

5.41 A gas at 772 mmHg and 35.0°C occupies a volume of 6.85 L. Calculate its volume at STP.

5.42 Dry ice is solid carbon dioxide. A 0.050-g sample of dry ice is placed in an evacuated 4.6-L vessel at 30°C. Calculate the pressure inside the vessel after all the dry ice has been converted to CO_2 gas.

5.43 A volume of 0.280 L of a gas at STP weighs 0.400 g. Calculate the molar mass of the gas.

5.44 A quantity of gas weighing 7.10 g at 741 torr and 44°C occupies a volume of 5.40 L. What is its molar mass?

5.45 The ozone molecules present in the stratosphere absorb much of the harmful radiation from the sun. Typically, the temperature and pressure of ozone in the stratosphere are 250 K and 1.0×10^{-3} atm, respectively. How many ozone molecules are present in 1.0 L of air under these conditions?

5.46 Assuming that air contains 78 percent N_2, 21 percent O_2, and 1 percent Ar, all by volume, how many molecules of each type of gas are present in 1.0 L of air at STP?

5.47 A 2.10-L vessel contains 4.65 g of a gas at 1.00 atm and 27.0°C. (a) Calculate the density of the gas in grams per liter. (b) What is the molar mass of the gas?

5.48 Calculate the density of hydrogen bromide (HBr) gas in grams per liter at 733 mmHg and 46°C.

5.49 A certain anesthetic contains 64.9 percent C, 13.5 percent H, and 21.6 percent O by mass. At 120°C and 750 mmHg, 1.00 L of the gaseous compound weighs 2.30 g. What is the molecular formula of the compound?

5.50 A compound has the empirical formula SF_4. At 20°C, 0.100 g of the gaseous compound occupies a volume of 22.1 mL and exerts a pressure of 1.02 atm. What is its molecular formula?

5.51 Dissolving 3.00 g of an impure sample of calcium carbonate in hydrochloric acid produced 0.656 L of carbon dioxide (measured at 20.0°C and 792 mmHg). Calculate the percent by mass of calcium carbonate in the sample. State any assumptions.

5.52 Calculate the mass in grams of hydrogen chloride produced when 5.6 L of molecular hydrogen measured at STP react with an excess of molecular chlorine gas.

5.53 A quantity of 0.225 g of a metal M (molar mass = 27.0 g/mol) liberated 0.303 L of molecular hydrogen (measured at 17°C and 741 mmHg) from an excess of hydrochloric acid. Deduce from these data the corresponding equation and write formulas for the oxide and sulfate of M.

5.54 A compound of P and F was analyzed as follows: Heating 0.2324 g of the compound in a 378-cm^3 container turned all of it to gas, which had a pressure of 97.3 mmHg at 77°C. Then the gas was mixed with calcium chloride solution, which turned all of the F to 0.2631 g of CaF_2. Determine the molecular formula of the compound.

Dalton's Law of Partial Pressures

Review Questions

5.55 Define Dalton's law of partial pressures and mole fraction. Does mole fraction have units?

5.56 A sample of air contains only nitrogen and oxygen gases whose partial pressures are 0.80 atm and 0.20 atm, respectively. Calculate the total pressure and the mole fractions of the gases.

Problems

5.57 A mixture of gases contains CH_4, C_2H_6, and C_3H_8. If the total pressure is 1.50 atm and the numbers of moles of the gases present are 0.31 mole for CH_4, 0.25 mole for C_2H_6, and 0.29 mole for C_3H_8, calculate the partial pressures of the gases.

5.58 A 2.5-L flask at 15°C contains a mixture of three gases, N_2, He, and Ne, at partial pressures of 0.32 atm for N_2, 0.15 atm for He, and 0.42 atm for Ne. (a) Calculate the total pressure of the mixture. (b) Calculate the volume in liters at STP occupied by He and Ne if the N_2 is removed selectively.

5.59 Dry air near sea level has the following composition by volume: N_2, 78.08 percent; O_2, 20.94 percent; Ar, 0.93 percent; CO_2, 0.05 percent. The atmospheric pressure is 1.00 atm. Calculate (a) the partial pressure of each gas in atm and (b) the concentration of each gas in moles per liter at 0°C. (*Hint:* Because volume is proportional to the number of moles present, mole fractions of gases can be expressed as ratios of volumes at the same temperature and pressure.)

5.60 A mixture of helium and neon gases is collected over water at 28.0°C and 745 mmHg. If the partial pressure of helium is 368 mmHg, what is the partial pressure of neon? (Vapor pressure of water at 28°C = 28.3 mmHg.)

5.61 Consider the three gas containers shown here. All of them have the same volume and are at the same temperature. (a) Which container has the smallest mole fraction of gas A (blue sphere)? (b) Which container has the highest partial pressure of gas B (green sphere)?

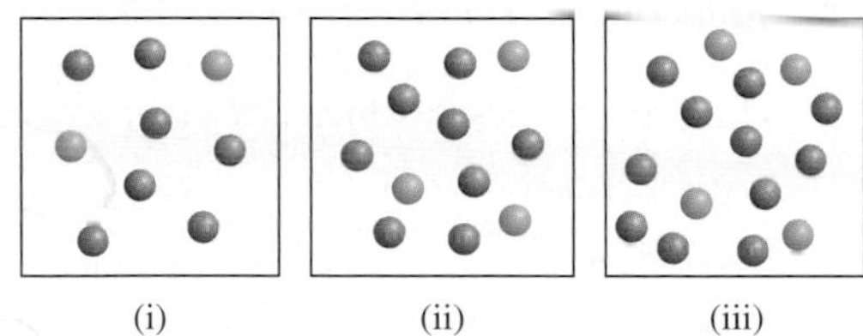

5.62 The volume of the box on the right is twice that of the box on the left. The boxes contain helium atoms (red) and hydrogen molecules (green) at the same temperature. (a) Which box has a higher total pressure? (b) Which box has a lower partial pressure of helium?

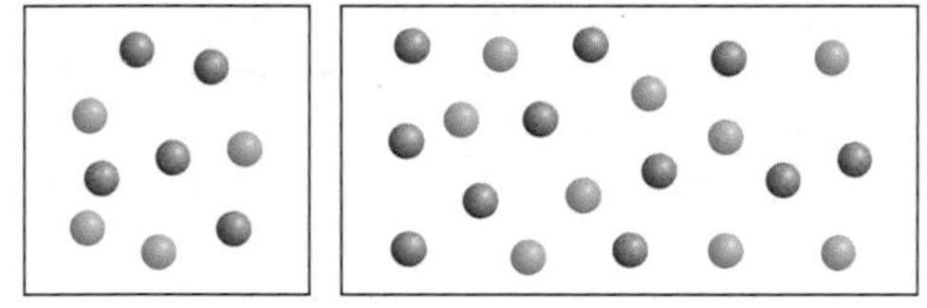

5.63 A piece of sodium metal undergoes complete reaction with water as follows:

$$2Na(s) + 2H_2O(l) \longrightarrow 2NaOH(aq) + H_2(g)$$

The hydrogen gas generated is collected over water at 25.0°C. The volume of the gas is 246 mL measured at 1.00 atm. Calculate the number of grams of sodium used in the reaction. (Vapor pressure of water at 25°C = 0.0313 atm.)

5.64 A sample of zinc metal is allowed to react completely with an excess of hydrochloric acid:

$$Zn(s) + 2HCl(aq) \longrightarrow ZnCl_2(aq) + H_2(g)$$

The hydrogen gas produced is collected over water at 25.0°C using an arrangement similar to that shown in Figure 5.14. The volume of the gas is 7.80 L, and the atmospheric pressure is 0.980 atm. Calculate the amount of zinc metal in grams consumed in the reaction. (Vapor pressure of water at 25°C = 23.8 mmHg.)

5.65 Helium is mixed with oxygen gas for deep sea divers. Calculate the percent by volume of oxygen gas in the mixture if the diver has to submerge to a depth where the total pressure is 4.2 atm. The partial pressure of oxygen is maintained at 0.20 atm at this depth.

5.66 A sample of ammonia (NH_3) gas is completely decomposed to nitrogen and hydrogen gases over heated iron wool. If the total pressure is 866 mmHg, calculate the partial pressures of N_2 and H_2.

Kinetic Molecular Theory of Gases

Review Questions

5.67 What are the basic assumptions of the kinetic molecular theory of gases?

5.68 What is thermal motion?

5.69 What does the Maxwell speed distribution curve tell us? Does Maxwell's theory work for a sample of 200 molecules? Explain.

5.70 Write the expression for the root-mean-square speed for a gas at temperature *T*. Define each term in the equation and show the units that are used in the calculations.

5.71 Which of the following two statements is correct? (a) Heat is produced by the collision of gas molecules against one another. (b) When a gas is heated, the molecules collide with one another more often.

5.72 Three gaseous compounds containing fluorine are illustrated here. Which of the three gases will have the highest root-mean-square speed? Which of the three gases will have the highest average kinetic energy at a given temperature?

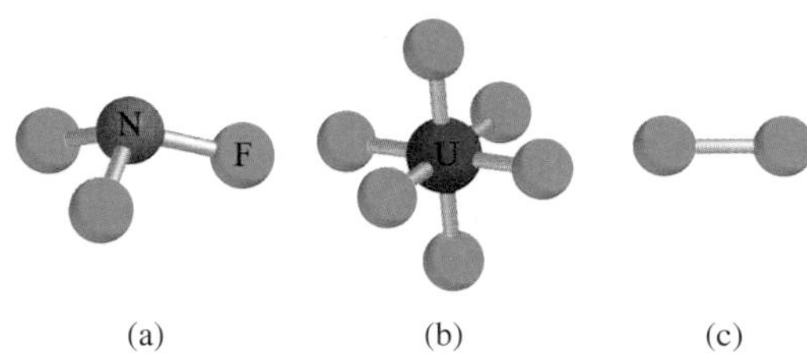

Problems

5.73 Compare the root-mean-square speeds of O_2 and UF_6 at 65°C.

5.74 The temperature in the stratosphere is −23°C. Calculate the root-mean-square speeds of N_2, O_2, and O_3 molecules in this region.

5.75 The average distance traveled by a molecule between successive collisions is called *mean free path*. For a given amount of a gas, how does the mean free path of a gas depend on (a) density, (b) temperature at constant volume, (c) pressure at constant temperature, (d) volume at constant temperature, and (e) size of the atoms?

5.76 At a certain temperature the speeds of six gaseous molecules in a container are 2.0 m/s, 2.2 m/s, 2.6 m/s, 2.7 m/s, 3.3 m/s, and 3.5 m/s. Calculate the root-mean-square speed and the average speed of the molecules. These two average values are close to each other, but the root-mean-square value is always the larger of the two. Why?

Deviation from Ideal Behavior

Review Questions

5.77 Give two pieces of evidence to show that gases do not behave ideally under all conditions.

5.78 Under what set of conditions would a gas be expected to behave most ideally? (a) High temperature and low pressure, (b) high temperature and high pressure, (c) low temperature and high pressure, (d) low temperature and low pressure.

5.79 Write the van der Waals equation for a real gas. Explain clearly the meaning of the corrective terms for pressure and volume.

5.80 The temperature of a real gas that is allowed to expand into a vacuum usually drops. Explain.

Problems

5.81 Using the data shown in Table 5.3, calculate the pressure exerted by 2.50 moles of CO_2 confined in a volume of 5.00 L at 450 K. Compare the pressure with that calculated using the ideal gas equation.

5.82 At 27°C, 10.0 moles of a gas in a 1.50-L container exert a pressure of 130 atm. Is this an ideal gas?

Additional Problems

5.83 Discuss the following phenomena in terms of the gas laws: (a) the pressure in an automobile tire increasing on a hot day, (b) the "popping" of a paper bag, (c) the expansion of a weather balloon as it rises in the air, (d) the loud noise heard when a lightbulb shatters.

5.84 Nitroglycerin, an explosive, decomposes according to the equation

$$4C_3H_5(NO_3)_3(s) \longrightarrow 12CO_2(g) + 10H_2O(g) + 6N_2(g) + O_2(g)$$

Calculate the total volume of gases produced when collected at 1.2 atm and 25°C from 2.6×10^2 g of nitroglycerin. What are the partial pressures of the gases under these conditions?

5.85 The empirical formula of a compound is CH. At 200°C, 0.145 g of this compound occupies 97.2 mL at a pressure of 0.74 atm. What is the molecular formula of the compound?

5.86 When ammonium nitrite (NH_4NO_2) is heated, it decomposes to give nitrogen gas. This property is used to inflate some tennis balls. (a) Write a balanced equation for the reaction. (b) Calculate the quantity (in grams) of NH_4NO_2 needed to inflate a tennis ball to a volume of 86.2 mL at 1.20 atm and 22°C.

5.87 The percent by mass of bicarbonate (HCO_3^-) in a certain Alka-Seltzer product is 32.5 percent. Calculate the volume of CO_2 generated (in milliliters) at 37°C and 1.00 atm when a person ingests a 3.29-g tablet. (*Hint:* The reaction is between HCO_3^- and HCl acid in the stomach.)

5.88 The boiling point of liquid nitrogen is −196°C. On the basis of this information alone, do you think nitrogen is an ideal gas?

5.89 In the metallurgical process of refining nickel, the metal is first combined with carbon monoxide to form tetracarbonylnickel, which is a gas at 43°C:

$$Ni(s) + 4CO(g) \longrightarrow Ni(CO)_4(g)$$

This reaction separates nickel from other solid impurities. (a) Starting with 86.4 g of Ni, calculate the pressure of $Ni(CO)_4$ in a container of volume 4.00 L. (Assume the above reaction goes to completion.) (b) On further heating the sample above 43°C, it is observed that the pressure of the gas increases much more rapidly than predicted based on the ideal gas equation. Explain.

5.90 The partial pressure of carbon dioxide varies with seasons. Would you expect the partial pressure in the Northern Hemisphere to be higher in the summer or winter? Explain.

5.91 A healthy adult exhales about 5.0×10^2 mL of a gaseous mixture with each breath. Calculate the number of molecules present in this volume at 37°C and 1.1 atm. List the major components of this gaseous mixture.

5.92 Sodium bicarbonate ($NaHCO_3$) is called baking soda because when heated, it releases carbon dioxide gas, which is responsible for the rising of cookies, doughnuts, and bread. (a) Calculate the volume (in liters) of CO_2 produced by heating 5.0 g of $NaHCO_3$ at 180°C and 1.3 atm. (b) Ammonium bicarbonate (NH_4HCO_3) has also been used for the same purpose. Suggest one advantage and one disadvantage of using NH_4HCO_3 instead of $NaHCO_3$ for baking.

5.93 A barometer having a cross-sectional area of 1.00 cm^2 at sea level measures a pressure of 76.0 cm of mercury. The pressure exerted by this column of mercury is equal to the pressure exerted by all the air on 1 cm^2 of Earth's surface. Given that the density of mercury is 13.6 g/mL, and the average radius of Earth is 6371 km, calculate the total mass of Earth's atmosphere in kilograms. (*Hint:* The surface area of a sphere is $4\pi r^2$, in which r is the radius of the sphere.)

5.94 Some commercial drain cleaners contain two components: sodium hydroxide and aluminum powder. When the mixture is poured down a clogged drain, the following reaction occurs:

$$2NaOH(aq) + 2Al(s) + 6H_2O(l) \longrightarrow 2NaAl(OH)_4(aq) + 3H_2(g)$$

The heat generated in this reaction helps melt away obstructions such as grease, and the hydrogen gas released stirs up the solids clogging the drain. Calculate the volume of H_2 formed at STP if 3.12 g of Al is treated with an excess of NaOH.

5.95 The volume of a sample of pure HCl gas was 189 mL at 25°C and 108 mmHg. It was completely dissolved in about 60 mL of water and titrated with an NaOH solution; 15.7 mL of the NaOH solution were required to neutralize the HCl. Calculate the molarity of the NaOH solution.

5.96 Propane (C_3H_8) burns in oxygen to produce carbon dioxide gas and water vapor. (a) Write a balanced equation for this reaction. (b) Calculate the number of liters of carbon dioxide measured at STP that could be produced from 7.45 g of propane.

5.97 Consider the apparatus shown here. When a small amount of water is introduced into the flask by squeezing the bulb of the medicine dropper, water is squirted upward out of the long glass tubing. Explain this observation. (*Hint:* Hydrogen chloride gas is soluble in water.)

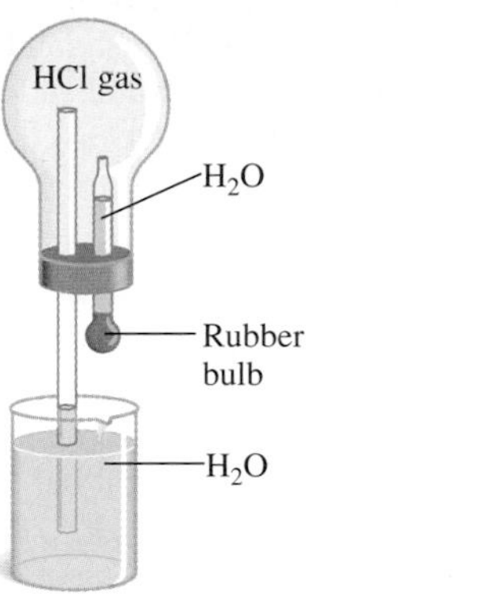

5.98 Nitric oxide (NO) reacts with molecular oxygen as follows:

$$2NO(g) + O_2(g) \longrightarrow 2NO_2(g)$$

Initially NO and O_2 are separated as shown in the figure. When the valve is opened, the reaction quickly goes to completion. Determine what gases remain at the end and calculate their partial pressures. Assume that the temperature remains constant at 25°C.

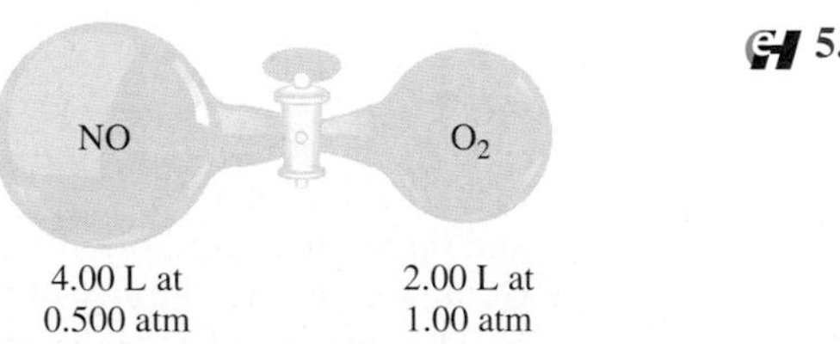

5.99 The apparatus shown in the diagram can be used to measure atomic and molecular speed. Suppose that a beam of metal atoms is directed at a rotating cylinder in a vacuum. A small opening in the cylinder allows the atoms to strike a target area. Because the cylinder is rotating, atoms traveling at different speeds will strike the target at different positions. In time, a layer of the metal will deposit on the target area, and the variation in its thickness is found to correspond to Maxwell's speed distribution. In one experiment it is found that at 850°C some bismuth (Bi) atoms struck the target at a point 2.80 cm from the spot directly opposite the slit. The diameter of the cylinder is 15.0 cm and it is rotating at 130 revolutions per second. (a) Calculate the speed (m/s) at which the target is moving. (*Hint:* The circumference of a circle is given by $2\pi r$, in which r is the radius.) (b) Calculate the time (in seconds) it takes for the target to travel 2.80 cm. (c) Determine the speed of the Bi atoms. Compare your result in (c) with the u_{rms} of Bi at 850°C. Comment on the difference.

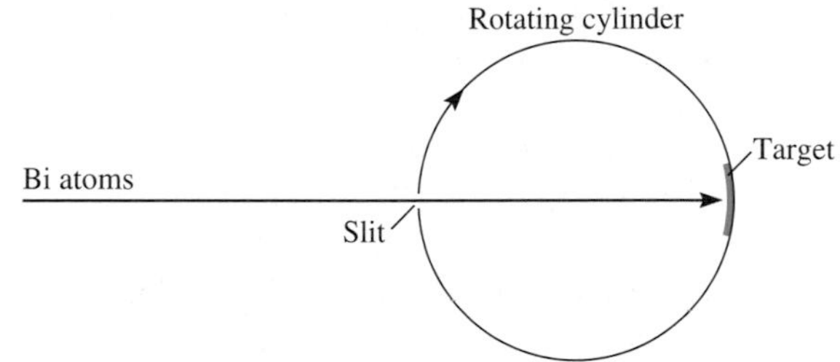

5.100 Acidic oxides such as carbon dioxide react with basic oxides like calcium oxide (CaO) and barium oxide (BaO) to form salts (metal carbonates). (a) Write equations representing these two reactions. (b) A student placed a mixture of BaO and CaO of combined mass 4.88 g in a 1.46-L flask containing carbon dioxide gas at 35°C and 746 mmHg. After the reactions were complete, she found that the CO_2 pressure had dropped to 252 mmHg. Calculate the percent composition of the mixture.

5.101 The running engine of an automobile produces carbon monoxide (CO), a toxic gas, at the rate of about 188 g CO per hour. A car is left idling at 20°C in a poorly ventilated garage that is 6.0 m long, 4.0 m wide, and 2.2 m high. (a) Calculate the rate of CO production in moles per minute. (b) How long would it take to build up a lethal concentration of CO of 1000 ppmv (parts per million by volume)?

5.102 Air entering the lungs ends up in tiny sacs called alveoli. It is from the alveoli that oxygen diffuses into the blood. The average radius of the alveoli is 0.0050 cm and the air inside contains 14 percent oxygen. Assuming that the pressure in the alveoli is 1.0 atm and the temperature is 37°C, calculate the number of oxygen molecules in one of the alveoli. (*Hint:* The volume of a sphere of radius r is $\frac{4}{3}\pi r^3$.)

5.103 It is said that every breath we take, on average, contains molecules that were once exhaled by Wolfgang Amadeus Mozart (1756–1791). These calculations demonstrate the validity of this statement. (a) Calculate the total number of molecules in the atmosphere. (*Hint:* Use the result in Problem 5.93 and 29.0 g/mol as the molar mass of air.) (b) Assuming the volume of every breath (inhale or exhale) is 500 mL, calculate the number of molecules exhaled in each breath at 37°C, which is human body temperature. (c) If Mozart's lifespan was exactly 35 years, what is the number of molecules exhaled in that period? (Given that an average person breathes 12 times per minute.) (d) Calculate the fraction of molecules in the atmosphere that were breathed out by Mozart. How many of Mozart's molecules do we breathe in with every inhalation of air? Round off your answer to one significant figure. (e) List three important assumptions in these calculations.

5.104 Under the same conditions of temperature and pressure, which of these gases would behave most ideally: Ne, N_2, or CH_4? Explain.

5.105 Based on your knowledge of the kinetic theory of gases, derive Graham's law of diffusion [Equation (5.17)].

5.106 A 6.11-g sample of a Cu-Zn alloy reacts with HCl acid to produce hydrogen gas. If the hydrogen gas has a volume of 1.26 L at 22°C and 728 mmHg, what is the percent of Zn in the alloy? (*Hint:* Cu does not react with HCl.)

5.107 Estimate the distance (in nanometers) between molecules of water vapor at 100°C and 1.0 atm. Assume ideal behavior. Repeat the calculation for liquid water at 100°C, given that the density of water is 0.96 g/cm^3 at that temperature. Comment on your results. (Assume water molecule to be a sphere with a diameter of 0.3 nm.) (*Hint:* First calculate the number density of water molecules. Next, convert the number density to linear density, that is, number of molecules in one direction.)

5.108 A stockroom supervisor measured the contents of a partially filled 25.0-gallon acetone drum on a day when the temperature was 18.0°C and atmospheric pressure was 750 mmHg, and found that 15.4 gallons of the solvent remained. After tightly sealing the drum, an assistant dropped the drum while carrying it upstairs to the organic laboratory. The drum was dented and its internal volume was decreased to 20.4 gallons. What is the total pressure inside the drum after the accident? The vapor pressure of acetone at 18.0°C is 400 mmHg. (*Hint:* At the time the drum was sealed, the pressure inside the drum, which is equal to the sum of the pressures of air and acetone, was equal to the atmospheric pressure.)

5.109 Lithium hydride reacts with water as follows:

$$LiH(s) + H_2O(l) \longrightarrow LiOH(aq) + H_2(g)$$

During World War II, U.S. pilots carried LiH tablets. In the event of a crash landing at sea, the LiH would react with the seawater and fill their life belts and lifeboats with hydrogen gas. How many grams of LiH are needed to fill a 4.1-L life belt at 0.97 atm and 12°C?

5.110 A sample of the gas discussed in Problem 5.38 is found to effuse through a porous barrier in 15.0 min. Under the same conditions of temperature and pressure, it takes N_2 12.0 min to effuse through the same barrier. Calculate the molar mass of the gas and suggest what gas it might be.

5.111 Nickel forms a gaseous compound of the formula $Ni(CO)_x$. What is the value of x given the fact that under the same conditions of temperature and pressure methane (CH_4) effuses 3.3 times faster than the compound?

Special Problems

5.112 Apply your knowledge of the kinetic theory of gases to these situations.

(a) Does a single molecule have a temperature?

(b) Two flasks of volumes V_1 and V_2 ($V_2 > V_1$) contain the same number of helium atoms at the same temperature. (i) Compare the root-mean-square (rms) speeds and average kinetic energies of the helium (He) atoms in the flasks. (ii) Compare the frequency and the force with which the He atoms collide with the walls of their containers.

(c) Equal numbers of He atoms are placed in two flasks of the same volume at temperatures T_1 and T_2 ($T_2 > T_1$). (i) Compare the rms speeds of the atoms in the two flasks. (ii) Compare the frequency and the force with which the He atoms collide with the walls of their containers.

(d) Equal numbers of He and neon (Ne) atoms are placed in two flasks of the same volume and the temperature of both gases is 74°C. Comment on the validity of these statements: (i) The rms speed of He is equal to that of Ne. (ii) The average kinetic energies of the two gases are equal. (iii) The rms speed of each He atom is 1.47×10^3 m/s.

5.113 Referring to the plot in Figure 5.19, (a) why do the plots of the gases dip before they rise? (b) Why do they all converge to 1 at very low P? (c) What is the meaning of the intercept on the ideal gas line? Does the intercept mean that the gas has become an ideal gas?

5.114 Referring to Figure 5.15, we see that the maximum of each speed distribution plot is called the most probable speed (u_{mp}) because it is the speed possessed by the largest number of molecules. It is given by $u_{mp} = \sqrt{2RT/\mathcal{M}}$. (a) Compare u_{mp} with u_{rms} for nitrogen at 25°C. (b) The following diagram shows the Maxwell speed distribution curves for an ideal gas at two different temperatures T_1 and T_2. Calculate the value of T_2.

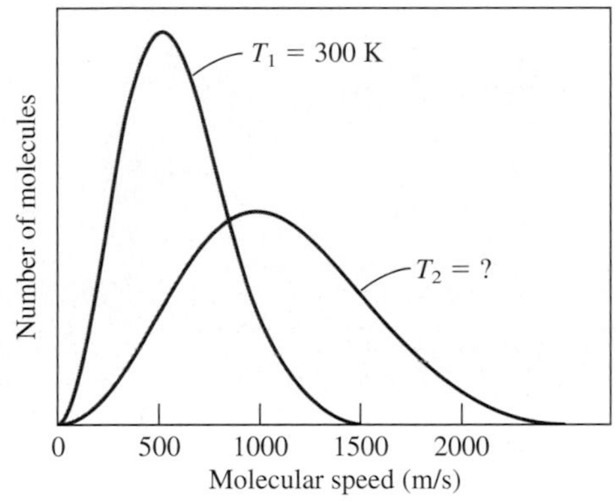

5.115 Use the kinetic theory of gases to explain why hot air rises.

5.116 One way to gain a physical understanding of b in the van der Waals equation is to calculate the "excluded volume." Assume that the distance of closest approach between two similar atoms is the sum of their radii ($2r$). (a) Calculate the volume around each atom into which the center of another atom cannot penetrate. (b) From your result in (a), calculate the excluded volume for 1 mole of the atoms, which is the constant b. How does this volume compare with the sum of the volumes of 1 mole of the atoms?

5.117 A 5.00-mole sample of NH_3 gas is kept in a 1.92 L container at 300 K. If the van der Waals equation is assumed to give the correct answer for the pressure of the gas, calculate the percent error made in using the ideal gas equation to calculate the pressure.

5.118 The root-mean-square speed of a certain gaseous oxide is 493 m/s at 20°C. What is the molecular formula of the compound?

5.119 In 2.00 min, 29.7 mL of He effuse through a small hole. Under the same conditions of pressure and temperature, 10.0 mL of a mixture of CO and CO_2 effuse through the hole in the same amount of time. Calculate the percent composition by volume of the mixture.

5.120 A gaseous reaction takes place at constant volume and constant pressure in a cylinder shown here. Which of the following equations best describes the reaction? The initial temperature (T_1) is twice that of the final temperature (T_2).

(a) A + B $\longrightarrow$ C

(b) AB $\longrightarrow$ C + D

(c) A + B $\longrightarrow$ C + D

(d) A + B $\longrightarrow$ 2C + D

Answers to Practice Exercises

5.1 0.986 atm. **5.2** 9.29 L. **5.3** 30.6 L. **5.4** 2.6 atm. **5.5** 44.1 g/mol. **5.6** 4.75 L. **5.7** CH_4: 1.29 atm; C_2H_6: 0.0657 atm; C_3H_8: 0.0181 atm. **5.8** 0.0653 g. **5.9** 321 m/s. **5.10** 146 g/mol. **5.11** 30.0 atm; 45.5 atm using the ideal gas equation.

CHAPTER

6

Energy Relationships in Chemical Reactions

Explosives such as dynamite and TNT undergo decomposition when detonated and release heat in exothermic reactions.

Chapter Outline

Student Interactive Activities

Animations
Heat Flow (6.2)

Electronic Homework
Example Practice Problems
End of Chapter Problems

Essential Concepts

Energy The many different forms of energy are, at least in principle, interconvertible.

First Law of Thermodynamics The first law of thermodynamics, which is based on the law of conservation of energy, relates the internal energy change of a system to the heat change and the work done. It can also be expressed to show the relationship between the internal energy change and enthalpy change of a process.

Thermochemistry Most chemical reactions involve the absorption or release of heat. At constant pressure, the heat change is equal to the enthalpy change. The heat change is measured by a calorimeter. Constant-pressure and constant-volume calorimeters are devices for measuring heat changes under the stated conditions.

Standard Enthalpy of Reaction Standard enthalpy of reaction is the enthalpy change when the reaction is carried out at 1 atm pressure. It can be calculated from the standard enthalpies of formation of reactants and products. Hess's law enables us to measure the standard enthalpy of formation of a compound in an indirect way.

6.1 The Nature of Energy and Types of Energy

"Energy" is a much-used term that represents a rather abstract concept. For instance, when we feel tired, we might say we haven't any *energy;* and we read about the need to find alternatives to nonrenewable *energy* sources. Unlike matter, energy is known and recognized by its effects. It cannot be seen, touched, smelled, or weighed.

Energy is usually defined as *the capacity to do work.* In Chapter 5 we defined work as "force × distance," but we will soon see that there are other kinds of work. All forms of energy are capable of doing work (that is, of exerting a force over a distance), but not all of them are equally relevant to chemistry. The energy contained in tidal waves, for example, can be harnessed to perform useful work, but the relationship between tidal waves and chemistry is minimal. Chemists define ***work*** as *directed energy change resulting from a process.* Kinetic energy—the energy produced by a moving object—is one form of energy that is of particular interest to chemists. Others include radiant energy, thermal energy, chemical energy, and potential energy.

Kinetic energy was first introduced in Chapter 5.

Radiant energy, or *solar energy, comes from the sun* and is Earth's primary energy source. Solar energy heats the atmosphere and Earth's surface, stimulates the growth of vegetation through the process known as photosynthesis, and influences global climate patterns.

Thermal energy is *the energy associated with the random motion of atoms and molecules.* In general, thermal energy can be calculated from temperature measurements. The more vigorous the motion of the atoms and molecules in a sample of matter, the hotter the sample is and the greater its thermal energy. However, we need to distinguish carefully between thermal energy and temperature. A cup of coffee at 70°C has a higher temperature than a bathtub filled with warm water at 40°C, but much more thermal energy is stored in the bathtub water because it has a much larger volume and greater mass than the coffee and therefore more water molecules and more molecular motion.

Chemical energy *is stored within the structural units of chemical substances;* its quantity is determined by the type and arrangement of constituent atoms. When substances participate in chemical reactions, chemical energy is released, stored, or converted to other forms of energy.

Potential energy is *energy available by virtue of an object's position.* For instance, because of its altitude, a rock at the top of a cliff has more potential energy and will make a bigger splash if it falls into the water below than a similar rock located partway down the cliff. Chemical energy can be considered a form of potential energy because it is associated with the relative positions and arrangements of atoms within a given substance.

As the water falls over the dam, its potential energy is converted to kinetic energy. Use of this energy to generate electricity is called hydroelectric power.

All forms of energy can be converted (at least in principle) from one form to another. We feel warm when we stand in sunlight because radiant energy is converted to thermal energy on our skin. When we exercise, chemical energy stored in our bodies is used to produce kinetic energy. When a ball starts to roll downhill, its potential energy is converted to kinetic energy. You can undoubtedly think of many other examples. Although energy can assume many different forms that are interconvertible, scientists have concluded that energy can be neither destroyed nor created. When one form of energy disappears, some other form of energy (of equal magnitude) must appear, and vice versa. This principle is summarized by the ***law of conservation of energy:*** *the total quantity of energy in the universe is assumed constant.*

This infrared photo shows where energy (heat) leaks through the house. The more red the color, the greater the energy is lost to the outside.

Animation: Heat Flow

6.2 Energy Changes in Chemical Reactions

Often the energy changes that take place during chemical reactions are of as much practical interest as the mass relationships we discussed in Chapter 3. For example, combustion reactions involving fuels such as natural gas and oil are carried out in daily life more for the thermal energy they release than for their products, which are water and carbon dioxide.

Almost all chemical reactions absorb or produce (release) energy, generally in the form of heat. It is important to understand the distinction between thermal energy and heat. ***Heat*** is *the transfer of thermal energy between two bodies that are at different temperatures.* Thus, we often speak of the "heat flow" from a hot object to a cold one. Although the term "heat" by itself implies the transfer of energy, we customarily talk of "heat absorbed" or "heat released" when describing the energy changes that occur during a process. ***Thermochemistry*** is *the study of heat change in chemical reactions.*

To analyze energy changes associated with chemical reactions we must first define the ***system,*** or *the specific part of the universe that is of interest to us.* For chemists, systems usually include substances involved in chemical and physical changes. For example, in an acid-base neutralization experiment, the system may be a beaker containing 50 mL of HCl to which 50 mL of NaOH is added. The ***surroundings*** are *the rest of the universe outside the system.*

There are three types of systems. An ***open system*** *can exchange mass and energy, usually in the form of heat with its surroundings.* For example, an open system may consist of a quantity of water in an open container, as shown in Figure 6.1(a). If we close the flask, as in Figure 6.1(b), so that no water vapor can escape from or condense into the container, we create a ***closed system,*** which *allows the transfer of energy (heat) but not mass.* By placing the water in a totally insulated container, we can construct an ***isolated system***, which *does not allow the transfer of either mass or energy,* as shown in Figure 6.1(c).

The combustion of hydrogen gas in oxygen is one of many chemical reactions that release considerable quantities of energy (Figure 6.2):

$$2H_2(g) + O_2(g) \longrightarrow 2H_2O(l) + \text{energy}$$

In this case, we label the reacting mixture (hydrogen, oxygen, and water molecules) the *system* and the rest of the universe the *surroundings.* Because energy cannot be created or destroyed, any energy lost by the system must be gained by the surroundings.

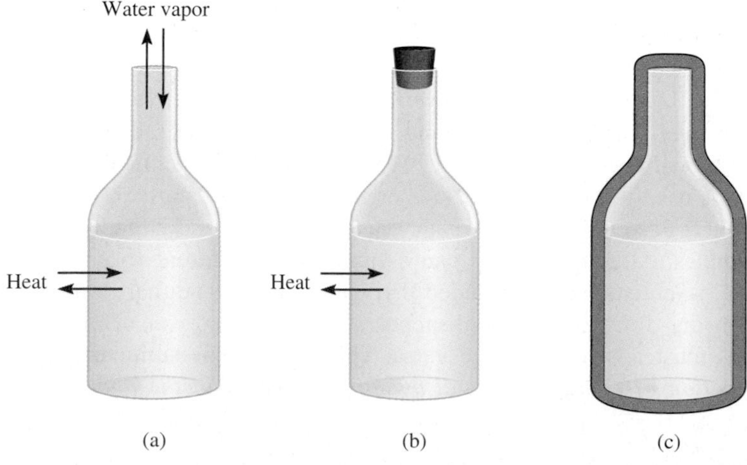

Figure 6.1
Three systems represented by water in a flask: (a) an open system, which allows exchange of both energy and mass with surroundings; (b) a closed system, which allows the exchange of energy but not mass; and (c) an isolated system, which allows neither energy nor mass to be exchanged (here the flask is enclosed by a vacuum jacket).

Figure 6.2
The Hindenburg disaster. The Hindenburg, a German airship filled with hydrogen gas, was destroyed in a spectacular fire at Lakehurst, New Jersey, in 1937.

Thus, the heat generated by the combustion process is transferred from the system to its surroundings. This reaction is an example of an ***exothermic process,*** which is *any process that gives off heat—*that is, *transfers thermal energy to the surroundings.*

Exo- comes from the Greek word meaning "outside"; ***endo-*** means "within."

Now consider another reaction, the decomposition of mercury(II) oxide (HgO) at high temperatures:

$$\text{energy} + 2\text{HgO}(s) \longrightarrow 2\text{Hg}(l) + \text{O}_2(g)$$

This reaction is an ***endothermic process,*** *in which heat has to be supplied to the system* (that is, to HgO) *by the surroundings.*

In exothermic reactions, the total energy of the products is less than the total energy of the reactants. The difference is the heat supplied by the system to the surroundings. Just the opposite happens in endothermic reactions. Here, the difference between the energy of the products and the energy of the reactants is equal to the heat supplied to the system by the surroundings.

On heating, HgO decomposes to give Hg and O_2.

REVIEW OF CONCEPTS

Classify each of the following as an open system, a closed system, or an isolated system.

(a) Soup kept in a closed thermo flask.
(b) A student reading in her dorm room.
(c) Air inside a tennis ball.

6.3 Introduction to Thermodynamics

Thermochemistry is part of a broader subject called ***thermodynamics,*** which is *the scientific study of the interconversion of heat and other kinds of energy.* The laws of thermodynamics provide useful guidelines for understanding the energetics and directions of processes. In this section we will concentrate on the first law of thermodynamics, which is particularly relevant to the study of thermochemistry. We will continue our discussion of thermodynamics in Chapter 18.

Figure 6.3
The gain in gravitational potential energy that occurs when a person climbs from the base to the top of a mountain is independent of the path taken.

In thermodynamics, we study changes in the ***state of a system,*** which is defined by *the values of all relevant macroscopic properties, for example, composition, energy, temperature, pressure, and volume.* Energy, pressure, volume, and temperature are said to be ***state functions****—properties that are determined by the state of the system, regardless of how that condition was achieved.* In other words, when the state of a system changes, the magnitude of change in any state function depends only on the initial and final states of the system and not on how the change is accomplished.

Changes in state functions do not depend on the pathway, but only on the initial and final states.

The state of a given amount of a gas is specified by its volume, pressure, and temperature. Consider a gas at 2 atm, 300 K, and 1 L (the initial state). Suppose a process is carried out at constant temperature such that the gas pressure decreases to 1 atm. According to Boyle's law, its volume must increase to 2 L. The final state then corresponds to 1 atm, 300 K, and 2 L. The change in volume (ΔV) is

The Greek letter *delta*, Δ, symbolizes change. We use Δ in this text always to mean final − initial.

$$\begin{aligned}\Delta V &= V_f - V_i \\ &= 2\text{ L} - 1\text{ L} \\ &= 1\text{ L}\end{aligned}$$

where V_i and V_f denote the initial and final volume, respectively. No matter how we arrive at the final state (for example, the pressure of the gas can be increased first and then decreased to 1 atm), the change in volume is always 1 L. Thus, the volume of a gas is a state function. In a similar manner, we can show that pressure and temperature are also state functions.

Energy is another state function. Using potential energy as an example, we find that the net increase in gravitational potential energy when we go from the same starting point to the top of a mountain is always the same, regardless of how we get there (Figure 6.3).

Recall that an object possesses potential energy by virtue of its position or chemical composition.

The First Law of Thermodynamics

The ***first law of thermodynamics,*** which is based on the law of conservation of energy, states that *energy can be converted from one form to another, but cannot be created or destroyed.*† How do we know this is so? It would be impossible to prove the validity of the first law of thermodynamics if we had to determine the total energy content of the universe. Even determining the total energy content of 1 g of iron, say, would be extremely difficult. Fortunately, we can test the validity of the first law by measuring only the *change* in the internal energy of a system between its *initial state* and its *final state* in a process. The change in internal energy ΔU is given by

$$\Delta U = U_f - U_i$$

†See footnote on p. 31 (Chapter 2) for a discussion of mass and energy relationship in chemical reactions.

where U_i and U_f are the internal energies of the system in the initial and final states, respectively.

The internal energy of a system has two components: kinetic energy and potential energy. The kinetic energy component consists of various types of molecular motion and the movement of electrons within molecules. Potential energy is determined by the attractive interactions between electrons and nuclei and by repulsive interactions between electrons and between nuclei in individual molecules, as well as by interaction between molecules. It is impossible to measure all these contributions accurately, so we cannot calculate the total energy of a system with any certainty. Changes in energy, on the other hand, can be determined experimentally.

Consider the reaction between 1 mole of sulfur and 1 mole of oxygen gas to produce 1 mole of sulfur dioxide:

$$S(s) + O_2(g) \longrightarrow SO_2(g)$$

Sulfur burning in air to form SO_2.

In this case, our system is composed of the reactant molecules S and O_2 (the initial state) and the product molecules SO_2 (the final state). We do not know the internal energy content of either the reactant molecules or the product molecules, but we can accurately measure the *change* in energy content, ΔU, given by

$$\begin{aligned}\Delta U &= U(\text{product}) - U(\text{reactants})\\ &= \text{energy content of 1 mol } SO_2(g) - \text{energy content of } [1 \text{ mol } S(s) + 1 \text{ mol } O_2(g)]\end{aligned}$$

We find that this reaction gives off heat. Therefore, the energy of the product is less than that of the reactants, and ΔU is negative.

Interpreting the release of heat in this reaction to mean that some of the chemical energy contained in the molecules has been converted to thermal energy, we conclude that the transfer of energy from the system to the surroundings does not change the total energy of the universe. That is, the sum of the energy changes must be zero:

$$\Delta U_{sys} + \Delta U_{surr} = 0$$

or

$$\Delta U_{sys} = -\Delta U_{surr}$$

where the subscripts "sys" and "surr" denote system and surroundings, respectively. Thus, if one system undergoes an energy change ΔU_{sys}, the rest of the universe, or the surroundings, must undergo a change in energy that is equal in magnitude but opposite in sign ($-\Delta U_{surr}$); energy gained in one place must have been lost somewhere else. Furthermore, because energy can be changed from one form to another, the energy lost by one system can be gained by another system in a different form. For example, the energy lost by burning oil in a power plant may ultimately turn up in our homes as electrical energy, heat, light, and so on.

In chemistry, we are normally interested in the energy changes associated with the system (which may be a flask containing reactants and products), not with its surroundings. Therefore, a more useful form of the first law is

$$\Delta U = q + w \qquad (6.1)$$

We use lowercase letters (such as w and q) to represent thermodynamic quantities that are not state functions.

(We drop the subscript "sys" for simplicity.) Equation (6.1) says that the change in the internal energy, ΔU, of a system is the sum of the heat exchange q between the system and the surroundings and the work done w on (or by) the system. The sign conventions for q and w are as follows: q is positive for an endothermic process and negative for an exothermic process and w is positive for work done on the system by the surroundings and negative for work done by the system on the surroundings. We can think of

For convenience, we sometimes omit the word "internal" when discussing the energy of a system.

Table 6.1 Sign Conventions for Work and Heat

Process	Sign
Work done by the system on the surroundings	−
Work done on the system by the surroundings	+
Heat absorbed by the system from the surroundings (endothermic process)	+
Heat absorbed by the surroundings from the system (exothermic process)	−

the first law of thermodynamics as an energy balance sheet, much like a money balance sheet kept in a bank that does currency exchange. You can withdraw or deposit money in either of two different currencies (like energy change due to heat exchange and work done). However, the value of your bank account depends only on the net amount of money left in it after these transactions, not on which currency you used.

Equation (6.1) may seem abstract, but it is actually quite logical. If a system loses heat to the surroundings or does work on the surroundings, we would expect its internal energy to decrease because those are energy-depleting processes. For this reason, both q and w are negative. Conversely, if heat is added to the system or if work is done on the system, then the internal energy of the system would increase. In this case, both q and w are positive. Table 6.1 summarizes the sign conventions for q and w.

Work and Heat

We will now look at the nature of work and heat in more detail.

Work

We have seen that work can be defined as force F multiplied by distance d:

$$w = Fd \tag{6.2}$$

In thermodynamics, work has a broader meaning that includes mechanical work (for example, a crane lifting a steel beam), electrical work (a battery supplying electrons to light the bulb of a flashlight), and surface work (blowing up a soap bubble). In this section we will concentrate on mechanical work; in Chapter 19 we will discuss the nature of electrical work.

One way to illustrate mechanical work is to study the expansion or compression of a gas. Many chemical and biological processes involve gas volume changes. Breathing and exhaling air involves the expansion and contraction of the tiny sacs called alveoli in the lungs. Another example is the internal combustion engine of the automobile. The successive expansion and compression of the cylinders due to the combustion of the gasoline-air mixture provide power to the vehicle. Figure 6.4 shows a gas in a cylinder fitted with a weightless, frictionless movable piston at a certain temperature, pressure, and volume. As it expands, the gas pushes the piston upward against a constant opposing external atmospheric pressure P. The work done by the gas on the surroundings is

$$w = -P\Delta V \tag{6.3}$$

where ΔV, the change in volume, is given by $V_f - V_i$. The minus sign in Equation (6.3) takes care of the sign convention for w. For gas expansion (work done by the system), $\Delta V > 0$, so $-P\Delta V$ is a negative quantity. For gas compression (work done on the system), $\Delta V < 0$, and $-P\Delta V$ is a positive quantity.

Note that "$-P\Delta V$" is often referred to as "P-V" work.

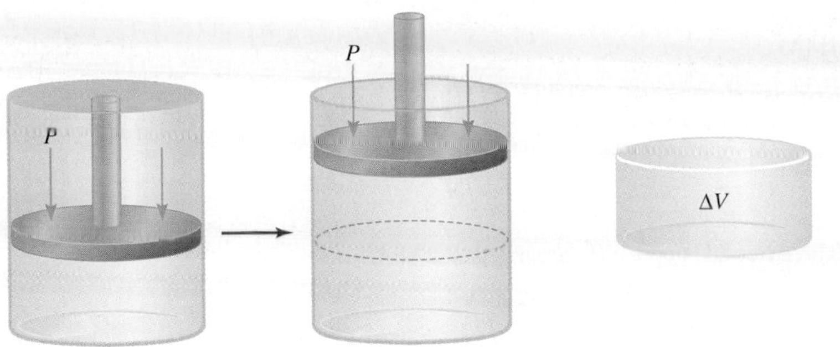

Figure 6.4
The expansion of a gas against a constant external pressure (such as atmospheric pressure). The gas is in a cylinder fitted with a weightless movable piston. The work done is given by $-P\Delta V$.

Equation (6.3) derives from the fact that pressure × volume can be expressed as (force/area) × volume; that is,

$$P \times V = \underbrace{\frac{F}{d^2}}_{\text{pressure}} \times \underbrace{d^3}_{\text{volume}} = Fd = w$$

where F is the opposing force and d has the dimension of length, d^2 has the dimensions of area, and d^3 has the dimensions of volume. Thus, the product of pressure and volume is equal to force times distance, or work. You can see that for a given increase in volume (that is, for a certain value of ΔV), the work done depends on the magnitude of the external, opposing pressure P. If P is zero (that is, if the gas is expanding against a vacuum), the work done must also be zero. If P is some positive, nonzero value, then the work done is given by $-P\Delta V$.

According to Equation (6.3), the units for work done by or on a gas are liters atmospheres. To express the work done in the more familiar unit of joules, we use the conversion factor (see Appendix 1).

$$1 \text{ L} \cdot \text{atm} = 101.3 \text{ J}$$

EXAMPLE 6.1

A certain gas expands in volume from 2.0 L to 6.0 L at constant temperature. Calculate the work done by the gas if it expands (a) against a vacuum and (b) against a constant pressure of 1.2 atm.

Strategy A simple sketch of the situation is helpful here:

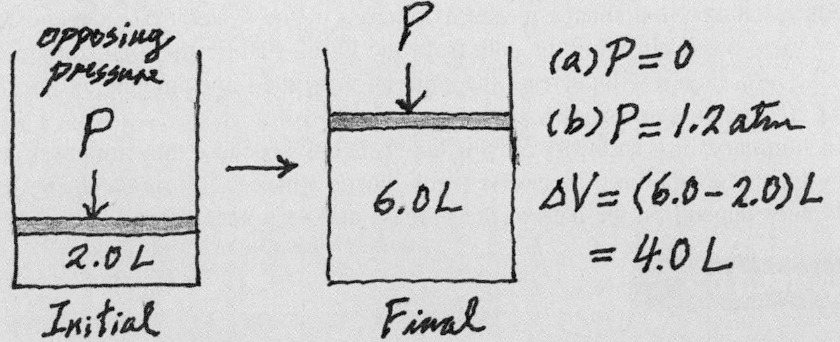

The work done in gas expansion is equal to the product of the external, opposing pressure and the change in volume. What is the conversion factor between L · atm and J?

(Continued)

Solution

(a) Because the external pressure is zero, no work is done in the expansion:

$$w = -P\Delta V$$
$$= -(0)(6.0 - 2.0)\ \text{L}$$
$$= 0$$

(b) The external, opposing pressure is 1.2 atm, so

$$w = -P\Delta V$$
$$= -(1.2\ \text{atm})(6.0 - 2.0)\ \text{L}$$
$$= -4.8\ \text{L} \cdot \text{atm}$$

To convert the answer to joules, we write

$$w = -4.8\ \text{L} \cdot \text{atm} \times \frac{101.3\ \text{J}}{1\ \text{L} \cdot \text{atm}}$$
$$= -4.9 \times 10^2\ \text{J}$$

Check Because this is gas expansion (work is done by the system on the surroundings), the work done has a negative sign.

Similar problems: 6.15, 6.16.

Practice Exercise A gas expands from 264 mL to 971 mL at constant temperature. Calculate the work done (in joules) by the gas if it expands (a) against a vacuum and (b) against a constant pressure of 4.00 atm.

Because temperature is kept constant, you can use Boyle's law to show that the final pressure is the same in (a) and (b).

Example 6.1 shows that work is not a state function. Although the initial and final states are the same in (a) and (b), the amount of work done is different because the external, opposing pressures are different. We *cannot* write $\Delta w = w_f - w_i$ for a change. Work done depends not only on the initial state and final state, but also on how the process is carried out, that is, on the path.

Heat

The other component of internal energy is heat, q. Like work, heat is not a state function. For example, it takes 4184 J of energy to raise the temperature of 100 g of water from 20°C to 30°C. This energy can be gained (a) directly as heat energy from a Bunsen burner, without doing any work on the water; (b) by doing work on the water without adding heat energy (for example, by stirring the water with a magnetic stir bar); or (c) by some combination of the procedures described in (a) and (b). This simple illustration shows that heat associated with a given process, like work, depends on how the process is carried out. It is important to note that regardless of which procedure is taken, the change in internal energy of the system, ΔU, depends on the sum of $(q + w)$. If changing the path from the initial state to the final state increases the value of q, then it will decrease the value of w by the same amount and vice versa, so that ΔU remains unchanged.

In summary, heat and work are not state functions because they are not properties of a system. They manifest themselves only during a process (during a change). Thus, their values depend on the path of the process and vary accordingly.

EXAMPLE 6.2

The work done when a gas is compressed in a cylinder like that shown in Figure 6.4 is 387 J. During this process, there is a heat transfer of 152 J from the gas to the surroundings. Calculate the energy change for this process.

(Continued)

Strategy Compression is work done on the gas, so what is the sign for w? Heat is released by the gas to the surroundings. Is this an endothermic or exothermic process? What is the sign for q?

Solution To calculate the energy change of the gas, we need Equation (6.1). Work of compression is positive and because heat is released by the gas, q is negative. Therefore, we have

$$\begin{aligned}\Delta U &= q + w \\ &= -152\ \text{J} + 387\ \text{J} \\ &= \boxed{235\ \text{J}}\end{aligned}$$

As a result, the energy of the gas increases by 235 J.

Similar problems: 6.17, 6.18.

Practice Exercise A gas expands and does P-V work on the surroundings equal to 279 J. At the same time, it absorbs 216 J of heat from the surroundings. What is the change in energy of the system?

Review of Concepts

Two ideal gases at the same temperature and pressure are placed in two equal-volume containers. One container has a fixed volume while the other is a cylinder fitted with a weightless movable piston like that shown in Figure 6.4. Initially, the gas pressures are equal to the external atmospheric pressure. The gases are then heated with a Bunsen burner. What are the signs of q and w for the gases under these conditions?

6.4 Enthalpy of Chemical Reactions

Our next step is to see how the first law of thermodynamics can be applied to processes carried out under different conditions. Specifically, we will consider a situation in which the volume of the system is kept constant and one in which the pressure applied on the system is kept constant. These are commonly encountered cases in the laboratory.

If a chemical reaction is run at constant volume, then $\Delta V = 0$ and no P-V work will result from this change. From Equation (6.1) it follows that

P-V work means work of gas expansion or gas compression.

$$\begin{aligned}\Delta U &= q - P\Delta V \\ &= q_v \end{aligned} \tag{6.4}$$

We add the subscript "v" to remind us that this is a constant-volume process. This equality may seem strange at first, for we showed earlier that q is not a state function. The process is carried out under constant-volume conditions, however, so that the heat change can have only a specific value, which is equal to ΔU.

Enthalpy

Constant-volume conditions are often inconvenient and sometimes impossible to achieve. Most reactions occur under conditions of constant pressure (usually atmospheric pressure). If such a reaction results in a net increase in the number of moles of a gas, then the system does work on the surroundings (expansion). This follows from the fact that for the gas formed to enter the atmosphere, it must push the surrounding air back.

Conversely, if more gas molecules are consumed than are produced, work is done on the system by the surroundings (compression). Finally, no work is done if there is no net change in the number of moles of gases from reactants to products.

In general, for a constant-pressure process we write

$$\begin{aligned} \Delta U &= q + w \\ &= q_p - P\Delta V \\ q_p &= \Delta U + P\Delta V \end{aligned} \tag{6.5}$$

where the subscript "p" denotes constant-pressure condition.

We now introduce a new thermodynamic function of a system called ***enthalpy (H),*** which is defined by the equation

$$H = U + PV \tag{6.6}$$

where U is the internal energy of the system and P and V are the pressure and volume of the system, respectively. Because U and PV have energy units, enthalpy also has energy units. Furthermore, U, P, and V are all state functions, that is, the changes in $(U + PV)$ depend only on the initial and final states. It follows, therefore, that the change in H, or ΔH, also depends only on the initial and final states. Thus, H is a state function.

For any process, the change in enthalpy according to Equation (6.6) is given by

$$\Delta H = \Delta U + \Delta(PV) \tag{6.7}$$

If the pressure is held constant, then

$$\Delta H = \Delta U + P\Delta V \tag{6.8}$$

Comparing Equation (6.8) with Equation (6.5), we see that for a constant-pressure process, $q_p = \Delta H$. Again, although q is not a state function, the heat change at constant pressure is equal to ΔH because the "path" is defined and therefore it can have only a specific value.

In Section 6.5, we will discuss ways to measure heat changes at constant volume and constant pressure.

We now have two quantities—ΔU and ΔH—that can be associated with a reaction. Both quantities measure energy changes, but under different conditions. If the reaction occurs under constant-volume conditions, then the heat change, q_v, is equal to ΔU. On the other hand, when the reaction is carried out at constant pressure, the heat change, q_p, is equal to ΔH.

Enthalpy of Reactions

Because most reactions are constant-pressure processes, we can equate the heat change in these cases to the change in enthalpy. For any reaction of the type

$$\text{reactants} \longrightarrow \text{products}$$

we define the change in enthalpy, called the ***enthalpy of reaction, ΔH,*** as *the difference between the enthalpies of the products and the enthalpies of the reactants:*

$$\Delta H = H(\text{products}) - H(\text{reactants}) \tag{6.9}$$

The enthalpy of reaction can be positive or negative, depending on the process. For an endothermic process (heat absorbed by the system from the surroundings), ΔH is positive (that is, $\Delta H > 0$). For an exothermic process (heat released by the system to the surroundings), ΔH is negative (that is, $\Delta H < 0$).

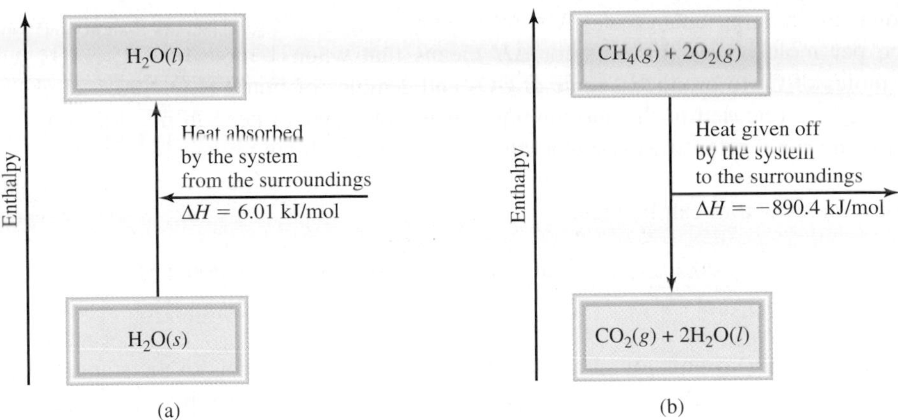

Figure 6.5
(a) Melting 1 mole of ice at 0°C (an endothermic process) results in an enthalpy increase in the system of 6.01 kJ. (b) Burning 1 mole of methane in oxygen gas (an exothermic process) results in an enthalpy decrease in the system of 890.4 kJ. Parts (a) and (b) are not drawn to the same scale.

An analogy for enthalpy change is a change in the balance in your bank account. Suppose your initial balance is \$100. After a transaction (deposit or withdrawal), the change in your bank balance, ΔX, is given by

$$\Delta X = X_{\text{final}} - X_{\text{initial}}$$

where X represents the bank balance. If you deposit \$80 into your account, then $\Delta X = \$180 - \$100 = \$80$. This corresponds to an endothermic reaction. (The balance increases and so does the enthalpy of the system.) On the other hand, a withdrawal of \$60 means $\Delta X = \$40 - \$100 = -\$60$. The negative sign of ΔX means your balance has decreased. Similarly, a negative value of ΔH reflects a decrease in enthalpy of the system as a result of an exothermic process. The difference between this analogy and Equation (6.9) is that while you always know your exact bank balance, there is no way to know the enthalpies of individual products and reactants. In practice, we can measure only the *difference* in their values.

This analogy assumes that you will not overdraw your bank account. The enthalpy of a substance *cannot* be negative.

Now let us apply the idea of enthalpy changes to two common processes, the first involving a physical change, the second a chemical change.

Thermochemical Equations

At 0°C and a pressure of 1 atm, ice melts to form liquid water. Measurements show that for every mole of ice converted to liquid water under these conditions, 6.01 kilojoules (kJ) of heat energy are absorbed by the system (ice). Because the pressure is constant, the heat change is equal to the enthalpy change, ΔH. Furthermore, this is an endothermic process, as expected for the energy-absorbing change of melting ice (Figure 6.5a). Therefore, ΔH is a positive quantity. The equation for this physical change is

$$H_2O(s) \longrightarrow H_2O(l) \qquad \Delta H = 6.01 \text{ kJ/mol}$$

The "per mole" in the unit for ΔH means that this is the enthalpy change *per mole of the reaction (or process) as it is written,* that is, when 1 mole of ice is converted to 1 mole of liquid water.

As another example, consider the combustion of methane (CH_4), the principal component of natural gas:

$$CH_4(g) + 2O_2(g) \longrightarrow CO_2(g) + 2H_2O(l) \qquad \Delta H = -890.4 \text{ kJ/mol}$$

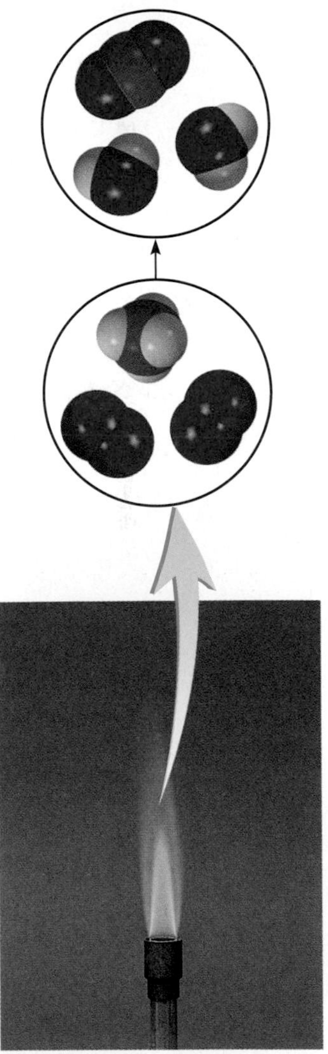

Methane gas burning from a Bunsen burner.

From experience we know that burning natural gas releases heat to the surroundings, so it is an exothermic process. Under constant-pressure condition, this heat change is equal to enthalpy change and ΔH must have a negative sign [Figure 6.5(b)]. Again, the per mole of reaction unit for ΔH means that when 1 mole of CH_4 reacts with 2 moles of O_2 to produce 1 mole of CO_2 and 2 moles of liquid H_2O, 890.4 kJ of heat energy are released to the surroundings. It is important to keep in mind that the ΔH value does not refer to a particular reactant or product. It simply means that the quoted ΔH value refers to all the reacting species in molar quantities. Thus, the following conversion factors can be created:

$$\frac{-890.4 \text{ kJ}}{1 \text{ mol } CH_4} \qquad \frac{-890.4 \text{ kJ}}{2 \text{ mol } O_2} \qquad \frac{-890.4 \text{ kJ}}{1 \text{ mol } CO_2} \qquad \frac{-890.4 \text{ kJ}}{2 \text{ mol } H_2O}$$

Expressing ΔH in units of kJ/mol (rather than just kJ) conforms to the standard convention; its merit will become apparent when we continue our study of thermodynamics in Chapter 18.

The equations for the melting of ice and the combustion of methane are examples of ***thermochemical equations,*** which *show the enthalpy changes as well as the mass relationships.* It is essential to specify a balanced equation when quoting the enthalpy change of a reaction. The following guidelines are helpful in writing and interpreting thermochemical equations:

1. When writing thermochemical equations, we must always specify the physical states of all reactants and products, because they help determine the actual enthalpy changes. For example, in the equation for the combustion of methane, if we show water vapor rather than liquid water as a product,

$$CH_4(g) + 2O_2(g) \longrightarrow CO_2(g) + 2H_2O(g) \quad \Delta H = -802.4 \text{ kJ/mol}$$

the enthalpy change is -802.4 kJ rather than -890.4 kJ because 88.0 kJ are needed to convert 2 moles of liquid water to water vapor; that is,

$$2H_2O(l) \longrightarrow 2H_2O(g) \qquad \Delta H = 88.0 \text{ kJ/mol}$$

Note that H is an extensive property.

2. If we multiply both sides of a thermochemical equation by a factor n, then ΔH must also change by the same factor. Returning to the melting of ice

$$H_2O(s) \longrightarrow H_2O(l) \qquad \Delta H = 6.01 \text{ kJ/mol}$$

If we multiply the equation throughout by 2; that is, if we set $n = 2$, then

$$2H_2O(s) \longrightarrow 2H_2O(l) \quad \Delta H = 2(6.01 \text{ kJ/mol}) = 12.0 \text{ kJ/mol}$$

3. When we reverse an equation, we change the roles of reactants and products. Consequently, the magnitude of ΔH for the equation remains the same, but its sign changes. For example, if a reaction consumes thermal energy from its surroundings (that is, if it is endothermic), then the reverse reaction must release thermal energy back to its surroundings (that is, it must be exothermic) and the enthalpy change expression must also change its sign. Thus, reversing the melting of ice and the combustion of methane, the thermochemical equations become

$$H_2O(l) \longrightarrow H_2O(s) \qquad \Delta H = -6.01 \text{ kJ/mol}$$

$$CO_2(g) + 2H_2O(l) \longrightarrow CH_4(g) + 2O_2(g) \qquad \Delta H = 890.4 \text{ kJ/mol}$$

and what was an endothermic process becomes exothermic, and vice versa.

EXAMPLE 6.3

Given the thermochemical equation

$$2SO_2(g) + O_2(g) \longrightarrow 2SO_3(g) \qquad \Delta H = -198.2 \text{ kJ/mol}$$

calculate the heat evolved when 93.4 g of SO_2 (molar mass = 64.07 g/mol) is converted to SO_3.

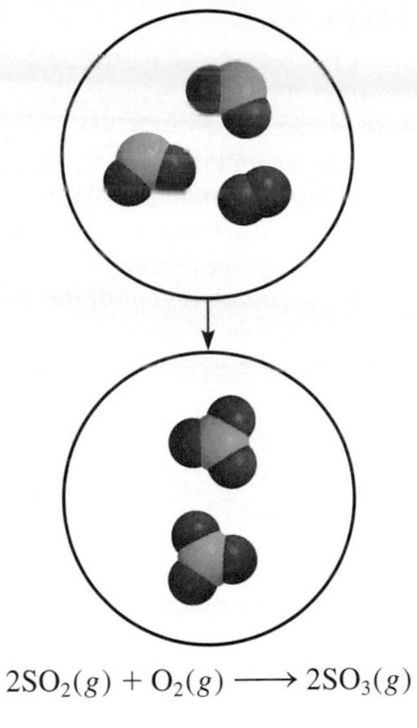

$2SO_2(g) + O_2(g) \longrightarrow 2SO_3(g)$

Strategy The thermochemical equation shows that for every 2 moles of SO_2 reacted, 198.2 kJ of heat are given off (note the negative sign). Therefore, the conversion factor is

$$\frac{-198.2 \text{ kJ}}{2 \text{ mol } SO_2}$$

How many moles of SO_2 are in 93.4 g of SO_2? What is the conversion factor between grams and moles?

Solution We need to first calculate the number of moles of SO_2 in 93.4 g of the compound and then find the number of kilojoules produced from the exothermic reaction. The sequence of conversions is as follows:

$$\text{grams of } SO_2 \longrightarrow \text{moles of } SO_2 \longrightarrow \text{kilojoules of heat generated}$$

Therefore, the enthalpy change for this reaction is given by

$$\Delta H = 93.4 \cancel{\text{g } SO_2} \times \frac{1 \cancel{\text{mol } SO_2}}{64.07 \cancel{\text{g } SO_2}} \times \frac{-198.2 \text{ kJ}}{2 \cancel{\text{mol } SO_2}} = -144 \text{ kJ}$$

and the heat released to the surroundings is 144 kJ.

Check Because 93.4 g is less than twice the molar mass of SO_2 (2 × 64.07 g) as shown in the preceding thermochemical equation, we expect the heat released to be smaller than 198.2 kJ.

Similar problem: 6.26.

Practice Exercise Calculate the heat evolved when 266 g of white phosphorus (P_4) burns in air according to the equation

$$P_4(s) + 5O_2(g) \longrightarrow P_4O_{10}(s) \qquad \Delta H = -3013 \text{ kJ/mol}$$

A Comparison of ΔH and ΔU

What is the relationship between ΔH and ΔU for a process? To find out, let us consider the reaction between sodium metal and water:

$$2Na(s) + 2H_2O(l) \longrightarrow 2NaOH(aq) + H_2(g) \qquad \Delta H = -367.5 \text{ kJ/mol}$$

This thermochemical equation says that when two moles of sodium react with an excess of water, 367.5 kJ of heat are given off. Note that one of the products is hydrogen gas, which must push back air to enter the atmosphere. Consequently, some of the energy produced by the reaction is used to do work of pushing back a volume of air (ΔV) against atmospheric pressure (P) (Figure 6.6). To calculate the change in internal energy, we rearrange Equation (6.8) as follows:

$$\Delta U = \Delta H - P\Delta V$$

Sodium reacting with water to form hydrogen gas.

If we assume the temperature to be 25°C and ignore the small change in the volume of the solution, we can show that the volume of 1 mole of H_2 gas at 1.0 atm and 298 K is 24.5 L, so that $-P\Delta V = -24.5 \text{ L} \cdot \text{atm}$ or -2.5 kJ. Finally,

Recall that 1 L · atm = 101.3 J.

$$\begin{aligned}\Delta U &= -367.5 \text{ kJ/mol} - 2.5 \text{ kJ/mol} \\ &= -370.0 \text{ kJ/mol}\end{aligned}$$

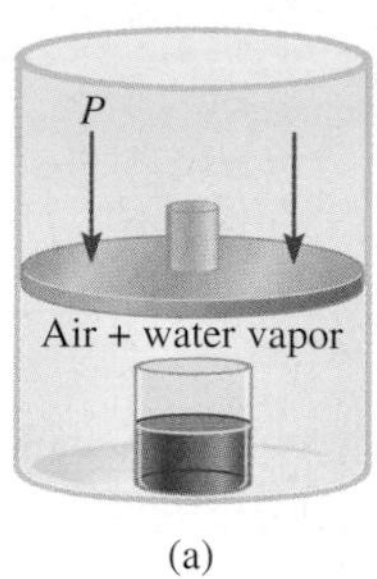

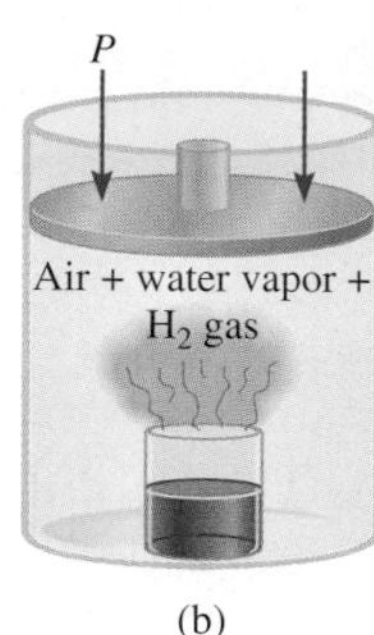

Figure 6.6
(a) A beaker of water inside a cylinder fitted with a movable piston. The pressure inside is equal to the atmospheric pressure. (b) As the sodium metal reacts with water, hydrogen gas pushes the piston upward (doing work on the surroundings) until the pressure inside is again equal to that of outside.

For reactions that do not result in a change in the number of moles of gases from reactants to products [for example, $H_2(g) + Cl_2(g) \longrightarrow 2HCl(g)$], $\Delta U = \Delta H$.

This calculation shows that ΔU and ΔH are approximately the same. The reason ΔH is smaller than ΔU in magnitude is that some of the internal energy released is used to do gas expansion work, so less heat is evolved. For reactions that do not involve gases, ΔV is usually very small and so ΔU is practically the same as ΔH.

Another way to calculate the internal energy change of a gaseous reaction is to assume ideal gas behavior and constant temperature. In this case,

$$\Delta U = \Delta H - \Delta(PV)$$
$$= \Delta H - \Delta(nRT)$$

or

$$\Delta U = \Delta H - RT\Delta n \qquad (6.10)$$

where Δn is defined as

$$\Delta n = \text{number of moles of product gases} - \text{number of moles of reactant gases}$$

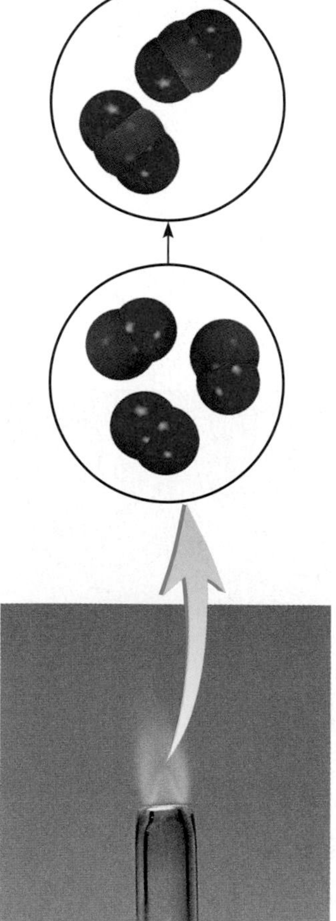

Carbon monoxide burns in air to form carbon dioxide.

EXAMPLE 6.4

Calculate the change in internal energy when 2 moles of CO are converted to 2 moles of CO_2 at 1 atm and 25°C:

$$2CO(g) + O_2(g) \longrightarrow 2CO_2(g) \qquad \Delta H = -566.0 \text{ kJ/mol}$$

Strategy We are given the enthalpy change, ΔH, for the reaction and are asked to calculate the change in internal energy, ΔU. Therefore, we need Equation (6.10). What is the change in the number of moles of gases? ΔH is given in kilojoules, so what units should we use for R?

Solution From the chemical equation we see that 3 moles of gases are converted to 2 moles of gases so that

$$\begin{aligned}\Delta n &= \text{number of moles of product gases} - \text{number of moles of reactant gases}\\ &= 2 - 3\\ &= -1\end{aligned}$$

Using 8.314 J/K · mol for R and T = 298 K in Equation (6.10), we write

$$\begin{aligned}\Delta U &= \Delta H - RT\Delta n\\ &= -566.0 \text{ kJ/mol} - (8.314 \text{ J/K} \cdot \text{mol})\left(\frac{1 \text{ kJ}}{1000 \text{ J}}\right)(298 \text{ K})(-1)\\ &= -563.5 \text{ kJ/mol}\end{aligned}$$

(Continued)

Check Knowing that the reacting gaseous system undergoes a compression (3 moles to 2 moles), is it reasonable to have $\Delta H > \Delta U$ in magnitude?

Practice Exercise What is ΔU for the formation of 1 mole of CO at 1 atm and 25°C?

$$C(graphite) + \frac{1}{2}O_2(g) \longrightarrow CO(g) \qquad \Delta H = -110.5 \text{ kJ/mol}$$

Similar problem: 6.27.

Review of Concepts

Which of the constant-pressure processes shown here has the smallest difference between ΔU and ΔH?

(a) water $\longrightarrow$ water vapor
(b) water $\longrightarrow$ ice
(c) ice $\longrightarrow$ water vapor

6.5 Calorimetry

In the laboratory, heat changes in physical and chemical processes are measured with a *calorimeter,* a closed container designed specifically for this purpose. Our discussion of ***calorimetry****, the measurement of heat changes,* will depend on an understanding of specific heat and heat capacity, so let us consider them first.

Specific Heat and Heat Capacity

The ***specific heat (s)*** of a substance is *the amount of heat required to raise the temperature of one gram of the substance by one degree Celsius.* The ***heat capacity (C)*** of a substance is *the amount of heat required to raise the temperature of a given quantity of the substance by one degree Celsius.* Specific heat is an intensive property, whereas heat capacity is an extensive property. The relationship between the heat capacity and specific heat of a substance is

$$C = ms \tag{6.11}$$

where m is the mass of the substance in grams. For example, the specific heat of water is 4.184 J/g · °C, and the heat capacity of 60.0 g of water is

$$(60.0 \text{ g})(4.184 \text{ J/g} \cdot °\text{C}) = 251 \text{ J/°C}$$

Note that specific heat has the units J/g · °C and heat capacity has the units J/°C. Table 6.2 shows the specific heat of some common substances.

If we know the specific heat and the amount of a substance, then the change in the sample's temperature (Δt) will tell us the amount of heat (q) that has been absorbed or released in a particular process. The equations for calculating the heat change are given by

$$q = ms\Delta t \tag{6.12}$$

$$q = C\Delta t \tag{6.13}$$

where Δt is the temperature change:

$$\Delta t = t_{\text{final}} - t_{\text{initial}}$$

The sign convention for q is the same as that for enthalpy change; q is positive for endothermic processes and negative for exothermic processes.

Table 6.2 The Specific Heats of Some Common Substances

Substance	Specific Heat (J/g · °C)
Al	0.900
Au	0.129
C (graphite)	0.720
C (diamond)	0.502
Cu	0.385
Fe	0.444
Hg	0.139
H_2O	4.184
C_2H_5OH (ethanol)	2.46

EXAMPLE 6.5

A 394-g sample of water is heated from 10.75°C to 83.20°C. Calculate the amount of heat absorbed (in kilojoules) by the water.

Strategy We know the quantity of water and the specific heat of water. With this information and the temperature rise, we can calculate the amount of heat absorbed (q).

Solution Using Equation (6.12), we write

$$\begin{aligned} q &= ms\Delta t \\ &= (394 \text{ g})(4.184 \text{ J/g} \cdot {}^\circ\text{C})(83.20^\circ\text{C} - 10.75^\circ\text{C}) \\ &= 1.19 \times 10^5 \text{ J} \times \frac{1 \text{ kJ}}{1000 \text{ J}} \\ &= 119 \text{ kJ} \end{aligned}$$

Check The units g and °C cancel, and we are left with the desired unit kJ. Because heat is absorbed by the water from the surroundings, it has a positive sign.

Similar problem: 6.34.

Practice Exercise An iron bar of mass 869 g cools from 94°C to 5°C. Calculate the heat released (in kilojoules) by the metal.

Constant-Volume Calorimetry

"Constant volume" refers to the volume of the container, which does not change during the reaction. Note that the container remains intact after the measurement. The term "bomb calorimeter" connotes the explosive nature of the reaction (on a small scale) in the presence of excess oxygen gas.

Heat of combustion is usually measured by placing a known mass of a compound in a steel container called a *constant-volume bomb calorimeter,* which is filled with oxygen at about 30 atm of pressure. The closed bomb is immersed in a known amount of water, as shown in Figure 6.7. The sample is ignited electrically, and the heat

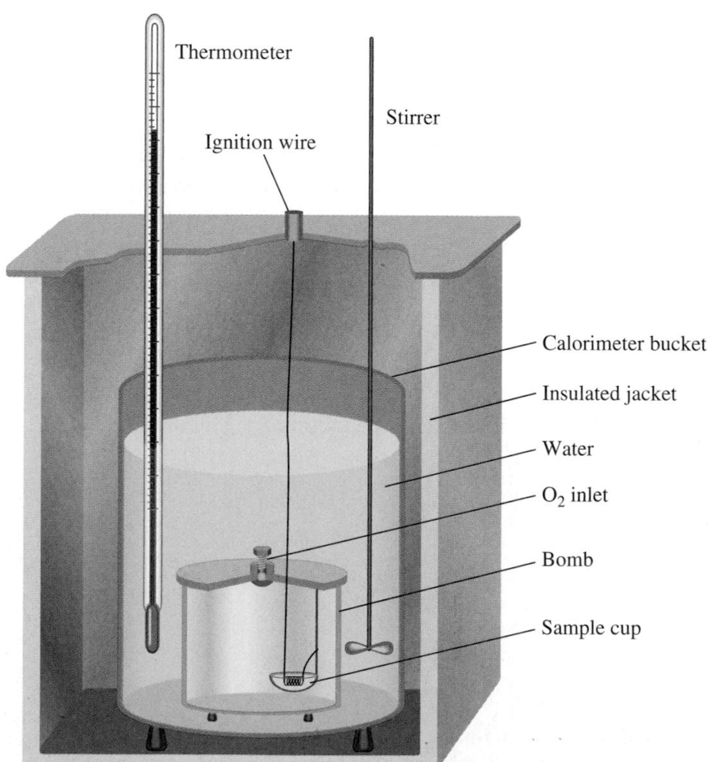

Figure 6.7
A constant-volume bomb calorimeter. The calorimeter is filled with oxygen gas before it is placed in the bucket. The sample is ignited electrically, and the heat produced by the reaction can be accurately determined by measuring the temperature increase in the known amount of surrounding water.

produced by the combustion reaction can be calculated accurately by recording the rise in temperature of the water. The heat given off by the sample is absorbed by the water and the bomb. The special design of the calorimeter enables us to assume that no heat (or mass) is lost to the surroundings during the time it takes to make measurements. Therefore, we can call the bomb and the water in which it is submerged an isolated system. Because no heat enters or leaves the system throughout the process, the heat change of the system (q_{sys}) must be zero and we can write

$$\begin{aligned} q_{sys} &= q_{cal} + q_{rxn} \\ &= 0 \end{aligned} \tag{6.14}$$

where q_{cal} and q_{rxn} are the heat changes for the calorimeter and the reaction, respectively. Thus,

$$q_{rxn} = -q_{cal} \tag{6.15}$$

To calculate q_{cal}, we need to know the heat capacity of the calorimeter (C_{cal}) and the temperature rise, that is

Note that C_{cal} comprises both the bomb and the surrounding water.

$$q_{cal} = C_{cal}\Delta t \tag{6.16}$$

The quantity C_{cal} is calibrated by burning a substance with an accurately known heat of combustion. For example, it is known that the combustion of 1 g of benzoic acid (C_6H_5COOH) releases 26.42 kJ of heat. If the temperature rise is 4.673°C, then the heat capacity of the calorimeter is given by

$$\begin{aligned} C_{cal} &= \frac{q_{cal}}{\Delta t} \\ &= \frac{26.42 \text{ kJ}}{4.673°\text{C}} = 5.654 \text{ kJ/°C} \end{aligned}$$

Note that although the combustion reaction is exothermic, q_{cal} is a positive quantity because it represents the heat absorbed by the calorimeter.

Once C_{cal} has been determined, the calorimeter can be used to measure the heat of combustion of other substances.

Note that because reactions in a bomb calorimeter occur under constant-volume rather than constant-pressure conditions, the heat changes *do not* correspond to the enthalpy change ΔH (see Section 6.4). It is possible to correct the measured heat changes so that they correspond to ΔH values, but the corrections usually are quite small so we will not concern ourselves with the details here. Finally, it is interesting to note that the energy contents of food and fuel (usually expressed in calories where 1 cal = 4.184 J) are measured with constant-volume calorimeters.

EXAMPLE 6.6

A quantity of 1.274 g of naphthalene ($C_{10}H_8$), a pungent-smelling substance used in moth repellants, was burned in a constant-volume bomb calorimeter. Consequently, the temperature of the water rose from 21.49°C to 26.52°C. If the heat capacity of the bomb plus water was 10.17 kJ/°C, calculate the heat of combustion of naphthalene on a molar basis; that is, find the molar heat of combustion.

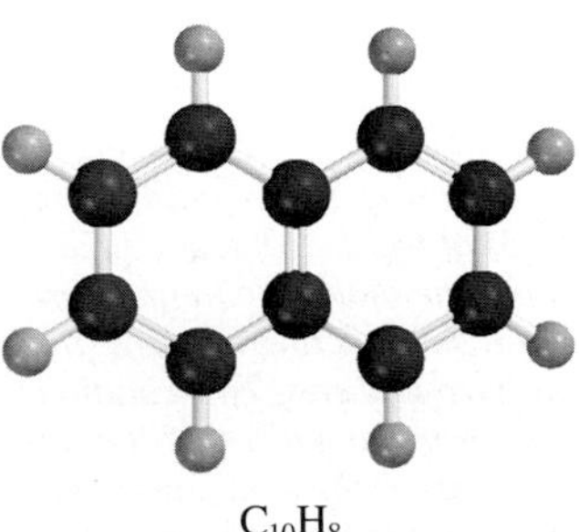

$C_{10}H_8$

Strategy Knowing the heat capacity and the temperature rise, how do we calculate the heat absorbed by the calorimeter? What is the heat generated by the combustion of 1.274 g of naphthalene? What is the conversion factor between grams and moles of naphthalene?

(Continued)

Solution The heat absorbed by the bomb and water is equal to the product of the heat capacity and the temperature change. From Equation (6.16), assuming no heat is lost to the surroundings, we write

$$\begin{aligned} q_{cal} &= C_{cal}\Delta t \\ &= (10.17\ \text{kJ/°C})(26.52°\text{C} - 21.49°\text{C}) \\ &= 51.16\ \text{kJ} \end{aligned}$$

Because $q_{sys} = q_{cal} + q_{rxn} = 0$, $q_{cal} = -q_{rxn}$. The heat change of the reaction is −51.56 kJ. This is the heat released by the combustion of 1.274 g of $C_{10}H_8$; therefore, we can write the conversion factor as

$$\frac{-51.16\ \text{kJ}}{1.274\ \text{g}\ C_{10}H_8}$$

The molar mass of naphthalene is 128.2 g, so the heat of combustion of 1 mole of naphthalene is

$$\begin{aligned} \text{molar heat of combustion} &= \frac{-51.16\ \text{kJ}}{1.274\ \cancel{\text{g}\ C_{10}H_8}} \times \frac{128.2\ \cancel{\text{g}\ C_{10}H_8}}{1\ \text{mol}\ C_{10}H_8} \\ &= -5.148 \times 10^3\ \text{kJ/mol} \end{aligned}$$

Similar problem: 6.37.

Check Knowing that the combustion reaction is exothermic and that the molar mass of naphthalene is much greater than 1.2 g, is the answer reasonable? Under these conditions, can the heat change (−51.16 kJ) be equated to the enthalpy change of the reaction?

Practice Exercise A quantity of 1.922 g of methanol (CH_3OH) was burned in a constant-volume bomb calorimeter. Consequently, the temperature of the water rose by 4.20°C. If the heat capacity of the bomb plus water was 10.4 kJ/°C, calculate the molar heat of combustion of methanol.

Constant-Pressure Calorimetry

A simpler device than the constant-volume calorimeter is the constant-pressure calorimeter, which is used to determine the heat changes for noncombustion reactions. A crude constant-pressure calorimeter can be constructed from two Styrofoam coffee cups, as shown in Figure 6.8. This device measures the heat effects of a variety of reactions, such as acid-base neutralization, as well as the heat of solution and heat of dilution. Because the pressure is constant, the heat change for the process (q_{rxn}) is equal to the enthalpy change (ΔH). As in the case of a constant-volume calorimeter, we treat the calorimeter as an isolated system. Furthermore, we neglect the small heat capacity of the coffee cups in our calculations. Table 6.3 lists some reactions that have been studied with the constant-pressure calorimeter.

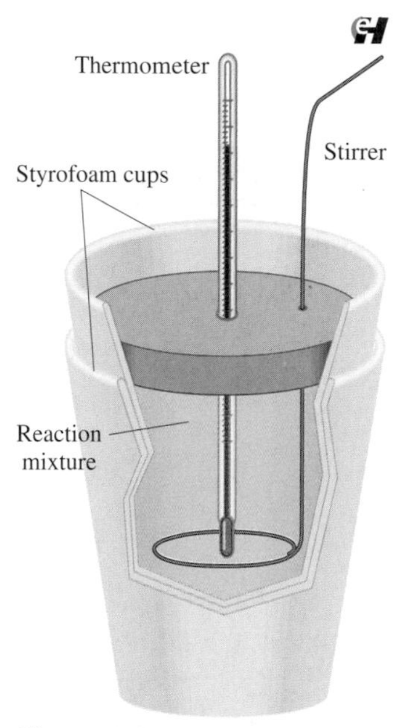

Figure 6.8
A constant-pressure calorimeter made of two Styrofoam coffee cups. The outer cup helps to insulate the reacting mixture from the surroundings. Two solutions of known volume containing the reactants at the same temperature are carefully mixed in the calorimeter. The heat produced or absorbed by the reaction can be determined by measuring the temperature change.

Table 6.3 Heats of Some Typical Reactions Measured at Constant Pressure

Type of Reaction	Example	ΔH (kJ)
Heat of neutralization	$HCl(aq) + NaOH(aq) \longrightarrow NaCl(aq) + H_2O(l)$	−56.2
Heat of ionization	$H_2O(l) \longrightarrow H^+(aq) + OH^-(aq)$	56.2
Heat of fusion	$H_2O(s) \longrightarrow H_2O(l)$	6.01
Heat of vaporization	$H_2O(l) \longrightarrow H_2O(g)$	44.0*
Heat of reaction	$MgCl_2(s) + 2Na(l) \longrightarrow 2NaCl(s) + Mg(s)$	−180.2

*Measured at 25°C. At 100°C, the value is 40.79 kJ.

EXAMPLE 6.7

A lead (Pb) pellet having a mass of 26.47 g at 89.98°C was placed in a constant-pressure calorimeter of negligible heat capacity containing 100.0 mL of water. The water temperature rose from 22.50°C to 23.17°C. What is the specific heat of the lead pellet?

Strategy A sketch of the initial and final situation is as follows:

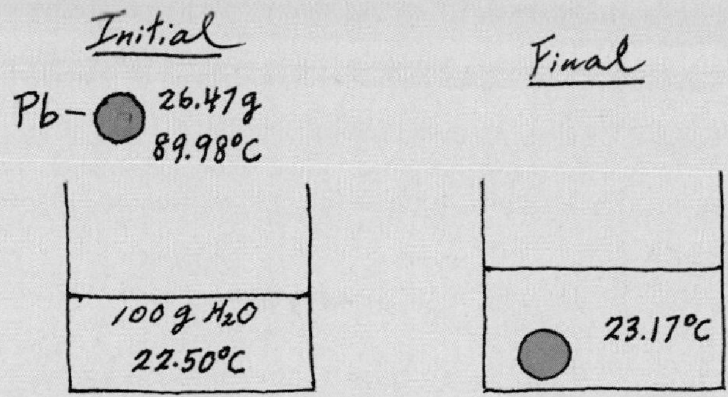

We know the masses of water and the lead pellet as well as the initial and final temperatures. Assuming no heat is lost to the surroundings, we can equate the heat lost by the lead pellet to the heat gained by the water. Knowing the specific heat of water, we can then calculate the specific heat of lead.

Solution Treating the calorimeter as an isolated system (no heat lost to the surroundings), we write

$$q_{Pb} + q_{H_2O} = 0$$

or

$$q_{Pb} = -q_{H_2O}$$

The heat gained by the water is given by

$$q_{H_2O} = ms\Delta t$$

where m and s are the mass and specific heat and $\Delta t = t_{final} - t_{initial}$. Therefore,

$$\begin{aligned} q_{H_2O} &= (100.0\ g)(4.184\ J/g \cdot °C)(23.17°C - 22.50°C) \\ &= 280.3\ g \end{aligned}$$

Because the heat lost by the lead pellet is equal to the heat gained by the water, so $q_{Pb} = -280.3$ J. Solving for the specific heat of Pb, we write

$$\begin{aligned} q_{Pb} &= ms\Delta t \\ -280.3\ J &= (26.47\ g)(s)(23.17°C - 89.98°C) \\ s &= \boxed{0.158\ J/g \cdot °C} \end{aligned}$$

Check The specific heat falls within the metals shown in Table 6.2.

Similar problem: 6.76.

Practice Exercise A 30.14-g stainless steel ball bearing at 117.82°C is placed in a constant-pressure calorimeter containing 120.0 mL of water at 18.44°C. If the specific heat of the ball bearing is 0.474 J/g · °C, calculate the final temperature of the water. Assume the calorimeter to have negligible heat capacity.

EXAMPLE 6.8

A quantity of 1.50×10^2 mL of 0.350 *M* HCl was mixed with 1.50×10^2 mL of 0.350 *M* NaOH in a constant-pressure calorimeter of negligible heat capacity. The initial temperature of the HCl and NaOH solutions was the same, 23.25°C, and the final temperature of the

(Continued)

mixed solution was 25.60°C. Calculate the heat change for the neutralization reaction on a molar basis

$$NaOH(aq) + HCl(aq) \longrightarrow NaCl(aq) + H_2O(l)$$

Assume that the densities and specific heats of the solutions are the same as for the water (1.00 g/mL and 4.184 J/g · °C, respectively).

Strategy Because the temperature rose, the neutralization reaction is exothermic. How do we calculate the heat absorbed by the combined solution? What is the heat of the reaction? What is the conversion factor for expressing the heat of reaction on a molar basis?

Solution Assuming no heat is lost to the surroundings, $q_{sys} = q_{soln} + q_{rxn} = 0$, so $q_{rxn} = -q_{soln}$, where q_{soln} is the heat absorbed by the combined solution. Because the density of the solution is 1.00 g/mL, the mass of a 150-mL solution is 150 g. Thus,

$$\begin{aligned} q_{soln} &= ms\Delta t \\ &= (1.50 \times 10^2 \text{ g} + 1.50 \times 10^2 \text{ g})(4.184 \text{ J/g} \cdot {}^\circ\text{C})(25.60^\circ\text{C} - 23.25^\circ\text{C}) \\ &= 2.95 \times 10^3 \text{ J} \\ &= 2.95 \text{ kJ} \end{aligned}$$

Because $q_{rxn} = -q_{soln}$, $q_{rxn} = -2.95$ kJ.

From the molarities given, the number of moles of both HCl and NaOH in 1.50×10^2 mL solution is

$$\frac{0.350 \text{ mol}}{1 \text{ L}} \times 0.150 \text{ L} = 0.0525 \text{ mol}$$

Therefore, the heat of neutralization when 1.00 mole of HCl reacts with 1.00 mole of NaOH is

$$\text{heat of neutralization} = \frac{-2.95 \text{ kJ}}{0.0525 \text{ mol}} = -56.2 \text{ kJ/mol}$$

Check Is the sign consistent with the nature of the reaction? Under the reaction condition, can the heat exchange be equated to the enthalpy change?

Similar problem: 6.38.

Practice Exercise A quantity of 4.00×10^2 mL of 0.600 M HNO_3 is mixed with 4.00×10^2 mL of 0.300 M $Ba(OH)_2$ in a constant-pressure calorimeter of negligible heat capacity. The initial temperature of both solutions is the same at 18.46°C. What is the final temperature of the solution? (Use the result in Example 6.8 for your calculation.)

REVIEW OF CONCEPTS

A 1-g sample of Al and a 1-g sample of Fe are heated from 50°C to 100°C. Which metal has absorbed a greater amount of heat?

6.6 Standard Enthalpy of Formation and Reaction

So far we have learned that we can determine the enthalpy change that accompanies a reaction by measuring the heat absorbed or released (at constant pressure). From Equation (6.9) we see that ΔH can also be calculated if we know the actual enthalpies of all reactants and products. However, as mentioned earlier, there is no way to measure the *absolute* value of the enthalpy of a substance. Only values *relative* to an arbitrary reference can be determined. This problem is similar to the one geographers face in expressing the elevations of specific mountains or valleys. Rather than trying to devise some type of "absolute" elevation scale (perhaps based on distance from the center of Earth?), by common agreement all geographic heights and depths are expressed relative to sea level, an arbitrary reference with a defined elevation of "zero" meters or feet. Similarly, chemists have agreed on an arbitrary reference point for enthalpy.

The "sea level" reference point for all enthalpy expressions is called the ***standard enthalpy of formation*** **(ΔH_f°)**. Substances are said to be in the ***standard state*** at 1 atm,† hence the term "standard enthalpy." The superscript "°" represents standard-state conditions (1 atm), and the subscript "f" stands for formation. By convention, *the standard enthalpy of formation of any element in its most stable form is zero.* Take the element oxygen as an example. Molecular oxygen (O_2) is more stable than the other allotropic form of oxygen, ozone (O_3), at 1 atm and 25°C. Thus, we can write $\Delta H_f^\circ(O_2) = 0$, but $\Delta H_f^\circ(O_3) = 142.2$ kJ/mol. Similarly, graphite is a more stable allotropic form of carbon than diamond at 1 atm and 25°C, so we have ΔH_f°(C, graphite) = 0 and ΔH_f°(C, diamond) = 1.90 kJ/mol. Based on this reference for elements, we can now define the standard enthalpy of formation of a compound as *the heat change that results when 1 mole of the compound is formed from its elements at a pressure of 1 atm.* Table 6.4 lists the standard

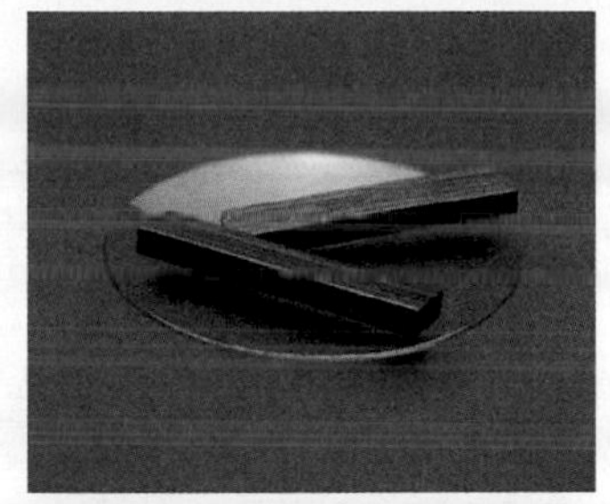

Graphite (top) and diamond (bottom).

Table 6.4 Standard Enthalpies of Formation of Some Inorganic Substances at 25°C

Substance	ΔH_f° (kJ/mol)	Substance	ΔH_f° (kJ/mol)
Ag(*s*)	0	$H_2O_2(l)$	−187.6
AgCl(*s*)	−127.04	Hg(*l*)	0
Al(*s*)	0	$I_2(s)$	0
$Al_2O_3(s)$	−1669.8	HI(*g*)	25.94
$Br_2(l)$	0	Mg(*s*)	0
HBr(*g*)	−36.2	MgO(*s*)	−601.8
C(graphite)	0	$MgCO_3(s)$	−1112.9
C(diamond)	1.90	$N_2(g)$	0
CO(*g*)	−110.5	$NH_3(g)$	−46.3
$CO_2(g)$	−393.5	NO(*g*)	90.4
Ca(*s*)	0	$NO_2(g)$	33.85
CaO(*s*)	−635.6	$N_2O_4(g)$	9.66
$CaCO_3(s)$	−1206.9	$N_2O(g)$	81.56
$Cl_2(g)$	0	O(*g*)	249.4
HCl(*g*)	−92.3	$O_2(g)$	0
Cu(*s*)	0	$O_3(g)$	142.2
CuO(*s*)	−155.2	S(rhombic)	0
$F_2(g)$	0	S(monoclinic)	0.30
HF(*g*)	−268.61	$SO_2(g)$	−296.1
H(*g*)	218.2	$SO_3(g)$	−395.2
$H_2(g)$	0	$H_2S(g)$	−20.15
$H_2O(g)$	−241.8	ZnO(*s*)	−347.98
$H_2O(l)$	−285.8	ZnS(s)	−202.9

†In thermodynamics, the standard pressure is defined as 1 bar, where 1 bar = 10^5 Pa = 0.987 atm. Because 1 bar differs from 1 atm by only 1.3 percent, we will continue to use 1 atm as the standard pressure. Note that the normal melting point and boiling point of a substance are defined in terms of 1 atm.

enthalpies of formation for a number of elements and compounds. (For a more complete list of ΔH_f° values, see Appendix 2.) Note that although the standard state does not specify a temperature, we will always use ΔH_f° values measured at 25°C for our discussion because most of the thermodynamic data are collected at this temperature.

The importance of the standard enthalpies of formation is that once we know their values, we can readily calculate the ***standard enthalpy of reaction,*** $\boldsymbol{\Delta H^\circ_{rxn}}$, defined as *the enthalpy of a reaction carried out at 1 atm.* For example, consider the hypothetical reaction

$$a\text{A} + b\text{B} \longrightarrow c\text{C} + d\text{D}$$

where a, b, c, and d are stoichiometric coefficients. For this reaction $\Delta H^\circ_{\text{rxn}}$ is given by

$$\Delta H^\circ_{\text{rxn}} = [c\Delta H_f^\circ(\text{C}) + d\Delta H_f^\circ(\text{D})] - [a\Delta H_f^\circ(\text{A}) + b\Delta H_f^\circ(\text{B})] \quad (6.17)$$

We can generalize Equation (6.17) as

$$\Delta H^\circ_{\text{rxn}} = \Sigma n\Delta H_f^\circ(\text{products}) - \Sigma m\Delta H_f^\circ(\text{reactants}) \quad (6.18)$$

where m and n denote the stoichiometric coefficients for the reactants and products, and Σ (sigma) means "the sum of." Note that in calculations, the stoichiometric coefficients are just numbers without units.

To use Equation (6.18) to calculate $\Delta H^\circ_{\text{rxn}}$ we must know the ΔH_f° values of the compounds that take part in the reaction. To determine these values we can apply the direct method or the indirect method.

The Direct Method

This method of measuring ΔH_f° works for compounds that can be readily synthesized from their elements. Suppose we want to know the enthalpy of formation of carbon dioxide. We must measure the enthalpy of the reaction when carbon (graphite) and molecular oxygen in their standard states are converted to carbon dioxide in its standard state:

$$\text{C(graphite)} + \text{O}_2(g) \longrightarrow \text{CO}_2(g) \quad \Delta H^\circ_{\text{rxn}} = -393.5 \text{ kJ/mol}$$

We know from experience that this combustion easily goes to completion. Thus, from Equation (6.17) we can write

$$\begin{aligned}\Delta H^\circ_{\text{rxn}} &= \Delta H_f^\circ(\text{CO}_2, g) - [\Delta H_f^\circ(\text{C, graphite}) + \Delta H_f^\circ(\text{O}_2, \text{g})] \\ &= -393.5 \text{ kJ/mol}\end{aligned}$$

Because both graphite and O_2 are stable allotropic forms of the elements, it follows that $\Delta H_f^\circ(\text{C, graphite})$ and $\Delta H_f^\circ(\text{O}_2, g)$ are zero. Therefore,

$$\Delta H^\circ_{\text{rxn}} = \Delta H_f^\circ(\text{CO}_2, \text{g}) = -393.5 \text{ kJ/mol}$$

or

$$\Delta H_f^\circ(\text{CO}_2, g) = -393.5 \text{ kJ/mol}$$

Note that arbitrarily assigning zero ΔH_f° for each element in its most stable form at the standard state does not affect our calculations in any way. Remember, in thermochemistry we are interested only in enthalpy *changes* because they can be determined experimentally whereas the absolute enthalpy values cannot. The choice of a zero "reference level" for enthalpy makes calculations easier to handle. Again referring to the terrestrial altitude analogy, we find that Mt. Everest is 8708 ft higher than Mt. McKinley. This difference in altitude is unaffected by the decision to set sea level at 0 ft or at 1000 ft.

Other compounds that can be studied by the direct method are SF_6, P_4O_{10}, and CS_2. The equations representing their syntheses are

$$S(\text{rhombic}) + 3F_2(g) \longrightarrow SF_6(g)$$
$$P_4(\text{white}) + 5O_2(g) \longrightarrow P_4O_{10}(s)$$
$$C(\text{graphite}) + 2S(\text{rhombic}) \longrightarrow CS_2(l)$$

Note that S(rhombic) and P(white) are the most stable allotropes of sulfur and phosphorus, respectively, at 1 atm and 25°C, so their ΔH_f° values are zero.

P_4

The Indirect Method

Many compounds cannot be directly synthesized from their elements. In some cases, the reaction proceeds too slowly, or side reactions produce substances other than the desired compound. In these cases, ΔH_f° can be determined by an indirect approach, which is based on Hess's law of heat summation, or simply Hess's law, named after the Swiss chemist Germain Hess. ***Hess's law*** can be stated as follows: *When reactants are converted to products, the change in enthalpy is the same whether the reaction takes place in one step or in a series of steps.* In other words, if we can break down the reaction of interest into a series of reactions for which ΔH_{rxn}° can be measured, we can calculate ΔH_{rxn}° for the overall reaction. Hess's law is based on the fact that because H is a state function, ΔH depends only on the initial and final state (that is, only on the nature of reactants and products). The enthalpy change would be the same whether the overall reaction takes place in one step or many steps.

White phosphorus burns in air to form P_4O_{10}.

An analogy for Hess's law is as follows. Suppose you go from the first floor to the sixth floor of a building by elevator. The gain in your gravitational potential energy (which corresponds to the enthalpy change for the overall process) is the same whether you go directly there or stop at each floor on your way up (breaking the trip into a series of steps).

Let's say we are interested in the standard enthalpy of formation of carbon monoxide (CO). We might represent the reaction as

$$C(\text{graphite}) + \tfrac{1}{2}O_2(g) \longrightarrow CO(g)$$

However, burning graphite also produces some carbon dioxide (CO_2), so we cannot measure the enthalpy change for CO directly as shown. Instead, we must employ an indirect route, based on Hess's law. It is possible to carry out the following two separate reactions, which do go to completion:

(a) $$C(\text{graphite}) + O_2(g) \longrightarrow CO_2(g) \quad \Delta H_{rxn}^\circ = -393.5 \text{ kJ/mol}$$
(b) $$CO(g) + \tfrac{1}{2}O_2(g) \longrightarrow CO_2(g) \quad \Delta H_{rxn}^\circ = -283.0 \text{ kJ/mol}$$

First, we reverse Equation (b) to get

(c) $$CO_2(g) \longrightarrow CO(g) + \tfrac{1}{2}O_2(g) \quad \Delta H_{rxn}^\circ = +283.0 \text{ kJ/mol}$$

Remember to reverse the sign of ΔH when you reverse an equation.

Because chemical equations can be added and subtracted just like algebraic equations, we carry out the operation (a) + (c) and obtain

(a) $$C(\text{graphite}) + O_2(g) \longrightarrow CO_2(g) \quad \Delta H_{rxn}^\circ = -393.5 \text{ kJ/mol}$$
(c) $$CO_2(g) \longrightarrow CO(g) + \tfrac{1}{2}O_2(g) \quad \Delta H_{rxn}^\circ = +283.0 \text{ kJ/mol}$$

(d) $$C(\text{graphite}) + \tfrac{1}{2}O_2(g) \longrightarrow CO(g) \quad \Delta H_{rxn}^\circ = -110.5 \text{ kJ/mol}$$

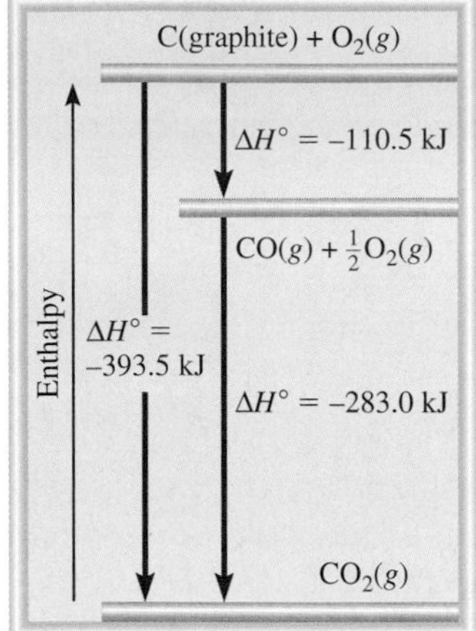

Figure 6.9
The enthalpy change for the formation of 1 mole of CO_2 from graphite and O_2 can be broken into two steps according to Hess's law.

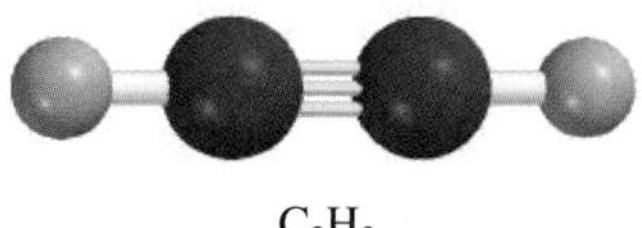

C_2H_2

Thus, $\Delta H_f^\circ(CO) = -110.5$ kJ/mol. Looking back, we see that the overall reaction is the formation of CO_2 [Equation (a)], which can be broken down into two parts [Equations (d) and (b)]. Figure 6.9 shows the overall scheme of our procedure.

The general rule in applying Hess's law is to arrange a series of chemical equations (corresponding to a series of steps) in such a way that, when added together, all species will cancel except for the reactants and products that appear in the overall reaction. This means that we want the elements on the left and the compound of interest on the right of the arrow. Further, we often need to multiply some or all of the equations representing the individual steps by the appropriate coefficients.

EXAMPLE 6.9

Calculate the standard enthalpy of formation of acetylene (C_2H_2) from its elements:

$$2C(\text{graphite}) + H_2(g) \longrightarrow C_2H_2(g)$$

The equations for each step and the corresponding enthalpy changes are

(a) $C(\text{graphite}) + O_2(g) \longrightarrow CO_2(g)$ $\quad \Delta H^\circ_{rxn} = -393.5$ kJ/mol

(b) $H_2(g) + \frac{1}{2}O_2(g) \longrightarrow H_2O(l)$ $\quad \Delta H^\circ_{rxn} = -285.8$ kJ/mol

(c) $2C_2H_2(g) + 5O_2(g) \longrightarrow 4CO_2(g) + 2H_2O(l)$ $\quad \Delta H^\circ_{rxn} = -2598.8$ kJ/mol

Strategy Our goal here is to calculate the enthalpy change for the formation of C_2H_2 from its elements C and H_2. The reaction does not occur directly, however, so we must use an indirect route using the information given by Equations (a), (b), and (c).

Solution Looking at the synthesis of C_2H_2, we need 2 moles of graphite as reactant. So we multiply Equation (a) by 2 to get

(d) $2C(\text{graphite}) + 2O_2(g) \longrightarrow 2CO_2(g)$ $\quad \Delta H^\circ_{rxn} = 2(-393.5 \text{ kJ/mol}) = -787.0$ kJ/mol

Next, we need 1 mole of H_2 as a reactant and this is provided by Equation (b). Last, we need 1 mole of C_2H_2 as a product. Equation (c) has 2 moles of C_2H_2 as a reactant so we need to reverse the equation and divide it by 2:

(e) $2CO_2(g) + H_2O(l) \longrightarrow C_2H_2(g) + \frac{5}{2}O_2(g)$ $\quad \Delta H^\circ_{rxn} = \frac{1}{2}(2598.8 \text{ kJ/mol}) = 1299.4$ kJ/mol

Adding Equations (d), (b), and (e) together, we get

$$\begin{array}{lr} 2C(\text{graphite}) + 2O_2(g) \longrightarrow CO_2(g) & \Delta H^\circ_{rxn} = -787.0 \text{ kJ/mol} \\ H_2(g) + \frac{1}{2}O_2(g) \longrightarrow H_2O(l) & \Delta H^\circ_{rxn} = -285.8 \text{ kJ/mol} \\ 2CO_2(g) + H_2O(l) \longrightarrow C_2H_2(g) + \frac{5}{2}O_2(g) & \Delta H^\circ_{rxn} = 1299.4 \text{ kJ/mol} \\ \hline 2C(\text{graphite}) + H_2(g) \longrightarrow C_2H_2(g) & \Delta H^\circ_{rxn} = 226.6 \text{ kJ/mol} \end{array}$$

Therefore, $\Delta H_f^\circ = \Delta H^\circ_{rxn} = 226.6$ kJ/mol. The ΔH_f° value means that when 1 mole of C_2H_2 is synthesized from 2 moles of C (graphite) and 1 mole of H_2, 226.6 kJ of heat are absorbed by the reacting system from the surroundings. Thus, this is an endothermic process.

An oxyacetylene torch has a high flame temperature (3000°C) and is used to weld metals.

Similar problems: 6.62, 6.63.

Practice Exercise Calculate the standard enthalpy of formation of carbon disulfide (CS_2) from its elements, given that

$$\begin{array}{lr} C(\text{graphite}) + O_2(g) \longrightarrow CO_2(g) & \Delta H^\circ_{rxn} = -393.5 \text{ kJ/mol} \\ S(\text{rhombic}) + O_2(g) \longrightarrow SO_2(g) & \Delta H^\circ_{rxn} = -296.4 \text{ kJ/mol} \\ CS_2(l) + 3O_2(g) \longrightarrow CO_2(g) + 2SO_2(g) & \Delta H^\circ_{rxn} = -1073.6 \text{ kJ/mol} \end{array}$$

EXAMPLE 6.10

The thermite reaction involves aluminum and iron(III) oxide

$$2Al(s) + Fe_2O_3(s) \longrightarrow Al_2O_3(s) + 2Fe(l)$$

This reaction is highly exothermic and the liquid iron formed is used to weld metals. Calculate the heat released in kilojoules per gram of Al reacted with Fe_2O_3. The ΔH_f° for Fe(l) is 12.40 kJ/mol.

The molten iron formed in a thermite reaction is run down into a mold between the ends of two railroad rails. On cooling, the rails are welded together.

Strategy The enthalpy of a reaction is the difference between the sum of the enthalpies of the products and the sum of the enthalpies of the reactants. The enthalpy of each species (reactant or product) is given by the product of the stoichiometric coefficient and the standard enthalpy of formation of the species.

Solution Using the given ΔH_f° value for Fe(l) and other ΔH_f° values in Appendix 2 and Equation (6.18), we write

$$\begin{aligned}\Delta H_{rxn}^\circ &= [\Delta H_f^\circ(Al_2O_3) + 2\Delta H_f^\circ(Fe)] - [2\Delta H_f^\circ(Al) + 2\Delta H_f^\circ(Fe_2O_3)] \\ &= [(-1669.8 \text{ kJ/mol}) + 2(12.40 \text{ kJ/mol})] - [2(0) + (-822.2 \text{ kJ/mol})] \\ &= -822.8 \text{ kJ/mol}\end{aligned}$$

This is the amount of heat released for two moles of Al reacted. We use the following ratio

$$\frac{-822.8 \text{ kJ}}{2 \text{ mol Al}}$$

to convert to kJ/g Al. The molar mass of Al is 26.98 g, so

$$\begin{aligned}\text{heat released per gram of Al} &= \frac{-822.8 \text{ kJ}}{2 \text{ mol Al}} \times \frac{1 \text{ mol Al}}{26.98 \text{ g Al}} \\ &= -15.25 \text{ kJ/g}\end{aligned}$$

Check Is the negative sign consistent with the exothermic nature of the reaction? As a quick check, we see that 2 moles of Al weigh about 54 g and give off about 823 kJ of heat when reacted with Fe_2O_3. Therefore, the heat given off per gram of Al reacted is approximately −830 kJ/54 g or −15.4 kJ/g.

Similar problems: 6.54, 6.57.

Practice Exercise Benzene (C_6H_6) burns in air to produce carbon dioxide and liquid water. Calculate the heat released (in kilojoules) per gram of the compound reacted with oxygen. The standard enthalpy of formation of benzene is 49.04 kJ/mol.

In general, the more negative the standard enthalpy of formation of a compound the more stable is the compound. Thus, in the thermite reaction the less stable Fe_2O_3 is converted to the more stable Al_2O_3, and we expect the reaction to release a large amount of heat.

Review of Concepts

Explain why reactions involving reactant compounds with positive ΔH_f° values are generally more exothermic than those with negative ΔH_f° values.

Key Equations

$\Delta U = q + w$ (6.1)	Mathematical statement of the first law of thermodynamics.
$w = -P\Delta V$ (6.3)	Calculating work done in gas expansion or compression.
$H = U + PV$ (6.6)	Definition of enthalpy.
$\Delta H = \Delta U + P\Delta V$ (6.8)	Calculating enthalpy (or energy) change for a constant-pressure process.
$\Delta U = \Delta H - RT\Delta n$ (6.10)	Calculating enthalpy (or energy) change for a constant-temperature process.
$C = ms$ (6.11)	Definition of heat capacity.
$q = ms\Delta t$ (6.12)	Calculating heat change in term of specific heat.
$q = C\Delta t$ (6.13)	Calculating heat change in terms of heat capacity.
$\Delta H^\circ_{rxn} = \Sigma n\Delta H^\circ_f(\text{products}) - \Sigma m\Delta H^\circ_f(\text{reactants})$ (6.18)	Calculating standard enthalpy of reaction.

Summary of Facts and Concepts

1. Energy is the capacity to do work. There are many forms of energy and they are all interconvertible. The law of conservation of energy states that the total amount of energy in the universe always stays the same.
2. Any process that gives off heat to the surroundings is called an exothermic process; any process that absorbs heat from the surroundings is an endothermic process.
3. The state of a system is defined by variables such as composition, volume, temperature, and pressure. The change in a state function for a system depends only on the initial and final states of the system, and not on the path by which the change is accomplished. Energy is a state function; work and heat are not.
4. Energy can be converted from one form to another, but it cannot be created or destroyed (first law of thermodynamics). In chemistry, we are concerned mainly with thermal energy, electrical energy, and mechanical energy, which is usually associated with pressure-volume work.
5. The change in enthalpy (ΔH, usually given in kilojoules) is a measure of the heat of a reaction (or any other process) at constant pressure. Enthalpy is a state function. A change in enthalpy ΔH is equal to $\Delta U + P\Delta V$ for a constant-pressure process. For chemical reactions at constant temperature, ΔH is given by $\Delta U + RT\Delta n$, where Δn is the difference between moles of gaseous products and moles of gaseous reactants.
6. Constant-volume and constant-pressure calorimeters are used to measure heat changes of physical and chemical processes.
7. Hess's law states that the overall enthalpy change in a reaction is equal to the sum of enthalpy changes for the individual steps that make up the overall reaction. The standard enthalpy of a reaction can be calculated from the standard enthalpies of formation of reactants and products.

Key Words

Calorimetry, p. 191
Chemical energy, p. 177
Closed system, p. 178
Endothermic process, p. 179
Energy, p. 177
Enthalpy (H), p. 186
Enthalpy of reaction (ΔH), p. 186
Exothermic process, p. 179
First law of thermodynamics, p. 180
Heat, p. 178
Heat capacity (C), p. 191
Hess's law, p. 199
Isolated system, p. 178
Law of conservation of energy, p. 177
Open system, p. 178
Potential energy, p. 177
Radiant energy, p. 177
Specific heat (s), p. 191
Standard enthalpy of formation (ΔH°_f), p. 197
Standard enthalpy of reaction (ΔH°_{rxn}), p. 198
Standard state, p. 197
State function, p. 180
State of a system, p. 180
Surroundings, p. 178
System, p. 178
Thermal energy, p. 177
Thermochemical equation, p. 188
Thermochemistry, p. 178
Thermodynamics, p. 179
Work, p. 177

Questions and Problems

Definitions

Review Questions

6.1 Define these terms: system, surroundings, open system, closed system, isolated system, thermal energy, chemical energy, potential energy, kinetic energy, law of conservation of energy.

6.2 What is heat? How does heat differ from thermal energy? Under what condition is heat transferred from one system to another?

6.3 What are the units for energy commonly employed in chemistry?

6.4 A truck initially traveling at 60 km per hour is brought to a complete stop at a traffic light. Does this change violate the law of conservation of energy? Explain.

6.5 These are various forms of energy: chemical, heat, light, mechanical, and electrical. Suggest ways of interconverting these forms of energy.

6.6 Describe the interconversions of forms of energy occurring in these processes: (a) You throw a softball up into the air and catch it. (b) You switch on a flashlight. (c) You ride the ski lift to the top of the hill and then ski down. (d) You strike a match and let it burn down.

Energy Changes in Chemical Reactions

Review Questions

6.7 Define these terms: thermochemistry, exothermic process, endothermic process.

6.8 Stoichiometry is based on the law of conservation of mass. On what law is thermochemistry based?

6.9 Describe two exothermic processes and two endothermic processes.

6.10 Decomposition reactions are usually endothermic, whereas combination reactions are usually exothermic. Give a qualitative explanation for these trends.

First Law of Thermodynamics

Review Questions

6.11 On what law is the first law of thermodynamics based? Explain the sign conventions in the equation $\Delta U = q + w$.

6.12 Explain what is meant by a state function. Give two examples of quantities that are state functions and two that are not.

6.13 The internal energy of an ideal gas depends only on its temperature. Do a first-law analysis of this process. A sample of an ideal gas is allowed to expand at constant temperature against atmospheric pressure. (a) Does the gas do work on its surroundings? (b) Is there heat exchange between the system and the surroundings? If so, in which direction? (c) What is ΔU for the gas for this process?

6.14 Consider these changes.

(a) $Hg(l) \longrightarrow Hg(g)$

(b) $3O_2(g) \longrightarrow 2O_3(g)$

(c) $CuSO_4 \cdot 5H_2O(s) \longrightarrow CuSO_4(s) + 5H_2O(g)$

(d) $H_2(g) + F_2(g) \longrightarrow 2HF(g)$

At constant pressure, in which of the reactions is work done by the system on the surroundings? By the surroundings on the system? In which of them is no work done?

Problems

6.15 A sample of nitrogen gas expands in volume from 1.6 L to 5.4 L at constant temperature. Calculate the work done in joules if the gas expands (a) against a vacuum, (b) against a constant pressure of 0.80 atm, and (c) against a constant pressure of 3.7 atm.

6.16 A gas expands in volume from 26.7 mL to 89.3 mL at constant temperature. Calculate the work done (in joules) if the gas expands (a) against a vacuum, (b) against a constant pressure of 1.5 atm, and (c) against a constant pressure of 2.8 atm.

6.17 A gas expands and does *P-V* work on the surroundings equal to 325 J. At the same time, it absorbs 127 J of heat from the surroundings. Calculate the change in energy of the gas.

6.18 The work done to compress a gas is 74 J. As a result, 26 J of heat is given off to the surroundings. Calculate the change in energy of the gas.

6.19 Calculate the work done when 50.0 g of tin dissolves in excess acid at 1.00 atm and 25°C:

$$Sn(s) + 2H^+(aq) \longrightarrow Sn^{2+}(aq) + H_2(g)$$

Assume ideal gas behavior.

6.20 Calculate the work done in joules when 1.0 mole of water vaporizes at 1.0 atm and 100°C. Assume that the volume of liquid water is negligible compared with that of steam at 100°C, and ideal gas behavior.

Enthalpy of Chemical Reactions

Review Questions

6.21 Define these terms: enthalpy, enthalpy of reaction. Under what condition is the heat of a reaction equal to the enthalpy change of the same reaction?

6.22 In writing thermochemical equations, why is it important to indicate the physical state (that is, gaseous, liquid, solid, or aqueous) of each substance?

6.23 Explain the meaning of this thermochemical equation:

$$4NH_3(g) + 5O_2(g) \longrightarrow 4NO(g) + 6H_2O(g) \quad \Delta H = -904 \text{ kJ/mol}$$

6.24 Consider this reaction:

$$2CH_3OH(l) + 3O_2(g) \longrightarrow 4H_2O(l) + 2CO_2(g) \quad \Delta H = -1452.8 \text{ kJ/mol}$$

What is the value of ΔH if (a) the equation is multiplied throughout by 2, (b) the direction of the reaction is reversed so that the products become the reactants and vice versa, (c) water vapor instead of liquid water is formed as the product?

Problems

6.25 The first step in the industrial recovery of copper from the copper sulfide ore is roasting, that is, the conversion of CuS to CuO by heating:

$$2CuS(s) + 3O_2(g) \longrightarrow 2CuO(s) + 2SO_2(g) \quad \Delta H = -805.6 \text{ kJ/mol}$$

Calculate the heat evolved (in kJ) per gram of CuS roasted.

6.26 Determine the amount of heat (in kJ) given off when 1.26×10^4 g of NO_2 are produced according to the equation

$$2NO(g) + O_2(g) \longrightarrow 2NO_2(g) \quad \Delta H = -114.6 \text{ kJ/mol}$$

6.27 Consider the reaction

$$2H_2O(g) \longrightarrow 2H_2(g) + O_2(g) \quad \Delta H = 483.6 \text{ kJ/mol}$$

If 2.0 moles of $H_2O(g)$ are converted to $H_2(g)$ and $O_2(g)$ against a pressure of 1.0 atm at 125°C, what is ΔU for this reaction?

6.28 Consider the reaction

$$H_2(g) + Cl_2(g) \longrightarrow 2HCl(g) \quad \Delta H = -184.6 \text{ kJ/mol}$$

If 3 moles of H_2 react with 3 moles of Cl_2 to form HCl, calculate the work done against a pressure of 1.0 atm at 25°C. What is ΔU for this reaction? Assume the reaction goes to completion.

Calorimetry

Review Questions

6.29 What is the difference between specific heat and heat capacity? What are the units for these two quantities? Which is the intensive property and which is the extensive property?

6.30 Define calorimetry and describe two commonly used calorimeters. In a calorimetric measurement, why is it important that we know the heat capacity of the calorimeter? How is this value determined?

Problems

6.31 Consider the following data:

Metal	Al	Cu
Mass (g)	10	30
Specific heat (J/g · °C)	0.900	0.385
Temperature (°C)	40	60

When these two metals are placed in contact, which of the following will take place?

(a) Heat will flow from Al to Cu because Al has a larger specific heat.

(b) Heat will flow from Cu to Al because Cu has a larger mass.

(c) Heat will flow from Cu to Al because Cu has a larger heat capacity.

(d) Heat will flow from Cu to Al because Cu is at a higher temperature.

(e) No heat will flow in either direction.

6.32 Consider two metals A and B, each having a mass of 100 g and an initial temperature of 20°C. The specific heat of A is larger than that of B. Under the same heating conditions, which metal would take longer to reach a temperature of 21°C?

6.33 A piece of silver of mass 362 g has a heat capacity of 85.7 J/°C. What is the specific heat of silver?

6.34 A 6.22-kg piece of copper metal is heated from 20.5°C to 324.3°C. Calculate the heat absorbed (in kJ) by the metal.

6.35 Calculate the amount of heat liberated (in kJ) from 366 g of mercury when it cools from 77.0°C to 12.0°C.

6.36 A sheet of gold weighing 10.0 g and at a temperature of 18.0°C is placed flat on a sheet of iron weighing 20.0 g and at a temperature of 55.6°C. What is the final temperature of the combined metals? Assume that no heat is lost to the surroundings. (*Hint:* The heat gained by the gold must be equal to the heat lost by the iron. The specific heats of the metals are given in Table 6.2.)

6.37 A 0.1375-g sample of solid magnesium is burned in a constant-volume bomb calorimeter that has a heat capacity of 3024 J/°C. The temperature increases by 1.126°C. Calculate the heat given off by the burning Mg, in kJ/g and in kJ/mol.

6.38 A quantity of 2.00×10^2 mL of 0.862 *M* HCl is mixed with 2.00×10^2 mL of 0.431 *M* $Ba(OH)_2$ in a constant-pressure calorimeter of negligible heat capacity. The initial temperature of the HCl and $Ba(OH)_2$ solutions is the same at 20.48°C. For the process

$$H^+(aq) + OH^-(aq) \longrightarrow H_2O(l)$$

the heat of neutralization is −56.2 kJ/mol. What is the final temperature of the mixed solution?

Standard Enthalpy of Formation and Reaction

Review Questions

6.39 What is meant by the standard-state condition?

6.40 How are the standard enthalpies of formation of an element and of a compound determined?

6.41 What is meant by the standard enthalpy of a reaction?

6.42 Write the equation for calculating the enthalpy of a reaction. Define all the terms.

6.43 State Hess's law. Explain, with one example, the usefulness of Hess's law in thermochemistry.

6.44 Describe how chemists use Hess's law to determine the ΔH_f° of a compound by measuring its heat (enthalpy) of combustion.

Problems

6.45 Which of the following standard enthalpy of formation values is not zero at 25°C? Na(*s*), Ne(*g*), $CH_4(g)$, $S_8(s)$, Hg(*l*), H(*g*).

6.46 The ΔH_f° values of the two allotropes of oxygen, O_2 and O_3, are 0 and 142.2 kJ/mol, respectively, at 25°C. Which is the more stable form at this temperature?

6.47 Which is the more negative quantity at 25°C: ΔH_f° for $H_2O(l)$ or ΔH_f° for $H_2O(g)$?

6.48 Predict the value of ΔH_f° (greater than, less than, or equal to zero) for these elements at 25°C: (a) $Br_2(g)$ and $Br_2(l)$, (b) $I_2(g)$ and $I_2(s)$.

6.49 In general, compounds with negative ΔH_f° values are more stable than those with positive ΔH_f° values. $H_2O_2(l)$ has a negative ΔH_f° (see Table 6.4). Why, then, does $H_2O_2(l)$ have a tendency to decompose to $H_2O(l)$ and $O_2(g)$?

6.50 Suggest ways (with appropriate equations) that would enable you to measure the ΔH_f° values of $Ag_2O(s)$ and $CaCl_2(s)$ from their elements. No calculations are necessary.

6.51 Calculate the heat of decomposition for this process at constant pressure and 25°C:

$$CaCO_3(s) \longrightarrow CaO(s) + CO_2(g)$$

(Look up the standard enthalpy of formation of the reactant and products in Table 6.4.)

6.52 The standard enthalpies of formation of ions in aqueous solutions are obtained by arbitrarily assigning a value of zero to H^+ ions; that is, $\Delta H_f^\circ[H^+(aq)] = 0$.

(a) For the following reaction

$$HCl(g) \xrightarrow{H_2O} H^+(aq) + Cl^-(aq) \qquad \Delta H^\circ = -74.9 \text{ kJ/mol}$$

calculate ΔH_f° for the Cl^- ions.

(b) Given that ΔH_f° for OH^- ions is −229.6 kJ/mol, calculate the enthalpy of neutralization when 1 mole of a strong monoprotic acid (such as HCl) is titrated by 1 mole of a strong base (such as KOH) at 25°C.

6.53 Calculate the heats of combustion for the following reactions from the standard enthalpies of formation listed in Appendix 2:

(a) $2H_2(g) + O_2(g) \longrightarrow 2H_2O(l)$

(b) $2C_2H_2(g) + 5O_2(g) \longrightarrow 4CO_2(g) + 2H_2O(l)$

6.54 Calculate the heats of combustion for the following reactions from the standard enthalpies of formation listed in Appendix 2:

(a) $C_2H_4(g) + 3O_2(g) \longrightarrow 2CO_2(g) + 2H_2O(l)$

(b) $2H_2S(g) + 3O_2(g) \longrightarrow 2H_2O(l) + 2SO_2(g)$

6.55 Methanol, ethanol, and *n*-propanol are three common alcohols. When 1.00 g of each of these alcohols is burned in air, heat is liberated as shown by the following data: (a) methanol (CH_3OH), −22.6 kJ; (b) ethanol (C_2H_5OH), −29.7 kJ; (c) *n*-propanol (C_3H_7OH), −33.4 kJ. Calculate the heats of combustion of these alcohols in kJ/mol.

6.56 The standard enthalpy change for the following reaction is 436.4 kJ/mol:

$$H_2(g) \longrightarrow H(g) + H(g)$$

Calculate the standard enthalpy of formation of atomic hydrogen (H).

6.57 From the standard enthalpies of formation, calculate ΔH_{rxn}° for the reaction

$$C_6H_{12}(l) + 9O_2(g) \longrightarrow 6CO_2(g) + 6H_2O(l)$$

For $C_6H_{12}(l)$, $\Delta H_f^\circ = -151.9$ kJ/mol.

6.58 The first step in the industrial recovery of zinc from the zinc sulfide ore is roasting, that is, the conversion of ZnS to ZnO by heating:

$$2ZnS(s) + 3O_2(g) \longrightarrow 2ZnO(s) + 2SO_2(g) \qquad \Delta H_{rxn}^\circ = -879 \text{ kJ/mol}$$

Calculate the heat evolved (in kJ) per gram of ZnS roasted.

6.59 Determine the amount of heat (in kJ) given off when 1.26×10^4 g of ammonia are produced according to the equation

$$N_2(g) + 3H_2(g) \longrightarrow 2NH_3(g) \qquad \Delta H_{rxn}^\circ = -92.6 \text{ kJ/mol}$$

Assume that the reaction takes place under standard-state conditions at 25°C.

6.60 At 850°C, $CaCO_3$ undergoes substantial decomposition to yield CaO and CO_2. Assuming that the ΔH_f° values of the reactant and products are the same at 850°C as they are at 25°C, calculate the enthalpy change (in kJ) if 66.8 g of CO_2 are produced in one reaction.

6.61 From these data,

$$S(\text{rhombic}) + O_2(g) \longrightarrow SO_2(g) \qquad \Delta H^\circ_{rxn} = -296.06 \text{ kJ/mol}$$

$$S(\text{monoclinic}) + O_2(g) \longrightarrow SO_2(g) \qquad \Delta H^\circ_{rxn} = -296.36 \text{ kJ/mol}$$

calculate the enthalpy change for the transformation

$$S(\text{rhombic}) \longrightarrow S(\text{monoclinic})$$

(Monoclinic and rhombic are different allotropic forms of elemental sulfur.)

6.62 From the following data,

$$C(\text{graphite}) + O_2(g) \longrightarrow CO(g) \qquad \Delta H^\circ_{rxn} = -393.5 \text{ kJ/mol}$$

$$H_2(g) + \tfrac{1}{2}O_2(g) \longrightarrow H_2O(l) \qquad \Delta H^\circ_{rxn} = -285.8 \text{ kJ/mol}$$

$$2C_2H_6(g) + 7O_2(g) \longrightarrow 4CO_2(g) + 6H_2O(l) \qquad \Delta H^\circ_{rxn} = -3119.6 \text{ kJ/mol}$$

calculate the enthalpy change for the reaction

$$2C(\text{graphite}) + 3H_2(g) \longrightarrow C_2H_6(g)$$

6.63 From the following heats of combustion,

$$CH_3OH(l) + \tfrac{3}{2}O_2(g) \longrightarrow CO_2(g) + 2H_2O(l) \qquad \Delta H^\circ_{rxn} = -726.4 \text{ kJ/mol}$$

$$C(\text{graphite}) + O_2(g) \longrightarrow CO_2(g) \qquad \Delta H^\circ_{rxn} = -393.5 \text{ kJ/mol}$$

$$H_2(g) + \tfrac{1}{2}O_2(g) \longrightarrow H_2O(l) \qquad \Delta H^\circ_{rxn} = -285.8 \text{ kJ/mol}$$

calculate the enthalpy of formation of methanol (CH_3OH) from its elements:

$$C(\text{graphite}) + 2H_2(g) + \tfrac{1}{2}O_2(g) \longrightarrow CH_3OH(l)$$

6.64 Calculate the standard enthalpy change for the reaction

$$2Al(s) + Fe_2O_3(s) \longrightarrow 2Fe(s) + Al_2O_3(s)$$

given that

$$2Al(s) + \tfrac{3}{2}O_2(g) \longrightarrow Al_2O_3(s) \qquad \Delta H^\circ_{rxn} = -1669.8 \text{ kJ/mol}$$

$$2Fe(s) + \tfrac{3}{2}O_2(g) \longrightarrow Fe_2O_3(s) \qquad \Delta H^\circ_{rxn} = -822.2 \text{ kJ/mol}$$

Additional Problems

6.65 The convention of arbitrarily assigning a zero enthalpy value for the most stable form of each element in the standard state at 25°C is a convenient way of dealing with enthalpies of reactions. Explain why this convention cannot be applied to nuclear reactions.

6.66 Consider the following two reactions:

$$A \longrightarrow 2B \qquad \Delta H^\circ_{rxn} = \Delta H_1$$

$$A \longrightarrow C \qquad \Delta H^\circ_{rxn} = \Delta H_2$$

Determine the enthalpy change for the process

$$2B \longrightarrow C$$

6.67 The standard enthalpy change ΔH° for the thermal decomposition of silver nitrate according to the following equation is +78.67 kJ:

$$AgNO_3(s) \longrightarrow AgNO_2(s) + \tfrac{1}{2}O_2(g)$$

The standard enthalpy of formation of $AgNO_3(s)$ is −123.02 kJ/mol. Calculate the standard enthalpy of formation of $AgNO_2(s)$.

6.68 Hydrazine, N_2H_4, decomposes according to the following reaction:

$$3N_2H_4(l) \longrightarrow 4NH_3(g) + N_2(g)$$

(a) Given that the standard enthalpy of formation of hydrazine is 50.42 kJ/mol, calculate ΔH° for its decomposition. (b) Both hydrazine and ammonia burn in oxygen to produce $H_2O(l)$ and $N_2(g)$. Write balanced equations for each of these processes and calculate ΔH° for each of them. On a mass basis (per kg), would hydrazine or ammonia be the better fuel?

6.69 Consider the reaction

$$N_2(g) + 3H_2(g) \longrightarrow 2NH_3(g) \qquad \Delta H^\circ_{rxn} = -92.6 \text{ kJ/mol}$$

If 2.0 moles of N_2 react with 6.0 moles of H_2 to form NH_3, calculate the work done (in joules) against a pressure of 1.0 atm at 25°C. What is ΔU for this reaction? Assume the reaction goes to completion.

6.70 Calculate the heat released when 2.00 L of $Cl_2(g)$ with a density of 1.88 g/L react with an excess of sodium metal as 25°C and 1 atm to form sodium chloride.

6.71 Photosynthesis produces glucose, $C_6H_{12}O_6$, and oxygen from carbon dioxide and water:

$$6CO_2 + 6H_2O \longrightarrow C_6H_{12}O_6 + 6O_2$$

(a) How would you determine experimentally the ΔH°_{rxn} value for this reaction? (b) Solar radiation produces about 7.0×10^{14} kg glucose a year on Earth. What is the corresponding ΔH° change?

6.72 A 2.10-mole sample of crystalline acetic acid, initially at 17.0°C, is allowed to melt at 17.0°C and is then heated to 118.1°C (its normal boiling point) at 1.00 atm. The sample is allowed to vaporize at 118.1°C and is then rapidly quenched to 17.0°C, so that it recrystallizes. Calculate $\Delta H°$ for the total process as described.

6.73 Calculate the work done in joules by the reaction

$$2Na(s) + 2H_2O(l) \longrightarrow 2NaOH(aq) + H_2(g)$$

when 0.34 g of Na reacts with water to form hydrogen gas at 0°C and 1.0 atm.

6.74 You are given the following data:

$$H_2(g) \longrightarrow 2H(g) \quad \Delta H° = 436.4 \text{ kJ/mol}$$

$$Br_2(g) \longrightarrow 2Br(g) \quad \Delta H° = 192.5 \text{ kJ/mol}$$

$$H_2(g) + Br_2(g) \longrightarrow 2HBr(g) \quad \Delta H° = -72.4 \text{ kJ/mol}$$

Calculate $\Delta H°$ for the reaction

$$H(g) + Br(g) \longrightarrow HBr(g)$$

6.75 Methanol (CH_3OH) is an organic solvent and is also used as a fuel in some automobile engines. From the following data, calculate the standard enthalpy of formation of methanol:

$$2CH_3OH(l) + 3O_2(g) \longrightarrow 2CO_2(g) + 4H_2O(l) \quad \Delta H°_{rxn} = -1452.8 \text{ kJ/mol}$$

6.76 A 44.0-g sample of an unknown metal at 99.0°C was placed in a constant-pressure calorimeter containing 80.0 g of water at 24.0°C. The final temperature of the system was found to be 28.4°C. Calculate the specific heat of the metal. (The heat capacity of the calorimeter is 12.4 J/°C.)

6.77 A 1.00-mole sample of ammonia at 14.0 atm and 25°C in a cylinder fitted with a movable piston expands against a constant external pressure of 1.00 atm. At equilibrium, the pressure and volume of the gas are 1.00 atm and 23.5 L, respectively. (a) Calculate the final temperature of the sample. (b) Calculate q, w, and ΔU for the process. The specific heat of ammonia is 0.0258 J/g · °C.

6.78 Producer gas (carbon monoxide) is prepared by passing air over red-hot coke:

$$C(s) + \tfrac{1}{2}O_2(g) \longrightarrow CO(g)$$

Water gas (mixture of carbon monoxide and hydrogen) is prepared by passing steam over red-hot coke:

$$C(s) + H_2O(g) \longrightarrow CO(g) + H_2(g)$$

For many years, both producer gas and water gas were used as fuels in industry and for domestic cooking. The large-scale preparation of these gases was carried out alternately, that is, first producer gas, then water gas, and so on. Using thermochemical reasoning, explain why this procedure was chosen.

6.79 If energy is conserved, how can there be an energy crisis?

6.80 The so-called hydrogen economy is based on hydrogen produced from water using solar energy. The gas is then burned as a fuel:

$$2H_2(g) + O_2(g) \longrightarrow 2H_2O(l)$$

A primary advantage of hydrogen as a fuel is that it is nonpolluting. A major disadvantage is that it is a gas and therefore is harder to store than liquids or solids. Calculate the volume of hydrogen gas at 25°C and 1.00 atm required to produce an amount of energy equivalent to that produced by the combustion of a gallon of octane (C_8H_{18}). The density of octane is 2.66 kg/gal, and its standard enthalpy of formation is −249.9 kJ/mol.

6.81 The combustion of what volume of ethane (C_2H_6), measured at 23.0°C and 752 mmHg, would be required to heat 855 g of water from 25.0°C to 98.0°C?

6.82 The heat of vaporization of a liquid (ΔH_{vap}) is the energy required to vaporize 1.00 g of the liquid at its boiling point. In one experiment, 60.0 g of liquid nitrogen (boiling point −196°C) are poured into a Styrofoam cup containing 2.00×10^2 g of water at 55.3°C. Calculate the molar heat of vaporization of liquid nitrogen if the final temperature of the water is 41.0°C.

6.83 Using the data in Appendix 2, calculate the enthalpy change for the gaseous reaction shown here. (*Hint:* First determine the limiting reagent.)

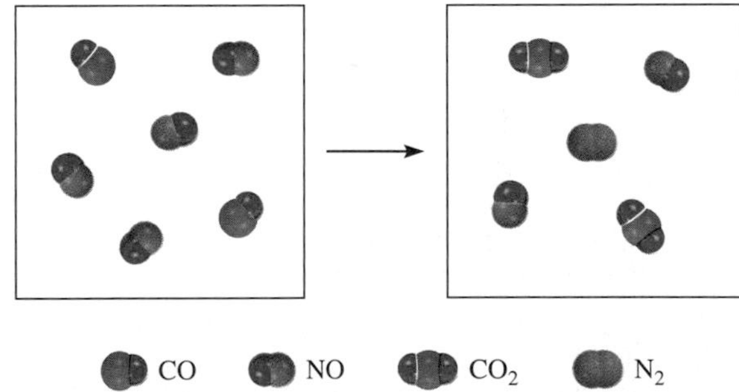

6.84 For which of the following reactions does $\Delta H°_{rxn} = \Delta H°_f$?

(a) $H_2(g) + S(\text{rhombic}) \longrightarrow H_2S(g)$
(b) $C(\text{diamond}) + O_2(g) \longrightarrow CO_2(g)$
(c) $H_2(g) + CuO(s) \longrightarrow H_2O(l) + Cu(s)$
(d) $O(g) + O_2(g) \longrightarrow O_3(g)$

6.85 A quantity of 0.020 mole of a gas initially at 0.050 L and 20°C undergoes a constant-temperature expansion until its volume is 0.50 L. Calculate the work done (in joules) by the gas if it expands (a) against a vacuum and (b) against a constant pressure of 0.20 atm. (c) If the gas in (b) is allowed to expand unchecked until its pressure is equal to the external pressure, what would its final volume be before it stopped expanding, and what would be the work done?

6.86 (a) For most efficient use, refrigerator freezer compartments should be fully packed with food. What is the thermochemical basis for this recommendation? (b) Starting at the same temperature, tea and coffee remain hot longer in a thermal flask than chicken noodle soup. Explain.

6.87 Calculate the standard enthalpy change for the fermentation process. (See Problem 3.72.)

6.88 Portable hot packs are available for skiers and people engaged in other outdoor activities in a cold climate. The air-permeable paper packet contains a mixture of powdered iron, sodium chloride, and other components, all moistened by a little water. The exothermic reaction that produces the heat is a very common one—the rusting of iron:

$$4Fe(s) + 3O_2(g) \longrightarrow 2Fe_2O_3(s)$$

When the outside plastic envelope is removed, O_2 molecules penetrate the paper, causing the reaction to begin. A typical packet contains 250 g of iron to warm your hands or feet for up to 4 h. How much heat (in kJ) is produced by this reaction? (*Hint:* See Appendix 2 for ΔH_f° values.)

6.89 A person ate 0.50 lb of cheese (an energy intake of 4000 kJ). Suppose that none of the energy was stored in his body. What mass (in grams) of water would he need to perspire in order to maintain his original temperature? (It takes 44.0 kJ to vaporize 1 mole of water.)

6.90 The total volume of the Pacific Ocean is estimated to be 7.2×10^8 km^3. A medium-sized atomic bomb produces 1.0×10^{15} J of energy upon explosion. Calculate the number of atomic bombs needed to release enough energy to raise the temperature of the water in the Pacific Ocean by 1°C.

6.91 A 19.2-g quantity of dry ice (solid carbon dioxide) is allowed to sublime (evaporate) in an apparatus like the one shown in Figure 6.4. Calculate the expansion work done against a constant external pressure of 0.995 atm and at a constant temperature of 22°C. Assume that the initial volume of dry ice is negligible and that CO_2 behaves like an ideal gas.

6.92 The enthalpy of combustion of benzoic acid (C_6H_5COOH) is commonly used as the standard for calibrating constant-volume bomb calorimeters; its value has been accurately determined to be −3226.7 kJ/mol. When 1.9862 g of benzoic acid are burned in a calorimeter, the temperature rises from 21.84°C to 25.67°C. What is the heat capacity of the bomb? (Assume that the quantity of water surrounding the bomb is exactly 2000 g.)

6.93 Lime is a term that includes calcium oxide (CaO, also called quicklime) and calcium hydroxide [$Ca(OH)_2$, also called slaked lime]. It is used in the steel industry to remove acidic impurities, in air-pollution control to remove acidic oxides such as SO_2, and in water treatment. Quicklime is made industrially by heating limestone ($CaCO_3$) above 2000°C:

$$CaCO_3(s) \longrightarrow CaO(s) + CO_2(g) \qquad \Delta H^\circ = 177.8 \text{ kJ/mol}$$

Slaked lime is produced by treating quicklime with water:

$$CaO(s) + H_2O(l) \longrightarrow Ca(OH)_2(s) \qquad \Delta H^\circ = -65.2 \text{ kJ/mol}$$

The exothermic reaction of quicklime with water and the rather small specific heats of both quicklime (0.946 J/g · °C) and slaked lime (1.20 J/g · °C) make it hazardous to store and transport lime in vessels made of wood. Wooden sailing ships carrying lime would occasionally catch fire when water leaked into the hold. (a) If a 500-g sample of water reacts with an equimolar amount of CaO (both at an initial temperature of 25°C), what is the final temperature of the product, $Ca(OH)_2$? Assume that the product absorbs all of the heat released in the reaction. (b) Given that the standard enthalpies of formation of CaO and H_2O are −635.6 kJ/mol and −285.8 kJ/mol, respectively, calculate the standard enthalpy of formation of $Ca(OH)_2$.

6.94 Calcium oxide (CaO) is used to remove sulfur dioxide generated by coal-burning power stations:

$$2CaO(s) + 2SO_2(g) + O_2(g) \longrightarrow 2CaSO_4(s)$$

Calculate the enthalpy change for this process if 6.6×10^5 g of SO_2 are removed by this process every day.

6.95 A balloon 16 m in diameter is inflated with helium at 18°C. (a) Calculate the mass of He in the balloon, assuming ideal behavior. (b) Calculate the work done (in joules) during the inflation process if the atmospheric pressure is 98.7 kPa.

6.96 (a) A person drinks four glasses of cold water (3.0°C) every day. The volume of each glass is 2.5×10^2 mL. How much heat (in kJ) does the body have to supply to raise the temperature of the water to 37°C, the body temperature? (b) How much heat would your body lose if you were to ingest 8.0×10^2 g of snow at 0°C to quench thirst? (The amount of heat necessary to melt snow is 6.01 kJ/mol.)

6.97 Determine the standard enthalpy of formation of ethanol (C_2H_5OH) from its standard enthalpy of combustion (−1367.4 kJ/mol).

6.98 Ice at 0°C is placed in a Styrofoam cup containing 361 g of a soft drink at 23°C. The specific heat of the drink is about the same as that of water. Some ice remains after the ice and soft drink reach an equilibrium temperature of 0°C. Determine the mass of ice that has melted. Ignore the heat capacity of the cup. (*Hint:* It takes 334 J to melt 1 g of ice at 0°C.)

6.99 A gas company in Massachusetts charges $1.30 for 15 ft^3 of natural gas (CH_4) measured at 20°C and 1.0 atm. Calculate the cost of heating 200 mL of water (enough to make a cup of coffee or tea) from 20°C to 100°C. Assume that only 50 percent of the heat generated by the combustion is used to heat the water; the rest of the heat is lost to the surroundings.

6.100 Calculate the internal energy of a Goodyear blimp filled with helium gas at 1.2×10^5 Pa. The volume of the blimp is 5.5×10^3 m^3. If all the energy were used to heat 10.0 tons of copper at 21°C, calculate the final temperature of the metal. (*Hint:* See Section 5.6 for help in calculating the internal energy of a gas. 1 ton = 9.072×10^5 g.)

6.101 Decomposition reactions are usually endothermic, whereas combination reactions are usually exothermic. Give a qualitative explanation for these trends.

6.102 Acetylene (C_2H_2) can be made by reacting calcium carbide (CaC_2) with water. (a) Write an equation for the reaction. (b) What is the maximum amount of heat (in kilojoules) that can be obtained from the combustion of acetylene, starting with 74.6 g of CaC_2?

6.103 When 1.034 g of naphthalene ($C_{10}H_8$) are burned in a constant-volume bomb calorimeter at 298 K, 41.56 kJ of heat are evolved. Calculate ΔU and ΔH for the reaction on a molar basis.

Special Problems

6.104 (a) A snowmaking machine contains a mixture of compressed air and water vapor at about 20 atm. When the mixture is sprayed into the atmosphere it expands so rapidly that, as a good approximation, no heat exchange occurs between the system (air and water) and its surroundings. (In thermodynamics, such a process is called an adiabatic process.) Do a first law of thermodynamics analysis to show how snow is formed under these conditions.

(b) If you have ever pumped air into a bicycle tire, you probably noticed a warming effect at the valve stem. The action of the pump compresses the air inside the pump and the tire. The process is rapid enough to be treated as an adiabatic process. Apply the first law of thermodynamics to account for the warming effect.

(c) A driver's manual states that the stopping distance quadruples as the speed doubles; that is, if it takes 30 ft to stop a car traveling at 25 mph, then it would take 120 ft to stop a car moving at 50 mph. Justify this statement by using the first law of thermodynamics. Assume that when a car is stopped, its kinetic energy ($\frac{1}{2}mu^2$) is totally converted to heat.

6.105 Why are cold, damp air and hot, humid air more uncomfortable than dry air at the same temperatures? (The specific heats of water vapor and air are approximately 1.9 J/g · °C and 1.0 J/g · °C, respectively.)

6.106 The average temperature in deserts is high during the day but quite cool at night, whereas that in regions along the coastline is more moderate. Explain.

6.107 From a thermochemical point of view, explain why a carbon dioxide fire extinguisher or water should not be used on a magnesium fire.

6.108 A 4.117-g impure sample of glucose ($C_6H_{12}O_6$) was burned in a constant-volume calorimeter having a heat capacity of 19.65 kJ/°C. If the rise in temperature is 3.134°C, calculate the percent by mass of the glucose in the sample. Assume that the impurities are unaffected by the combustion process. See Appendix 2 for thermodynamic data.

6.109 Construct a table with the headings q, w, ΔU, and ΔH. For each of the following processes, deduce whether each of the quantities listed is positive (+), negative (−), or zero (0). (a) Freezing of benzene. (b) Compression of an ideal gas at constant temperature. (c) Reaction of sodium with water. (d) Boiling liquid ammonia. (e) Heating a gas at constant volume. (f) Melting of ice.

6.110 The combustion of 0.4196 g of a hydrocarbon releases 17.55 kJ of heat. The masses of the products are CO_2 = 1.419 g and H_2O = 0.290 g. (a) What is the empirical formula of the compound? (b) If the approximate molar mass of the compound is 76 g, calculate its standard enthalpy of formation.

6.111 Metabolic activity in the human body releases approximately 1.0×10^4 kJ of heat per day. Assuming the body is 50 kg of water, how much would the body temperature rise if it were an isolated system? How much water must the body eliminate as perspiration to maintain the normal body temperature (98.6°F)? Comment on your results. The heat of vaporization of water may be taken as 2.41 kJ/g.

6.112 Starting at A, an ideal gas undergoes a cyclic process involving expansion and compression, as shown here. Calculate the total work done. Does your result support the notion that work is not a state function?

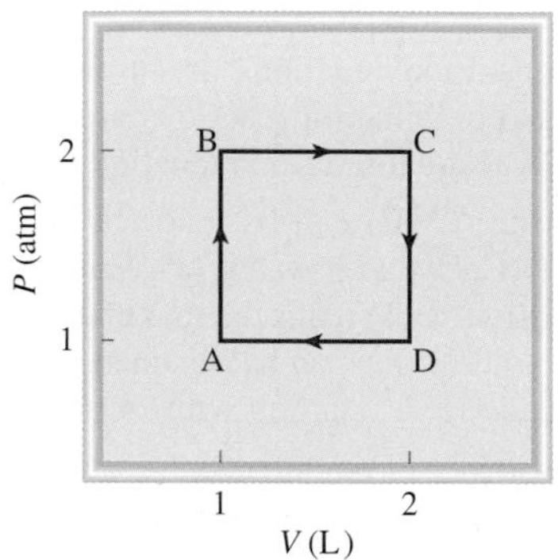

6.113 Give an example for each of the following situations: (a) Adding heat to a system raises its temperature, (b) adding heat to a system does not change (raise) its temperature, and (c) a system's temperature is changed even though no heat is added or removed from it.

6.114 For reactions in condensed phases (liquids and solids), the difference between ΔH and ΔU is usually quite small. This statement holds for reactions carried out under atmospheric conditions. For certain geochemical processes, however, the external pressure may be so great that ΔH and ΔU can differ by a significant amount. A well-known example is the slow conversion of graphite to diamond under Earth's surface. Calculate ($\Delta H - \Delta U$) for the conversion of 1 mole of graphite to 1 mole of diamond at a pressure of 50,000 atm. The densities of graphite and diamond are 2.25 g/cm^3 and 3.52 g/cm^3, respectively.

Answers to Practice Exercises

6.1 (a) 0, (b) −286 J. **6.2** −63 J. **6.3** -6.47×10^3 kJ. **6.4** −111.7 kJ/mol. **6.5** −34.3 kJ. **6.6** −728 kJ/mol. **6.7** 21.19°C. **6.8** 22.49°C. **6.9** 87.3 kJ/mol. **6.10** −41.83 kJ/g.

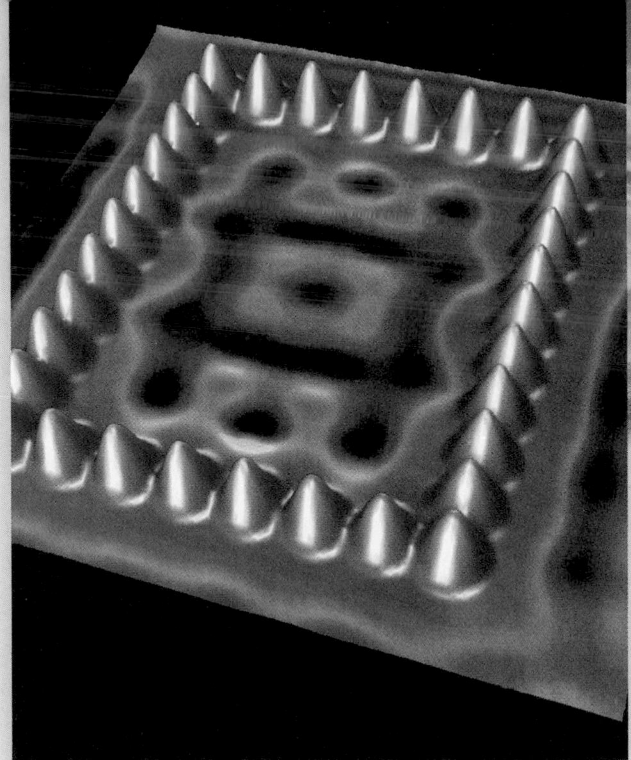

Iron atoms arranged on a copper surface create a quantum corral where the wave properties of electrons can be observed.

CHAPTER

7

The Electronic Structure of Atoms

Chapter Outline

Student Interactive Activities

Animations
Emission Spectra (7.3)
Electron Configurations (7.8)

Electronic Homework
Example Practice Problems
End of Chapter Problems

Essential Concepts

Planck's Quantum Theory To explain the dependence of radiation emitted by objects on wavelength, Planck proposed that atoms and molecules could emit (or absorb) energy in discrete quantities called quanta. Planck's theory revolutionized physics.

The Advent of Quantum Mechanics Planck's work led to the explanation of the photoelectric effect by Einstein, who postulated that light consists of particles called photons, and the emission spectrum of the hydrogen atom by Bohr. Further advancements to quantum theory were made by de Broglie, who demonstrated that an electron possesses both particle and wave properties, and Heisenberg, who derived an inherent limitation to measuring submicroscopic systems. These developments culminated in the Schrödinger equation, which describes the behavior and energy of electrons, atoms, and molecules.

The Hydrogen Atom The solution to the Schrödinger equation for the hydrogen atom shows quantized energies for the electron and a set of wave functions called atomic orbitals. The atomic orbitals are labeled with specific quantum numbers; the orbitals tell us the regions in which an electron can be located. The results obtained for hydrogen, with minor modifications, can be applied to more complex atoms.

The Building-Up Principle The periodic table can be constructed by increasing atomic number and adding electrons in a stepwise fashion. Specific guidelines (the Pauli exclusion principle and Hund's rule) help us write ground-state electron configurations of the elements, which tell us how electrons are distributed among the atomic orbitals.

7.1 From Classical Physics to Quantum Theory

Early attempts to understand atoms and molecules met with only limited success. By assuming that molecules behave like rebounding balls, physicists were able to predict and explain some macroscopic phenomena, such as the pressure exerted by a gas. However, their model did not account for the stability of molecules; that is, it could not explain the forces that hold atoms together. It took a long time to realize—and an even longer time to accept—that the properties of atoms and molecules are *not* governed by the same laws that work so well for larger objects.

The new era in physics started in 1900 with a young German physicist named Max Planck. While analyzing the data on the radiation emitted by solids heated to various temperatures, Planck discovered that atoms and molecules emit energy only in certain discrete quantities, or *quanta.* Physicists had always assumed that energy is continuous, which meant that any amount of energy could be released in a radiation process, so Planck's *quantum theory* turned physics upside down. Indeed, the flurry of research that ensued altered our concept of nature forever.

Ocean water waves.

To understand quantum theory, we must know something about the nature of waves. A ***wave*** can be thought of as *a vibrating disturbance by which energy is transmitted.* The speed of a wave depends on the type of wave and the nature of the medium through which the wave is traveling (for example, air, water, or a vacuum). *The distance between identical points on successive waves* is called the ***wavelength*** λ (lambda). The ***frequency*** ν (nu) of the wave is *the number of waves that pass through a particular point in one second.* The ***amplitude*** is *the vertical distance from the midline of a wave to the peak or trough* (Figure 7.1a). Figure 7.1b shows two waves that have the same amplitude but different wavelengths and frequencies.

An important property of a wave traveling through space is its speed (u), which is given by the product of its wavelength and its frequency:

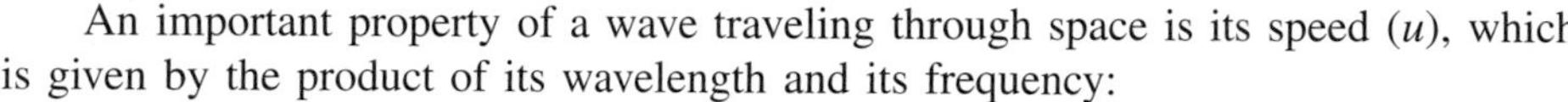

$$u = \lambda\nu \qquad (7.1)$$

The inherent "sensibility" of Equation (7.1) becomes apparent if we analyze the physical dimensions involved in the three terms. The wavelength (λ) expresses the length

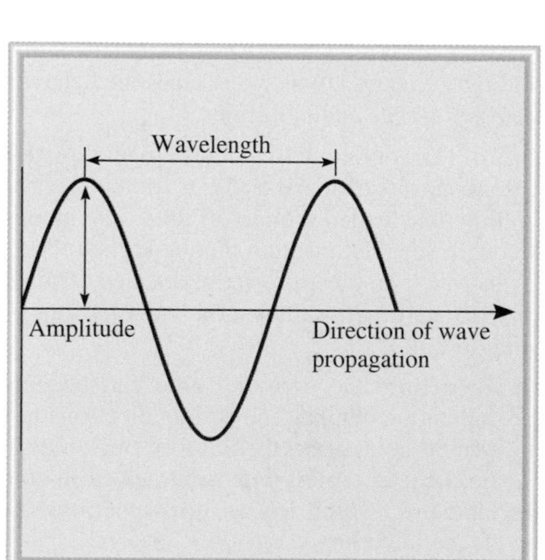

(a)

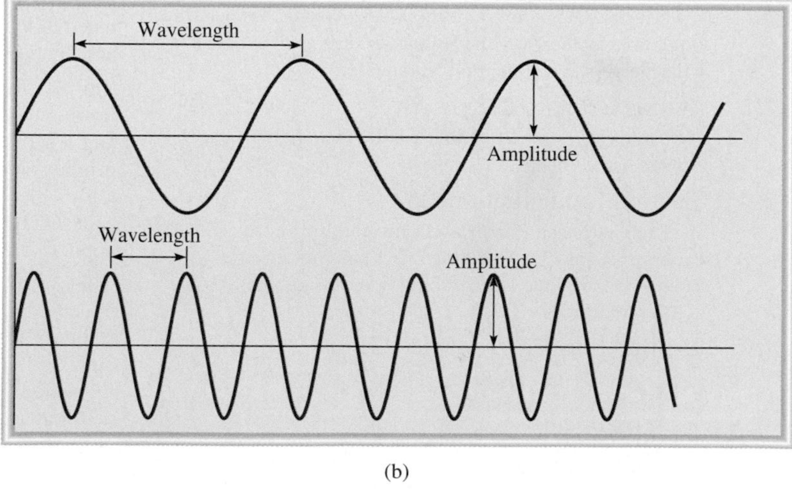

(b)

Figure 7.1

(a) Wavelength and amplitude. (b) Two waves having different wavelengths and frequencies. The wavelength of the top wave is three times that of the lower wave, but its frequency is only one-third that of the lower wave. Both waves have the same amplitude and speed.

of a wave, or distance/wave. The frequency (ν) indicates the number of these waves that pass any reference point per unit of time, or waves/time. Thus, the product of these terms results in dimensions of distance/time, which is speed:

$$\frac{\text{distance}}{\text{time}} = \frac{\text{distance}}{\cancel{\text{wave}}} \times \frac{\cancel{\text{waves}}}{\text{time}}$$

Wavelength is usually expressed in units of meters, centimeters, or nanometers, and frequency is measured in hertz (Hz), where

$$1 \text{ Hz} = 1 \text{ cycle/s}$$

The word "cycle" may be left out and the frequency expressed as, for example, 25/s (read as "25 per second").

Electromagnetic Radiation

There are many kinds of waves, such as water waves, sound waves, and light waves. In 1873 James Clerk Maxwell proposed that visible light consists of electromagnetic waves. According to Maxwell's theory, an ***electromagnetic wave*** *has an electric field component and a magnetic field component.* These two components have the same wavelength and frequency, and hence the same speed, but they travel in mutually perpendicular planes (Figure 7.2). The significance of Maxwell's theory is that it provides a mathematical description of the general behavior of light. In particular, his model accurately describes how energy in the form of radiation can be propagated through space as vibrating electric and magnetic fields. ***Electromagnetic radiation*** is *the emission and transmission of energy in the form of electromagnetic waves.*

Sound waves and water waves are not electromagnetic waves, but X rays and radio waves are.

Electromagnetic waves travel 3.00×10^8 meters per second (rounded off), or 186,000 miles per second, in a vacuum. This speed differs from one medium to another, but not enough to distort our calculations significantly. By convention, we use the symbol c for the speed of electromagnetic waves, or as it is more commonly called, the *speed of light.* The wavelength of electromagnetic waves is usually given in nanometers (nm).

A more accurate value for the speed of light is given on the inside back cover of the book.

EXAMPLE 7.1

The wavelength of the green light from a traffic signal is centered at 522 nm. What is the frequency of this radiation?

Strategy We are given the wavelength of an electromagnetic wave and asked to calculate its frequency. Rearranging Equation (7.1) and replacing u with c (the speed of light) gives

$$\nu = \frac{c}{\lambda}$$

Solution Because the speed of light is given in meters per second, it is convenient to first convert wavelength to meters. Recall that $1 \text{ nm} = 1 \times 10^{-9}$ m (see Table 1.3). We write

$$\lambda = 522 \cancel{\text{nm}} \times \frac{1 \times 10^{-9} \text{ m}}{1 \cancel{\text{nm}}} = 522 \times 10^{-9} \text{ m}$$

$$= 5.22 \times 10^{-7} \text{ m}$$

(Continued)

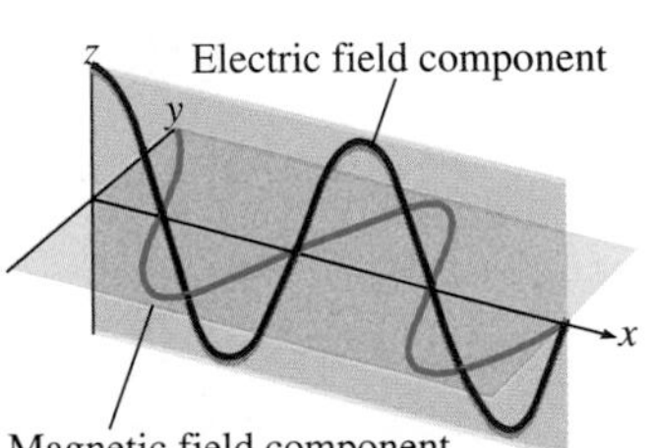

Figure 7.2
The electric field and magnetic field components of an electromagnetic wave. These two components have the same wavelength, frequency, and amplitude, but they vibrate in two mutually perpendicular planes.

Substituting in the wavelength and the speed of light (3.00×10^8 m/s), the frequency is

$$\nu = \frac{3.00 \times 10^8 \text{ m/s}}{5.22 \times 10^{-7} \text{ m}}$$
$$= 5.75 \times 10^{14}\text{/s, or } 5.75 \times 10^{14} \text{ Hz}$$

Check The answer shows that 5.75×10^{14} waves pass a fixed point every second. This very high frequency is in accordance with the very high speed of light.

Similar problem: 7.7.

Practice Exercise What is the wavelength (in meters) of an electromagnetic wave whose frequency is 3.64×10^7 Hz?

Figure 7.3 shows various types of electromagnetic radiation, which differ from one another in wavelength and frequency. The long radio waves are emitted by large antennas, such as those used by broadcasting stations. The shorter, visible light waves are produced by the motions of electrons within atoms and molecules. The shortest waves, which also have the highest frequency, are associated with γ (gamma) rays, which result from changes within the nucleus of the atom (see Chapter 2). As we will

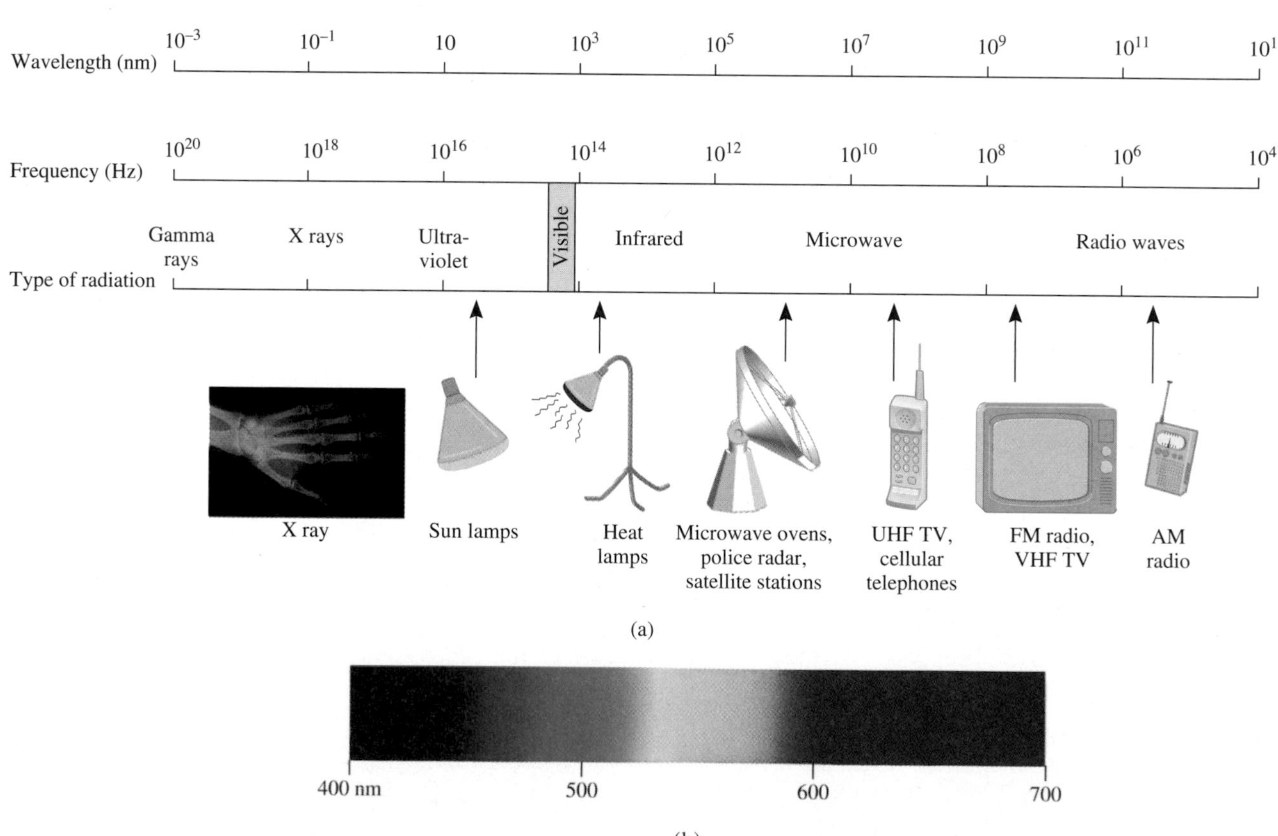

Figure 7.3
(a) Types of electromagnetic radiation. Gamma rays have the shortest wavelength and highest frequency; radio waves have the longest wavelength and the lowest frequency. Each type of radiation is spread over a specific range of wavelengths (and frequencies). (b) Visible light ranges from a wavelength of 400 nm (violet) to 700 nm (red).

see shortly, the higher the frequency, the more energetic the radiation. Thus, ultraviolet radiation, X rays, and γ rays are high-energy radiation.

Planck's Quantum Theory

When solids are heated, they emit electromagnetic radiation over a wide range of wavelengths. The dull red glow of an electric heater and the bright white light of a tungsten lightbulb are examples of radiation from heated solids.

Measurements taken in the latter part of the nineteenth century showed that the amount of radiant energy emitted by an object at a certain temperature depends on its wavelength. Attempts to account for this dependence in terms of established wave theory and thermodynamic laws were only partially successful. One theory explained short-wavelength dependence but failed to account for the longer wavelengths. Another theory accounted for the longer wavelengths but failed for short wave lengths. It seemed that something fundamental was missing from the laws of classical physics.

The failure in the short-wavelength region is called the ultraviolet catastrophe.

Planck solved the problem with an assumption that departed drastically from accepted concepts. Classical physics assumed that atoms and molecules could emit (or absorb) any arbitrary amount of radiant energy. Planck said that atoms and molecules could emit (or absorb) energy only in discrete quantities, like small packages or bundles. Planck gave the name ***quantum*** to *the smallest quantity of energy that can be emitted (or absorbed) in the form of electromagnetic radiation.* The energy E of a single quantum of energy is given by

$$E = h\nu \tag{7.2}$$

where h is called *Planck's constant* and ν is the frequency of radiation. The value of Planck's constant is 6.63×10^{-34} J · s. Because $\nu = c/\lambda$, Equation (7.2) can also be expressed as

$$E = h\frac{c}{\lambda} \tag{7.3}$$

According to quantum theory, energy is always emitted in integral multiples of $h\nu$; for example, $h\nu$, 2 $h\nu$, 3 $h\nu$, . . . , but never, for example, 1.67 $h\nu$ or 4.98 $h\nu$. At the time Planck presented his theory, he could not explain why energies should be fixed or quantized in this manner. Starting with this hypothesis, however, he had no trouble correlating the experimental data for emission by solids over the *entire* range of wavelengths; they all supported the quantum theory.

The idea that energy should be quantized or "bundled" may seem strange, but the concept of quantization has many analogies. For example, an electric charge is also quantized; there can be only whole-number multiples of e, the charge of one electron. Matter itself is quantized because the numbers of electrons, protons, and neutrons and the numbers of atoms in a sample of matter must also be integers. Our money system is based on a "quantum" of value called a penny. Even processes in living systems involve quantized phenomena. The eggs laid by hens are quantized, and a pregnant cat gives birth to an integral number of kittens, not to one-half or three-quarters of a kitten.

Review of Concepts

From the following colors of visible light, which is most energetic? (a) green, (b) red, (c) yellow.

7.2 The Photoelectric Effect

The equation for the energy of the photon has the same form as Equation (7.2) because, as we will see shortly, electromagnetic radiation is emitted as well as absorbed in the form of photons.

In 1905, only 5 years after Planck presented his quantum theory, the German-American physicist Albert Einstein used the theory to solve another mystery in physics, the ***photoelectric effect,*** a phenomenon in which *electrons are ejected from the surface of certain metals exposed to light of at least a certain minimum frequency,* called *the threshold frequency* (Figure 7.4). The number of electrons ejected was proportional to the intensity (or brightness) of the light, but the energies of the ejected electrons were not. Below the threshold frequency no electrons were ejected no matter how intense the light.

The photoelectric effect could not be explained by the wave theory of light. Einstein, however, made an extraordinary assumption. He suggested that a beam of light is really a stream of particles. These *particles of light* are now called ***photons.*** Using Planck's quantum theory of radiation as a starting point, Einstein deduced that each photon must possess energy E, given by the equation

$$E = h\nu$$

where ν is the frequency of light.

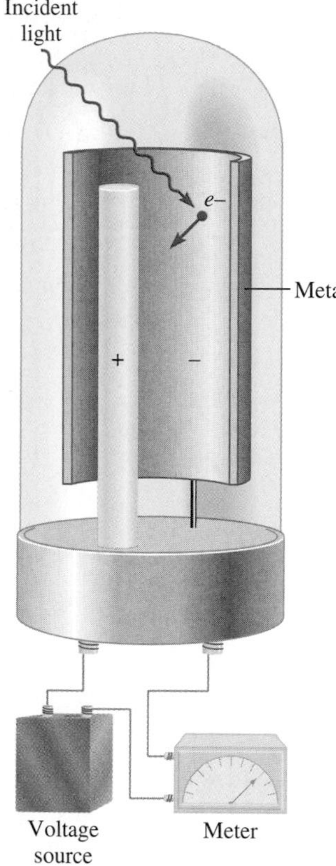

Figure 7.4
An apparatus for studying the photoelectric effect. Light of a certain frequency falls on a clean metal surface. Ejected electrons are attracted toward the positive electrode. The flow of electrons is registered by a detecting meter.

EXAMPLE 7.2

Calculate the energy (in joules) of (a) a photon with a wavelength of 3.75×10^4 nm (infrared region) and (b) a photon with a wavelength of 3.75×10^{-2} nm (X ray region).

Strategy In both (a) and (b), we are given the wavelength of a photon and asked to calculate its energy. We need to use Equation (7.3) to calculate the energy. Planck's constant is given in the text and also on the back inside cover.

Solution (a) From Equation (7.3),

$$E = h\frac{c}{\lambda}$$

$$= \frac{(6.63 \times 10^{-34}\text{ J}\cdot\text{s})(3.00 \times 10^8\text{ m/s})}{(3.75 \times 10^4\text{ nm})\dfrac{1 \times 10^{-9}\text{ m}}{1\text{ nm}}}$$

$$= 5.30 \times 10^{-21}\text{ J}$$

This is the energy of a single photon with a 3.75×10^4 nm wavelength.

(b) Following the same procedure as in (a), we can show that the energy of the photon that has a wavelength of 3.75×10^{-2} nm is 5.30×10^{-15} J.

Check Because the energy of a photon increases with decreasing wavelength, we see that an "X ray" photon is 1×10^6, or a million times, more energetic than an "infrared" photon.

Similar problem: 7.15.

Practice Exercise The energy of a photon is 5.87×10^{-20} J. What is its wavelength (in nanometers)?

Electrons are held in a metal by attractive forces, and so removing them from the metal requires light of a sufficiently high frequency (which corresponds to sufficiently high energy) to break them free. Shining a beam of light onto a metal surface can be thought of as shooting a beam of particles—photons—at the metal atoms. If the frequency of photons is such that $h\nu$ is exactly equal to the energy that binds the electrons in the metal, then the light will have just enough energy to knock the electrons loose. If we use light of a higher frequency, then not only will the electrons be knocked loose, but they will also acquire some kinetic energy. This situation is summarized by the equation

$$h\nu = \text{KE} + W \tag{7.4}$$

where KE is the kinetic energy of the ejected electron and W is the work function, which is a measure of how strongly the electrons are held in the metal. Rewriting Equation (7.4) as

$$\text{KE} = h\nu - W$$

shows that the more energetic the photon (that is, the higher the frequency), the greater the kinetic energy of the ejected electron.

Now consider two beams of light having the same frequency (which is greater than the threshold frequency) but different intensities. The more intense beam of light consists of a larger number of photons; consequently, it ejects more electrons from the metal's surface than the weaker beam of light. Thus, the more intense the light, the greater the number of electrons emitted by the target metal; the higher the frequency of the light, the greater the kinetic energy of the ejected electrons.

EXAMPLE 7.3

The work function of magnesium metal is 5.86×10^{-19} J. (a) Calculate the minimum frequency of light required to release electrons from the metal. (b) Calculate the kinetic energy of the ejected electron if light of frequency $2.00 \times 10^{15}\ \text{s}^{-1}$ is used for irradiating the metal.

Strategy (a) The relationship between the work function of an element and the frequency of light is given by Equation (7.4). The minimum frequency of light needed to dislodge an electron is the point where the kinetic energy of the ejected electron is zero. (b) Knowing both the work function and the frequency of light, we can solve for the kinetic energy of the ejected electron.

Solution (a) Setting KE = 0 in Equation (7.4), we write

$$h\nu = W$$

Thus,

$$\nu = \frac{W}{h} = \frac{5.86 \times 10^{-19}\ \text{J}}{6.63 \times 10^{-34}\ \text{J} \cdot \text{s}} = 8.84 \times 10^{14}\ \text{s}^{-1}$$

(b) Rearranging Equation (7.4) gives

$$\begin{aligned}\text{KE} &= h\nu - W \\ &= (6.63 \times 10^{-34}\ \text{J} \cdot \text{s})(2.00 \times 10^{15}\ \text{s}^{-1}) - 5.86 \times 10^{-19}\ \text{J} \\ &= 7.40 \times 10^{-19}\ \text{J}\end{aligned}$$

Check The kinetic energy of the ejected electron (7.40×10^{-19} J) is smaller than the energy of the photon (1.33×10^{-18} J). Therefore, the answer is reasonable.

Similar problem: 7.21.

Practice Exercise The work function of titanium metal is 6.93×10^{-19} J. Calculate the kinetic energy of the ejected electrons if light of frequency $2.50 \times 10^{15}\ \text{s}^{-1}$ is used to irradiate the metal.

Einstein's theory of light posed a dilemma for scientists. On the one hand, it explains the photoelectric effect satisfactorily. On the other hand, the particle theory of light is not consistent with the known wave behavior of light. The only way to resolve the dilemma is to accept the idea that light possesses *both* particlelike and wavelike properties. Depending on the experiment, light behaves either as a wave or as a stream of particles. This concept was totally alien to the way physicists had thought about matter and radiation, and it took a long time for them to accept it. We will see in Section 7.4 that a dual nature (particles and waves) is not unique to light but is characteristic of all matter, including electrons.

Review of Concepts

A clean metal surface is irradiated with light of three different wavelengths λ_1, λ_2, and λ_3. The kinetic energies of the ejected electrons are as follows: λ_1: 7.2×10^{-20} J; λ_2: approximately zero; λ_3: 5.8×10^{-19} J. Which light has the shortest wavelength and which has the longest wavelength?

7.3 Bohr's Theory of the Hydrogen Atom

Einstein's work paved the way for the solution of yet another nineteenth-century "mystery" in physics: the emission spectra of atoms.

Emission Spectra

Animation:
Emission Spectra

Since the seventeenth century, when Newton showed that sunlight is composed of various color components that can be recombined to produce white light, chemists and physicists have studied the characteristics of ***emission spectra,*** that is, *either continuous or line spectra of radiation emitted by substances.* The emission spectrum of a substance can be seen by energizing a sample of material either with thermal energy or with some other form of energy (such as a high-voltage electrical discharge if the substance is gaseous). A "red-hot" or "white-hot" iron bar freshly removed from

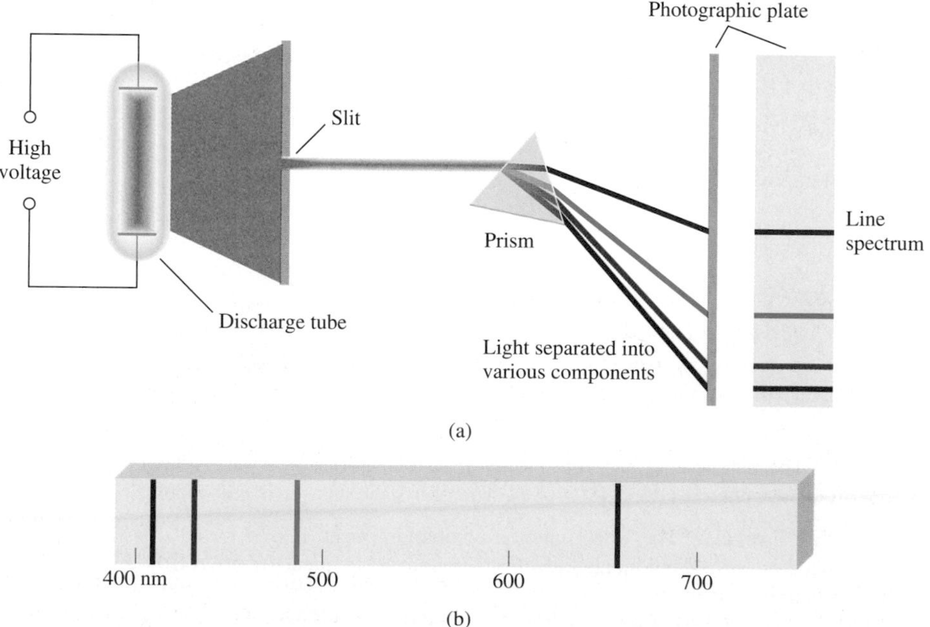

Figure 7.5
(a) An experimental arrangement for studying the emission spectra of atoms and molecules. The gas under study is in a discharge tube containing two electrodes. As electrons flow from the negative electrode to the positive electrode, they collide with the gas. This collision process eventually leads to the emission of light by the atoms (or molecules). The emitted light is separated into its components by a prism. Each component color is focused at a definitive position, according to its wavelength, and forms a colored image of the slit on the photographic plate. The colored images are called spectral lines. (b) The line emission spectrum of hydrogen atoms.

a high-temperature source produces a characteristic glow. This visible glow is the portion of its emission spectrum that is sensed by eye. The warmth of the same iron bar represents another portion of its emission spectrum—the infrared region. A feature common to the emission spectra of the sun and of a heated solid is that both are continuous; that is, all wavelengths of visible light are represented in the spectra (see the visible region in Figure 7.3).

The emission spectra of atoms in the gas phase, on the other hand, do not show a continuous spread of wavelengths from red to violet; rather, the atoms produce bright lines in different parts of the visible spectrum. These ***line spectra*** are *the light emission only at specific wavelengths.* Figure 7.5 is a schematic diagram of a discharge tube that is used to study emission spectra, and Figure 7.6 shows the color emitted by hydrogen atoms in a discharge tube.

Every element has a unique emission spectrum. The characteristic lines in atomic spectra can be used in chemical analysis to identify unknown atoms, much as fingerprints are used to identify people. When the lines of the emission spectrum of a known element exactly match the lines of the emission spectrum of an unknown sample, the identity of the sample is established. Although the utility of this procedure was recognized some time ago in chemical analysis, the origin of these lines was unknown until early in the twentieth century.

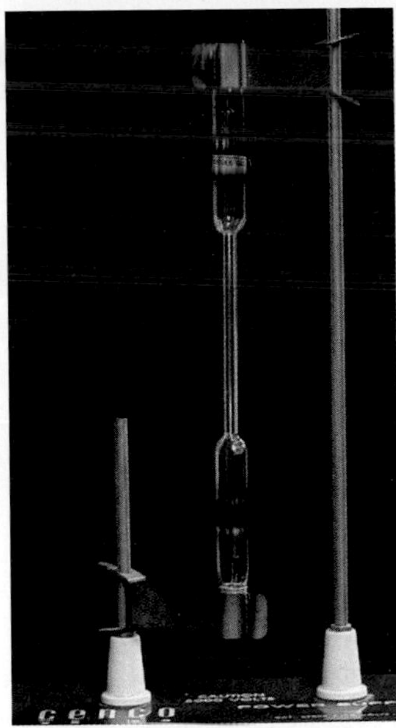

Figure 7.6
Color emitted by hydrogen atoms in a discharge tube. The color observed results from the combination of the colors emitted in the visible spectrum.

Emission Spectrum of the Hydrogen Atom

In 1913, not too long after Planck's and Einstein's discoveries, a theoretical explanation of the emission spectrum of the hydrogen atom was presented by the Danish physicist Niels Bohr. Bohr's treatment is very complex and is no longer considered to be correct in all its details. Thus, we will concentrate only on his important assumptions and final results, which do account for the spectral lines.

According to the laws of classical physics, an electron moving in an orbit of a hydrogen atom would experience an acceleration toward the nucleus by radiating away energy in the form of electromagnetic waves. Thus, such an electron would quickly spiral into the nucleus and annihilate itself with the proton. To explain why this does not happen, Bohr postulated that the electron is allowed to occupy only certain orbits of specific energies. In other words, the energies of the electron are quantized. An electron in any of the allowed orbits will not radiate energy and therefore will not spiral into the nucleus. Bohr attributed the emission of radiation by an energized hydrogen atom to the electron dropping from a higher-energy orbit to a lower one and giving up a quantum of energy (a photon) in the form of light (Figure 7.7). Bohr showed that the energies that an electron in hydrogen atom can occupy are given by

$$E_n = -R_{\mathrm{H}}\left(\frac{1}{n^2}\right) \tag{7.5}$$

where R_{H}, the Rydberg constant for the hydrogen atom (after the Swedish physicist Johannes Rydberg), has the value 2.18×10^{-18} J. The number n is an integer called the principal quantum number; it has the values $n = 1, 2, 3, \ldots$.

The negative sign in Equation (7.5) is an arbitrary convention, signifying that the energy of the electron in the atom is *lower* than the energy of a *free electron,* which is an electron that is infinitely far from the nucleus. The energy of a free electron is arbitrarily assigned a value of zero. Mathematically, this corresponds to setting n equal to infinity in Equation (7.5), so that $E_\infty = 0$. As the electron gets closer to the nucleus (as n decreases), E_n becomes larger in absolute value, but also more negative. The most negative value, then, is reached when $n = 1$, which corresponds to the most stable energy state. We call this the ***ground state,*** or the ***ground level,*** which refers to *the*

When a high voltage is applied between the forks, some of the sodium ions in the pickle are converted to sodium atoms in an excited state. These atoms emit the characteristic yellow light as they relax to the ground state.

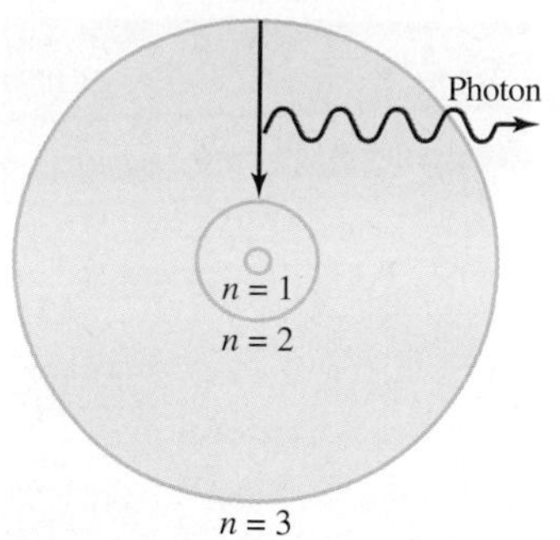

Figure 7.7
The emission process in an excited hydrogen atom, according to Bohr's theory. An electron originally in a higher-energy orbit (n = 3) falls back to a lower-energy orbit (n = 2). As a result, a photon with energy hν is given off. The value of hν is equal to the difference in energies of the two orbits occupied by the electron in the emission process. For simplicity, only three orbits are shown.

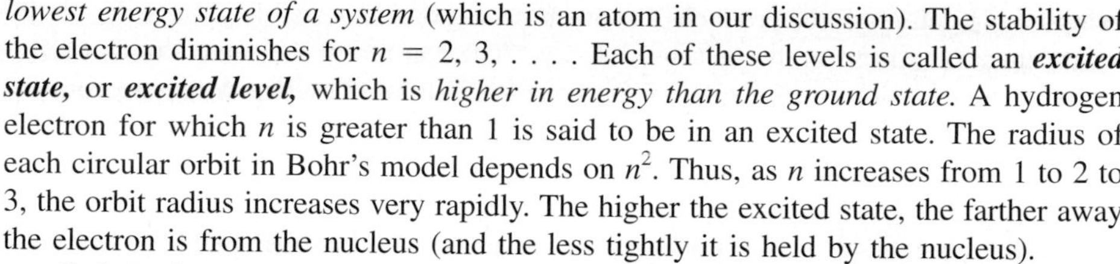

lowest energy state of a system (which is an atom in our discussion). The stability of the electron diminishes for $n = 2, 3, \ldots$. Each of these levels is called an ***excited state,*** or ***excited level,*** which is *higher in energy than the ground state.* A hydrogen electron for which n is greater than 1 is said to be in an excited state. The radius of each circular orbit in Bohr's model depends on n^2. Thus, as n increases from 1 to 2 to 3, the orbit radius increases very rapidly. The higher the excited state, the farther away the electron is from the nucleus (and the less tightly it is held by the nucleus).

Bohr's theory enables us to explain the line spectrum of the hydrogen atom. Radiant energy absorbed by the atom causes the electron to move from a lower-energy state (characterized by a smaller n value) to a higher-energy state (characterized by a larger n value). Conversely, radiant energy (in the form of a photon) is emitted when the electron moves from a higher-energy state to a lower-energy state. The quantized movement of the electron from one energy state to another is analogous to the movement of a tennis ball either up or down a set of stairs (Figure 7.8). The ball can be on any of several steps but never between steps. The journey from a lower step to a higher one is an energy-requiring process, whereas movement from a higher step to a lower step is an energy-releasing process. The quantity of energy involved in either type of change is determined by the distance between the beginning and ending steps. Similarly, the amount of energy needed to move an electron in the Bohr atom depends on the difference in energy levels between the initial and final states.

To apply Equation (7.5) to the emission process in a hydrogen atom, let us suppose that the electron is initially in an excited state characterized by the principal quantum number n_i. During emission, the electron drops to a lower energy state characterized by the principal quantum number n_f (the subscripts i and f denote the initial and final states, respectively). This lower energy state may be either a less excited state or the ground state. The difference between the energies of the initial and final states is

$$\Delta E = E_f - E_i$$

From Equation (7.5),

$$E_f = -R_H\left(\frac{1}{n_f^2}\right)$$

and

$$E_i = -R_H\left(\frac{1}{n_i^2}\right)$$

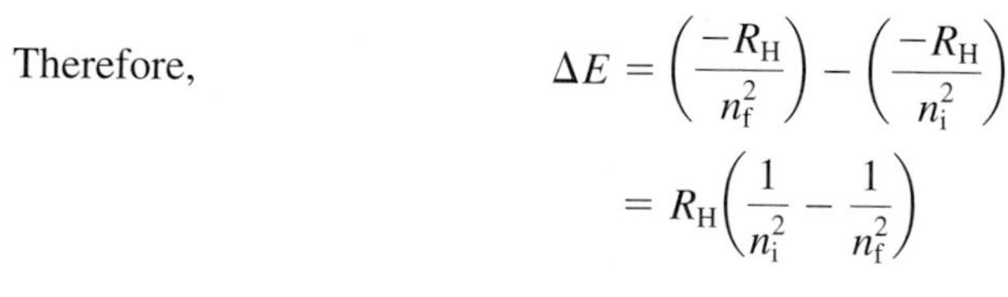

Therefore,

$$\Delta E = \left(\frac{-R_H}{n_f^2}\right) - \left(\frac{-R_H}{n_i^2}\right)$$

$$= R_H\left(\frac{1}{n_i^2} - \frac{1}{n_f^2}\right)$$

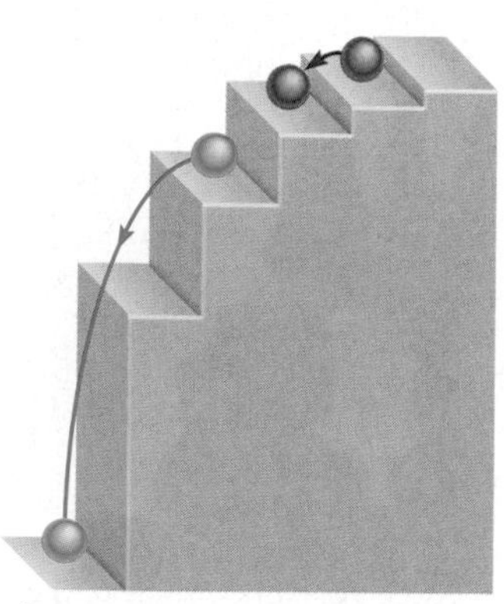

Figure 7.8
A mechanical analogy for the emission process. The ball can rest on any step but not between steps.

Because this transition results in the emission of a photon of frequency ν and energy $h\nu$, we can write

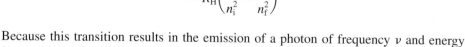

$$\Delta E = h\nu = R_H\left(\frac{1}{n_i^2} - \frac{1}{n_f^2}\right) \quad (7.6)$$

When a photon is emitted, $n_i > n_f$. Consequently the term in parentheses is negative and ΔE is negative (energy is lost to the surroundings). When energy is absorbed, $n_i < n_f$ and the term in parentheses is positive, so ΔE is positive. Each spectral line

Table 7.1 The Various Series in Atomic Hydrogen Emission Spectrum

Series	n_f	n_i	Spectrum Region
Lyman	1	2, 3, 4, . . .	Ultraviolet
Balmer	2	3, 4, 5, . . .	Visible and ultraviolet
Paschen	3	4, 5, 6, . . .	Infrared
Brackett	4	5, 6, 7, . . .	Infrared

in the emission spectrum corresponds to a particular transition in a hydrogen atom. When we study a large number of hydrogen atoms, we observe all possible transitions and hence the corresponding spectral lines. The brightness of a spectral line depends on how many photons of the same wavelength are emitted.

The emission spectrum of hydrogen includes a wide range of wavelengths from the infrared to the ultraviolet. Table 7.1 lists the series of transitions in the hydrogen spectrum; they are named after their discoverers. The Balmer series was particularly easy to study because a number of its lines fall in the visible range.

Figure 7.7 shows a single transition. However, it is more informative to express transitions as shown in Figure 7.9. Each horizontal line represents an allowed energy level for the electron in a hydrogen atom. The energy levels are labeled with their principal quantum numbers.

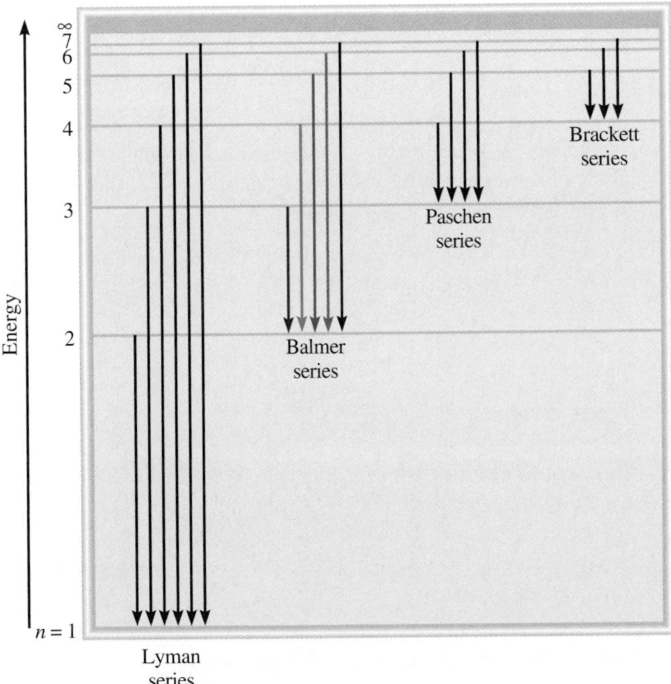

Figure 7.9
The energy levels in the hydrogen atom and the various emission series. Each energy level corresponds to the energy associated with an allowed energy state for an orbit, as postulated by Bohr and shown in Figure 7.7. The emission lines are labeled according to the scheme in Table 7.1.

EXAMPLE 7.4

What is the wavelength of a photon (in nanometers) emitted during a transition from the $n_i = 5$ state to the $n_f = 2$ state in the hydrogen atom?

Strategy We are given the initial and final states in the emission process. We can calculate the energy of the emitted photon using Equation (7.6). Then from Equations (7.2) and (7.1) we can solve for the wavelength of the photon. The value of Rydberg's constant is given in the text.

Solution From Equation (7.6) we write

$$\begin{aligned}\Delta E &= R_H\left(\frac{1}{n_i^2} - \frac{1}{n_f^2}\right)\\ &= 2.18 \times 10^{-18}\,\text{J}\left(\frac{1}{5^2} - \frac{1}{2^2}\right)\\ &= -4.58 \times 10^{-19}\,\text{J}\end{aligned}$$

The negative sign is in accord with our convention that energy is given off to the surroundings.

The negative sign indicates that this is energy associated with an emission process. To calculate the wavelength, we will omit the minus sign for ΔE because the wavelength of the photon must be positive. Because $\Delta E = h\nu$ or $\nu = \Delta E/h$, we can calculate the wavelength of the photon by writing

$$\begin{aligned}\lambda &= \frac{c}{\nu}\\ &= \frac{ch}{\Delta E}\\ &= \frac{(3.00 \times 10^8\,\text{m/s})(6.63 \times 10^{-34}\,\text{J}\cdot\text{s})}{4.58 \times 10^{-19}\,\text{J}}\\ &= 4.34 \times 10^{-7}\,\text{m}\\ &= 4.34 \times 10^{-7}\,\text{m} \times \left(\frac{1\,\text{nm}}{1 \times 10^{-9}\,\text{m}}\right) = 434\,\text{nm}\end{aligned}$$

Check The wavelength is in the visible region of the electromagnetic region (see Figure 7.3). This is consistent with the fact that because $n_f = 2$, this transition gives rise to a spectral line in the Balmer series (see Table 7.1).

Similar problems: 7.31, 7.32.

Practice Exercise What is the wavelength (in nanometers) of a photon emitted during a transition from $n_i = 6$ to $n_f = 4$ state in the H atom?

REVIEW OF CONCEPTS

Which transition would require more energy in a hydrogen atom? (a) $n_i = 1$ to $n_f = 2$, or (b) $n_i = 2$ to $n_f = 4$.

7.4 The Dual Nature of the Electron

Physicists were both mystified and intrigued by Bohr's theory. They questioned why the energies of the hydrogen electron are quantized. Or, phrasing the question in a more concrete way, Why is the electron in a Bohr atom restricted to orbiting the nucleus at certain fixed distances? For a decade no one, not even Bohr himself, had

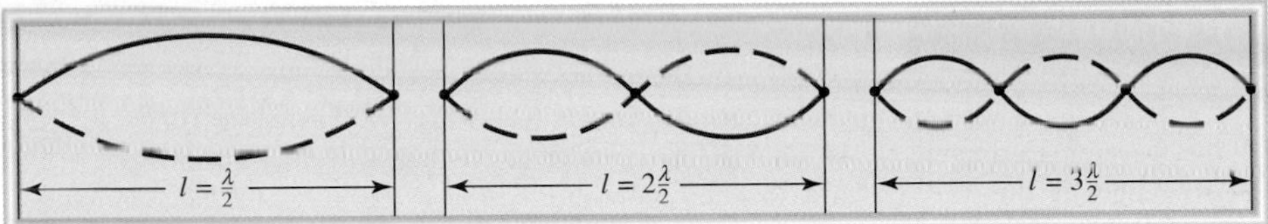

Figure 7.10
The standing waves generated by plucking a guitar string. Each dot represents a node. The length of the string (l) must be equal to a whole number times one-half the wavelength (λ/2).

a logical explanation. In 1924 the French physicist Louis de Broglie provided a solution to this puzzle. De Broglie reasoned that if light waves can behave like a stream of particles (photons), then perhaps particles such as electrons can possess wave properties. According to de Broglie, an electron bound to the nucleus behaves like a *standing wave.* Standing waves can be generated by plucking, say, a guitar string (Figure 7.10). The waves are described as standing, or stationary, because they do not travel along the string. Some points on the string, called ***nodes,*** do not move at all; that is, *the amplitude of the wave at these points is zero.* There is a node at each end, and there may be nodes between the ends. The greater the frequency of vibration, the shorter the wavelength of the standing wave and the greater the number of nodes. As Figure 7.10 shows, there can be only certain wavelengths in any of the allowed motions of the string.

De Broglie argued that if an electron does behave like a standing wave in the hydrogen atom, the length of the wave must fit the circumference of the orbit exactly (Figure 7.11). Otherwise the wave would partially cancel itself on each successive orbit. Eventually the amplitude of the wave would be reduced to zero, and the wave would not exist.

The relation between the circumference of an allowed orbit ($2\pi r$) and the wavelength (λ) of the electron is given by

$$2\pi r = n\lambda \tag{7.7}$$

where r is the radius of the orbit, λ is the wavelength of the electron wave, and $n = 1, 2, 3, \ldots$. Because n is an integer, it follows that r can have only certain values as n increases from 1 to 2 to 3 and so on. And because the energy of the electron depends on the size of the orbit (or the value of r), its value must be quantized.

De Broglie's reasoning led to the conclusion that waves can behave like particles and particles can exhibit wavelike properties. De Broglie deduced that the particle and wave properties are related by the expression

$$\lambda = \frac{h}{mu} \tag{7.8}$$

where λ, m, and u are the wavelengths associated with a moving particle, its mass, and its velocity, respectively. Equation (7.8) implies that a particle in motion can be treated as a wave, and a wave can exhibit the properties of a particle. Note that the left side of Equation (7.8) involves the wavelike property of wavelength, whereas the right side makes references to mass, a distinctly particlelike property.

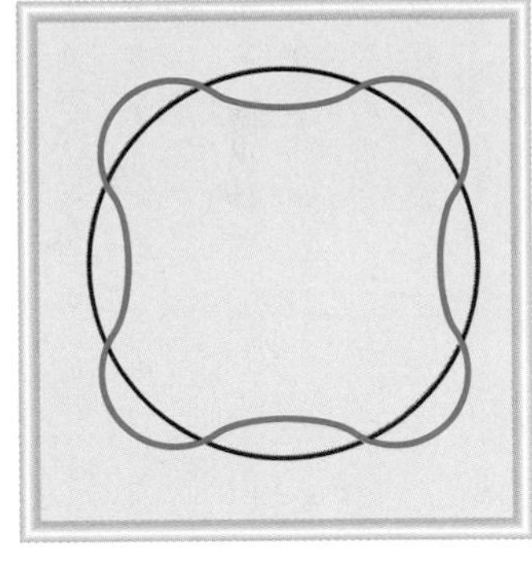

(a)

(b)

Figure 7.11
(a) The circumference of the orbit is equal to an integral number of wavelengths. This is an allowed orbit. (b) The circumference of the orbit is not equal to an integral number of wavelengths. As a result, the electron wave does not close in on itself. This is a nonallowed orbit.

EXAMPLE 7.5

Calculate the wavelength of the "particle" in the following two cases: (a) Calculate the wavelength associated with a 5.50×10^{-2}-kg tennis ball traveling at 45 m/s. (b) Calculate the wavelength associated with an electron (9.1094×10^{-31} kg) moving at 45 m/s.

Strategy We are given the mass and the speed of the particle in (a) and (b) and asked to calculate the wavelength, so we need Equation (7.8). Note that because the units of Planck's constants are J · s, m and u must be in kg and m/s (1 J = 1 kg m^2/s^2), respectively.

Solution (a) Using Equation (7.8) we write

$$\lambda = \frac{h}{mu}$$

$$= \frac{6.63 \times 10^{-34}\ \text{J} \cdot \text{s}}{(5.50 \times 10^{-2}\ \text{kg}) \times 45\ \text{m/s}}$$

$$= 2.7 \times 10^{-34}\ \text{m}$$

Comment This is an exceedingly small wavelength considering that the size of an atom itself is on the order of 1×10^{-10} m. For this reason, no existing measuring device can detect the wave properties of a tennis ball.

(b) In this case,

$$\lambda = \frac{h}{mu}$$

$$= \frac{6.63 \times 10^{-34}\ \text{J} \cdot \text{s}}{(9.1094 \times 10^{-31}\ \text{kg}) \times 45\ \text{m/s}}$$

$$= 1.6 \times 10^{-5}\ \text{m}$$

Comment This wavelength (1.6×10^{-5} m or 1.6×10^{4} nm) is in the infrared region. This calculation shows that only electrons (and other submicroscopic particles) have measurable wavelengths.

Similar problems: 7.40, 7.41.

Practice Exercise Calculate the wavelength (in nanometers) of a H atom (mass = 1.674×10^{-27} kg) moving at 7.00×10^{2} cm/s.

Example 7.5 shows that although de Broglie's equation can be applied to diverse systems, the wave properties become observable only for submicroscopic objects.

Shortly after de Broglie introduced his equation, Clinton Davisson and Lester Germer in the United States and G. P. Thomson in England demonstrated that electrons do indeed possess wavelike properties. By directing a beam of electrons through a thin piece of gold foil, Thomson obtained a set of concentric rings on a screen, similar to the pattern observed when X rays (which are waves) were used. Figure 7.12 shows such a pattern for aluminum.

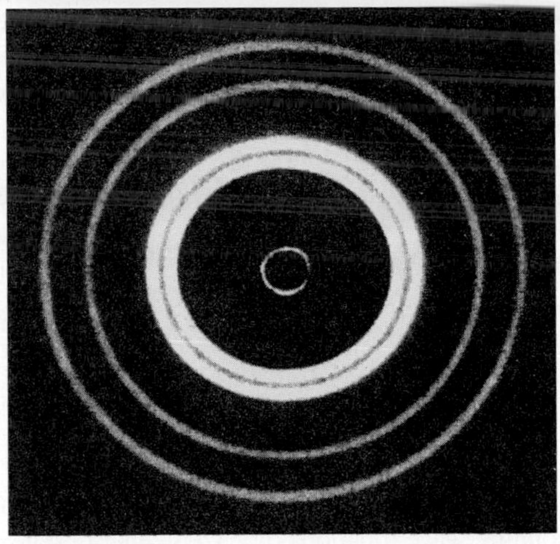

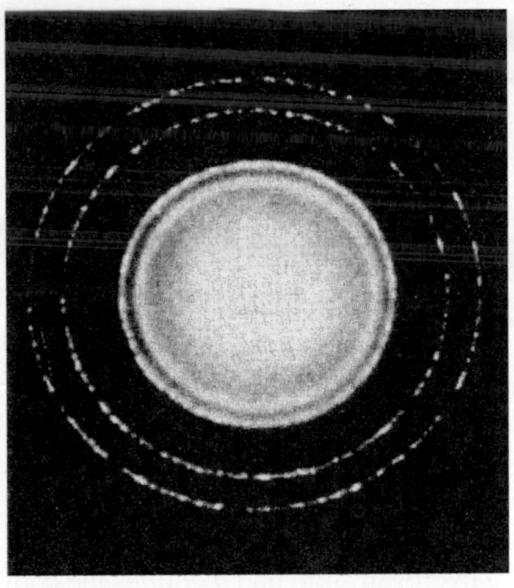

Figure 7.12
(a) X-ray diffraction pattern of aluminum foil. (b) Electron diffraction of aluminum foil. The similarity of these two patterns shows that electrons can behave like X rays and display wave properties.

REVIEW OF CONCEPTS

Which quantity in Equation (7.8) is responsible for the fact that macroscopic objects do not show observable wave properties?

7.5 Quantum Mechanics

The spectacular success of Bohr's theory was followed by a series of disappointments. Bohr's approach could not account for the emission spectra of atoms containing more than one electron, such as atoms of helium and lithium. Nor did it explain why extra lines appear in the hydrogen emission spectrum when a magnetic field is applied. Another problem arose with the discovery that electrons are wavelike: How can the "position" of a wave be specified? We cannot define the precise location of a wave because a wave extends in space.

In reality, Bohr's theory accounted for the observed emission spectra of He^+ and Li^{2+} ions, as well as that of hydrogen. However, all three systems have one feature in common—each contains a single electron. Thus, the Bohr model worked successfully only for the hydrogen atom and for "hydrogenlike ions."

The dual nature of electrons was particularly troublesome because of the electron's exceedingly small mass. To describe the problem of trying to locate a subatomic particle that behaves like a wave, the German physicist Werner Heisenberg formulated what is now known as the ***Heisenberg uncertainty principle:*** *It is impossible to know simultaneously both the momentum (mass times velocity) and the position of a particle with certainty.* In other words, to get a precise measurement of the momentum of a particle we must settle for less precise knowledge of the particle's position, and vice versa. Applying the Heisenberg uncertainty principle to the hydrogen atom, we see that it is inherently impossible to know simultaneously both the precise location and precise momentum of the electron. Thus, it is not appropriate to imagine the electron circling the nucleus in well-defined orbits.

To be sure, Bohr made a significant contribution to our understanding of atoms, and his suggestion that the energy of an electron in an atom is quantized remains unchallenged. But his theory did not provide a complete description of electronic behavior in atoms. In 1926 the Austrian physicist Erwin Schrödinger, using a complicated mathematical technique, formulated an equation that describes the behavior and energies of submicroscopic particles in general, an equation analogous to Newton's laws of motion for macroscopic objects. The *Schrödinger equation* requires advanced calculus to solve, and we will not discuss it here. It is important to know, however, that the equation incorporates both particle behavior, in terms of mass m, and wave behavior, in terms of a *wave function* ψ (psi), which depends on the location in space of the system (such as an electron in an atom).

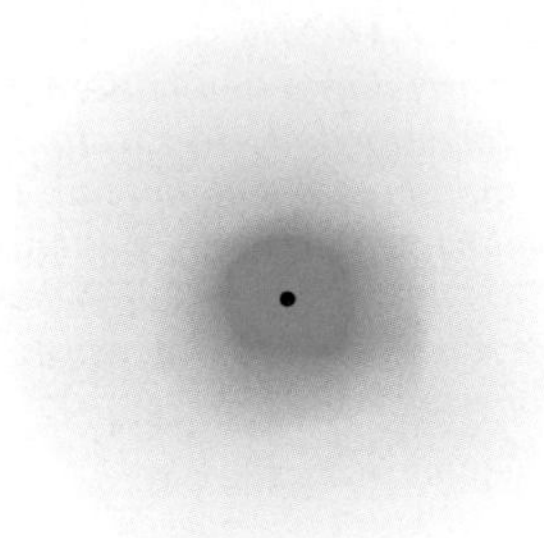

Figure 7.13
A representation of the electron density distribution surrounding the nucleus in the hydrogen atom. It shows a high probability of finding the electron closer to the nucleus.

The wave function itself has no direct physical meaning. However, the probability of finding the electron in a certain region in space is proportional to the square of the wave function, ψ^2. The idea of relating ψ^2 to probability stemmed from a wave theory analogy. According to wave theory, the intensity of light is proportional to the square of the amplitude of the wave, or ψ^2. The most likely place to find a photon is where the intensity is greatest, that is, where the value of ψ^2 is greatest. A similar argument associates ψ^2 with the likelihood of finding an electron in regions surrounding the nucleus.

Schrödinger's equation began a new era in physics and chemistry, for it launched a new field, *quantum mechanics* (also called *wave mechanics*). We now refer to the developments in quantum theory from 1913—the time Bohr presented his analysis for the hydrogen atom—to 1926 as "old quantum theory."

The Quantum Mechanical Description of the Hydrogen Atom

The Schrödinger equation specifies the possible energy states the electron can occupy in a hydrogen atom and identifies the corresponding wave functions (ψ). These energy states and wave functions are characterized by a set of quantum numbers (to be discussed shortly), with which we can construct a comprehensive model of the hydrogen atom.

Although quantum mechanics tells us that we cannot pinpoint an electron in an atom, it does define the region where the electron might be at a given time. The concept of ***electron density*** *gives the probability that an electron will be found in a particular region of an atom.* The square of the wave function, ψ^2, defines the distribution of electron density in three-dimensional space around the nucleus. Regions of high electron density represent a high probability of locating the electron, whereas the opposite holds for regions of low electron density (Figure 7.13).

To distinguish the quantum mechanical description of an atom from Bohr's model, we speak of an atomic orbital, rather than an orbit. An ***atomic orbital*** can be thought of as *the wave function of an electron in an atom.* When we say that an electron is in a certain orbital, we mean that the distribution of the electron density or the probability of locating the electron in space is described by the square of the wave function associated with that orbital. An atomic orbital, therefore, has a characteristic energy, as well as a characteristic distribution of electron density.

The Schrödinger equation works nicely for the simple hydrogen atom with its one proton and one electron, but it turns out that it cannot be solved exactly for any atom containing more than one electron! Fortunately, chemists and physicists have learned to get around this kind of difficulty by approximation. For example, although the behavior of electrons in ***many-electron atoms*** (that is, *atoms containing two or more electrons*) is not the same as in the hydrogen atom, we assume that the difference is probably not too great. Thus, we can use the energies and wave functions obtained from the hydrogen atom as good approximations of the behavior of electrons in more complex atoms. In fact, this approach provides fairly reliable descriptions of electronic behavior in many-electron atoms.

Although the helium atom has only two electrons, in quantum mechanics it is regarded as a many-electron atom.

Review of Concepts

What does the square of the wave function, ψ^2, for an orbital describe?

7.6 Quantum Numbers

In quantum mechanics, three ***quantum numbers*** are required to *describe the distribution of electrons in hydrogen and other atoms.* These numbers are derived from the mathematical solution of the Schrödinger equation for the hydrogen atom. They are called the *principal quantum number,* the *angular momentum quantum number,* and the *magnetic quantum number.* These quantum numbers will be used to describe atomic orbitals and to label electrons that reside in them. A fourth quantum number—the *spin quantum number*—describes the behavior of a specific electron and completes the description of electrons in atoms.

The Principal Quantum Number (n)

The principal quantum number (n) can have integral values 1, 2, 3, and so forth; it corresponds to the quantum number in Equation (7.5). In a hydrogen atom, the value of n determines the energy of an orbital. As we will see shortly, this is not the case for a many-electron atom. The principal quantum number also relates to the average distance of the electron from the nucleus in a particular orbital. The larger n is, the greater the average distance of an electron in the orbital from the nucleus and therefore the larger the orbital.

Equation (7.5) holds only for the hydrogen atom.

The Angular Momentum Quantum Number (ℓ)

The angular momentum quantum number (ℓ) tells us the "shape" of the orbitals (see Section 7.7). The values of ℓ depend on the value of the principal quantum number, n. For a given value of n, ℓ has possible integral values from 0 to $(n - 1)$. If $n = 1$, there is only one possible value of ℓ; that is, $\ell = n - 1 = 1 - 1 = 0$. If $n = 2$, there are two values of ℓ, given by 0 and 1. If $n = 3$, there are three values of ℓ, given by 0, 1, and 2. The value of ℓ is generally designated by the letters s, p, d, . . . as follows:

The value of ℓ is fixed based on the type of the orbital.

ℓ	0	1	2	3	4	5
Name of orbital	s	p	d	f	g	h

Thus, if $\ell = 0$, we have an s orbital; if $\ell = 1$, we have a p orbital; and so on.

A collection of orbitals with the same value of n is frequently called a shell. One or more orbitals with the same n and ℓ values are referred to as a subshell. For example, the shell with $n = 2$ is composed of two subshells, $\ell = 0$ and 1 (the allowed values for $n = 2$). These subshells are called the $2s$ and $2p$ subshells where 2 denotes the value of n, and s and p denote the values of ℓ.

Remember that the "2" in 2s refers to the value of n and the "s" symbolizes the value of ℓ.

The Magnetic Quantum Number (m_ℓ)

The magnetic quantum number (m_ℓ) describes the orientation of the orbital in space (to be discussed in Section 7.7). Within a subshell, the value of m_ℓ depends on the value of the angular momentum quantum number, ℓ. For a certain value of ℓ, there are $(2\ell + 1)$ integral values of m_ℓ as follows:

$$-\ell, (-\ell + 1), \ldots 0, \ldots (+\ell - 1), +\ell$$

If $\ell = 0$, then $m_\ell = 0$. If $\ell = 1$, then there are $[(2 \times 1) + 1]$, or three values of m_ℓ, namely, -1, 0, and 1. If $\ell = 2$, there are $[(2 \times 2) + 1]$, or five values of m_ℓ, namely, -2, -1, 0, 1, and 2. The number of m_ℓ values indicates the number of orbitals in a subshell with a particular ℓ value.

To conclude our discussion of these three quantum numbers, let us consider a situation in which $n = 2$ and $\ell = 1$. The values of n and ℓ indicate that we have a $2p$ subshell, and in this subshell we have *three* $2p$ orbitals (because there are three values of m_ℓ, given by -1, 0, and 1).

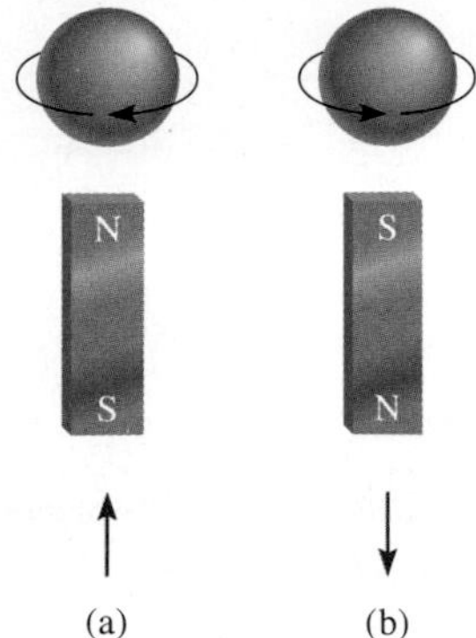

Figure 7.14
The (a) clockwise and (b) counterclockwise spins of an electron. The magnetic fields generated by these two spinning motions are analogous to those from the two magnets. The upward and downward arrows are used to denote the direction of spin.

The Electron Spin Quantum Number (m_s)

Experiments on the emission spectra of hydrogen and sodium atoms indicated that lines in the emission spectra could be split by the application of an external magnetic field. The only way physicists could explain these results was to assume that electrons act like tiny magnets. If electrons are thought of as spinning on their own axes, as Earth does, their magnetic properties can be accounted for. According to electromagnetic theory, a spinning charge generates a magnetic field, and it is this motion that causes an electron to behave like a magnet. Figure 7.14 shows the two possible spinning motions of an electron, one clockwise and the other counterclockwise. To take the electron spin into account, it is necessary to introduce a fourth quantum number, called the electron spin quantum number (m_s), which has a value of $+\frac{1}{2}$ or $-\frac{1}{2}$.

REVIEW OF CONCEPTS

Give the four quantum numbers for each of the two electrons in a 6*s* orbital.

7.7 Atomic Orbitals

Table 7.2 shows the relation between quantum numbers and atomic orbitals. We see that when $\ell = 0$, $(2\ell + 1) = 1$ and there is only one value of m_ℓ, thus, we have an *s* orbital. When $\ell = 1$, $(2\ell + 1) = 3$, so there are three values of m_ℓ or three *p* orbitals, labeled p_x, p_y, and p_z. When $\ell = 2$, $(2\ell + 1) = 5$ and there are five values of m_ℓ, and the corresponding five *d* orbitals are labeled with more elaborate subscripts. In the following sections we will consider the *s*, *p*, and *d* orbitals separately.

s Orbitals

That the wave function for an orbital theoretically has no outer limit as one moves outward from the nucleus raises interesting philosophical questions regarding the sizes of atoms. Chemists have agreed on an operational definition of atomic size, as we will see in Chapter 8.

One of the important questions we ask when studying the properties of atomic orbitals is, What are the shapes of the orbitals? Strictly speaking, an orbital does not have a well-defined shape because the wave function characterizing the orbital extends from the nucleus to infinity. In that sense, it is difficult to say what an orbital looks like. On the other hand, it is certainly convenient to think of orbitals as having specific shapes, particularly in discussing the formation of chemical bonds between atoms, as we will do in Chapters 9 and 10.

An *s* subshell has one orbital, a *p* subshell has three orbitals, and a *d* subshell has five orbitals.

Table 7.2 Relation Between Quantum Numbers and Atomic Orbitals

n	ℓ	m_ℓ	Number of Orbitals	Atomic Orbital Designations
1	0	0	1	1*s*
2	0	0	1	2*s*
	1	−1, 0, 1	3	$2p_x$, $2p_y$, $2p_z$
3	0	0	1	3*s*
	1	−1, 0, 1	3	$3p_x$, $3p_y$, $3p_z$
	2	−2, −1, 0, 1, 2	5	$3d_{xy}$, $3d_{yz}$, $3d_{xz}$, $3d_{x^2-y^2}$, $3d_{z^2}$
⋮	⋮	⋮	⋮	⋮

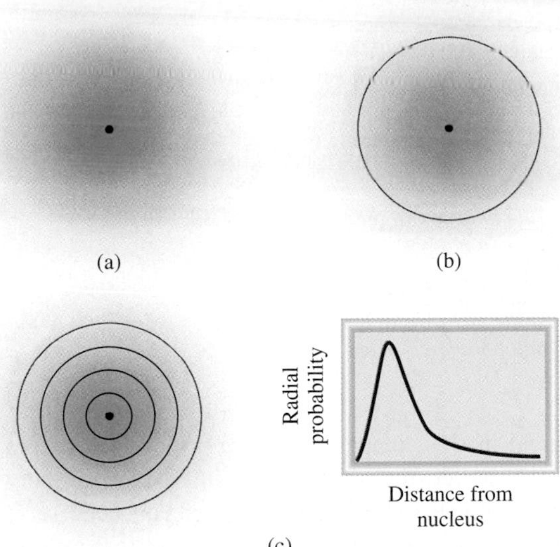

Figure 7.15
(a) Plot of electron density in the hydrogen 1s orbital as a function of the distance from the nucleus. The electron density falls off rapidly as the distance from the nucleus increases. (b) Boundary surface diagram of the hydrogen 1s orbital. (c) A more realistic way of viewing electron density distribution is to divide the 1s orbital into successive spherical thin shells. A plot of the probability of finding the electron in each shell, called radial probability, as a function of distance shows a maximum at 52.9 pm from the nucleus. Interestingly, this is equal to the radius of the innermost orbit in the Bohr model.

Although in principle an electron can be found anywhere, we know that most of the time it is quite close to the nucleus. Figure 7.15(a) shows the distribution of electron density in a hydrogen 1*s* orbital moving outward from the nucleus. As you can see, the electron density falls off rapidly as the distance from the nucleus increases. Roughly speaking, there is about a 90 percent probability of finding the electron within a sphere of radius 100 pm (1 pm $= 1 \times 10^{-12}$ m) surrounding the nucleus. Thus, we can represent the 1*s* orbital by drawing a ***boundary surface diagram*** that *encloses about 90 percent of the total electron density in an orbital,* as shown in Figure 7.15(b). A 1*s* orbital represented in this manner is merely a sphere.

Figure 7.16 shows boundary surface diagrams for the 1*s*, 2*s*, and 3*s* hydrogen atomic orbitals. All *s* orbitals are spherical in shape but differ in size, which increases as the principal quantum number increases. Although the details of electron density variation within each boundary surface are lost, there is no serious disadvantage. For us the most important features of atomic orbitals are their shapes and *relative* sizes, which are adequately represented by boundary surface diagrams.

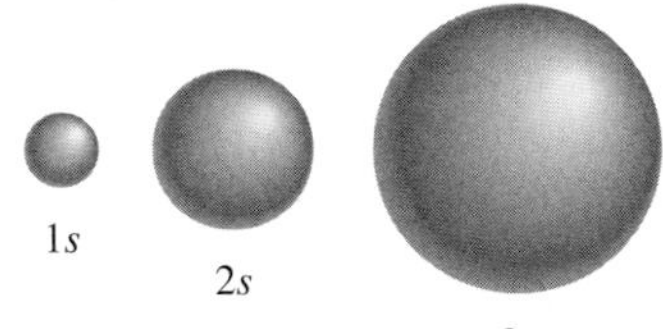

Figure 7.16
Boundary surface diagrams of the hydrogen 1s, 2s, and 3s orbitals. Each sphere contains about 90 percent of the total electron density. All s orbitals are spherical. Roughly speaking, the size of an orbital is proportional to n^2, where n is the principal quantum number.

p Orbitals

It should be clear that the *p* orbitals start with the principal quantum number $n = 2$. If $n = 1$, then the angular momentum quantum number ℓ can assume only the value of zero; therefore, there is only a 1*s* orbital. As we saw earlier, when $\ell = 1$, the magnetic quantum number m_ℓ can have values of -1, 0, and 1. Starting with $n = 2$ and $\ell = 1$, we therefore have three 2*p* orbitals: $2p_x$, $2p_y$, and $2p_z$ (Figure 7.17). The letter subscripts indicate the axes along which the orbitals are oriented. These three

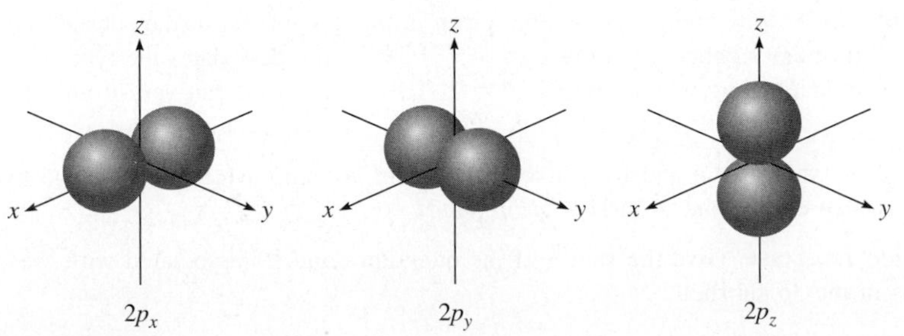

Figure 7.17
The boundary surface diagrams of the three 2p orbitals. These orbitals are identical in shape and energy, but their orientations are different. The p orbitals of higher principal quantum numbers have a similar shape.

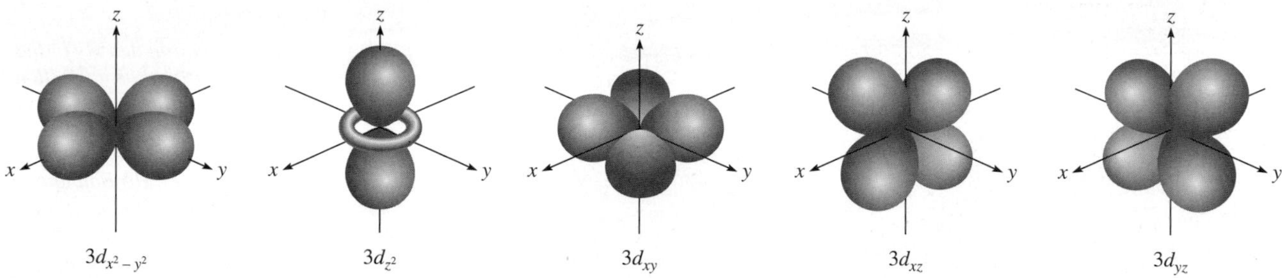

Figure 7.18
Boundary surface diagrams of the five 3d orbitals. Although the $3d_{z^2}$ orbital looks different, it is equivalent to the other four orbitals in all other respects. The d orbitals of higher principal quantum numbers have similar shapes.

Orbitals that have the same energy are said to be degenerate orbitals.

p orbitals are identical in size, shape, and energy; they differ from one another only in orientation. Note, however, that there is no simple relation between the values of m_ℓ and the *x*, *y*, and *z* directions. For our purpose, you need to remember only that because there are three possible values of m_ℓ, there are three *p* orbitals with different orientations.

The boundary surface diagrams of *p* orbitals in Figure 7.17 show that each *p* orbital can be thought of as two lobes on opposite sides of the nucleus. Like *s* orbitals, *p* orbitals increase in size from 2*p* to 3*p* to 4*p* orbital and so on.

d Orbitals and Other Higher-Energy Orbitals

When $\ell = 2$, there are five values of m_ℓ, which correspond to five *d* orbitals. The lowest value of *n* for a *d* orbital is 3. Because ℓ can never be greater than $n - 1$, when $n = 3$ and $\ell = 2$, we have five 3*d* orbitals ($3d_{xy}$, $3d_{yz}$, $3d_{xz}$, $3d_{x^2-y^2}$, and $3d_{z^2}$), as shown in Figure 7.18. As in the case of the *p* orbitals, the different orientations of the *d* orbitals correspond to the different values of m_ℓ, but again there is no direct correspondence between a given orientation and a particular m_ℓ value. All the 3*d* orbitals in an atom are identical in energy. The *d* orbitals for which *n* is greater than 3 (4*d*, 5*d*, . . .) have similar shapes.

Orbitals having higher energy than *d* orbitals are labeled *f*, *g*, . . . and so on. The *f* orbitals are important in accounting for the behavior of elements with atomic numbers greater than 57, but their shapes are difficult to represent. In general chemistry we are not concerned with orbitals having ℓ values greater than 3 (the *g* orbitals and beyond).

EXAMPLE 7.6

List the values of *n*, ℓ, and m_ℓ for orbitals in the 4*d* subshell.

Strategy What are the relationships among *n*, ℓ, and m_ℓ? What do "4" and "*d*" represent in 4*d*?

Solution As we saw earlier, the number given in the designation of the subshell is the principal quantum number, so in this case $n = 4$. The letter designates the type of orbital. Because we are dealing with *d* orbitals, $\ell = 2$. The values of m_ℓ can vary from $-\ell$ to ℓ. Therefore, m_ℓ can be $-2, -1, 0, 1,$ or 2.

Check The values of *n* and ℓ are fixed for 4*d*, but m_ℓ can have any one of the five values, which correspond to the five *d* orbitals.

Similar problem: 7.55.

Practice Exercise Give the values of the quantum numbers associated with the orbitals in the 3*p* subshell.

EXAMPLE 7.7

What is the total number of orbitals associated with the principal quantum number $n = 3$?

Strategy To calculate the total number of orbitals for a given n value, we need to first write the possible values of ℓ. We then determine how many m_ℓ values are associated with each value of ℓ. The total number of orbitals is equal to the sum of all the m_ℓ values.

Solution For $n = 3$, the possible values of ℓ are 0, 1, and 2. Thus, there is one 3s orbital ($n = 3$, $\ell = 0$, and $m_\ell = 0$); there are three 3p orbitals ($n = 3$, $\ell = 1$, and $m_\ell = -1, 0, 1$); there are five 3d orbitals ($n = 3$, $\ell = 2$, and $m_\ell = -2, -1, 0, 1, 2$). The total number of orbitals is $1 + 3 + 5 = 9$.

Check The total number of orbitals for a given value of n is n^2. So here we have $3^2 = 9$. Can you prove the validity of this relationship?

Similar problem: 7.60.

Practice Exercise What is the total number of orbitals associated with the principal quantum number $n = 4$?

The Energies of Orbitals

Now that we have some understanding of the shapes and sizes of atomic orbitals, we are ready to inquire into their relative energies and look at how energy levels affect the actual arrangement of electrons in atoms.

According to Equation (7.5), the energy of an electron in a hydrogen atom is determined solely by its principal quantum number. Thus, the energies of hydrogen orbitals increase as follows (Figure 7.19):

$$1s < 2s = 2p < 3s = 3p = 3d < 4s = 4p = 4d = 4f < \cdots$$

Although the electron density distributions are different in the 2s and 2p orbitals, hydrogen's electron has the same energy whether it is in the 2s orbital or a 2p orbital. The 1s orbital in a hydrogen atom corresponds to the most stable condition, the ground state. An electron residing in this orbital is most strongly held by the nucleus because it is closest to the nucleus. An electron in the 2s, 2p, or higher orbitals in a hydrogen atom is in an excited state.

The energy picture is more complex for many-electron atoms than for hydrogen. The energy of an electron in such an atom depends on its angular momentum quantum

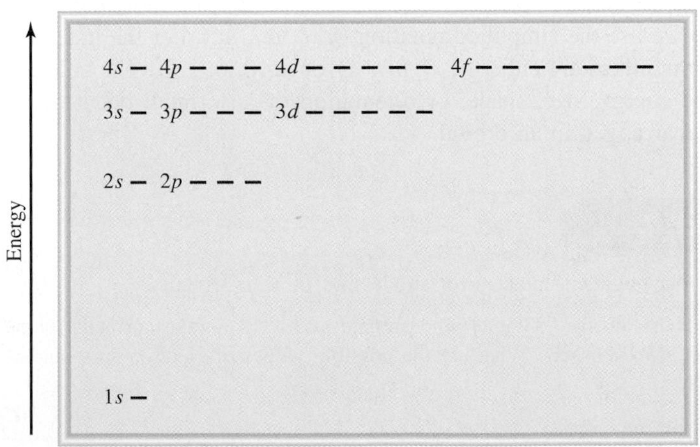

Figure 7.19
Orbital energy levels in the hydrogen atom. Each short horizontal line represents one orbital. Orbitals with the same principal quantum number (n) all have the same energy.

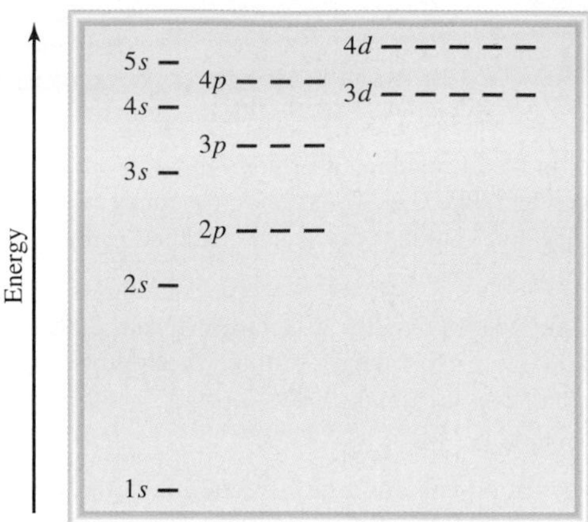

Figure 7.20
Orbital energy levels in a many-electron atom. Note that the energy level depends on both n and ℓ values.

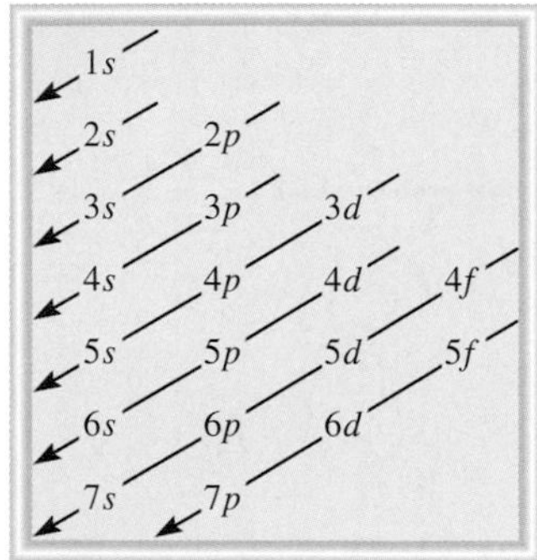

Figure 7.21
The order in which atomic subshells are filled in a many-electron atom. Start with the 1s orbital and move downward, following the direction of the arrows. Thus, the order goes as follows: 1s < 2s < 2p < 3s < 3p < 4s < 3d < · · · .

number as well as on its principal quantum number (Figure 7.20). For many-electron atoms, the 3*d* energy level is very close to the 4*s* energy level. The total energy of an atom, however, depends not only on the sum of the orbital energies but also on the energy of repulsion between the electrons in these orbitals (each orbital can accommodate up to two electrons, as we will see in Section 7.8). It turns out that the total energy of an atom is lower when the 4*s* subshell is filled before a 3*d* subshell. Figure 7.21 depicts the order in which atomic orbitals are filled in a many-electron atom. We will consider specific examples in Section 7.8.

REVIEW OF CONCEPTS

Why is it not possible to have a 2*d* orbital but a 3*d* orbital is allowed?

7.8 Electron Configuration

Animation:
Electron Configurations

The four quantum numbers n, ℓ, m_ℓ, and m_s enable us to label completely an electron in any orbital in any atom. In a sense, we can regard the set of four quantum numbers as the "address" of an electron in an atom, somewhat in the same way that a street address, city, state, and postal ZIP code specify the address of an individual. For example, the four quantum numbers for a 2*s* orbital electron are $n = 2$, $\ell = 0$, $m_\ell = 0$, and $m_s = +\frac{1}{2}$ or $-\frac{1}{2}$. It is inconvenient to write out all the individual quantum numbers, and so we use the simplified notation (n, ℓ, m_ℓ, m_s). For the preceding example, the quantum numbers are either $(2, 0, 0, +\frac{1}{2})$ or $(2, 0, 0, -\frac{1}{2})$. The value of m_s has no effect on the energy, size, shape, or orientation of an orbital, but it determines how electrons are arranged in an orbital.

EXAMPLE 7.8

Write the four quantum numbers for an electron in a 3*p* orbital.

Strategy What do the "3" and "*p*" designate in 3*p*? How many orbitals (values of m_ℓ) are there in a 3*p* subshell? What are the possible values of electron spin quantum number?

(Continued)

Solution To start with, we know that the principal quantum number n is 3 and the angular momentum quantum number ℓ must be 1 (because we are dealing with a p orbital). For $\ell = 1$, there are three values of m_ℓ given by -1, 0, and 1. Because the electron spin quantum number m_s can be either $+\frac{1}{2}$ or $-\frac{1}{2}$, we conclude that there are six possible ways to designate the electron using the (n, ℓ, m_ℓ, m_s) notation:

$$(3, 1, -1, +\tfrac{1}{2}) \qquad (3, 1, -1, -\tfrac{1}{2})$$
$$(3, 1, 0, +\tfrac{1}{2}) \qquad (3, 1, 0, -\tfrac{1}{2})$$
$$(3, 1, 1, +\tfrac{1}{2}) \qquad (3, 1, 1, -\tfrac{1}{2})$$

Check In these six designations we see that the values of n and ℓ are constant, but the values of m_ℓ and m_s can vary.

Similar problem: 7.56.

Practice Exercise Write the four quantum numbers for an electron in a 4d orbital.

The hydrogen atom is a particularly simple system because it contains only one electron. The electron may reside in the 1s orbital (the ground state), or it may be found in some higher-energy orbital (an excited state). For many-electron atoms, however, we must know the ***electron configuration*** of the atom, that is, *how the electrons are distributed among the various atomic orbitals,* in order to understand electronic behavior. We will use the first 10 elements (hydrogen to neon) to illustrate the rules for writing electron configurations for atoms in the *ground state.* (Section 7.9 will describe how these rules can be applied to the remainder of the elements in the periodic table.) For this discussion, recall that the number of electrons in an atom is equal to its atomic number Z.

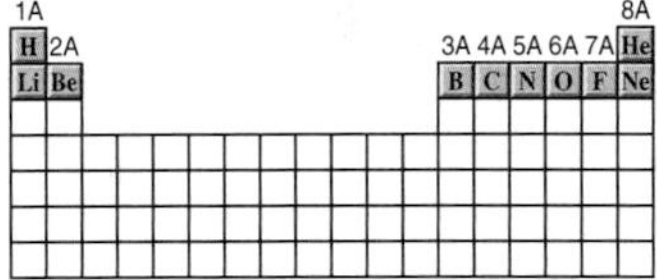

Figure 7.19 indicates that the electron in a ground-state hydrogen atom must be in the 1s orbital, so its electron configuration is $1s^1$:

$1s^1$

denotes the number of electrons in the orbital or subshell

denotes the principal quantum number n

denotes the angular momentum quantum number ℓ

The electron configuration can also be represented by an *orbital diagram* that shows the spin of the electron (see Figure 7.14):

H [↑]
$1s^1$

The upward arrow denotes one of the two possible spinning motions of the electron. (Alternatively, we could have represented the electron with a downward arrow.) The box represents an atomic orbital.

Remember that the direction of electron spin has no effect on the energy of the electron.

The Pauli Exclusion Principle

For many-electron atoms we use the ***Pauli exclusion principle*** (after the Austrian physicist Wolfgang Pauli) to determine electron configurations. This principle states that *no two electrons in an atom can have the same four quantum numbers.* If two electrons in an atom should have the same n, ℓ, and m_ℓ values (that is, these two electrons are in the *same* atomic orbital), then they must have different values of m_s. In other words, only two electrons may occupy the same atomic orbital, and these electrons

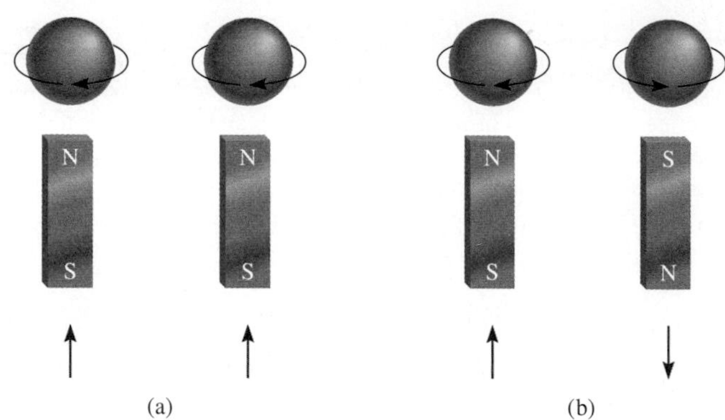

Figure 7.22
The (a) parallel and (b) antiparallel spins of two electrons. In (a), the two magnetic fields reinforce each other. In (b), the two magnetic fields cancel each other.

must have opposite spins. Consider the helium atom, which has two electrons. The three possible ways of placing two electrons in the 1*s* orbital are as follows:

He	↑↑	↓↓	↑↓
	$1s^2$	$1s^2$	$1s^2$
	(a)	(b)	(c)

Diagrams (a) and (b) are ruled out by the Pauli exclusion principle. In (a), both electrons have the same upward spin and would have the quantum numbers $(1, 0, 0, +\frac{1}{2})$; in (b), both electrons have downward spins and would have the quantum numbers $(1, 0, 0, -\frac{1}{2})$. Only the configuration in (c) is physically acceptable, because one electron has the quantum numbers $(1, 0, 0, +\frac{1}{2})$ and the other has $(1, 0, 0, -\frac{1}{2})$. Thus, the helium atom has the following configuration:

Electrons that have opposite spins are said to be paired. In helium, $m_s = +\frac{1}{2}$ for one electron; $m_s = -\frac{1}{2}$ for the other.

He	↑↓
	$1s^2$

Note that $1s^2$ is read "one *s* two," not "one *s* squared."

Diamagnetism and Paramagnetism

The Pauli exclusion principle is one of the fundamental principles of quantum mechanics. It can be tested by a simple observation. If the two electrons in the 1*s* orbital of a helium atom had the same, or parallel, spins (↑↑ or ↓↓), their net magnetic fields would reinforce each other [Figure 7.22(a)]. Such an arrangement would make the helium gas paramagnetic. ***Paramagnetic*** substances are those that *contain net unpaired spins and are attracted by a magnet.* On the other hand, if the electron spins are paired, or antiparallel to each other (↑↓ or ↓↑), the magnetic effects cancel out [Figure 7.22(b)]. ***Diamagnetic*** substances *do not contain net unpaired spins and are slightly repelled by a magnet.*

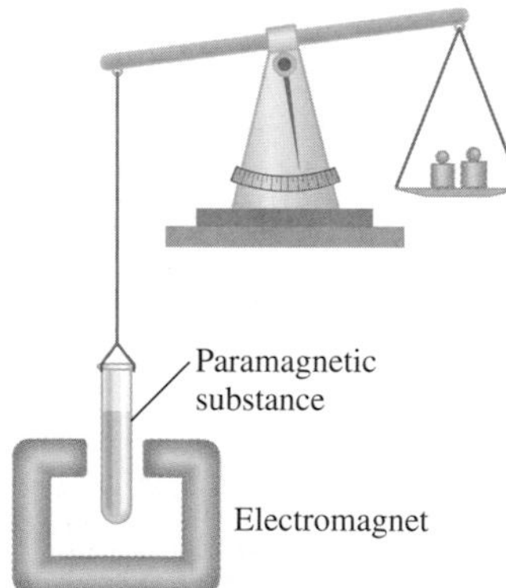

Figure 7.23
Initially the paramagnetic substance was weighed on a balance. When the electromagnet is turned on, the balance is offset because the sample tube is drawn into the magnetic field. Knowing the concentration and the additional mass needed to reestablish balance, it is possible to calculate the number of unpaired electrons in the substance.

Measurements of magnetic properties provide the most direct evidence for specific electron configurations of elements. Advances in instrument design during the last 30 years or so enable us to determine the number of unpaired electrons in an atom (Figure 7.23). By experiment we find that the helium atom in its ground state has no net magnetic field. Therefore, the two electrons in the 1*s* orbital must be paired in accord with the Pauli exclusion principle, and the helium gas is diamagnetic. A useful rule to keep in mind is that any atom with an *odd* number of electrons will always contain one or more unpaired spins because we need an even number of electrons for complete pairing. On the other hand, atoms containing an even number of electrons may or may not contain unpaired spins. We will see the reason for this behavior shortly.

As another example, consider the lithium atom ($Z = 3$), which has three electrons. The third electron cannot go into the 1*s* orbital because it would inevitably have the same set of four quantum numbers as one of the first two electrons. Therefore, this electron "enters" the next (energetically) higher orbital, which is the 2*s* orbital (see Figure 7.20). The electron configuration of lithium is $1s^2 2s^1$, and its orbital diagram is

Li [↑↓] [↑]
$1s^2$ $2s^1$

The lithium atom contains one unpaired electron and the lithium metal is therefore paramagnetic.

The Shielding Effect in Many-Electron Atoms

Experimentally we find that the 2*s* orbital lies at a lower energy level than the 2*p* orbital in a many-electron atom. Why? In comparing the electron configurations of $1s^2 2s^1$ and $1s^2 2p^1$, we note that, in both cases, the 1*s* orbital is filled with two electrons. Figure 7.24 shows the radial probability plots for the 1*s*, 2*s*, and 2*p* orbitals. Because the 2*s* and 2*p* orbitals are larger than the 1*s* orbital, an electron in either of these orbitals will spend more time away from the nucleus than an electron in the 1*s* orbital. Thus, we can speak of a 2*s* or 2*p* electron being partly "shielded" from the attractive force of the nucleus by the 1*s* electrons. The important consequence of the shielding effect is that it *reduces* the electrostatic attraction between the protons in the nucleus and the electron in the 2*s* or 2*p* orbital.

The manner in which the electron density varies as we move from the nucleus outward depends on the type of orbital. Although a 2*s* electron spends most of its time (on average) slightly farther from the nucleus than a 2*p* electron, the electron density near the nucleus is actually greater for the 2*s* electron (see the small maximum for the 2*s* orbital in Figure 7.24). For this reason, the 2*s* orbital is said to be more "penetrating" than the 2*p* orbital. Therefore, a 2*s* electron is less shielded by the 1*s* electrons and is more strongly held by the nucleus. In fact, for the same principal quantum number *n*, the penetrating power decreases as the angular momentum quantum number ℓ increases, or

$$s > p > d > f > \cdots$$

Because the stability of an electron is determined by the strength of its attraction to the nucleus, it follows that a 2*s* electron will be lower in energy than a 2*p* electron. To put it another way, less energy is required to remove a 2*p* electron than a 2*s* electron because a 2*p* electron is not held quite as strongly by the nucleus. The hydrogen atom has only one electron and, therefore, is without such a shielding effect.

Continuing our discussion of atoms of the first 10 elements, we go next to beryllium ($Z = 4$). The ground-state electron configuration of beryllium is $1s^2 2s^2$, or

Be [↑↓] [↑↓]
$1s^2$ $2s^2$

Beryllium is diamagnetic, as we would expect.

The electron configuration of boron ($Z = 5$) is $1s^2 2s^2 2p^1$, or

B [↑↓] [↑↓] [↑| |]
$1s^2$ $2s^2$ $2p^1$

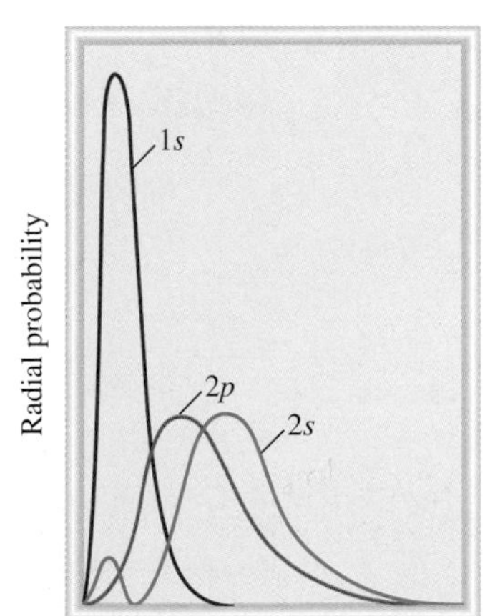

Figure 7.24
Radial probability plots for the 1s, 2s, and 2p orbitals. The 1s electrons effectively shield both the 2s and 2p electrons from the nucleus. The 2s orbital is more penetrating than the 2p orbital.

Note that the unpaired electron can be in the $2p_x$, $2p_y$, or $2p_z$ orbital. The choice is completely arbitrary because the three *p* orbitals are equivalent in energy. As the diagram shows, boron is paramagnetic.

Hund's Rule

The electron configuration of carbon ($Z = 6$) is $1s^22s^22p^2$. The following are different ways of distributing two electrons among three *p* orbitals:

↑↓		
$2p_x$	$2p_y$	$2p_z$

(a)

↑	↓	
$2p_x$	$2p_y$	$2p_z$

(b)

↑	↑	
$2p_x$	$2p_y$	$2p_z$

(c)

None of the three arrangements violates the Pauli exclusion principle, so we must determine which one will give the greatest stability. The answer is provided by ***Hund's rule*** (after the German physicist Frederick Hund), which states that *the most stable arrangement of electrons in subshells is the one with the greatest number of parallel spins.* The arrangement shown in (c) satisfies this condition. In both (a) and (b) the two spins cancel each other. Thus, the orbital diagram for carbon is

C ↑↓ ($1s^2$) ↑↓ ($2s^1$) ↑ ↑ □ ($2p^2$)

Qualitatively, we can understand why (c) is preferred to (a). In (a), the two electrons are in the same $2p_x$ orbital, and their proximity results in a greater mutual repulsion than when they occupy two separate orbitals, say $2p_x$ and $2p_y$. The choice of (c) over (b) is more subtle but can be justified on theoretical grounds. The fact that carbon atoms contain two unpaired electrons is in accord with Hund's rule.

The electron configuration of nitrogen ($Z = 7$) is $1s^22s^22p^3$:

N ↑↓ ($1s^2$) ↑↓ ($2s^2$) ↑ ↑ ↑ ($2p^3$)

Again, Hund's rule dictates that all three 2*p* electrons have spins parallel to one another; the nitrogen atom contains three unpaired electrons.

The electron configuration of oxygen ($Z = 8$) is $1s^22s^22p^4$. An oxygen atom has two unpaired electrons:

O ↑↓ ($1s^2$) ↑↓ ($2s^2$) ↑↓ ↑ ↑ ($2p^4$)

The electron configuration of fluorine (Z = 9) is $1s^22s^22p^5$. The nine electrons are arranged as follows:

F ↑↓ ($1s^2$) ↑↓ ($2s^2$) ↑↓ ↑↓ ↑ ($2p^5$)

The fluorine atom has one unpaired electron.

In neon ($Z = 10$), the 2*p* subshell is completely filled. The electron configuration of neon is $1s^22s^22p^6$, and *all* the electrons are paired, as follows:

Ne ↑↓ ($1s^2$) ↑↓ ($2s^2$) ↑↓ ↑↓ ↑↓ ($2p^6$)

The neon gas should be diamagnetic, and experimental observation bears out this prediction.

General Rules for Assigning Electrons to Atomic Orbitals

Based on the preceding examples we can formulate some general rules for determining the maximum number of electrons that can be assigned to the various subshells and orbitals for a given value of n:

1. Each shell or principal level of quantum number n contains n subshells. For example, if $n = 2$, then there are two subshells (two values of ℓ) of angular momentum quantum numbers 0 and 1.
2. Each subshell of quantum number ℓ contains $(2\ell + 1)$ orbitals. For example, if $\ell = 1$, then there are three p orbitals.
3. No more than two electrons can be placed in each orbital. Therefore, the maximum number of electrons is simply twice the number of orbitals that are employed.
4. A quick way to determine the maximum number of electrons that an atom can have in a principal level n is to use the formula $2n^2$.

EXAMPLE 7.9

What is the maximum number of electrons that can be present in the principal level for which $n = 3$?

Strategy We are given the principal quantum number (n) so we can determine all the possible values of the angular momentum quantum number (ℓ). The preceding rule shows that the number of orbitals for each value of ℓ is $(2\ell + 1)$. Thus, we can determine the total number of orbitals. How many electrons can each orbital accommodate?

Solution When $n = 3$, $\ell = 0$, 1, and 2. The number of orbitals for each value of ℓ is given by

Value of ℓ	Number of Orbitals $(2\ell + 1)$
0	1
1	3
2	5

The total number of orbitals is nine. Because each orbital can accommodate two electrons, the maximum number of electrons that can reside in the orbitals is 2×9, or 18.

Check If we use the formula (n^2) in Example 7.7, we find that the total number of orbitals is 3^2 and the total number of electrons is $2(3^2)$ or 18. In general, the number of electrons in a given principal energy level n is $2n^2$.

Practice Exercise Calculate the total number of electrons that can be present in the principal level for which $n = 4$.

Similar problems: 7.62, 7.63.

EXAMPLE 7.10

An oxygen atom has a total of eight electrons. Write the four quantum numbers for each of the eight electrons in the ground state.

(Continued)

Strategy We start with $n = 1$ and proceed to fill orbitals in the order shown in Figure 7.21. For each value of n we determine the possible values of ℓ. For each value of ℓ, we assign the possible values of m_ℓ. We can place electrons in the orbitals according to the Pauli exclusion principle and Hund's rule.

Solution We start with $n = 1$, so $\ell = 0$, a subshell corresponding to the $1s$ orbital. This orbital can accommodate a total of two electrons. Next, $n = 2$, and ℓ may be either 0 or 1. The $\ell = 0$ subshell contains one $2s$ orbital, which can accommodate two electrons. The remaining four electrons are placed in the $\ell = 1$ subshell, which contains three $2p$ orbitals. The orbital diagram is

O	↑↓	↑↓	↑↓ ↑ ↑
	$1s^2$	$2s^2$	$2p^4$

The results are summarized in the following table:

Electron	n	ℓ	m_ℓ	m_s	Orbital
1	1	0	0	$+\frac{1}{2}$	$1s$
2	1	0	0	$-\frac{1}{2}$	
3	2	0	0	$+\frac{1}{2}$	$2s$
4	2	0	0	$-\frac{1}{2}$	
5	2	1	-1	$+\frac{1}{2}$	$2p_x, 2p_y, 2p_z$
6	2	1	0	$+\frac{1}{2}$	
7	2	1	1	$+\frac{1}{2}$	
8	2	1	1	$-\frac{1}{2}$	

Of course, the placement of the eighth electron in the orbital labeled $m_\ell = 1$ is completely arbitrary. It would be equally correct to assign it to $m_\ell = 0$ or $m_\ell = -1$.

Similar problem: 7.85.

Practice Exercise Write a complete set of quantum numbers for each of the electrons in boron (B).

At this point let's summarize what our examination has revealed about ground-state electron configurations and the properties of electrons in atoms:

1. No two electrons in the same atom can have the same four quantum numbers. This is the Pauli exclusion principle.
2. Each orbital can be occupied by a maximum of two electrons. They must have opposite spins, or different electron spin quantum numbers.
3. The most stable arrangement of electrons in a subshell is the one that has the greatest number of parallel spins. This is Hund's rule.
4. Atoms in which one or more electrons are unpaired are paramagnetic. Atoms in which all the electron spins are paired are diamagnetic.
5. In a hydrogen atom, the energy of the electron depends only on its principal quantum number n. In a many-electron atom, the energy of an electron depends on both n and its angular momentum quantum number ℓ.
6. In a many-electron atom the subshells are filled in the order shown in Figure 7.21.
7. For electrons of the same principal quantum number, their penetrating power, or proximity to the nucleus, decreases in the order $s > p > d > f$. This means that, for example, more energy is required to separate a $3s$ electron from a many-electron atom than is required to remove a $3p$ electron.

Review of Concepts

Given the ground state electron configuration $1s^22s^22p^63s^23p^3$, which of the four quantum numbers would be the same for the three 3*p* electrons?

7.9 The Building-Up Principle

Here we will extend the rules used in writing electron configurations for the first 10 elements to the rest of the elements. This process is based on the Aufbau principle. The ***Aufbau principle*** dictates that *as protons are added one by one to the nucleus to build up the elements, electrons are similarly added to the atomic orbitals.* Through this process we gain a detailed knowledge of the ground-state electron configurations of the elements. As we will see later, knowledge of electron configurations helps us to understand and predict the properties of the elements; it also explains why the periodic table works so well.

The German word "Aufbau" means "building up."

Table 7.3 gives the ground-state electron configurations of elements from H ($Z = 1$) through Rg ($Z = 111$). The electron configurations of all elements except hydrogen and helium are represented by a ***noble gas core,*** which *shows in brackets the noble gas element that most nearly precedes the element being considered,* followed by the symbol for the highest filled subshells in the outermost shells. Notice that the electron configurations of the highest filled subshells in the outermost shells for the elements sodium ($Z = 11$) through argon ($Z = 18$) follow a pattern similar to those of lithium ($Z = 3$) through neon ($Z = 10$).

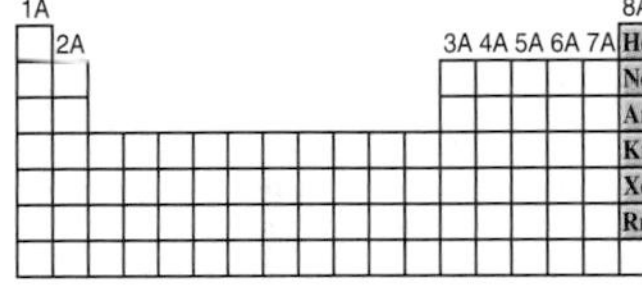

The noble gases.

As mentioned in Section 7.7, the 4*s* subshell is filled before the 3*d* subshell in a many-electron atom (see Figure 7.21). Thus, the electron configuration of potassium ($Z = 19$) is $1s^22s^22p^63s^23p^64s^1$. Because $1s^22s^22p^63s^23p^6$ is the electron configuration of argon, we can simplify the electron configuration of potassium by writing $[Ar]4s^1$, where [Ar] denotes the "argon core." Similarly, we can write the electron configuration of calcium ($Z = 20$) as $[Ar]4s^2$. The placement of the outermost electron in the 4*s* orbital (rather than in the 3*d* orbital) of potassium is strongly supported by experimental evidence. The following comparison also suggests that this is the correct configuration. The chemistry of potassium is very similar to that of lithium and sodium, the first two alkali metals. The outermost electron of both lithium and sodium is in an *s* orbital (there is no ambiguity in assigning their electron configurations); therefore, we expect the last electron in potassium to occupy the 4*s* rather than the 3*d* orbital.

The elements from scandium ($Z = 21$) to copper ($Z = 29$) are transition metals. ***Transition metals*** either *have incompletely filled d subshells or readily give rise to cations that have incompletely filled d subshells.* Consider the first transition metal series, from scandium through copper. In this series additional electrons are placed in the 3*d* orbitals, according to Hund's rule. However, there are two irregularities. The electron configuration of chromium ($Z = 24$) is $[Ar]4s^13d^5$ and not $[Ar]4s^23d^4$, as we might expect. A similar break in the pattern is observed for copper, whose electron configuration is $[Ar]4s^13d^{10}$ rather than $[Ar]4s^23d^9$. The reason for these irregularities is that a slightly greater stability is associated with the half-filled ($3d^5$) and completely filled ($3d^{10}$) subshells. Electrons in the same subshell (in this case, the *d* orbitals) have equal energy but different spatial distributions. Consequently, their shielding of one another is relatively small, and the electrons are more strongly attracted by the nucleus when they have the $3d^5$ configuration. According to Hund's rule, the orbital diagram for Cr is

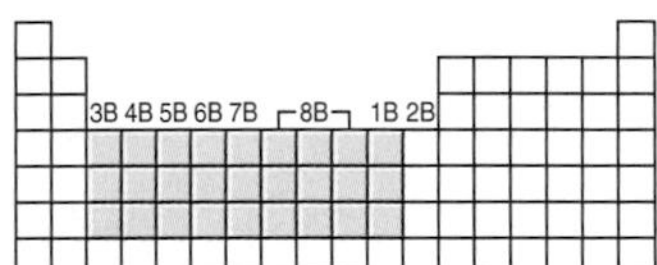

The transition metals.

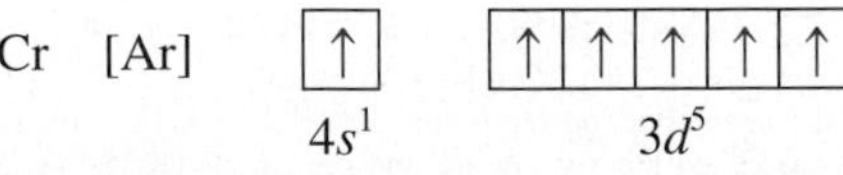

Table 7.3 The Ground-State Electron Configurations of the Elements*

Atomic Number	Symbol	Electron Configuration	Atomic Number	Symbol	Electron Configuration	Atomic Number	Symbol	Electron Configuration
1	H	$1s^1$	38	Sr	$[Kr]5s^2$	75	Re	$[Xe]6s^24f^{14}5d^5$
2	He	$1s^2$	39	Y	$[Kr]5s^24d^1$	76	Os	$[Xe]6s^24f^{14}5d^6$
3	Li	$[He]2s^1$	40	Zr	$[Kr]5s^24d^2$	77	Ir	$[Xe]6s^24f^{14}5d^7$
4	Be	$[He]2s^2$	41	Nb	$[Kr]5s^14d^4$	78	Pt	$[Xe]6s^14f^{14}5d^9$
5	B	$[He]2s^22p^1$	42	Mo	$[Kr]5s^14d^5$	79	Au	$[Xe]6s^14f^{14}5d^{10}$
6	C	$[He]2s^22p^2$	43	Tc	$[Kr]5s^24d^5$	80	Hg	$[Xe]6s^24f^{14}5d^{10}$
7	N	$[He]2s^22p^3$	44	Ru	$[Kr]5s^14d^7$	81	Tl	$[Xe]6s^24f^{14}5d^{10}6p^1$
8	O	$[He]2s^22p^4$	45	Rh	$[Kr]5s^14d^8$	82	Pb	$[Xe]6s^24f^{14}5d^{10}6p^2$
9	F	$[He]2s^22p^5$	46	Pd	$[Kr]4d^{10}$	83	Bi	$[Xe]6s^24f^{14}5d^{10}6p^3$
10	Ne	$[He]2s^22p^6$	47	Ag	$[Kr]5s^14d^{10}$	84	Po	$[Xe]6s^24f^{14}5d^{10}6p^4$
11	Na	$[Ne]3s^1$	48	Cd	$[Kr]5s^24d^{10}$	85	At	$[Xe]6s^24f^{14}5d^{10}6p^5$
12	Mg	$[Ne]3s^2$	49	In	$[Kr]5s^24d^{10}5p^1$	86	Rn	$[Xe]6s^24f^{14}5d^{10}6p^6$
13	Al	$[Ne]3s^23p^1$	50	Sn	$[Kr]5s^24d^{10}5p^2$	87	Fr	$[Rn]7s^1$
14	Si	$[Ne]3s^23p^2$	51	Sb	$[Kr]5s^24d^{10}5p^3$	88	Ra	$[Rn]7s^2$
15	P	$[Ne]3s^23p^3$	52	Te	$[Kr]5s^24d^{10}5p^4$	89	Ac	$[Rn]7s^26d^1$
16	S	$[Ne]3s^23p^4$	53	I	$[Kr]5s^24d^{10}5p^5$	90	Th	$[Rn]7s^26d^2$
17	Cl	$[Ne]3s^23p^5$	54	Xe	$[Kr]5s^24d^{10}5p^6$	91	Pa	$[Rn]7s^25f^26d^1$
18	Ar	$[Ne]3s^23p^6$	55	Cs	$[Xe]6s^1$	92	U	$[Rn]7s^25f^36d^1$
19	K	$[Ar]4s^1$	56	Ba	$[Xe]6s^2$	93	Np	$[Rn]7s^25f^46d^1$
20	Ca	$[Ar]4s^2$	57	La	$[Xe]6s^25d^1$	94	Pu	$[Rn]7s^25f^6$
21	Sc	$[Ar]4s^23d^1$	58	Ce	$[Xe]6s^24f^15d^1$	95	Am	$[Rn]7s^25f^7$
22	Ti	$[Ar]4s^23d^2$	59	Pr	$[Xe]6s^24f^3$	96	Cm	$[Rn]7s^25f^76d^1$
23	V	$[Ar]4s^23d^3$	60	Nd	$[Xe]6s^24f^4$	97	Bk	$[Rn]7s^25f^9$
24	Cr	$[Ar]4s^13d^5$	61	Pm	$[Xe]6s^24f^5$	98	Cf	$[Rn]7s^25f^{10}$
25	Mn	$[Ar]4s^23d^5$	62	Sm	$[Xe]6s^24f^6$	99	Es	$[Rn]7s^25f^{11}$
26	Fe	$[Ar]4s^23d^6$	63	Eu	$[Xe]6s^24f^7$	100	Fm	$[Rn]7s^25f^{12}$
27	Co	$[Ar]4s^23d^7$	64	Gd	$[Xe]6s^24f^75d^1$	101	Md	$[Rn]7s^25f^{13}$
28	Ni	$[Ar]4s^23d^8$	65	Tb	$[Xe]6s^24f^9$	102	No	$[Rn]7s^25f^{14}$
29	Cu	$[Ar]4s^13d^{10}$	66	Dy	$[Xe]6s^24f^{10}$	103	Lr	$[Rn]7s^25f^{14}6d^1$
30	Zn	$[Ar]4s^23d^{10}$	67	Ho	$[Xe]6s^24f^{11}$	104	Rf	$[Rn]7s^25f^{14}6d^2$
31	Ga	$[Ar]4s^23d^{10}4p^1$	68	Er	$[Xe]6s^24f^{12}$	105	Db	$[Rn]7s^25f^{14}6d^3$
32	Ge	$[Ar]4s^23d^{10}4p^2$	69	Tm	$[Xe]6s^24f^{13}$	106	Sg	$[Rn]7s^25f^{14}6d^4$
33	As	$[Ar]4s^23d^{10}4p^3$	70	Yb	$[Xe]6s^24f^{14}$	107	Bh	$[Rn]7s^25f^{14}6d^5$
34	Se	$[Ar]4s^23d^{10}4p^4$	71	Lu	$[Xe]6s^24f^{14}5d^1$	108	Hs	$[Rn]7s^25f^{14}6d^6$
35	Br	$[Ar]4s^23d^{10}4p^5$	72	Hf	$[Xe]6s^24f^{14}5d^2$	109	Mt	$[Rn]7s^25f^{14}6d^7$
36	Kr	$[Ar]4s^23d^{10}4p^6$	73	Ta	$[Xe]6s^24f^{14}5d^3$	110	Ds	$[Rn]7s^25f^{14}6d^8$
37	Rb	$[Kr]5s^1$	74	W	$[Xe]6s^24f^{14}5d^4$	111	Rg	$[Rn]7s^25f^{14}6d^9$

*The symbol [He] is called the helium core and represents $1s^2$. [Ne] is called the neon core and represents $1s^22s^22p^6$. [Ar] is called the argon core and represents $[Ne]3s^23p^6$. [Kr] is called the krypton core and represents $[Ar]4s^23d^{10}4p^6$. [Xe] is called the xenon core and represents $[Kr]5s^24d^{10}5p^6$. [Rn] is called the radon core and represents $[Xe]6s^24f^{14}5d^{10}6p^6$.

Thus, Cr has a total of six unpaired electrons. The orbital diagram for copper is

$$\text{Cu}\quad[\text{Ar}]\quad \underset{4s^1}{\boxed{\uparrow}}\quad \underset{3d^{10}}{\boxed{\uparrow\downarrow}\boxed{\uparrow\downarrow}\boxed{\uparrow\downarrow}\boxed{\uparrow\downarrow}\boxed{\uparrow\downarrow}}$$

Again, extra stability is gained in this case by having the 3*d* subshell completely filled.

For elements Zn ($Z = 30$) through Kr ($Z = 36$), the 4*s* and 4*p* subshells fill in a straightforward manner. With rubidium ($Z = 37$), electrons begin to enter the $n = 5$ energy level.

The electron configurations in the second transition metal series [yttrium ($Z = 39$) to silver ($Z = 47$)] are also irregular, but we will not be concerned with the details here.

The sixth period of the periodic table begins with cesium ($Z = 55$) and barium ($Z = 56$), whose electron configurations are $[\text{Xe}]6s^1$ and $[\text{Xe}]6s^2$, respectively. Next we come to lanthanum ($Z = 57$). From Figure 7.21 we would expect that after filling the 6*s* orbital we would place the additional electrons in 4*f* orbitals. In reality, the energies of the 5*d* and 4*f* orbitals are very close; in fact, for lanthanum 4*f* is slightly higher in energy than 5*d*. Thus, lanthanum's electron configuration is $[\text{Xe}]6s^25d^1$ and not $[\text{Xe}]6s^24f^1$.

Following lanthanum are the 14 elements known as the ***lanthanides,*** or ***rare earth series*** [cerium ($Z = 58$) to lutetium ($Z = 71$)]. The rare earth metals *have incompletely filled 4f subshells or readily give rise to cations that have incompletely filled 4f subshells.* In this series, the added electrons are placed in 4*f* orbitals. After the 4*f* subshell is completely filled, the next electron enters the 5*d* subshell of lutetium. Note that the electron configuration of gadolinium ($Z = 64$) is $[\text{Xe}]6s^24f^75d^1$ rather than $[\text{Xe}]6s^24f^8$. Like chromium, gadolinium gains extra stability by having a half-filled subshell ($4f^7$).

The third transition metal series, including lanthanum and hafnium ($Z = 72$) and extending through gold ($Z = 79$), is characterized by the filling of the 5*d* subshell. With Hg ($Z = 80$), both the 6*s* and 5*d* orbitals are now filled. The 6*p* subshell is filled next, which takes us to radon ($Z = 86$).

The *last row of elements* is the ***actinide series,*** which starts at thorium ($Z = 90$). *Most of these elements are not found in nature but have been synthesized.*

With few exceptions, you should be able to write the electron configuration of any element, using Figure 7.21 as a guide. Elements that require particular care are the transition metals, the lanthanides, and the actinides. As we noted earlier, at larger values of the principal quantum number *n*, the order of subshell filling may reverse from one element to the next. Figure 7.25 groups the elements according to the type of subshell in which the outermost electrons are placed.

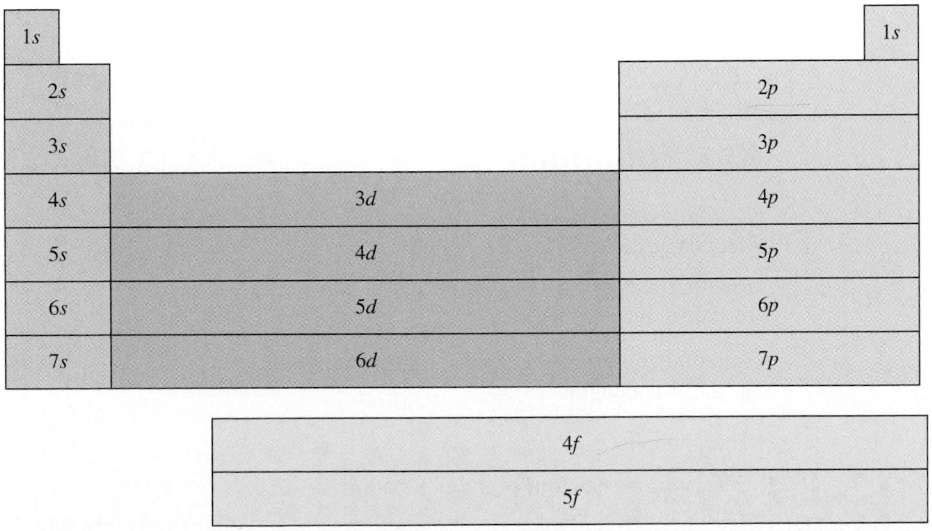

Figure 7.25
Classification of groups of elements in the periodic table according to the type of subshell being filled with electrons.

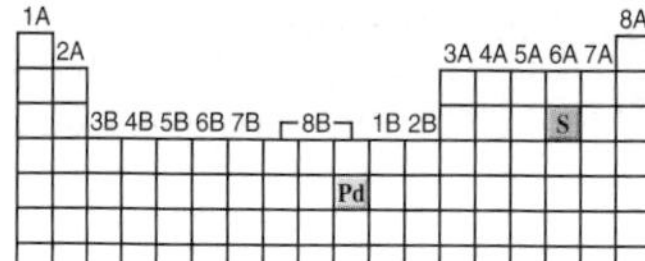

EXAMPLE 7.11

Write the ground-state electron configurations for (a) sulfur (S) and (b) palladium (Pd), which is diamagnetic.

(a) Strategy How many electrons are in the S ($Z = 16$) atom? We start with $n = 1$ and proceed to fill orbitals in the order shown in Figure 7.21. For each value of ℓ, we assign the possible values of m_ℓ. We can place electrons in the orbitals according to the Pauli exclusion principle and Hund's rule and then write the electron configuration. The task is simplified if we use the noble-gas core preceding S for the inner electrons.

Solution Sulfur has 16 electrons. The noble gas core in this case is [Ne]. (Ne is the noble gas in the period preceding sulfur.) [Ne] represents $1s^22s^22p^6$. This leaves us 6 electrons to fill the 3*s* subshell and partially fill the 3*p* subshell. Thus, the electron configuration of S is $1s^22s^22p^63s^23p^4$ or $[Ne]3s^23p^4$.

(b) Strategy We use the same approach as that in (a). What does it mean to say that Pd is a diamagnetic element?

Solution Palladium has 46 electrons. The noble-gas core in this case is [Kr]. (Kr is the noble gas in the period preceding palladium.) [Kr] represents

$$1s^22s^22p^63s^23p^64s^23d^{10}4p^6$$

The remaining 10 electrons are distributed among the 4*d* and 5*s* orbitals. The three choices are (1) $4d^{10}$, (2) $4d^95s^1$, and (3) $4d^85s^2$. Because palladium is diamagnetic, all the electrons are paired and its electron configuration must be

$$1s^22s^22p^63s^23p^64s^23d^{10}4p^64d^{10}$$

or simply $[Kr]4d^{10}$. The configurations in (2) and (3) both represent paramagnetic elements.

Check To confirm the answer, write the orbital diagrams for (1), (2), and (3).

Similar problems: 7.91, 7.92.

Practice Exercise Write the ground-state electron configuration for phosphorus (P).

Review of Concepts

Identify the atom that has the following ground-state electron configuration: $[Ar]4s^23d^8$.

Key Equations

$u = \lambda\nu$ (7.1)	Relating speed of a wave to its wavelength and frequency.
$E = h\nu$ (7.2)	Relating energy of a quantum (and of a photon) to the frequency.
$E = h\frac{c}{\lambda}$ (7.3)	Relating energy of a quantum (and of a photon) to the wavelength.
$h\nu = \text{KE} + W$ (7.4)	The photoelectric effect
$E_n = -R_H\left(\frac{1}{n^2}\right)$ (7.5)	Energy of an electron in the *n*th state in a hydrogen atom.

$$\Delta E = h\nu = R_H\left(\frac{1}{n_i^2} - \frac{1}{n_f^2}\right) \quad (7.6)$$

Energy of a photon emitted (or absorbed) as the electron undergoes a transition from the n_i level to the n_f level.

$$\lambda = \frac{h}{mu} \quad (7.8)$$

Relating wavelength of a particle to its mass m and velocity u.

Summary of Facts and Concepts

1. The quantum theory developed by Planck successfully explains the emission of radiation by heated solids. The quantum theory states that radiant energy is emitted by atoms and molecules in small discrete amounts (quanta), rather than over a continuous range. This behavior is governed by the relationship $E = h\nu$, where E is the energy of the radiation, h is Planck's constant, and ν is the frequency of the radiation. Energy is always emitted in whole-number multiples of $h\nu$ ($1\ h\nu$, $2\ h\nu$, $3\ h\nu$, . . .).
2. Using quantum theory, Einstein solved another mystery of physics—the photoelectric effect. Einstein proposed that light can behave like a stream of particles (photons).
3. The line spectrum of hydrogen, yet another mystery to nineteenth-century physicists, was also explained by applying the quantum theory. Bohr developed a model of the hydrogen atom in which the energy of its single electron is quantized—limited to certain energy values determined by an integer, the principal quantum number.
4. An electron in its most stable energy state is said to be in the ground state, and an electron at an energy level higher than its most stable state is said to be in an excited state. In the Bohr model, an electron emits a photon when it drops from a higher-energy state (an excited state) to a lower-energy state (the ground state or another, less excited state). The release of specific amounts of energy in the form of photons accounts for the lines in the hydrogen emission spectrum.
5. De Broglie extended Einstein's wave-particle description of light to all matter in motion. The wavelength of a moving particle of mass m and velocity u is given by the de Broglie equation $\lambda = h/mu$.
6. The Schrödinger equation describes the motions and energies of submicroscopic particles. This equation launched quantum mechanics and a new era in physics.
7. The Schrödinger equation tells us the possible energy states of the electron in a hydrogen atom and the probability of its location in a particular region surrounding the nucleus. These results can be applied with reasonable accuracy to many-electron atoms.
8. An atomic orbital is a function (ψ) that defines the distribution of electron density (ψ^2) in space. Orbitals are represented by electron density diagrams or boundary surface diagrams.
9. Four quantum numbers characterize each electron in an atom: the principal quantum number n identifies the main energy level, or shell, of the orbital; the angular momentum quantum number ℓ indicates the shape of the orbital; the magnetic quantum number m_ℓ specifies the orientation of the orbital in space; and the electron spin quantum number m_s indicates the direction of the electron's spin on its own axis.
10. The single s orbital for each energy level is spherical and centered on the nucleus. The three p orbitals present at $n = 2$ and higher; each has two lobes, and the pairs of lobes are arranged at right angles to one another. Starting with $n = 3$, there are five d orbitals, with more complex shapes and orientations.
11. The energy of the electron in a hydrogen atom is determined solely by its principal quantum number. In many-electron atoms, the principal quantum number and the angular momentum quantum number together determine the energy of an electron.
12. No two electrons in the same atom can have the same four quantum numbers (the Pauli exclusion principle).
13. The most stable arrangement of electrons in a subshell is the one that has the greatest number of parallel spins (Hund's rule). Atoms with one or more unpaired electron spins are paramagnetic. Atoms in which all electrons are paired are diamagnetic.
14. The Aufbau principle provides the guideline for building up the elements. The periodic table classifies the elements according to their atomic numbers and thus also by the electronic configurations of their atoms.

Key Words

Actinide series, p. 241
Amplitude, p. 212
Atomic orbital, p. 226
Aufbau principle, p. 239
Boundary surface diagram, p. 229
Diamagnetic, p. 234
Electromagnetic radiation, p. 213
Electromagnetic wave, p. 213
Electron configuration, p. 233
Electron density, p. 226

Questions and Problems

Quantum Theory and Electromagnetic Radiation

Review Questions

7.1 What is a wave? Explain the following terms associated with waves: wavelength, frequency, amplitude.

7.2 What are the units for wavelength and frequency of electromagnetic waves? What is the speed of light in meters per second and miles per hour?

7.3 List the types of electromagnetic radiation, starting with the radiation having the longest wavelength and ending with the radiation having the shortest wavelength.

7.4 Give the high and low wavelength values that define the visible region of the electromagnetic spectrum.

7.5 Briefly explain Planck's quantum theory and explain what a quantum is. What are the units for Planck's constant?

7.6 Give two everyday examples that illustrate the concept of quantization.

Problems

7.7 (a) What is the wavelength (in nanometers) of light having a frequency of 8.6×10^{13} Hz? (b) What is the frequency (in Hz) of light having a wavelength of 566 nm?

7.8 (a) What is the frequency of light having a wavelength of 456 nm? (b) What is the wavelength (in nanometers) of radiation having a frequency of 2.45×10^{9} Hz? (This is the type of radiation used in microwave ovens.)

7.9 The average distance between Mars and Earth is about 1.3×10^{8} miles. How long would it take TV pictures transmitted from the *Viking* space vehicle on Mars' surface to reach Earth? (1 mile = 1.61 km.)

7.10 How many minutes would it take a radio wave to travel from the planet Venus to Earth? (Average distance from Venus to Earth = 28 million miles.)

7.11 The SI unit of time is the second, which is defined as 9,192,631,770 cycles of radiation associated with a certain emission process in the cesium atom. Calculate the wavelength of this radiation (to three significant figures). In which region of the electromagnetic spectrum is this wavelength found?

7.12 The SI unit of length is the meter, which is defined as the length equal to 1,650,763.73 wavelengths of the light emitted by a particular energy transition in krypton atoms. Calculate the frequency of the light to three significant figures.

The Photoelectric Effect

Review Questions

7.13 Explain what is meant by the photoelectric effect.

7.14 What are photons? What role did Einstein's explanation of the photoelectric effect play in the development of the particle-wave interpretation of the nature of electromagnetic radiation?

Problems

7.15 A photon has a wavelength of 624 nm. Calculate the energy of the photon in joules.

7.16 The blue color of the sky results from the scattering of sunlight by air molecules. The blue light has a frequency of about 7.5×10^{14} Hz. (a) Calculate the wavelength, in nm, associated with this radiation, and (b) calculate the energy, in joules, of a single photon associated with this frequency.

7.17 A photon has a frequency of 6.0×10^{14} Hz. (a) Convert this frequency into wavelength (nm). Does this frequency fall in the visible region? (b) Calculate the energy (in joules) of this photon. (c) Calculate the energy (in joules) of 1 mole of photons all with this frequency.

7.18 What is the wavelength, in nm, of radiation that has an energy content of 1.0×10^{3} kJ/mol? In which region of the electromagnetic spectrum is this radiation found?

7.19 When copper is bombarded with high-energy electrons, X rays are emitted. Calculate the energy (in joules) associated with the photons if the wavelength of the X rays is 0.154 nm.

7.20 A particular form of electromagnetic radiation has a frequency of 8.11×10^{14} Hz. (a) What is its wavelength in nanometers? In meters? (b) To what region of the electromagnetic spectrum would you assign it? (c) What is the energy (in joules) of one quantum of this radiation?

7.21 The work function of potassium is 3.68×10^{-19} J. (a) What is the minimum frequency of light needed to eject electrons from the metal? (b) Calculate the kinetic energy of the ejected electrons when light of frequency equal to 8.62×10^{14} s^{-1} is used for irradiation.

7.22 When light of frequency equal to 2.11×10^{15} s^{-1} shines on the surface of gold metal, the kinetic energy of ejected electrons is found to be 5.83×10^{-19} J. What is the work function of gold?

Bohr's Theory of the Hydrogen Atom

Review Questions

7.23 What is an energy level? Explain the difference between ground state and excited state.

7.24 Briefly describe Bohr's theory of the hydrogen atom and how it explains the appearance of an emission spectrum. How does Bohr's theory differ from concepts of classical physics?

Problems

7.25 Explain why elements produce their own characteristic colors when they emit photons.

7.26 Some copper compounds emit green light when they are heated in a flame. How would you determine whether the light is of one wavelength or a mixture of two or more wavelengths?

7.27 Is it possible for a fluorescent material to emit radiation in the ultraviolet region after absorbing visible light? Explain your answer.

7.28 Explain how astronomers are able to tell which elements are present in distant stars by analyzing the electromagnetic radiation emitted by the stars.

7.29 Consider the following energy levels of a hypothetical atom:

E_4 ________ -1.0×10^{-19} J
E_3 ________ -5.0×10^{-19} J
E_2 ________ -10×10^{-19} J
E_1 ________ -15×10^{-19} J

(a) What is the wavelength of the photon needed to excite an electron from E_1 to E_4? (b) What is the energy (in joules) a photon must have in order to excite an electron from E_2 to E_3? (c) When an electron drops from the E_3 level to the E_1 level, the atom is said to undergo emission. Calculate the wavelength of the photon emitted in this process.

7.30 The first line of the Balmer series occurs at a wavelength of 656.3 nm. What is the energy difference between the two energy levels involved in the emission that results in this spectral line?

7.31 Calculate the wavelength (in nanometers) of a photon emitted by a hydrogen atom when its electron drops from the $n = 5$ state to the $n = 3$ state.

7.32 Calculate the frequency (Hz) and wavelength (nm) of the emitted photon when an electron drops from the $n = 4$ to the $n = 2$ level in a hydrogen atom.

7.33 Careful spectral analysis shows that the familiar yellow light of sodium lamps (such as street lamps) is made up of photons of two wavelengths, 589.0 nm and 589.6 nm. What is the difference in energy (in joules) between photons with these wavelengths?

7.34 An electron in the hydrogen atom makes a transition from an energy state of principal quantum numbers n_i to the $n = 2$ state. If the photon emitted has a wavelength of 434 nm, what is the value of n_i?

Particle-Wave Duality

Review Questions

7.35 Explain the statement, Matter and radiation have a "dual nature."

7.36 How does de Broglie's hypothesis account for the fact that the energies of the electron in a hydrogen atom are quantized?

7.37 Why is Equation (7.8) meaningful only for submicroscopic particles, such as electrons and atoms, and not for macroscopic objects?

7.38 Does a baseball in flight possess wave properties? If so, why can we not determine its wave properties?

Problems

7.39 Thermal neutrons are neutrons that move at speeds comparable to those of air molecules at room temperature. These neutrons are most effective in initiating a nuclear chain reaction among ^{235}U isotopes. Calculate the wavelength (in nm) associated with a beam of neutrons moving at 7.00×10^2 m/s. (Mass of a neutron $= 1.675 \times 10^{-27}$ kg.)

7.40 Protons can be accelerated to speeds near that of light in particle accelerators. Estimate the wavelength (in nm) of such a proton moving at 2.90×10^8 m/s. (Mass of a proton $= 1.673 \times 10^{-27}$ kg.)

7.41 What is the de Broglie wavelength, in cm, of a 12.4-g hummingbird flying at 1.20×10^2 mph? (1 mile = 1.61 km.)

7.42 What is the de Broglie wavelength (in nm) associated with a 2.5-g Ping-Pong ball traveling 35 mph?

Quantum Mechanics

Review Questions

7.43 What are the inadequacies of Bohr's theory?

7.44 What is the Heisenberg uncertainty principle? What is the Schrödinger equation?

7.45 What is the physical significance of the wave function?

7.46 How is the concept of electron density used to describe the position of an electron in the quantum mechanical treatment of an atom?

7.47 What is an atomic orbital? How does an atomic orbital differ from an orbit?

7.48 Describe the characteristics of an *s* orbital, a *p* orbital, and a *d* orbital. Which of the following orbitals do not exist: 1*p*, 2*s*, 2*d*, 3*p*, 3*d*, 3*f*, 4*g*?

7.49 Why is a boundary surface diagram useful in representing an atomic orbital?

7.50 Describe the four quantum numbers used to characterize an electron in an atom.

7.51 Which quantum number defines a shell? Which quantum numbers define a subshell?

7.52 Which of the four quantum numbers (n, ℓ, m_ℓ, m_s) determine (a) the energy of an electron in a hydrogen atom and in a many-electron atom, (b) the size of an orbital, (c) the shape of an orbital, (d) the orientation of an orbital in space?

Problems

7.53 An electron in a certain atom is in the $n = 2$ quantum level. List the possible values of ℓ and m_ℓ that it can have.

7.54 An electron in an atom is in the $n = 3$ quantum level. List the possible values of ℓ and m_ℓ that it can have.

7.55 Give the values of the quantum numbers associated with the following orbitals: (a) 2*p*, (b) 3*s*, (c) 5*d*.

7.56 Give the values of the four quantum numbers of an electron in the following orbitals: (a) 3*s*, (b) 4*p*, (c) 3*d*.

7.57 Discuss the similarities and differences between a 1*s* and a 2*s* orbital.

7.58 What is the difference between a $2p_x$ and a $2p_y$ orbital?

7.59 List all the possible subshells and orbitals associated with the principal quantum number n, if $n = 5$.

7.60 List all the possible subshells and orbitals associated with the principal quantum number n, if $n = 6$.

7.61 Calculate the total number of electrons that can occupy (a) one *s* orbital, (b) three *p* orbitals, (c) five *d* orbitals, (d) seven *f* orbitals.

7.62 What is the total number of electrons that can be held in all orbitals having the same principal quantum number n?

7.63 Determine the maximum number of electrons that can be found in each of the following subshells: 3*s*, 3*d*, 4*p*, 4*f*, 5*f*.

7.64 Indicate the total number of (a) *p* electrons in N ($Z = 7$); (b) *s* electrons in Si ($Z = 14$); and (c) 3*d* electrons in S ($Z = 16$).

7.65 Make a chart of all allowable orbitals in the first four principal energy levels of the hydrogen atom. Designate each by type (for example, *s*, *p*) and indicate how many orbitals of each type there are.

7.66 Why do the 3*s*, 3*p*, and 3*d* orbitals have the same energy in a hydrogen atom but different energies in a many-electron atom?

7.67 For each of the following pairs of hydrogen orbitals, indicate which is higher in energy: (a) 1*s*, 2*s*; (b) 2*p*, 3*p*; (c) $3d_{xy}$, $3d_{yz}$; (d) 3*s*, 3*d*; (e) 4*f*, 5*s*.

7.68 Which orbital in each of the following pairs is lower in energy in a many-electron atom? (a) 2*s*, 2*p*; (b) 3*p*, 3*d*; (c) 3*s*, 4*s*; (d) 4*d*, 5*f*.

Atomic Orbitals

Review Questions

7.69 Describe the shapes of *s*, *p*, and *d* orbitals. How are these orbitals related to the quantum numbers n, ℓ, and m_ℓ?

7.70 List the hydrogen orbitals in increasing order of energy.

Electron Configuration

Review Questions

7.71 What is electron configuration? Describe the roles that the Pauli exclusion principle and Hund's rule play in writing the electron configuration of elements.

7.72 Explain the meaning of the symbol $4d^6$.

7.73 Explain the meaning of diamagnetic and paramagnetic. Give an example of an element that is diamagnetic and one that is paramagnetic. What does it mean when we say that electrons are paired?

7.74 What is meant by the term "shielding of electrons" in an atom? Using the Li atom as an example, describe the effect of shielding on the energy of electrons in an atom.

7.75 Define the following terms and give an example of each: transition metals, lanthanides, actinides.

7.76 Explain why the ground-state electron configurations of Cr and Cu are different from what we might expect.

7.77 Explain what is meant by a noble gas core. Write the electron configuration of a xenon core.

7.78 Comment on the correctness of the following statement: The probability of finding two electrons with the same four quantum numbers in an atom is zero.

Problems

7.79 Indicate which of the following sets of quantum numbers in an atom are unacceptable and explain why: (a) $(1, 0, \frac{1}{2}, \frac{1}{2})$, (b) $(3, 0, 0, +\frac{1}{2})$, (c) $(2, 2, 1, +\frac{1}{2})$, (d) $(4, 3, -2, +\frac{1}{2})$, (e) $(3, 2, 1, 1)$.

7.80 The ground-state electron configurations listed here are incorrect. Explain what mistakes have been made in each and write the correct electron configurations.

Al: $1s^22s^22p^43s^23p^3$

B: $1s^22s^22p^5$

F: $1s^22s^22p^6$

7.81 The atomic number of an element is 73. Is this element diamagnetic or paramagnetic?

7.82 Indicate the number of unpaired electrons present in each of the following atoms: B, Ne, P, Sc, Mn, Se, Kr, Fe, Cd, I, Pb.

7.83 Write the ground-state electron configurations for the following elements: B, V, Ni, As, I, Au.

7.84 Write the ground-state electron configurations for the following elements: Ge, Fe, Zn, Ni, W, Tl.

7.85 The electron configuration of a neutral atom is $1s^22s^22p^63s^2$. Write a complete set of quantum numbers for each of the electrons. Name the element.

7.86 Which of the following species has the most unpaired electrons? S^+, S, or S^-. Explain how you arrive at your answer.

The Aufbau Principle

Review Questions

7.87 State the Aufbau principle and explain the role it plays in classifying the elements in the periodic table.

7.88 Describe the characteristics of the following groups of elements: transition metals, lanthanides, actinides.

7.89 What is the noble gas core? How does it simplify the writing of electron configurations?

7.90 What are the group and period of the element osmium?

Problems

7.91 Use the Aufbau principle to obtain the ground-state electron configuration of selenium.

7.92 Use the Aufbau principle to obtain the ground-state electron configuration of technetium.

Additional Problems

7.93 When a compound containing cesium ion is heated in a Bunsen burner flame, photons with an energy of 4.30×10^{-19} J are emitted. What color is the cesium flame?

7.94 What is the maximum number of electrons in an atom that can have the following quantum numbers? Specify the orbitals in which the electrons would be found. (a) $n = 2$, $m_s = +\frac{1}{2}$; (b) $n = 4$, $m_\ell = +1$; (c) $n = 3$, $\ell = 2$; (d) $n = 2$, $\ell = 0$, $m_s = -\frac{1}{2}$; (e) $n = 4$, $\ell = 3$, $m_\ell = -2$.

7.95 Identify the following individuals and their contributions to the development of quantum theory: Bohr, de Broglie, Einstein, Planck, Heisenberg, Schrödinger.

7.96 What properties of electrons are used in the operation of an electron microscope?

7.97 How many photons at 660 nm must be absorbed to melt 5.0×10^2 g of ice? On average, how many H_2O molecules does one photon convert from ice to water? (*Hint:* It takes 334 J to melt 1 g of ice at 0°C.)

7.98 A certain pitcher's fastballs have been clocked at about 100 mph. (a) Calculate the wavelength of a 0.141-kg baseball (in nm) at this speed. (b) What is the wavelength of a hydrogen atom at the same speed? (1 mile = 1609 m.)

7.99 Considering only the ground-state electron configuration, are there more diamagnetic or paramagnetic elements? Explain.

7.100 A ruby laser produces radiation of wavelength 633 nm in pulses whose duration is 1.00×10^{-9} s. (a) If the laser produces 0.376 J of energy per pulse, how many photons are produced in each pulse? (b) Calculate the power (in watts) delivered by the laser per pulse. (1 W = 1 J/s.)

7.101 A 368-g sample of water absorbs infrared radiation at 1.06×10^4 nm from a carbon dioxide laser. Suppose all the absorbed radiation is converted to heat. Calculate the number of photons at this wavelength required to raise the temperature of the water by 5.00°C.

7.102 Photodissociation of water

$$H_2O(l) + h\nu \longrightarrow H_2(g) + \tfrac{1}{2}O_2(g)$$

has been suggested as a source of hydrogen. The ΔH°_{rxn} for the reaction, calculated from thermochemical data, is 285.8 kJ per mole of water decomposed. Calculate the maximum wavelength (in nm) that would provide the necessary energy. In principle, is it feasible to use sunlight as a source of energy for this process?

7.103 Spectral lines of the Lyman and Balmer series do not overlap. Verify this statement by calculating the longest wavelength associated with the Lyman series and the shortest wavelength associated with the Balmer series (in nm).

7.104 Only a fraction of the electrical energy supplied to a tungsten lightbulb is converted to visible light. The rest of the energy shows up as infrared radiation (that is, heat). A 75-W lightbulb converts 15.0 percent of the energy supplied to it into visible light (assume the wavelength to be 550 nm). How many photons are emitted by the lightbulb per second? (1 W = 1 J/s.)

7.105 A microwave oven operating at 1.22×10^8 nm is used to heat 150 mL of water (roughly the volume of a tea cup) from 20°C to 100°C. Calculate the number of photons needed if 92.0 percent of microwave energy is converted to the thermal energy of water.

7.106 The He^+ ion contains only one electron and is therefore a hydrogen-like ion. Calculate the wavelengths, in increasing order, of the first four transitions in the Balmer series of the He^+ ion. Compare these wavelengths with the same transitions in a H atom. Comment on the differences. (The Rydberg constant for He^+ is 8.72×10^{-18} J.)

7.107 Ozone (O_3) in the stratosphere absorbs the harmful radiation from the sun by undergoing decomposition: $O_3 \longrightarrow O + O_2$. (a) Referring to Table 6.4, calculate the $\Delta H°$ for this process. (b) Calculate the maximum wavelength of photons (in nm) that possess this energy to cause the decomposition of ozone photochemically.

7.108 The retina of a human eye can detect light when radiant energy incident on it is at least 4.0×10^{-17} J. For light of 600-nm wavelength, how many photons does this correspond to?

7.109 An electron in an excited state in a hydrogen atom can return to the ground state in two different ways: (a) via a direct transition in which a photon of wavelength λ_1 is emitted and (b) via an intermediate excited state reached by the emission of a photon of wavelength λ_2. This intermediate excited state then decays to the ground state by emitting another photon of wavelength λ_3. Derive an equation that relates λ_1 to λ_2 and λ_3.

7.110 A photoelectric experiment was performed by separately shining a laser at 450 nm (blue light) and a laser at 560 nm (yellow light) on a clean metal surface and measuring the number and kinetic energy of the ejected electrons. Which light would generate more electrons? Which light would eject electrons with greater kinetic energy? Assume that the same amount of energy is delivered to the metal surface by each laser and that the frequencies of the laser lights exceed the threshold frequency.

7.111 The UV light that is responsible for tanning the skin falls in the 320- to 400-nm region. Calculate the total energy (in joules) absorbed by a person exposed to this radiation for 2.0 h, given that there are 2.0×10^{16} photons hitting Earth's surface per square centimeter per second over a 80-nm (320 nm to 400 nm) range and that the exposed body area is 0.45 m^2. Assume that only half of the radiation is absorbed and the other half is reflected by the body. (*Hint:* Use an average wavelength of 360 nm in calculating the energy of a photon.)

7.112 Calculate the wavelength of a helium atom whose speed is equal to the root-mean-square speed at 20°C.

7.113 The sun is surrounded by a white circle of gaseous material called the corona, which becomes visible during a total eclipse of the sun. The temperature of the corona is in the millions of degrees Celsius, which is high enough to break up molecules and remove some or all of the electrons from atoms. One way astronomers have been able to estimate the temperature of the corona is by studying the emission lines of ions of certain elements. For example, the emission spectrum of Fe^{14+} ions has been recorded and analyzed. Knowing that it takes 3.5×10^4 kJ/mol to convert Fe^{13+} to Fe^{14+}, estimate the temperature of the sun's corona. (*Hint:* The average kinetic energy of one mole of a gas is $\frac{3}{2}RT$.)

7.114 The radioactive Co-60 isotope is used in nuclear medicine to treat certain types of cancer. Calculate the wavelength and frequency of an emitted gamma particle having the energy of 1.29×10^{11} J/mol.

Special Problems

7.115 An electron in a hydrogen atom is excited from the ground state to the $n = 4$ state. Comment on the correctness of the following statements (true or false).

(a) $n = 4$ is the first excited state.

(b) It takes more energy to ionize (remove) the electron from $n = 4$ than from the ground state.

(c) The electron is farther from the nucleus (on average) in $n = 4$ than from the ground state.

(d) The wavelength of light emitted when the electron drops from $n = 4$ to $n = 1$ is longer than that from $n = 4$ to $n = 2$.

(e) The wavelength the atom absorbs in going from $n = 1$ to $n = 4$ is the same as that emitted as it goes from $n = 4$ to $n = 1$.

7.116 When an electron makes a transition between energy levels of a hydrogen atom, there are no restrictions on the initial and final values of the principal quantum number n. However, there is a quantum mechanical rule that restricts the initial and final values of the orbital angular momentum ℓ. This is the *selection rule,* which states that $\Delta\ell = \pm 1$, that is, in a transition, the value of ℓ can only increase or decrease by one. According to this rule, which of the following transitions are allowed: (a) $1s \longrightarrow 2s$, (b) $2p \longrightarrow 1s$, (c) $1s \longrightarrow 3d$, (d) $3d \longrightarrow 4f$, (e) $4d \longrightarrow 3s$?

7.117 For hydrogenlike ions, that is, ions containing only one electron, Equation (7.5) is modified as follows: $E_n = -R_H Z^2(1/n^2)$, where Z is the atomic number of the parent atom. The figure here represents the emission spectrum of such a hydrogenlike ion in the gas phase. All the lines result from the electronic transitions from the excited states to the $n = 2$ state. (a) What electronic transitions correspond to lines B and C? (b) If the wavelength of line C is 27.1 nm, calculate the wavelengths of lines A and B. (c) Calculate the energy needed to remove the electron from the ion in the $n = 4$ state. (d) What is the physical significance of the continuum?

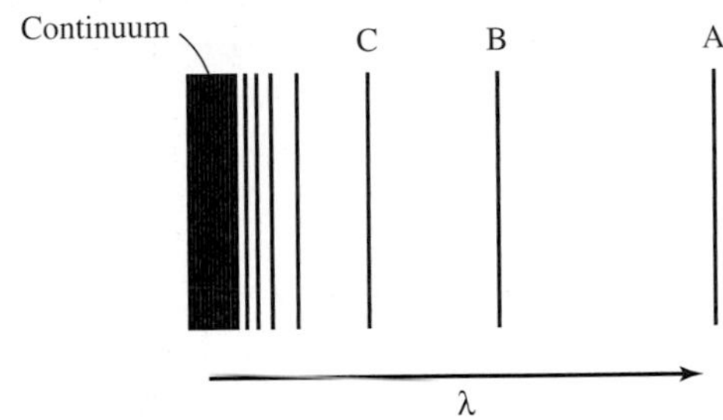

7.118 Calculate the energies needed to remove an electron from the $n = 1$ state and the $n = 5$ state in the Li^{2+} ion. What is the wavelength (in nm) of the emitted photon in a transition from $n = 5$ to $n = 1$? The Rydberg constant for hydrogenlike ions is $(2.18 \times 10^{-18}\ J)Z^2$, where Z is the atomic number.

7.119 According to Einstein's special theory of relativity, the mass of a moving particle, m_{moving}, is related to its mass at rest, m_{rest}, by the following equation

$$m_{moving} = \frac{m_{rest}}{\sqrt{1 - \left(\frac{u}{c}\right)^2}}$$

where u and c are the speeds of the particle and light, respectively. (a) In particle accelerators, protons, electrons, and other charged particles are often accelerated to speeds close to the speed of light. Calculate the wavelength (in nm) of a proton moving at 50.0 percent the speed of light. The mass of a proton is 1.673×10^{-27} kg. (b) Calculate the mass of a 6.0×10^{-2} kg tennis ball moving at 63 m/s. Comment on your results.

7.120 The mathematical equation for studying the photoelectric effect is

$$h\nu = W + \tfrac{1}{2}m_e u^2$$

where ν is the frequency of light shining on the metal, W is the work function (see p. 217), m_e and u are the mass and speed of the ejected electron. In an experiment, a student found that a maximum wavelength of 351 nm is needed to just dislodge electrons from a zinc metal surface. Calculate the velocity (in m/s) of an ejected electron when she employed light with a wavelength of 313 nm.

7.121 Blackbody radiation is the term used to describe the dependence of the radiation energy emitted by an object on wavelength at a certain temperature. Planck proposed the quantum theory to account for this dependence. Shown in the figure is a plot of the radiation energy emitted by our sun versus wavelength. This curve is characteristic of the temperature at the surface of the sun. At a higher temperature, the curve has a similar shape but the maximum will shift to a shorter wavelength. What does this curve reveal about two consequences of great biological significance on Earth?

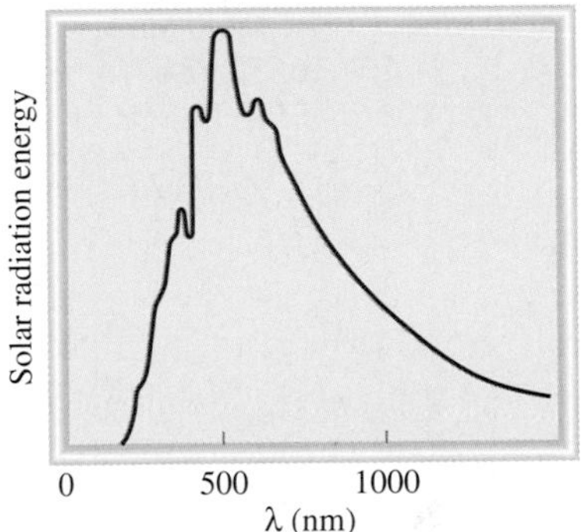

7.122 All molecules undergo vibrational motions. Quantum mechanical treatment shows that the vibrational energy, E_{vib}, of a diatomic molecule like HCl is given by

$$E_{vib} = \left(n + \frac{1}{2}\right)h\nu$$

where n is a quantum given by $n = 0, 1, 2, 3, \ldots$ and ν is the fundamental frequency of vibration. (a) Sketch the first three vibrational energy levels for HCl. (b) Calculate the energy required to excite a HCl molecule from the ground level to the first excited level. The fundamental frequency of vibration for HCl is $8.66 \times 10^{13}\ s^{-1}$. (c) The fact that the lowest vibrational energy in the ground level is not zero but equal to $\frac{1}{2}h\nu$ means that molecules will vibrate at all temperatures, including the absolute zero. Use the Heisenberg uncertainty principle to justify this prediction. (*Hint:* Consider a nonvibrating molecule and predict the uncertainty in the momentum and hence the uncertainty in the position.)

7.123 According to Wien's law, the wavelength of maximum intensity in blackbody radiation, λ_{max}, is given by

$$\lambda_{max} = \frac{b}{T}$$

where b is a constant (2.898×10^6 nm · K) and T is the temperature of the radiating body in kelvins. (a) Estimate the temperature at the surface of the sun. (b) How are astronomers able to determine the temperature of stars in general? (*Hint:* See Problem 7.121.)

7.124 The wave function for the 2*s* orbital in the hydrogen atom is

$$\psi_{2s} = \frac{1}{\sqrt{2a_0^3}}\left(1 - \frac{\rho}{2}\right)e^{-\rho/2}$$

where a_0 is the value of the radius of the first Bohr orbit, equal to 0.529 nm, ρ is $Z(r/a_0)$, and r is the distance from the nucleus in meters. Calculate the location of the node of the 2*s* wave function from the nucleus.

Answers to Practice Exercises

7.1 8.24 m. **7.2** 3.39×10^3 nm. **7.3** 9.65×10^{-19} J. **7.4** 2.63×10^3 nm. **7.5** 56.6 nm. **7.6** $n = 3$, $\ell = 1$, $m_\ell = -1, 0, 1$. **7.7** 16. **7.8** $(4, 2, -2, +\frac{1}{2})$, $(4, 2, -1, +\frac{1}{2})$, $(4, 2, 0, +\frac{1}{2})$, $(4, 2, 1, +\frac{1}{2})$, $(4, 2, 2, +\frac{1}{2})$, $(4, 2, -2, -\frac{1}{2})$, $(4, 2, -1, -\frac{1}{2})$, $(4, 2, 0 -\frac{1}{2})$, $(4, 2, 1, -\frac{1}{2})$, $(4, 2, 2, -\frac{1}{2})$. **7.9** 32. **7.10** $(1, 0, 0, +\frac{1}{2})$, $(1, 0, 0, -\frac{1}{2})$, $(2, 0, 0, +\frac{1}{2})$, $(2, 0, 0, -\frac{1}{2})$, $(2, 1, -1, -\frac{1}{2})$. There are five other acceptable ways to write the quantum numbers for the last electron (in the 2*p* orbital). **7.11** $[Ne]3s^23p^3$.

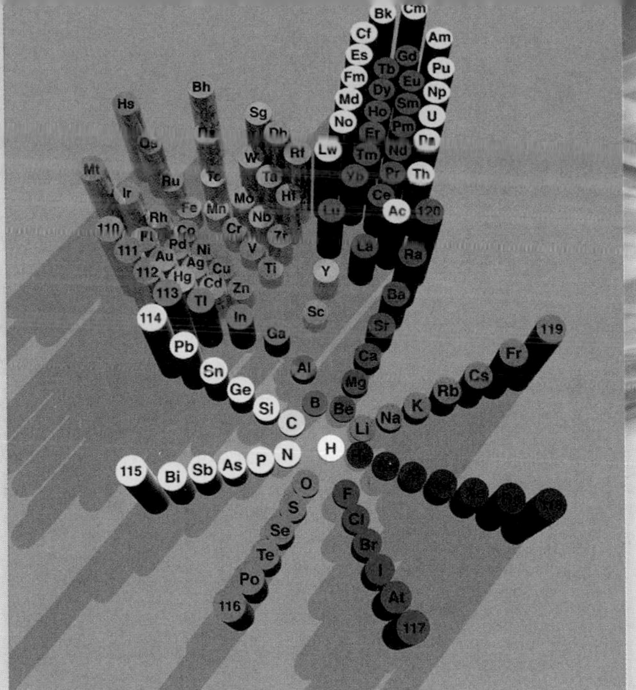

While the recurring or "periodic" trends in the properties of elements are most commonly illustrated in tabular form, alternative geometric arrangements are possible.

CHAPTER

8

The Periodic Table

Chapter Outline

Student Interactive Activities

Animations
Atomic and Ionic Radius (8.3)

Electronic Homework
Example Practice Problems
End of Chapter Problems

Essential Concepts

Development of the Periodic Table In the nineteenth century, chemists noticed a regular, periodic recurrence of chemical and physical properties of elements. In particular, the periodic table drawn up by Mendeleev grouped the elements accurately and was able to predict the properties of several elements that had not yet been discovered.

Periodic Classification of the Elements Elements are grouped according to their outer-shell electron configurations, which account for their similar chemical behavior. Special names are assigned to these various groups.

Periodic Variation in Properties Overall, physical properties such as atomic and ionic radii of the elements vary in a regular and periodic fashion. Similar variation is also noted in their chemical properties. Chemical properties of special importance are ionization energy, which measures the tendency of an atom of an element to lose an electron, and electron affinity, which measures the tendency of an atom to accept an electron. Ionization energy and electron affinity form the basis for understanding chemical bond formation.

8.1 Development of the Periodic Table

In the nineteenth century, when chemists had only a vague idea of atoms and molecules and did not know of the existence of electrons and protons, they devised the periodic table using their knowledge of atomic masses. Accurate measurements of the atomic masses of many elements had already been made. Arranging elements according to their atomic masses in a periodic table seemed logical to chemists, who felt that chemical behavior should somehow be related to atomic mass.

In 1864 the English chemist John Newlands noticed that when the known elements were arranged in order of atomic mass, every eighth element had similar properties. Newlands referred to this peculiar relationship as the *law of octaves*. However, this "law" turned out to be inadequate for elements beyond calcium, and Newlands's work was not accepted by the scientific community.

Five years later the Russian chemist Dmitri Mendeleev and the German chemist Lothar Meyer independently proposed a much more extensive tabulation of the elements, based on the regular, periodic recurrence of properties. Mendeleev's classification was a great improvement over Newlands's for two reasons. First, it grouped the elements together more accurately, according to their properties. Equally important, it made possible the prediction of the properties of several elements that had not yet been discovered. For example, Mendeleev proposed the existence of an unknown element that he called eka-aluminum. (*Eka* is a Sanskrit word meaning "first"; thus, eka-aluminum would be the first element under aluminum in the same group.) When gallium was discovered 4 years later, its properties closely matched the predicted properties of eka-aluminum as shown here.

	Eka-Aluminum (Ea)	Gallium (Ga)
Atomic mass	68 amu	69.9 amu
Melting point	Low	29.78°C
Density	5.9 g/cm^3	5.94 g/cm^3
Formula of oxide	Ea_2O_3	Ga_2O_3

Gallium melts in a person's hand (body temperature is about 37°C).

Nevertheless, the early versions of the periodic table had some glaring inconsistencies. For example, the atomic mass of argon (39.95 amu) is greater than that of potassium (39.10 amu). If elements were arranged solely according to increasing atomic mass, argon would appear in the position occupied by potassium in our modern periodic table (see the inside front cover). But no chemist would place argon, an inert gas, in the same group as lithium and sodium, two very reactive metals. This and other discrepancies suggested that some fundamental property other than atomic mass is the basis of the observed periodicity. This property turned out to be associated with atomic number.

Using data from α-scattering experiments (see Section 2.2), Rutherford was able to estimate the number of positive charges in the nucleus of a few elements, but until 1913 there was no general procedure for determining atomic numbers. In that year a young English physicist, Henry Moseley, discovered a correlation between atomic number and the frequency of X rays generated by the bombardment of the element under study with high-energy electrons. With a few exceptions, Moseley found that the order of increasing atomic number is the same as the order of increasing atomic mass. For example, calcium is the twentieth element in increasing order of atomic mass, and it has an atomic number of 20. The discrepancies that bothered scientists now made sense. The atomic number of argon is 18 and that of potassium is 19, so potassium should follow argon in the periodic table.

Appendix 4 explains the names and symbols of the elements.

A modern periodic table usually shows the atomic number along with the element symbol. As you already know, the atomic number also indicates the number of electrons in the atoms of an element. Electron configurations of elements help explain the recurrence of physical and chemical properties. The importance and usefulness of the periodic table lie in the fact that we can use our understanding of the general properties and trends within a group or a period to predict with considerable accuracy the properties of any element, even though that element may be unfamiliar to us.

8.2 Periodic Classification of the Elements

Figure 8.1 shows the periodic table together with the outermost ground-state electron configurations of the elements. (The electron configurations of the elements are also given in Table 7.3.) Starting with hydrogen, we see that the subshells are filled in the order shown in Figure 7.21. According to the type of subshell being filled, the elements can be divided into categories—the representative elements, the noble gases, the transition elements (or transition metals), the lanthanides, and the actinides. Referring to Figure 8.1, the ***representative elements*** (also called *main group elements*) are *the elements in Groups 1A through 7A, all of which have incompletely filled s or p subshells of the highest principal quantum number.* With the exception of helium, the *noble gases* (the Group 8A elements) all have a completely filled *p* subshell. (The electron configurations are $1s^2$ for helium and ns^2np^6 for the other noble gases, in which *n* is the principal quantum number for the outermost shell.) The transition metals are the elements in Groups 1B and 3B through 8B, which have incompletely filled *d* subshells or readily produce cations with incompletely filled *d*

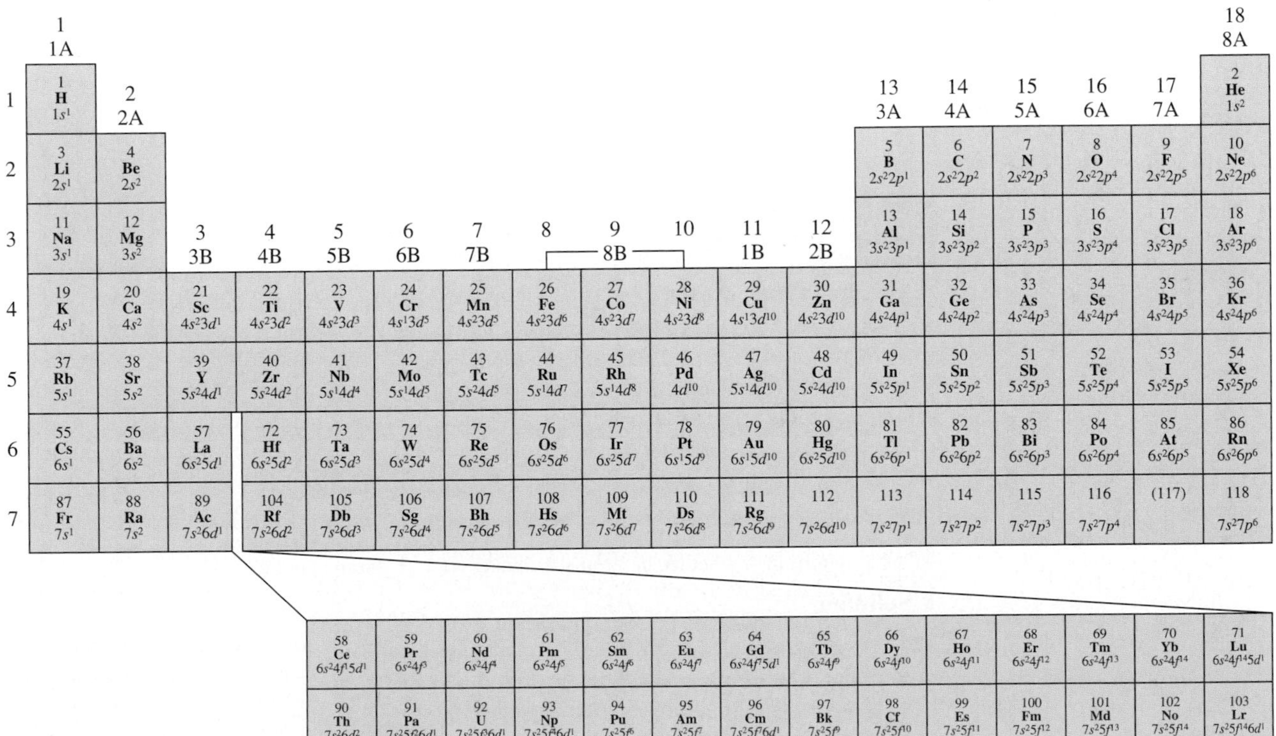

Figure 8.1
The ground-state electron configurations of the elements. For simplicity, only the configurations of the outer electrons are shown.

Table 8.1

Electron Configurations of Group 1A and Group 2A Elements

Group 1A	Group 2A
Li [He]$2s^1$	Be [He]$2s^2$
Na [Ne]$3s^1$	Mg [Ne]$3s^2$
K [Ar]$4s^1$	Ca [Ar]$4s^2$
Rb [Kr]$5s^1$	Sr [Kr]$5s^2$
Cs [Xe]$6s^1$	Ba [Xe]$6s^2$
Fr [Rn]$7s^1$	Ra [Rn]$7s^2$

For the representative elements, the valence electrons are simply those electrons at the highest principal energy level *n*.

subshells. (These metals are sometimes referred to as the *d*-block transition elements.) The Group 2B elements are Zn, Cd, and Hg, which are neither representative elements nor transition metals. The lanthanides and actinides are sometimes called *f*-block transition elements because they have incompletely filled *f* subshells.

A clear pattern emerges when we examine the electron configurations of the elements in a particular group. The electron configurations for Groups 1A and 2A elements are shown in Table 8.1. We see that all members of the Group 1A alkali metals have similar outer electron configurations; each has a noble gas core and an ns^1 configuration of the outer electron. Similarly, the Group 2A alkaline earth metals have a noble gas core and an ns^2 configuration of the outer electrons. *The outer electrons of an atom, which are those involved in chemical bonding,* are often called the ***valence electrons.*** Having the same number of valence electrons accounts for similarities in chemical behavior among the elements within each of these groups. This observation holds true also for the halogens (the Group 7A elements), which have outer electron configurations of ns^2np^5 and exhibit very similar properties. We must be careful, however, in predicting properties for Groups 3A through 6A. For example, the elements in Group 4A all have the same outer electron configuration, ns^2np^2, but there is much variation in chemical properties among these elements: Carbon is a nonmetal, silicon and germanium are metalloids, and tin and lead are metals.

As a group, the noble gases behave very similarly. With the exception of krypton and xenon, these elements are totally inert chemically. The reason is that these elements all have completely filled outer ns^2np^6 subshells, a condition that represents great stability. Although the outer electron configuration of the transition metals is not always the same within a group and there is no regular pattern in the change of the electron configuration from one metal to the next in the same period, all transition metals share many characteristics that set them apart from other elements. The reason is that these metals all have an incompletely filled *d* subshell. Likewise, the lanthanide (and the actinide) elements resemble one another within the series because they have incompletely filled *f* subshells. Figure 8.2 distinguishes the groups of elements discussed here.

EXAMPLE 8.1

An atom of a certain element has 14 electrons. Without consulting a periodic table, answer the following questions: (a) What is the ground-state configuration of the element? (b) How should the element be classified? (c) Are the atoms of the element diamagnetic or paramagnetic?

Strategy (a) We refer to the building-up principle (the Aufbau principle) discussed in Section 7.9 and start writing the electron configuration with principal quantum number $n = 1$ and continuing upward until all the electrons are accounted for. (b) What are the electron configuration characteristics of representative elements? transition elements? noble gases? (c) Examine the pairing scheme of the electrons in the outermost shell. What determines whether an atom is diamagnetic or paramagnetic?

Solution

(a) We know that for $n = 1$, we have a 1*s* orbital (2 electrons); for $n = 2$, we have a 2*s* orbital (2 electrons) and three 2*p* orbitals (6 electrons); for $n = 3$, we have a 3*s* orbital (2 electrons). The number of electrons left is $14 - 12 = 2$ and these two electrons are placed in the 3*p* orbitals. The electron configuration is $1s^22s^22p^63s^23p^2$.

(Continued)

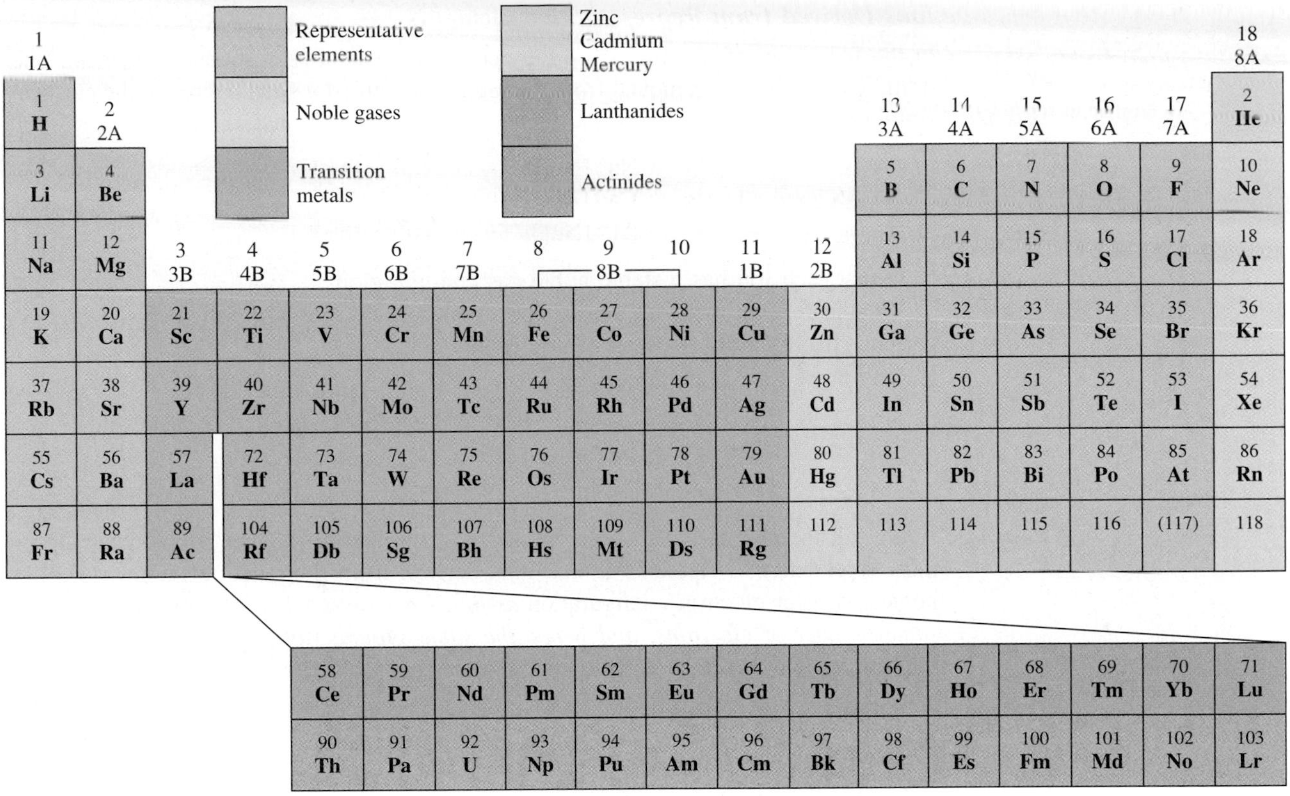

1 1A	2 2A	3 3B	4 4B	5 5B	6 6B	7 7B	8	9 8B	10	11 1B	12 2B	13 3A	14 4A	15 5A	16 6A	17 7A	18 8A
1 H																	2 He
3 Li	4 Be											5 B	6 C	7 N	8 O	9 F	10 Ne
11 Na	12 Mg											13 Al	14 Si	15 P	16 S	17 Cl	18 Ar
19 K	20 Ca	21 Sc	22 Ti	23 V	24 Cr	25 Mn	26 Fe	27 Co	28 Ni	29 Cu	30 Zn	31 Ga	32 Ge	33 As	34 Se	35 Br	36 Kr
37 Rb	38 Sr	39 Y	40 Zr	41 Nb	42 Mo	43 Tc	44 Ru	45 Rh	46 Pd	47 Ag	48 Cd	49 In	50 Sn	51 Sb	52 Te	53 I	54 Xe
55 Cs	56 Ba	57 La	72 Hf	73 Ta	74 W	75 Re	76 Os	77 Ir	78 Pt	79 Au	80 Hg	81 Tl	82 Pb	83 Bi	84 Po	85 At	86 Rn
87 Fr	88 Ra	89 Ac	104 Rf	105 Db	106 Sg	107 Bh	108 Hs	109 Mt	110 Ds	111 Rg	112	113	114	115	116	(117)	118

58 Ce	59 Pr	60 Nd	61 Pm	62 Sm	63 Eu	64 Gd	65 Tb	66 Dy	67 Ho	68 Er	69 Tm	70 Yb	71 Lu
90 Th	91 Pa	92 U	93 Np	94 Pu	95 Am	96 Cm	97 Bk	98 Cf	99 Es	100 Fm	101 Md	102 No	103 Lr

Figure 8.2
Classification of the elements. Note that the Group 2B elements are often classified as transition metals even though they do not exhibit the characteristics of the transition metals.

(b) Because the $3p$ subshell is not completely filled, this is a representative element. Based on the information given, we cannot say whether it is a metal, a nonmetal, or a metalloid.

(c) According to Hund's rule, the two electrons in the $3p$ orbitals have parallel spins (two unpaired electrons). Therefore, the atoms of the element are paramagnetic.

Check For (b), note that a transition metal possesses an incompletely filled d subshell and a noble gas has a completely filled outer shell. For (c), recall that if the atoms of an element contain an even number of electrons, then they may or may not be paramagnetic.

Practice Exercise An atom of a certain element has 20 electrons. (a) Write the ground-state electron configuration of the element, (b) classify the element and, (c) determine whether the atoms of the element are diamagnetic or paramagnetic.

Similar problem: 8.16.

Electron Configurations of Cations and Anions

Because many ionic compounds are made up of monatomic anions and/or cations, it is helpful to know how to write the electron configurations of these ionic species. The procedure for writing the electron configurations of ions requires only a slight extension of the method used for neutral atoms. We will group the ions in two categories for discussion.

Ions Derived from Representative Elements

In the formation of a cation from the neutral atom of a representative element, one or more electrons are removed from the highest occupied n shell. Here are the electron configurations of some atoms and their corresponding cations:

$$\begin{array}{ll} \text{Na: [Ne]}3s^1 & \text{Na}^+\text{: [Ne]} \\ \text{Ca: [Ar]}4s^2 & \text{Ca}^{2+}\text{: [Ar]} \\ \text{Al: [Ne]}3s^23p^1 & \text{Al}^{3+}\text{: [Ne]} \end{array}$$

Note that each ion has a stable noble gas configuration.

In the formation of an anion, one or more electrons are added to the highest partially filled n shell. Consider these examples:

$$\begin{array}{ll} \text{H: }1s^1 & \text{H}^-\text{: }1s^2 \text{ or [He]} \\ \text{F: }1s^22s^22p^5 & \text{F}^-\text{: }1s^22s^22p^6 \text{ or [Ne]} \\ \text{O: }1s^22s^22p^4 & \text{O}^{2-}\text{: }1s^22s^22p^6 \text{ or [Ne]} \\ \text{N: }1s^22s^22p^3 & \text{N}^{3-}\text{: }1s^22s^22p^6 \text{ or [Ne]} \end{array}$$

Again, all the anions have stable noble gas configurations. Thus, a characteristic of most representative elements is that ions derived from their neutral atoms have the noble gas outer electron configuration ns^2np^6. *Ions, or atoms and ions, that have the same number of electrons, and hence the same ground-state electron configuration, are said to be* ***isoelectronic***. Thus, H^- and He are isoelectronic; F^-, Na^+, and Ne are isoelectronic; and so on.

Cations Derived from Transition Metals

In Section 7.9 we saw that in the first-row transition metals (Sc to Cu), the $4s$ orbital is always filled before the $3d$ orbitals. Consider manganese, whose electron configuration is $[Ar]4s^23d^5$. When the Mn^{2+} ion is formed, we might expect the two electrons to be removed from the $3d$ orbitals to yield $[Ar]4s^23d^3$. In fact, the electron configuration of Mn^{2+} is $[Ar]3d^5$! The reason is that the electron-electron and electron-nucleus interactions in a neutral atom can be quite different from those in its ion. Thus, whereas the $4s$ orbital is always filled before the $3d$ orbital in Mn, electrons are removed from the $4s$ orbital in forming Mn^{2+} because the $3d$ orbital is more stable than the $4s$ orbital in transition metal ions. Therefore, when a cation is formed from an atom of a transition metal, electrons are always removed first from the ns orbital and then from the $(n - 1)d$ orbitals.

Bear in mind that the order of electron filling does not determine or predict the order of electron removal for transition metals. For these metals, the ns electrons are lost before the $(n - 1)d$ electrons.

Keep in mind that most transition metals can form more than one cation and that frequently the cations are not isoelectronic with the preceding noble gases.

REVIEW OF CONCEPTS

What must the charge be on a Fe ion that is isoelectronic with Co^{3+}?

8.3 Periodic Variation in Physical Properties

As we have seen, the electron configurations of the elements show a periodic variation with increasing atomic number. Consequently, there are also periodic variations in physical and chemical behavior. In this section and Sections 8.4 and 8.5, we will examine some physical properties of elements that are in the same group or period and additional properties that influence the chemical behavior of the elements.

First, let's look at the concept of effective nuclear charge, which has a direct bearing on many atomic properties.

Effective Nuclear Charge

In Chapter 7, we discussed the shielding effect that electrons close to the nucleus have on outer-shell electrons in many-electron atoms. The presence of other electrons in an atom reduces the electrostatic attraction between a given electron and the positively charged protons in the nucleus. The ***effective nuclear charge*** **(Z_{eff})** *is the nuclear charge felt by an electron when both the actual nuclear charge (Z) and the repulsive effects (shielding) of the other electrons are taken into account.* In general, Z_{eff} is given by

$$Z_{eff} = Z - \sigma \tag{8.1}$$

where σ (sigma) is called the *shielding constant* (also called the *screening constant*). The shielding constant is greater than zero but smaller than Z.

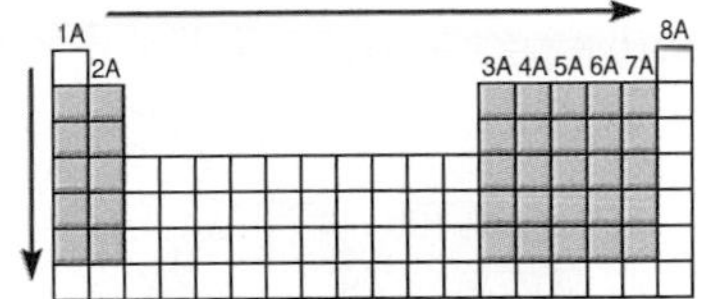

The increase in effective nuclear charge from left to right across a period and from top to bottom in a group for representative elements.

One way to illustrate how electrons in an atom shield one another is to consider the amounts of energy required to remove the two electrons from a helium atom. Experiments show that it takes 3.94×10^{-18} J to remove the first electron and 8.72×10^{-18} J to remove the second electron. There is no shielding once the first electron is removed, so the second electron feels the full effect of the +2 nuclear charge.

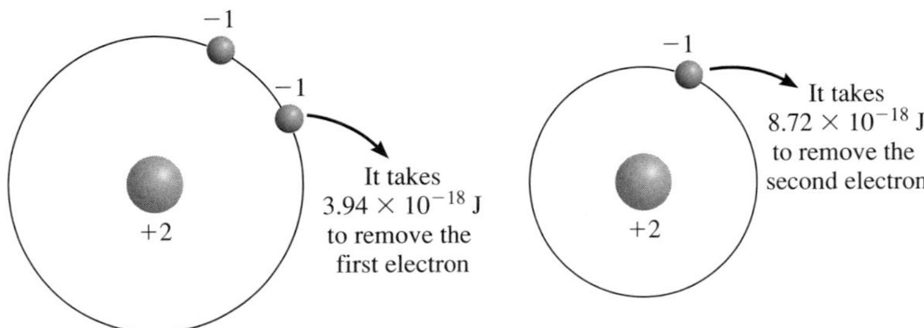

Because the core electrons are, on average, closer to the nucleus than valence electrons, core electrons shield valence electrons much more than valence electrons shield one another. Consider the second-period elements from Li to Ne. Moving from left to right, we find the number of core electrons ($1s^2$) remains constant while the nuclear charge increases. However, because the added electron is a valence electron and valence electrons do not shield each other well, the net effect of moving across the period is a greater effective nuclear charge felt by the valence electrons, as shown here.

See Figure 7.24 for radial probability plots of 1*s* and 2*s* orbitals.

	Li	**Be**	**B**	**C**	**N**	**O**	**F**	**Ne**
Z	3	4	5	6	7	8	9	10
Z_{eff}	1.28	1.91	2.42	3.14	3.83	4.45	5.10	5.76

The effective nuclear charge also increases as we go down a particular periodic group. However, because the valence electrons are now added to increasingly large shells as *n* increases, the electrostatic attraction between the nucleus and the valence electrons actually decreases.

Atomic Radius

A number of physical properties, including density, melting point, and boiling point, are related to the sizes of atoms, but atomic size is difficult to define. As we saw in Chapter 7, the electron density in an atom extends far beyond the nucleus, but we normally think of atomic size as the volume containing about 90 percent of the total electron density around the nucleus. When we must be even more specific, we define the size of an atom in terms of its ***atomic radius,*** which is *one-half the distance between the two nuclei in two adjacent metal atoms.*

For atoms linked together to form an extensive three-dimensional network, atomic radius is simply one-half the distance between the nuclei in two neighboring atoms [Figure 8.3(a)]. For elements that exist as simple diatomic molecules, the atomic radius is one-half the distance between the nuclei of the two atoms in a particular molecule [Figure 8.3(b)].

Figure 8.4 shows the atomic radius of many elements according to their positions in the periodic table, and Figure 8.5 plots the atomic radii of these elements against their atomic numbers. Periodic trends are clearly evident. Consider the second-period elements. Because the effective nuclear charge increases from left to right, the added valence electron at each step is more strongly attracted by the nucleus than the one before. Therefore, we expect and indeed find the atomic radius decreases from Li to Ne. Within a group, we find that atomic radius increases with atomic number. For the alkali metals in Group 1A, the valence electron resides in the *ns* orbital. Because orbital

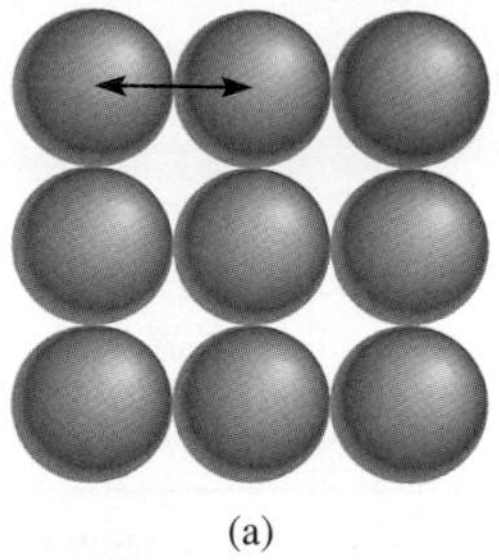

(a)

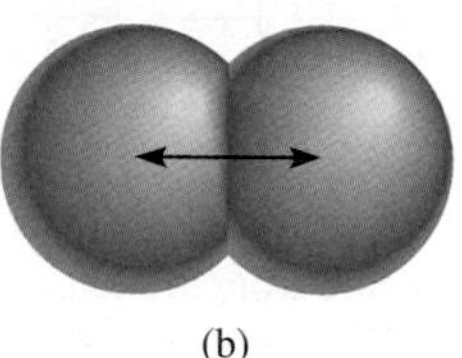

(b)

Figure 8.3

(a) In metals such as beryllium, the atomic radius is defined as one-half the distance between the centers of two adjacent atoms. (b) For elements that exist as diatomic molecules, such as iodine, the radius of the atom is defined as one-half the distance between the centers of the atoms in the molecules.

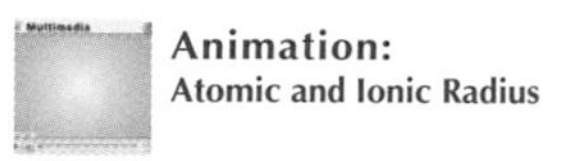

Animation:
Atomic and Ionic Radius

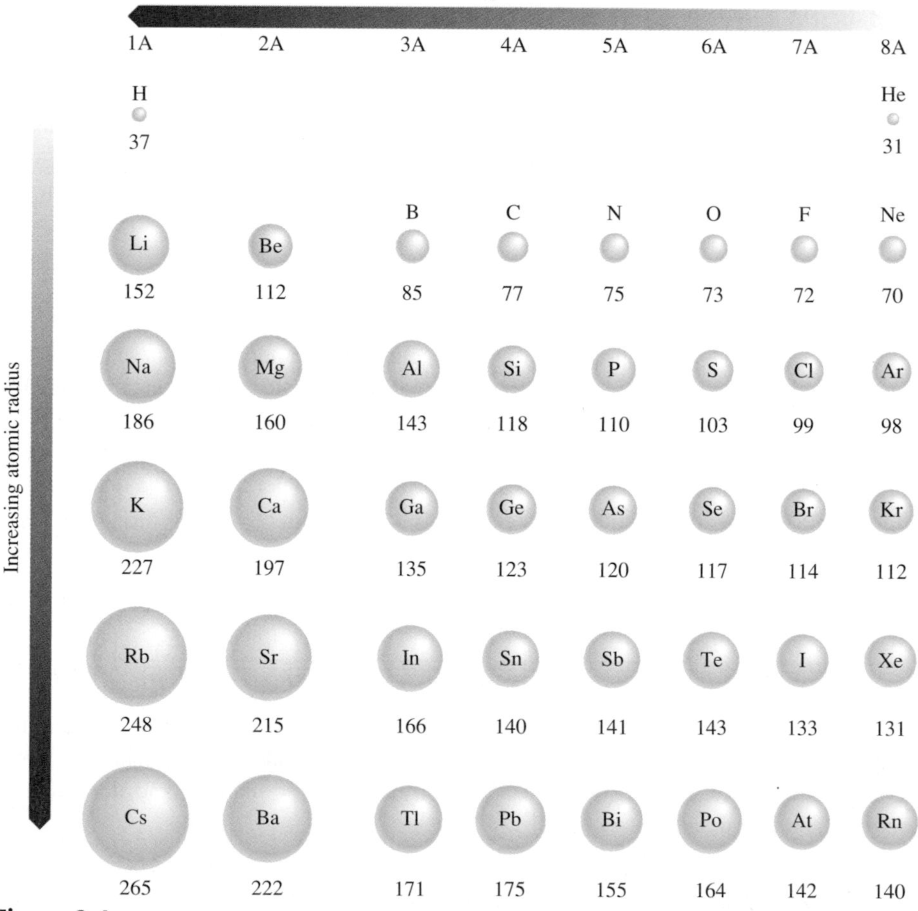

Figure 8.4

Atomic radii (in picometers) of representative elements according to their positions in the periodic table. Note that there is no general agreement on the size of atomic radii. We focus only on the trends in atomic radii, not on their precise values.

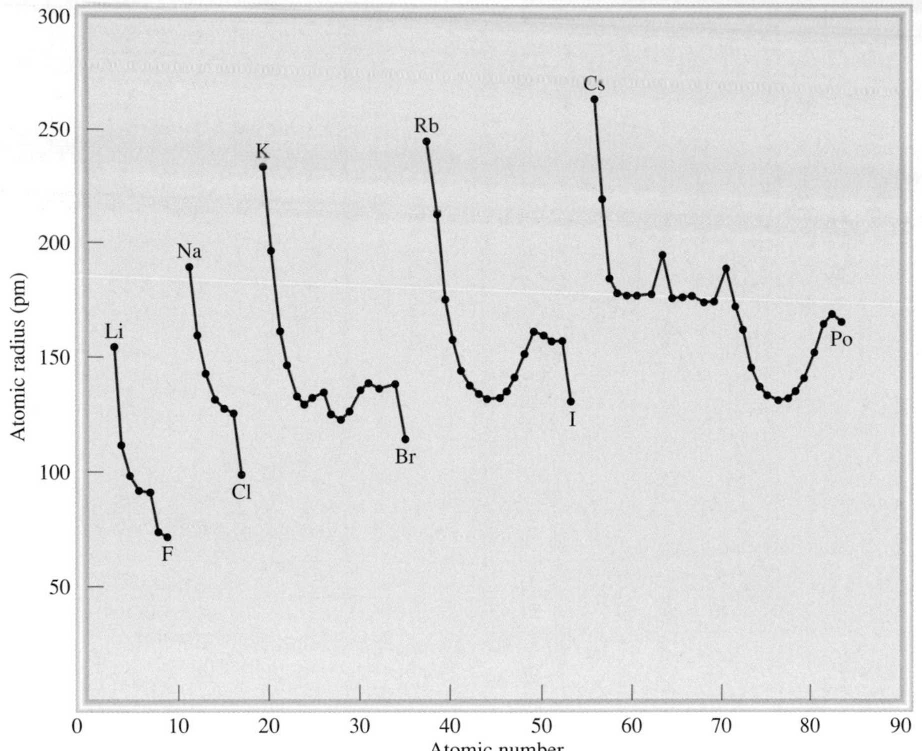

Figure 8.5
Plot of atomic radii (in picometers) of elements against their atomic numbers.

size increases with the increasing principal quantum number *n,* the size of the atomic radius increases from Li to Cs.

EXAMPLE 8.2

Referring to a periodic table, arrange the following atoms in order of increasing atomic radius: P, Si, N.

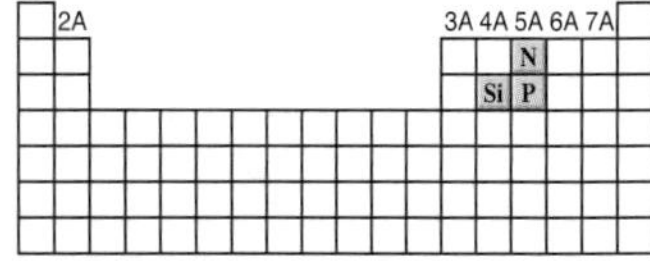

Strategy What are the trends in atomic radii in a periodic group and in a particular period? Which of the preceding elements are in the same group? in the same period?

Solution From Figure 8.2 we see that N and P are in the same group (Group 5A). Therefore, the radius of N is smaller than that of P (atomic radius increases as we go down a group). Both Si and P are in the third period, and Si is to the left of P. Therefore, the radius of P is smaller than that of Si (atomic radius decreases as we move from left to right across a period). Thus, the order of increasing radius is $N < P < Si$.

Similar problems: 8.37, 8.38.

Practice Exercise Arrange the following atoms in order of decreasing radius: C, Li, Be.

Ionic Radius

Ionic radius is *the radius of a cation or an anion.* Ionic radius affects the physical and chemical properties of an ionic compound. For example, the three-dimensional structure of an ionic compound depends on the relative sizes of its cations and anions.

When a neutral atom is converted to an ion, we expect a change in size. If the atom forms an anion, its size (or radius) increases, because the nuclear charge remains the same but the repulsion resulting from the additional electron(s) enlarges the domain of the electron cloud. On the other hand, removing one or more electrons from an atom reduces electron-electron repulsion but the nuclear charge

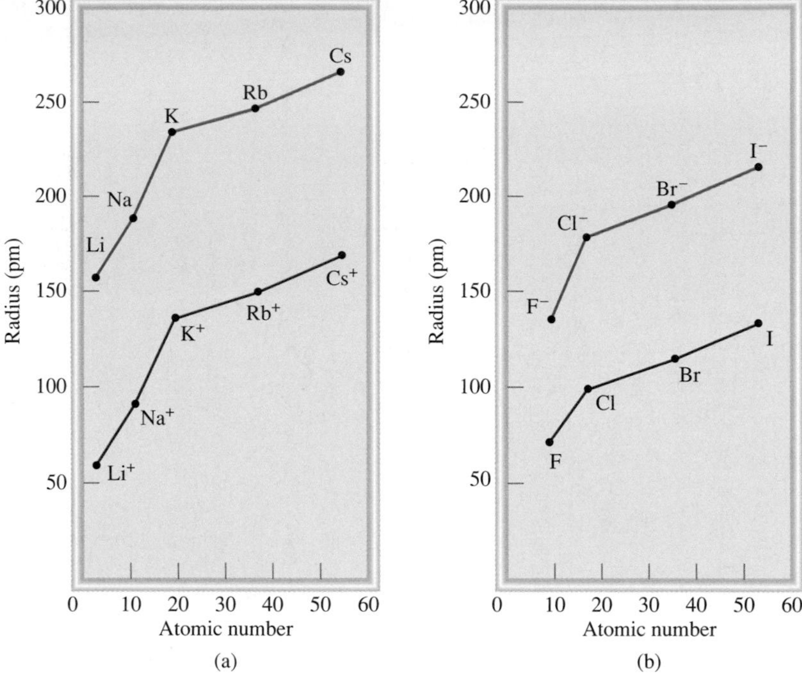

Figure 8.6
Comparison of atomic radii with ionic radii. (a) Alkali metals and alkali metal cations. (b) Halogens and halide ions.

remains the same, so the electron cloud shrinks, and the cation is smaller than the atom. Figure 8.6 shows the changes in size that result when alkali metals are converted to cations and halogens are converted to anions; Figure 8.7 shows the changes in size that occur when a lithium atom reacts with a fluorine atom to form a LiF unit.

Figure 8.8 shows the radii of ions derived from the familiar elements, arranged according to elements' positions in the periodic table. We can see parallel trends between atomic radii and ionic radii. For example, from top to bottom both the atomic radius and the ionic radius increase within a group. For ions derived from elements in different groups, a size comparison is meaningful only if the ions are isoelectronic. If we examine isoelectronic ions, we find that cations are smaller than anions. For example, Na^+ is smaller than F^-. Both ions have the same number of electrons, but Na ($Z = 11$) has more protons than F ($Z = 9$). The larger effective nuclear charge of Na^+ results in a smaller radius.

For isoelectronic ions, the size of the ion is based on the size of the electron cloud, not on the number of protons in the nucleus.

Focusing on isoelectronic cations, we see that the radii of *tripositive ions* (ions that bear three positive charges) are smaller than those of *dipositive ions* (ions that bear two positive charges), which in turn are smaller than *unipositive ions* (ions that bear one positive charge). This trend is nicely illustrated by the sizes of three isoelectronic ions in the third period: Al^{3+}, Mg^{2+}, and Na^+ (see Figure 8.8). The Al^{3+} ion has the same number of electrons as Mg^{2+}, but it has one more proton. Thus, the electron cloud in Al^{3+} is pulled inward more than that in Mg^{2+}. The smaller radius of Mg^{2+} compared with that of Na^{2+} can be similarly explained. Turning to isoelectronic anions,

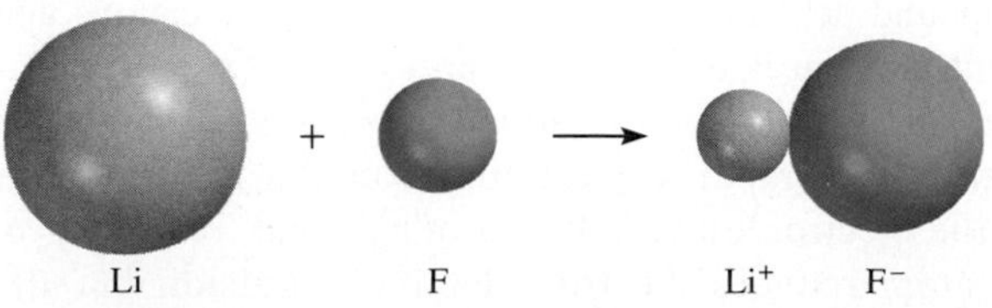

Figure 8.7
Changes in the sizes of Li and F when they react to form LiF.

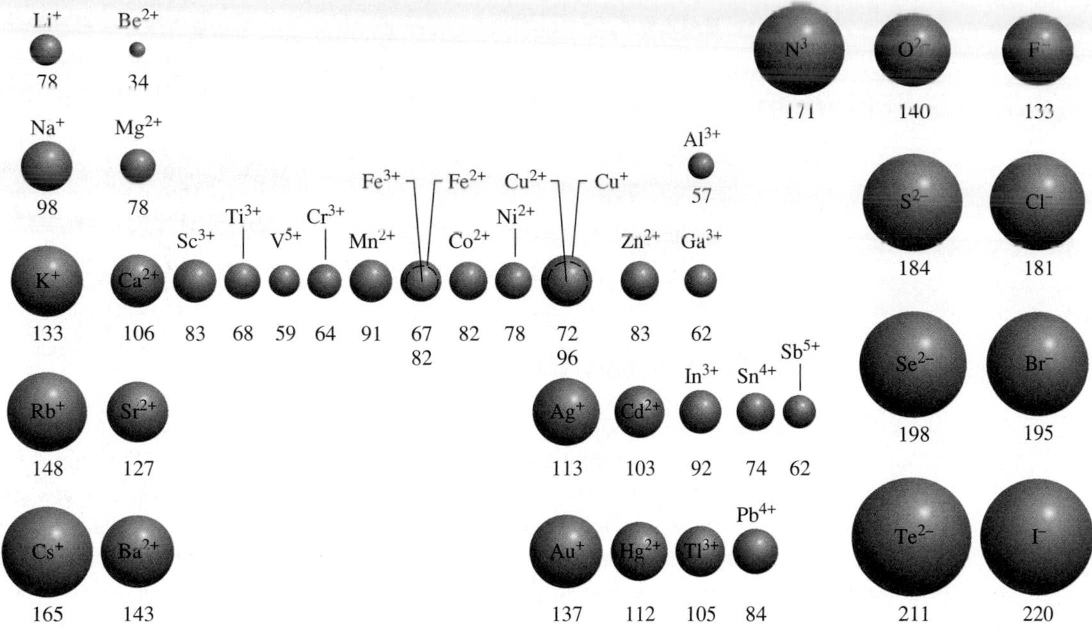

Figure 8.8

The radii (in picometers) of ions of familiar elements arranged according to the elements' positions in the periodic table.

we find that the radius increases as we go from ions with uninegative charge (−) to those with dinegative charge (2−), and so on. Thus, the oxide ion is larger than the fluoride ion because oxygen has one fewer proton than fluorine; the electron cloud is spread out more in O^{2-}.

EXAMPLE 8.3

For each of the following pairs, indicate which one of the two species is larger: (a) O^{2-} or F^-; (b) Sr^{2+} or Ca^{2+}; (c) Co^{3+} or Co^{2+}.

Strategy In comparing ionic radii, it is useful to classify the ions into three categories: (1) isoelectronic ions, (2) ions that carry the same charges and are generated from atoms of the same periodic group, and (3) ions carrying different charges but are generated from the same atom. In case (1), ions carrying a greater negative charge are always larger; in case (2), ions from atoms having a greater atomic number are always larger; in case (3), ions having a smaller positive charge are always larger.

Solution

(a) O^{2-} and F^- are isoelectronic anions, both containing 10 electrons. Because O^{2-} has only eight protons and F^- has nine, the smaller attraction exerted by the nucleus on the electrons results in a larger O^{2-} ion.

(b) Both Ca and Sr belong to Group 2A (the alkaline earth metals). Thus, the Sr^{2+} ion is larger than Ca^{2+} because Sr's valence electrons are in a larger shell ($n = 5$) than are Ca's ($n = 4$).

(c) Both ions have the same nuclear charge, but Co^{2+} has one more electron (25 electrons compared to 24 electrons for Co^{3+}) and hence greater electron-electron repulsion. The radius of Co^{2+} is larger.

Similar problems: 8.43, 8.45.

Practice Exercise Select the smaller ion in each of the following pairs: (a) Li^+, K^+; (b) Au^+, Au^{3+}; (c) N^{3-}, P^{3-}.

Review of Concepts

Identify the spheres shown here with each of the following: S^{2-}, Mg^{2+}, F^-, Na^+.

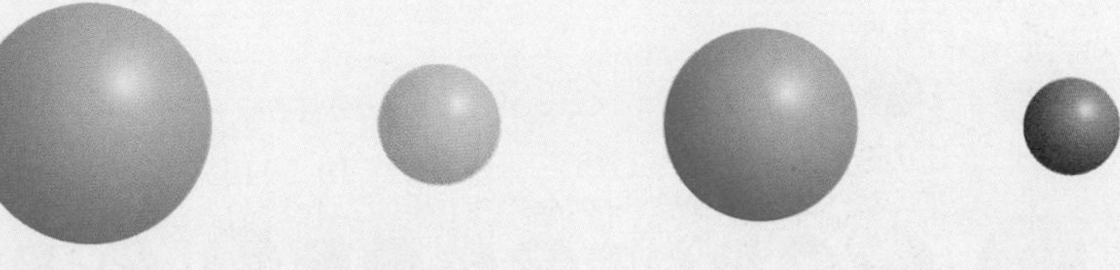

8.4 Ionization Energy

As we will see throughout this book, the chemical properties of any atom are determined by the configuration of the atom's valence electrons. The stability of these outermost electrons is reflected directly in the atom's ionization energies. ***Ionization energy*** is *the minimum energy (in kJ/mol) required to remove an electron from a gaseous atom in its ground state*. In other words, ionization energy is the amount of energy in kilojoules needed to strip 1 mole of electrons from 1 mole of gaseous atoms. Gaseous atoms are specified in this definition because an atom in the gas phase is virtually uninfluenced by its neighbors and so there are no intermolecular forces (that is, forces between molecules) to take into account when measuring ionization energy.

Note that although valence electrons are relatively easy to remove from the atom, core electrons are much harder to remove. Thus, there is a large jump in ionization energy between the last valence electron and the first core electron.

The magnitude of ionization energy is a measure of how "tightly" the electron is held in the atom. The higher the ionization energy, the more difficult it is to remove the electron. For a many-electron atom, the amount of energy required to remove the first electron from the atom in its ground state,

$$\text{energy} + X(g) \longrightarrow X^+(g) + e^- \qquad (8.2)$$

is called the *first ionization energy* (I_1). In Equation (8.2), X represents an atom of any element and e^- is an electron. The second ionization energy (I_2) and the third ionization energy (I_3) are shown in the following equations:

$$\text{energy} + X^+(g) \longrightarrow X^{2+}(g) + e^- \quad \text{second ionization}$$
$$\text{energy} + X^{2+}(g) \longrightarrow X^{3+}(g) + e^- \quad \text{third ionization}$$

The pattern continues for the removal of subsequent electrons.

When an electron is removed from an atom, the repulsion among the remaining electrons decreases. Because the nuclear charge remains constant, more energy is needed to remove another electron from the positively charged ion. Thus, ionization energies always increase in the following order:

$$I_1 < I_2 < I_3 < \cdots$$

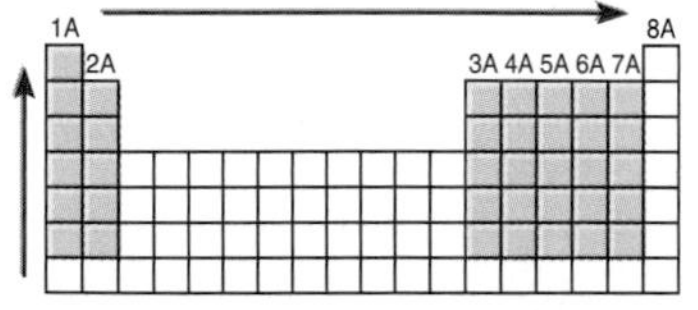

The increase in first ionization energy from left to right across a period and from bottom to top in a group for representative elements.

Table 8.2 lists the ionization energies of the first 20 elements. Ionization is always an endothermic process. By convention, energy absorbed by atoms (or ions) in the ionization process has a positive value. Thus, ionization energies are all positive quantities. Figure 8.9 shows the variation of the first ionization energy with atomic number. The plot clearly exhibits the periodicity in the stability of the most loosely held electron. Note that, apart from small irregularities, the first ionization energies of elements in a period increase with increasing atomic number. This trend is due to the increase in effective nuclear charge from left to right (as in the case of atomic radii variation). A larger effective nuclear charge means a more tightly held valence electron, and hence a higher first ionization energy. A notable feature of Figure 8.9 is the peaks, which correspond to the

Table 8.2 The Ionization Energies (kJ/mol) of the First 20 Elements

Z	Element	First	Second	Third	Fourth	Fifth	Sixth
1	H	1,312					
2	He	2,373	5,251				
3	Li	520	7,300	11,815			
4	Be	899	1,757	14,850	21,005		
5	B	801	2,430	3,660	25,000	32,820	
6	C	1,086	2,350	4,620	6,220	38,000	47,261
7	N	1,400	2,860	4,580	7,500	9,400	53,000
8	O	1,314	3,390	5,300	7,470	11,000	13,000
9	F	1,680	3,370	6,050	8,400	11,000	15,200
10	Ne	2,080	3,950	6,120	9,370	12,200	15,000
11	Na	495.9	4,560	6,900	9,540	13,400	16,600
12	Mg	738.1	1,450	7,730	10,500	13,600	18,000
13	Al	577.9	1,820	2,750	11,600	14,800	18,400
14	Si	786.3	1,580	3,230	4,360	16,000	20,000
15	P	1,012	1,904	2,910	4,960	6,240	21,000
16	S	999.5	2,250	3,360	4,660	6,990	8,500
17	Cl	1,251	2,297	3,820	5,160	6,540	9,300
18	Ar	1,521	2,666	3,900	5,770	7,240	8,800
19	K	418.7	3,052	4,410	5,900	8,000	9,600
20	Ca	589.5	1,145	4,900	6,500	8,100	11,000

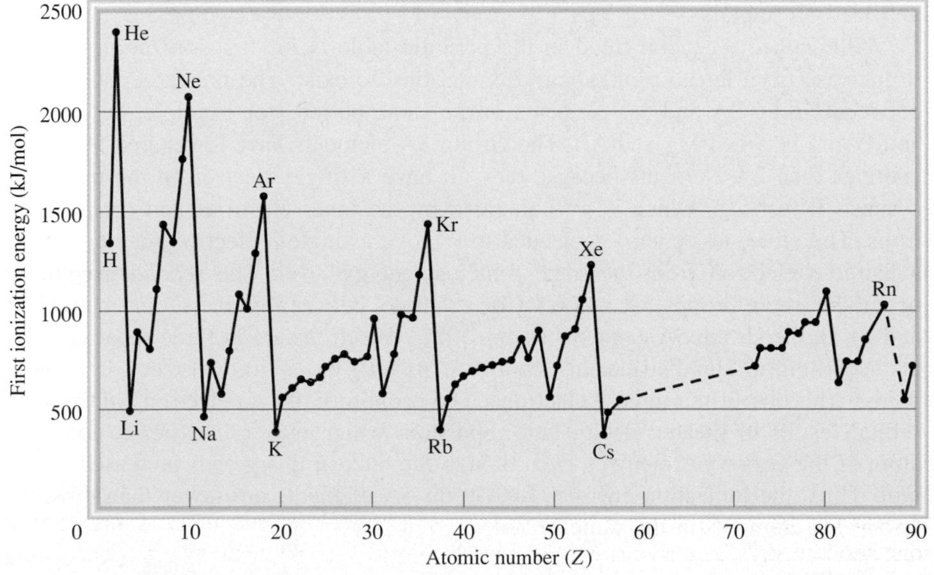

Figure 8.9
Variation of the first ionization energy with atomic number. Note that the noble gases have high ionization energies, whereas the alkali metals and alkaline earth metals have low ionization energies.

noble gases. The high ionization energies of the noble gases, stemming from their large effective nuclear charge, comprise one of the reasons for this stability. In fact, helium ($1s^2$) has the highest first ionization energy of all the elements.

At the bottom of the graph in Figure 8.9 are the Group 1A elements (the alkali metals), which have the lowest first ionization energies. Each of these metals has one valence electron (the outermost electron configuration is ns^1), which is effectively shielded by the completely filled inner shells. Consequently, it is energetically easy to remove an electron from the atom of an alkali metal to form a unipositive ion (Li^+, Na^+, K^+, . . .). Significantly, the electron configurations of these cations are isoelectronic with those noble gases just preceding them in the periodic table.

The Group 2A elements (the alkaline earth metals) have higher first ionization energies than the alkali metals do. The alkaline earth metals have two valence electrons (the outermost electron configuration is ns^2). Because these two s electrons do not shield each other well, the effective nuclear charge for an alkaline earth metal atom is larger than that for the preceding alkali metal. Most alkaline earth compounds contain dipositive ions (Mg^{2+}, Ca^{2+}, Sr^{2+}, Ba^{2+}). The Be^{2+} ion is isoelectronic with Li^+ and with He, Mg^{2+}, is isoelectronic with Na^+ and with Ne, and so on.

As Figure 8.9 shows, metals have relatively low ionization energies compared to nonmetals. The ionization energies of the metalloids generally fall between those of metals and nonmetals. The difference in ionization energies suggests why metals always form cations and nonmetals form anions in ionic compounds. (The only important nonmetallic cation is the ammonium ion, NH_4^+.) For a given group, ionization energy decreases with increasing atomic number (that is, as we move down the group). Elements in the same group have similar outer electron configurations. However, as the principal quantum number n increases, so does the average distance of a valence electron from the nucleus. A greater separation between the electron and the nucleus means a weaker attraction, so that it becomes easier to remove the first electron as we go from element to element down a group, even though the effective nuclear charge also increases in the same direction. Thus, the metallic character of the elements within a group increases from top to bottom. This trend is particularly noticeable for elements in Groups 3A to 7A. For example, in Group 4A, carbon is a nonmetal, silicon and germanium are metalloids, and tin and lead are metals.

Although the general trend in the periodic table is for first ionization energies to increase from left to right, some irregularities do exist. The first exception occurs between Group 2A and 3A elements in the same period (for example, between Be and B and between Mg and Al). The Group 3A elements have lower first ionization energies than 2A elements because they all have a single electron in the outermost p subshell (ns^2np^1), which is well shielded by the inner electrons and the ns^2 electrons. Therefore, less energy is needed to remove a single p electron than to remove a paired s electron from the same principal energy level. The second irregularity occurs between Groups 5A and 6A (for example, between N and O and between P and S). In the Group 5A elements (ns^2np^3) the p electrons are in three separate orbitals according to Hund's rule. In Group 6A (ns^2np^4) the additional electron must be paired with one of the three p electrons. The proximity of two electrons in the same orbital results in greater electrostatic repulsion, which makes it easier to ionize an atom of the Group 6A element, even though the nuclear charge has increased by one unit. Thus, the ionization energies for Group 6A elements are lower than those for Group 5A elements in the same period.

EXAMPLE 8.4

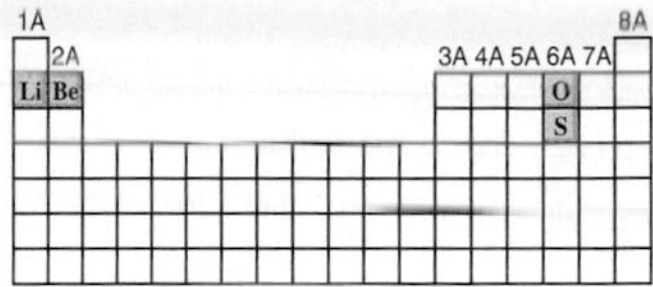

(a) Which atom should have a smaller first ionization energy: oxygen or sulfur?
(b) Which atom should have a higher second ionization energy: lithium or beryllium?

Strategy (a) First ionization energy decreases as we go down a group because the outermost electron is farther away from the nucleus and feels less attraction. (b) Removal of the outermost electron requires less energy if it is shielded by a filled inner shell.

Solution

(a) Oxygen and sulfur are members of Group 6A. They have the same valence electron configuration (ns^2np^4), but the $3p$ electron in sulfur is farther from the nucleus and experiences less nuclear attraction than the $2p$ electron in oxygen. Thus, we predict that sulfur should have a smaller first ionization energy.

(b) The electron configurations of Li and Be are $1s^22s^1$ and $1s^22s^2$, respectively. The second ionization energy is the minimum energy required to remove an electron from a gaseous unipositive ion in its ground state. For the second ionization process we write

$$\underset{1s^2}{Li^+(g)} \longrightarrow \underset{1s^1}{Li^{2+}(g)} + e^-$$

$$\underset{1s^22s^1}{Be^+(g)} \longrightarrow \underset{1s^2}{Be^{2+}(g)} + e^-$$

Because $1s$ electrons shield $2s$ electrons much more effectively than they shield each other, we predict that it should be easier to remove a $2s$ electron from Be^+ than to remove a $1s$ electron from Li^+.

Check Compare your result with the data shown in Table 8.2. In (a), is your prediction consistent with the fact that the metallic character of the elements increases as we move down a periodic group? In (b), does your prediction account for the fact that alkali metals form +1 ions while alkaline earth metals form +2 ions?

Similar problem: 8.53.

Practice Exercise (a) Which of the following atoms should have a larger first ionization energy: N or P? (b) Which of the following atoms should have a smaller second ionization energy: Na or Mg?

REVIEW OF CONCEPTS

Label the plots shown here for the first, second, and third ionization energies for Mg, Al, and K.

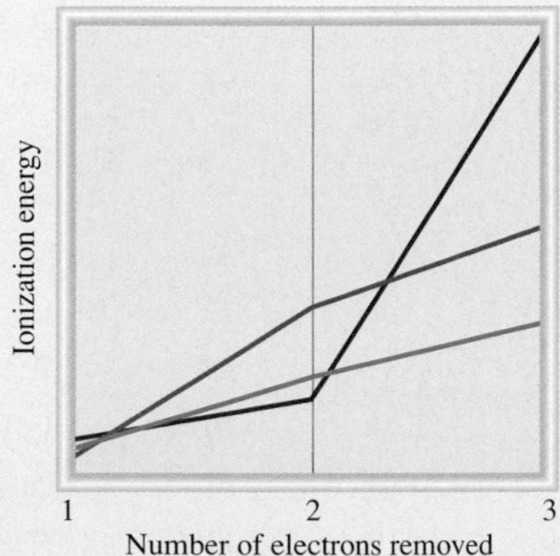

8.5 Electron Affinity

Another property that greatly influences the chemical behavior of atoms is their ability to accept one or more electrons. This property is called ***electron affinity,*** which is *the negative of the energy change that occurs when an electron is accepted by an atom in the gaseous state to form an anion.*

$$X(g) + e^- \longrightarrow X^-(g) \tag{8.3}$$

Consider the process in which a gaseous fluorine atom accepts an electron:

$$F(g) + e^- \longrightarrow F^-(g) \qquad \Delta H = -328 \text{ kJ/mol}$$

Electron affinity is positive if the reaction is exothermic and negative if the reaction is endothermic.

The electron affinity of fluorine is therefore assigned a value of +328 kJ/mol. The more positive is the electron affinity of an element, the greater is the affinity of an atom of the element to accept an electron. Another way of viewing electron affinity is to think of it as the energy that must be supplied to remove an electron from the anion. For fluorine, we write

$$F^-(g) \longrightarrow F(g) + e^- \qquad \Delta H = +328 \text{ kJ/mol}$$

Thus, a large positive electron affinity means that the negative ion is very stable (that is, the atom has a great tendency to accept an electron), just as a high ionization energy of an atom means that the electron in the atom is very stable.

Experimentally, electron affinity is determined by removing the additional electron from an anion. In contrast to ionization energies, however, electron affinities are difficult to measure because the anions of many elements are unstable. Table 8.3 shows the electron affinities of some representative elements and the noble gases. The overall trend is an increase in the tendency to accept electrons (electron affinity values become more positive) from left to right across a period. The electron

Table 8.3 Electron Affinities (kJ/mol) of Some Representative Elements and the Noble Gases*

1A	2A	3A	4A	5A	6A	7A	8A
H							He
73							<0
Li	Be	B	C	N	O	F	Ne
60	≤0	27	122	0	141	328	<0
Na	Mg	Al	Si	P	S	Cl	Ar
53	≤0	44	134	72	200	349	<0
K	Ca	Ga	Ge	As	Se	Br	Kr
48	2.4	29	118	77	195	325	<0
Rb	Sr	In	Sn	Sb	Te	I	Xe
47	4.7	29	121	101	190	295	<0
Cs	Ba	Tl	Pb	Bi	Po	At	Rn
45	14	30	110	110	?	?	<0

*The electron affinities of the noble gases, Be, and Mg have not been determined experimentally, but are believed to be close to zero or negative.

affinities of metals are generally lower than those of nonmetals. The values vary little within a given group. The halogens (Group 7A) have the highest electron affinity values.

There is a general correlation between electron affinity and effective nuclear charge, which also increases from left to right in a given period (see p. 257). However, as in the case of ionization energies, there are some irregularities. For example, the electron affinity of a Group 2A element is lower than that for the corresponding Group 1A element, and the electron affinity of a Group 5A element is lower that that for the corresponding Group 4A element. These exceptions are due to the valence electron configurations of the elements involved. An electron added to a Group 2A element must end up in a higher-energy *np* orbital, where it is effectively shielded by the ns^2 electrons and therefore experiences a weaker attraction to the nucleus. Therefore, it has a lower electron affinity than the corresponding Group 1A element. Likewise, it is harder to add an electron to a Group 5A element (ns^2np^3) than to the corresponding Group 4A element (ns^2np^2) because the electron added to the Group 5A element must be placed in a *np* orbital that already contains an electron and will therefore experience a greater electrostatic repulsion. Finally, in spite of the fact that noble gases have high effective nuclear charge, they have extremely low electron affinities (zero or negative values). The reason is that an electron added to an atom with an ns^2np^6 configuration has to enter an $(n + 1)s$ orbital, where it is well shielded by the core electrons and will only be very weakly attracted by the nucleus. This analysis also explains why species with complete valence shells tend to be chemically stable.

There is a much less regular variation in electron affinities from top to bottom within a group (see Table 8.3)

EXAMPLE 8.5

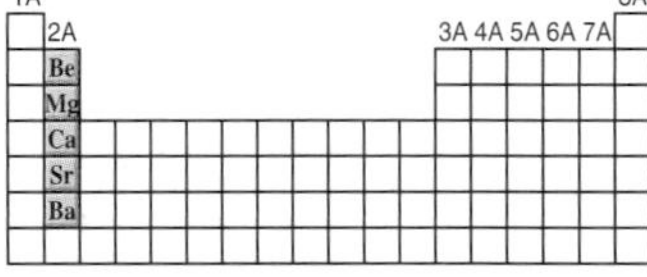

Why are the electron affinities of the alkaline earth metals, shown in Table 8.3, either negative or small positive values?

Strategy What are the electron configurations of alkaline earth metals? Would the added electron to such an atom be held strongly by the nucleus?

Solution The valence electron configuration of the alkaline earth metals is ns^2, where n is the highest principal quantum number. For the process

$$\underset{ns^2}{\mathrm{M}(g)} + e^- \longrightarrow \underset{ns^2np^1}{\mathrm{M}^-(g)}$$

where M denotes a member of the Group 2A family, the extra electron must enter the *np* subshell, which is effectively shielded by the two *ns* electrons (the *ns* electrons are more penetrating than the *np* electrons) and the inner electrons. Consequently, alkaline earth metals have little tendency to pick up an extra electron.

Similar problem: 8.62.

Practice Exercise Is it likely that Ar will form the anion Ar^-?

Review of Concepts

Why is it possible to measure the successive ionization energies of an atom until all the electrons are removed, but it becomes increasingly difficult and often impossible to measure the electron affinity of an atom beyond the first stage?

8.6 Variation in Chemical Properties of the Representative Elements

Ionization energy and electron affinity help chemists understand the types of reactions that elements undergo and the nature of the elements' compounds. On a conceptual level, these two measures are related in a simple way: Ionization energy indexes the attraction of an atom for its own electrons, whereas electron affinity expresses the attraction of an atom for an additional electron from some other source. Together they give us insight into the general attraction of an atom for electrons. With these concepts we can survey the chemical behavior of the elements systematically, paying particular attention to the relationship between chemical properties and electron configuration.

We have seen that the metallic character of the elements *decreases* from left to right across a period and *increases* from top to bottom within a group. On the basis of these trends and the knowledge that metals usually have low ionization energies while nonmetals usually have high electron affinities, we can frequently predict the outcome of a reaction involving some of these elements.

General Trends in Chemical Properties

Before we study the elements in individual groups, let us look at some overall trends. We have said that elements in the same group resemble one another in chemical behavior because they have similar valence electron configurations. This statement, although correct in the general sense, must be applied with caution. Chemists have long known that the first member of each group (the element in the second period from lithium to fluorine) differs from the rest of the members of the same group. Lithium, for example, exhibits many, but not all, of the properties characteristic of the alkali metals. Similarly, beryllium is a somewhat atypical member of Group 2A, and so on. The difference can be attributed to the unusually small size of the first element in each group (see Figure 8.4).

Another trend in the chemical behavior of the representative elements is the diagonal relationship. ***Diagonal relationships*** are *similarities between pairs of elements in different groups and periods of the periodic table*. Specifically, the first three members of the second period (Li, Be, and B) exhibit many similarities to those elements located diagonally below them in the periodic table (Figure 8.10). The reason for this phenomenon is the closeness of the charge densities of their cations. (*Charge density* is the charge of an ion divided by its volume.) Cations with comparable charge densities react similarly with anions and therefore form the same type of compounds. Thus, the chemistry of lithium resembles that of magnesium in some ways; the same holds for beryllium and aluminum and for boron and silicon. Each of these pairs is said to exhibit a diagonal relationship. We will see a number of examples of this relationship later.

Bear in mind that a comparison of the properties of elements in the same group is most valid if we are dealing with elements of the same type with respect to their metallic character. This guideline applies to the elements in Groups 1A and 2A, which are all metals, and to the elements in Groups 7A and 8A, which are all nonmetals. In Groups 3A through 6A, where the elements change either from nonmetals to metals or from nonmetals to metalloids, it is natural to expect greater variation in chemical properties even though the members of the same group have similar outer electron configurations.

Now let us take a closer look at the chemical properties of the representative elements and the noble gases. (We will consider the chemistry of the transition metals in Chapter 20.)

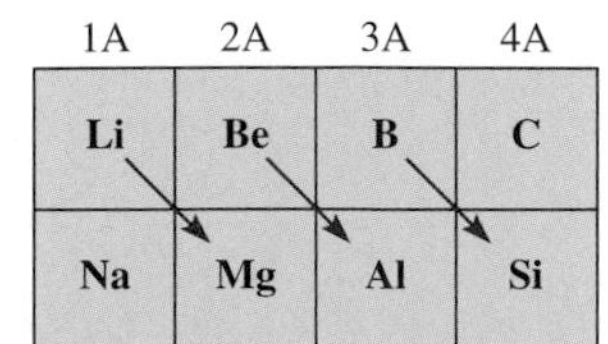

Figure 8.10
Diagonal relationships in the periodic table.

Hydrogen (1s[1])

There is no totally suitable position for hydrogen in the periodic table. Traditionally hydrogen is shown in Group 1A, but it really could be a class by itself. Like the alkali metals, it has a single *s* valence electron and forms a unipositive ion (H^+), which is hydrated in solution. On the other hand, hydrogen also forms the hydride ion (H^-) in ionic compounds such as NaH and CaH_2. In this respect, hydrogen resembles the halogens, all of which form uninegative ions (F^-, Cl^-, Br^-, and I^-) in ionic compounds. Ionic hydrides react with water to produce hydrogen gas and the corresponding metal hydroxides:

$$2NaH(s) + 2H_2O(l) \longrightarrow 2NaOH(aq) + H_2(g)$$
$$CaH_2(s) + 2H_2O(l) \longrightarrow Ca(OH)_2(s) + 2H_2(g)$$

Of course, the most important compound of hydrogen is water, which forms when hydrogen burns in air:

$$2H_2(g) + O_2(g) \longrightarrow 2H_2O(l)$$

Group 1A Elements (ns^1, $n \geq 2$)

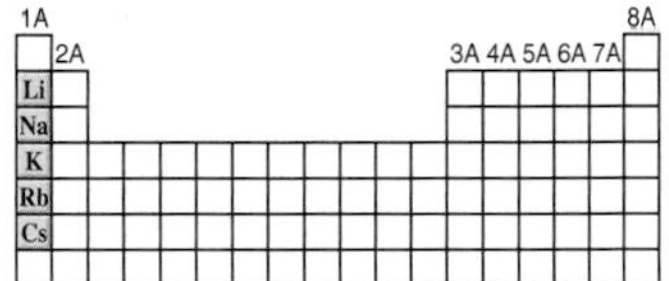

Figure 8.11 shows the Group 1A elements, the alkali metals. All of these elements have low ionization energies and therefore a great tendency to lose the single valence electron. In fact, in the vast majority of their compounds they are unipositive ions. These metals are so reactive that they are never found in the pure state in nature. They react with water to produce hydrogen gas and the corresponding metal hydroxide:

$$2M(s) + 2H_2O(l) \longrightarrow 2MOH(aq) + H_2(g)$$

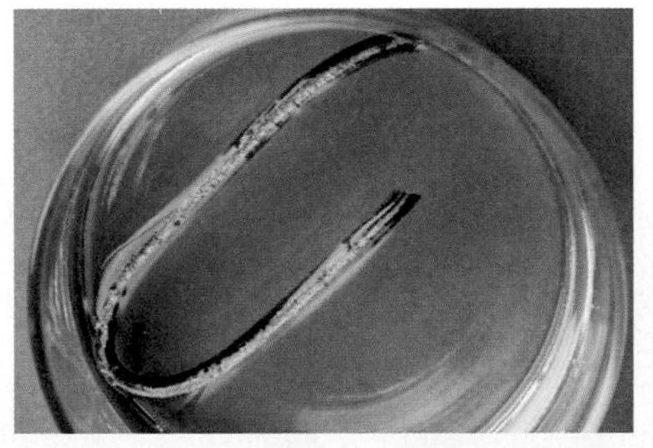

Lithium (Li)

Sodium (Na)

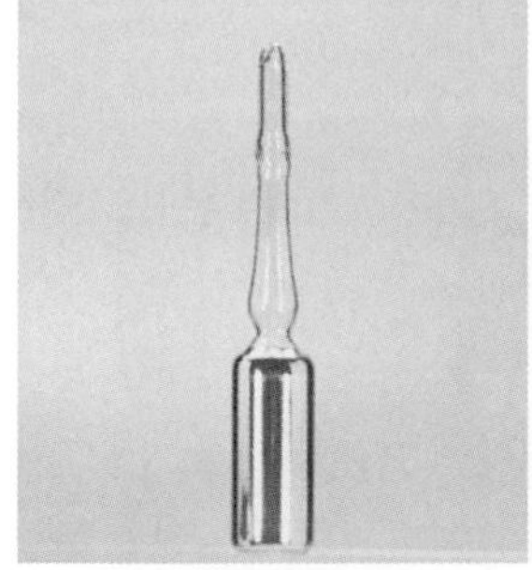

Potassium (K)

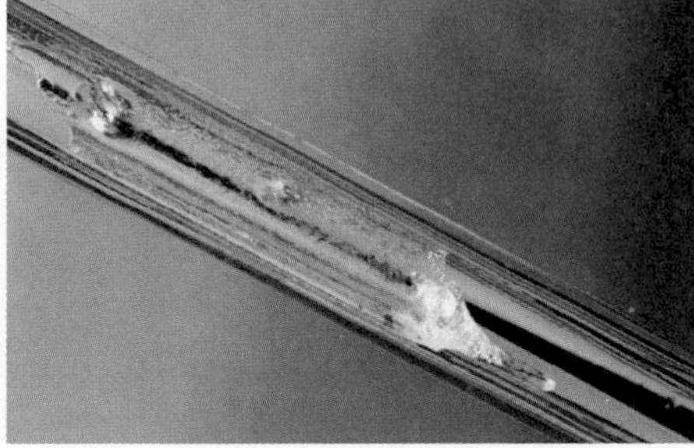

Rubidium (Rb)

Cesium (Cs)

Figure 8.11

The Group 1A elements: the alkali metals. Francium (not shown) is radioactive.

where M denotes an alkali metal. When exposed to air, they gradually lose their shiny appearance as they combine with oxygen gas to form oxides. Lithium forms lithium oxide (containing the O^{2-} ion):

$$4Li(s) + O_2(s) \longrightarrow 2Li_2O(s)$$

The other alkali metals all form oxides and *peroxides* (containing the O_2^{2-} ion). For example,

$$2Na(s) + O_2(g) \longrightarrow Na_2O_2(s)$$

Potassium, rubidium, and cesium also form *superoxides* (containing the O_2^- ion):

$$K(s) + O_2(g) \longrightarrow KO_2(s)$$

The reason that different types of oxides are formed when alkali metals react with oxygen has to do with the stability of the oxides in the solid state. Because these oxides are all ionic compounds, their stability depends on how strongly the cations and anions attract one another. Lithium tends to form predominantly lithium oxide because this compound is more stable than lithium peroxide. The formation of other alkali metal oxides can be explained similarly.

Group 2A Elements (ns^2, $n \geq 2$)

Figure 8.12 shows the Group 2A elements. As a group, the alkaline earth metals are somewhat less reactive than the alkali metals. Both the first and the second ionization energies decrease from beryllium to barium. Thus, the tendency is to form M^{2+} ions (where M denotes an alkaline earth metal atom), and hence the metallic character increases from top to bottom. Most beryllium compounds (BeH_2 and beryllium halides, such as $BeCl_2$) and some magnesium compounds (MgH_2, for example) are molecular rather than ionic in nature.

Figure 8.12
The Group 2A elements: the alkaline earth metals. Radium is a radioactive element.

The reactivities of alkaline earth metals with water vary quite markedly. Beryllium does not react with water; magnesium reacts slowly with steam; calcium, strontium, and barium are reactive enough to attack cold water:

$$Ba(s) + 2H_2O(l) \longrightarrow Ba(OH)_2(aq) + H_2(g)$$

The reactivities of the alkaline earth metals toward oxygen also increase from Be to Ba. Beryllium and magnesium form oxides (BeO and MgO) only at elevated temperatures, whereas CaO, SrO, and BaO form at room temperature.

Magnesium reacts with acids in aqueous solution, liberating hydrogen gas:

$$Mg(s) + 2H^+(aq) \longrightarrow Mg^{2+}(aq) + H_2(g)$$

Calcium, strontium, and barium also react with aqueous acid solutions to produce hydrogen gas. However, because these metals also attack water, two different reactions will occur simultaneously.

The chemical properties of calcium and strontium provide an interesting example of periodic group similarity. Strontium-90, a radioactive isotope, is a major product of an atomic bomb explosion. If an atomic bomb is exploded in the atmosphere, the strontium-90 formed will eventually settle on land and water, and it will reach our bodies via a relatively short food chain. For example, if cows eat contaminated grass and drink contaminated water, they will pass along strontium-90 in their milk. Because calcium and strontium are chemically similar, Sr^{2+} ions can replace Ca^{2+} ions in our bones. Constant exposure of the body to the high-energy radiation emitted by the strontium-90 isotopes can lead to anemia, leukemia, and other chronic illnesses.

Group 3A Elements (ns^2np^1, $n \geq 2$)

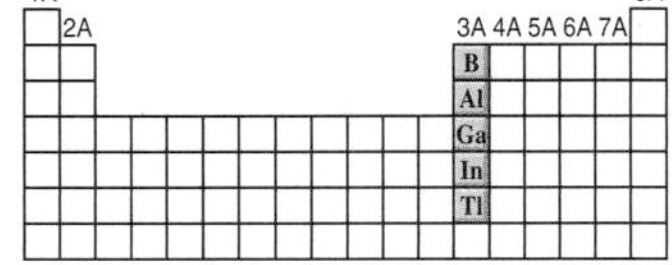

The first member of Group 3A, boron, is a metalloid; the rest are metals (Figure 8.13). Boron does not form binary ionic compounds and is unreactive toward oxygen gas and water. The next element, aluminum, readily forms aluminum oxide when exposed to air:

$$4Al(s) + 3O_2(g) \longrightarrow 2Al_2O_3(s)$$

Boron (B)

Aluminum (Al)

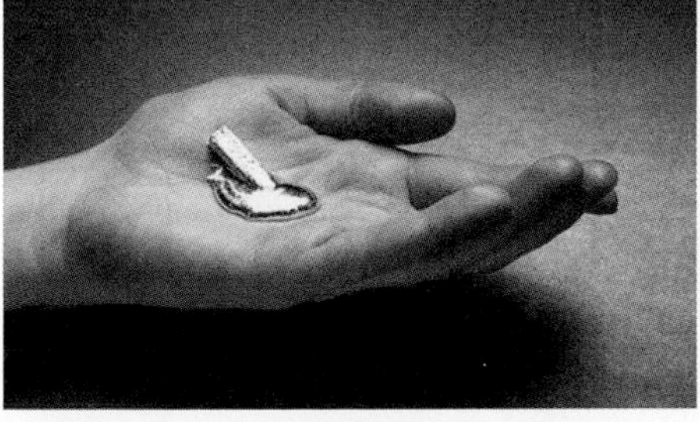

Gallium (Ga)

Indium (In)

Figure 8.13
The Group 3A elements. The low melting point of gallium (29.8°C) causes it to melt when held in hand.

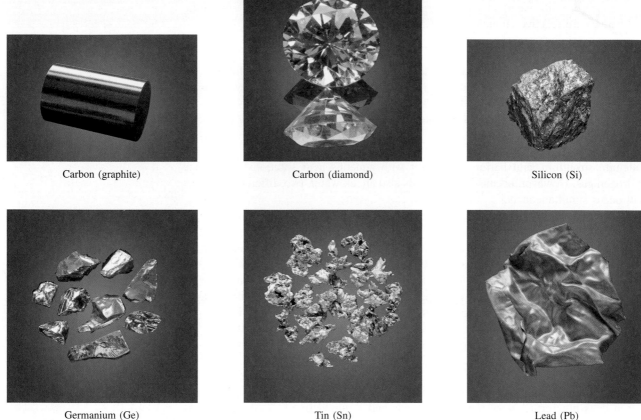

Carbon (graphite)

Carbon (diamond)

Silicon (Si)

Germanium (Ge)

Tin (Sn)

Lead (Pb)

Figure 8.14
The Group 4A elements.

Aluminum that has a protective coating of aluminum oxide is less reactive than elemental aluminum. Aluminum forms only tripositive ions. It reacts with hydrochloric acid as follows:

$$2\text{Al}(s) + 6\text{H}^+(aq) \longrightarrow 2\text{Al}^{3+}(aq) + 3\text{H}_2(g)$$

The other Group 3A metallic elements form both unipositive and tripositive ions. Moving down the group, we find that the unipositive ion becomes more stable than the tripositive ion.

The metallic elements in Group 3A also form many molecular compounds. For example, aluminum reacts with hydrogen to form AlH_3, which resembles BeH_2 in its properties. (Here is an example of the diagonal relationship.) Thus, from left to right across the periodic table, we are seeing a gradual shift from metallic to nonmetallic character in the representative elements.

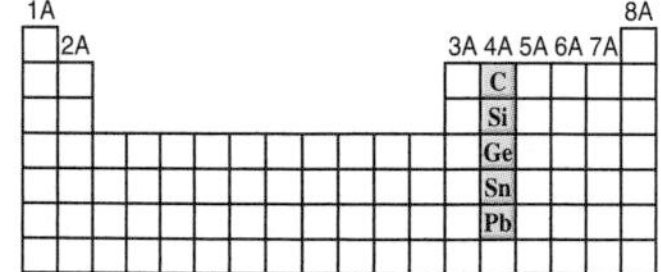

Group 4A Elements (ns^2np^2, $n \geq 2$)

The first member of Group 4A, carbon, is a nonmetal, and the next two members, silicon and germanium, are metalloids (Figure 8.14). The metallic elements of this group, tin and lead, do not react with water, but they do react with acids (hydrochloric acid, for example) to liberate hydrogen gas:

$$\text{Sn}(s) + 2\text{H}^+(aq) \longrightarrow \text{Sn}^{2+}(aq) + \text{H}_2(g)$$
$$\text{Pb}(s) + 2\text{H}^+(aq) \longrightarrow \text{Pb}^{2+}(aq) + \text{H}_2(g)$$

Liquid nitrogen (N_2)

White and red phosphorus (P)

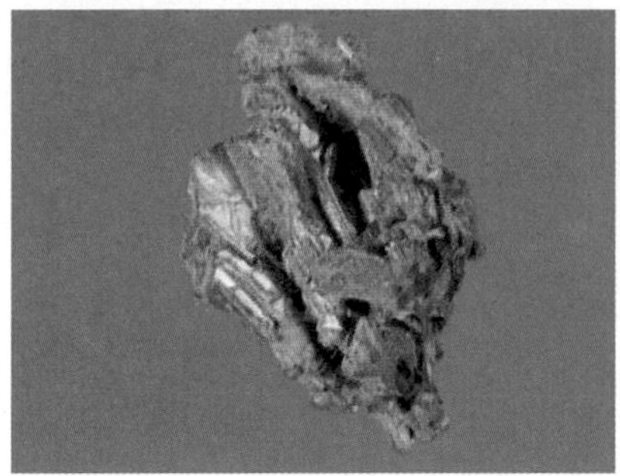

Arsenic (As)

Antimony (Sb)

Bismuth (Bi)

Figure 8.15
The Group 5A elements. Molecular nitrogen is a colorless, odorless gas.

The Group 4A elements form compounds in both the +2 and +4 oxidation states. For carbon and silicon, the +4 oxidation state is the more stable one. For example, CO_2 is more stable than CO, and SiO_2 is a stable compound, but SiO does not exist under normal conditions. As we move down the group, however, the trend in stability is reversed. In tin compounds, the +4 oxidation state is only slightly more stable than the +2 oxidation state. In lead compounds, the +2 oxidation state is unquestionably the more stable one. The outer electron configuration of lead is $6s^26p^2$, and lead tends to lose only the $6p$ electrons (to form Pb^{2+}) rather than both the $6p$ and $6s$ electrons (to form Pb^{4+}).

Group 5A Elements (ns^2np^3, $n \geq 2$)

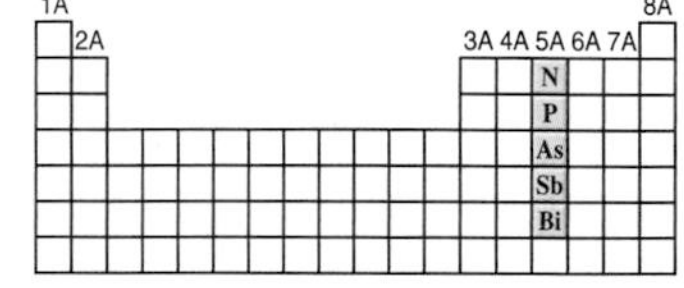

In Group 5A, nitrogen and phosphorus are nonmetals, arsenic and antimony are metalloids, and bismuth is a metal (Figure 8.15). Thus, we expect a greater variation in properties within the group.

Elemental nitrogen is a diatomic gas (N_2). It forms a number of oxides (NO, N_2O, NO_2, N_2O_4, and N_2O_5), of which only N_2O_5 is a solid; the others are gases. Nitrogen has a tendency to accept three electrons to form the nitride ion, N^{3-} (thus, achieving the electron configuration $1s^22s^22p^6$, which is isoelectronic with neon). Most metallic nitrides (Li_3N and Mg_3N_2, for example) are ionic compounds. Phosphorus exists as P_4 molecules. It forms two solid oxides with the formulas P_4O_6 and P_4O_{10}. The important oxoacids HNO_3 and H_3PO_4 are formed when the following oxides react with water:

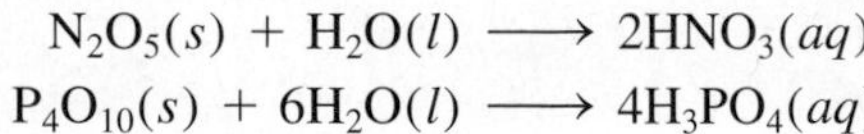

$$N_2O_5(s) + H_2O(l) \longrightarrow 2HNO_3(aq)$$
$$P_4O_{10}(s) + 6H_2O(l) \longrightarrow 4H_3PO_4(aq)$$

Sulfur (S_8)

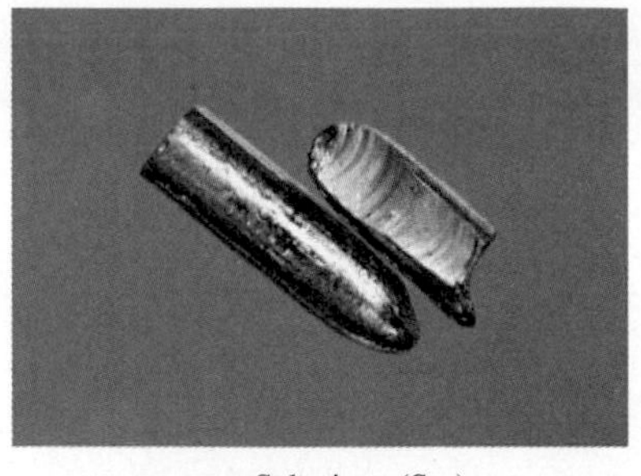

Selenium (Se_8)

Tellurium (Te)

Figure 8.16

The Group 6A elements sulfur, selenium, and tellurium. Molecular oxygen is a colorless, odorless gas. Polonium (not shown) is radioactive.

Arsenic, antimony, and bismuth have extensive three-dimensional structures. Bismuth is a far less reactive metal than those in the preceding groups.

Group 6A Elements (ns^2np^4, $n \geq 2$)

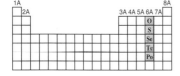

The first three members of Group 6A (oxygen, sulfur, and selenium) are nonmetals, and the last two (tellurium and polonium) are metalloids (Figure 8.16). Oxygen is a diatomic gas; elemental sulfur and selenium have the molecular formulas S_8 and Se_8, respectively; tellurium and polonium have more extensive three-dimensional structures. (Polonium is a radioactive element that is difficult to study in the laboratory.) Oxygen has a tendency to accept two electrons to form the oxide ion (O^{2-}) in many ionic compounds. Sulfur, selenium, and tellurium also form dinegative anions (S^{2-}, Se^{2-}, and Te^{2-}). The elements in this group (especially oxygen) form a large number of molecular compounds with nonmetals. The important compounds of sulfur are SO_2, SO_3, and H_2S. Sulfuric acid is formed when sulfur trioxide reacts with water:

$$SO_3(g) + H_2O(l) \longrightarrow H_2SO_4(aq)$$

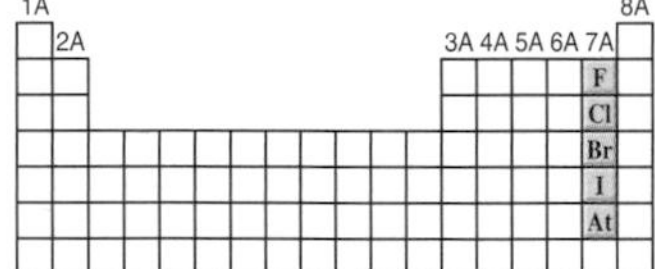

Group 7A Elements (ns^2np^5, $n \geq 2$)

All the halogens are nonmetals with the general formula X_2, where X denotes a halogen element (Figure 8.17). Because of their great reactivity, the halogens are never found in the elemental form in nature. (The last member of Group 7A, astatine, is a

Figure 8.17

The Group 7A elements chlorine, bromine, and iodine. Fluorine is a greenish-yellow gas that attacks ordinary glassware. Astatine is radioactive.

radioactive element. Little is known about its properties.) Fluorine is so reactive that it attacks water to generate oxygen:

$$2F_2(g) + 2H_2O(l) \longrightarrow 4HF(aq) + O_2(g)$$

Actually the reaction between molecular fluorine and water is quite complex; the products formed depend on reaction conditions. The reaction just shown is one of several possible changes.

The halogens have high ionization energies and large positive electron affinities. Anions derived from the halogens (F^-, Cl^-, Br^-, and I^-) are called *halides*. They are isoelectronic with the noble gases immediately to their right in the periodic table. For example, F^- is isoelectronic with Ne, Cl^- with Ar, and so on. The vast majority of the alkali metal halides and alkaline earth metal halides are ionic compounds. The halogens also form many molecular compounds among themselves (such as ICl and BrF_3) and with nonmetallic elements in other groups (such as NF_3, PCl_5, and SF_6). The halogens react with hydrogen to form hydrogen halides:

$$H_2(g) + X_2(g) \longrightarrow 2HX(g)$$

When this reaction involves fluorine, it is explosive, but it becomes less and less violent as we substitute chlorine, bromine, and iodine. The hydrogen halides dissolve in water to form hydrohalic acids. Hydrofluoric acid (HF) is a weak acid (that is, it is a weak electrolyte), but the other hydrohalic acids (HCl, HBr, and HI) are all strong acids (strong electrolytes).

Group 8A Elements (ns^2np^6, $n \geq 2$)

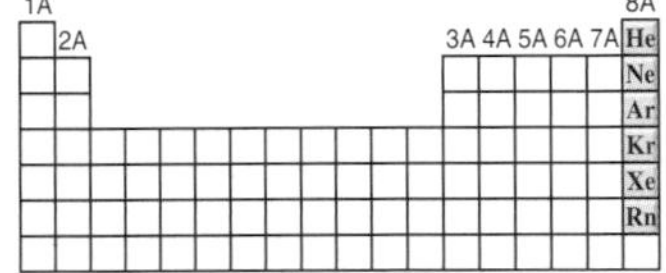

All noble gases exist as monatomic species (Figure 8.18). Their atoms have completely filled outer *ns* and *np* subshells, which give them great stability. (Helium is $1s^2$.) The Group 8A ionization energies are among the highest of all elements, and these gases have no tendency to accept extra electrons. For years these elements were called inert gases, and rightly so. Until 1962 no one had been able to prepare a compound containing any of these elements. The British chemist Neil Bartlett shattered

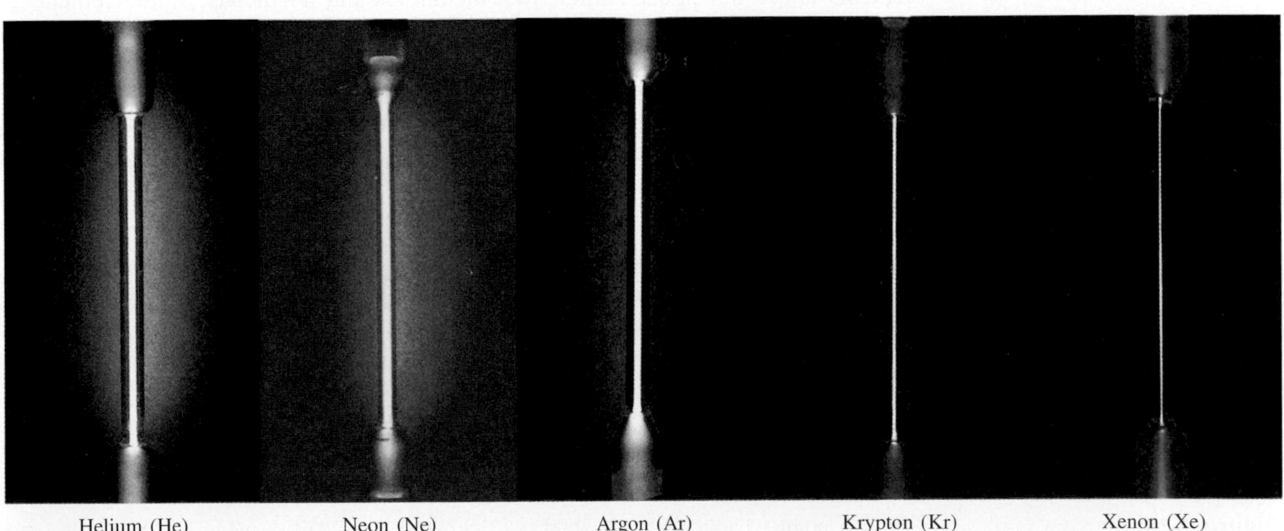

Figure 8.18
All noble gases are colorless and odorless. These pictures show the colors emitted by the gases from a discharge tube.

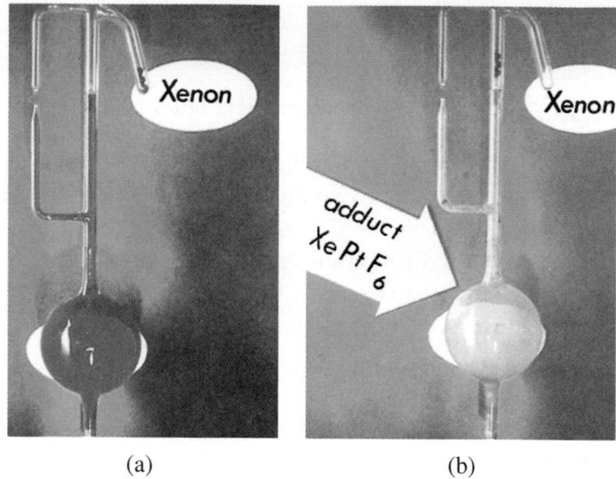

Figure 8.19
(a) Xenon gas (colorless) and PtF_6 (red gas) separated from each other. (b) When the two gases are allowed to mix, a yellow-orange solid compound is formed. Note that the product was initially erroneously given the formula $XePtF_6$.

chemists' long-held views of these elements when he exposed xenon to platinum hexafluoride, a strong oxidizing agent, and brought about the following reaction (Figure 8.19):

$$\mathrm{Xe}(g) + 2\mathrm{PtF_6}(g) \longrightarrow \mathrm{XeF^+Pt_2F_{11}^-}(s)$$

In 2000, chemists prepared a compound containing argon (HArF), which is stable only at very low temperatures.

Since then, a number of xenon compounds (XeF_4, XeO_3, XeO_4, $XeOF_4$) and a few krypton compounds (KrF_2, for example) have been prepared (Figure 8.20). Despite the immense interest in the chemistry of the noble gases, however, their compounds do not have any large-scale commercial applications, and they are not involved in natural biological processes. No compounds of helium and neon are known.

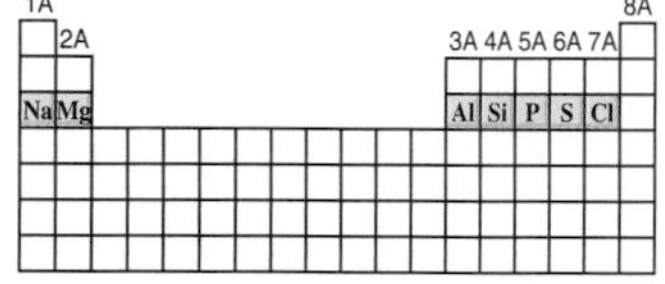

Properties of Oxides Across a Period

One way to compare the properties of the representative elements across a period is to examine the properties of a series of similar compounds. Because oxygen combines with almost all elements, we will compare the properties of oxides of the third-period elements to see how metals differ from metalloids and nonmetals. Some elements in the third period (P, S, and Cl) form several types of oxides, but for simplicity we will consider only those oxides in which the elements have the highest oxidation number. Table 8.4 lists a few general characteristics of these oxides. We observed earlier that oxygen has a tendency to form the oxide ion. This tendency is greatly favored when oxygen combines with metals that have low ionization energies, namely, those in

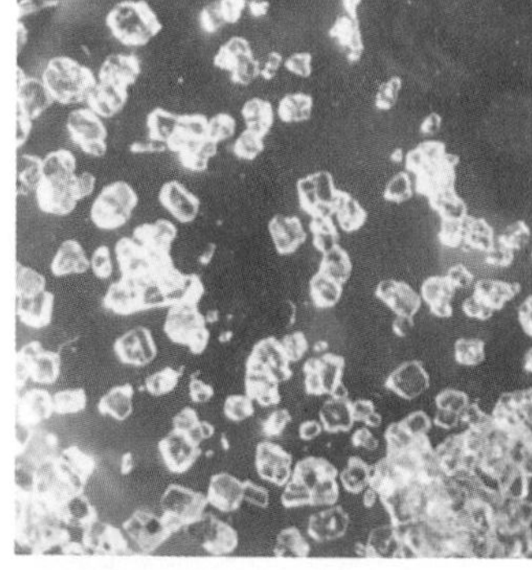

Figure 8.20
Crystals of xenon tetrafluoride (XeF_4).

Table 8.4 Some Properties of Oxides of the Third-Period Elements

	Na_2O	MgO	Al_2O_3	SiO_2	P_4O_{10}	SO_3	Cl_2O_7
Type of compound	⟵ Ionic ⟶			⟵ Molecular ⟶			
Structure	⟵ Extensive three-dimensional ⟶				⟵ Discrete molecular units ⟶		
Melting point (°C)	1275	2800	2045	1610	580	16.8	−91.5
Boiling point (°C)	?	3600	2980	2230	?	44.8	82
Acid-base nature	Basic	Basic	Amphoteric	⟵ Acidic ⟶			

Groups 1A and 2A, plus aluminum. Thus, Na_2O, MgO, and Al_2O_3 are ionic compounds, as indicated by their high melting points and boiling points. They have extensive three-dimensional structures in which each cation is surrounded by a specific number of anions, and vice versa. As the ionization energies of the elements increase from left to right, so does the molecular nature of the oxides that are formed. Silicon is a metalloid; its oxide (SiO_2) also has a huge three-dimensional network, although no ions are present. The oxides of phosphorus, sulfur, and chlorine are molecular compounds composed of small discrete units. The weak attractions among these molecules result in relatively low melting points and boiling points.

Most oxides can be classified as acidic or basic depending on whether they produce acids or bases when dissolved in water or react as acids or bases in certain processes. Some oxides are ***amphoteric,*** which means that they *display both acidic and basic properties*. The first two oxides of the third period, Na_2O and MgO, are basic oxides. For example, Na_2O reacts with water to form the base sodium hydroxide:

$$Na_2O(s) + H_2O(l) \longrightarrow 2NaOH(aq)$$

Magnesium oxide is quite insoluble; it does not react with water to any appreciable extent. However, it does react with acids in a manner that resembles an acid-base reaction:

$$MgO(s) + 2HCl(aq) \longrightarrow MgCl_2(aq) + H_2O(l)$$

Note that the products of this reaction are a salt ($MgCl_2$) and water, the usual products of an acid-base neutralization.

Aluminum oxide is even less soluble than magnesium oxide; it too does not react with water. However, it shows basic properties by reacting with acids:

$$Al_2O_3(s) + 6HCl(aq) \longrightarrow 2AlCl_3(aq) + 3H_2O(l)$$

It also exhibits acidic properties by reacting with bases:

$$Al_2O_3(s) + 2NaOH(aq) + 3H_2O(l) \longrightarrow 2NaAl(OH)_4(aq)$$

Note that this acid-base neutralization produces a salt but no water.

Thus, Al_2O_3 is classified as an amphoteric oxide because it has properties of both acids and bases. Other amphoteric oxides are ZnO, BeO, and Bi_2O_3.

Silicon dioxide is insoluble and does not react with water. It has acidic properties, however, because it reacts with very concentrated bases:

$$SiO_2(s) + 2NaOH(aq) \longrightarrow Na_2SiO_3(aq) + H_2O(l)$$

For this reason, concentrated aqueous, strong bases such as NaOH(*aq*) should not be stored in Pyrex glassware, which is made of SiO_2.

The remaining third-period oxides are acidic. They react with water to form phosphoric acid (H_3PO_4), sulfuric acid (H_2SO_4), and perchloric acid ($HClO_4$):

$$P_4O_{10}(s) + 6H_2O(l) \longrightarrow 4H_3PO_4(aq)$$
$$SO_3(g) + H_2O(l) \longrightarrow H_2SO_4(aq)$$
$$Cl_2O_7(l) + H_2O(l) \longrightarrow 2HClO_4(aq)$$

Certain oxides such as CO and NO are neutral; that is, they do not react with water to produce an acidic or basic solution. In general, oxides containing nonmetallic elements are not basic.

This brief examination of oxides of the third-period elements shows that as the metallic character of the elements decreases from left to right across the period, their oxides change from basic to amphoteric to acidic. Metallic oxides are usually basic, and most oxides of nonmetals are acidic. The intermediate properties of the oxides (as shown by the amphoteric oxides) are exhibited by elements whose positions are intermediate within the period. Note also that because the metallic character of the elements increases from top to bottom within a group of representative elements, we would expect oxides of elements with higher atomic numbers to be more basic than the lighter elements. This is indeed the case.

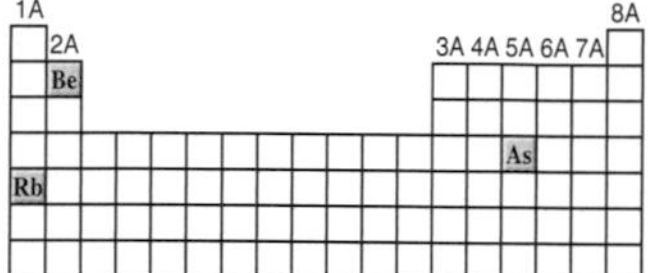

EXAMPLE 8.6

Classify the following oxides as acidic, basic, or amphoteric: (a) Rb_2O, (b) BeO, (c) As_2O_5.

Strategy What type of elements form acidic oxides? basic oxides? amphoteric oxides?

Solution

(a) Because rubidium is an alkali metal, we would expect Rb_2O to be a basic oxide.

(b) Beryllium is an alkaline earth metal. However, because it is the first member of Group 2A, we expect that it may differ somewhat from the other members of the group. In the text we saw that Al_2O_3 is amphoteric. Because beryllium and aluminum exhibit a diagonal relationship, BeO may resemble Al_2O_3 in properties. It turns out that BeO is also an amphoteric oxide.

(c) Because arsenic is a nonmetal, we expect As_2O_5 to be an acidic oxide.

Similar problem: 8.70.

Practice Exercise Classify the following oxides as acidic, basic, or amphoteric: (a) ZnO, (b) P_4O_{10}, (c) CaO.

Review of Concepts

An oxide of an element was determined to be basic. Which of the following could be that element? (a) Ba, (b) Al, and (c) Sb.

Key Equation

$Z_{eff} = Z - \sigma$ (8.1) Definition of effective nuclear charge.

Summary of Facts and Concepts

1. Nineteenth-century chemists developed the periodic table by arranging elements in the increasing order of their atomic masses. Discrepancies in early versions of the periodic table were resolved by arranging the elements in order of their atomic numbers.
2. Electron configuration determines the properties of an element. The modern periodic table classifies the elements according to their atomic numbers, and thus also by their electron configurations. The configuration of the valence electrons directly affects the properties of the atoms of the representative elements.
3. Periodic variations in the physical properties of the elements reflect differences in atomic structure. The metallic character of elements decreases across a period from metals through the metalloids to nonmetals and increases from top to bottom within a particular group of representative elements.
4. Atomic radius varies periodically with the arrangement of the elements in the periodic table. It decreases from left to right and increases from top to bottom.
5. Ionization energy is a measure of the tendency of an atom to resist the loss of an electron. The higher the ionization energy, the stronger the attraction between the nucleus and an electron. Electron affinity is a measure of the tendency of an atom to gain an electron. The more positive the electron affinity, the greater the tendency for

the atom to gain an electron. Metals usually have low ionization energies, and nonmetals usually have high electron affinities.

6. Noble gases are very stable because their outer ns and np subshells are completely filled. The metals among the representative elements (in Groups 1A, 2A, and 3A) tend to lose electrons until their cations become isoelectronic with the noble gases that precede them in the periodic table. The nonmetals in Groups 5A, 6A, and 7A tend to accept electrons until their anions become isoelectronic with the noble gases that follow them in the periodic table.

Key Words

Amphoteric oxide, p. 277
Atomic radius, p. 258
Diagonal relationship, p. 268
Effective nuclear charge (Z_{eff}), p. 257
Electron affinity, p. 266
Ionic radius, p. 259
Ionization energy, p. 262
Isoelectronic, p. 256
Representative elements, p. 253
Valence electrons, p. 254

Questions and Problems

Development of the Periodic Table

Review Questions

8.1 Briefly describe the significance of Mendeleev's periodic table.

8.2 What is Moseley's contribution to the modern periodic table?

8.3 Describe the general layout of a modern periodic table.

8.4 What is the most important relationship among elements in the same group in the periodic table?

Periodic Classification of the Elements

Review Questions

8.5 Which of these elements are metals, nonmetals, and metalloids: As, Xe, Fe, Li, B, Cl, Ba, P, I, Si?

8.6 Compare the physical and chemical properties of metals and nonmetals.

8.7 Draw a rough sketch of a periodic table (no details are required). Indicate where metals, nonmetals, and metalloids are located.

8.8 What is a representative element? Give names and symbols of four representative elements.

8.9 Without referring to a periodic table, write the name and symbol for an element in each of these groups: 1A, 2A, 3A, 4A, 5A, 6A, 7A, 8A, transition metals.

8.10 Indicate whether these elements exist as atomic species, molecular species, or extensive three-dimensional structures in their most stable state at 25°C and 1 atm, and write the molecular or empirical formula for the elements: phosphorus, iodine, magnesium, neon, arsenic, sulfur, carbon, selenium, and oxygen.

8.11 You are given a dark shiny solid and asked to determine whether it is iodine or a metallic element. Suggest a nondestructive test that would enable you to arrive at the correct answer.

8.12 Define valence electrons. For representative elements, the number of valence electrons of an element is equal to its group number. Show that this is true for the following elements: Al, Sr, K, Br, P, S, C.

8.13 Write the outer electron configurations for (a) the alkali metals, (b) the alkaline earth metals, (c) the halogens, (d) the noble gases.

8.14 Use the first-row transition metals (Sc to Cu) as an example to illustrate the characteristics of the electron configurations of transition metals.

Problems

8.15 In the periodic table, the element hydrogen is sometimes grouped with the alkali metals (as in this book) and sometimes with the halogens. Explain why hydrogen can resemble both the Group 1A and the Group 7A elements.

8.16 A neutral atom of a certain element has 17 electrons. Without consulting a periodic table, (a) write the ground-state electron configuration of the element, (b) classify the element, (c) determine whether the atoms of this element are diamagnetic or paramagnetic.

8.17 Group these electron configurations in pairs that would represent similar chemical properties of their atoms:

(a) $1s^22s^22p^63s^2$
(b) $1s^22s^22p^3$
(c) $1s^22s^22p^63s^23p^64s^23d^{10}4p^6$
(d) $1s^22s^2$
(e) $1s^22s^22p^6$
(f) $1s^22s^22p^63s^23p^3$

8.18 Group these electron configurations in pairs that would represent similar chemical properties of their atoms:
(a) $1s^22s^22p^5$
(b) $1s^22s^1$
(c) $1s^22s^22p^6$
(d) $1s^22s^22p^63s^23p^5$
(e) $1s^22s^22p^63s^23p^64s^1$
(f) $1s^22s^22p^63s^23p^64s^23d^{10}4p^6$

8.19 Without referring to a periodic table, write the electron configurations of elements with these atomic numbers: (a) 9, (b) 20, (c) 26, (d) 33. Classify the elements.

8.20 Specify in what group of the periodic table each of these elements is found: (a) [Ne]$3s^1$, (b) [Ne]$3s^23p^3$, (c) [Ne]$3s^23p^6$, (d) [Ar]$4s^23d^8$.

8.21 An ion M^{2+} derived from a metal in the first transition metal series has four and only four electrons in the $3d$ subshell. What element might M be?

8.22 A metal ion with a net +3 charge has five electrons in the $3d$ subshell. Identify the metal.

Electron Configurations of Ions

Review Questions

8.23 What is the characteristic of the electron configuration of stable ions derived from representative elements?

8.24 What do we mean when we say that two ions or an atom and an ion are isoelectronic?

8.25 What is wrong with the statement "The atoms of element X are isoelectronic with the atoms of element Y"?

8.26 Give three examples of first-row transition metal (Sc to Cu) ions whose electron configurations are represented by the argon core.

Problems

8.27 Write ground-state electron configurations for these ions: (a) Li^+, (b) H^-, (c) N^{3-}, (d) F^-, (e) S^{2-}, (f) Al^{3+}, (g) Se^{2-}, (h) Br^-, (i) Rb^+, (j) Sr^{2+}, (k) Sn^{2+}.

8.28 Write ground-state electron configurations for these ions, which play important roles in biochemical processes in our bodies: (a) Na^+, (b) Mg^{2+}, (c) Cl^-, (d) K^+, (e) Ca^{2+}, (f) Fe^{2+}, (g) Cu^{2+}, (h) Zn^{2+}.

8.29 Write ground-state electron configurations for these transition metal ions: (a) Sc^{3+}, (b) Ti^{4+}, (c) V^{5+}, (d) Cr^{3+}, (e) Mn^{2+}, (f) Fe^{2+}, (g) Fe^{3+}, (h) Co^{2+}, (i) Ni^{2+}, (j) Cu^+, (k) Cu^{2+}, (l) Ag^+, (m) Au^+, (n) Au^{3+}, (o) Pt^{2+}.

8.30 Name the ions with +3 charges that have these electron configurations: (a) [Ar]$3d^3$, (b) [Ar], (c) [Kr]$4d^6$, (d) [Xe]$4f^{14}5d^6$.

8.31 Which of these species are isoelectronic with each other: C, Cl^-, Mn^{2+}, B^-, Ar, Zn, Fe^{3+}, Ge^{2+}?

8.32 Group the species that are isoelectronic: Be^{2+}, F^-, Fe^{2+}, N^{3-}, He, S^{2-}, Co^{3+}, Ar.

Periodic Variation in Physical Properties

Review Questions

8.33 Define atomic radius. Does the size of an atom have a precise meaning?

8.34 How does atomic radius change as we move (a) from left to right across the period and (b) from top to bottom in a group?

8.35 Define ionic radius. How does the size change when an atom is converted to (a) an anion and (b) a cation?

8.36 Explain why, for isoelectronic ions, the anions are larger than the cations.

Problems

8.37 On the basis of their positions in the periodic table, select the atom with the larger atomic radius in each of these pairs: (a) Na, Cs; (b) Be, Ba; (c) N, Sb; (d) F, Br; (e) Ne, Xe.

8.38 Arrange the following atoms in order of decreasing atomic radius: Na, Al, P, Cl, Mg.

8.39 Which is the largest atom in Group 4A?

8.40 Which is the smallest atom in Group 7A?

8.41 Why is the radius of the lithium atom considerably larger than the radius of the hydrogen atom?

8.42 Use the second period of the periodic table as an example to show that the sizes of atoms decrease as we move from left to right. Explain the trend.

8.43 In each of the following pairs, indicate which one of the two species is smaller: (a) Cl or Cl^-, (b) Na or Na^+, (c) O^{2-} or S^{2-}, (d) Mg^{2+} or Al^{3+}, (e) Au^+ or Au^{3+}.

8.44 List these ions in order of increasing ionic radius: N^{3-}, Na^+, F^-, Mg^{2+}, O^{2-}.

8.45 Explain which of these ions is larger, and why: Cu^+ or Cu^{2+}.

8.46 Explain which of these anions is larger, and why: Se^{2-} or Te^{2-}.

8.47 Give the physical states (gas, liquid, or solid) of the representative elements in the fourth period at 1 atm and 25°C: K, Ca, Ga, Ge, As, Se, Br.

8.48 The boiling points of neon and krypton are −245.9°C and −152.9°C, respectively. Using these data, estimate the boiling point of argon. (*Hint:* The properties of argon are intermediate between those of neon and krypton.)

Ionization Energy

Review Questions

8.49 Define ionization energy. Ionization energy is usually measured in the gaseous state. Why? Why is the second ionization energy always greater than the first ionization energy for any element?

8.50 Sketch an outline of the periodic table and show group and period trends in the first ionization energy of the elements. What types of elements have the highest ionization energies and what types the lowest ionization energies?

Problems

8.51 Use the third period of the periodic table as an example to illustrate the change in first ionization energies of the elements as we move from left to right. Explain the trend.

8.52 Ionization energy usually increases from left to right across a given period. Aluminum, however, has a lower ionization energy than magnesium. Explain.

8.53 The first and second ionization energies of K are 419 kJ/mol and 3052 kJ/mol, and those of Ca are 590 kJ/mol and 1145 kJ/mol, respectively. Compare their values and comment on the differences.

8.54 Two atoms have the electron configurations $1s^22s^22p^6$ and $1s^22s^22p^63s^1$. The first ionization energy of one is 2080 kJ/mol, and that of the other is 496 kJ/mol. Pair each ionization energy with one of the given electron configurations. Justify your choice.

8.55 A hydrogenlike ion is an ion containing only one electron. The energies of the electron in a hydrogenlike ion are given by

$$E_n = -(2.18 \times 10^{-18}\text{ J})Z^2\left(\frac{1}{n^2}\right)$$

in which n is the principal quantum number and Z is the atomic number of the element. Calculate the ionization energy (in kilojoules per mole) of the He^+ ion.

8.56 Plasma is a state of matter in which a gaseous system consists of positive ions and electrons. In the plasma state, a mercury atom would be stripped of its 80 electrons and exist as Hg^{80+}. Use the equation in Problem 8.55 to calculate the energy required for the last step of ionization; that is,

$$Hg^{79+}(g) \longrightarrow Hg^{80+}(g) + e^-$$

Electron Affinity

Review Questions

8.57 (a) Define electron affinity. Electron affinity is usually measured with atoms in the gaseous state. Why? (b) Ionization energy is always a positive quantity, whereas electron affinity may be either positive or negative. Explain.

8.58 Explain the trends in electron affinity from aluminum to chlorine (see Table 8.3).

Problems

8.59 Arrange the elements in each of these groups in order of increasing electron affinity: (a) Li, Na, K; (b) F, Cl, Br, I.

8.60 Which of these elements would you expect to have the greatest electron affinity? He, K, Co, S, Cl.

8.61 From the electron-affinity values for the alkali metals, do you think it is possible for these metals to form an anion like M^-, where M represents an alkali metal?

8.62 Explain why alkali metals have a greater affinity for electrons than alkaline earth metals do.

Variation in Chemical Properties

Review Questions

8.63 Explain what is meant by the diagonal relationship. List two pairs of elements that show this relationship.

8.64 Which elements are more likely to form acidic oxides? basic oxides? amphoteric oxides?

Problems

8.65 Use the alkali metals and alkaline earth metals as examples to show how we can predict the chemical properties of elements simply from their electron configurations.

8.66 Based on your knowledge of the chemistry of the alkali metals, predict some of the chemical properties of francium, the last member of the group.

8.67 As a group, the noble gases are very stable chemically (only Kr and Xe are known to form some compounds). Why?

8.68 Why are the Group 1B elements more stable than the Group 1A elements even though they seem to have the same outer electron configuration ns^1 in which n is the principal quantum number of the outermost shell?

8.69 How do the chemical properties of oxides change as we move across a period from left to right? as we move down a particular group?

8.70 Predict (and give balanced equations for) the reactions between each of these oxides and water: (a) Li_2O, (b) CaO, (c) CO_2.

8.71 Write formulas and give names for the binary hydrogen compounds of the second-period elements (Li to F). Describe the changes in physical and chemical properties of these compounds as we move across the period from left to right.

8.72 Which oxide is more basic, MgO or BaO? Why?

Additional Problems

8.73 State whether each of these properties of the representative elements generally increases or decreases (a) from left to right across a period and (b) from top to bottom in a group: metallic character, atomic size, ionization energy, acidity of oxides.

8.74 With reference to the periodic table, name (a) a halogen element in the fourth period, (b) an element similar to phosphorus in chemical properties, (c) the most reactive metal in the fifth period, (d) an element that has an atomic number smaller than 20 and is similar to strontium.

8.75 Why do elements that have high ionization energies usually have more positive electron affinities?

8.76 Arrange the following isoelectronic species in order of (a) increasing ionic radius and (b) increasing ionization energy: O^{2-}, F^-, Na^+, Mg^{2+}.

8.77 Write the empirical (or molecular) formulas of compounds that the elements in the third period (sodium to chlorine) are expected to form with (a) molecular oxygen and (b) molecular chlorine. In each case indicate whether you expect the compound to be ionic or molecular.

8.78 Element M is a shiny and highly reactive metal (melting point 63°C), and element X is a highly reactive nonmetal (melting point −7.2°C). They react to form a compound with the empirical formula MX, a colorless, brittle solid that melts at 734°C. When dissolved in water or when molten, the substance conducts electricity. When chlorine gas is bubbled through an aqueous solution containing MX, a reddish-brown liquid appears and Cl^- ions are formed. From these observations, identify M and X. (You may need to consult a handbook of chemistry for the melting-point values.)

8.79 Match each of the elements on the right with its description on the left:

(a) A dark-red liquid	Calcium (Ca)
(b) A colorless gas that burns in oxygen gas	Gold (Au)
	Hydrogen (H_2)
(c) A reactive metal that attacks water	Neon (Ne)
	Bromine (Br_2)
(d) A shiny metal that is used in jewelry	
(e) A totally inert gas	

8.80 Arrange these species in isoelectronic pairs: O^+, Ar, S^{2-}, Ne, Zn, Cs^+, N^{3-}, As^{3+}, N, Xe.

8.81 In which of these are the species written in decreasing radius? (a) Be, Mg, Ba, (b) N^{3-}, O^{2-}, F^-, (c) Tl^{3+}, Tl^{2+}, Tl^+.

8.82 Which of these properties show a clear periodic variation? (a) first ionization energy, (b) molar mass of the elements, (c) number of isotopes of an element, (d) atomic radius.

8.83 When carbon dioxide is bubbled through a clear calcium hydroxide solution, the solution becomes milky. Write an equation for the reaction and explain how this reaction illustrates that CO_2 is an acidic oxide.

8.84 You are given four substances: a fuming red liquid, a dark metallic-looking solid, a pale-yellow gas, and a yellow-green gas that attacks glass. You are told that these substances are the first four members of Group 7A, the halogens. Name each one.

8.85 For each pair of elements listed here, give three properties that show their chemical similarity: (a) sodium and potassium and (b) chlorine and bromine.

8.86 Name the element that forms compounds, under appropriate conditions, with every other element in the periodic table except He and Ne.

8.87 Explain why the first electron affinity of sulfur is 200 kJ/mol but the second electron affinity is −649 kJ/mol.

8.88 The H^- ion and the He atom have two 1*s* electrons each. Which of the two species is larger? Explain.

8.89 Acidic oxides are those that react with water to produce acid solutions, whereas the reactions of basic oxides with water produce basic solutions. Nonmetallic oxides are usually acidic, whereas those of metals are basic. Predict the products of the reactions of these oxides with water: Na_2O, BaO, CO_2, N_2O_5, P_4O_{10}, SO_3. Write an equation for each of the reactions.

8.90 Write the formulas and names of the oxides of the second-period elements (Li to N). Identify the oxides as acidic, basic, or amphoteric.

8.91 State whether each of the elements listed here is a gas, a liquid, or a solid under atmospheric conditions. Also state whether it exists in the elemental form as atoms, as molecules, or as a three-dimensional network: Mg, Cl, Si, Kr, O, I, Hg, Br.

8.92 What factors account for the unique nature of hydrogen?

8.93 The formula for calculating the energies of an electron in a hydrogen-like ion is

$$E_n = -(2.18 \times 10^{-18}\ \text{J})Z^2\left(\frac{1}{n^2}\right)$$

This equation cannot be applied to many-electron atoms. One way to modify it for the more complex atoms is to replace *Z* with $(Z - \sigma)$, in which Z is the atomic number and σ is a positive dimensionless quantity called the shielding constant. Consider the helium atom as an example. The physical significance of σ is that it represents the extent of shielding that the two 1*s* electrons exert on each other. Thus, the quantity $(Z - \sigma)$ is appropriately called the "effective nuclear charge." Calculate the value of σ if the first ionization energy of helium is 3.94×10^{-18} J per atom. (Ignore the minus sign in the given equation in your calculation.)

8.94 A technique called photoelectron spectroscopy is used to measure the ionization energy of atoms. A sample is irradiated with ultraviolet (UV) light, and electrons are ejected from the valence shell. The kinetic energies of the ejected electrons are measured. Because the energy of the UV photon and the kinetic energy of the ejected electron are known, we can write

$$h\nu = \text{IE} + \tfrac{1}{2}mu^2$$

in which ν is the frequency of the UV light, and m and u are the mass and velocity of the electron, respectively. In one experiment the kinetic energy of the ejected electron from potassium is found to be 5.34×10^{-19} J using a UV source of wavelength 162 nm. Calculate the ionization energy of potassium. How can you be sure that this ionization energy corresponds to the electron in the valence shell (that is, the most loosely held electron)?

8.95 A student is given samples of three elements, X, Y, and Z, which could be an alkali metal, a member of Group 4A, and a member of Group 5A. She makes the following observations: Element X has a metallic luster and conducts electricity. It reacts slowly with hydrochloric acid to produce hydrogen gas. Element Y is a light-yellow solid that does not conduct electricity. Element Z has a metallic luster and conducts electricity. When exposed to air, it slowly forms a white powder. A solution of the white powder in water is basic. What can you conclude about the elements from these observations?

8.96 Using these melting-point data, estimate the melting point of francium, which is a radioactive element:

Metal	Li	Na	K	Rb	Cs
melting point (°C)	180.5	97.8	63.3	38.9	28.4

(*Hint:* Plot melting point versus atomic number.)

8.97 Experimentally, the electron affinity of an element can be determined by using a laser light to ionize the anion of the element in the gas phase:

$$X^-(g) + h\nu \longrightarrow X(g) + e^-$$

Referring to Table 8.3, calculate the photon wavelength (in nanometers) corresponding to the electron affinity for chlorine. In what region of the electromagnetic spectrum does this wavelength fall?

8.98 Name an element in Group 1A or Group 2A that is an important constituent of each of these substances: (a) remedy for acid indigestion, (b) coolant in nuclear reactors, (c) Epsom salt, (d) baking powder, (e) gunpowder, (f) a light alloy, (g) fertilizer that also neutralizes acid rain, (h) cement, and (i) grit for icy roads. You may need to ask your instructor about some of the items.

8.99 Explain why the electron affinity of nitrogen is approximately zero, although the elements on either side, carbon and oxygen, have substantial positive electron affinities.

8.100 Little is known of the chemistry of astatine, the last member of Group 7A. Describe the physical characteristics that you would expect this halogen to have. Predict the products of the reaction between sodium astatide (NaAt) and sulfuric acid. (*Hint:* Sulfuric acid is an oxidizing agent.)

8.101 The ionization energies of sodium (in kJ/mol), starting with the first and ending with the eleventh, are 495.9, 4560, 6900, 9540, 13,400, 16,600, 20,120, 25,490, 28,930, 141,360, 170,000. Plot the log of ionization energy (y axis) versus the number of ionization (x axis); for example, log 495.9 is plotted versus 1 (labeled I_1, the first ionization energy), log 4560 is plotted versus 2 (labeled I_2, the second ionization energy), and so on. (a) Label I_1 through I_{11} with the electrons in orbitals such as $1s$, $2s$, $2p$, and $3s$. (b) What can you deduce about electron shells from the breaks in the curve?

8.102 Calculate the maximum wavelength of light (in nanometers) required to ionize a single sodium atom.

8.103 The first four ionization energies of an element are approximately 738 kJ/mol, 1450 kJ/mol, 7.7×10^3 kJ/mol, and 1.1×10^4 kJ/mol. To which periodic group does this element belong? Why?

8.104 Match each of the elements on the right with its description on the left:

(a) A greenish-yellow gas that reacts with water	Nitrogen (N_2)
(b) A soft metal that reacts with water to produce hydrogen	Boron (B)
(c) A metalloid that is hard and has a high melting point	Aluminum (Al)
(d) A colorless, odorless gas	Fluorine (F_2)
(e) A more reactive metal than iron, which does not corrode in air	Sodium (Na)

8.105 When magnesium metal is burned in air, it forms two products A and B. A reacts with water to form a basic solution. B reacts with water to form a similar solution as that of A plus a gas with a pungent odor. Identify A and B and write equations for the reactions.

Special Problems

8.106 In the late 1800s the British physicist Lord Rayleigh accurately determined the atomic masses of a number of elements, but he obtained a puzzling result with nitrogen. One of his methods of preparing nitrogen was by the thermal decomposition of ammonia:

$$2NH_3(g) \longrightarrow N_2(g) + 3H_2(g)$$

Another method was to start with air and remove oxygen, carbon dioxide, and water vapor from it. Invariably, the nitrogen from air was a little denser (by about 0.5 percent) than the nitrogen from ammonia.

Later the English chemist Sir William Ramsay carried out an experiment in which he passed nitrogen, which he had obtained from air by Raleigh's procedure, over red-hot magnesium to convert it to magnesium nitride:

$$3Mg(s) + N_2(g) \longrightarrow Mg_3N_2(s)$$

After all of the nitrogen had reacted with magnesium, Ramsay was left with an unknown gas that would not combine with anything. The atomic mass of this gas was determined to be 39.95 amu. Ramsay called the gas *argon,* which means "the lazy one" in Greek.

(a) Later Rayleigh and Ramsay, with the help of Sir William Crookes, the inventor of the discharge tube, showed that argon was a new element. Describe the type of experiment performed that led them to the conclusion.

(b) Why did it take so long to discover argon?

(c) Once argon had been discovered, why did it take relatively little time to discover the rest of the noble gases?

(d) Why was helium the last noble gas to be discovered on Earth?

(e) The only confirmed compound of radon is radon fluoride, RnF. Give two reasons why there are so few known radon compounds.

8.107 On the same graph, plot the effective nuclear charge (in parentheses) and atomic radius (see Figure 8.4) versus atomic number for the second-period elements: Li(1.30), Be(1.95), B(2.60), C(3.25), N(3.90), O(4.55), F(5.20), Ne(5.85). Comment on the trends.

8.108 The ionization energy of a certain element is 412 kJ/mol. When the atoms of this element are in the first excited state, however, the ionization energy is only 126 kJ/mol. Based on this information, calculate the wavelength of light emitted in a transition from the first excited state to the ground state.

8.109 Referring to Table 8.2, explain why the first ionization energy of helium is less than twice the ionization energy of hydrogen, but the second ionization energy of helium is greater than twice the ionization energy of hydrogen. [*Hint*: According to Coulomb's law, the energy between two charges Q_1 and Q_2 separated by distance r is proportional to (Q_1Q_2/r).]

8.110 Ammonium nitrate (NH_4NO_3) is the most important nitrogen-containing fertilizer in the world. Describe how you would prepare this compound, given only air and water as the starting materials. You may have any device at your disposal for this task.

8.111 To prevent the formation of oxides, peroxides, and superoxides, alkali metals are sometimes stored in an inert atmosphere. Which of the following gases should not be used for lithium: Ne, Ar, N_2, Kr? Explain. (*Hint:* As mentioned in the chapter, Li and Mg exhibit a diagonal relationship. Compare the common compounds of these two elements.)

Answers to Practice Exercises

8.1 (a) $1s^22s^22p^63s^23p^64s^2$, (b) it is a representative element, (c) diamagnetic. **8.2** Li > Be > C. **8.3** (a) Li^+, (b) Au^{3+}, (c) N^{3-}. **8.4** (a) N, (b) Mg. **8.5** No. **8.6** (a) amphoteric, (b) acidic, (c) basic.

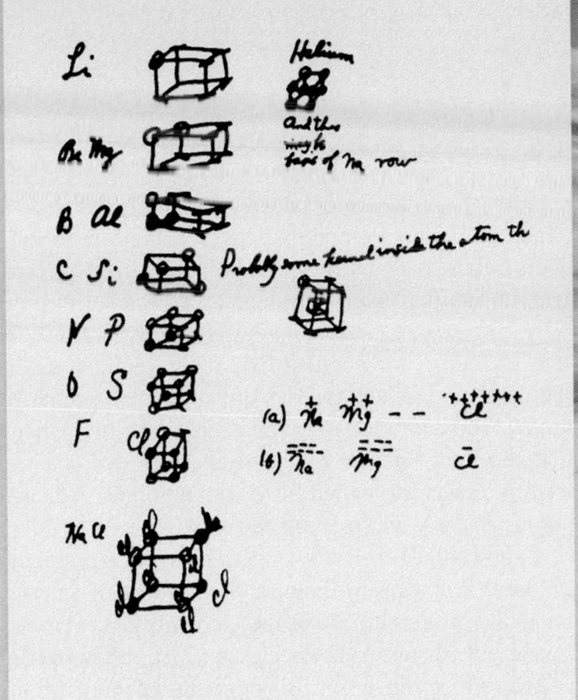

Lewis first sketched his idea about the octet rule on the back of an envelope.

CHAPTER

9

Chemical Bonding I: The Covalent Bond

Chapter Outline

Student Interactive Activities

Animations
Ionic versus Covalent Bonding (9.4)
Resonance (9.8)

Electronic Homework
Example Practice Problems
End of Chapter Problems

Essential Concepts

The Ionic Bond An ionic bond is the electrostatic force that holds the cations and anions in an ionic compound. The stability of ionic compounds is determined by their lattice energies.

The Covalent Bond Lewis postulated the formation of a covalent bond in which atoms share one or more pairs of electrons. The octet rule was formulated to predict the correctness of Lewis structures. This rule says that an atom other than hydrogen tends to form bonds until it is surrounded by eight valence electrons.

Characteristics of Lewis Structures In addition to covalent bonds, a Lewis structure also shows lone pairs, which are pairs of electrons not involved in bonding, on atoms and formal charges, which are the result of bookkeeping of electrons used in bonding. A resonance structure is one of two or more Lewis structures for a single molecule that cannot be described fully with only one Lewis structure.

Exceptions to the Octet Rule The octet rule applies mainly to the second-period elements. The three categories of exceptions to the octet rule are the incomplete octet, in which an atom in a molecule has fewer than eight valence electrons, the odd-electron molecules, which have an odd number of valence electrons, and the expanded octet, in which an atom has more than eight valence electrons. These exceptions can be explained by more refined theories of chemical bonding.

Thermochemistry Based on Bond Enthalpy From a knowledge of the strength of covalent bonds or bond enthalpies, it is possible to estimate the enthalpy change of a reaction.

9.1 Lewis Dot Symbols

The development of the periodic table and concept of electron configuration gave chemists a rationale for molecule and compound formation. This explanation, formulated by the American chemist Gilbert Lewis, is that atoms combine to achieve a more stable electron configuration. Maximum stability results when an atom is isoelectronic with a noble gas.

When atoms interact to form a chemical bond, only their outer regions are in contact. For this reason, when we study chemical bonding, we are concerned primarily with the valence electrons of the atoms. To keep track of valence electrons in a chemical reaction, and to make sure that the total number of electrons does not change, chemists use a system of dots devised by Lewis and called Lewis dot symbols. A ***Lewis dot symbol*** *consists of the symbol of an element and one dot for each valence electron in an atom of the element.* Figure 9.1 shows the Lewis dot symbols for the representative elements and the noble gases. Note that, except for helium, the number of valence electrons each atom has is the same as the group number of the element. For example, Li is a Group 1A element and has one dot for one valence electron; Be, a Group 2A element, has two valence electrons (two dots); and so on. Elements in the same group have similar outer electron configurations and hence similar Lewis dot symbols. The transition metals, lanthanides, and actinides all have incompletely filled inner shells, and in general, we cannot write simple Lewis dot symbols for them.

In this chapter, we will learn to use electron configurations and the periodic table to predict the type of bond atoms will form, as well as the number of bonds an atom of a particular element can form and the stability of the product.

REVIEW OF CONCEPTS

What is the maximum number of dots that can be drawn around the atom of a representative element?

1 1A	2 2A	3 3B	4 4B	5 5B	6 6B	7 7B	8	9 8B	10	11 1B	12 2B	13 3A	14 4A	15 5A	16 6A	17 7A	18 8A
·H																	He:
·Li	·Be·											·B·	·C·	·N·	·O·	:F·	:Ne:
·Na	·Mg·											·Al·	·Si·	·P·	·S·	:Cl·	:Ar:
·K	·Ca·											·Ga·	·Ge·	·As·	·Se·	:Br·	:Kr:
·Rb	·Sr·											·In·	·Sn·	·Sb·	·Te·	:I·	:Xe:
·Cs	·Ba·											·Tl·	·Pb·	·Bi·	·Po·	:At·	:Rn:
·Fr	·Ra·																

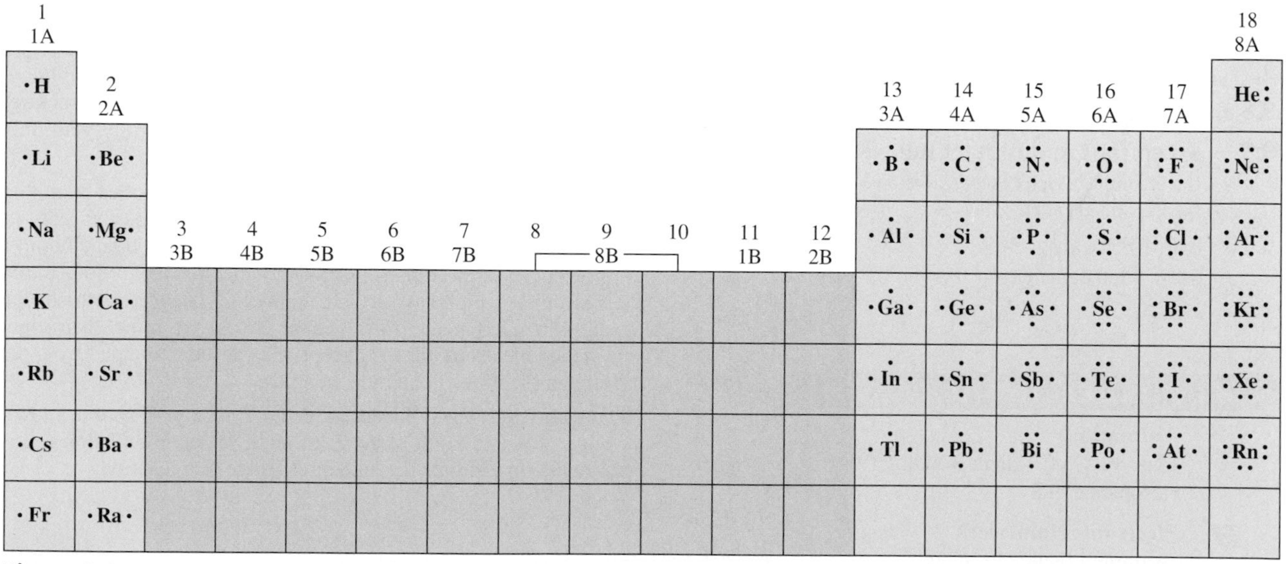

Figure 9.1

Lewis dot symbols for the representative elements and the noble gases. The number of unpaired dots corresponds to the number of bonds an atom of the element can form in a compound.

9.2 The Ionic Bond

In Chapter 8 we saw that atoms of elements with low ionization energies tend to form cations, while those with high electron affinities tend to form anions. As a rule, the elements most likely to form cations in ionic compounds are the alkali metals and alkaline earth metals, and the elements most likely to form anions are the halogens and oxygen. Consequently, a wide variety of ionic compounds combine a Group 1A or Group 2A metal with a halogen or oxygen. An ***ionic bond*** is *the electrostatic force that holds ions together in an ionic compound.* Consider, for example, the reaction between lithium and fluorine to form lithium fluoride, a poisonous white powder used in lowering the melting point of solders and in manufacturing ceramics. The electron configuration of lithium is $1s^22s^1$, and that of fluorine is $1s^22s^22p^5$. When lithium and fluorine atoms come in contact with each other, the outer $2s^1$ valence electron of lithium is transferred to the fluorine atom. Using Lewis dot symbols, we represent the reaction like this:

Lithium fluoride. Industrially, LiF (like most other ionic compounds) is obtained by purifying minerals containing the compound.

$$\underset{1s^22s^1}{\cdot\text{Li}} + \underset{1s^22s^22p^5}{:\ddot{\underset{..}{\text{F}}}\cdot} \longrightarrow \underset{1s^2}{\text{Li}^+}\ \underset{1s^22s^22p^6}{:\ddot{\underset{..}{\text{F}}}:^-} \quad \text{(or LiF)} \tag{9.1}$$

For convenience, imagine that this reaction occurs in separate steps—first the ionization of Li:

$$\cdot\text{Li} \longrightarrow \text{Li}^+ + e^-$$

and then the acceptance of an electron by F:

$$:\ddot{\underset{..}{\text{F}}}\cdot + e^- \longrightarrow :\ddot{\underset{..}{\text{F}}}:^-$$

Next, imagine the two separate ions joining to form a LiF unit:

$$\text{Li}^+ + :\ddot{\underset{..}{\text{F}}}:^- \longrightarrow \text{Li}^+:\ddot{\underset{..}{\text{F}}}:^-$$

Note that the sum of these three equations is

$$\cdot\text{Li} + :\ddot{\underset{..}{\text{F}}}\cdot \longrightarrow \text{Li}^+:\ddot{\underset{..}{\text{F}}}:^-$$

We normally write the empirical formulas of ionic compounds without showing the charges. The + and − are shown here to emphasize the transfer of electrons.

which is the same as Equation (9.1). The ionic bond in LiF is the electrostatic attraction between the positively charged lithium ion and the negatively charged fluoride ion. The compound itself is electrically neutral.

Many other common reactions lead to the formation of ionic bonds. For instance, calcium burns in oxygen to form calcium oxide:

$$2\text{Ca}(s) + \text{O}_2(g) \longrightarrow 2\text{CaO}(s)$$

Assuming that the diatomic O_2 molecule first splits into separate oxygen atoms (we will look at the energetics of this step later), we can represent the reaction with Lewis symbols:

$$\underset{[\text{Ar}]4s^2}{\cdot\text{Ca}\cdot} + \underset{1s^22s^22p^4}{\cdot\ddot{\underset{..}{\text{O}}}\cdot} \longrightarrow \underset{[\text{Ar}]}{\text{Ca}^{2+}}\ \underset{[\text{Ne}]}{:\ddot{\underset{..}{\text{O}}}:^{2-}}$$

There is a transfer of two electrons from the calcium atom to the oxygen atom. Note that the resulting calcium ion (Ca^{2+}) has the argon electron configuration, the oxide ion (O^{2-}) is isoelectronic with neon, and the compound (CaO) is electrically neutral.

In many cases, the cation and the anion in a compound do not carry the same charges. For instance, when lithium burns in air to form lithium oxide (Li_2O), the balanced equation is

$$4Li(s) + O_2(g) \longrightarrow 2Li_2O(s)$$

Using Lewis dot symbols, we write

$$\begin{array}{ccccc} 2 \cdot \text{Li} + & \cdot\ddot{\underset{\cdot\cdot}{\text{O}}}\cdot & \longrightarrow 2\text{Li}^+ & :\ddot{\underset{\cdot\cdot}{\text{O}}}:^{2-} & (\text{or Li}_2\text{O}) \\ 1s^22s^1 & 1s^22s^22p^4 & [\text{He}] & [\text{Ne}] & \end{array}$$

In this process, the oxygen atom receives two electrons (one from each of the two lithium atoms) to form the oxide ion. The Li^+ ion is isoelectronic with helium.

When magnesium reacts with nitrogen at elevated temperatures, a white solid compound, magnesium nitride (Mg_3N_2), forms:

$$3Mg(s) + N_2(g) \longrightarrow Mg_3N_2(s)$$

or

$$\begin{array}{ccccc} 3 \cdot \text{Mg} \cdot + & 2 \cdot\ddot{\underset{\cdot}{\text{N}}}\cdot & \longrightarrow 3\text{Mg}^{2+} & 2\ :\ddot{\underset{\cdot\cdot}{\text{N}}}:^{3-} & (\text{or Mg}_3\text{N}_2) \\ [\text{Ne}]3s^2 & 1s^22s^22p^3 & [\text{Ne}] & [\text{Ne}] & \end{array}$$

The reaction involves the transfer of six electrons (two from each Mg atom) to two nitrogen atoms. The resulting magnesium ion (Mg^{2+}) and the nitride ion (N^{3-}) are both isoelectronic with neon. Because there are three +2 ions and two −3 ions, the charges balance and the compound is electrically neutral.

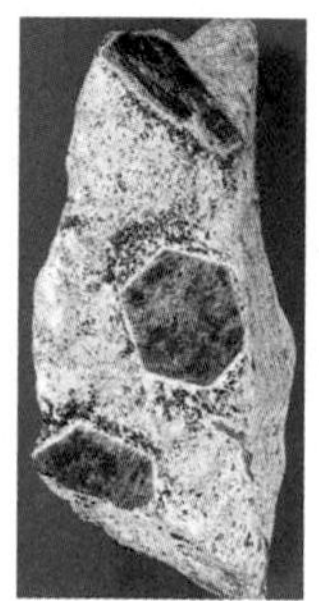

The mineral corundum (Al_2O_3).

EXAMPLE 9.1

Use Lewis dot symbols to show the formation of aluminum oxide (Al_2O_3).

Strategy We use electroneutrality as our guide in writing formulas for ionic compounds, that is, the total positive charges on the cations must be equal to the total negative charges on the anions.

Solution According to Figure 9.1, the Lewis dot symbols of Al and O are

$$\cdot\dot{\text{Al}}\cdot \qquad \cdot\ddot{\underset{\cdot\cdot}{\text{O}}}\cdot$$

Because aluminum tends to form the cation (Al^{3+}) and oxygen the anion (O^{2-}) in ionic compounds, the transfer of electrons is from Al to O. There are three valence electrons in each Al atom; each O atom needs two electrons to form the O^{2-} ion, which is isoelectronic with neon. Thus, the simplest neutralizing ratio of Al^{3+} to O^{2-} is 2:3; two Al^{3+} ions have a total charge of +6, and three O^{2-} ions have a total charge of −6. So the empirical formula of aluminum oxide is Al_2O_3, and the reaction is

$$\begin{array}{ccccc} 2 \cdot\dot{\text{Al}}\cdot + & 3 \cdot\ddot{\underset{\cdot\cdot}{\text{O}}}\cdot & \longrightarrow 2\text{Al}^{3+} & 3\ :\ddot{\underset{\cdot\cdot}{\text{O}}}:^{2-} & (\text{or Al}_2\text{O}_3) \\ [\text{Ne}]3s^23p^1 & 1s^22s^22p^4 & [\text{Ne}] & [\text{Ne}] & \end{array}$$

Check Make sure that the number of valence electrons (24) is the same on both sides of the equation. Are the subscripts in Al_2O_3 reduced to the smallest possible whole numbers?

Similar problems: 9.17, 9.18.

Practice Exercise Use Lewis dot symbols to represent the formation of barium hydride.

Review of Concepts

Use Lewis dot symbols to represent the formation of rubidium sulfide.

9.3 Lattice Energy of Ionic Compounds

We can predict which elements are likely to form ionic compounds based on ionization energy and electron affinity, but how do we evaluate the stability of an ionic compound? Ionization energy and electron affinity are defined for processes occurring in the gas phase, but at 1 atm and 25°C all ionic compounds are solids. The solid state is a very different environment because each cation in a solid is surrounded by a specific number of anions, and vice versa. Thus, the overall stability of a solid ionic compound depends on the interactions of all these ions and not merely on the interaction of a single cation with a single anion. A quantitative measure of the stability of any ionic solid is its ***lattice energy,*** defined as *the energy required to completely separate one mole of a solid ionic compound into gaseous ions.*

Lattice energy is determined by the charge of the ions and the distance between the ions.

The Born-Haber Cycle for Determining Lattice Energies

Lattice energy cannot be measured directly. However, if we know the structure and composition of an ionic compound, we can calculate the compound's lattice energy by using ***Coulomb's law,*** which states that *the potential energy (E) between two ions is directly proportional to the product of their charges and inversely proportional to the distance of separation between them.* For a single Li^+ ion and a single F^- ion separated by distance r, the potential energy of the system is given by

$$E \propto \frac{Q_{Li^+}Q_{F^-}}{r}$$

$$= k\frac{Q_{Li^+}Q_{F^-}}{r} \tag{9.2}$$

Because energy = force × distance, Coulomb's law can also be stated as

$$F = k\frac{Q_{Li^+}Q_{F^-}}{r^2}$$

where F is the force between the ions.

where Q_{Li^+} and Q_{F^-} are the charges on the Li^+ and F^- ions and k is the proportionality constant. Because Q_{Li^+} is positive and Q_{F^-} is negative, E is a negative quantity, and the formation of an ionic bond from Li^+ and F^- is an exothermic process. Consequently, energy must be supplied to reverse the process (in other words, the lattice energy of LiF is positive), and so a bonded pair of Li^+ and F^- ions is more stable than separate Li^+ and F^- ions.

We can also determine lattice energy indirectly, by assuming that the formation of an ionic compound takes place in a series of steps. This procedure, known as the ***Born-Haber cycle,*** *relates lattice energies of ionic compounds to ionization energies, electron affinities, and other atomic and molecular properties.* It is based on Hess's law (see Section 6.6). Developed by the German physicist Max Born and the German chemist Fritz Haber, the Born-Haber cycle defines the various steps that precede the formation of an ionic solid. We will illustrate its use to find the lattice energy of lithium fluoride.

Consider the reaction between lithium and fluorine:

$$Li(s) + \tfrac{1}{2}F_2(g) \longrightarrow LiF(s)$$

The standard enthalpy change for this reaction is −594.1 kJ/mol. (Because the reactants and product are in their standard states, that is, at 1 atm, the enthalpy change is also the standard enthalpy of formation for LiF.) Keeping in mind that the sum of enthalpy changes for the steps is equal to the enthalpy change for the overall reaction (−594.1 kJ/mol), we can trace the formation of LiF from its elements through five separate steps. The process may not occur exactly this way, but this pathway enables us to analyze the energy changes of ionic compound formation, with the application of Hess's law.

1. Convert solid lithium to lithium vapor (the direct conversion of a solid to a gas is called sublimation):

$$\text{Li}(s) \longrightarrow \text{Li}(g) \qquad \Delta H_1^\circ = 155.2 \text{ kJ/mol}$$

   The energy of sublimation for lithium is 155.2 kJ/mol.

2. Dissociate $\frac{1}{2}$ mole of F_2 gas into separate gaseous F atoms:

The F atoms in a F_2 molecule are held together by a covalent bond. The energy required to break the bond is called the bond enthalpy (see Section 9.10).

$$\tfrac{1}{2}\text{F}_2(g) \longrightarrow \text{F}(g) \qquad \Delta H_2^\circ = 75.3 \text{ kJ/mol}$$

   The energy needed to break the bonds in 1 mole of F_2 molecules is 150.6 kJ. Here we are breaking the bonds in half a mole of F_2, so the enthalpy change is 150.6/2, or 75.3, kJ.

3. Ionize 1 mole of gaseous Li atoms (see Table 8.3):

$$\text{Li}(g) \longrightarrow \text{Li}^+(g) + e^- \qquad \Delta H_3^\circ = 520 \text{ kJ/mol}$$

   This process corresponds to the first ionization of lithium.

4. Add 1 mole of electrons to 1 mole of gaseous F atoms. As discussed on page 266, the energy change for this process is just the opposite of electron affinity (see Table 8.3):

$$\text{F}(g) + e^- \longrightarrow \text{F}^-(g) \qquad \Delta H_4^\circ = -328 \text{ kJ/mol}$$

5. Combine 1 mole of gaseous Li^+ and 1 mole of F^- to form 1 mole of solid LiF:

$$\text{Li}^+(g) + \text{F}^-(g) \longrightarrow \text{LiF}(s) \qquad \Delta H_5^\circ = ?$$

The reverse of step 5,

$$\text{energy} + \text{LiF}(s) \longrightarrow \text{Li}^+(g) + \text{F}^-(g)$$

defines the lattice energy of LiF. Thus, the lattice energy must have the same magnitude as ΔH_5° but an opposite sign. Although we cannot determine ΔH_5° directly, we can calculate its value by the following procedure:

$$\begin{array}{llll}
1. & \text{Li}(s) & \longrightarrow \text{Li}(g) & \Delta H_1^\circ = 155.2 \text{ kJ/mol} \\
2. & \tfrac{1}{2}\text{F}_2(g) & \longrightarrow \text{F}(g) & \Delta H_2^\circ = 75.3 \text{ kJ/mol} \\
3. & \text{Li}(g) & \longrightarrow \text{Li}^+(g) + e^- & \Delta H_3^\circ = 520 \text{ kJ/mol} \\
4. & \text{F}(g) + e^- & \longrightarrow \text{F}^-(g) & \Delta H_4^\circ = -328 \text{ kJ/mol} \\
5. & \text{Li}^+(g) + \text{F}^-(g) & \longrightarrow \text{LiF}(s) & \Delta H_5^\circ = ? \\
\hline
 & \text{Li}(s) + \tfrac{1}{2}\text{F}_2(g) & \longrightarrow \text{LiF}(s) & \Delta H^\circ_{\text{overall}} = -594.1 \text{ kJ/mol}
\end{array}$$

According to Hess's law, we can write

$$\Delta H^\circ_{\text{overall}} = \Delta H_1^\circ + \Delta H_2^\circ + \Delta H_3^\circ + \Delta H_4^\circ + \Delta H_5^\circ$$

or

$$-594.1 \text{ kJ/mol} = 155.2 \text{ kJ/mol} + 75.3 \text{ kJ/mol} + 520 \text{ kJ/mol} - 328 \text{ kJ/mol} + \Delta H_5^\circ$$

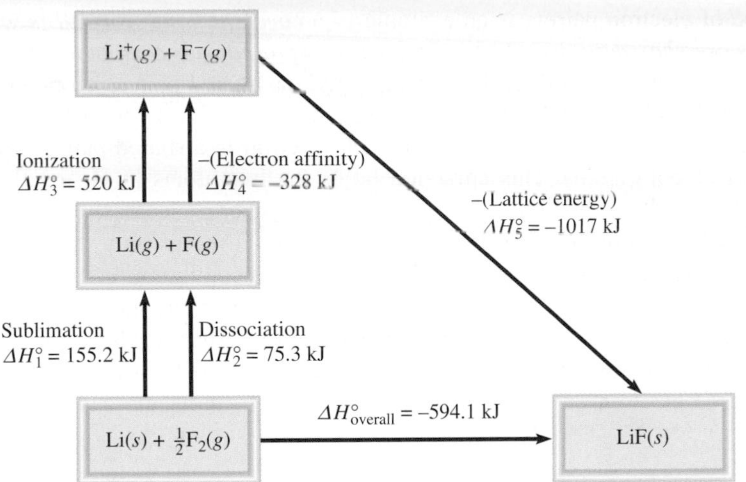

Figure 9.2
The Born-Haber cycle for the formation of 1 mole of solid LiF.

Hence,

$$\Delta H_5^\circ = -1017 \text{ kJ/mol}$$

and the lattice energy of LiF is +1017 kJ/mol.

Figure 9.2 summarizes the Born-Haber cycle for LiF. Steps 1, 2, and 3 all require the input of energy. On the other hand, steps 4 and 5 release energy. Because ΔH_5° is a large negative quantity, the lattice energy of LiF is a large positive quantity, which accounts for the stability of solid LiF. The greater the lattice energy, the more stable the ionic compound. Keep in mind that lattice energy is *always* a positive quantity because the separation of ions in a solid into ions in the gas phase is, by Coulomb's law, an endothermic process.

Table 9.1 lists the lattice energies and the melting points of several common ionic compounds. There is a rough correlation between lattice energy and melting point. The larger the lattice energy, the more stable the solid and the more tightly held the ions. It takes more energy to melt such a solid, and so the solid has a higher melting point than one with a smaller lattice energy. Note that $MgCl_2$, MgO, and CaO have unusually high lattice energies. The first of these ionic compounds has a doubly charged cation (Mg^{2+}) and in the second and third compounds there is an interaction between two doubly charged species (Mg^{2+} or Ca^{2+} and O^{2-}). The coulombic attractions between two doubly charged species, or between a doubly charged ion and a singly charged ion, are much stronger than those between singly charged anions and cations.

Table 9.1

Lattice Energies and Melting Points of Some Ionic Compounds

	Lattice Energy (kJ/mol)	Melting Point (°C)
LiF	1017	845
LiCl	828	610
NaCl	788	801
NaBr	736	750
$MgCl_2$	2527	714
MgO	3890	2800
CaO	3414	2580

Review of Concepts

Which of the following compounds has a larger lattice energy, LiCl or CsBr?

9.4 The Covalent Bond

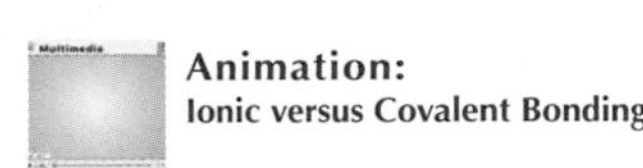

Although the concept of molecules goes back to the seventeenth century, it was not until early in the twentieth century that chemists began to understand how and why molecules form. The first major breakthrough was Gilbert Lewis's suggestion that a chemical bond involves electron sharing by atoms. He depicted the formation of a chemical bond in H_2 as

$$\text{H}\cdot + \cdot\text{H} \longrightarrow \text{H:H}$$

This type of electron pairing is an example of a ***covalent bond,*** *a bond in which two electrons are shared by two atoms.* ***Covalent compounds*** are *compounds that contain only covalent bonds.* For the sake of simplicity, the shared pair of electrons is often represented by a single line. Thus, the covalent bond in the hydrogen molecule can be written as H—H. In a covalent bond, each electron in a shared pair is attracted to the nuclei of both atoms. This attraction holds the two atoms in H_2 together and is responsible for the formation of covalent bonds in other molecules.

The discussion applies only to representative elements. Remember that for these elements, the number of valence electrons is equal to the group number (Groups 1A–7A).

Covalent bonding between many-electron atoms involves only the valence electrons. Consider the fluorine molecule, F_2. The electron configuration of F is $1s^22s^22p^5$. The $1s$ electrons are low in energy and stay near the nucleus most of the time. For this reason they do not participate in bond formation. Thus, each F atom has seven valence electrons (the $2s$ and $2p$ electrons). According to Figure 9.1, there is only one unpaired electron on F, so the formation of the F_2 molecule can be represented as follows:

$$:\ddot{\underset{..}{F}}\cdot + \cdot\ddot{\underset{..}{F}}: \longrightarrow :\ddot{\underset{..}{F}}:\ddot{\underset{..}{F}}: \quad \text{or} \quad :\ddot{\underset{..}{F}}—\ddot{\underset{..}{F}}:$$

Note that only two valence electrons participate in the formation of F_2. The other, nonbonding electrons, are called ***lone pairs***—*pairs of valence electrons that are not involved in covalent bond formation.* Thus, each F in F_2 has three lone pairs of electrons:

$$\text{lone pairs} \longrightarrow :\ddot{\underset{..}{F}}—\ddot{\underset{..}{F}}: \longleftarrow \text{lone pairs}$$

The structures we use to represent covalent compounds, such as H_2 and F_2, are called Lewis structures. A ***Lewis structure*** is *a representation of covalent bonding in which shared electron pairs are shown either as lines or as pairs of dots between two atoms, and lone pairs are shown as pairs of dots on individual atoms.* Only valence electrons are shown in a Lewis structure.

Let us consider the Lewis structure of the water molecule. Figure 9.1 shows the Lewis dot symbol for oxygen with two unpaired dots or two unpaired electrons, as we expect that O might form two covalent bonds. Because hydrogen has only one electron, it can form only one covalent bond. Thus, the Lewis structure for water is

$$H:\ddot{\underset{..}{O}}:H \quad \text{or} \quad H—\ddot{\underset{..}{O}}—H$$

In this case, the O atom has two lone pairs. The hydrogen atom has no lone pairs because its only electron is used to form a covalent bond.

In the F_2 and H_2O molecules, the F and O atoms achieve a noble gas configuration by sharing electrons:

$8e^-$ $8e^-$ $\quad$ $2e^-$ $8e^-$ $2e^-$

The formation of these molecules illustrates the ***octet rule,*** formulated by Lewis: *An atom other than hydrogen tends to form bonds until it is surrounded by eight valence electrons.* In other words, a covalent bond forms when there are not enough electrons for each individual atom to have a complete octet. By sharing electrons in a covalent bond, the individual atoms can complete their octets. The requirement for hydrogen is that it attain the electron configuration of helium, or a total of two electrons.

The octet rule works mainly for elements in the second period of the periodic table. These elements have only 2*s* and 2*p* subshells, which can hold a total of eight electrons. When an atom of one of these elements forms a covalent compound, it can attain the noble gas electron configuration [Ne] by sharing electrons with other atoms in the same compound. Later, we will discuss a number of important exceptions to the octet rule that give us further insight into the nature of chemical bonding.

Atoms can form different types of covalent bonds. In a ***single bond,*** *two atoms are held together by one electron pair.* Many compounds are held together by ***multiple bonds,*** that is, bonds formed when *two atoms share two or more pairs of electrons.* If *two atoms share two pairs of electrons,* the covalent bond is called a ***double bond.*** Double bonds are found in molecules of carbon dioxide (CO_2) and ethylene (C_2H_4):

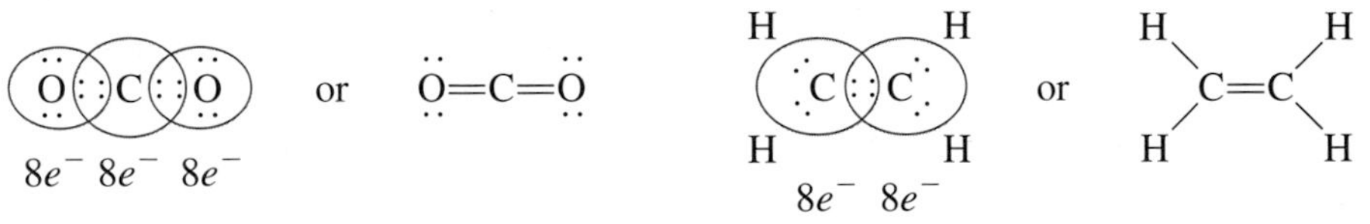

$$\ddot{O}::C::\ddot{O} \quad \text{or} \quad \ddot{O}=C=\ddot{O}$$

8e^- 8e^- 8e^-

$$\begin{matrix} H & & H \\ & C::C & \\ H & & H \end{matrix} \quad \text{or} \quad \begin{matrix} H & & H \\ & C=C & \\ H & & H \end{matrix}$$

8e^- 8e^-

A ***triple bond*** arises when *two atoms share three pairs of electrons,* as in the nitrogen molecule (N_2):

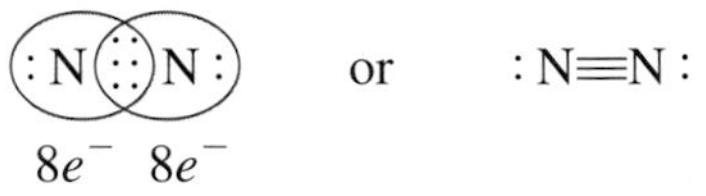

$$:N:::N: \quad \text{or} \quad :N\equiv N:$$

8e^- 8e^-

The acetylene molecule (C_2H_2) also contains a triple bond, in this case between two carbon atoms:

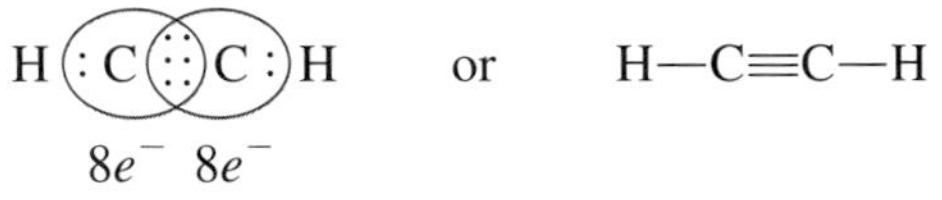

$$H:C:::C:H \quad \text{or} \quad H{-}C\equiv C{-}H$$

8e^- 8e^-

Note that in ethylene and acetylene all the valence electrons are used in bonding; there are no lone pairs on the carbon atoms. In fact, with the exception of carbon monoxide, the vast majority of stable molecules containing carbon do not have lone pairs on the carbon atoms.

For the same pair of atoms, multiple bonds are shorter than single covalent bonds. ***Bond length*** is defined as the *distance between the nuclei of two covalently bonded atoms in a molecule* (Figure 9.3). Table 9.2 shows some experimentally determined bond lengths. For a given pair of atoms, such as carbon and nitrogen, triple bonds are shorter than double bonds, which, in turn, are shorter than single bonds. The shorter multiple bonds are also more stable than single bonds, as we will see later.

74 pm 161 pm

H_2 HI

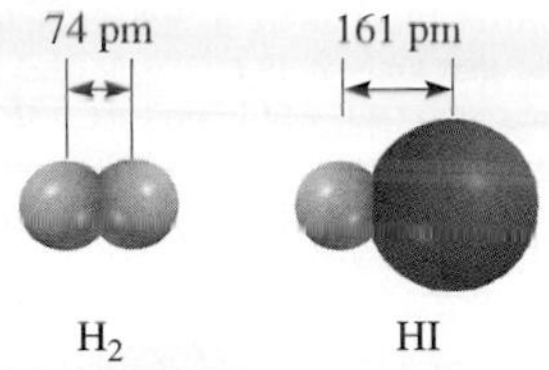

Figure 9.3
Bond length (in pm) in H_2 and HI.

Shortly, you will be introduced to the rules for writing proper Lewis structures. Here we simply want to become familiar with the language associated with them.

Table 9.2
Average Bond Lengths of Some Common Single, Double, and Triple Bonds

Bond Type	Bond Length (pm)
C—H	107
C—O	143
C=O	121
C—C	154
C=C	133
C≡C	120
C—N	143
C=N	138
C≡N	116
N—O	136
N=O	122
O—H	96

REVIEW OF CONCEPTS

Why is it not possible for hydrogen to form double or triple bonds in covalently bonded compounds?

9.5 Electronegativity

A covalent bond, as we have said, is the sharing of an electron pair by two atoms. In a molecule like H_2, in which the atoms are identical, we expect the electrons to be equally shared—that is, the electrons spend the same amount of time in the vicinity

Hydrogen fluoride is a clear, fuming liquid that boils at 19.8°C. It is used to make refrigerants and to prepare hydrofluoric acid.

of each atom. However, in the covalently bonded HF molecule, the H and F atoms do not share the bonding electrons equally because H and F are different atoms:

$$\mathrm{H{-}\ddot{\underset{\cdot\cdot}{F}}:}$$

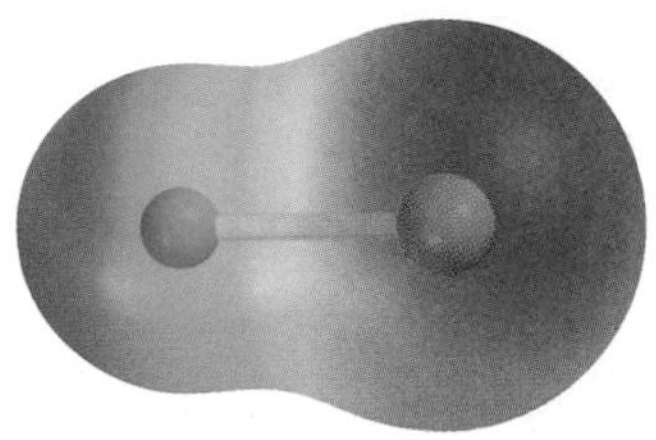

Figure 9.4
Electrostatic potential map of the HF molecule. The distribution varies according to the colors of the rainbow. The most electron-rich region is red; the most electron-poor region is blue.

The bond in HF is called a ***polar covalent bond,*** or simply a *polar bond,* because *the electrons spend more time in the vicinity of one atom than the other.* Experimental evidence indicates that in the HF molecule the electrons spend more time near the F atom. We can think of this unequal sharing of electrons as a partial electron transfer or a shift in electron density, as it is more commonly described, from H to F (Figure 9.4). This "unequal sharing" of the bonding electron pair results in a relatively greater electron density near the fluorine atom and a correspondingly lower electron density near hydrogen. The HF bond and other polar bonds can be thought of as being intermediate between a (nonpolar) covalent bond, in which the sharing of electrons is exactly equal, and an ionic bond, in which the transfer of the electron(s) is nearly complete.

A property that helps us distinguish a nonpolar covalent bond from a polar covalent bond is ***electronegativity,*** *the ability of an atom to attract toward itself the electrons in a chemical bond.* Elements with high electronegativity have a greater tendency to attract electrons than do elements with low electronegativity. As we might expect, electronegativity is related to electron affinity and ionization energy. Thus, an atom such as fluorine, which has a high electron affinity (tends to pick up electrons easily) and a high ionization energy (does not lose electrons easily), has a high electronegativity. On the other hand, sodium has a low electron affinity, a low ionization energy, and a low electronegativity.

Electronegativity values have no units.

Electronegativity is a relative concept, meaning that an element's electronegativity can be measured only in relation to the electronegativity of other elements. The American chemist Linus Pauling devised a method for calculating *relative* electronegativities of most elements. These values are shown in Figure 9.5. A careful examination of this chart reveals trends and relationships among electronegativity values of

Increasing electronegativity (→ across a period)

Increasing electronegativity (↑ up a group)

1A	2A	3B	4B	5B	6B	7B	8B	8B	8B	1B	2B	3A	4A	5A	6A	7A	8A
H 2.1																	
Li 1.0	**Be** 1.5											**B** 2.0	**C** 2.5	**N** 3.0	**O** 3.5	**F** 4.0	
Na 0.9	**Mg** 1.2											**Al** 1.5	**Si** 1.8	**P** 2.1	**S** 2.5	**Cl** 3.0	
K 0.8	**Ca** 1.0	**Sc** 1.3	**Ti** 1.5	**V** 1.6	**Cr** 1.6	**Mn** 1.5	**Fe** 1.8	**Co** 1.9	**Ni** 1.9	**Cu** 1.9	**Zn** 1.6	**Ga** 1.6	**Ge** 1.8	**As** 2.0	**Se** 2.4	**Br** 2.8	**Kr** 3.0
Rb 0.8	**Sr** 1.0	**Y** 1.2	**Zr** 1.4	**Nb** 1.6	**Mo** 1.8	**Tc** 1.9	**Ru** 2.2	**Rh** 2.2	**Pd** 2.2	**Ag** 1.9	**Cd** 1.7	**In** 1.7	**Sn** 1.8	**Sb** 1.9	**Te** 2.1	**I** 2.5	**Xe** 2.6
Cs 0.7	**Ba** 0.9	**La-Lu** 1.0-1.2	**Hf** 1.3	**Ta** 1.5	**W** 1.7	**Re** 1.9	**Os** 2.2	**Ir** 2.2	**Pt** 2.2	**Au** 2.4	**Hg** 1.9	**Tl** 1.8	**Pb** 1.9	**Bi** 1.9	**Po** 2.0	**At** 2.2	
Fr 0.7	**Ra** 0.9																

Figure 9.5
The electronegativities of common elements.

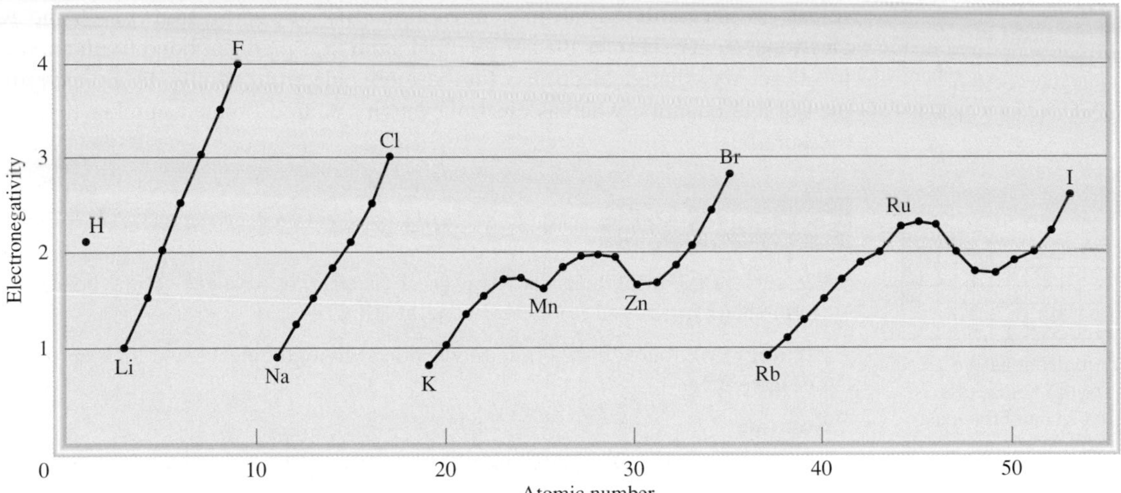

Figure 9.6
Variation of electronegativity with atomic number. The halogens have the highest electronegativities, and the alkali metals the lowest.

different elements. In general, electronegativity increases from left to right across a period in the periodic table, as the metallic character of the elements decreases. Within each group, electronegativity decreases with increasing atomic number, and increasing metallic character. Note that the transition metals do not follow these trends. The most electronegative elements—the halogens, oxygen, nitrogen, and sulfur—are found in the upper right-hand corner of the periodic table, and the least electronegative elements (the alkali and alkaline earth metals) are clustered near the lower left-hand corner. These trends are readily apparent on a graph, as shown in Figure 9.6.

Atoms of elements with widely different electronegativities tend to form ionic bonds (such as those that exist in NaCl and CaO compounds) with each other because the atom of the less electronegative element gives up its electron(s) to the atom of the more electronegative element. An ionic bond generally joins an atom of a metallic element and an atom of a nonmetallic element. Atoms of elements with comparable electronegativities tend to form polar covalent bonds with each other because the shift in electron density is usually small. Most covalent bonds involve atoms of nonmetallic elements. Only atoms of the same element, which have the same electronegativity, can be joined by a pure covalent bond. These trends and characteristics are what we would expect, given our knowledge of ionization energies and electron affinities.

There is no sharp distinction between a polar covalent bond and an ionic bond, but the following rules are helpful as a rough guide. An ionic bond forms when the electronegativity difference between the two bonding atoms is 2.0 or more. This rule applies to most but not all ionic compounds. A polar covalent bond forms when the electronegativity difference between the atoms is in the range of 0.3–2.0. If the electronegativity difference is below 0.3, the bond is normally classified as a covalent bond, with little or no polarity. Sometimes chemists use the quantity *percent ionic character* to describe the nature of a bond. A purely ionic bond would have 100 percent ionic character, although no such bond is known, whereas a purely covalent bond (such as that in H_2) has 0 percent ionic character. As Figure 9.7 shows there is a correlation between the percent ionic character of a bond and the electronegativity difference between the bonding atoms.

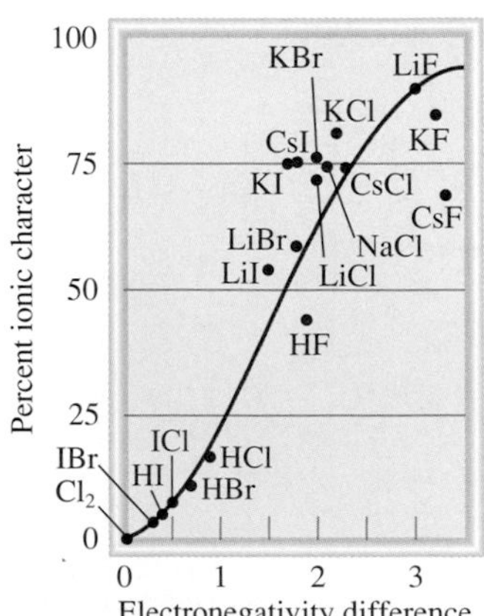

Figure 9.7
Relation between percent ionic character and electronegativity difference.

Electronegativity and electron affinity are related but different concepts. Both indicate the tendency of an atom to attract electrons. However, electron affinity refers to an isolated atom's attraction for an additional electron, whereas electronegativity signifies the ability of an atom in a chemical bond (with another atom) to attract the shared electrons. Furthermore, electron affinity is an experimentally measurable quantity, whereas electronegativity is an estimated number that cannot be measured.

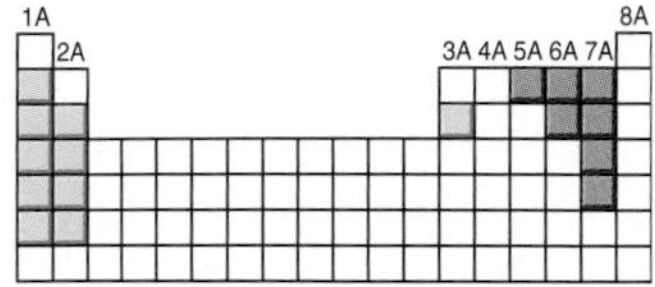

The most electronegative elements are the nonmetals (Groups 5A–7A) and the least electronegative elements are the alkali and alkaline earth metals (Groups 1A–2A) and aluminum. Beryllium, the first member of Group 2A, forms mostly covalent compounds.

EXAMPLE 9.2

Classify the following bonds as ionic, polar covalent, or covalent: (a) the bond in HCl, (b) the bond in KF, and (c) the CC bond in H_3CCH_3.

Strategy We follow the 2.0 rule of electronegativity difference and look up the values in Figure 9.5.

Solution

(a) The electronegativity difference between H and Cl is 0.9, which is appreciable but not large enough (by the 2.0 rule) to qualify HCl as an ionic compound. Therefore, the bond between H and Cl is polar covalent.

(b) The electronegativity difference between K and F is 3.2, which is well above the 2.0 mark; therefore, the bond between K and F is ionic.

(c) The two C atoms are identical in every respect—they are bonded to each other and each is bonded to three other H atoms. Therefore, the bond between them is purely covalent.

Similar problems: 9.37, 9.38.

Practice Exercise Which of the following bonds is covalent, which is polar covalent, and which is ionic? (a) the bond in CsCl, (b) the bond in H_2S, (c) the NN bond in H_2NNH_2.

Electronegativity and Oxidation Number

In Chapter 4 we introduced the rules for assigning oxidation numbers of elements in their compounds. The concept of electronegativity is the basis for these rules. In essence, oxidation number refers to the number of charges an atom would have if electrons were transferred completely to the more electronegative of the bonded atoms in a molecule.

Consider the NH_3 molecule, in which the N atom forms three single bonds with the H atoms. Because N is more electronegative than H, electron density will be shifted from H to N. If the transfer were complete, each H would donate an electron to N, which would have a total charge of -3 while each H would have a charge of $+1$. Thus, we assign an oxidation number of -3 to N and an oxidation number of $+1$ to H in NH_3.

Oxygen usually has an oxidation number of -2 in its compounds, except in hydrogen peroxide (H_2O_2), whose Lewis structure is

$$\mathrm{H{-}\ddot{\underset{..}{O}}{-}\ddot{\underset{..}{O}}{-}H}$$

A bond between identical atoms makes no contribution to the oxidation number of those atoms because the electron pair of that bond is *equally* shared. Because H has an oxidation number of $+1$, each O atom has an oxidation number of -1.

Can you see now why fluorine always has an oxidation number of -1? It is the most electronegative element known, and it *usually* forms a single bond in its compounds. Therefore, it would bear a -1 charge if electron transfer were complete.

Review of Concepts

Identify the electrostatic potential maps shown here with HCl and LiH. In both diagrams, the H atom is on the left.

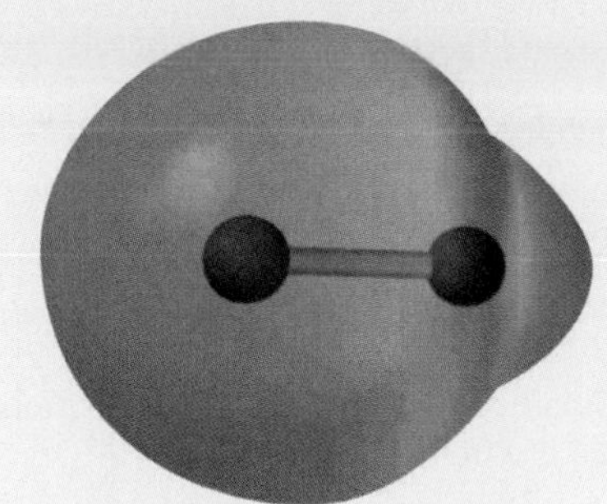

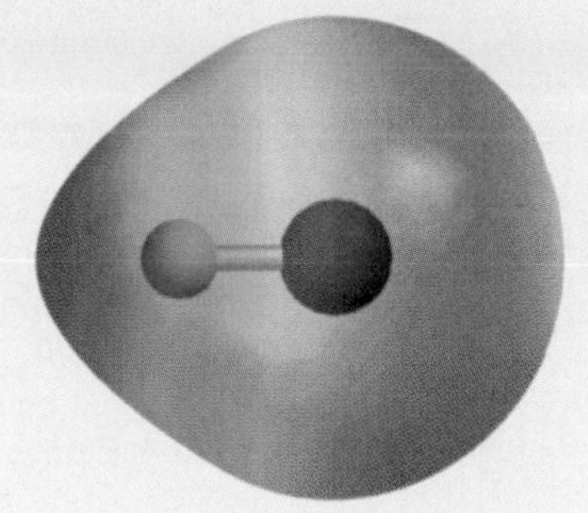

9.6 Writing Lewis Structures

Although the octet rule and Lewis structures do not present a complete picture of covalent bonding, they do help to explain the bonding scheme in many compounds and account for the properties and reactions of molecules. For this reason, you should practice writing Lewis structures of compounds. The basic steps are as follows:

1. Write the skeletal structure of the compound, using chemical symbols and placing bonded atoms next to one another. For simple compounds, this task is fairly easy. For more complex compounds, we must either be given the information or make an intelligent guess about it. In general, the least electronegative atom occupies the central position. Hydrogen and fluorine usually occupy the terminal (end) positions in the Lewis structure.
2. Count the total number of valence electrons present, referring, if necessary, to Figure 9.1. For polyatomic anions, add the number of negative charges to that total. (For example, for the CO_3^{2-} ion we add two electrons because the 2− charge indicates that there are two more electrons than are provided by the atoms.) For polyatomic cations, we subtract the number of positive charges from this total. (Thus, for NH_4^+ we subtract one electron because the 1+ charge indicates a loss of one electron from the group of atoms.)
3. Draw a single covalent bond between the central atom and each of the surrounding atoms. Complete the octets of the atoms bonded to the central atom. (Remember that the valence shell of a hydrogen atom is complete with only two electrons.) Electrons belonging to the central or surrounding atoms must be shown as lone pairs if they are not involved in bonding. The total number of electrons to be used is that determined in step 2.
4. After completing steps 1–3, if the central atom has fewer than eight electrons, try adding double or triple bonds between the surrounding atoms and the central atom, using lone pairs from the surrounding atoms to complete the octet of the central atom.

Hydrogen follows a "duet rule" when drawing Lewis structures.

NF_3 is a colorless, odorless, unreactive gas.

EXAMPLE 9.3

Write the Lewis structure for nitrogen trifluoride (NF_3) in which all three F atoms are bonded to the N atom.

Solution We follow the preceding procedure for writing Lewis structures.

Step 1: The N atom is less electronegative than F, so the skeletal structure of NF_3 is

F N F
F

Step 2: The outer-shell electron configurations of N and F are $2s^22p^3$ and $2s^22p^5$, respectively. Thus, there are $5 + (3 \times 7)$, or 26, valence electrons to account for in NF_3.

Step 3: We draw a single covalent bond between N and each F, and complete the octets for the F atoms. We place the remaining two electrons on N:

$$:\ddot{\underset{..}{F}}-\ddot{N}-\ddot{\underset{..}{F}}: \quad \text{with } :\underset{..}{F}: \text{ bonded below N}$$

Because this structure satisfies the octet rule for all the atoms, step 4 is not required.

Check Count the valence electrons in NF_3 (in bonds and in lone pairs). The result is 26, the same as the total number of valence electrons on three F atoms ($3 \times 7 = 21$) and one N atom (5).

Similar problem: 9.41.

Practice Exercise Write the Lewis structure for carbon disulfide (CS_2).

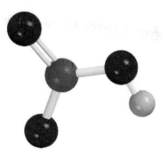

HNO_3 is a strong electrolyte.

EXAMPLE 9.4

Write the Lewis structure for nitric acid (HNO_3) in which the three O atoms are bonded to the central N atom and the ionizable H atom is bonded to one of the O atoms.

Solution We follow the procedure already outlined for writing Lewis structures.

Step 1: The skeletal structure of HNO_3 is

O N O H
O

Step 2: The outer-shell electron configurations of N, O, and H are $2s^22p^3$, $2s^22p^4$, and $1s^1$, respectively. Thus, there are $5 + (3 \times 6) + 1$, or 24, valence electrons to account for in HNO_3.

Step 3: We draw a single covalent bond between N and each of the three O atoms and between one O atom and the H atom. Then we fill in electrons to comply with the octet rule for the O atoms:

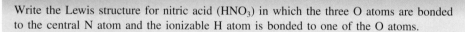

$$:\ddot{\underset{..}{O}}-N-\ddot{\underset{..}{O}}-H \quad \text{with } :\underset{..}{\ddot{O}}: \text{ bonded below N}$$

Step 4: We see that this structure satisfies the octet rule for all the O atoms but not for the N atom. The N atom has only six electrons. Therefore, we move a lone pair

(Continued)

from one of the end O atoms to form another bond with N. Now the octet rule is also satisfied for the N atom:

```
 ..      ..
 O=N—O—H
 ..      ..
   |
  :O:
   ..
```

Check Make sure that all the atoms (except H) satisfy the octet rule. Count the valence electrons in HNO_3 (in bonds and in lone pairs). The result is 24, the same as the total number of valence electrons on three O atoms ($3 \times 6 = 18$), one N atom (5), and one H atom (1).

Similar problem: 9.41.

Practice Exercise Write the Lewis structure for formic acid (HCOOH).

EXAMPLE 9.5

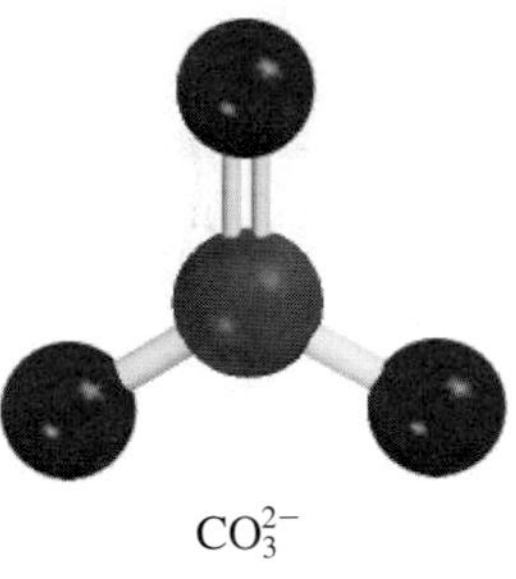

CO_3^{2-}

Write the Lewis structure for the carbonate ion (CO_3^{2-}).

Solution We follow the preceding procedure for writing Lewis structures and note that this is an anion with two negative charges.

Step 1: We can deduce the skeletal structure of the carbonate ion by recognizing that C is less electronegative than O. Therefore, it is most likely to occupy a central position as follows:

```
   O
O  C  O
```

Step 2: The outer-shell electron configurations of C and O are $2s^22p^2$ and $2s^22p^4$, respectively, and the ion itself has two negative charges. Thus, the total number of electrons is $4 + (3 \times 6) + 2$, or 24.

Step 3: We draw a single covalent bond between C and each O and comply with the octet rule for the O atoms:

```
      ..
     :O:
 ..   |   ..
:O—C—O:
 ..       ..
```

This structure shows all 24 electrons.

Step 4: Although the octet rule is satisfied for the O atoms, it is not for the C atom. Therefore we move a lone pair from one of the O atoms to form another bond with C. Now the octet rule is also satisfied for the C atom:

```
⌈     :O:      ⌉2−
|  ..  ‖   ..  |
| :O—C—O:  |
⌊  ..       ..  ⌋
```

We use the brackets to indicate that the 2− charge is on the whole molecule.

Check Make sure that all the atoms satisfy the octet rule. Count the valence electrons in CO_3^{2-} (in chemical bonds and in lone pairs). The result is 24, the same as the total number of valence electrons on three O atoms ($3 \times 6 = 18$), one C atom (4), and two negative charges (2).

Similar problem: 9.42.

Practice Exercise Write the Lewis structure for the nitrite ion (NO_2^-).

Review of Concepts

The molecular model shown here represents guanine, a component of a DNA molecule. Only the connections between the atoms are shown in this model. Draw a complete Lewis structure of the molecule, showing all the multiple bonds and lone pairs. (For color code, see inside back endpaper.)

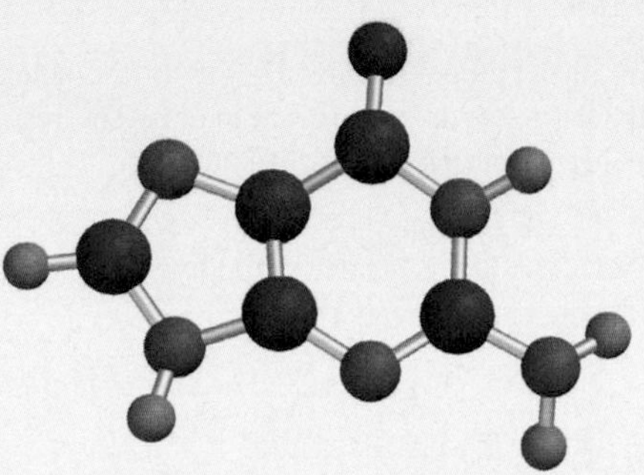

9.7 Formal Charge and Lewis Structure

By comparing the number of electrons in an isolated atom with the number of electrons that are associated with the same atom in a Lewis structure, we can determine the distribution of electrons in the molecule and draw the most plausible Lewis structure. The bookkeeping procedure is as follows: In an isolated atom, the number of electrons associated with the atom is simply the number of valence electrons. (As usual, we need not be concerned with the inner electrons.) In a molecule, electrons associated with the atom are the nonbonding electrons plus the electrons in the bonding pair(s) between the atom and other atom(s). However, because electrons are shared in a bond, we must divide the electrons in a bonding pair equally between the atoms forming the bond. An atom's ***formal charge*** is *the electrical charge difference between the valence electrons in an isolated atom and the number of electrons assigned to that atom in a Lewis structure.*

To assign the number of electrons on an atom in a Lewis structure, we proceed as follows:

- All the atom's nonbonding electrons are assigned to the atom.
- We break the bond(s) between the atom and other atom(s) and assign half of the bonding electrons to the atom.

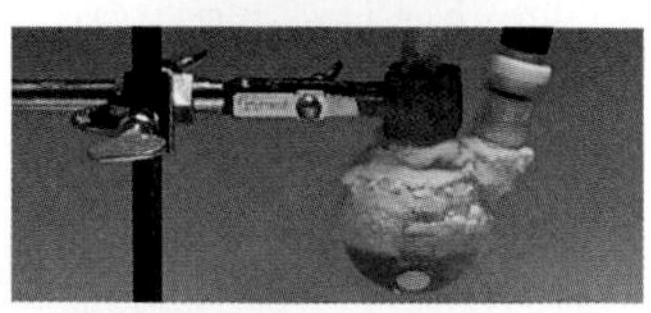

Liquid ozone below its boiling point (−111.3°C). Ozone is a toxic, light-blue gas with a pungent odor.

Let us illustrate the concept of formal charge using the ozone molecule (O_3). Proceeding by steps, as we did in Examples 9.3 and 9.4, we draw the skeletal structure of O_3 and then add bonds and electrons to satisfy the octet rule for the two end atoms:

$$:\underset{\cdot\cdot}{\ddot{O}}-\ddot{O}-\underset{\cdot\cdot}{\ddot{O}}:$$

You can see that although all available electrons are used, the octet rule is not satisfied for the central atom. To remedy this, we convert a lone pair on one of the end atoms to a second bond between that end atom and the central atom, as follows:

$$\underset{\cdot\cdot}{\ddot{O}}=\ddot{O}-\underset{\cdot\cdot}{\ddot{O}}:$$

The formal charge on each atom in O_3 can now be calculated according to the following scheme:

	O	O	O
Valence e^-	6	6	6
e^- assigned to atom	6	5	7
Difference (formal charge)	0	+1	−1

Assign half of the bonding electrons to each atom.

where the wavy red lines denote the breaking of the bonds. Note that the breaking of a single bond results in the transfer of an electron, the breaking of a double bond results in a transfer of two electrons to each of the bonding atoms, and so on. Thus, the formal charges of the atoms in O_3 are

By the same token, the breaking of a triple bond transfers three electrons to each of the bonding atoms.

$$\ddot{\underset{..}{O}}=\overset{+}{\ddot{O}}-\ddot{\underset{..}{O}}:^{-}$$

For single positive and negative charges, we normally omit the numeral 1.

When you write formal charges, these rules are helpful:

1. For molecules, the sum of the formal charges must add up to zero because molecules are electrically neutral species. (This rule applies, for example, to the O_3 molecule.)
2. For cations, the sum of the formal charges must equal the positive charge.
3. For anions, the sum of the formal charges must equal the negative charge.

In determining formal charges, does the atom in the molecule (or ion) have more electrons than its valence electrons (negative formal charge), or does the atom have fewer electrons than its valence electrons (positive formal charge)?

Keep in mind that formal charges do not represent actual charge separation within the molecule. In the O_3 molecule, for example, there is no evidence that the central atom bears a net +1 charge or that one of the end atoms bears a −1 charge. Writing these charges on the atoms in the Lewis structure merely helps us keep track of the valence electrons in the molecule.

EXAMPLE 9.6

Write formal charges for the carbonate ion.

Solution The Lewis structure for the carbonate ion was developed in Example 9.5:

$$\left[\begin{matrix} & :O: & \\ & \| & \\ :\ddot{\underset{..}{O}}- & C & -\ddot{\underset{..}{O}}: \end{matrix}\right]^{2-}$$

The formal charges on the atoms can be calculated using the given procedure.

The C atom: The C atom has four valence electrons and there are no nonbonding electrons on the atom in the Lewis structure. The breaking of the double bond and two single bonds results in the transfer of four electrons to the C atom. Therefore, the formal charge is 4 − 4 = 0.

The O atom in C=O: The O atom has six valence electrons and there are four nonbonding electrons on the atom. The breaking of the double bond results in the transfer of two electrons to the O atom. Here the formal charge is 6 − 4 − 2 = 0.

The O atom in C—O: This atom has six nonbonding electrons and the breaking of the single bond transfers another electron to it. Therefore, the formal charge is 6 − 6 − 1 = −1.

(Continued)

Thus, the Lewis structure for CO_3^{2-} with formal charges is

$$\begin{array}{c} :\ddot{O}: \\ \| \\ ^{-}:\ddot{\underset{..}{O}}-C-\ddot{\underset{..}{O}}:^{-} \end{array}$$

Check Note that the sum of the formal charges is −2, the same as the charge on the carbonate ion.

Similar problem: 9.42.

Practice Exercise Write formal charges for the nitrite ion (NO_2^-).

Sometimes there is more than one acceptable Lewis structure for a given species. In such cases, we can often select the most plausible Lewis structure by using formal charges and the following guidelines:

- For molecules, a Lewis structure in which there are no formal charges is preferable to one in which formal charges are present.
- Lewis structures with large formal charges (+2, +3, and/or −2, −3, and so on) are less plausible than those with small formal charges.
- Among Lewis structures having similar distributions of formal charges, the most plausible structure is the one in which negative formal charges are placed on the more electronegative atoms.

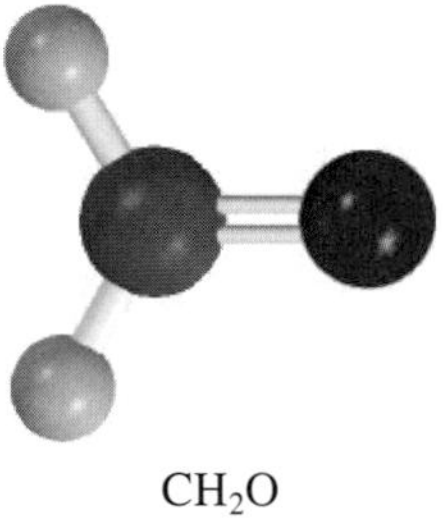

CH_2O

EXAMPLE 9.7

Formaldehyde (CH_2O), a liquid with a disagreeable odor, traditionally has been used to preserve laboratory specimens. Draw the most likely Lewis structure for the compound.

Strategy A plausible Lewis structure should satisfy the octet rule for all the elements, except H, and have the formal charges (if any) distributed according to electronegativity guidelines.

Solution The two possible skeletal structures are

$$\text{H C O H} \qquad \begin{array}{c} \text{H} \\ \text{C O} \\ \text{H} \end{array}$$

(a) (b)

First we draw the Lewis structures for each of these possibilities

$$\text{H}-\overset{-}{\underset{}{\ddot{C}}}=\overset{+}{\ddot{O}}-\text{H} \qquad \begin{array}{l} \text{H} \\ \ \ \diagdown \\ \ \ \ \text{C}=\ddot{\underset{..}{O}} \\ \ \ \diagup \\ \text{H} \end{array}$$

(a) (b)

To show the formal charges, we follow the procedure given in Example 9.6. In (a), the C atom has a total of five electrons (one lone pair plus three electrons from the breaking of a single and a double bond). Because C has four valence electrons, the formal charge on the atom is 4 − 5 = −1. The O atom has a total of five electrons (one lone pair and three electrons from the breaking of a single and a double bond). Because O has six valence electrons, the formal charge on the atom is 6 − 5 = +1.

(Continued)

In (b) the C atom has a total of four electrons from the breaking of two single bonds and a double bond, so its formal charge is 4 − 4 = 0. The O atom has a total of six electrons (two lone pairs and two electrons from the breaking of the double bond). Therefore, the formal charge on the atom is 6 − 6 = 0. Although both structures satisfy the octet rule, (b) is the more likely structure because it carries no formal charges.

Check In each case, make sure that the total number of valence electrons is 12. Can you suggest two other reasons why (a) is less plausible?

Similar problem: 9.43.

Practice Exercise Draw the most reasonable Lewis structure of a molecule that contains an N atom, a C atom, and an H atom.

REVIEW OF CONCEPTS

Consider three of the possible atomic arrangements for cyanamide (CH_2N_2): (a) H_2CNN, (b) H_2NCN, (c) HNNCH. Using formal charges as a guide, determine the actual atomic arrangement of cyanamide.

9.8 The Concept of Resonance

Our drawing of the Lewis structure for ozone (O_3) satisfied the octet rule for the central atom because we placed a double bond between it and one of the two end O atoms. In fact, we can put the double bond at either end of the molecule, as shown by these two equivalent Lewis structures:

$$\ddot{\underset{\cdot\cdot}{O}}=\overset{+}{\ddot{O}}-\ddot{\underset{\cdot\cdot}{O}}:^{-} \qquad {}^{-}:\ddot{\underset{\cdot\cdot}{O}}-\overset{+}{\ddot{O}}=\ddot{\underset{\cdot\cdot}{O}}$$

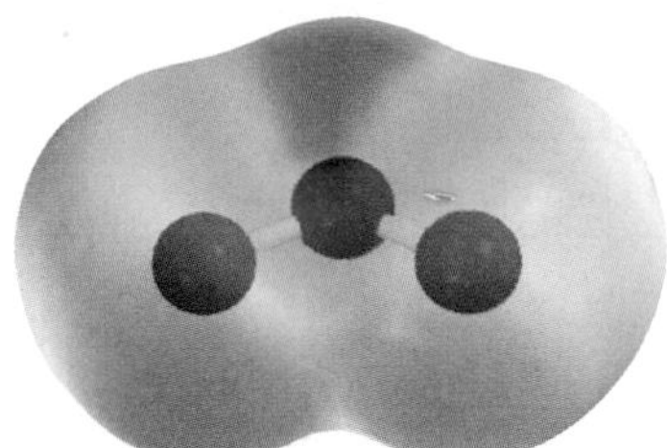

Electrostatic potential map of O_3. The electron density is evenly distributed between the two end O atoms.

However, neither one of these two Lewis structures accounts for the known bond lengths in O_3.

We would expect the O—O bond in O_3 to be longer than the O=O bond because double bonds are known to be shorter than single bonds. Yet experimental evidence shows that both oxygen-to-oxygen bonds are equal in length (128 pm). We resolve this discrepancy by using *both* Lewis structures to represent the ozone molecule:

$$\ddot{\underset{\cdot\cdot}{O}}=\overset{+}{\ddot{O}}-\ddot{\underset{\cdot\cdot}{O}}:^{-} \longleftrightarrow {}^{-}:\ddot{\underset{\cdot\cdot}{O}}-\overset{+}{\ddot{O}}=\ddot{\underset{\cdot\cdot}{O}}$$

Each of these structures is called a resonance structure. A ***resonance structure,*** then, is *one of two or more Lewis structures for a single molecule that cannot be represented accurately by only one Lewis structure.* The double-headed arrow indicates that the structures shown are resonance structures.

The term ***resonance*** itself means *the use of two or more Lewis structures to represent a particular molecule.* Like the medieval European traveler to Africa who described a rhinoceros as a cross between a griffin and a unicorn, two familiar but imaginary animals, we describe ozone, a real molecule, in terms of two familiar but nonexistent structures.

Animation:
Resonance

A common misconception about resonance is the notion that a molecule such as ozone somehow shifts quickly back and forth from one resonance structure to the other. Keep in mind that *neither* resonance structure adequately represents the actual molecule, which has its own unique, stable structure. "Resonance" is a human invention, designed to address the limitations in these simple bonding models. To extend the animal analogy, a rhinoceros is a distinct creature, not some oscillation between mythical griffin and unicorn!

The carbonate ion provides another example of resonance:

$$^{-}\!:\!\ddot{\underset{\cdot\cdot}{O}}\!-\!\overset{\overset{:O:}{\|}}{C}\!-\!\ddot{\underset{\cdot\cdot}{O}}\!:^{-} \longleftrightarrow \ddot{O}\!=\!\overset{\overset{:\ddot{O}:^{-}}{|}}{C}\!-\!\ddot{\underset{\cdot\cdot}{O}}\!:^{-} \longleftrightarrow {}^{-}\!:\!\ddot{\underset{\cdot\cdot}{O}}\!-\!\overset{\overset{:\ddot{O}:^{-}}{|}}{C}\!=\!\ddot{O}$$

According to experimental evidence, all carbon-to-oxygen bonds in CO_3^{2-} are equivalent. Therefore, the properties of the carbonate ion are best explained by considering its resonance structures together.

The concept of resonance applies equally well to organic systems. A good example is the benzene molecule (C_6H_6):

The hexagonal structure of benzene was first proposed by the German chemist August Kekulé (1829–1896).

If one of these resonance structures corresponded to the actual structure of benzene, there would be two different bond lengths between adjacent C atoms, one characteristic of the single bond and the other of the double bond. In fact, the distance between all adjacent C atoms in benzene is 140 pm, which is shorter than a C—C bond (154 pm) and longer than a C═C bond (133 pm).

A simpler way of drawing the structure of the benzene molecule and other compounds containing the "benzene ring" is to show only the skeleton and not the carbon and hydrogen atoms. By this convention the resonance structures are represented by

Note that the C atoms at the corners of the hexagon and the H atoms are all omitted, although they are understood to exist. Only the bonds between the C atoms are shown.

Remember this important rule for drawing resonance structures: The positions of electrons (that is, bonds), but not those of atoms, can be rearranged in different resonance structures. In other words, the same atoms must be bonded to one another in all the resonance structures for a given species.

So far, the resonance structures shown in the examples all contribute equally to the real structure of the molecules and ion. This is not always the case, as we will see in Example 9.8.

EXAMPLE 9.8

Draw three resonance structures for the molecule nitrous oxide, N_2O (the atomic arrangement is NNO). Indicate formal charges. Rank the structures in their relative importance to the overall properties of the molecule.

Strategy The skeletal structure for N_2O is

N N O

We follow the procedure used for drawing Lewis structure and calculating formal charges in Examples 9.5 and 9.6.

Solution The three resonance structures are

$$\overset{-}{\ddot{\underset{..}{N}}}=\overset{+}{N}=\ddot{\underset{..}{O}} \qquad :\!N\equiv\overset{+}{N}-\overset{-}{\ddot{\underset{..}{O}}}\!: \qquad :\!\overset{2-}{\ddot{\underset{..}{N}}}-\overset{+}{N}\equiv\overset{+}{O}\!:$$

(a) (b) (c)

We see that all three structures show formal charges. Structure (b) is the most important one because the negative charge is on the more electronegative oxygen atom. Structure (c) is the least important one because it has a larger separation of formal charges. Also, the positive charge is on the more electronegative oxygen atom.

Check Make sure there is no change in the positions of the atoms in the structures. Because N has five valence electrons and O has six valence electrons, the total number of valence electrons is $5 \times 2 + 6 = 16$. The sum of formal charges is zero in each structure.

Practice Exercise Draw three resonance structures for the thiocyanate ion, SCN^-. Rank the structures in decreasing order of importance.

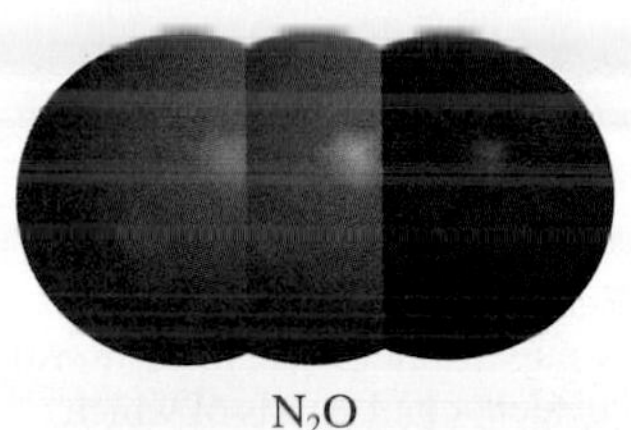

N_2O

Resonance structures with formal charges greater than +2 or −2 are usually considered highly implausible and can be discarded.

Similar problems: 9.49, 9.54.

REVIEW OF CONCEPTS

The molecular model shown here represents acetamide, which is used as an organic solvent. Only the connections between the atoms are shown in this model. Draw two resonance structures for the molecule, showing the positions of multiple bonds and formal charges. (For color code, see inside back endpaper.)

9.9 Exceptions to the Octet Rule

As mentioned earlier, the octet rule applies mainly to the second-period elements. Exceptions to the octet rule fall into three categories characterized by an incomplete octet, an odd number of electrons, or more than eight valence electrons around the central atom.

The Incomplete Octet

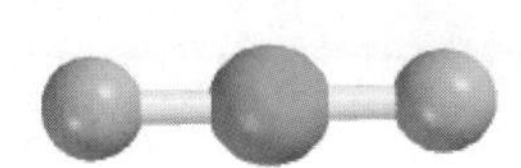

Beryllium, unlike the other Group 2A elements, forms mostly covalent compounds of which BeH_2 is an example.

In some compounds, the number of electrons surrounding the central atom in a stable molecule is fewer than eight. Consider, for example, beryllium, which is a Group 2A (and a second-period) element. The electron configuration of beryllium is $1s^22s^2$; it has two valence electrons in the 2*s* orbital. In the gas phase, beryllium hydride (BeH_2) exists as discrete molecules. The Lewis structure of BeH_2 is

$$\mathrm{H—Be—H}$$

As you can see, only four electrons surround the Be atom, and there is no way to satisfy the octet rule for beryllium in this molecule.

Elements in Group 3A, particularly boron and aluminum, also tend to form compounds in which they are surrounded by fewer than eight electrons. Take boron as an example. Because its electron configuration is $1s^22s^22p^1$, it has a total of three valence electrons. Boron reacts with the halogens to form a class of compounds having the general formula BX_3, where X is a halogen atom. Thus, in boron trifluoride there are only six electrons around the boron atom:

```
        ..
       :F:
   ..   |
  :F—B
   ..   |
       :F:
        ..
```

The following resonance structures all contain a double bond between B and F and satisfy the octet rule for boron:

```
      ..             ..+             ..
     :F:            :F:             :F:
 ..   |         ..   ||         ..   |
+F=B⁻   ⟷    :F—B⁻    ⟷    :F—B⁻
 ..   |         ..   |          ..   ||
     :F:            :F:             :F:+
      ..             ..              ..
```

The fact that the B—F bond length in BF_3 (130.9 pm) is shorter than a single bond (137.3 pm) lends support to the resonance structures even though in each case the negative formal charge is placed on the B atom and the positive formal charge on the F atom.

Although boron trifluoride is stable, it readily reacts with ammonia. This reaction is better represented by using the Lewis structure in which boron has only six valence electrons around it:

```
      ..                          ..
     :F:     H                   :F:   H
 ..   |      |               ..   |    |
:F—B  +  :N—H  ⟶   :F—B⁻—N⁺—H
 ..   |      |               ..   |    |
     :F:     H                   :F:   H
      ..                          ..
```

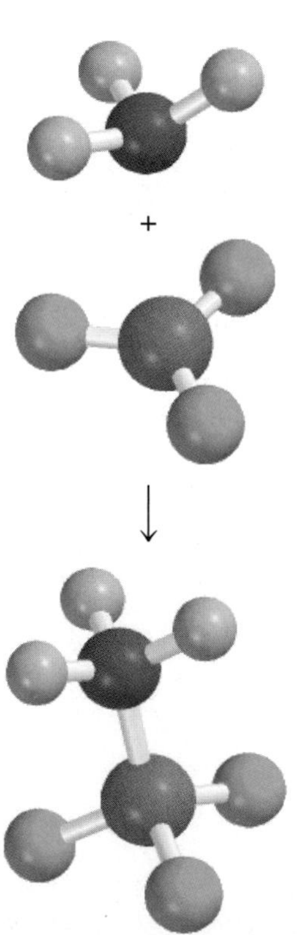

$NH_3 + BF_3 \longrightarrow H_3N—BF_3$

It seems that the properties of the BF_3 molecule are best explained by all four resonance structures.

The B—N bond in the preceding compound is different from the covalent bonds discussed so far in the sense that both electrons are contributed by the N atom. This type of bond is called a ***coordinate covalent bond*** (also referred to as a *dative bond*), defined as *a covalent bond in which one of the atoms donates both electrons.* Although the properties of a coordinate covalent bond do not differ from those of a normal covalent bond (because all electrons are alike no matter what their source), the distinction is useful for keeping track of valence electrons and assigning formal charges.

Odd-Electron Molecules

Some molecules contain an *odd* number of electrons. Among them are nitric oxide (NO) and nitrogen dioxide (NO_2):

$$\dot{\ddot{\text{N}}}=\ddot{\text{O}} \qquad \ddot{\text{O}}=\dot{\text{N}}^{+}-\ddot{\text{O}}:^{-}$$

Because we need an even number of electrons for complete pairing (to reach eight), the octet rule clearly cannot be satisfied for all the atoms in any of these molecules.

Odd-electron molecules are sometimes called *radicals.* Many radicals are highly reactive. The reason is that there is a tendency for the unpaired electron to form a covalent bond with an unpaired electron on another molecule. For example, when two nitrogen dioxide molecules collide, they form dinitrogen tetroxide in which the octet rule is satisfied for both the N and O atoms:

$$\text{O}_2\dot{\text{N}} + \dot{\text{N}}\text{O}_2 \longrightarrow \text{O}_2\text{N}-\text{NO}_2$$

The Expanded Octet

Atoms of the second-period elements cannot have more than eight valence electrons around the central atom, but atoms of elements in and beyond the third period of the periodic table form some compounds in which more than eight electrons surround the central atom. In addition to the 3*s* and 3*p* orbitals, elements in the third period also have 3*d* orbitals that can be used in bonding. These orbitals enable an atom to form an *expanded octet.* One compound in which there is an expanded octet is sulfur hexafluoride, a very stable compound. The electron configuration of sulfur is $[\text{Ne}]3s^23p^4$. In SF_6, each of sulfur's six valence electrons forms a covalent bond with a fluorine atom, so there are twelve electrons around the central sulfur atom:

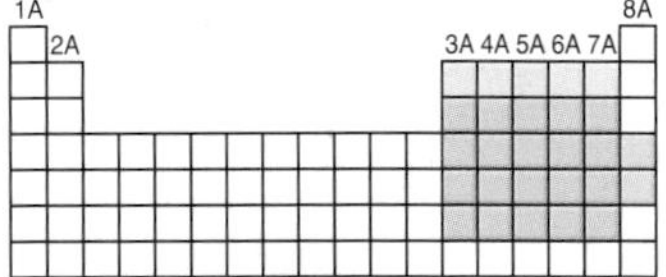

Yellow: second-period elements cannot have an expanded octet. Blue: third-period elements and beyond can have an expanded octet. Green: the noble gases usually only have an expanded octet.

$$\text{SF}_6:\ \text{S bonded to six }:\ddot{\text{F}}:\text{ atoms}$$

In Chapter 10 we will see that these 12 electrons, or six bonding pairs, are accommodated in six orbitals that originate from the one 3*s*, the three 3*p*, and two of the five 3*d* orbitals. Sulfur also forms many compounds in which it obeys the octet rule. In sulfur dichloride, for instance, S is surrounded by only eight electrons:

$$:\ddot{\text{Cl}}-\ddot{\text{S}}-\ddot{\text{Cl}}:$$

Sulfur dichloride is a toxic, foul-smelling cherry-red liquid (boiling point: 59°C).

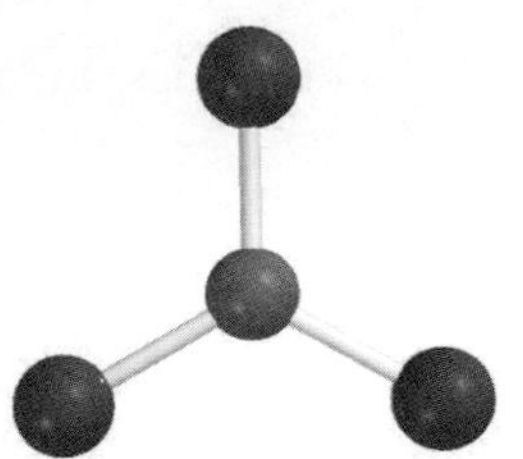

AlI_3 has a tendency to dimerize or form two units as Al_2I_6.

EXAMPLE 9.9

Draw the Lewis structure for aluminum triiodide (AlI_3).

Strategy We follow the procedures used in Examples 9.5 and 9.6 to draw the Lewis structure and calculate formal charges.

Solution The outer-shell electron configurations of Al and I are $3s^23p^1$ and $5s^25p^5$, respectively. The total number of valence electrons is $3 + (3 \times 7)$ or 24. Because Al is less electronegative than I, it occupies a central position and forms three bonds with the I atoms:

```
     ..
    :I:
 ..  |
:I—Al
 ..  |
    :I:
     ..
```

Note that there are no formal charges on the Al and I atoms.

Similar problem: 9.60.

Check Although the octet rule is satisfied for the I atoms, there are only six valence electrons around the Al atom. Thus, AlI_3 is an example of the incomplete octet.

Practice Exercise Draw the Lewis structure for BeF_2.

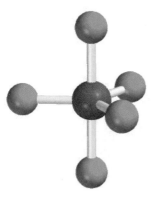

PF_5 is a reactive gaseous compound.

EXAMPLE 9.10

Draw the Lewis structure for phosphorus pentafluoride (PF_5), in which all five F atoms are bonded to the central P atom.

Strategy Note that P is a third-period element. We follow the procedures given in Examples 9.5 and 9.6 to draw the Lewis structure and calculate formal charges.

Solution The outer-shell electron configurations for P and F are $3s^23p^3$ and $2s^22p^5$, respectively, and so the total number of valence electrons is $5 + (5 \times 7)$, or 40. Phosphorus, like sulfur, is a third-period element, and therefore it can have an expanded octet. The Lewis structure of PF_5 is

```
     ..
    :F: ..
     | F:
 ..  |/ ..
:F—P\ ..
 ..  | F:
    :F: ..
     ..
```

Note that there are no formal charges on the P and F atoms.

Similar problem: 9.62.

Check Although the octet rule is satisfied for the F atoms, there are 10 valence electrons around the P atom, giving it an expanded octet.

Practice Exercise Draw the Lewis structure for arsenic pentafluoride (AsF_5).

A final note about the expanded octet: In drawing Lewis structures of compounds containing a central atom from the third period and beyond, sometimes we find that the octet rule is satisfied for all the atoms but there are still valence electrons left to place. In such cases, the extra electrons should be placed as lone pairs on the central atom.

EXAMPLE 9.11

Draw a Lewis structure of the noble gas compound xenon tetrafluoride (XeF_4) in which all F atoms are bonded to the central Xe atom.

Strategy Note that Xe is a fifth-period element. We follow the procedures in Examples 9.5 and 9.6 for drawing the Lewis structure and calculating formal charges.

Solution *Step 1:* The skeletal structure of XeF_4 is

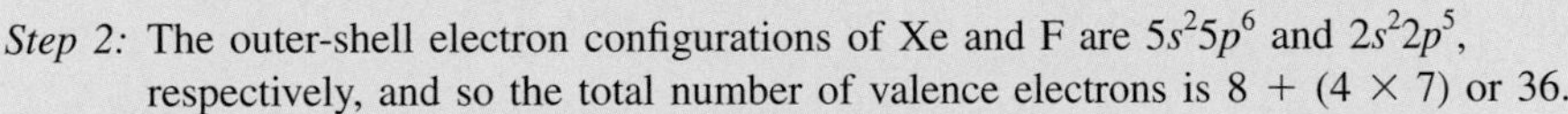

Step 2: The outer-shell electron configurations of Xe and F are $5s^2 5p^6$ and $2s^2 2p^5$, respectively, and so the total number of valence electrons is $8 + (4 \times 7)$ or 36.

Step 3: We draw a single covalent bond between all the bonding atoms. The octet rule is satisfied for the F atoms, each of which has three lone pairs. The sum of the lone pair electrons on the four F atoms (4×6) and the four bonding pairs (4×2) is 32. Therefore, the remaining four electrons are shown as two lone pairs on the Xe atom:

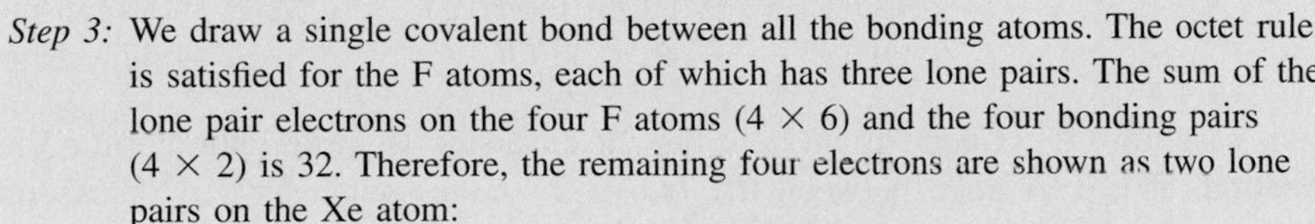

We see that the Xe atom has an expanded octet. There are no formal charges on the Xe and F atoms.

Practice Exercise Write the Lewis structure of sulfur tetrafluoride (SF_4).

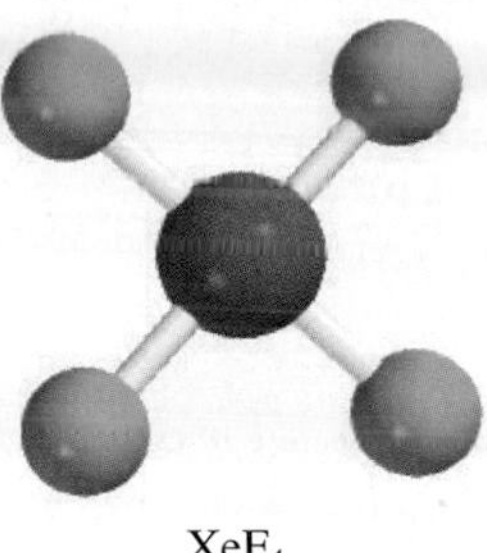

XeF_4

Similar problem: 9.61.

REVIEW OF CONCEPTS

Both boron and aluminum tend to form compounds in which they are surrounded with fewer than eight electrons. However, aluminum is able to form compounds and polyatomic ions where it is surrounded by more than eight electrons (e.g., AlF_6^-). Why is it possible for aluminum, but not boron, to expand the octet?

9.10 Bond Enthalpy

A measure of the stability of a molecule is its ***bond enthalpy,*** which is *the enthalpy change required to break a particular bond in 1 mole of gaseous molecules.* (Bond enthalpies in solids and liquids are affected by neighboring molecules.) The experimentally determined bond enthalpy of the diatomic hydrogen molecule, for example, is

$$H_2(g) \longrightarrow H(g) + H(g) \qquad \Delta H° = 436.4 \text{ kJ/mol}$$

Remember that it takes energy to break a bond so that energy is released when a bond is formed.

This equation tells us that breaking the covalent bonds in 1 mole of gaseous H_2 molecules requires 436.4 kJ of energy. For the less stable chlorine molecule,

$$Cl_2(g) \longrightarrow Cl(g) + Cl(g) \qquad \Delta H° = 242.7 \text{ kJ/mol}$$

Bond enthalpies can also be directly measured for diatomic molecules containing unlike elements, such as HCl,

$$HCl(g) \longrightarrow H(g) + Cl(g) \qquad \Delta H^\circ = 431.9 \text{ kJ/mol}$$

as well as for molecules containing double and triple bonds:

The Lewis structure of O_2 is $\ddot{\underset{..}{O}}{=}\ddot{\underset{..}{O}}$ and that for N_2 is :N≡N:.

$$O_2(g) \longrightarrow O(g) + O(g) \qquad \Delta H^\circ = 498.7 \text{ kJ/mol}$$
$$N_2(g) \longrightarrow N(g) + N(g) \qquad \Delta H^\circ = 941.4 \text{ kJ/mol}$$

Measuring the strength of covalent bonds in polyatomic molecules is more complicated. For example, measurements show that the energy needed to break the first O—H bond in H_2O is different from that needed to break the second O—H bond:

$$H_2O(g) \longrightarrow H(g) + OH(g) \qquad \Delta H^\circ = 502 \text{ kJ/mol}$$
$$OH(g) \longrightarrow H(g) + O(g) \qquad \Delta H^\circ = 427 \text{ kJ/mol}$$

In each case, an O—H bond is broken, but the first step is more endothermic than the second. The difference between the two ΔH° values suggests that the second O—H bond itself has undergone change, because of the changes in the chemical environment.

Now we can understand why the bond enthalpy of the same O—H bond in two different molecules such as methanol (CH_3OH) and water (H_2O) will not be the same: their environments are different. Thus, for polyatomic molecules we speak of the *average* bond enthalpy of a particular bond. For example, we can measure the energy of the O—H bond in 10 different polyatomic molecules and obtain the average O—H bond enthalpy by dividing the sum of the bond enthalpies by 10. Table 9.3 lists the average bond enthalpies of a number of diatomic and polyatomic molecules. As stated earlier, triple bonds are stronger than double bonds, which, in turn, are stronger than single bonds.

Use of Bond Enthalpies in Thermochemistry

A comparison of the thermochemical changes that take place during a number of reactions (Chapter 6) reveals a strikingly wide variation in the enthalpies of different reactions. For example, the combustion of hydrogen gas in oxygen gas is fairly exothermic:

$$H_2(g) + \tfrac{1}{2}O_2(g) \longrightarrow H_2O(l) \qquad \Delta H^\circ = -285.8 \text{ kJ/mol}$$

On the other hand, the formation of glucose ($C_6H_{12}O_6$) from water and carbon dioxide, best achieved by photosynthesis, is highly endothermic:

$$6CO_2(g) + 6H_2O(l) \longrightarrow C_6H_{12}O_6(s) + 6O_2(g) \qquad \Delta H^\circ = 2801 \text{ kJ/mol}$$

We can account for such variations by looking at the stability of individual reactant and product molecules. After all, most chemical reactions involve the making and breaking of bonds. Therefore, knowing the bond enthalpies and hence the stability of molecules tells us something about the thermochemical nature of reactions that molecules undergo.

In many cases, it is possible to predict the approximate enthalpy of reaction by using the average bond enthalpies. Because energy is always required to break chemical bonds and chemical bond formation is always accompanied by a release of energy,

Table 9.3 Some Bond Enthalpies of Diatomic Molecules* and Average Bond Enthalpies for Bonds in Polyatomic Molecules

Bond	Bond Enthalpy (kJ/mol)	Bond	Bond Enthalpy (kJ/mol)
H—H	436.4	C—S	255
H—N	393	C=S	477
H—O	460	N—N	193
H—S	368	N=N	418
H—P	326	N≡N	941.4
H—F	568.2	N—O	176
H—Cl	431.9	N=O	607
H—Br	366.1	O—O	142
H—I	298.3	O=O	498.7
C—H	414	O—P	502
C—C	347	O=S	469
C=C	620	P—P	197
C≡C	812	P=P	489
C—N	276	S—S	268
C=N	615	S=S	352
C≡N	891	F—F	156.9
C—O	351	Cl—Cl	242.7
C=O†	745	Br—Br	192.5
C—P	263	I—I	151.0

*Bond enthalpies for diatomic molecules (in color) have more significant figures than bond enthalpies for bonds in polyatomic molecules because the bond enthalpies of diatomic molecules are directly measurable quantities and not averaged over many compounds.
†The C=O bond enthalpy in CO_2 is 799 kJ/mol.

we can estimate the enthalpy of a reaction by counting the total number of bonds broken and formed in the reaction and recording all the corresponding energy changes. The enthalpy of reaction in the *gas phase* is given by

$$\begin{aligned}\Delta H° &= \Sigma BE(\text{reactants}) - \Sigma BE(\text{products}) \\ &= \text{total energy input} - \text{total energy released}\end{aligned} \tag{9.3}$$

where BE stands for average bond enthalpy and Σ is the summation sign. As written, Equation (9.3) takes care of the sign convention for $\Delta H°$. Thus, if the total energy input is greater than the total energy released, $\Delta H°$ is positive and the reaction is endothermic. On the other hand, if more energy is released than absorbed, $\Delta H°$ is negative and the reaction is exothermic (Figure 9.8). If reactants and products are all diatomic molecules, then Equation (9.3) will yield accurate results because the bond enthalpies of diatomic molecules are accurately known. If some or all of the reactants and products are polyatomic molecules, Equation (9.3) will yield only approximate results because the bond enthalpies used will be averages.

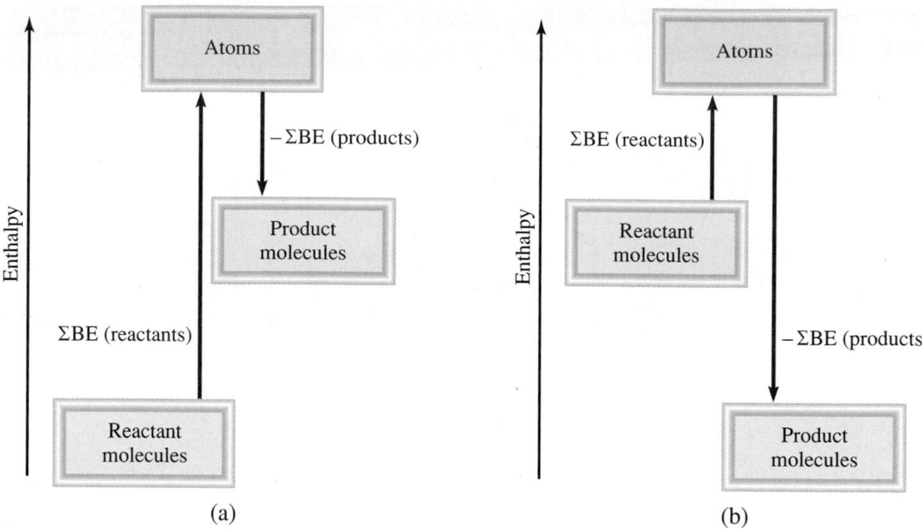

Figure 9.8
Bond enthalpy changes in (a) an endothermic reaction and (b) an exothermic reaction.

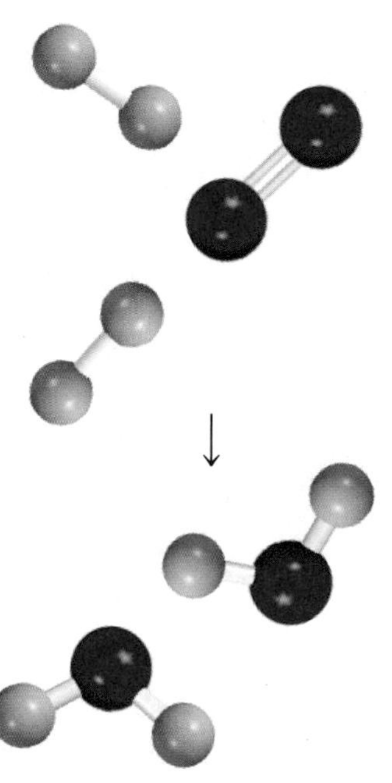

EXAMPLE 9.12

Estimate the enthalpy change for the combustion of hydrogen gas:

$$2H_2(g) + O_2(g) \longrightarrow 2H_2O(g)$$

Strategy Note that H_2O is a polyatomic molecule, and so we need to use the average bond enthalpy value for the O—H bond.

Solution We construct the following table:

Type of bonds broken	*Number of bonds broken*	*Bond enthalpy (kJ/mol)*	*Energy change (kJ/mol)*
H—H (H_2)	2	436.4	872.8
O═O (O_2)	1	498.7	498.7
Type of bonds formed	*Number of bonds formed*	*Bond enthalpy (kJ/mol)*	*Energy change (kJ/mol)*
O—H (H_2O)	4	460	1840

Next, we obtain the total energy input and total energy released:

$$\text{total energy input} = 872.8 \text{ kJ/mol} + 498.7 \text{ kJ/mol} = 1371.5 \text{ kJ/mol}$$
$$\text{total energy released} = 1840 \text{ kJ/mol}$$

Using Equation (9.3), we write

$$\Delta H° = 1371.5 \text{ kJ/mol} - 1840 \text{ kJ/mol} = -469 \text{ kJ/mol}$$

This result is only an estimate because the bond enthalpy of O—H is an average quantity. Alternatively, we can use Equation (6.18) and the data in Appendix 2 to calculate the enthalpy of reaction:

$$\begin{aligned}\Delta H° &= 2\Delta H_f°(H_2O) - [2\Delta H_f°(H_2) + \Delta H_f°(O_2)] \\ &= 2(-241.8 \text{ kJ/mol}) - 0 - 0 \\ &= -483.6 \text{ kJ/mol}\end{aligned}$$

(Continued)

Check Note that the estimated value based on average bond enthalpies is quite close to the value calculated using ΔH_f° data. In general, Equation (9.3) works best for reactions that are either quite endothermic or quite exothermic, that is, reactions for which $\Delta H_{rxn}^\circ >$ 100 kJ/mol or for which $\Delta H_{rxn}^\circ < -100$ kJ/mol.

Similar problem: 9.70.

Practice Exercise For the reaction

$$H_2(g) + C_2H_4(g) \longrightarrow C_2H_6(g)$$

(a) Estimate the enthalpy of reaction, using the bond enthalpy values in Table 9.3.
(b) Calculate the enthalpy of reaction, using standard enthalpies of formation. (ΔH_f° for H_2, C_2H_4, and C_2H_6 are 0, 52.3 kJ/mol, and -84.7 kJ/mol, respectively.)

REVIEW OF CONCEPTS

Why does ΔH_{rxn}° calculated using bond enthalpies not always agree with that calculated using the ΔH_f° values?

Key Equation

$\Delta H^\circ = \Sigma BE(\text{reactants}) - \Sigma BE(\text{products})$ (9.3) Calculating enthalpy change of a reaction from bond enthalpies.

Summary of Facts and Concepts

1. A Lewis dot symbol shows the number of valence electrons possessed by an atom of a given element. Lewis dot symbols are useful mainly for the representative elements.
2. In a covalent bond, two electrons (one pair) are shared by two atoms. In multiple covalent bonds, two or three electron pairs are shared by two atoms. Some bonded atoms possess lone pairs, that is, pairs of valence electrons not involved in bonding. The arrangement of bonding electrons and lone pairs around each atom in a molecule is represented by the Lewis structure.
3. Electronegativity is a measure of the ability of an atom to attract electrons in a chemical bond.
4. The octet rule predicts that atoms form enough covalent bonds to surround themselves with eight electrons each. When one atom in a covalently bonded pair donates two electrons to the bond, the Lewis structure can include the formal charge on each atom as a means of keeping track of the valence electrons. There are exceptions to the octet rule, particularly for covalent beryllium compounds, elements in Group 3A, and elements in the third period and beyond in the periodic table.
5. For some molecules or polyatomic ions, two or more Lewis structures based on the same skeletal structure satisfy the octet rule and appear chemically reasonable. Such resonance structures taken together represent the molecule or ion.
6. The strength of a covalent bond is measured in terms of its bond enthalpy. Bond enthalpies can be used to estimate the enthalpy of reactions.

Key Words

Bond enthalpy, p. 309
Bond length, p. 293
Born-Haber cycle, p. 289
Coordinate covalent bond, p. 306
Coulomb's law, p. 289
Covalent bond, p. 292
Covalent compound, p. 292
Double bond, p. 293
Electronegativity, p. 294
Formal charge, p. 300
Ionic bond, p. 287
Lattice energy, p. 289
Lewis dot symbol, p. 286
Lewis structure, p. 292
Lone pair, p. 292
Multiple bond, p. 293
Octet rule, p. 292
Polar covalent bond, p. 294
Resonance, p. 303
Resonance structure, p. 303
Single bond, p. 293
Triple bond, p. 293

Questions and Problems

Lewis Dot Symbols

Review Questions

9.1 What is a Lewis dot symbol? To what elements does the symbol mainly apply?

9.2 Use the second member of each group from Group 1A to Group 7A to show that the number of valence electrons on an atom of the element is the same as its group number.

9.3 Without referring to Figure 9.1, write Lewis dot symbols for atoms of the following elements: (a) Be, (b) K, (c) Ca, (d) Ga, (e) O, (f) Br, (g) N, (h) I, (i) As, (j) F.

9.4 Write Lewis dot symbols for the following ions: (a) Li^+, (b) Cl^-, (c) S^{2-}, (d) Sr^{2+}, (e) N^{3-}.

9.5 Write Lewis dot symbols for the following atoms and ions: (a) I, (b) I^-, (c) S, (d) S^{2-}, (e) P, (f) P^{3-}, (g) Na, (h) Na^+, (i) Mg, (j) Mg^{2+}, (k) Al, (l) Al^{3+}, (m) Pb, (n) Pb^{2+}.

The Ionic Bond

Review Questions

9.6 Explain what an ionic bond is.

9.7 Explain how ionization energy and electron affinity determine whether atoms of elements will combine to form ionic compounds.

9.8 Name five metals and five nonmetals that are very likely to form ionic compounds. Write formulas for compounds that might result from the combination of these metals and nonmetals. Name these compounds.

9.9 Name one ionic compound that contains only nonmetallic elements.

9.10 Name one ionic compound that contains a polyatomic cation and a polyatomic anion (see Table 2.3).

9.11 Explain why ions with charges greater than 3 are seldom found in ionic compounds.

9.12 The term "molar mass" was introduced in Chapter 3. What is the advantage of using the term "molar mass" when we discuss ionic compounds?

9.13 In which of the following states would NaCl be electrically conducting? (a) solid, (b) molten (that is, melted), (c) dissolved in water. Explain your answers.

9.14 Beryllium forms a compound with chlorine that has the empirical formula $BeCl_2$. How would you determine whether it is an ionic compound? (The compound is not soluble in water.)

Problems

9.15 An ionic bond is formed between a cation A^+ and an anion B^-. How would the energy of the ionic bond [see Equation (9.2)] be affected by the following changes? (a) doubling the radius of A^+, (b) tripling the charge on A^+, (c) doubling the charges on A^+ and B^-, (d) decreasing the radii of A^+ and B^- to half their original values.

9.16 Give the empirical formulas and names of the compounds formed from the following pairs of ions: (a) Rb^+ and I^-, (b) Cs^+ and SO_4^{2-}, (c) Sr^{2+} and N^{3-}, (d) Al^{3+} and S^{2-}.

9.17 Use Lewis dot symbols to show the transfer of electrons between the following atoms to form cations and anions: (a) Na and F, (b) K and S, (c) Ba and O, (d) Al and N.

9.18 Write the Lewis dot symbols of the reactants and products in the following reactions. (First balance the equations.)

(a) $Sr + Se \longrightarrow SrSe$

(b) $Ca + H_2 \longrightarrow CaH_2$

(c) $Li + N_2 \longrightarrow Li_3N$

(d) $Al + S \longrightarrow Al_2S_3$

9.19 For each of the following pairs of elements, state whether the binary compound they form is likely to be ionic or covalent. Write the empirical formula and name of the compound: (a) I and Cl, (b) Mg and F.

9.20 For each of the following pairs of elements, state whether the binary compound they form is likely to be ionic or covalent. Write the empirical formula and name of the compound: (a) B and F, (b) K and Br.

Lattice Energy of Ionic Compounds

Review Questions

9.21 What is lattice energy and what role does it play in the stability of ionic compounds?

9.22 Explain how the lattice energy of an ionic compound such as KCl can be determined using the Born-Haber cycle. On what law is this procedure based?

9.23 Specify which compound in the following pairs of ionic compounds has the higher lattice energy: (a) KCl or MgO, (b) LiF or LiBr, (c) Mg_3N_2 or NaCl. Explain your choice.

9.24 Compare the stability (in the solid state) of the following pairs of compounds: (a) LiF and LiF_2 (containing the Li^{2+} ion), (b) Cs_2O and CsO (containing the O^- ion), (c) $CaBr_2$ and $CaBr_3$ (containing the Ca^{3+} ion).

Problems

9.25 Use the Born-Haber cycle outlined in Section 9.3 for LiF to calculate the lattice energy of NaCl. [The heat of sublimation of Na is 108 kJ/mol and $\Delta H_f^\circ(NaCl) = -411$ kJ/mol. Energy needed to dissociate $\frac{1}{2}$ mole of Cl_2 into Cl atoms = 121.4 kJ].

9.26 Calculate the lattice energy of calcium chloride given that the heat of sublimation of Ca is 121 kJ/mol and $\Delta H_f^\circ(CaCl_2) = -795$ kJ/mol. (See Tables 8.2 and 8.3 for other data.)

The Covalent Bond

Review Questions

9.27 What is Lewis's contribution to our understanding of the covalent bond?

9.28 What is the difference between a Lewis dot symbol and a Lewis structure?

9.29 How many lone pairs are on the underlined atoms in these compounds? H<u>Br</u>, H_2<u>S</u>, <u>C</u>H_4.

9.30 Distinguish among single, double, and triple bonds in a molecule, and give an example of each.

Electronegativity and Bond Type

Review Questions

9.31 Define electronegativity, and explain the difference between electronegativity and electron affinity. Describe in general how the electronegativities of the elements change according to position in the periodic table.

9.32 What is a polar covalent bond? Name two compounds that contain one or more polar covalent bonds.

Problems

9.33 List these bonds in order of increasing ionic character: the lithium-to-fluorine bond in LiF, the potassium-to-oxygen bond in K_2O, the nitrogen-to-nitrogen bond in N_2, the sulfur-to-oxygen bond in SO_2, the chlorine-to-fluorine bond in ClF_3.

9.34 Arrange these bonds in order of increasing ionic character: carbon to hydrogen, fluorine to hydrogen, bromine to hydrogen, sodium to chlorine, potassium to fluorine, lithium to chlorine.

9.35 Four atoms are arbitrarily labeled D, E, F, and G. Their electronegativities are: D = 3.8, E = 3.3, F = 2.8, and G = 1.3. If the atoms of these elements form the molecules DE, DG, EG, and DF, how would you arrange these molecules in order of increasing covalent bond character?

9.36 List these bonds in order of increasing ionic character: cesium to fluorine, chlorine to chlorine, bromine to chlorine, silicon to carbon.

9.37 Classify these bonds as ionic, polar covalent, or covalent, and give your reasons: (a) the CC bond in H_3CCH_3, (b) the KI bond in KI, (c) the NB bond in H_3NBCl_3, (d) the ClO bond in ClO_2.

9.38 Classify these bonds as ionic, polar covalent, or covalent, and give your reasons: (a) the SiSi bond in $Cl_3SiSiCl_3$, (b) the SiCl bond in $Cl_3SiSiCl_3$, (c) the CaF bond in CaF_2, (d) the NH bond in NH_3.

Lewis Structure and the Octet Rule

Review Questions

9.39 Summarize the essential features of the Lewis octet rule. The octet rule applies mainly to the second-period elements. Explain.

9.40 Explain the concept of formal charge. Do formal charges on a molecule represent actual separation of charges?

Problems

9.41 Write Lewis structures for these molecules: (a) ICl, (b) PH_3, (c) P_4 (each P is bonded to three other P atoms), (d) H_2S, (e) N_2H_4, (f) $HClO_3$, (g) $COBr_2$ (C is bonded to O and Br atoms).

9.42 Write Lewis structures for these ions: (a) O_2^{2-}, (b) C_2^{2-}, (c) NO^+, (d) NH_4^+. Show formal charges.

9.43 The following Lewis structures for (a) HCN, (b) C_2H_2, (c) SnO_2, (d) BF_3, (e) HOF, (f) HCOF, and (g) NF_3 are incorrect. Explain what is wrong with each one and give a correct structure for the molecule. (Relative positions of atoms are shown correctly.)

(a) H—C̈═N̈

(b) H═C═C═H

(c) Ö—Sn—Ö

(d) :F̤—B(—:F̤:)—F̤: (B bonded to three F atoms)

(e) H—Ö═F̤:

(f) H—C(—F̤:)—Ö (C bonded to H, F, and O)

(g) :F̤—N(—:F̤:)—F̤: (N bonded to three F atoms)

9.44 The skeletal structure of acetic acid in this structure is correct, but some of the bonds are wrong. (a) Identify the incorrect bonds and explain what is wrong with them. (b) Write the correct Lewis structure for acetic acid.

```
   H :O:
   |  |
H═C—C—Ö—H
   |
   H
```

Resonance

Review Questions

9.45 Define bond length, resonance, and resonance structure.

9.46 Is it possible to "trap" a resonance structure of a compound for study? Explain.

Problems

9.47 The resonance concept is sometimes described by analogy to a mule, which is a cross between a horse and a donkey. Compare this analogy with that used in this chapter, that is, the description of a rhinoceros as a cross between a griffin and a unicorn. Which description is more appropriate? Why?

9.48 What are the other two reasons for choosing (b) in Example 9.7?

9.49 Write Lewis structures for these species, including all resonance forms, and show formal charges: (a) HCO_2^-, (b) $CH_2NO_2^-$. Relative positions of the atoms are as follows:

```
      O          H       O
H  C                 C — N
      O          H       O
```

9.50 Draw three resonance structures for the chlorate ion, ClO_3^-. Show formal charges.

9.51 Write three resonance structures for hydrazoic acid, HN_3. The atomic arrangement is HNNN. Show formal charges.

9.52 Draw two resonance structures for diazomethane, CH_2N_2. Show formal charges. The skeletal structure of the molecule is

```
 H
    C  N  N
 H
```

9.53 Draw three reasonable resonance structures for the OCN^- ion. Show formal charges.

9.54 Draw three resonance structures for the molecule OCS in which the atoms are arranged in the order OCS. Indicate formal charges.

9.55 Draw three resonance structures for the molecule N_2O_3 (atomic arrangement is $ONNO_2$). Show formal charges.

Exceptions to the Octet Rule

Review Questions

9.56 Why does the octet rule not hold for many compounds containing elements in the third period of the periodic table and beyond?

9.57 Because fluorine has seven valence electrons ($2s^22p^5$), seven covalent bonds in principle could form around the atom. Such a compound might be FH_7 or FCl_7. These compounds have never been prepared. Why?

9.58 What is a coordinate covalent bond? Is it different from a normal covalent bond?

Problems

9.59 The BCl_3 molecule has an incomplete octet around B. Draw three resonance structures of the molecule in which the octet rule is satisfied for both the B and the Cl atoms. Show formal charges.

9.60 In the vapor phase, beryllium chloride consists of discrete molecular units $BeCl_2$. Is the octet rule satisfied for Be in this compound? If not, can you form an octet around Be by drawing another resonance structure? How plausible is this structure?

9.61 Of the noble gases, only Kr, Xe, and Rn are known to form a few compounds with O and/or F. Write Lewis structures for these molecules: (a) XeF_2, (b) XeF_4, (c) XeF_6, (d) $XeOF_4$, (e) XeO_2F_2. In each case Xe is the central atom.

9.62 Write a Lewis structure for $SbCl_5$. Is the octet rule obeyed in this molecule?

9.63 Write Lewis structures for SeF_4 and SeF_6. Is the octet rule satisfied for Se?

9.64 Write Lewis structures for the reaction

$$AlCl_3 + Cl^- \longrightarrow AlCl_4^-$$

What kind of bond is between Al and Cl in the product?

Bond Enthalpies

Review Questions

9.65 Define bond enthalpy. Bond enthalpies of polyatomic molecules are average values. Why?

9.66 Explain why the bond enthalpy of a molecule is usually defined in terms of a gas-phase reaction. Why are bond-breaking processes always endothermic and bond-forming processes always exothermic?

Problems

9.67 From these data, calculate the average bond enthalpy for the N—H bond:

$$NH_3(g) \longrightarrow NH_2(g) + H(g) \quad \Delta H° = 435 \text{ kJ/mol}$$
$$NH_2(g) \longrightarrow NH(g) + H(g) \quad \Delta H° = 381 \text{ kJ/mol}$$
$$NH(g) \longrightarrow N(g) + H(g) \quad \Delta H° = 360 \text{ kJ/mol}$$

9.68 For the reaction

$$O(g) + O_2(g) \longrightarrow O_3(g) \quad \Delta H° = -107.2 \text{ kJ/mol}$$

calculate the average bond enthalpy in O_3.

9.69 The bond enthalpy of $F_2(g)$ is 156.9 kJ/mol. Calculate $\Delta H_f°$ for F(g).

9.70 (a) For the reaction

$$2C_2H_6(g) + 7O_2(g) \longrightarrow 4CO_2(g) + 6H_2O(g)$$

predict the enthalpy of reaction from the average bond enthalpies in Table 9.3. (b) Calculate the enthalpy of reaction from the standard enthalpies of formation (see Appendix 2) of the reactant and product molecules, and compare the result with your answer for part (a).

Additional Problems

9.71 Match each of these energy changes with one of the processes given: ionization energy, electron affinity, bond enthalpy, standard enthalpy of formation.

(a) $F(g) + e^- \longrightarrow F^-(g)$
(b) $F_2(g) \longrightarrow 2F(g)$
(c) $Na(g) \longrightarrow Na^+(g) + e^-$
(d) $Na(s) + \frac{1}{2}F_2(g) \longrightarrow NaF(s)$

9.72 The formulas for the fluorides of the third-period elements are NaF, MgF_2, AlF_3, SiF_4, PF_5, SF_6, and ClF_3. Classify these compounds as covalent or ionic.

9.73 Use the ionization energy (see Table 8.2) and electron affinity values (see Table 8.3) to calculate the energy change (in kilojoules) for these reactions:

(a) $Li(g) + I(g) \longrightarrow Li^+(g) + I^-(g)$
(b) $Na(g) + F(g) \longrightarrow Na^+(g) + F^-(g)$
(c) $K(g) + Cl(g) \longrightarrow K^+(g) + Cl^-(g)$

9.74 Describe some characteristics of an ionic compound such as KF that would distinguish it from a covalent compound such as CO_2.

9.75 Write Lewis structures for BrF_3, ClF_5, and IF_7. Identify those in which the octet rule is not obeyed.

9.76 Write three reasonable resonance structures of the azide ion N_3^- in which the atoms are arranged as NNN. Show formal charges.

9.77 The amide group plays an important role in determining the structure of proteins:

```
        :O:
  ..    ||
 —N—C—
   |
   H
```

Draw another resonance structure of this group. Show formal charges.

9.78 Give an example of an ion or molecule containing Al that (a) obeys the octet rule, (b) has an expanded octet, and (c) has an incomplete octet.

9.79 Draw four reasonable resonance structures for the PO_3F^{2-} ion. The central P atom is bonded to the three O atoms and to the F atom. Show formal charges.

9.80 Attempts to prepare these as stable species under atmospheric conditions have failed. Suggest reasons for the failure.

CF_2 CH_5 FH_2^- PI_5

9.81 Draw reasonable resonance structures for these sulfur-containing ions: (a) HSO_4^-, (b) SO_4^{2-}, (c) HSO_3^-, (d) SO_3^{2-}.

9.82 True or false: (a) Formal charges represent actual separation of charges; (b) ΔH°_{rxn} can be estimated from bond enthalpies of reactants and products; (c) all second-period elements obey the octet rule in their compounds; (d) the resonance structures of a molecule can be separated from one another.

9.83 A rule for drawing plausible Lewis structures is that the central atom is invariably less electronegative than the surrounding atoms. Explain why this is so. Why does this rule not apply to compounds like H_2O and NH_3?

9.84 Using this information:

$$C(s) \longrightarrow C(g) \quad \Delta H^\circ_{rxn} = 716 \text{ kJ/mol}$$
$$2H_2(g) \longrightarrow 4H(g) \quad \Delta H^\circ_{rxn} = 872.8 \text{ kJ/mol}$$

and the fact that the average C—H bond enthalpy is 414 kJ/mol, estimate the standard enthalpy of formation of methane (CH_4).

9.85 Based on energy considerations, which of these two reactions will occur more readily?

(a) $Cl(g) + CH_4(g) \longrightarrow CH_3Cl(g) + H(g)$
(b) $Cl(g) + CH_4(g) \longrightarrow CH_3(g) + HCl(g)$

(*Hint:* Refer to Table 9.3, and assume that the average bond enthalpy of the C—Cl bond is 338 kJ/mol.)

9.86 Which of these molecules has the shortest nitrogen-to-nitrogen bond? Explain.

N_2H_4 N_2O N_2 N_2O_4

9.87 Most organic acids can be represented as RCOOH, in which COOH is the carboxyl group and R is the rest of the molecule. (For example, R is CH_3 in acetic acid, CH_3COOH.) (a) Draw a Lewis structure of the carboxyl group. (b) Upon ionization, the carboxyl group is converted to the carboxylate group, COO^-. Draw resonance structures of the carboxylate group.

9.88 Which of these molecules or ions are isoelectronic: NH_4^+, C_6H_6, CO, CH_4, N_2, $B_3N_3H_6$?

9.89 These species have been detected in interstellar space: (a) CH, (b) OH, (c) C_2, (d) HNC, (e) HCO. Draw Lewis structures of these species and indicate whether they are diamagnetic or paramagnetic.

9.90 The amide ion, NH_2^-, is a Brønsted base. Represent the reaction between the amide ion and water in terms of Lewis structures.

9.91 Draw Lewis structures of these organic molecules: (a) tetrafluoroethylene (C_2F_4), (b) propane (C_3H_8), (c) butadiene ($CH_2CHCHCH_2$), (d) propyne (CH_3CCH), (e) benzoic acid (C_6H_5COOH). (*Hint:* To draw C_6H_5COOH, replace an H atom in benzene with a COOH group.)

9.92 The triiodide ion (I_3^-) in which the I atoms are arranged as III is stable, but the corresponding F_3^- ion does not exist. Explain.

9.93 Compare the bond enthalpy of F_2 with the energy change for this process:

$$F_2(g) \longrightarrow F^+(g) + F^-(g)$$

Which is the preferred dissociation for F_2, energetically speaking?

9.94 Methyl isocyanate, CH_3NCO, is used to make certain pesticides. In December 1984, water leaked into a tank containing this substance at a chemical plant to produce a toxic cloud that killed thousands of people in Bhopal, India. Draw Lewis structures for this compound, showing formal charges.

9.95 The chlorine nitrate molecule ($ClONO_2$) is believed to be involved in the destruction of ozone in the Antarctic stratosphere. Draw a plausible Lewis structure for the molecule.

9.96 Several resonance structures of the molecule CO_2 are given here. Explain why some of them are likely to be of little importance in describing the bonding in this molecule.

(a) $\ddot{\underset{\cdot\cdot}{O}}{=}C{=}\ddot{\underset{\cdot\cdot}{O}}$ (c) $:\overset{+}{O}{\equiv}\overset{-}{\ddot{C}}\;\; \ddot{\underset{\cdot\cdot}{O}}:$

(b) $:\overset{+}{O}{\equiv}C{-}\overset{-}{\ddot{\underset{\cdot\cdot}{O}}}:$ (d) $:\overset{-}{\ddot{\underset{\cdot\cdot}{O}}}{-}\overset{2+}{C}{-}\overset{-}{\ddot{\underset{\cdot\cdot}{O}}}:$

9.97 Draw a Lewis structure for each of these organic molecules in which the carbon atoms are bonded to each other by single bonds: C_2H_6, C_4H_{10}, C_5H_{12}.

9.98 Draw Lewis structures for these chlorofluorocarbons (CFCs), which are partly responsible for the depletion of ozone in the stratosphere: $CFCl_3$, CF_2Cl_2, CHF_2Cl, CF_3CHF_2.

9.99 Draw Lewis structures for these organic molecules, in each of which there is one C=C bond and the rest of the carbon atoms are joined by C—C bonds: C_2H_3F, C_3H_6, C_4H_8.

9.100 Calculate $\Delta H°$ for the reaction

$$H_2(g) + I_2(g) \longrightarrow 2HI(g)$$

using (a) Equation (9.3) and (b) Equation (6.18), given that ΔH_f° for $I_2(g)$ is 61.0 kJ/mol.

9.101 Draw Lewis structures of these organic molecules: (a) methanol (CH_3OH); (b) ethanol (CH_3CH_2OH); (c) tetraethyllead [$Pb(CH_2CH_3)_4$], which was used in "leaded" gasoline; (d) methylamine (CH_3NH_2); (e) mustard gas ($ClCH_2CH_2SCH_2CH_2Cl$), a poisonous gas used in World War I; (f) urea [$(NH_2)_2CO$], a fertilizer; (g) glycine (NH_2CH_2COOH), an amino acid.

9.102 Write Lewis structures for these four isoelectronic species: (a) CO, (b) NO^+, (c) CN^-, (d) N_2. Show formal charges.

9.103 Oxygen forms three types of ionic compounds in which the anions are oxide (O^{2-}), peroxide (O_2^{2-}), and superoxide (O_2^-). Draw Lewis structures of these ions.

9.104 Comment on the correctness of this statement: All compounds containing a noble gas atom violate the octet rule.

9.105 (a) From these data:

$$F_2(g) \longrightarrow 2F(g) \qquad \Delta H^\circ_{rxn} = 156.9 \text{ kJ/mol}$$
$$F^-(g) \longrightarrow F(g) + e^- \qquad \Delta H^\circ_{rxn} = 333 \text{ kJ/mol}$$
$$F_2^-(g) \longrightarrow F_2(g) + e^- \qquad \Delta H^\circ_{rxn} = 290 \text{ kJ/mol}$$

calculate the bond enthalpy of the F_2^- ion. (b) Explain the difference between the bond enthalpies of F_2 and F_2^-.

9.106 Write three resonance structures for the isocyanate ion (CNO^-). Rank them in importance.

9.107 The only known argon-containing compound is HArF, which was prepared in 2000. Draw a Lewis structure of the compound.

9.108 Experiments show that it takes 1656 kJ/mol to break all the bonds in methane (CH_4) and 4006 kJ/mol to break all the bonds in propane (C_3H_8). Based on these data, calculate the average bond enthalpy of the C—C bond.

9.109 Among the common inhaled anesthetics are

halothane: $CF_3CHClBr$

enflurane: $CHFClCF_2OCHF_2$

isoflurane: $CF_3CHClOCHF_2$

methoxyflurane: $CHCl_2CF_2OCH_3$

Draw Lewis structures of these molecules.

9.110 Industrially, ammonia is synthesized by the Haber process at high pressures and temperatures:

$$N_2(g) + 3H_2(g) \longrightarrow 2NH_3(g)$$

Calculate the enthalpy change for the reaction using (a) bond enthalpies and Equation (9.3) and (b) the ΔH_f° values in Appendix 2.

Special Problems

9.111 The neutral hydroxyl radical (OH) plays an important role in atmospheric chemistry. It is highly reactive and has a tendency to combine with an H atom from other compounds, causing them to break up. Thus, it is sometimes called a "detergent" radical because it helps to clean up the atmosphere.

(a) Write the Lewis structure for the radical.

(b) Refer to Table 9.3 and explain why the radical has a high affinity for H atoms.

(c) Estimate the enthalpy change for the following reaction:

$$OH(g) + CH_4(g) \longrightarrow CH_3(g) + H_2O(g)$$

(d) The radical is generated when sunlight hits water vapor. Calculate the maximum wavelength (in nanometers) required to break up an O—H bond in H_2O.

9.112 Ethylene dichloride ($C_2H_4Cl_2$) is used to make vinyl chloride (C_2H_3Cl), which, in turn, is used to manufacture the plastic poly(vinyl chloride) (PVC), found in piping, siding, floor tiles, and toys.

(a) Write the Lewis structures of ethylene dichloride and vinyl chloride. Classify the bonds as covalent or polar.

(b) Poly(vinyl chloride) is a polymer; that is, it is a molecule with very high molar mass (on the order of thousands to millions of grams). It is formed by joining many vinyl chloride molecules together. The repeating unit in poly(vinyl chloride) is $—CH_2—CHCl—$. Draw a portion of the molecule showing three such repeating units.

(c) Calculate the enthalpy change when 1.0×10^3 kg of vinyl chloride react to form poly(vinyl chloride). Comment on your answer in relation to industrial design for such a process.

9.113 Sulfuric acid (H_2SO_4), the most important industrial chemical in the world, is prepared by oxidizing sulfur to sulfur dioxide and then to sulfur trioxide. Although sulfur trioxide reacts with water to form sulfuric acid, it forms a mist of fine droplets of H_2SO_4 with water vapor that is hard to condense. Instead, sulfur trioxide is first dissolved in 98 percent sulfuric acid to form oleum ($H_2S_2O_7$). On treatment with water, concentrated sulfuric acid can be generated. Write equations for all the steps and draw Lewis structures of oleum.

9.114 The species H_3^+ is the simplest polyatomic ion. The geometry of the ion is that of an equilateral triangle. (a) Draw three resonance structures to represent the ion. (b) Given the following information

$$2H + H^+ \longrightarrow H_3^+ \quad \Delta H° = -849 \text{ kJ/mol}$$

and

$$H_2 \longrightarrow 2H \quad \Delta H° = 436.4 \text{ kJ/mol}$$

calculate $\Delta H°$ for the reaction

$$H^+ + H_2 \longrightarrow H_3^+$$

9.115 The bond enthalpy of the C—N bond in the amide group of proteins (see Problem 9.77) can be treated as an average of C—N and C=N bonds. Calculate the maximum wavelength of light needed to break the bond.

9.116 In 1999 an unusual cation containing only nitrogen (N_5^+) was prepared. Draw three resonance structures of the ion, showing formal charges. (*Hint:* The N atoms are joined in a linear fashion.)

9.117 Give a brief description of the medical uses of the following ionic compounds: $AgNO_3$, $BaSO_4$, $CaSO_4$, KI, Li_2CO_3, $Mg(OH)_2$, $MgSO_4$, $NaHCO_3$, Na_2CO_3, NaF, TiO_2, ZnO. You would need to do a Web search of some of these compounds.

9.118 Use Table 9.3 to estimate the bond enthalpy of the C—C, N—N, and O—O bonds in C_2H_6, N_2H_4, and H_2O_2, respectively. What effect do lone pairs on adjacent atoms have on the strength of the particular bonds?

Answers to Practice Exercises

9.1 $\cdot Ba \cdot + 2 \cdot H \longrightarrow Ba^{2+}\ 2H:^-$ (or BaH_2)
$[Xe]6s^2 \quad 1s^1 \qquad [Xe]\ [He]$

9.2 (a) Ionic, (b) polar covalent, (c) covalent.

9.3 S=C=S **9.4** H—C(=O)—O—H **9.5** [O=N—O:]$^-$

9.6 O=N—O:$^-$ **9.7** H—C≡N:

9.8 S=C=N$^-$ ⟷ $^-$:S—C≡N: ⟷ $^+$:S≡C—N:$^{2-}$

The first structure is the most important; the last structure is the least important.

9.9 :F—Be—F:

9.10 AsF_5 (:F—As with four additional F atoms)

9.11 SF_4 (S bonded to four F atoms, with one lone pair on S)

9.12 (a) −119 kJ/mol, (b) −137.0 kJ/mol.

CHAPTER

10

Chemical Bonding II: Molecular Geometry and Hybridization of Atomic Orbitals

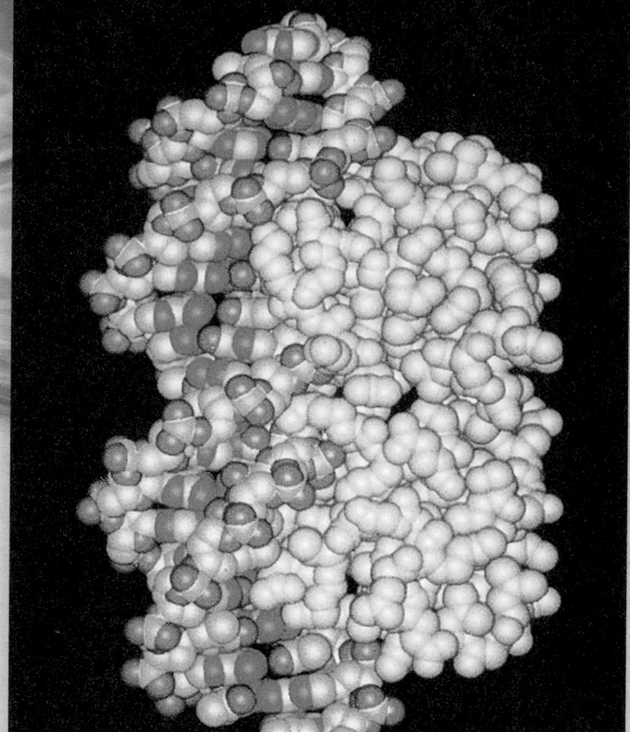

Molecular models are used to study complex biochemical reactions such as those between protein and DNA molecules.

Chapter Outline

Student Interactive Activities

Animations
VSEPR (10.1)
Polarity of Molecules (10.2)
Hybridization (10.4)
Sigma and Pi Bonds (10.5)

Electronic Homework
Example Practice Problems
End of Chapter Problems

Essential Concepts

Molecular Geometry Molecular geometry refers to the three-dimensional arrangement of atoms in a molecule. For relatively small molecules, in which the central atom contains two to six bonds, geometries can be reliably predicted by the valence-shell electron-pair repulsion (VSEPR) model. This model is based on the assumption that chemical bonds and lone pairs tend to remain as far apart as possible to minimize repulsion.

Dipole Moments In a diatomic molecule, the difference in the electronegativities of bonding atoms results in a polar bond and a dipole moment. The dipole moment of a molecule made up of three or more atoms depends on both the polarity of the bonds and molecular geometry. Dipole moment measurements can help us distinguish between different possible geometries of a molecule.

Hybridization of Atomic Orbitals Hybridization is the quantum mechanical description of chemical bonding. Atomic orbitals are hybridized, or mixed, to form hybrid orbitals. These orbitals then interact with other atomic orbitals to form chemical bonds. Various molecular geometries can be generated by different hybridizations. The hybridization concept accounts for the exception to the octet rule and also explains the formation of double and triple bonds.

Molecular Orbital Theory Molecular orbital theory describes bonding in terms of the combination of atomic orbitals to form orbitals that are associated with the molecule as a whole. Molecules are stable if the number of electrons in bonding molecular orbitals is greater than that in antibonding molecular orbitals. We write electron configurations for molecular orbitals as we do for atomic orbitals, using the Pauli exclusion principle and Hund's rule.

10.1 Molecular Geometry

Molecular geometry is the three-dimensional arrangement of atoms in a molecule. A molecule's geometry affects its physical and chemical properties, such as melting point, boiling point, density, and the types of reactions it undergoes. In general, bond lengths and bond angles must be determined by experiment. However, there is a simple procedure that enables us to predict with considerable success the overall geometry of a molecule or ion if we know the number of electrons surrounding a central atom in its Lewis structure. The basis of this approach is the assumption that electron pairs in the valence shell of an atom repel one another. The ***valence shell*** is *the outermost electron-occupied shell of an atom; it holds the electrons that are usually involved in bonding.* In a covalent bond, a pair of electrons, often called the *bonding pair,* is responsible for holding two atoms together. However, in a polyatomic molecule, where there are two or more bonds between the central atom and the surrounding atoms, the repulsion between electrons in different bonding pairs causes them to remain as far apart as possible. The geometry that the molecule ultimately assumes (as defined by the positions of all the atoms) minimizes the repulsion. This approach to the study of molecular geometry is called the ***valence-shell electron-pair repulsion (VSEPR) model,*** because *it accounts for the geometric arrangements of electron pairs around a central atom in terms of the electrostatic repulsion between electron pairs.*

The term "central atom" means an atom that is not a terminal atom in a polyatomic molecule.

VSEPR is pronounced "vesper."

Animation: VSEPR

Two general rules govern the use of the VSEPR model:

1. As far as electron-pair repulsion is concerned, double bonds and triple bonds can be treated like single bonds. This approximation is good for qualitative purposes. However, you should realize that in reality multiple bonds are "larger" than single bonds; that is, because there are two or three bonds between two atoms, the electron density occupies more space.
2. If a molecule has two or more resonance structures, we can apply the VSEPR model to any one of them. Formal charges are usually not shown.

With this model in mind, we can predict the geometry of molecules (and ions) in a systematic way. For this purpose, it is convenient to divide molecules into two categories, according to whether or not the central atom has lone pairs.

Molecules in Which the Central Atom Has No Lone Pairs

For simplicity we will consider molecules that contain atoms of only two elements, A and B, of which A is the central atom. These molecules have the general formula AB_x, where x is an integer 2, 3, ... (If $x = 1$, we have the diatomic molecule AB, which is linear by definition.) In the vast majority of cases, x is between 2 and 6.

Table 10.1 shows five possible arrangements of electron pairs around the central atom A. As a result of mutual repulsion, the electron pairs stay as far from one another as possible. Note that the table shows arrangements of the electron pairs but not the positions of the atoms that surround the central atom. Molecules in which the central atom has no lone pairs have one of these five arrangements of bonding pairs. Using Table 10.1 as a reference, let us take a close look at the geometry of molecules with the formulas AB_2, AB_3, AB_4, AB_5, and AB_6.

AB_2: Beryllium Chloride ($BeCl_2$)

The Lewis structure of beryllium chloride in the gaseous state is

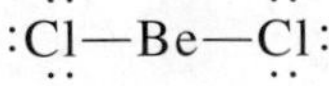

Table 10.1 Arrangement of Electron Pairs About a Central Atom (A) in a Molecule and Geometry of Some Simple Molecules and Ions in Which the Central Atom Has No Lone Pairs

Number of Electron Pairs	Arrangement of Electron Pairs*	Molecular Geometry*	Examples
2	180° Linear	B—A—B Linear	$BeCl_2$, $HgCl_2$
3	120° Trigonal planar	Trigonal planar	BF_3
4	109.5° Tetrahedral	Tetrahedral	CH_4, NH_4^+
5	90°, 120° Trigonal bipyramidal	Trigonal bipyramidal	PCl_5
6	90°, 90° Octahedral	Octahedral	SF_6

*The colored lines are used only to show the overall shapes; they do not represent bonds.

Because the bonding pairs repel each other, they must be at opposite ends of a straight line in order for them to be as far apart as possible. Thus, the ClBeCl angle is predicted to be 180°, and the molecule is linear (see Table 10.1). The "ball-and-stick" model of $BeCl_2$ is

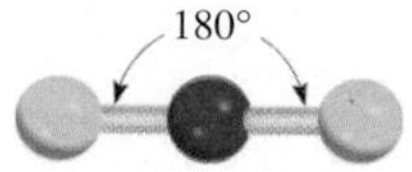

The blue and yellow spheres are for atoms in general.

AB_3: *Boron Trifluoride* (BF_3)

Boron trifluoride contains three covalent bonds, or bonding pairs. In the most stable arrangement, the three BF bonds point to the corners of an equilateral triangle with B in the center of the triangle:

```
      ..
     :F:
      |
      B
  .. / \ ..
 :F      F:
  ..      ..
```

According to Table 10.1, the geometry of BF_3 is *trigonal planar* because the three end atoms are at the corners of an equilateral triangle that is planar:

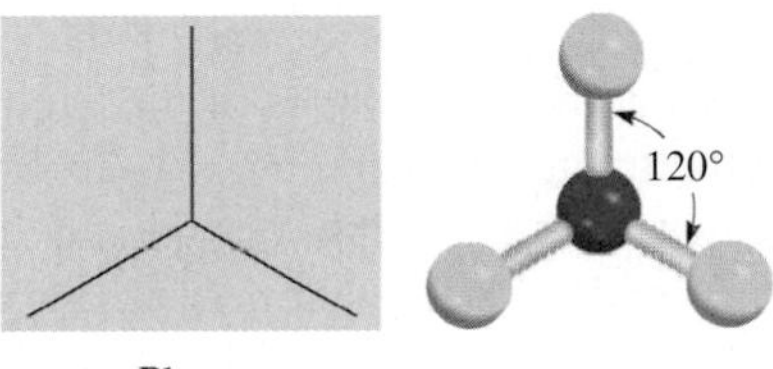

Planar

Thus, each of the three FBF angles is 120°, and all four atoms lie in the same plane.

AB_4: *Methane* (CH_4)

The Lewis structure of methane is

```
     H
     |
 H—C—H
     |
     H
```

Because there are four bonding pairs, the geometry of CH_4 is tetrahedral (see Table 10.1). A *tetrahedron* has four sides (the prefix *tetra* means "four"), or faces, all of which are equilateral triangles. In a tetrahedral molecule, the central atom (C in this case) is located at the center of the tetrahedron and the other four atoms are at the corners. The bond angles are all 109.5°.

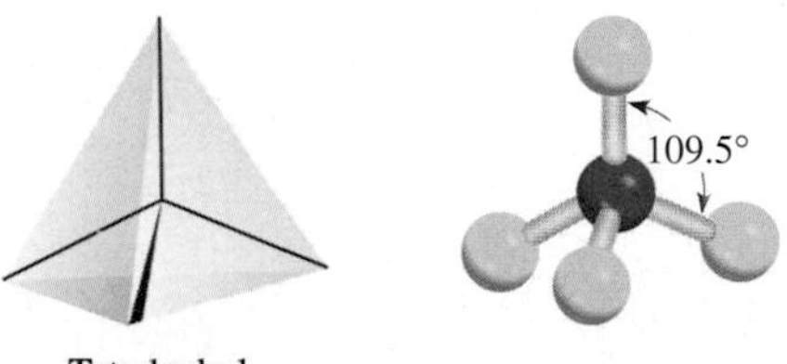

Tetrahedral

AB_5: *Phosphorus Pentachloride* (PCl_5)

The Lewis structure of phosphorus pentachloride (in the gas phase) is

```
   ..   ..
  :Cl  :Cl:
   ..\  |   ..
      P—Cl:
   ../  |   ..
  :Cl  :Cl:
   ..   ..
```

The only way to minimize the repulsive forces among the five bonding pairs is to arrange the PCl bonds in the form of a trigonal bipyramid (see Table 10.1). A trigonal bipyramid can be generated by joining two tetrahedrons along a common triangular base:

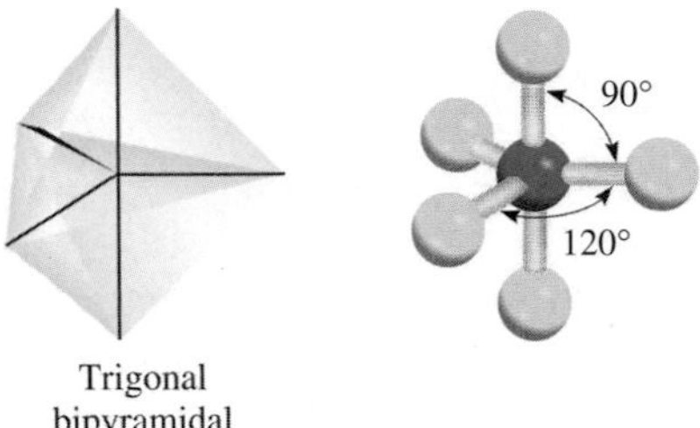

Trigonal
bipyramidal

The central atom (P in this case) is at the center of the common triangle with the surrounding atoms positioned at the five corners of the trigonal bipyramid. The atoms that are above and below the triangular plane are said to occupy *axial* positions, and those that are in the triangular plane are said to occupy *equatorial* positions. The angle between any two equatorial bonds is 120°; that between an axial bond and an equatorial bond is 90°, and that between the two axial bonds is 180°.

AB_6: Sulfur Hexafluoride (SF_6)

The Lewis structure of sulfur hexafluoride is

:F: (top), :F: (bottom), and four :F: atoms each bonded to a central S

The most stable arrangement of the six SF bonding pairs is in the shape of an octahedron, shown in Table 10.1. An octahedron has eight sides (the prefix *octa* means "eight"). It can be generated by joining two square pyramids on a common base. The central atom (S in this case) is at the center of the square base and the surrounding atoms are at the six corners. All bond angles are 90° except the one made by the bonds between the central atom and the pairs of atoms that are diametrically opposite each other. That angle is 180°. Because the six bonds are equivalent in an octahedral molecule, we cannot use the terms "axial" and "equatorial" as in a trigonal bipyramidal molecule.

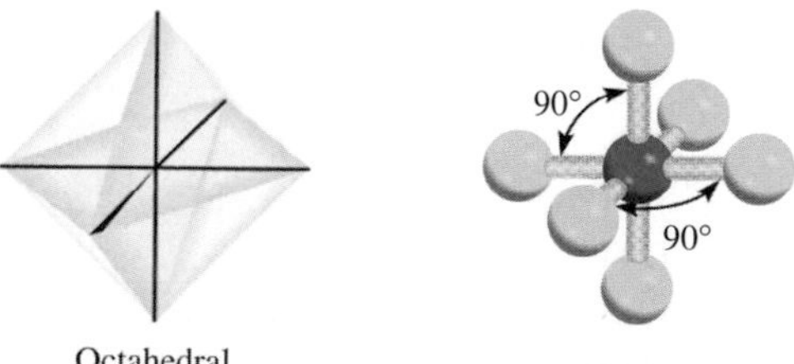

Octahedral

Molecules in Which the Central Atom Has One or More Lone Pairs

Determining the geometry of a molecule is more complicated if the central atom has both lone pairs and bonding pairs. In such molecules there are three types of repulsive forces—those between bonding pairs, those between lone pairs, and those between a bonding pair and a lone pair. In general, according to the VSEPR model, the repulsive forces decrease in the following order:

lone-pair vs. lone-pair repulsion	>	lone-pair vs. bonding-pair repulsion	>	bonding-pair vs. bonding-pair repulsion

Electrons in a bond are held by the attractive forces exerted by the nuclei of the two bonded atoms. These electrons have less "spatial distribution" than lone pairs; that is, they take up less space than lone-pair electrons, which are associated with only one particular atom. Because lone-pair electrons in a molecule occupy more space, they experience greater repulsion from neighboring lone pairs and bonding pairs. To keep track of the total number of bonding pairs and lone pairs, we designate molecules with lone pairs as AB_xE_y, where A is the central atom, B is a surrounding atom, and E is a lone pair on A. Both x and y are integers; $x = 2, 3, \ldots,$ and $y = 1, 2, \ldots.$ Thus, the values of x and y indicate the number of surrounding atoms and number of lone pairs on the central atom, respectively. The simplest such molecule would be a triatomic molecule with one lone pair on the central atom and the formula is AB_2E.

As the following examples show, in most cases the presence of lone pairs on the central atom makes it difficult to predict the bond angles accurately.

AB_2E: Sulfur Dioxide (SO_2)

The Lewis structure of sulfur dioxide is

O=S=O

Because VSEPR treats double bonds as though they were single, the SO_2 molecule can be viewed as consisting of three electron pairs on the central S atom. Of these, two are bonding pairs and one is a lone pair. In Table 10.1 we see that the overall arrangement of three electron pairs is trigonal planar. But because one of the electron pairs is a lone pair, the SO_2 molecule has a "bent" shape.

S
O O

Because the lone-pair versus bonding-pair repulsion is greater than the bonding-pair versus bonding-pair repulsion, the two sulfur-to-oxygen bonds are pushed together slightly and the OSO angle is less than 120°.

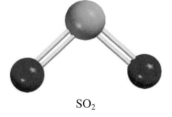

SO_2

AB_3E: Ammonia (NH_3)

The ammonia molecule contains three bonding pairs and one lone pair:

H—N—H
|
H

As Table 10.1 shows, the overall arrangement of four electron pairs is tetrahedral. But in NH_3 one of the electron pairs is a lone pair, so the geometry of NH_3 is trigonal pyramidal (so called because it looks like a pyramid, with the N atom at the apex). Because the lone pair repels the bonding pairs more strongly, the three NH bonding pairs are pushed closer together:

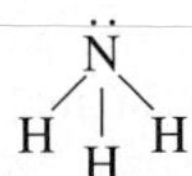

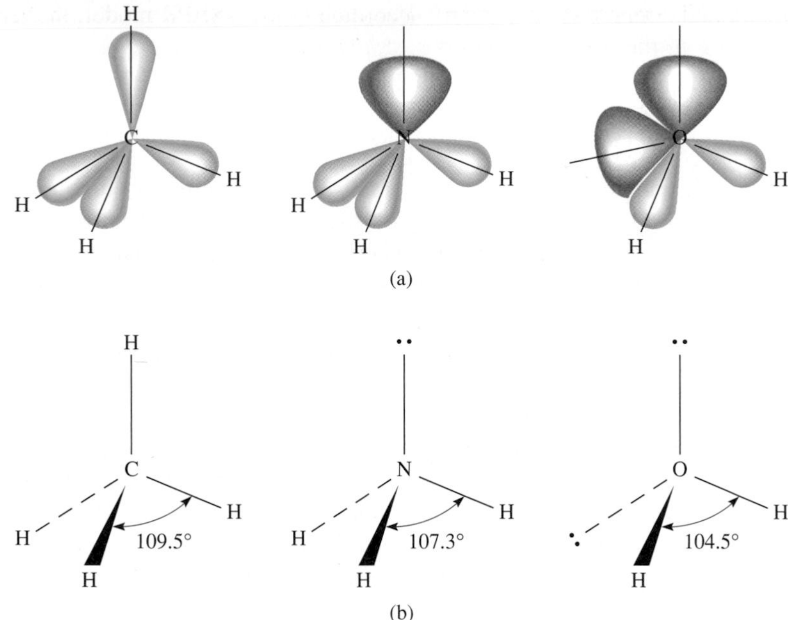

Figure 10.1
(a) The relative sizes of bonding pairs and lone pairs in CH_4, NH_3, and H_2O. (b) The bond angles in CH_4, NH_3, and H_2O. Note that the dashed lines represent a bond axes behind the plane of the paper, the wedged lines represent a bond axes in front of the plane of the paper, and the thin solid lines represent bonds in the plane of the paper.

Thus, the HNH angle in ammonia is smaller than the ideal tetrahedral angle of 109.5° (Figure 10.1).

AB_2E_2: Water (H_2O)

A water molecule contains two bonding pairs and two lone pairs:

H—Ö—H (with a lone pair above and below O)

The overall arrangement of the four electron pairs in water is tetrahedral, the same as in ammonia. However, unlike ammonia, water has two lone pairs on the central O atom. These lone pairs tend to be as far from each other as possible. Consequently, the two O—H bonding pairs are pushed toward each other, and we predict an even greater deviation from the tetrahedral angle than in NH_3. As Figure 10.1 shows, the HOH angle is 104.5°. The geometry of H_2O is bent:

:O: bonded to H and H (bent)

AB_4E: Sulfur Tetrafluoride (SF_4)

The Lewis structure of SF_4 is

:F̤: and :F̤: bonded to S̤ (with one lone pair), bonded to :F̤: and :F̤:

The central sulfur atom has five electron pairs whose arrangement, according to Table 10.1, is trigonal bipyramidal. In the SF_4 molecule, however, one of the electron pairs is a lone pair, so the molecule must have one of the following geometries:

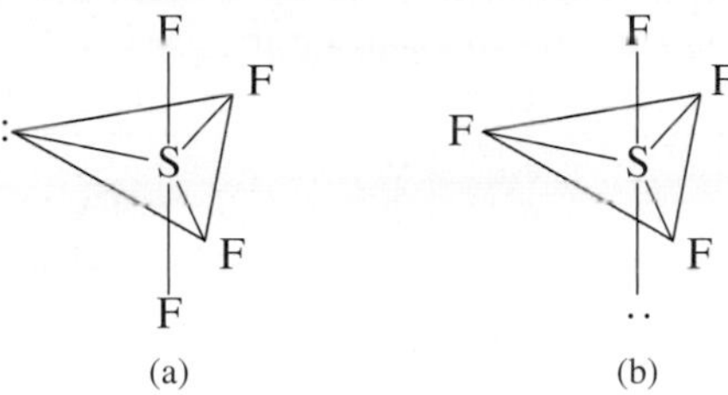

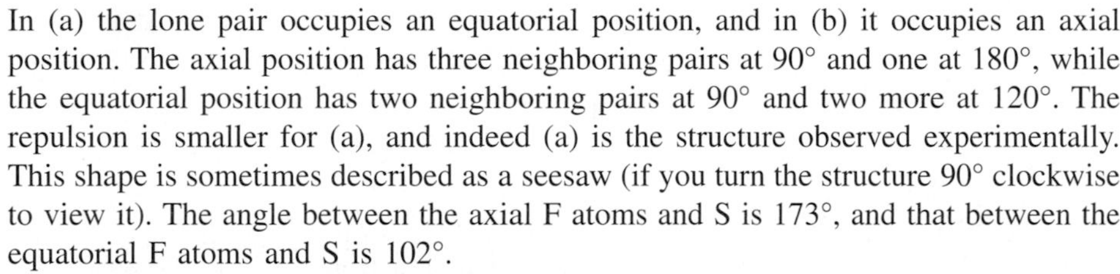

In (a) the lone pair occupies an equatorial position, and in (b) it occupies an axial position. The axial position has three neighboring pairs at 90° and one at 180°, while the equatorial position has two neighboring pairs at 90° and two more at 120°. The repulsion is smaller for (a), and indeed (a) is the structure observed experimentally. This shape is sometimes described as a seesaw (if you turn the structure 90° clockwise to view it). The angle between the axial F atoms and S is 173°, and that between the equatorial F atoms and S is 102°.

Table 10.2 shows the geometries of simple molecules in which the central atom has one or more lone pairs, including some that we have not discussed.

Geometry of Molecules with More Than One Central Atom

So far we have discussed the geometry of molecules having only one central atom. The overall geometry of molecules with more than one central atom is difficult to define in most cases. Often we can describe only the shape around each of the central atoms. For example, consider methanol, CH_3OH, whose Lewis structure is shown next:

```
     H
     |   ..
  H—C—O—H
     |   ..
     H
```

The two central (nonterminal) atoms in methanol are C and O. We can say that the three CH and the CO bonding pairs are tetrahedrally arranged about the C atom. The HCH and OCH bond angles are approximately 109°. The O atom here is like the one in water in that it has two lone pairs and two bonding pairs. Therefore, the HOC portion of the molecule is bent, and the angle HOC is approximately equal to 105° (Figure 10.2).

Guidelines for Applying the VSEPR Model

Having studied the geometries of molecules in two categories (central atoms with and without lone pairs), let us consider some rules for applying the VSEPR model to all types of molecules:

1. Write the Lewis structure of the molecule, considering only the electron pairs around the central atom (that is, the atom that is bonded to more than one other atom).
2. Count the number of electron pairs around the central atom (bonding pairs and lone pairs). Treat double and triple bonds as though they were single bonds. Refer to Table 10.1 to predict the overall arrangement of the electron pairs.

Figure 10.2
The geometry of CH_3OH.

Table 10.2 Geometry of Simple Molecules and Ions in Which the Central Atom Has One or More Lone Pairs

Class of molecule	Total number of electron pairs	Number of bonding pairs	Number of lone pairs	Arrangement of electron pairs*	Geometry	Examples
AB_2E	3	2	1	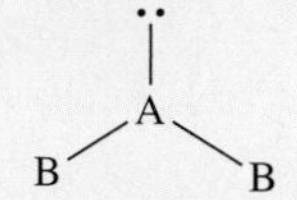 Trigonal planar	Bent	 SO_2
AB_3E	4	3	1	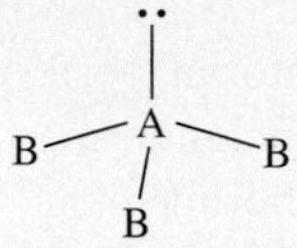 Tetrahedral	Trigonal pyramidal	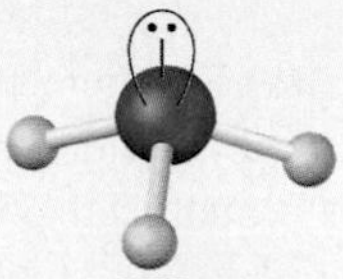 NH_3
AB_2E_2	4	2	2	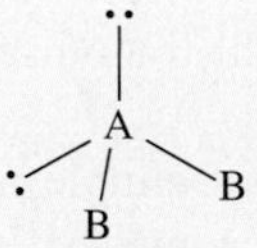 Tetrahedral	Bent	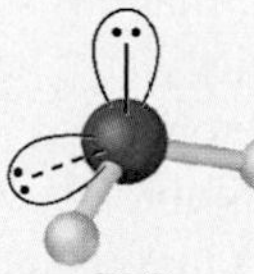 H_2O
AB_4E	5	4	1	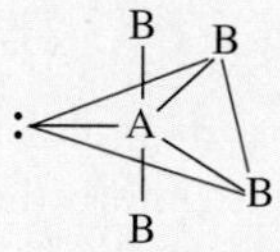 Trigonal bipyramidal	Distorted tetrahedron (or seesaw)	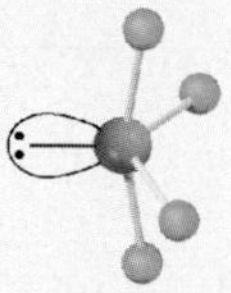 SF_4
AB_3E_2	5	3	2	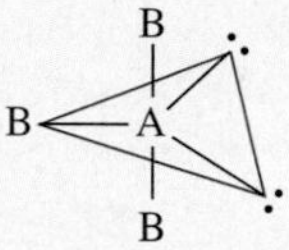 Trigonal bipyramidal	T-shaped	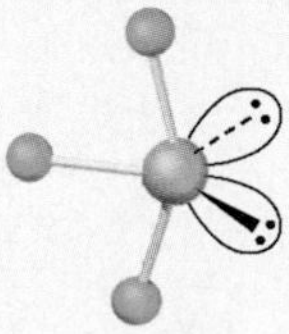 ClF_3
AB_2E_3	5	2	3	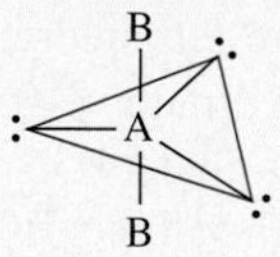 Trigonal bipyramidal	Linear	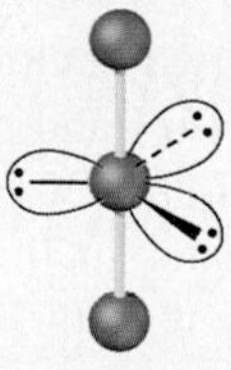 I_3^-
AB_5E	6	5	1	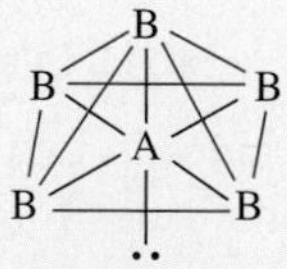 Octahedral	Square pyramidal	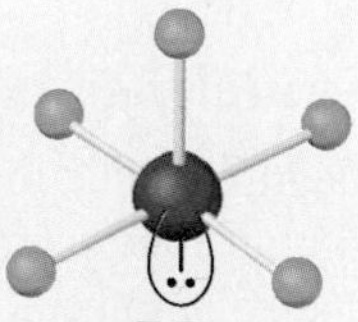 BrF_5
AB_4E_2	6	4	2	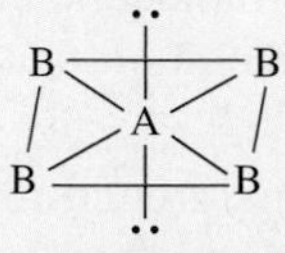 Octahedral	Square planar	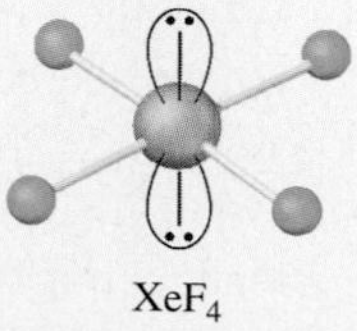 XeF_4

*The colored lines are used to show the overall shape, not bonds.

3. Use Tables 10.1 and 10.2 to predict the geometry of the molecule.
4. In predicting bond angles, note that a lone pair repels another lone pair or a bonding pair more strongly than a bonding pair repels another bonding pair. Remember that in general there is no easy way to predict bond angles accurately when the central atom possesses one or more lone pairs.

The VSEPR model generates reliable predictions of the geometries of a variety of molecular structures. Chemists use the VSEPR approach because of its simplicity. Although there are some theoretical concerns about whether "electron-pair repulsion" actually determines molecular shapes, the assumption that it does leads to useful (and generally reliable) predictions. We need not ask more of any model at this stage in the study of chemistry. Example 10.1 illustrates the application of VSEPR.

EXAMPLE 10.1

Use the VSEPR model to predict the geometry of the following molecules and ions: (a) AsH_3, (b) OF_2, (c) $AlCl_4^-$, (d) I_3^-, (e) C_2H_4.

Strategy The sequence of steps in determining molecular geometry is as follows:

draw Lewis structure ⟶ find arrangement of electron pairs ⟶ find arrangement of bonding pairs ⟶ determine geometry based on bonding pairs

Solution

(a) The Lewis structure of AsH_3 is

$$\begin{array}{c} \text{H—}\ddot{\text{As}}\text{—H} \\ | \\ \text{H} \end{array}$$

AsH_3

There are four electron pairs around the central atom; therefore, the electron pair arrangement is tetrahedral (see Table 10.1). Recall that the geometry of a molecule is determined only by the arrangement of atoms (in this case the As and H atoms). Thus, removing the lone pair leaves us with three bonding pairs and a trigonal pyramidal geometry, like NH_3. We cannot predict the HAsH angle accurately, but we know that it is less than 109.5° because the repulsion of the bonding electron pairs in the As—H bonds by the lone pair on As is greater than the repulsion between the bonding pairs.

(b) The Lewis structure of OF_2 is

$$:\ddot{\underset{\cdot\cdot}{\text{F}}}\text{—}\ddot{\underset{\cdot\cdot}{\text{O}}}\text{—}\ddot{\underset{\cdot\cdot}{\text{F}}}:$$

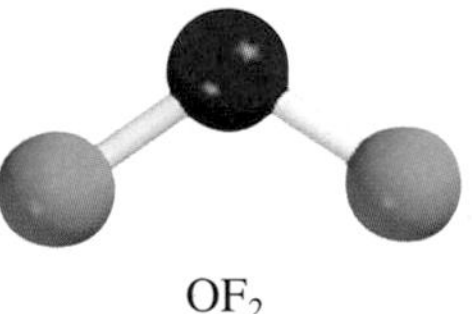

OF_2

There are four electron pairs around the central atom; therefore, the electron pair arrangement is tetrahedral (see Table 10.1). Recall that the geometry of a molecule is determined only by the arrangement of atoms (in this case the O and F atoms). Thus, removing the two lone pairs leaves us with two bonding pairs and a bent geometry, like H_2O. We cannot predict the FOF angle accurately, but we know that it must be less than 109.5° because the repulsion of the bonding electron pairs in the O—F bonds by the lone pairs on O is greater than the repulsion between the bonding pairs.

(c) The Lewis structure of $AlCl_4^-$ is

$$\left[\begin{array}{c} :\ddot{\text{Cl}}: \\ | \\ :\ddot{\underset{\cdot\cdot}{\text{Cl}}}\text{—Al—}\ddot{\underset{\cdot\cdot}{\text{Cl}}}: \\ | \\ :\underset{\cdot\cdot}{\text{Cl}}: \end{array}\right]^-$$

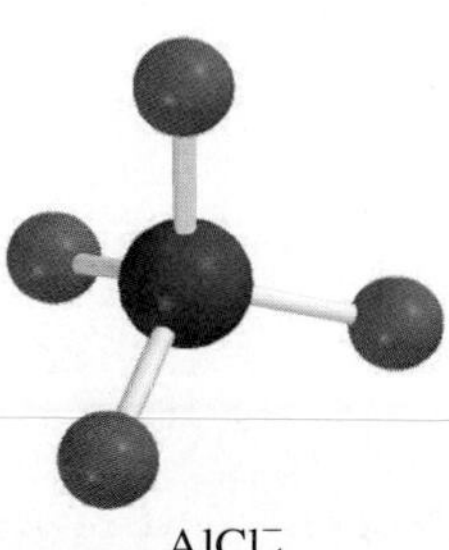

$AlCl_4^-$

(Continued)

There are four electron pairs around the central I atom; therefore, the electron pair arrangement is tetrahedral. Because there are no lone pairs present, the arrangement of the bonding pairs is the same as the electron pair arrangement. Therefore, $AlCl_4^-$ has a tetrahedral geometry and the ClAlCl angles are all 109.5°.

(d) The Lewis structure of I_3^- is

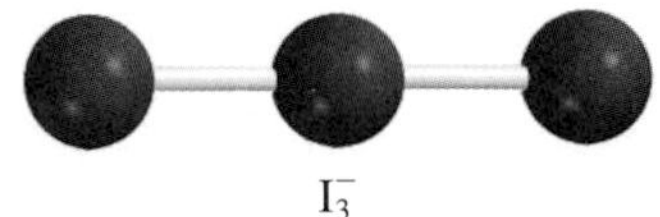

I_3^-

$$\left[:\ddot{\underset{..}{I}}-\ddot{\underset{..}{I}}-\ddot{\underset{..}{I}}:\right]^-$$

There are five electron pairs around the central I atom; therefore, the electron pair arrangement is trigonal bipyramidal. Of the five electron pairs, three are lone pairs and two are bonding pairs. Recall that the lone pairs preferentially occupy the equatorial positions in a trigonal bipyramid (see Table 10.2). Thus, removing the lone pairs leaves us with a linear geometry for I_3^-, that is, all three I atoms lie in a straight line.

(e) The Lewis structure of C_2H_4 is

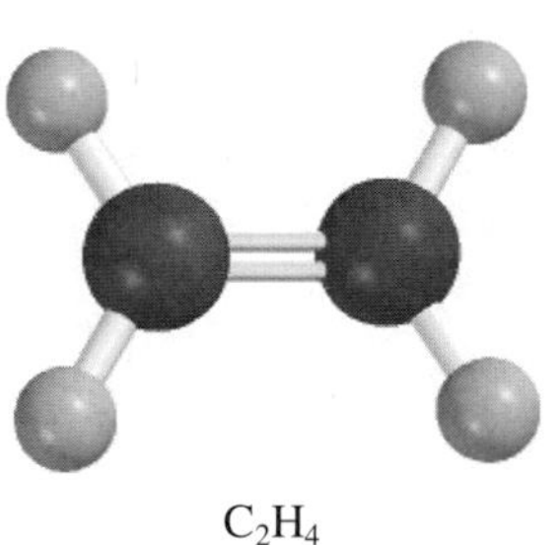

C_2H_4

```
H       H
 \     /
  C = C
 /     \
H       H
```

The C═C bond is treated as though it were a single bond in the VSEPR model. Because there are three electron pairs around each C atom and there are no lone pairs present, the arrangement around each C atom has a trigonal planar shape like BF_3, discussed earlier. Thus, the predicted bond angles in C_2H_4 are all 120°.

```
H  120°  H
 \      /
  C = C    120°
 /      \
H  120°  H
```

Comment (1) The I_3^- ion is one of the few structures for which the bond angle (180°) can be predicted accurately even though the central atom contains lone pairs. (2) In C_2H_4, all six atoms lie in the same plane. The overall planar geometry is not predicted by the VSEPR model, but we will see why the molecule prefers to be planar later. In reality, the angles are close, but not equal, to 120° because the bonds are not all equivalent.

Similar problems: 10.7, 10.8, 10.9.

Practice Exercise Use the VSEPR model to predict the geometry of (a) $SiBr_4$, (b) CS_2, and (c) NO_3^-.

Review of Concepts

Which of the following geometries has a greater stability for tin(IV) hydride?

10.2 Dipole Moments

In Section 9.2, we learned that hydrogen fluoride is a covalent compound with a polar bond. There is a shift of electron density from H to F because the F atom is more electronegative than the H atom (see Figure 9.3). The shift of electron density is symbolized by placing a crossed arrow (⟼) above the Lewis structure to indicate the direction of the shift. For example,

$$\overset{\longmapsto}{\text{H—}\ddot{\underset{\cdot\cdot}{\text{F}}}\text{:}}$$

The consequent charge separation can be represented as

$$\overset{\delta+\quad\delta-}{\text{H—}\ddot{\underset{\cdot\cdot}{\text{F}}}\text{:}}$$

where δ (delta) denotes a partial charge. This separation of charges can be confirmed in an electric field (Figure 10.3). When the field is turned on, HF molecules orient their negative ends toward the positive plate and their positive ends toward the negative plate. This alignment of molecules can be detected experimentally.

A quantitative measure of the polarity of a bond is its ***dipole moment*** **(μ)**, which is *the product of the charge Q and the distance r between the charges:*

$$\mu = Q \times r \tag{10.1}$$

To maintain electrical neutrality, the charges on both ends of an electrically neutral diatomic molecule must be equal in magnitude and opposite in sign. However, in Equation (10.1), Q refers only to the magnitude of the charge and not to its sign, so μ is always positive. Dipole moments are usually expressed in debye units (D), named for the Dutch-American chemist and physicist Peter Debye. The conversion factor is

In a diatomic molecule like HF, the charge Q is equal to $\delta+$ and $\delta-$.

$$1 \text{ D} = 3.336 \times 10^{-30} \text{ C m}$$

where C is coulomb and m is meter.

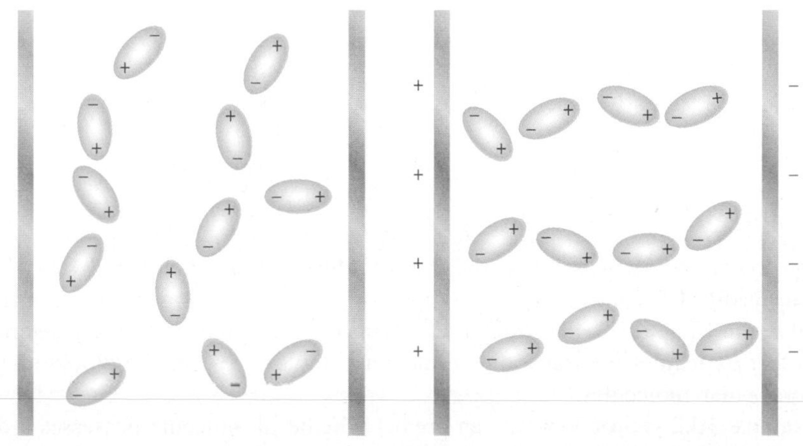

Figure 10.3
Behavior of polar molecules (a) in the absence of an external electric field and (b) when the electric field is turned on. Nonpolar molecules are not affected by an electric field.

Animation:
Polarity of Molecules

Diatomic molecules containing atoms of *different* elements (for example, HCl, CO, and NO) *have dipole moments* and are called ***polar molecules.*** Diatomic molecules containing atoms of the *same* element (for example, H_2, O_2, and F_2) are examples of ***nonpolar molecules*** because they *do not have dipole moments.* For a molecule made up of three or more atoms, both the polarity of the bonds and the molecular geometry determine whether there is a dipole moment. Even if polar bonds are present, the molecule will not necessarily have a dipole moment. Carbon dioxide (CO_2), for example, is a triatomic molecule, so its geometry is either linear or bent:

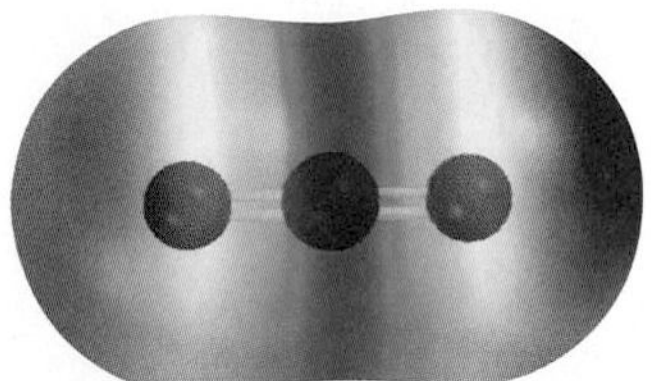

Each carbon-to-oxygen bond is polar, with the electron density shifted toward the more electronegative oxygen atom. However, the linear geometry of the molecule results in the cancellation of the two bond moments.

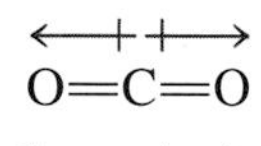

linear molecule
(no dipole moment)

O C O
resultant dipole moment

bent molecule
(would have a dipole moment)

The arrows show the shift of electron density from the less electronegative carbon atom to the more electronegative oxygen atom. In each case, the dipole moment of the entire molecule is made up of two *bond moments,* that is, individual dipole moments in the polar C=O bonds. The bond moment is a *vector quantity,* which means that it has both magnitude and direction. The measured dipole moment is equal to the vector sum of the bond moments. The two bond moments in CO_2 are equal in magnitude. Because they point in opposite directions in a linear CO_2 molecule, the sum or resultant dipole moment would be zero. On the other hand, if the CO_2 molecule were bent, the two bond moments would partially reinforce each other, so that the molecule would have a dipole moment. Experimentally it is found that carbon dioxide has no dipole moment. Therefore, we conclude that the carbon dioxide molecule is linear. The linear nature of carbon dioxide has been confirmed through other experimental measurements.

The VSEPR model predicts that CO_2 is a linear molecule.

Dipole moments can be used to distinguish between molecules that have the same formula but different structures. For example, the following molecules both exist; they have the same molecular formula ($C_2H_2Cl_2$), the same number and type of bonds, but different molecular structures:

resultant dipole moment

Cl Cl
C=C
H H

cis-dichloroethylene
$\mu = 1.89$ D

Cl H
C=C
H Cl

trans-dichloroethylene
$\mu = 0$

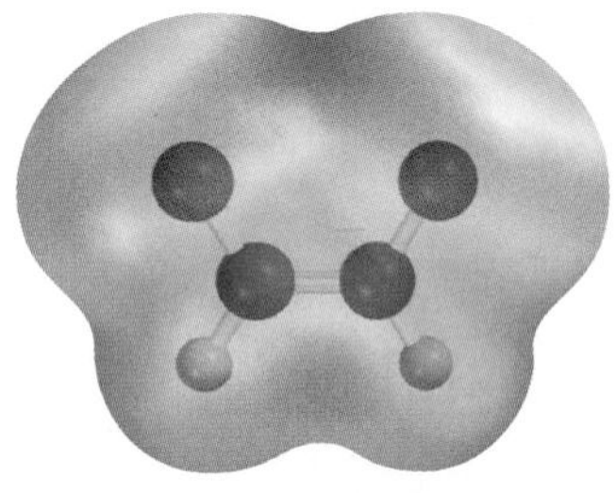

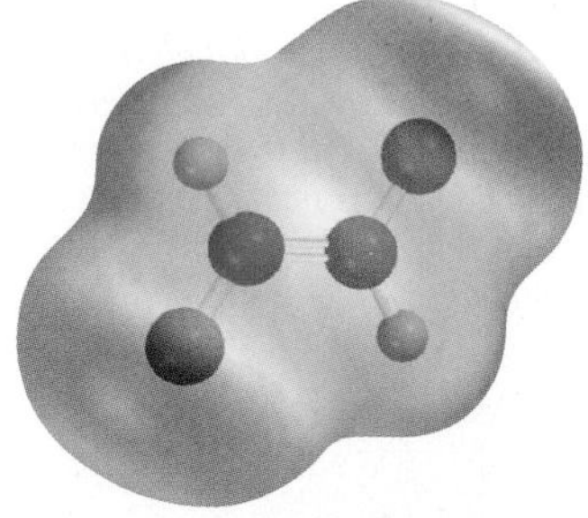

In *cis*-dichloroethylene (top), the bond moments reinforce one another and the molecule is polar. The opposite holds for *trans*-dichloroethylene (bottom) and the molecule is nonpolar.

Because *cis*-dichloroethylene is a polar molecule but *trans*-dichloroethylene is not, they can readily be distinguished by a dipole moment measurement. Additionally, as we will see in Chapter 11, the strength of intermolecular forces is partially determined by whether molecules possess a dipole moment. Table 10.3 lists the dipole moments of several polar molecules.

Example 10.2 shows how we can predict whether a molecule possesses a dipole moment if we know its molecular geometry.

Table 10.3 Dipole Moments of Some Polar Molecules

Molecule	Geometry	Dipole Moment (D)
HF	Linear	1.92
HCl	Linear	1.08
HBr	Linear	0.78
HI	Linear	0.38
H_2O	Bent	1.87
H_2S	Bent	1.10
NH_3	Trigonal pyramidal	1.46
SO_2	Bent	1.60

EXAMPLE 10.2

Predict whether each of the following molecules has a dipole moment: (a) BrCl, (b) BF_3 (trigonal planar), (c) CH_2Cl_2 (tetrahedral).

Strategy Keep in mind that the dipole moment of a molecule depends on both the difference in electronegativities of the elements present and its geometry. A molecule can have polar bonds (if the bonded atoms have different electronegativities), but it may not possess a dipole moment if it has a highly symmetrical geometry.

Solution

(a) Because bromine chloride is diatomic, it has a linear geometry. Chlorine is more electronegative than bromine (see Figure 9.5), so BrCl is polar with chlorine at the negative end

$$\overset{+\!\!\longrightarrow}{\text{Br—Cl}}$$

Thus, the molecule does have a dipole moment. In fact, all diatomic molecules containing different elements possess a dipole moment.

(b) Because fluorine is more electronegative than boron, each B—F bond in BF_3 (boron trifluoride) is polar and the three bond moments are equal. However, the symmetry of a trigonal planar shape means that the three bond moments exactly cancel one another:

F
|
B
F F

An analogy is an object that is pulled in the directions shown by the three bond moments. If the forces are equal, the object will not move. Consequently, BF_3 has no dipole moment; it is a nonpolar molecule.

(c) The Lewis structure of CH_2Cl_2 (methylene chloride) is

Cl
|
H—C—H
|
Cl

(Continued)

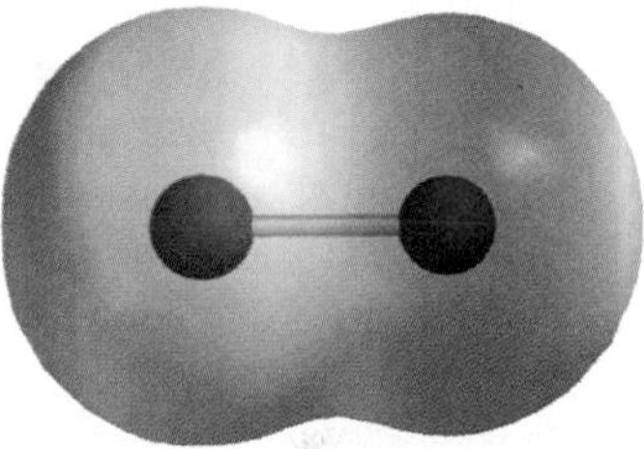

Electrostatic potential map of BrCl shows that the electron density is shifted toward the Cl atom.

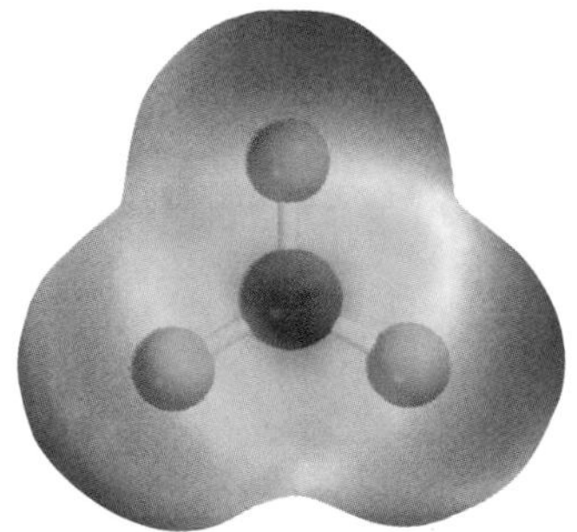

Electrostatic potential map shows that the electron density is symmetrically distributed in the BF_3 molecule.

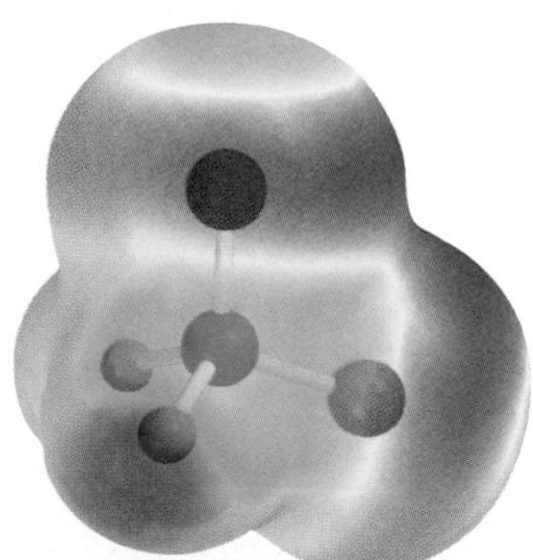

Electrostatic potential map of CH_2Cl_2. The electron density is shifted toward the electronegative Cl atoms.

This molecule is similar to CH_4 in that it has an overall tetrahedral shape. However, because not all the bonds are identical, there are three different bond angles: HCH, HCCl, and ClCCl. These bond angles are close to, but not equal to, 109.5°. Because chlorine is more electronegative than carbon, which is more electronegative than hydrogen, the bond moments do not cancel and the molecule possesses a dipole moment:

resultant dipole moment

Similar problems: 10.19, 10.21, 10.22.

Thus, CH_2Cl_2 is a polar molecule.

Practice Exercise Does the $AlCl_3$ molecule have a dipole moment?

REVIEW OF CONCEPTS

Carbon dioxide has a linear geometry and is nonpolar. Yet we know that the molecule executes bending and stretching motions that create a dipole moment. How would you reconcile these conflicting descriptions about CO_2?

10.3 Valence Bond Theory

The VSEPR model, based largely on Lewis structures, provides a relatively simple and straightforward method for predicting the geometry of molecules. But as we noted earlier, the Lewis theory of chemical bonding does not clearly explain why chemical bonds exist. Relating the formation of a covalent bond to the pairing of electrons was a step in the right direction, but it did not go far enough. For example, the Lewis theory describes the single bond between the H atoms in H_2 and that between the F atoms in F_2 in essentially the same way—as the pairing of two electrons. Yet these two molecules have quite different bond enthalpies and bond lengths (436.4 kJ/mol and 74 pm for H_2 and 150.6 kJ/mol and 142 pm for F_2). These and many other facts cannot be explained by the Lewis theory. For a more complete explanation of chemical bond formation we look to quantum mechanics. In fact, the quantum mechanical study of chemical bonding also provides a means for understanding molecular geometry.

At present, two quantum mechanical theories are used to describe covalent bond formation and the electronic structure of molecules. *Valence bond (VB) theory* assumes that the electrons in a molecule occupy atomic orbitals of the individual atoms. It enables us to retain a picture of individual atoms taking part in the bond formation. The second theory, called *molecular orbital (MO) theory,* assumes the formation of molecular orbitals from the atomic orbitals. Neither theory perfectly explains all aspects of bonding, but each has contributed something to our understanding of many observed molecular properties.

Let us start our discussion of valence bond theory by considering the formation of a H_2 molecule from two H atoms. The Lewis theory describes the H—H bond in terms of the pairing of the two electrons on the H atoms. In the framework of valence bond theory, the covalent H—H bond is formed by the *overlap* of the two 1*s* orbitals in the H atoms. By overlap, we mean that the two orbitals share a common region in space.

What happens to two H atoms as they move toward each other and form a bond? Initially, when the two atoms are far apart, there is no interaction. We say that the

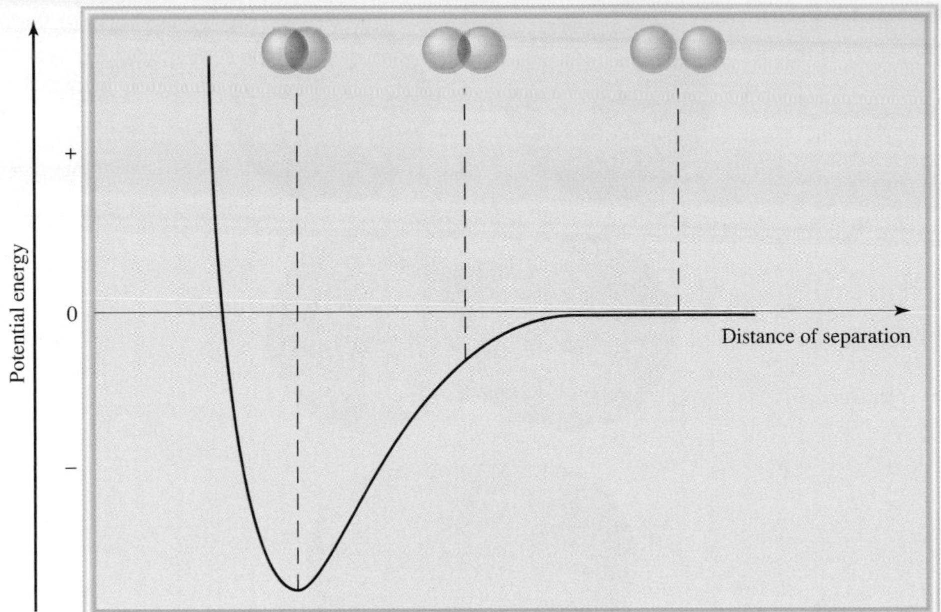

Figure 10.4
Change in potential energy of two H atoms with their distance of separation. At the point of minimum potential energy, the H_2 molecule is in its most stable state and the bond length is 74 pm. The spheres represent the 1s orbitals.

potential energy of this system (that is, the two H atoms) is zero. As the atoms approach each other, each electron is attracted by the nucleus of the other atom; at the same time, the electrons repel each other, as do the nuclei. While the atoms are still separated, attraction is stronger than repulsion, so that the potential energy of the system *decreases* (that is, it becomes negative) as the atoms approach each other (Figure 10.4). This trend continues until the potential energy reaches a minimum value. At this point, when the system has the lowest potential energy, it is most stable. This condition corresponds to substantial overlap of the 1*s* orbitals and the formation of a stable H_2 molecule. If the distance between nuclei were to decrease further, the potential energy would rise steeply and finally become positive as a result of the increased electron-electron and nuclear-nuclear repulsions. In accord with the law of conservation of energy, the decrease in potential energy as a result of H_2 formation must be accompanied by a release of energy. Experiments show that as a H_2 molecule is formed from two H atoms, heat is given off. The converse is also true. To break a H—H bond, energy must be supplied to the molecule. Figure 10.5 is another way of viewing the formation of an H_2 molecule.

Recall that an object has potential energy by virtue of its position.

Thus, valence bond theory gives a clearer picture of chemical bond formation than the Lewis theory does. Valence bond theory states that a stable molecule forms from reacting atoms when the potential energy of the system has decreased to a minimum; the Lewis theory ignores energy changes in chemical bond formation.

The concept of overlapping atomic orbitals applies equally well to diatomic molecules other than H_2. Thus, a stable F_2 molecule forms when the 2*p* orbitals (containing the unpaired electrons) in the two F atoms overlap to form a covalent bond. Similarly, the formation of the HF molecule can be explained by the overlap of the 1*s* orbital in H with the 2*p* orbital in F. In each case, VB theory accounts for the changes in potential energy as the distance between the reacting atoms changes. Because the orbitals involved are not the same kind in all cases, we can see why the bond enthalpies and bond lengths in H_2, F_2, and HF might be different. As we stated earlier, Lewis theory treats *all* covalent bonds the same way and offers no explanation for the differences among covalent bonds.

The orbital diagram of the F atom is shown on p. 236.

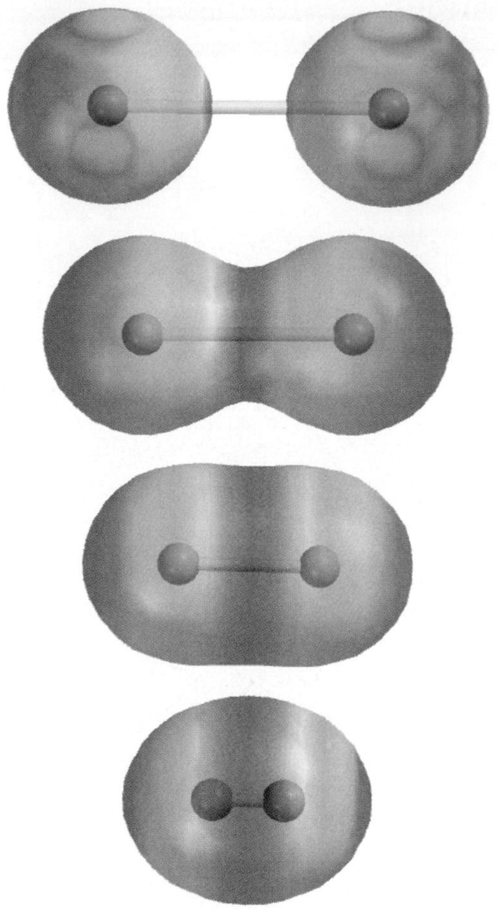

Figure 10.5
Top to bottom: As two H atoms approach each other, their 1s orbitals begin to interact and each electron begins to feel the attraction of the other proton. Gradually, the electron density builds up in the region between the two nuclei (red color). Eventually, a stable H_2 molecule is formed when the internuclear distance is 74 pm.

REVIEW OF CONCEPTS

How does valence bond theory treat covalent bonds differently from Lewis theory, where all covalent bonds are the same?

10.4 Hybridization of Atomic Orbitals

The concept of atomic orbital overlap should apply also to polyatomic molecules. However, a satisfactory bonding scheme must account for molecular geometry. We will discuss three examples of VB treatment of bonding in polyatomic molecules.

Animation: Hybridization

sp^3 Hybridization

Consider the CH_4 molecule. Focusing only on the valence electrons, we can represent the orbital diagram of C as

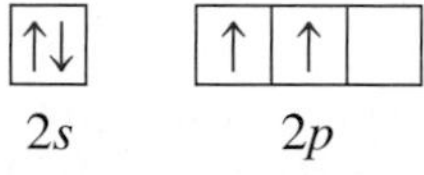

Because the carbon atom has two unpaired electrons (one in each of the two $2p$ orbitals), it can form only two bonds with hydrogen in its ground state. Although the

species CH_2 is known, it is very unstable. To account for the four C—H bonds in methane, we can try to promote (that is, energetically excite) an electron from the 2*s* orbital to the 2*p* orbital:

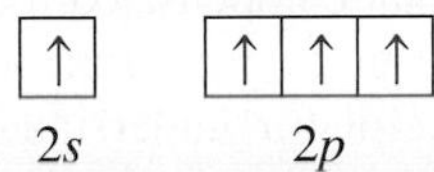

Now there are four unpaired electrons on C that could form four C—H bonds. However, the geometry is wrong, because three of the HCH bond angles would have to be 90° (remember that the three 2*p* orbitals on carbon are mutually perpendicular), and yet *all* HCH angles are 109.5°.

To explain the bonding in methane, VB theory uses hypothetical ***hybrid orbitals***, which are *atomic orbitals obtained when two or more nonequivalent orbitals of the same atom combine in preparation for covalent bond formation.* ***Hybridization*** is the term applied to *the mixing of atomic orbitals in an atom (usually a central atom) to generate a set of hybrid orbitals.* We can generate four equivalent hybrid orbitals for carbon by mixing the 2*s* orbital and the three 2*p* orbitals:

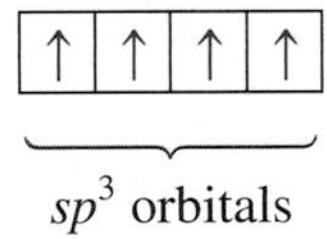

Because the new orbitals are formed from one *s* and three *p* orbitals, they are called sp^3 hybrid orbitals. Figure 10.6 shows the shape and orientations of the sp^3 orbitals. These four hybrid orbitals are directed toward the four corners of a regular tetrahedron.

sp^3 is pronounced "*s-p* three."

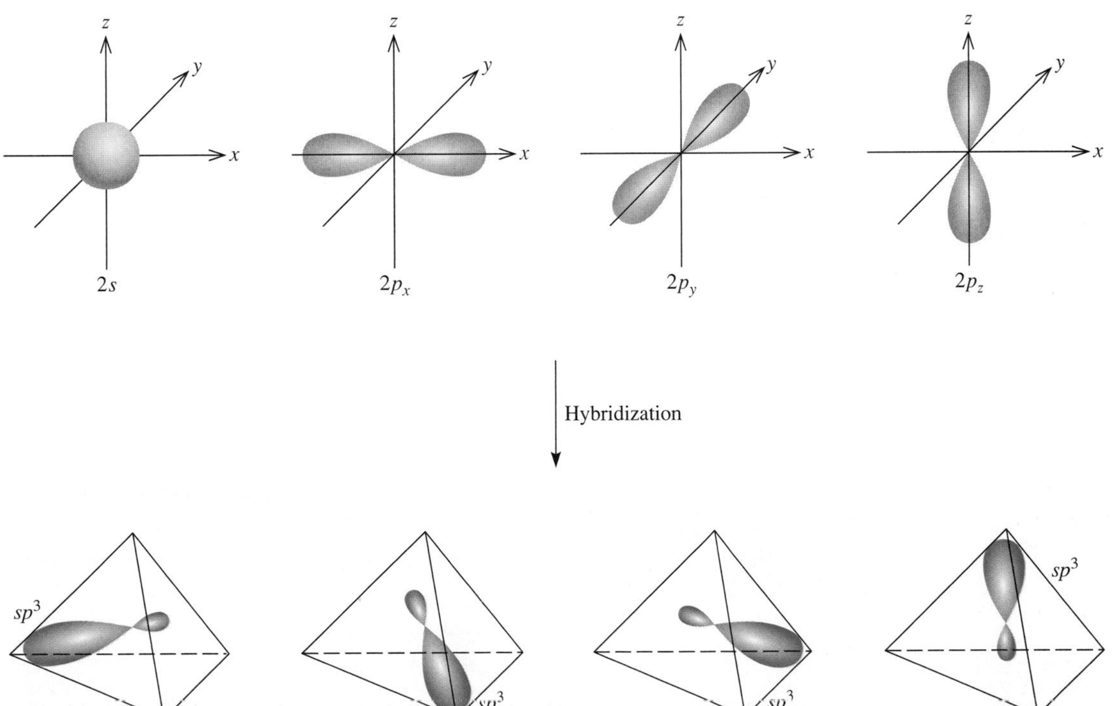

Figure 10.6
Formation of four sp^3 hybrid orbitals from one 2s and three 2p orbitals. The sp^3 orbitals point to the corners of a tetrahedron.

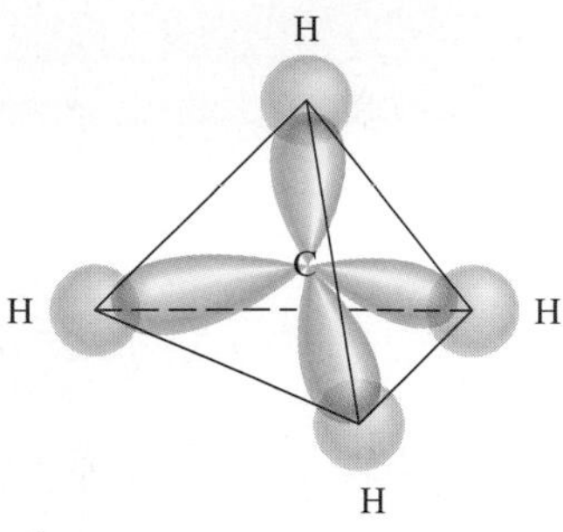

Figure 10.7
Formation of four bonds between the carbon sp^3 hybrid orbitals and the hydrogen 1s orbitals in CH_4. The smaller lobes are not shown.

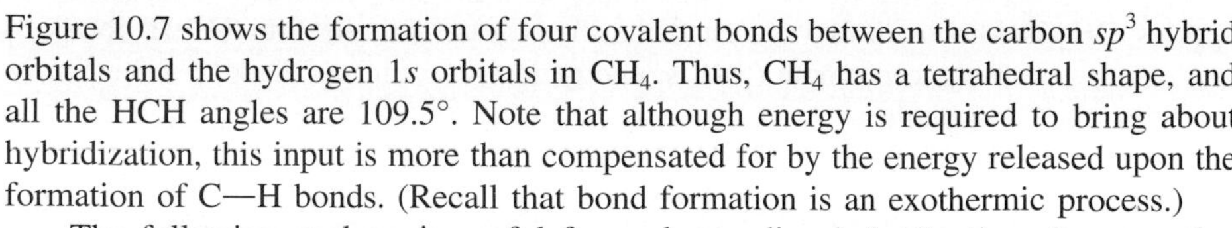

Figure 10.7 shows the formation of four covalent bonds between the carbon sp^3 hybrid orbitals and the hydrogen 1*s* orbitals in CH_4. Thus, CH_4 has a tetrahedral shape, and all the HCH angles are 109.5°. Note that although energy is required to bring about hybridization, this input is more than compensated for by the energy released upon the formation of C—H bonds. (Recall that bond formation is an exothermic process.)

The following analogy is useful for understanding hybridization. Suppose that we have a beaker of a red solution and three beakers of blue solutions and that the volume of each is 50 mL. The red solution corresponds to one 2*s* orbital, the blue solutions represent three 2*p* orbitals, and the four equal volumes symbolize four separate orbitals. By mixing the solutions we obtain 200 mL of a purple solution, which can be divided into four 50-mL portions (that is, the hybridization process generates four sp^3 orbitals). Just as the purple color is made up of the red and blue components of the original solutions, the sp^3 hybrid orbitals possess both *s* and *p* orbital characteristics.

Another example of sp^3 hybridization is ammonia (NH_3). Table 10.1 shows that the arrangement of four electron pairs is tetrahedral, so that the bonding in NH_3 can be explained by assuming that N, like C in CH_4, is sp^3-hybridized. The ground-state electron configuration of N is $1s^22s^22p^3$, so that the orbital diagram for the sp^3 hybridized N atom is

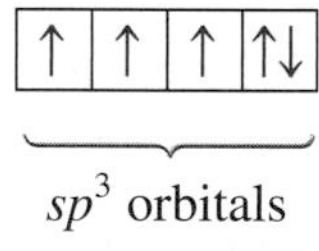

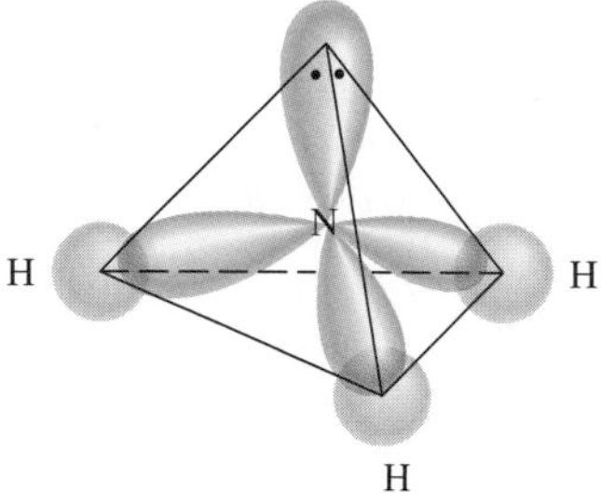

Figure 10.8
The sp^3-hybridized N atom in NH_3. Three sp^3 hybrid orbitals form bonds with the H atoms. The fourth is occupied by nitrogen's lone pair.

Three of the four hybrid orbitals form covalent N—H bonds, and the fourth hybrid orbital accommodates the lone pair on nitrogen (Figure 10.8). Repulsion between the lone-pair electrons and electrons in the bonding orbitals decreases the HNH bond angles from 109.5° to 107.3°.

It is important to understand the relationship between hybridization and the VSEPR model. We use hybridization to describe the bonding scheme only when the arrangement of electron pairs has been predicted using VSEPR. If the VSEPR model predicts a tetrahedral arrangement of electron pairs, then we assume that one *s* and three *p* orbitals are hybridized to form four sp^3 hybrid orbitals. The following are examples of other types of hybridization.

sp Hybridization

The beryllium chloride ($BeCl_2$) molecule is predicted to be linear by VSEPR. The orbital diagram for the valence electrons in Be is

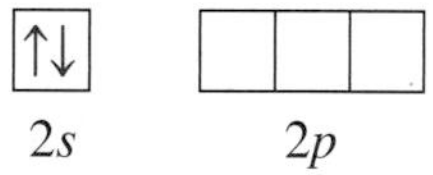

We know that in its ground state Be does not form covalent bonds with Cl because its electrons are paired in the 2*s* orbital. So we turn to hybridization for an explanation of Be's bonding behavior. First, we promote a 2*s* electron to a 2*p* orbital, resulting in

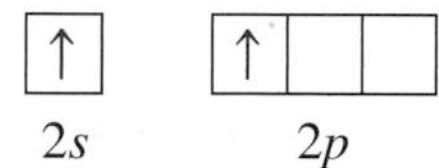

Now there are two Be orbitals available for bonding, the 2*s* and 2*p*. However, if two Cl atoms were to combine with Be in this excited state, one Cl atom would share a 2*s* electron and the other Cl would share a 2*p* electron, making two

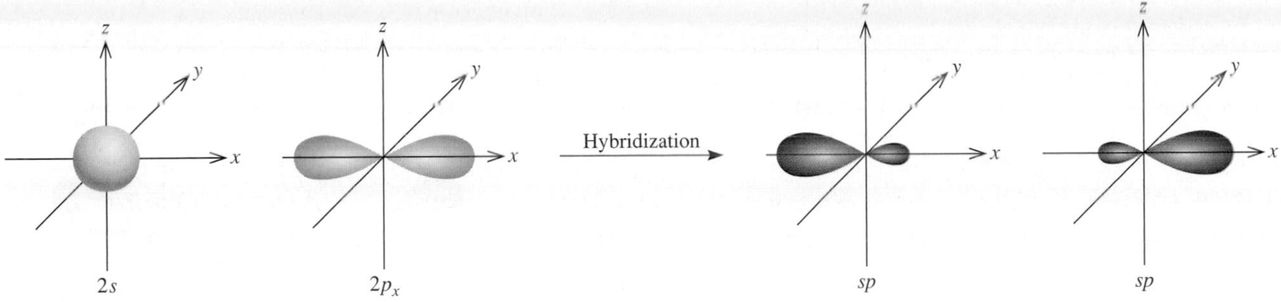

Figure 10.9
Formation of two sp hybrid orbitals from one 2s and one 2p orbital.

nonequivalent BeCl bonds. This scheme contradicts experimental evidence. In the actual $BeCl_2$ molecule, the two BeCl bonds are identical in every respect. Thus, the 2*s* and 2*p* orbitals must be mixed, or hybridized, to form two equivalent *sp* hybrid orbitals:

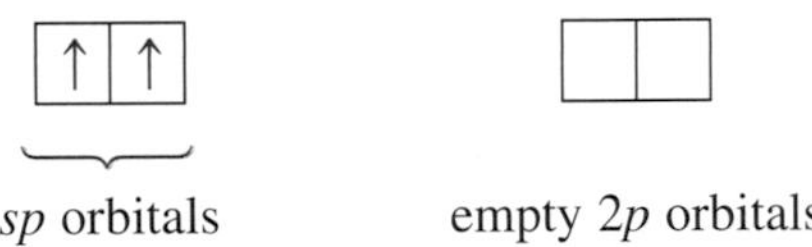

Figure 10.10
The linear geometry of $BeCl_2$ can be explained by assuming that Be is sp-hybridized. The two sp hybrid orbitals overlap with the two chlorine 3p orbitals to form two covalent bonds.

Figure 10.9 shows the shape and orientation of the *sp* orbitals. These two hybrid orbitals lie on the same line, the *x*-axis, so that the angle between them is 180°. Each of the BeCl bonds is then formed by the overlap of a Be *sp* hybrid orbital and a Cl 3*p* orbital, and the resulting $BeCl_2$ molecule has a linear geometry (Figure 10.10).

sp^2 Hybridization

Next we will look at the BF_3 (boron trifluoride) molecule, known to have planar geometry based on VSEPR. Considering only the valence electrons, the orbital diagram of B is

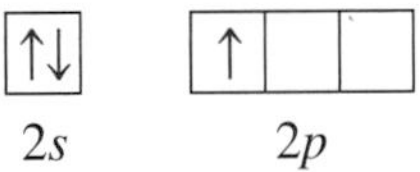

First, we promote a 2*s* electron to an empty 2*p* orbital:

↑	↑ ↑ □
2s	2p

Mixing the 2*s* orbital with the two 2*p* orbitals generates three sp^2 hybrid orbitals:

sp^2 is pronounced "s-p two."

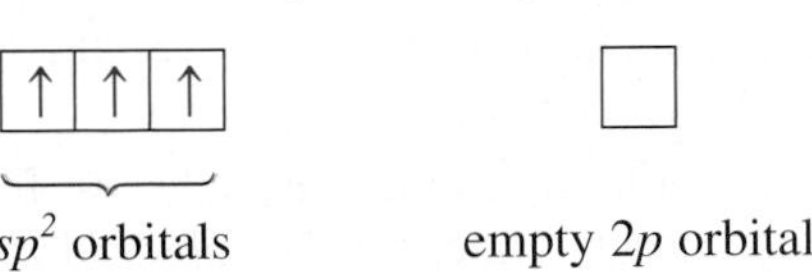

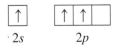

These three sp^2 orbitals lie in the same plane, and the angle between any two of them is 120° (Figure 10.11). Each of the BF bonds is formed by the overlap of a boron sp^2 hybrid orbital and a fluorine 2*p* orbital (Figure 10.12). The BF_3 molecule is planar with all the FBF angles equal to 120°. This result conforms to experimental findings and also to VSEPR predictions.

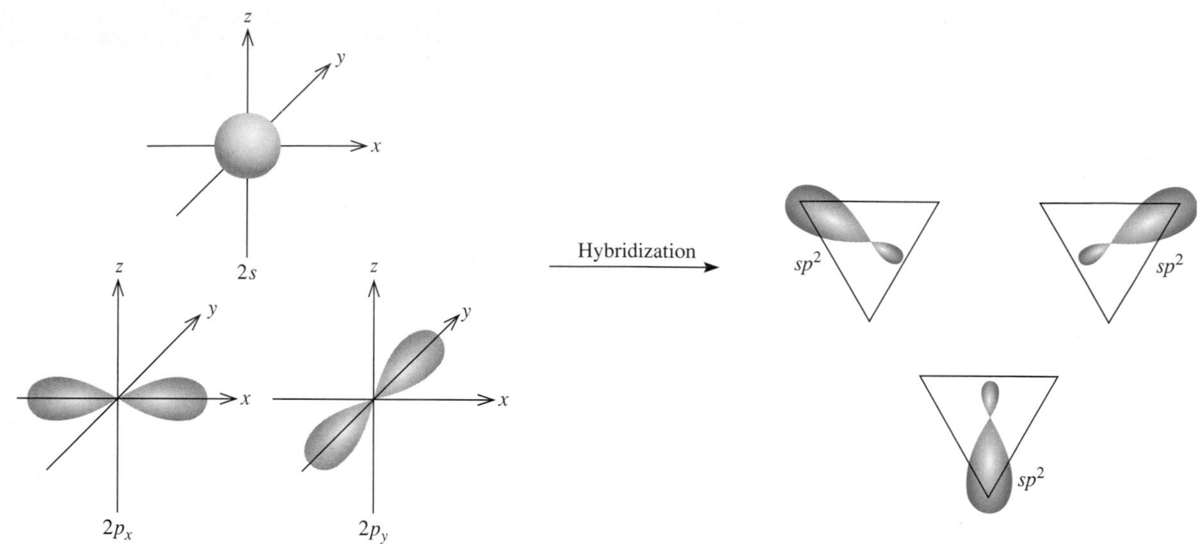

Figure 10.11
Formation of three sp^2 hybrid orbitals from one 2s and two 2p orbitals. The sp^2 orbitals point to the corners of an equilateral triangle.

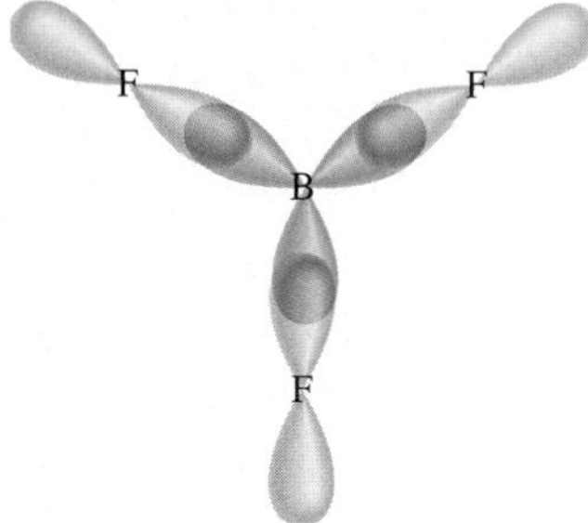

Figure 10.12
The sp^2 hybrid orbitals of boron overlap with the 2p orbitals of fluorine. The BF_3 molecule is planar, and all the FBF angles are 120°.

You may have noticed an interesting connection between hybridization and the octet rule. Regardless of the type of hybridization, an atom starting with one *s* and three *p* orbitals would still possess four orbitals, enough to accommodate a total of eight electrons in a compound. For elements in the second period of the periodic table, eight is the maximum number of electrons that an atom of any of these elements can accommodate in the valence shell. This is the reason that the octet rule is usually obeyed by the second-period elements.

The situation is different for an atom of a third-period element. If we use only the 3*s* and 3*p* orbitals of the atom to form hybrid orbitals in a molecule, then the octet rule applies. However, in some molecules the same atom may use one or more 3*d* orbitals, in addition to the 3*s* and 3*p* orbitals, to form hybrid orbitals. In these cases, the octet rule does not hold. We will see specific examples of the participation of the 3*d* orbital in hybridization shortly.

To summarize our discussion of hybridization, we note that

1. The concept of hybridization is not applied to isolated atoms. It is a theoretical model used only to explain covalent bonding.
2. Hybridization is the mixing of at least two nonequivalent atomic orbitals, for example, *s* and *p* orbitals. Therefore, a hybrid orbital is not a pure atomic orbital. Hybrid orbitals and pure atomic orbitals have very different shapes.
3. The number of hybrid orbitals generated is equal to the number of pure atomic orbitals that participate in the hybridization process.
4. Hybridization requires an input of energy; however, the system more than recovers this energy during bond formation.
5. Covalent bonds in polyatomic molecules and ions are formed by the overlap of hybrid orbitals, or of hybrid orbitals with unhybridized ones. Therefore, the hybridization bonding scheme is still within the framework of valence bond theory; electrons in a molecule are assumed to occupy hybrid orbitals of the individual atoms.

Table 10.4 summarizes *sp*, sp^2, and sp^3 hybridization (as well as other types that we will discuss shortly).

Table 10.4 Important Hybrid Orbitals and Their Shapes

Pure Atomic Orbitals of the Central Atom	Hybridization of the Central Atom	Number of Hybrid Orbitals	Shape of Hybrid Orbitals	Examples
s, p	*sp*	2	 180° Linear	$BeCl_2$
s, p, p	sp^2	3	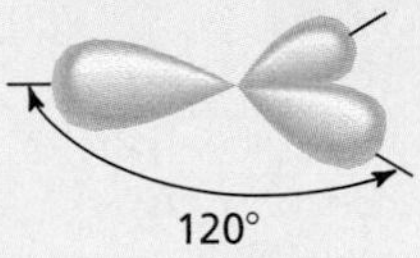 120° Trigonal planar	BF_3
s, p, p, p	sp^3	4	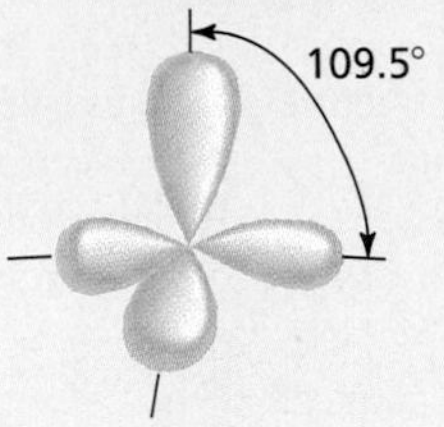 109.5° Tetrahedral	CH_4, NH_4^+
s, p, p, p, d	sp^3d	5	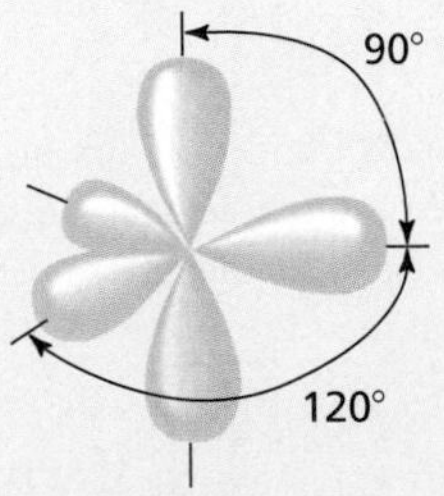 90° 120° Trigonal bipyramidal	PCl_5
s, p, p, p, d, d	sp^3d^2	6	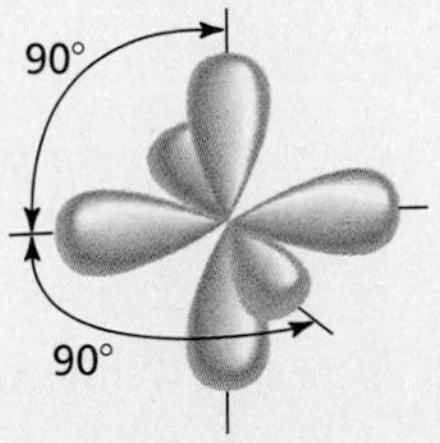 90° 90° Octahedral	SF_6

Procedure for Hybridizing Atomic Orbitals

Before going on to discuss the hybridization of *d* orbitals, let us specify what we need to know to apply hybridization to bonding in polyatomic molecules in general. In essence, hybridization simply extends Lewis theory and the VSEPR model. To assign a suitable state of hybridization to the central atom in a molecule, we must have some idea about the geometry of the molecule. The steps are as follows:

1. Draw the Lewis structure of the molecule.
2. Predict the overall arrangement of the electron pairs (both bonding pairs and lone pairs) using the VSEPR model (see Table 10.1).
3. Deduce the hybridization of the central atom by matching the arrangement of the electron pairs with those of the hybrid orbitals shown in Table 10.4.

EXAMPLE 10.3

Determine the hybridization state of the central (underlined) atom in each of the following molecules: (a) $\underline{Be}H_2$, (b) $\underline{Al}I_3$, and (c) $\underline{P}F_3$. Describe the hybridization process and determine the molecular geometry in each case.

Strategy The steps for determining the hybridization of the central atom in a molecule are:

draw Lewis structure of the molecule ⟶ use VSEPR to determine the electron pair arrangement surrounding the central atom (Table 10.1) ⟶ use Table 10.4 to determine the hybridization state of the central atom

Solution

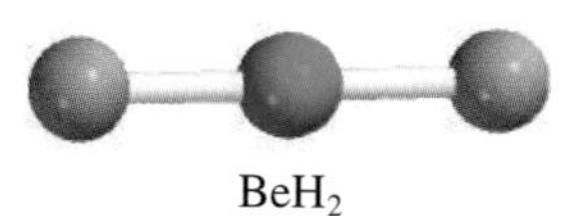

BeH_2

(a) The ground-state electron configuration of Be is $1s^22s^2$ and the Be atom has two valence electrons. The Lewis structure of BeH_2 is

$$H{-}Be{-}H$$

There are two bonding pairs around Be; therefore, the electron pair arrangement is linear. We conclude that Be uses *sp* hybrid orbitals in bonding with H, because *sp* orbitals have a linear arrangement (see Table 10.4). The hybridization process can be imagined as follows. First we draw the orbital diagram for the ground state of Be:

↑↓				
2s		2p		

By promoting a 2*s* electron to the 2*p* orbital, we get the excited state:

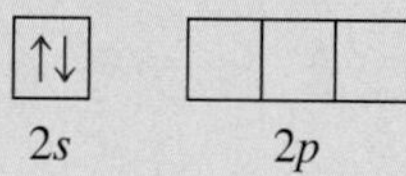

The 2*s* and 2*p* orbitals then mix to form two hybrid orbitals:

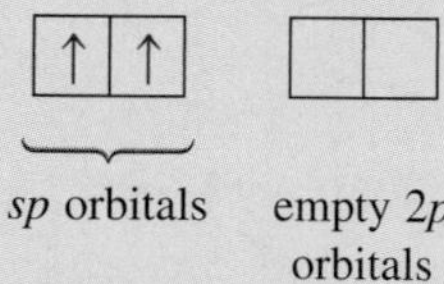

The two Be—H bonds are formed by the overlap of the Be *sp* orbitals with the 1*s* orbitals of the H atoms. Thus, BeH_2 is a linear molecule.

(Continued)

(b) The ground-state electron configuration of Al is $[Ne]3s^23p^1$. Therefore, the Al atom has three valence electrons. The Lewis structure of AlI_3 is

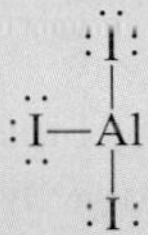

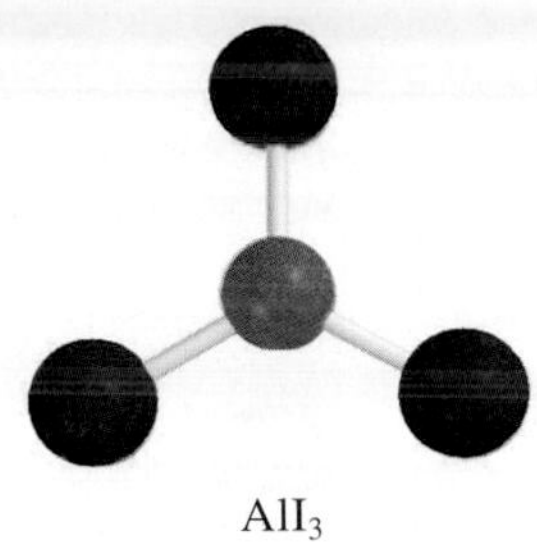
AlI_3

There are three pairs of electrons around Al; therefore, the electron pair arrangement is trigonal planar. We conclude that Al uses sp^2 hybrid orbitals in bonding with I because sp^2 orbitals have a trigonal planar arrangement (see Table 10.4). The orbital diagram of the ground-state Al atom is

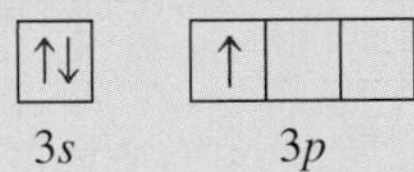

By promoting a 3s electron into the 3p orbital we obtain the following excited state:

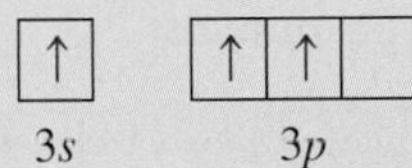

The 3s and two 3p orbitals then mix to form three sp^2 hybrid orbitals:

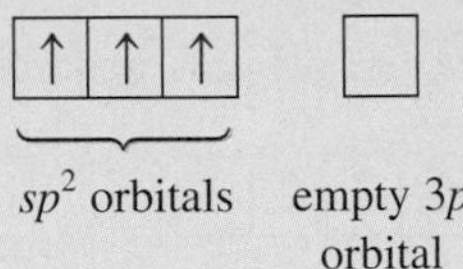

The sp^2 hybrid orbitals overlap with the 5p orbitals of I to form three covalent Al—I bonds. We predict that the AlI_3 molecule is trigonal planar and all the IAlI angles are 120°.

(c) The ground-state electron configuration of P is $[Ne]3s^23p^3$. Therefore, the P atom has five valence electrons. The Lewis structure of PF_3 is

:F—P—F:
:F:

PF_3

There are four pairs of electrons around P; therefore, the electron pair arrangement is tetrahedral. We conclude that P uses sp^3 hybrid orbitals in bonding to F, because sp^3 orbitals have a tetrahedral arrangement (see Table 10.4). The hybridization process can be imagined to take place as follows. The orbital diagram of the ground-state P atom is

↑↓ ↑ ↑ ↑
3s 3p

By mixing the 3s and 3p orbitals, we obtain four sp^3 hybrid orbitals.

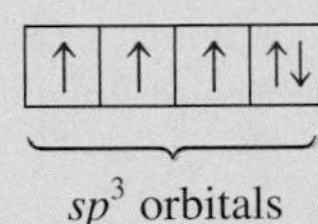

As in the case of NH_3, one of the sp^3 hybrid orbitals is used to accommodate the lone pair on P. The other three sp^3 hybrid orbitals form covalent P—F bonds with the 2p orbitals of F. We predict the geometry of the molecule to be trigonal pyramidal; the FPF angle should be somewhat less than 109.5°.

Similar problems: 10.31, 10.32.

Practice Exercise Determine the hybridization state of the underlined atoms in the following compounds: (a) <u>Si</u>Br_4 and (b) <u>B</u>Cl_3.

Hybridization of *s*, *p*, and *d* Orbitals

We have seen that hybridization neatly explains bonding that involves *s* and *p* orbitals. For elements in the third period and beyond, however, we cannot always account for molecular geometry by assuming that only *s* and *p* orbitals hybridize. To understand the formation of molecules with trigonal bipyramidal and octahedral geometries, for instance, we must include *d* orbitals in the hybridization concept.

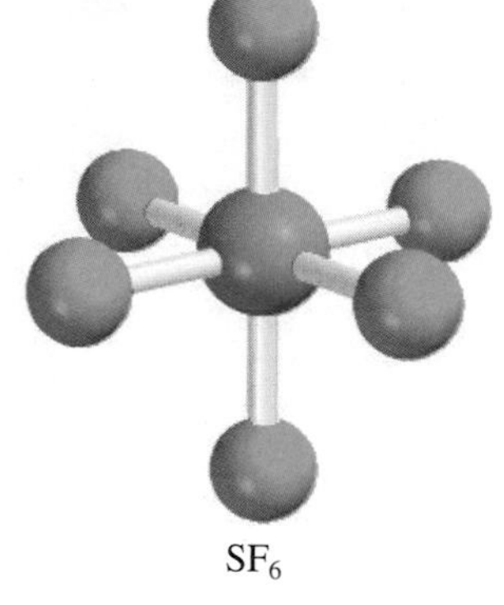

SF_6

Consider the SF_6 molecule as an example. In Section 10.1 we saw that this molecule has octahedral geometry, which is also the arrangement of the six electron pairs. Table 10.4 shows that the S atom is sp^3d^2-hybridized in SF_6. The ground-state electron configuration of S is [Ne]$3s^23p^4$:

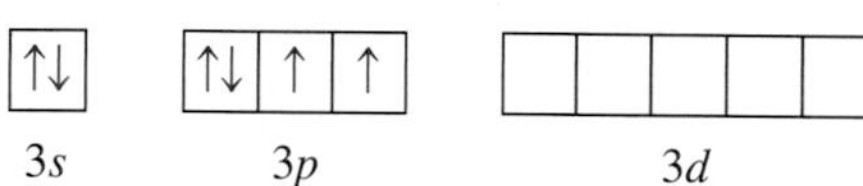

Because the 3*d* level is quite close in energy to the 3*s* and 3*p* levels, we can promote 3*s* and 3*p* electrons to two of the 3*d* orbitals:

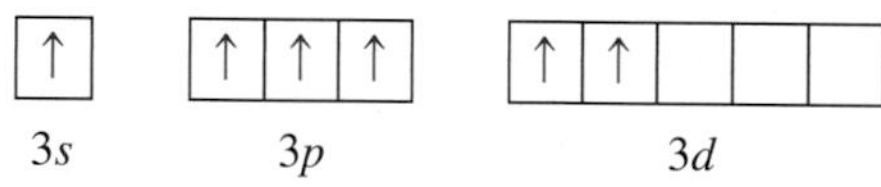

sp^3d^2 is pronounced "s-p three d two."

Mixing the 3*s*, three 3*p*, and two 3*d* orbitals generates six sp^3d^2 hybrid orbitals:

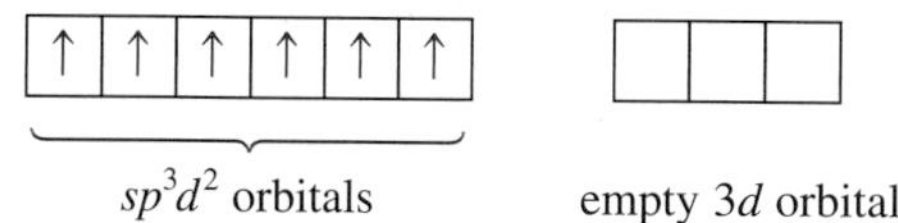

The six S—F bonds are formed by the overlap of the hybrid orbitals of the S atom with the 2*p* orbitals of the F atoms. Because there are 12 electrons around the S atom, the octet rule is violated. The use of *d* orbitals in addition to *s* and *p* orbitals to form an expanded octet (see Section 9.9) is an example of *valence-shell expansion.* Second-period elements, unlike third-period elements, do not have 2*d* energy levels, so they can never expand their valence shells. (Recall that when $n = 2$, $l = 0$ and 1. Thus, we can only have 2*s* and 2*p* orbitals.) Hence, atoms of second-period elements can never be surrounded by more than eight electrons in any of their compounds.

PBr_5

EXAMPLE 10.4

Describe the hybridization state of phosphorus in phosphorus pentabromide (PBr_5).

Strategy Follow the same procedure shown in Example 10.3.

Solution The ground-state electron configuration of P is [Ne]$3s^23p^3$. Therefore, the P atom has five valence electrons. The Lewis structure of PBr_5 is

```
        :Br:
  :Br\   |
       P—Br:
  :Br/   |
        :Br:
```

There are five pairs of electrons around P; therefore, the electron pair arrangement is trigonal bipyramidal. We conclude that P uses sp^3d hybrid orbitals in bonding to Br,

(Continued)

because sp^3d hybrid orbitals have a trigonal bipyramidal arrangement (see Table 10.4). The hybridization process can be imagined as follows. The orbital diagram of the ground-state P atom is

↑↓ ↑ ↑ ↑ □□□□□

3*s* 3*p* 3*d*

Promoting a 3*s* electron into a 3*d* orbital results in the following excited state:

↑ ↑ ↑ ↑ ↑□□□□

3*s* 3*p* 3*d*

Mixing the one 3*s*, three 3*p*, and one 3*d* orbitals generates five sp^3d hybrid orbitals:

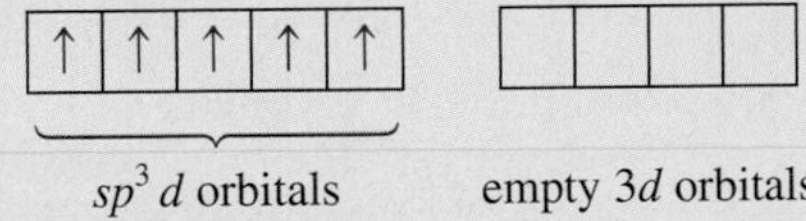

sp^3d orbitals empty 3*d* orbitals

These hybrid orbitals overlap the 4*p* orbitals of Br to form five covalent P—Br bonds. Because there are no lone pairs on the P atom, the geometry of PBr_5 is trigonal bipyramidal.

Practice Exercise Describe the hybridization state of Se in SeF_6.

Similar problem: 10.40.

Review of Concepts

Why is it not possible to generate sp^4 hybrid orbitals?

10.5 Hybridization in Molecules Containing Double and Triple Bonds

The concept of hybridization is useful also for molecules with double and triple bonds. Consider the ethylene molecule, C_2H_4, as an example. In Example 10.1 we saw that C_2H_4 contains a carbon-carbon double bond and has planar geometry. Both the geometry and the bonding can be understood if we assume that each carbon atom is sp^2-hybridized. Figure 10.13 shows orbital diagrams of this hybridization process. We assume that only the $2p_x$ and $2p_y$ orbitals combine with the 2*s* orbital, and that the $2p_z$ orbital remains unchanged. Figure 10.14 shows that the $2p_z$ orbital is perpendicular to the plane of the hybrid orbitals. Now, how do we account for the bonding of the C atoms? As Figure 10.15(a) shows, each carbon atom uses the three sp^2 hybrid orbitals to form two bonds with the two hydrogen 1*s* orbitals and one bond with the sp^2 hybrid orbital of the adjacent C atom. In addition, the two unhybridized $2p_z$ orbitals of the C atoms form another bond by overlapping sideways [Figure 10.15(b)].

A distinction is made between the two types of covalent bonds in C_2H_4. The three bonds formed by each C atom in Figure 10.15(a) are all ***sigma bonds (σ bonds),*** *covalent bonds formed by orbitals overlapping end-to-end, with the electron density concentrated between the nuclei of the bonding atoms.* The second type is called a ***pi bond (π bond),*** which is defined as *a covalent bond formed by sideways overlapping orbitals with electron density concentrated above and below the plane of the nuclei of the bonding atoms.* The two C atoms form a pi bond, as shown in Figure 10.15(b). This pi bond

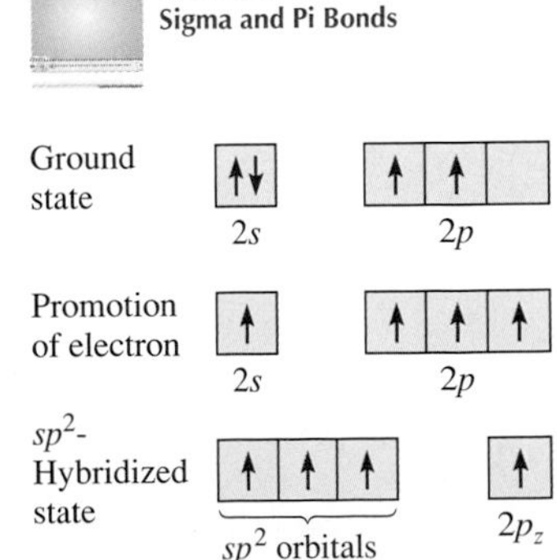

Figure 10.13
The sp^2 hybridization of a carbon atom. The 2s orbital is mixed with only two 2p orbitals to form three equivalent sp^2 hybrid orbitals. This process leaves an electron in the unhybridized orbital, the $2p_z$ orbital.

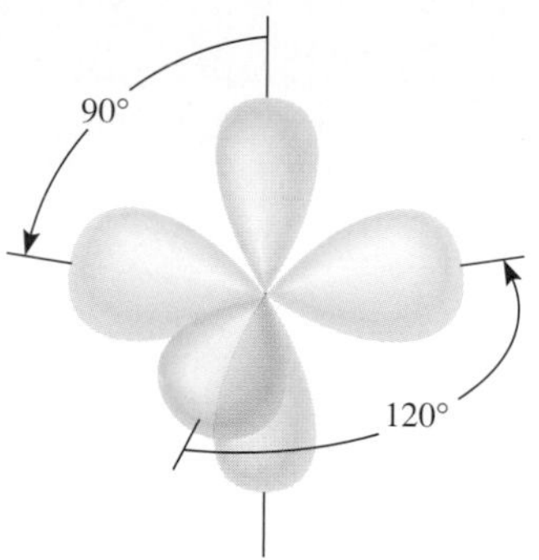

Figure 10.14
Each carbon atom in the C_2H_4 molecule has three sp^2 hybrid orbitals (green) and one unhybridized $2p_z$ orbital (gray), which is perpendicular to the plane of the hybrid orbitals.

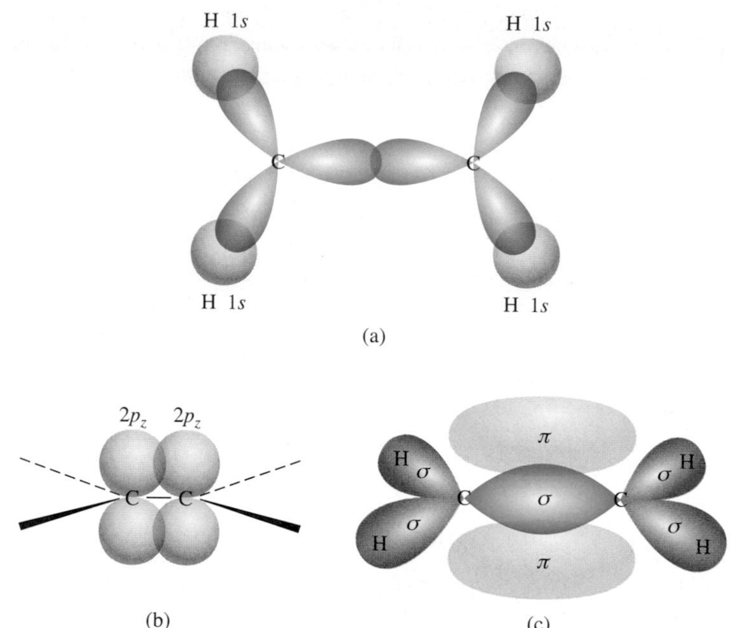

Figure 10.15
Bonding in ethylene, C_2H_4. (a) Top view of the sigma bonds between carbon atoms and between carbon and hydrogen atoms. All the atoms lie in the same plane, making C_2H_4 a planar molecule. (b) Side view showing how the two $2p_z$ orbitals on the two carbon atoms overlap, leading to the formation of a pi bond. The solid, dashed, and wedged lines show the directions of the sigma bonds. (c) The interactions in (a) and (b) lead to the formation of the sigma bonds and the pi bond in ethylene. Note that the pi bond lies above and below the plane of the molecule.

formation gives ethylene its planar geometry. Figure 10.15(c) shows the orientation of the sigma and pi bonds. Figure 10.16 is yet another way of looking at the planar C_2H_4 molecule and the formation of the pi bond. Although we normally represent the carbon-carbon double bond as C═C (as in a Lewis structure), it is important to keep in mind that the two bonds are different types: One is a sigma bond and the other is a pi bond. In fact, the bond enthalpies of the carbon-carbon pi and sigma bonds are about 270 kJ/mol and 350 kJ/mol, respectively.

The acetylene molecule (C_2H_2) contains a carbon-carbon triple bond. Because the molecule is linear, we can explain its geometry and bonding by assuming that each

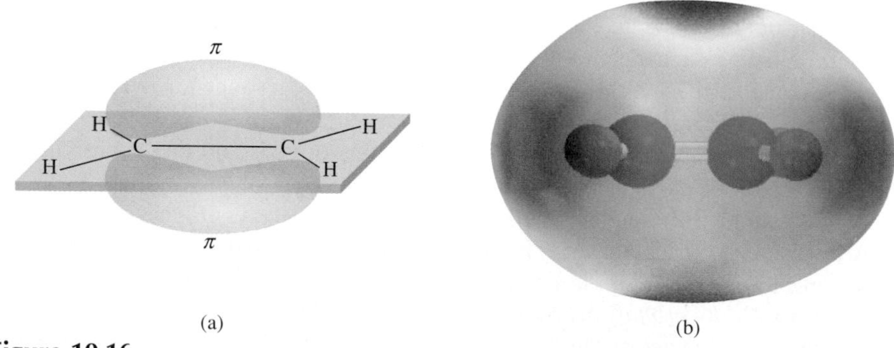

Figure 10.16
(a) Another view of the pi bond formation in the C_2H_4 molecule. Note that all six atoms are in the same plane. It is the overlap of the $2p_z$ orbitals that causes the molecule to assume a planar structure. (b) Electrostatic potential map of C_2H_4.

C atom is *sp*-hybridized by mixing the 2*s* with the $2p_x$ orbital (Figure 10.17). As Figure 10.18 shows, the two *sp* hybrid orbitals of each C atom form one sigma bond with a hydrogen 1*s* orbital and another sigma bond with the other C atom. In addition, two pi bonds are formed by the sideways overlap of the unhybridized $2p_y$ and $2p_z$ orbitals. Thus, the C≡C bond is made up of one sigma bond and two pi bonds.

The following rule helps us predict hybridization in molecules containing multiple bonds: If the central atom forms a double bond, it is sp^2-hybridized; if it forms two double bonds or a triple bond, it is *sp*-hybridized. Note that this rule applies only to atoms of the second-period elements. Atoms of third-period elements and beyond that form multiple bonds present a more complicated picture and will not be dealt with here.

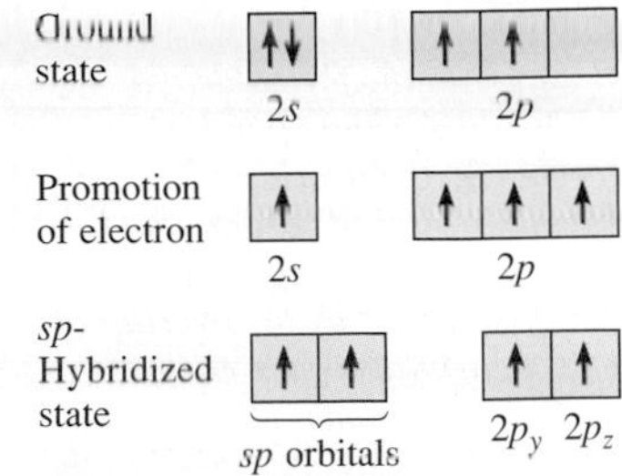

Figure 10.17
The sp hybridization of a carbon atom. The 2s orbital is mixed with only one 2p orbital to form two sp hybrid orbitals. This process leaves an electron in each of the two unhybridized 2p orbitals, namely, the $2p_y$ and $2p_z$ orbitals.

EXAMPLE 10.5

Describe the bonding in the formaldehyde molecule whose Lewis structure is

$$\begin{array}{l} \text{H} \\ \quad \diagdown \\ \quad\ \text{C}=\ddot{\underset{\cdot\cdot}{\text{O}}} \\ \quad \diagup \\ \text{H} \end{array}$$

Assume that the O atom is sp^2-hybridized.

Strategy Follow the procedure shown in Example 10.3.

Solution There are three pairs of electrons around the C atom; therefore, the electron pair arrangement is trigonal planar. (Recall that a double bond is treated as a single bond in the VSEPR model.) We conclude that C uses sp^2 hybrid orbitals in bonding, because

(Continued)

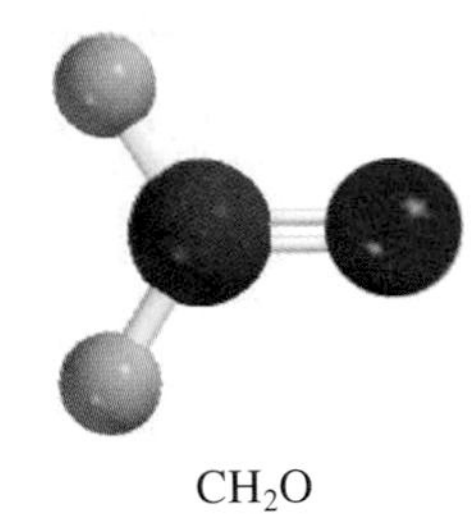

CH_2O

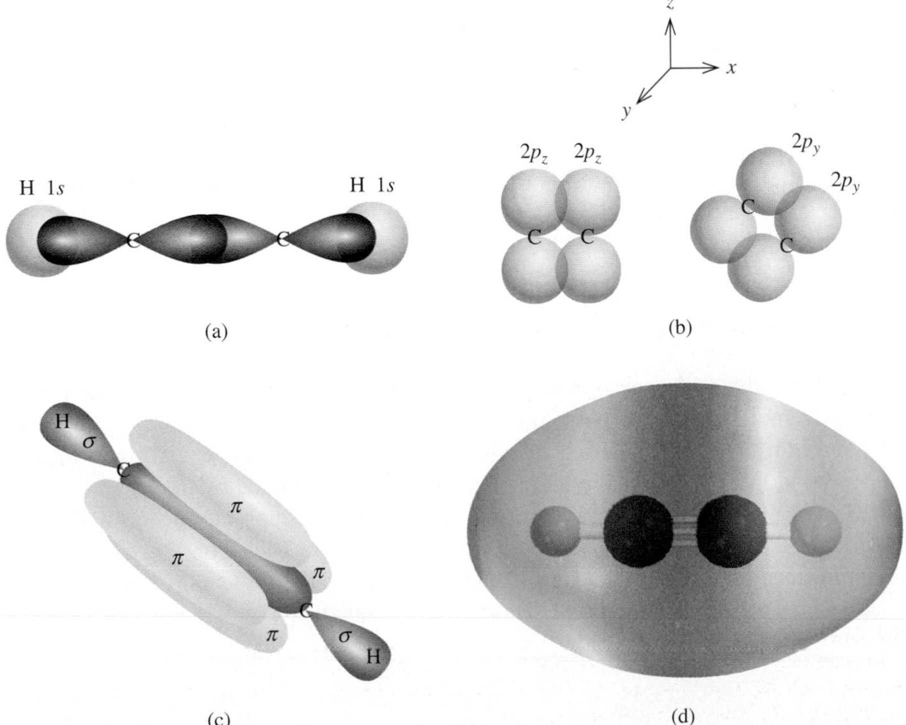

Figure 10.18
Bonding in acetylene, C_2H_2. (a) Top view showing the overlap of the sp orbitals between the C atoms and the overlap of the sp orbital with the 1s orbital between the C and H atoms. All the atoms lie along a straight line; therefore acetylene is a linear molecule. (b) Side view showing the overlap of the two $2p_y$ orbitals and of the two $2p_z$ orbitals of the two carbon atoms, which leads to the formation of two pi bonds. (c) Formation of the sigma and pi bonds as a result of the interactions in (a) and (b). (d) Electrostatic potential map of C_2H_2.

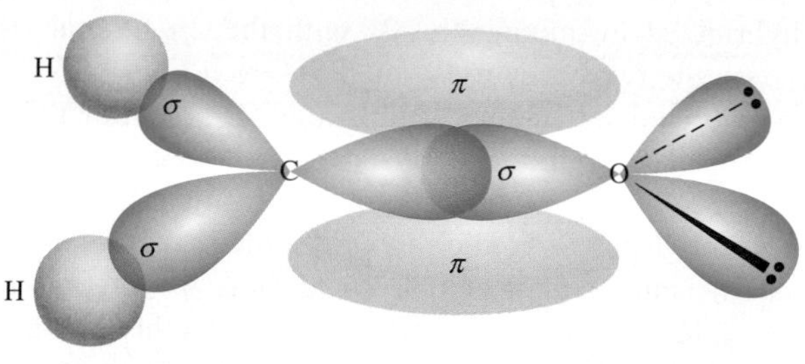

Figure 10.19
Bonding in the formaldehyde molecule. A sigma bond is formed by the overlap of the sp^2 hybrid orbital of carbon and the sp^2 hybrid orbital of oxygen; a pi bond is formed by the overlap of the $2p_z$ orbitals of the carbon and oxygen atoms. The two lone pairs on oxygen are placed in the other two sp^2 orbitals of oxygen.

sp^2 hybrid orbitals have a trigonal planar arrangement (see Table 10.4). We can imagine the hybridization processes for C and O as follows:

	2s	2p		sp^2 orbitals	$2p_z$
C	↑↓	↑ ↑ □	→	↑ ↑ ↑	↑
O	↑↓	↑↓ ↑ ↑	→	↑ ↑↓ ↑↓	↑

Carbon has one electron in each of the three sp^2 orbitals, which are used to form sigma bonds with the H atoms and the O atom. There is also an electron in the $2p_z$ orbital, which forms a pi bond with oxygen. Oxygen has two electrons in two of its sp^2 hybrid orbitals. These are the lone pairs on oxygen. Its third sp^2 hybrid orbital with one electron is used to form a sigma bond with carbon. The $2p_z$ orbital (with one electron) overlaps with the $2p_z$ orbital of C to form a pi bond (Figure 10.19).

Similar problems: 10.36, 10.39.

Practice Exercise Describe the bonding in the hydrogen cyanide molecule, HCN. Assume that N is *sp*-hybridized.

REVIEW OF CONCEPTS

Which of the following pairs of atomic orbitals on adjacent nuclei can overlap to form a sigma bond? a pi bond? Which cannot overlap (no bond)? Consider the *x*-axis to be the internuclear axis. (a) 1*s* and 2*s*, (b) 1*s* and $2p_x$, (c) $2p_y$ and $2p_y$, (d) $3p_y$ and $3p_z$, (e) $2p_x$ and $3p_x$.

10.6 Molecular Orbital Theory

Valence bond theory is one of the two quantum mechanical approaches that explain bonding in molecules. It accounts, at least qualitatively, for the stability of the covalent bond in terms of overlapping atomic orbitals. Using the concept of hybridization, valence bond theory can explain molecular geometries predicted by the VSEPR model. However, the assumption that electrons in a molecule occupy atomic orbitals of the individual atoms can be only an approximation, because each bonding electron in a molecule must be in an orbital that is characteristic of the molecule as a whole.

In some cases, valence bond theory cannot satisfactorily account for observed properties of molecules. Consider the oxygen molecule, whose Lewis structure is

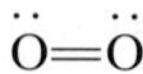

According to this description, all the electrons in O_2 are paired and oxygen should therefore be diamagnetic. But experiments have shown that the oxygen molecule has two unpaired electrons (Figure 10.20). This finding suggests a fundamental deficiency in valence bond theory, one that justifies searching for an alternative bonding approach that accounts for the properties of O_2 and other molecules that do not match the predictions of valence bond theory.

Magnetic and other properties of molecules are sometimes better explained by another quantum mechanical approach called *molecular orbital (MO) theory.* Molecular orbital theory describes covalent bonds in terms of ***molecular orbitals,*** which *result from interaction of the atomic orbitals of the bonding atoms and are associated with the entire molecule.* The difference between a molecular orbital and an atomic orbital is that an atomic orbital is associated with only one atom.

Figure 10.20
Liquid oxygen caught between the poles of a magnet, because the O_2 molecules are paramagnetic, having two parallel spins.

Bonding and Antibonding Molecular Orbitals

According to MO theory, the overlap of the 1*s* orbitals of two hydrogen atoms leads to the formation of two molecular orbitals: one bonding molecular orbital and one antibonding molecular orbital. A ***bonding molecular orbital*** has *lower energy and greater stability than the atomic orbitals from which it was formed.* An ***antibonding molecular orbital*** has *higher energy and lower stability than the atomic orbitals from which it was formed.* As the names "bonding" and "antibonding" suggest, placing electrons in a bonding molecular orbital yields a stable covalent bond, whereas placing electrons in an antibonding molecular orbital results in an unstable bond.

In the bonding molecular orbital, the electron density is greatest between the nuclei of the bonding atoms. In the antibonding molecular orbital, on the other hand, the electron density decreases to zero between the nuclei. We can understand this distinction if we recall that electrons in orbitals have wave characteristics. A property unique to waves enables waves of the same type to interact in such a way that the resultant wave has either an enhanced amplitude or a diminished amplitude. In the former case, we call the interaction *constructive interference;* in the latter case, it is *destructive interference* (Figure 10.21).

The formation of bonding molecular orbitals corresponds to constructive interference (the increase in amplitude is analogous to the buildup of electron density between the two nuclei). The formation of antibonding molecular orbitals corresponds to destructive interference (the decrease in amplitude is analogous to the decrease in electron density between the two nuclei). The constructive and destructive interactions between the two 1*s* orbitals in the H_2 molecule, then, lead to the formation of a sigma bonding molecular orbital (σ_{1s}) and a sigma antibonding molecular orbital $\sigma^{\star}_{1s}$:

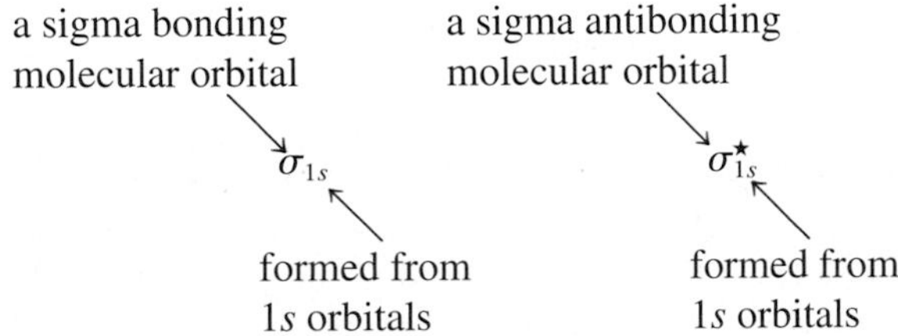

where the star denotes an antibonding moleculer orbital.

In a ***sigma molecular orbital*** (bonding or antibonding) *the electron density is concentrated symmetrically around a line between the two nuclei of the bonding atoms.* Two electrons in a sigma molecular orbital form a sigma bond (see Section 10.5). Remember that a single covalent bond (such as H—H or F—F) is almost always a sigma bond.

The two electrons in the sigma molecular orbital are paired. The Pauli exclusion principle applies to molecules as well as to atoms.

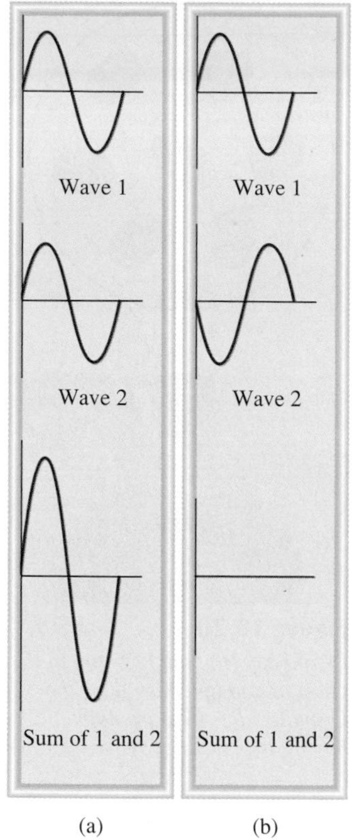

Figure 10.21
Constructive interference (a) and destructive interference (b) of two waves of the same wavelength and amplitude.

Figure 10.22 shows the *molecular orbital energy level diagram*—that is, the relative energy levels of the orbitals produced in the formation of the H_2 molecule—and the constructive and destructive interactions between the two 1*s* orbitals. Notice that in the antibonding molecular orbital there is a *node* between the nuclei that signifies zero electron density. The nuclei are repelled by each other's positive charges, rather than held together. Electrons in the antibonding molecular orbital have higher energy (and less stability) than they would have in the isolated atoms. On the other hand, electrons in the bonding molecular orbital have less energy (and hence greater stability) than they would have in the isolated atoms.

Although we have used the hydrogen molecule to illustrate molecular orbital formation, the concept is equally applicable to other molecules. In the H_2 molecule, we consider only the interaction between 1*s* orbitals; with more complex molecules, we need to consider additional atomic orbitals as well. Nevertheless, for all *s* orbitals, the process is the same as for 1*s* orbitals. Thus, the interaction between two 2*s* or 3*s* orbitals can be understood in terms of the molecular orbital energy level diagram and the formation of bonding and antibonding molecular orbitals shown in Figure 10.22.

For *p* orbitals, the process is more complex because they can interact with each other in two different ways. For example, two 2*p* orbitals can approach each other end-to-end to produce a sigma bonding and a sigma antibonding molecular orbital, as shown in Figure 10.23(a). Alternatively, the two *p* orbitals can overlap sideways to generate a bonding and an antibonding pi molecular orbital [Figure 10.23(b)].

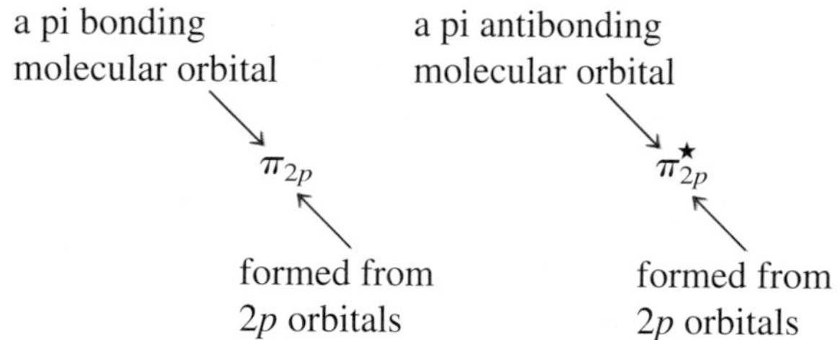

In a ***pi molecular orbital*** (bonding or antibonding), *the electron density is concentrated above and below a line joining the two nuclei of the bonding atoms.* Two electrons in

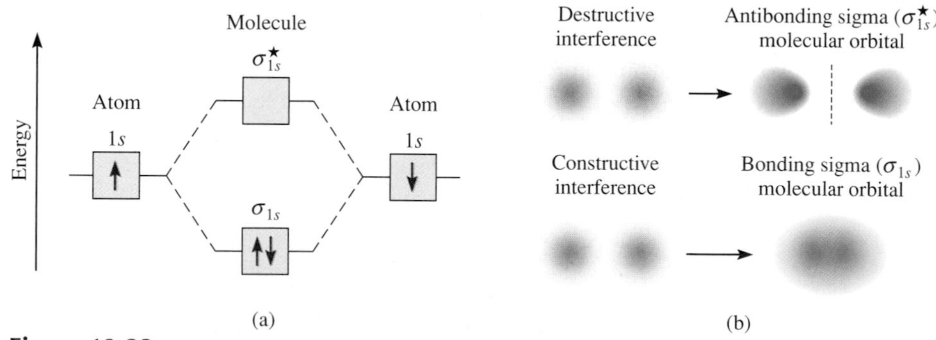

Figure 10.22
(a) Energy levels of bonding and antibonding molecular orbitals in the H_2 molecule. Note that the two electrons in the σ_{1s} orbital must have opposite spins in accord with the Pauli exclusion principle. Keep in mind that the higher the energy of the molecular orbital, the less stable the electrons in that molecular orbital. (b) Constructive and destructive interferences between the two hydrogen 1s orbitals lead to the formation of a bonding and an antibonding molecular orbital. In the bonding molecular orbital, there is a buildup between the nuclei of electron density, which acts as a negatively charged "glue" to hold the positively charged nuclei together. In the antibonding molecular orbital, there is a nodal plane between the nuclei, where the electron density is zero.

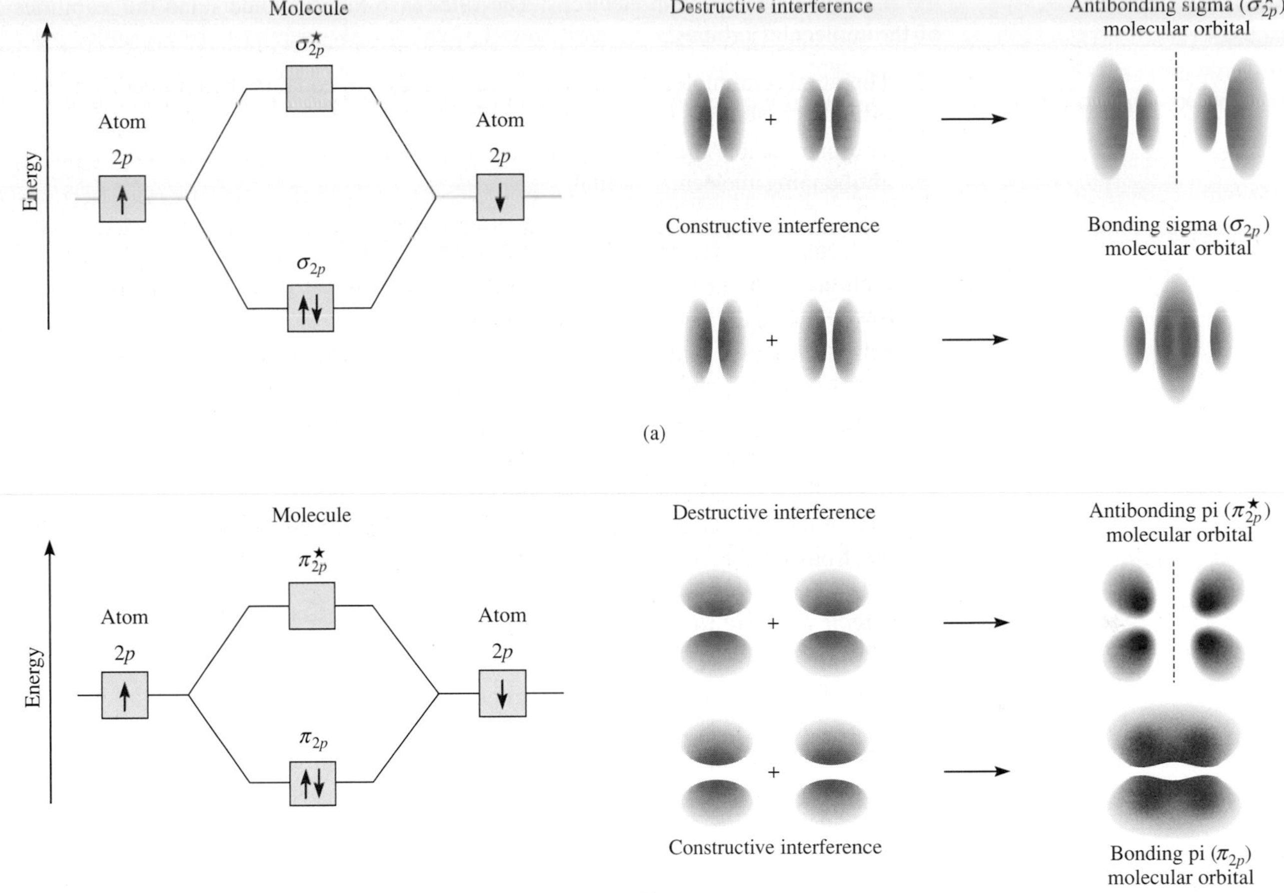

Figure 10.23

Two possible interactions between two equivalent p orbitals and the corresponding molecular orbitals. (a) When the p orbitals overlap end-to-end, a sigma bonding and a sigma antibonding molecular orbital form. (b) When the p orbitals overlap side-to-side, a pi bonding and a pi antibonding molecular orbital form. Normally, a sigma bonding molecular orbital is more stable than a pi bonding molecular orbital, because side-to-side interaction leads to a smaller overlap of the p orbitals than does end-to-end interaction. We assume that the $2p_x$ orbitals take part in the sigma molecular orbital formation. The $2p_y$ and $2p_z$ orbitals can interact to form only π molecular orbitals. The behavior shown in (b) represents the interaction between the $2p_y$ orbitals or the $2p_z$ orbitals. In both cases, the dashed line represents a nodal plane between the nuclei, where the electron density is zero.

a pi molecular orbital form a pi bond (see Section 10.5). A double bond is almost always composed of a sigma bond and a pi bond; a triple bond is always a sigma bond plus two pi bonds.

Molecular Orbital Configurations

To understand properties of molecules, we must know how electrons are distributed among molecular orbitals. The procedure for determining the electron configuration of a molecule is analogous to the one we use to determine the electron configurations of atoms (see Section 7.8).

Rules Governing Molecular Electron Configuration and Stability

To write the electron configuration of a molecule, we must first arrange the molecular orbitals in order of increasing energy. Then we can use the following guidelines to fill the molecular orbitals with electrons. The rules also help us understand the stabilities of the molecular orbitals.

1. The number of molecular orbitals formed is always equal to the number of atomic orbitals combined.
2. The more stable the bonding molecular orbital, the less stable the corresponding antibonding molecular orbital.
3. The filling of molecular orbitals proceeds from low to high energies. In a stable molecule, the number of electrons in bonding molecular orbitals is always greater than that in antibonding molecular orbitals because we place electrons first in the lower-energy bonding molecular orbitals.
4. Like an atomic orbital, each molecular orbital can accommodate up to two electrons with opposite spins in accordance with the Pauli exclusion principle.
5. When electrons are added to molecular orbitals of the same energy, the most stable arrangement is predicted by Hund's rule; that is, electrons enter these molecular orbitals with parallel spins.
6. The number of electrons in the molecular orbitals is equal to the sum of all the electrons on the bonding atoms.

Hydrogen and Helium Molecules

Later in this section, we will study molecules formed by atoms of the second-period elements. Before we do, it will be instructive to predict the relative stabilities of the simple species H_2^+, H_2, He_2^+, and He_2, using the energy-level diagrams shown in Figure 10.24. The σ_{1s} and $\sigma_{1s}^{\star}$ orbitals can accommodate a maximum of four electrons. The total number of electrons increases from one for H_2^+ to four for He_2. The Pauli exclusion principle stipulates that each molecular orbital can accommodate a maximum of two electrons with opposite spins. We are concerned only with the ground-state electron configurations in these cases.

To evaluate the stabilities of these species we determine their ***bond order,*** defined as

$$\text{bond order} = \frac{1}{2}\left(\begin{array}{c}\text{number of electrons}\\ \text{in bonding MOs}\end{array} - \begin{array}{c}\text{number of electrons}\\ \text{in antibonding MOs}\end{array}\right) \qquad (10.2)$$

The quantitative measure of the strength of a bond is bond enthalpy (Section 9.10).

The bond order indicates the approximate strength of a bond. For example, if there are two electrons in the bonding molecular orbital and none in the antibonding molecular orbital, the bond order is one, which means that there is one covalent bond and that the molecule is stable. Note that the bond order can be a fraction, but a bond order of zero (or a negative value) means the bond has no stability and the molecule

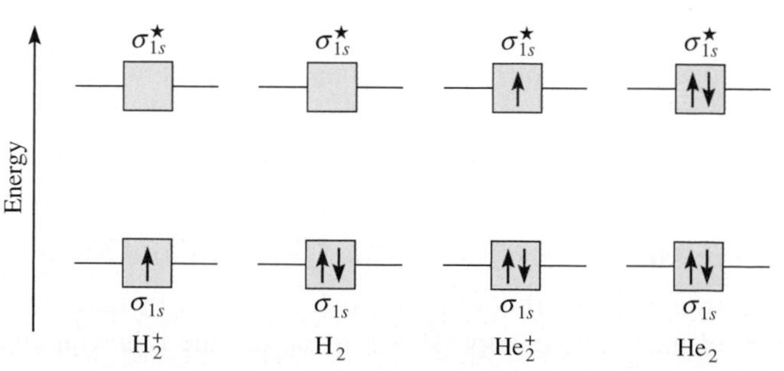

Figure 10.24
Energy levels of the bonding and antibonding molecular orbitals in H_2^+, H_2, He_2^+, and He_2. In all these species, the molecular orbitals are formed by the interaction of two 1s orbitals.

cannot exist. Bond order can be used only qualitatively for purposes of comparison. For example, a bonding sigma molecular orbital with two electrons and a bonding pi molecular orbital with two electrons would each have a bond order of one. Yet, these two bonds must differ in bond strength (and bond length) because of the differences in the extent of atomic orbital overlap.

We are ready now to make predictions about the stability of H_2^+, H_2, He_2^+, and He_2 (see Figure 10.24). The H_2^+ molecular ion has only one electron in the σ_{1s} orbital. Because a covalent bond consists of two electrons in a bonding molecular orbital, H_2^+ has only half of one bond, or a bond order of $\frac{1}{2}$. Thus, we predict that the H_2^+ molecule may be a stable species. The electron configuration of H_2^+ is written as $(\sigma_{1s})^1$.

The superscript in $(\sigma_{1s})^1$ indicates that there is one electron in the sigma bonding molecular orbital.

The H_2 molecule has two electrons, both of which are in the σ_{1s} orbital. According to our scheme, two electrons equal one full bond; therefore, the H_2 molecule has a bond order of one, or one full covalent bond. The electron configuration of H_2 is $(\sigma_{1s})^2$.

As for the He_2^+ molecular ion, we place the first two electrons in the σ_{1s} orbital and the third electron in the $\sigma_{1s}^\star$ orbital. Because the antibonding molecular orbital is destabilizing, we expect He_2^+ to be less stable than H_2. Roughly speaking, the instability resulting from the electron in the $\sigma_{1s}^\star$ orbital is balanced by one of the σ_{1s} electrons. The bond order is $\frac{1}{2}(2 - 1) = \frac{1}{2}$ and the overall stability of He_2^+ is similar to that of the H_2^+ molecule. The electron configuration of He_2^+ is $(\sigma_{1s})^2(\sigma_{1s}^\star)^1$.

In He_2 there would be two electrons in the σ_{1s} orbital and two electrons in the $\sigma_{1s}^\star$ orbital, so the molecule would have a bond order of zero and no net stability. The electron configuration of He_2 would be $(\sigma_{1s})^2(\sigma_{1s}^\star)^2$.

To summarize, we can arrange our examples in order of decreasing stability:

$$H_2 > H_2^+, He_2^+ > He_2$$

We know that the hydrogen molecule is a stable species. Our simple molecular orbital method predicts that H_2^+ and He_2^+ also possess some stability, because both have bond orders of $\frac{1}{2}$. Indeed, their existence has been confirmed by experiment. It turns out that H_2^+ is somewhat more stable than He_2^+, because there is only one electron in the hydrogen molecular ion and therefore it has no electron-electron repulsion. Furthermore, H_2^+ also has less nuclear repulsion than He_2^+. Our prediction about He_2 is that it would have no stability, but in 1993 He_2 gas was found to exist. The "molecule" is extremely unstable and has only a transient existence under specially created conditions.

Homonuclear Diatomic Molecules of Second-Period Elements

We are now ready to study the ground-state electron configuration of molecules containing second-period elements. We will consider only the simplest case, that of ***homonuclear diatomic molecules,*** or *diatomic molecules containing atoms of the same elements.*

Figure 10.25 shows the molecular orbital energy level diagram for the first member of the second period, Li_2. These molecular orbitals are formed by the overlap of $1s$ and $2s$ orbitals. We will use this diagram to build up all the diatomic molecules, as we will see shortly.

The situation is more complex when the bonding also involves p orbitals. Two p orbitals can form either a sigma bond or a pi bond. Because there are three p orbitals for each atom of a second-period element, we know that one sigma and two pi molecular orbitals will result from the constructive interaction. The sigma molecular orbital is formed by the overlap of the $2p_x$ orbitals along the internuclear axis, that is, the x-axis. The $2p_y$ and $2p_z$ orbitals are perpendicular to the x-axis, and they will overlap sideways

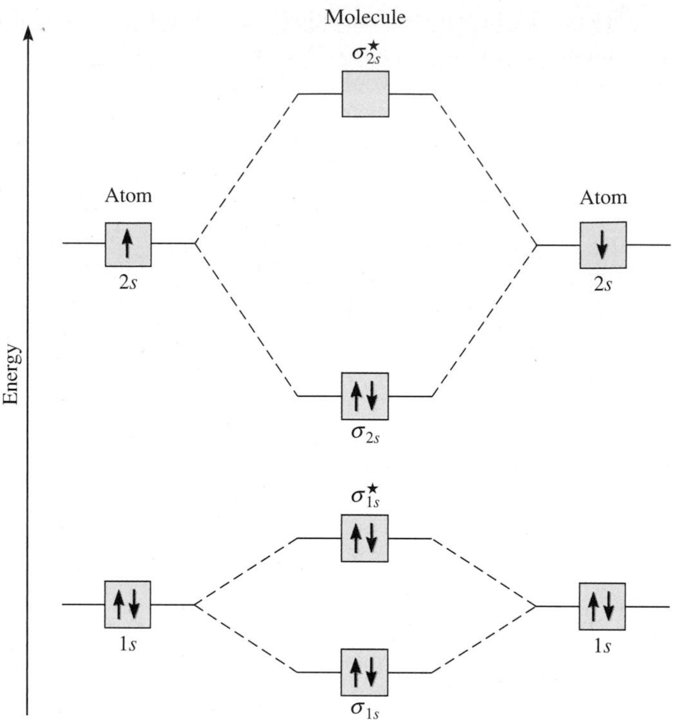

Figure 10.25
Molecular orbital energy level diagram for the Li_2 molecule. The six electrons in Li_2 (Li's electron configuration is $1s^2 2s^1$) are in the σ_{1s}, $\sigma^\star_{1s}$, and σ_{2s} orbitals. Because there are two electrons in each σ_{1s} and $\sigma^\star_{1s}$ (just as in He_2), there is no net bonding or antibonding effect. Therefore, the single covalent bond in Li_2 is formed by the two electrons in the bonding molecular orbital σ_{2s}. Note that although the antibonding orbital ($\sigma^\star_{1s}$) has greater energy and is thus less stable than the bonding orbital (σ_{1s}), this antibonding orbital has less energy and greater stability than the σ_{2s} bonding orbital.

to give two pi molecular orbitals. The molecular orbitals are called σ_{2p_x}, π_{2p_y}, and π_{2p_z} orbitals, where the subscripts indicate which atomic orbitals take part in forming the molecular orbitals. As shown in Figure 10.23, overlap of the two p orbitals is normally greater in a σ molecular orbital than in a π molecular orbital, so we would expect the former to be lower in energy. However, the energies of molecular orbitals actually increase as follows:

$$\sigma_{1s} < \sigma^\star_{1s} < \sigma_{2s} < \sigma^\star_{2s} < \pi_{2p_y} = \pi_{2p_z} < \sigma_{2p_x} < \pi^\star_{2p_y} = \pi^\star_{2p_z} < \sigma^\star_{2p_x}$$

The inversion of the σ_{2p_x} orbital and the π_{2p_y} and π_{2p_z} orbitals is due to the interaction between the 2s orbital on one atom with the 2p orbital on the other. In MO terminology, we say there is mixing between these orbitals. The condition for mixing is that the 2s and 2p orbitals must be close in energy. This condition is met for the lighter molecules B_2, C_2, and N_2 with the result that the σ_{2p_x} orbital is raised in energy relative to the π_{2p_y} and π_{2p_z} orbitals as already shown. The mixing is less pronounced for O_2 and F_2 so the σ_{2p_x} orbital lies lower in energy than the π_{2p_y} and π_{2p_z} orbitals in these molecules.

With these concepts and Figure 10.26, which shows the order of increasing energies for 2p molecular orbitals, we can write the electron configurations and predict the magnetic properties and bond orders of second-period homonuclear diatomic molecules.

The Carbon Molecule (C_2)

The carbon atom has the electron configuration $1s^2 2s^2 2p^2$; thus, there are 12 electrons in the C_2 molecule. From the bonding scheme for Li_2, we place four additional carbon electrons in the π_{2p_y} and π_{2p_z} orbitals. Therefore, C_2 has the electron configuration

$$(\sigma_{1s})^2(\sigma^\star_{1s})^2(\sigma_{2s})^2(\sigma^\star_{2s})^2(\pi_{2p_y})^2(\pi_{2p_z})^2$$

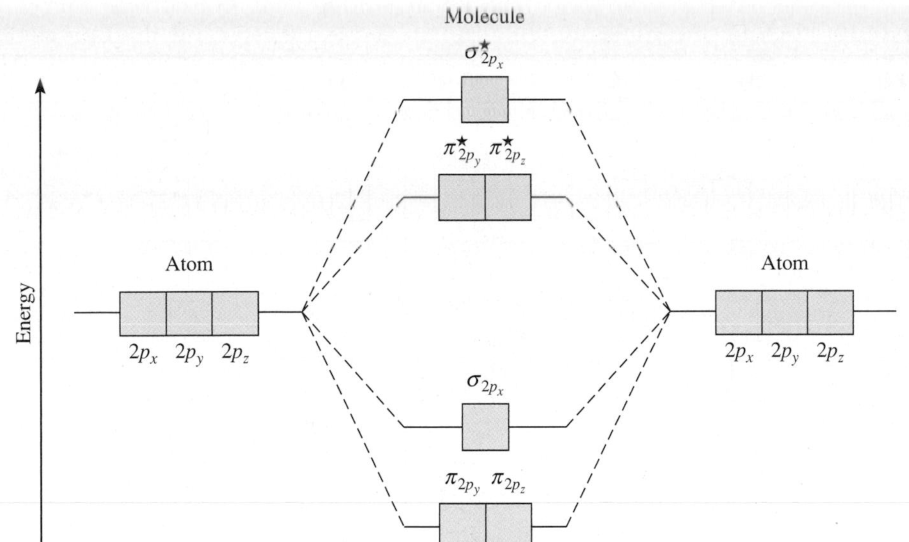

Figure 10.26 *General molecular orbital energy level diagram for the second-period homonuclear diatomic molecules Li_2, Be_2, B_2, C_2, and N_2. For simplicity, the σ_{1s} and σ_{2s} orbitals have been omitted. Note that in these molecules the σ_{2p_x} orbital is higher in energy than either the π_{2p_y} or the π_{2p_z} orbitals. This means that electrons in the σ_{2p_x} orbitals are less stable than those in π_{2p_y} and π_{2p_z}. For O_2 and F_2, the σ_{2p_x} orbital is lower in energy than π_{2p_y} and π_{2p_z}.*

Its bond order is 2, and the molecule has no unpaired electrons. Again, diamagnetic C_2 molecules have been detected in the vapor state. Note that the double bonds in C_2 are both pi bonds because of the four electrons in the two pi molecular orbitals. In most other molecules, a double bond is made up of a sigma bond and a pi bond.

The Oxygen Molecule (O_2)

As we stated earlier, valence bond theory does not account for the magnetic properties of the oxygen molecule. To show the two unpaired electrons on O_2, we need to draw an alternative to the resonance structure shown on p. 348:

$$\cdot\ddot{\underset{\cdot\cdot}{\text{O}}}\text{—}\ddot{\underset{\cdot\cdot}{\text{O}}}\cdot$$

This structure is unsatisfactory on at least two counts. First, it implies the presence of a single covalent bond, but experimental evidence strongly suggests that there is a double bond in this molecule. Second, it places seven valence electrons around each oxygen atom, a violation of the octet rule.

The ground-state electron configuration of O is $1s^22s^22p^4$; thus, there are 16 electrons in O_2. Using the order of increasing energies of the molecular orbitals discussed above, we write the ground-state electron configuration of O_2 as

$$(\sigma_{1s})^2(\sigma^{\star}_{1s})^2(\sigma_{2s})^2(\sigma^{\star}_{2s})^2(\sigma_{2p_x})^2(\pi_{2p_y})^2(\pi_{2p_z})^2(\pi^{\star}_{2p_y})^1(\pi^{\star}_{2p_z})^1$$

According to Hund's rule, the last two electrons enter the $\pi^{\star}_{2p_y}$ and $\pi^{\star}_{2p_z}$ orbitals with parallel spins. Ignoring the σ_{1s} and σ_{2s} orbitals (because their net effects on bonding are zero), we calculate the bond order of O_2 using Equation (10.2):

$$\text{bond order} = \tfrac{1}{2}(6 - 2) = 2$$

Therefore, the O_2 molecule has a bond order of 2 and oxygen is paramagnetic, a prediction that corresponds to experimental observations.

Table 10.5 Properties of Homonuclear Diatomic Molecules of the Second-Period Elements*

	Li_2	B_2	C_2	N_2	O_2	F_2	
$\sigma^\star_{2p_x}$	☐	☐	☐	☐	☐	☐	$\sigma^\star_{2p_x}$
$\pi^\star_{2p_y}, \pi^\star_{2p_z}$	☐☐	☐☐	☐☐	☐☐	↑ ↑	↑↓ ↑↓	$\pi^\star_{2p_y}, \pi^\star_{2p_z}$
σ_{2p_x}	☐	☐	☐	↑↓	↑↓ ↑↓	↑↓ ↑↓	π_{2p_y}, π_{2p_z}
π_{2p_y}, π_{2p_z}	☐☐	↑ ↑	↑↓ ↑↓	↑↓ ↑↓	↑↓	↑↓	σ_{2p_x}
$\sigma^\star_{2s}$	☐	↑↓	↑↓	↑↓	↑↓	↑↓	$\sigma^\star_{2s}$
σ_{2s}	↑↓	↑↓	↑↓	↑↓	↑↓	↑↓	σ_{2s}
Bond order	1	1	2	3	2	1	
Bond length (pm)	267	159	131	110	121	142	
Bond enthalpy (kJ/mol)	104.6	288.7	627.6	941.4	498.7	156.9	
Magnetic properties	Diamagnetic	Paramagnetic	Diamagnetic	Diamagnetic	Paramagnetic	Diamagnetic	

*For simplicity the σ_{1s} and $\sigma^\star_{1s}$ orbitals are omitted. These two orbitals hold a total of four electrons. Remember that for O_2 and F_2, σ_{2p_x} is lower in energy than π_{2p_y} and π_{2p_z}.

Table 10.5 summarizes the general properties of the stable diatomic molecules of the second period.

EXAMPLE 10.6

The N_2^+ ion can be prepared by bombarding the N_2 molecule with fast-moving electrons. Predict the following properties of N_2^+: (a) electron configuration, (b) bond order, (c) magnetic properties, and (d) bond length relative to the bond length of N_2 (is it longer or shorter?).

Strategy From Table 10.5 we can deduce the properties of ions generated from the homonuclear molecules. How does the stability of a molecule depend on the number of electrons in bonding and antibonding molecular orbitals? From what molecular orbital is an electron removed to form the N_2^+ ion from N_2? What properties determine whether a species is diamagnetic or paramagnetic?

Solution From Table 10.5 we can deduce the properties of ions generated from the homonuclear diatomic molecules.

(a) Because N_2^+ has one fewer electron than N_2, its electron configuration is

$$(\sigma_{1s})^2(\sigma^\star_{1s})^2(\sigma_{2s})^2(\sigma^\star_{2s})^2(\pi_{2p_y})^2(\pi_{2p_z})^2(\sigma_{2p_x})^1$$

(b) The bond order of N_2^+ is found by using Equation (10.2):

$$\text{bond order} = \tfrac{1}{2}(9 - 4) = 2.5$$

(c) N_2^+ has one unpaired electron, so it is paramagnetic.

(d) Because the electrons in the bonding molecular orbitals are responsible for holding the atoms together, N_2^+ should have a weaker and, therefore, longer bond than N_2. (In fact, the bond length of N_2^+ is 112 pm, compared with 110 pm for N_2.)

(Continued)

Check Because an electron is removed from a bonding molecular orbital, we expect the bond order to decrease. The N_2^+ ion has an odd number of electrons (13), so it should be paramagnetic.

Similar problems: 10.57, 10.59.

Practice Exercise Which of the following species has a longer bond length: F_2, F_2^+, or F_2^-?

REVIEW OF CONCEPTS

Why is a fractional bond order such as 1.5 possible according to molecular orbital theory?

Key Equations

$\mu = Q \times r$ (10.1) Expressing dipole moment in terms of charge (Q) and distance of separation (r) between charges.

$$\text{bond order} = \frac{1}{2}\left(\begin{matrix}\text{number of electrons} \\ \text{in bonding MOs}\end{matrix} - \begin{matrix}\text{number of electrons} \\ \text{in antibonding MOs}\end{matrix}\right) \quad (10.2)$$

Summary of Facts and Concepts

1. The VSEPR model for predicting molecular geometry is based on the assumption that valence-shell electron pairs repel one another and tend to stay as far apart as possible. According to the VSEPR model, molecular geometry can be predicted from the number of bonding electron pairs and lone pairs. Lone pairs repel other pairs more strongly than bonding pairs do and thus distort bond angles from those of the ideal geometry.
2. The dipole moment is a measure of the charge separation in molecules containing atoms of different electronegativities. The dipole moment of a molecule is the resultant of whatever bond moments are present in a molecule. Information about molecular geometry can be obtained from dipole moment measurements.
3. In valence bond theory, hybridized atomic orbitals are formed by the combination and rearrangement of orbitals of the same atom. The hybridized orbitals are all of equal energy and electron density, and the number of hybridized orbitals is equal to the number of pure atomic orbitals that combine. Valence-shell expansion can be explained by assuming hybridization of *s*, *p*, and *d* orbitals.
4. In *sp* hybridization, the two hybrid orbitals lie in a straight line; in sp^2 hybridization, the three hybrid orbitals are directed toward the corners of a triangle; in sp^3 hybridization, the four hybrid orbitals are directed toward the corners of a tetrahedron; in sp^3d hybridization, the five hybrid orbitals are directed toward the corners of a trigonal bipyramid; in sp^3d^2 hybridization, the six hybrid orbitals are directed toward the corners of an octahedron.
5. In an sp^2-hybridized atom (for example, carbon), the one unhybridized *p* orbital can form a pi bond with another *p* orbital. A carbon-carbon double bond consists of a sigma bond and a pi bond. In an *sp*-hybridized carbon atom, the two unhybridized *p* orbitals can form two pi bonds with two *p* orbitals on another atom (or atoms). A carbon-carbon triple bond consists of one sigma bond and two pi bonds.
6. Molecular orbital theory describes bonding in terms of the combination and rearrangement of atomic orbitals to form orbitals that are associated with the molecule as a whole. Bonding molecular orbitals increase electron density between the nuclei and are lower in energy than individual atomic orbitals. Antibonding molecular orbitals have a region of zero electron density between the nuclei, and an energy level higher than that of the individual atomic orbitals. Molecules are stable if the number of electrons in bonding molecular orbitals is greater than that in antibonding molecular orbitals.
7. We write electron configurations for molecular orbitals as we do for atomic orbitals, referring to the Pauli exclusion principle and Hund's rule.

Key Words

Antibonding molecular orbital, p. 349
Bond order, p. 352
Bonding molecular orbital, p. 349
Dipole moment (μ), p. 331
Homonuclear diatomic molecule, p. 353
Hybrid orbital, p. 337
Hybridization, p. 337
Molecular orbital, p. 349
Nonpolar molecule, p. 332
Pi bond (π bond), p. 345
Pi molecular orbital, p. 350
Polar molecule, p. 332
Sigma bond (σ bond), p. 345
Sigma molecular orbital, p. 349
Valence shell, p. 321
Valence-shell electron-pair repulsion (VSEPR) model, p. 321

Questions and Problems

Molecular Geometry

Review Questions

10.1 How is the geometry of a molecule defined and why is the study of molecular geometry important?

10.2 Sketch the shape of a linear triatomic molecule, a trigonal planar molecule containing four atoms, a tetrahedral molecule, a trigonal bipyramidal molecule, and an octahedral molecule. Give the bond angles in each case.

10.3 How many atoms are directly bonded to the central atom in a tetrahedral molecule, a trigonal bipyramidal molecule, and an octahedral molecule?

10.4 Discuss the basic features of the VSEPR model. Explain why the magnitude of repulsion decreases in this order: lone pair-lone pair > lone pair-bonding pair > bonding pair-bonding pair.

10.5 In the trigonal bipyramidal arrangement, why does a lone pair occupy an equatorial position rather than an axial position?

10.6 The geometry of CH_4 could be square planar, with the four H atoms at the corners of a square and the C atom at the center of the square. Sketch this geometry and compare its stability with that of a tetrahedral CH_4 molecule.

Problems

10.7 Predict the geometries of these species using the VSEPR method: (a) PCl_3, (b) $CHCl_3$, (c) SiH_4, (d) $TeCl_4$.

10.8 Predict the geometries of these species: (a) $AlCl_3$, (b) $ZnCl_2$, (c) $ZnCl_4^{2-}$.

10.9 Predict the geometry of these molecules and ion using the VSEPR method: (a) $HgBr_2$, (b) N_2O (arrangement of atoms is NNO), (c) SCN^- (arrangement of atoms is SCN).

10.10 Predict the geometries of these ions: (a) NH_4^+, (b) NH_2^-, (c) CO_3^{2-}, (d) ICl_2^-, (e) ICl_4^-, (f) AlH_4^-, (g) $SnCl_5^-$, (h) H_3O^+, (i) BeF_4^{2-}.

10.11 Describe the geometry around each of the three central atoms in the CH_3COOH molecule.

10.12 Which of these species are tetrahedral? $SiCl_4$, SeF_4, XeF_4, CI_4, $CdCl_4^{2-}$.

Dipole Moments

Review Questions

10.13 Define dipole moment. What are the units and symbol for dipole moment?

10.14 What is the relationship between the dipole moment and bond moment? How is it possible for a molecule to have bond moments and yet be nonpolar?

10.15 Explain why an atom cannot have a permanent dipole moment.

10.16 The bonds in beryllium hydride (BeH_2) molecules are polar, and yet the dipole moment of the molecule is zero. Explain.

Problems

10.17 Referring to Table 10.3, arrange the following molecules in order of increasing dipole moment: H_2O, H_2S, H_2Te, H_2Se.

10.18 The dipole moments of the hydrogen halides decrease from HF to HI (see Table 10.3). Explain this trend.

10.19 List these molecules in order of increasing dipole moment: H_2O, CBr_4, H_2S, HF, NH_3, CO_2.

10.20 Does the molecule OCS have a higher or lower dipole moment than CS_2?

10.21 Which of these molecules has a higher dipole moment?

Br\C=C/H, H/ \Br (a) — Br(H)C=C(H)Br

Br\C=C/Br, H/ \H (b) — Br(H)C=C(H)Br

(a) (b)

10.22 Arrange these compounds in order of increasing dipole moment:

Cl Cl Cl Cl Cl Cl Cl Cl Cl Cl Cl

(a) (b) (c) (d)

10.34 (continued) Describe the changes in hybridization (if any) of the B and N atoms as a result of this reaction.

Valence Bond Theory

Review Questions

10.23 What is valence bond theory? How does it differ from the Lewis concept of chemical bonding?

10.24 Use valence bond theory to explain the bonding in Cl_2 and HCl. Show how the atomic orbitals overlap when a bond is formed.

10.25 Draw a potential energy curve for the bond formation in F_2.

Hybridization

Review Questions

10.26 What is the hybridization of atomic orbitals? Why is it impossible for an isolated atom to exist in the hybridized state?

10.27 How does a hybrid orbital differ from a pure atomic orbital? Can two 2*p* orbitals of an atom hybridize to give two hybridized orbitals?

10.28 What is the angle between these two hybrid orbitals on the same atom? (a) *sp* and *sp* hybrid orbitals, (b) sp^2 and sp^2 hybrid orbitals, (c) sp^3 and sp^3 hybrid orbitals.

10.29 How would you distinguish between a sigma bond and a pi bond?

10.30 Which of these pairs of atomic orbitals of adjacent nuclei can overlap to form a sigma bond? Which overlap to form a pi bond? Which cannot overlap (no bond)? Consider the *x*-axis to be the internuclear axis, that is, the line joining the nuclei of the two atoms. (a) 1*s* and 1*s*, (b) 1*s* and $2p_x$, (c) $2p_x$ and $2p_y$, (d) $3p_y$ and $3p_y$, (e) $2p_x$ and $2p_x$, (f) 1*s* and 2*s*.

Problems

10.31 Describe the bonding scheme of the AsH_3 molecule in terms of hybridization.

10.32 What is the hybridization state of Si in SiH_4 and in $H_3Si—SiH_3$?

10.33 Describe the change in hybridization (if any) of the Al atom in this reaction:

$$AlCl_3 + Cl^- \longrightarrow AlCl_4^-$$

10.34 Consider the reaction

$$BF_3 + NH_3 \longrightarrow F_3B—NH_3$$

Describe the changes in hybridization (if any) of the B and N atoms as a result of this reaction.

10.35 What hybrid orbitals are used by nitrogen atoms in these species? (a) NH_3, (b) $H_2N—NH_2$, (c) NO_3^-.

10.36 What are the hybrid orbitals of the carbon atoms in these molecules?

(a) $H_3C—CH_3$
(b) $H_3C—CH{=}CH_2$
(c) $CH_3—C{\equiv}C—CH_2OH$
(d) $CH_3CH{=}O$
(e) CH_3COOH.

10.37 Specify which hybrid orbitals are used by carbon atoms in these species: (a) CO, (b) CO_2, (c) CN^-.

10.38 What is the hybridization state of the central N atom in the azide ion, N_3^-? (Arrangement of atoms: NNN.)

10.39 The allene molecule $H_2C{=}C{=}CH_2$ is linear (the three C atoms lie on a straight line). What are the hybridization states of the carbon atoms? Draw diagrams to show the formation of sigma bonds and pi bonds in allene.

10.40 Describe the hybridization of phosphorus in PF_5.

10.41 How many sigma bonds and pi bonds are there in each of these molecules?

(a) $Cl—CH_2—Cl$ (H above and below C)
(b) $HHC{=}CClH$ (H, H on left C; Cl, H on right C)
(c) $H_3C—CH{=}CH—C{\equiv}C—H$

10.42 How many pi bonds and sigma bonds are there in the tetracyanoethylene molecule?

$$(N{\equiv}C)_2C{=}C(C{\equiv}N)_2$$

10.43 Give the formula of a cation comprised of iodine and fluorine in which the iodine atom is sp^3d-hybridized.

10.44 Give the formula of an anion comprised of iodine and fluorine in which the iodine atom is sp^3d^2-hybridized.

Molecular Orbital Theory

Review Questions

10.45 What is molecular orbital theory? How does it differ from valence bond theory?

10.46 Define these terms: bonding molecular orbital, antibonding molecular orbital, pi molecular orbital, sigma molecular orbital.

10.47 Sketch the shapes of these molecular orbitals: σ_{1s}, $\sigma^{\star}_{1s}$, π_{2p}, and $\pi^{\star}_{2p}$. How do their energies compare?

10.48 Explain the significance of bond order. Can bond order be used for quantitative comparisons of the strengths of chemical bonds?

Problems

10.49 Explain in molecular orbital terms the changes in H—H internuclear distance that occur as the molecular H_2 is ionized first to H_2^+ and then to H_2^{2+}.

10.50 The formation of H^+ from two H atoms is an energetically favorable process. Yet statistically there is less than a 100 percent chance that any two H atoms will undergo the reaction. Apart from energy considerations, how would you account for this observation based on the electron spins in the two H atoms?

10.51 Draw a molecular orbital energy level diagram for each of these species: He_2, HHe, He_2^+. Compare their relative stabilities in terms of bond orders. (Treat HHe as a diatomic molecule with three electrons.)

10.52 Arrange these species in order of increasing stability: Li_2, Li_2^+, Li_2^-. Justify your choice with a molecular orbital energy level diagram.

10.53 Use molecular orbital theory to explain why the Be_2 molecule does not exist.

10.54 Which of these species has a longer bond, B_2 or B_2^+? Explain in terms of molecular orbital theory.

10.55 Acetylene (C_2H_2) has a tendency to lose two protons (H^+) and form the carbide ion (C_2^{2-}), which is present in a number of ionic compounds, such as CaC_2 and MgC_2. Describe the bonding scheme in the C_2^{2-} ion in terms of molecular orbital theory. Compare the bond order in C_2^{2-} with that in C_2.

10.56 Compare the Lewis and molecular orbital treatments of the oxygen molecule.

10.57 Explain why the bond order of N_2 is greater than that of N_2^+, but the bond order of O_2 is less than that of O_2^+.

10.58 Compare the relative stability of these species and indicate their magnetic properties (that is, diamagnetic or paramagnetic): O_2, O_2^+, O_2^- (superoxide ion), O_2^{2-} (peroxide ion).

10.59 Use molecular orbital theory to compare the relative stabilities of F_2 and F_2^+.

10.60 A single bond is almost always a sigma bond, and a double bond is almost always made up of a sigma bond and a pi bond. There are very few exceptions to this rule. Show that the B_2 and C_2 molecules are examples of the exceptions.

Additional Problems

10.61 Which of these species is not likely to have a tetrahedral shape? (a) $SiBr_4$, (b) NF_4^+, (c) SF_4, (d) $BeCl_4^{2-}$, (e) BF_4^-, (f) $AlCl_4^-$.

10.62 Draw the Lewis structure of mercury(II) bromide. Is this molecule linear or bent? How would you establish its geometry?

10.63 Sketch the bond moments and resultant dipole moments for these molecules: H_2O, PCl_3, XeF_4, PCl_5, SF_6.

10.64 Although both carbon and silicon are in Group 4A, very few Si=Si bonds are known. Account for the instability of silicon-to-silicon double bonds in general. (*Hint:* Compare the atomic radii of C and Si in Figure 8.4. What effect would the larger size have on pi bond formation?)

10.65 Predict the geometry of sulfur dichloride (SCl_2) and the hybridization of the sulfur atom.

10.66 Antimony pentafluoride, SbF_5, reacts with XeF_4 and XeF_6 to form ionic compounds, $XeF_3^+SbF_6^-$ and $XeF_5^+SbF_6^-$. Describe the geometries of the cations and anion in these two compounds.

10.67 Draw Lewis structures and give the other information requested for the following molecules: (a) BF_3. Shape: planar or nonplanar? (b) ClO_3^-. Shape: planar or nonplanar? (c) H_2O. Show the direction of the resultant dipole moment. (d) OF_2. Polar or nonpolar molecule? (e) SeO_2. Estimate the OSeO bond angle.

10.68 Predict the bond angles for these molecules: (a) $BeCl_2$, (b) BCl_3, (c) CCl_4, (d) CH_3Cl, (e) Hg_2Cl_2 (arrangement of atoms: ClHgHgCl), (f) $SnCl_2$, (g) H_2O_2, (h) SnH_4.

10.69 Briefly compare the VSEPR and hybridization approaches to the study of molecular geometry.

10.70 Describe the hybridization state of arsenic in arsenic pentafluoride (AsF_5).

10.71 Draw Lewis structures and give the other information requested for these: (a) SO_3. Polar or nonpolar molecule? (b) PF_3. Polar or nonpolar molecule? (c) F_3SiH. Show the direction of the resultant dipole moment. (d) SiH_3^-. Planar or pyramidal shape? (e) Br_2CH_2. Polar or nonpolar molecule?

10.72 Which of these molecules and ions are linear? ICl_2^-, IF_2^+, OF_2, SnI_2, $CdBr_2$.

10.73 Draw the Lewis structure for the $BeCl_4^{2-}$ ion. Predict its geometry and describe the hybridization state of the Be atom.

10.74 The N_2F_2 molecule can exist in either of these two forms:

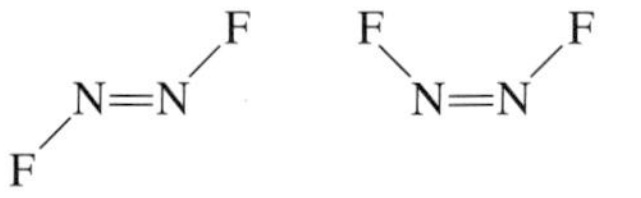

(a) What is the hybridization of N in the molecule?

(b) Which structure has a dipole moment?

10.75 Cyclopropane (C_3H_6) has the shape of a triangle in which a C atom is bonded to two H atoms and two other C atoms at each corner. Cubane (C_8H_8) has the shape of a cube in which a C atom is bonded to one H atom and three other C atoms at each corner. (a) Draw Lewis structures of these molecules. (b) Compare the CCC angles in these molecules with those predicted for an sp^3-hybridized C atom. (c) Would you expect these molecules to be easy to make?

10.76 The compound 1,2-dichloroethane ($C_2H_4Cl_2$) is nonpolar, while *cis*-dichloroethylene ($C_2H_2Cl_2$) has a dipole moment:

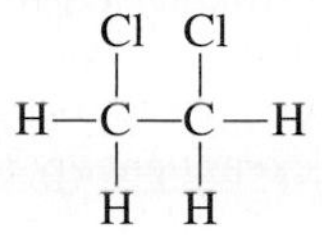

1,2-dichloroethane 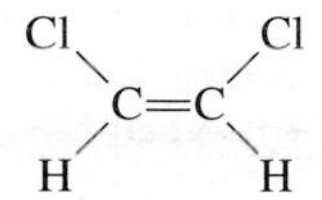   *cis*-dichloroethylene

The reason for the difference is that groups connected by a single bond can rotate with respect to each other, but no rotation occurs when a double bond connects the groups. On the basis of bonding considerations, explain why rotation occurs in 1,2-dichloroethane but not in *cis*-dichloroethylene.

10.77 Does the following molecule have a dipole moment?

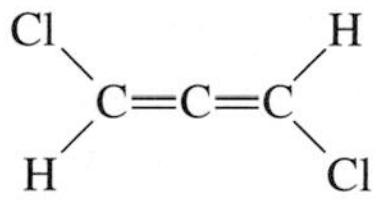

(*Hint:* See the answer to Problem 10.39.)

10.78 The compounds carbon tetrachloride (CCl_4) and silicon tetrachloride ($SiCl_4$) are similar both in geometry and hybridization. However, CCl_4 does not react with water but $SiCl_4$ does. Explain the difference in their chemical reactivities. (*Hint:* The first step of the reaction is believed to be the addition of a water molecule to the Si atom in $SiCl_4$.)

10.79 Write the ground-state electron configuration for B_2. Is the molecule diamagnetic or paramagnetic?

10.80 What are the hybridization states of the C and N atoms in this molecule?

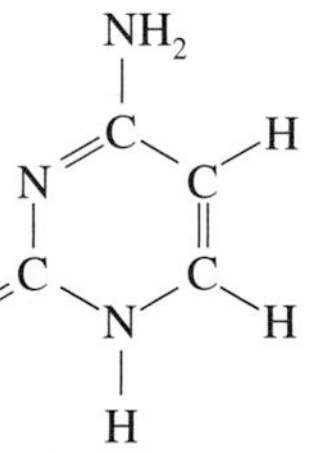

10.81 Use molecular orbital theory to explain the difference between the bond enthalpies of F_2 and F_2^- (see Problem 9.105).

10.82 The ionic character of the bond in a diatomic molecule can be estimated by the formula

$$\frac{\mu}{ed} \times 100\%$$

where μ is the experimentally measured dipole moment (in C m), e is the electronic charge (1.6022×10^{-19} C), and d is the bond length in meters. (The quantity ed is the hypothetical dipole moment for the case in which the transfer of an electron from the less electronegative to the more electronegative atom is complete.) Given that the dipole moment and bond length of HF are 1.92 D and 91.7 pm, respectively, calculate the percent ionic character of the molecule.

10.83 The geometries discussed in this chapter all lend themselves to fairly straightforward elucidation of bond angles. The exception is the tetrahedron, because its bond angles are hard to visualize. Consider the CCl_4 molecule, which has a tetrahedral geometry and is nonpolar. By equating the bond moment of a particular C—Cl bond to the resultant bond moments of the other three C—Cl bonds in opposite directions, show that the bond angles are all equal to 109.5°.

10.84 Aluminum trichloride ($AlCl_3$) is an electron-deficient molecule. It has a tendency to form a dimer (a molecule made of two $AlCl_3$ units):

$$AlCl_3 + AlCl_3 \longrightarrow Al_2Cl_6$$

(a) Draw a Lewis structure for the dimer. (b) Describe the hybridization state of Al in $AlCl_3$ and Al_2Cl_6. (c) Sketch the geometry of the dimer. (d) Do these molecules possess a dipole moment?

10.85 Assume that the third-period element phosphorus forms a diatomic molecule, P_2, in an analogous way as nitrogen does to form N_2. (a) Write the electronic configuration for P_2. Use $[Ne_2]$ to represent the electron configuration for the first two periods. (b) Calculate its bond order. (c) What are its magnetic properties (diamagnetic or paramagnetic)?

Special Problems

10.86 Progesterone is a hormone responsible for female sex characteristics. In the usual shorthand structure, each point where lines meet represents a C atom, and most H atoms are not shown. Draw the complete structure of the molecule, showing all C and H atoms. Indicate which C atoms are sp^2- and sp^3-hybridized.

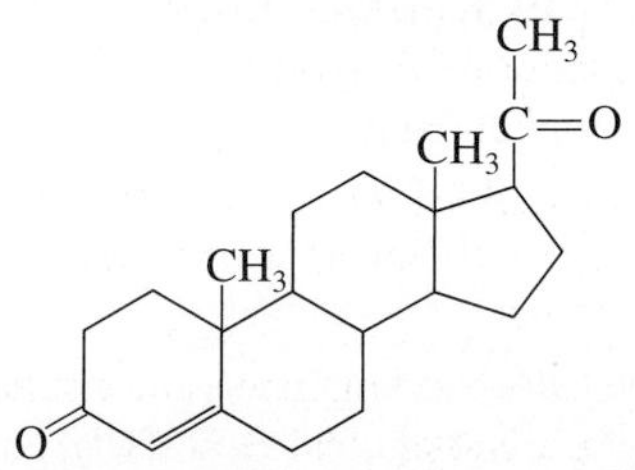

10.87 Greenhouse gases absorb (and trap) outgoing infared radiation (heat) from Earth and contribute to global warming. The molecule of a greenhouse gas either possesses a permanent dipole moment or has a changing dipole moment during its vibrational motions. Consider three of the vibrational modes of carbon dioxide

$$\overleftarrow{\text{O}}{=}\text{C}{=}\overrightarrow{\text{O}} \qquad \overrightarrow{\text{O}}{=}\overleftarrow{\text{C}}{=}\overrightarrow{\text{O}} \qquad \overset{\uparrow}{\text{O}}{=}\underset{\downarrow}{\text{C}}{=}\overset{\uparrow}{\text{O}}$$

where the arrows indicate the movement of the atoms. (During a complete cycle of vibration, the atoms move toward one extreme position and then reverse their direction to the other extreme position.) Which of the preceding vibrations are responsible for CO_2 behaving as a greenhouse gas? Which of the following molecules can act as a greenhouse gas: N_2, O_2, CO, NO_2, and N_2O?

10.88 The molecules *cis*-dichloroethylene and *trans*-dichloroethylene shown on p. 332 can be interconverted by heating or irradiation. (a) Starting with *cis*-dichloroethylene, show that rotating the C═C bond by 180° will break only the pi bond but will leave the sigma bond intact. Explain the formation of *trans*-dichloroethylene from this process. (Treat the rotation as two, stepwise 90° rotations.) (b) Account for the difference in the bond enthalpies for the pi bond (about 270 kJ/mol) and the sigma bond (about 350 kJ/mol). (c) Calculate the longest wavelength of light needed to bring about this conversion.

10.89 For each pair listed here, state which one has a higher first ionization energy and explain your choice: (a) H or H_2, (b) N or N_2, (c) O or O_2, (d) F or F_2.

10.90 The molecule benzyne (C_6H_4) is a very reactive species. It resembles benzene in that it has a six-membered ring of carbon atoms. Draw a Lewis structure of the molecule and account for the molecule's high reactivity.

10.91 Consider a N_2 molecule in its first excited electronic state; that is, when an electron in the highest occupied molecular orbital is promoted to the lowest empty molecular obital. (a) Identify the molecular orbitals involved and sketch a diagram to show the transition. (b) Compare the bond order and bond length of N_2^* with N_2, where the asterisk denotes the excited molecule. (c) Is N_2^* diamagnetic or paramagnetic? (d) When N_2^* loses its excess energy and converts to the ground state N_2, it emits a photon of wavelength 470 nm, which makes up part of the auroras lights. Calculate the energy difference between these levels.

10.92 As mentioned in the chapter, the Lewis structure for O_2 is

$$\ddot{\underset{\cdot\cdot}{\text{O}}}{=}\ddot{\underset{\cdot\cdot}{\text{O}}}$$

Use the molecular orbital theory to show that the structure actually corresponds to an excited state of the oxygen molecule.

10.93 Draw the Lewis structure of ketene (C_2H_2O) and describe the hybridization states of the C atoms. The molecule does not contain O—H bonds. On separate diagrams, sketch the formation of sigma and pi bonds.

10.94 TCDD, or 2,3,7,8-tetrachlorodibenzo-p-dioxin, is a highly toxic compound

Cl O Cl
Cl O Cl

It gained considerable notoriety in 2004 when it was implicated in the murder plot of a Ukrainian politician. (a) Describe its geometry and state whether the molecule has a dipole moment. (b) How many pi bonds and sigma bonds are there in the molecule?

10.95 Carbon suboxide (C_3O_2) is a colorless pungent-smelling gas. Does it possess a dipole moment?

10.96 Which of the following ions possess a dipole moment? (a) ClF_2^+, (b) ClF_2^-, (c) IF_4^+, (d) IF_4^-.

Answers to Practice Exercises

10.1 (a) Tetrahedral, (b) linear, (c) trigonal planar. **10.2** No. **10.3** (a) sp^3, (b) sp^2. **10.4** sp^3d^2. **10.5** The C atom is *sp*-hybridized. It forms a sigma bond with the H atom and another sigma bond with the N atom. The two unhybridized *p* orbitals on the C atom are used to form two pi bonds with the N atom. The lone pair on the N atom is placed in the *sp* orbital. **10.6** F_2^-.

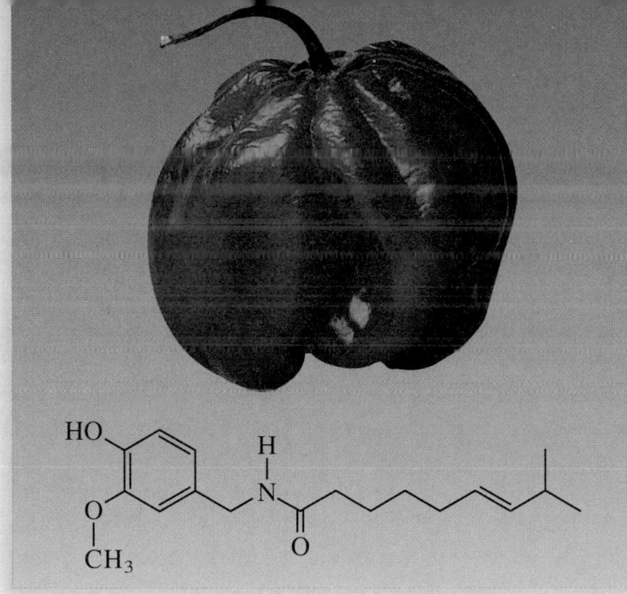

The burning sensation of chili peppers such as habaneros is mostly due to the organic compound capsaicin (illustrated by its skeletal structure).

CHAPTER

11

Introduction to Organic Chemistry

Chapter Outline

Student Interactive Activities

Animations
Chirality (11.5)

Electronic Homework
Example Practice Problems
End of Chapter Problems

Essential Concepts

Organic Compounds Organic compounds contain primarily carbon and hydrogen atoms, plus nitrogen, oxygen, sulfur, and atoms of other elements. The parent compounds of all organic compounds are the hydrocarbons—the alkanes (containing only single bonds), the alkenes (containing carbon-carbon double bonds), the alkynes (containing carbon-carbon triple bonds), and the aromatic hydrocarbons (containing the benzene ring).

Functional Groups The reactivity of organic compounds can be reliably predicted by the presence of functional groups, which are groups of atoms that are largely responsible for the chemical behavior of the compounds.

Chirality Certain organic compounds can exist as nonsuperimposable mirror-image twins. These compounds are said to be chiral. The pure enantiomer of a compound can rotate plane-polarized light. Enantiomers have identical physical properties but exhibit different chemical properties toward another chiral substance.

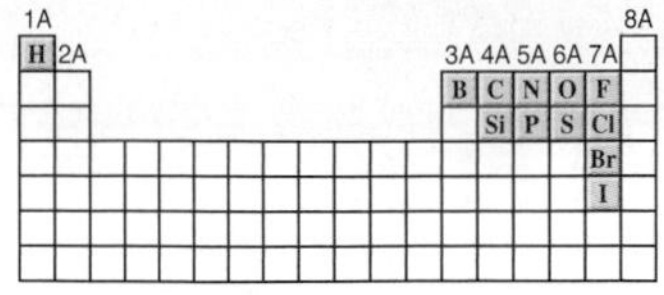

Common elements in organic compounds.

11.1 Classes of Organic Compounds

Carbon can form more compounds than most other elements because carbon atoms are able not only to form single, double, and triple carbon-carbon bonds, but also to link up with each other in chains and ring structures. *The branch of chemistry that deals with carbon compounds* is ***organic chemistry.***

Classes of organic compounds can be distinguished according to functional groups they contain. A ***functional group*** is *a group of atoms that is largely responsible for the chemical behavior of the parent molecule.* Different molecules containing the same kind of functional group or groups undergo similar reactions. Thus, by learning the characteristic properties of a few functional groups, we can study and understand the properties of many organic compounds. In the second half of this chapter we will discuss the functional groups known as alcohols, ethers, aldehydes and ketones, carboxylic acids, and amines.

All organic compounds are derived from a group of compounds known as ***hydrocarbons*** because they are *made up of only hydrogen and carbon.* On the basis of structure, hydrocarbons are divided into two main classes—aliphatic and aromatic. ***Aliphatic hydrocarbons*** *do not contain the benzene group, or the benzene ring,* whereas ***aromatic hydrocarbons*** *contain one or more benzene rings.*

11.2 Aliphatic Hydrocarbons

Aliphatic hydrocarbons are divided into alkanes, alkenes, and alkynes, discussed in this section (Figure 11.1).

Alkanes

Alkanes *are hydrocarbons that have the general formula* C_nH_{2n+2}, *where* $n = 1, 2, \ldots$. The essential characteristic of alkanes is that *only single covalent bonds are present.* The alkanes are known as ***saturated hydrocarbons*** because they *contain the maximum number of hydrogen atoms that can bond with the number of carbon atoms present.*

For a given number of carbon atoms, the saturated hydrocarbon contains the largest number of hydrogen atoms.

The simplest alkane (that is, with $n = 1$) is methane CH_4, which is a natural product of the anaerobic bacterial decomposition of vegetable matter under water. Because it was first collected in marshes, methane became known as "marsh gas." A rather improbable

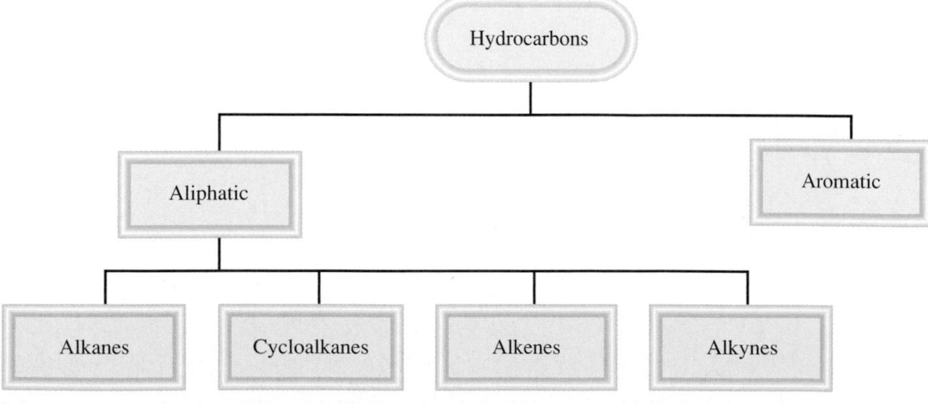

Figure 11.1
Classification of hydrocarbons.

```
     H              H  H              H  H  H
     |              |  |              |  |  |
   H-C-H          H-C--C-H          H-C--C--C-H
     |              |  |              |  |  |
     H              H  H              H  H  H

  Methane          Ethane            Propane

                                         H
                                         |
                                       H-C-H
     H  H  H  H                        H | H
     |  |  |  |                        | | |
   H-C--C--C--C-H                    H-C-C-C-H
     |  |  |  |                        | | |
     H  H  H  H                        H H H

      n-Butane                       Isobutane
```

Figure 11.2
Structures of the first four alkanes. Note that butane can exist in two structurally different forms, called structural isomers.

but proven source of methane is termites. When these voracious insects consume wood, the microorganisms that inhabit their digestive system break down cellulose (the major component of wood) into methane, carbon dioxide, and other compounds. An estimated 170 million tons of methane are produced annually by termites! It is also produced in some sewage treatment processes. Commercially, methane is obtained from natural gas.

Termites are a natural source of methane.

Figure 11.2 shows the structures of the first four alkanes ($n = 1$ to $n = 4$). Natural gas is a mixture of methane, ethane, and a small amount of propane. We discussed the bonding scheme of methane in Chapter 10. The carbon atoms in all the alkanes can be assumed to be sp^3-hybridized. The structures of ethane and propane are straightforward, for there is only one way to join the carbon atoms in these molecules. Butane, however, has two possible bonding schemes resulting in different compounds called *n*-butane (*n* stands for normal) and isobutane. *n*-Butane is a straight-chain alkane because the carbon atoms are joined in a continuous chain. In a branched-chain alkane like isobutane, one or more carbon atoms are bonded to a nonterminal carbon atom. *Isomers that differ in the order in which atoms are connected* are called ***structural isomers.***

In the alkane series, as the number of carbon atoms increases, the number of structural isomers increases rapidly. For example, C_4H_{10} has two isomers; $C_{10}H_{22}$ has 75 isomers; and $C_{30}H_{62}$ has over 400 million possible isomers! Obviously, most of these isomers do not exist in nature nor have they been synthesized. Nevertheless, the numbers help to explain why carbon is found in so many more compounds than any other element.

EXAMPLE 11.1

How many structural isomers can be identified for pentane, C_5H_{12}?

Strategy For small hydrocarbon molecules (eight or fewer C atoms), it is relatively easy to determine the number of structural isomers by trial and error.

Solution The first step is to write the straight-chain structure:

```
     H  H  H  H  H
     |  |  |  |  |
   H-C--C--C--C--C-H
     |  |  |  |  |
     H  H  H  H  H
```

n-pentane
(b.p. 36.1°C)

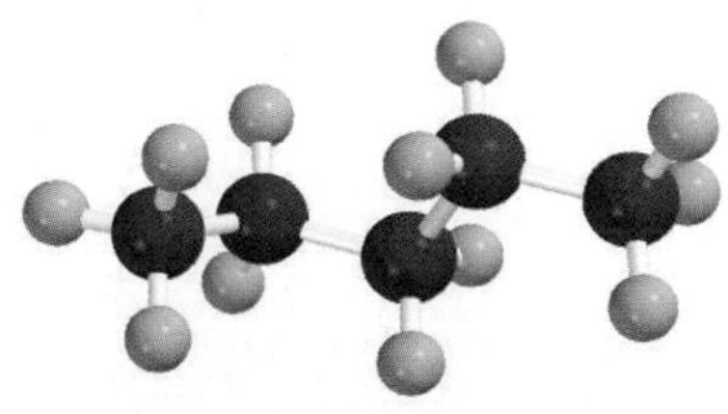

n-pentane

(Continued)

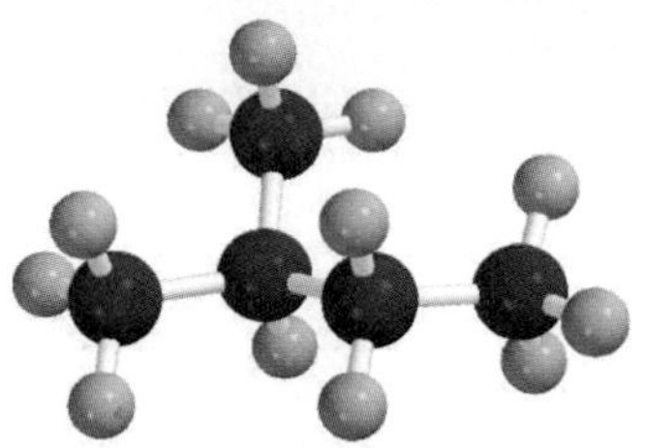

2-methylbutane

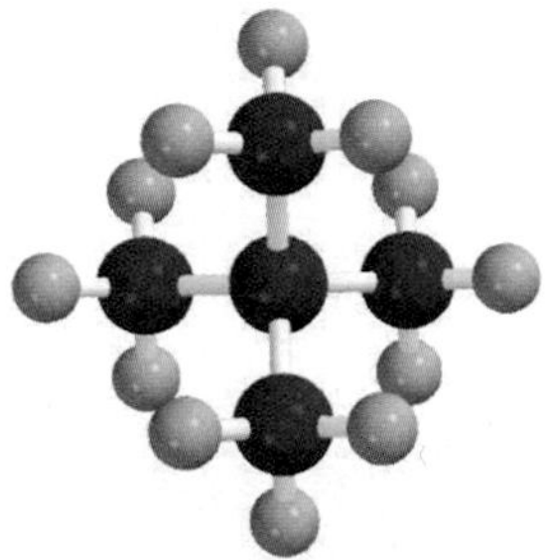

2,2-dimethylpropane

Similar problems: 11.11, 11.12.

The second structure, by necessity, must be a branched chain:

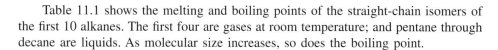

```
   H  CH3  H   H
   |   |   |   |
H—C—C————C—C—H
   |   |   |   |
   H   H   H   H
```

2-methylbutane
(b.p. 27.9°C)

Yet another branched-chain structure is possible:

```
   H  CH3  H
   |   |   |
H—C—C————C—H
   |   |   |
   H  CH3  H
```

2,2-dimethylpropane
(b.p. 9.5°C)

We can draw no other structure for an alkane having the molecular formula C_5H_{12}. Thus, pentane has three structural isomers, in which the numbers of carbon and hydrogen atoms remain unchanged despite the differences in structure.

Practice Exercise How many structural isomers are there in the alkane C_6H_{14}?

Table 11.1 shows the melting and boiling points of the straight-chain isomers of the first 10 alkanes. The first four are gases at room temperature; and pentane through decane are liquids. As molecular size increases, so does the boiling point.

Drawing Chemical Structures

Shortly we will discuss the nomenclature of alkanes.

Before proceeding further, it is useful to learn different ways of drawing the structure of organic compounds. Consider the alkane 2-methylbutane (C_5H_{12}). To see how atoms are connected in this molecule, we need to first write a more detailed molecular formula,

Crude oil is the source of many hydrocarbons.

Table 11.1 The First 10 Straight-Chain Alkanes

Name of Hydrocarbon	Molecular Formula	Number of Carbon Atoms	Melting Point (°C)	Boiling Point (°C)
Methane	CH_4	1	−182.5	−161.6
Ethane	$CH_3—CH_3$	2	−183.3	−88.6
Propane	$CH_3—CH_2—CH_3$	3	−189.7	−42.1
Butane	$CH_3—(CH_2)_2—CH_3$	4	−138.3	−0.5
Pentane	$CH_3—(CH_2)_3—CH_3$	5	−129.8	36.1
Hexane	$CH_3—(CH_2)_4—CH_3$	6	−95.3	68.7
Heptane	$CH_3—(CH_2)_5—CH_3$	7	−90.6	98.4
Octane	$CH_3—(CH_2)_6—CH_3$	8	−56.8	125.7
Nonane	$CH_3—(CH_2)_7—CH_3$	9	−53.5	150.8
Decane	$CH_3—(CH_2)_8—CH_3$	10	−29.7	174.0

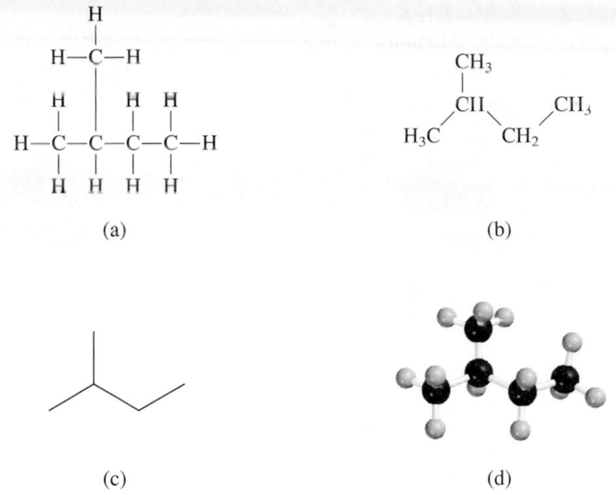

(a) (b)

(c) (d)

Figure 11.3
Different representations of 2-methylbutane. (a) Structural formula. (b) Abbreviated formula. (c) Skeletal formula. (d) Molecular model.

$CH_3CH(CH_3)CH_2CH_3$, and then draw its structural formula, shown in Figure 11.3(a). While informative, this structure is time-consuming to draw. Therefore, chemists have devised ways to simplify the representation. Figure 11.3(b) is an abbreviated version and the structure shown in Figure 11.3(c) is called the *skeletal structure* in which all the C and H letters are omitted. A carbon atom is assumed to be at each intersection of two lines (bonds) and at the end of each line. Because every C atom forms four bonds, we can always deduce the number of H atoms bonded to any C atom. One of the two end CH_3 groups is represented by a vertical line. What is lacking in these structures, however, is the three-dimensionality of the molecule, which is shown by the molecular model in Figure 11.3(d). Depending on the purpose of discussion, any of these representations can be used to describe the properties of the molecule.

Skeletal structure is the simplest structure. Atoms other than C and H must be shown explicitly in a skeletal structure.

Conformation of Ethane

Molecular geometry gives the spatial arrangement of atoms in a molecule. However, atoms are not held rigidly in position because of internal molecular motions. For this reason, even a simple molecule like ethane may be structurally more complicated than we think.

The two C atoms in ethane are sp^3-hybridized and they are joined by a sigma bond (see Section 10.5). Sigma bonds have cylindrical symmetry, that is, the overlap of the sp^3 orbitals is the same regardless of the rotation of the C—C bond. Yet this bond rotation is not totally free because of the interactions between the H atoms on different C atoms. Figure 11.4 shows the two extreme *conformations* of ethane. ***Conformations*** are *different spatial arrangements of a molecule that are generated by rotation about single bonds.* In the staggered conformation, the three H atoms on one C atom are pointing away from the three H atoms on the other C atom, whereas in the eclipsed conformation the two groups of H atoms are aligned parallel to one another.

A simpler and effective way of viewing these two conformations is by using the Newman projection, also shown in Figure 11.4. Look at the C—C bond end-on. The two C atoms are represented by a circle. The C—H bonds attached to the front C atom are the lines going to the center of the circle, and the C—H bonds attached to the rear C atom appear as lines going to the edge of the circle. The eclipsed form of ethane is less stable than the staggered form. Figure 11.5 shows the variation in the potential energy of ethane as a function of rotation. The rotation of one CH_3 group relative to the other is described in terms of the angle between the C—H bonds on

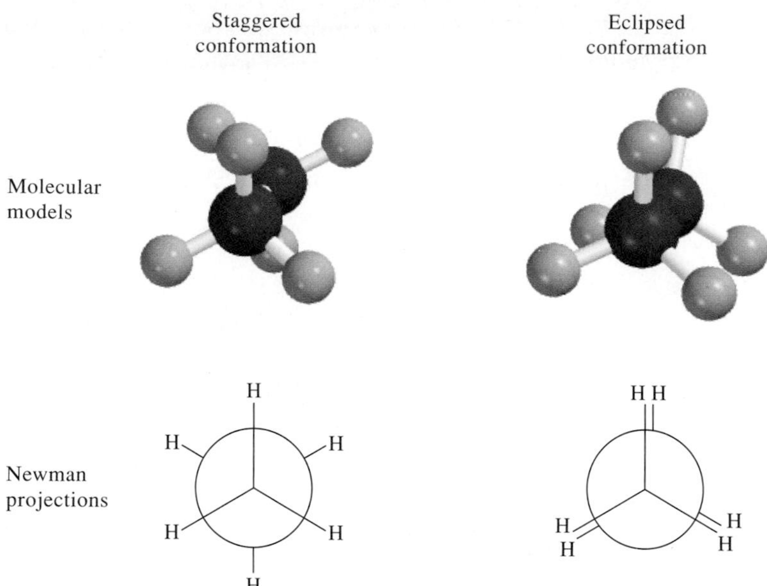

Figure 11.4
Molecular models and Newman projections of the staggered and eclipsed conformations of ethane. The dihedral angle in the staggered form is 60° and that in the eclipsed form is 0°. The C—C bond is rotated slightly in the Newman projection of the eclipsed form in order to show the H atoms attached to the back C atom. The proximity of the H atoms on the two C atoms in the eclipsed form results in a greater repulsion, and hence its instability relative to the staggered form.

front and back carbons, called the *dihedral* angle. The dihedral angle for the first eclipsed conformation is zero. A clockwise rotation of 60° about the C—C bond generates a staggered conformation, which is converted to another eclipsed conformation by a similar rotation and so on.

Conformational analysis of molecules is of great importance in understanding the details of reactions ranging from simple hydrocarbons to proteins and DNAs.

Alkane Nomenclature

The nomenclature of alkanes and all other organic compounds is based on the recommendations of the International Union of Pure and Applied Chemistry (IUPAC). The first four alkanes (methane, ethane, propane, and butane) have nonsystematic names. As Table 11.1 shows, the number of carbon atoms is reflected in the Greek

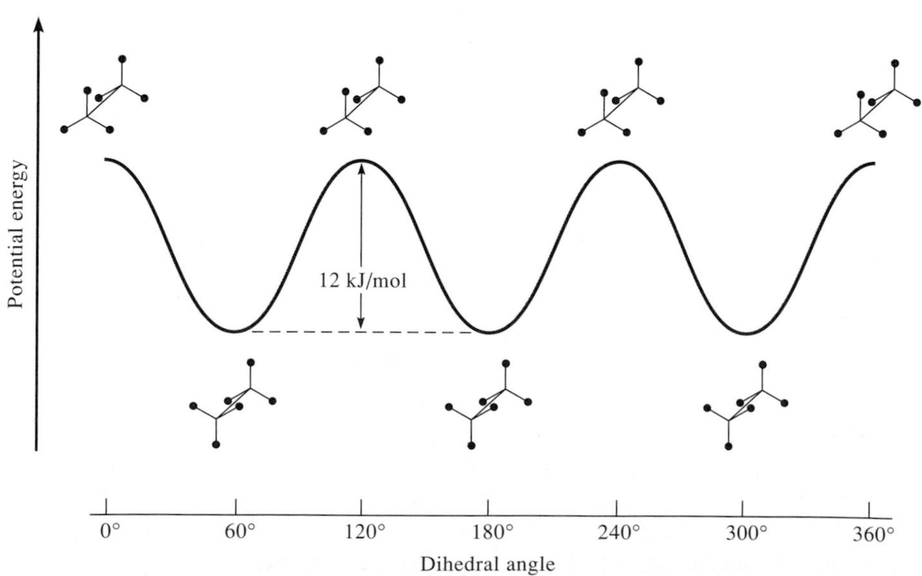

Figure 11.5
Potential energy diagram for the internal rotation in ethane. Here the dihedral angle is defined by the angle between the two C—H bonds (with the red spheres representing the H atoms). Dihedral angles of 0°, 120°, 240°, and 360° represent the eclipsed conformation, while those of 60°, 180°, and 300° represent the staggered conformation. Thus, a rotation of 60° changes the eclipsed conformation to the staggered one and vice versa. The staggered conformation is more stable than the eclipsed conformation by 12 kJ/mol. However, these two forms interconvert rapidly and cannot be separated from each other.

prefixes for the alkanes containing 5 to 10 carbons. We now apply the IUPAC rules to the following examples:

1. The parent name of the hydrocarbon is that given to the longest continuous chain of carbon atoms in the molecule. Thus, the parent name of the following compound is heptane because there are seven carbon atoms in the longest chain

$$\overset{1}{CH_3}-\overset{2}{CH_2}-\overset{3}{CH_2}-\overset{\overset{CH_3}{|}}{\underset{4}{CH}}-\overset{5}{CH_2}-\overset{6}{CH_2}-\overset{7}{CH_3}$$

2. An alkane less one hydrogen atom is an *alkyl* group. For example, when a hydrogen atom is removed from methane, we are left with the CH_3 fragment, which is called a *methyl* group. Similarly, removing a hydrogen atom from the ethane molecule gives an *ethyl* group, or C_2H_5. Table 11.2 lists the names of several common alkyl groups. Any chain branching off the longest chain is named as an alkyl group.
3. When one or more hydrogen atoms are replaced by other groups, the name of the compound must indicate the locations of carbon atoms where replacements are made. The procedure is to number each carbon atom on the longest chain in the direction that gives the smaller numbers for the locations of all branches. Consider the two different systems for the *same* compound shown below:

$$\overset{1}{CH_3}-\overset{\overset{CH_3}{|}}{\underset{2}{CH}}-\overset{3}{CH_2}-\overset{4}{CH_2}-\overset{5}{CH_3} \qquad \overset{1}{CH_3}-\overset{2}{CH_2}-\overset{3}{CH_2}-\overset{\overset{CH_3}{|}}{\underset{4}{CH}}-\overset{5}{CH_3}$$

2-methylpentane 4-methylpentane

The compound on the left is numbered correctly because the methyl group is located at carbon 2 of the pentane chain; in the compound on the right, the methyl group is located at carbon 4. Thus, the name of the compound is 2-methylpentane, and not 4-methylpentane. Note that the branch name and the parent name are written as a single word, and a hyphen follows the number.

4. When there is more than one alkyl branch of the same kind present, we use a prefix such as *di-*, *tri-*, or *tetra-* with the name of the alkyl group. Consider the following examples:

$$\overset{1}{CH_3}-\overset{\overset{CH_3}{|}}{\underset{2}{CH}}-\overset{\overset{CH_3}{|}}{\underset{3}{CH}}-\overset{4}{CH_2}-\overset{5}{CH_2}-\overset{6}{CH_3} \qquad \overset{1}{CH_3}-\overset{2}{CH_2}-\underset{\underset{CH_3}{|}}{\overset{\overset{CH_3}{|}}{\underset{3}{C}}}-\overset{4}{CH_2}-\overset{5}{CH_2}-\overset{6}{CH_3}$$

2,3-dimethylhexane 3,3-dimethylhexane

When there are two or more different alkyl groups, the names of the groups are listed alphabetically. For example,

$$\overset{1}{CH_3}-\overset{2}{CH_2}-\overset{\overset{CH_3}{|}}{\underset{3}{CH}}-\overset{\overset{C_2H_5}{|}}{\underset{4}{CH}}-\overset{5}{CH_2}-\overset{6}{CH_2}-\overset{7}{CH_3}$$

4-ethyl-3-methylheptane

5. Of course, alkanes can have many different types of substituents. Table 11.3 lists the names of some substituents, including bromo and nitro. Thus, the compound

$$\overset{1}{CH_3}-\overset{\overset{NO_2}{|}}{\underset{2}{CH}}-\overset{\overset{Br}{|}}{\underset{3}{CH}}-\overset{4}{CH_2}-\overset{5}{CH_2}-\overset{6}{CH_3}$$

Table 11.2

Common Alkyl Groups

Name	Formula
Methyl	$-CH_3$
Ethyl	$-CH_2-CH_3$
n-Propyl	$-(CH_2)_2-CH_3$
n-Butyl	$-(CH_2)_3-CH_3$
Isopropyl	$-\underset{\underset{CH_3}{\vert}}{\overset{\overset{CH_3}{\vert}}{C}}-H$
t-Butyl*	$-\underset{\underset{CH_3}{\vert}}{\overset{\overset{CH_3}{\vert}}{C}}-CH_3$

*The letter *t* stands for tertiary.

Table 11.3

Names of Common Substituent Groups

Functional Group	Name
$-NH_2$	Amino
$-F$	Fluoro
$-Cl$	Chloro
$-Br$	Bromo
$-I$	Iodo
$-NO_2$	Nitro
$-CH{=}CH_2$	Vinyl

is called 3-bromo-2-nitrohexane. Note that the substituent groups are listed alphabetically in the name, and the chain is numbered in the direction that gives the lowest number to the first substituted carbon atom.

EXAMPLE 11.2

Give the IUPAC name of the following compound:

$$CH_3{-}\underset{\displaystyle CH_3}{\overset{\displaystyle CH_3}{\underset{|}{\overset{|}{C}}}}{-}CH_2{-}\overset{\displaystyle CH_3}{\overset{|}{CH}}{-}CH_2{-}CH_3$$

Strategy We follow the IUPAC rules and use the information in Table 11.2 to name the compound. How many C atoms are there in the longest chain?

Solution The longest chain has six C atoms so the parent compound is called hexane. Note that there are two methyl groups attached to carbon number 2 and one methyl group attached to carbon number 4.

$$\overset{1}{CH_3}{-}\underset{\displaystyle CH_3}{\overset{\displaystyle CH_3}{\underset{|}{\overset{2|}{C}}}}{-}\overset{3}{CH_2}{-}\overset{\displaystyle CH_3}{\overset{4|}{CH}}{-}\overset{5}{CH_2}{-}\overset{6}{CH_3}$$

Similar problems: 11.28(a), (b), (c).

Therefore, we call the compound 2,2,4-trimethylhexane.

Practice Exercise Give the IUPAC name of the following compound:

$$CH_3{-}\overset{\displaystyle CH_3}{\overset{|}{CH}}{-}CH_2{-}\overset{\displaystyle C_2H_5}{\overset{|}{CH}}{-}CH_2{-}\overset{\displaystyle C_2H_5}{\overset{|}{CH}}{-}CH_2{-}CH_3$$

Example 11.3 shows that prefixes such as di-, tri-, and tetra- are used as needed, but are ignored when alphabetizing.

EXAMPLE 11.3

Write the structural formula of 3-ethyl-2,2-dimethylpentane.

Strategy We follow the preceding procedure and the information in Table 11.2 to write the structural formula of the compound. How many C atoms are there in the longest chain?

Solution The parent compound is pentane, so the longest chain has five C atoms. There are two methyl groups attached to carbon number 2 and one ethyl group attached to carbon number 3. Therefore, the structure of the compound is

$$\overset{1}{CH_3}{-}\underset{\displaystyle CH_3}{\overset{\displaystyle CH_3}{\underset{|}{\overset{2|}{C}}}}{-}\overset{\displaystyle C_2H_5}{\overset{3|}{CH}}{-}\overset{4}{CH_2}{-}\overset{5}{CH_3}$$

Similar problems: 11.27(a), (c), (e).

Practice Exercise Write the structural formula of 5-ethyl-2,6-dimethyloctane.

Reactions of Alkanes

Alkanes are generally not considered to be very reactive substances. However, under suitable conditions they do react. For example, natural gas, gasoline, and fuel oil are alkanes that undergo highly exothermic combustion reactions:

$$CH_4(g) + 2O_2(g) \longrightarrow CO_2(g) + 2H_2O(l) \qquad \Delta H^\circ = -890.4\ \text{kJ/mol}$$
$$2C_2H_6(g) + 7O_2(g) \longrightarrow 4CO_2(g) + 6H_2O(l) \qquad \Delta H^\circ = -3119\ \text{kJ/mol}$$

These, and similar combustion reactions, have long been utilized in industrial processes and in domestic heating and cooking.

Halogenation of alkanes—that is, the replacement of one or more hydrogen atoms by halogen atoms—is another type of reaction that alkanes undergo. When a mixture of methane and chlorine is heated above 100°C or irradiated with light of a suitable wavelength, methyl chloride is produced:

$$CH_4(g) + Cl_2(g) \longrightarrow \underset{\text{methyl chloride}}{CH_3Cl(g)} + HCl(g)$$

If an excess of chlorine gas is present, the reaction can proceed further:

$$CH_3Cl(g) + Cl_2(g) \longrightarrow \underset{\text{methylene chloride}}{CH_2Cl_2(l)} + HCl(g)$$
$$CH_2Cl_2(l) + Cl_2(g) \longrightarrow \underset{\text{chloroform}}{CHCl_3(l)} + HCl(g)$$
$$CHCl_3(l) + Cl_2(g) \longrightarrow \underset{\text{carbon tetrachloride}}{CCl_4(l)} + HCl(g)$$

A great deal of experimental evidence suggests that the initial step of the first halogenation reaction occurs as follows:

$$Cl_2 + \text{energy} \longrightarrow Cl\cdot + Cl\cdot$$

Thus, the covalent bond in Cl_2 breaks and two chlorine atoms form. We know it is the Cl—Cl bond that breaks when the mixture is heated or irradiated because the bond enthalpy of Cl_2 is 242.7 kJ/mol, whereas about 414 kJ/mol are needed to break C—H bonds in CH_4.

A chlorine atom is a ***radical,*** which *contains an unpaired electron* (shown by a single dot). Chlorine atoms are highly reactive and attack methane molecules according to the equation

$$CH_4 + Cl\cdot \longrightarrow \cdot CH_3 + HCl$$

This reaction produces hydrogen chloride and the methyl radical $\cdot CH_3$. The methyl radical is another reactive species; it combines with molecular chlorine to give methyl chloride and a chlorine atom:

$$\cdot CH_3 + Cl_2 \longrightarrow CH_3Cl + Cl\cdot$$

The production of methylene chloride from methyl chloride and any further reactions can be explained in the same way. The actual mechanism is more complex than the scheme we have shown because "side reactions" that do not lead to the desired products often take place, such as

$$Cl\cdot + Cl\cdot \longrightarrow Cl_2$$
$$\cdot CH_3 + \cdot CH_3 \longrightarrow C_2H_6$$

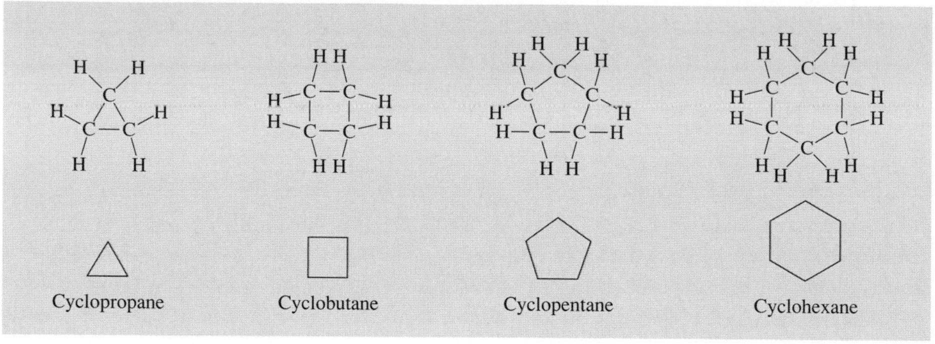

Figure 11.6
Structures of the first four cycloalkanes and their simplified forms.

Cyclopropane Cyclobutane Cyclopentane Cyclohexane

The systematic names of methyl chloride, methylene chloride, and chloroform are chloromethane, dichloromethane, and trichloromethane, respectively.

Alkanes in which one or more hydrogen atoms have been replaced by a halogen atom are called *alkyl halides.* Among the large number of alkyl halides, the best known are chloroform ($CHCl_3$), carbon tetrachloride (CCl_4), methylene chloride (CH_2Cl_2), and the chlorofluorohydrocarbons.

Chloroform is a volatile, sweet-tasting liquid that was used for many years as an anesthetic. However, because of its toxicity (it can severely damage the liver, kidneys, and heart) it has been replaced by other compounds. Carbon tetrachloride, also a toxic substance, serves as a cleaning liquid, for it removes grease stains from clothing. Methylene chloride is used as a solvent to decaffeinate coffee and as a paint remover.

Cycloalkanes

In addition to C, atoms such as N, O, and S may also occupy the ring positions in these compounds.

Alkanes whose carbon atoms are joined in rings are known as ***cycloalkanes.*** They have the general formula C_nH_{2n}, where $n = 3, 4, \ldots$. The simplest cycloalkane is cyclopropane, C_3H_6 (Figure 11.6). Many biologically significant substances such as antibiotics, sugars, cholesterol, and hormones contain one or more such ring systems. Cyclohexane can assume two different conformations called the chair and boat that are relatively free of angle strain (Figure 11.7). By "angle strain" we mean that the bond angles at each carbon atom deviate from the tetrahedral value of 109.5° required for sp^3 hybridization.

Alkenes

The ***alkenes*** (also called *olefins*) *contain at least one carbon-carbon double bond. Alkenes have the general formula* C_nH_{2n}, *where* $n = 2, 3, \ldots$. The simplest alkene

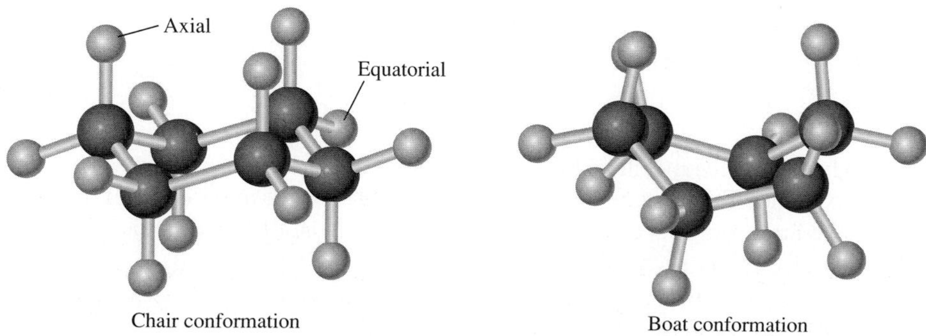

Figure 11.7
The cyclohexane molecule can exist in various shapes. The most stable shape is the chair conformation and a less stable one is the boat conformation. Two types of H atoms are labeled axial and equatorial, respectively.

is C_2H_4, ethylene, in which both carbon atoms are sp^2-hybridized and the double bond is made up of a sigma bond and a pi bond (see Section 10.5).

Geometric Isomers of Alkenes

In a compound such as ethane, C_2H_6, the rotation of the two methyl groups about the carbon-carbon single bond (which is a sigma bond) is quite free. The situation is different for molecules that contain carbon-carbon double bonds, such as ethylene, C_2H_4. In addition to the sigma bond, there is a pi bond between the two carbon atoms. Rotation about the carbon-carbon linkage does not affect the sigma bond, but it does move the two $2p_z$ orbitals out of alignment for overlap and, hence, partially or totally destroys the pi bond (see Figure 10.15). This process requires an input of energy on the order of 270 kJ/mol. For this reason, the rotation of a carbon-carbon double bond is considerably restricted, but not impossible. Consequently, molecules containing carbon-carbon double bonds (that is, the alkenes) may have ***geometric isomers,*** which *have the same type and number of atoms and the same chemical bonds but different spatial arrangements. Such isomers cannot be interconverted without breaking a chemical bond.*

The molecule dichloroethylene, ClHC=CHCl, can exist as one of the two geometric isomers called *cis*-1,2-dichloroethylene and *trans*-1,2-dichloroethylene:

resultant dipole moment

cis-1,2-dichloroethylene
μ = 1.89 D
b.p. 60.3°C

trans-1,2-dichloroethylene
μ = 0
b.p. 47.5°C

where the term *cis* means that two particular atoms (or groups of atoms) are adjacent to each other, and *trans* means that the two atoms (or groups of atoms) are across from each other. Generally, *cis* and *trans* isomers have distinctly different physical and chemical properties. Heat or irradiation with light is commonly used to bring about the conversion of one geometric isomer to another, a process called *cis-trans isomerization,* or geometric isomerization (Figure 11.8).

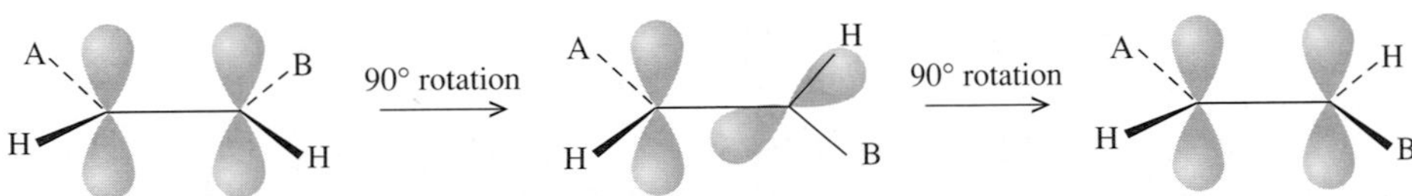

Figure 11.8
Breaking and remaking the pi bond. When a compound containing a C=C bond is heated or excited by light, the weaker pi bond is broken. This allows the free rotation of the single carbon-to-carbon sigma bond. A rotation of 180° converts a cis isomer to a trans isomer or the other way around. Note that a dashed line represents a bond axis behind the plane of the paper, the wedged line represents a bond axis in front of the paper, and the solid line represents bonds in the plane of the paper. The letters A and B represent atoms (other than H) or groups of atoms. Here we have a cis-to-trans isomerization.

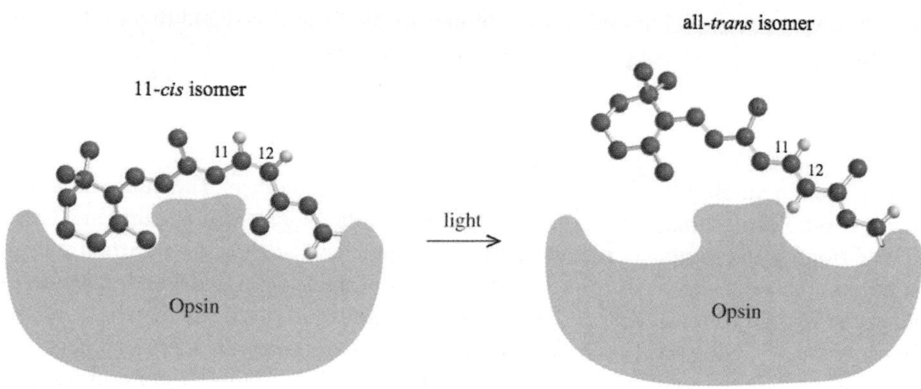

Figure 11.9
The primary event in the vision process is the conversion of 11-cis retinal to the all-trans isomer on rhodopsin. The double bond at which the isomerization occurs is between carbon-11 and carbon-12. For simplicity, most of the H atoms are omitted. In the absence of light, this transformation takes place about once in a thousand years!

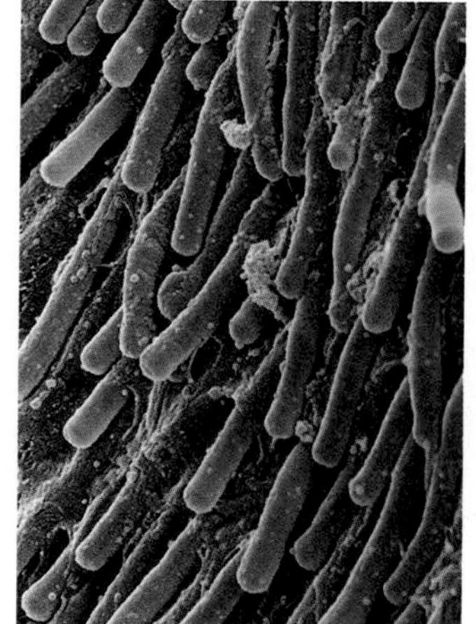

An electron micrograph of rod-shaped cells (containing rhodopsins) in the retina.

Cis-trans Isomerization in the Vision Process

The molecules in the retina that respond to light are rhodopsin, which has two components called 11-*cis* retinal and opsin (Figure 11.9). Retinal is the light-sensitive component and opsin is a protein molecule. Upon receiving a photon in the visible region the 11-*cis* retinal isomerizes to the all-*trans* retinal by breaking a carbon-carbon pi bond. With the pi bond broken, the remaining carbon-carbon sigma bond is free to rotate and transforms into the all-*trans* retinal. At this point an electrical impulse is generated and transmitted to the brain, which forms a visual image. The all-*trans* retinal does not fit into the binding site on opsin and eventually separates from the protein. In time, the *trans* isomer is converted back to 11-*cis* retinal by an enzyme (in the absence of light) and rhodopsin is regenerated by the binding of the *cis* isomer to opsin and the visual cycle can begin again.

Alkene Nomenclature

In naming alkenes we indicate the positions of the carbon-carbon double bonds. The names of compounds containing C=C bonds end with *-ene.* As with the alkanes, the name of the parent compound is determined by the number of carbon atoms in the longest chain (see Table 11.1), as shown here:

$$CH_2{=}CH{-}CH_2{-}CH_3 \qquad H_3C{-}CH{=}CH{-}CH_3$$

1-butene   2-butene

The numbers in the names of alkenes refer to the lowest numbered carbon atom in the chain that is part of the C=C bond of the alkene. The name "butene" means that there are four carbon atoms in the longest chain. Alkene nomenclature must specify whether a given molecule is *cis* or *trans* if it is a geometric isomer, such as

In the *cis* isomer, the two H atoms are on the same side of the C=C bond; in the *trans* isomer, the two H atoms are across from each other.

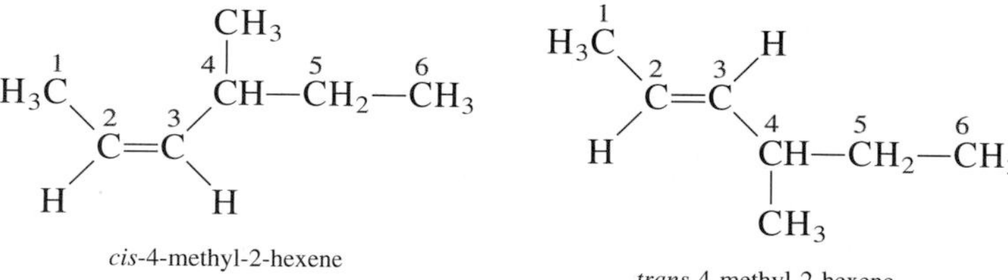

cis-4-methyl-2-hexene   *trans*-4-methyl-2-hexene

Properties and Reactions of Alkenes

Ethylene is an extremely important substance because it is used in large quantities in manufacturing organic polymers (very large molecules) and in preparing many other organic chemicals. Ethylene and other alkenes are prepared industrially by the *cracking* process, that is, the thermal decomposition of a large hydrocarbon into smaller molecules. When ethane is heated to about 800°C in the presence of platinum, it undergoes the following reaction:

$$C_2H_6(g) \xrightarrow[\text{catalyst}]{\text{Pt}} CH_2{=}CH_2(g) + H_2(g)$$

The platinum acts as a *catalyst,* a substance that speeds up a reaction without being used up in the process and therefore does not appear on either side of the equation. Other alkenes can be prepared by cracking the higher members of the alkane family.

Alkenes are classified as ***unsaturated hydrocarbons,*** *compounds with double or triple carbon-carbon bonds.* Unsaturated hydrocarbons commonly undergo ***addition reactions*** in which *one molecule adds to another to form a single product.* An example of an addition reaction is ***hydrogenation,*** which is *the addition of molecular hydrogen to compounds containing C=C and C≡C bonds*

$$H_2 + \begin{matrix} H & & H \\ & C{=}C & \\ H & & H \end{matrix} \longrightarrow \begin{matrix} & H & H & \\ & | & | & \\ H- & C- & C & -H \\ & | & | & \\ & H & H & \end{matrix}$$

Hydrogenation is an important process in the food industry. Vegetable oils have considerable nutritional value, but many oils must be hydrogenated to eliminate some of the C=C bonds before they can be used to prepare food. Upon exposure to air, *polyunsaturated* molecules—molecules with many C=C bonds—undergo oxidation to yield unpleasant-tasting products (vegetable oil that has oxidized is said to be rancid). In the hydrogenation process, a small amount of nickel catalyst is added to the oil and the mixture is exposed to hydrogen gas at high temperature and pressure. Afterward, the nickel is removed by filtration. Hydrogenation reduces the number of double bonds in the molecule but does not completely eliminate them. If all the double bonds are eliminated, the oil becomes hard and brittle (Figure 11.10). Under controlled

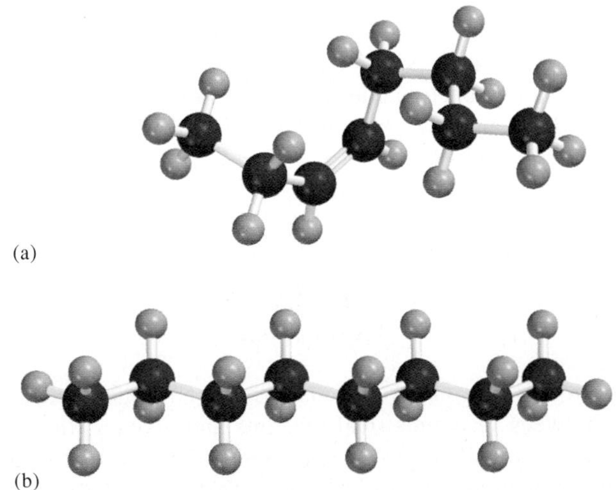

Figure 11.10
Oils and fats have side chains that resemble hydrocarbons. (a) The side chains of oils contain one or more C=C bonds. The cis form of the hydrocarbon chains prevents close packing of the molecules. Therefore, oils are liquids. (b) Upon hydrogenation, the saturated hydrocarbon chains stack well together. As a result, fats have higher density than oils and are solids at room temperature.

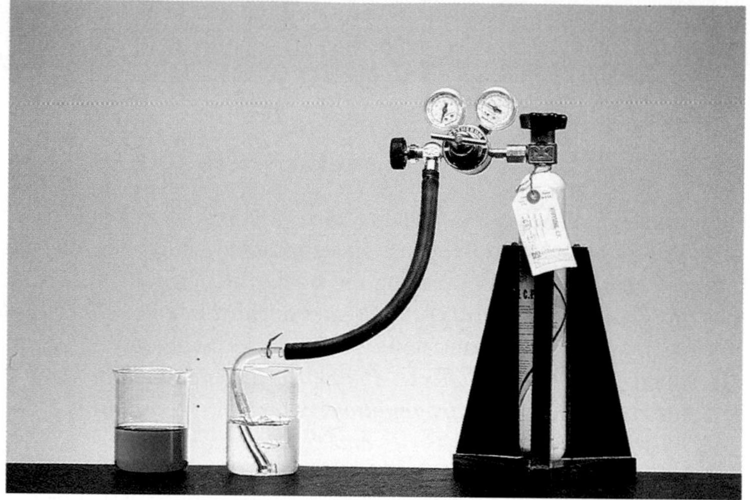

Figure 11.11
When ethylene gas is bubbled through an aqueous bromine solution, the reddish brown color gradually disappears due to the formation of 1,2-dibromoethane, which is colorless.

conditions, suitable cooking oils and margarine may be prepared by hydrogenation from vegetable oils extracted from cottonseed, corn, and soybeans.

Other addition reactions to the C=C bond involve the hydrogen halides and halogens (Figure 11.11):

$$H_2C{=}CH_2 + X_2 \longrightarrow CH_2X{-}CH_2X$$
$$H_2C{=}CH_2 + HX \longrightarrow CH_3{-}CH_2X$$

in which X represents a halogen atom. Figure 11.12 shows the electron density maps of HCl and ethylene. When the two molecules react, the interaction is between the electron-rich region (pi electrons of the double bond) and the electron-poor region of HCl, which is the H atom. The steps are

$$H_2C{=}CH_2 + HCl \longrightarrow H_2\overset{\overset{\displaystyle H}{|}}{C}{-}\overset{+}{C}H_2 + Cl^- \longrightarrow CH_3{-}CH_2Cl$$

The addition of a hydrogen halide to an unsymmetrical alkene such as propene is more complicated because two products are possible:

$$\underset{\text{propene}}{(H)_2C{=}C(H)(CH_3)} + HBr \longrightarrow \underset{\text{1-bromopropane}}{H{-}\underset{Br}{\overset{H}{C}}{-}\underset{H}{\overset{H}{C}}{-}CH_3} \quad \text{and/or} \quad \underset{\text{2-bromopropane}}{H{-}\underset{H}{\overset{H}{C}}{-}\underset{Br}{\overset{H}{C}}{-}CH_3}$$

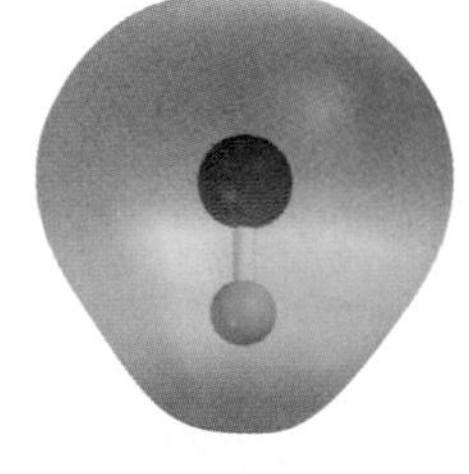

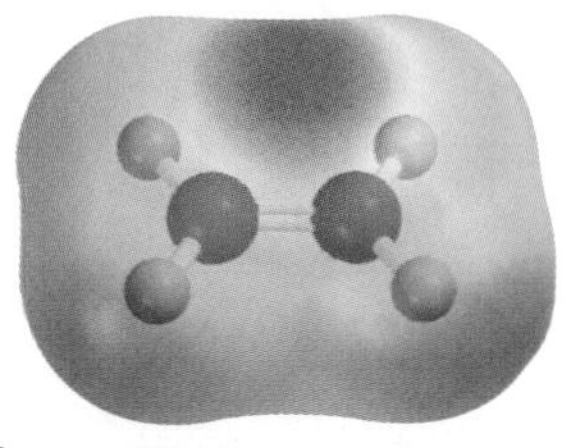

Figure 11.12
The addition reaction between HCl and ethylene. The initial interaction is between the positive end of HCl (blue) and the electron-rich region of ethylene (red), which is associated with the pi electrons of the C=C bond.

In reality, however, only 2-bromopropane is formed. This phenomenon was observed in all reactions between unsymmetrical reagents and alkenes. In 1871 the Russian chemist Vladimir Markovnikov postulated a generalization that enables us to predict the outcome of such an addition reaction. This generalization, now known as *Markovnikov's rule,* states that in the addition of unsymmetrical (that is, polar) reagents to alkenes, the positive portion of the reagent (usually hydrogen) adds to the carbon atom in the double bond that already has the most hydrogen atoms. As the marginal figure on p. 377 shows, the C atom to which the two H atoms are

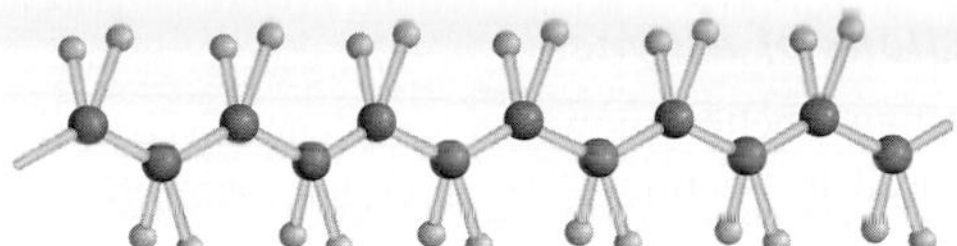

Figure 11.13
Structure of polyethylene. Each carbon atom is sp^3-hybridized.

attached has a higher electron density. Therefore, this is the site for the H^+ ion (from HBr) to form a C—H bond, followed by the formation of the C—Br bond on the other C atom.

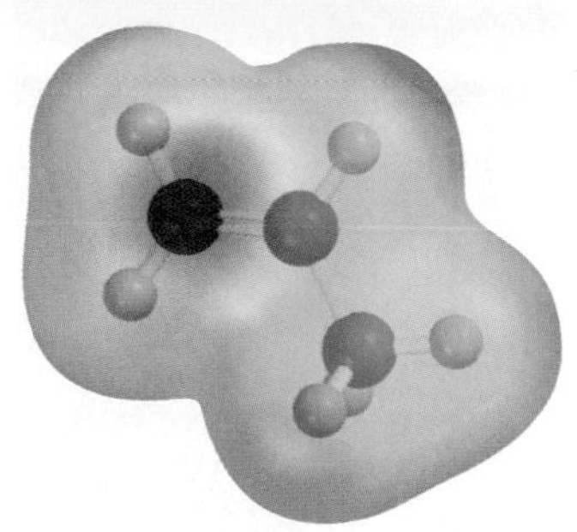

The electron density is higher on the carbon atom in the CH_2 group in propene. This is the site of hydrogen addition by hydrogen halides.

Finally we note that ethylene undergoes a different type of addition reaction that leads to the formation of a polymer. In this process, first an *initiator* molecule (R_2) is heated to produce two radicals:

$$R_2 \longrightarrow 2R\cdot$$

The reactive radical attacks an ethylene molecule to generate a new radical (it is the pi bond that is broken in the polymerization of ethylene):

$$R\cdot + CH_2{=}CH_2 \longrightarrow R{-}CH_2{-}CH_2\cdot$$

which further reacts with another ethylene molecule, and so on:

$$R{-}CH_2{-}CH_2\cdot + CH_2{=}CH_2 \longrightarrow R{-}CH_2{-}CH_2{-}CH_2{-}CH_2\cdot$$

Very quickly a long chain of CH_2 groups is built. Eventually, this process is terminated by the combination of two long-chain radicals to give the polymer called polyethylene (Figure 11.13):

$$R{+}CH_2{-}CH_2{+}_nCH_2CH_2\cdot + R{+}CH_2{-}CH_2{+}_nCH_2CH_2\cdot \longrightarrow$$
$$R{+}CH_2{-}CH_2{+}_nCH_2CH_2{-}CH_2CH_2{+}CH_2{-}CH_2{+}_nR$$

where ${+}CH_2{-}CH_2{+}_n$ is a convenient shorthand convention for representing the repeating unit in the polymer. The value of n is understood to be very large, on the order of thousands.

Under different conditions, it is possible to prepare polyethylene with branched chains. Today, many different forms of polyethylene with widely different physical properties are known. Polyethylene is mainly used in films in frozen food packaging and other product wrappings. A specially treated type of polyethylene, called Tyvek, is used in home construction and mailer envelopes.

Common mailing envelopes made of Tyvek.

Alkynes

***Alkynes** contain at least one carbon-carbon triple bond.* They have the *general formula C_nH_{2n-2}, where $n = 2, 3, \ldots$.*

Alkyne Nomenclature

Names of compounds containing C≡C bonds end with *-yne*. Again the name of the parent compound is determined by the number of carbon atoms in the longest chain (see Table 11.1 for names of alkane counterparts). As in the case of alkenes, the names of alkynes indicate the position of the carbon-carbon triple bond, as, for example, in

$$\underset{\text{1-butyne}}{HC{\equiv}C{-}CH_2{-}CH_3} \qquad \underset{\text{2-butyne}}{H_3C{-}C{\equiv}C{-}CH_3}$$

Properties and Reactions of Alkynes

The reaction of calcium carbide with water produces acetylene, a flammable gas.

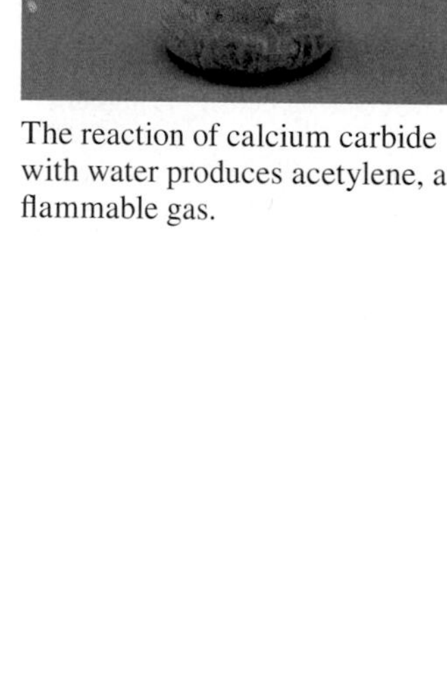

The simplest alkyne is ethyne, better known as acetylene (C_2H_2). The structure and bonding of C_2H_2 were discussed in Section 10.5. Acetylene is a colorless gas (b.p. $-84°C$) prepared in the laboratory by the reaction between calcium carbide and water:

$$CaC_2(s) + 2H_2O(l) \longrightarrow C_2H_2(g) + Ca(OH)_2(aq)$$

Industrially, it is prepared by the thermal decomposition of ethylene at about 1100°C:

$$C_2H_4(g) \longrightarrow C_2H_2(g) + H_2(g)$$

Acetylene has many important uses in industry. Because of its high heat of combustion

$$2C_2H_2(g) + 5O_2(g) \longrightarrow 4CO_2(g) + 2H_2O(l) \quad \Delta H° = -2599.2 \text{ kJ/mol}$$

acetylene burned in an "oxyacetylene torch" gives an extremely hot flame (about 3000°C). Thus, oxyacetylene torches are used to weld metals (see p. 200).

Acetylene is unstable and has a tendency to decompose:

$$C_2H_2(g) \longrightarrow 2C(s) + H_2(g)$$

In the presence of a suitable catalyst or when the gas is kept under pressure, this reaction can occur with explosive violence. To be transported safely, it must be dissolved in an inert organic solvent such as acetone at moderate pressure. In the liquid state, acetylene is very sensitive to shock and is highly explosive.

Being an unsaturated hydrocarbon, acetylene can be hydrogenated to yield ethylene:

$$C_2H_2(g) + H_2(g) \longrightarrow C_2H_4(g)$$

It undergoes these addition reactions with hydrogen halides and halogens:

$$CH{\equiv}CH(g) + HX(g) \longrightarrow CH_2{=}CHX(g)$$
$$CH{\equiv}CH(g) + X_2(g) \longrightarrow CHX{=}CHX(g)$$
$$CH{\equiv}CH(g) + 2X_2(g) \longrightarrow CHX_2{-}CHX_2(l)$$

Methylacetylene (propyne), $CH_3{-}C{\equiv}C{-}H$, is the next member in the alkyne family. It undergoes reactions similar to those of acetylene. The addition reactions of propyne also obey Markovnikov's rule:

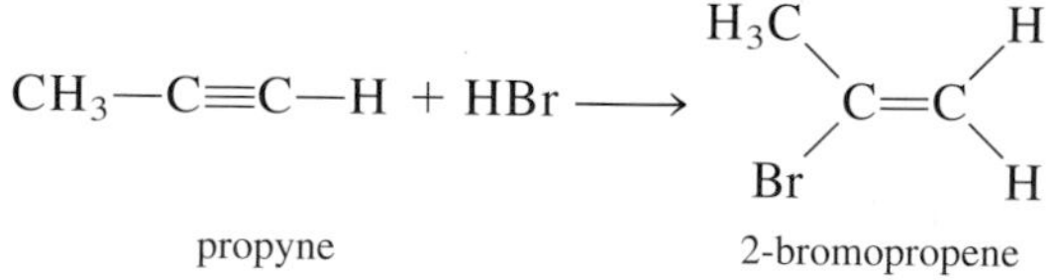

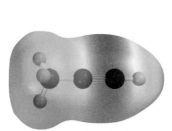

Propyne. Can you account for Markovnikov's rule in this molecule?

REVIEW OF CONCEPTS

How could an alkene and an alkyne be distinguished by using only a hydrogenation reaction?

11.3 Aromatic Hydrocarbons

Benzene (C_6H_6) is the parent compound of this large family of organic substances. As we saw in Section 9.8, the properties of benzene are best represented by both of the following resonance structures (p. 304):

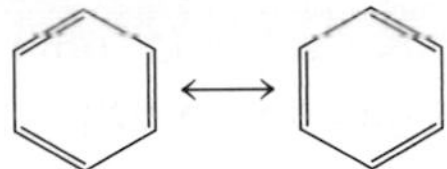

An electron micrograph of benzene molecules, which shows clearly the ring structure.

Benzene is a planar hexagonal molecule with carbon atoms situated at the six corners. All carbon-carbon bonds are equal in length and strength, as are all carbon-hydrogen bonds, and the CCC and HCC angles are all 120°. Therefore, each carbon atom is sp^2-hybridized; it forms three sigma bonds with two adjacent carbon atoms and a hydrogen atom (Figure 11.14). This arrangement leaves an unhybridized $2p_z$ orbital on each carbon atom, perpendicular to the plane of the benzene molecule, or *benzene ring,* as it is often called. So far the description resembles the configuration of ethylene (C_2H_4), discussed in Section 10.5, except that in this case there are six unhybridized $2p_z$ orbitals in a cyclic arrangement.

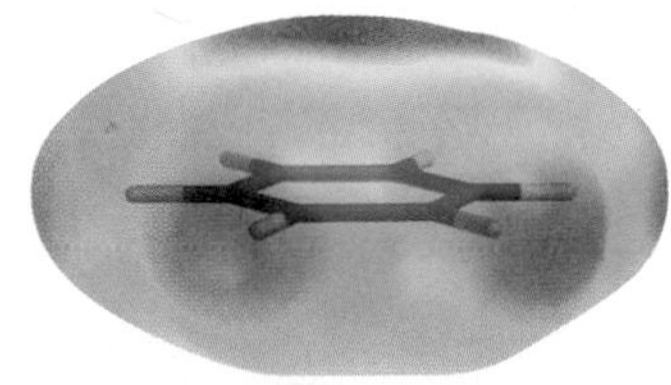

Electrostatic potential map of benzene shows the electron density (red color) above and below the plane of the molecule. For simplicity, only the framework of the molecule is shown.

Because of their similar shape and orientation, each $2p_z$ orbital overlaps two others, one on each adjacent carbon atom. According to the rules listed on p. 351, the interaction of six $2p_z$ orbitals leads to the formation of six pi molecular orbitals, of which three are bonding and three antibonding. A benzene molecule in the ground state therefore has six electrons in the three pi bonding molecular orbitals, two electrons with paired spins in each orbital (Figure 11.15).

In the ethylene molecule, the overlap of the two $2p_z$ orbitals gives rise to a bonding and an antibonding molecular orbital, which are localized over the two C atoms. The interaction of the $2p_z$ orbitals in benzene, however, leads to the formation of ***delocalized molecular orbitals,*** which *are not confined between two adjacent bonding atoms, but actually extend over three or more atoms.* Therefore, electrons residing in any of these orbitals are free to move around the benzene ring. For this reason, the structure of benzene is sometimes represented as

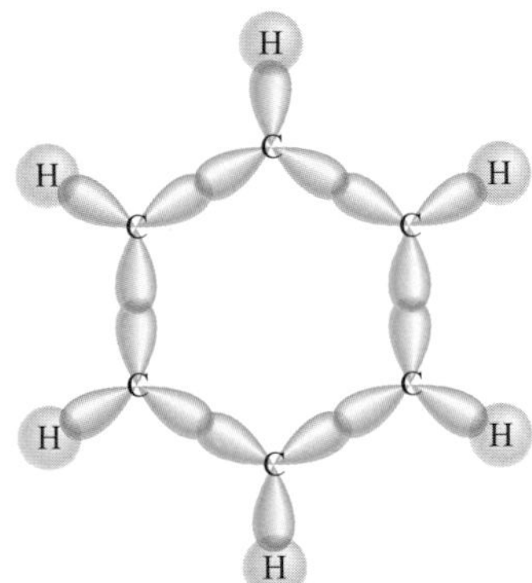

Figure 11.14
The sigma bond framework of the benzene molecule. Each C atom is sp^2-hybridized and forms sigma bonds with two adjacent C atoms and another sigma bond with an H atom.

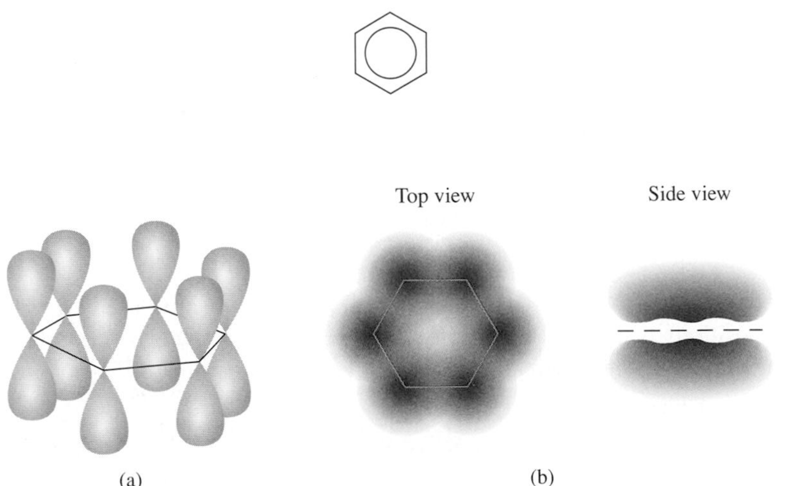

Figure 11.15
(a) The six $2p_z$ orbitals on the carbon atoms in benzene. (b) The delocalized molecular orbital formed by the overlap of the $2p_z$ orbitals. The delocalized molecular orbital possesses pi symmetry and lies above and below the plane of the benzene ring. Actually, these $2p_z$ orbitals can combine in six different ways to yield three bonding molecular orbitals and three antibonding molecular orbitals. The one shown here is the most stable.

in which the circle indicates that the pi bonds between carbon atoms are not confined to individual pairs of atoms; rather, the pi electron densities are evenly distributed throughout the benzene molecule. As we will see shortly, electron delocalization imparts extra stability to aromatic hydrocarbons.

We can now state that each carbon-to-carbon linkage in benzene contains a sigma bond and a "partial" pi bond. The bond order between any two adjacent carbon atoms is therefore between 1 and 2. Thus, molecular orbital theory offers an alternative to the resonance approach, which is based on valence bond theory.

Nomenclature of Aromatic Compounds

The naming of monosubstituted benzenes, that is, benzenes in which one H atom has been replaced by another atom or a group of atoms, is quite straightforward, as shown next:

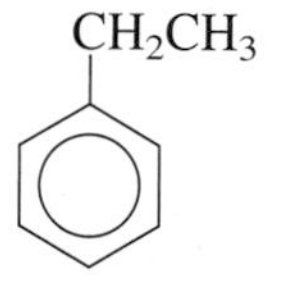

CH_2CH_3 Cl NH_2 NO_2

ethylbenzene 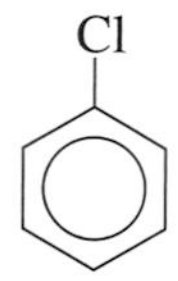 chlorobenzene 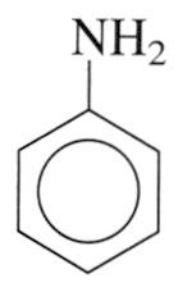 aminobenzene (aniline) 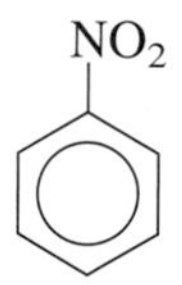 nitrobenzene

If more than one substituent is present, we must indicate the location of the second group relative to the first. The systematic way to accomplish this is to number the carbon atoms as follows:

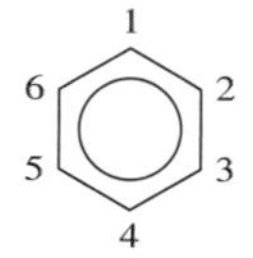

1 2 3 4 5 6

Three different dibromobenzenes are possible:

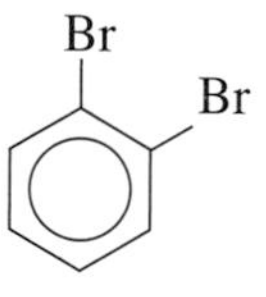

Br Br Br Br Br Br

1,2-dibromobenzene (*o*-dibromobenzene) 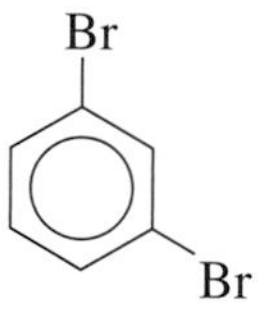 1,3-dibromobenzene (*m*-dibromobenzene) 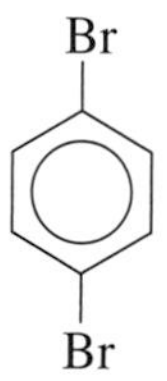 1,4-dibromobenzene (*p*-dibromobenzene)

The prefixes *o*- (*ortho*-), *m*- (*meta*-), and *p*- (*para*-) are also used to denote the relative positions of the two substituted groups, as just shown for the dibromobenzenes. Compounds in which the two substituted groups are different are named accordingly. Thus,

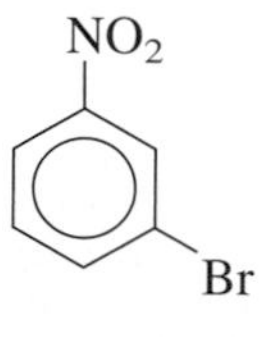

NO_2 Br

is named 3-bromonitrobenzene, or *m*-bromonitrobenzene.

Finally we note that the group containing benzene minus a hydrogen atom (C_6H_5) is called the *phenyl* group. Thus, the following molecule is called 2-phenylpropane:

$CH_3—CH—CH_3$

This compound is also called isopropylbenzene (see Table 11.2).

Properties and Reactions of Aromatic Compounds

Benzene is a colorless, flammable liquid obtained chiefly from petroleum and coal tar. Perhaps the most remarkable chemical property of benzene is its relative inertness. Although it has the same empirical formula as acetylene (CH) and a high degree of unsaturation, it is much less reactive than either ethylene or acetylene. The stability of benzene is the result of electron delocalization. In fact, benzene can be hydrogenated, but only with difficulty. The following reaction is carried out at significantly higher temperatures and pressures than are similar reactions for the alkenes:

$$C_6H_6 + 3H_2 \xrightarrow[\text{catalyst}]{\text{Pt}} C_6H_{12}$$

cyclohexane

A catalyst is a substance that can speed up the rate of a reaction without itself being used up. More on this topic in Chapter 14.

We saw earlier that alkenes react readily with halogens and hydrogen halides to form addition products, because the pi bond in C=C can be broken more easily. The most common reaction of halogens with benzene is substitution. For example,

$$C_6H_6 + Br_2 \xrightarrow[\text{catalyst}]{FeBr_3} C_6H_5Br + HBr$$

bromobenzene

Note that if the reaction were addition, electron delocalization would be destroyed in the product

$$C_6H_6Br_2$$

and the molecule would not have the aromatic characteristic of chemical unreactivity.

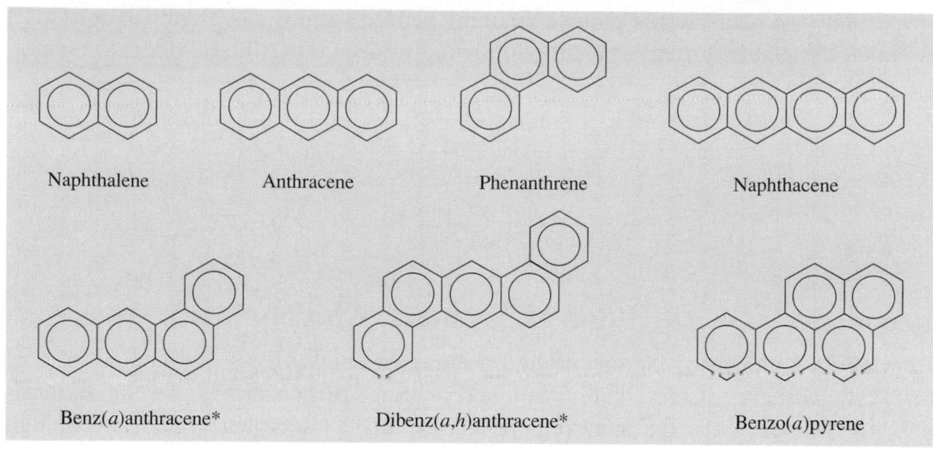

Figure 11.16
*Some polycyclic aromatic hydrocarbons. Compounds denoted by * are potent carcinogens. An enormous number of such compounds exist in nature.*

Alkyl groups can be introduced into the ring system by allowing benzene to react with an alkyl halide using $AlCl_3$ as the catalyst:

$$C_6H_6 + \underset{\text{ethyl chloride}}{CH_3CH_2Cl} \xrightarrow[\text{catalyst}]{AlCl_3} \underset{\text{ethylbenzene}}{C_6H_5CH_2CH_3} + HCl$$

An enormously large number of compounds can be generated from substances in which benzene rings are fused together. Some of these *polycyclic* aromatic hydrocarbons are shown in Figure 11.16. The best known of these compounds is naphthalene, which is used in mothballs. These and many other similar compounds are present in coal tar. Some of the compounds with several rings are powerful carcinogens—they can cause cancer in humans and other animals.

Review of Concepts

Benzene has sp^2-hybridized carbon atoms and multiple bonds. However, unlike ethylene, geometric isomerism is not possible in benzene. Explain.

11.4 Chemistry of the Functional Groups

We now examine some organic functional groups, groups that are responsible for most of the reactions of the parent compounds. In particular, we focus on oxygen-containing and nitrogen-containing compounds.

Alcohols

All ***alcohols*** contain *the hydroxyl functional group, —OH.* Some common alcohols are shown in Figure 11.17. Ethyl alcohol, or ethanol, is by far the best known. It is produced biologically by the fermentation of sugar or starch. In the absence of oxygen the enzymes present in bacterial cultures or yeast catalyze the reaction

$$C_6H_{12}O_6(aq) \xrightarrow{\text{enzymes}} \underset{\text{ethanol}}{2CH_3CH_2OH(aq)} + 2CO_2(g)$$

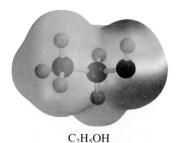

C_2H_5OH

```
     H                 H  H                 H  H  H
     |                 |  |                 |  |  |
  H—C—OH            H—C—C—OH             H—C—C—C—H
     |                 |  |                 |  |  |
     H                 H  H                 H  OH H
```

Methanol (methyl alcohol) — Ethanol (ethyl alcohol) — 2-Propanol (isopropyl alcohol)

```
                     H  H
                     |  |
                  H—C—C—H
                     |  |
                     OH OH
```

Phenol — Ethylene glycol

Figure 11.17
Common alcohols. Note that all the compounds contain the OH group. The properties of phenol are quite different from those of the aliphatic alcohols.

This process gives off energy, which microorganisms, in turn, use for growth and other functions.

Commercially, ethanol is prepared by an addition reaction in which water is combined with ethylene at about 280°C and 300 atm:

$$CH_2{=}CH_2(g) + H_2O(g) \xrightarrow{H_2SO_4} CH_3CH_2OH(g)$$

Ethanol has countless applications as a solvent for organic chemicals and as a starting compound for the manufacture of dyes, synthetic drugs, cosmetics, and explosives. It is also a constituent of alcoholic beverages. Ethanol is the only nontoxic (more properly, the least toxic) of the straight-chain alcohols; our bodies produce an enzyme, called *alcohol dehydrogenase,* which helps metabolize ethanol by oxidizing it to acetaldehyde:

$$CH_3CH_2OH \xrightarrow[\text{dehydrogenase}]{\text{alcohol}} \underset{\text{acetaldehyde}}{CH_3CHO} + H_2$$

This equation is a simplified version of what actually takes place; the H atoms are taken up by other molecules, so that no H_2 gas is evolved.

Ethanol can also be oxidized by inorganic oxidizing agents, such as acidified potassium dichromate, to acetic acid:

$$3CH_3CH_2OH + \underset{\text{orange-yellow}}{2K_2Cr_2O_7} + 8H_2SO_4 \longrightarrow 3CH_3COOH + \underset{\text{green}}{2Cr_2(SO_4)_3} + 2K_2SO_4 + 11H_2O$$

Left: A $K_2Cr_2O_7$ solution. Right: A $Cr_2(SO_4)_3$ solution.

This reaction has been employed by law enforcement agencies to test drivers suspected of being drunk. A sample of the driver's breath is drawn into a device called a breath analyzer, where it is reacted with an acidic potassium dichromate solution. From the color change (orange-yellow to green) it is possible to determine the alcohol content in the driver's blood.

Ethanol is called an aliphatic alcohol because it is derived from an alkane (ethane). The simplest aliphatic alcohol is methanol, CH_3OH. Called *wood alcohol,* it was prepared at one time by the dry distillation of wood. It is now synthesized industrially by the reaction of carbon monoxide and molecular hydrogen at high temperatures and pressures:

$$CO(g) + 2H_2(g) \xrightarrow[\text{catalyst}]{Fe_2O_3} \underset{\text{methanol}}{CH_3OH(l)}$$

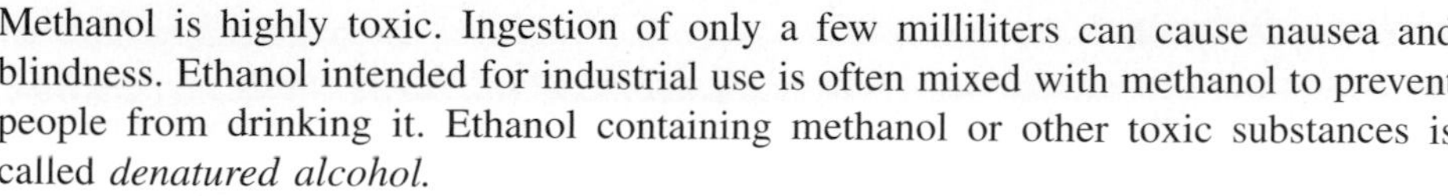

Methanol is highly toxic. Ingestion of only a few milliliters can cause nausea and blindness. Ethanol intended for industrial use is often mixed with methanol to prevent people from drinking it. Ethanol containing methanol or other toxic substances is called *denatured alcohol.*

Alcohols react more slowly with sodium metal than water does.

The alcohols are very weakly acidic; they do not react with strong bases, such as NaOH. The alkali metals react with alcohols to produce hydrogen:

$$2CH_3OH + 2Na \longrightarrow \underset{\text{sodium methoxide}}{2CH_3ONa} + H_2$$

However, the reaction is much less violent than that between Na and water:

$$2H_2O + 2Na \longrightarrow 2NaOH + H_2$$

Two other familiar aliphatic alcohols are 2-propanol (or isopropyl alcohol), commonly known as rubbing alcohol, and ethylene glycol, which is used as an antifreeze. Most alcohols—especially those with low molar masses—are highly flammable.

Ethers

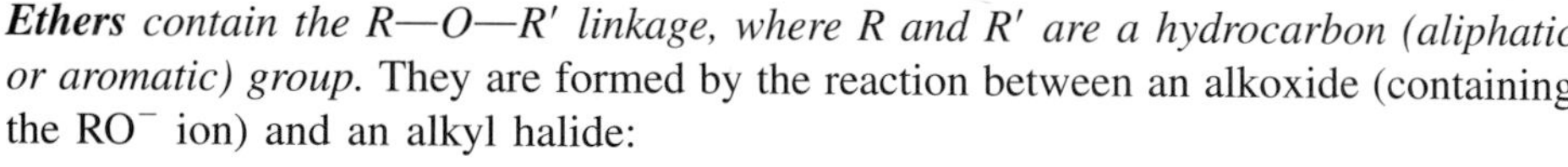

Ethers *contain the R—O—R′ linkage, where R and R′ are a hydrocarbon (aliphatic or aromatic) group.* They are formed by the reaction between an alkoxide (containing the RO^- ion) and an alkyl halide:

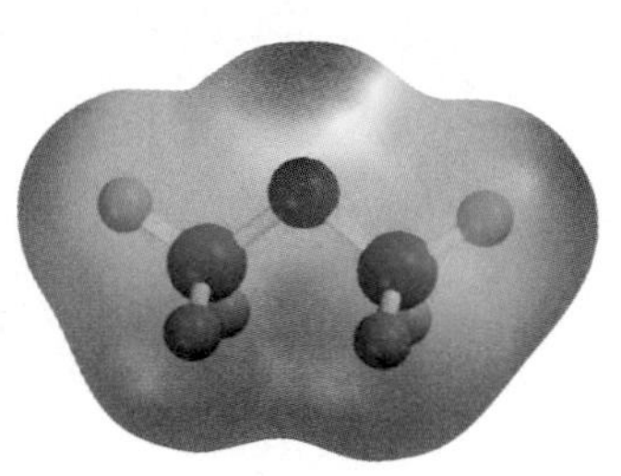

CH_3OCH_3

$$\underset{\text{sodium methoxide}}{NaOCH_3} + \underset{\text{methyl bromide}}{CH_3Br} \longrightarrow \underset{\text{dimethyl ether}}{CH_3OCH_3} + NaBr$$

Diethyl ether is prepared on an industrial scale by heating ethanol with sulfuric acid at 140°C

$$C_2H_5OH + C_2H_5OH \longrightarrow C_2H_5OC_2H_5 + H_2O$$

This reaction is an example of a ***condensation reaction,*** which is characterized by *the joining of two molecules and the elimination of a small molecule, usually water.*

Like alcohols, ethers are extremely flammable. When left standing in air, they have a tendency to slowly form explosive peroxides:

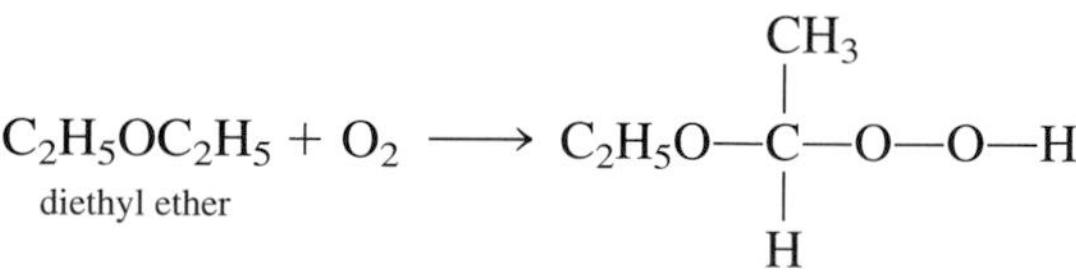

$$\underset{\text{diethyl ether}}{C_2H_5OC_2H_5} + O_2 \longrightarrow C_2H_5O{-}\overset{\displaystyle CH_3}{\underset{\displaystyle H}{C}}{-}O{-}O{-}H$$

1-ethyoxyethyl hydroperoxide

Peroxides contain the —O—O— linkage; the simplest peroxide is hydrogen peroxide, H_2O_2. Diethyl ether, commonly known as "ether," was used as an anesthetic for many years. It produces unconsciousness by depressing the activity of the central nervous system. The major disadvantages of diethyl ether are its irritating effects on the respiratory system and the occurrence of postanesthetic nausea and vomiting. "Neothyl," or methyl propyl ether, $CH_3OCH_2CH_2CH_3$, is currently favored as an anesthetic because it is relatively free of side effects.

Aldehydes and Ketones

Under mild oxidation conditions, it is possible to convert alcohols to aldehydes and ketones:

$$CH_3OH + \tfrac{1}{2}O_2 \longrightarrow \underset{\text{formaldehyde}}{H_2C{=}O} + H_2O$$

$$C_2H_5OH + \tfrac{1}{2}O_2 \longrightarrow \underset{\text{acetaldehyde}}{(H_3C)(H)C{=}O} + H_2O$$

$$CH_3{-}\overset{H}{\underset{OH}{C}}{-}CH_3 + \tfrac{1}{2}O_2 \longrightarrow \underset{\text{acetone}}{(H_3C)(H_3C)C{=}O} + H_2O$$

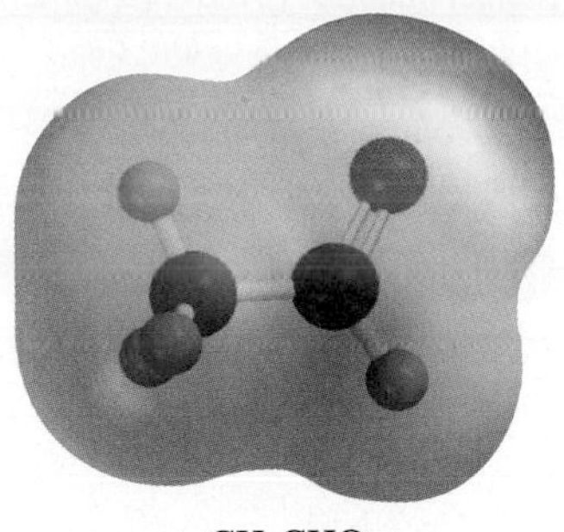
CH_3CHO

The functional group in these compounds is the *carbonyl group*, >C=O. In an ***aldehyde*** *at least one hydrogen atom is bonded to the carbon in the carbonyl group.* In a ***ketone,*** *the carbon atom in the carbonyl group is bonded to two hydrocarbon groups.*

The simplest aldehyde, formaldehyde ($H_2C{=}O$) has a tendency to *polymerize;* that is, the individual molecules join together to form a compound of high molar mass. This action gives off much heat and is often explosive, so formaldehyde is usually prepared and stored in aqueous solution (to reduce the concentration). This rather disagreeable-smelling liquid is used as a starting material in the polymer industry and in the laboratory as a preservative for animal specimens. Interestingly, the higher molar mass aldehydes, such as cinnamic aldehyde

$$C_6H_5{-}CH{=}CH{-}C(H){=}O$$

Cinnamic aldehyde gives cinnamon its characteristic aroma.

have a pleasant odor and are used in the manufacture of perfumes.

Ketones generally are less reactive than aldehydes. The simplest ketone is acetone, a pleasant-smelling liquid that is used mainly as a solvent for organic compounds and nail polish remover.

Carboxylic Acids

Under appropriate conditions both alcohols and aldehydes can be oxidized to ***carboxylic acids,*** *acids that contain the carboxyl group, —COOH:*

$$CH_3CH_2OH + O_2 \longrightarrow CH_3COOH + H_2O$$
$$CH_3CHO + \tfrac{1}{2}O_2 \longrightarrow CH_3COOH$$

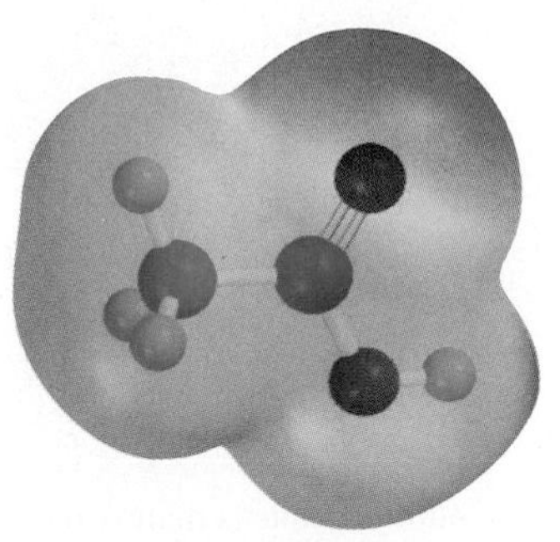
CH_3COOH

These reactions occur so readily, in fact, that wine must be protected from atmospheric oxygen while in storage. Otherwise, it would soon turn to vinegar due to the formation of acetic acid. Figure 11.18 shows the structure of some of the common carboxylic acids.

Carboxylic acids are widely distributed in nature; they are found in both the plant and animal kingdoms. All protein molecules are made of amino acids, a special kind of carboxylic acid containing an amino group ($—NH_2$) and a carboxyl group (—COOH).

Figure 11.18
Some common carboxylic acids. Note that they all contain the COOH group. (Glycine is one of the amino acids found in proteins.)

Formic acid   Acetic acid   Butyric acid   Benzoic acid

Glycine   Oxalic acid   Citric acid

Unlike the inorganic acids HCl, HNO_3, and H_2SO_4, carboxylic acids are usually weak. They react with alcohols to form pleasant-smelling esters:

This is a condensation reaction.

$$\underset{\text{acetic acid}}{CH_3COOH} + \underset{\text{ethanol}}{HOCH_2CH_3} \longrightarrow \underset{\text{ethyl acetate}}{CH_3{-}\overset{\displaystyle O}{\overset{\|}{C}}{-}O{-}CH_2CH_3} + H_2O$$

Other common reactions of carboxylic acids are neutralization

$$CH_3COOH + NaOH \longrightarrow CH_3COONa + H_2O$$

and formation of acid halides, such as acetyl chloride

$$CH_3COOH + PCl_5 \longrightarrow \underset{\text{acetyl chloride}}{CH_3COCl} + HCl + \underset{\text{phosphoryl chloride}}{POCl_3}$$

Acid halides are reactive compounds used as intermediates in the preparation of many other organic compounds.

Esters

***Esters** have the general formula R′COOR, in which R′ can be H, an alkyl, or an aromatic hydrocarbon group and R is an alkyl or an aromatic hydrocarbon group.* Esters are used in the manufacture of perfumes and as flavoring agents in the confectionery and soft-drink industries. Many fruits owe their characteristic smell and flavor to the presence of esters. For example, bananas contain isopentyl acetate [$CH_3COOCH_2CH_2CH(CH_3)_2$], oranges contain octyl acetate ($CH_3COOC_8H_{17}$), and apples contain methyl butyrate ($CH_3CH_2CH_2COOCH_3$).

The odor of fruits is mainly due to the ester compounds in them.

The functional group in esters is —COOR. In the presence of an acid catalyst, such as HCl, esters undergo a reaction with water (a *hydrolysis* reaction) to regenerate a carboxylic acid and an alcohol. For example, in acid solution, ethyl acetate is converted to acetic acid:

$$\underset{\text{ethyl acetate}}{CH_3COOC_2H_5} + H_2O \rightleftharpoons \underset{\text{acetic acid}}{CH_3COOH} + \underset{\text{ethanol}}{C_2H_5OH}$$

However, this reaction does not go to completion because the reverse reaction, that is, the formation of an ester from an alcohol and an acid, also occurs to an appreciable

(a)

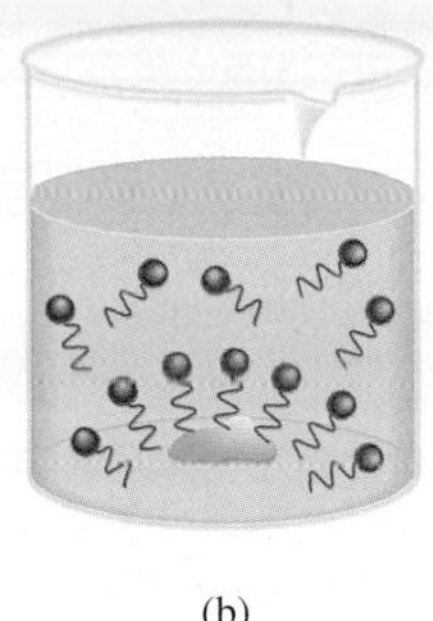
(b)

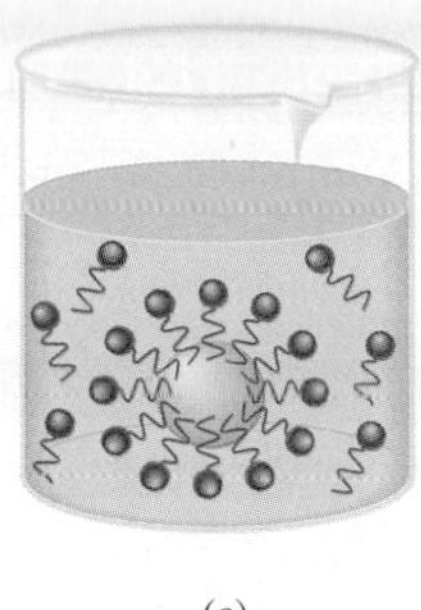
(c)

Figure 11.19
The cleansing action of soap. The soap molecule is represented by a polar head and zigzag hydrocarbon tail. An oily spot (a) can be removed by soap (b) because the nonpolar tail dissolves in the oil, and (c) the entire system becomes soluble in water because the exterior portion is now ionic.

extent. On the other hand, when the hydrolysis reaction is run in aqueous NaOH solution, ethyl acetate is converted to sodium acetate, which does not react with ethanol, so this reaction goes to completion from left to right:

$$\underset{\text{ethyl acetate}}{CH_3COOC_2H_5} + NaOH \longrightarrow \underset{\text{sodium acetate}}{CH_3COO^-Na^+} + \underset{\text{ethanol}}{C_2H_5OH}$$

The term ***saponification*** (meaning *soapmaking*) was originally used to describe the reaction between an ester and sodium hydroxide to yield soap (sodium stearate):

$$\underset{\text{ethyl stearate}}{C_{17}H_{35}COOC_2H_5} + NaOH \longrightarrow \underset{\text{sodium stearate}}{C_{17}H_{35}COO^-Na^+} + C_2H_5OH$$

Saponification is now *a general term for alkaline hydrolysis of any type of ester.* Soaps are characterized by a long nonpolar hydrocarbon chain and a polar head (the —COO^- group). The hydrocarbon chain is readily soluble in oily substances, while the ionic carboxylate group (—COO^-) remains outside the oily nonpolar surface. Figure 11.19 shows the action of soap.

Amines

Amines are *organic bases that have the general formula R_3N, in which one of the R groups must be an alkyl group or an aromatic hydrocarbon group.* Like ammonia, amines are weak Brønsted bases that react with water as follows:

$$RNH_2 + H_2O \longrightarrow RNH_3^+ + OH^-$$

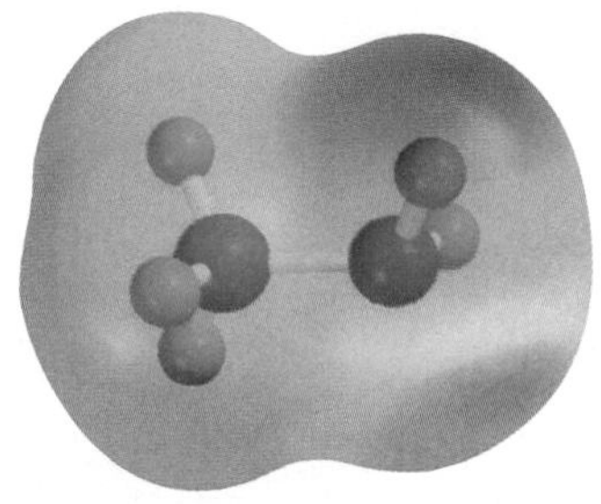
CH_3NH_2

Like all bases, the amines form salts when allowed to react with acids:

$$\underset{\text{methylamine}}{CH_3NH_2} + HCl \longrightarrow \underset{\text{methylammonium chloride}}{CH_3NH_3^+Cl^-}$$

These salts are usually colorless, odorless solids that are soluble in water. Many of the aromatic amines are carcinogenic.

Summary of Functional Groups

Table 11.4 summarizes the common functional groups, including the C=C and C≡C groups. Organic compounds commonly contain more than one functional group. Generally, the reactivity of a compound is determined by the number and types of functional groups in its makeup.

Table 11.4 Important Functional Groups and Their Reactions

Functional Group	Name	Typical Reactions
$>C{=}C<$	Carbon-carbon double bond	Addition reactions with halogens, hydrogen halides, and water; hydrogenation to yield alkanes
$-C{\equiv}C-$	Carbon-carbon triple bond	Addition reactions with halogens, hydrogen halides; hydrogenation to yield alkenes and alkanes
$-\ddot{\underset{..}{X}}:$ (X = F, Cl, Br, I)	Halogen	Exchange reactions: $CH_3CH_2Br + KI \longrightarrow CH_3CH_2I + KBr$
$-\ddot{\underset{..}{O}}-H$	Hydroxyl	Esterification (formation of an ester) with carboxylic acids; oxidation to aldehydes, ketones, and carboxylic acids
$>C{=}\ddot{\underset{..}{O}}$	Carbonyl	Reduction to yield alcohols; oxidation of aldehydes to yield carboxylic acids
$-C(=\ddot{O}:)-\ddot{\underset{..}{O}}-H$	Carboxyl	Esterification with alcohols; reaction with phosphorus pentachloride to yield acid chlorides
$-C(=\ddot{O}:)-\ddot{\underset{..}{O}}-R$ (R = hydrocarbon)	Ester	Hydrolysis to yield acids and alcohols
$-\ddot{N}(R)R$ (R = H or hydrocarbon)	Amine	Formation of ammonium salts with acids

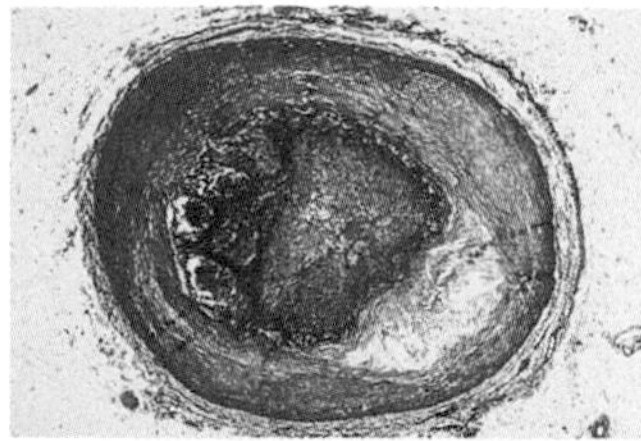
An artery becoming blocked by cholesterol.

EXAMPLE 11.4

Cholesterol is a major component of gallstones, and it is believed that the cholesterol level in the blood is a contributing factor in certain types of heart disease. From the following structure of the compound, predict its reaction with (a) Br_2, (b) H_2 (in the presence of a Pt catalyst), (c) CH_3COOH.

[Structure of cholesterol: steroid ring system with HO, two CH_3 groups, C_8H_{17} side chain, and one C=C double bond]

Strategy To predict the type of reactions a molecule may undergo, we must first identify the functional groups present (see Table 11.4).

Solution There are two functional groups in cholesterol: the hydroxyl group and the carbon-carbon double bond.

(a) The reaction with bromine results in the addition of bromine to the double-bonded carbons, which become single-bonded.

(Continued)

Figure 11.20 *The products formed by the reaction of cholesterol with (a) molecular bromine, (b) molecular hydrogen, and (c) acetic acid.*

(b) This is a hydrogenation reaction. Again, the carbon-carbon double bond is converted to a carbon-carbon single bond.

(c) The acetic acid (CH_3COOH) reacts with the hydroxyl group to form an ester and water. Figure 11.20 shows the products of these reactions.

Practice Exercise Predict the products of the following reaction:

$$CH_3OH + CH_3CH_2COOH \longrightarrow ?$$

Similar problem: 11.41.

Review of Concepts

Identify all of the functional groups in vanillin, the primary component in vanilla bean extract.

11.5 Chirality—The Handedness of Molecules

Many organic compounds can exist as mirror-image twins, in which one partner may cure disease, quell a headache, or smell good, whereas its mirror-reversed counterpart may be poisonous, smell repugnant, or simply be inert. Compounds that come as mirror image pairs are sometimes compared with the left and right hands and are referred to as ***chiral,*** or *handed,* molecules. Although every molecule can have a mirror image, the difference between chiral and *achiral* (meaning nonchiral) molecules is that only the twins of the former are nonsuperimposable.

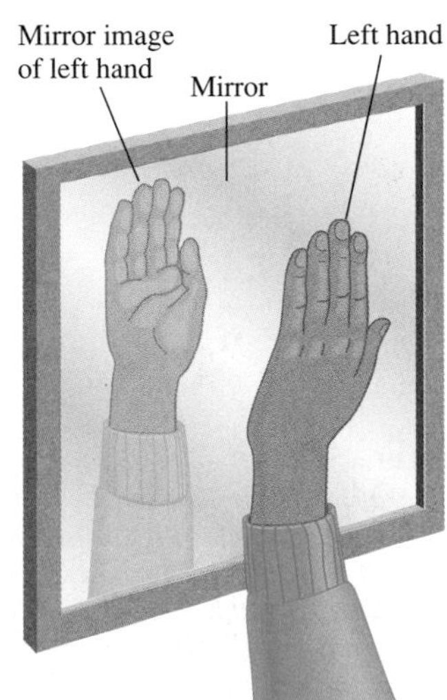

A left hand and its mirror image which looks the same as the right hand.

Consider the substituted methanes CH_2ClBr and CHFClBr. Figure 11.21 shows perspective drawings of these two molecules and their mirror images. The two mirror images of Figure 11.21(a) are superimposable, but those of Figure 11.21(b) are not, no matter how we rotate the molecules. Thus, the CHFClBr molecule is chiral. Careful observation shows that most simple chiral molecules contain at least one *asymmetric* carbon atom—that is, a carbon atom bonded to four different atoms or groups of atoms.

Animations: Chirality

The nonsuperimposable mirror images of a chiral compound are called ***enantiomers.*** Like geometric isomers, enantiomers come in pairs. However, the enantiomers of a compound have identical physical and chemical properties, such as melting point, boiling point, and chemical reactivity toward molecules that are not chiral themselves. Each enantiomer of a chiral molecule is said to be optically active because of its ability to rotate the plane of polarization of polarized light.

An older term for enantiomers is optical isomers.

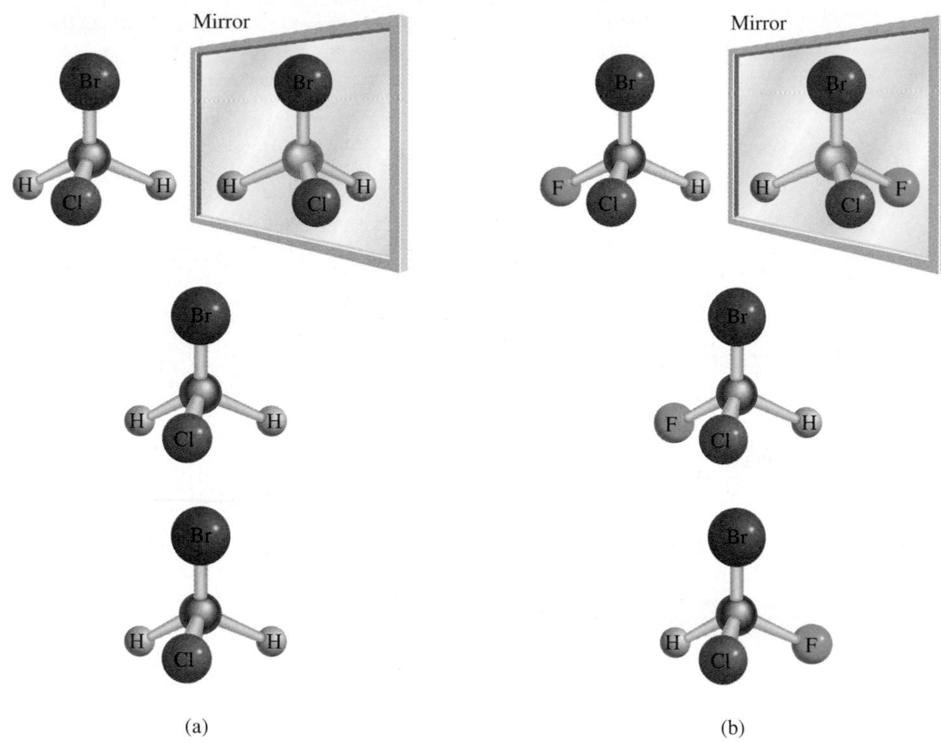

Figure 11.21
(a) The CH_2ClBr molecule and its mirror image. Because the molecule and its mirror image are superimposable, the molecule is said to be achiral. (b) The CHFClBr molecule and its mirror image. Because the molecule and its mirror image are not superimposable, no matter how we rotate one with respect to the other, the molecule is said to be chiral.

Unlike ordinary light, which vibrates in all directions, plane-polarized light vibrates only in a single plane. To *study the interaction between plane-polarized light and chiral molecules* we use a ***polarimeter,*** shown schematically in Figure 11.22. A beam of unpolarized light first passes through a polarizer, and then through a sample tube containing a solution of a chiral compound. As the polarized light passes through the sample tube, its plane of polarization is rotated either to the right or to

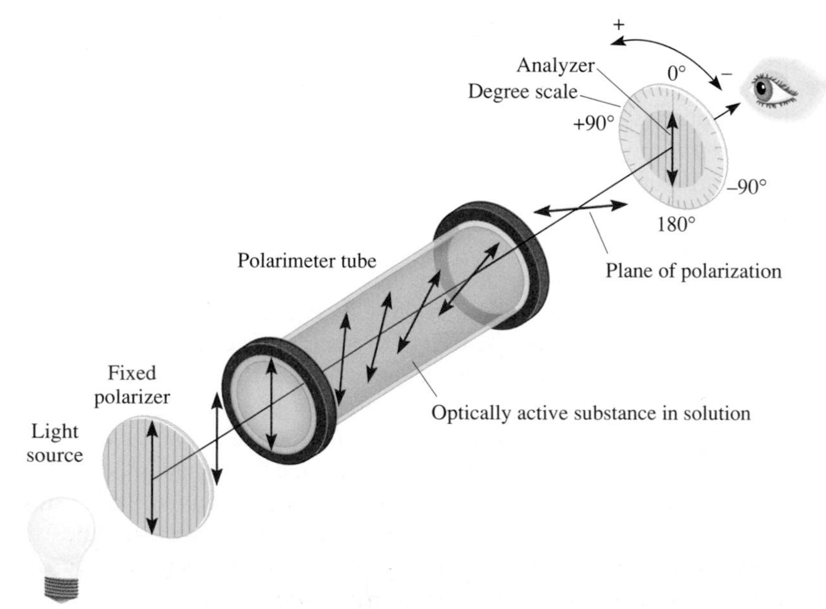

Figure 11.22
Operation of a polarimeter. Initially, the tube is filled with an achiral compound. The analyzer is rotated so that its plane of polarization is perpendicular to that of the polarizer. Under this condition, no light reaches the observer. Next, a chiral compound is placed in the tube as shown. The plane of polarization of the polarized light is rotated as it travels through the tube so that some light reaches the observer. Rotating the analyzer (either to the left or to the right) until no light reaches the observer again allows the angle of optical rotation to be measured.

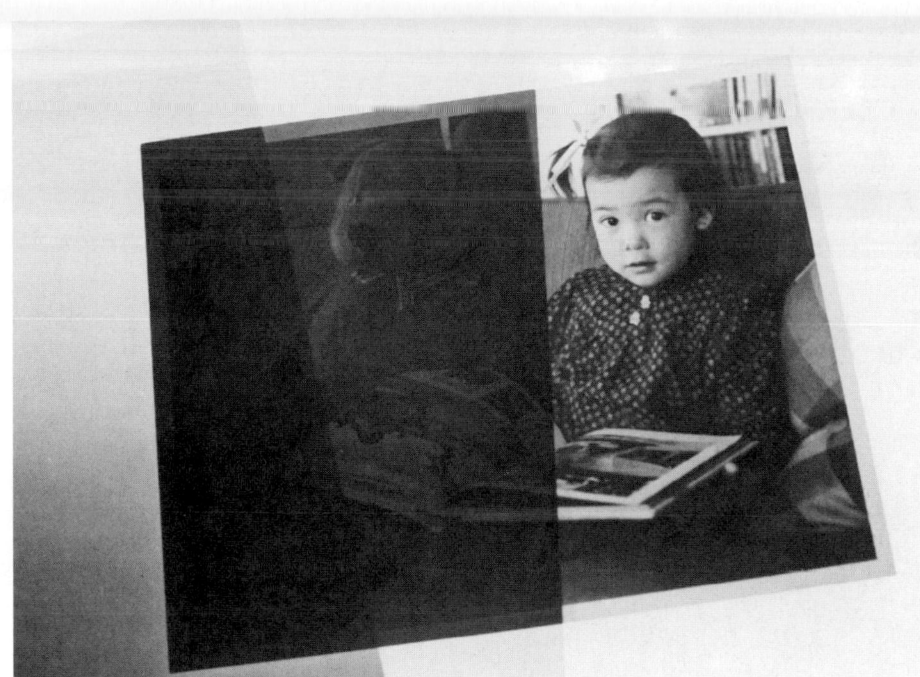

Figure 11.23
With one Polaroid sheet over a picture, light passes through. With a second sheet of Polaroid placed over the first so that the axes of polarization of the sheets are perpendicular, little or no light passes through. If the axes of polarization of the two sheets were parallel, light would pass through.

the left. This rotation can be measured directly by turning the analyzer in the appropriate direction until minimal light transmission is achieved (Figure 11.23). If the plane of polarization is rotated to the right, the isomer is said to be dextrorotatory (+); it is levorotatory (−) if the rotation is to the left. Enantiomers of a chiral substance always rotate the light by the same amount, but in opposite directions. Thus, in *an equimolar mixture of two enantiomers,* called a ***racemic mixture,*** the net rotation is zero.

Chirality plays an important role in biological systems. Protein molecules have many asymmetric carbon atoms and their functions are often influenced by their chirality. Because the enantiomers of a chiral compound usually behave very differently from each other in the body, chiral twins are coming under increasing scrutiny among pharmaceutical manufacturers. More than half of the most prescribed drugs in 2008 are chiral. In most of these cases only one enantiomer of the drug works as a medicine, whereas the other form is useless or less effective or may even cause serious side effects. The best-known case in which the use of a racemic mixture of a drug had tragic consequences occurred in Europe in the late 1950s. The drug thalidomide was prescribed for pregnant women there as an antidote to morning sickness. But by 1962, the drug had to be withdrawn from the market after thousands of deformed children had been born to mothers who had taken it. Only later did researchers discover that the sedative properties of thalidomide belong to (+)-thalidomide and that (−)-thalidomide is a potent mutagen. (A *mutagen* is a substance that causes gene mutation, usually leading to deformed offspring.)

Figure 11.24 shows the two enantiomeric forms of another drug, ibuprofen. This popular pain reliever is sold as a racemic mixture, but only the one on the left is potent. The other form is ineffective but also harmless. Organic chemists today are actively researching ways to synthesize enantiomerically pure drugs, or "chiral drugs." Chiral drugs contain only one enantiomeric form both for efficiency and for protection against possible side effects from its mirror-image twin.

As of 2009, one of the best-selling chiral drugs, Lipitor, which controls cholesterol levels, is sold as a pure enantiomer.

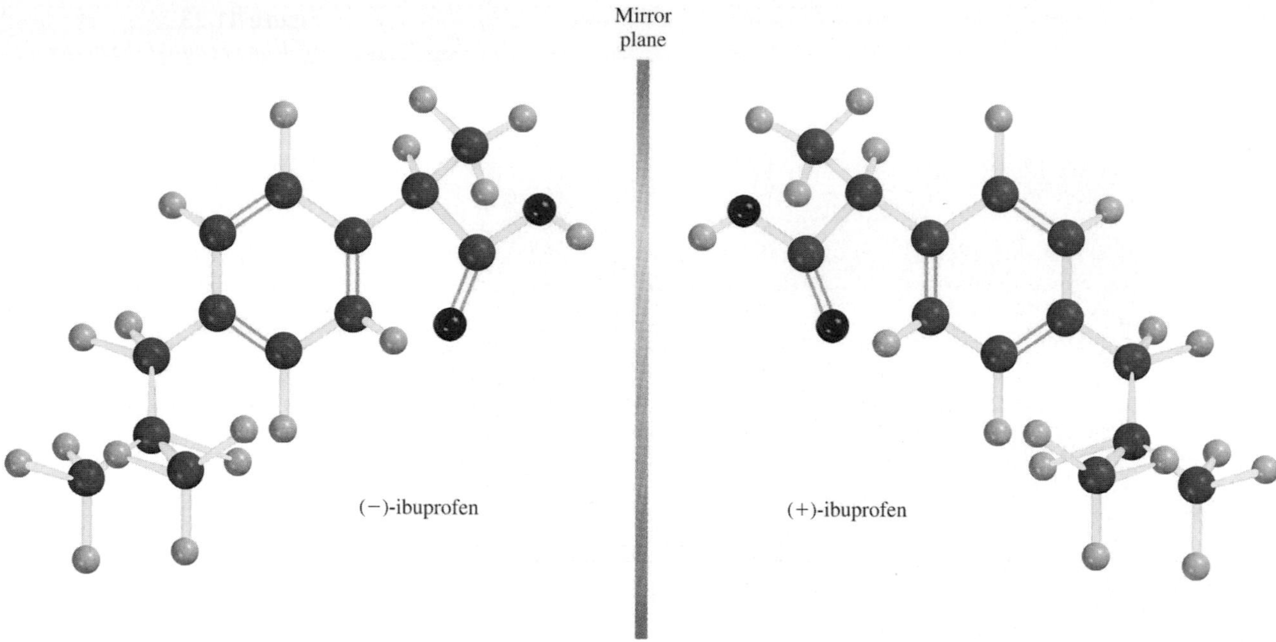

Figure 11.24
The enantiomers of ibuprofen are mirror images of each other. There is only one asymmetric C atom in the molecule. Can you spot it?

EXAMPLE 11.5

Is the following molecule chiral?

$$\mathrm{H{-}\overset{\displaystyle Cl}{\underset{\displaystyle CH_3}{C}}{-}CH_2{-}CH_3}$$

Strategy Recall the condition for chirality. Is the central C atom asymmetric; that is, does it have four different atoms or different groups attached to it?

Solution We note that the central carbon atom is bonded to a hydrogen atom, a chlorine atom, a —CH_3 group, and a —CH_2—CH_3 group. Therefore, the central carbon atom is asymmetric and the molecule is chiral.

Similar problems: 11.45, 11.46.

Practice Exercise Is the following molecule chiral?

$$\mathrm{I{-}\overset{\displaystyle Br}{\underset{\displaystyle Br}{C}}{-}CH_2{-}CH_3}$$

Review of Concepts

How many chiral carbon centers are in ephedrine, a drug used as a stimulant and decongestant?

$$\mathrm{C_6H_5{-}\overset{\displaystyle H}{\underset{\displaystyle OH}{C}}{-}\overset{\displaystyle H}{\underset{\displaystyle CH_3}{C}}{-}NHCH_3}$$

Summary of Facts and Concepts

1. Because carbon atoms can link up with other carbon atoms in straight and branched chains, carbon can form more compounds than most other elements.
2. Alkanes and cycloalkanes are saturated hydrocarbons. Methane, CH_4, is the simplest of the alkanes, a family of hydrocarbons with the general formula C_nH_{2n+2}. The cycloalkanes are a subfamily of alkanes whose carbon atoms are joined in a ring. Ethylene, $CH_2{=}CH_2$, is the simplest of alkenes, a class of hydrocarbons containing carbon-carbon double bonds and having the general formula C_nH_{2n}. Unsymmetrical alkenes can exist as *cis* and *trans* isomers. Acetylene, $CH{\equiv}CH$, is the simplest of the alkynes, which are compounds that have the general formula C_nH_{2n-2} and contain carbon-carbon triple bonds. Compounds that contain one or more benzene rings are called aromatic hydrocarbons. The stability of the benzene molecule is the result of electron delocalization.
3. Functional groups determine the chemical reactivity of molecules in which they are found. Classes of compounds characterized by their functional groups include alcohols, ethers, aldehydes and ketones, carboxylic acids and esters, and amines.
4. Chirality refers to molecules that have nonsuperimposable mirror images. Most chiral molecules contain one or more asymmetric carbon atoms. Chiral molecules are widespread in biological systems and are important in drug design.

Key Words

Addition reaction, p. 375
Alcohol, p. 382
Aldehyde, p. 385
Aliphatic hydrocarbon, p. 364
Alkane, p. 364
Alkene, p. 372
Alkyne, p. 377
Amine, p. 387
Aromatic hydrocarbon, p. 364
Carboxylic acid, p. 385
Chiral, p. 389
Condensation reaction, 384
Conformations, p. 367
Cycloalkane, p. 372
Delocalized molecular orbitals, p. 379
Enantiomer, p. 389
Ester, p. 386
Ether, p. 384
Functional group, p. 364
Geometric isomers, p. 373
Hydrocarbon, p. 364
Hydrogenation, p. 375
Ketone, p. 385
Organic chemistry, p. 364
Polarimeter, p. 390
Racemic mixture, p. 391
Radical, p. 371
Saponification, p. 387
Saturated hydrocarbon, p. 364
Structural isomer, p. 365
Unsaturated hydrocarbon, p. 375

Questions and Problems

Aliphatic Hydrocarbons

Review Questions

11.1 Explain why carbon is able to form so many more compounds than most other elements.

11.2 What is the difference between aliphatic and aromatic hydrocarbons?

11.3 What do "saturated" and "unsaturated" mean when applied to hydrocarbons? Give examples of a saturated hydrocarbon and an unsaturated hydrocarbon.

11.4 What are structural isomers?

11.5 Use ethane as an example to explain the meaning of conformations. What are Newman projections? How do the conformations of a molecule differ from structural isomers?

11.6 Draw skeletal structures of the boat and chair forms of cyclohexane.

11.7 Alkenes exhibit geometric isomerism because rotation about the C=C bond is restricted. Explain.

11.8 Why is it that alkanes and alkynes, unlike alkenes, have no geometric isomers?

11.9 What is Markovnikov's rule?

11.10 Describe reactions that are characteristic of alkanes, alkenes, and alkynes.

Problems

11.11 Draw all possible structural isomers for this alkane: C_7H_{16}.

11.12 How many distinct chloropentanes, $C_5H_{11}Cl$, could be produced in the direct chlorination of *n*-pentane, $CH_3(CH_2)_3CH_3$? Draw the structure of each molecule.

11.13 Draw all possible isomers for the molecule C_4H_8.

11.14 Draw all possible isomers for the molecule C_3H_5Br.

11.15 The structural isomers of pentane, C_5H_{12}, have quite different boiling points (see Example 11.1). Explain the observed variation in boiling point, in terms of structure.

11.16 Discuss how you can determine which of these compounds might be alkanes, cycloalkanes, alkenes, or alkynes, without drawing their formulas: (a) C_6H_{12}, (b) C_4H_6, (c) C_5II_{12}, (d) C_7H_{14}, (e) C_3H_4.

11.17 Draw Newman projections of the staggered and eclipsed conformations of propane. Rank them in stability.

11.18 Draw Newman projections of four different conformations of butane. Rank them in stability. (*Hint:* Two of the conformations represent the most stable forms and the other two the least stable forms.)

11.19 Draw the structures of *cis*-2-butene and *trans*-2-butene. Which of the two compounds would give off more heat on hydrogenation to butane? Explain.

11.20 Would you expect cyclobutadiene to be a stable molecule? Explain.

H—C=C—H (ring: C—C / ‖ ‖ / C—C, each C bearing H)

11.21 How many different isomers can be derived from ethylene if two hydrogen atoms are replaced by a fluorine atom and a chlorine atom? Draw their structures and name them. Indicate which are structural isomers and which are geometric isomers.

11.22 Suggest two chemical tests that would help you distinguish between these two compounds:

(a) $CH_3CH_2CH_2CH_2CH_3$

(b) $CH_3CH_2CH_2CH{=}CH_2$

11.23 Sulfuric acid (H_2SO_4) adds to the double bond of alkenes as H^+ and $^-OSO_3H$. Predict the products when sulfuric acid reacts with (a) ethylene and (b) propene.

11.24 Acetylene is an unstable compound. It has a tendency to form benzene as follows:

$$3C_2H_2(g) \longrightarrow C_6H_6(l)$$

Calculate the standard enthalpy change in kJ/mol for this reaction at 25°C.

11.25 Predict products when HBr is added to (a) 1-butene and (b) 2-butene.

11.26 Geometric isomers are not restricted to compounds containing the C=C bond. For example, certain disubstituted cycloalkanes can exist in the *cis* and the *trans* forms. Label the following molecules as the *cis* and *trans* isomer, of the same compound:

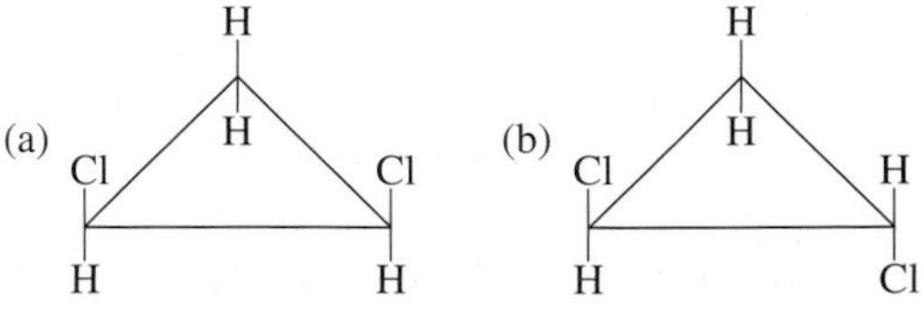

11.27 Write the structural formulas for these organic compounds: (a) 3-methylhexane, (b) 1,3,5-trichlorocyclohexane, (c) 2,3-dimethylpentane, (d) 2-bromo-4-phenylpentane, (e) 3,4,5-trimethyloctane.

11.28 Name these compounds:

(a) $CH_3-CH(CH_3)-CH_2-CH_2-CH_3$

(b) $CH_3-CH(C_2H_5)-CH(CH_3)-CH(CH_3)-CH_3$

(c) $CH_3-CH_2-CH(CH_2-CH_2-CH_3)-CH_2-CH_3$

(d) $CH_2{=}CH-CH(Br)-CH_2-CH_3$

(e) $CH_3-C{\equiv}C-CH_2-CH_3$

Aromatic Hydrocarbons

Review Questions

11.29 Comment on the extra stability of benzene compared to ethylene. Why does ethylene undergo addition reactions while benzene usually undergoes substitution reactions?

11.30 Benzene and cyclohexane both contain six-membered rings. Benzene is planar and cyclohexane is nonplanar. Explain.

Problems

11.31 Write structures for the compounds shown below: (a) 1-bromo-3-methylbenzene, (b) 1-chloro-2-propylbenzene, (c) 1,2,4,5-tetramethylbenzene.

11.32 Name these compounds:

(a) benzene ring with Cl and CH_3 (para)

(b) benzene ring with NO_2 and CH_2CH_3 (meta)

(c) benzene ring with CH_3, CH_3, H_3C, CH_3

(d) benzene ring with $CH_3-CH-CH{=}CH_2$

Chemistry of the Functional Groups

Review Questions

11.33 What are functional groups? Why is it logical and useful to classify organic compounds according to their functional groups?

11.34 Draw the Lewis structure for each of these functional groups: alcohol, ether, aldehyde, ketone, carboxylic acid, ester, amine.

Problems

11.35 Draw one possible structure for molecules with these formulas: (a) CH_4O, (b) C_2H_6O, (c) $C_3H_6O_2$, (d) C_3H_8O.

11.36 Classify each of these molecules as alcohol, aldehyde, ketone, carboxylic acid, amine, or ether:

(a) $CH_3{-}O{-}CH_2{-}CH_3$ (b) $CH_3{-}CH_2{-}NH_2$

(c) $CH_3{-}CH_2{-}C(=O){-}H$ (d) $CH_3{-}C(=O){-}CH_2{-}CH_3$

(e) $H{-}C(=O){-}OH$ (f) $H_3C{-}CH_2CH_2{-}OH$

(g) $C_6H_5{-}CH_2{-}CH(NH_2){-}C(=O){-}OH$

11.37 Generally aldehydes are more susceptible to oxidation in air than are ketones. Use acetaldehyde and acetone as examples and show why ketones such as acetone are more stable than aldehydes in this respect.

11.38 Complete this equation and identify the products:

$$HCOOH + CH_3OH \longrightarrow$$

11.39 A compound has the empirical formula $C_5H_{12}O$. Upon controlled oxidation, it is converted into a compound of empirical formula $C_5H_{10}O$, which behaves as a ketone. Draw possible structures for the original compound and the final compound.

11.40 A compound having the molecular formula $C_4H_{10}O$ does not react with sodium metal. In the presence of light, the compound reacts with Cl_2 to form three compounds having the formula C_4H_9OCl. Draw a structure for the original compound that is consistent with this information.

11.41 Predict the product or products of each of these reactions:

(a) $CH_3CH_2OH + HCOOH \longrightarrow$

(b) $H{-}C{\equiv}C{-}CH_3 + H_2 \longrightarrow$

(c) $(C_2H_5)(H)C{=}C(H)(H) + HBr \longrightarrow$

11.42 Identify the functional groups in each of these molecules:

(a) $CH_3CH_2COCH_2CH_2CH_3$

(b) $CH_3COOC_2H_5$

(c) $CH_3CH_2OCH_2CH_2CH_2CH_3$

Chirality

Review Questions

11.43 What factor determines whether a carbon atom in a compound is asymmetric?

11.44 Give examples of a chiral substituted alkane and an achiral substituted alkane.

Problems

11.45 Which of these amino acids are chiral:

(a) $CH_3CH(NH_2)COOH$,

(b) $CH_2(NH_2)COOH$,

(c) $CH_2(OH)CH(NH_2)COOH$?

11.46 Indicate the asymmetric carbon atoms in these compounds:

(a) $CH_3{-}CH_2{-}CH(CH_3){-}CH(NH_2){-}C(=O){-}NH_2$

(b)

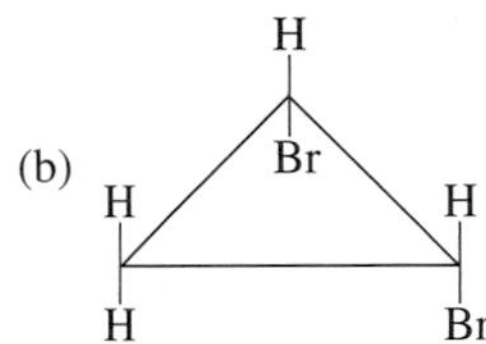

Additional Problems

11.47 Draw all the possible structural isomers for the molecule having the formula C_7H_7Cl. The molecule contains one benzene ring.

11.48 Given these data

$$C_2H_4(g) + 3O_2(g) \longrightarrow 2CO_2(g) + 2H_2O(l) \quad \Delta H^\circ = -1411 \text{ kJ/mol}$$

$$2C_2H_2(g) + 5O_2(g) \longrightarrow 4CO_2(g) + 2H_2O(l) \quad \Delta H^\circ = -2599 \text{ kJ/mol}$$

$$H_2(g) + \tfrac{1}{2}O_2(g) \longrightarrow H_2O(l) \quad \Delta H^\circ = -285.8 \text{ kJ/mol}$$

calculate the heat of hydrogenation for acetylene:

$$C_2H_2(g) + H_2(g) \longrightarrow C_2H_4(g)$$

11.49 State which member of each of these pairs of compounds is the more reactive and explain why: (a) propane and cyclopropane, (b) ethylene and methane, (c) acetaldehyde and acetone.

11.50 Like ethylene, tetrafluoroethylene (C_2F_4) undergoes polymerization reaction to form polytetrafluoroethylene (Teflon). Draw a repeating unit of the polymer.

11.51 An organic compound is found to contain 37.5 percent carbon, 3.2 percent hydrogen, and 59.3 percent fluorine by mass. These pressure and volume data were obtained for 1.00 g of this substance at 90°C:

P **(atm)**	***V*** **(L)**
2.00	0.332
1.50	0.409
1.00	0.564
0.50	1.028

The molecule is known to have no dipole moment. (a) What is the empirical formula of this substance? (b) Does this substance behave as an ideal gas? (c) What is its molecular formula? (d) Draw the Lewis structure of this molecule and describe its geometry. (e) What is the systematic name of this compound?

11.52 State at least one commercial use for each of the following compounds: (a) 2-propanol, (b) acetic acid, (c) naphthalene, (d) methanol, (e) ethanol, (f) ethylene glycol, (g) methane, (h) ethylene.

11.53 How many liters of air (78 percent N_2, 22 percent O_2 by volume) at 20°C and 1.00 atm are needed for the complete combustion of 1.0 L of octane, C_8H_{18}, a typical gasoline component that has a density of 0.70 g/mL?

11.54 How many carbon-carbon sigma bonds are present in each of these molecules? (a) 2-butyne, (b) anthracene (see Figure 11.16), (c) 2,3-dimethylpentane.

11.55 How many carbon-carbon sigma bonds are present in each of these molecules? (a) benzene, (b) cyclobutane, (c) 3-ethyl-2-methylpentane.

11.56 The combustion of 20.63 mg of compound Y, which contains only C, H, and O, with excess oxygen gave 57.94 mg of CO_2 and 11.85 mg of H_2O. (a) Calculate how many milligrams of C, H, and O were present in the original sample of Y. (b) Derive the empirical formula of Y. (c) Suggest a plausible structure for Y if the empirical formula is the same as the molecular formula.

11.57 Draw all the structural isomers of compounds with the formula $C_4H_8Cl_2$. Indicate which isomers are chiral and give them systematic names.

11.58 The combustion of 3.795 mg of liquid B, which contains only C, H, and O, with excess oxygen gave 9.708 mg of CO_2 and 3.969 mg of H_2O. In a molar mass determination, 0.205 g of B vaporized at 1.00 atm and 200.0°C and occupied a volume of 89.8 mL. Derive the empirical formula, molar mass, and molecular formula of B and draw three plausible structures.

11.59 Beginning with 3-methyl-1-butyne, show how you would prepare these compounds:

```
              Br CH3
              |  |
(a) CH2=C—CH—CH3
```

```
                     CH3
                     |
(b) BrCH2—CBr2—CH—CH3
```

```
          Br   CH3
          |    |
(c) CH3—CH—CH—CH3
```

11.60 Write structural formulas for these compounds: (a) *trans*-2-pentene, (b) 2-ethyl-1-butene, (c) 4-ethyl-*trans*-2-heptene, (d) 3-phenyl-1-butyne.

11.61 Suppose benzene contained three distinct single bonds and three distinct double bonds. How many different structural isomers would there be for dichlorobenzene ($C_6H_4Cl_2$)? Draw all your proposed structures.

11.62 Write the structural formula of an aldehyde that is an isomer of acetone.

11.63 Draw structures for these compounds: (a) cyclopentane, (b) *cis*-2-butene, (c) 2-hexanol, (d) 1,4-dibromobenzene, (e) 2-butyne.

11.64 Name the classes to which these compounds belong:

(a) C_4H_9OH (b) $CH_3OC_2H_5$
(c) C_2H_5CHO (d) C_6H_5COOH
(e) CH_3NH_2

11.65 Ethanol, C_2H_5OH, and dimethyl ether, CH_3OCH_3, are structural isomers. Compare their melting points, boiling points, and solubilities in water.

11.66 Amines are Brønsted bases. The unpleasant smell of fish is due to the presence of certain amines. Explain why cooks often add lemon juice to suppress the odor of fish (in addition to enhancing the flavor).

11.67 You are given two bottles, each containing a colorless liquid. You are told that one liquid is cyclohexane and the other is benzene. Suggest one chemical test that would enable you to distinguish between these two liquids.

11.68 Give the chemical names of these organic compounds and write their formulas: marsh gas, grain alcohol, wood alcohol, rubbing alcohol, antifreeze, mothballs, chief ingredient of vinegar.

11.69 The compound $CH_3—C\equiv C—CH_3$ is hydrogenated to an alkene using platinum as the catalyst. If the product is the pure *cis* isomer, what can you deduce about the mechanism?

11.70 How many asymmetric carbon atoms are present in each of these compounds?

```
      H  H  H
      |  |  |
(a) H—C—C—C—Cl
      |  |  |
      H  Cl H
```

```
        OH   CH3
        |    |
(b) CH3—C————C—CH2OH
        |    |
        H    H
```

(c)

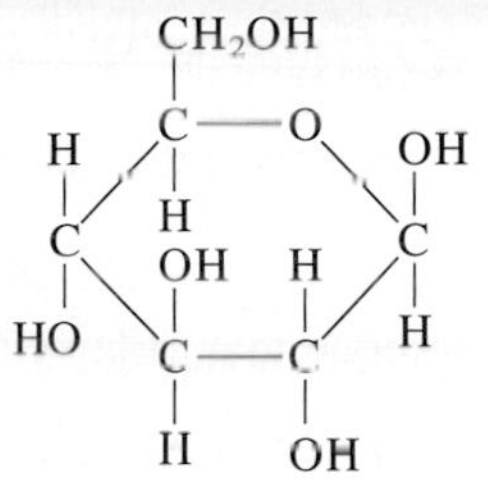

```
          CH2OH
          |
      H   C———O    OH
      |  /|     \  |
      C   H      C
      |  OH  H   |
     HO \ |   | / H
          C———C
          |   |
          H   OH
```

11.71 Isopropyl alcohol is prepared by reacting propene (CH_3CHCH_2) with sulfuric acid, followed by treatment with water. (a) Show the sequence of steps leading to the product. What is the role of sulfuric acid? (b) Draw the structure of an alcohol that is an isomer of isopropyl alcohol. (c) Is isopropyl alcohol a chiral molecule? (d) What property of isopropyl alcohol makes it useful as a rubbing alcohol?

11.72 When a mixture of methane and bromine vapor is exposed to light, this reaction occurs slowly:

$$CH_4(g) + Br_2(g) \longrightarrow CH_3Br(g) + HBr(g)$$

Suggest a mechanism for this reaction. (*Hint:* Bromine vapor is deep red; methane is colorless.)

Special Problems

11.73 Octane number is assigned to gasoline to indicate the tendency of "knocking" in the automobile's engine. The higher the octane number, the more smoothly the fuel will burn without knocking. Branched-chain aliphatic hydrocarbons have higher octane numbers than straight-chain aliphatic hydrocarbons, and aromatic hydrocarbons have the highest octane numbers.

(a) Arrange these compounds in the order of decreasing octane numbers: 2,2,4-trimethylpentane, toluene (methylbenzene), *n*-heptane, and 2-methylhexane.

(b) Oil refineries carry out *catalytic reforming* in which a straight-chain hydrocarbon, in the presence of a catalyst, is converted to an aromatic molecule and a useful by-product. Write an equation for the conversion from *n*-heptane to toluene.

(c) Until 2000, *tert*-butylmethyl ether had been widely used as an antiknocking agent to enhance the octane number of gasoline. Write the structural formula of the compound.

11.74 Fats and oils are names for the same class of compounds, called triglycerides, which contain three ester groups

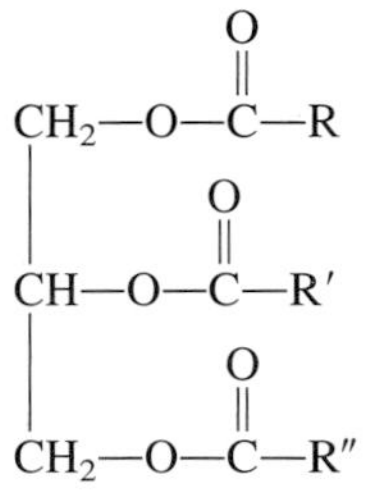

```
             O
             ‖
CH2—O—C—R
|
|            O
|            ‖
CH—O—C—R′
|
|            O
|            ‖
CH2—O—C—R″
```

A fat or oil

in which R, R′, and R″ represent long hydrocarbon chains.

(a) Suggest a reaction that leads to the formation of a triglyceride molecule, starting with glycerol and carboxylic acids (see p. 407 for structure of glycerol).

(b) In the old days, soaps were made by hydrolyzing animal fat with lye (a sodium hydroxide solution). Write an equation for this reaction.

(c) The difference between fats and oils is that at room temperature, the former are solid and the latter are liquids. Fats are usually produced by animals, whereas oils are commonly found in plants. The melting points of these substances are determined by the number of C=C bonds (or the extent of unsaturation) present—the larger the number of C=C bonds, the lower the melting point and the more likely the substance is a liquid. Explain.

(d) One way to convert liquid oil to solid fat is to hydrogenate the oil, a process by which some or all of the C=C bonds are converted to C—C bonds. This procedure prolongs shelf life of the oil by removing the more reactive C=C group and facilitates packaging. How would you carry out such a process (that is, what reagents and catalyst would you employ)?

(e) The degree of unsaturation of oil can be determined by reacting the oil with iodine, which reacts with the C=C as follows:

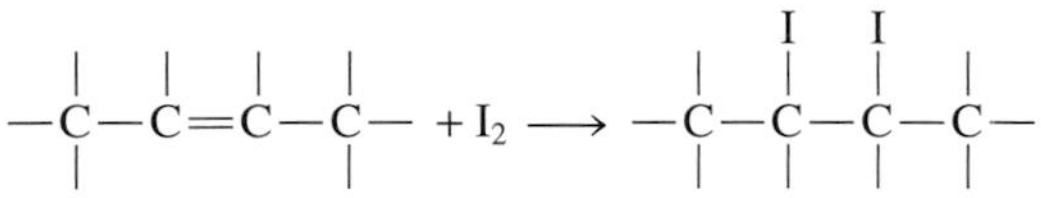

$$-\overset{|}{\underset{|}{C}}-\overset{|}{C}=\overset{|}{C}-\overset{|}{\underset{|}{C}}- + I_2 \longrightarrow -\overset{|}{\underset{|}{C}}-\overset{I}{\underset{|}{C}}-\overset{I}{\underset{|}{C}}-\overset{|}{\underset{|}{C}}-$$

The procedure is to add a known amount of iodine to the oil and allow the reaction to go to completion. The amount of excess (unreacted) iodine is determined by titrating the remaining iodine with a standard sodium thiosulfate ($Na_2S_2O_3$) solution:

$$I_2 + 2Na_2S_2O_3 \longrightarrow Na_2S_4O_6 + 2NaI$$

The number of grams of iodine that reacts with 100 g of oil is called the *iodine number*. In one case, 43.8 g of I_2 were treated with 35.3 g of corn oil. The excess iodine required 20.6 mL of 0.142 *M* $Na_2S_2O_3$ for neutralization. Calculate the iodine number of the corn oil.

11.75 2-Butanone can be reduced to 2-butanol by reagents such as lithium aluminum hydride ($LiAlH_4$). (a) Write the formula of the product. Is it chiral? (b) In reality, the product does not exhibit optical activity. Explain.

11.76 Write the structures of three alkenes that yield 2-methylbutane on hydrogenation.

11.77 Write the structural formulas of the alcohols with the formula $C_6H_{13}O$ and indicate those that are chiral. Show only the C atoms and the —OH groups.

11.78 An alcohol was converted to a carboxylic acid with acidic potassium dichromate. A 4.46-g sample of the acid was added to 50.0 mL of 2.27 *M* NaOH and the excess NaOH required 28.7 mL of 1.86 *M* HCl for neutralization. What is the molecular formula of the alcohol?

Answers to Practice Exercises

11.1 5. **11.2** 4,6-diethyl-2-methyloctane

11.3
$$\begin{array}{c} \quad\quad\;\; CH_3 \quad\quad\quad\quad\quad\quad\quad\;\; C_2H_5\;CH_3 \\ \quad\quad\;\; | \quad\quad\quad\quad\quad\quad\quad\quad\quad | \quad\;\; | \\ CH_3-CH-CH_2-CH_2-CH-CH-CH_2-CH_3 \end{array}$$

11.4 $CH_3CH_2COOCH_3$ and H_2O. **11.5** No.

Under atmospheric conditions, solid carbon dioxide (dry ice) does not melt; it only sublimes.

CHAPTER 12

Intermolecular Forces and Liquids and Solids

Chapter Outline

Student Interactive Activities

Animations
Packing Spheres (12.4)
Equilibrium Vapor Pressure (12.6)

Electronic Homework
Example Practice Problems
End of Chapter Problems

Essential Concepts

Intermolecular Forces Intermolecular forces, which are responsible for the nonideal behavior of gases, also account for the existence of the condensed states of matter—liquids and solids. They exist between polar molecules, between ions and polar molecules, and between nonpolar molecules. A special type of intermolecular force, called the hydrogen bond, describes the interaction between the hydrogen atom in a polar bond and an electronegative atom such as O, N, or F.

The Liquid State Liquids tend to assume the shapes of their containers. The surface tension of a liquid is the energy required to increase its surface area. It manifests itself in capillary action, which is responsible for the rise (or depression) of a liquid in a narrow tubing. Viscosity is a measure of a liquid's resistance to flow. It always decreases with increasing temperature. The structure of water is unique in that its solid state (ice) is less dense than its liquid state.

The Crystalline State A crystalline solid possesses rigid and long-range order. Different crystal structures can be generated by packing identical spheres in three dimensions.

Bonding in Solids Atoms, molecules, or ions are held in a solid by different types of bonding. Electrostatic forces are responsible for ionic solids, intermolecular forces are responsible for molecular solids, covalent bonds are responsible for covalent solids, and a special type of interaction, which involves electrons being delocalized over the entire crystal, accounts for the existence of metals.

Phase Transitions The states of matter can be interconverted by heating or cooling. Two phases are in equilibrium at the transition temperature such as boiling or freezing. Solids can also be directly converted to vapor by sublimation. Above a certain temperature, called the critical temperature, the gas of a substance cannot be made to liquefy. The pressure-temperature relationships of solid, liquid, and vapor phases are best represented by a phase diagram.

12.1 The Kinetic Molecular Theory of Liquids and Solids

In Chapter 5 we used the kinetic molecular theory to explain the behavior of gases in terms of the constant, random motion of gas molecules. In gases, the distances between molecules are so great (compared with their diameters) that at ordinary temperatures and pressures (say, 25°C and 1 atm), there is no appreciable interaction between the molecules. Because there is a great deal of empty space in a gas—that is, space that is not occupied by molecules—gases can be readily compressed. The lack of strong forces between molecules also allows a gas to expand to fill the volume of its container. Furthermore, the large amount of empty space explains why gases have very low densities under normal conditions.

Liquids and solids are quite a different story. The principal difference between the condensed states (liquids and solids) and the gaseous state is the distance between molecules. In a liquid, the molecules are so close together that there is very little empty space. Thus, liquids are much more difficult to compress than gases, and they are also much denser under normal conditions. Molecules in a liquid are held together by one or more types of attractive forces, which will be discussed in Section 12.2. A liquid also has a definite volume, because molecules in a liquid do not break away from the attractive forces. The molecules can, however, move past one another freely, and so a liquid can flow, can be poured, and assumes the shape of its container.

In a solid, molecules are held rigidly in position with virtually no freedom of motion. Many solids are characterized by long-range order; that is, the molecules are arranged in regular configurations in three dimensions. There is even less empty space in a solid than in a liquid. Thus, solids are almost incompressible and possess definite shape and volume. With very few exceptions (water being the most important), the density of the solid form is higher than that of the liquid form for a given substance. It is not uncommon for two states of a substance to coexist. An ice cube (solid) floating in a glass of water (liquid) is a familiar example. Chemists refer to the different states of a substance that are present in a system as *phases*. Thus, our glass of ice water contains both the solid phase and the liquid phase of water. In this chapter we will use the term "phase" when talking about changes of state involving one substance, as well as systems containing more than one phase of a substance. Table 12.1 summarizes some of the characteristic properties of the three phases of matter.

Table 12.1 Characteristic Properties of Gases, Liquids, and Solids

State of Matter	Volume/Shape	Density	Compressibility	Motion of Molecules
Gas	Assumes the volume and shape of its container	Low	Very compressible	Very free motion
Liquid	Has a definite volume but assumes the shape of its container	High	Only slightly compressible	Slide past one another freely
Solid	Has a definite volume and shape	High	Virtually incompressible	Vibrate about fixed positions

12.2 Intermolecular Forces

Intermolecular forces are *attractive forces between molecules.* Intermolecular forces are responsible for the nonideal behavior of gases described in Chapter 5. They exert even more influence in the condensed phases of matter—liquids and solids. As the temperature of a gas drops, the average kinetic energy of its molecules decreases. Eventually, at a sufficiently low temperature, the molecules no longer have enough energy to break away from the attraction of neighboring molecules. At this point, the molecules aggregate to form small drops of liquid. This transition from the gaseous to the liquid phase is known as *condensation.*

For simplicity we use the term "intermolecular forces" for both atoms and molecules.

In contrast to intermolecular forces, ***intramolecular forces*** *hold atoms together in a molecule.* (Chemical bonding, discussed in Chapters 9 and 10, involves intramolecular forces.) Intramolecular forces stabilize individual molecules, whereas intermolecular forces are primarily responsible for the bulk properties of matter (for example, melting point and boiling point).

Generally, intermolecular forces are much weaker than intramolecular forces. Much less energy is usually required to evaporate a liquid than to break the bonds in the molecules of the liquid. For example, it takes about 41 kJ of energy to vaporize 1 mole of water at its boiling point; but about 930 kJ of energy are necessary to break the two O—H bonds in 1 mole of water molecules. The boiling points of substances often reflect the strength of the intermolecular forces operating among the molecules. At the boiling point, enough energy must be supplied to overcome the attractive forces among molecules before they can enter the vapor phase. If it takes more energy to separate molecules of substance A than of substance B because A molecules are held together by stronger intermolecular forces, then the boiling point of A is higher than that of B. The same principle applies also to the melting points of the substances. In general, the melting points of substances increase with the strength of the intermolecular forces.

To discuss the properties of condensed matter, we must understand the different types of intermolecular forces. *Dipole-dipole, dipole-induced dipole,* and *dispersion forces* make up what chemists commonly refer to as ***van der Waals forces,*** after the Dutch physicist Johannes van der Waals (see Section 5.7). Ions and dipoles are attracted to one another by electrostatic forces called *ion-dipole forces,* which are *not* van der Waals forces. *Hydrogen bonding* is a particularly strong type of dipole-dipole interaction. Because only a few elements can participate in hydrogen bond formation, it is treated as a separate category. Depending on the phase of a substance, the nature of chemical bonds, and the types of elements present, more than one type of interaction may contribute to the total attraction between molecules, as we will see below.

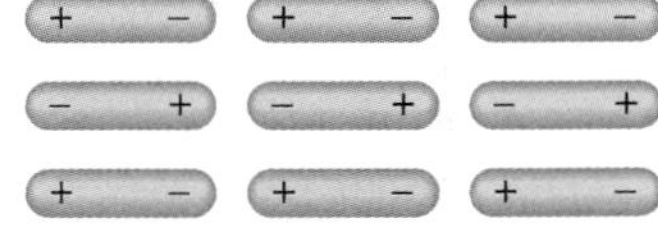

Figure 12.1
Molecules that have a permanent dipole moment tend to align with opposite polarities in the solid phase for maximum attractive information.

Dipole-Dipole Forces

Dipole-dipole forces are *attractive forces between polar molecules,* that is, between molecules that possess dipole moments (see Section 10.2). Their origin is electrostatic, and they can be understood in terms of Coulomb's law. The larger the dipole moment, the greater the force. Figure 12.1 shows the orientation of polar molecules in a solid. In liquids, polar molecules are not held as rigidly as in a solid, but they tend to align in a way that, on average, maximizes the attractive interaction.

Ion-Dipole Forces

Coulomb's law also explains ***ion-dipole forces,*** which *attract an ion (either a cation or an anion) and a polar molecule to each other* (Figure 12.2). The strength of this interaction depends on the charge and size of the ion and on the magnitude of the dipole

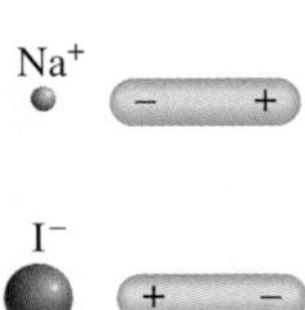

Figure 12.2
Two types of ion-dipole interaction.

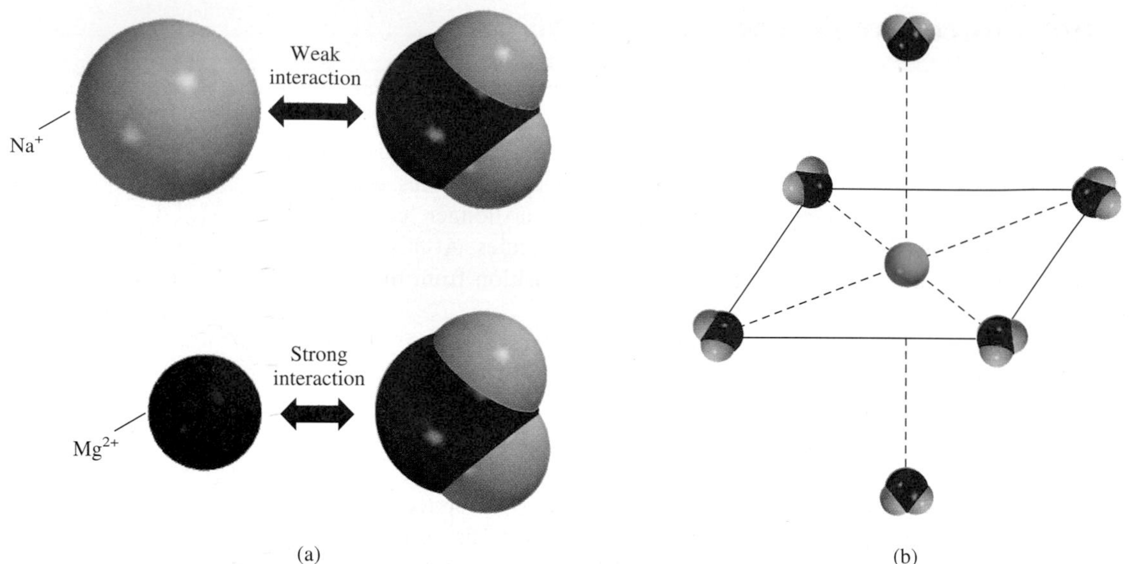

Figure 12.3
(a) Interaction of a water molecule with a Na^+ ion and a Mg^{2+} ion. (b) In aqueous solutions, metal ions are usually surrounded by six water molecules in an octahedral arrangement.

moment and size of the molecule. The charges on cations are generally more concentrated, because cations are usually smaller than anions. Therefore, a cation interacts more strongly with dipoles than does an anion having a charge of the same magnitude.

Hydration, discussed in Section 4.1, is one example of ion-dipole interaction. Figure 12.3 shows the ion-dipole interaction between the Na^+ and Mg^{2+} ions with a water molecule, which has a large dipole moment (1.87 D). Because the Mg^{2+} ion has a higher charge and a smaller ionic radius (78 pm) than that of the Na^+ ion (98 pm), it interacts more strongly with water molecules. (In reality, each ion is surrounded by a number of water molecules in solution.) Similar differences exist for anions of different charges and sizes.

Dispersion Forces

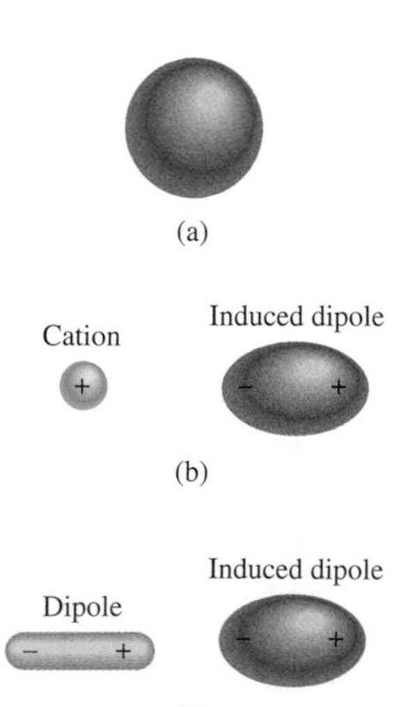

Figure 12.4
(a) Spherical charge distribution in a helium atom. (b) Distortion caused by the approach of a cation. (c) Distortion caused by the approach of a dipole.

What attractive interaction occurs in nonpolar substances? To learn the answer to this question, consider the arrangement shown in Figure 12.4. If we place an ion or a polar molecule near an atom (or a nonpolar molecule), the electron distribution of the atom (or molecule) is distorted by the force exerted by the ion or the polar molecule, resulting in a kind of dipole. The dipole in the atom (or nonpolar molecule) is said to be an ***induced dipole*** because *the separation of positive and negative charges in the atom (or nonpolar molecule) is due to the proximity of an ion or a polar molecule.* The attractive interaction between an ion and the induced dipole is called *ion-induced dipole interaction,* and the attractive interaction between a polar molecule and the induced dipole is called *dipole-induced dipole interaction.*

The likelihood of a dipole moment being induced depends not only on the charge on the ion or the strength of the dipole but also on the ***polarizability*** of the atom or molecule—that is, *the ease with which the electron distribution in the atom (or molecule) can be distorted.* Generally, the larger the number of electrons and the more diffuse the electron cloud in the atom or molecule, the greater its polarizability. By *diffuse cloud* we mean an electron cloud that is spread over an appreciable volume, so that the electrons are not held tightly by the nucleus.

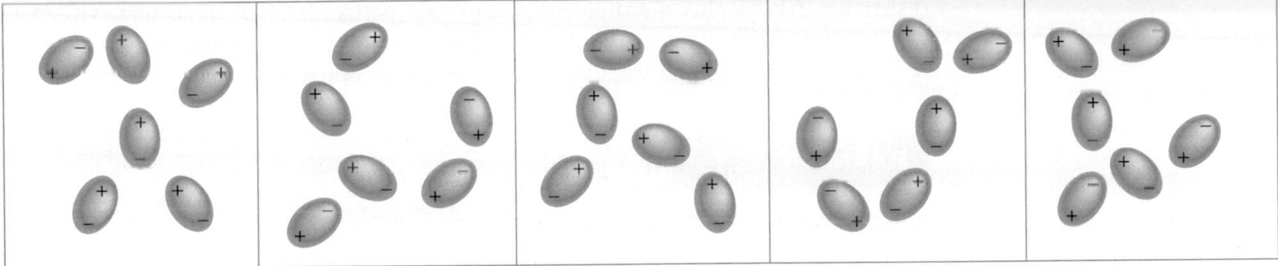

Figure 12.5
Induced dipoles interacting with each other. Such patterns exist only momentarily; new arrangements are formed in the next instant. This type of interaction is responsible for the condensation of nonpolar gases.

Polarizability allows gases containing atoms or nonpolar molecules (for example, He and N_2) to condense. In a helium atom, the electrons are moving at some distance from the nucleus. At any instant it is likely that the atom has a dipole moment created by the specific positions of the electrons. This dipole moment is called an *instantaneous dipole* because it lasts for just a tiny fraction of a second. In the next instant, the electrons are in different locations and the atom has a new instantaneous dipole, and so on. Averaged over time (that is, the time it takes to make a dipole moment measurement), however, the atom has no dipole moment because the instantaneous dipoles all cancel one another. In a collection of He atoms, an instantaneous dipole of one He atom can induce a dipole in each of its nearest neighbors (Figure 12.5). At the next moment, a different instantaneous dipole can create temporary dipoles in the surrounding He atoms. The important point is that this kind of interaction produces ***dispersion forces,*** *attractive forces that arise as a result of temporary dipoles induced in atoms or molecules*. At very low temperatures (and reduced atomic speeds), dispersion forces are strong enough to hold He atoms together, causing the gas to condense. The attraction between nonpolar molecules can be explained similarly.

A quantum mechanical interpretation of temporary dipoles was provided by the German physicist Fritz London in 1930. London showed that the magnitude of this attractive interaction is directly proportional to the polarizability of the atom or molecule. As we might expect, dispersion forces may be quite weak. This is certainly true for helium, which has a boiling point of only 4.2 K, or −269°C. (Note that helium has only two electrons, which are tightly held in the 1*s* orbital. Therefore, the helium atom has a low polarizability.)

Dispersion forces, which are also called London forces, usually increase with molar mass because molecules with larger molar mass tend to have more electrons, and dispersion forces increase in strength with the number of electrons. Furthermore, larger molar mass often means a bigger atom whose electron distribution is more easily disturbed because the outer electrons are less tightly held by the nuclei. Table 12.2 compares the melting points of similar substances that consist of nonpolar molecules. As expected, the melting point increases as the number of electrons in the molecule increases. Because these are all nonpolar molecules, the only attractive intermolecular forces present are the dispersion forces.

In many cases, dispersion forces are comparable to or even greater than the dipole-dipole forces between polar molecules. For a dramatic illustration, let us compare the boiling points of CH_3F (−78.4°C) and CCl_4 (76.5°C). Although CH_3F has a dipole moment of 1.8 D, it boils at a much lower temperature than CCl_4, a nonpolar molecule. CCl_4 boils at a higher temperature simply because it contains more electrons. As a result, the dispersion forces between CCl_4 molecules are stronger than the

Table 12.2
Melting Points of Similar Nonpolar Compounds

Compound	Melting Point (°C)
CH_4	−182.5
CF_4	−150.0
CCl_4	−23.0
CBr_4	90.0
CI_4	171.0

dispersion forces plus the dipole-dipole forces between CH_3F molecules. (Keep in mind that dispersion forces exist among species of all types, whether they are neutral or bear a net charge and whether they are polar or nonpolar.)

EXAMPLE 12.1

What type(s) of intermolecular forces exist between the following pairs: (a) HBr and H_2S, (b) Cl_2 and CBr_4, (c) I_2 and NO_3^-, (d) NH_3 and C_6H_6?

Strategy Classify the species into three categories: ionic, polar (possessing a dipole moment), and nonpolar. Keep in mind that dispersion forces exist between *all* species.

Solution

(a) Both HBr and H_2S are polar molecules.

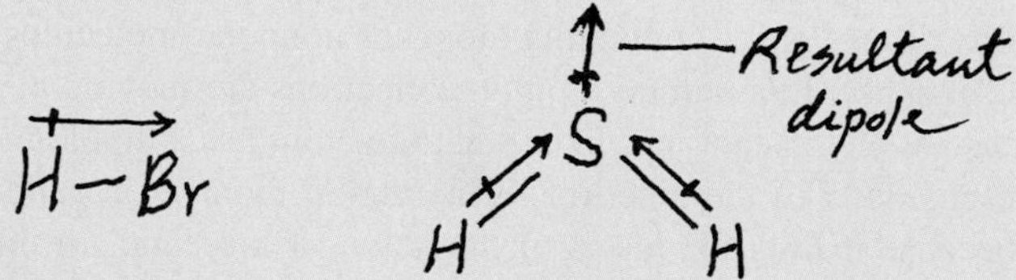

Therefore, the intermolecular forces present are dipole-dipole forces, as well as dispersion forces.

(b) Both Cl_2 and CBr_4 are nonpolar, so there are only dispersion forces between these molecules.

Cl–Cl
$\mu = 0$

$\mu = 0$

(c) I_2 is a homonuclear diatomic molecule and therefore nonpolar, so the forces between it and the ion NO_3^- are ion-induced dipole forces and dispersion forces.

(d) NH_3 is polar, and C_6H_6 is nonpolar. The forces are dipole-induced dipole forces and dispersion forces.

Similar problem: 12.10.

Practice Exercise Name the type(s) of intermolecular forces that exists between molecules (or basic units) in each of the following species: (a) LiF, (b) CH_4, (c) SO_2.

The Hydrogen Bond

Normally, the boiling points of a series of similar compounds containing elements in the same periodic group increase with increasing molar mass. This increase in boiling point is due to the increase in dispersion forces for molecules with more electrons. Hydrogen compounds of Group 4A follow this trend, as Figure 12.6 shows. The lightest compound, CH_4, has the lowest boiling point, and the heaviest compound, SnH_4, has the highest boiling point. However, hydrogen compounds of the elements in Groups 5A, 6A, and 7A do not follow this trend. In each of these series, the lightest compound (NH_3, H_2O, and HF) has the highest boiling point, contrary to our

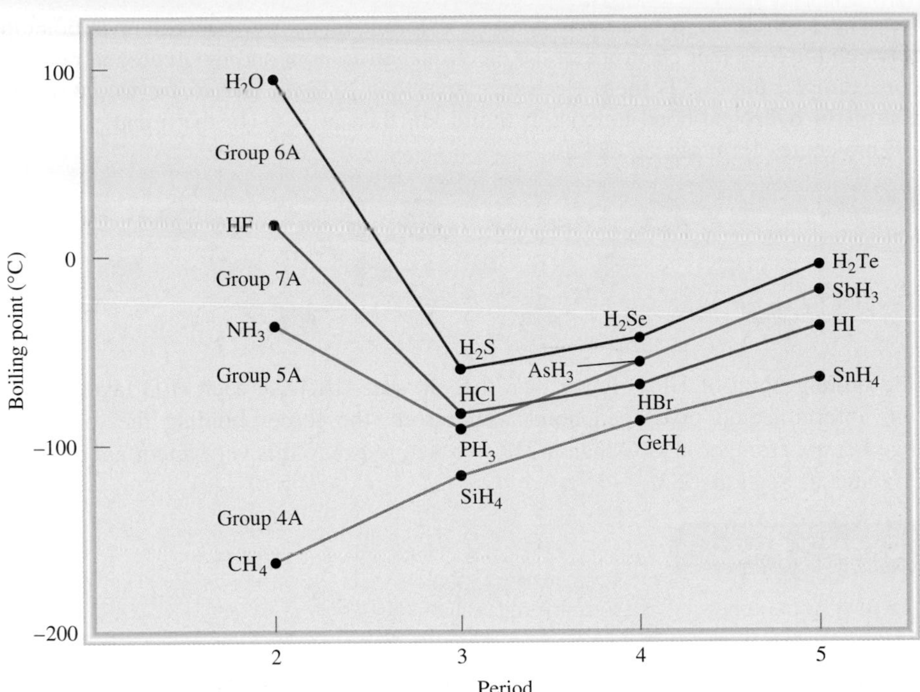

Figure 12.6
Boiling points of the hydrogen compounds of Groups 4A, 5A, 6A, and 7A elements. Although normally we expect the boiling point to increase as we move down a group, we see that three compounds (NH_3, H_2O, and HF) behave differently. The anomaly can be explained in terms of intermolecular hydrogen bonding.

expectations based on molar mass. This observation must mean that there are stronger intermolecular attractions in NH_3, H_2O, and HF, compared to other molecules in the same groups. In fact, this particularly strong type of intermolecular attraction is called the ***hydrogen bond,*** which is *a special type of dipole-dipole interaction between the hydrogen atom in a polar bond, such as N—H, O—H, or F—H, and an electronegative O, N, or F atom.* The interaction is written

$$\text{A—H} \cdots \text{B} \qquad \text{or} \qquad \text{A—H} \cdots \text{A}$$

A and B represent O, N, or F; A—H is one molecule or part of a molecule and B is a part of another molecule; and the dotted line represents the hydrogen bond. The three atoms usually lie in a straight line, but the angle AHB (or AHA) can deviate as much as 30° from linearity. Note that the O, N, and F atoms all possess at least one lone pair that can interact with the hydrogen atom in hydrogen bonding.

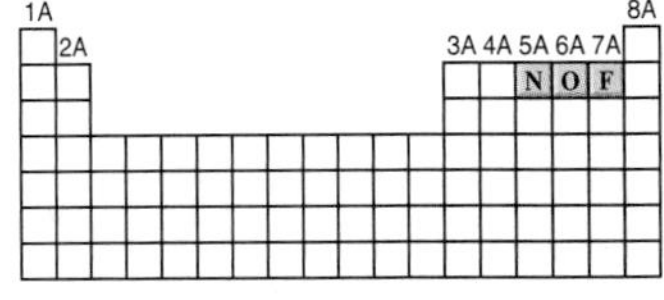

The three most electronegative elements that take part in hydrogen bonding.

The average energy of a hydrogen bond is quite large for a dipole-dipole interaction (up to 40 kJ/mol). Thus, hydrogen bonds have a powerful effect on the structures and properties of many compounds. Figure 12.7 shows several examples of hydrogen bonding.

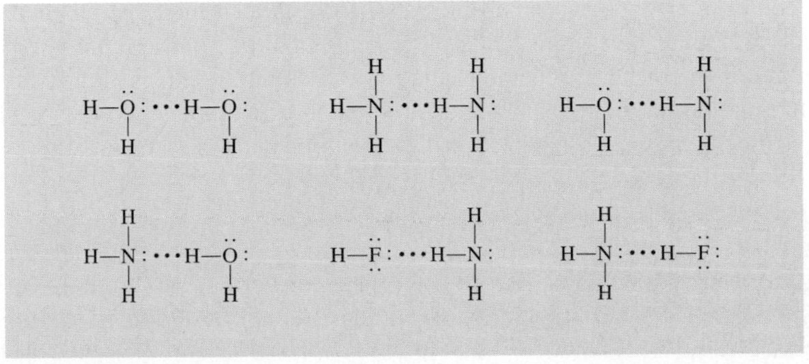

Figure 12.7
Hydrogen bonding in water, ammonia, and hydrogen fluoride. Solid lines represent covalent bonds, and dotted lines represent hydrogen bonds.

The strength of a hydrogen bond is determined by the coulombic interaction between the lone-pair electrons of the electronegative atom and the hydrogen nucleus. For example, fluorine is more electronegative than oxygen, and so we would expect a stronger hydrogen bond to exist in liquid HF than in H_2O. In the liquid phase, the HF molecules form zigzag chains:

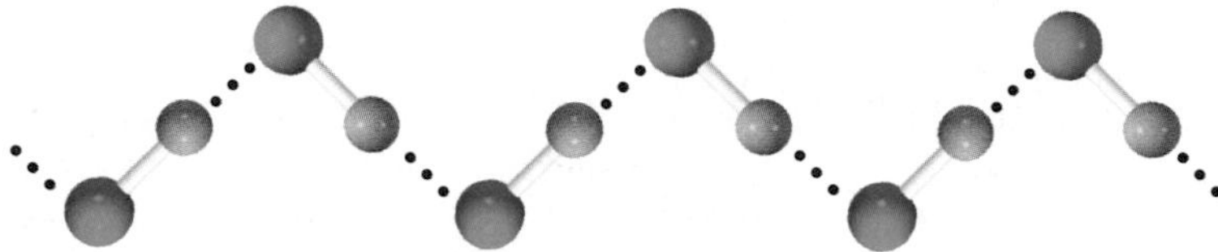

The boiling point of HF is lower than that of water because each H_2O takes part in *four* intermolecular hydrogen bonds. Therefore, the forces holding the molecules together are stronger in H_2O than in HF. We will return to this very important property of water in Section 12.3.

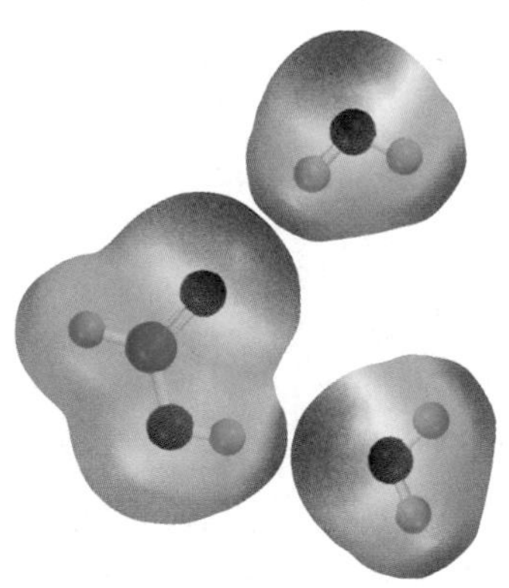

HCOOH forms hydrogen bonds with two H_2O molecules.

EXAMPLE 12.2

Which of the following can form hydrogen bonds with water? CH_3OCH_3, CH_4, F^-, HCOOH, Na^+.

Strategy A species can form hydrogen bonds with water if it contains one of the three electronegative elements (F, O, or N) or it has an H atom bonded to one of these three elements.

Solution There are no electronegative elements (F, O, or N) in either CH_4 or Na^+. Therefore, only CH_3OCH_3, F^-, and HCOOH can form hydrogen bonds with water.

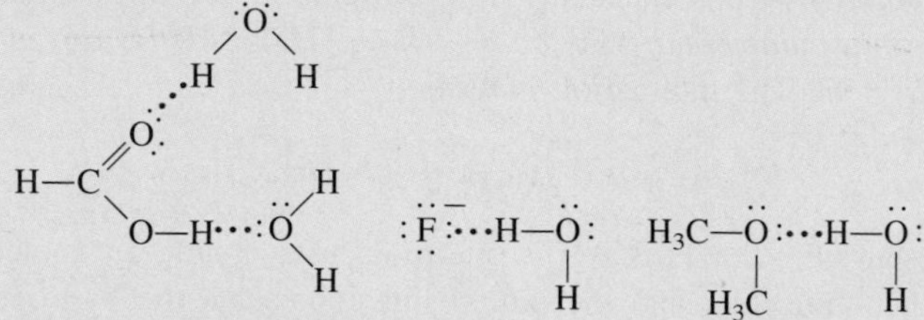

Check Note that HCOOH (formic acid) can form hydrogen bonds with water in two different ways.

Similar problem: 12.12.

Practice Exercise Which of the following species are capable of hydrogen bonding among themselves? (a) H_2S, (b) C_6H_6, (c) CH_3OH.

The intermolecular forces discussed so far are all attractive in nature. Keep in mind, though, that molecules also exert repulsive forces on one another. Thus, when two molecules approach each other, the repulsion between the electrons and between the nuclei in the molecules comes into play. The magnitude of the repulsive force rises very steeply as the distance separating the molecules in a condensed phase decreases. This is the reason that liquids and solids are so hard to compress. In these phases, the molecules are already in close contact with one another, and so they greatly resist being compressed further.

Review of Concepts

Which of the following compounds is most likely to exist as a liquid at room temperature: ethane (C_2H_6), hydrazine (N_2H_4), fluoromethane (CH_3F)?

12.3 Properties of Liquids

Intermolecular forces give rise to a number of structural features and properties of liquids. In this section we will look at two such phenomena associated with liquids in general: surface tension and viscosity. Then we will discuss the structure and properties of water.

Surface Tension

Molecules within a liquid are pulled in all directions by intermolecular forces; there is no tendency for them to be pulled in any one way. However, molecules at the surface are pulled downward and sideways by other molecules, but not upward away from the surface (Figure 12.8). These intermolecular attractions thus tend to pull the molecules into the liquid and cause the surface to tighten like an elastic film. Because there is little or no attraction between polar water molecules and, say, the nonpolar wax molecules on a freshly waxed car, a drop of water assumes the shape of a small round bead, because a sphere minimizes the surface area of a liquid. The waxy surface of a wet apple also produces this effect (Figure 12.9).

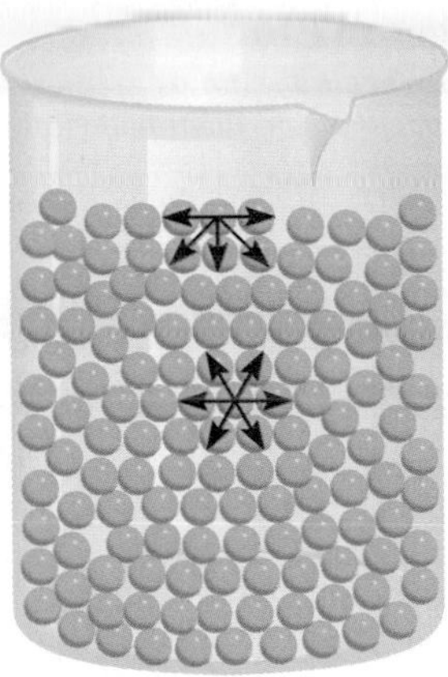

Figure 12.8
Intermolecular forces acting on a molecule in the surface layer of a liquid and in the interior region of the liquid.

A measure of the elastic force in the surface of a liquid is surface tension. The ***surface tension*** is *the amount of energy required to stretch or increase the surface of a liquid by a unit area* (for example, by 1 cm^2). Liquids that have strong intermolecular forces also have high surface tensions. Thus, because of hydrogen bonding, water has a considerably greater surface tension than most other liquids.

Surface tension enables the water strider to "walk" on water.

Another example of surface tension is *capillary action.* Figure 12.10(a) shows water rising spontaneously in a capillary tube. A thin film of water adheres to the wall of the glass tube. The surface tension of water causes this film to contract, and as it does, it pulls the water up the tube. Two types of forces bring about capillary action. One is ***cohesion,*** which is *the intermolecular attraction between like molecules* (in this case, the water molecules). The second force, called ***adhesion,*** is *an attraction between unlike molecules,* such as those in water and in the sides of a glass tube. If adhesion is stronger than cohesion, as it is in Figure 12.10(a), the contents of the tube will be pulled upward. This process continues until the adhesive force is balanced by the weight of the water in the tube. This action is by no means universal among liquids, as Figure 12.10(b) shows. In mercury, cohesion is greater than the adhesion between mercury and glass, so that when a capillary tube is dipped in mercury, the result is a depression or lowering, at the mercury level—that is, the height of the liquid in the capillary tube is below the surface of the mercury.

Viscosity

The expression "slow as molasses in January" owes its truth to another physical property of liquids called viscosity. ***Viscosity*** is *a measure of a fluid's resistance to flow.* The greater the viscosity, the more slowly the liquid flows. The viscosity of a liquid usually decreases as temperature increases; thus, hot molasses flows much faster than cold molasses.

Liquids that have strong intermolecular forces have higher viscosities than those that have weak intermolecular forces (Table 12.3). Water has a higher viscosity than many other liquids because of its ability to form hydrogen bonds. Interestingly, the viscosity of glycerol is significantly higher than that of all the other liquids listed in Table 12.3. Glycerol has the structure

$$\begin{array}{l} CH_2{-}OH \\ | \\ CH{-}OH \\ | \\ CH_2{-}OH \end{array}$$

Figure 12.9
Water beads on an apple, which has a waxy surface.

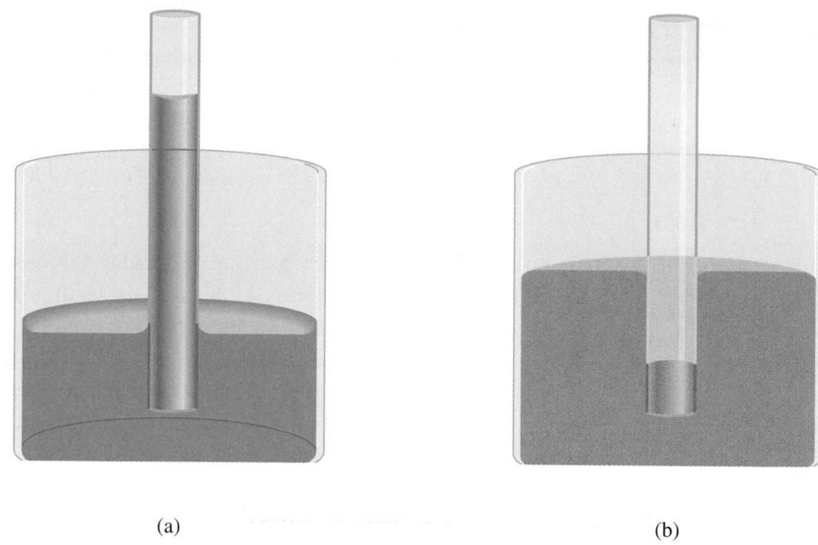

Figure 12.10
(a) When adhesion is greater than cohesion, the liquid (for example, water) rises in the capillary tube. (b) When cohesion is greater than adhesion, as it is for mercury, a depression of the liquid in the capillary tube results. Note that the meniscus in the tube of water is concave, or rounded downward, whereas that in the tube of mercury is convex, or rounded upward.

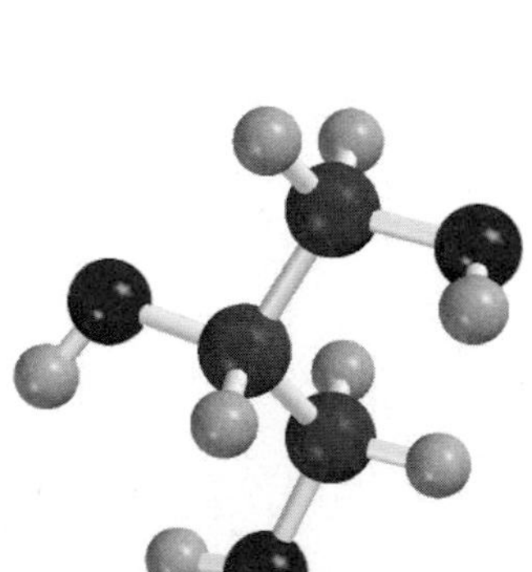

Glycerol is a clear, odorless, syrupy liquid used to make explosives, ink, and lubricants.

Like water, glycerol can form hydrogen bonds. Each glycerol molecule has three —OH groups that can participate in hydrogen bonding with other glycerol molecules. Furthermore, because of their shape, the molecules have a great tendency to become entangled rather than to slip past one another as the molecules of less viscous liquids do. These interactions contribute to its high viscosity.

The Structure and Properties of Water

Water is so common a substance on Earth that we often overlook its unique nature. All life processes involve water. Water is an excellent solvent for many ionic compounds, as well as for other substances capable of forming hydrogen bonds with water.

If water did not have the ability to form hydrogen bonds, it would be a gas at room temperature.

As Table 6.2 shows, water has a high specific heat. The reason is that to raise the temperature of water (that is, to increase the average kinetic energy of water molecules), we must first break the many intermolecular hydrogen bonds. Thus, water can absorb a substantial amount of heat while its temperature rises only slightly. The converse is also true: Water can give off much heat with only a slight decrease in its temperature. For this reason, the huge quantities of water that are present in our lakes

Table 12.3 Viscosity of Some Common Liquids at 20°C

Liquid	Viscosity (N s/m^2)*
Acetone (C_3H_6O)	3.16×10^{-4}
Benzene (C_6H_6)	6.25×10^{-4}
Blood	4×10^{-3}
Carbon tetrachloride (CCl_4)	9.69×10^{-4}
Diethyl ether ($C_2H_5OC_2H_5$)	2.33×10^{-4}
Ethanol (C_2H_5OH)	1.20×10^{-3}
Glycerol ($C_3H_8O_3$)	1.49
Mercury (Hg)	1.55×10^{-3}
Water (H_2O)	1.01×10^{-3}

*The SI units of viscosity are newton-second per meter squared.

Figure 12.11
Left: Ice cubes float on water. Right: Solid benzene sinks to the bottom of liquid benzene.

and oceans can effectively moderate the climate of adjacent land areas by absorbing heat in the summer and giving off heat in the winter, with only small changes in the temperature of the body of water.

The most striking property of water is that its solid form is less dense than its liquid form: ice floats at the surface of liquid water. The density of almost all other substances is greater in the solid state than in the liquid state (Figure 12.11).

To understand why water is different, we have to examine the electronic structure of the H_2O molecule. As we saw in Chapter 9, there are two pairs of nonbonding electrons, or two lone pairs, on the oxygen atom:

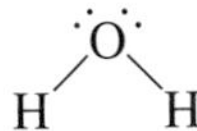

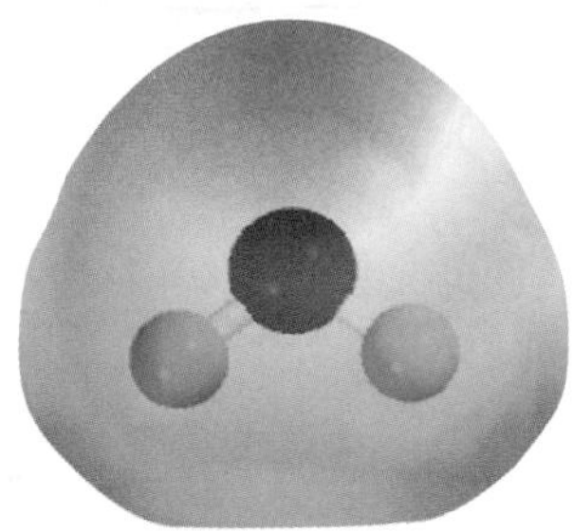

Electrostatic potential map of water.

Although many compounds can form intermolecular hydrogen bonds, the difference between H_2O and other polar molecules, such as NH_3 and HF, is that each oxygen atom can form *two* hydrogen bonds, the same as the number of lone electron pairs on the oxygen atom. Thus, water molecules are joined together in an extensive three-dimensional network in which each oxygen atom is approximately tetrahedrally bonded to four hydrogen atoms, two by covalent bonds and two by hydrogen bonds. This equality in the number of hydrogen atoms and lone pairs is not characteristic of NH_3 or HF or, for that matter, of any other molecule capable of forming hydrogen bonds. Consequently, these other molecules can form rings or chains, but not three-dimensional structures.

The highly ordered three-dimensional structure of ice (Figure 12.12) prevents the molecules from getting too close to one another. But consider what happens when ice melts. At the melting point, a number of water molecules have enough kinetic energy to break free of the intermolecular hydrogen bonds. These molecules become trapped in the cavities of the three-dimensional structure, which is broken down into smaller clusters. As a result, there are more molecules per unit volume in liquid water than in ice. Thus, because density = mass/volume, the density of water is greater than that of ice. With further heating, more water molecules are released from intermolecular hydrogen bonding, so that the density of water tends to increase with rising temperature just above the melting point. Of course, at the same time, water expands as it is being heated so that its density is decreased. These two processes—the trapping of free water molecules in cavities and thermal expansion—act in opposite directions. From 0°C to 4°C, the trapping prevails and water becomes progressively denser. Beyond 4°C, however, thermal expansion predominates and the density of water decreases with increasing temperature (Figure 12.13).

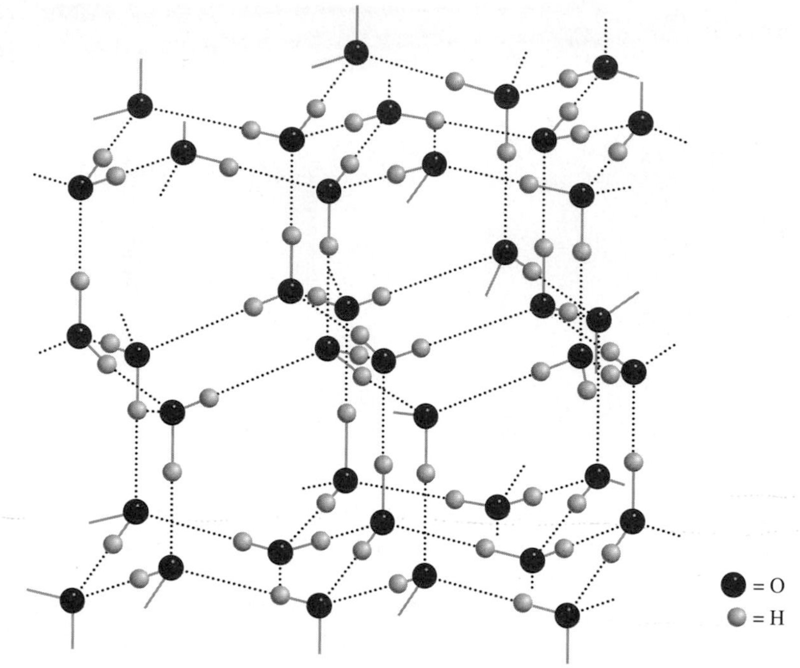

Figure 12.12
The three-dimensional structure of ice. Each O atom is bonded to four H atoms. The covalent bonds are shown by short solid lines and the weaker hydrogen bonds by long dotted lines between O and H. The empty space in the structure accounts for the low density of ice.

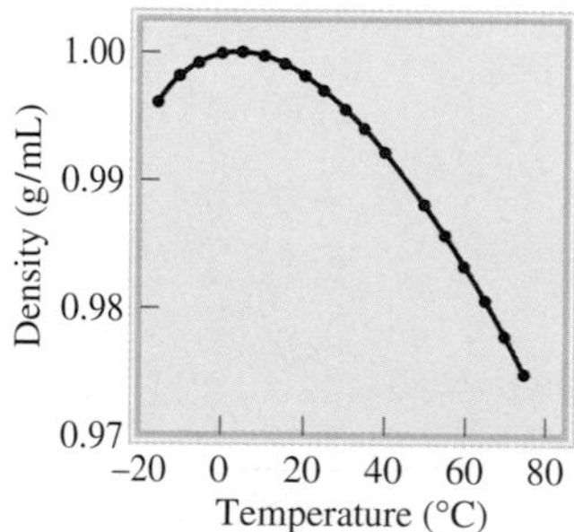

Figure 12.13
Plot of density versus temperature for liquid water. The maximum density of water is reached at 4°C. The density of ice at 0°C is about 0.92 g/cm^3.

Review of Concepts

Why are motorists advised to use more viscous oils for their engines in the summer and less viscous oils in the winter?

12.4 Crystal Structure

Solids can be divided into two categories: crystalline and amorphous. Ice is a ***crystalline solid,*** which *possesses rigid and long-range order; its atoms, molecules, or ions occupy specific positions.* The arrangement of atoms, molecules, or ions in a crystalline solid is such that the net attractive intermolecular forces are at their maximum. The forces responsible for the stability of any crystal can be ionic forces, covalent bonds, van der Waals forces, hydrogen bonds, or a combination of these forces. *Amorphous solids,* such as glass, lack a well-defined arrangement and long-range molecular order. In this section we will concentrate on the structure of crystalline solids.

Virtually all we know about crystal structure has been learned from X-ray diffraction studies. ***X-ray diffraction*** refers to *the scattering of X rays by the units of a crystalline solid.* The scattering, or diffraction, patterns produced are used to deduce the arrangement of particles in the crystalline solid.

The basic repeating structural unit of a crystalline solid is a ***unit cell.*** Figure 12.14 shows a unit cell and its extension in three dimensions. Each sphere *represents an atom, an ion, or a molecule* and is called a ***lattice point.*** In many crystals, the lattice point does not actually contain an atom, ion, or molecule. Rather, there may be several atoms, ions, or molecules identically arranged about each lattice point. For simplicity, however, we can assume that each lattice point is occupied by an atom. Every crystalline solid can be described in terms of one of the seven types of

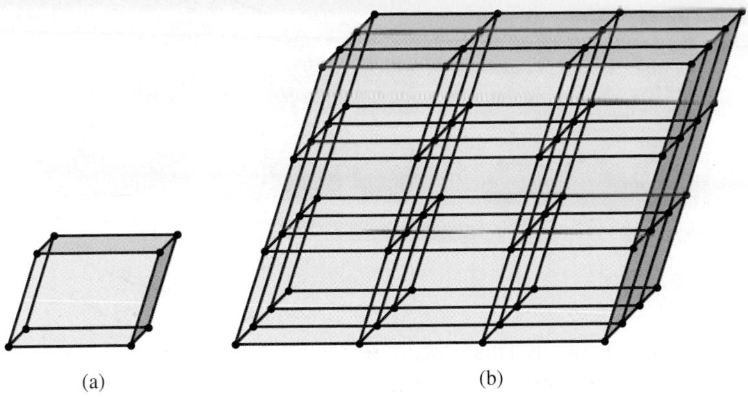

Figure 12.14
(a) A unit cell and (b) its extension in three dimensions. The black spheres represent either atoms or molecules.

unit cells shown in Figure 12.15. The geometry of the cubic unit cell is particularly simple because all sides and all angles are equal. Any of the unit cells, when repeated in space in all three dimensions, forms the lattice structure characteristic of a crystalline solid.

Packing Spheres

Animation: Packing Spheres

We can understand the general geometric requirements for crystal formation by considering the different ways of packing a number of identical spheres (Ping-Pong balls, for example) to form an ordered three-dimensional structure. The way the spheres are arranged in layers determines what type of unit cell we have.

In the simplest case, a layer of spheres can be arranged as shown in Figure 12.16(a). The three-dimensional structure can be generated by placing a layer above and below this layer in such a way that spheres in one layer are directly over the

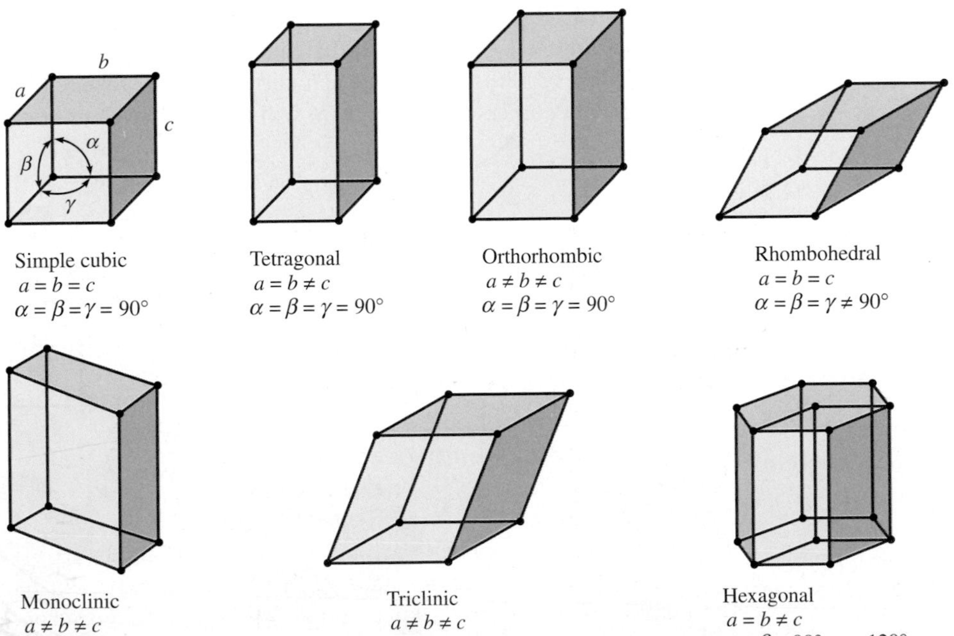

Figure 12.15
The seven types of unit cells. Angle α is defined by edges b and c, angle β by edges a and c, and angle γ by edges a and b.

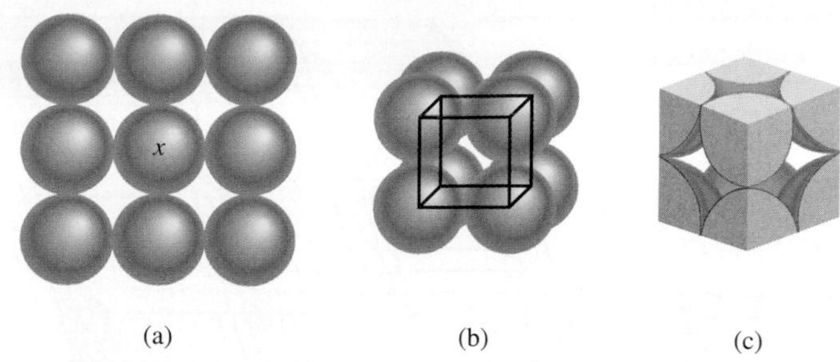

Figure 12.16
Arrangement of identical spheres in a simple cubic cell. (a) Top view of one layer of spheres. (b) Definition of a simple cubic cell. (c) Because each sphere is shared by eight unit cells and there are eight corners in a cube, there is the equivalent of one complete sphere inside a simple cubic unit cell.

spheres in the layer below it. This procedure can be extended to generate many, many layers, as in the case of a crystal. Focusing on the sphere marked with x, we see that it is in contact with four spheres in its own layer, one sphere in the layer above, and one sphere in the layer below. Each sphere in this arrangement is said to have a coordination number of 6 because it has six immediate neighbors. The ***coordination number*** is defined as *the number of atoms (or ions) surrounding an atom (or ion) in a crystal lattice.* The basic, repeating unit in this array of spheres is called a *simple cubic cell* (scc) [Figure 12.16(b)].

The other types of cubic cells are the *body-centered cubic cell* (bcc) and the *face-centered cubic cell* (fcc) (Figure 12.17). A body-centered cubic arrangement differs from a simple cube in that the second layer of spheres fits into the depressions of the first layer and the third layer into the depressions of the second layer. The coordination number of each sphere in this structure is 8 (each sphere is in contact with four spheres in the layer above and four spheres in the layer below). In the face-centered cubic cell there are spheres at the center of each of the six faces of the cube in addition to the eight corner spheres and the coordination number of each sphere is 12.

Because every unit cell in a crystalline solid is adjacent to other unit cells, most of a cell's atoms are shared by neighboring cells. For example, in all types of cubic cells, each corner atom belongs to eight unit cells [Figure 12.18(a)]; an edge atom is shared by four unit cells [Figure 12.18(b)], and a face-centered atom is shared by two unit cells [Figure 12.18(c)]. Because each corner sphere is shared by eight unit cells and there are eight corners in a cube, there will be the equivalent of only one complete

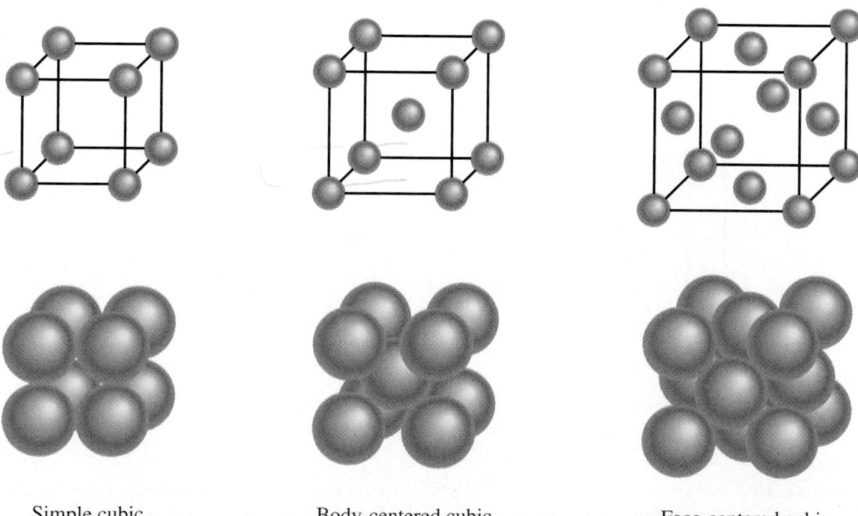

Figure 12.17
Three types of cubic cells. In reality, the spheres representing atoms, molecules, or ions are in contact with one another in these cubic cells.

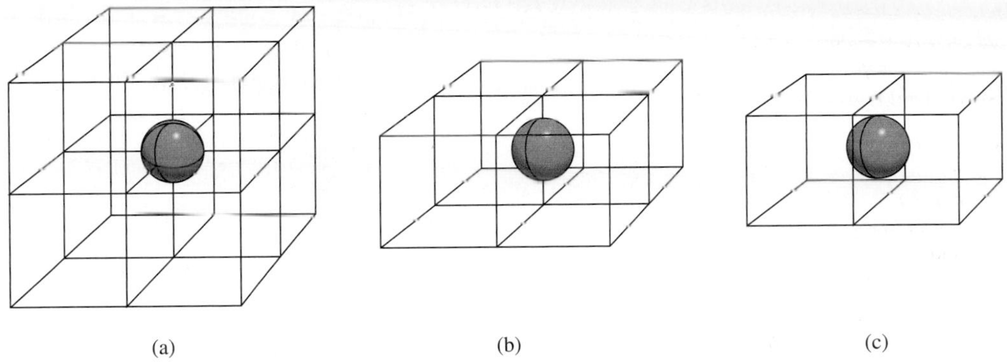

Figure 12.18
(a) A corner atom in any cell is shared by eight unit cells. (b) An edge atom is shared by four unit cells. (c) A face-centered atom in a cubic cell is shared by two unit cells.

sphere inside a simple cubic unit cell (Figure 12.19). A body-centered cubic cell contains the equivalent of two complete spheres, one in the center and eight shared corner spheres. A face-centered cubic cell contains four complete spheres—three from the six face-centered atoms and one from the eight shared corner spheres.

Closest Packing

Clearly there is more empty space in the simple cubic and body-centered cubic cells than in the face-centered cubic cell. ***Closest packing,*** *the most efficient arrangement of spheres,* starts with the structure shown in Figure 12.20(a), which we call layer A. Focusing on the only enclosed sphere, we see that it has six immediate neighbors in that layer. In the second layer (which we call layer B), spheres are packed into the depressions between the spheres in the first layer so that all the spheres are as close together as possible [Figure 12.20(b)].

These oranges are in a closest packed arrangement as shown in Figure 12.20(a).

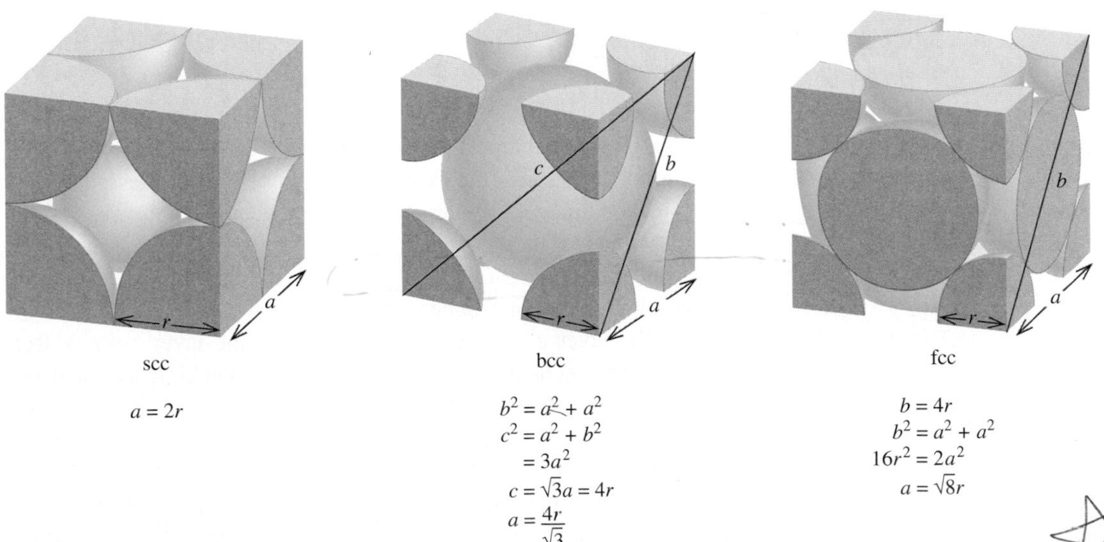

Figure 12.19
The relationship between the edge length (a) and radius (r) of atoms in the simple cubic cell (scc), body-centered cubic cell (bcc), and face-centered cubic cell (fcc).

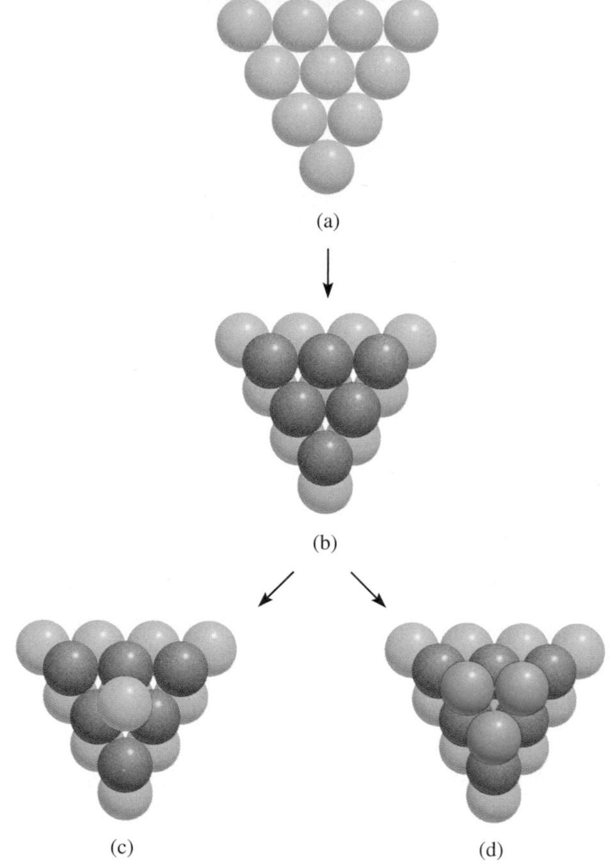

Figure 12.20
(a) In a close-packed layer, each sphere is in contact with six others. (b) Spheres in the second layer fit into the depressions between the first-layer spheres. (c) In the hexagonal close-packed structure, each third layer sphere is directly over a first-layer sphere. (d) In the cubic close-packed structure, each third layer sphere fits into a depression that is directly over a depression in the first layer.

There are two ways that a third-layer sphere may cover the second layer to achieve closest packing. The spheres may fit into the depressions so that each third-layer sphere is directly over a first-layer sphere [Figure 12.20(c)]. Because there is no difference between the arrangement of the first and third layers, we also call the third layer layer A. Alternatively, the third-layer spheres may fit into the depressions that lie directly over the depressions in the first layer [Figure 12.20(d)]. In this case, we call the third layer layer C. Figure 12.21 shows the "exploded views" and the structures resulting from these two arrangements. The ABA arrangement is known as the *hexagonal close-packed (hcp) structure,* and the ABC arrangement is the *cubic close-packed (ccp) structure,* which corresponds to the face-centered cube already described. Note that in the hcp structure, the spheres in every other layer occupy the same vertical position (ABABAB . . .), while in the ccp structure, the spheres in every fourth layer occupy the same vertical position (ABCABCA . . .). In both structures, each sphere has a coordination number of 12 (each sphere is in contact with six spheres in its own layer, three spheres in the layer above, and three sphere in the layer below). Both the hcp and ccp structures represent the most efficient way of packing identical spheres in a unit cell, and there is no way to increase the coordination number to beyond 12.

Many metals and noble gases, which are monatomic, form crystals with hcp or ccp structures. For example, magnesium, titanium, and zinc crystallize with their atoms in a hcp array, while aluminum, nickel, and silver crystallize in the ccp arrangement. All solid noble gases have the ccp structure except helium, which crystallizes in the hcp structure. It is natural to ask why a series of related substances, such as the transition metals or the noble gases, would form different crystal structures. The answer lies in the relative stability of a particualar crystal structure, which is governed

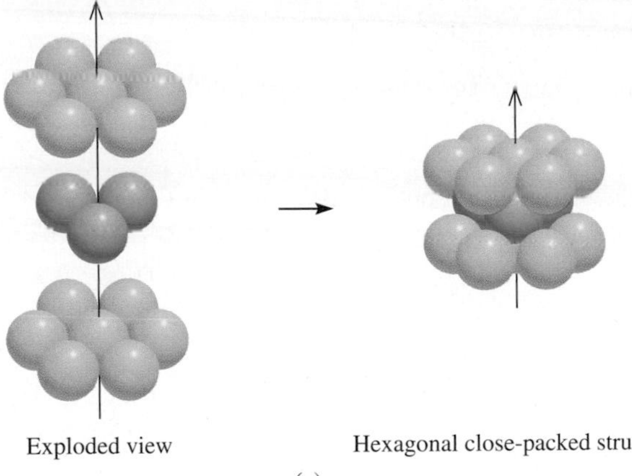

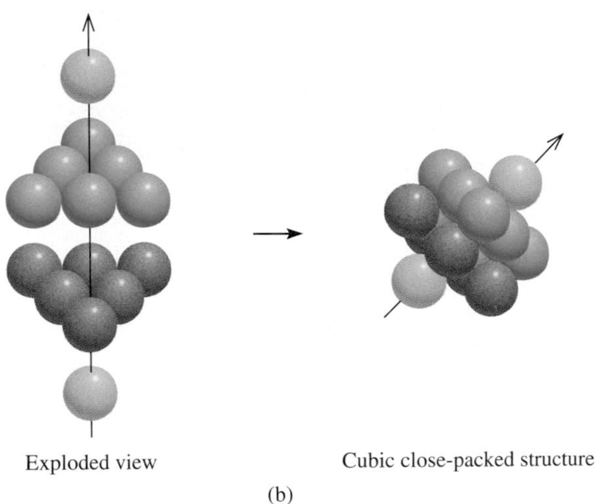

Figure 12.21 *Exploded views of (a) a hexagonal close-packed structure and (b) a cubic close-packed structure. The arrow is tilted to show the face-centered cubic unit cell more clearly. Note that this arrangement is the same as the face-centered unit cell.*

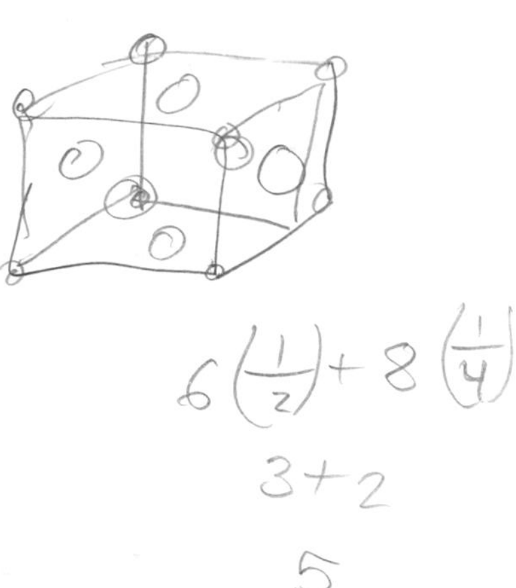

by intermolecular forces. Thus, magnesium metal has the hcp structure because this arrangement of Mg atoms results in the greatest stability of the solid.

Figure 12.19 also summarizes the relationship between the atomic radius r and the edge length a of a simple cubic cell, a body-centered cubic cell, and a face-centered cubic cell. This relationship can be used to determine the density of a crystal or the radius of an atom, as Example 12.3 shows.

EXAMPLE 12.3

Gold (Au) crystallizes in a cubic close-packed structure (the face-centered cubic unit cell) and has a density of 19.3 g/cm^3. Calculate the atomic radius of gold in picometers.

Strategy We want to calculate the radius of a gold atom. For a face-centered cubic unit cell, the relationship between radius (r) and edge length (a), according to Figure 12.19, is $a = \sqrt{8}r$. Therefore, to determine r of a Au atom, we need to find a. The volume of a cube is $V = a^3$ or $a = \sqrt[3]{V}$. Thus, if we can determine the volume of the unit cell, we can calculate a. We are given the density in the problem.

$$\underset{\text{given}}{\text{density}} = \frac{\overset{\text{need to find}}{\text{mass}}}{\underset{\text{want to calculate}}{\text{volume}}}$$

(Continued)

The sequence of steps is summarized as follows:

density of unit cell ⟶ volume of unit cell ⟶ edge length of unit cell ⟶ radius of Au atom

Solution

Step 1: We know the density, so in order to determine the volume, we find the mass of the unit cell. Each unit cell has eight corners and six faces. The total number of atoms within such a cell, according to Figure 12.18, is

$$\left(8 \times \frac{1}{8}\right) + \left(6 \times \frac{1}{2}\right) = 4$$

The mass of a unit cell in grams is

$$m = \frac{4 \text{ atoms}}{1 \text{ unit cell}} \times \frac{1 \text{ mol}}{6.022 \times 10^{23} \text{ atoms}} \times \frac{197.0 \text{ g Au}}{1 \text{ mol Au}}$$
$$= 1.31 \times 10^{-21} \text{ g/unit cell}$$

Remember that density is an intensive property, so that it is the same for one unit cell and 1 cm^3 of the substance.

From the definition of density ($d = m/V$), we calculate the volume of the unit cell as follows:

$$V = \frac{m}{d} = \frac{1.31 \times 10^{-21} \text{ g}}{19.3 \text{ g/cm}^3} = 6.79 \times 10^{-23} \text{ cm}^3$$

Step 2: Because volume is length cubed, we take the cubic root of the volume of the unit cell to obtain the edge length (a) of the cell

$$a = \sqrt[3]{V}$$
$$= \sqrt[3]{6.79 \times 10^{-23} \text{ cm}^3}$$
$$= 4.08 \times 10^{-8} \text{ cm}$$

Step 3: From Figure 12.19 we see that the radius of an Au sphere (r) is related to the edge length by

$$a = \sqrt{8}r$$

Therefore,

$$r = \frac{a}{\sqrt{8}} = \frac{4.08 \times 10^{-8} \text{ cm}}{\sqrt{8}}$$
$$= 1.44 \times 10^{-8} \text{ cm}$$
$$= 1.44 \times 10^{-8} \text{ cm} \times \frac{1 \times 10^{-2} \text{ m}}{1 \text{ cm}} \times \frac{1 \text{ pm}}{1 \times 10^{-12} \text{ m}}$$
$$= 144 \text{ pm}$$

Similar problem: 12.48.

Practice Exercise When silver crystallizes, it forms face-centered cubic cells. The unit cell edge length is 408.7 pm. Calculate the density of silver.

Review of Concepts

Tungsten crystallizes in a body-centered cubic lattice (the W atoms occupy only the lattice points). How many W atoms are present in a unit cell?

12.5 Bonding in Solids

The structure and properties of crystalline solids, such as melting point, density, and hardness, are determined by the attractive forces that hold the particles together. We can classify crystals according to the types of forces between particles: ionic, molecular, covalent, and metallic (Table 12.4).

Table 12.4 Types of Crystals and General Properties

Type of Crystal	Force(s) Holding the Units Together	General Properties	Examples
Ionic	Electrostatic attraction	Hard, brittle, high melting point, poor conductor of heat and electricity	NaCl, LiF, MgO, $CaCO_3$
Molecular*	Dispersion forces, dipole-dipole forces, hydrogen bonds	Soft, low melting point, poor conductor of heat and electricity	Ar, CO_2, I_2, H_2O, $C_{12}H_{22}O_{11}$ (sucrose)
Covalent	Covalent bond	Hard, high melting point, poor conductor of heat and electricity	C (diamond),† SiO_2 (quartz)
Metallic	Metallic bond	Soft to hard, low to high melting point, good conductor of heat and electricity	All metallic elements; for example, Na, Mg, Fe, Cu

*Included in this category are crystals made up of individual atoms.
†Diamond is a good thermal conductor.

Ionic Crystals

Ionic crystals consist of ions held together by ionic bonds. The structure of an ionic crystal depends on the charges on the cation and anion and on their radii. We have already discussed the structure of sodium chloride, which has a face-centered cubic lattice (see Figure 2.12). Figure 12.22 shows the structures of three other ionic crystals: CsCl, ZnS, and CaF_2. Because Cs^+ is considerably larger than Na^+, CsCl has the simple cubic lattice structure. ZnS has the *zincblende* structure, which is based on the face-centered cubic lattice. If the S^{2-} ions are located at the lattice points, the Zn^{2+} ions are located one-fourth of the distance along each body diagonal. Other ionic compounds that have the zincblende structure include CuCl, BeS, CdS, and HgS. CaF_2 has the *fluorite* structure. The Ca^{2+} ions are located at the lattice points, and each F^- ion is tetrahedrally surrounded by four Ca^{2+} ions. The compounds SrF_2, BaF_2, $BaCl_2$, and PbF_2 also have the fluorite structure.

These giant ionic potassium dihydrogen phosphate crystals were grown in the laboratory. The largest one weighs 701 lb!

Ionic solids have high melting points, an indication of the strong cohesive force holding the ions together. These solids do not conduct electricity because the ions are fixed in position. However, in the molten state (that is, when melted) or dissolved in water, the ions are free to move and the resulting liquid is electrically conducting.

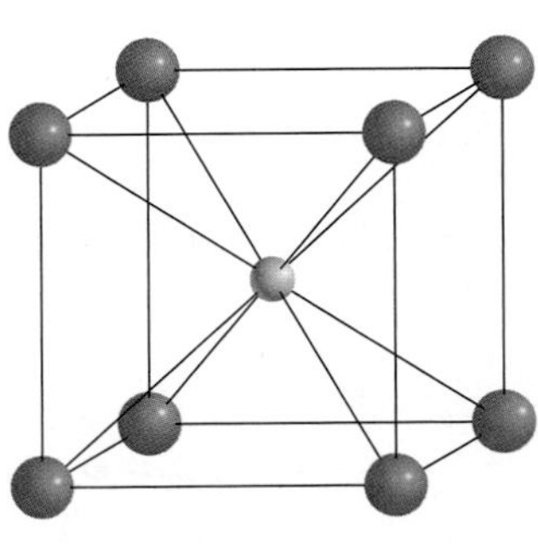

(a)

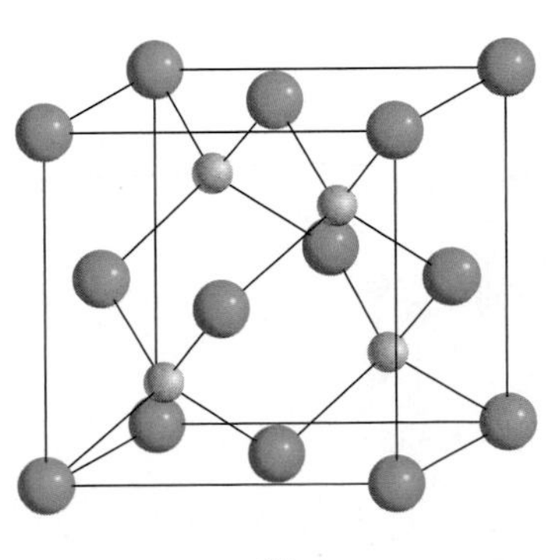

(b)

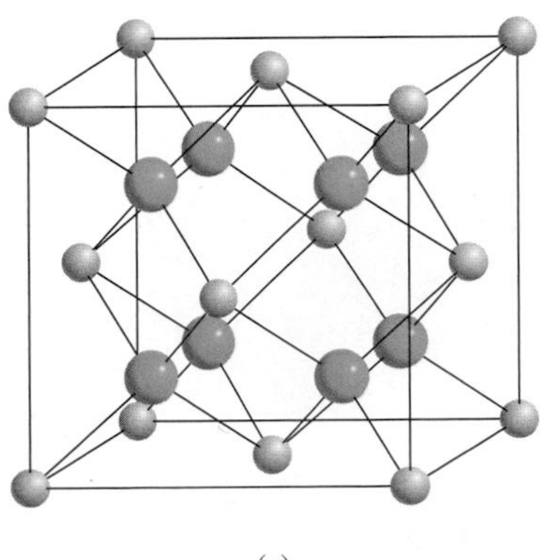

(c)

Figure 12.22
Crystal structures of (a) CsCl, (b) ZnS, and (c) CaF_2. In each case, the cation is the smaller sphere.

EXAMPLE 12.4

How many Na^+ and Cl^- ions are in each NaCl unit cell?

Solution NaCl has a structure based on a face-centered cubic lattice. As Figure 2.12 shows, one whole Na^+ ion is at the center of the unit cell, and there are 12 Na^+ ions at the edges. Because each edge Na^+ ion is shared by four unit cells, the total number of Na^+ ions is $1 + (12 \times \frac{1}{4}) = 4$. Similarly, there are six Cl^- ions at the face centers and eight Cl^- ions at the corners. Each face-centered ion is shared by two unit cells, and each corner ion is shared by eight unit cells (see Figure 12.18), so the total number of Cl^- ions is $(6 \times \frac{1}{2}) + (8 \times \frac{1}{8}) = 4$. Thus, there are four Na^+ ions and four Cl^- ions in each NaCl unit cell. Figure 12.23 shows the portions of the Na^+ and Cl^- ions *within* a unit cell.

Similar problem: 12.47.

Practice Exercise How many atoms are in a body-centered cube, assuming that all atoms occupy lattice points?

Molecular Crystals

Molecular crystals consist of atoms or molecules held together by van der Waals forces and/or hydrogen bonding. An example of a molecular crystal is solid sulfur dioxide (SO_2), in which the predominant attractive force is dipole-dipole interaction. Intermolecular hydrogen bonding is mainly responsible for the three-dimensional ice lattice (see Figure 12.12). Other examples of molecular crystals are I_2, P_4, and S_8.

In general, except in ice, molecules in molecular crystals are packed together as closely as their size and shape allow. Because van der Waals forces and hydrogen bonding are generally quite weak compared with covalent and ionic bonds, molecular crystals are more easily broken apart than ionic and covalent crystals. Indeed, most molecular crystals melt below 200°C.

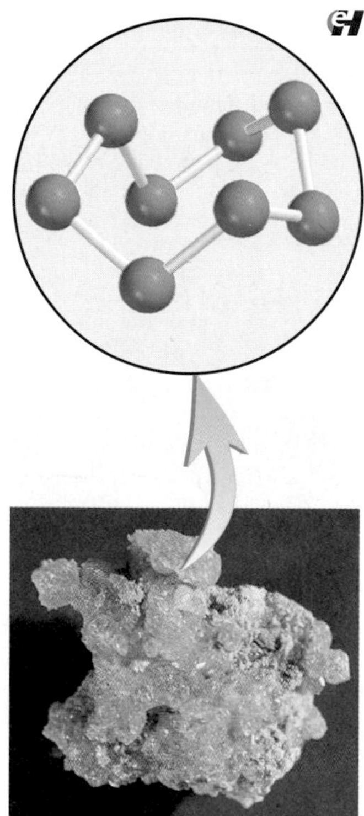

Sulfur.

Covalent Crystals

In covalent crystals (sometimes called covalent network crystals), atoms are held together entirely by covalent bonds in an extensive three-dimensional network. No discrete molecules are present, as in the case of molecular solids. Well-known examples are the two allotropes of carbon: diamond and graphite (see Figure 8.14). In diamond each carbon atom is tetrahedrally bonded to four other atoms (Figure 12.24). The strong covalent bonds in three dimensions contribute to diamond's unusual

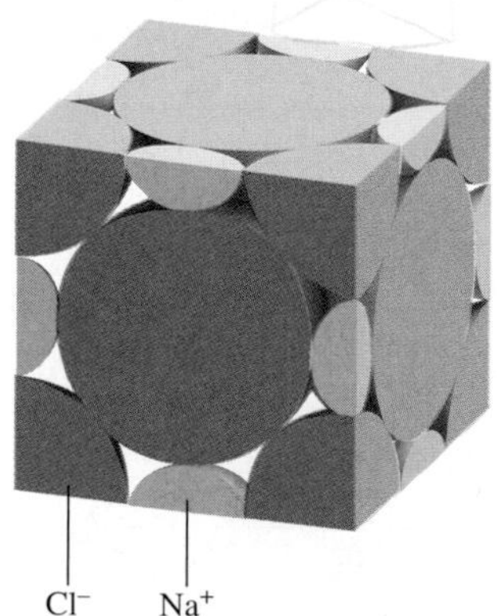

Figure 12.23
Portions of Na^+ and Cl^- ions within a face-centered cubic unit cell.

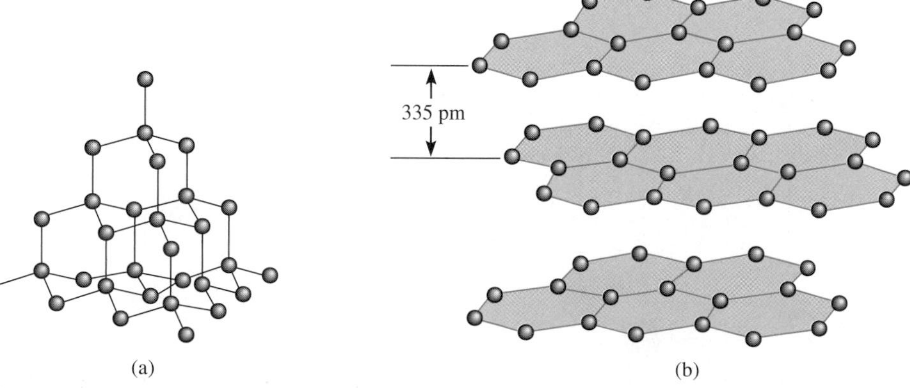

Figure 12.24
(a) The structure of diamond. Each carbon is tetrahedrally bonded to four other carbon atoms. (b) The structure of graphite. The distance between successive layers is 335 pm.

hardness (it is the hardest material known) and high melting point (3550°C). In graphite, carbon atoms are arranged in six-membered rings. The atoms are all sp^2-hybridized; each atom is covalently bonded to three other atoms. The unhybridized $2p$ orbital is used in pi bonding. In fact, the electrons in these $2p$ orbitals are free to move around, making graphite a good conductor of electricity in the planes of the bonded carbon atoms. The layers are held together by the weak van der Waals forces. The covalent bonds in graphite account for its hardness; however, because the layers can slide over one another, graphite is slippery to the touch and is effective as a lubricant. It is also used in pencils and in ribbons made for computer printers and typewriters.

The central electrode in flashlight batteries is made of graphite.

Another type of covalent crystal is quartz (SiO_2). The arrangement of silicon atoms in quartz is similar to that of carbon in diamond, but in quartz there is an oxygen atom between each pair of Si atoms. Because Si and O have different electronegativities (see Figure 9.4), the Si—O bond is polar. Nevertheless, SiO_2 is similar to diamond in many respects, such as hardness and high melting point (1610°C).

Quartz.

Metallic Crystals

In a sense, the structure of metallic crystals is the simplest to deal with, because every lattice point in a crystal is occupied by an atom of the same metal. The bonding in metals is quite different from that in other types of crystals. In a metal, the bonding electrons are spread (or *delocalized*) over the entire crystal. In fact, metal atoms in a crystal can be imagined as an array of positive ions immersed in a sea of delocalized valence electrons (Figure 12.25). The great cohesive force resulting from delocalization is responsible for a metal's strength, which increases as the number of electrons available for bonding increases. For example, the melting point of sodium, with one valence electron, is 97.6°C, whereas that of aluminum, with three valence electrons, is 660°C. The mobility of the delocalized electrons makes metals good conductors of heat and electricity.

This comparison applies only to representative elements.

Solids are most stable in crystalline form. However, if a solid is formed rapidly (for example, when a liquid is cooled quickly), its atoms or molecules do not have time to align themselves and may become locked in positions other than those of a regular crystal. The resulting solid is said to be *amorphous*. ***Amorphous solids***, such as glass, *lack a regular three-dimensional arrangement of atoms.*

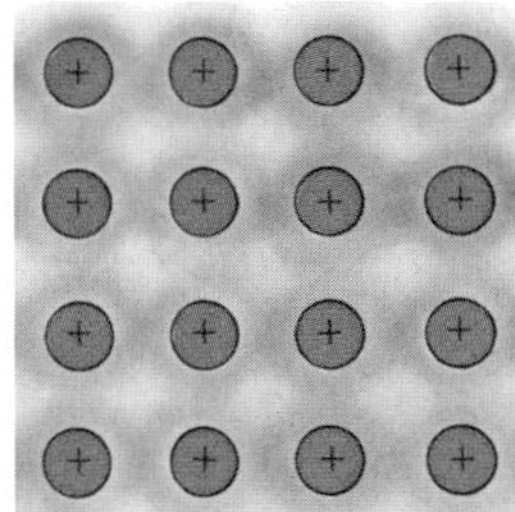

Figure 12.25

A cross section of a metallic crystal. Each circled positive charge represents the nucleus and inner electrons of a metal atom. The gray area surrounding the positive metal ions indicates the mobile sea of electrons.

Review of Concepts

Shown here is a zinc oxide unit cell. What is the formula of the compound?

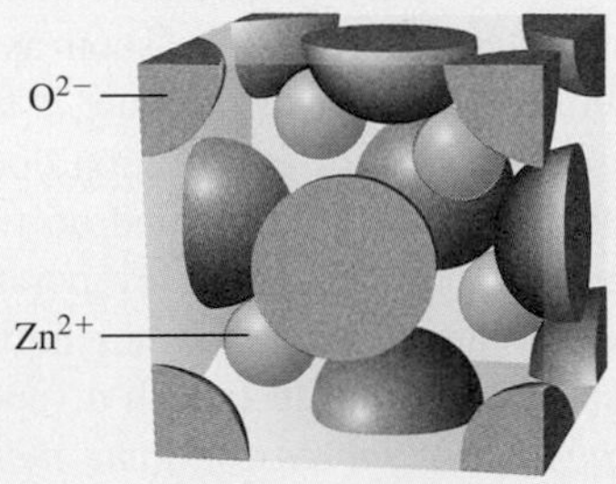

12.6 Phase Changes

The discussions in Chapter 5 and in this chapter have given us an overview of the properties of the three states of matter: gas, liquid, and solid. Each of these states is often referred to as a ***phase***, which is *a homogeneous part of the system in contact*

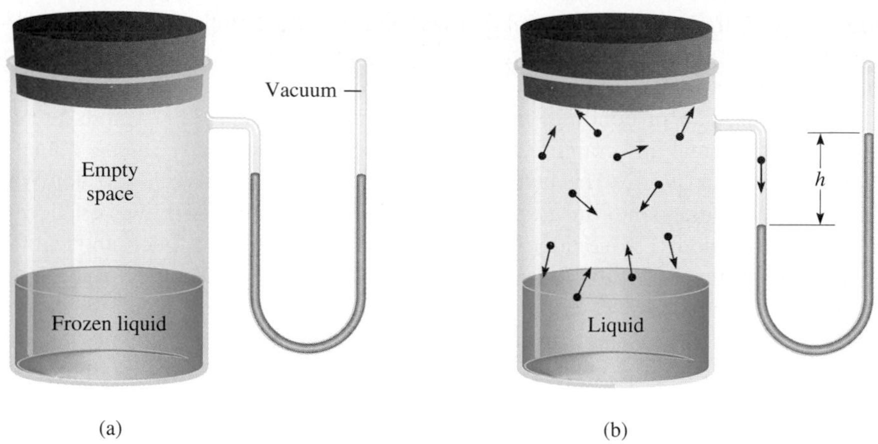

Figure 12.26
Apparatus for measuring the vapor pressure of a liquid (a) Initially the liquid is frozen so there are no molecules in the vapor phase. (b) On heating, a liquid phase is formed and evaporization begins. At equilibrium, the number of molecules leaving the liquid is equal to the number of molecules returning to the liquid. The difference in the mercury levels (h) gives the equilibrium vapor pressure of the liquid at the specified temperature.

with other parts of the system but separated from them by a well-defined boundary. An ice cube floating in water makes up two phases of water—the solid phase (ice) and the liquid phase (water). ***Phase changes,*** *transformations from one phase to another,* occur when energy (usually in the form of heat) is added or removed from a substance. Phase changes are physical changes that are characterized by changes in molecular order; molecules in the solid state have the most order, and those in the gas phase have the greatest randomness. Keeping in mind the relationship between energy change and the increase or decrease in molecular order will help us understand the nature of phase changes.

Liquid-Vapor Equilibrium

Vapor Pressure

The difference between a gas and a vapor is explained on p. 137.

Molecules in a liquid are not fixed in a rigid lattice. Although they lack the total freedom of gaseous molecules, these molecules are in constant motion. Because liquids are denser than gases, the collision rate among molecules is much higher in the liquid phase than in the gas phase. *At any given temperature, a certain number of the molecules in a liquid possess sufficient kinetic energy to escape from the surface.* This process is called ***evaporation,*** or ***vaporization.***

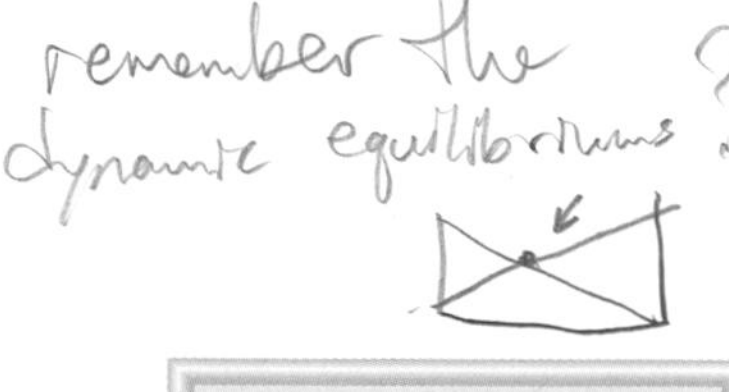

When a liquid evaporates, its gaseous molecules exert a vapor pressure. Consider the apparatus shown in Figure 12.26. Before the evaporation process starts, the mercury levels in the U-shaped manometer are equal. As soon as some molecules leave the liquid, a vapor phase is established. The vapor pressure is measurable only when a fair amount of vapor is present. The process of evaporation does not continue indefinitely, however. Eventually, the mercury levels stabilize and no further changes are seen.

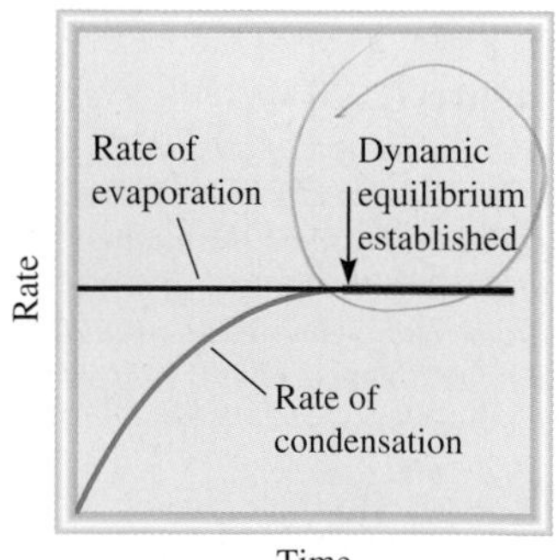

Figure 12.27
Comparison of the rates of evaporation and condensation at constant temperature.

What happens at the molecular level during evaporation? In the beginning, the traffic is only one way: Molecules are moving from the liquid to the empty space. Soon the molecules in the space above the liquid establish a vapor phase. *As the concentration of molecules in the vapor phase increases, some molecules return to the liquid phase,* a process called ***condensation.*** Condensation occurs because a molecule striking the liquid surface becomes trapped by intermolecular forces in the liquid.

The rate of evaporation is constant at any given temperature, and the rate of condensation increases with increasing concentration of molecules in the vapor phase. A state of ***dynamic equilibrium,*** in which *the rate of a forward process is exactly balanced by the rate of the reverse process,* is reached when the rates of condensation and evaporation become equal (Figure 12.27). *The vapor pressure measured under* *dynamic equilibrium of condensation and evaporation* is called the ***equilibrium vapor pressure.*** We often use the simpler term "vapor pressure" when we talk about the equilibrium vapor pressure of a liquid. This practice is acceptable as long as we know the meaning of the abbreviated term.

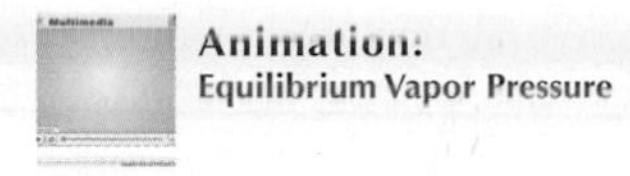

It is important to note that the equilibrium vapor pressure is the *maximum* vapor pressure a liquid exerts at a given temperature and that it is constant at constant temperature. Vapor pressure does change with temperature, however. Plots of vapor pressure versus temperature for three different liquids are shown in Figure 12.28. We know that the number of molecules with higher kinetic energies is greater at the higher temperature and therefore so is the evaporation rate. For this reason, the vapor pressure of a liquid always increases with temperature. For example, the vapor pressure of water is 17.5 mmHg at 20°C, but it rises to 760 mmHg at 100°C.

Equilibrium vapor pressure is independent of the amount of liquid as long as there is some liquid present.

Heat of Vaporization and Boiling Point

A measure of how strongly molecules are held in a liquid is its ***molar heat of vaporization*** (***ΔH_{vap}***), defined as *the energy* (usually in kilojoules) *required to vaporize one mole of a liquid.* The molar heat of vaporization is directly related to the strength of intermolecular forces that exist in the liquid. If the intermolecular attraction is strong, it takes a lot of energy to free the molecules from the liquid phase. Consequently, the liquid has a relatively low vapor pressure and a high molar heat of vaporization.

The previous discussion predicts that the equilibrium vapor pressure (P) of a liquid should increase with increasing temperature as shown in Figure 12.28. Analysis of this behavior reveals that the quantitative relationship between the vapor pressure P of a liquid and the absolute temperature T is given by the Clausius-Clapeyron equation:

$$\ln P = -\frac{\Delta H_{\text{vap}}}{RT} + C \qquad (12.1)$$

in which ln is the natural logarithm, R is the gas constant (8.314 J/K · mol), and C is a constant. The Clausius-Clapeyron equation has the form of the linear equation $y = mx + b$:

$$\begin{array}{ccccc} \ln P = & \left(-\dfrac{\Delta H_{\text{vap}}}{R}\right) & \left(\dfrac{1}{T}\right) & + C \\ \updownarrow & \updownarrow & \updownarrow & \updownarrow \\ y = & m & x & + b \end{array}$$

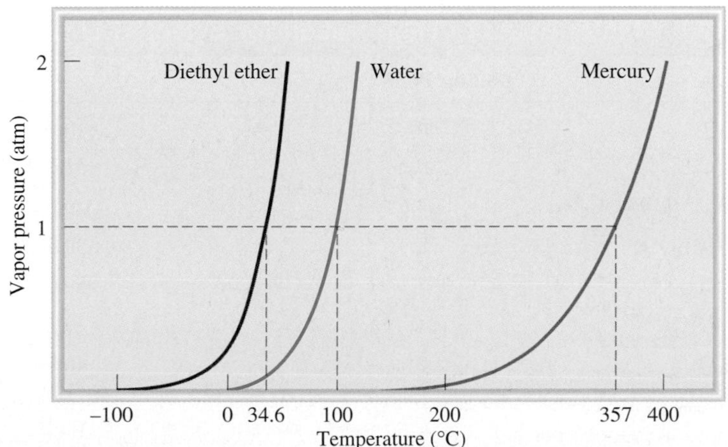

Figure 12.28
The increase in vapor pressure with temperature for three liquids. The normal boiling points of the liquids (at 1 atm) are shown on the horizontal axis. The strong metallic bonding in mercury results in a much lower vapor pressure of the liquid at room temperature.

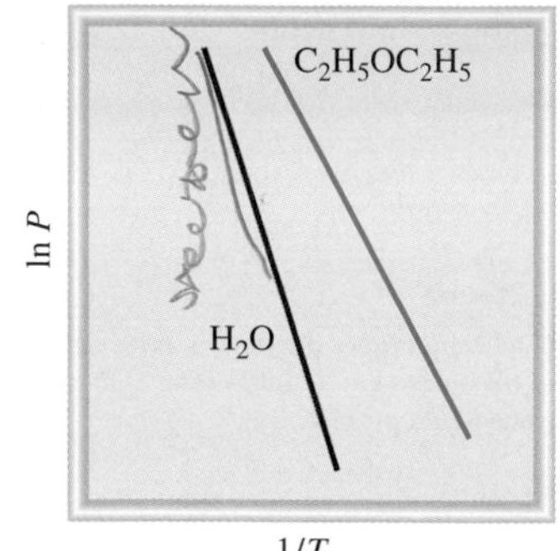

Figure 12.29
Plots of ln P versus 1/T for water and diethyl ether. The slope in each case is equal to $-\Delta H_{vap}/R$.

By measuring the vapor pressure of a liquid at different temperatures and plotting ln P versus $1/T$, we determine the slope of the line described by the equation, which is equal to $-\Delta H_{vap}/R$. (ΔH_{vap} is assumed to be independent of temperature.) This is the method used to determine heats of vaporization. Figure 12.29 shows plots of ln P versus $1/T$ for water and diethyl ether ($C_2H_5OC_2H_5$). Note that the straight line for water has a steeper slope because water has a larger ΔH_{vap} (Table 12.5).

If we know the values of ΔH_{vap} and P of a liquid at one temperature, we can use the Clausius-Clapeyron equation to calculate the vapor pressure of the liquid at a different temperature. At temperatures T_1 and T_2 the vapor pressures are P_1 and P_2. From Equation (12.1) we can write

$$\ln P_1 = -\frac{\Delta H_{vap}}{RT_1} + C \tag{12.2}$$

$$\ln P_2 = -\frac{\Delta H_{vap}}{RT_2} + C \tag{12.3}$$

Subtracting Equation (12.3) from Equation (12.2) we obtain

$$\begin{aligned}\ln P_1 - \ln P_2 &= -\frac{\Delta H_{vap}}{RT_1} - \left(-\frac{\Delta H_{vap}}{RT_2}\right)\\ &= \frac{\Delta H_{vap}}{R}\left(\frac{1}{T_2} - \frac{1}{T_1}\right)\end{aligned}$$

Hence,

$$\ln \frac{P_1}{P_2} = \frac{\Delta H_{vap}}{R}\left(\frac{1}{T_2} - \frac{1}{T_1}\right)$$

or

$$\ln \frac{P_1}{P_2} = \frac{\Delta H_{vap}}{R}\left(\frac{T_1 - T_2}{T_1T_2}\right) \tag{12.4}$$

Table 12.5 Molar Heats of Vaporization for Selected Liquids

Substance	Boiling Point* (°C)	ΔH_{vap} (kJ/mol)
Argon (Ar)	−186	6.3
Benzene (C_6H_6)	80.1	31.0
Diethyl ether ($C_2H_5OC_2H_5$)	34.6	26.0
Ethanol (C_2H_5OH)	78.3	39.3
Mercury (Hg)	357	59.0
Methane (CH_4)	−164	9.2
Water (H_2O)	100	40.79

*Measured at 1 atm.

EXAMPLE 12.5

Diethyl ether is a volatile, highly flammable organic liquid that is used mainly as a solvent. The vapor pressure of diethyl ether is 401 mmHg at 18°C. Calculate its vapor pressure at 29°C.

$C_2H_5OC_2H_5$

Strategy We are given the vapor pressure of diethyl ether at one temperature and asked to find the pressure at another temperature. Therefore, we need Equation (12.4).

Solution Table 12.5 tells us that $\Delta H_{vap} = 26.0$ kJ/mol. The data are

$$P_1 = 401 \text{ mmHg} \qquad P_2 = ?$$
$$T_1 = 18°\text{C} = 291 \text{ K} \qquad T_2 = 29°\text{C} = 302 \text{ K}$$

From Equation (12.4) we have

$$\ln \frac{401}{P_2} = \frac{26{,}000 \text{ J/mol}}{8.314 \text{ J/K} \cdot \text{mol}} \left[\frac{291 \text{ K} - 302 \text{ K}}{(291 \text{ K})(302 \text{ K})} \right]$$

Taking the antilog of both sides (see Appendix 3), we obtain

$$\frac{401}{P_2} = e^{-0.391} = 0.676$$

Hence,

$$P_2 = 593 \text{ mmHg}$$

Check We expect the vapor pressure to be greater at the higher temperature. Therefore, the answer is reasonable.

Practice Exercise The vapor pressure of ethanol is 100 mmHg at 34.9°C. What is its vapor pressure at 63.5°C? (ΔH_{vap} for ethanol is 39.3 kJ/mol.)

Similar problem: 12.80.

A practical way to demonstrate the molar heat of vaporization is by rubbing alcohol on your hands. The heat from your hands increases the kinetic energy of the alcohol molecules. The alcohol evaporates rapidly, extracting heat from your hands and cooling them. The process is similar to perspiration, which is one means by which the human body maintains a constant temperature. Because of the strong intermolecular hydrogen bonding that exists in water, a considerable amount of energy is needed to vaporize the water in perspiration from the body's surface. This energy is supplied by the heat generated in various metabolic processes.

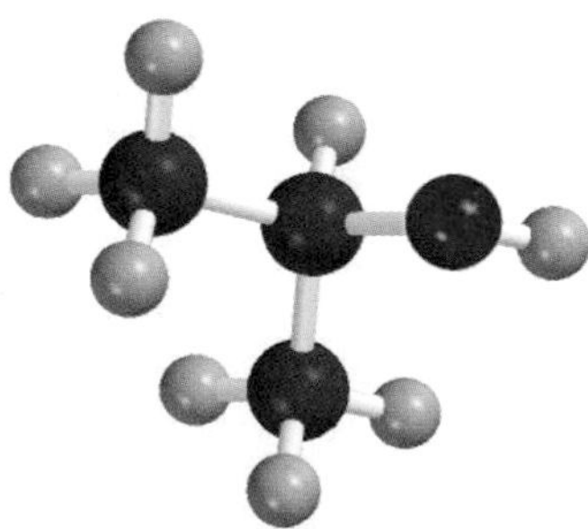

Isopropanol (rubbing alcohol)

You have already seen that the vapor pressure of a liquid increases with temperature. For every liquid there exists a temperature at which the liquid begins to boil. The ***boiling point*** is *the temperature at which the vapor pressure of a liquid is equal to the external pressure.* The *normal* boiling point of a liquid is the boiling point when the external pressure is 1 atm.

At the boiling point, bubbles form within the liquid. When a bubble forms, the liquid originally occupying that space is pushed aside, and the level of the liquid in the container is forced to rise. The pressure exerted *on* the bubble is largely atmospheric pressure, plus some *hydrostatic pressure* (that is, pressure caused by the presence of liquid). The pressure *inside* the bubble is due solely to the vapor pressure of the liquid. When the vapor pressure equals the external pressure, the bubble rises to the surface of the liquid and bursts. If the vapor pressure in the bubble were lower than the external pressure, the bubble would collapse before it could rise. We can thus conclude that the boiling point of a liquid depends on the external pressure. (We usually ignore the small contribution caused by the hydrostatic pressure.) For example, at 1 atm water boils at 100°C, but if the pressure is reduced to 0.5 atm, water boils at only 82°C.

Because the boiling point is defined in terms of the vapor pressure of the liquid, we expect the boiling point to be related to the molar heat of vaporization: The higher ΔH_{vap}, the higher the boiling point. The data in Table 12.5 roughly confirm our prediction. Ultimately, both the boiling point and ΔH_{vap} are determined by the strength of intermolecular forces. For example, argon (Ar) and methane (CH_4), which have weak dispersion forces, have low boiling points and small molar heats of vaporization. Diethyl ether ($C_2H_5OC_2H_5$) has a dipole moment, and the dipole-dipole forces account for its moderately high boiling point and ΔH_{vap}. Both ethanol (C_2H_5OH) and water have strong hydrogen bonding, which accounts for their high boiling points and large ΔH_{vap} values. Strong metallic bonding causes mercury to have the highest boiling point and ΔH_{vap} of this group of liquids. Interestingly, the boiling point of benzene, which is nonpolar, is comparable to that of ethanol. Benzene has a large polarizability, and the dispersion forces among benzene molecules can be as strong as or even stronger than dipole-dipole forces and/or hydrogen bonds.

Critical Temperature and Pressure

The opposite of evaporation is condensation. In principle, a gas can be made to liquefy by either one of two techniques. By cooling a sample of gas we decrease the kinetic energy of its molecules, and eventually molecules aggregate to form small drops of liquid. Alternatively, we may apply pressure to the gas. Under compression, the average distance between molecules is reduced so that they are held together by mutual attraction. Industrial liquefaction processes combine these two methods.

Every substance has a ***critical temperature*** (T_c), *above which its gas form cannot be made to liquefy, no matter how great the applied pressure.* This is also *the highest temperature at which a substance can exist as a liquid. The minimum pressure that must be applied to bring about liquefaction at the critical temperature* is called the ***critical pressure*** (P_c). The existence of the critical temperature can be qualitatively explained as follows. The intermolecular attraction is a finite quantity for any given substance. Below T_c, this force is sufficiently strong to hold the molecules together (under some appropriate pressure) in a liquid. Above T_c, molecular motion becomes so energetic that the molecules can always break away from this attraction. Figure 12.30 shows what happens when sulfur hexafluoride is heated above its critical temperature (45.5°C) and then cooled down to below 45.5°C.

Intermolecular forces are independent of temperature; the kinetic energy of molecules increases with temperature.

Table 12.6 lists the critical temperatures and critical pressures of a number of common substances. Benzene, ethanol, mercury, and water, which have strong intermolecular forces, also have high critical temperatures compared with the other substances listed in the table.

Liquid-Solid Equilibrium

"Fusion" refers to the process of melting. Thus, a "fuse" breaks an electrical circuit when a metallic strip melts due to the heat generated by excessively high electrical current.

The transformation of liquid to solid is called *freezing,* and the reverse process is called *melting* or *fusion.* The ***melting point*** of a solid (or the freezing point of a liquid) is *the temperature at which solid and liquid phases coexist in equilibrium.* The *normal* melting point (or the *normal* freezing point) of a substance is the melting point (or freezing point) measured at 1 atm pressure. We generally omit the word "normal" in referring to the melting point of a substance at 1 atm.

The most familiar liquid-solid equilibrium is that of water and ice. At 0°C and 1 atm, the dynamic equilibrium is represented by

$$\text{ice} \rightleftharpoons \text{water}$$

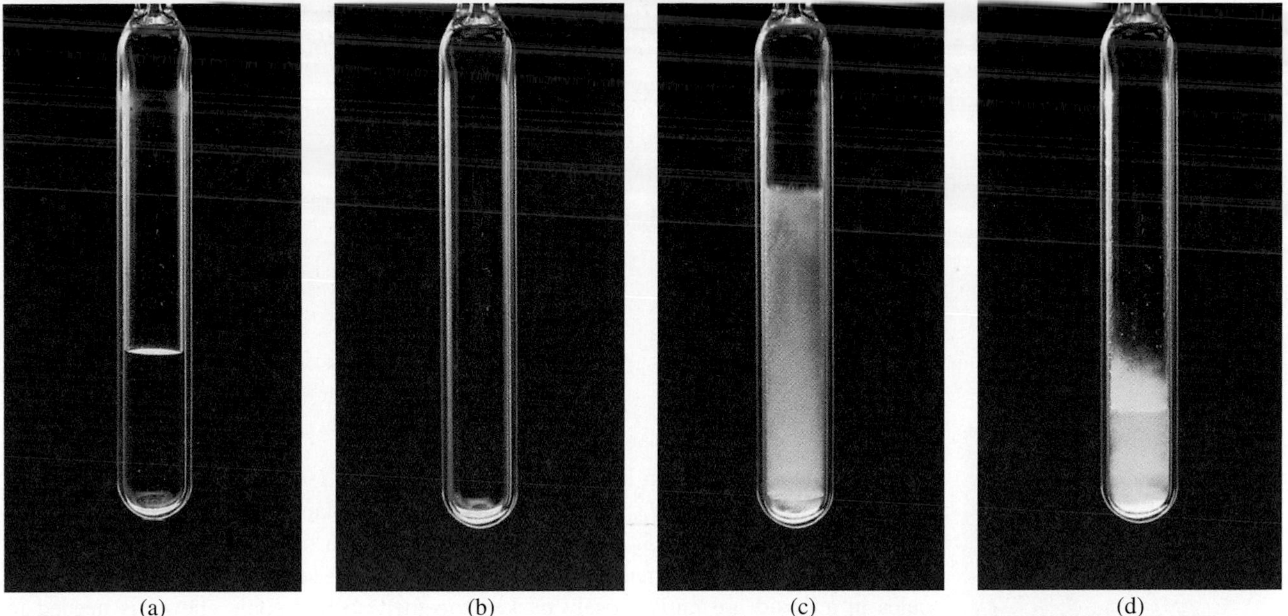

Figure 12.30
The critical phenomenon of sulfur hexafluoride. (a) Below the critical temperature the clear liquid phase is visible. (b) Above the critical temperature the liquid phase has disappeared. (c) The substance is cooled just below its critical temperature. The fog represents the condensation of vapor. (d) Finally, the liquid phase reappears.

A practical illustration of this dynamic equilibrium is provided by a glass of ice water. As the ice cubes melt to form water, some of the water between the ice cubes may freeze, thus joining the cubes together. This is not a true dynamic equilibrium; however, because the glass is not kept at 0°C, all the ice cubes will eventually melt away.

Table 12.6 Critical Temperatures and Critical Pressures of Selected Substances

Substance	T_c(°C)	P_c(atm)
Ammonia (NH_3)	132.4	111.5
Argon (Ar)	−186	6.3
Benzene (C_6H_6)	288.9	47.9
Carbon dioxide (CO_2)	31.0	73.0
Diethyl ether ($C_2H_5OC_2H_5$)	192.6	35.6
Ethanol (C_2H_5OH)	243	63.0
Mercury (Hg)	1462	1036
Methane (CH_4)	−83.0	45.6
Molecular hydrogen (H_2)	−239.9	12.8
Molecular nitrogen (N_2)	−147.1	33.5
Molecular oxygen (O_2)	−118.8	49.7
Sulfur hexafluoride (SF_6)	45.5	37.6
Water (H_2O)	374.4	219.5

Table 12.7 Molar Heats of Fusion for Selected Substances

Substance	Melting Point* (°C)	ΔH_{fus} (kJ/mol)
Argon (Ar)	−190	1.3
Benzene (C_6H_6)	5.5	10.9
Diethyl ether ($C_2H_5OC_2H_5$)	−116.2	6.90
Ethanol (C_2H_5OH)	−117.3	7.61
Mercury (Hg)	−39	23.4
Methane (CH_4)	−183	0.84
Water (H_2O)	0	6.01

*Measured at 1 atm.

Solid iodine in equilibrium with its vapor.

The energy (usually in kilojoules) *required to melt 1 mole of a solid* is called the ***molar heat of fusion*** **(ΔH_{fus})**. Table 12.7 shows the molar heats of fusion for the substances listed in Table 12.5. A comparison of the data in the two tables shows that for each substance ΔH_{fus} is smaller than ΔH_{vap}. This is consistent with the fact that molecules in a liquid are fairly closely packed together, so that some energy is needed to bring about the rearrangement from solid to liquid. On the other hand, when a liquid evaporates, its molecules are completely separated from one another and considerably more energy is required to overcome the attractive forces.

Solid-Vapor Equilibrium

Solids, too, undergo evaporation and therefore possess a vapor pressure. Consider the following dynamic equilibrium:

$$\text{solid} \rightleftharpoons \text{vapor}$$

The process in which molecules go directly from the solid into the vapor phase is called ***sublimation,*** and the reverse process (that is, *from vapor directly to solid*) is called ***deposition.*** Naphthalene (the substance used to make mothballs) has a fairly high vapor pressure for a solid (1 mmHg at 53°C); thus its pungent vapor quickly permeates an enclosed space. Generally, because molecules are more tightly held in a solid, the vapor pressure of a solid is much less than that of the corresponding liquid. *The energy* (usually in kilojoules) *required to sublime 1 mole of a solid,* called the ***molar heat of sublimation*** **(ΔH_{sub})**, is given by the sum of the molar heats of fusion and vaporization:

$$\Delta H_{sub} = \Delta H_{fus} + \Delta H_{vap} \tag{12.5}$$

Strictly speaking, Equation (12.5), which is an illustration of Hess's law, holds if all the phase changes occur at the same temperature. The enthalpy, or heat change, for the overall process is the same whether the substance changes directly from the solid to the vapor form or goes from solid to liquid and then to vapor. Figure 12.31 summarizes the types of phase changes discussed in this section.

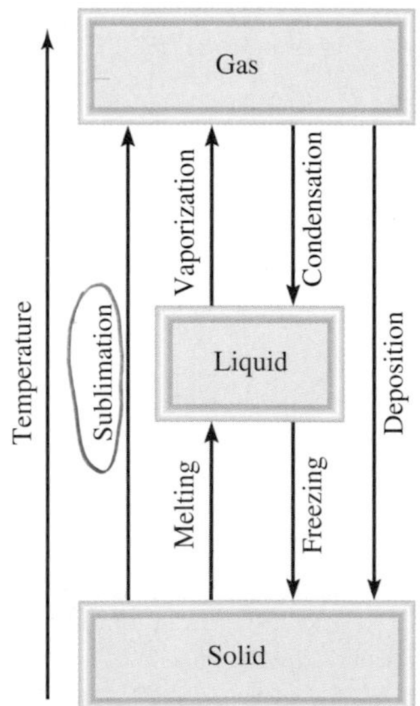

Figure 12.31
The various phase changes that a substance can undergo.

Review of Concepts

A student studies the ln P versus $1/T$ plots for two organic liquids: methanol (CH_3OH) and dimethyl ether (CH_3OCH_3), such as those in Figure 12.29. The slopes are -2.32×10^3 K and -4.50×10^3 K, respectively. How should she assign the ΔH_{vap} values to these two compounds?

12.7 Phase Diagrams

The overall relationships among the solid, liquid, and vapor phases are best represented in a single graph known as a phase diagram. A ***phase diagram*** *summarizes the conditions under which a substance exists as a solid, liquid, or gas.* In this section we will briefly discuss the phase diagrams of water and carbon dioxide.

Water

Figure 12.32(a) shows the phase diagram of water. The graph is divided into three regions, each of which represents a pure phase. The line separating any two regions indicates conditions under which these two phases can exist in equilibrium. For example, the curve between the liquid and vapor phases shows the variation of vapor pressure with temperature. The other two curves similarly indicate conditions for equilibrium between ice and liquid water and between ice and water vapor. (Note that the solid-liquid boundary line has a negative slope.) The point at which all three curves meet is called the ***triple point.*** For water, this point is at 0.01°C and 0.006 atm. This is *the only temperature and pressure at which all three phases can be in equilibrium with one another.*

Phase diagrams enable us to predict changes in the melting point and boiling point of a substance as a result of changes in the external pressure; we can also anticipate directions of phase transitions brought about by changes in temperature and pressure. The normal melting point and boiling point of water, measured at 1 atm, are 0°C and 100°C, respectively. What would happen if melting and boiling were carried out at some other pressure? Figure 12.32(b) shows clearly that increasing the pressure above 1 atm will raise the boiling point and lower the melting point. A decrease in pressure will lower the boiling point and raise the melting point.

Carbon Dioxide

The phase diagram of carbon dioxide (Figure 12.33) is similar to that of water, with one important exception—the slope of the curve between solid and liquid is positive. In fact, this holds true for almost all other substances. Water behaves differently

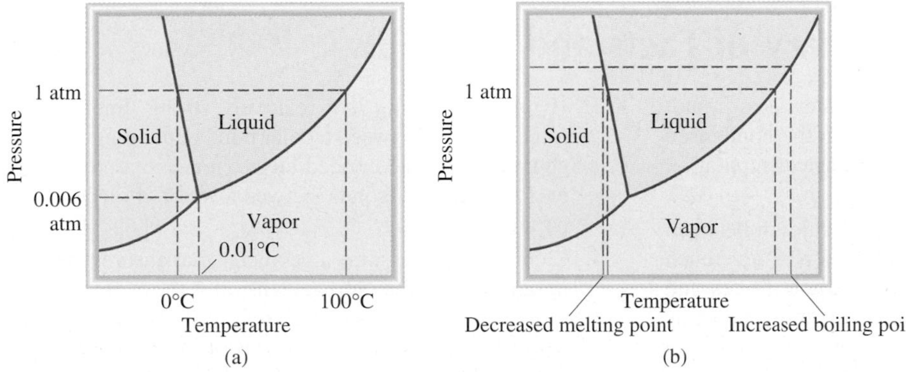

Figure 12.32
(a) The phase diagram of water. Each solid line between two phases specifies the conditions of pressure and temperature under which the two phases can exist in equilibrium. The point at which all three phases can exist in equilibrium (0.006 atm and 0.01°C) is called the triple point. (b) This phase diagram tells us that increasing the pressure on ice lowers its melting point and that increasing the pressure of liquid water raises its boiling point.

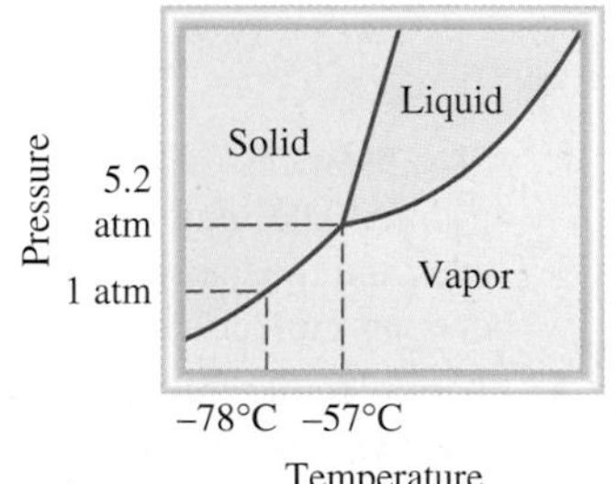

Figure 12.33
The phase diagram of carbon dioxide. Note that the solid-liquid boundary line has a positive slope. The liquid phase is not stable below 5.2 atm, so that only the solid and vapor phases can exist under atmospheric conditions.

Figure 12.34
Under atmospheric conditions, solid carbon dioxide does not melt; it can only sublime. The cold carbon dioxide gas causes nearby water vapor to condense and form a fog.

because ice is *less* dense than liquid water. The triple point of carbon dioxide is at 5.2 atm and −57°C.

An interesting observation can be made about the phase diagram in Figure 12.33. As you can see, the entire liquid phase lies well above atmospheric pressure; therefore, it is impossible for solid carbon dioxide to melt at 1 atm. Instead, when solid CO_2 is heated to −78°C at 1 atm, it sublimes. In fact, solid carbon dioxide is called dry ice because it looks like ice and *does not melt* (Figure 12.34). Because of this property, dry ice is useful as a refrigerant.

Review of Concepts

Which phase diagram corresponds to a substance that will sublime rather than melt as it is heated at 1 atm?

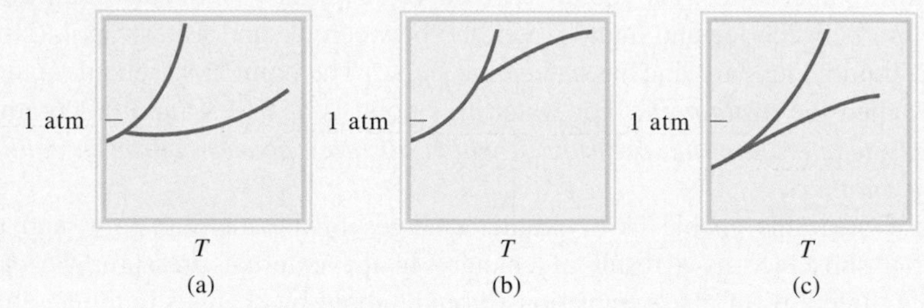

Key Equation

$\ln P = -\frac{\Delta H_{vap}}{RT} + C$	(12.1)	Clausius-Clapeyron equation for determining ΔH_{vap} of a liquid.
$\ln\frac{P_1}{P_2} = \frac{\Delta H_{vap}}{R}\left(\frac{T_1 - T_2}{T_1T_2}\right)$	(12.4)	For calculating ΔH_{vap}, vapor pressure, or boiling point of a liquid.
$\Delta H_{sub} = \Delta H_{fus} + \Delta H_{vap}$	(12.5)	Application of Hess's law.

Summary of Facts and Concepts

1. All substances exist in one of three states: gas, liquid, or solid. The major difference between the condensed states and the gaseous state is the distance of separation between molecules.
2. Intermolecular forces act between molecules or between molecules and ions. Generally, these forces are much weaker than bonding forces. Dipole-dipole forces and ion-dipole forces attract molecules with dipole moments to other polar molecules or ions. Dispersion forces are the result of temporary dipole moments induced in ordinarily nonpolar molecules. The extent to which a dipole moment can be induced in a molecule is determined by its polarizability. The term "van der Waals forces" refers to dipole-dipole, dipole-induced dipole, and dispersion forces.
3. Hydrogen bonding is a relatively strong dipole-dipole force that acts between a polar bond containing a hydrogen atom and the bonded electronegative atoms, N, O, or F. Hydrogen bonds between water molecules are particularly strong.
4. Liquids tend to assume a geometry that ensures the minimum surface area. Surface tension is the energy needed to expand a liquid surface area; strong intermolecular forces lead to greater surface tension. Viscosity is a measure of the resistance of a liquid to flow; it decreases with increasing temperature.
5. Water molecules in the solid state form a three-dimensional network in which each oxygen atom is covalently bonded to two hydrogen atoms and is hydrogen-bonded to two hydrogen atoms. This unique structure accounts

for the fact that ice is less dense than liquid water. Water is also ideally suited for its ecological role by its high specific heat, another property imparted by its strong hydrogen bonding. Large bodies of water are able to moderate the climate by giving off and absorbing substantial amounts of heat with only small changes in the water temperature.

6. All solids are either crystalline (with a regular structure of atoms, ions, or molecules) or amorphous (without a regular structure). The basic structural unit of a crystalline solid is the unit cell, which is repeated to form a three-dimensional crystal lattice.
7. The four types of crystals and the forces that hold their particles together are ionic crystals, held together by ionic bonding; molecular crystals, van der Waals forces and/or hydrogen bonding; covalent crystals, covalent bonding; and metallic crystals, metallic bonding.
8. A liquid in a closed vessel eventually establishes a dynamic equilibrium between evaporation and condensation. The vapor pressure over the liquid under these conditions is the equilibrium vapor pressure, which is often referred to simply as vapor pressure. At the boiling point, the vapor pressure of a liquid equals the external pressure. The molar heat of vaporization of a liquid is the energy required to vaporize 1 mole of the liquid. It can be determined by measuring the vapor pressure of the liquid as a function of temperature and using Equation (12.1). The molar heat of fusion of a solid is the energy required to melt 1 mole of the solid.
9. For every substance there is a temperature, called the critical temperature, above which its gas form cannot be made to liquefy.
10. The relationships among the three phases of a single substance are represented by a phase diagram, in which each region represents a pure phase and the boundaries between the regions show the temperatures and pressures at which the two phases are in equilibrium. At the triple point, all three phases are in equilibrium.

Key Words

Adhesion, p. 407
Amorphous solid, p. 419
Boiling point, p. 423
Closest packing, p. 413
Cohesion, p. 407
Condensation, p. 420
Coordination number, p. 412
Critical pressure (P_c), p. 424
Critical temperature (T_c), p. 424
Crystalline solid, p. 410
Deposition, p. 426
Dipole-dipole forces, p. 401
Dispersion forces, p. 403
Dynamic equilibrium, p. 420
Equilibrium vapor pressure, p. 421
Evaporation, p. 420
Hydrogen bond, p. 405
Induced dipole, p. 402
Intermolecular forces, p. 401
Intramolecular forces, p. 401
Ion-dipole forces, p. 401
Lattice point, p. 410
Melting point, p. 424
Molar heat of fusion (ΔH_{fus}), p. 426
Molar heat of sublimation (ΔH_{sub}), p. 426
Molar heat of vaporization (ΔH_{vap}), p. 421
Phase, p. 419
Phase changes, p. 420
Phase diagram, p. 427
Polarizability, p. 402
Sublimation, p. 426
Surface tension, p. 407
Triple point, p. 427
Unit cell, p. 410
van der Waals forces, p. 401
Vaporization, p. 420
Viscosity, p. 407
X-ray diffraction, p. 410

Questions and Problems

Intermolecular Forces

Review Questions

12.1 Define these terms and give an example for each category: (a) dipole-dipole interaction, (b) dipole-induced dipole interaction, (c) ion-dipole interaction, (d) dispersion forces, (e) van der Waals forces.

12.2 Explain the term "polarizability." What kind of molecules tend to have high polarizabilities? What is the relationship between polarizability and intermolecular forces?

12.3 Explain the difference between the temporary dipole moment induced in a molecule and the permanent dipole moment in a polar molecule.

12.4 Give some evidence that all molecules exert attractive forces on one another.

12.5 What type of physical properties would you need to consider in comparing the strength of intermolecular forces in solids and in liquids?

12.6 Which elements can take part in hydrogen bonding?

Problems

12.7 The compounds Br_2 and ICl have the same number of electrons, yet Br_2 melts at $-7.2°C$, whereas ICl melts at 27.2°C. Explain.

12.8 If you lived in Alaska, state which of these natural gases you would keep in an outdoor storage tank in winter and explain why: methane (CH_4), propane (C_3H_8), or butane (C_4H_{10}).

12.9 The binary hydrogen compounds of the Group 4A elements are CH_4 (−162°C), SiH_4 (−112°C), GeH_4 (−88°C), and SnH_4 (−52°C). The temperatures in parentheses are the corresponding boiling points. Explain the increase in boiling points from CH_4 to SnH_4.

12.10 List the types of intermolecular forces that exist in each of these species: (a) benzene (C_6H_6), (b) CH_3Cl, (c) PF_3, (d) NaCl, (e) CS_2.

12.11 Ammonia is both a donor and an acceptor of hydrogen in hydrogen bond formation. Draw a diagram to show the hydrogen bonding of an ammonia molecule with two other ammonia molecules.

12.12 Which of these species are capable of hydrogen bonding among themselves: (a) C_2H_6, (b) HI, (c) KF, (d) BeH_2, (e) CH_3COOH?

12.13 Arrange the following compounds in order of increasing boiling point: RbF, CO_2, CH_3OH, CH_3Br. Explain your arrangement.

12.14 Diethyl ether has a boiling point of 34.5°C, and 1-butanol has a boiling point of 117°C.

```
   H  H     H  H
   |  |     |  |
H—C—C—O—C—C—H
   |  |     |  |
   H  H     H  H
```
diethyl ether

```
   H  H  H  H
   |  |  |  |
H—C—C—C—C—OH
   |  |  |  |
   H  H  H  H
```
1-butanol

Both of these compounds have the same numbers and types of atoms. Explain the difference in their boiling points.

12.15 Which member of each of these pairs of substances would you expect to have a higher boiling point: (a) O_2 or N_2, (b) SO_2 or CO_2, (c) HF or HI?

12.16 State which substance in each of these pairs you would expect to have the higher boiling point and explain why: (a) Ne or Xe, (b) CO_2 or CS_2, (c) CH_4 or Cl_2, (d) F_2 or LiF, (e) NH_3 or PH_3.

12.17 Explain in terms of intermolecular forces why (a) NH_3 has a higher boiling point than CH_4 and (b) KCl has a higher melting point than I_2.

12.18 What kind of attractive forces must be overcome to (a) melt ice, (b) boil molecular bromine, (c) melt solid iodine, and (d) dissociate F_2 into F atoms?

12.19 These nonpolar molecules have the same number and type of atoms. Which one would you expect to have a higher boiling point?

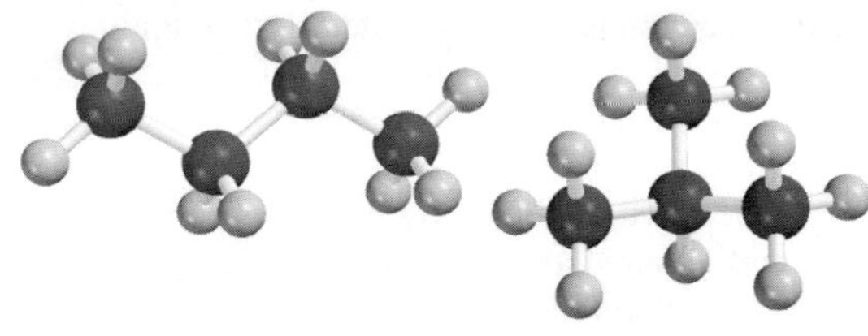

(*Hint:* Molecules that can be stacked together more easily have greater intermolecular attraction.)

12.20 Explain the difference in the melting points of these compounds:

(*Hint:* Only one of the two can form intramolecular hydrogen bonds.)

The Liquid State

Review Questions

12.21 Explain why liquids, unlike gases, are virtually incompressible.

12.22 Define surface tension. What is the relationship between the intermolecular forces that exist in a liquid and its surface tension?

12.23 Despite the fact that stainless steel is much denser than water, a stainless-steel razor blade can be made to float on water. Why?

12.24 Use water and mercury as examples to explain adhesion and cohesion.

12.25 A glass can be filled slightly above the rim with water. Explain why the water does not overflow.

12.26 Draw diagrams showing the capillary action of (a) water and (b) mercury in three tubes of different radii.

12.27 What is viscosity? What is the relationship between the intermolecular forces that exist in a liquid and its viscosity?

12.28 Why does the viscosity of a liquid decrease with increasing temperature?

12.29 Why is ice less dense than water?

12.30 Outdoor water pipes have to be drained or insulated in winter in a cold climate. Why?

Problems

12.31 Predict which of these liquids has the greater surface tension: ethanol (C_2H_5OH) or dimethyl ether (CH_3OCH_3).

12.32 Predict the viscosity of ethylene glycol

$$\begin{array}{c} CH_2-OH \\ | \\ CH_2-OH \end{array}$$

relative to that of ethanol and glycerol (see Table 12.3).

Crystalline Solids

Review Questions

12.33 Define these terms: crystalline solid, lattice point, unit cell, coordination number.

12.34 Describe the geometries of these cubic cells: simple cubic cell, body-centered cubic cell, face-centered cubic cell. Which of these cells would give the highest density for the same type of atoms?

Problems

12.35 Describe, with examples, these types of crystals: (a) ionic crystals, (b) covalent crystals, (c) molecular crystals, (d) metallic crystals.

12.36 A solid is hard, brittle, and electrically nonconducting. Its melt (the liquid form of the substance) and an aqueous solution containing the substance do conduct electricity. Classify the solid.

12.37 A solid is soft and has a low melting point (below 100°C). The solid, its melt, and a solution containing the substance are all nonconductors of electricity. Classify the solid.

12.38 A solid is very hard and has a high melting point. Neither the solid nor its melt conducts electricity. Classify the solid.

12.39 Why are metals good conductors of heat and electricity? Why does the ability of a metal to conduct electricity decrease with increasing temperature?

12.40 Classify the solid states of the elements in the second period of the periodic table.

12.41 The melting points of the oxides of the third-period elements are given in parentheses: Na_2O (1275°C), MgO (2800°C), Al_2O_3 (2045°C), SiO_2 (1610°C), P_4O_{10} (580°C), SO_3 (16.8°C), Cl_2O_7 (−91.5°C). Classify these solids.

12.42 Which of these are molecular solids and which are covalent solids: Se_8, HBr, Si, CO_2, C, P_4O_6, B, SiH_4?

12.43 What is the coordination number of each sphere in (a) a simple cubic lattice, (b) a body-centered cubic lattice, and (c) a face-centered cubic lattice? Assume the spheres to be of equal size.

12.44 Calculate the number of spheres in these unit cells: simple cubic, body-centered cubic, and face-centered cubic cells. Assume that the spheres are of equal size and that they are only at the lattice points.

12.45 Metallic iron crystallizes in a cubic lattice. The unit cell edge length is 287 pm. The density of iron is 7.87 g/cm^3. How many iron atoms are there within a unit cell?

12.46 Barium metal crystallizes in a body-centered cubic lattice (the Ba atoms are at the lattice points only). The unit cell edge length is 502 pm, and the density of Ba is 3.50 g/cm^3. Using this information, calculate Avogadro's number. (*Hint:* First calculate the volume occupied by 1 mole of Ba atoms in the unit cells. Next calculate the volume occupied by one of the Ba atoms in the unit cell.)

12.47 Vanadium crystallizes in a body-centered cubic lattice (the V atoms occupy only the lattice points). How many V atoms are in a unit cell?

12.48 Europium crystallizes in a body-centered cubic lattice (the Eu atoms occupy only the lattice points). The density of Eu is 5.26 g/cm^3. Calculate the unit cell edge length in picometers.

12.49 Crystalline silicon has a cubic structure. The unit cell edge length is 543 pm. The density of the solid is 2.33 g/cm^3. Calculate the number of Si atoms in one unit cell.

12.50 A face-centered cubic cell contains 8 X atoms at the corners of the cell and 6 Y atoms at the faces. What is the empirical formula of the solid?

12.51 Classify the crystalline form of these substances as ionic crystals, covalent crystals, molecular crystals, or metallic crystals: (a) CO_2, (b) B, (c) S_8, (d) KBr, (e) Mg, (f) SiO_2, (g) LiCl, (h) Cr.

12.52 Explain why diamond is harder than graphite. Why is graphite an electrical conductor but diamond is not?

Phase Changes

Review Questions

12.53 Define phase change. Name all possible changes that can occur among the vapor, liquid, and solid states of a substance.

12.54 What is the equilibrium vapor pressure of a liquid? How does it change with temperature?

12.55 Use any one of the phase changes to explain what is meant by dynamic equilibrium.

12.56 Define these terms: (a) molar heat of vaporization, (b) molar heat of fusion, (c) molar heat of sublimation. What are their units?

12.57 How is the molar heat of sublimation related to the molar heats of vaporization and fusion? On what law is this relation based?

12.58 What can we learn about the strength of intermolecular forces in a liquid from its molar heat of vaporization?

12.59 The greater the molar heat of vaporization of a liquid, the greater its vapor pressure. True or false?

12.60 Define boiling point. How does the boiling point of a liquid depend on external pressure? Referring to Table 5.2, what is the boiling point of water when the external pressure is 187.5 mmHg?

12.61 As a liquid is heated at constant pressure, its temperature rises. This trend continues until the boiling point of the liquid is reached. No further rise in the temperature of the liquid can be induced by heating. Explain.

12.62 Define critical temperature. What is the significance of critical temperature in the liquefaction of gases?

12.63 What is the relationship between intermolecular forces in a liquid and the liquid's boiling point and critical temperature? Why is the critical temperature of water greater than that of most other substances?

12.64 How do the boiling points and melting points of water and carbon tetrachloride vary with pressure? Explain any difference in behavior of these two substances.

12.65 Why is solid carbon dioxide called dry ice?

12.66 The vapor pressure of a liquid in a closed container depends on which of these: (a) the volume above the liquid, (b) the amount of liquid present, (c) temperature?

12.67 Referring to Figure 12.28, estimate the boiling points of diethyl ether, water, and mercury at 0.5 atm.

12.68 Wet clothes dry more quickly on a hot, dry day than on a hot, humid day. Explain.

12.69 Which of the following phase transitions gives off more heat: (a) 1 mole of steam to 1 mole of water at 100°C or (b) 1 mole of water to 1 mole of ice at 0°C?

12.70 A beaker of water is heated to boiling by a Bunsen burner. Would adding another burner raise the boiling point of water? Explain.

Problems

12.71 Calculate the amount of heat (in kilojoules) required to convert 74.6 g of water to steam at 100°C.

12.72 How much heat (in kilojoules) is needed to convert 866 g of ice at −10°C to steam at 126°C? (The specific heats of ice and steam are 2.03 J/g · °C and 1.99 J/g · °C, respectively.)

12.73 How is the rate of evaporation of a liquid affected by (a) temperature, (b) the surface area of liquid exposed to air, (c) intermolecular forces?

12.74 The molar heats of fusion and sublimation of molecular iodine are 15.27 kJ/mol and 62.30 kJ/mol, respectively. Estimate the molar heat of vaporization of liquid iodine.

12.75 These compounds are liquid at −10°C; their boiling points are given: butane, −0.5°C; ethanol, 78.3°C; toluene, 110.6°C. At −10°C, which of these liquids would you expect to have the highest vapor pressure? Which the lowest?

12.76 Freeze-dried coffee is prepared by freezing a sample of brewed coffee and then removing the ice component by vacuum-pumping the sample. Describe the phase changes taking place during these processes.

12.77 A student hangs wet clothes outdoors on a winter day when the temperature is −15°C. After a few hours, the clothes are found to be fairly dry. Describe the phase changes in this drying process.

12.78 Steam at 100°C causes more serious burns than water at 100°C. Why?

12.79 Vapor pressure measurements at several different temperatures are shown here for mercury. Determine graphically the molar heat of vaporization for mercury.

t (°C)	200	250	300	320	340
P (mmHg)	17.3	74.4	246.8	376.3	557.9

12.80 The vapor pressure of benzene, C_6H_6, is 40.1 mmHg at 7.6°C. What is its vapor pressure at 60.6°C? The molar heat of vaporization of benzene is 31.0 kJ/mol.

12.81 The vapor pressure of liquid X is lower than that of liquid Y at 20°C, but higher at 60°C. What can you deduce about the relative magnitude of the molar heats of vaporization of X and Y?

Phase Diagrams

Review Questions

12.82 What is a phase diagram? What useful information can be obtained from a phase diagram?

12.83 Explain how water's phase diagram differs from those of most substances. What property of water causes the difference?

Problems

12.84 The phase diagram of sulfur is shown here. (a) How many triple points are there? (b) Monoclinic and rhombic are two allotropes of sulfur. Which is more stable under atmospheric conditions? (c) Describe what happens when sulfur at 1 atm is heated from 80°C to 300°C?

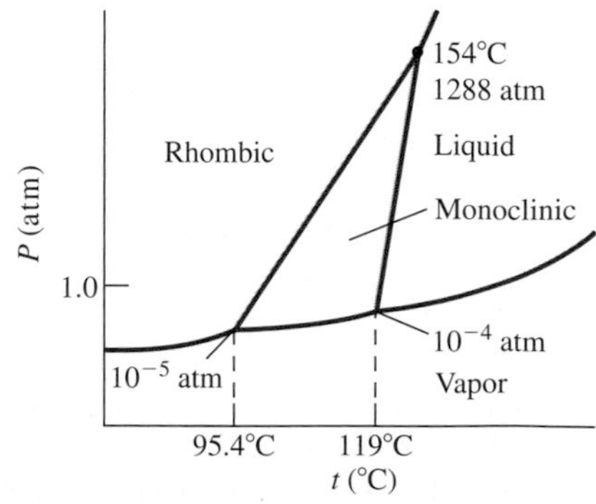

12.85 A length of wire is placed on top of a block of ice. The ends of the wire extend over the edges of the ice, and a heavy weight is attached to each end. It is found that the ice under the wire gradually melts, so that the wire slowly moves through the ice block. At the same time, the water above the wire refreezes. Explain the phase changes that accompany this phenomenon.

12.86 Consider the phase diagram of water shown here. Label the regions. Predict what would happen if we did the following: (a) Starting at A, we raise the temperature at constant pressure. (b) Starting at C, we lower the temperature at constant pressure. (c) Starting at B, we lower the pressure at constant temperature.

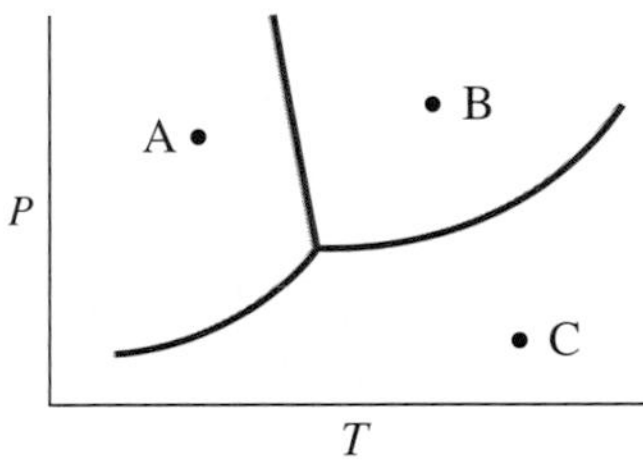

12.87 The boiling point and freezing point of sulfur dioxide are −10°C and −72.7°C (at 1 atm), respectively. The triple point is −75.5°C and 1.65×10^{-3} atm, and its critical point is at 157°C and 78 atm. On the basis of this information, draw a rough sketch of the phase diagram of SO_2.

Additional Problems

12.88 Name the kinds of attractive forces that must be overcome to (a) boil liquid ammonia, (b) melt solid phosphorus (P_4), (c) dissolve CsI in liquid HF, (d) melt potassium metal.

12.89 Which of these indicates very strong intermolecular forces in a liquid: (a) a very low surface tension, (b) a very low critical temperature, (c) a very low boiling point, (d) a very low vapor pressure?

12.90 At −35°C, liquid HI has a higher vapor pressure than liquid HF. Explain.

12.91 From these properties of elemental boron, classify it as one of the crystalline solids discussed in Section 12.5: high melting point (2300°C), poor conductor of heat and electricity, insoluble in water, very hard substance.

12.92 Referring to Figure 12.33, determine the stable phase of CO_2 at (a) 4 atm and −60°C and (b) 0.5 atm and −20°C.

12.93 A solid contains X, Y, and Z atoms in a cubic lattice with X atoms in the corners, Y atoms in the body-centered positions, and Z atoms on the faces of the cell. What is the empirical formula of the compound?

12.94 A CO_2 fire extinguisher is located on the outside of a building in Massachusetts. During the winter months, one can hear a sloshing sound when the extinguisher is gently shaken. In the summertime the sound is often absent. Explain. Assume that the extinguisher has no leaks and that it has not been used.

12.95 What is the vapor pressure of mercury at its normal boiling point (357°C)?

12.96 A flask containing water is connected to a powerful vacuum pump. When the pump is turned on, the water begins to boil. After a few minutes, the same water begins to freeze. Eventually, the ice disappears. Explain what happens at each step.

12.97 The liquid-vapor boundary line in the phase diagram of any substance always stops abruptly at a certain point. Why?

12.98 Given the phase diagram of carbon shown here, answer these questions: (a) How many triple points are there and what are the phases that can coexist at each triple point? (b) Which has a higher density, graphite or diamond? (c) Synthetic diamond can be made from graphite. Using the phase diagram, how would you go about making diamond?

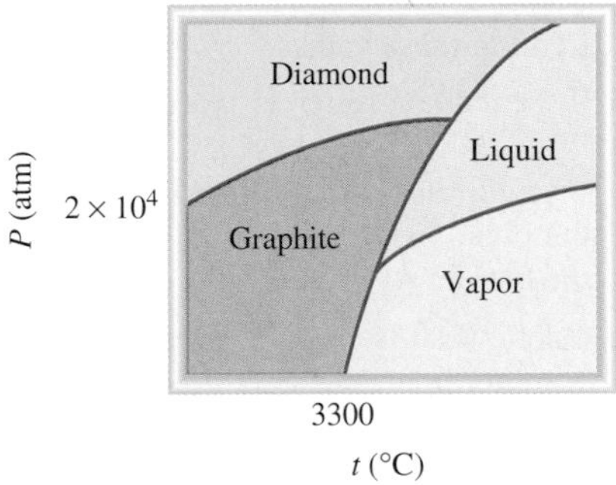

12.99 Estimate the molar heat of vaporization of a liquid whose vapor pressure doubles when the temperature is raised from 85°C to 95°C.

12.100 A student is given four samples of solids W, X, Y, and Z, all of which have a metallic luster. She is told that the solids are gold, lead sulfide, mica (which is quartz, or SiO_2), and iodine. The results of her investigation are (a) W is a good electrical conductor; X, Y, and Z are poor electrical conductors; (b) when the solids are hit with a hammer, W flattens out, X shatters into many pieces, Y is smashed into a powder, and Z is not affected; (c) when the solids are heated with a Bunsen burner, Y melts with some sublimation, but X, W, and Z do not melt; (d) in treatment with 6 *M* HNO_3, X dissolves; there is no effect on W, Y, or Z. On the basis of her studies, identify the solids.

12.101 Which of these statements are false: (a) Dipole-dipole interactions between molecules are greatest if the molecules possess only temporary dipole moments. (b) All compounds containing hydrogen atoms can participate in hydrogen bond formation. (c) Dispersion forces exist between all atoms, molecules, and ions. (d) The extent of ion-induced dipole interaction depends only on the charge on the ion.

12.102 The south pole of Mars is covered with dry ice, which partly sublimes during the summer. The CO_2 vapor recondenses in the winter when the temperature drops to 150 K. Given that the heat of sublimation of CO_2 is 25.9 kJ/mol, calculate the atmospheric pressure on the surface of Mars. [*Hint:* Use Figure 12.33 to determine the normal sublimation temperature of dry ice and Equation (12.4), which also applies to sublimation.]

12.103 The standard enthalpy of formation of gaseous molecular bromine is 30.7 kJ/mol. Use this information to calculate the molar heat of vaporization of molecular bromine at 25°C.

12.104 Heats of hydration, that is, heat changes that occur when ions become hydrated in solution, are largely due to ion-dipole interactions. The heats of hydration for the alkali metal ions are Li^+, −520 kJ/mol; Na^+, −405 kJ/mol; K^+, −321 kJ/mol. Account for the trend in these values.

12.105 A beaker of water is placed in a closed container. Predict the effect on the vapor pressure of the water when (a) its temperature is lowered, (b) the volume of the container is doubled, (c) more water is added to the beaker.

12.106 Ozone (O_3) is a strong agent that can oxidize all the common metals except gold and platinum. A convenient test for ozone is based on its action on mercury. When exposed to ozone, mercury becomes dull looking and sticks to glass tubing (instead of flowing freely through it). Write a balanced equation for the reaction. What property of mercury is altered by its interaction with ozone?

12.107 The phase diagram of helium is shown here. Helium is the only known substance that has two different liquid phases called helium-I and helium-II. (a) What is the maximum temperature at which helium-II can exist? (b) What is the minimum pressure at which solid helium can exist? (c) What is the normal boiling point of helium-I? (d) Can solid helium sublime? (e) How many triple points are there?

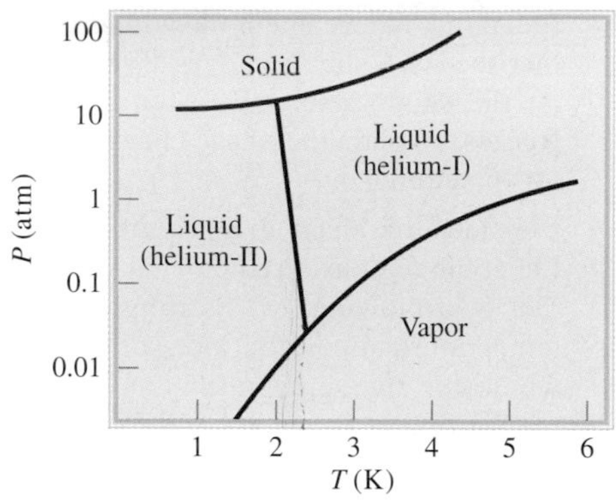

12.108 A 1.20-g sample of water is injected into an evacuated 5.00-L flask at 65°C. What percentage of the water will be vapor when the system reaches equilibrium? Assume ideal behavior of water vapor and that the volume of liquid water is negligible. The vapor pressure of water at 65°C is 187.5 mmHg.

12.109 Swimming coaches sometimes suggest that a drop of alcohol (ethanol) placed in an ear plugged with water "draws out the water." Explain this action from a molecular point of view.

12.110 Argon crystallizes in the face-centered cubic arrangement at 40 K. Given that the atomic radius of argon is 191 pm, calculate the density of solid argon.

12.111 Use the concept of intermolecular forces to explain why the far end of a walking cane rises when one raises the handle.

12.112 Why do citrus growers spray their trees with water to protect them from freezing?

12.113 What is the origin of dark spots on the inner glass walls of an old tungsten lightbulb? What is the purpose of filling these lightbulbs with argon gas?

12.114 A student heated a beaker of cold water (on a tripod) with a Bunsen burner. When the gas is ignited, she noticed that there was water condensed on the outside of the beaker. Explain what happened.

Special Problems

12.115 A quantitative measure of how efficiently spheres pack into unit cells is called *packing efficiency,* which is the percentage of the cell space occupied by the spheres. Calculate the packing efficiencies of a simple cubic cell, a body-centered cubic cell, and a face-centered cubic cell. (*Hint:* Refer to Figure 12.19 and use the relationship that the volume of a sphere is $\frac{4}{3}\pi r^3$, in which r is the radius of the sphere.)

12.116 A chemistry instructor performed the following mystery demonstration. Just before the students arrived in class, she heated some water to boiling in an Erlenmeyer flask. She then removed the flask from the flame and closed the flask with a rubber stopper. After the class commenced, she held the flask in front of the students and announced that she could make the water boil simply by rubbing an ice cube on the outside walls of the flask. To the amazement of everyone, it worked. Can you give the explanation for this phenomenon?

12.117 Silicon used in computer chips must have an impurity level below 10^{-9} (that is, fewer than one impurity atom for every 10^9 Si atoms). Silicon is prepared by the reduction of quartz (SiO_2) with coke (a form of carbon made by the destructive distillation of coal) at about 2000°C:

$$SiO_2 + 2C(s) \longrightarrow Si(l) + 2CO(g)$$

Next, solid silicon is separated from other solid impurities by treatment with hydrogen chloride at 350°C to form gaseous trichlorosilane ($SiCl_3H$):

$$Si(s) + 3HCl(g) \longrightarrow SiCl_3H(g) + H_2(g)$$

Finally, ultrapure Si can be obtained by reversing the above reaction at 1000°C:

$$SiCl_3H(g) + H_2(g) \longrightarrow Si(s) + 3HCl(g)$$

(a) Trichlorosilane has a vapor pressure of 0.258 atm at −2°C. What is its normal boiling point? Is trichlorosilane's boiling point consistent with the type of intermolecular forces that exist among its molecules? (The molar heat of vaporization of trichlorosilane is 28.8 kJ/mol.)

(b) What types of crystals do Si and SiO_2 form?

(c) Silicon has a diamond crystal structure (see Figure 12.24). Each cubic unit cell (edge length a = 543 pm) contains eight Si atoms. If there are 1.0×10^{13} boron atoms per cubic centimeter in a sample of pure silicon, how many Si atoms are there for every B atom in the sample? Does this sample satisfy the 10^{-9} purity requirement for the electronic grade silicon?

12.118 Iron crystallizes in a body-centered cubic lattice. The cell length as determined by X-ray diffraction is 286.7 pm. Given that the density of iron is 7.874 g/cm^3, calculate Avogadro's number.

12.119 The boiling point of methanol is 65.0°C and the standard enthalpy of formation of methanol vapor is −210.2 kJ/mol. Calculate the vapor pressure of methanol (in mmHg) at 25°C. (*Hint:* See Appendix 2 for other thermodynamic data of methanol.)

12.120 An alkali metal in the form of a cube of edge length 0.171 cm is vaporized in a 0.843-L container at 1235 K. The vapor pressure is 19.2 mmHg. Identify the metal by calculating the atomic radius in picometers and the density. (*Hint:* You need to consult Figures 8.4, 12.19, and a chemistry handbook. All alkali metals form body-centered cubic lattices.)

12.121 A sample of water shows the following behavior as it is heated at a constant rate:

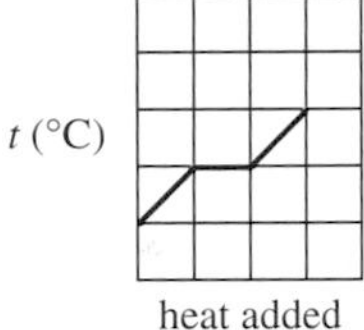

If twice the mass of water has the same amount of heat transferred to it, which of the following graphs best describes the temperature variation? Note that the scales for all the graphs are the same.

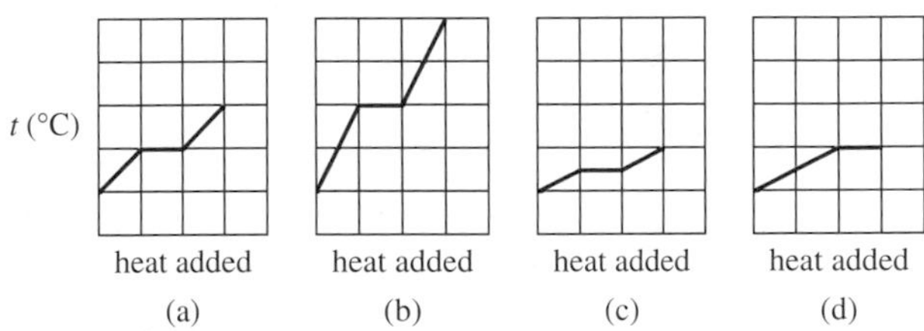

(Used with permission from the *Journal of Chemical Education,* Vol. 79, No. 7, 2002, pp. 889–895; © 2002, Division of Chemical Education, Inc.)

Answers to Practice Exercises

12.1 (a) Ionic and dispersion forces, (b) dispersion forces, (c) dipole-dipole and dispersion forces. **12.2** CH_3OH. **12.3** 10.50 g/cm^3. **12.4** Two. **12.5** 369 mmHg.

CHAPTER

13

Physical Properties of Solutions

A sugar cube dissolving in water. The properties of a solution are markedly different from those of its solvent.

Chapter Outline

Student Interactive Activities

Animations
Dissolution of an Ionic and a Covalent Compound (13.2)
Osmosis (13.6)

Electronic Homework
Example Practice Problems
End of Chapter Problems

Essential Concepts

Solutions There are many types of solutions; the most common is the liquid solution in which the solvent is a liquid and the solute is a solid or a liquid. Molecules that possess similar types of intermolecular forces readily mix with each other. Solubility is a quantitative measure of the amount of a solute dissolved in a solvent at a specific temperature.

Concentration Units The four common concentration units for solutions are percent by mass, mole fraction, molarity, and molality. Each one has its advantages and limitations.

Effect of Temperature and Pressure on Solubility Temperature generally has a marked influence on the solubility of a substance. Pressure can affect the solubility of a gas in a liquid but has little effect if the solute is a solid or liquid.

Colligative Properties The presence of a solute affects the vapor pressure, boiling point, and freezing point of a solvent. In addition, when a solution is separated from the solvent by a semipermeable membrane, osmosis, the passage of solvent molecules from the solvent to the solution, occurs. Equations have been derived that relate the extent of the changes in these properties to the concentration of the solution.

13.1 Types of Solutions

Most chemical reactions take place not between pure solids, liquids, or gases, but among ions and molecules dissolved in water or other solvents. In Section 4.1 we noted that a solution is a homogeneous mixture of two or more substances. Because this definition places no restriction on the nature of the substances involved, we can distinguish six types of solutions, depending on the original states (solid, liquid, or gas) of the solution components. Table 13.1 gives examples of each of these types.

Our focus here will be on solutions involving at least one liquid component—that is, gas-liquid, liquid-liquid, and solid-liquid solutions. And, perhaps not too surprisingly, the liquid solvent in most of the solutions we will study is water.

Chemists also characterize solutions by their capacity to dissolve a solute. *A solution that contains the maximum amount of a solute in a given solvent, at a specific temperature,* is called a ***saturated solution.*** Before the saturation point is reached, the solution is said to be ***unsaturated;*** it *contains less solute than it has the capacity to dissolve.* A third type, a ***supersaturated solution,*** *contains more solute than is present in a saturated solution.* Supersaturated solutions are not very stable. In time, some of the solute will come out of a supersaturated solution as crystals. *The process in which dissolved solute comes out of solution and forms crystals* is called ***crystallization.*** Note that both precipitation and crystallization describe the separation of excess solid substance from a supersaturated solution. However, solids formed by the two processes differ in appearance. We normally think of precipitates as being made up of small particles, whereas crystals may be large and well formed (Figure 13.1).

Figure 13.1
In a supersaturated sodium acetate solution (top), sodium acetate crystals rapidly form when a small seed crystal is added.

13.2 A Molecular View of the Solution Process

In liquids and solids, molecules are held together by intermolecular attractions. These forces also play a central role in the formation of solutions. When one substance (the solute) dissolves in another (the solvent), particles of the solute disperse throughout the solvent. The solute particles occupy positions that are normally taken by solvent molecules. The ease with which a solute particle replaces a solvent molecule depends on the relative strengths of three types of interactions:

- solvent-solvent interaction
- solute-solute interaction
- solvent-solute interaction

Table 13.1 Types of Solutions

Solute	Solvent	State of Resulting Solution	Examples
Gas	Gas	Gas	Air
Gas	Liquid	Liquid	Soda water (CO_2 in water)
Gas	Solid	Solid	H_2 gas in palladium
Liquid	Liquid	Liquid	Ethanol in water
Solid	Liquid	Liquid	NaCl in water
Solid	Solid	Solid	Brass (Cu/Zn), solder (Sn/Pb)

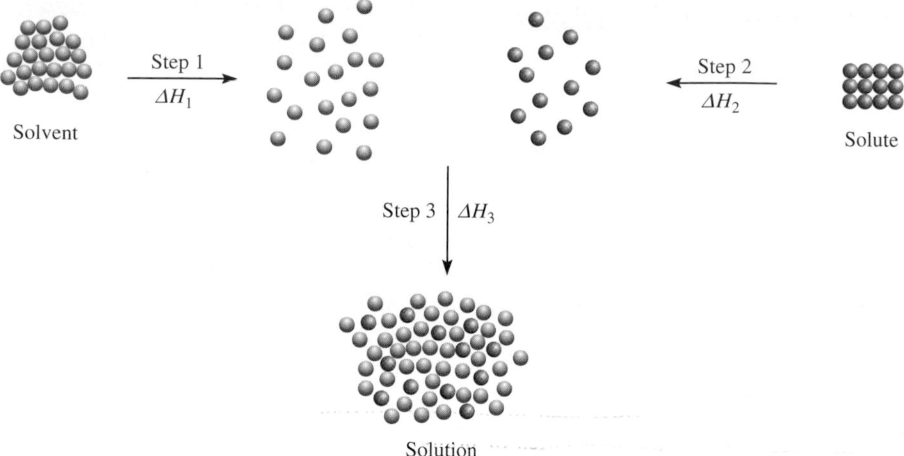

Figure 13.2
A molecular view of the solution process portrayed as taking in three steps: First the solvent and solute molecules are separated (steps 1 and 2). Then the solvent and solute molecules mix (step 3).

Animation:
Dissolution of an Ionic and a Covalent Compound

For simplicity, we can imagine the solution process taking place in three distinct steps (Figure 13.2). Step 1 is the separation of solvent molecules, and step 2 entails the separation of solute molecules. These steps require energy input to break attractive intermolecular forces; therefore, they are endothermic. In step 3 the solvent and solute molecules mix. This step may be exothermic or endothermic. The heat of solution ΔH_{soln} is given by

This equation is an application of Hess's law.

$$\Delta H_{soln} = \Delta H_1 + \Delta H_2 + \Delta H_3 \tag{13.1}$$

If the solute-solvent attraction is stronger than the solvent-solvent attraction and solute-solute attraction, the solution process is favorable; that is, it is exothermic ($\Delta H_{soln} < 0$). If the solute-solvent interaction is weaker than the solvent-solvent and solute-solute interactions, the solution process is endothermic ($\Delta H_{soln} > 0$).

You may wonder why a solute dissolves in a solvent at all if the attraction among its own molecules is stronger than that between its molecules and the solvent molecules. The solution process, like all physical and chemical processes, is governed by two factors. One is energy, which determines whether a solution process is exothermic or endothermic. The second factor is an inherent tendency toward disorder in all natural events. In much the same way that a deck of new playing cards becomes mixed up after it has been shuffled a few times, when solute and solvent molecules mix to form a solution, there is an increase in randomness or disorder. In the pure state, the solvent and solute possess a fair degree of order, characterized by the more or less regular arrangement of atoms, molecules, or ions in three-dimensional space. Much of this order is destroyed when the solute dissolves in the solvent (see Figure 13.2). Therefore, the solution process is accompanied by an increase in disorder or randomness. It is the increase in disorder of the system that favors the solubility of any substance, even if the solution process is endothermic.

Solubility is a measure of the amount of a solute that will dissolve in a solvent at a specific temperature. The saying "like dissolves like" helps in predicting the solubility of a substance in a solvent. What this expression means is that two substances with intermolecular forces of similar type and magnitude are likely to be soluble in each other. For example, both carbon tetrachloride (CCl_4) and benzene (C_6H_6) are nonpolar liquids. The only intermolecular forces present in these substances are dispersion forces (see Section 12.2). When these two liquids are mixed, they readily dissolve in each other, because the attraction between CCl_4 and C_6H_6 molecules is comparable in magnitude to that between CCl_4 molecules and between

C_6H_6 molecules. When *two liquids are completely soluble in each other in all proportions,* as in this case, they are said to be ***miscible.*** Alcohols such as methanol, ethanol, and ethylene glycol are miscible with water because of their ability to form hydrogen bonds with water molecules:

```
     H              H  H               H  H
     |              |  |               |  |
  H—C—O—H       H—C—C—O—H       H—O—C—C—O—H
     |              |  |               |  |
     H              H  H               H  H
  methanol         ethanol        1,2-ethylene glycol
```

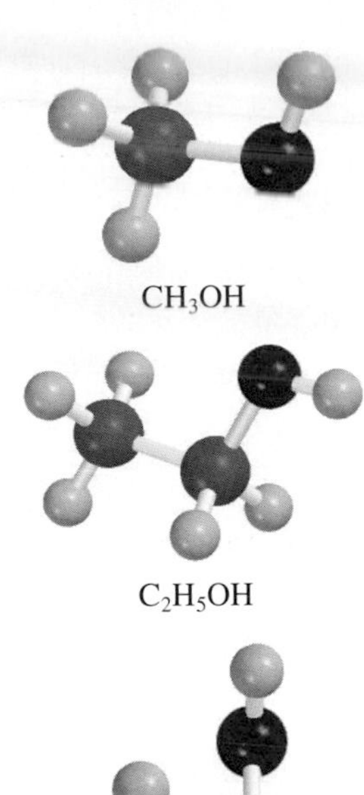

CH_3OH

C_2H_5OH

$CH_2(OH)CH_2(OH)$

When sodium chloride dissolves in water, the ions are stabilized in solution by hydration, which involves ion-dipole interaction. In general, we predict that ionic compounds should be much more soluble in polar solvents, such as water, liquid ammonia, and liquid hydrogen fluoride, than in nonpolar solvents, such as benzene and carbon tetrachloride. Because the molecules of nonpolar solvents lack a dipole moment, they cannot effectively solvate the Na^+ and Cl^- ions. (***Solvation*** is *the process in which an ion or a molecule is surrounded by solvent molecules arranged in a specific manner.* When the solvent is water, the process is called *hydration.*) The predominant intermolecular interaction between ions and nonpolar compounds is ion-induced dipole interaction, which is much weaker than ion-dipole interaction. Consequently, ionic compounds usually have extremely low solubility in nonpolar solvents.

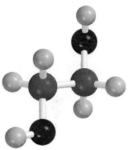

EXAMPLE 13.1

Predict the relative solubilities in the following cases: (a) Bromine (Br_2) in benzene (C_6H_6, $\mu = 0$ D) and in water ($\mu = 1.87$ D), (b) KCl in carbon tetrachloride (CCl_4, $\mu = 0$ D) and in liquid ammonia (NH_3, $\mu = 1.46$ D), (c) formaldehyde (CH_2O) in carbon disulfide (CS_2, $\mu = 0$) and in water.

Strategy In predicting solubility, remember the saying: Like dissolves like. A nonpolar solute will dissolve in a nonpolar solvent; ionic compounds will generally dissolve in polar solvents due to favorable ion-dipole interaction; solutes that can form hydrogen bonds with the solvent will have high solubility in the solvent.

Solution

(a) Br_2 is a nonpolar molecule and therefore should be more soluble in C_6H_6, which is also nonpolar, than in water. The only intermolecular forces between Br_2 and C_6H_6 are dispersion forces.

(b) KCl is an ionic compound. For it to dissolve, the individual K^+ and Cl^- ions must be stabilized by ion-dipole interaction. Because CCl_4 has no dipole moment, KCl should be more soluble in liquid NH_3, a polar molecule with a large dipole moment.

(c) Because CH_2O is a polar molecule and CS_2 (a linear molecule) is nonpolar, the forces between molecules of CH_2O and CS_2 are dipole-induced dipole and dispersion. On the other hand, CH_2O can form hydrogen bonds with water, so it should be more soluble in that solvent.

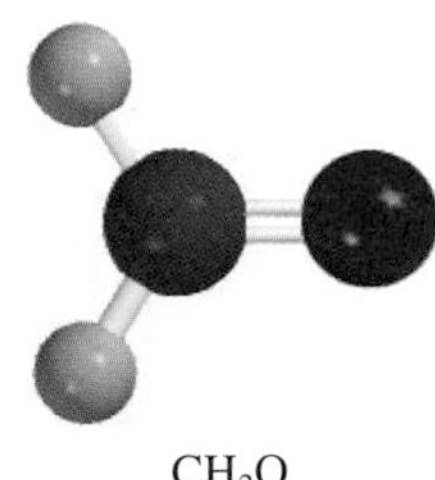

CH_2O

Similar problem: 13.9.

Practice Exercise Is iodine (I_2) more soluble in water or in carbon disulfide (CS_2)?

Review of Concepts

Which of the following would you expect to be more soluble in benzene than in water: C_4H_{10}, HBr, CS_2, I_2?

13.3 Concentration Units

Quantitative study of a solution requires that we know its *concentration,* that is, the amount of solute present in a given amount of solution. Chemists use several different concentration units, each of which has advantages as well as limitations. Let us examine the three most common units of concentration: percent by mass, molarity, and molality.

Types of Concentration Units

Percent by Mass

The ***percent by mass*** (also called the *percent by weight* or the *weight percent*) is defined as

$$\text{percent by mass of solute} = \frac{\text{mass of solute}}{\text{mass of solute} + \text{mass of solvent}} \times 100\%$$

$$= \frac{\text{mass of solute}}{\text{mass of soln}} \times 100\% \tag{13.2}$$

The percent by mass is a unitless number because it is a ratio of two similar quantities.

Molarity (M)

For calculations involving molarity, see Example 4.5 on p. 119.

The molarity unit was defined in Section 4.5 as the number of moles of solute in 1 L of solution; that is,

$$\text{molarity} = \frac{\text{moles of solute}}{\text{liters of soln}} \tag{13.3}$$

Thus, molarity has the units of mole per liter (mol/L).

Molality (m)

Molality is *the number of moles of solute dissolved in 1 kg (1000 g) of solvent*—that is,

$$\text{molality} = \frac{\text{moles of solute}}{\text{mass of solvent (kg)}} \tag{13.4}$$

For example, to prepare a 1 *molal,* or 1 *m,* sodium sulfate (Na_2SO_4) aqueous solution, we need to dissolve 1 mole (142.0 g) of the substance in 1000 g (1 kg) of water. Depending on the nature of the solute-solvent interaction, the final volume of the solution will be either greater or less than 1000 mL. It is also possible, though very unlikely, that the final volume could be equal to 1000 mL.

EXAMPLE 13.2

Calculate the molality of a sulfuric acid solution containing 35.2 g of sulfuric acid in 237 g of water. The molar mass of sulfuric acid is 98.09 g.

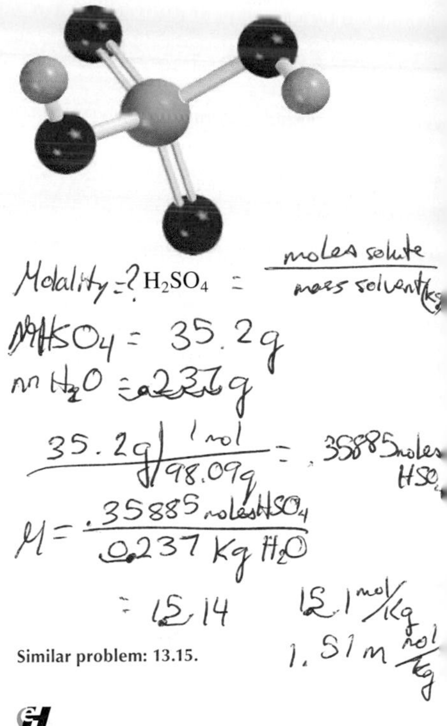

Strategy To calculate the molality of a solution, we need to know the number of moles of solute and the mass of the solvent in kilograms.

Solution The definition of molality (m) is

$$m = \frac{\text{moles of solute}}{\text{mass of solvent (kg)}}$$

First, we find the number of moles of sulfuric acid in 35.2 g of the acid, using its molar mass as the conversion factor:

$$\text{moles of } H_2SO_4 = 35.2 \cancel{\text{g } H_2SO_4} \times \frac{1 \text{ mol } H_2SO_4}{98.09 \cancel{\text{g } H_2SO_4}}$$

$$= 0.359 \text{ mol } H_2SO_4$$

The mass of water is 237 g, or 0.237 kg. Therefore,

$$\text{molality} = \frac{0.359 \text{ mol } H_2SO_4}{0.237 \text{ kg } H_2O}$$

$$= 1.51\ m$$

Similar problem: 13.15.

Practice Exercise What is the molality of a solution containing 7.78 g of urea $[(NH_2)_2CO]$ in 203 g of water?

Comparison of Concentration Units

The choice of a concentration unit is based on the purpose of the experiment. The advantage of molarity is that it is generally easier to measure the volume of a solution, using precisely calibrated volumetric flasks, than to weigh the solvent, as we saw in Section 4.5. For this reason, molarity is often preferred over molality. On the other hand, molality is independent of temperature, because the concentration is expressed in number of moles of solute and mass of solvent. The volume of a solution typically increases with increasing temperature, so that a solution that is 1.0 M at 25°C may become 0.97 M at 45°C because of the increase in volume on warming. This concentration dependence on temperature can significantly affect the accuracy of an experiment. Therefore, it is sometimes preferable to use molality instead of molarity.

Percent by mass is similar to molality in that it is independent of temperature. Furthermore, because it is defined in terms of ratio of mass of solute to mass of solution, we do not need to know the molar mass of the solute to calculate the percent by mass.

Sometimes it is desirable to convert one concentration unit of a solution to another; for example, the same solution may be employed for different experiments that require different concentration units for calculations. Suppose we want to express the concentration of a 0.396 m glucose ($C_6H_{12}O_6$) solution in molarity. We know there is 0.396 mole of glucose in 1000 g of the solvent and we need to determine the volume of this solution to calculate molarity. First, we calculate the mass of the solution from the molar mass of glucose:

$$\left(0.396 \cancel{\text{mol } C_6H_{12}O_6} \times \frac{180.2 \text{ g}}{1 \cancel{\text{mol } C_6H_{12}O_6}}\right) + 1000 \text{ g } H_2O \text{ soln} = 1071 \text{ g}$$

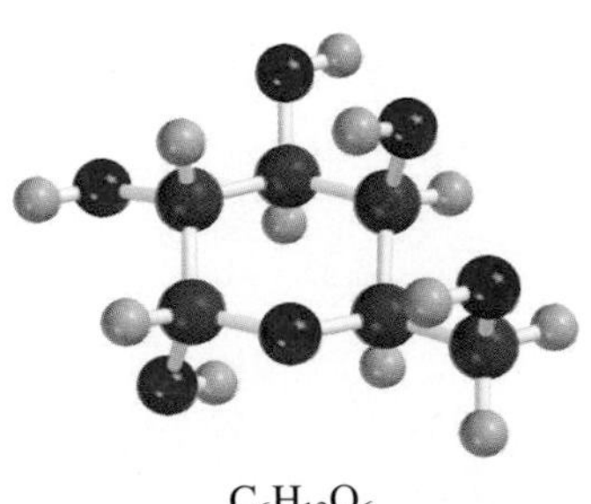

$C_6H_{12}O_6$

The next step is to experimentally determine the density of the solution, which is found to be 1.16 g/mL. We can now calculate the volume of the solution in liters by writing

$$\begin{aligned}\text{volume} &= \frac{\text{mass}}{\text{density}} \\ &= \frac{1071\text{ g}}{1.16\text{ g/mL}} \times \frac{1\text{ L}}{1000\text{ mL}} \\ &= 0.923\text{ L}\end{aligned}$$

Finally, the molarity of the solution is given by

$$\begin{aligned}\text{molarity} &= \frac{\text{moles of solute}}{\text{liters of soln}} \\ &= \frac{0.396\text{ mol}}{0.923\text{ L}} \\ &= 0.429\text{ mol/L} = 0.429\ M\end{aligned}$$

As you can see, the density of the solution serves as a conversion factor between molality and molarity.

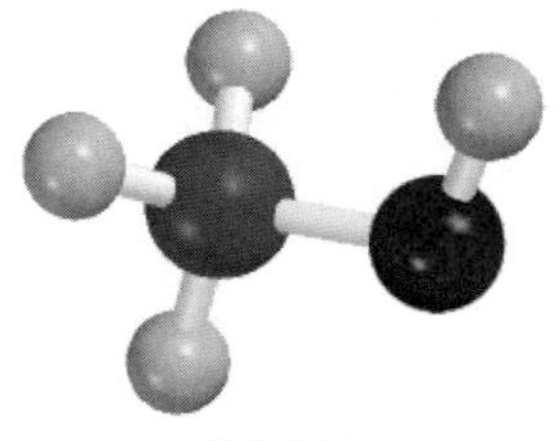

CH_3OH

EXAMPLE 13.3

The density of a 4.86 M aqueous solution of methanol (CH_3OH) is 0.973 g/mL. What is the molality of the solution? The molar mass of methanol is 32.04 g.

Strategy To calculate the molality, we need to know the number of moles of methanol and the mass of solvent in kilograms. We assume 1 L of solution, so the number of moles of methanol is 4.86 mol.

$$m = \frac{\text{moles of solute}}{\text{mass of solvent (kg)}}$$

(want to calculate: m; given: moles of solute; need to find: mass of solvent (kg))

Solution Our first step is to calculate the mass of water in 1 L of the solution, using density as a conversion factor. The total mass of 1 L of a 4.86 M solution of methanol is

$$1\text{ L soln} \times \frac{1000\text{ mL soln}}{1\text{ L soln}} \times \frac{0.973\text{ g}}{1\text{ mL soln}} = 973\text{ g}$$

Because this solution contains 4.86 moles of methanol, the amount of water (solvent) in the solution is

$$\begin{aligned}\text{mass of } H_2O &= \text{mass of soln} - \text{mass of solute} \\ &= 973\text{ g} - \left(4.86\text{ mol } CH_3OH \times \frac{32.04\text{ g } CH_3OH}{1\text{ mol } CH_3OH}\right) \\ &= 817\text{ g}\end{aligned}$$

The molality of the solution can be calculated by converting 817 g to 0.817 kg:

$$\begin{aligned}\text{molality} &= \frac{4.86\text{ mol } CH_3OH}{0.817\text{ kg } H_2O} \\ &= 5.95\ m\end{aligned}$$

(Continued)

Check For very dilute aqueous solutions, molarity and molality have nearly identical values. However, for more concentrated aqueous solutions, the mass of the solute is significant, which in turn decreases the mass of the solvent. Thus, it is not uncommon for molality to be considerably larger than molarity in such cases.

Practice Exercise Calculate the molality of a 5.86 *M* ethanol (C_2H_5OH) aqueous solution whose density is 0.927 g/mol.

Similar problem: 13.16(a).

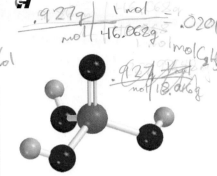

EXAMPLE 13.4

Calculate the molality of a 29.7 percent (by mass) aqueous solution of phosphoric acid (H_3PO_4). The molar mass of phosphoric acid is 97.99 g.

Strategy In solving this type of problem, it is convenient to assume that we start with 100.0 g of the solution. If the mass of phosphoric acid is 29.7 percent, or 29.7 g, the percent by mass and mass of water must be 100.0% − 29.7% = 70.3% and 70.3 g.

Solution From the known molar mass of phosphoric acid, we can calculate the molality in two steps, as shown in Example 13.2. First we calculate the number of moles of phosphoric acid in 29.7 g of the acid:

$$\text{moles of } H_3PO_4 = 29.7 \text{ g } \cancel{H_3PO_4} \times \frac{1 \text{ mol } H_3PO_4}{97.99 \text{ g } \cancel{H_3PO_4}}$$

$$= 0.303 \text{ mol } H_3PO_4$$

The mass of water is 70.3 g, or 0.0703 kg. Therefore, the molality is given by

$$\text{molality} = \frac{0.303 \text{ mol } H_3PO_4}{0.0703 \text{ kg } H_2O}$$

$$= 4.31\ m$$

H_3PO_4

Practice Exercise Calculate the molality of a 44.6 percent (by mass) aqueous solution of sodium chloride.

Similar problem: 13.16(b).

REVIEW OF CONCEPTS

A solution is prepared at 20°C and its concentration is expressed in three different units: percent by mass, molality, and molarity. The solution is then heated to 75°C. Which of the concentration units will change (increase or decrease)?

13.4 Effect of Temperature on Solubility

Recall that solubility is defined as the maximum amount of a solute that will dissolve in a given quantity of solvent *at a specific temperature.* For most substances, temperature affects solubility. In this section we will consider the effects of temperature on the solubility of solids and gases.

Solid Solubility and Temperature

Figure 13.3 shows the temperature dependence of the solubility of some ionic compounds in water. In most but certainly not all cases, the solubility of a solid substance increases with temperature. However, there is no clear correlation between the sign of ΔH_{soln} and the variation of solubility with temperature. For example, the solution process of $CaCl_2$ is exothermic and that of NH_4NO_3 is endothermic. But the solubility

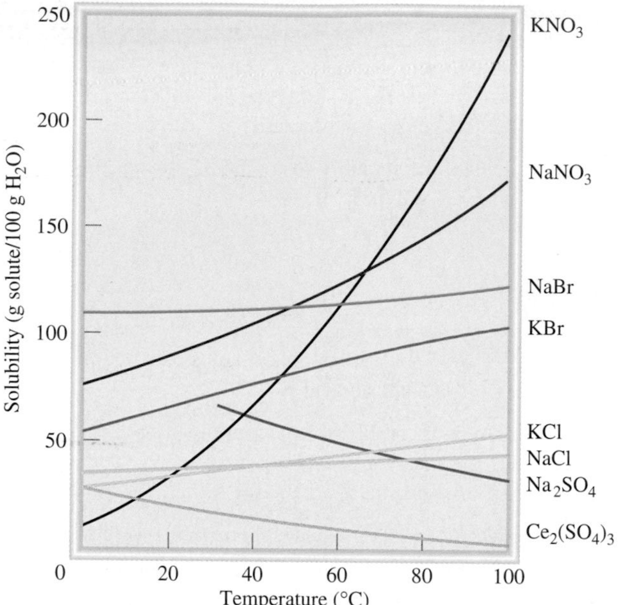

Figure 13.3
Dependence on temperature of the solubility of some ionic compounds in water.

of both compounds increases with increasing temperature. In general, the effect of temperature on solubility is best determined experimentally.

Gas Solubility and Temperature

The solubility of gases in water usually decreases with increasing temperature (Figure 13.4). When water is heated in a beaker, you can see bubbles of air forming on the side of the glass before the water boils. As the temperature rises, the dissolved air molecules begin to "boil out" of the solution long before the water itself boils.

The reduced solubility of molecular oxygen in hot water has a direct bearing on ***thermal pollution***, that is, *the heating of the environment*—usually waterways—*to temperatures that are harmful to its living inhabitants.* It is estimated that every year in the United States some 100,000 billion gallons of water are used for industrial cooling, mostly in electric power and nuclear power production. This process heats up the water, which is then returned to the rivers and lakes from which it was taken. Ecologists have become increasingly concerned about the effect of thermal pollution on aquatic life. Fish, like all other cold-blooded animals, have much more difficulty coping with rapid temperature fluctuation in the environment than humans do. An increase in water temperature accelerates their rate of metabolism, which generally doubles with each 10°C rise. The speedup of metabolism increases the fish's need for oxygen at the same time that the supply of oxygen decreases because of its lower solubility in heated water. Effective ways to cool power plants while doing only minimal damage to the biological environment are being sought.

On the lighter side, a knowledge of the variation of gas solubility with temperature can improve one's performance in a popular recreational sport—fishing. On a hot summer day, an experienced fisherman usually picks a deep spot in the river or lake to cast the bait. Because the oxygen content is greater in the deeper, cooler region, most fish will be found there.

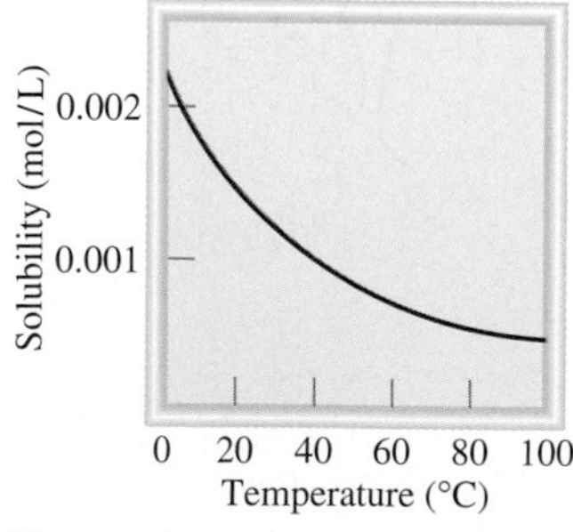

Figure 13.4
Dependence on temperature of the solubility of O_2 gas in water. Note that the solubility decreases as temperature increases. The pressure of the gas over the solution is 1 atm.

Review of Concepts

Using Figure 13.3, rank the potassium salts in increasing order of solubility at 40°C.

13.5 Effect of Pressure on the Solubility of Gases

For all practical purposes, external pressure has no influence on the solubilities of liquids and solids, but it does greatly affect the solubility of gases. The quantitative relationship between gas solubility and pressure is given by ***Henry's law,*** which states that *the solubility of a gas in a liquid is proportional to the pressure of the gas over the solution:*

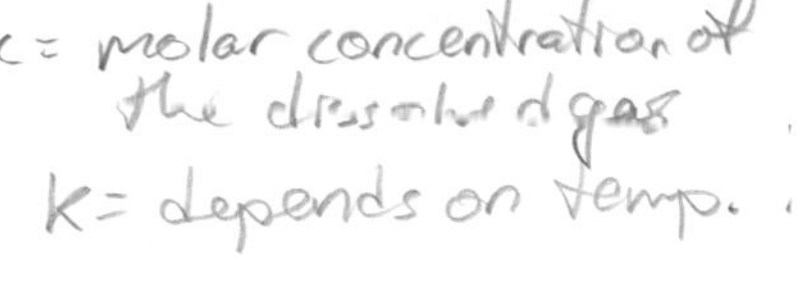

c = molar concentration of the dissolved gas
k = depends on temp.

$$c \propto P$$

$$c = kP \tag{13.5}$$

Here c is the molar concentration (moles per liter) of the dissolved gas; P is the pressure (in atmospheres) of the gas over the solution at equilibrium; and, for a given gas, k is a constant that depends only on temperature. The constant k has the units mol/L · atm. You can see that when the pressure of the gas is 1 atm, c is *numerically* equal to k.

Each gas has a different k value at a given temperature.

Henry's law can be understood qualitatively in terms of the kinetic molecular theory. The amount of gas that will dissolve in a solvent depends on how frequently the molecules in the gas phase collide with the liquid surface and become trapped by the condensed phase. Suppose we have a gas in dynamic equilibrium with a solution [Figure 13.5(a)]. At every instant, the number of gas molecules entering the solution is equal to the number of dissolved molecules moving into the gas phase. When the partial pressure is increased, more molecules dissolve in the liquid because more molecules are striking the surface of the liquid. This process continues until the concentration of the solution is again such that the number of molecules leaving the solution per second equals the number entering the solution [Figure 13.5(b)]. Because of the increased concentration of molecules in both the gas and solution phases, this number is greater in (b) than in (a), where the partial pressure is lower.

A practical demonstration of Henry's law is the effervescence of a soft drink when the cap of the bottle is removed. Before the beverage bottle is sealed, it is pressurized with a mixture of air and CO_2 saturated with water vapor. Because of the high partial pressure of CO_2 in the pressurizing gas mixture, the amount dissolved in the soft drink is many times the amount that would dissolve under normal

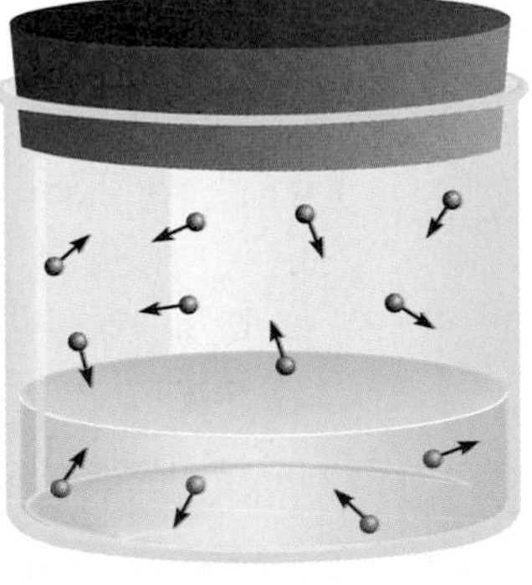

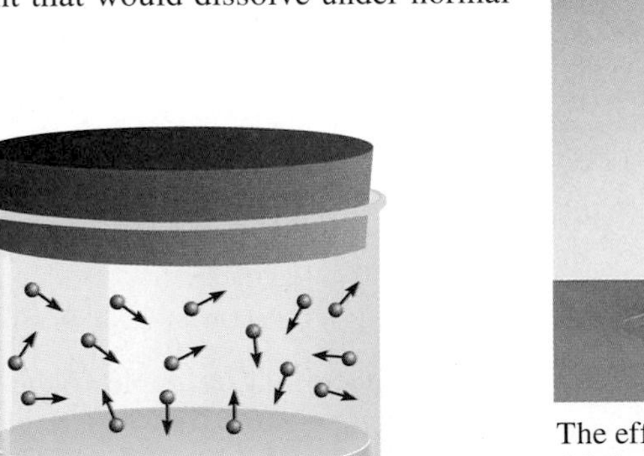

(a) (b)

Figure 13.5
A molecular interpretation of Henry's law. When the partial pressure of the gas over the solution increases from (a) to (b), the concentration of the dissolved gas also increases according to Equation (13.5).

The effervescence of a soft drink. The bottle was shaken before being opened to dramatize the escape of CO_2.

atmospheric conditions. When the cap is removed, the pressurized gases escape, eventually the pressure in the bottle falls to atmospheric pressure, and the amount of CO_2 remaining in the beverage is determined only by the normal atmospheric partial pressure of CO_2, 0.0003 atm. The excess dissolved CO_2 comes out of solution, causing the effervescence.

EXAMPLE 13.5

The solubility of nitrogen gas at 25°C and 1 atm is 6.8×10^{-4} mol/L. What is the concentration of nitrogen dissolved in water under atmospheric conditions? The partial pressure of nitrogen gas in the atmosphere is 0.78 atm.

Strategy The given solubility enables us to calculate Henry's law constant (k), which can then be used to determine the concentration of the solution.

Solution The first step is to calculate the quantity k in Equation (13.5):

$$c = kP$$
$$6.8 \times 10^{-4} \text{ mol/L} = k\,(1 \text{ atm})$$
$$k = 6.8 \times 10^{-4} \text{ mol/L} \cdot \text{atm}$$

Therefore, the solubility of nitrogen gas in water is

$$c = (6.8 \times 10^{-4} \text{ mol/L} \cdot \text{atm})(0.78 \text{ atm})$$
$$= 5.3 \times 10^{-4} \text{ mol/L}$$
$$= 5.3 \times 10^{-4}\, M$$

The decrease in solubility is the result of lowering the pressure from 1 atm to 0.78 atm.

Check The ratio of the concentrations $[(5.3 \times 10^{-4}\, M/6.8 \times 10^{-4}\, M) = 0.78]$ should be equal to the ratio of the pressures (0.78 atm/1.0 atm = 0.78).

Similar problem: 13.35.

Practice Exercise Calculate the molar concentration of oxygen in water at 25°C for a partial pressure of 0.22 atm. The Henry's law constant for oxygen is 1.3×10^{-3} mol/L · atm.

Most gases obey Henry's law, but there are some important exceptions. For example, if the dissolved gas *reacts* with water, higher solubilities can result. The solubility of ammonia is much higher than expected because of the reaction

$$NH_3 + H_2O \rightleftharpoons NH_4^+ + OH^-$$

Carbon dioxide also reacts with water, as follows:

$$CO_2 + H_2O \rightleftharpoons H_2CO_3$$

Another interesting example is the dissolution of molecular oxygen in blood. Normally, oxygen gas is only sparingly soluble in water (see the Practice Exercise in Example 13.5). However, its solubility in blood is dramatically greater because of the high content of hemoglobin (Hb) molecules. Each hemoglobin molecule can bind up to four oxygen molecules, which are eventually delivered to the tissues for use in metabolism:

$$Hb + 4O_2 \rightleftharpoons Hb(O_2)_4$$

This is the process that accounts for the high solubility of molecular oxygen in blood.

REVIEW OF CONCEPTS

Which of the following gases has the greatest Henry's law constant in water at 25°C: CH_4, Ne, HCl, H_2?

13.6 Colligative Properties

Several important properties of solutions depend on the number of solute particles in solution and not on the nature of the solute particles. These properties are called ***colligative properties*** (or collective properties) because they are bound together by a common origin; that is, they all depend on the number of solute particles present, whether these particles are atoms, ions, or molecules. The colligative properties are vapor-pressure lowering, boiling-point elevation, freezing-point depression, and osmotic pressure. We will first discuss the colligative properties of nonelectrolyte solutions. It is important to keep in mind that we are talking about relatively dilute solutions, that is, solutions whose concentrations are $\lesssim 0.2\ M$.

Vapor-Pressure Lowering

To review the concept of equilibrium vapor pressure as it applies to pure liquids, see Section 12.6.

If a solute is ***nonvolatile*** (that is, it *does not have a measurable vapor pressure*), the vapor pressure of its solution is always less than that of the pure solvent. Thus, the relationship between solution vapor pressure and solvent vapor pressure depends on the concentration of the solute in the solution. This relationship is given by ***Raoult's law*** (after the French chemist Francois Raoult), which states that *the vapor pressure of a solvent over a solution, P_1, is given by the vapor pressure of the pure solvent, P_1°, times the mole fraction of the solvent in the solution, X_1:*

$$P_1 = X_1 P_1^\circ \tag{13.6}$$

In a solution containing only one solute, $X_1 = 1 - X_2$, in which X_2 is the mole fraction of the solute (see Section 5.5). Equation (13.6) can therefore be rewritten as

$$P_1 = (1 - X_2)P_1^\circ$$

$$P_1^\circ - P_1 = \Delta P = X_2 P_1^\circ \tag{13.7}$$

We see that the *decrease* in vapor pressure, ΔP, is directly proportional to the concentration (measured in mole fraction) of the solute present.

EXAMPLE 13.6

Calculate the vapor pressure of a solution made by dissolving 198 g of glucose (molar mass = 180.2 g/mol) in 435 mL of water at 35°C. What is the vapor-pressure lowering? The vapor pressure of pure water at 35°C is given in Table 5.2 (p. 156). Assume the density of the solvent is 1.00 g/mL.

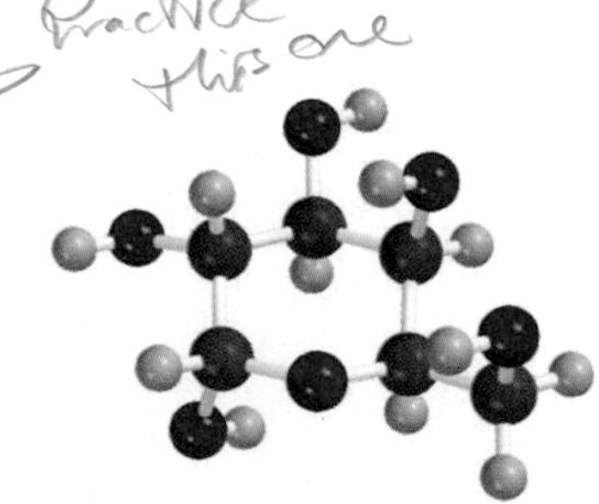

$C_6H_{12}O_6$

Strategy We need Raoult's law [Equation (13.6)] to determine the vapor pressure of a solution. Note that glucose is a nonvolatile solute.

Solution The vapor pressure of a solution (P_1) is

$$\underset{\text{want to calculate}}{P_1} = X_1 \underset{\text{given}}{\overset{\text{need to find}}{}} P_1^\circ$$

(need to find → X_1; want to calculate → P_1; given → P_1°)

(Continued)

First we calculate the number of moles of glucose and water in the solution:

$$n_1(\text{water}) = 435\ \cancel{\text{mL}} \times \frac{1.00\ \cancel{\text{g}}}{1\ \cancel{\text{mL}}} \times \frac{1\ \text{mol}}{18.02\ \cancel{\text{g}}} = 24.1\ \text{mol}$$

$$n_2(\text{glucose}) = 198\ \cancel{\text{g}} \times \frac{1\ \text{mol}}{180.2\ \cancel{\text{g}}} = 1.10\ \text{mol}$$

The mole fraction of water, X_1, is given by

$$X_1 = \frac{n_1}{n_1 + n_2}$$

$$= \frac{24.1\ \text{mol}}{24.1\ \text{mol} + 1.10\ \text{mol}} = 0.956$$

From Table 5.2, we find the vapor pressure of water at 35°C to be 42.18 mmHg. Therefore, the vapor pressure of the glucose solution is

$$P_1 = 0.956 \times 42.18\ \text{mmHg}$$
$$= 40.3\ \text{mmHg}$$

Finally, the vapor-pressure lowering is (42.18 – 40.3) mmHg, or 1.9 mmHg.

Check We can also calculate the vapor pressure lowering using Equation (13.7). Because the mole fraction of glucose is (1 − 0.956), or 0.044, the vapor pressure lowering is given by (0.044)(42.18 mmHg) or 1.9 mmHg.

Similar problems: 13.47, 13.48.

Practice Exercise Calculate the vapor pressure of a solution made by dissolving 82.4 g of urea (molar mass = 60.06 g/mol) in 212 mL of water at 35°C. What is the vapor pressure lowering?

Why is the vapor pressure of a solution less than that of its pure solvent? As was mentioned in Section 13.2, one driving force in physical and chemical processes is the increase in disorder—the greater the disorder created, the more favorable the process. Vaporization increases the disorder of a system because molecules in a vapor have less order than those in a liquid. Because a solution is more disordered than a pure solvent, the difference in disorder between a solution and a vapor is less than that between a pure solvent and a vapor. Thus, solvent molecules have less of a tendency to leave a solution than to leave the pure solvent to become vapor, and the vapor pressure of a solution is less than that of the solvent.

If both components of a solution are ***volatile*** (that is, *have measurable vapor pressure*), the vapor pressure of the solution is the sum of the individual partial pressures. Raoult's law holds equally well in this case:

$$P_A = X_A P_A^\circ$$
$$P_B = X_B P_B^\circ$$

in which P_A and P_B are the partial pressures over the solution for components A and B; P_A° and P_B° are the vapor pressures of the pure substances; and X_A and X_B are their mole fractions. The total pressure is given by Dalton's law of partial pressures (see Section 5.5):

$$P_T = P_A + P_B$$

Benzene and toluene have similar structures and therefore similar intermolecular forces:

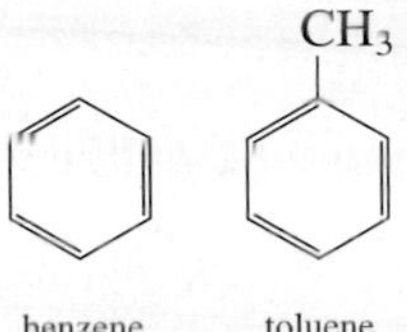

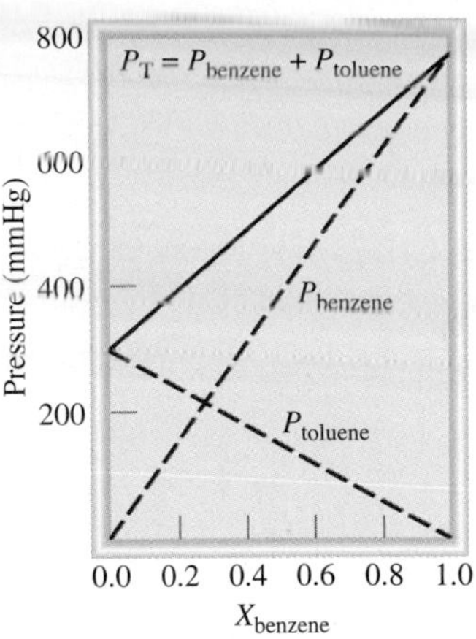

Figure 13.6
The dependence of the partial pressures of benzene and toluene on their mole fractions in a benzene-toluene solution ($X_{toluene} = 1 - X_{benzene}$) at 80°C. This solution is said to be ideal because the vapor pressures obey Raoult's law.

In a solution of benzene and toluene, the vapor pressure of each component obeys Raoult's law. Figure 13.6 shows the dependence of the total vapor pressure (P_T) in a benzene-toluene solution on the composition of the solution. Note that we need only express the composition of the solution in terms of the mole fraction of one component. For every value of $X_{benzene}$, the mole fraction of toluene is given by $(1 - X_{benzene})$. The benzene-toluene solution is one of the few examples of an ***ideal solution,*** which is *any solution that obeys Raoult's law.* One characteristic of an ideal solution is that the intermolecular forces between solute and solvent molecules are equal to those between solute molecules and between solvent molecules. Consequently, the heat of solution, ΔH_{soln}, is always zero.

Boiling-Point Elevation

Because the presence of a *nonvolatile* solute lowers the vapor pressure of a solution, it must also affect the boiling point of the solution. The boiling point of a solution is the temperature at which its vapor pressure equals the external atmospheric pressure (see Section 12.6). Figure 13.7 shows the phase diagram of water and the changes that occur in an aqueous solution. Because at any temperature the vapor pressure of the solution is lower than that of the pure solvent, the liquid-vapor curve for the solution lies below that for the pure solvent. Consequently, the solution curve (dotted line) intersects the horizontal line that marks $P = 1$ atm at a *higher* temperature than the normal boiling point of the pure solvent. This graphical analysis shows that the boiling point of the solution is higher than that of water. The *boiling-point elevation,* ΔT_b, is defined as

$$\Delta T_b = T_b - T_b^\circ$$

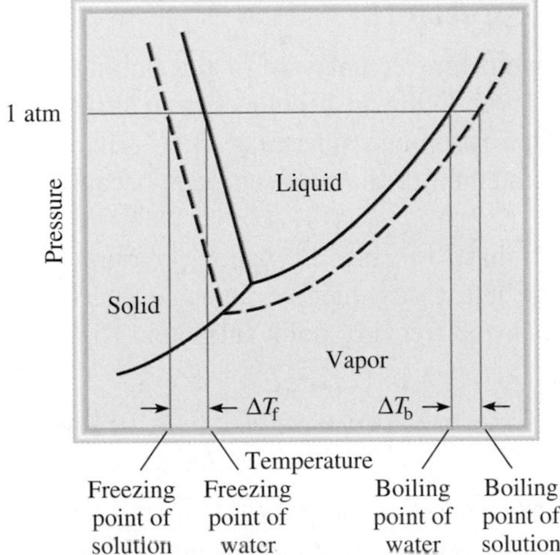

Figure 13.7
Phase diagram illustrating the boiling-point elevation and freezing-point depression of aqueous solutions. The dashed curves pertain to the solution, and the solid curves to the pure solvent. As you can see, the boiling point of the solution is higher than that of water, and the freezing point of the solution is lower than that of water.

Table 13.2 Molal Boiling-Point Elevation and Freezing-Point Depression Constants of Several Common Liquids

Solvent	Normal Freezing Point (°C)*	K_f (°C/m)	Normal Boiling Point (°C)*	K_b (°C/m)
Water	0	1.86	100	0.52
Benzene	5.5	5.12	80.1	2.53
Ethanol	−117.3	1.99	78.4	1.22
Acetic acid	16.6	3.90	117.9	2.93
Cyclohexane	6.6	20.0	80.7	2.79

*Measured at 1 atm.

In calculating the new boiling point, add ΔT_b to the normal boiling point of the solvent.

in which T_b is the boiling point of the solution and T_b° the boiling point of the pure solvent. Because ΔT_b is proportional to the vapor-pressure lowering, it is also proportional to the concentration (molality) of the solution. That is,

$$\Delta T_b \propto m$$

$$\Delta T_b = K_b m \tag{13.8}$$

in which m is the molality of the solution and K_b is the *molal boiling-point elevation constant*. The units of K_b are °C/m.

It is important to understand the choice of concentration unit here. We are dealing with a system (the solution) whose temperature is *not* kept constant, so we cannot express the concentration units in molarity because molarity changes with temperature.

Table 13.2 lists the value of K_b for several common solvents. Using the boiling-point elevation constant for water and Equation (13.8), you can see that if the molality of an aqueous solution is 1.00 m, the boiling point will be 100.52°C.

Freezing-Point Depression

De-icing of airplanes is based on freezing-point depression.

A nonscientist may remain forever unaware of the boiling-point elevation phenomenon, but a careful observer living in a cold climate is familiar with freezing-point depression. Ice on frozen roads and sidewalks melts when sprinkled with salts such as NaCl or $CaCl_2$. This method of thawing succeeds because it depresses the freezing point of water.

Figure 13.7 shows that lowering the vapor pressure of the solution shifts the solid-liquid curve to the left. Consequently, this line intersects the horizontal line at a temperature *lower* than the freezing point of water. The *freezing-point depression*, ΔT_f, is defined as

$$\Delta T_f = T_f^\circ - T_f$$

In calculating the new freezing point, subtract ΔT_f from the normal freezing point of the solvent.

in which T_f° is the freezing point of the pure solvent, and T_f the freezing point of the solution. Again, ΔT_f is proportional to the concentration of the solution:

$$\Delta T_f \propto m$$

$$\Delta T_f = K_f m \tag{13.9}$$

in which m is the concentration of the solute in molality units, and K_f is the *molal freezing-point depression constant* (see Table 13.2). Like K_b, K_f has the units °C/m.

A qualitative explanation of the freezing-point depression phenomenon is as follows. Freezing involves a transition from the disordered state to the ordered state. For this to happen, energy must be removed from the system. Because a solution has greater disorder than the solvent, more energy needs to be removed from it to create order than in the case of a pure solvent. Therefore, the solution has a lower freezing point than the solvent. Note that when a solution freezes, the solid that separates is the solvent component.

Whereas the solute must be nonvolatile in the case of boiling-point elevation, no such restriction applies to freezing-point depression. For example, methanol (CH_3OH), a fairly volatile liquid that boils at only 65°C, has sometimes been used as an antifreeze in automobile radiators.

EXAMPLE 13.7

Ethylene glycol (EG), $CH_2(OH)CH_2(OH)$, is a common automobile antifreeze. It is water soluble and fairly nonvolatile (bp 197°C). Calculate the freezing point of a solution containing 724 g of this substance in 2603 g of water. Would you keep this substance in your car radiator during the summer? The molar mass of ethylene glycol is 62.07 g.

In cold climate regions, antifreeze must be used in car radiators in winter.

Strategy This question asks for the depression in freezing point of the solution.

$$\Delta T_f = K_f m$$

(constant: K_f; want to calculate: ΔT_f; need to find: m)

The information given enables us to calculate the molality of the solution and we refer to Table 13.2 for the K_f of water.

Solution To solve for the molality of the solution, we need to know the number of moles of EG and the mass of the solvent in kilograms. We find the molar mass of EG, and convert the mass of the solvent to 2.603 kg, and calculate the molality as follows:

$$724 \text{ g EG} \times \frac{1 \text{ mol EG}}{62.07 \text{ g EG}} = 11.7 \text{ mol EG}$$

$$\text{molality} = \frac{\text{moles of solute}}{\text{mass of solvent (kg)}}$$

$$= \frac{11.7 \text{ mol EG}}{2.603 \text{ kg H}_2\text{O}} = 4.49 \text{ mol EG/kg H}_2\text{O}$$

$$= 4.49\ m$$

From Equation (13.9) and Table 13.2 we write

$$\begin{aligned}\Delta T_f &= K_f m \\ &= (1.86\text{°C}/m)(4.49\ m) \\ &= 8.35\text{°C}\end{aligned}$$

Because pure water freezes at 0°C, the solution will freeze at −8.35°C. We can calculate boiling-point elevation in the same way as follows:

$$\begin{aligned}\Delta T_b &= K_b m \\ &= (0.52\text{°C}/m)(4.49\ m) \\ &= 2.3\text{°C}\end{aligned}$$

(Continued)

Because the solution will boil at (100 + 2.3)°C, or 102.3°C, it would be preferable to leave the antifreeze in your car radiator in summer to prevent the solution from boiling.

Similar problems: 13.54, 13.57.

Practice Exercise Calculate the boiling point and freezing point of a solution containing 478 g of ethylene glycol in 3202 g of water.

Review of Concepts

The diagram here shows the vapor pressure curves for pure benzene and a solution of a nonvolatile solute in benzene. Estimate the molality of the benzene solution.

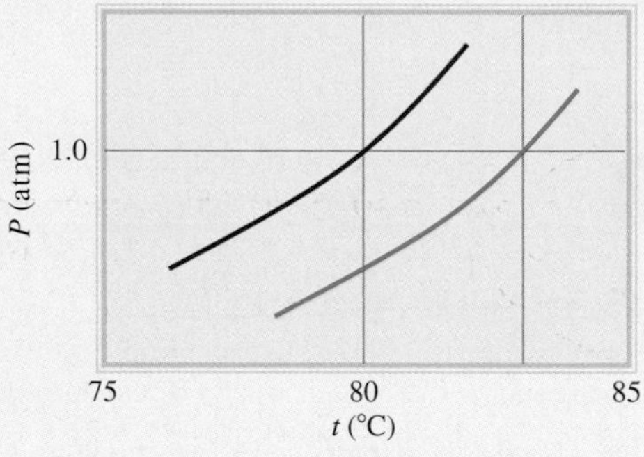

Osmotic Pressure

Animation: Osmosis

Many chemical and biological processes depend on the selective passage of solvent molecules through a porous membrane from a dilute solution to a more concentrated one. Figure 13.8 illustrates this phenomenon. The left compartment of the apparatus contains pure solvent; the right compartment contains a solution. The two compartments are separated by a ***semipermeable membrane,*** which *allows solvent molecules to pass through but blocks the passage of solute molecules.* At the start, the water levels in the two tubes are equal [see Figure 13.8(a)]. After some time, the level in

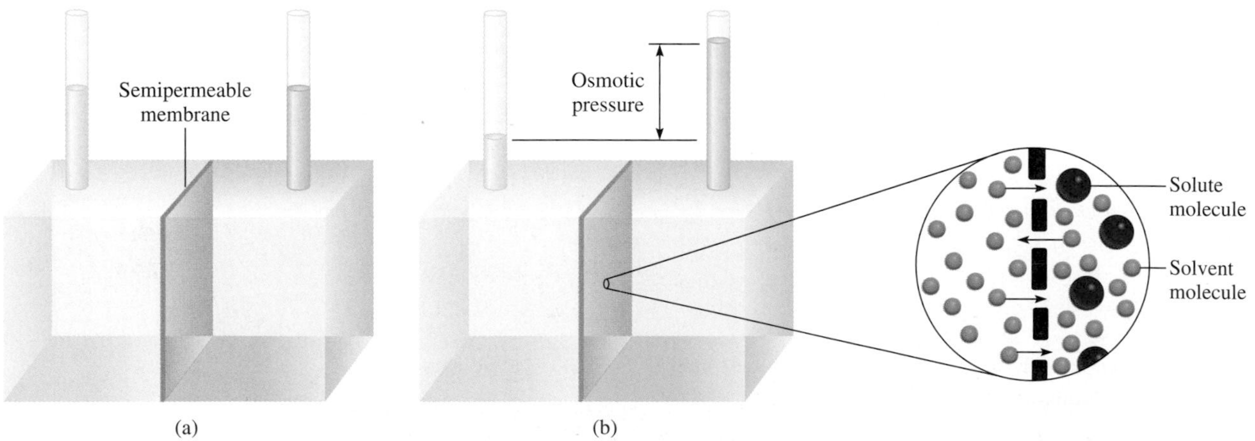

Figure 13.8
Osmotic pressure. (a) The levels of the pure solvent (left) and of the solution (right) are equal at the start. (b) During osmosis, the level on the solution side rises as a result of the net flow of solvent from left to right. The osmotic pressure is equal to the hydrostatic pressure exerted by the column of fluid in the right tube at equilibrium. Basically the same effect occurs when the pure solvent is replaced by a more dilute solution than that on the right.

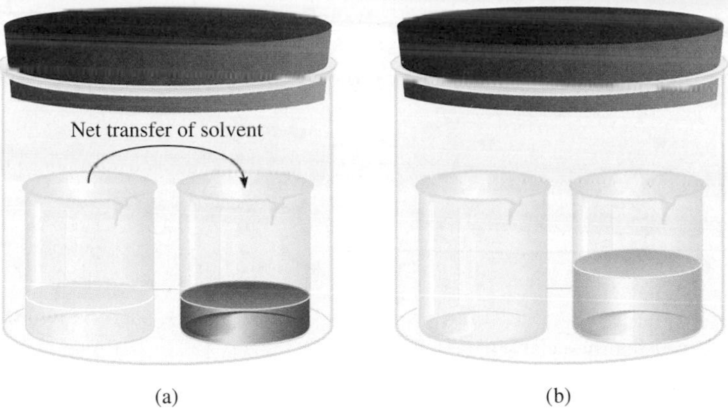

Figure 13.9
(a) Unequal vapor pressures inside the container lead to a net transfer of water from the left beaker (which contains pure water) to the right one (which contains a solution). (b) At equilibrium, all the water in the left beaker has been transferred to the right beaker. This driving force for solvent transfer is analogous to the osmotic phenomenon that is shown in Figure 13.8.

the right tube begins to rise; this continues until equilibrium is reached. *The net movement of solvent molecules through a semipermeable membrane from a pure solvent or from a dilute solution to a more concentrated solution* is called **osmosis.** The ***osmotic pressure*** (π) of a solution is *the pressure required to stop osmosis.* As shown in Figure 13.8(b), this pressure can be measured directly from the difference in the final fluid levels.

What causes water to move spontaneously from left to right in this case? Compare the vapor pressure of pure water and that of water from a solution (Figure 13.9). Because the vapor pressure of pure water is higher, there is a net transfer of water from the left beaker to the right one. Given enough time, the transfer will continue to completion. A similar force causes water to move into the solution during osmosis.

Although osmosis is a common and well-studied phenomenon, relatively little is known about how the semipermeable membrane stops some molecules yet allows others to pass. In some cases, it is simply a matter of size. A semipermeable membrane may have pores small enough to let only the solvent molecules through. In other cases, a different mechanism may be responsible for the membrane's selectivity—for example, the solvent's greater "solubility" in the membrane.

The osmotic pressure of a solution is given by

$$\pi = MRT \tag{13.10}$$

in which M is the molarity of solution, R is the gas constant (0.0821 L · atm/K · mol), and T is the absolute temperature. The osmotic pressure, π, is expressed in atmospheres. Because osmotic pressure measurements are carried out at constant temperature, we express the concentration here in terms of the more convenient units of molarity rather than molality.

Like boiling-point elevation and freezing-point depression, osmotic pressure is directly proportional to the concentration of solution. This is what we would expect, bearing in mind that all colligative properties depend only on the number of solute particles in solution. If two solutions are of equal concentration and, hence, of the same osmotic pressure, they are said to be *isotonic*. If two solutions are of unequal osmotic pressures, the more concentrated solution is said to be *hypertonic* and the more dilute solution is described as *hypotonic* (Figure 13.10).

The osmotic pressure phenomenon manifests itself in many interesting applications. To study the contents of red blood cells, which are protected from the external environment by a semipermeable membrane, biochemists use a technique called

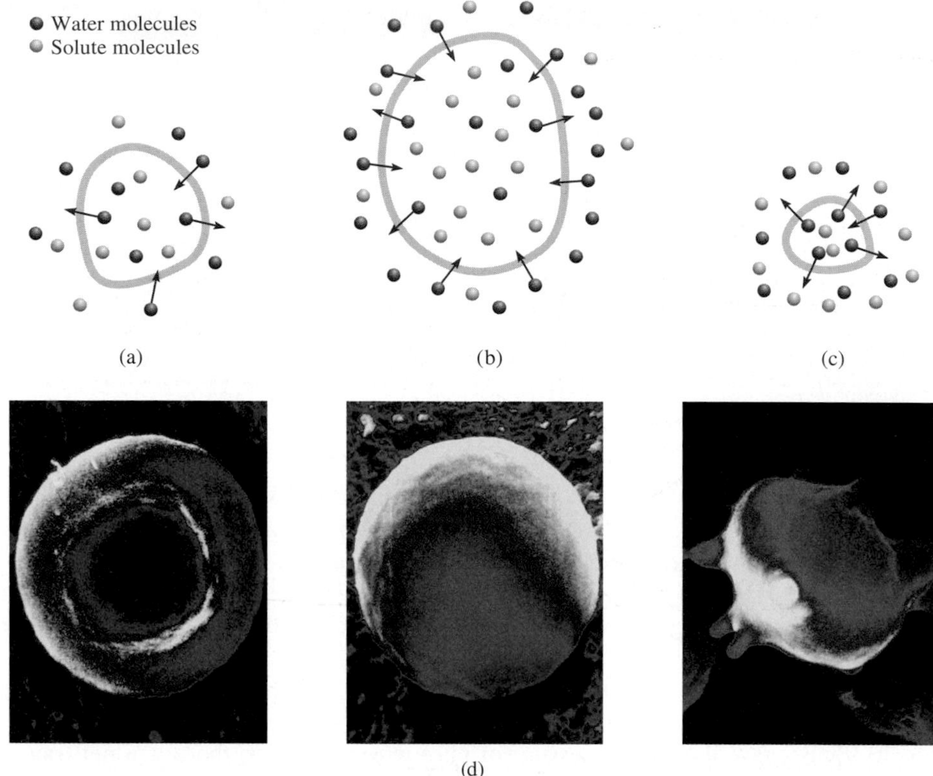

Figure 13.10
A cell in (a) an isotonic solution, (b) a hypotonic solution, and (c) a hypertonic solution. The cell remains unchanged in (a), swells in (b), and shrinks in (c). (d) From left to right: a red blood cell in an isotonic solution, in a hypotonic solution, and in a hypertonic solution.

hemolysis. The red blood cells are placed in a hypotonic solution. Because the hypotonic solution is less concentrated than the interior of the cell, water moves into the cells, as shown in Figure 13.10(b). The cells swell and eventually burst, releasing hemoglobin and other molecules.

Home preserving of jam and jelly provides another example of the use of osmotic pressure. A large quantity of sugar is actually essential to the preservation process because the sugar helps to kill bacteria that may cause botulism. As Figure 13.10(c) shows, when a bacterial cell is in a hypertonic (high-concentration) sugar solution, the intracellular water tends to move out of the bacterial cell to the more concentrated solution by osmosis. This process, known as *crenation*, causes the cell to shrink and, eventually, to cease functioning. The natural acidity of fruits also inhibits bacterial growth.

Osmotic pressure also is the major mechanism for transporting water upward in plants. Because leaves constantly lose water to the air, in a process called *transpiration,* the solute concentrations in leaf fluids increase. Water is pushed up through the trunk, branches, and stems of trees by osmotic pressure. Up to 10 to 15 atm pressure is necessary to transport water to the leaves at the tops of California's redwoods, which reach about 120 m in height. (The capillary action discussed in Section 12.3 is responsible for the rise of water only up to a few centimeters.)

California redwoods.

Review of Concepts

What does it mean when we say that the osmotic pressure of a sample of seawater is 25 atm at a certain temperature?

Using Colligative Properties to Determine Molar Mass

The colligative properties of nonelectrolyte solutions provide a means of determining the molar mass of a solute. Theoretically, any of the four colligative properties are suitable for this purpose. In practice, however, only freezing-point depression and osmotic pressure are used because they show the most pronounced changes.

EXAMPLE 13.8

A 9.66-g sample of a compound with the empirical formula C_5H_4 is dissolved in 284 g of benzene. The freezing point of the solution is 1.37°C below that of pure benzene. What are the molar mass and molecular formula of this compound?

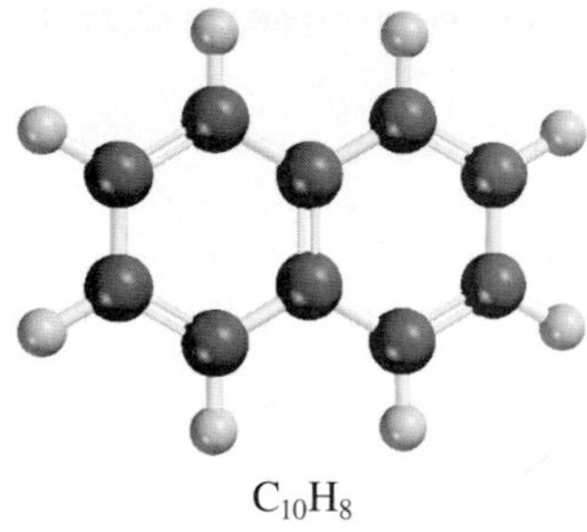

$C_{10}H_8$

Strategy Solving this problem requires three steps. First, we calculate the molality of the solution from the depression in freezing point. Next, from the molality we determine the number of moles in 9.66 g of the compound and hence its molar mass. Finally, comparing the experimental molar mass with the empirical molar mass enables us to write the molecular formula.

Solution The sequence of conversions for calculating the molar mass of the compound is

$$\text{freezing-point depression} \longrightarrow \text{molality} \longrightarrow \text{number of moles} \longrightarrow \text{molar mass}$$

Our first step is to calculate the molality of the solution. From Equation (13.9) and Table 13.2 we write

$$\text{molality} = \frac{\Delta T_f}{K_f} = \frac{1.37°\text{C}}{5.12°\text{C}/m} = 0.268\ m$$

Because there is 0.268 mole of the solute in 1 kg of solvent, the number of moles of solute in 284 g, or 0.284 kg, of solvent is

$$0.284\ \text{kg} \times \frac{0.268\ \text{mol}}{1\ \text{kg}} = 0.0761\ \text{mol}$$

Thus, the molar mass of the solute is

$$\text{molar mass} = \frac{\text{grams of compound}}{\text{moles of compound}}$$

$$= \frac{9.66\ \text{g}}{0.0761\ \text{mol}} = 127\ \text{g/mol}$$

Now we can determine the ratio

$$\frac{\text{molar mass}}{\text{empirical molar mass}} = \frac{127\ \text{g/mol}}{64\ \text{g/mol}} \approx 2$$

Therefore, the molecular formula is $(C_5H_4)_2$ or $C_{10}H_8$ (naphthalene).

Similar problem: 13.55.

Practice Exercise A solution of 0.85 g of an organic compound in 100.0 g of benzene has a freezing point of 5.16°C. What are the molality of the solution and the molar mass of the solute?

EXAMPLE 13.9

A solution is prepared by dissolving 44.1 g of hemoglobin (Hb) in enough water to make up 1 L in volume. If the osmotic pressure of the solution is found to be 12.6 mmHg at 25°C, calculate the molar mass of hemoglobin.

(Continued)

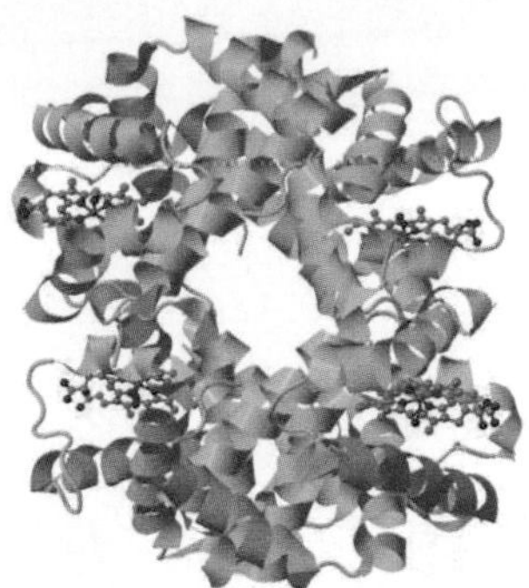

Ribbon representation of the structure of hemoglobin. There are four heme groups in each molecule that bind oxygen molecules.

Strategy We are asked to calculate the molar mass of Hb. The steps are similar to those outlined in Example 13.8. From the osmotic pressure of the solution, we calculate the molarity of the solution. Then, from its molarity, we determine the number of moles in 44.1 g of Hb and hence its molar mass. What units should we use for π and temperature?

Solution The sequence of conversions is as follows:

$$\text{osmotic pressure} \longrightarrow \text{molarity} \longrightarrow \text{number of moles} \longrightarrow \text{molar mass}$$

First, we calculate the molarity using Equation (13.10)

$$\pi = MRT$$

$$M = \frac{\pi}{RT}$$

$$= \frac{12.6 \text{ mmHg} \times \dfrac{1 \text{ atm}}{760 \text{ mmHg}}}{(0.0821 \text{ L} \cdot \text{atm/K} \cdot \text{mol})(298 \text{ K})}$$

$$= 6.78 \times 10^{-4}\, M$$

The volume of the solution is 1 L, so it must contain 6.78×10^{-4} mole of Hb. We use this quantity to calculate the molar mass:

$$\text{moles of Hb} = \frac{\text{mass of Hb}}{\text{molar mass of Hb}}$$

$$\text{molar mass of Hb} = \frac{\text{mass of Hb}}{\text{moles of Hb}}$$

$$= \frac{44.1 \text{ g}}{6.78 \times 10^{-4} \text{ mol}}$$

$$= 6.50 \times 10^{4} \text{ g/mol}$$

Similar problems: 13.62, 13.64.

Practice Exercise A 202-mL benzene solution containing 2.47 g of an organic polymer has an osmotic pressure of 8.63 mmHg at 21°C. Calculate the molar mass of the polymer.

The density of mercury is 13.6 g/mL. Therefore, 12.6 mmHg corresponds to a column of water 17.1 cm in height.

A pressure of 12.6 mmHg, as in Example 13.9, can be measured easily and accurately. For this reason, osmotic pressure measurements are very useful for determining the molar masses of large molecules, such as proteins. To see how much more practical the osmotic pressure technique is than freezing-point depression would be, let us estimate the change in freezing point of the same hemoglobin solution. If an aqueous solution is quite dilute, we can assume that molarity is roughly equal to molality. (Molarity would be equal to molality if the density of the aqueous solution were 1 g/mL.) Hence, from Equation (13.9) we write

$$\Delta T_f = (1.86°\text{C}/m)(6.78 \times 10^{-4}\, m)$$
$$= 1.26 \times 10^{-3}°\text{C}$$

The freezing-point depression of one-thousandth of a degree is too small a temperature change to measure accurately. For this reason, the freezing-point depression technique is more suitable for determining the molar mass of smaller and more soluble molecules, those having molar masses of 500 g or less, because the freezing-point depressions of their solutions are much greater.

Colligative Properties of Electrolyte Solutions

The colligative properties of electrolytes require a slightly different approach than the one used for the colligative properties of nonelectrolytes. The reason is that electrolytes dissociate into ions in solution, and so one unit of an electrolyte compound separates into two or more particles when it dissolves. (Remember, it is the number of solute particles that determines the colligative properties of a solution.) For example, each unit of NaCl dissociates into two ions—Na^+ and Cl^-. Thus, the colligative properties of a 0.1 *m* NaCl solution should be twice as great as those of a 0.1 *m* solution containing a nonelectrolyte, such as sucrose. Similarly, we would expect a 0.1 *m* $CaCl_2$ solution to depress the freezing point by three times as much as a 0.1 *m* sucrose solution. To account for this effect we must modify the equations for colligative properties as follows:

$$\Delta T_b = iK_bm \qquad (13.11)$$

$$\Delta T_f = iK_fm \qquad (13.12)$$

$$\pi = iMRT \qquad (13.13)$$

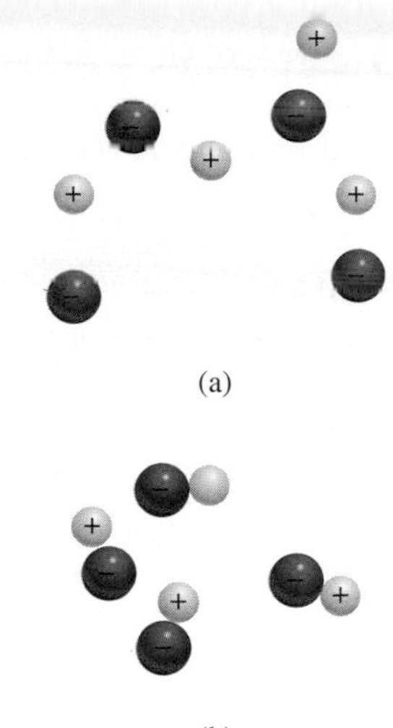

Figure 13.11
(a) Free ions and (b) ion pairs in solution. Such an ion pair bears no net charge and therefore cannot conduct electricity in solution.

The variable *i* is the *van't Hoff factor,* which is defined as

$$i = \frac{\text{actual number of particles in soln after dissociation}}{\text{number of formula units initially dissolved in soln}} \qquad (13.14)$$

Thus, *i* should be 1 for all nonelectrolytes. For strong electrolytes such as NaCl and KNO_3, *i* should be 2, and for strong electrolytes such as Na_2SO_4 and $MgCl_2$, *i* should be 3.

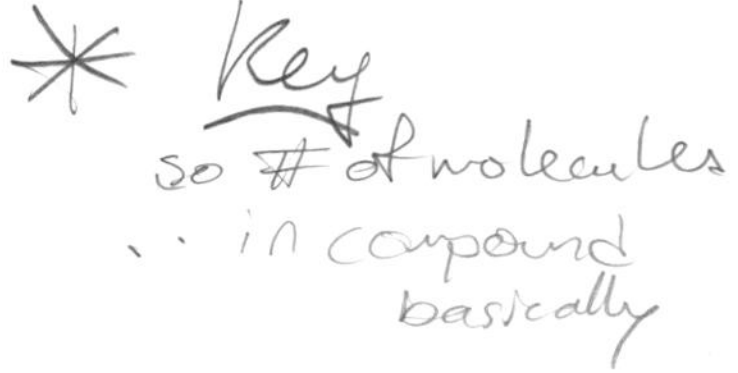

In reality, the colligative properties of electrolyte solutions are usually smaller than anticipated because at higher concentrations, electrostatic forces come into play, drawing cations and anions together. *A cation and an anion held together by electrostatic forces* is called an ***ion pair.*** The formation of an ion pair reduces the number of particles in solution by one, causing a reduction in the colligative properties (Figure 13.11). Table 13.3 shows the experimentally measured values of *i* and those calculated assuming complete dissociation. As you can see, the agreement is close but not perfect, indicating that the extent of ion-pair formation in these solutions is appreciable.

Review of Concepts

Indicate which compound in each of the following groups has a greater tendency to form ion pairs in water: (a) NaCl or Na_2SO_4, (b) $MgCl_2$ or $MgSO_4$, (c) LiBr or KBr.

Table 13.3 The van't Hoff Factor of 0.0500 *M* Electrolyte Solutions at 25°C

Electrolyte	*i* (measured)	*i* (calculated)
Sucrose*	1.0	1.0
HCl	1.9	2.0
NaCl	1.9	2.0
$MgSO_4$	1.3	2.0
$MgCl_2$	2.7	3.0
$FeCl_3$	3.4	4.0

*Sucrose is a nonelectrolyte. It is listed here for comparison only.

EXAMPLE 13.10

The osmotic pressure of a 0.010 *M* potassium iodide (KI) solution at 25°C is 0.465 atm. Calculate the van't Hoff factor for KI at this concentration.

Strategy Note that KI is a strong electrolyte, so we expect it to dissociate completely in solution. If so, its osmotic pressure would be

$$2(0.010\,M)(0.0821\text{ L}\cdot\text{atm/K}\cdot\text{mol})(298\text{ K}) = 0.489\text{ atm}$$

However, the measured osmotic pressure is only 0.465 atm. The smaller than predicted osmotic pressure means that there is ion-pair formation, which reduces the number of solute particles (K^+ and I^- ions) in solution.

Solution From Equation (13.13) we have

$$\begin{aligned} i &= \frac{\pi}{MRT} \\ &= \frac{0.465\text{ atm}}{(0.010\,M)(0.0821\text{ L}\cdot\text{atm/K}\cdot\text{mol})(298\text{ K})} \\ &= 1.90 \end{aligned}$$

Similar problem: 13.75.

Practice Exercise The freezing-point depression of a 0.100 *m* $MgSO_4$ solution is 0.225°C. Calculate the van't Hoff factor of $MgSO_4$ at this concentration.

Review of Concepts

The osmotic pressure of blood is about 7.4 atm. What is the approximate concentration of a saline solution (NaCl solution) a physician should use for intravenous injection? Use 37°C for physiological temperature.

Key Equations

$\text{molality } (m) = \dfrac{\text{moles of solute}}{\text{mass of solvent (kg)}}$	(13.4)	Calculating the molality of a solution.
$c = kP$	(13.5)	Henry's law for calculating solubility of gases.
$P_1 = X_1 P_1^\circ$	(13.6)	Raoult's law relating the vapor pressure of a liquid to its vapor pressure in a solution.
$\Delta P = X_2 P_1^\circ$	(13.7)	Vapor pressure lowering in terms of the concentration of solution.
$\Delta T_b = K_b m$	(13.8)	Boiling-point elevation.
$\Delta T_f = K_f m$	(13.9)	Freezing-point depression.
$\pi = MRT$	(13.10)	Osmotic pressure of a solution.
$i = \dfrac{\text{actual number of particles in soln after dissociation}}{\text{number of formula units initially dissolved in soln}}$	(13.14)	Calculating the van't Hoff factor for an electrolyte solution.

Summary of Facts and Concepts

1. Solutions are homogeneous mixtures of two or more substances, which may be solids, liquids, or gases. The ease of dissolution of a solute in a solvent is governed by intermolecular forces. Energy and the increase in disorder that result when molecules of the solute and solvent mix to form a solution are the forces driving the solution process.
2. The concentration of a solution can be expressed as percent by mass, mole fraction, molarity, and molality. The circumstances dictate which units are appropriate.
3. A rise in temperature usually increases the solubility of solid and liquid substances and decreases the solubility of gases. According to Henry's law, the solubility of a gas in a liquid is directly proportional to the partial pressure of the gas over the solution.
4. Raoult's law states that the partial pressure of a substance A over a solution is related to the mole fraction (X_A) of A and to the vapor pressure (P_A°) of pure A as: $P_A = X_A P_A^\circ$. An ideal solution obeys Raoult's law over the entire range of concentration. In practice, very few solutions exhibit ideal behavior.
5. Vapor-pressure lowering, boiling-point elevation, freezing-point depression, and osmotic pressure are colligative properties of solutions; that is, they are properties that depend only on the number of solute particles that are present and not on their nature. In electrolyte solutions, the interaction between ions leads to the formation of ion pairs. The van't Hoff factor provides a measure of the extent of ion-pair formation in solution.

Key Words

Questions and Problems

The Solution Process

Review Questions

13.1 Briefly describe the solution process at the molecular level. Use the dissolution of a solid in a liquid as an example.

13.2 Basing your answer on intermolecular force considerations, explain what "like dissolves like" means.

13.3 What is solvation? What are the factors that influence the extent to which solvation occurs? Give two examples of solvation, including one that involves ion-dipole interaction and another in which dispersion forces come into play.

13.4 As you know, some solution processes are endothermic and others are exothermic. Provide a molecular interpretation for the difference.

13.5 Explain why the solution process invariably leads to an increase in disorder.

13.6 Describe the factors that affect the solubility of a solid in a liquid. What does it mean to say that two liquids are miscible?

Problems

13.7 Why is naphthalene ($C_{10}H_8$) more soluble than CsF in benzene?

13.8 Explain why ethanol (C_2H_5OH) is not soluble in cyclohexane (C_6H_{12}).

13.9 Arrange these compounds in order of increasing solubility in water: O_2, LiCl, Br_2, CH_3OH (methanol).

13.10 Explain the variations in solubility in water of the alcohols listed here:

Compound	**Solubility in Water g/100 g, 20°C**
CH_3OH	∞
CH_3CH_2OH	∞
$CH_3CH_2CH_2OH$	∞
$CH_3CH_2CH_2CH_2OH$	9
$CH_3CH_2CH_2CH_2CH_2OH$	2.7

Note: ∞ means the alcohol and water are completely miscible in all proportions.

Concentration Units

Review Questions

13.11 Define these concentration terms and give their units: percent by mass, molarity, molality. Compare their advantages and disadvantages.

13.12 Outline the steps required for conversion among molarity, molality, and percent by mass.

Problems

13.13 Calculate the percent by mass of the solute in each of these aqueous solutions: (a) 5.50 g of NaBr in 78.2 g of solution, (b) 31.0 g of KCl in 152 g of water, (c) 4.5 g of toluene in 29 g of benzene.

13.14 Calculate the amount of water (in grams) that must be added to (a) 5.00 g of urea $[(NH_2)_2CO]$ in the preparation of a 16.2 percent by mass solution and (b) 26.2 g of $MgCl_2$ in the preparation of a 1.5 percent by mass solution.

13.15 Calculate the molality of each of these solutions: (a) 14.3 g of sucrose ($C_{12}H_{22}O_{11}$) in 676 g of water, (b) 7.20 moles of ethylene glycol ($C_2H_6O_2$) in 3546 g of water.

13.16 Calculate the molality of each of the following aqueous solutions: (a) 2.50 *M* NaCl solution (density of solution = 1.08 g/mL), (b) 48.2 percent by mass KBr solution.

13.17 Calculate the molalities of these aqueous solutions: (a) 1.22 *M* sugar ($C_{12}H_{22}O_{11}$) solution (density of solution = 1.12 g/mL), (b) 0.87 *M* NaOH solution (density of solution = 1.04 g/mL), (c) 5.24 *M* $NaHCO_3$ solution (density of solution = 1.19 g/mL).

13.18 For dilute aqueous solutions in which the density of the solution is roughly equal to that of the pure solvent, the molarity of the solution is equal to its molality. Show that this statement is correct for a 0.010 *M* urea $[(NH_2)_2CO]$ solution.

13.19 The alcohol content of hard liquor is normally given in terms of the "proof," which is defined as twice the percentage by volume of ethanol (C_2H_5OH) present. Calculate the number of grams of alcohol present in 1.00 L of 75 proof gin. The density of ethanol is 0.798 g/mL.

13.20 The concentrated sulfuric acid we use in the laboratory is 98.0 percent H_2SO_4 by mass. Calculate the molality and molarity of the acid solution. The density of the solution is 1.83 g/mL.

13.21 Calculate the molarity and the molality of NH_3 for a solution of 30.0 g of NH_3 in 70.0 g of water. The density of the solution is 0.982 g/mL.

13.22 The density of an aqueous solution containing 10.0 percent ethanol (C_2H_5OH) by mass is 0.984 g/mL. (a) Calculate the molality of this solution. (b) Calculate its molarity. (c) What volume of the solution would contain 0.125 mole of ethanol?

Effect of Temperature and Pressure on Solubility

Review Questions

13.23 How do the solubilities of most ionic compounds in water change with temperature?

13.24 What is the effect of pressure on the solubility of a liquid in liquid and of a solid in liquid?

Problems

13.25 A 3.20-g sample of a salt dissolves in 9.10 g of water to give a saturated solution at 25°C. What is the solubility (in g salt/100 g of H_2O) of the salt?

13.26 The solubility of KNO_3 is 155 g per 100 g of water at 75°C and 38.0 g at 25°C. What mass (in grams) of KNO_3 will crystallize out of solution if exactly 100 g of its saturated solution at 75°C are cooled to 25°C?

Gas Solubility

Review Questions

13.27 Discuss the factors that influence the solubility of a gas in a liquid. Explain why the solubility of a gas in a liquid usually decreases with increasing temperature.

13.28 What is thermal pollution? Why is it harmful to aquatic life?

13.29 What is Henry's law? Define each term in the equation, and give its units. Explain the law in terms of the kinetic molecular theory of gases.

13.30 Give two exceptions to Henry's law.

Problems

13.31 A student is observing two beakers of water. One beaker is heated to 30°C, and the other is heated to 100°C. In each case, bubbles form in the water. Are these bubbles of the same origin? Explain.

13.32 A man bought a goldfish in a pet shop. Upon returning home, he put the goldfish in a bowl of recently boiled water that had been cooled quickly. A few minutes later the fish was dead. Explain what happened to the fish.

13.33 A beaker of water is initially saturated with dissolved air. Explain what happens when He gas at 1 atm is bubbled through the solution for a long time.

13.34 A miner working 260 m below sea level opened a carbonated soft drink during a lunch break. To his surprise, the soft drink tasted rather "flat." Shortly afterward, the miner took an elevator to the surface. During the trip up, he could not stop belching. Why?

13.35 The solubility of CO_2 in water at 25°C and 1 atm is 0.034 mol/L. What is its solubility under atmospheric conditions? (The partial pressure of CO_2 in air is 0.0003 atm.) Assume that CO_2 obeys Henry's law.

13.36 The solubility of N_2 in blood at 37°C and at a partial pressure of 0.80 atm is 5.6×10^{-4} mol/L. A deep-sea diver breathes compressed air with the partial pressure of N_2 equal to 4.0 atm. Assume that the total volume of blood in the body is 5.0 L. Calculate the amount of N_2 gas released (in liters) when the diver returns to the surface of the water, where the partial pressure of N_2 is 0.80 atm.

Colligative Properties of Nonelectrolyte Solutions

Review Questions

13.37 What are colligative properties? What is the meaning of the word "colligative" in this context?

13.38 Define Raoult's law. Define each term in the equation representing Raoult's law, and give its units. What is an ideal solution?

13.39 Define boiling-point elevation and freezing-point depression. Write the equations relating boiling-point elevation and freezing-point depression to the concentration of the solution. Define all the terms, and give their units.

13.40 How is the lowering in vapor pressure related to a rise in the boiling point of a solution?

13.41 Use a phase diagram to show the difference in freezing point and boiling point between an aqueous urea solution and pure water.

13.42 What is osmosis? What is a semipermeable membrane?

13.43 Write the equation relating osmotic pressure to the concentration of a solution. Define all the terms and give their units.

13.44 Explain why molality is used for boiling-point elevation and freezing-point depression calculations and molarity is used in osmotic pressure calculations.

13.45 Describe how you would use the freezing-point depression and osmotic pressure measurements to determine the molar mass of a compound. Why is the boiling-point elevation phenomenon normally not used for this purpose?

13.46 Explain why it is essential that fluids used in intravenous injections have approximately the same osmotic pressure as blood.

Problems

13.47 A solution is prepared by dissolving 396 g of sucrose ($C_{12}H_{22}O_{11}$) in 624 g of water. What is the vapor pressure of this solution at 30°C? (The vapor pressure of water is 31.8 mmHg at 30°C.)

13.48 How many grams of sucrose ($C_{12}H_{22}O_{11}$) must be added to 552 g of water to give a solution with a vapor pressure 2.0 mmHg less than that of pure water at 20°C? (The vapor pressure of water at 20°C is 17.5 mmHg.)

13.49 The vapor pressure of benzene is 100.0 mmHg at 26.1°C. Calculate the vapor pressure of a solution containing 24.6 g of camphor ($C_{10}H_{16}O$) dissolved in 98.5 g of benzene. (Camphor is a low-volatility solid.)

13.50 The vapor pressures of ethanol (C_2H_5OH) and 1-propanol (C_3H_7OH) at 35°C are 100 mmHg and 37.6 mmHg, respectively. Assume ideal behavior and calculate the partial pressures of ethanol and 1-propanol at 35°C over a solution of ethanol in 1-propanol, in which the mole fraction of ethanol is 0.300.

13.51 The vapor pressure of ethanol (C_2H_5OH) at 20°C is 44 mmHg, and the vapor pressure of methanol (CH_3OH) at the same temperature is 94 mmHg. A mixture of 30.0 g of methanol and 45.0 g of ethanol is prepared (and may be assumed to behave as an ideal solution). (a) Calculate the vapor pressure of methanol and ethanol above this solution at 20°C. (b) Calculate the mole fraction of methanol and ethanol in the vapor above this solution at 20°C.

13.52 How many grams of urea [$(NH_2)_2CO$] must be added to 450 g of water to give a solution with a vapor pressure 2.50 mmHg less than that of pure water at 30°C? (The vapor pressure of water at 30°C is 31.8 mmHg.)

13.53 What are the boiling point and freezing point of a 2.47 *m* solution of naphthalene in benzene? (The boiling point and freezing point of benzene are 80.1°C and 5.5°C, respectively.)

13.54 An aqueous solution contains the amino acid glycine (NH_2CH_2COOH). Assuming no ionization of the acid, calculate the molality of the solution if it freezes at −1.1°C.

13.55 Pheromones are compounds secreted by the females of many insect species to attract males. One of these compounds contains 80.78% C, 13.56% H, and 5.66% O. A solution of 1.00 g of this pheromone in 8.50 g of benzene freezes at 3.37°C. What are the molecular formula and molar mass of the compound? (The normal freezing point of pure benzene is 5.50°C).

13.56 The elemental analysis of an organic solid extracted from gum arabic showed that it contained 40.0% C, 6.7% H, and 53.3% O. A solution of 0.650 g of the solid in 27.8 g of the solvent diphenyl gave a freezing-point depression of 1.56°C. Calculate the molar mass and molecular formula of the solid. (K_f for diphenyl is 8.00°C/*m*.)

13.57 How many liters of the antifreeze ethylene glycol [$CH_2(OH)CH_2(OH)$] would you add to a car radiator containing 6.50 L of water if the coldest winter temperature in your area is −20°C? Calculate the boiling point of this water-ethylene glycol mixture. The density of ethylene glycol is 1.11 g/mL.

13.58 A solution is prepared by condensing 4.00 L of a gas, measured at 27°C and 748 mmHg pressure, into 58.0 g of benzene. Calculate the freezing point of this solution.

13.59 The molar mass of benzoic acid (C_6H_5COOH) determined by measuring the freezing-point depression in benzene is twice that expected for the molecular formula, $C_7H_6O_2$. Explain this apparent anomaly.

13.60 A solution of 2.50 g of a compound of empirical formula C_6H_5P in 25.0 g of benzene is observed to freeze at 4.3°C. Calculate the molar mass of the solute and its molecular formula.

13.61 What is the osmotic pressure (in atmospheres) of a 12.36 *M* aqueous urea solution at 22.0°C?

13.62 A solution containing 0.8330 g of a protein of unknown structure in 170.0 mL of aqueous solution was found to have an osmotic pressure of 5.20 mmHg at 25°C. Determine the molar mass of the protein.

13.63 A quantity of 7.480 g of an organic compound is dissolved in water to make 300.0 mL of solution. The solution has an osmotic pressure of 1.43 atm at 27°C. The analysis of this compound shows it to contain 41.8% C, 4.7% H, 37.3% O, and 16.3% N. Calculate the molecular formula of the organic compound.

13.64 A solution of 6.85 g of a carbohydrate in 100.0 g of water has a density of 1.024 g/mL and an osmotic pressure of 4.61 atm at 20.0°C. Calculate the molar mass of the carbohydrate.

Colligative Properties of Electrolyte Solutions

Review Questions

13.65 Define ion pairs. What effect does ion-pair formation have on the colligative properties of a solution? How does the ease of ion-pair formation depend on (a) charges on the ions, (b) size of the ions, (c) nature of the solvent (polar versus nonpolar), (d) concentration?

13.66 Define the van't Hoff factor. What information does this quantity provide?

Problems

13.67 Which of these two aqueous solutions has (a) the higher boiling point, (b) the higher freezing point, and (c) the lower vapor pressure: 0.35 *m* $CaCl_2$ or 0.90 *m* urea? State your reasons.

13.68 Consider two aqueous solutions, one of sucrose ($C_{12}H_{22}O_{11}$) and the other of nitric acid (HNO_3), both of which freeze at −1.5°C. What other properties do these solutions have in common?

13.69 Arrange these solutions in order of decreasing freezing point: (a) 0.10 *m* Na_3PO_4, (b) 0.35 *m* NaCl, (c) 0.20 *m* $MgCl_2$, (d) 0.15 *m* $C_6H_{12}O_6$, (e) 0.15 *m* CH_3COOH.

13.70 Arrange these aqueous solutions in order of decreasing freezing point and explain your reasons: (a) 0.50 *m* HCl, (b) 0.50 *m* glucose, (c) 0.50 *m* acetic acid.

13.71 What are the normal freezing points and boiling points of the following solutions: (a) 21.2 g NaCl in 135 mL of water, (b) 15.4 g of urea in 66.7 mL of water?

13.72 At 25°C the vapor pressure of pure water is 23.76 mmHg and that of seawater is 22.98 mmHg. Assuming that seawater contains only NaCl, estimate its concentration in molality units.

13.73 Both NaCl and $CaCl_2$ are used to melt ice on roads in winter. What advantages do these substances have over sucrose or urea in lowering the freezing point of water?

13.74 A 0.86 percent by mass solution of NaCl is called "physiological saline" because its osmotic pressure is equal to that of the solution in blood cells. Calculate the osmotic pressure of this solution at normal body temperature (37°C). Note that the density of the saline solution is 1.005 g/mL.

13.75 The osmotic pressure of 0.010 *M* solutions of $CaCl_2$ and urea at 25°C are 0.605 atm and 0.245 atm, respectively. Calculate the van't Hoff factor for the $CaCl_2$ solution.

13.76 Calculate the osmotic pressure of a 0.0500 *M* $MgSO_4$ solution at 22°C. (*Hint:* See Table 13.3.)

Additional Problems

13.77 Lysozyme is an enzyme that cleaves bacterial cell walls. A sample of lysozyme extracted from chicken egg white has a molar mass of 13,930 g. A quantity of 0.100 g of this enzyme is dissolved in 150 g of water at 25°C. Calculate the vapor-pressure lowering, the depression in freezing point, the elevation in boiling point, and the osmotic pressure of this solution. (The vapor pressure of water at 25°C = 23.76 mmHg.

13.78 Solutions A and B containing the same solute have osmotic pressures of 2.4 atm and 4.6 atm, respectively, at a certain temperature. What is the osmotic pressure of a solution prepared by mixing equal volumes of A and B at the same temperature?

13.79 A cucumber placed in concentrated brine (saltwater) shrivels into a pickle. Explain.

13.80 Two liquids A and B have vapor pressures of 76 mmHg and 132 mmHg, respectively, at 25°C. What is the total vapor pressure of the ideal solution made up of (a) 1.00 mole of A and 1.00 mole of B and (b) 2.00 moles of A and 5.00 moles of B?

13.81 Calculate the van't Hoff factor of Na_3PO_4 in a 0.40 *m* aqueous solution whose boiling point is 100.78°C.

13.82 A 262-mL sample of a sugar solution containing 1.22 g of the sugar has an osmotic pressure of 30.3 mmHg at 35°C. What is the molar mass of the sugar?

13.83 Consider these three mercury manometers. One of them has 1 mL of water placed on top of the mercury, another has 1 mL of a 1 *m* urea solution placed on top of the mercury, and the third one has 1 mL of a 1 *m*

NaCl solution placed on top of the mercury. Identify X, Y, and Z with these solutions.

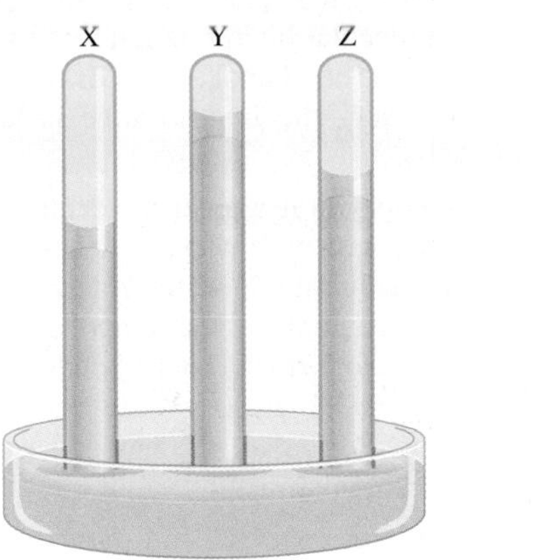

13.84 A forensic chemist is given a white powder for analysis. She dissolves 0.50 g of the substance in 8.0 g of benzene. The solution freezes at 3.9°C. Can the chemist conclude that the compound is cocaine ($C_{17}H_{21}NO_4$)? What assumptions are made in the analysis?

13.85 "Time-release" drugs have the advantage of releasing the drug to the body at a constant rate so that the drug concentration at any time is not so high as to have harmful side effects or so low as to be ineffective. A schematic diagram of a pill that works on this basis is shown here. Explain how it works.

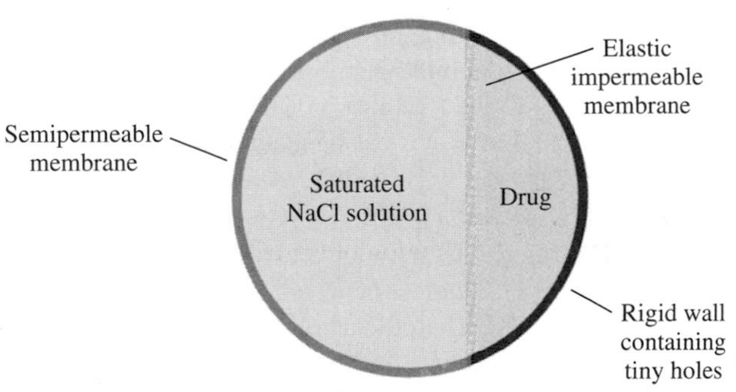

13.86 Concentrated hydrochloric acid is usually available at 37.7 percent by mass. What is its concentration in molarity? (The density of the solution is 1.19 g/mL.)

13.87 A protein has been isolated as a salt with the formula $Na_{20}P$ (this notation means that there are 20 Na^+ ions associated with a negatively charged protein P^{20-}). The osmotic pressure of a 10.0-mL solution containing 0.225 g of the protein is 0.257 atm at 25.0°C. (a) Calculate the molar mass of the protein from these data. (b) What is the actual molar mass of the protein?

13.88 A nonvolatile organic compound Z was used to make two solutions. Solution A contains 5.00 g of Z dissolved in 100 g of water, and solution B contains 2.31 g of Z dissolved in 100 g of benzene. Solution A has a vapor pressure of 754.5 mmHg at the normal boiling point of water, and solution B has the same vapor pressure at the normal boiling point of benzene. Calculate the molar mass of Z in solutions A and B and account for the difference.

13.89 Hydrogen peroxide with a concentration of 3.0 percent (3.0 g of H_2O_2 in 100 mL of solution) is sold in drugstores for use as an antiseptic. For a 10.0-mL 3.0 percent H_2O_2 solution, calculate (a) the oxygen gas produced (in liters) at STP when the compound undergoes complete decomposition and (b) the ratio of the volume of O_2 collected to the initial volume of the H_2O_2 solution.

13.90 Before a carbonated beverage bottle is sealed, it is pressurized with a mixture of air and carbon dioxide. (a) Explain the effervescence that occurs when the cap of the bottle is removed. (b) What causes the fog to form near the mouth of the bottle right after the cap is removed?

13.91 Two beakers, one containing a 50-mL aqueous 1.0 *M* glucose solution and the other a 50-mL aqueous 2.0 *M* glucose solution, are placed under a tightly sealed bell jar as that shown in Figure 13.9 at room temperature. What are the volumes in these two beakers at equilibrium? Assume ideal behavior.

13.92 Explain each of these statements: (a) The boiling point of seawater is higher than that of pure water. (b) Carbon dioxide escapes from the solution when the cap is removed from a soft-drink bottle. (c) Molal concentrations and molar concentrations of dilute aqueous solutions are approximately equal. (d) In discussing the colligative properties of a solution (other than osmotic pressure), it is preferable to express the concentration in units of molality rather than in molarity. (e) Methanol (b.p. 65°C) is useful as an auto antifreeze, but it should be removed from the car radiator during the summer season.

13.93 Acetic acid is a weak acid that ionizes in solution as follows:

$$CH_3COOH(aq) \rightleftharpoons CH_3COO^-(aq) + H^+(aq)$$

If the freezing point of a 0.106 *m* CH_3COOH solution is −0.203°C, calculate the percent of the acid that has undergone ionization.

13.94 A 1.32-g sample of a mixture of cyclohexane (C_6H_{12}) and naphthalene ($C_{10}H_8$) is dissolved in 18.9 g of benzene (C_6H_6). The freezing point of the solution is 2.2°C. Calculate the mass percent of the mixture.

13.95 How does each of the following affect the solubility of an ionic compound: (a) lattice energy, (b) solvent (polar versus nonpolar), (c) enthalpies of hydration of cation and anion?

13.96 A solution contains two volatile liquids A and B. Complete the following table, in which the symbol $\longleftrightarrow$ indicates attractive intermolecular forces.

Attractive Forces	Deviation from Raoult's Law	ΔH_{soln}
$A \longleftrightarrow A$, $B \longleftrightarrow B > A \longleftrightarrow B$		
	Negative	
		Zero

A negative deviation means the vapor pressure of the solution is less than that expected from Raoult's law. The opposite holds for a positive deviation.

13.97 A mixture of ethanol and 1-propanol behaves ideally at 36°C and is in equilibrium with its vapor. If the mole fraction of ethanol in the solution is 0.62, calculate its mole fraction in the vapor phase at this temperature. (The vapor pressures of pure ethanol and 1-propanol at 36°C are 108 mmHg and 40.0 mmHg, respectively.)

13.98 For ideal solutions, the volumes are additive. This means that if 5 mL of A and 5 mL of B form an ideal solution, the volume of the solution is 10 mL. Provide a molecular interpretation for this observation. When 500 mL of ethanol (C_2H_5OH) are mixed with 500 mL of water, the final volume is less than 1000 mL. Why?

13.99 Acetic acid is a polar molecule and can form hydrogen bonds with water molecules. Therefore, it has a high solubility in water. Yet acetic acid is also soluble in benzene (C_6H_6), a nonpolar solvent that lacks the ability to form hydrogen bonds. A solution of 3.8 g of CH_3COOH in 80 g C_6H_6 has a freezing point of 3.5°C. Calculate the molar mass of the solute and explain your result.

13.100 A mixture of NaCl and sucrose ($C_{12}H_{22}O_{11}$) of combined mass 10.2 g is dissolved in enough water to make up a 250 mL solution. The osmotic pressure of the solution is 7.32 atm at 23°C. Calculate the mass percent of NaCl in the mixture.

Special Problems

13.101 Desalination is a process by which salts are removed from seawater. Three major ways to accomplish desalination are distillation, freezing, and reverse osmosis. The freezing method is based on the fact that when an aqueous solution freezes, the solid that separates from the solution is almost pure water. Reverse osmosis uses water movement from a more concentrated solution to a less concentrated one through a semipermeable membrane.

(a) With reference to Figure 13.8, draw a diagram showing how reverse osmosis can be carried out.

(b) What are the advantages and disadvantages of reverse osmosis compared to the freezing and boiling methods?

(c) What minimum pressure (in atm) must be applied to seawater at 25°C for reverse osmosis to occur? (Treat seawater as an 0.70 *M* NaCl solution.)

13.102 Liquids A (molar mass 100 g/mol) and B (molar mass 110 g/mol) form an ideal solution. At 55°C, A has a vapor pressure of 95 mmHg and B has a vapor pressure of 42 mmHg. A solution is prepared by mixing equal masses of A and B. (a) Calculate the mole fraction of each component in the solution. (b) Calculate the partial pressures of A and B over the solution at 55°C. (c) Suppose that some of the vapor described in (b) is condensed to a liquid. Calculate the mole fraction of each component in this liquid and the vapor pressure of each component above this liquid at 55°C.

13.103 A very long pipe is capped at one end with a semipermeable membrane. How deep (in meters) must the pipe be immersed into the sea for fresh water to begin to pass through the membrane? Assume the water to be at 20°C and treat it as a 0.70 *M* NaCl solution. The density of seawater is 1.03 g/cm^3 and the acceleration due to gravity is 9.81 m/s^2.

13.104 A mixture of liquids A and B exhibits ideal behavior. At 84°C, the total vapor pressure of a solution containing 1.2 moles of A and 2.3 moles of B is 331 mmHg. Upon the addition of another mole of B to the solution, the vapor pressure increases to 347 mmHg. Calculate the vapor pressures of pure A and B at 84°C.

13.105 Using Henry's law and the ideal gas equation to prove the statement that the volume of a gas that dissolves in a given amount of solvent is *independent* of the pressure of the gas. (*Hint:* Henry's law can be modified as $n = kP$, where n is the number of moles of the gas dissolved in the solvent.)

13.106 At 298 K, the osmotic pressure of a glucose solution is 10.50 atm. Calculate the freezing point of the solution. The density of the solution is 1.16 g/mL.

13.107 A student carried out the following procedure to measure the pressure of carbon dioxide in a soft drink bottle. First, she weighed the bottle (853.5 g). Next, she carefully removed the cap to let the CO_2 gas escape. She then reweighed the bottle with the cap (851.3 g). Finally, she measured the volume of the soft drink (452.4 mL). Given that Henry's law constant for CO_2 in water at 25°C is 3.4×10^{-2} mol/L · atm, calculate the pressure of CO_2 in the original bottle. Why is this pressure only an estimate of the true value?

13.108 Valinomycin is an antibiotic. It functions by binding K^+ ions and transporting them across the membrane into cells to offset the ionic balance. The molecule is represented here by its *skeletal* structure in which the end of each straight line corresponds to a carbon atom (unless a N or an O atom is shown at the end of the line). There are as many H atoms attached to each C atom as necessary to give each C atom a total of four bonds. Use the "like dissolves like" guideline to explain its function. (*Hint:* The $—CH_3$ groups at the two ends of the Y shape are nonpolar.)

Answers to Practice Exercises

13.1 Carbon disulfide. **13.2** 0.638 *m*. **13.3** 8.92 *m*. **13.4** 13.8 *m*. **13.5** 2.9×10^{-4} *M*. **13.6** 37.8 mmHg; 4.4 mmHg. **13.7** 101.3°C; −4.48°C. **13.8** 0.066 *m*; 1.3×10^2 g/mol. **13.9** 2.60×10^4 g/mol. **13.10** 1.21.

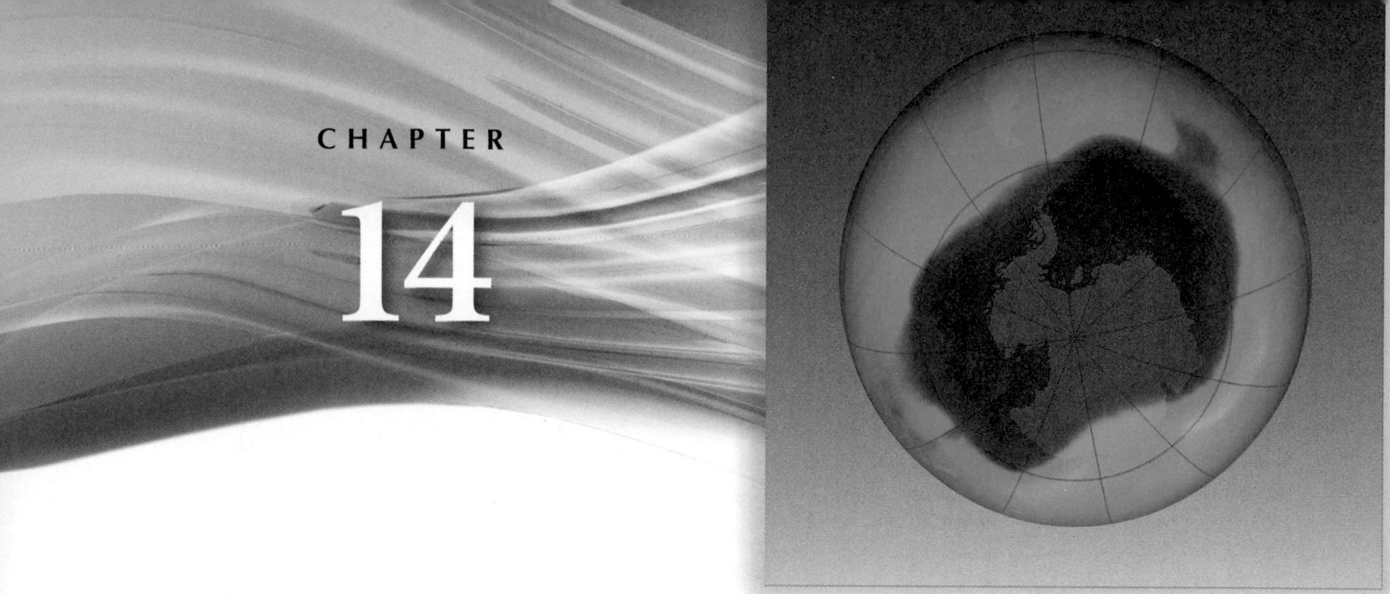

CHAPTER 14

Chemical Kinetics

The study of chemical kinetics and reaction mechanisms proved critical in determining the cause of the Antarctic ozone hole, shown using satellite imaging.

Chapter Outline

Student Interactive Activities

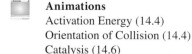

Animations
Activation Energy (14.4)
Orientation of Collision (14.4)
Catalysis (14.6)

Electronic Homework
Example Practice Problems
End of Chapter Problems

Essential Concepts

Rate of a Reaction The rate of a reaction measures how fast a reactant is consumed or how fast a product is formed. The rate is expressed as a ratio of the change in concentration to elapsed time.

Rate Laws Experimental measurement of the rate leads to the rate law for the reaction, which expresses the rate in terms of the rate constant and the concentrations of the reactants. The dependence of rate on concentrations gives the order of a reaction. A reaction can be described as zero order if the rate does not depend on the concentration of the reactant, or first order if it depends on the reactant raised to the first power. Higher orders and fractional orders are also known. An important characteristic of reaction rates is the time required for the concentration of a reactant to decrease to half of its initial concentration, called the half-life. For first-order reactions, the half-life is independent of the initial concentration.

Temperature Dependence of Rate Constants To react, molecules must possess energy equal to or greater than the activation energy. The rate constant generally increases with increasing temperature. The Arrhenius equation relates the rate constant to activation energy and temperature.

Reaction Mechanism The progress of a reaction can be broken into a series of elementary steps at the molecular level, and the sequence of such steps is the mechanism of the reaction. Elementary steps can be unimolecular, involving one molecule, bimolecular, where two molecules react, or in rare cases, termolecular, involving the simultaneous encounter of three molecules. The rate of a reaction having more than one elementary step is governed by the slowest step, called the rate-determining step.

Catalysis A catalyst speeds up the rate of a reaction without itself being consumed. In heterogeneous catalysis, the reactants and catalyst are in different phases. In homogeneous catalysis, the reactants and catalyst are dispersed in a single phase. Enzymes, which are highly efficient catalysts, play a central role in all living systems.

14.1 The Rate of a Reaction

The area of chemistry concerned with the speed, or rate, at which a chemical reaction occurs is called ***chemical kinetics.*** The word "kinetic" suggests movement or change; in Chapter 5 we defined kinetic energy as the energy available because of the motion of an object. Here kinetics refers to the rate of a reaction, or the ***reaction rate,*** which is *the change in concentration of a reactant or a product with time* (M/s).

We know that any reaction can be represented by the general equation

$$\text{reactants} \longrightarrow \text{products}$$

This equation tells us that, during the course of a reaction, reactant molecules are consumed while product molecules are formed. As a result, we can follow the progress of a reaction by monitoring either the decrease in concentration of the reactants or the increase in concentration of the products.

Figure 14.1 shows the progress of a simple reaction in which A molecules are converted to B molecules (for example, the conversion of *cis*-1,2-dichloroethylene to *trans*-1,2-dichloroethylene shown on p. 373):

$$\text{A} \longrightarrow \text{B}$$

The decrease in the number of A molecules and the increase in the number of B molecules with time are shown in Figure 14.2. In general, it is more convenient to express the rate in terms of change in concentration with time. Thus, for the preceding reaction we can express the rate as

$$\text{rate} = -\frac{\Delta[\text{A}]}{\Delta t} \quad \text{or} \quad \text{rate} = \frac{\Delta[\text{B}]}{\Delta t}$$

Recall that Δ denotes the difference between the final and initial state.

in which $\Delta[\text{A}]$ and $\Delta[\text{B}]$ are the changes in concentration (in molarity) over a period Δt. Because the concentration of A *decreases* during the time interval, $\Delta[\text{A}]$ is a negative quantity. The rate of a reaction is a positive quantity, so a minus sign is needed in the rate expression to make the rate positive. On the other hand, the rate of product formation does not require a minus sign because $\Delta[\text{B}]$ is a positive quantity (the concentration of B *increases* with time).

For more complex reactions, we must be careful in writing the rate expression. Consider, for example, the reaction

$$2\text{A} \longrightarrow \text{B}$$

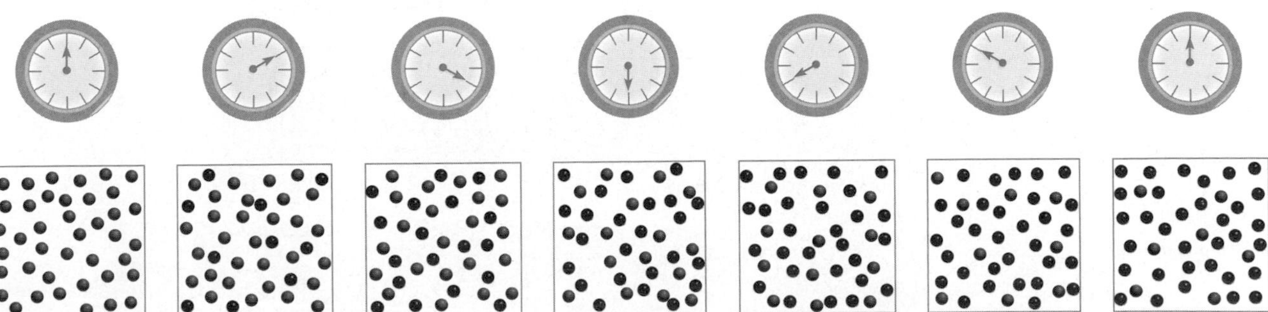

Figure 14.1
The progress of reaction A ⟶ B at 10-s intervals over a period of 60 s. Initially, only A molecules (gray spheres) are present. As time progresses, B molecules (red spheres) are formed.

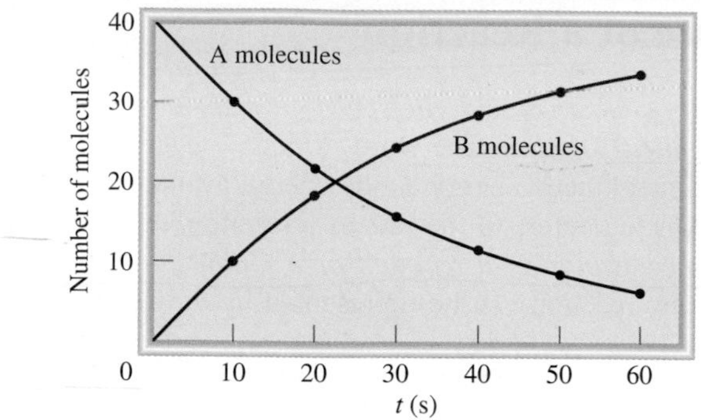

Figure 14.2
The rate of reaction A $\longrightarrow$ B, represented as the decrease of A molecules with time and as the increase of B molecules with time.

Two moles of A disappear for each mole of B that forms—that is, the rate at which B forms is one half the rate at which A disappears. We write the rate as either

$$\text{rate} = -\frac{1}{2}\frac{\Delta[\text{A}]}{\Delta t} \quad \text{or} \quad \text{rate} = \frac{\Delta[\text{B}]}{\Delta t}$$

For the reaction

$$a\text{A} + b\text{B} \longrightarrow c\text{C} + d\text{D}$$

the rate is given by

$$\text{rate} = -\frac{1}{a}\frac{\Delta[\text{A}]}{\Delta t} = -\frac{1}{b}\frac{\Delta[\text{B}]}{\Delta t} = \frac{1}{c}\frac{\Delta[\text{C}]}{\Delta t} = \frac{1}{d}\frac{\Delta[\text{D}]}{\Delta t}$$

EXAMPLE 14.1

Write the rate expressions for the following reactions in terms of the disappearance of the reactants and the appearance of the products:

(a) $I^-(aq) + OCl^-(aq) \longrightarrow Cl^-(aq) + OI^-(aq)$
(b) $3O_2(g) \longrightarrow 2O_3(g)$
(c) $4NH_3(g) + 5O_2(g) \longrightarrow 4NO(g) + 6H_2O(g)$

Solution

(a) Because each of the stoichiometric coefficients equals 1,

$$\text{rate} = -\frac{\Delta[\text{I}^-]}{\Delta t} = -\frac{\Delta[\text{OCl}^-]}{\Delta t} = \frac{\Delta[\text{Cl}^-]}{\Delta t} = \frac{\Delta[\text{OI}^-]}{\Delta t}$$

(b) Here the coefficients are 3 and 2, so

$$\text{rate} = -\frac{1}{3}\frac{\Delta[\text{O}_2]}{\Delta t} = \frac{1}{2}\frac{\Delta[\text{O}_3]}{\Delta t}$$

(c) In this reaction

Similar problem: 14.5.

$$\text{rate} = -\frac{1}{4}\frac{\Delta[\text{NH}_3]}{\Delta t} = -\frac{1}{5}\frac{\Delta[\text{O}_2]}{\Delta t} = \frac{1}{4}\frac{\Delta[\text{NO}]}{\Delta t} = \frac{1}{6}\frac{\Delta[\text{H}_2\text{O}]}{\Delta t}$$

Practice Exercise Write the rate expression for the following reaction:

$$CH_4(g) + 2O_2(g) \longrightarrow CO_2(g) + 2H_2O(g)$$

EXAMPLE 14.2

Consider the reaction

$$4NO_2(g) + O_2(g) \longrightarrow 2N_2O_5(g)$$

Suppose that, at a particular moment during the reaction, molecular oxygen is reacting at the rate of 0.037 *M*/s. (a) At what rate is N_2O_5 being formed? (b) At what rate is NO_2 reacting?

Strategy To calculate the rate of formation of N_2O_5 and disappearance of NO_2, we need to express the rate of the reaction in terms of the stoichiometric coefficients as in Example 14.1:

$$\text{rate} = -\frac{1}{4}\frac{\Delta[NO_2]}{\Delta t} = -\frac{\Delta[O_2]}{\Delta t} = \frac{1}{2}\frac{\Delta[N_2O_5]}{\Delta t}$$

We are given

$$\frac{\Delta[O_2]}{\Delta t} = -0.037\ M/\text{s}$$

where the minus sign shows that the concentration of O_2 is decreasing with time.

Solution

(a) From the preceding rate expression, we have

$$-\frac{\Delta[O_2]}{\Delta t} = \frac{1}{2}\frac{\Delta[N_2O_5]}{\Delta t}$$

Therefore,

$$\frac{\Delta[N_2O_5]}{\Delta t} = -2(-0.037\ M/\text{s}) = 0.074\ M/\text{s}$$

(b) Here we have

$$-\frac{1}{4}\frac{\Delta[NO_2]}{\Delta t} = -\frac{\Delta[O_2]}{\Delta t}$$

so

$$\frac{\Delta[NO_2]}{\Delta t} = 4(-0.037\ M/\text{s}) = -0.15\ M/\text{s}$$

Similar problem: 14.6.

Practice Exercise Consider the reaction

$$4PH_3(g) \longrightarrow P_4(g) + 6H_2(g)$$

Suppose that, at a particular moment during the reaction, molecular hydrogen is being formed at the rate of 0.078 *M*/s. (a) At what rate is P_4 being formed? (b) At what rate is PH_3 reacting?

Depending on the nature of the reaction, there are a number of ways in which to measure reaction rate. For example, in aqueous solution, molecular bromine reacts with formic acid (HCOOH) as

$$Br_2(aq) + HCOOH(aq) \longrightarrow 2H^+(aq) + 2Br^-(aq) + CO_2(g)$$

Molecular bromine is reddish brown. All other species in the reaction are colorless. As the reaction progresses, the concentration of Br_2 steadily decreases and its color

Figure 14.3
The decrease in bromine concentration as time elapses shows up as a loss of color (from left to right).

$2H_2O_2 \longrightarrow 2H_2O + O_2$

fades (Figure 14.3). Thus, the change in concentration (which is evident by the intensity of the color) with time can be followed with a spectrometer (Figure 14.4). We can determine the reaction rate graphically by plotting the concentration of bromine versus time, as Figure 14.5 shows. The rate of the reaction at a particular instant is given by the slope of the tangent (which is $\Delta[Br_2]/\Delta t$) at that instant. In a certain experiment, we find that the rate is 2.96×10^{-5} M/s at 100 s after the start of the reaction, 2.09×10^{-5} M/s at 200 s, and so on. Because generally the rate is proportional to the concentration of the reactant, it is not surprising that its value falls as the concentration of bromine decreases.

If one of the products or reactants of a reaction is a gas, we can use a manometer to find the reaction rate. To illustrate this method, let us consider the decomposition of hydrogen peroxide:

$$2H_2O_2(l) \longrightarrow 2H_2O(l) + O_2(g)$$

In this case, the rate of decomposition can be conveniently determined by measuring the rate of oxygen evolution with a manometer (Figure 14.6). The oxygen

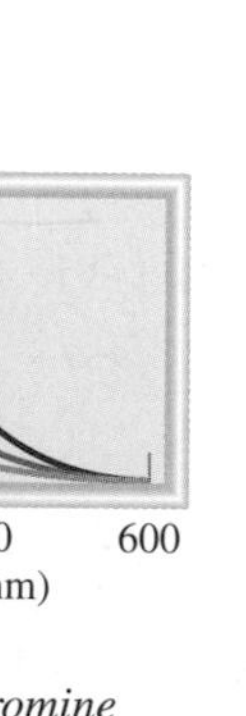

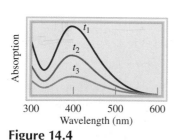

Figure 14.4
Plot of absorption of bromine versus wavelength. The maximum absorption of visible light by bromine occurs at 393 nm. As the reaction progresses (t_1 to t_3), the absorption, which is proportional to $[Br_2]$, decreases.

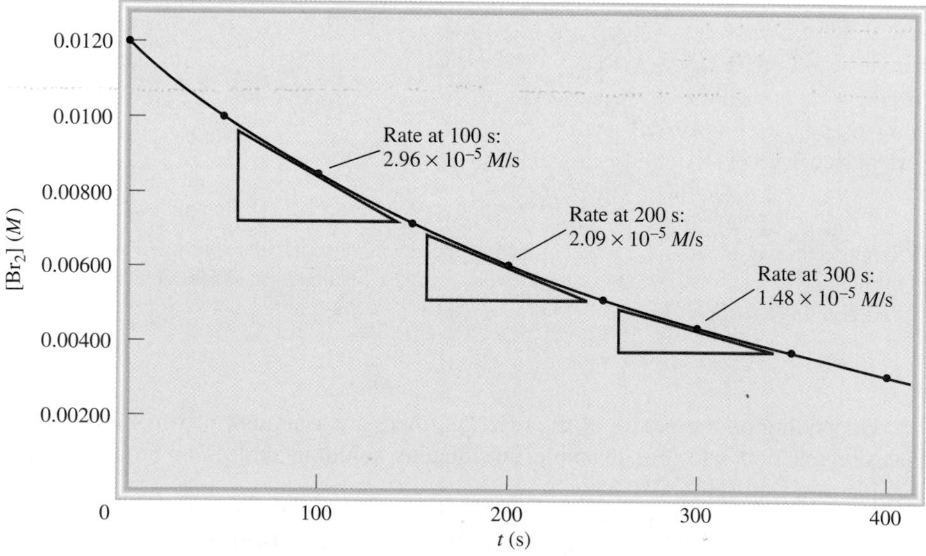

Figure 14.5
The instantaneous rates of the reaction between molecular bromine and formic acid at t = 100 s, 200 s, and 300 s are given by the slopes of the tangents at these times.

pressure can be readily converted to concentration by using the ideal gas equation [Equation (5.8)]:

$$PV = nRT$$

or

$$P = \frac{n}{V}RT = MRT$$

in which n/V gives the molarity (M) of oxygen gas. Rearranging the equation, we get

$$M = \frac{1}{RT}P$$

The reaction rate, which is given by the rate of oxygen production, can now be written as

$$\text{rate} = \frac{\Delta[O_2]}{\Delta t} = \frac{1}{RT}\frac{\Delta P}{\Delta t}$$

If a reaction either consumes or generates ions, its rate can be measured by monitoring electrical conductance. If H^+ ion is the reactant or product, we can determine the reaction rate by measuring the solution's pH as a function of time.

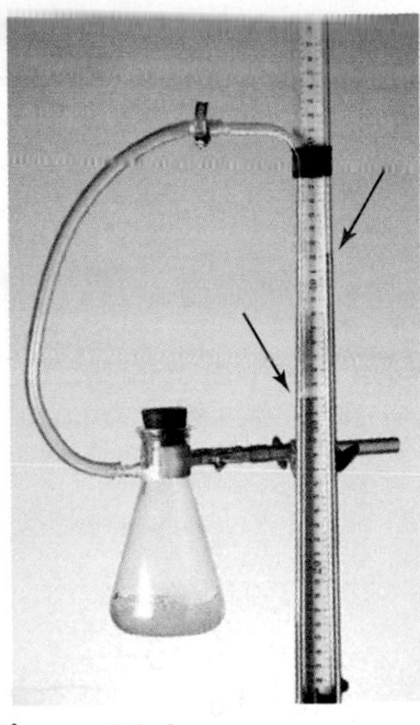

Figure 14.6
The rate of hydrogen peroxide decomposition can be measured with a manometer, which shows the increase in the oxygen gas pressure with time. The arrows show the mercury levels in the U tube.

REVIEW OF CONCEPTS

Write a balanced equation for a gas-phase reaction whose rate is given by

$$\text{rate} = -\frac{\Delta[SO_2]}{\Delta t} = -\frac{1}{3}\frac{\Delta[CO]}{\Delta t} = \frac{1}{2}\frac{\Delta[CO_2]}{\Delta t} = \frac{\Delta[COS]}{\Delta t}$$

14.2 The Rate Laws

One way to study the effect of reactant concentration on reaction rate is to determine how the initial rate depends on the starting concentrations. In general, it is preferable to measure the initial rate because as the reaction proceeds, the concentrations of the reactants decrease and it may become difficult to measure the changes accurately. Also, there may be a reverse reaction such that

$$\text{products} \longrightarrow \text{reactants}$$

which would introduce error in the rate measurement. Both of these complications are virtually absent during the early stages of the reaction.

Table 14.1 shows three experimental rate measurements for the reaction

$$F_2(g) + 2ClO_2(g) \longrightarrow 2FClO_2(g)$$

Table 14.1 Rate Data for the Reaction Between F_2 and ClO_2

$[F_2]$ (M)	$[ClO_2]$ (M)	Initial Rate (M/s)
0.10	0.010	1.2×10^{-3}
0.10	0.040	4.8×10^{-3}
0.20	0.010	2.4×10^{-3}

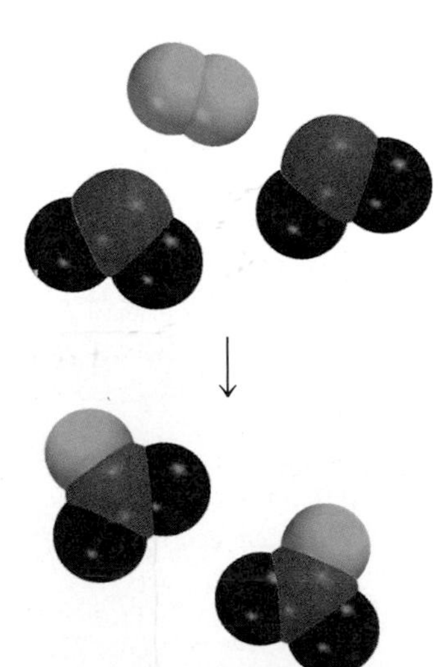

$F_2 + 2ClO_2 \longrightarrow 2FClO_2$

Looking at table entries 1 and 3, we see that if we double $[F_2]$ while holding $[ClO_2]$ constant, the rate doubles. Thus, the rate is directly proportional to $[F_2]$. Similarly, the data in entries 1 and 2 show that when we quadruple $[ClO_2]$ at constant $[F_2]$, the rate increases by four times, so that the rate is also directly proportional to $[ClO_2]$. We can summarize these observations by writing

$$\text{rate} \propto [F_2][ClO_2]$$
$$\text{rate} = k[F_2][ClO_2]$$

As we will see, for a given reaction, k is affected only by a change in temperature.

The term k is the ***rate constant,*** *a constant of proportionality between the reaction rate and the concentrations of the reactants.* This equation is known as the ***rate law,*** *an expression relating the rate of a reaction to the rate constant and the concentrations of the reactants.* From the reactant concentrations and the initial rate, we can also calculate the rate constant. Using the first entry of data in Table 14.1, we can write

what K is equal to →

$$\begin{aligned} k &= \frac{\text{rate}}{[F_2][ClO_2]} \\ &= \frac{1.2 \times 10^{-3}\ M/\text{s}}{(0.10\ M)(0.010\ M)} \\ &= 1.2/M \cdot S \end{aligned}$$

For a general reaction of the type

$$a\text{A} + b\text{B} \longrightarrow c\text{C} + d\text{D}$$

the rate law takes the form

Note that x and y are *not* related to a and b. They must be determined experimentally.

$$\text{rate} = k[\text{A}]^x[\text{B}]^y \tag{14.1}$$

If we know the values of k, x, and y, as well as the concentrations of A and B, we can use the rate law to calculate the rate of the reaction. Like k, x and y must be determined experimentally. *The sum of the powers to which all reactant concentrations appearing in the rate law are raised* is called the overall ***reaction order.*** In the rate law expression shown, the overall reaction order is given by $x + y$. For the reaction involving F_2 and ClO_2, the overall order is 1 + 1, or 2. We say that the reaction is first order in F_2 and first order in ClO_2, or second order overall. Note that reaction order is *always* determined by reactant concentrations and never by product concentrations.

Reaction order enables us to appreciate better the dependence of rate on reactant concentrations. Suppose, for example, that, for a certain reaction, $x = 1$ and $y = 2$. The rate law for the reaction from Equation (14.1) is

$$\text{rate} = k[\text{A}][\text{B}]^2$$

This reaction is first order in A, second order in B, and third order overall (1 + 2 = 3). Let us assume that initially [A] = 1.0 M and [B] = 1.0 M. The rate law tells us that if we double the concentration of A from 1.0 M to 2.0 M at constant [B], we also double the reaction rate:

$$\text{for [A]} = 1.0\ M \qquad \text{rate}_1 = k(1.0\ M)(1.0\ M)^2 = k(1.0\ M^3)$$
$$\text{for [A]} = 2.0\ M \qquad \text{rate}_2 = k(2.0\ M)(1.0\ M)^2 = k(2.0\ M^3)$$

Hence,

$$\text{rate}_2 = 2(\text{rate}_1)$$

On the other hand, if we double the concentration of B from 1.0 *M* to 2.0 *M* at constant [A], the reaction rate will increase by a factor of 4 because of the power 2 in the exponent:

$$\text{for } [B] = 1.0\,M \qquad \text{rate}_1 = k(1.0\,M)(1.0\,M)^2$$
$$= k(1.0\,M^3)$$
$$\text{for } [B] = 2.0\,M \qquad \text{rate}_2 = k(1.0\,M)(2.0\,M)^2$$
$$= k(4.0\,M^3)$$

Hence,

$$\text{rate}_2 = 4(\text{rate}_1)$$

If, for a certain reaction, $x = 0$ and $y = 1$, then the rate law is

$$\text{rate} = k[A]^0[B]$$
$$= k[B]$$

This reaction is zero order in A, first order in B, and first order overall. Thus, the rate of this reaction is *independent* of the concentration of A.

Zero order does not mean that the rate is zero. It just means that the rate is independent of the concentration of A present.

Experimental Determination of Rate Laws

If a reaction involves only one reactant, the rate law can be readily determined by measuring the initial rate of the reaction as a function of the reactant's concentration. For example, if the rate doubles when the concentration of the reactant doubles, then the reaction is first order in the reactant. If the rate quadruples when the concentration doubles, the reaction is second order in the reactant.

For a reaction involving more than one reactant, we can find the rate law by measuring the dependence of the reaction rate on the concentration of each reactant, one at a time. We fix the concentrations of all but one reactant and record the rate of the reaction as a function of the concentration of that reactant. Any changes in the rate must be due only to changes in that substance. The dependence thus observed gives us the order in that particular reactant. The same procedure is then applied to the next reactant, and so on. This approach is known as the *isolation method.*

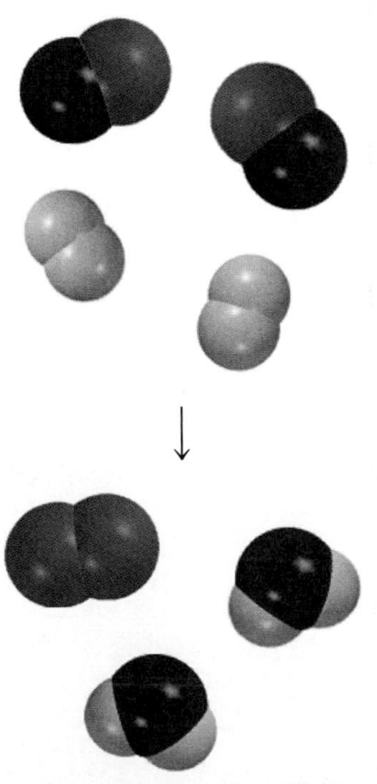

$2NO + 2H_2 \longrightarrow N_2 + 2H_2O$

EXAMPLE 14.3

The reaction of nitric oxide with hydrogen at 1280°C is

$$2NO(g) + 2H_2(g) \longrightarrow N_2(g) + 2H_2O(g)$$

From the following data collected at this temperature, determine (a) the rate law, (b) the rate constant, and (c) the rate of the reaction when $[NO] = 12.0 \times 10^{-3}\,M$ and $[H_2] = 6.0 \times 10^{-3}\,M$.

Experiment	[NO] (*M*)	$[H_2]$ (*M*)	Initial Rate (*M*/s)
1	5.0×10^{-3}	2.0×10^{-3}	1.3×10^{-5}
2	10.0×10^{-3}	2.0×10^{-3}	5.0×10^{-5}
3	10.0×10^{-3}	4.0×10^{-3}	10.0×10^{-5}

Strategy We are given a set of concentration and reaction rate data and asked to determine the rate law and the rate constant. We assume that the rate law takes the form

$$\text{rate} = k[NO]^x[H_2]^y$$

(Continued)

How do we use the data to determine x and y? Once the orders of the reactants are known, we can calculate k from any set of rate and concentrations. Finally, the rate law enables us to calculate the rate at any concentrations of NO and H_2.

Solution

(a) Experiments 1 and 2 show that when we double the concentration of NO at constant concentration of H_2, the rate quadruples. Taking the ratio of the rates from these two experiments

$$\frac{\text{rate}_2}{\text{rate}_1} = \frac{5.0 \times 10^{-5}\ M/\text{s}}{1.3 \times 10^{-5}\ M/\text{s}} \approx 4 = \frac{k(10.0 \times 10^{-3}\ M)^x(2.0 \times 10^{-3}\ M)^y}{k(5.0 \times 10^{-3}\ M)^x(2.0 \times 10^{-3}\ M)^y}$$

Therefore,

$$\frac{(10.0 \times 10^{-3}\ M)^x}{(5.0 \times 10^{-3}\ M)^x} = 2^x = 4$$

or $x = 2$, that is, the reaction is second order in NO. Experiments 2 and 3 indicate that doubling $[H_2]$ at constant [NO] doubles the rate. Here we write the ratio as

$$\frac{\text{rate}_3}{\text{rate}_2} = \frac{10.0 \times 10^{-5}\ M/\text{s}}{5.0 \times 10^{-5}\ M/\text{s}} = 2 = \frac{k(10.0 \times 10^{-3}\ M)^x(4.0 \times 10^{-3}\ M)^y}{k(10.0 \times 10^{-3}\ M)^x(2.0 \times 10^{-3}\ M)^y}$$

Therefore,

$$\frac{(4.0 \times 10^{-3}\ M)^y}{(2.0 \times 10^{-3}\ M)^y} = 2^y = 2$$

or $y = 1$, that is, the reaction is first order in H_2. Hence, the rate law is given by

$$\text{rate} = k[\text{NO}]^2[\text{H}_2]$$

which shows that it is a (2 + 1) or third-order reaction overall.

(b) The rate constant k can be calculated using the data from any one of the experiments. Rearranging the rate law, we get

$$k = \frac{\text{rate}}{[\text{NO}]^2[\text{H}_2]}$$

The data from experiment 2 give us

$$k = \frac{5.0 \times 10^{-5}\ M/\text{s}}{(10.0 \times 10^{-3}\ M)^2(2.0 \times 10^{-3}\ M)}$$
$$= 2.5 \times 10^2/M^2 \cdot \text{s}$$

(c) Using the known rate constant and concentrations of NO and H_2, we write

$$\text{rate} = (2.5 \times 10^2/M^2 \cdot \text{s})(12.0 \times 10^{-3}\ M)^2(6.0 \times 10^{-3}\ M)$$
$$= 2.2 \times 10^{-4}\ M/\text{s}$$

Comment Note that the reaction is first order in H_2, whereas the stoichiometric coefficient for H_2 in the balanced equation is 2. The order of a reactant is not related to the stoichiometric coefficient of the reactant in the overall balanced equation.

Similar problem: 14.15.

Practice Exercise The reaction of peroxydisulfate ion ($S_2O_8^{2-}$) with iodide ion (I^-) is

$$S_2O_8^{2-}(aq) + 3I^-(aq) \longrightarrow 2SO_4^{2-}(aq) + I_3^-(aq)$$

From the following data collected at a certain temperature, determine the rate law and calculate the rate constant.

Experiment	$[S_2O_8^{2-}]$ (M)	$[I^-]$ (M)	Initial Rate (M/s)
1	0.080	0.034	2.2×10^{-4}
2	0.080	0.017	1.1×10^{-4}
3	0.16	0.017	2.2×10^{-4}

REVIEW OF CONCEPTS

The relative rates of the reaction 2A + B ⟶ products shown in diagrams (a)–(c) are 1:2:4. The red spheres represent A molecules and the green spheres represent B molecules. Write a rate law for this reaction.

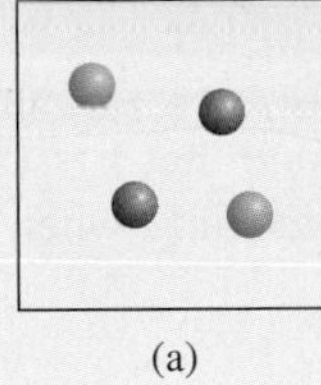
(a)

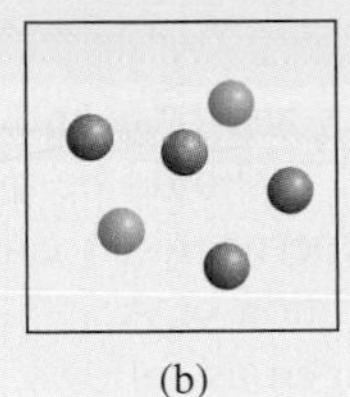
(b)

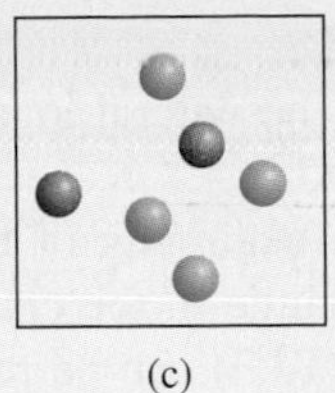
(c)

14.3 Relation Between Reactant Concentrations and Time

Rate laws enable us to calculate the rate of a reaction from the rate constant and reactant concentrations. They can also be converted into equations that enable us to determine the concentrations of reactants at any time during the course of a reaction. We will illustrate this application by considering first one of the simplest kind of rate laws—that applying to reactions that are first order overall.

First-Order Reactions

A ***first-order reaction*** is *a reaction whose rate depends on the reactant concentration raised to the first power.* In a first-order reaction of the type

$$\text{A} \longrightarrow \text{product}$$

the rate is

$$\text{rate} = -\frac{\Delta[\text{A}]}{\Delta t}$$

From the rate law, we also know that

$$\text{rate} = k[\text{A}]$$

For a first-order reaction, doubling the concentration of the reactant doubles the rate.

Thus,

$$-\frac{\Delta[\text{A}]}{\Delta t} = k[\text{A}] \quad (14.2)$$

We can determine the units of the first-order rate constant k by transposing:

$$k = -\frac{\Delta[\text{A}]}{[\text{A}]}\frac{1}{\Delta t}$$

Because the unit for Δ[A] and [A] is M and that for Δt is s, the unit for k is

$$\frac{M}{M\,\text{s}} = \frac{1}{\text{s}} = \text{s}^{-1}$$

In differential form, Equation (14.2) becomes

$$-\frac{d[A]}{dt} = k[A]$$

Rearranging, we get

$$\frac{d[A]}{[A]} = -kdt$$

Integrating between $t = 0$ and $t = t$ gives

$$\int_{[A]_0}^{[A]_t} \frac{d[A]}{[A]} = -k\int_0^t dt$$

$$\ln[A]_t - \ln[A]_0 = -kt$$

or

$$\ln\frac{[A]_t}{[A]_0} = -kt$$

(The minus sign does not enter into the evaluation of units.) Using calculus, we can show from Equation (14.2) that

$$\ln\frac{[A]_t}{[A]_0} = -kt \qquad (14.3)$$

in which ln is the natural logarithm, and $[A]_0$ and $[A]_t$ are the concentrations of A at times $t = 0$ and $t = t$, respectively. It should be understood that $t = 0$ need not correspond to the beginning of the experiment; it can be any time when we choose to monitor the change in the concentration of A.

Equation (14.3) can be rearranged as follows:

$$\ln [A]_t - \ln [A]_0 = -kt$$

or

$$\ln [A]_t = -kt + \ln [A]_0 \qquad (14.4)$$

Equation (14.4) has the form of the linear equation $y = mx + b$, in which m is the slope of the line that is the graph of the equation:

$$\begin{array}{ccccccc} \ln [A]_t & = & (-k) & (t) & + & \ln [A]_0 \\ \updownarrow & & \updownarrow & \updownarrow & & \updownarrow \\ y & = & m & x & + & b \end{array}$$

Thus, a plot of $\ln [A]_t$ versus t (or y versus x) gives a straight line with a slope of $-k$ (or m). This enables us to calculate the rate constant k. Figure 14.7 shows the characteristics of a first-order reaction.

There are many known first-order reactions. All nuclear decay processes are first order (see Chapter 21). Another example is the decomposition of ethane (C_2H_6) into highly reactive methyl radicals (CH_3):

$$C_2H_6 \longrightarrow 2CH_3$$

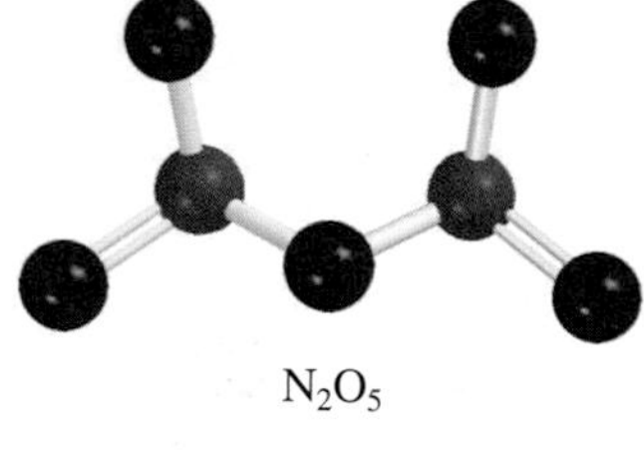

N_2O_5

Now let us determine graphically the order and rate constant of the decomposition of dinitrogen pentoxide in carbon tetrachloride (CCl_4) solvent at 45°C:

$$2N_2O_5(CCl_4) \longrightarrow 4NO_2(g) + O_2(g)$$

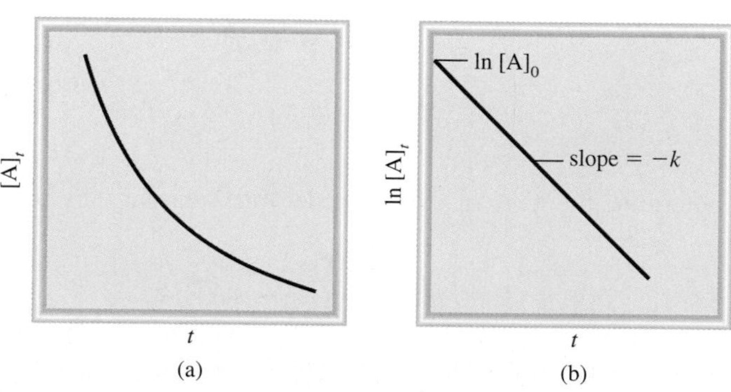

Figure 14.7
First-order reaction characteristics: (a) The exponential decrease of reactant concentration with time. (b) A plot of ln $[A]_t$ versus t. The slope of the line is equal to $-k$.

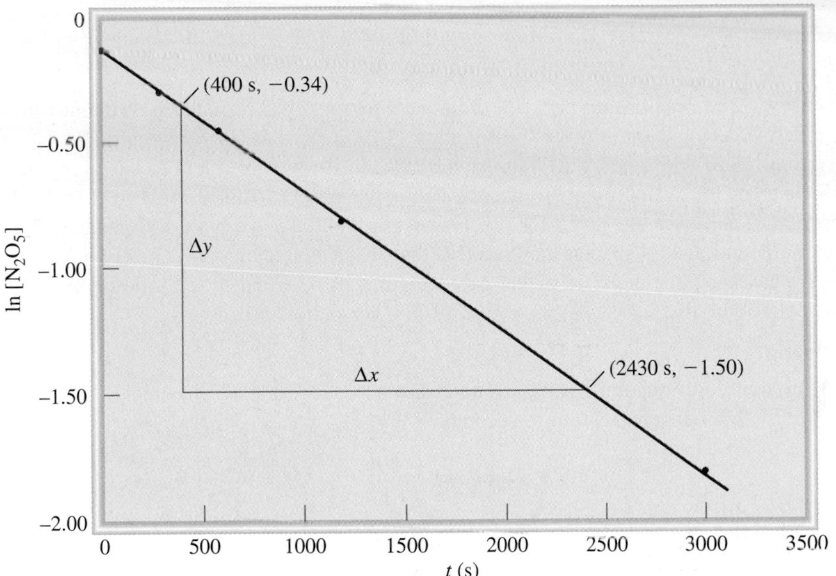

Figure 14.8
Plot of ln $[N_2O_5]$ versus time. The rate constant can be determined from the slope of the straight line.

This table shows the variation of N_2O_5 concentration with time, and the corresponding ln $[N_2O_5]$ values

t(s)	$[N_2O_5]$	ln $[N_2O_5]$
0	0.91	−0.094
300	0.75	−0.29
600	0.64	−0.45
1200	0.44	−0.82
3000	0.16	−1.83

N_2O_5 decomposes to give NO_2 (brown color) and colorless O_2 gases.

Applying Equation (14.4) we plot ln $[N_2O_5]$ versus t, as shown in Figure 14.8. The fact that the points lie on a straight line shows that the rate law is first order. Next, we determine the rate constant from the slope. We select two points far apart on the line and subtract their y and x values as

$$\begin{aligned}\text{slope } (m) &= \frac{\Delta y}{\Delta x}\\ &= \frac{-1.50 - (-0.34)}{(2430 - 400)\text{ s}}\\ &= -5.7 \times 10^{-4}\text{ s}^{-1}\end{aligned}$$

Because $m = -k$, we get $k = 5.7 \times 10^{-4}\text{ s}^{-1}$.

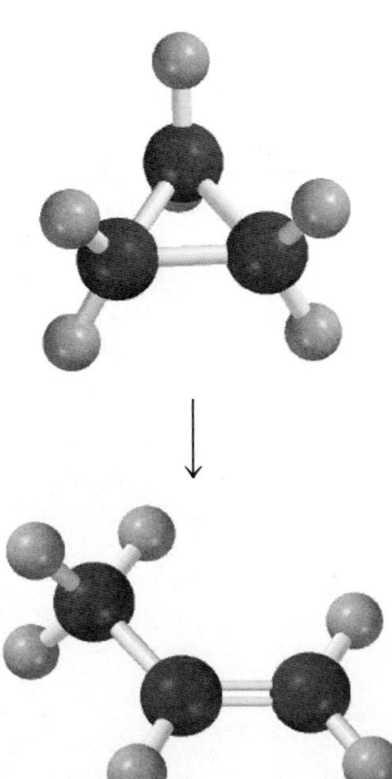

EXAMPLE 14.4

The conversion of cyclopropane to propene in the gas phase is a first-order reaction with a rate constant of $6.7 \times 10^{-4}\text{ s}^{-1}$ at 500°C.

$$\underset{\text{cyclopropane}}{\overset{CH_2}{\overset{/\ \backslash}{CH_2{-}CH_2}}} \longrightarrow \underset{\text{propene}}{CH_3{-}CH{=}CH_2}$$

(a) If the initial concentration of cyclopropane was 0.25 M, what is the concentration after 8.8 min? (b) How long (in minutes) will it take for the concentration of

(Continued)

cyclopropane to decrease from 0.25 M to 0.15 M? (c) How long (in minutes) will it take to convert 74 percent of the starting material?

Strategy The relationship between the concentrations of a reactant at different times in a first-order reaction is given by Equation (14.3) or (14.4). In (a) we are given $[A]_0 = 0.25\ M$ and asked for $[A]_t$ after 8.8 min. In (b) we are asked to calculate the time it takes for cyclopropane to decrease in concentration from 0.25 M to 0.15 M. No concentration values are given for (c). However, if initially we have 100 percent of the compound and 74 percent has reacted, then what is left must be (100% − 74%), or 26 percent. Thus, the ratio of the percentages will be equal to the ratio of the actual concentrations; that is, $[A]_t/[A]_0 = 26\%/100\%$, or 0.26/1.00.

Solution

(a) In applying Equation (14.4), we note that because k is given in units of s^{-1}, we must first convert 8.8 min to seconds:

$$8.8\ \cancel{\text{min}} \times \frac{60\ \text{s}}{1\ \cancel{\text{min}}} = 528\ \text{s}$$

We write

$$\begin{aligned}\ln [A]_t &= -kt + \ln [A]_0 \\ &= -(6.7 \times 10^{-4}\ \text{s}^{-1})(528\ \text{s}) + \ln (0.25) \\ &= -1.74\end{aligned}$$

Hence,

$$[A]_t = e^{-1.74} = 0.18\ M$$

Note that in the ln $[A]_0$ term, $[A]_0$ is expressed as a dimensionless quantity (0.25) because we cannot take the logarithm of units.

(b) Using Equation (14.3),

$$\ln \frac{0.15\ M}{0.25\ M} = -(6.7 \times 10^{-4}\ \text{s}^{-1})t$$

$$t = 7.6 \times 10^2\ \cancel{\text{s}} \times \frac{1\ \text{min}}{60\ \cancel{\text{s}}}$$

$$= 13\ \text{min}$$

(c) From Equation (14.3),

$$\ln \frac{0.26}{1.00} = -(6.7 \times 10^{-4}\ \text{s}^{-1})t$$

Similar problems: 14.22(b), 14.23(a).

$$t = 2.0 \times 10^3\ \cancel{\text{s}} \times \frac{1\ \text{min}}{60\ \cancel{\text{s}}} = 33\ \text{min}$$

Practice Exercise The reaction $2A \longrightarrow B$ is first order in A with a rate constant of $2.8 \times 10^{-2}\ \text{s}^{-1}$ at 80°C. How long (in seconds) will it take for A to decrease from 0.88 M to 0.14 M?

Half-Life

The ***half-life*** of a reaction, $t_{\frac{1}{2}}$, is *the time required for the concentration of a reactant to decrease to half of its initial concentration.* We can obtain an expression for $t_{\frac{1}{2}}$ for a first-order reaction as shown next. Rearranging Equation (14.3) we get

$$t = \frac{1}{k} \ln \frac{[A]_0}{[A]_t}$$

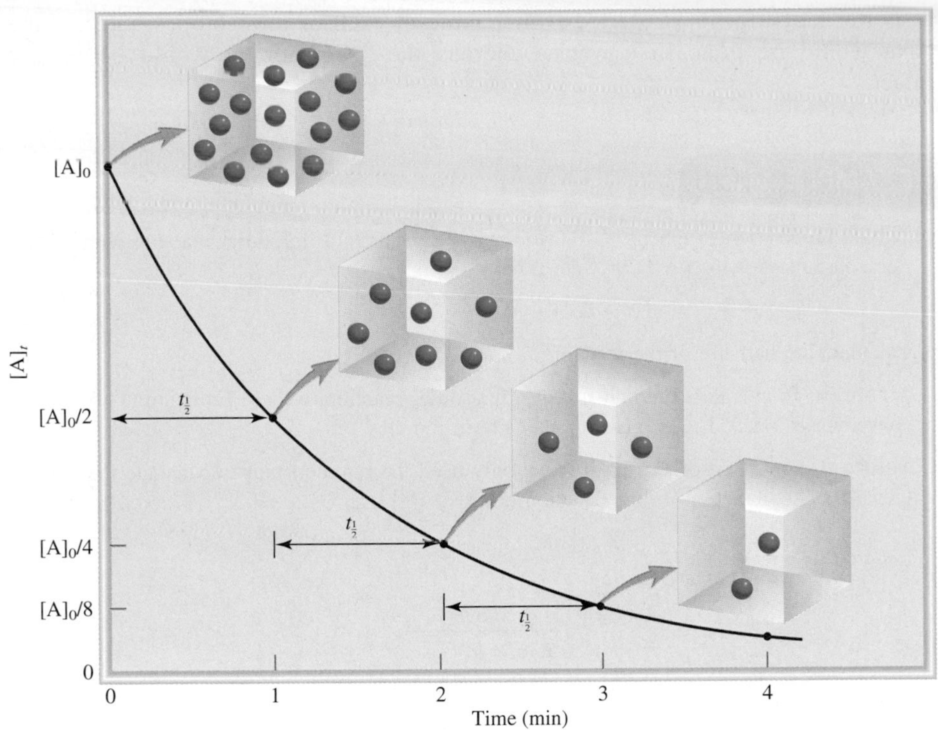

Figure 14.9
A plot of $[A]_t$ versus time for the first-order reaction A $\longrightarrow$ products. The half-life of the reaction is 1 min. After the elapse of each half-life, the concentration of A is halved.

By the definition of half-life, when $t = t_{\frac{1}{2}}$, $[A]_t = [A]_0/2$, so

$$t_{\frac{1}{2}} = \frac{1}{k} \ln \frac{[A]_0}{[A]_0/2}$$

or

$$t_{\frac{1}{2}} = \frac{1}{k} \ln 2 = \frac{0.693}{k} \qquad (14.5)$$

Equation (14.5) tells us that the half-life of a first-order reaction is *independent* of the initial concentration of the reactant. Thus, it takes the same time for the concentration of the reactant to decrease from 1.0 *M* to 0.50 *M*, say, as it does for a decrease in concentration from 0.10 *M* to 0.050 *M* (Figure 14.9). Measuring the half-life of a reaction is one way to determine the rate constant of a first-order reaction.

This analogy is helpful in understanding Equation (14.5). The duration of a college undergraduate's career, assuming the student does not take any time off, is 4 years. Thus, the half-life of his or her stay at the college is 2 years. This half-life is not affected by how many other students are present. Similarly, the half-life of a first-order reaction is concentration independent.

The usefulness of $t_{\frac{1}{2}}$ is that it gives us an estimate of the magnitude of the rate constant—the shorter the half-life, the larger the k. Consider, for example, two radioactive isotopes that are used in nuclear medicine: ^{24}Na ($t_{\frac{1}{2}}$ = 14.7 h) and ^{60}Co ($t_{\frac{1}{2}}$ = 5.3 yr). It is obvious that the ^{24}Na isotope decays faster because it has a shorter half-life. If we started with 1 mole of each of the isotopes, most of the ^{24}Na would be gone in a week, whereas the ^{60}Co sample would be mostly intact.

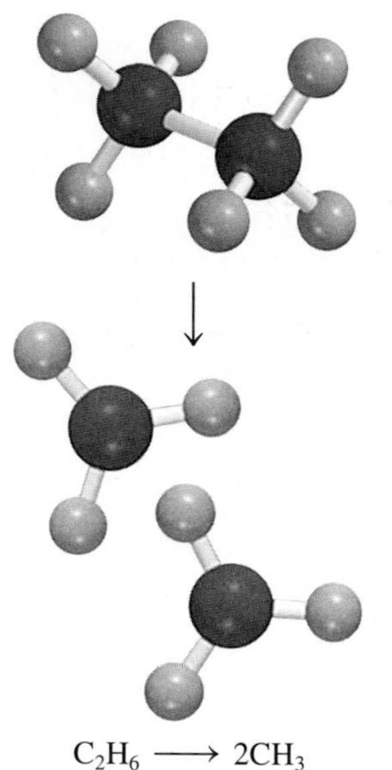

$C_2H_6 \longrightarrow 2CH_3$

Similar problem: 14.22(a).

EXAMPLE 14.5

The decomposition of ethane (C_2H_6) to methyl radicals is a first-order reaction with a rate constant of $5.36 \times 10^{-4}\ s^{-1}$ at 700°C:

$$C_2H_6(g) \longrightarrow 2CH_3(g)$$

Calculate the half-life of the reaction in minutes.

Strategy To calculate the half-life of a first-order reaction, we use Equation (14.5). A conversion is needed to express the half-life in minutes.

Solution For a first-order reaction, we only need the rate constant to calculate the half-life of the reaction. From Equation (14.5)

$$\begin{aligned} t_{\frac{1}{2}} &= \frac{0.693}{k} \\ &= \frac{0.693}{5.36 \times 10^{-4}\ s^{-1}} \\ &= 1.29 \times 10^3\ \cancel{s} \times \frac{1\ \text{min}}{60\ \cancel{s}} \\ &= 21.5\ \text{min} \end{aligned}$$

Practice Exercise Calculate the half-life of the decomposition of N_2O_5, discussed on p. 477.

REVIEW OF CONCEPTS

Consider the first-order reaction A $\longrightarrow$ B in which A molecules (blue spheres) are converted to B molecules (yellow spheres). (a) What are the half-life and rate constant for the reaction? (b) How many molecules of A and B are present at $t = 20$ s and $t = 30$ s?

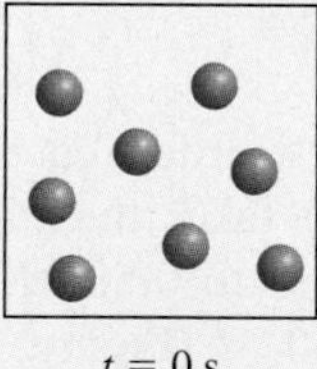

$t = 0$ s

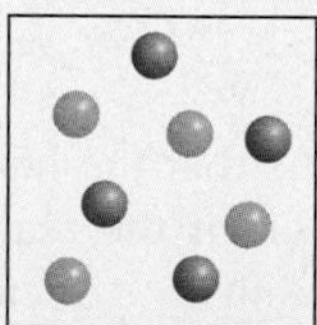

$t = 10$ s

Second-Order Reactions

A ***second-order reaction*** is *a reaction whose rate depends on the concentration of one reactant raised to the second power or on the concentrations of two different reactants, each raised to the first power.* The simpler type involves only one kind of reactant molecule:

$$A \longrightarrow \text{product}$$

for which

$$\text{rate} = -\frac{\Delta[\text{A}]}{\Delta t}$$

From the rate law,

$$\text{rate} = k[\text{A}]^2$$

As before, we can determine the units of k by writing

$$k = \frac{\text{rate}}{[\text{A}]^2} = \frac{M/\text{s}}{M^2} = 1/M \cdot \text{s}$$

Another type of second-order reaction is

$$\text{A} + \text{B} \longrightarrow \text{product}$$

and the rate law is given by

$$\text{rate} = k[\text{A}][\text{B}]$$

The reaction is first order in A and first order in B, so it has an overall reaction order of 2.

Using calculus, we can obtain the following expressions for "A ⟶ product" second-order reactions:

$$\frac{1}{[\text{A}]_t} = \frac{1}{[\text{A}]_0} + kt \qquad (14.6)$$

Equation 14.6 is the result of

$$\int_{[\text{A}]_0}^{[\text{A}]_t} \frac{d[\text{A}]}{[\text{A}]^2} = -k\int_0^t dt$$

Equation (14.6) has the form of a linear equation. As Figure 14.10 shows, a plot of $1/[\text{A}]_t$ versus t gives a straight line with slope $= k$ and y intercept $= 1/[\text{A}]_0$. (The corresponding equation for "A + B ⟶ product" reactions is too complex for our discussion.)

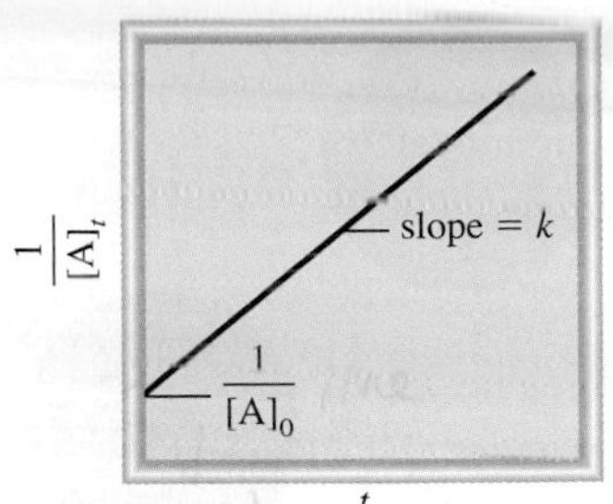

Figure 14.10
A plot of $1/[\text{A}]_t$ versus t for a second-order reaction. The slope of the line is equal to k.

We can obtain an equation for the half-life of a second-order reaction by setting $[\text{A}]_t = [\text{A}]_0/2$ in Equation (14.6):

$$\frac{1}{[\text{A}]_0/2} = \frac{1}{[\text{A}]_0} + kt_{\frac{1}{2}}$$

Solving for $t_{\frac{1}{2}}$ we obtain

$$t_{\frac{1}{2}} = \frac{1}{k[\text{A}]_0} \qquad (14.7)$$

Note that the half-life of a second-order reaction is inversely proportional to the initial reactant concentration. This result makes sense because the half-life should be shorter in the early stage of the reaction when more reactant molecules are present to collide with each other. Measuring the half-lives at different initial concentrations is one way to distinguish between a first-order and a second-order reaction.

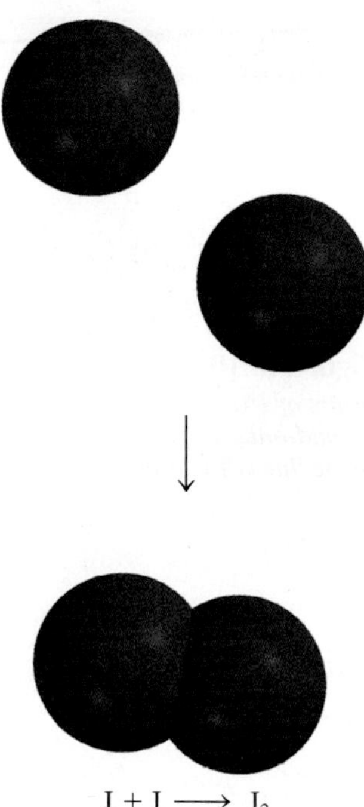

EXAMPLE 14.6

Iodine atoms combine to form molecular iodine in the gas phase:

$$I(g) + I(g) \longrightarrow I_2(g)$$

This reaction follows second-order kinetics and has the high rate constant $7.0 \times 10^9/M \cdot s$ at 23°C. (a) If the initial concentration of I was 0.068 *M*, calculate the concentration after 3.5 min. (b) Calculate the half-life of the reaction if the initial concentration of I is 0.53 *M* and if it is 0.39 *M*.

Strategy

(a) The relationship between the concentrations of a reactant at different times is given by the integrated rate law. Because this is a second-order reaction, we use Equation (14.6). (b) We are asked to calculate the half-life. The half-life for a second-order reaction is given by Equation (14.7).

Solution

(a) To calculate the concentration of a species at a later time of a second-order reaction, we need the initial concentration and the rate constant. Applying Equation (14.6),

$$\frac{1}{[A]_t} = kt + \frac{1}{[A]_0}$$

$$\frac{1}{[A]_t} = (7.0 \times 10^9 /M \cdot s)\left(3.5 \text{ min} \times \frac{60 \text{ s}}{1 \text{ min}}\right) + \frac{1}{0.068\ M}$$

where $[A]_t$ is the concentration at t = 3.5 min. Solving the equation, we get

$$[A]_t = 6.8 \times 10^{-13}\ M$$

This is such a low concentration that it is virtually undetectable. The very large rate constant for the reaction means that practically all the I atoms combine after only 3.5 min of reaction time.

(b) We need Equation (14.7) for this part.

For $[I]_0 = 0.53\ M$,

$$t_{\frac{1}{2}} = \frac{1}{k[A]_0} = \frac{1}{(7.0 \times 10^9/M \cdot s)(0.53\ M)} = 2.7 \times 10^{-10}\ s$$

For $[I]_0 = 0.39\ M$,

$$t_{\frac{1}{2}} = \frac{1}{k[A]_0} = \frac{1}{(7.0 \times 10^9/M \cdot s)(0.39\ M)} = 3.7 \times 10^{-10}\ s$$

Check These results confirm that the half-life of a second-order reaction, unlike that of a first-order reaction, is not a constant but depends on the initial concentration of the reactant(s).

Similar problem: 14.24.

Practice Exercise The reaction $2A \longrightarrow B$ is second order with a rate constant of $51/M \cdot \text{min}$ at 24°C. (a) Starting with $[A]_0 = 0.0092\ M$, how long will it take for $[A]_t = 3.7 \times 10^{-3}\ M$? (b) Calculate the half-life of the reaction.

Zero-Order Reactions

First- and second-order reactions are the most common reaction types. Reactions whose order is zero are rare. For a zero-order reaction

$$A \longrightarrow \text{product}$$

Table 14.2 Summary of the Kinetics of Zero-Order, First-Order, and Second-Order Reactions

Order	Rate Law	Concentration-Time Equation	Half-Life
0	Rate $= k$	$[A]_t = -kt + [A]_0$	$\frac{[A]_0}{2k}$
1	Rate $= k[A]$	$\ln\frac{[A]_t}{[A]_0} = -kt$	$\frac{0.693}{k}$
2	Rate $= k[A]^2$	$\frac{1}{[A]_t} = kt + \frac{1}{[A]_0}$	$\frac{1}{k[A]_0}$

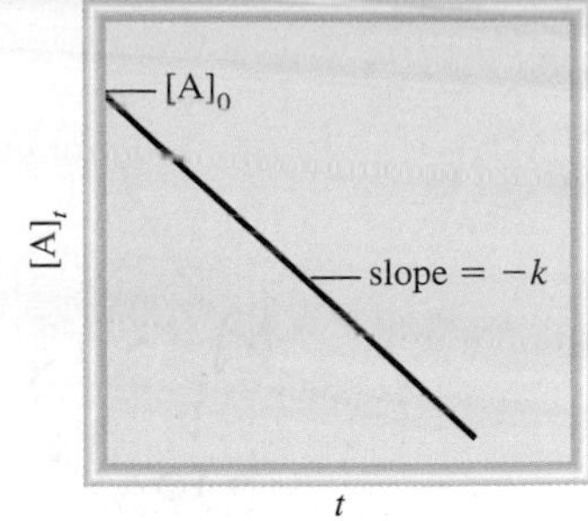

Figure 14.11 *A plot of $[A]_t$ versus t for a zero-order reaction. The slope of the line is equal to $-k$.*

the rate law is given by

$$\text{rate} = k[A]^0 = k$$

Recall that any number raised to the power of zero is equal to one.

Thus, the rate of a zero-order reaction is a *constant,* independent of reactant concentration. Using calculus, we can show that

$$[A]_t = -kt + [A]_0 \quad (14.8)$$

Equation 14.8 is the result of

$$\int_{[A]_0}^{[A]_t} d[A] = -k\int_0^t dt$$

Equation (14.8) has the form of a linear equation. As Figure 14.11 shows, a plot of $[A]_t$ versus t gives a straight line with slope $= -k$ and y intercept $= [A]_0$. To calculate the half-life of a zero-order reaction, we set $[A]_t = [A]_0/2$ in Equation (14.8) and obtain

$$t_{\frac{1}{2}} = \frac{[A]_0}{2k} \quad (14.9)$$

Many of the known zero-order reactions take place on a metal surface. An example is the decomposition of nitrous oxide (N_2O) to nitrogen and oxygen in the presence of platinum (Pt):

$$2N_2O(g) \longrightarrow 2N_2(g) + O_2(g)$$

When all the binding sites on Pt are occupied, the rate becomes constant regardless of the amount of N_2O present in the gas phase. As we will see in Section 14.6, another well-studied zero-order reaction occurs in enzyme catalysis.

Keep in mind that $[A]_0$ and $[A]_t$ in Equation (14.8) refer to the concentration of N_2O in the gas phase.

Third-order and higher order reactions are quite complex; they are not presented in this book. Table 14.2 summarizes the kinetics of zero-order, first-order, and second-order reactions.

14.4 Activation Energy and Temperature Dependence of Rate Constants

With very few exceptions, reaction rates increase with increasing temperature. For example, much less time is required to hard-boil an egg at 100°C (about 10 min) than at 80°C (about 30 min). Conversely, an effective way to preserve foods is to store

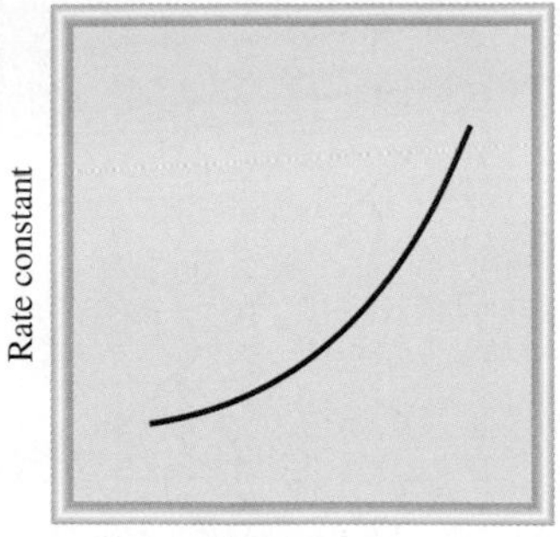

Figure 14.12
Dependence of rate constant on temperature. The rate constants of most reactions increase with increasing temperature.

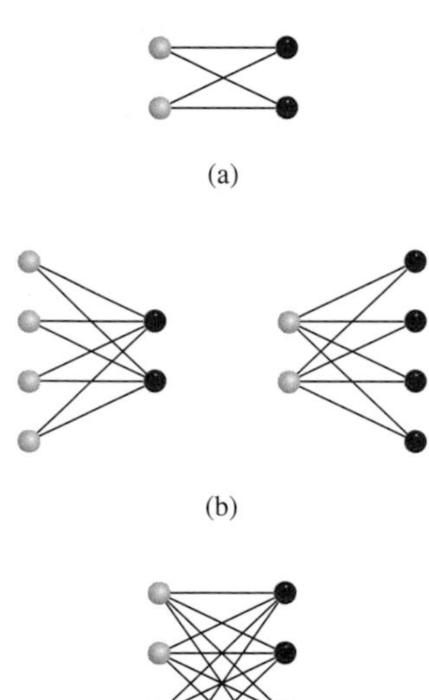

Figure 14.13
Dependence of number of collisions on concentration. We consider here only A-B collisions, which can lead to formation of products. (a) There are four possible collisions among two A and two B molecules. (b) Doubling the number of either type of molecule (but not both) increases the number of collisions to eight. (c) Doubling both the A and B molecules increases the number of collisions to sixteen.

Animation: Activation Energy

them at subzero temperatures, thereby slowing the rate of bacterial decay. Figure 14.12 shows a typical example of the relationship between the rate constant of a reaction and temperature. To explain this behavior, we must ask how reactions get started in the first place.

The Collision Theory of Chemical Kinetics

The kinetic molecular theory of gases (p. 157) states that gas molecules frequently collide with one another. Therefore it seems logical to assume—and it is generally true—that chemical reactions occur as a result of collisions between reacting molecules. In terms of the *collision theory* of chemical kinetics, then, we expect the rate of a reaction to be directly proportional to the number of molecular collisions per second, or to the frequency of molecular collisions:

$$\text{rate} \propto \frac{\text{number of collisions}}{\text{s}}$$

This simple relationship explains the dependence of reaction rate on concentration.

Consider the reaction of A molecules with B molecules to form some product. Suppose that each product molecule is formed by the direct combination of an A molecule and a B molecule. If we doubled the concentration of A, say, then the number of A-B collisions would also double, because, in any given volume, there would be twice as many A molecules that could collide with B molecules (Figure 14.13). Consequently, the rate would increase by a factor of 2. Similarly, doubling the concentration of B molecules would increase the rate twofold. Thus, we can express the rate law as

$$\text{rate} = k[\text{A}][\text{B}]$$

The reaction is first order in both A and B and obeys second-order kinetics.

The collision theory is intuitively appealing, but the relationship between rate and molecular collisions is more complicated than you might expect. The implication of the collision theory is that a reaction always occurs when an A and a B molecule collide. However, not all collisions lead to reactions. Calculations based on the kinetic molecular theory show that, at ordinary pressures (say, 1 atm) and temperatures (say, 298 K), there are about 1×10^{27} binary collisions (collisions between two molecules) in 1 mL of volume every second, in the gas phase. Even more collisions per second occur in liquids. If every binary collision led to a product, then most reactions would be complete almost instantaneously. In practice, we find that the rates of reactions differ greatly. This means that, in many cases, collisions alone do not guarantee that a reaction will take place.

Any molecule in motion possesses kinetic energy; the faster it moves, the greater the kinetic energy. When molecules collide, part of their kinetic energy is converted to vibrational energy. If the initial kinetic energies are large, then the colliding molecules will vibrate so strongly as to break some of the chemical bonds. This bond fracture is the first step toward product formation. If the initial kinetic energies are small, the molecules will merely bounce off each other intact. Energetically speaking, there is some minimum collision energy below which no reaction occurs.

We postulate that, to react, the colliding molecules must have a total kinetic energy equal to or greater than the ***activation energy*** **(E_a)**, which is *the minimum amount of energy required to initiate a chemical reaction*. Lacking this energy, the molecules remain intact, and no change results from the collision. *The species temporarily formed by the reactant molecules as a result of the collision before they form the product* is called the ***activated complex*** (also called the ***transition state***).

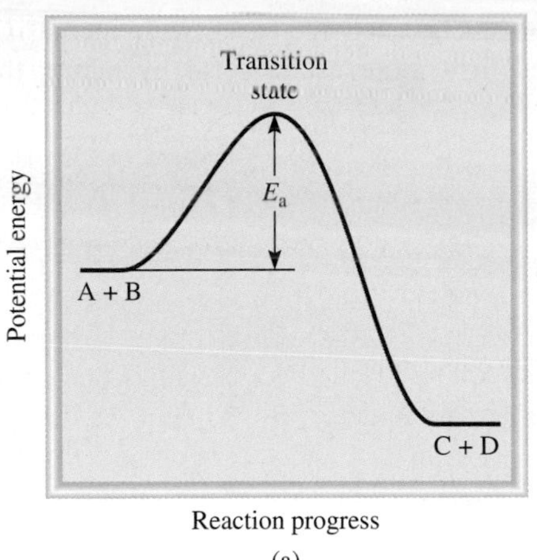

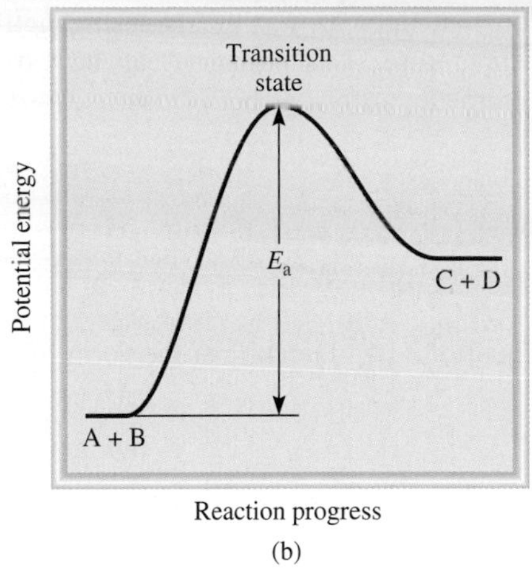

Figure 14.14
Potential energy profiles for (a) exothermic and (b) endothermic reactions. These plots show the change in potential energy as reactants A and B are converted to products C and D. The transition state is a highly unstable species with a high potential energy. The activation energy is defined for the forward reaction in both (a) and (b). Note that the products C and D are more stable than the reactants in (a) and less stable than those in (b).

Figure 14.14 shows two different potential energy profiles for the reaction

$$\text{A} + \text{B} \longrightarrow \text{C} + \text{D}$$

If the products are more stable than the reactants, then the reaction will be accompanied by a release of heat; that is, the reaction is exothermic [Figure 14.14(a)]. On the other hand, if the products are less stable than the reactants, then heat will be absorbed by the reacting mixture from the surroundings and we have an endothermic reaction [Figure 14.14(b)]. In both cases, we plot the potential energy of the reacting system versus the progress of the reaction. Qualitatively, these plots show the potential energy changes as reactants are converted to products.

We can think of activation energy as a barrier that prevents less energetic molecules from reacting. Because the number of reactant molecules in an ordinary reaction is very large, the speeds, and hence also the kinetic energies of the molecules, vary greatly. Normally, only a small fraction of the colliding molecules—the fastest-moving ones—have enough kinetic energy to exceed the activation energy. These molecules can therefore take part in the reaction. The increase in the rate (or the rate constant) with temperature can now be explained: The speeds of the molecules obey the Maxwell distributions shown in Figure 5.15. Compare the speed distributions at two different temperatures. Because more high-energy molecules are present at the higher temperature, the rate of product formation is also greater at the higher temperature.

The Arrhenius Equation

The dependence of the rate constant of a reaction on temperature can be expressed by this equation, now known as the *Arrhenius equation:*

$$k = Ae^{-E_a/RT} \tag{14.10}$$

in which E_a is the activation energy of the reaction (in kilojoules per mole), R is the gas constant (8.314 J/K · mol), T is the absolute temperature, and e is the base of the natural logarithm scale (see Appendix 3). The quantity A represents the collision frequency and is called the *frequency factor*. It can be treated as a constant for a given reacting system over a fairly wide temperature range. Equation (14.10) shows that the rate constant is directly proportional to A and, therefore, to the collision frequency. Further, because of the minus sign associated with the exponent E_a/RT, the rate constant decreases with increasing activation energy and increases with increasing temperature. This equation can be expressed in a more useful form by taking the natural logarithm of both sides:

$$\ln k = \ln Ae^{-E_a/RT}$$

$$\ln k = \ln A - \frac{E_a}{RT} \tag{14.11}$$

Equation (14.11) can take the form of a linear equation:

$$\ln k = \left(-\frac{E_a}{R}\right)\left(\frac{1}{T}\right) + \ln A \tag{14.12}$$

$$\begin{array}{ccccc} \updownarrow & & \updownarrow\ \updownarrow & & \updownarrow \\ y & = & m\quad x & + & b \end{array}$$

Thus, a plot of ln k versus $1/T$ gives a straight line whose slope m is equal to $-E_a/R$ and whose intercept b with the ordinate (the y-axis) is ln A.

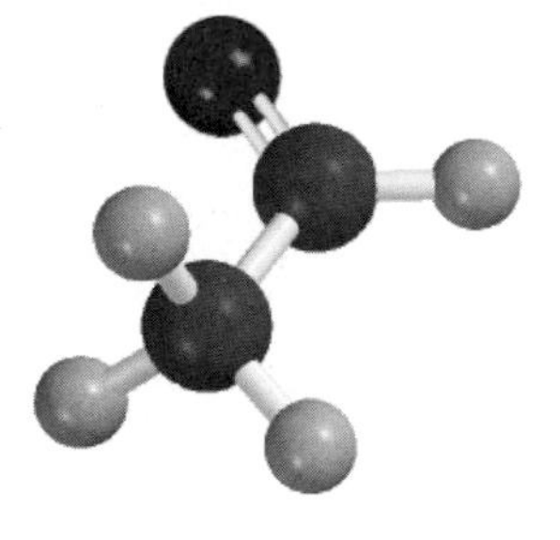

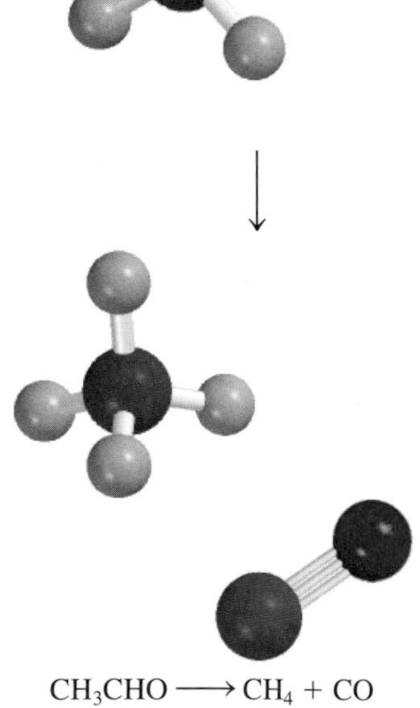

$CH_3CHO \longrightarrow CH_4 + CO$

EXAMPLE 14.7

The rate constants for the decomposition of acetaldehyde

$$CH_3CHO(g) \longrightarrow CH_4(g) + CO(g)$$

were measured at five different temperatures. The data are shown in the table. Plot ln k versus $1/T$, and determine the activation energy (in kJ/mol) for the reaction. This reaction has been experimentally shown to be "$\frac{3}{2}$" order in CH_3CHO, so k has the units of $1/M^{\frac{1}{2}} \cdot s$.

$k\ (1/M^{\frac{1}{2}} \cdot s)$	T (K)
0.011	700
0.035	730
0.105	760
0.343	790
0.789	810

Strategy Consider the Arrhenius equation written as a linear equation

$$\ln k = \left(-\frac{E_a}{R}\right)\left(\frac{1}{T}\right) + \ln A$$

A plot of ln k versus $1/T$ (y versus x) will produce a straight line with a slope equal to $-E_a/R$. Thus, the activation energy can be determined from the slope of the plot.

Solution First, we convert the data to the following table:

ln k	$1/T$ (K^{-1})
−4.51	1.43×10^{-3}
−3.35	1.37×10^{-3}
−2.254	1.32×10^{-3}
−1.070	1.27×10^{-3}
−0.237	1.23×10^{-3}

(Continued)

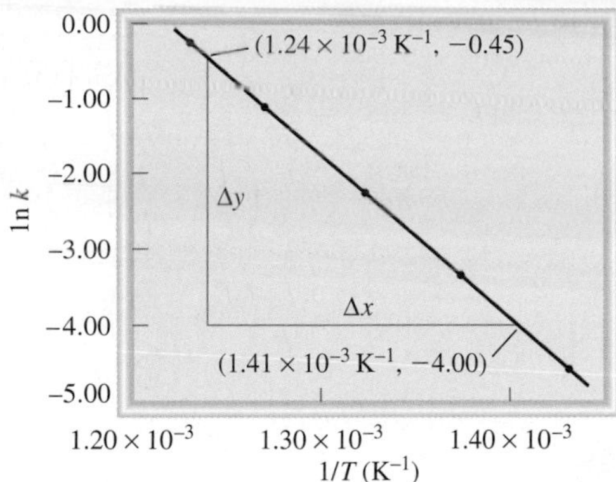

Figure 14.15
Plot of ln k versus 1/T. The slope of the line is equal to $-E_a/R$.

A plot of these data yields the graph in Figure 14.15. The slope of the line is calculated from two pairs of coordinates:

$$\text{slope} = \frac{-4.00 - (-0.45)}{(1.41 - 1.24) \times 10^{-3}\ \text{K}^{-1}} = -2.09 \times 10^4\ \text{K}$$

From the linear form of Equation (14.12)

$$\text{slope} = -\frac{E_a}{R} = -2.09 \times 10^4\ \text{K}$$

$$\begin{aligned} E_a &= (8.314\ \text{J/K} \cdot \text{mol})(2.09 \times 10^4\ \text{K}) \\ &= 1.74 \times 10^5\ \text{J/mol} \\ &= 1.74 \times 10^2\ \text{kJ/mol} \end{aligned}$$

Check It is important to note that although the rate constant itself has the units $1/M^{\frac{1}{2}} \cdot \text{s}$, the quantity ln *k* has no units (we cannot take the logarithm of a unit).

Similar problem: 14.33.

Practice Exercise The second-order rate constant for the decomposition of nitrous oxide (N_2O) into nitrogen molecule and oxygen atom has been measured at different temperatures:

k (1/$M \cdot$ s)	t (°C)
1.87×10^{-3}	600
0.0113	650
0.0569	700

Determine graphically the activation energy for the reaction.

An equation relating the rate constants k_1 and k_2 at temperatures T_1 and T_2 can be used to calculate the activation energy or to find the rate constant at another temperature if the activation energy is known. To derive such an equation we start with Equation (14.11):

$$\ln k_1 = \ln A - \frac{E_a}{RT_1}$$

$$\ln k_2 = \ln A - \frac{E_a}{RT_2}$$

Subtracting ln k_2 from ln k_1 gives

$$\ln k_1 - \ln k_2 = \frac{E_a}{R}\left(\frac{1}{T_2} - \frac{1}{T_1}\right)$$

$$\ln \frac{k_1}{k_2} = \frac{E_a}{R}\left(\frac{1}{T_2} - \frac{1}{T_1}\right)$$

$$\ln \frac{k_1}{k_2} = \frac{E_a}{R}\left(\frac{T_1 - T_2}{T_1 T_2}\right) \tag{14.13}$$

EXAMPLE 14.8

The rate constant of a first-order reaction is $4.68 \times 10^{-2}\ s^{-1}$ at 298 K. What is the rate constant at 375 K if the activation energy for the reaction is 33.1 kJ/mol?

Strategy A modified form of the Arrhenius equation relates two rate constants at two different temperatures [see Equation (14.13)]. Make sure the units of R and E_a are consistent.

Solution The data arc

$$k_1 = 4.68 \times 10^{-2}\ s^{-1} \qquad k_2 = ?$$
$$T_1 = 298\ K \qquad T_2 = 375\ K$$

Substituting in Equation (14.13),

$$\ln \frac{4.68 \times 10^{-2}\ s^{-1}}{k_2} = \frac{33.1 \times 10^3\ J/mol}{8.314\ J/K \cdot mol}\left[\frac{298\ K - 375\ K}{(298\ K)(375\ K)}\right]$$

We convert E_a to units of J/mol to match the units of R. Solving the equation gives

$$\ln \frac{4.68 \times 10^{-2}\ s^{-1}}{k_2} = -2.74$$

$$\frac{4.68 \times 10^{-2}\ s^{-1}}{k_2} = e^{-2.74} = 0.0646$$

$$k_2 = 0.724\ s^{-1}$$

Check The rate constant is expected to be greater at a higher temperature. Therefore, the answer is reasonable.

Similar problem: 14.36.

Practice Exercise The first-order rate constant for the reaction of methyl chloride (CH_3Cl) with water to produce methanol (CH_3OH) and hydrochloric acid (HCl) is $3.32 \times 10^{-10}\ s^{-1}$ at 25°C. Calculate the rate constant at 40°C if the activation energy is 116 kJ/mol.

Animation: Orientation of Collision

For simple reactions (for example, reactions between atoms), we can equate the frequency factor (A) in the Arrhenius equation with the frequency of collision between the reacting species. For more complex reactions, we must also consider the "orientation factor," that is, how reacting molecules are oriented relative to each other. The reaction between carbon monoxide (CO) and nitrogen dioxide (NO_2) to form carbon dioxide (CO_2) and nitric oxide (NO) illustrates this point:

$$CO(g) + NO_2(g) \longrightarrow CO_2(g) + NO(g)$$

This reaction is most favorable when the reacting molecules approach each other according to that shown in Figure 14.16(a). Otherwise, few or no products are formed

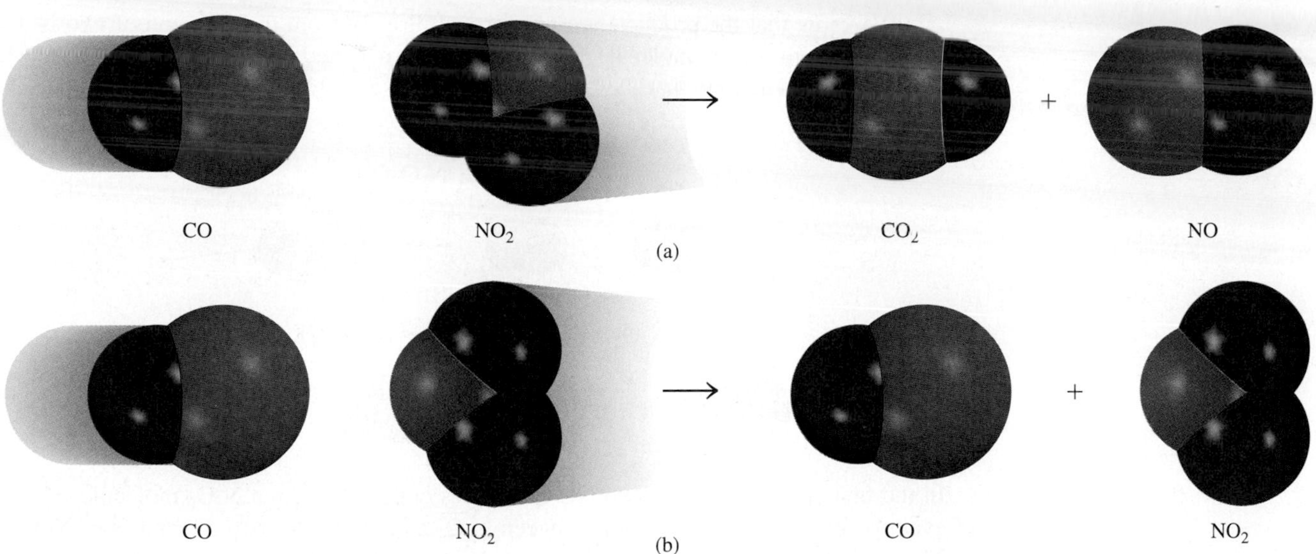

Figure 14.16
The orientations of the molecules shown in (a) are effective and will likely lead to formation of products. The orientations shown in (b) are ineffective and no products will be formed.

[Figure 14.16(b)]. The quantitative treatment of orientation factor is to modify Equation (14.10) as follows:

$$k = pAe^{-E_a/RT} \qquad (14.14)$$

where p is the orientation factor. The orientation factor is a unitless quantity; its value ranges from 1 for reactions involving atoms such as $I + I \longrightarrow I_2$ to 10^{-6} or smaller for reactions involving molecules.

REVIEW OF CONCEPTS

(a) What can you deduce about the magnitude of the activation energy of a reaction if its rate constant changes appreciably with a small change in temperature? (b) If a reaction occurs every time two reacting molecules collide, what can you say about the orientation factor and the activation energy of the reaction?

14.5 Reaction Mechanisms

As we mentioned earlier, an overall balanced chemical equation does not tell us much about how a reaction actually takes place. In many cases, it merely represents the sum of several ***elementary steps,*** or *elementary reactions, a series of simple reactions that represent the progress of the overall reaction at the molecular level.* The term for *the sequence of elementary steps that leads to product formation* is ***reaction mechanism.*** The reaction mechanism is comparable to the route of travel followed during a trip; the overall chemical equation specifies only the origin and destination.

As an example of a reaction mechanism, let us consider the reaction between nitric oxide and oxygen:

$$2NO(g) + O_2(g) \longrightarrow 2NO_2(g)$$

We know that the products are not formed directly from the collision of two NO molecules with an O_2 molecule because N_2O_2 is detected during the course of the reaction. Let us assume that the reaction actually takes place via two elementary steps as follows:

$$2NO(g) \longrightarrow N_2O_2(g)$$

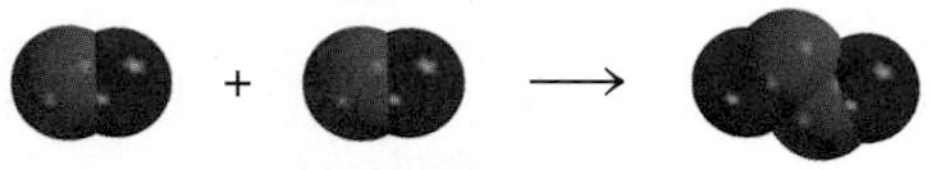

$$N_2O_2(g) + O_2(g) \longrightarrow 2NO_2(g)$$

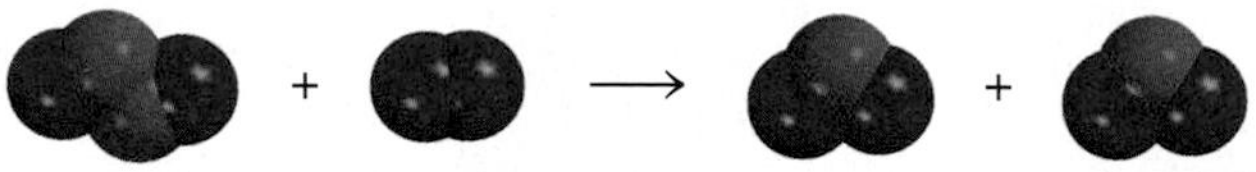

In the first elementary step, two NO molecules collide to form a N_2O_2 molecule. This event is followed by the reaction between N_2O_2 and O_2 to give two molecules of NO_2. The net chemical equation, which represents the overall change, is given by the sum of the elementary steps:

The sum of the elementary steps must give the overall balanced equation.

Step 1:	$NO + NO \longrightarrow N_2O_2$
Step 2:	$N_2O_2 + O_2 \longrightarrow 2NO_2$
Overall reaction:	$2NO + \cancel{N_2O_2} + O_2 \longrightarrow \cancel{N_2O_2} + 2NO_2$

Species such as N_2O_2 are called ***intermediates*** because they *appear in the mechanism of the reaction (that is, the elementary steps) but not in the overall balanced equation.* Keep in mind that an intermediate is always formed in an early elementary step and consumed in a later elementary step.

The ***molecularity of a reaction*** is *the number of molecules reacting in an elementary step.* These molecules may be of the same or different types. Each of the elementary steps just discussed is called a ***bimolecular reaction,*** *an elementary step that involves two molecules.* An example of a ***unimolecular reaction,*** *an elementary step in which only one reacting molecule participates,* is the conversion of cyclopropane to propene discussed in Example 14.4. Very few ***termolecular reactions,*** *reactions that involve the participation of three molecules in one elementary step,* are known, because the simultaneous encounter of three molecules is a far less likely event than a bimolecular collision.

Rate Laws and Elementary Steps

Knowing the elementary steps of a reaction enables us to deduce the rate law. Suppose we have the following elementary reaction:

$$A \longrightarrow \text{products}$$

Because there is only one molecule present, this is a unimolecular reaction. It follows that the larger the number of A molecules present, the faster the rate of product formation. Thus, the rate of a unimolecular reaction is directly proportional to the concentration of A, or is first order in A:

$$\text{rate} = k[A]$$

For a bimolecular elementary reaction involving A and B molecules,

$$A + B \longrightarrow \text{product}$$

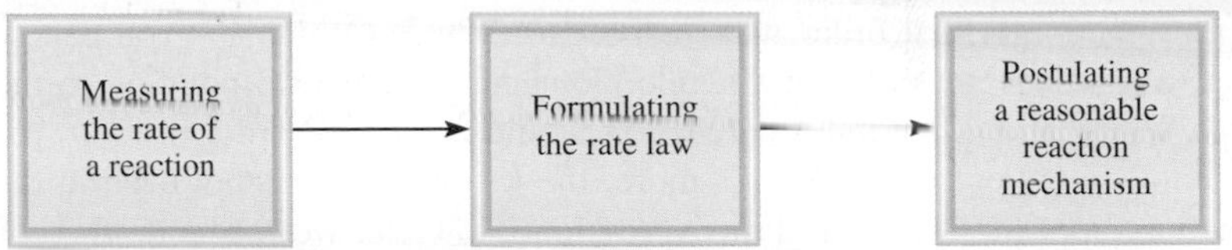

Figure 14.17
Sequence of steps in the study of a reaction mechanism.

the rate of product formation depends on how frequently A and B collide, which in turn depends on the concentrations of A and B. Thus, we can express the rate as

$$\text{rate} = k[\text{A}][\text{B}]$$

Similarly, for a bimolecular elementary reaction of the type

$$\text{A} + \text{A} \longrightarrow \text{products}$$

or

$$2\text{A} \longrightarrow \text{products}$$

the rate becomes

$$\text{rate} = k[\text{A}]^2$$

The preceding examples show that the reaction order for each reactant in an elementary reaction is equal to its stoichiometric coefficient in the chemical equation for that step. In general, we cannot tell by merely looking at the overall balanced equation whether the reaction occurs as shown or in a series of steps. This determination is made in the laboratory.

Note that the rate law can be written directly from the coefficients of an elementary step.

When we study a reaction that has more than one elementary step, the rate law for the overall process is given by the ***rate-determining step,*** which is *the slowest step in the sequence of steps leading to product formation.*

An analogy for the rate-determining step is the flow of traffic along a narrow road. Assuming the cars cannot pass one another on the road, the rate at which the cars travel is governed by the slowest-moving car.

Experimental studies of reaction mechanisms begin with the collection of data (rate measurements). Next, we analyze the data to determine the rate constant and order of the reaction, and we write the rate law. Finally, we suggest a plausible mechanism for the reaction in terms of elementary steps (Figure 14.17). The elementary steps must satisfy two requirements:

- The sum of the elementary steps must give the overall balanced equation for the reaction.
- The rate-determining step should predict the same rate law as is determined experimentally.

Remember that for a proposed reaction scheme, we must be able to detect the presence of any intermediate(s) formed in one or more elementary steps.

The decomposition of hydrogen peroxide illustrates the elucidation of reaction mechanisms by experimental studies. This reaction is facilitated by iodide ions (I^-) (Figure 14.18). The overall reaction is

$$2H_2O_2(aq) \longrightarrow 2H_2O(l) + O_2(g)$$

By experiment, the rate law is found to be

$$\text{rate} = k[H_2O_2][I^-]$$

Figure 14.18
The decomposition of hydrogen peroxide is catalyzed by the iodide ion. A few drops of liquid soap have been added to the solution to dramatize the evolution of oxygen gas. (Some of the iodide ions are oxidized to molecular iodine, which then reacts with iodide ions to form the brown triiodide ion, I_3^-.)

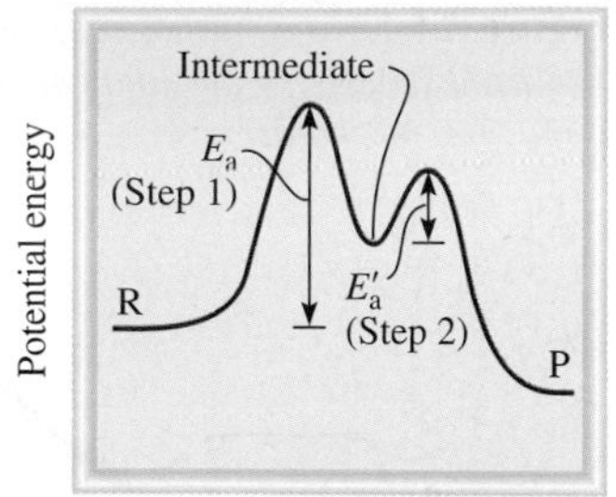

Figure 14.19
Potential energy profile for a two-step reaction in which the first step is rate-determining. R and P represent reactants and products, respectively.

Thus, the reaction is first order with respect to both H_2O_2 and I^-. You can see that decomposition does not occur in a single elementary step corresponding to the overall balanced equation. If it did, the reaction would be second order in H_2O_2 (note the coefficient 2 in the equation). What's more, the I^- ion, which is not even in the overall equation, appears in the rate law expression. How can we reconcile these facts?

We can account for the observed rate law by assuming that the reaction takes place in two separate elementary steps, each of which is bimolecular:

$$\textit{Step 1:} \quad H_2O_2 + I^- \xrightarrow{k_1} H_2O + IO^-$$

$$\textit{Step 2:} \quad H_2O_2 + IO^- \xrightarrow{k_2} H_2O + O_2 + I^-$$

If we further assume that step 1 is the rate-determining step, then the rate of the reaction can be determined from the first step alone:

$$\text{rate} = k_1[H_2O_2][I^-]$$

where $k_1 = k$. Note that the IO^- ion is an intermediate because it does not appear in the overall balanced equation. Although the I^- ion also does not appear in the overall equation, I^- differs from IO^- in that the former is present at the start of the reaction and at its completion. The function of I^- is to speed up the reaction—that is, it is a *catalyst*. We will discuss catalysis in Section 14.6. Figure 14.19 shows the potential energy profile for a reaction like the decomposition of H_2O_2. We see that the first step, which is rate determining, has a larger activation energy than the second step. The intermediate, although stable enough to be observed, reacts quickly to form the products.

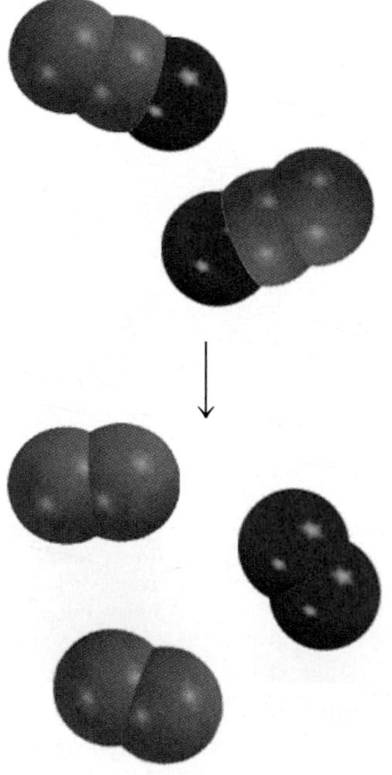

$2N_2O \longrightarrow 2N_2 + O_2$

EXAMPLE 14.9

The gas-phase decomposition of nitrous oxide (N_2O) is believed to occur via two elementary steps:

$$\textit{Step 1:} \quad N_2O \xrightarrow{k_1} N_2 + O$$

$$\textit{Step 2:} \quad N_2O + O \xrightarrow{k_2} N_2 + O_2$$

Experimentally the rate law is found to be rate $= k[N_2O]$. (a) Write the equation for the overall reaction. (b) Identify the intermediates. (c) What can you say about the relative rates of steps 1 and 2?

Strategy (a) Because the overall reaction can be broken down into elementary steps, knowing the elementary steps would enable us to write the overall reaction. (b) What are the characteristics of an intermediate? Does it appear in the overall reaction? (c) What determines which elementary step is rate determining? How does a knowledge of the rate-determining step help us write the rate law of a reaction?

Solution

(a) Adding the equations for steps 1 and 2 gives the overall reaction:

$$2N_2O \longrightarrow 2N_2 + O_2$$

(b) Because the O atom is produced in the first elementary step and it does not appear in the overall balanced equation, it is an intermediate.

(c) If we assume that step 1 is the rate-determining step (that is, if $k_2 \gg k_1$), then the rate of the overall reaction is given by

$$\text{rate} = k_1[N_2O]$$

and $k = k_1$.

(Continued)

Check Step 1 must be the rate-determining step because the rate law written from this step matches the experimentally determined rate law, that is, rate $= k[N_2O]$.

Similar problem: 14.47.

Practice Exercise The reaction between NO_2 and CO to produce NO and CO_2 is believed to occur via two steps:

$$\textit{Step 1:} \quad NO_2 + NO_2 \longrightarrow NO + NO_3$$

$$\textit{Step 2:} \quad NO_3 + CO \longrightarrow NO_2 + CO_2$$

The experimental rate law is rate $= k[NO_2]^2$. (a) Write the equation for the overall reaction. (b) Identify the intermediate. (c) What can you say about the relative rates of steps 1 and 2?

Review of Concepts

Compounds A and B react according to the equation $2A + B \longrightarrow C$. The mechanism of this reaction, which involves the intermediate E, occurs in two elementary steps—a slow step followed by a fast step. Which of the two elementary steps includes the intermediate E as a reactant?

14.6 Catalysis

Animation:
Catalysis

We saw in studying the decomposition of hydrogen peroxide that the reaction rate depends on the concentration of iodide ions even though I^- does not appear in the overall equation. We noted there that I^- acts as a catalyst for that reaction. A ***catalyst*** is *a substance that increases the rate of a reaction by lowering the activation energy.* It does so by providing an alternate reaction pathway. The catalyst may react to form an intermediate with the reactant, but it is regenerated in a subsequent step so it is not consumed in the reaction.

A rise in temperature also increases the rate of a reaction. However, at high temperatures, the products formed may undergo other reactions, thereby reducing the yield.

In the laboratory preparation of molecular oxygen, a sample of potassium chlorate is heated; the reaction is (see p. 155)

$$2KClO_3(s) \longrightarrow 2KCl(s) + 3O_2(g)$$

However, this thermal decomposition is very slow in the absence of a catalyst. The rate of decomposition can be increased dramatically by adding a small amount of the catalyst manganese dioxide (MnO_2), a black powdery substance. All the MnO_2 can be recovered at the end of the reaction, just as all the I^- ions remain following H_2O_2 decomposition.

A catalyst speeds up a reaction by providing a set of elementary steps with more favorable kinetics than those that exist in its absence. From Equation (14.10) we know that the rate constant k (and hence the rate) of a reaction depends on the frequency factor A and the activation energy E_a—the larger the A or the smaller the E_a, the greater the rate. In many cases, a catalyst increases the rate by lowering the activation energy for the reaction.

To extend the traffic analogy, adding a catalyst can be compared with building a tunnel through a mountain to connect two towns that were previously linked by a winding road over the mountain.

Let us assume that the following reaction has a certain rate constant k and an activation energy E_a:

$$A + B \xrightarrow{k} C + D$$

In the presence of a catalyst, however, the rate constant is k_c (called the *catalytic rate constant*):

$$A + B \xrightarrow{k_c} C + D$$

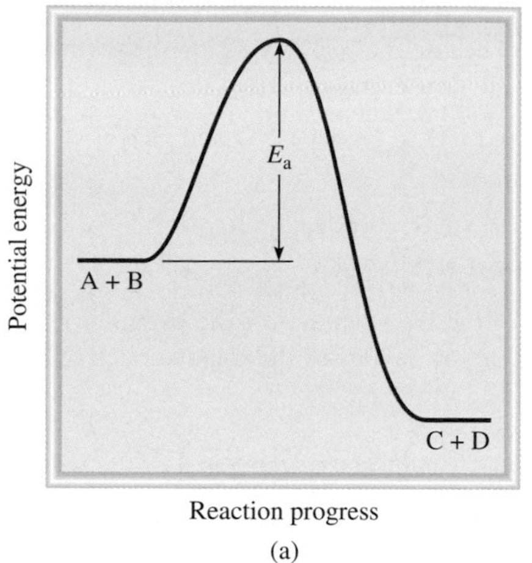

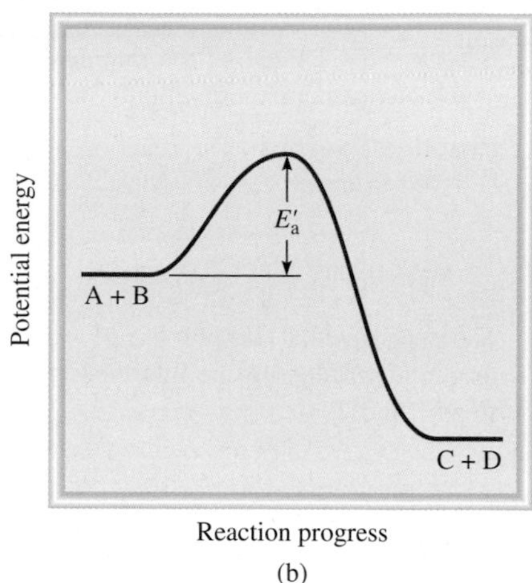

Figure 14.20
Comparison of the activation energy barriers of an uncatalyzed reaction and the same reaction with a catalyst. The catalyst lowers the energy barrier but does not affect the actual energies of the reactants or products. Although the reactants and products are the same in both cases, the reaction mechanisms and rate laws are different in (a) and (b).

By the definition of a catalyst,

$$\text{rate}_{\text{catalyzed}} > \text{rate}_{\text{uncatalyzed}}$$

A catalyst lowers the activation energy for both the forward and reverse reactions.

Figure 14.20 shows the potential energy profiles for both reactions. Note that the total energies of the reactants (A and B) and those of the products (C and D) are unaffected by the catalyst; the only difference between the two is a lowering of the activation energy from E_a to E'_a. Because the activation energy for the reverse reaction is also lowered, a catalyst enhances the rate of the reverse reaction to the same extent as it does the forward reaction rate.

There are three general types of catalysis, depending on the nature of the rate-increasing substance: heterogeneous catalysis, homogeneous catalysis, and enzyme catalysis.

Heterogeneous Catalysis

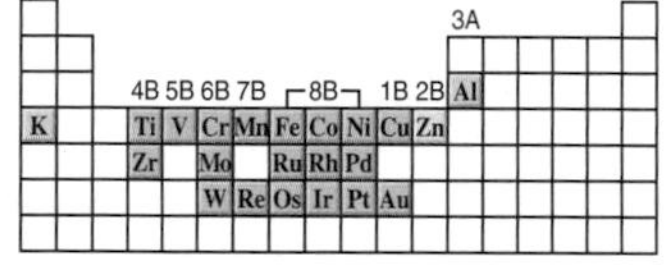

Metals and compounds of metals that are most frequently used in heterogeneous catalysis.

In *heterogeneous catalysis,* the reactants and the catalyst are in different phases. Usually the catalyst is a solid and the reactants are either gases or liquids. Heterogeneous catalysis is by far the most important type of catalysis in industrial chemistry, especially in the synthesis of many key chemicals. Here we describe three specific examples of heterogeneous catalysis.

The Haber Synthesis of Ammonia

Ammonia is an extremely valuable inorganic substance used in the fertilizer industry, the manufacture of explosives, and many other applications. Around the turn of the twentieth century, many chemists strove to synthesize ammonia from nitrogen and hydrogen. The supply of atmospheric nitrogen is virtually inexhaustible, and hydrogen gas can be produced readily by passing steam over heated coal:

$$H_2O(g) + C(s) \longrightarrow CO(g) + H_2(g)$$

Hydrogen is also a by-product of petroleum refining.

The formation of NH_3 from N_2 and H_2 is exothermic:

$$N_2(g) + 3H_2(g) \longrightarrow 2NH_3(g) \qquad \Delta H° = -92.6 \text{ kJ/mol}$$

But the reaction rate is extremely slow at room temperature. To be practical on a large scale, a reaction must occur at an appreciable rate *and* it must have a high yield of the desired product. Raising the temperature does accelerate the preceding reaction,

Figure 14.21
The catalytic action in the synthesis of ammonia. First the H_2 and N_2 molecules bind to the surface of the catalyst. This interaction weakens the covalent bonds within the molecules and eventually causes the molecules to dissociate. The highly reactive H and N atoms combine to form NH_3 molecules, which then leave the surface.

but at the same time it promotes the decomposition of NH_3 molecules into N_2 and H_2, thus lowering the yield of NH_3.

In 1905, after testing literally hundreds of compounds at various temperatures and pressures, the German chemist Fritz Haber discovered that iron plus a few percent of oxides of potassium and aluminum catalyze the reaction of hydrogen with nitrogen to yield ammonia at about 500°C. This procedure is known as the *Haber process.*

In heterogeneous catalysis, the surface of the solid catalyst is usually the site of the reaction. The initial step in the Haber process involves the dissociation of N_2 and H_2 on the metal surface (Figure 14.21). Although the dissociated species are not truly free atoms because they are bonded to the metal surface, they are highly reactive. The two reactant molecules behave very differently on the catalyst surface. Studies show that H_2 dissociates into atomic hydrogen at temperatures as low as −196°C (the boiling point of liquid nitrogen). Nitrogen molecules, on the other hand, dissociate at about 500°C. The highly reactive N and H atoms combine rapidly at high temperatures to produce the desired NH_3 molecules:

$$N + 3H \longrightarrow NH_3$$

The Manufacture of Nitric Acid

Nitric acid is one of the most important inorganic acids. It is used in the production of fertilizers, dyes, drugs, and explosives. The major industrial method of producing nitric acid is the *Ostwald process,* after the German chemist Wilhelm Ostwald. The starting materials, ammonia and molecular oxygen, are heated in the presence of a platinum-rhodium catalyst (Figure 14.22) to about 800°C:

$$4NH_3(g) + 5O_2(g) \longrightarrow 4NO(g) + 6H_2O(g)$$

The nitric oxide formed readily oxidizes (without catalysis) to nitrogen dioxide:

$$2NO(g) + O_2(g) \longrightarrow 2NO_2(g)$$

When dissolved in water, NO_2 forms both nitrous acid and nitric acid:

$$2NO_2(g) + H_2O(l) \longrightarrow HNO_2(aq) + HNO_3(aq)$$

On heating, nitrous acid is converted to nitric acid as follows:

$$3HNO_2(aq) \longrightarrow HNO_3(aq) + H_2O(l) + 2NO(g)$$

The NO generated can be recycled to produce NO_2 in the second step.

Figure 14.22
Platinum-rhodium catalyst used in the Ostwald process.

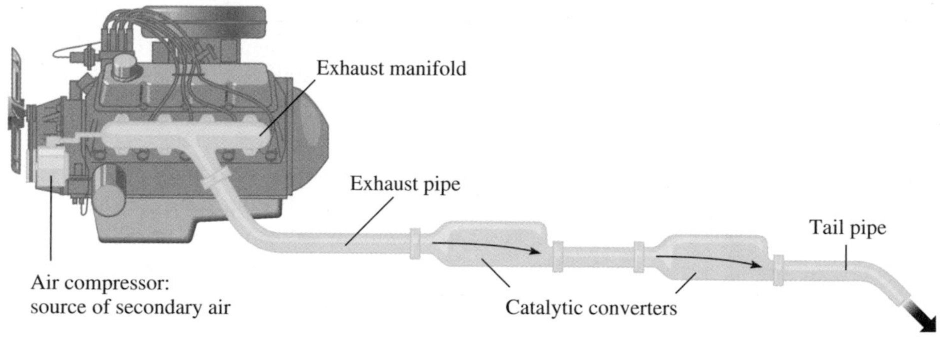

Figure 14.23
A two-stage catalytic converter for an automobile.

Catalytic Converters

At high temperatures inside a running car's engine, nitrogen and oxygen gases react to form nitric oxide:

$$N_2(g) + O_2(g) \rightleftharpoons 2NO(g)$$

When released into the atmosphere, NO rapidly combines with O_2 to form NO_2. Nitrogen dioxide and other gases emitted by an automobile, such as carbon monoxide (CO) and various unburned hydrocarbons, make automobile exhaust a major source of air pollution.

Figure 14.24
A cross-sectional view of a catalytic converter. The beads contain platinum, palladium, and rhodium, which catalyze the combustion of CO and hydrocarbons.

Most new cars are equipped with catalytic converters (Figure 14.23). An efficient catalytic converter serves two purposes: It oxidizes CO and unburned hydrocarbons to CO_2 and H_2O, and it reduces NO and NO_2 to N_2 and O_2. Hot exhaust gases into which air has been injected are passed through the first chamber of one converter to accelerate the complete burning of hydrocarbons and to decrease CO emission. (A cross section of the catalytic converter, containing Pt or Pd or a transition metal oxide such as CuO or Cr_2O_3, is shown in Figure 14.24.) However, because high temperatures increase NO production, a second chamber containing a different catalyst (a transition metal or a transition metal oxide) and operating at a lower temperature is required to dissociate NO into N_2 and O_2 before the exhaust is discharged through the tailpipe.

Homogeneous Catalysis

In *homogeneous catalysis* the reactants and catalyst are dispersed in a single phase, usually liquid. Acid and base catalyses are the most important type of homogeneous catalysis in liquid solution. For example, the reaction of ethyl acetate with water to form acetic acid and ethanol normally occurs too slowly to be measured.

$$\underset{\text{ethyl acetate}}{CH_3{-}\overset{\overset{\displaystyle O}{\|}}{C}{-}O{-}C_2H_5} + H_2O \longrightarrow \underset{\text{acetic acid}}{CH_3{-}\overset{\overset{\displaystyle O}{\|}}{C}{-}OH} + \underset{\text{ethanol}}{C_2H_5OH}$$

In the absence of the catalyst, the rate law is given by

$$\text{rate} = k[CH_3COOC_2H_5]$$

This reaction is zero order in water because water's concentration is very high and therefore it is unaffected by the reaction.

However, the reaction can be catalyzed by an acid. In the presence of hydrochloric acid, the rate is given by

$$\text{rate} = k_c[CH_3COOC_2H_5][H^+]$$

Enzyme Catalysis

Of all the intricate processes that have evolved in living systems, none is more striking or more essential than enzyme catalysis. ***Enzymes*** are *biological catalysts*. The amazing fact about enzymes is that not only can they increase the rate of biochemical reactions by factors ranging from 10^6 to 10^{18}, but they are also highly specific. An enzyme acts only on certain molecules, called *substrates* (that is, reactants), while leaving the rest of the system unaffected. It has been estimated that an average living cell may contain some 3000 different enzymes, each of them catalyzing a specific reaction in which a substrate is converted into the appropriate products. Enzyme catalyses are usually homogeneous with the substrate and enzyme present in the same aqueous solution.

An enzyme is typically a large protein molecule that contains one or more *active sites* where interactions with substrates take place. These sites are structurally compatible with specific molecules, in much the same way as a key fits a particular lock (Figure 14.25). However, an enzyme molecule (or at least its active site) has a fair amount of structural flexibility and can modify its shape to accommodate different kinds of substrates (Figure 14.26).

The mathematical treatment of enzyme kinetics is quite complex, even when we know the basic steps involved in the reaction. A simplified scheme is

$$E + S \rightleftharpoons ES$$
$$ES \xrightarrow{k} P + E$$

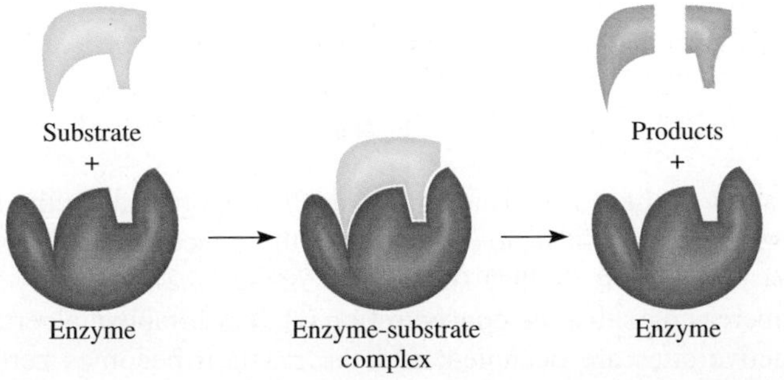

Figure 14.25
The lock-and-key model of an enzyme's specificity for substrate molecules.

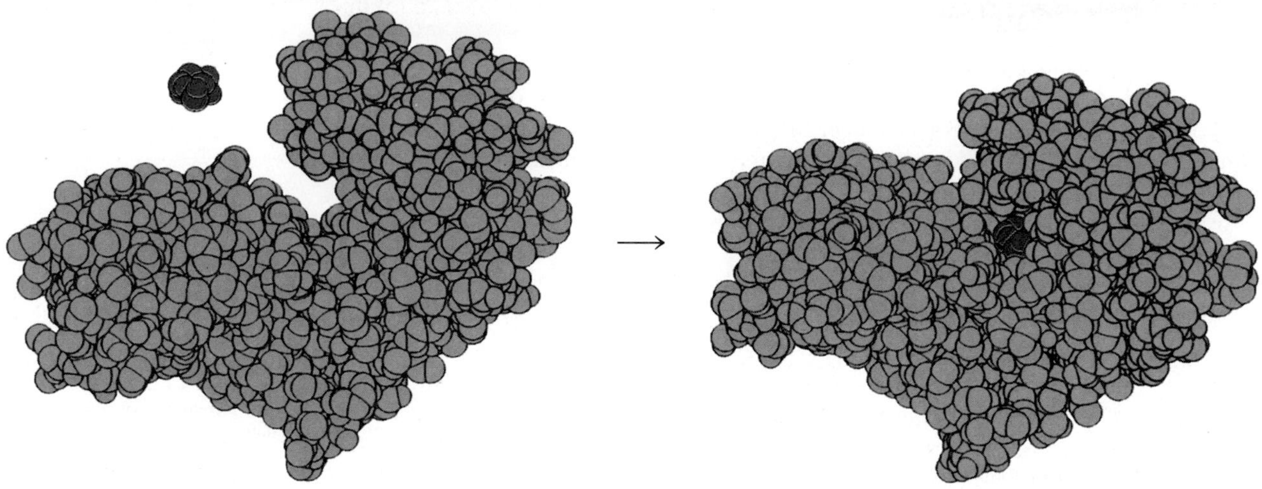

Figure 14.26
Left to right: The binding of glucose molecule (red) to hexokinase (an enzyme in the metabolic pathway). Note how the region at the active site closes around glucose after binding. Frequently, the geometries of both the substrate and the active site are altered to fit each other.

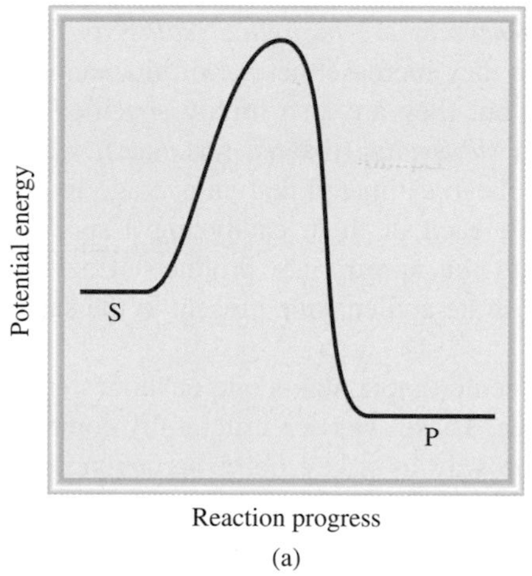

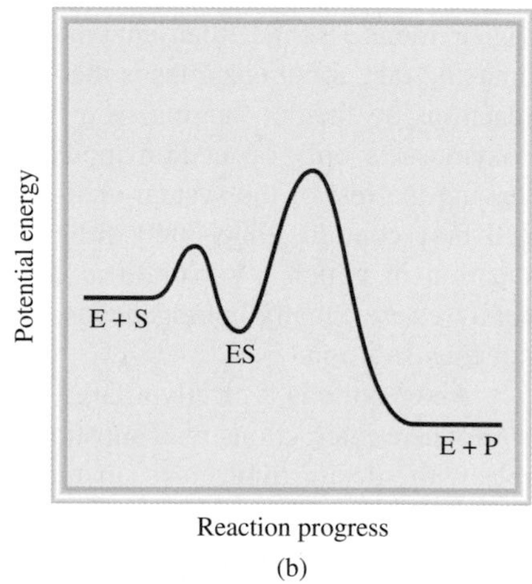

Figure 14.27
Comparison of (a) an uncatalyzed reaction and (b) the same reaction catalyzed by an enzyme. The plot in (b) assumes that the catalyzed reaction has a two-step mechanism, in which the second step (ES $\longrightarrow$ E + P) is rate-determining.

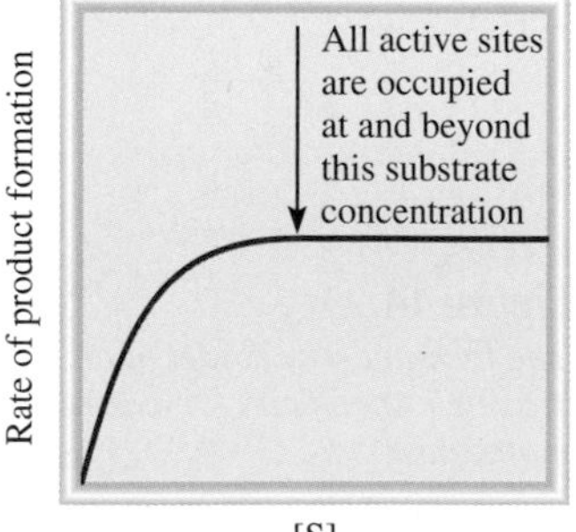

Figure 14.28
Plot of the rate of product formation versus substrate concentration in an enzyme–catalyzed reaction.

in which E, S, and P represent enzyme, substrate, and product, and ES is the enzyme-substrate intermediate. Figure 14.27 shows the potential energy profile for the reaction. It is often assumed that the formation of ES and its decomposition back to enzyme and substrate molecules occur rapidly and that the rate-determining step is the formation of product. In general, the rate of such a reaction is given by the equation

$$\text{rate} = \frac{\Delta[\text{P}]}{\Delta t} = k[\text{ES}]$$

The concentration of the ES intermediate is itself proportional to the amount of the substrate present, and a plot of the rate versus the concentration of substrate typically yields a curve such as that shown in Figure 14.28. Initially the rate rises rapidly with increasing substrate concentration. However, above a certain concentration all the active sites are occupied, and the reaction becomes zero order in the substrate. That is, the rate remains the same even though the substrate concentration increases. At and beyond this point, the rate of formation of product depends only on how fast the ES intermediate breaks down, not on the number of substrate molecules present.

Review of Concepts

Consider the following two-step reaction mechanism.

Step 1:	$NO(g) + O_3(g) \longrightarrow NO_2(g) + O_2(g)$
Step 2:	$O(g) + NO_2(g) \longrightarrow NO(g) + O_2(g)$
Overall:	$O_3(g) + O(g) \longrightarrow 2O_2(g)$

Identify the intermediate and catalyst in this reaction. Also classify the catalyst as heterogeneous or homogeneous.

Key Equations

$\text{rate} = k[A]^x[B]^y$	(14.1)	Rate law expressions. The sum $(x + y)$ gives the overall order of the reaction.
$\ln \frac{[A]_t}{[A]_0} = -kt$	(14.3)	Relationship between concentration and time for a first-order reaction.
$\ln [A]_t = -kt + \ln [A]_0$	(14.4)	Equation for the graphical determination of k for a first-order reaction.
$t_{\frac{1}{2}} = \frac{0.693}{k}$	(14.5)	Half-life for a first-order reaction.
$\frac{1}{[A]_t} = kt + \frac{1}{[A]_0}$	(14.6)	Relationship between concentration and time for a second-order reaction.
$[A]_t = -kt + [A]_0$	(14.8)	Relationship between concentration and time for a zero-order reaction.
$k = Ae^{-E_a/RT}$	(14.10)	The Arrhenius equation expressing the dependence of the rate constant on activation energy and temperature.
$\ln k = \left(-\frac{E_a}{R}\right)\left(\frac{1}{T}\right) + \ln A$	(14.12)	Equation for the graphical determination of activation energy.
$\ln \frac{k_1}{k_2} = \frac{E_a}{R}\left(\frac{T_1 - T_2}{T_1T_2}\right)$	(14.13)	Relationship of rate constants at two different temperatures.

Summary of Facts and Concepts

1. The rate of a chemical reaction is the change in the concentration of reactants or products over time. The rate is not constant, but varies continuously as concentrations change.
2. The rate law expresses the relationship of the rate of a reaction to the rate constant and the concentrations of the reactants raised to appropriate powers. The rate constant k for a given reaction changes only with temperature.
3. Reaction order is the power to which the concentration of a given reactant is raised in the rate law. Overall reaction order is the sum of the powers to which reactant concentrations are raised in the rate law. The rate law and the reaction order cannot be determined from the stoichiometry of the overall equation for a reaction; they must be determined by experiment. For a zero-order reaction, the reaction rate is equal to the rate constant.

4. The half-life of a reaction (the time it takes for the concentration of a reactant to decrease by one-half) can be used to determine the rate constant of a first-order reaction.
5. In terms of collision theory, a reaction occurs when molecules collide with sufficient energy, called the activation energy, to break the bonds and initiate the reaction. The rate constant and the activation energy are related by the Arrhenius equation.
6. The overall balanced equation for a reaction may be the sum of a series of simple reactions, called elementary steps. The complete series of elementary steps for a reaction is the reaction mechanism.
7. If one step in a reaction mechanism is much slower than all other steps, it is the rate-determining step.
8. A catalyst speeds up a reaction usually by lowering the value of E_a. A catalyst can be recovered unchanged at the end of a reaction.
9. In heterogeneous catalysis, which is of great industrial importance, the catalyst is a solid and the reactants are gases or liquids. In homogeneous catalysis, the catalyst and the reactants are in the same phase. Enzymes are catalysts in living systems.

Key Words

Activated complex, p. 484
Activation energy (E_a), p. 484
Bimolecular reaction, p. 490
Catalyst, p. 493
Chemical kinetics, p. 467
Elementary step, p. 489
Enzyme, p. 497
First-order reaction, p. 475
Half-life ($t_{\frac{1}{2}}$), p. 478
Intermediate, p. 490
Molecularity of a reaction, p. 490
Rate constant (k), p. 472
Rate-determining step, p. 491
Rate law, p. 472
Reaction mechanism, p. 489
Reaction order, p. 472
Reaction rate, p. 467
Second-order reaction, p. 480
Termolecular reaction, p. 490
Transition state, p. 484
Unimolecular reaction, p. 490

Questions and Problems

Reaction Rate

Review Questions

14.1 What is meant by the rate of a chemical reaction?

14.2 What are the units of the rate of a reaction?

14.3 What are the advantages of measuring the initial rate of a reaction?

14.4 Can you suggest two reactions that are very slow (take days or longer to complete) and two reactions that are very fast (are over in minutes or seconds)?

Problems

14.5 Write the reaction rate expressions for these reactions in terms of the disappearance of the reactants and the appearance of products:
(a) $H_2(g) + I_2(g) \longrightarrow 2HI(g)$
(b) $2H_2(g) + O_2(g) \longrightarrow 2H_2O(g)$
(c) $5Br^-(aq) + BrO_3^-(aq) + 6H^+(aq) \longrightarrow 3Br_2(aq) + 3H_2O(l)$

14.6 Consider the reaction

$$N_2(g) + 3H_2(g) \longrightarrow 2NH_3(g)$$

Suppose that at a particular moment during the reaction molecular hydrogen is reacting at the rate of 0.074 M/s. (a) At what rate is ammonia being formed? (b) At what rate is molecular nitrogen reacting?

Rate Laws

Review Questions

14.7 Explain what is meant by the rate law of a reaction.

14.8 What are the units for the rate constants of first-order and second-order reactions?

14.9 Write an equation relating the concentration of a reactant A at $t = 0$ to that at $t = t$ for a first-order reaction. Define all the terms and give their units.

14.10 Consider the zero-order reaction A $\longrightarrow$ product. (a) Write the rate law for the reaction. (b) What are the units for the rate constant? (c) Plot the rate of the reaction versus [A].

14.11 On which of these quantities does the rate constant of a reaction depend: (a) concentrations of reactants, (b) nature of reactants, (c) temperature?

14.12 For each of these pairs of reaction conditions, indicate which has the faster rate of formation of hydrogen gas: (a) sodium or potassium with water, (b) magnesium or iron with 1.0 M HCl, (c) magnesium rod or magnesium powder with 1.0 M HCl,

(d) magnesium with 0.10 *M* HCl or magnesium with 1.0 *M* HCl.

Problems

14.13 The rate law for the reaction

$$NH_4^+(aq) + NO_2^-(aq) \longrightarrow N_2(g) + 2H_2O(l)$$

is given by rate $= k[NH_4^+][NO_2^-]$. At 25°C, the rate constant is $3.0 \times 10^{-4}/M \cdot s$. Calculate the rate of the reaction at this temperature if $[NH_4^+] = 0.26\ M$ and $[NO_2^-] = 0.080\ M$.

14.14 Starting with the data in Table 14.1, (a) deduce the rate law for the reaction, (b) calculate the rate constant, and (c) calculate the rate of the reaction at the time when $[F_2] = 0.010\ M$ and $[ClO_2] = 0.020\ M$.

14.15 Consider the reaction

$$A + B \longrightarrow \text{products}$$

From these data obtained at a certain temperature, determine the order of the reaction and calculate the rate constant:

[A] (*M*)	[B] (*M*)	Rate (*M*/s)
1.50	1.50	3.20×10^{-1}
1.50	2.50	3.20×10^{-1}
3.00	1.50	6.40×10^{-1}

14.16 Consider the reaction

$$X + Y \longrightarrow Z$$

These data are obtained at 360 K:

Initial Rate of Disappearance of X (*M*/s)	[X]	[Y]
0.147	0.10	0.50
0.127	0.20	0.30
4.064	0.40	0.60
1.016	0.20	0.60
0.508	0.40	0.30

(a) Determine the order of the reaction. (b) Determine the initial rate of disappearance of X when the concentration of X is 0.30 *M* and that of Y is 0.40 *M*.

14.17 Determine the overall orders of the reactions to which these rate laws apply: (a) rate $= k[NO_2]^2$; (b) rate $= k$; (c) rate $= k[H_2][Br_2]^{\frac{1}{2}}$; (d) rate $= k[NO]^2[O_2]$.

14.18 Consider the reaction

$$A \longrightarrow B$$

The rate of the reaction is $1.6 \times 10^{-2}\ M/s$ when the concentration of A is 0.35 *M*. Calculate the rate constant if the reaction is (a) first order in A, (b) second order in A.

Relationship Between Reactant Concentration and Time

Review Questions

14.19 Define the half-life of a reaction. Write the equation relating the half-life of a first-order reaction to the rate constant.

14.20 For a first-order reaction, how long will it take for the concentration of reactant to fall to one-eighth its original value? Express your answer in terms of the half-life ($t_{\frac{1}{2}}$) and in terms of the rate constant *k*.

Problems

14.21 What is the half-life of a compound if 75 percent of a given sample of the compound decomposes in 60 min? Assume first-order kinetics.

14.22 The thermal decomposition of phosphine (PH_3) into phosphorus and molecular hydrogen is a first-order reaction:

$$4PH_3(g) \longrightarrow P_4(g) + 6H_2(g)$$

The half-life of the reaction is 35.0 s at 680°C. Calculate (a) the first-order rate constant for the reaction and (b) the time required for 95 percent of the phosphine to decompose.

14.23 The rate constant for the second-order reaction

$$2NOBr(g) \longrightarrow 2NO(g) + Br_2(g)$$

is $0.80/M \cdot s$ at 10°C. (a) Starting with a concentration of 0.086 *M*, calculate the concentration of NOBr after 22 s. (b) Calculate the half-lives when $[NOBr]_0 = 0.072\ M$ and $[NOBr]_0 = 0.054\ M$.

14.24 The rate constant for the second-order reaction

$$2NO_2(g) \longrightarrow 2NO(g) + O_2(g)$$

is $0.54/M \cdot s$ at 300°C. (a) How long (in seconds) would it take for the concentration of NO_2 to decrease from 0.62 *M* to 0.28 *M*? (b) Calculate the half-lives at these two concentrations.

14.25 Consider the first-order reaction $A \longrightarrow B$ shown here. (a) What is the rate constant of the reaction? (b) How many A (yellow) and B (blue) molecules are present at $t = 20$ s and 30 s.

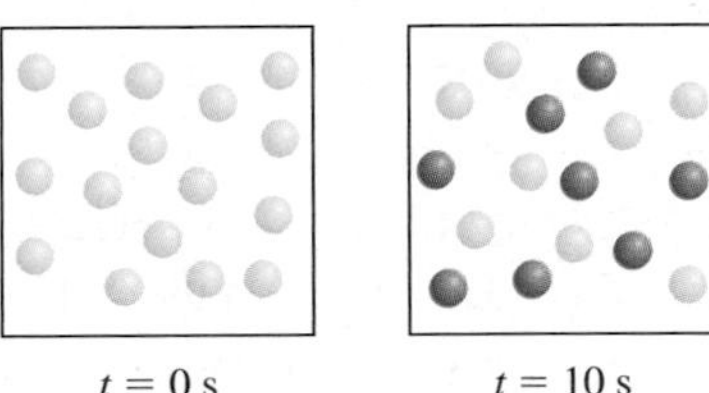

14.26 The reaction $X \longrightarrow Y$ shown here follows first-order kinetics. Initially different amounts of X molecules are placed in three equal-volume containers at the same temperature. (a) What are the relative rates of the reaction in these three containers? (b) How would the relative rates be affected if the volume of each container were doubled? (c) What are the relative half-lives of the reactions in (i) to (iii)?

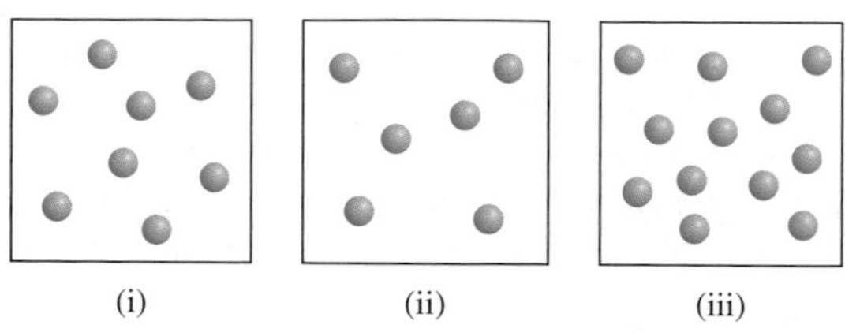

Activation Energy

Review Questions

14.27 Define activation energy. What role does activation energy play in chemical kinetics?

14.28 Write the Arrhenius equation and define all terms.

14.29 Use the Arrhenius equation to show why the rate constant of a reaction (a) decreases with increasing activation energy and (b) increases with increasing temperature.

14.30 As we know, methane burns readily in oxygen in a highly exothermic reaction. Yet a mixture of methane and oxygen gas can be kept indefinitely without any apparent change. Explain.

14.31 Sketch a potential-energy-versus-reaction-progress plot for the following reactions:

(a) $S(s) + O_2(g) \longrightarrow SO_2(g)$
$\Delta H° = -296.06$ kJ/mol

(b) $Cl_2(g) \longrightarrow Cl(g) + Cl(g)$
$\Delta H° = 242.7$ kJ/mol

14.32 The reaction $H + H_2 \longrightarrow H_2 + H$ has been studied for many years. Sketch a potential-energy-versus-reaction-progress diagram for this reaction.

Problems

14.33 Variation of the rate constant with temperature for the first-order reaction

$$2N_2O_5(g) \longrightarrow 2N_2O_4(g) + O_2(g)$$

is given in the following table. Determine graphically the activation energy for the reaction.

T (K)	*k* (s^{-1})
273	7.87×10^3
298	3.46×10^5
318	4.98×10^6
338	4.87×10^7

14.34 Given the same concentrations, the reaction

$$CO(g) + Cl_2(g) \longrightarrow COCl_2(g)$$

at 250°C is 1.50×10^3 times as fast as the same reaction at 150°C. Calculate the energy of activation for this reaction. Assume that the frequency factor is constant.

14.35 For the reaction

$$NO(g) + O_3(g) \longrightarrow NO_2(g) + O_2(g)$$

the frequency factor *A* is $8.7 \times 10^{12}\ s^{-1}$ and the activation energy is 63 kJ/mol. What is the rate constant for the reaction at 75°C?

14.36 The rate constant of a first-order reaction is $4.60 \times 10^{-4}\ s^{-1}$ at 350°C. If the activation energy is 104 kJ/mol, calculate the temperature at which its rate constant is $8.80 \times 10^{-4}\ s^{-1}$.

14.37 The rate constants of some reactions double with every 10-degree rise in temperature. Assume a reaction takes place at 295 K and 305 K. What must the activation energy be for the rate constant to double as described?

14.38 The rate at which tree crickets chirp is 2.0×10^2 per minute at 27°C but only 39.6 per minute at 5°C. From these data, calculate the "energy of activation" for the chirping process. (*Hint:* The ratio of rates is equal to the ratio of rate constants.)

14.39 The diagram here describes the initial state of the reaction $A_2 + B_2 \longrightarrow 2AB$.

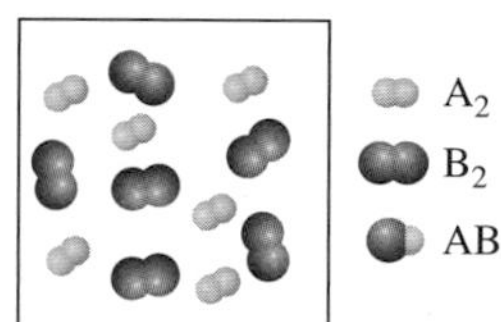

Suppose the reaction is carried out at two temperatures as shown below. Which picture represents the result at the higher temperature? (The reaction proceeds for the same amount of time at both temperatures.)

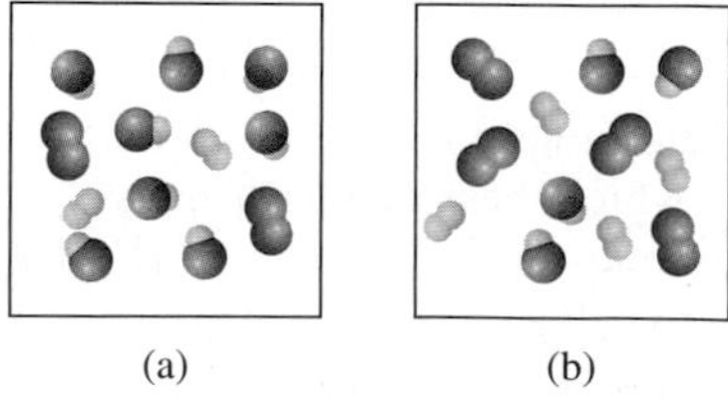

Reaction Mechanisms

Review Questions

14.40 What do we mean by the mechanism of a reaction? What is an elementary step?

14.41 Classify each of the following elementary steps as unimolecular, bimolecular, or termolecular.

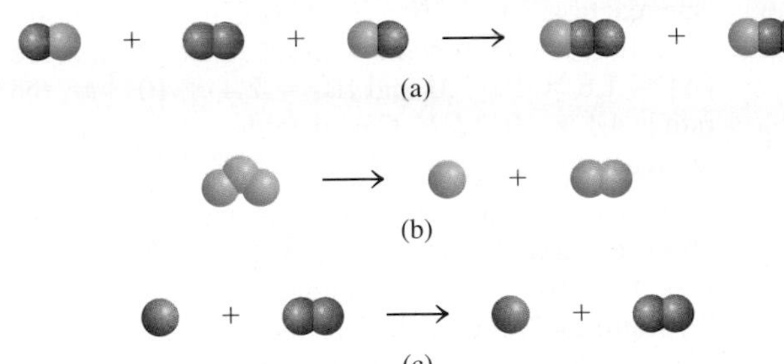

14.42 Reactions can be classified as unimolecular, bimolecular, and so on. Why are there no zero-molecular reactions?

14.43 Explain why termolecular reactions are rare.

14.44 What is the rate-determining step of a reaction? Give an everyday analogy to illustrate the meaning of the term "rate determining."

14.45 The equation for the combustion of ethane (C_2H_6) is

$$2C_2H_6 + 7O_2 \longrightarrow 4CO_2 + 6H_2O$$

Explain why it is unlikely that this equation also represents the elementary step for the reaction.

14.46 Which of these species cannot be isolated in a reaction: activated complex, product, intermediate?

Problems

14.47 The rate law for the reaction

$$2NO(g) + Cl_2(g) \longrightarrow 2NOCl(g)$$

is given by rate $= k[NO][Cl_2]$. (a) What is the order of the reaction? (b) A mechanism involving these steps has been proposed for the reaction

$$NO(g) + Cl_2(g) \longrightarrow NOCl_2(g)$$
$$NOCl_2(g) + NO(g) \longrightarrow 2NOCl(g)$$

If this mechanism is correct, what does it imply about the relative rates of these two steps?

14.48 For the reaction $X_2 + Y + Z \longrightarrow XY + XZ$ it is found that doubling the concentration of X_2 doubles the reaction rate, tripling the concentration of Y triples the rate, and doubling the concentration of Z has no effect. (a) What is the rate law for this reaction? (b) Why is it that the change in the concentration of Z has no effect on the rate? (c) Suggest a mechanism for the reaction that is consistent with the rate law.

Catalysis

Review Questions

14.49 How does a catalyst increase the rate of a reaction?

14.50 What are the characteristics of a catalyst?

14.51 A certain reaction is known to proceed slowly at room temperature. Is it possible to make the reaction proceed at a faster rate without changing the temperature?

14.52 Distinguish between homogeneous catalysis and heterogeneous catalysis. Describe some important industrial processes that utilize heterogeneous catalysis.

14.53 Are enzyme-catalyzed reactions examples of homogeneous or heterogeneous catalysis?

14.54 The concentrations of enzymes in cells are usually quite small. What is the biological significance of this fact?

Problems

14.55 Most reactions, including enzyme-catalyzed reactions, proceed faster at higher temperatures. However, for a given enzyme, the rate drops off abruptly at a certain temperature. Account for this behavior.

14.56 Consider this mechanism for the enzyme-catalyzed reaction

$$E + S \underset{k_{-1}}{\overset{k_1}{\rightleftharpoons}} ES \qquad \text{(fast equilibrium)}$$
$$ES \xrightarrow{k_2} E + P \qquad \text{(slow)}$$

Derive an expression for the rate law of the reaction in terms of the concentrations of E and S. (*Hint:* To solve for [ES], make use of the fact that, at equilibrium, the rate of the forward reaction is equal to the rate of the reverse reaction.)

Additional Problems

14.57 Suggest experimental means by which the rates of the following reactions could be followed:

(a) $CaCO_3(s) \longrightarrow CaO(s) + CO_2(g)$
(b) $Cl_2(g) + 2Br^-(aq) \longrightarrow Br_2(aq) + 2Cl^-(aq)$
(c) $C_2H_6(g) \longrightarrow C_2H_4(g) + H_2(g)$

14.58 The following diagrams represent the progress of the reaction $A \longrightarrow B$, where the red spheres represent A molecules and the green spheres represent B molecules. Calculate the rate constant of the reaction.

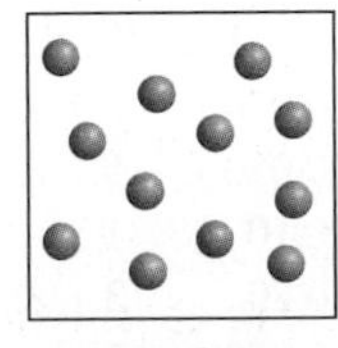
$t = 0$ s

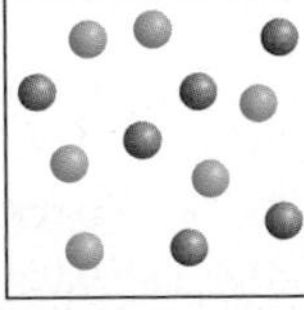
$t = 20$ s

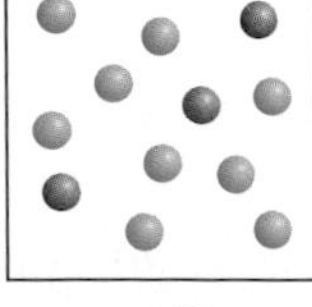
$t = 40$ s

14.59 The following diagrams show the progress of the reaction $2A \longrightarrow A_2$. Determine whether the reaction is first order or second order and calculate the rate constant.

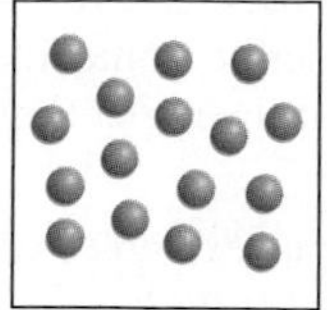
$t = 0$ min

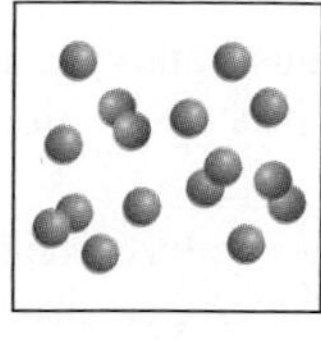
$t = 15$ min

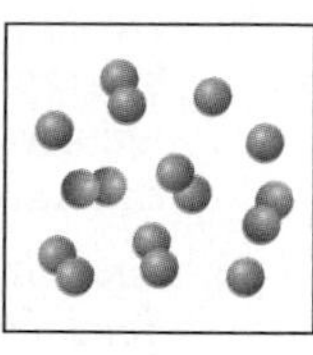
$t = 30$ min

14.60 In a certain industrial process using a heterogeneous catalyst, the volume of the catalyst (in the shape of a sphere) is 10.0 cm^3. Calculate the surface area of the catalyst. If the sphere is broken down into eight spheres, each of which has a volume of 1.25 cm^3, what is the total surface area of the spheres? Which of the two geometric configurations of the catalyst is more effective? Explain. (The surface area of a sphere is $4\pi r^2$, in which r is the radius of the sphere.)

14.61 When methyl phosphate is heated in acid solution, it reacts with water:

$$CH_3OPO_3H_2 + H_2O \longrightarrow CH_3OH + H_3PO_4$$

If the reaction is carried out in water enriched with ^{18}O, the oxygen-18 isotope is found in the phosphoric acid product but not in the methanol. What does this tell us about the bond-breaking scheme in the reaction?

14.62 The rate of the reaction

$$CH_3COOC_2H_5(aq) + H_2O(l) \longrightarrow CH_3COOH(aq) + C_2H_5OH(aq)$$

shows first-order characteristics—that is, rate = $k[CH_3COOC_2H_5]$—even though this is a second-order reaction (first order in $CH_3COOC_2H_5$ and first order in H_2O). Explain.

14.63 Explain why most metals used in catalysis are transition metals.

14.64 The bromination of acetone is acid-catalyzed:

$$CH_3COCH_3 + Br_2 \xrightarrow[\text{catalyst}]{H^+} CH_3COCH_2Br + H^+ + Br^-$$

The rate of disappearance of bromine was measured for several different concentrations of acetone, bromine, and H^+ ions at a certain temperature:

	$[CH_3COCH_3]$	$[Br_2]$	$[H^+]$	Rate of Disappearance of Br_2 (M/s)
(a)	0.30	0.050	0.050	5.7×10^{-5}
(b)	0.30	0.10	0.050	5.7×10^{-5}
(c)	0.30	0.050	0.10	1.2×10^{-4}
(d)	0.40	0.050	0.20	3.1×10^{-4}
(e)	0.40	0.050	0.050	7.6×10^{-5}

(a) What is the rate law for the reaction? (b) Determine the rate constant.

14.65 The reaction $2A + 3B \longrightarrow C$ is first order with respect to A and B. When the initial concentrations are $[A] = 1.6 \times 10^{-2}$ M and $[B] = 2.4 \times 10^{-3}$ M, the rate is 4.1×10^{-4} M/s. Calculate the rate constant of the reaction.

14.66 The decomposition of N_2O to N_2 and O_2 is a first-order reaction. At 730°C the half-life of the reaction is 3.58×10^3 min. If the initial pressure of N_2O is 2.10 atm at 730°C, calculate the total gas pressure after one half-life. Assume that the volume remains constant.

14.67 The reaction $S_2O_8^{2-} + 2I^- \longrightarrow 2SO_4^{2-} + I_2$ proceeds slowly in aqueous solution, but it can be catalyzed by the Fe^{3+} ion. Given that Fe^{3+} can oxidize I^- and Fe^{2+} can reduce $S_2O_8^{2-}$, write a plausible two-step mechanism for this reaction. Explain why the uncatalyzed reaction is slow.

14.68 What are the units of the rate constant for a third-order reaction?

14.69 Consider the zero-order reaction $A \longrightarrow B$. Sketch the following plots: (a) rate versus [A] and (b) [A] versus t.

14.70 A flask contains a mixture of compounds A and B. Both compounds decompose by first-order kinetics. The half-lives are 50.0 min for A and 18.0 min for B. If the concentrations of A and B are equal initially, how long will it take for the concentration of A to be four times that of B?

14.71 Referring to the decomposition of N_2O_5 in Problem 14.33, explain how you would measure the partial pressure of N_2O_5 as a function of time.

14.72 The rate law for the reaction $2NO_2(g) \longrightarrow N_2O_4(g)$ is rate = $k[NO_2]^2$. Which of these changes will change the value of k? (a) The pressure of NO_2 is doubled. (b) The reaction is run in an organic solvent. (c) The volume of the container is doubled. (d) The temperature is decreased. (e) A catalyst is added to the container.

14.73 The reaction of G_2 with E_2 to form 2EG is exothermic, and the reaction of G_2 with X_2 to form 2XG is endothermic. The activation energy of the exothermic reaction is greater than that of the endothermic reaction. Sketch the potential energy profile diagrams for these two reactions on the same graph.

14.74 In the nuclear industry, workers use a rule of thumb that the radioactivity from any sample will be relatively harmless after 10 half-lives. Calculate the fraction of a radioactive sample that remains after this time. (*Hint:* Radioactive decays obey first-order kinetics.)

14.75 Briefly comment on the effect of a catalyst on each of the following: (a) activation energy, (b) reaction mechanism, (c) enthalpy of reaction, (d) rate of forward step, (e) rate of reverse step.

14.76 A quantity of 6 g of granulated Zn is added to a solution of 2 *M* HCl in a beaker at room temperature. Hydrogen gas is generated. For each of the following changes (at constant volume of the acid) state whether the rate of hydrogen gas evolution will be increased, decreased, or unchanged: (a) 6 g of powdered Zn is used; (b) 4 g of granulated Zn is used; (c) 2 *M* acetic acid is used instead of 2 *M* HCl; (d) temperature is raised to 40°C.

14.77 These data were collected for the reaction between hydrogen and nitric oxide at 700°C:

$$2H_2(g) + 2NO(g) \longrightarrow 2H_2O(g) + N_2(g)$$

Experiment	$[H_2]$	[NO]	Initial Rate (*M*/s)
1	0.010	0.025	2.4×10^{-6}
2	0.0050	0.025	1.2×10^{-6}
3	0.010	0.0125	0.60×10^{-6}

(a) Determine the order of the reaction. (b) Calculate the rate constant. (c) Suggest a plausible mechanism that is consistent with the rate law. (*Hint:* Assume the oxygen atom is the intermediate.)

14.78 A certain first-order reaction is 35.5 percent complete in 4.90 min at 25°C. What is its rate constant?

14.79 The decomposition of dinitrogen pentoxide has been studied in carbon tetrachloride solvent (CCl_4) at a certain temperature:

$$2N_2O_5 \longrightarrow 4NO_2 + O_2$$

$[N_2O_5]$ (*M*)	Initial Rate (*M*/s)
0.92	0.95×10^{-5}
1.23	1.20×10^{-5}
1.79	1.93×10^{-5}
2.00	2.10×10^{-5}
2.21	2.26×10^{-5}

Determine graphically the rate law for the reaction and calculate the rate constant.

14.80 The thermal decomposition of N_2O_5 obeys first-order kinetics. At 45°C, a plot of ln $[N_2O_5]$ versus t gives a slope of -6.18×10^{-4} min^{-1}. What is the half-life of the reaction?

14.81 When a mixture of methane and bromine is exposed to light, the following reaction occurs slowly:

$$CH_4(g) + Br_2(g) \longrightarrow CH_3Br(g) + HBr(g)$$

Suggest a reasonable mechanism for this reaction. (*Hint:* Bromine vapor is deep red; methane is colorless.)

14.82 Consider this elementary step:

$$X + 2Y \longrightarrow XY_2$$

(a) Write a rate law for this reaction. (b) If the initial rate of formation of XY_2 is 3.8×10^{-3} *M*/s and the initial concentrations of X and Y are 0.26 *M* and 0.88 *M*, what is the rate constant of the reaction?

14.83 Consider the reaction

$$C_2H_5I(aq) + H_2O(l) \longrightarrow C_2H_5OH(aq) + H^+(aq) + I^-(aq)$$

How could you follow the progress of the reaction by measuring the electrical conductance of the solution?

14.84 A compound X undergoes two *simultaneous* first-order reactions as follows: X $\longrightarrow$ Y with rate constant k_1 and X $\longrightarrow$ Z with rate constant k_2. The ratio of k_1/k_2 at 40°C is 8.0. What is the ratio at 300°C? Assume that the frequency factor of the two reactions is the same.

14.85 In recent years ozone in the stratosphere has been depleted at an alarmingly fast rate by chlorofluorocarbons (CFCs). A CFC molecule such as $CFCl_3$ is first decomposed by UV radiation:

$$CFCl_3 \longrightarrow CFCl_2 + Cl$$

The chlorine radical then reacts with ozone as follows:

$$Cl + O_3 \longrightarrow ClO + O_2$$
$$ClO + O \longrightarrow Cl + O_2$$

(a) Write the overall reaction for the last two steps. (b) What are the roles of Cl and ClO? (c) Why is the fluorine radical not important in this mechanism? (d) One suggestion to reduce the concentration of chlorine radicals is to add hydrocarbons such as ethane (C_2H_6) to the stratosphere. How will this work?

14.86 Consider a car fitted with a catalytic converter. The first 10 min or so after it is started are the most polluting. Why?

14.87 Strontium-90, a radioactive isotope, is a major product of an atomic bomb explosion. It has a half-life of 28.1 yr. (a) Calculate the first-order rate constant for the nuclear decay. (b) Calculate the fraction of ^{90}Sr that remains after 10 half-lives. (c) Calculate the number of years required for 99.0 percent of ^{90}Sr to disappear.

14.88 The following mechanism has been proposed for the reaction described in Problem 14.64:

$$CH_3{-}C({=}O){-}CH_3 + H_3O^+ \rightleftharpoons CH_3{-}C({=}{}^+OH){-}CH_3 + H_2O \quad \text{(fast equilibrium)}$$

$$CH_3{-}C({=}{}^+OH){-}CH_3 + H_2O \longrightarrow CH_3{-}C(OH){=}CH_2 + H_3O^+ \quad \text{(slow)}$$

$$CH_3{-}C(OH){=}CH_2 + Br_2 \longrightarrow CH_3{-}C({=}O){-}CH_2Br + HBr \quad \text{(fast)}$$

Show that the rate law deduced from the mechanism is consistent with that shown in (a) of Problem 14.64.

14.89 The integrated rate law for the zero-order reaction $A \longrightarrow B$ is $[A]_t = [A]_0 - kt$. (a) Sketch the following plots: (i) rate versus $[A]_t$ and (ii) $[A]_t$ versus t. (b) Derive an expression for the half-life of the reaction. (c) Calculate the time in half-lives when the integrated rate law is no longer valid, that is, when $[A]_t = 0$.

14.90 Strictly speaking, the rate law derived for the reaction in Problem 14.77 applies only to certain concentrations of H_2. The general rate law for the reaction takes the form

$$\text{rate} = \frac{k_1[NO]^2[H_2]}{1 + k_2[H_2]}$$

in which k_1 and k_2 are constants. Derive rate law expressions under the conditions of very high and very low hydrogen concentrations. Does the result from Problem 14.77 agree with one of the rate expressions here?

14.91 (a) What can you deduce about the activation energy of a reaction if its rate constant changes significantly with a small change in temperature? (b) If a bimolecular reaction occurs every time an A and a B molecule collide, what can you say about the orientation factor and activation energy of the reaction?

14.92 The rate law for this reaction

$$CO(g) + NO_2(g) \longrightarrow CO_2(g) + NO(g)$$

is rate $= k[NO_2]^2$. Suggest a plausible mechanism for the reaction, given that the unstable species NO_3 is an intermediate.

14.93 Radioactive plutonium-239 ($t_{\frac{1}{2}} = 2.44 \times 10^5$ yr) is used in nuclear reactors and atomic bombs. If there are 5.0×10^2 g of the isotope in a small atomic bomb, how long will it take for the substance to decay to 1.0×10^2 g, too small an amount for an effective bomb? (*Hint:* Radioactive decays follow first-order kinetics.)

14.94 Many reactions involving heterogeneous catalysts are zero order; that is, rate $= k$. An example is the decomposition of phosphine (PH_3) over tungsten (W):

$$4PH_3(g) \longrightarrow P_4(g) + 6H_2(g)$$

It is found that the reaction is independent of $[PH_3]$ as long as phosphine's pressure is sufficiently high (≥ 1 atm). Explain.

14.95 Thallium(I) is oxidized by cerium(IV) as follows:

$$Tl^+ + 2Ce^{4+} \longrightarrow Tl^{3+} + 2Ce^{3+}$$

The elementary steps, in the presence of Mn(II), are as follows:

$$Ce^{4+} + Mn^{2+} \longrightarrow Ce^{3+} + Mn^{3+}$$
$$Ce^{4+} + Mn^{3+} \longrightarrow Ce^{3+} + Mn^{4+}$$
$$Tl^{+} + Mn^{4+} \longrightarrow Tl^{3+} + Mn^{2+}$$

(a) Identify the catalyst, intermediates, and the rate-determining step if the rate law is given by rate $= k[Ce^{4+}][Mn^{2+}]$. (b) Explain why the reaction is slow without the catalyst. (c) Classify the type of catalysis (homogeneous or heterogeneous).

14.96 Consider the following elementary steps for a consecutive reaction

$$A \xrightarrow{k_1} B \xrightarrow{k_2} C$$

(a) Write an expression for the rate of change of B. (b) Derive an expression for the concentration of B under steady-state conditions; that is, when B is decomposing to C at the same rate as it is formed from A.

14.97 For gas-phase reactions, we can replace the concentration terms in Equation (14.3) with the pressures of the gaseous reactant. (a) Derive the equation

$$\ln \frac{P_t}{P_0} = -kt$$

where P_t and P_0 are the pressures at $t = t$ and $t = 0$, respectively. (b) Consider the decomposition of azomethane

$$CH_3{-}N{=}N{-}CH_3(g) \longrightarrow N_2(g) + C_2H_6(g)$$

The data obtained at 300°C are shown in the following table:

Time (s)	Partial Pressure of Azomethane (mmHg)
0	284
100	220
150	193
200	170
250	150
300	132

Are these values consistent with first-order kinetics? If so, determine the rate constant by plotting the data as shown in Figure 14.7(b). (c) Determine the rate constant by the half-life method.

14.98 The hydrolysis of methyl acetate

$$\underset{\text{methyl acetate}}{CH_3{-}\overset{\overset{\displaystyle O}{\|}}{C}{-}O{-}CH_3} + H_2O \longrightarrow \underset{\text{acetic acid}}{CH_3{-}\overset{\overset{\displaystyle O}{\|}}{C}{-}OH} + \underset{\text{methanol}}{CH_3OH}$$

involves the breaking of a C—O bond. The two possibilities are

$$\underset{\text{(a)}}{CH_3-\overset{\overset{\displaystyle O}{\|}}{C}+O-CH_3} \qquad \underset{\text{(b)}}{CH_3-\overset{\overset{\displaystyle O}{\|}}{C}-O+CH_3}$$

Suggest an experiment that would enable you to distinguish between these two possibilities.

14.99 The following gas-phase reaction was studied at 290°C by observing the change in pressure as a function of time in a constant-volume vessel:

$$ClCO_2CCl_3(g) \longrightarrow 2COCl_2(g)$$

Determine the order of the reaction and the rate constant based on the following data:

Time (s)	*P* (mmHg)
0	15.76
181	18.88
513	22.79
1164	27.08

where *P* is the total pressure.

14.100 Consider the potential energy profiles for the following three reactions (from left to right). (1) Rank the rates (slowest to fastest) of the reactions. (2) Calculate ΔH for each reaction and determine which reaction(s) are exothermic and which reaction(s) are endothermic. Assume the reactions have roughly the same frequency factors.

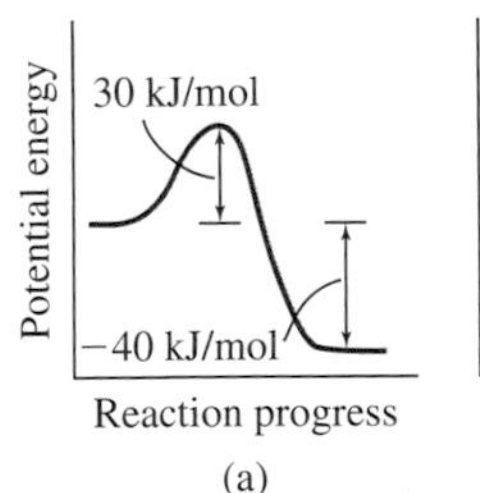

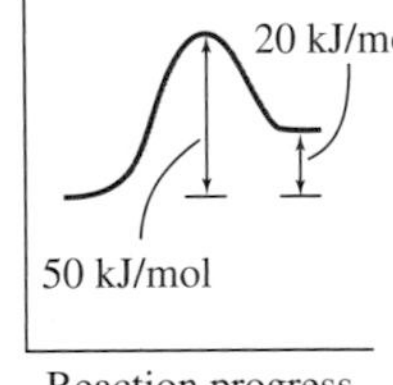

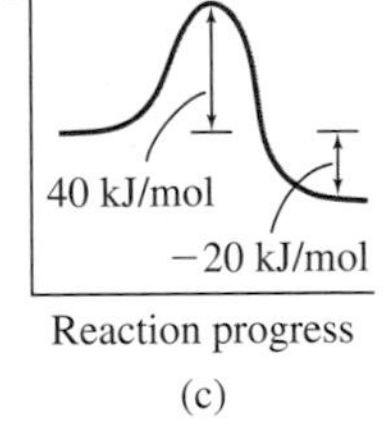

14.101 Consider the following potential energy profile for the A ⟶ D reaction. (a) How many elementary steps are there? (b) How many intermediates are formed? (c) Which step is rate determining? (d) Is the overall reaction exothermic or endothermic?

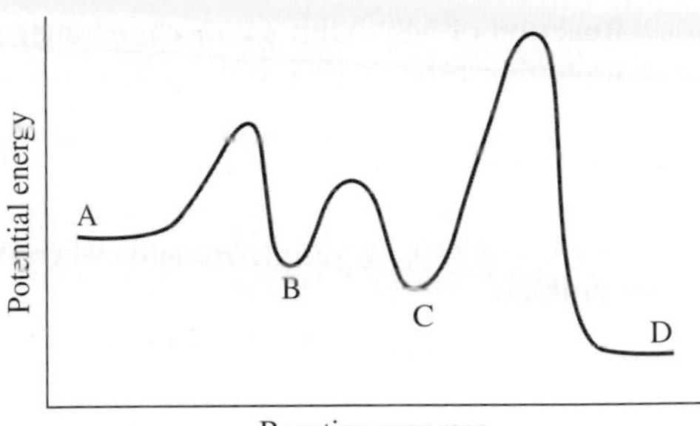

14.102 Hydrogen and iodine monochloride react as follows:

$$H_2(g) + 2ICl(g) \longrightarrow 2HCl(g) + I_2(g)$$

The rate law for the reaction is rate $= k[H_2][ICl]$. Suggest a possible mechanism for the reaction.

14.103 The activation energy for the decomposition of hydrogen peroxide

$$2H_2O_2(aq) \longrightarrow 2H_2O(l) + O_2(g)$$

is 42 kJ/mol, whereas when the reaction is catalyzed by the enzyme catalase, it is 7.0 kJ/mol. Calculate the temperature that would cause the nonenzymatic catalysis to proceed as rapidly as the enzyme-catalyzed decomposition at 20°C. Assume the frequency factor A to be the same in both cases.

14.104 To carry out metabolism, oxygen is taken up by hemoglobin (Hb) to form oxyhemoglobin (HbO_2) according to the simplified equation

$$Hb(aq) + O_2(aq) \xrightarrow{k} HbO_2(aq)$$

where the second-order rate constant is $2.1 \times 10^6/M \cdot s$ at 37°C. (The reaction is first order in Hb and O_2.) For an average adult, the concentrations of Hb and O_2 in the blood at the lungs are $8.0 \times 10^{-6}\ M$ and $1.5 \times 10^{-6}\ M$, respectively. (a) Calculate the rate of formation of HbO_2. (b) Calculate the rate of consumption of O_2. (c) The rate of formation of HbO_2 increases to $1.4 \times 10^{-4}\ M/s$ during exercise to meet the demand of increased metabolism rate. Assuming the Hb concentration to remain the same, what must be the oxygen concentration to sustain this rate of HbO_2 formation?

Special Problems

14.105 Polyethylene is used in many items such as water pipes, bottles, electrical insulation, toys, and mailer envelopes. It is a *polymer,* a molecule with a very high molar mass made by joining many ethylene molecules (the basic unit is called a monomer) together (see p. 377). The initiation step is

$$R_2 \xrightarrow{k_1} 2R\cdot \qquad \text{initiation}$$

The R · species (called a radical) reacts with an ethylene molecule (M) to generate another radical

$$R\cdot + M \longrightarrow M_1\cdot$$

Reaction of $M_1 \cdot$ with another monomer leads to the growth or propagation of the polymer chain:

$$M_1 \cdot + M \xrightarrow{k_p} M_2 \cdot \qquad \text{propagation}$$

This step can be repeated with hundreds of monomer units. The propagation terminates when two radicals combine

$$M' \cdot + M'' \cdot \xrightarrow{k_t} M'\text{—}M'' \qquad \text{termination}$$

(a) The initiator used in the polymerization of ethylene is benzoyl peroxide [$(C_6H_5COO)_2$]:

$$(C_6H_5COO)_2 \longrightarrow 2C_6H_5COO \cdot$$

This is a first-order reaction. The half-life of benzoyl peroxide at 100°C is 19.8 min. (a) Calculate the rate constant (in min^{-1}) of the reaction. (b) If the half-life of benzoyl peroxide is 7.30 h or 438 min, at 70°C, what is the activation energy (in kJ/mol) for the decomposition of benzoyl peroxide? (c) Write the rate laws for the elementary steps in the above polymerization process and identify the reactant, product, and intermediates. (d) What condition would favor the growth of long high-molar-mass polyethylenes?

14.106 Ethanol is a toxic substance that, when consumed in excess, can impair respiratory and cardiac functions by interference with the neurotransmitters of the nervous system. In the human body, ethanol is metabolized by the enzyme alcohol dehydrogenase to acetaldehyde, which causes "hangovers." (a) Based on your knowledge of enzyme kinetics, explain why binge drinking (that is, consuming too much alcohol too fast) can prove fatal. (b) Methanol is even more toxic than ethanol. It is also metabolized by alcohol dehydrogenase, and the product, formaldehyde, can cause blindness or death. An antidote to methanol poisoning is ethanol. Explain how this procedure works.

14.107 At a certain elevated temperature, ammonia decomposes on the surface of tungsten metal as follows:

$$2NH_3 \longrightarrow N_2 + 3H_2$$

From the following plot of the rate of the reaction versus the pressure of NH_3, describe the mechanism of the reaction.

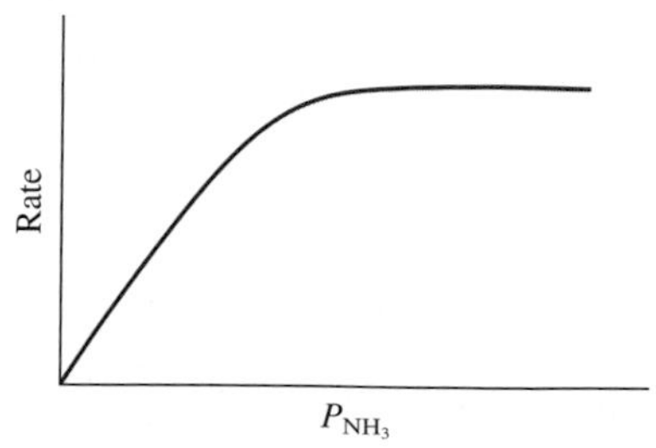

14.108 The following expression shows the dependence of the half-life of a reaction ($t_{\frac{1}{2}}$) on the initial reactant concentration $[A]_0$:

$$t_{\frac{1}{2}} \propto \frac{1}{[A]_0^{n-1}}$$

where n is the order of the reaction. Verify this dependence for zero-, first-, and second-order reactions.

14.109 The rate constant for the gaseous reaction

$$H_2(g) + I_2(g) \longrightarrow 2HI(g)$$

is $2.42 \times 10^{-2}/M \cdot s$ at 400°C. Initially an equimolar sample of H_2 and I_2 is placed in a vessel at 400°C and the total pressure is 1658 mmHg. (a) What is the initial rate (M/min) of formation of HI? (b) What are the rate of formation of HI and the concentration of HI (in molarity) after 10.0 min?

14.110 When the concentration of A in the reaction A $\longrightarrow$ B was changed from 1.20 M to 0.60 M, the half-life increased from 2.0 min to 4.0 min at 25°C. Calculate the order of the reaction and the rate constant. (*Hint:* Use the equation in Problem 14.108.)

14.111 The activation energy for the reaction

$$N_2O(g) \longrightarrow N_2(g) + O(g)$$

is 2.4×10^2 kJ/mol at 600 K. Calculate the percentage of the increase in rate from 600 K to 606 K. Comment on your results.

14.112 The rate of a reaction was followed by the absorption of light by the reactants and products as a function of wavelengths (λ_1, λ_2, λ_3) as time progresses. Which of the following mechanisms is consistent with the experimental data?

(a) A $\longrightarrow$ B, A $\longrightarrow$ C
(b) A $\longrightarrow$ B + C
(c) A $\longrightarrow$ B, B $\longrightarrow$ C + D
(d) A $\longrightarrow$ B, B $\longrightarrow$ C

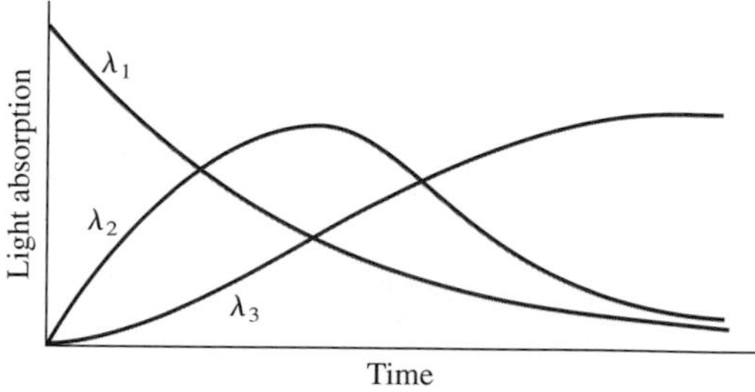

14.113 A gas mixture containing CH_3 fragments, C_2H_6 molecules, and an inert gas (He) was prepared at 600 K with a total pressure of 5.42 atm. The elementary reaction

$$CH_3 + C_2H_6 \longrightarrow CH_4 + C_2H_5$$

has a second-order rate constant of $3.0 \times 10^4/M \cdot s$. Given that the mole fractions of CH_3 and C_2H_6 are 0.00093 and 0.00077, respectively, calculate the initial rate of the reaction at this temperature.

14.114 To prevent brain damage, a drastic medical procedure is to lower the body temperature of someone who has suffered cardiac arrest. What is the physiochemical basis for this treatment?

14.115 The activation energy (E_a) for the reaction

$$2N_2O(g) \longrightarrow 2N_2(g) + O_2(g) \quad \Delta H° = -164 \text{ kJ/mol}$$

is 240 kJ/mol. What is E_a for the reverse reaction?

Answers to Practice Exercises

14.1 rate $= -\frac{\Delta[CH_4]}{\Delta t} = -\frac{1}{2}\frac{\Delta[O_2]}{\Delta t} = \frac{\Delta[CO_2]}{\Delta t} = \frac{1}{2}\frac{\Delta[H_2O]}{\Delta t}$. **14.2** (a) 0.013 M/s, (b) -0.052 M/s. **14.3** rate $= k[S_2O_8^{2-}][I^-]$; $k = 8.1 \times 10^{-2}/M \cdot s$. **14.4** 66 s. **14.5** 1.2×10^3 s. **14.6** (a) 3.2 min, (b) 2.1 min. **14.7** 240 kJ/mol. **14.8** $3.13 \times 10^{-9}\ s^{-1}$. **14.9** (a) $NO_2 + CO \longrightarrow NO + CO_2$, (b) NO_3, (c) the first step is rate-determining.

Chemical Equilibrium

The brown haze of Los Angeles is caused by high levels of nitrogen dioxide (NO_2), which participates in a chemical equilibrium with nitric oxide (NO) and ozone (O_3), among other air pollutants.

Chapter Outline

Student Interactive Activities

Animations
Chemical Equilibrium (15.1)
Le Châtelier's Principle (15.4)

Electronic Homework
Example Practice Problems
End of Chapter Problems

Essential Concepts

Chemical Equilibrium Chemical equilibrium describes the state in which the rates of forward and reverse reactions are equal and the concentrations of the reactants and products remain unchanged with time. This state of dynamic equilibrium is characterized by an equilibrium constant. Depending on the nature of reacting species, the equilibrium constant can be expressed in terms of molarities (for solutions) or partial pressures (for gases). The equilibrium constant provides information about the net direction of a reversible reaction and the concentrations of the equilibrium mixture.

Factors That Affect Chemical Equilibrium Changes in concentration can affect the position of an equilibrium state—that is, the relative amounts of reactants and products. Changes in pressure and volume may have the same effect for gaseous systems at equilibrium. Only a change in temperature can alter the value of equilibrium constant. A catalyst can establish the equilibrium state faster by speeding the forward and reverse reactions, but it can change neither the equilibrium position nor the equilibrium constant.

15.1 The Concept of Equilibrium

Animation: Chemical Equilibrium

Few chemical reactions proceed in only one direction. Most are, at least to some extent, reversible. At the start of a reversible process, the reaction proceeds toward the formation of products. As soon as some product molecules are formed, the reverse process—that is, the formation of reactant molecules from product molecules—begins to take place. *When the rates of the forward and reverse reactions are equal and the concentrations of the reactants and products no longer change with time,* ***chemical equilibrium*** is reached.

Chemical equilibrium is a dynamic process. As such, it can be likened to the movement of skiers at a busy ski resort, where the number of skiers carried up the mountain on the chair lift is equal to the number coming down the slopes. Thus, although there is a constant transfer of skiers, the number of people at the top and the number at the bottom of the slope do not change.

Note that a chemical equilibrium reaction involves different substances as reactants and products. Equilibrium between two phases of the same substance is called ***physical equilibrium*** because *the changes that occur are physical processes.* The vaporization of water in a closed container at a given temperature is an example of physical equilibrium. In this instance, the number of H_2O molecules leaving and the number returning to the liquid phase are equal:

$$H_2O(l) \rightleftharpoons H_2O(g)$$

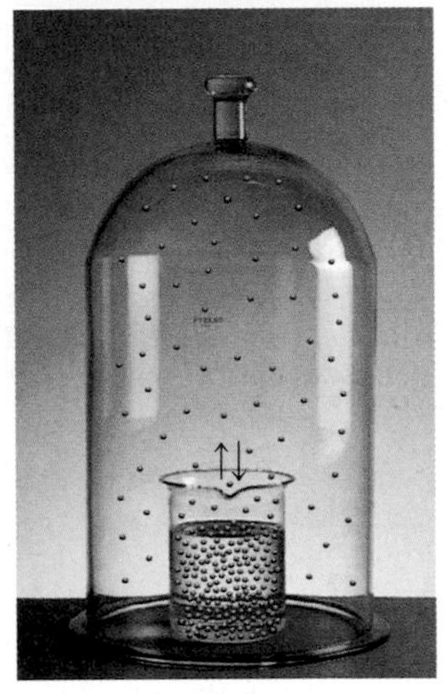

Liquid water in equilibrium with its vapor in a closed system at room temperature.

(Recall from Chapter 4 that the double arrow means that the reaction is reversible.) The study of physical equilibrium yields useful information, such as the equilibrium vapor pressure (see Section 12.6). However, chemists are particularly interested in chemical equilibrium processes, such as the reversible reaction involving nitrogen dioxide (NO_2) and dinitrogen tetroxide (N_2O_4). The progress of the reaction

$$N_2O_4(g) \rightleftharpoons 2NO_2(g)$$

can be monitored easily because N_2O_4 is a colorless gas, whereas NO_2 has a dark-brown color that makes it sometimes visible in polluted air. Suppose that a known amount of N_2O_4 is injected into an evacuated flask. Some brown color appears immediately, indicating the formation of NO_2 molecules. The color intensifies as the dissociation of N_2O_4 continues until eventually equilibrium is reached. Beyond that point, no further change in color is observed. By experiment we find that we can also reach the equilibrium state by starting with pure NO_2 or with a mixture of NO_2 and N_2O_4. In each case, we observe an initial change in color, caused either by the formation of NO_2 (if the color intensifies) or by the depletion of NO_2 (if the color fades), and then the final state in which the color of NO_2 no longer changes. Depending on the temperature of the reacting system and on the initial amounts of NO_2 and N_2O_4, the concentrations of NO_2 and N_2O_4 at equilibrium differ from system to system (Figure 15.1).

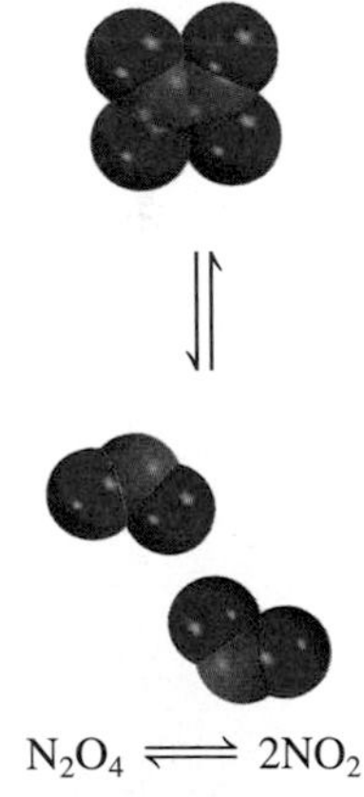

$N_2O_4 \rightleftharpoons 2NO_2$

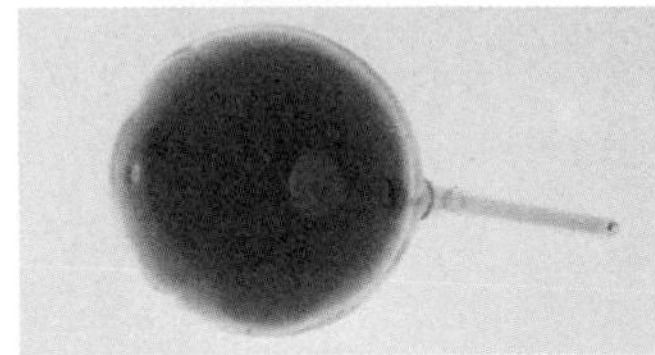

NO_2 and N_2O_4 gases at equilibrium.

The Equilibrium Constant

Table 15.1 shows some experimental data for this reaction at 25°C. The gas concentrations are expressed in molarity, which can be calculated from the number of moles of gases present initially and at equilibrium and the volume of the flask in liters. Note that the equilibrium concentrations of NO_2 and N_2O_4 vary, depending on the starting concentrations. We can look for relationships between $[NO_2]$ and $[N_2O_4]$ present at equilibrium by comparing the ratios of their concentrations. The simplest

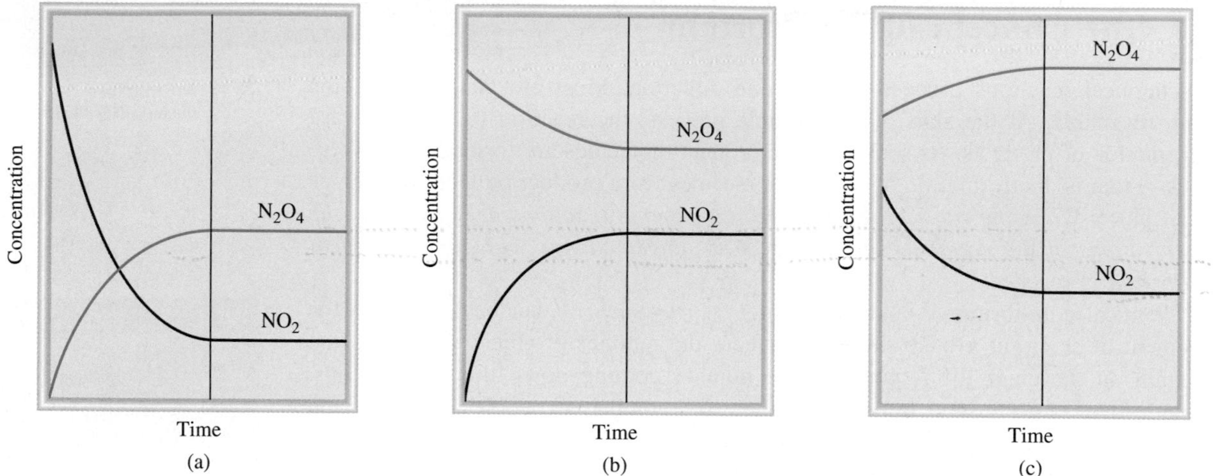

Figure 15.1
Change in the concentrations of NO_2 and N_2O_4 with time, in three situations. (a) Initially only NO_2 is present. (b) Initially only N_2O_4 is present. (c) Initially a mixture of NO_2 and N_2O_4 is present. In each case, equilibrium is established to the right of the vertical line.

ratio, that is, $[NO_2]/[N_2O_4]$, gives scattered values. But if we examine other possible mathematical relationships, we find that the ratio $[NO_2]^2/[N_2O_4]$ at equilibrium gives a nearly constant value that averages 4.63×10^{-3}, regardless of the initial concentrations present:

$$K = \frac{[NO_2]^2}{[N_2O_4]} = 4.63 \times 10^{-3} \tag{15.1}$$

Note that the exponent 2 in $[NO_2]^2$ is the same as the stoichiometric coefficient for NO_2 in the reversible equation. It turns out that for every reversible reaction, there is a specific mathematical ratio between the equilibrium concentrations of products and reactants that yields a constant value.

We can generalize this discussion by considering the following reversible reaction:

$$aA + bB \rightleftharpoons cC + dD$$

Table 15.1 The NO_2–N_2O_4 System at 25°C

Initial Concentrations (M)		Equilibrium Concentrations (M)		Ratio of Concentrations at Equilibrium	
$[NO_2]$	$[N_2O_4]$	$[NO_2]$	$[N_2O_4]$	$\frac{[NO_2]}{[N_2O_4]}$	$\frac{[NO_2]^2}{[N_2O_4]}$
0.000	0.670	0.0547	0.643	0.0851	4.65×10^{-3}
0.0500	0.446	0.0457	0.448	0.102	4.66×10^{-3}
0.0300	0.500	0.0475	0.491	0.0967	4.60×10^{-3}
0.0400	0.600	0.0523	0.594	0.0880	4.60×10^{-3}
0.200	0.000	0.0204	0.0898	0.227	4.63×10^{-3}

in which a, b, c, and d are the stoichiometric coefficients for the reacting species A, B, C, and D. The equilibrium constant for the reaction at a particular temperature is

$$K = \frac{[C]^c[D]^d}{[A]^a[B]^b} \tag{15.2}$$

Equilibrium concentrations must be used in this equation.

Equation (15.2) is the mathematical form of the *law of mass action.* It *relates the concentrations of reactants and products at equilibrium* in terms of a quantity called the ***equilibrium constant.*** The equilibrium constant is defined by a quotient. The numerator is obtained by multiplying together the equilibrium concentrations of the *products,* each raised to a power equal to its stoichiometric coefficient in the balanced equation. The same procedure is applied to the equilibrium concentrations of *reactants* to obtain the denominator. This formulation is based on purely empirical evidence, such as the study of reactions like NO_2–N_2O_4.

In keeping with the convention, we refer to substances on the left of the equilibrium arrows as "reactants" and those on the right as "products."

The equilibrium constant has its origin in thermodynamics, to be discussed in Chapter 18. However, we can gain some insight into K by considering the kinetics of chemical reactions. Let us suppose that this reversible reaction occurs via a mechanism of a single *elementary step* in both the forward and reverse directions:

To review reaction mechanisms, see Section 14.5.

$$A + 2B \underset{k_r}{\overset{k_f}{\rightleftharpoons}} AB_2$$

The forward rate is given by

$$\text{rate}_f = k_f[A][B]^2$$

and the reverse rate is

$$\text{rate}_r = k_r[AB_2]$$

in which k_f and k_r are the rate constants for the forward and reverse directions, respectively. At equilibrium, when no net changes occur, the two rates must be equal:

$$\text{rate}_f = \text{rate}_r$$

or

$$k_f[A][B]^2 = k_r[AB_2]$$

$$\frac{k_f}{k_r} = \frac{[AB_2]}{[A][B]^2}$$

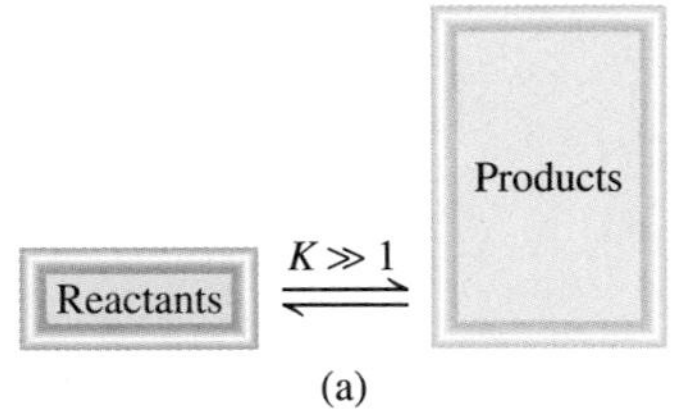

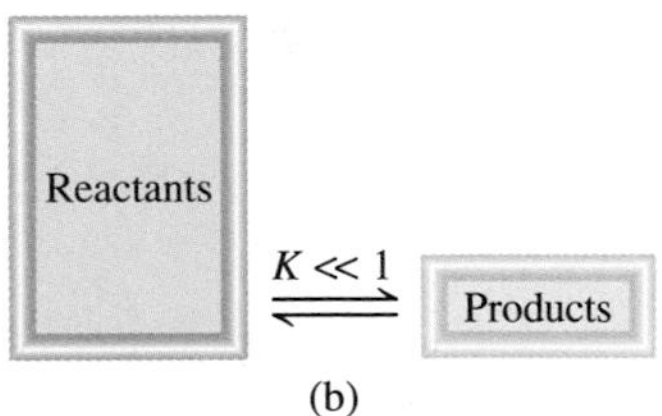

Figure 15.2
(a) At equilibrium, there are more products than reactants, and the equilibrium is said to lie to the right. (b) In the opposite situation, when there are more reactants than products, the equilibrium is said to lie to the left.

Because both k_f and k_r are constants at a given temperature, their ratio is also a constant, which is equal to the equilibrium constant K_c.

$$\frac{k_f}{k_r} = K_c = \frac{[AB_2]}{[A][B]^2}$$

So K_c is a constant regardless of the equilibrium concentrations of the reacting species because it is always equal to k_f/k_r, the quotient of two quantities that are themselves constant at a given temperature. Because rate constants are temperature-dependent [see Equation (14.9)], it follows that the equilibrium constant must also change with temperature.

Finally, we note that if the equilibrium constant is much greater than 1 (that is, $K \gg 1$), the equilibrium will lie to the right of the reaction arrows and favor the products. Conversely, if the equilibrium constant is much smaller than 1 (that is, $K \ll 1$), the equilibrium will lie to the left and favor the reactants (Figure 15.2).

The signs $\gg$ and $\ll$ mean "much greater than" and "much smaller than," respectively.

Review of Concepts

Consider the equilibrium $X \rightleftharpoons Y$, where the forward reaction rate constant is greater than the reverse reaction rate constant. Which of the following is true about the equilibrium constant? (a) $K_c > 1$, (b) $K_c < 1$, (c) $K_c = 1$.

15.2 Ways of Expressing Equilibrium Constants

To use equilibrium constants, we must express them in terms of the reactant and product concentrations. Our only guidance is the law of mass action [Equation (15.2)]. However, because the concentrations of the reactants and products can be expressed in different units and because the reacting species are not always in the same phase, there may be more than one way to express the equilibrium constant for the *same* reaction. To begin with, we will consider reactions in which the reactants and products are in the same phase.

Homogeneous Equilibria

The term ***homogeneous equilibrium*** applies to reactions in which *all reacting species are in the same phase.* An example of homogeneous gas-phase equilibrium is the dissociation of N_2O_4. The equilibrium constant, as given in Equation (15.1), is

$$K_c = \frac{[NO_2]^2}{[N_2O_4]}$$

Note that the subscript in K_c denotes that the concentrations of the reacting species are expressed in moles per liter. The concentrations of reactants and products in gaseous reactions can also be expressed in terms of their partial pressures. From Equation (5.8) we see that at constant temperature the pressure P of a gas is directly related to the concentration in moles per liter of the gas; that is, $P = (n/V)RT$. Thus, for the equilibrium process

$$N_2O_4(g) \rightleftharpoons 2NO_2(g)$$

we can write

$$K_P = \frac{P^2_{NO_2}}{P_{N_2O_4}} \tag{15.3}$$

in which P_{NO_2} and $P_{N_2O_4}$ are the equilibrium partial pressures (in atmospheres) of NO_2 and N_2O_4, respectively. The subscript in K_P tells us that equilibrium concentrations are expressed in terms of pressure.

In general, K_c is not equal to K_P, because the partial pressures of reactants and products are not equal to their concentrations expressed in moles per liter. A simple relationship between K_P and K_c can be derived as follows. Let us consider this equilibrium in the gas phase:

$$aA(g) \rightleftharpoons bB(g)$$

in which a and b are stoichiometric coefficients. The equilibrium constant K_c is

$$K_c = \frac{[B]^b}{[A]^a}$$

and the expression for K_P is

$$K_P = \frac{P_B^b}{P_A^a}$$

in which P_A and P_B are the partial pressures of A and B. Assuming ideal gas behavior,

$$P_A V = n_A RT$$
$$P_A = \frac{n_A RT}{V}$$

in which V is the volume of the container in liters. Also,

$$P_B V = n_B RT$$
$$P_B = \frac{n_B RT}{V}$$

Substituting these relations into the expression for K_P, we obtain

$$K_P = \frac{\left(\frac{n_B RT}{V}\right)^b}{\left(\frac{n_A RT}{V}\right)^a} = \frac{\left(\frac{n_B}{V}\right)^b}{\left(\frac{n_A}{V}\right)^a}(RT)^{b-a}$$

Now both n_A/V and n_B/V have the units of moles per liter and can be replaced by [A] and [B], so that

$$K_P = \frac{[B]^b}{[A]^a}(RT)^{\Delta n}$$
$$= K_c(RT)^{\Delta n} \qquad (15.4)$$

in which

$$\Delta n = b - a$$
$$= \text{moles of gaseous products} - \text{moles of gaseous reactants}$$

Because pressure is usually expressed in atmospheres, the gas constant R is given by 0.0821 L · atm/K · mol, and we can write the relationship between K_P and K_c as

$$K_P = K_c(0.0821T)^{\Delta n} \qquad (15.5)$$

To use this equation, the pressures in K_P must be in atm.

In general, $K_P \neq K_c$ except in the special case when $\Delta n = 0$. In that case, Equation (15.5) can be written as

$$K_P = K_c(0.0821T)^0$$
$$= K_c$$

Any number raised to the zero power is equal to 1.

As another example of homogeneous equilibrium, let us consider the ionization of acetic acid (CH_3COOH) in water:

$$CH_3COOH(aq) + H_2O(l) \rightleftharpoons CH_3COO^-(aq) + H_3O^+(aq)$$

The equilibrium constant is

$$K'_c = \frac{[CH_3COO^-][H_3O^+]}{[CH_3COOH][H_2O]}$$

(We use the prime for K_c here to distinguish it from the final form of equilibrium constant to be derived shortly.) However, in 1 L, or 1000 g, of water, there are 1000 g/(18.02 g/mol), or 55.5 mol, of water. Therefore, the "concentration" of water, or $[H_2O]$, is 55.5 mol/L, or 55.5 *M*. This is a large quantity compared with the concentrations of other species in solution (usually 1 *M* or smaller), and we can assume that it does not change appreciably during the course of a reaction. Thus, we can treat $[H_2O]$ as a constant and rewrite the equilibrium constant as

$$K_c = \frac{[CH_3COO^-][H_3O^+]}{[CH_3COOH]}$$

in which

$$K_c = K'_c[H_2O]$$

Equilibrium Constant and Units

For nonideal systems, the activities are not exactly numerically equal to concentrations. In some cases, the differences can be appreciable. Unless otherwise noted, we will treat all systems as ideal.

Note that it is general practice not to include units for the equilibrium constant. In thermodynamics, the equilibrium constant is defined in terms of *activities* rather than concentrations. For an ideal system, the activity of a substance is the ratio of its concentration (or partial pressure) to a standard value, which is 1 *M* (or 1 atm). This procedure eliminates all units but does not alter the numerical parts of the concentration or pressure. Consequently, *K* has no units. We will extend this practice to acid-base equilibria and solubility equilibria in Chapters 16 and 17.

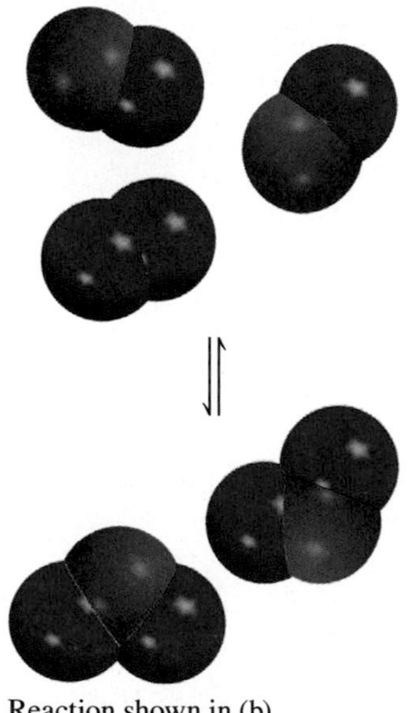
Reaction shown in (b).

EXAMPLE 15.1

Write expressions for K_c, and K_P if applicable, for the following reversible reactions at equilibrium:

(a) $HF(aq) + H_2O(l) \rightleftharpoons H_3O^+(aq) + F^-(aq)$

(b) $2NO(g) + O_2(g) \rightleftharpoons 2NO_2(g)$

(c) $CH_3COOH(aq) + C_2H_5OH(aq) \rightleftharpoons CH_3COOC_2H_5(aq) + H_2O(l)$

Strategy Keep in mind the following facts: (1) the K_P expression applies only to gaseous reactions and (2) the concentration of solvent (usually water) does not appear in the equilibrium constant expression.

Solution

(a) Because there are no gases present, K_P does not apply and we have only K_c.

$$K'_c = \frac{[H_3O^+][F^-]}{[HF][H_2O]}$$

HF is a weak acid, so that the amount of water consumed in acid ionizations is negligible compared with the total amount of water present as solvent. Thus, we can rewrite the equilibrium constant as

$$K_c = \frac{[H_3O^+][F^-]}{[HF]}$$

(Continued)

(b) $$K_c = \frac{[NO_2]^2}{[NO]^2[O_2]} \qquad K_P = \frac{P^2_{NO_2}}{P^2_{NO}P_{O_2}}$$

(c) The equilibrium constant K'_c is given by

$$K'_c = \frac{[CH_3COOC_2H_5][H_2O]}{[CH_3COOH][C_2H_5OH]}$$

Because the water produced in the reaction is negligible compared with the water solvent, the concentration of water does not change. Thus, we can write the new equilibrium constant as

$$K_c = \frac{[CH_3COOC_2H_5]}{[CH_3COOH][C_2H_5OH]}$$

Similar problem: 15.7.

Practice Exercise Write K_c and K_P for the decomposition of dinitrogen pentoxide:

$$2N_2O_5(g) \rightleftharpoons 4NO_2(g) + O_2(g)$$

EXAMPLE 15.2

The equilibrium constant K_P for the decomposition of phosphorus pentachloride to phosphorus trichloride and molecular chlorine

$$PCl_5(g) \rightleftharpoons PCl_3(g) + Cl_2(g)$$

is found to be 1.05 at 250°C. If the equilibrium partial pressures of PCl_5 and PCl_3 are 0.973 atm and 0.548 atm, respectively, what is the equilibrium partial pressure of Cl_2 at 250°C.

Strategy The concentrations of the reacting gases are given in atm, so we can express the equilibrium constant in K_P. From the known K_P value and the equilibrium pressures of PCl_3 and PCl_5, we can solve for P_{Cl_2}.

Solution First, we write K_P in terms of the partial pressures of the reacting species

$$K_P = \frac{P_{PCl_3}P_{Cl_2}}{P_{Cl_5}}$$

Knowing the partial pressures, we write

$$1.05 = \frac{(0.548)(P_{Cl_2})}{(0.973)}$$

or

$$P_{Cl_2} = \frac{(1.05)(0.973)}{(0.548)} = 1.86 \text{ atm}$$

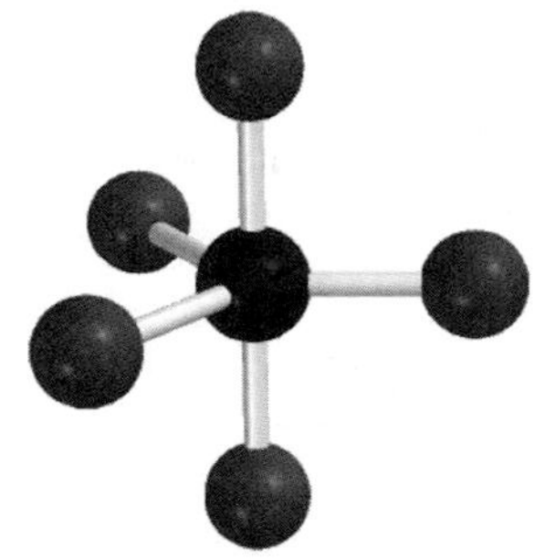

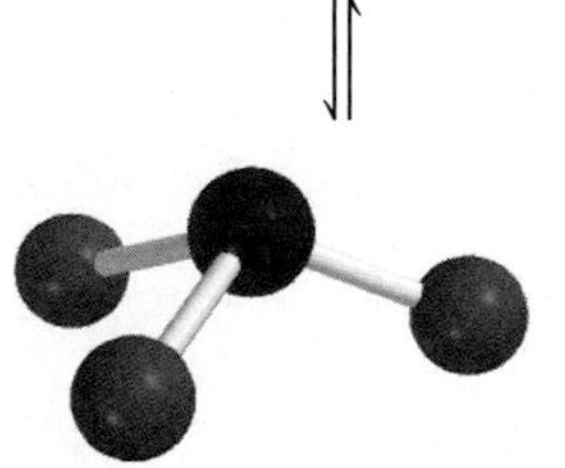

$PCl_5 \rightleftharpoons PCl_3 + Cl_2$

Check Note that we have added atm as the unit for P_{Cl_2}.

Similar problem: 15.17.

Practice Exercise The equilibrium constant K_P for the reaction

$$2NO_2(g) \rightleftharpoons 2NO(g) + O_2(g)$$

is 158 at 1000 K. Calculate P_{O_2} if $P_{NO_2} = 0.400$ atm and $P_{NO} = 0.270$ atm.

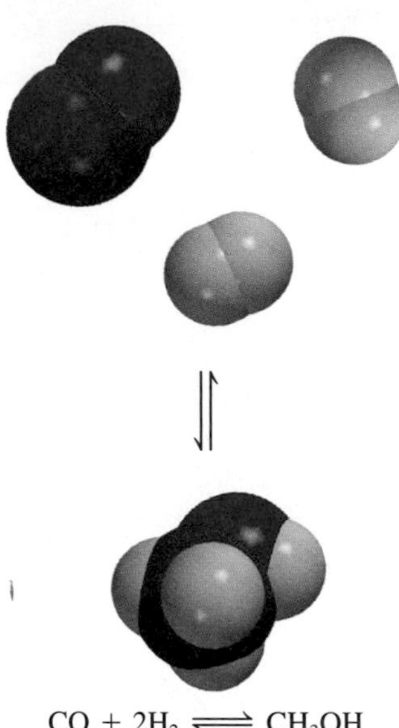

$CO + 2H_2 \rightleftharpoons CH_3OH$

EXAMPLE 15.3

Methanol (CH_3OH) is manufactured industrially by the reaction

$$CO(g) + 2H_2(g) \rightleftharpoons CH_3OH(g)$$

The equilibrium constant (K_c) for the reaction is 10.5 at 220°C. What is the value of K_P at this temperature?

Strategy The relationship between K_c and K_P is given by Equation (15.5). What is the change in the number of moles of gases from reactants to product? Recall that

$$\Delta n = \text{moles of gaseous products} - \text{moles of gaseous reactants}$$

What unit of temperature should we use?

Solution The relationship between K_c and K_P is

$$K_P = K_c(0.0821T)^{\Delta n}$$

Because $T = 273 + 220 = 493$ K and $\Delta n = 1 - 3 = -2$, we have

$$K_P = (10.5)(0.0821 \times 493)^{-2}$$
$$= 6.41 \times 10^{-3}$$

Check Note that K_P, like K_c, is a dimensionless quantity. This example shows that we can get a quite different value for the equilibrium constant for the same reaction, depending on whether we express the concentrations in moles per liter or in atmospheres.

Similar problem: 15.15.

Practice Exercise For the reaction

$$N_2(g) + 3H_2(g) \rightleftharpoons 2NH_3(g)$$

K_P is 4.3×10^{-4} at 375°C. Calculate K_c for the reaction.

Heterogeneous Equilibria

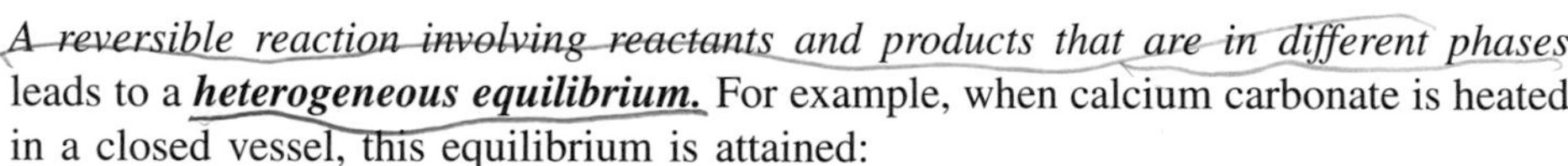

A reversible reaction involving reactants and products that are in different phases leads to a ***heterogeneous equilibrium.*** For example, when calcium carbonate is heated in a closed vessel, this equilibrium is attained:

$$CaCO_3(s) \rightleftharpoons CaO(s) + CO_2(g)$$

The mineral calcite is made of calcium carbonate, as are chalk and marble.

The two solids and one gas constitute three separate phases. At equilibrium, we might write the equilibrium constant as

$$K'_c = \frac{[CaO][CO_2]}{[CaCO_3]} \tag{15.6}$$

However, the "concentration" of a solid, like its density, is an intensive property and does not depend on how much of the substance is present. [Note that the units of concentration (moles per liter) can be converted to units of density (grams per cubic centimeters) and vice versa.] For this reason, the terms $[CaCO_3]$ and $[CaO]$ are themselves constants and can be combined with the equilibrium constant. We can simplify the equilibrium expression by writing

$$\frac{[CaCO_3]}{[CaO]}K'_c = K_c = [CO_2] \tag{15.7}$$

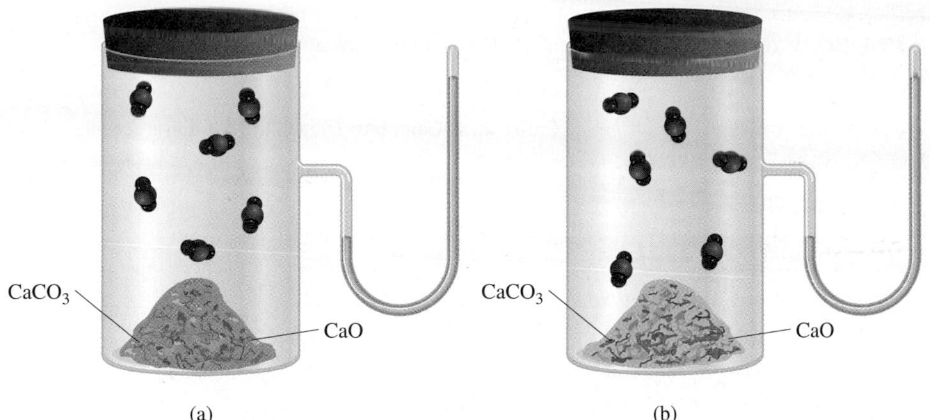

Figure 15.3
In (a) and (b) the equilibrium pressure of CO_2 is the same at the same temperature, despite the presence of different amounts of $CaCO_3$ (represented by orange color) and CaO (represented by green color).

in which K_c, the "new" equilibrium constant, is now conveniently expressed in terms of a single concentration, that of CO_2. Keep in mind that the value of K_c does not depend on how much $CaCO_3$ and CaO are present, as long as some of each is present at equilibrium (Figure 15.3).

The situation becomes simpler if we replace concentrations with activities. In thermodynamics, the activity of a pure solid is 1. Thus, the concentration terms for $CaCO_3$ and CaO are both unity, and from the preceding equilibrium equation, we can immediately write $K_c = [CO_2]$. Similarly, the activity of a pure liquid is also 1. Thus, if a reactant or a product is a liquid, we can omit it in the equilibrium constant expression.

Alternatively, we can express the equilibrium constant as

$$K_P = P_{CO_2} \tag{15.8}$$

The equilibrium constant in this case is numerically equal to the pressure of CO_2 gas, an easily measurable quantity.

EXAMPLE 15.4

Consider the following heterogeneous equilibrium:

$$CaCO_3(s) \rightleftharpoons CaO(s) + CO_2(g)$$

At 800°C, the pressure of CO_2 is 0.236 atm. Calculate (a) K_P and (b) K_c for the reaction at this temperature.

Strategy Remember that pure solids do not appear in the equilibrium constant expression. The relationship between K_P and K_c is given by Equation (15.5).

Solution

(a) Using Equation (15.8) we write

$$K_P = P_{CO_2}$$
$$= 0.236$$

(b) From Equation (15.5), we know

$$K_P = K_c(0.0821T)^{\Delta n}$$

In this case, $T = 800 + 273 = 1073$ K and $\Delta n = 1$, so we substitute these values in the equation and obtain

$$0.236 = K_c(0.0821 \times 1073)$$
$$K_c = 2.68 \times 10^{-3}$$

Similar problem: 15.20.

(Continued)

Practice Exercise Consider the following equilibrium at 395 K:

$$NH_4HS(s) \rightleftharpoons NH_3(g) + H_2S(g)$$

The partial pressure of each gas is 0.265 atm. Calculate K_P and K_c for the reaction.

REVIEW OF CONCEPTS

For which of the following reactions is K_c equal to K_p?

(a) $4NH_3(g) + 5O_2(g) \rightleftharpoons 4NO(g) + 6H_2O(g)$

(b) $2H_2O_2(aq) \rightleftharpoons 2H_2O(l) + O_2(g)$

(c) $PCl_3(g) + 3NH_3(g) \rightleftharpoons 3HCl(g) + P(NH_2)_3(g)$

The Form of *K* and the Equilibrium Equation

Before closing this section, we note these two important rules about writing equilibrium constants:

The reciprocal of *x* is 1/*x*.

1. When the equation for a reversible reaction is written in the opposite direction, the equilibrium constant becomes the reciprocal of the original equilibrium constant. Thus, if we write the NO_2–N_2O_4 equilibrium at 25°C as

$$N_2O_4(g) \rightleftharpoons 2NO_2(g)$$

then

$$K_c = \frac{[NO_2]^2}{[N_2O_4]} = 4.63 \times 10^{-3}$$

However, we can represent the equilibrium equally well as

$$2NO_2(g) \rightleftharpoons N_2O_4(g)$$

and the equilibrium constant is now given by

$$K_c' = \frac{[N_2O_4]}{[NO_2]^2} = \frac{1}{K_c} = \frac{1}{4.63 \times 10^{-3}} = 216$$

You can see that $K_c = 1/K_c'$ or $K_c K_c' = 1.00$. Either K_c or K_c' is a valid equilibrium constant, but it is meaningless to say that the equilibrium constant for the NO_2–N_2O_4 system is 4.63×10^{-3}, or 216, unless we also specify how the equilibrium equation is written.

2. The value of *K* also depends on how the equilibrium equation is balanced. Consider the following two ways of describing the same equilibrium:

$$\tfrac{1}{2}N_2O_4(g) \rightleftharpoons NO_2(g) \qquad K_c' = \frac{[NO_2]}{[N_2O_4]^{1/2}}$$

$$N_2O_4(g) \rightleftharpoons 2NO_2(g) \qquad K_c = \frac{[NO_2]^2}{[N_2O_4]}$$

Looking at the exponents we see that $K_c' = \sqrt{K_c}$. In Table 15.1 we find the average value of $K_c = 4.63 \times 10^{-3}$; therefore, $K_c' = 0.0680$.

Thus, if you double a chemical equation throughout, the corresponding equilibrium constant will be the square of the original value; if you triple the equation, the equilibrium constant will be the cube of the original value, and so on. The NO_2–N_2O_4 example illustrates once again the need to write the particular chemical equation when quoting the numerical value of an equilibrium constant.

Summary of the Rules for Writing Equilibrium Constant Expressions

1. The concentrations of the reacting species in the condensed phase are expressed in moles per liter; in the gaseous phase, the concentrations can be expressed in moles per liter or in atmospheres. K_c is related to K_P by a simple equation [Equation (15.5)].
2. The concentrations of pure solids, pure liquids (in heterogeneous equilibria), and solvents (in homogeneous equilibria) do not appear in the equilibrium constant expressions.
3. The equilibrium constant (K_c or K_P) is dimensionless.
4. In quoting a value for the equilibrium constant, we must specify the balanced equation and the temperature.

REVIEW OF CONCEPTS

From the following equilibrium constant expression, write a balanced chemical equation for the gas-phase reaction. Does $K_c = K_P$ for this reaction?

$$K_c = \frac{[NH_3]^2[H_2O]^4}{[NO_2]^2[H_2]^7}$$

15.3 What Does the Equilibrium Constant Tell Us?

We have seen that the equilibrium constant for a given reaction can be calculated from known equilibrium concentrations. Once we know the value of the equilibrium constant, we can use Equation (15.2) to calculate unknown equilibrium concentrations—remembering, of course, that the equilibrium constant has a constant value only if the temperature does not change. In general, the equilibrium constant helps us to predict the direction in which a reaction mixture will proceed to achieve equilibrium and to calculate the concentrations of reactants and products once equilibrium has been reached. These uses of the equilibrium constant will be explored in this section.

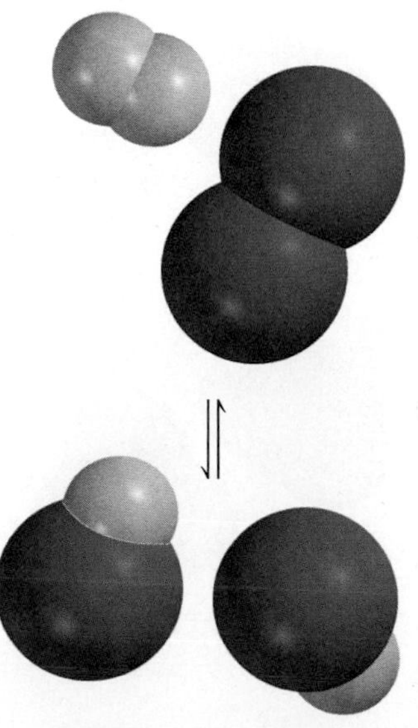

$H_2 + I_2 \rightleftharpoons 2HI$

Predicting the Direction of a Reaction

The equilibrium constant K_c for the reaction

$$H_2(g) + I_2(g) \rightleftharpoons 2HI(g)$$

is 54.3 at 430°C. Suppose that in a certain experiment we place 0.243 mole of H_2, 0.146 mole of I_2, and 1.98 moles of HI all in a 1.00-L container at 430°C. Will there be a net reaction to form more H_2 and I_2 or more HI? Inserting the starting concentrations in the equilibrium constant expression, we write

$$\frac{[HI]_0^2}{[H_2]_0[I_2]_0} = \frac{(1.98)^2}{(0.243)(0.146)} = 111$$

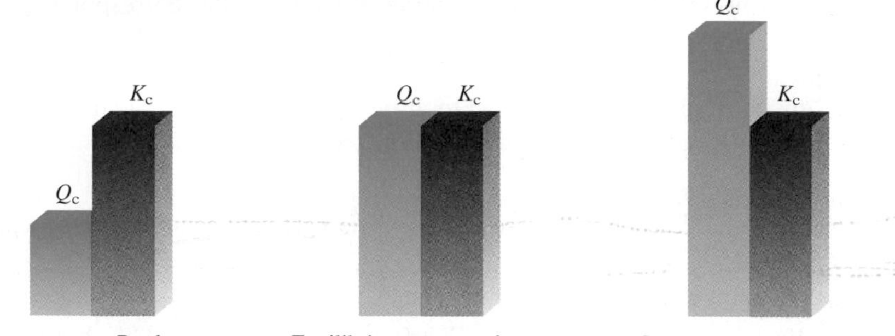

Figure 15.4
The direction of a reversible reaction to reach equilibrium depends on the relative magnitudes of Q_c and K_c. Note that K_c is a constant at a given temperature, but Q_c varies according to the relative amounts of reactants and products present.

where the subscript 0 indicates initial concentrations. Because the calculated quotient $[HI]_0^2/[H_2]_0[I_2]_0$ is greater than K_c, this system is not at equilibrium. Consequently, some of the HI will react to form more H_2 and I_2 (decreasing the value of the quotient). Thus, the net reaction proceeds from right to left to reach equilibrium.

Keep in mind that the method for calculating Q is the same as that for K, except that nonequilibrium concentrations are used.

The quantity obtained by substituting the initial concentrations into the equilibrium constant expression is called the ***reaction quotient*** **(Q_c)**. To determine in which direction the net reaction will proceed to achieve equilibrium, we compare the values of Q_c and K_c. The three possible cases are as follows (Figure 15.4):

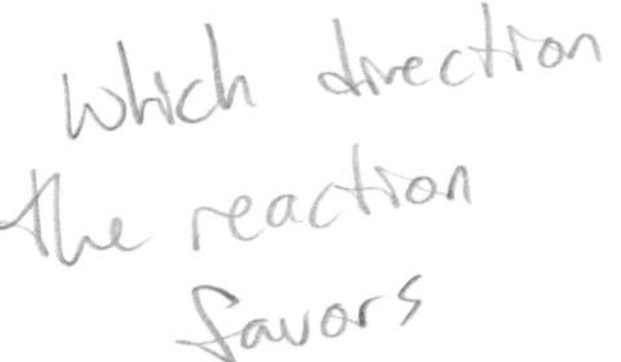

- $Q_c < K_c$ The ratio of initial concentrations of products to reactants is too small. To reach equilibrium, reactants must be converted to products. The system proceeds from left to right (consuming reactants, forming products) to reach equilibrium.
- $Q_c = K_c$ The initial concentrations are equilibrium concentrations. The system is at equilibrium.
- $Q_c > K_c$ The ratio of initial concentrations of products to reactants is too large. To reach equilibrium, products must be converted to reactants. The system proceeds from right to left (consuming products, forming reactants) to reach equilibrium.

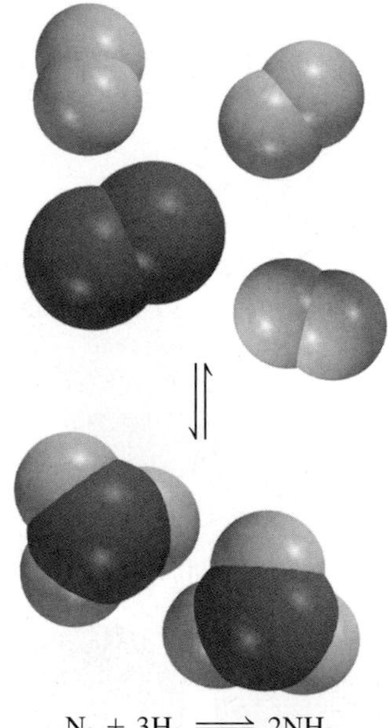

$N_2 + 3H_2 \rightleftharpoons 2NH_3$

EXAMPLE 15.5

At the start of a reaction, there are 0.351 mol N_2, 3.67×10^{-2} mol H_2, and 7.51×10^{-4} mol NH_3 in a 3.75-L reaction vessel at 375°C. If the equilibrium constant (K_c) for the reaction

$$N_2(g) + 3H_2(g) \rightleftharpoons 2NH_3(g)$$

is 1.2 at this temperature, decide whether the system is at equilibrium. If it is not, predict which way the net reaction will proceed to reach equilibrium.

Strategy We are given the initial amounts of gases (in moles) in a vessel of known volume (in liters), so we can calculate their molar concentrations and hence the reaction quotient (Q_c). How does a comparison of Q_c with K_c enable us to determine if the system is at equilibrium or, if not, in which direction the net reaction will proceed to reach equilibrium?

Solution The initial concentrations of the reacting species are

$$[N_2]_0 = \frac{0.351 \text{ mol}}{3.75 \text{ L}} = 0.0936\ M$$

$$[H_2]_0 = \frac{3.67 \times 10^{-2} \text{ mol}}{3.75 \text{ L}} = 9.79 \times 10^{-3}\ M$$

$$[NH_3]_0 = \frac{7.51 \times 10^{-4} \text{ mol}}{3.75 \text{ L}} = 2.00 \times 10^{-4}\ M$$

(Continued)

Next we write

$$Q_c = \frac{[NH_3]_0^2}{[N_2]_0[H_2]_0^3} = \frac{(2.00 \times 10^{-4})^2}{(0.0936)(9.79 \times 10^{-3})^3} = 0.455$$

Because Q_c is smaller than K_c (1.2), the system is not at equilibrium. The net result will be an increase in the concentration of NH_3 and a decrease in the concentrations of N_2 and H_2. That is, the net reaction will proceed from left to right until equilibrium is reached.

Similar problem: 15.21.

Practice Exercise The equilibrium constant (K_c) for the formation of nitrosyl chloride, an orange-yellow compound, from nitric oxide and molecular chlorine

$$2NO(g) + Cl_2(g) \rightleftharpoons 2NOCl(g)$$

is 6.5×10^4 at 35°C. In a certain experiment, 2.0×10^{-2} mole of NO, 8.3×10^{-3} mole of Cl_2, and 6.8 moles of NOCl are mixed in a 2.0-L flask. In which direction will the system proceed to reach equilibrium?

Calculating Equilibrium Concentrations

If we know the equilibrium constant for a particular reaction, we can calculate the concentrations in the equilibrium mixture from a knowledge of the initial concentrations. Depending on the information given, the calculation may be straightforward or complex. In the most common situation, only the initial reactant concentrations are given. Let us consider this system involving a pair of geometric isomers in an organic solvent (Figure 15.5), which has an equilibrium constant (K_c) of 24.0 at 200°C:

$$cis\text{-stilbene} \rightleftharpoons trans\text{-stilbene}$$

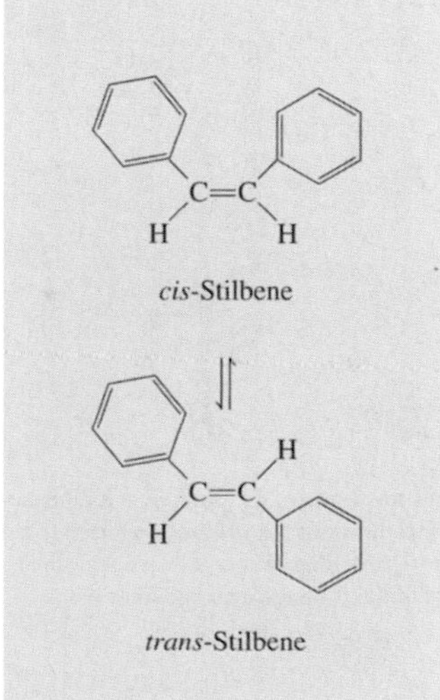

Figure 15.5
The equilibrium between cis-stilbene and trans-stilbene. Note that both molecules have the same molecular formula ($C_{14}H_{12}$) and also the same type of bonds. However, in cis-stilbene, the benzene rings are on one side of the C=C bond and the H atoms are on the other side whereas in trans-stilbene the benzene rings (and the H atoms) are across from the C=C bond. These compounds have different melting points and dipole moments.

Suppose that only *cis*-stilbene is initially present at a concentration of 0.850 mol/L. How do we calculate the concentrations of *cis*- and *trans*-stilbene at equilibrium? From the stoichiometry of the reaction we see that for every mole of *cis*-stilbene converted, 1 mole of *trans*-stilbene is formed. Let x be the equilibrium concentration of *trans*-stilbene in moles per liter; therefore, the equilibrium concentration of *cis*-stilbene must be $(0.850 - x)$ mol/L. It is useful to summarize the changes in concentration as

	cis-stilbene ⇌	*trans*-stilbene
Initial (M):	0.850	0
Change (M):	$-x$	$+x$
Equilibrium (M):	$(0.850 - x)$	x

A positive (+) change represents an increase and a negative (−) change indicates a decrease in concentration at equilibrium. Next, we set up the equilibrium constant expression

$$K_c = \frac{[trans\text{-stilbene}]}{[cis\text{-stilbene}]}$$

$$24.0 = \frac{x}{0.850 - x}$$

$$x = 0.816\,M$$

Having solved for x, we calculate the equilibrium concentrations of *cis*-stilbene and *trans*-stilbene as

$$[cis\text{-stilbene}] = (0.850 - 0.816)\ M = 0.034\ M$$
$$[trans\text{-stilbene}] = 0.816\ M$$

We summarize our approach to solving equilibrium constant problems as

1. Express the equilibrium concentrations of all species in terms of the initial concentrations and a single unknown quantity x, which represents the change in concentration.
2. Write the equilibrium constant expression in terms of the equilibrium concentrations. Knowing the value of the equilibrium constant, solve for x.
3. Having solved for x, calculate the equilibrium concentrations of all species.

EXAMPLE 15.6

A mixture of 0.750 mol H_2 and 0.750 mol I_2 was placed in a 1.00-L stainless-steel flask at 430°C. The equilibrium constant K_c for the reaction $H_2(g) + I_2(g) \rightleftharpoons 2HI(g)$ is 54.3 at this temperature. Calculate the concentrations of H_2, I_2, and HI at equilibrium.

Strategy We are given the initial amounts of gases (in moles) in a vessel of known volume (in liters), so we can calculate their molar concentrations. Because initially no HI was present, the system could not be at equilibrium. Therefore, some H_2 would react with the same amount of I_2 (why?) to form HI until equilibrium was established.

Solution We follow the preceding procedure to calculate the equilibrium concentrations.

Step 1: The stoichiometry of the reaction is 1 mol H_2 reacting with 1 mol I_2 to yield 2 mol HI. Let x be the depletion in concentration (mol/L) of H_2 and I_2 at equilibrium. It follows that the equilibrium concentration of HI must be $2x$. We summarize the changes in concentrations as follows:

This procedure for solving equilibrium concentrations is sometimes referred to as the ICE method, where the acronym stands for Initial, Change, and Equilibrium.

	H_2	+	I_2	$\rightleftharpoons$	$2HI$
Initial (M):	0.750		0.750		0.000
Change (M):	$-x$		$-x$		$+2x$
Equilibrium (M):	$(0.750 - x)$		$(0.750 - x)$		$2x$

Step 2: The equilibrium constant is given by

$$K_c = \frac{[HI]^2}{[H_2][I_2]}$$

Substituting, we get

$$54.3 = \frac{(2x)^2}{(0.750 - x)(0.750 - x)}$$

Taking the square root of both sides, we get

$$7.37 = \frac{2x}{0.750 - x}$$
$$x = 0.590\ M$$

(Continued)

Step 3: At equilibrium, the concentrations are

$$[H_2] = (0.750 - 0.590)\ M = 0.160\ M$$
$$[I_2] = (0.750 - 0.590)\ M = 0.160\ M$$
$$[HI] = 2 \times 0.590\ M = 1.18\ M$$

Check You can check your answers by calculating K_c using the equilibrium concentrations. Remember that K_c is a constant for a particular reaction at a given temperature.

Similar problem: 15.33.

Practice Exercise Consider the reaction in Example 15.6. Starting with a concentration of 0.040 M for HI, calculate the concentrations of HI, H_2, and I_2 at equilibrium.

EXAMPLE 15.7

For the same reaction and temperature as in Example 15.6, suppose that the initial concentrations of H_2, I_2, and HI are 0.00623 M, 0.00414 M, and 0.0224 M, respectively. Calculate the concentrations of these species at equilibrium.

Strategy From the initial concentrations we can calculate the reaction quotient (Q_c) to see if the system is at equilibrium or, if not, in which direction the net reaction will proceed to reach equilibrium. A comparison of Q_c with K_c also enables us to determine if there will be a depletion in H_2 and I_2 or HI as equilibrium is established.

Solution First we calculate Q_c as follows:

$$Q_c = \frac{[HI]_0^2}{[H_2]_0[I_2]_0} = \frac{(0.0224)^2}{(0.00623)(0.00414)} = 19.5$$

Because Q_c (19.5) is smaller than K_c (54.3), we conclude that the net reaction will proceed from left to right until equilibrium is reached (see Figure 15.4); that is, there will be a depletion of H_2 and I_2 and a gain in HI.

Step 1: Let x be the depletion in concentration (mol/L) of H_2 and I_2 at equilibrium. From the stoichiometry of the reaction it follows that the increase in concentration for HI must be $2x$. Next we write

	H_2	+	I_2	$\rightleftharpoons$	2HI
Initial (M):	0.00623		0.00414		0.0224
Change (M):	$-x$		$-x$		$+2x$
Equilibrium (M):	$(0.00623 - x)$		$(0.00414 - x)$		$(0.0224 + 2x)$

Step 2: The equilibrium constant is

$$K_c = \frac{[HI]^2}{[H_2][I_2]}$$

Substituting, we get

$$54.3 = \frac{(0.0224 + 2x)^2}{(0.00623 - x)(0.00414 - x)}$$

It is not possible to solve this equation by the square root shortcut, as the starting concentrations $[H_2]$ and $[I_2]$ are unequal. Instead, we must first carry out the multiplications

$$54.3(2.58 \times 10^{-5} - 0.0104x + x^2) = 5.02 \times 10^{-4} + 0.0896x + 4x^2$$

Collecting terms, we get

$$50.3x^2 - 0.654x + 8.98 \times 10^{-4} = 0$$

(Continued)

This is a quadratic equation of the form $ax^2 + bx + c = 0$. The solution for a quadratic equation (see Appendix 3) is

$$x = \frac{-b \pm \sqrt{b^2 - 4ac}}{2a}$$

Here we have $a = 50.3$, $b = -0.654$, and $c = 8.98 \times 10^{-4}$, so that

$$x = \frac{0.654 \pm \sqrt{(-0.654)^2 - 4(50.3)(8.98 \times 10^{-4})}}{2 \times 50.3}$$

$$x = 0.0114\,M \qquad \text{or} \qquad x = 0.00156\,M$$

The first solution is physically impossible because the amounts of H_2 and I_2 reacted would be more than those originally present. The second solution gives the correct answer. Note that in solving quadratic equations of this type, one answer is always physically impossible, so choosing a value for x is easy.

Step 3: At equilibrium, the concentrations are

$$[H_2] = (0.00623 - 0.00156)\,M = 0.00467\,M$$
$$[I_2] = (0.00414 - 0.00156)\,M = 0.00258\,M$$
$$[HI] = (0.0224 + 2 \times 0.00156)\,M = 0.0255\,M$$

Check You can check the answers by calculating K_c using the equilibrium concentrations. Remember that K_c is a constant for a particular reaction at a given temperature.

Similar problem: 15.78.

Practice Exercise At 1280°C the equilibrium constant (K_c) for the reaction

$$Br_2(g) \rightleftharpoons 2Br(g)$$

is 1.1×10^{-3}. If the initial concentrations are $[Br_2] = 6.3 \times 10^{-2}\,M$ and $[Br] = 1.2 \times 10^{-2}\,M$, calculate the concentrations of these species at equilibrium.

Review of Concepts

The equilibrium constant (K_c) for the $A_2 + B_2 \rightleftharpoons 2AB$ reaction is 3 at a certain temperature. Which of the diagrams shown here corresponds to the reaction at equilibrium? For those mixtures that are not at equilibrium, will the net reaction move in the forward or reverse direction to reach equilibrium?

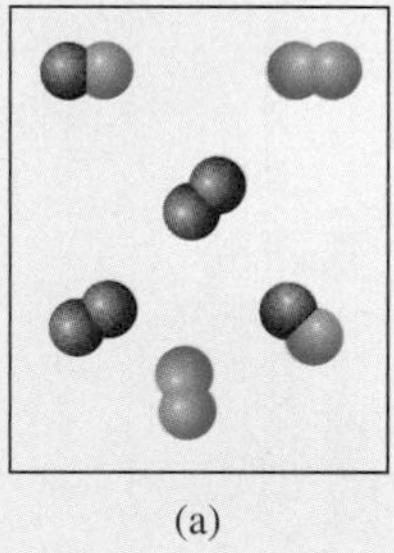

(a)

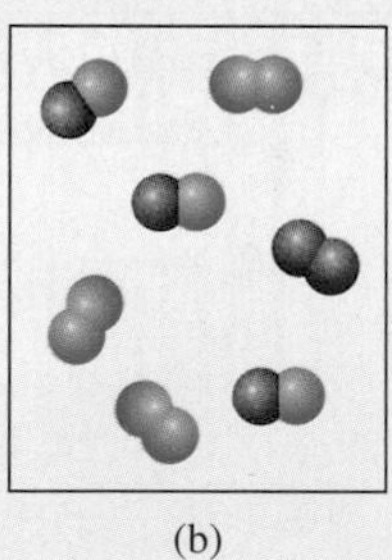

(b)

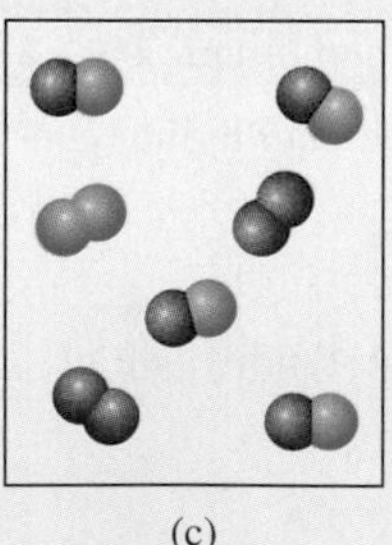

(c)

15.4 Factors That Affect Chemical Equilibrium

Chemical equilibrium represents a balance between forward and reverse reactions. In most cases, this balance is quite delicate. Changes in experimental conditions may disturb the balance and shift the equilibrium position so that more or less of the desired product is formed. When we say that an equilibrium position shifts to the right, for example, we mean that the net reaction is now from left to right. At our disposal are the following experimentally controllable variables: concentration, pressure, volume, and temperature. Here we will examine how each of these variables affects a reacting system at equilibrium. In addition, we will examine the effect of a catalyst on equilibrium.

Le Châtelier's Principle

There is a general rule that helps us to predict the direction in which an equilibrium reaction will move when a change in concentration, pressure, volume, or temperature occurs. The rule, known as ***Le Châtelier's principle*** (after the French chemist Henri Le Châtelier), states that *if an external stress is applied to a system at equilibrium, the system adjusts in such a way that the stress is partially offset as it tries to reestablish equilibrium.* The word "stress" here means a change in concentration, pressure, volume, or temperature that removes a system from the equilibrium state. We will use Le Châtelier's principle to assess the effects of such changes.

Changes in Concentrations

Iron(III) thiocyanate $[Fe(SCN)_3]$ dissolves readily in water to give a red solution. The red color is due to the presence of hydrated $FeSCN^{2+}$ ion. The equilibrium between undissociated $FeSCN^{2+}$ and the Fe^{3+} and SCN^- ions is given by

$$\underset{\text{red}}{FeSCN^{2+}(aq)} \rightleftharpoons \underset{\text{pale yellow}}{Fe^{3+}(aq)} + \underset{\text{colorless}}{SCN^-(aq)}$$

What happens if we add some sodium thiocyanate (NaSCN) to this solution? In this case, the stress applied to the equilibrium system is an increase in the concentration of SCN^- (from the dissociation of NaSCN). To offset this stress, some Fe^{3+} ions react with the added SCN^- ions, and the equilibrium shifts from right to left:

$$FeSCN^{2+}(aq) \longleftarrow Fe^{3+}(aq) + SCN^-(aq)$$

Consequently, the red color of the solution deepens (Figure 15.6). Similarly, if we added iron(III) nitrate $[Fe(NO_3)_3]$ to the original solution, the red color would also deepen because the additional Fe^{3+} ions [from $Fe(NO_3)_3$] would shift the equilibrium from right to left. Both Na^+ and NO_3^- are colorless spectator ions.

(a) (b) (c) (d)

Figure 15.6
Effect of concentration change on the position of equilibrium. (a) An aqueous $Fe(SCN)_3$ solution. The color of the solution is due to both the red $FeSCN^{2+}$ and the yellow Fe^{3+} ions. (b) After the addition of some NaSCN to the solution in (a), the equilibrium shifts to the left. (c) After the addition of some $Fe(NO_3)_3$ to the solution in (a), the equilibrium shifts to the left. (d) After the addition of some $H_2C_2O_4$ to the solution in (a), the equilibrium shifts to the right. The yellow color is due to the $Fe(C_2O_4)_3^{3-}$ ions.

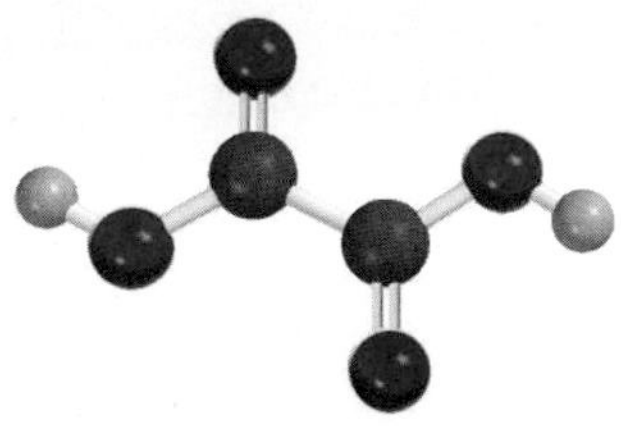

Oxalic acid is sometimes used to remove bathtub rings that consist of rust, or Fe_2O_3.

Now suppose we add some oxalic acid ($H_2C_2O_4$) to the original solution. Oxalic acid ionizes in water to form the oxalate ion, $C_2O_4^{2-}$, which binds strongly to the Fe^{3+} ions. The formation of the stable yellow ion $Fe(C_2O_4)_3^{3-}$ removes free Fe^{3+} ions from solution. Consequently, more $FeSCN^{2+}$ units dissociate and the equilibrium shifts from left to right:

$$FeSCN^{2+}(aq) \longrightarrow Fe^{3+}(aq) + SCN^-(aq)$$

The red solution will turn yellow because of the formation of $Fe(C_2O_4)_3^{3-}$ ions.

This experiment demonstrates that at equilibrium all reactants and products are present in the reacting system. Second, increasing the concentrations of the products (Fe^{3+} or SCN^-) shifts the equilibrium to the left, and decreasing the concentration of the product Fe^{3+} shifts the equilibrium to the right. These results are just as predicted by Le Châtelier's principle.

Le Châtelier's principle simply summarizes the observed behavior of equilibrium systems; therefore, it is incorrect to say that a given equilibrium shift occurs "because of" Le Châtelier's principle.

EXAMPLE 15.8

At 720°C, the equilibrium constant K_c for the reaction

$$N_2(g) + 3H_2(g) \rightleftharpoons 2NH_3(g)$$

is 2.37×10^{-3}. In a certain experiment, the equilibrium concentrations are $[N_2] = 0.683\ M$, $[H_2] = 8.80\ M$, and $[NH_3] = 1.05\ M$. Suppose some NH_3 is added to the mixture so that its concentration is increased to 3.65 M. (a) Use Le Châtelier's principle to predict the shift in direction of the net reaction to reach a new equilibrium.(b) Confirm your prediction by calculating the reaction quotient Q_c and comparing its value with K_c.

Strategy (a) What is the stress applied to the system? How does the system adjust to offset the stress? (b) At the instant when some NH_3 is added, the system is no longer at equilibrium. How do we calculate the Q_c for the reaction at this point? How does a comparison of Q_c with K_c tell us the direction of the net reaction to reach equilibrium?

Solution

(a) The stress applied to the system is the addition of NH_3. To offset this stress, some NH_3 reacts to produce N_2 and H_2 until a new equilibrium is established. The net reaction therefore shifts from right to left; that is,

$$N_2(g) + 3H_2(g) \longleftarrow 2NH_3(g)$$

(b) At the instant when some of the NH_3 is added, the system is no longer at equilibrium. The reaction quotient is given by

$$Q_c = \frac{[NH_3]_0^2}{[N_2]_0[H_2]_0^3} = \frac{(3.65)^2}{(0.683)(8.80)^3} = 2.86 \times 10^{-2}$$

Because this value is greater than 2.37×10^{-3}, the net reaction shifts from right to left until Q_c equals K_c.

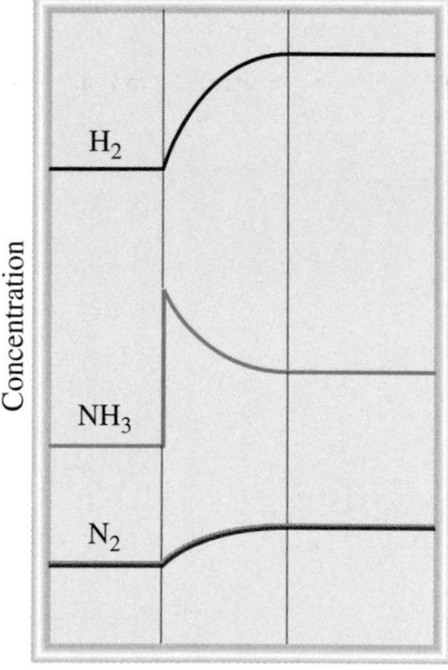

Figure 15.7
Changes in concentration of H_2, N_2, and NH_3 after the addition of NH_3 to the equilibrium mixture. When the new equilibrium is established, all the concentrations are changed but K_c remains the same because temperature remains constant.

Figure 15.7 shows qualitatively the changes in concentrations of the reacting species.

Similar problem: 15.36.

Practice Exercise At 430°C, the equilibrium constant (K_P) for the reaction

$$2NO(g) + O_2(g) \rightleftharpoons 2NO_2(g)$$

is 1.5×10^5. In one experiment, the initial pressures of NO, O_2, and NO_2 are 2.1×10^{-3} atm, 1.1×10^{-2} atm, and 0.14 atm, respectively. Calculate Q_P and predict the direction that the net reaction will shift to reach equilibrium.

Changes in Pressure and Volume

Changes in pressure ordinarily do not affect the concentrations of reacting species in condensed phases (say, in an aqueous solution) because liquids and solids are virtually incompressible. On the other hand, concentrations of gases are greatly affected by changes in pressure. Let us look again at Equation (5.8):

$$PV = nRT$$

$$P = \left(\frac{n}{V}\right)RT$$

Thus, P and V are related to each other inversely: The greater the pressure, the smaller the volume, and vice versa. Note, too, that the term (n/V) is the concentration of the gas in moles per liter, and it varies directly with pressure.

Suppose that the equilibrium system

$$N_2O_4(g) \rightleftharpoons 2NO_2(g)$$

is in a cylinder fitted with a movable piston. What happens if we increase the pressure on the gases by pushing down on the piston at constant temperature? As the volume decreases, the concentration (n/V) of both NO_2 and N_2O_4 increases. Because the concentration of NO_2 is squared, the increase in pressure increases the numerator more than the denominator. The system is no longer at equilibrium, so we write

$$Q_c = \frac{[NO_2]_0^2}{[N_2O_4]_0}$$

The shift in equilibrium can also be predicted using Le Châtelier's principle.

Thus, $Q_c > K_c$, and the net reaction will shift to the left until $Q_c = K_c$ (Figure 15.8). Conversely, a decrease in pressure (increase in volume) would result in $Q_c < K_c$; the net reaction would shift to the right until $Q_c = K_c$.

In general, an increase in pressure (decrease in volume) favors the net reaction that decreases the total number of moles of gases (the reverse reaction, in the preceding case), and a decrease in pressure (increase in volume) favors the net reaction that increases the total number of moles of gases (here, the forward reaction). For reactions in which there is no change in the number of moles of gases, for example, $H_2(g) + Cl_2(g) \rightleftharpoons 2HCl(g)$, a pressure (or volume) change has no effect on the position of equilibrium.

It is possible to change the pressure of a system without changing its volume. Suppose the NO_2–N_2O_4 system is contained in a stainless-steel vessel whose volume is constant. We can increase the total pressure in the vessel by adding an inert gas (helium, for example) to the equilibrium system. Adding helium to the equilibrium mixture at constant volume increases the total gas pressure and decreases the mole fractions of both NO_2 and N_2O_4; but the partial pressure of each gas, given by the product of its mole fraction and total pressure (see Section 5.5), does not change. Thus, the presence of an inert gas in such a case does not affect the equilibrium.

Figure 15.8
The effect of an increase in pressure on the $N_2O_4(g) \rightleftharpoons 2NO_2(g)$ equilibrium.

EXAMPLE 15.9

Consider the following equilibrium systems:

(a) $2PbS(s) + 3O_2(g) \rightleftharpoons 2PbO(s) + 2SO_2(g)$
(b) $PCl_5(g) \rightleftharpoons PCl_3(g) + Cl_2(g)$
(c) $H_2(g) + CO_2(g) \rightleftharpoons H_2O(g) + CO(g)$

(Continued)

Predict the direction of the net reaction in each case as a result of increasing the pressure (decreasing the volume) on the system at constant temperature.

Strategy A change in pressure can affect only the volume of a gas, but not that of a solid because solids (and liquids) are much less compressible. The stress applied is an increase in pressure. According to Le Châtelier's principle, the system will adjust to partially offset this stress. In other words, the system will adjust to decrease the pressure. This can be achieved by shifting to the side of the equation that has fewer moles of gas. Recall that pressure is directly proportional to moles of gas: $PV = nRT$ so $P \propto n$.

Solution

(a) Consider only the gaseous molecules. In the balanced equation there are 3 moles of gaseous reactants and 2 moles of gaseous products. Therefore, the net reaction will shift toward the products (to the right) when the pressure is increased.

(b) The number of moles of products is 2 and that of reactants is 1; therefore, the net reaction will shift to the left, toward the reactant.

(c) The number of moles of products is equal to the number of moles of reactants, so a change in pressure has no effect on the equilibrium.

Similar problem: 15.46.

Check In each case, the prediction is consistent with Le Châtelier's principle.

Practice Exercise Consider the equilibrium reaction involving nitrosyl chloride, nitric oxide, and molecular chlorine

$$2NOCl(g) \rightleftharpoons 2NO(g) + Cl_2(g)$$

Predict the direction of the net reaction as a result of decreasing the pressure (increasing the volume) on the system at constant temperature.

REVIEW OF CONCEPTS

The diagram here shows the gaseous reaction $2A \rightleftharpoons A_2$ at equilibrium. If the pressure is decreased by increasing the volume at constant temperature, how would the concentrations of A and A_2 change when a new equilibrium is established?

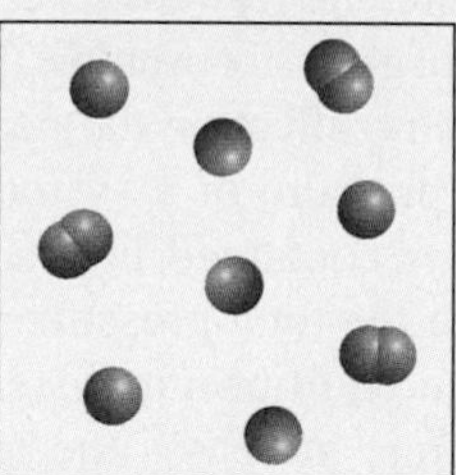

Changes in Temperature

A change in concentration, pressure, or volume may alter the equilibrium position, that is, the relative amounts of reactants and products, but it does not change the value of the equilibrium constant. Only a change in temperature can alter the equilibrium constant. To see why, let us consider the reaction

$$N_2O_4(g) \rightleftharpoons 2NO_2(g)$$

The forward reaction is endothermic (absorbs heat, $\Delta H° > 0$):

$$\text{heat} + N_2O_4(g) \longrightarrow 2NO_2(g) \qquad \Delta H° = 58.0 \text{ kJ/mol}$$

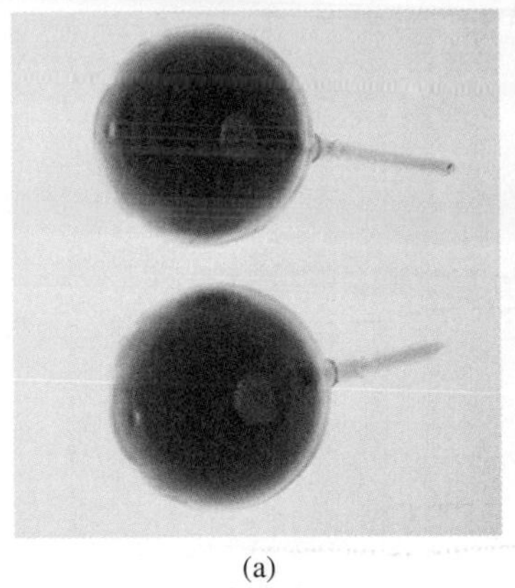

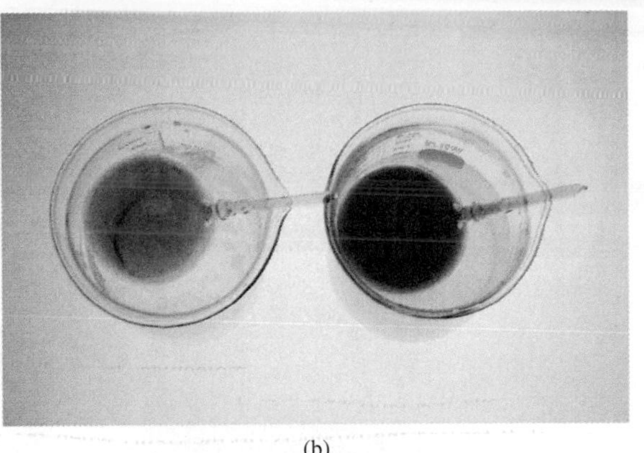

(a) (b)

Figure 15.9
(a) Two bulbs containing a mixture of NO_2 and N_2O_4 gases at equilibrium at room temperature. (b) When one bulb is immersed in ice water (left), its color becomes lighter, indicating the formation of colorless N_2O_4 gas. When the other bulb is immersed in hot water, its color darkens, indicating an increase in NO_2.

so the reverse reaction is exothermic (releases heat, $\Delta H° < 0$):

$$2NO_2(g) \longrightarrow N_2O_4(g) + \text{heat} \qquad \Delta H° = -58.0 \text{ kJ/mol}$$

At equilibrium at a certain temperature, the heat effect is zero because there is no net reaction. If we treat heat as though it were a chemical reagent, then a rise in temperature "adds" heat to the system and a drop in temperature "removes" heat from the system. As with a change in any other parameter (concentration, pressure, or volume), the system shifts to reduce the effect of the change. Therefore, a temperature increase favors the endothermic direction (from left to right of the equilibrium equation), which decreases $[N_2O_4]$ and increases $[NO_2]$. A temperature decrease favors the exothermic direction (from right to left of the equilibrium equation), which decreases $[NO_2]$ and increases $[N_2O_4]$. Consequently, the equilibrium constant, given by

$$K_c = \frac{[NO_2]^2}{[N_2O_4]}$$

increases when the system is heated and decreases when the system is cooled (Figure 15.9).

As another example, consider the equilibrium between the following ions:

$$\underset{\text{blue}}{CoCl_4^{2-}} + 6H_2O \rightleftharpoons \underset{\text{pink}}{Co(H_2O)_6^{2+}} + 4Cl^-$$

The formation of $CoCl_4^{2-}$ is endothermic. On heating, the equilibrium shifts to the left and the solution turns blue. Cooling favors the exothermic reaction [the formation of $Co(H_2O)_6^{2+}$] and the solution turns pink (Figure 15.10).

In summary, *a temperature increase favors an endothermic reaction, and a temperature decrease favors an exothermic reaction.*

The Effect of a Catalyst

We know that a catalyst enhances the rate of a reaction by lowering the reaction's activation energy (Section 14.4). However, as Figure 14.17 shows, a catalyst lowers the activation energy of the forward reaction and the reverse reaction to the same extent. We can therefore conclude that the presence of a catalyst does not alter the equilibrium constant, nor does it shift the position of an equilibrium system. Adding a catalyst to a reaction mixture that is not at equilibrium will simply cause the mixture to reach equilibrium sooner. The same equilibrium mixture could be obtained without the catalyst, but we might have to wait much longer for it to happen.

Figure 15.10
(Left) Heating favors the formation of the blue $CoCl_4^{2-}$ ion. (Right) Cooling favors the formation of the pink $Co(H_2O)_6^{2+}$ ion.

Summary of Factors That May Affect the Equilibrium Position

We have considered four ways to affect a reacting system at equilibrium. It is important to remember that, of the four, *only a change in temperature changes the value of the equilibrium constant.* Changes in concentration, pressure, and volume can alter the equilibrium concentrations of the reacting mixture, but they cannot change the equilibrium constant as long as the temperature does not change. A catalyst can speed up the process, but it has no effect on the equilibrium constant or on the equilibrium concentrations of the reacting species.

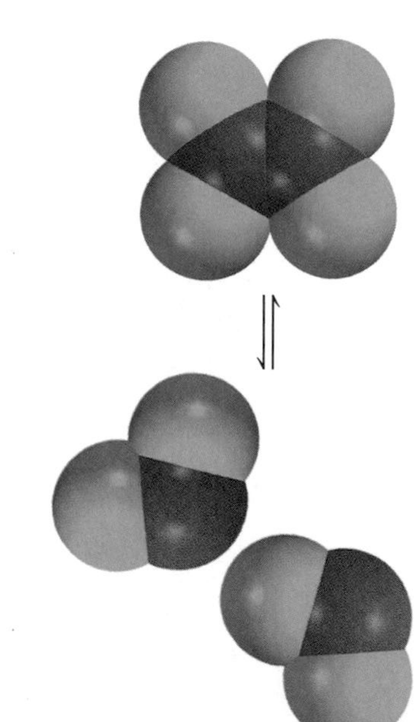

$N_2F_4 \rightleftharpoons 2NF_2$

EXAMPLE 15.10

Consider the following equilibrium process between dinitrogen tetrafluoride (N_2F_4) and nitrogen difluoride (NF_2):

$$N_2F_4(g) \rightleftharpoons 2NF_2(g) \qquad \Delta H° = 38.5 \text{ kJ/mol}$$

Predict the changes in the equilibrium if (a) the reacting mixture is heated at constant volume; (b) some N_2F_4 gas is removed from the reacting mixture at constant temperature and volume; (c) the pressure on the reacting mixture is decreased at constant temperature; and (d) a catalyst is added to the reacting mixture.

Strategy (a) What does the sign of $\Delta H°$ indicate about the heat change (endothermic or exothermic) for the forward reaction? (b) Would the removal of some N_2F_4 increase or decrease the Q_c of the reaction? (c) How would the decrease in pressure change the volume of the system? (d) What is the function of a catalyst? How does it affect a reacting system not at equilibrium? at equilibrium?

Solution

(a) The stress applied is the heat added to the system. Note that the $N_2F_4 \longrightarrow 2NF_2$ reaction is an endothermic process ($\Delta H° > 0$), which absorbs heat from the surroundings. Therefore, we can think of heat as a reactant

$$\text{heat} + N_2F_4(g) \rightleftharpoons 2NF_2(g)$$

The system will adjust to remove some of the added heat by undergoing a decomposition reaction (from left to right). The equilibrium constant

$$K_c = \frac{[NF_2]^2}{[N_2F_4]}$$

(Continued)

will therefore increase with increasing temperature because the concentration of NF_2 has increased and that of N_2F_4 has decreased. Recall that the equilibrium constant is a constant only at a particular temperature. If the temperature is changed, then the equilibrium constant will also change.

(b) The stress here is the removal of N_2F_4 gas. The system will shift to replace some of the N_2F_4 removed. Therefore, the system shifts from right to left until equilibrium is reestablished. As a result, some NF_2 combines to form N_2F_4.

Comment The equilibrium constant remains unchanged in this case because temperature is held constant. It might seem that K_c should change because NF_2 combines to produce N_2F_4. Remember, however, that initially some N_2F_4 was removed. The system adjusts to replace only some of the N_2F_4 that was removed, so that overall the amount of N_2F_4 has decreased. In fact, by the time the equilibrium is reestablished, the amounts of both NF_2 and N_2F_4 have decreased. Looking at the equilibrium constant expression, we see that dividing a smaller numerator by a smaller denominator gives the same value of K_c.

(c) The stress applied is a decrease in pressure (which is accompanied by an increase in gas volume). The system will adjust to remove the stress by increasing the pressure. Recall that pressure is directly proportional to the number of moles of a gas. In the balanced equation, we see that the formation of NF_2 from N_2F_4 will increase the total number of moles of gases and hence the pressure. Therefore, the system will shift from left to right to reestablish equilibrium. The equilibrium constant will remain unchanged because temperature is held constant.

(d) The function of a catalyst is to increase the rate of a reaction. If a catalyst is added to a reacting system not at equilibrium, the system will reach equilibrium faster than if left undisturbed. If a system is already at equilibrium, as in this case, the addition of a catalyst will not affect either the concentrations of NF_2 and N_2F_4 or the equilibrium constant.

Similar problems: 15.47, 15.48.

Practice Exercise Consider the equilibrium between molecular oxygen and ozone

$$3O_2(g) \rightleftharpoons 2O_3(g) \qquad \Delta H^\circ = 284 \text{ kJ/mol}$$

What would be the effect of (a) increasing the pressure on the system by decreasing the volume, (b) adding O_2 to the system, (c) decreasing the temperature, and (d) adding a catalyst?

Review of Concepts

The diagrams shown here represent the reaction $X_2 + Y_2 \rightleftharpoons 2XY$ at equilibrium at two temperatures ($T_2 > T_1$). Is the reaction endothermic or exothermic?

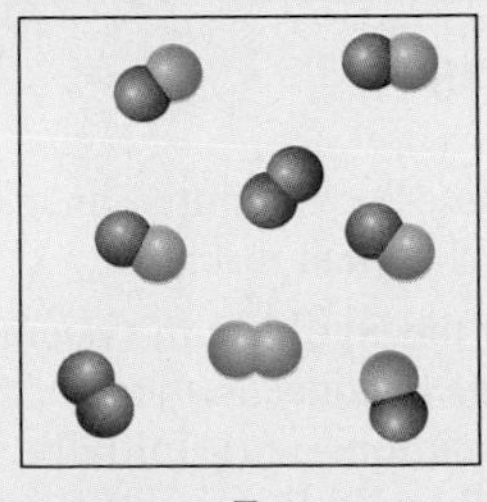

T_1

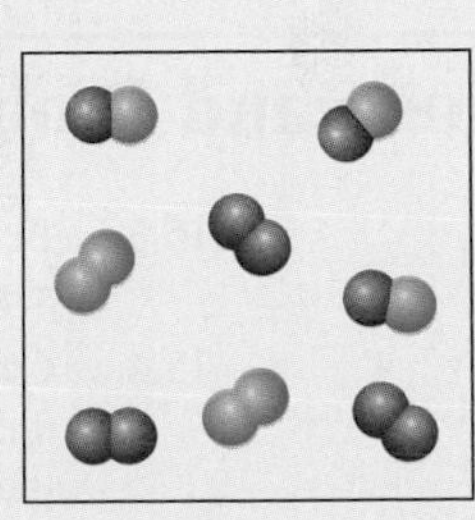

T_2

Key Equations

$K = \frac{[C]^c[D]^d}{[A]^a[B]^b}$	(15.2)	Law of mass action. General expression of equilibrium constant.
$K_P = K_c(0.0821T)^{\Delta n}$	(15.5)	Relationship between K_P and K_c.

Summary of Facts and Concepts

1. Dynamic equilibria between phases are called physical equilibria. Chemical equilibrium is a reversible process in which the rates of the forward and reverse reactions are equal and the concentrations of reactants and products do not change with time.
2. For the general chemical reaction

$$aA + bB \rightleftharpoons cC + dD$$

the concentrations of reactants and products at equilibrium (in moles per liter) are related by the equilibrium constant expression [Equation (15.2)].
3. The equilibrium constant for gases, K_P, expresses the relationship of the equilibrium partial pressures (in atm) of reactants and products.
4. A chemical equilibrium process in which all reactants and products are in the same phase is homogeneous. If the reactants and products are not all in the same phase, the equilibrium is heterogeneous. The concentrations of pure solids, pure liquids, and solvents are constant and do not appear in the equilibrium constant expression of a reaction.
5. The value of K depends on how the chemical equation is balanced, and the equilibrium constant for the reverse of a particular reaction is the reciprocal of the equilibrium constant of that reaction.
6. The equilibrium constant is the ratio of the rate constant for the forward reaction to that for the reverse reaction.
7. The reaction quotient Q has the same form as the equilibrium constant, but it applies to a reaction that may not be at equilibrium. If $Q > K$, the reaction will proceed from right to left to achieve equilibrium. If $Q < K$, the reaction will proceed from left to right to achieve equilibrium.
8. Le Châtelier's principle states that if an external stress is applied to a system at chemical equilibrium, the system will adjust to partially offset the stress.
9. Only a change in temperature changes the value of the equilibrium constant for a particular reaction. Changes in concentration, pressure, or volume may change the equilibrium concentrations of reactants and products. The addition of a catalyst hastens the attainment of equilibrium but does not affect the equilibrium concentrations of reactants and products.

Key Words

Chemical equilibrium, p. 511
Equilibrium constant, p. 513
Heterogeneous equilibrium, p. 518
Homogeneous equilibrium, p. 514
Le Châtelier's principle, p. 527
Physical equilibrium, p. 511
Reaction quotient (Q_c), p. 522

Questions and Problems

Concept of Equilibrium

Review Questions

15.1 Define equilibrium. Give two examples of a dynamic equilibrium.

15.2 Explain the difference between physical equilibrium and chemical equilibrium. Give two examples of each.

15.3 Briefly describe the importance of equilibrium in the study of chemical reactions.

15.4 Consider the equilibrium system $3A \rightleftharpoons B$. Sketch the change in concentrations of A and B with time for these situations: (a) Initially only A is present; (b) initially only B is present; (c) initially both A and B are present (with A in higher concentration).

In each case, assume that the concentration of B is higher than that of A at equilibrium.

Equilibrium Constant Expressions

Review Questions

15.5 Define homogeneous equilibrium and heterogeneous equilibrium. Give two examples of each.

15.6 What do the symbols K_c and K_P represent?

Problems

15.7 Write equilibrium constant expressions for K_c and for K_P, if applicable, for these processes:
(a) $2CO_2(g) \rightleftharpoons 2CO(g) + O_2(g)$
(b) $3O_2(g) \rightleftharpoons 2O_3(g)$
(c) $CO(g) + Cl_2(g) \rightleftharpoons COCl_2(g)$
(d) $H_2O(g) + C(s) \rightleftharpoons CO(g) + H_2(g)$
(e) $HCOOH(aq) \rightleftharpoons H^+(aq) + HCOO^-(aq)$
(f) $2HgO(s) \rightleftharpoons 2Hg(l) + O_2(g)$

15.8 Write the expressions for the equilibrium constants K_P of these thermal decompositions:
(a) $2NaHCO_3(s) \rightleftharpoons Na_2CO_3(s) + CO_2(g) + H_2O(g)$
(b) $2CaSO_4(s) \rightleftharpoons 2CaO(s) + 2SO_2(g) + O_2(g)$

15.9 Write the equilibrium constant expressions for K_c and K_P, if applicable, for these reactions:
(a) $2NO_2(g) + 7H_2(g) \rightleftharpoons 2NH_3(g) + 4H_2O(l)$
(b) $2ZnS(s) + 3O_2(g) \rightleftharpoons 2ZnO(s) + 2SO_2(g)$
(c) $C(s) + CO_2(g) \rightleftharpoons 2CO(g)$
(d) $C_6H_5COOH(aq) \rightleftharpoons C_6H_5COO^-(aq) + H^+(aq)$

Calculating Equilibrium Constants

Review Question

15.10 Write the equation relating K_c and K_P and define all the terms.

Problems

15.11 The equilibrium constant for the reaction $A \rightleftharpoons B$ is $K_c = 10$ at a certain temperature. (1) Starting with only reactant A, which of the diagrams best represents the system at equilibrium? (2) Which of the diagrams best represents the system at equilibrium if $K_c = 0.10$? Explain why you can calculate K_c in each case without knowing the volume of the container. The gray spheres represent the A molecules and the green spheres represent the B molecules.

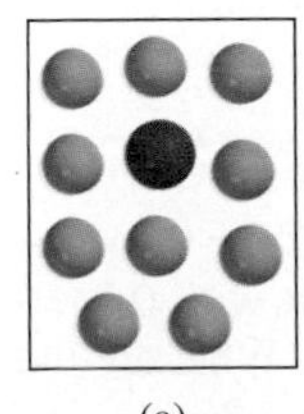
(a)

(b)

(c)

(d)

15.12 The following diagrams represent the equilibrium state for three different reactions of the type $A + X \rightleftharpoons AX$ (X = B, C, or D):

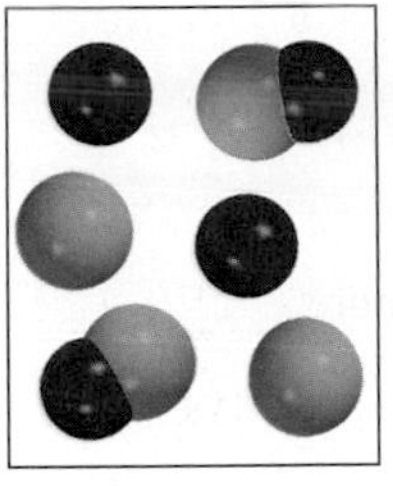
$A + B \rightleftharpoons AB$

$A + C \rightleftharpoons AC$

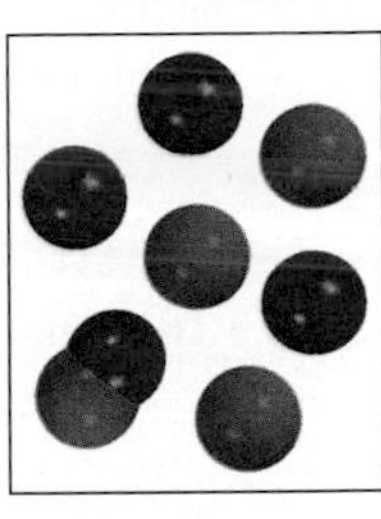
$A + D \rightleftharpoons AD$

(a) Which reaction has the largest equilibrium constant? (b) Which reaction has the smallest equilibrium constant?

15.13 The equilibrium constant (K_c) for the reaction

$$2HCl(g) \rightleftharpoons H_2(g) + Cl_2(g)$$

is 4.17×10^{-34} at 25°C. What is the equilibrium constant for the reaction

$$H_2(g) + Cl_2(g) \rightleftharpoons 2HCl(g)$$

at the same temperature?

15.14 Consider the following equilibrium process at 700°C:

$$2H_2(g) + S_2(g) \rightleftharpoons 2H_2S(g)$$

Analysis shows that there are 2.50 moles of H_2, 1.35×10^{-5} mole of S_2, and 8.70 moles of H_2S present in a 12.0-L flask at equilibrium. Calculate the equilibrium constant K_c for the reaction.

15.15 What is the K_P at 1273°C for the reaction

$$2CO(g) + O_2(g) \rightleftharpoons 2CO_2(g)$$

if K_c is 2.24×10^{22} at the same temperature?

15.16 The equilibrium constant K_P for the reaction

$$2SO_3(g) \rightleftharpoons 2SO_2(g) + O_2(g)$$

is 5.0×10^{-4} at 302°C. What is K_c for this reaction?

15.17 Consider this reaction:

$$N_2(g) + O_2(g) \rightleftharpoons 2NO(g)$$

If the equilibrium partial pressures of N_2, O_2, and NO are 0.15 atm, 0.33 atm, and 0.050 atm, respectively, at 2200°C, what is K_P?

15.18 A reaction vessel contains NH_3, N_2, and H_2 at equilibrium at a certain temperature. The equilibrium concentrations are $[NH_3] = 0.25\ M$, $[N_2] = 0.11\ M$, and $[H_2] = 1.91\ M$. Calculate the equilibrium constant K_c for the synthesis of ammonia if the reaction is represented as
(a) $N_2(g) + 3H_2(g) \rightleftharpoons 2NH_3(g)$
(b) $\frac{1}{2}N_2(g) + \frac{3}{2}H_2(g) \rightleftharpoons NH_3(g)$

15.19 The equilibrium constant K_c for the reaction

$$I_2(g) \rightleftharpoons 2I(g)$$

is 3.8×10^{-5} at 727°C. Calculate K_c and K_P for the equilibrium

$$2I(g) \rightleftharpoons I_2(g)$$

at the same temperature.

15.20 The pressure of the reacting mixture

$$CaCO_3(s) \rightleftharpoons CaO(s) + CO_2(g)$$

at equilibrium is 0.105 atm at 350°C. Calculate K_P and K_c for this reaction.

15.21 The equilibrium constant K_P for the reaction

$$PCl_5(g) \rightleftharpoons PCl_3(g) + Cl_2(g)$$

is 1.05 at 250°C. The reaction starts with a mixture of PCl_5, PCl_3, and Cl_2 at pressures of 0.177 atm, 0.223 atm, and 0.111 atm, respectively, at 250°C. When the mixture comes to equilibrium at that temperature, which pressures will have decreased and which will have increased? Explain why.

15.22 Ammonium carbamate, $NH_4CO_2NH_2$, decomposes as

$$NH_4CO_2NH_2(s) \rightleftharpoons 2NH_3(g) + CO_2(g)$$

Starting with only the solid, it is found that at 40°C the total gas pressure (NH_3 and CO_2) is 0.363 atm. Calculate the equilibrium constant K_P.

15.23 Consider the following reaction at 1600°C:

$$Br_2(g) \rightleftharpoons 2Br(g)$$

When 1.05 moles of Br_2 are put in a 0.980-L flask, 1.20 percent of the Br_2 undergoes dissociation. Calculate the equilibrium constant K_c for the reaction.

15.24 Pure phosgene gas ($COCl_2$), 3.00×10^{-2} mol, was placed in a 1.50-L container. It was heated to 800 K, and at equilibrium the pressure of CO was found to be 0.497 atm. Calculate the equilibrium constant K_P for the reaction

$$CO(g) + Cl_2(g) \rightleftharpoons COCl_2(g)$$

15.25 Consider the equilibrium

$$2NOBr(g) \rightleftharpoons 2NO(g) + Br_2(g)$$

If nitrosyl bromide, NOBr, is 34 percent dissociated at 25°C and the total pressure is 0.25 atm, calculate K_P and K_c for the dissociation at this temperature.

15.26 A 2.50-mol quantity of NOCl was initially placed in a 1.50-L reaction chamber at 400°C. After equilibrium was established, it was found that 28.0 percent of the NOCl had dissociated:

$$2NOCl(g) \rightleftharpoons 2NO(g) + Cl_2(g)$$

Calculate the equilibrium constant K_c for the reaction.

Calculating Equilibrium Concentrations

Review Questions

15.27 Define reaction quotient. How does it differ from equilibrium constant?

15.28 Outline the steps for calculating the concentrations of reacting species in an equilibrium reaction.

Problems

15.29 The equilibrium constant K_P for the reaction

$$2SO_2(g) + O_2(g) \rightleftharpoons 2SO_3(g)$$

is 5.60×10^4 at 350°C. SO_2 and O_2 are mixed initially at 0.350 atm and 0.762 atm, respectively, at 350°C. When the mixture equilibrates, is the total pressure less than or greater than the sum of the initial pressures, 1.112 atm?

15.30 For the synthesis of ammonia

$$N_2(g) + 3H_2(g) \rightleftharpoons 2NH_3(g)$$

the equilibrium constant K_c at 375°C is 1.2. Starting with $[H_2]_0 = 0.76\ M$, $[N_2]_0 = 0.60\ M$, and $[NH_3]_0 = 0.48\ M$, when this mixture comes to equilibrium, which gases will have increased in concentration and which will have decreased in concentration?

15.31 For the reaction

$$H_2(g) + CO_2(g) \rightleftharpoons H_2O(g) + CO(g)$$

at 700°C, $K_c = 0.534$. Calculate the number of moles of H_2 formed at equilibrium if a mixture of 0.300 mole of CO and 0.300 mole of H_2O is heated to 700°C in a 10.0-L container.

15.32 A sample of pure NO_2 gas heated to 1000 K decomposes:

$$2NO_2(g) \rightleftharpoons 2NO(g) + O_2(g)$$

The equilibrium constant K_P is 158. Analysis shows that the partial pressure of O_2 is 0.25 atm at equilibrium. Calculate the pressure of NO and NO_2 in the mixture.

15.33 The equilibrium constant K_c for the reaction

$$H_2(g) + Br_2(g) \rightleftharpoons 2HBr(g)$$

is 2.18×10^6 at 730°C. Starting with 3.20 moles HBr in a 12.0-L reaction vessel, calculate the concentrations of H_2, Br_2, and HBr at equilibrium.

15.34 The dissociation of molecular iodine into iodine atoms is represented as

$$I_2(g) \rightleftharpoons 2I(g)$$

At 1000 K, the equilibrium constant K_c for the reaction is 3.80×10^{-5}. Suppose you start with 0.0456 mole of I_2 in a 2.30-L flask at 1000 K. What are the concentrations of the gases at equilibrium?

15.35 The equilibrium constant K_c for the decomposition of phosgene, $COCl_2$, is 4.63×10^{-3} at 527°C:

$$COCl_2(g) \rightleftharpoons CO(g) + Cl_2(g)$$

Calculate the equilibrium partial pressure of all the components, starting with pure phosgene at 0.760 atm.

15.36 Consider this equilibrium process at 686°C:

$$CO_2(g) + H_2(g) \rightleftharpoons CO(g) + H_2O(g)$$

The equilibrium concentrations of the reacting species are $[CO] = 0.050\ M$, $[H_2] = 0.045\ M$, $[CO_2] = 0.086\ M$, and $[H_2O] = 0.040\ M$. (a) Calculate K_c for the reaction at 686°C. (b) If the concentration of CO_2 were raised to 0.50 mol/L by the addition of CO_2, what would be the concentrations of all the gases when equilibrium is reestablished?

15.37 Consider the heterogeneous equilibrium process:

$$C(s) + CO_2(g) \rightleftharpoons 2CO(g)$$

At 700°C, the total pressure of the system is found to be 4.50 atm. If the equilibrium constant K_P is 1.52, calculate the equilibrium partial pressures of CO_2 and CO.

15.38 The equilibrium constant K_c for the reaction

$$H_2(g) + CO_2(g) \rightleftharpoons H_2O(g) + CO(g)$$

is 4.2 at 1650°C. Initially 0.80 mol H_2 and 0.80 mol CO_2 are injected into a 5.0-L flask. Calculate the concentration of each species at equilibrium.

Le Châtelier's Principle

Review Questions

15.39 Explain Le Châtelier's principle. How can this principle help us maximize the yields of reactions?

15.40 Use Le Châtelier's principle to explain why the equilibrium vapor pressure of a liquid increases with increasing temperature.

15.41 List four factors that can shift the position of an equilibrium. Which one can alter the value of the equilibrium constant?

15.42 What is meant by "the position of an equilibrium"? Does the addition of a catalyst have any effects on the position of an equilibrium?

Problems

15.43 Consider this equilibrium system:

$$SO_2(g) + Cl_2(g) \rightleftharpoons SO_2Cl_2(g)$$

Predict how the equilibrium position would change if (a) Cl_2 gas were added to the system, (b) SO_2Cl_2 were removed from the system, (c) SO_2 were removed from the system. The temperature remains constant.

15.44 Heating solid sodium bicarbonate in a closed vessel established this equilibrium:

$$2NaHCO_3(s) \rightleftharpoons Na_2CO_3(s) + H_2O(g) + CO_2(g)$$

What would happen to the equilibrium position if (a) some of the CO_2 were removed from the system, (b) some solid Na_2CO_3 were added to the system, (c) some of the solid $NaHCO_3$ were removed from the system? The temperature remains constant.

15.45 Consider these equilibrium systems:

(a) $A \rightleftharpoons 2B$ $\qquad \Delta H° = 20.0$ kJ/mol

(b) $A + B \rightleftharpoons C$ $\qquad \Delta H° = -5.4$ kJ/mol

(c) $A \rightleftharpoons B$ $\qquad \Delta H° = 0.0$ kJ/mol

Predict the change in the equilibrium constant K_c that would occur in each case if the temperature of the reacting system were raised.

15.46 What effect does an increase in pressure have on each of these systems at equilibrium?

(a) $A(s) \rightleftharpoons 2B(s)$

(b) $2A(l) \rightleftharpoons B(l)$

(c) $A(s) \rightleftharpoons B(g)$

(d) $A(g) \rightleftharpoons B(g)$

(e) $A(g) \rightleftharpoons 2B(g)$

The temperature is kept constant. In each case, the reacting mixture is in a cylinder fitted with a movable piston.

15.47 Consider the equilibrium

$$2I(g) \rightleftharpoons I_2(g)$$

What would be the effect on the position of equilibrium of (a) increasing the total pressure on the system by decreasing its volume, (b) adding I_2 to the reaction mixture, (c) decreasing the temperature?

15.48 Consider this equilibrium process:

$$PCl_5(g) \rightleftharpoons PCl_3(g) + Cl_2(g) \qquad \Delta H° = 92.5 \text{ kJ/mol}$$

Predict the direction of the shift in equilibrium when (a) the temperature is raised, (b) more chlorine gas is added to the reaction mixture, (c) some PCl_3 is removed from the mixture, (d) the pressure on the gases is increased, (e) a catalyst is added to the reaction mixture.

15.49 Consider the reaction

$$2SO_2(g) + O_2(g) \rightleftharpoons 2SO_3(g) \qquad \Delta H° = -198.2 \text{ kJ/mol}$$

Comment on the changes in the concentrations of SO_2, O_2, and SO_3 at equilibrium if we were to (a) increase the temperature, (b) increase the pressure, (c) increase SO_2, (d) add a catalyst, (e) add helium at constant volume.

15.50 In the uncatalyzed reaction

$$N_2O_4(g) \rightleftharpoons 2NO_2(g)$$

at 100°C the pressures of the gases at equilibrium are $P_{N_2O_4} = 0.377$ atm and $P_{NO_2} = 1.56$ atm. What would happen to these pressures if a catalyst were present?

15.51 Consider the gas-phase reaction

$$2CO(g) + O_2(g) \rightleftharpoons 2CO_2(g)$$

Predict the shift in the equilibrium position when helium gas is added to the equilibrium mixture (a) at constant pressure and (b) at constant volume.

15.52 Consider this reaction at equilibrium in a closed container:

$$CaCO_3(s) \rightleftharpoons CaO(s) + CO_2(g)$$

What would happen if (a) the volume is increased, (b) some CaO is added to the mixture, (c) some $CaCO_3$ is removed, (d) some CO_2 is added to the mixture, (e) a few drops of an NaOH solution are added to the mixture, (f) a few drops of an HCl solution are added to the mixture (ignore the reaction between CO_2 and water), (g) the temperature is increased?

Additional Problems

15.53 Consider the statement: The equilibrium constant of a reacting mixture of solid NH_4Cl and gaseous NH_3 and HCl is 0.316. List three important pieces of information that are missing from this statement.

15.54 Pure NOCl gas was heated at 240°C in a 1.00-L container. At equilibrium the total pressure was 1.00 atm and the NOCl pressure was 0.64 atm.

$$2NOCl(g) \rightleftharpoons 2NO(g) + Cl_2(g)$$

(a) Calculate the partial pressures of NO and Cl_2 in the system. (b) Calculate the equilibrium constant K_P.

15.55 Consider this reaction:

$$N_2(g) + O_2(g) \rightleftharpoons 2NO(g)$$

The equilibrium constant K_P for the reaction is 1.0×10^{-15} at 25°C and 0.050 at 2200°C. Is the formation of nitric oxide endothermic or exothermic? Explain your answer.

15.56 Baking soda (sodium bicarbonate) undergoes thermal decomposition as

$$2NaHCO_3(s) \rightleftharpoons Na_2CO_3(s) + CO_2(g) + H_2O(g)$$

Would we obtain more CO_2 and H_2O by adding extra baking soda to the reaction mixture in (a) a closed vessel or (b) an open vessel?

15.57 Consider the following reaction at equilibrium:

$$A(g) \rightleftharpoons 2B(g)$$

From the following data, calculate the equilibrium constant (both K_P and K_c) at each temperature. Is the reaction endothermic or exothermic?

Temperature (°C)	[A]	[B]
200	0.0125	0.843
300	0.171	0.764
400	0.250	0.724

15.58 The equilibrium constant K_P for the reaction

$$2H_2O(g) \rightleftharpoons 2H_2(g) + O_2(g)$$

is found to be 2×10^{-42} at 25°C. (a) What is K_c for the reaction at the same temperature? (b) The very small value of K_P (and K_c) indicates that the reaction overwhelmingly favors the formation of water molecules. Explain why, despite this fact, a mixture of hydrogen and oxygen gases can be kept at room temperature without any change.

15.59 Consider the following reacting system:

$$2NO(g) + Cl_2(g) \rightleftharpoons 2NOCl(g)$$

What combination of temperature and pressure (high or low) would maximize the yield of NOCl? [*Hint:* $\Delta H_f^\circ(NOCl) = 51.7$ kJ/mol. You will also need to consult Appendix 2.]

15.60 At a certain temperature and a total pressure of 1.2 atm, the partial pressures of an equilibrium mixture

$$2A(g) \rightleftharpoons B(g)$$

are $P_A = 0.60$ atm and $P_B = 0.60$ atm. (a) Calculate the K_P for the reaction at this temperature. (b) If the total pressure were increased to 1.5 atm, what would be the partial pressures of A and B at equilibrium?

15.61 The decomposition of ammonium hydrogen sulfide

$$NH_4HS(s) \rightleftharpoons NH_3(g) + H_2S(g)$$

is an endothermic process. A 6.1589-g sample of the solid is placed in an evacuated 4.000-L vessel at exactly 24°C. After equilibrium has been established, the total pressure inside is 0.709 atm. Some solid NH_4HS remains in the vessel. (a) What is the K_P for the reaction? (b) What percentage of the solid has decomposed? (c) If the volume of the vessel were doubled at constant temperature, what would happen to the amount of solid in the vessel?

15.62 Consider the reaction

$$2NO(g) + O_2(g) \rightleftharpoons 2NO_2(g)$$

At 430°C, an equilibrium mixture consists of 0.020 mole of O_2, 0.040 mole of NO, and 0.96 mole of NO_2. Calculate K_P for the reaction, given that the total pressure is 0.20 atm.

15.63 When heated, ammonium carbamate decomposes as

$$NH_4CO_2NH_2(s) \rightleftharpoons 2NH_3(g) + CO_2(g)$$

At a certain temperature the equilibrium pressure of the system is 0.318 atm. Calculate K_P for the reaction.

15.64 A mixture of 0.47 mole of H_2 and 3.59 moles of HCl is heated to 2800°C. Calculate the equilibrium partial pressures of H_2, Cl_2, and HCl if the total pressure is 2.00 atm. The K_P for the reaction $H_2(g) + Cl_2(g) \rightleftharpoons 2HCl(g)$ is 193 at 2800°C.

15.65 Consider the reaction in a closed container:

$$N_2O_4(g) \rightleftharpoons 2NO_2(g)$$

Initially, 1 mole of N_2O_4 is present. At equilibrium, α mole of N_2O_4 has dissociated to form NO_2. (a) Derive an expression for K_P in terms of α and P, the total pressure. (b) How does the expression in (a) help you predict the shift in equilibrium caused by an increase in P? Does your prediction agree with Le Châtelier's principle?

15.66 One mole of N_2 and 3 moles of H_2 are placed in a flask at 397°C. Calculate the total pressure of the system at equilibrium if the mole fraction of NH_3 is found to be 0.21. The K_P for the reaction is 4.31×10^{-4}.

15.67 At 1130°C the equilibrium constant (K_c) for the reaction

$$2H_2S(g) \rightleftharpoons 2H_2(g) + S_2(g)$$

is 2.25×10^{-4}. If $[H_2S] = 4.84 \times 10^{-3}$ M and $[H_2] = 1.50 \times 10^{-3}$ M, calculate $[S_2]$.

15.68 A quantity of 6.75 g of SO_2Cl_2 was placed in a 2.00-L flask. At 648 K, there is 0.0345 mole of SO_2 present. Calculate K_c for the reaction

$$SO_2Cl_2(g) \rightleftharpoons SO_2(g) + Cl_2(g)$$

15.69 The formation of SO_3 from SO_2 and O_2 is an intermediate step in the manufacture of sulfuric acid, and it is also responsible for the acid rain phenomenon. The equilibrium constant (K_P) for the reaction

$$2SO_2(g) + O_2(g) \rightleftharpoons 2SO_3(g)$$

is 0.13 at 830°C. In one experiment 2.00 moles of SO_2 and 2.00 moles of O_2 were initially present in a flask. What must the total pressure be at equilibrium to have an 80.0 percent yield of SO_3?

15.70 Consider the dissociation of iodine:

$$I_2(g) \rightleftharpoons 2I(g)$$

A 1.00-g sample of I_2 is heated at 1200°C in a 500-mL flask. At equilibrium the total pressure is 1.51 atm. Calculate K_P for the reaction. [*Hint:* Use the result in problem 15.65(a). The degree of dissociation α can be obtained by first calculating the ratio of observed pressure over calculated pressure, assuming no dissociation.]

15.71 Eggshells are composed mostly of calcium carbonate ($CaCO_3$) formed by the reaction

$$Ca^{2+}(aq) + CO_3^{2-}(aq) \longrightarrow CaCO_3(s)$$

The carbonate ions are supplied by carbon dioxide produced as a result of metabolism. Explain why eggshells are thinner in the summer, when the rate of chicken panting is greater. Suggest a remedy for this situation.

15.72 The equilibrium constant K_P for the following reaction is found to be 4.31×10^{-4} at 375°C:

$$N_2(g) + 3H_2(g) \rightleftharpoons 2NH_3(g)$$

In a certain experiment a student starts with 0.862 atm of N_2 and 0.373 atm of H_2 in a constant-volume vessel at 375°C. Calculate the partial pressures of all species when equilibrium is reached.

15.73 A quantity of 0.20 mole of carbon dioxide was heated at a certain temperature with an excess of graphite in a closed container until the following equilibrium was reached:

$$C(s) + CO_2(g) \rightleftharpoons 2CO(g)$$

Under this condition, the average molar mass of the gases was found to be 35 g/mol. (a) Calculate the mole fractions of CO and CO_2. (b) What is the K_P for the equilibrium if the total pressure was 11 atm? (*Hint:* The average molar mass is the sum of the products of the mole fraction of each gas and its molar mass.)

15.74 When dissolved in water, glucose (corn sugar) and fructose (fruit sugar) exist in equilibrium as follows:

$$\text{fructose} \rightleftharpoons \text{glucose}$$

A chemist prepared a 0.244 M fructose solution at 25°C. At equilibrium, it was found that its concentration had decreased to 0.113 M. (a) Calculate the equilibrium constant for the reaction. (b) At equilibrium, what percentage of fructose was converted to glucose?

15.75 At room temperature, solid iodine is in equilibrium with its vapor through sublimation and deposition (see Figure 8.17). Describe how you would use radioactive iodine, in either solid or vapor form, to show that there is a dynamic equilibrium between these two phases.

15.76 At 1024°C, the pressure of oxygen gas from the decomposition of copper(II) oxide (CuO) is 0.49 atm:

$$4CuO(s) \rightleftharpoons 2Cu_2O(s) + O_2(g)$$

(a) What is the K_P for the reaction? (b) Calculate the fraction of CuO decomposed if 0.16 mole of it is placed in a 2.0-L flask at 1024°C. (c) What would be the fraction if a 1.0-mole sample of CuO were used? (d) What is the smallest amount of CuO (in moles) that would establish the equilibrium?

15.77 A mixture containing 3.9 moles of NO and 0.88 mole of CO_2 was allowed to react in a flask at a certain temperature according to the equation

$$NO(g) + CO_2(g) \rightleftharpoons NO_2(g) + CO(g)$$

At equilibrium, 0.11 mole of CO_2 was present. Calculate the equilibrium constant K_c of this reaction.

15.78 The equilibrium constant K_c for the reaction

$$H_2(g) + I_2(g) \rightleftharpoons 2HI(g)$$

is 54.3 at 430°C. At the start of the reaction there are 0.714 mole of H_2, 0.984 mole of I_2, and 0.886 mole of HI in a 2.40-L reaction chamber. Calculate the concentrations of the gases at equilibrium.

15.79 On heating, a gaseous compound A dissociates as follows:

$$A(g) \rightleftharpoons B(g) + C(g)$$

In an experiment A was heated at a certain temperature until its equilibrium pressure reached 0.14P, in which P is the total pressure. Calculate the equilibrium constant (K_P) of this reaction.

15.80 When a gas was heated under atmospheric conditions, its color was found to deepen. Heating above 150°C caused the color to fade, and at 550°C the color was barely detectable. However, at 550°C, the color was partially restored by increasing the pressure of the system. Which of these best fits this description? Justify your choice. (a) A mixture of hydrogen and bromine, (b) pure bromine, (c) a mixture of nitrogen dioxide and dinitrogen tetroxide. (*Hint:* Bromine has a reddish color and nitrogen dioxide is a brown gas. The other gases are colorless.)

15.81 The equilibrium constant K_c for the following reaction is 0.65 at 395°C.

$$N_2(g) + 3H_2(g) \rightleftharpoons 2NH_3(g)$$

(a) What is the value of K_P for this reaction?
(b) What is the value of the equilibrium constant K_c for $2NH_3(g) \rightleftharpoons N_2(g) + 3H_2(g)$?
(c) What is K_c for $\frac{1}{2}N_2(g) + \frac{3}{2}H_2(g) \rightleftharpoons NH_3(g)$?
(d) What are the values of K_P for the reactions described in (b) and (c)?

15.82 A sealed glass bulb contains a mixture of NO_2 and N_2O_4 gases. When the bulb is heated from 20°C to 40°C, what happens to these properties of the gases: (a) color, (b) pressure, (c) average molar mass, (d) degree of dissociation (from N_2O_4 to NO_2), (e) density? Assume that volume remains constant. (*Hint:* NO_2 is a brown gas; N_2O_4 is colorless.)

15.83 At 20°C, the vapor pressure of water is 0.0231 atm. Calculate K_P and K_c for the process

$$H_2O(l) \rightleftharpoons H_2O(g)$$

15.84 Industrially, sodium metal is obtained by electrolyzing molten sodium chloride. The reaction at the cathode is $Na^+ + e^- \longrightarrow Na$. We might expect that potassium metal would also be prepared by electrolyzing molten potassium chloride. However, potassium metal is soluble in molten potassium chloride and therefore is hard to recover. Furthermore, potassium vaporizes readily at the operating temperature, creating hazardous conditions. Instead, potassium is prepared by the distillation of molten potassium chloride in the presence of sodium vapor at 892°C:

$$Na(g) + KCl(l) \rightleftharpoons NaCl(l) + K(g)$$

In view of the fact that potassium is a stronger reducing agent than sodium, explain why this approach works. (The boiling points of sodium and potassium are 892°C and 770°C, respectively.)

15.85 In the gas phase, nitrogen dioxide is actually a mixture of nitrogen dioxide (NO_2) and dinitrogen tetroxide (N_2O_4). If the density of such a mixture at 74°C and 1.3 atm is 2.9 g/L, calculate the partial pressures of the gases and K_P.

15.86 About 75 percent of hydrogen for industrial use is produced by the *steam-reforming* process. This process is carried out in two stages called primary and secondary reforming. In the primary stage, a mixture of steam and methane at about 30 atm is heated over a nickel catalyst at 800°C to give hydrogen and carbon monoxide:

$$CH_4(g) + H_2O(g) \rightleftharpoons CO(g) + 3H_2(g) \qquad \Delta H° = 206 \text{ kJ/mol}$$

The secondary stage is carried out at about 1000°C, in the presence of air, to convert the remaining methane to hydrogen:

$$CH_4(g) + \tfrac{1}{2}O_2(g) \rightleftharpoons CO(g) + 2H_2(g) \qquad \Delta H° = 35.7 \text{ kJ/mol}$$

(a) What conditions of temperature and pressure would favor the formation of products in both the primary and secondary stages? (b) The equilibrium constant K_c for the primary stage is 18 at 800°C. (i) Calculate K_P for the reaction. (ii) If the partial pressures of methane and steam were both 15 atm at the start, what are the pressures of all the gases at equilibrium?

15.87 Photosynthesis can be represented by

$$6CO_2(g) + 6H_2O(l) \rightleftharpoons C_6H_{12}O_6(s) + 6O_2(g) \qquad \Delta H° = 2801 \text{ kJ/mol}$$

Explain how the equilibrium would be affected by the following changes: (a) partial pressure of CO_2 is increased, (b) O_2 is removed from the mixture, (c) $C_6H_{12}O_6$ (sucrose) is removed from the mixture, (d) more water is added, (e) a catalyst is added, (f) temperature is decreased, (g) more sunlight falls on the plants.

15.88 Consider the decomposition of ammonium chloride at a certain temperature:

$$NH_4Cl(s) \rightleftharpoons NH_3(g) + HCl(g)$$

Calculate the equilibrium constant K_P if the total pressure is 2.2 atm at that temperature.

15.89 At 25°C, the equilibrium partial pressures of NO_2 and N_2O_4 are 0.15 atm and 0.20 atm, respectively. If the volume is doubled at constant temperature, calculate the partial pressures of the gases when a new equilibrium is established.

15.90 In 1899 the German chemist Ludwig Mond developed a process for purifying nickel by converting it to the volatile nickel tetracarbonyl [$Ni(CO)_4$] (b.p. = 42.2°C):

$$Ni(s) + 4CO(g) \rightleftharpoons Ni(CO)_4(g)$$

(a) Describe how you can separate nickel and its solid impurities. (b) How would you recover nickel? [ΔH_f° for $Ni(CO)_4$ is −602.9 kJ/mol.]

15.91 Consider the equilibrium reaction in Problem 15.21. A quantity of 2.50 g of PCl_5 is placed in an evacuated flask of volume 0.500 L and heated to 250°C. (a) Calculate the pressure of PCl_5 if it did not dissociate. (b) Calculate the partial pressure of PCl_5 at equilibrium. (c) What is the total pressure at equilibrium? (d) What is the degree of dissociation of PCl_5? (The degree of dissociation is given by the fraction of PCl_5 that has undergone dissociation.)

15.92 The vapor pressure of mercury is 0.0020 mmHg at 26°C. (a) Calculate K_c and K_P for the process $Hg(l) \rightleftharpoons Hg(g)$. (b) A chemist breaks a thermometer and spills mercury onto the floor of a laboratory measuring 6.1 m long, 5.3 m wide, and 3.1 m high. Calculate the mass of mercury (in grams) vaporized at equilibrium and the concentration of mercury vapor in mg/m^3. Does this concentration exceed the safety limit of 0.050 mg/m^3? (Ignore the volume of furniture and other objects in the laboratory.)

15.93 Consider the potential energy diagrams for two types of reactions $A \rightleftharpoons B$. In each case, answer the following questions for the system at equilibrium. (a) How would a catalyst affect the forward and reverse rates of the reaction? (b) How would a catalyst affect the energies of the reactant and product? (c) How would an increase in temperature affect the equilibrium constant? (d) If the only effect of a catalyst is to lower the activation energies for the forward and reverse reactions, show that the equilibrium constant remains unchanged if a catalyst is added to the reacting mixture.

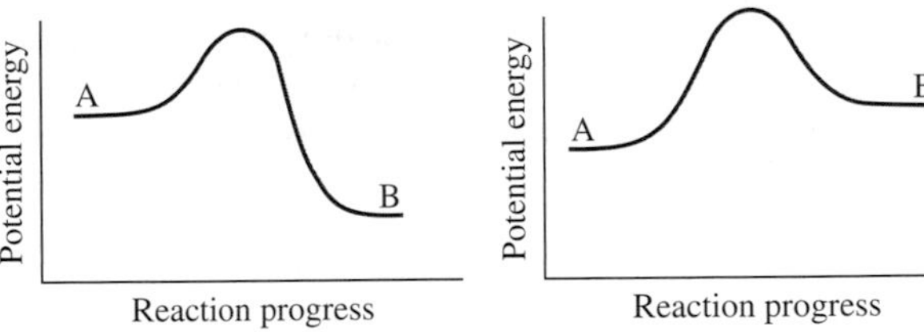

15.94 The equilibrium constant K_c for the reaction $2NH_3(g) \rightleftharpoons N_2(g) + 3H_2(g)$ is 0.83 at 375°C. A 14.6-g sample of ammonia is placed in a 4.00-L flask and heated to 375°C. Calculate the concentrations of all the gases when equilibrium is reached.

15.95 A quantity of 1.0 mole of N_2O_4 was introduced into an evacuated vessel and allowed to attain equilibrium at a certain temperature

$$N_2O_4(g) \rightleftharpoons 2NO_2(g)$$

The average molar mass of the reacting mixture was 70.6 g/mol. (a) Calculate the mole fractions of the gases. (b) Calculate K_P for the reaction if the total pressure was 1.2 atm. (c) What would be the mole fractions if the pressure were increased to 4.0 atm by reducing the volume at the same temperature?

15.96 The equilibrium constant (K_P) for the reaction

$$C(s) + CO_2(g) \rightleftharpoons 2CO(g)$$

is 1.9 at 727°C. What total pressure must be applied to the reacting system to obtain 0.012 mole of CO_2 and 0.025 mole of CO?

Special Problems

15.97 In this chapter, we learned that a catalyst has no effect on the position of an equilibrium because it speeds up both the forward and reverse rates to the same extent. To test this statement, consider a situation in which an equilibrium of the type

$$2A(g) \rightleftharpoons B(g)$$

is established inside a cylinder fitted with a weightless piston. The piston is attached by a string to the cover of a box containing a catalyst. When the piston moves upward (expanding against atmospheric pressure), the cover is lifted and the catalyst is exposed to the gases. When the piston moves downward, the box is closed. Assume that the catalyst speeds up the forward reaction ($2A \longrightarrow B$) but does not affect the reverse process ($B \longrightarrow 2A$). Suppose the catalyst is suddenly exposed to the equilibrium system as shown here. Describe what would happen subsequently. How does this "thought" experiment convince you that no such catalyst can exist?

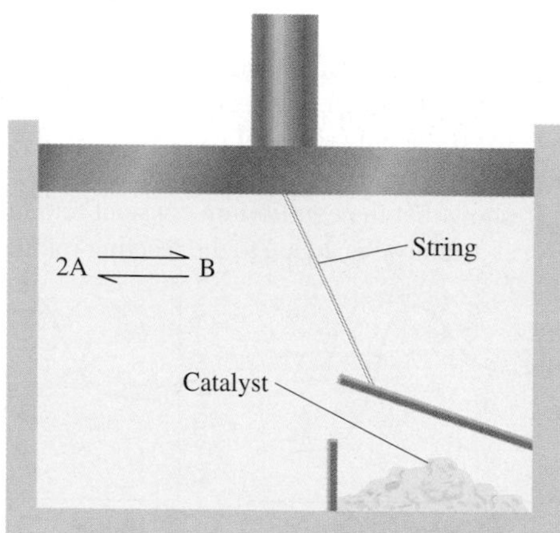

15.98 At 25°C, a mixture of NO_2 and N_2O_4 gases are in equilibrium in a cylinder fitted with a movable piston. The concentrations are: $[NO_2] = 0.0475\ M$ and $[N_2O_4] = 0.491\ M$. The volume of the gas mixture is halved by pushing down on the piston at constant temperature. Calculate the concentrations of the gases when equilibrium is reestablished. Will the color become darker or lighter after the change? [*Hint:* K_c for the dissociation of N_2O_4 is 4.63×10^{-3}. $N_2O_4(g)$ is colorless and $NO_2(g)$ has a brown color.]

15.99 The dependence of the equilibrium constant of a reaction on temperature is given by the van't Hoff equation:

$$\ln K = -\frac{\Delta H°}{RT} + C$$

where C is a constant. The following table gives the equilibrium constant (K_P) for the reaction at various temperatures

$$2NO(g) + O_2(g) \rightleftharpoons 2NO_2(g)$$

K_P	138	5.12	0.436	0.0626	0.0130
T(K)	600	700	800	900	1000

Determine graphically the $\Delta H°$ for the reaction.

15.100 (a) Use the van't Hoff equation in Problem 15.99 to derive the following expression, which relates the equilibrium constants at two different temperatures

$$\ln\frac{K_1}{K_2} = \frac{\Delta H°}{R}\left(\frac{1}{T_2} - \frac{1}{T_1}\right)$$

How does this equation support the prediction based on Le Châtelier's principle about the shift in equilibrium with temperature? (b) The vapor pressures of water are 31.82 mmHg at 30°C and 92.51 mmHg at 50°C. Calculate the molar heat of vaporization of water.

15.101 The "boat" form and "chair" form of cyclohexane (C_6H_{12}) interconverts as shown here:

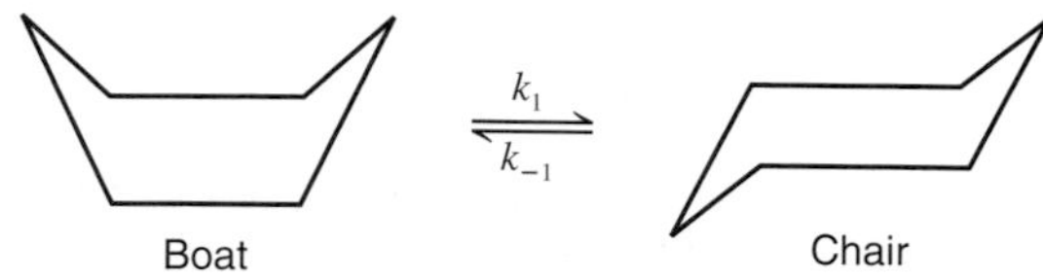

In this representation, the H atoms are omitted and a C atom is assumed to be at each intersection of two lines (bonds). The conversion is first order in each direction. The activation energy for the chair $\longrightarrow$ boat conversion is 41 kJ/mol. If the frequency factor is $1.0 \times 10^{12}\ s^{-1}$, what is k_1 at 298 K? The equilibrium constant K_c for the reaction is 9.83×10^3 at 298 K.

15.102 Consider the following reaction at a certain temperature

$$A_2 + B_2 \rightleftharpoons 2AB$$

The mixing of 1 mole of A_2 with 3 moles of B_2 gives rise to x mole of AB at equilibrium. The addition of 2 more moles of A_2 produces another x mole of AB. What is the equilibrium constant for the reaction?

15.103 Iodine is sparingly soluble in water but much more so in carbon tetrachloride (CCl_4). The equilibrium constant, also called the partition coefficient, for the distribution of I_2 between these two phases

$$I_2(aq) \rightleftharpoons I_2(CCl_4)$$

is 83 at 20°C. (a) A student adds 0.030 L of CCl_4 to 0.200 L of an aqueous solution containing 0.032 g I_2. The mixture is shaken and the two phases are then allowed to separate. Calculate the fraction of I_2 remaining in the aqueous phase. (b) The student now repeats the extraction of I_2 with another 0.030 L of CCl_4. Calculate the fraction of the I_2 from the original solution that remains in the aqueous phase. (c) Compare the result in (b) with a single extraction using 0.060 L of CCl_4. Comment on the difference.

15.104 Consider the following equilibrium system:

$$N_2O_4(g) \rightleftharpoons 2NO_2(g) \quad \Delta H° = 58.0\ \text{kJ/mol}$$

(a) If the volume of the reacting system is changed at constant temperature, describe what a plot of P versus $1/V$ would look like for the system. (*Hint:* See Figure 5.6.) (b) If the temperatures of the reacting system is changed at constant pressure, describe what a plot of V versus T would look like for the system. (*Hint:* See Figure 5.8.)

Answers to Practice Exercises

15.1 $K_c = \frac{[NO_2]^4[O_2]}{[N_2O_5]^2}$ $K_P = \frac{P^4_{NO_2}P_{O_2}}{P^2_{N_2O_5}}$. **15.2** 347 atm. **15.3** 1.2. **15.4** $K_P = 0.0702$; $K_c = 6.68 \times 10^{-5}$. **15.5** From right to left. **15.6** $[HI] = 0.031\ M$, $[H_2] = 4.3 \times 10^{-3}\ M$, $[I_2] = 4.3 \times 10^{-3}\ M$.

15.7 $[Br_2] = 0.065\ M$, $[Br] = 8.4 \times 10^{-3}\ M$. **15.8** $Q_P = 4.0 \times 10^5$, the net reaction will shift from right to left. **15.9** Left to right. **15.10** The equilibrium will shift from (a) left to right, (b) left to right, and (c) right to left. (d) A catalyst has no effect on the equilibrium.

CHAPTER 16

Acids and Bases

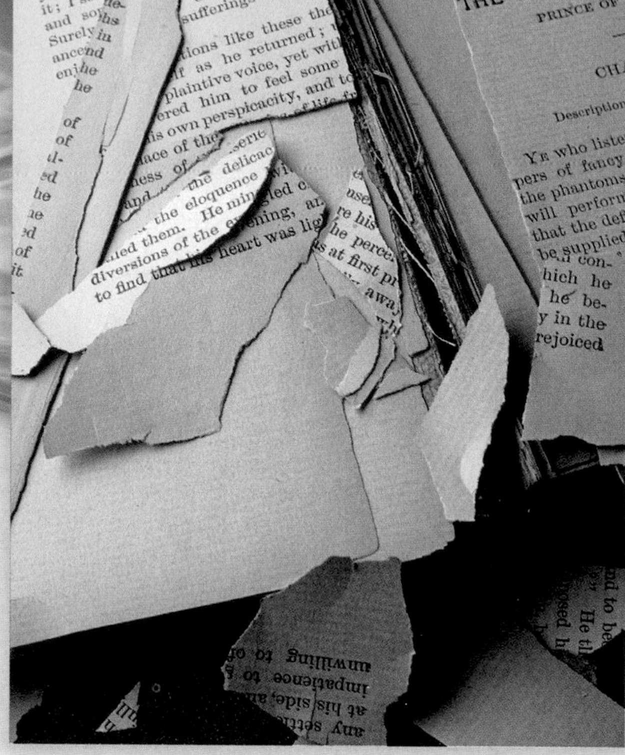

Without neutralization and careful storage, many old books will succumb to the effects of acids generated during the manufacturing process and the decomposition of the cellulose fibers.

Chapter Outline

Student Interactive Activities

Animations
Acid Ionization (16.5)
Base Ionization (16.6)

Electronic Homework
Example Practice Problems
End of Chapter Problems

Essential Concepts

Brønsted Acids and Bases A Brønsted acid can donate a proton and a Brønsted base can accept a proton. For every Brønsted acid, there exists a conjugate Brønsted base and vice versa.

Acid-Base Properties of Water and the pH Scale Water acts both as a Brønsted acid and as a Brønsted base. At 25°C, the concentrations of H^+ and OH^- ions are both at 10^{-7} *M*. The pH scale is established to express the acidity of a solution—the smaller the pH, the higher the H^+ concentration and the greater the acidity.

Acid and Base Ionization Constants Strong acids and strong bases are assumed to ionize completely. Most weak acids and bases ionize to a small extent. The concentrations of the acid, conjugate base, and H^+ ion at equilibrium can be calculated from the acid ionization constant, which is the equilibrium constant for the reaction.

Molecular Structure and Acid Strength The strength of a series of structurally similar acids can be compared using parameters such as bond enthalpy, bond polarity, and oxidation number.

Acid-Base Properties of Salts and Oxides Many salts react with water in a process called hydrolysis. From the nature of the cation and anion present in the salt, it is possible to predict the pH of the resulting solution. Most oxides also react with water to produce acidic or basic solutions.

Lewis Acids and Bases A more general definition of acids and bases characterizes an acid as a substance that can accept a pair of electrons and a base as a substance that can donate a pair of electrons. All Brønsted acids and bases are Lewis acids and bases.

16.1 Brønsted Acids and Bases

In Chapter 4 we defined a Brønsted acid as a substance capable of donating a proton, and a Brønsted base as a substance capable of accepting a proton. These definitions are generally suitable for discussion of the properties and reactions of acids and bases.

Conjugate Acid-Base Pairs

An extension of the Brønsted definition of acids and bases is the concept of the ***conjugate acid-base pair,*** which can be defined as *an acid and its conjugate base or a base and its conjugate acid.* The conjugate base of a Brønsted acid is the species that remains when one proton has been removed from the acid. Conversely, a conjugate acid results from the addition of a proton to a Brønsted base.

Conjugate means "joined together."

Every Brønsted acid has a conjugate base, and every Brønsted base has a conjugate acid. For example, the chloride ion (Cl^-) is the conjugate base formed from the acid HCl, and H_2O is the conjugate base of the acid H_3O^+. Similarly, the ionization of acetic acid can be represented as

$$\text{H}_3\text{C}{-}\text{C}({=}\ddot{\text{O}}{:}){-}\ddot{\text{O}}{-}\text{H} + \text{H}{-}\ddot{\text{O}}{:}{-}\text{H} \rightleftharpoons \text{H}_3\text{C}{-}\text{C}({=}\ddot{\text{O}}{:}){-}\ddot{\text{O}}{:}^- + \left[\text{H}{-}\ddot{\text{O}}{-}\text{H} \atop \text{H}\right]^+$$

$$\underset{\text{acid}_1}{CH_3COOH(aq)} + \underset{\text{base}_2}{H_2O(l)} \rightleftharpoons \underset{\text{base}_1}{CH_3COO^-(aq)} + \underset{\text{acid}_2}{H_3O^+(aq)}$$

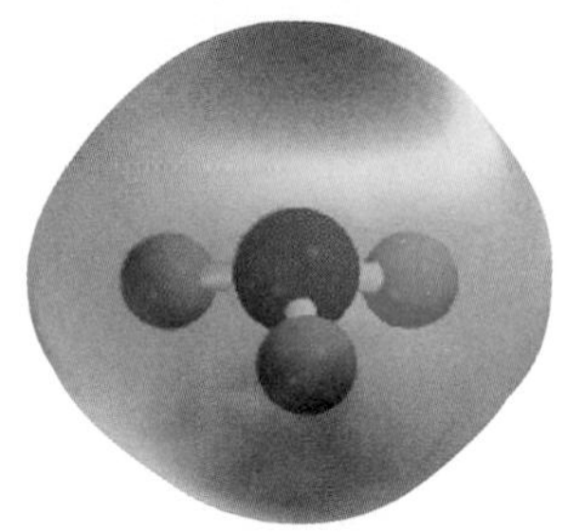

Electrostatic potential map of the hydronium ion. The proton is always associated with water molecules in aqueous solution. The H_3O^+ ion is the simplest formula of a hydrated proton.

The subscripts 1 and 2 designate the two conjugate acid-base pairs. Thus, the acetate ion (CH_3COO^-) is the conjugate base of the acid CH_3COOH. Both the ionization of HCl (see Section 4.3) and the ionization of CH_3COOH are examples of Brønsted acid-base reactions.

The Brønsted definition also enables us to classify ammonia as a base because of its ability to accept a proton:

$$\text{H}{-}\ddot{\text{N}}(\text{H}){-}\text{H} + \text{H}{-}\ddot{\text{O}}{:}{-}\text{H} \rightleftharpoons \left[\text{H}{-}\text{N}(\text{H})_2{-}\text{H}\right]^+ + \text{H}{-}\ddot{\text{O}}{:}^-$$

$$\underset{\text{base}_1}{NH_3(aq)} + \underset{\text{acid}_2}{H_2O(l)} \rightleftharpoons \underset{\text{acid}_1}{NH_4^+(aq)} + \underset{\text{base}_2}{OH^-(aq)}$$

In this case, NH_4^+ is the conjugate acid of the base NH_3, and OH^- is the conjugate base of the acid H_2O. Note that the atom in the Brønsted base that accepts a H^+ ion must have a lone pair.

EXAMPLE 16.1

Identify the conjugate acid-base pairs in the reaction between ammonia and hydrofluoric acid in aqueous solution

$$NH_3(aq) + HF(aq) \rightleftharpoons NH_4^+(aq) + F^-(aq)$$

Strategy Remember that a conjugate base always has one fewer H atom and one more negative charge (or one fewer positive charge) than the formula of the corresponding acid.

(Continued)

Solution NH_3 has one fewer H atom and one fewer positive charge than NH_4^+. F^- has one fewer H atom and one more negative charge than HF. Therefore, the conjugate acid-base pairs are (1) NH_4^+ and NH_3 and (2) HF and F^-.

Similar problem: 16.5.

Practice Exercise Identify the conjugate acid-base pairs for the reaction

$$CN^- + H_2O \rightleftharpoons HCN + OH^-$$

It is acceptable to represent the proton in aqueous solution either as H^+ or as H_3O^+. The formula H^+ is less cumbersome in calculations involving hydrogen ion concentrations and in calculations involving equilibrium constants, whereas H_3O^+ is more representative of the proton in aqueous solution.

REVIEW OF CONCEPTS

Which of the following does not constitute a conjugate acid-base pair? (a) HNO_2, NO_2^-; (b) H_2CO_3, CO_3^{2-}; (c) $CH_3NH_3^+$, CH_3NH_2.

16.2 The Acid-Base Properties of Water

Water, as we know, is a unique solvent. One of its special properties is its ability to act either as an acid or as a base. Water functions as a base in reactions with acids such as HCl and CH_3COOH, and it functions as an acid in reactions with bases such as NH_3. Water is a very weak electrolyte and therefore a poor conductor of electricity, but it does undergo ionization to a small extent:

Tap water and water from underground sources do conduct electricity because they contain many dissolved ions.

$$H_2O(l) \rightleftharpoons H^+(aq) + OH^-(aq) \tag{16.1}$$

This reaction is sometimes called the *autoionization* of water. To describe the acid-base properties of water in the Brønsted framework, we express its autoionization as follows (also shown in Figure 16.1):

$$\mathrm{H{-}\ddot{\underset{|}{\underset{H}{O}}}{:}} + \mathrm{H{-}\ddot{\underset{|}{\underset{H}{O}}}{:}} \rightleftharpoons \left[\mathrm{H{-}\ddot{\underset{|}{\underset{H}{O}}}{-}H}\right]^+ + \mathrm{H{-}\ddot{\underset{\cdot\cdot}{O}}{:}^-}$$

or

$$\underset{\text{acid}_1}{H_2O} + \underset{\text{base}_2}{H_2O} \rightleftharpoons \underset{\text{acid}_2}{H_3O^+} + \underset{\text{base}_1}{OH^-} \tag{16.2}$$

The acid-base conjugate pairs are (1) H_2O (acid) and OH^- (base) and (2) H_3O^+ (acid) and H_2O (base).

The Ion Product of Water

In the study of acid-base reactions, the hydrogen ion concentration is key; its value indicates the acidity or basicity of the solution. Because only a very small fraction of water molecules are ionized, the concentration of water, $[H_2O]$, remains virtually

Recall that in pure water, $[H_2O] = 55.5\ M$ (see p. 516).

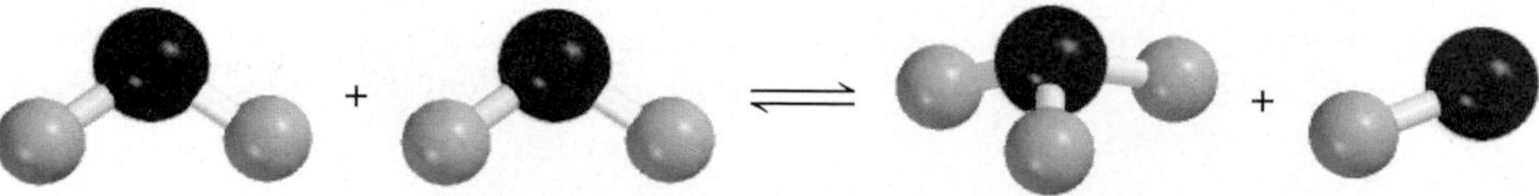

Figure 16.1
Reaction between two water molecules to form hydronium and hydroxide ions.

unchanged. Therefore, the equilibrium constant for the autoionization of water, according to Equation (16.2), is

$$K_c = [H_3O^+][OH^-]$$

Because we use $H^+(aq)$ and $H_3O^+(aq)$ interchangeably to represent the hydrated proton, the equilibrium constant can also be expressed as

$$K_c = [H^+][OH^-]$$

To indicate that the equilibrium constant refers to the autoionization of water, we replace K_c by K_w

$$K_w = [H_3O^+][OH^-] = [H^+][OH^-] \quad (16.3)$$

Kw refers to the autoionization of water. "Ion-product constant"

where K_w is called the ***ion-product constant,*** which is *the product of the molar concentrations of H^+ and OH^- ions at a particular temperature.*

In pure water at 25°C, the concentrations of H^+ and OH^- ions are equal and found to be $[H^+] = 1.0 \times 10^{-7}$ *M* and $[OH^-] = 1.0 \times 10^{-7}$ *M*. Thus, from Equation (16.3), at 25°C

If you could randomly remove and examine 10 particles (H_2O, H^+, or OH^-) per second from a liter of water, it would take you 2 years, working nonstop, to find one H^+ ion!

$$K_w = (1.0 \times 10^{-7})(1.0 \times 10^{-7}) = 1.0 \times 10^{-14}$$

Whether we have pure water or an aqueous solution of dissolved species, the following relation *always* holds at 25°C:

$$K_w = [H^+][OH^-] = 1.0 \times 10^{-14} \quad (16.4)$$

Whenever $[H^+] = [OH^-]$ the aqueous solution is said to be neutral. In an acidic solution, there is an excess of H^+ ions and $[H^+] > [OH^-]$. In a basic solution, there is an excess of hydroxide ions, so $[H^+] < [OH^-]$. In practice, we can change the concentration of either H^+ or OH^- ions in solution, but we cannot vary both of them independently. If we adjust the solution so that $[H^+] = 1.0 \times 10^{-6}$ *M*, the OH^- concentration *must* change to

$$[OH^-] = \frac{K_w}{[H^+]} = \frac{1.0 \times 10^{-14}}{1.0 \times 10^{-6}} = 1.0 \times 10^{-8}\ M$$

EXAMPLE 16.2

The concentration of OH^- ions in a certain household ammonia cleaning solution is 0.0030 *M*. Calculate the concentration of H^+ ions.

Strategy We are given the concentration of OH^- ions and are asked to calculate $[H^+]$. The relationship between $[H^+]$ and $[OH^-]$ in water or an aqueous solution is given by the ion-product of water, K_w [Equation (16.4)].

Solution Rearranging Equation (16.4), we write

$$[H^+] = \frac{K_w}{[OH^-]} = \frac{1.0 \times 10^{-14}}{0.0030} = 3.3 \times 10^{-12}\ M$$

Check Because $[H^+] < [OH^-]$, the solution is basic, as we would expect from the earlier discussion of the reaction of ammonia with water.

Similar problem: 16.16(c).

Practice Exercise Calculate the concentration of OH^- ions in a HCl solution whose hydrogen ion concentration is 1.3 *M*.

Review of Concepts

If the H^+ ion concentration in an aqueous solution is 0.0010 *M*, why is it not possible for the OH^- ion concentration to be 1.0×10^{-10} *M*?

16.3 pH—A Measure of Acidity

Because the concentrations of H^+ and OH^- ions in aqueous solutions are frequently very small numbers and therefore inconvenient to work with, the Danish chemist Soren Sorensen in 1909 proposed a more practical measure called pH. The ***pH*** of a solution is defined as *the negative logarithm of the hydrogen ion concentration (in mol/L):*

Note that a unit pH change corresponds to a 10-fold change in $[H^+]$.

$$\mathrm{pH} = -\log [H_3O^+] \quad \text{or} \quad \mathrm{pH} = -\log [H^+] \tag{16.5}$$

Keep in mind that Equation (16.5) is simply a definition designed to give us convenient numbers to work with. The negative logarithm gives us a positive number for pH, which otherwise would be negative due to the small value of $[H^+]$. Furthermore, the term $[H^+]$ in Equation (16.5) pertains only to the *numerical part* of the expression for hydrogen ion concentration, for we cannot take the logarithm of units. Thus, like the equilibrium constant, the pH of a solution is a dimensionless quantity.

Because pH is simply a way to express hydrogen ion concentration, acidic and basic solutions at 25°C can be distinguished by their pH values, as follows:

The pH of concentrated acid solutions can be negative. For example, the pH of a 2.0 *M* HCl solution is −0.30.

Acidic solutions: $[H^+] > 1.0 \times 10^{-7}\,M$, pH < 7.00
Basic solutions: $[H^+] < 1.0 \times 10^{-7}\,M$, pH > 7.00
Neutral solutions: $[H^+] = 1.0 \times 10^{-7}\,M$, pH $= 7.00$

Notice that pH increases as $[H^+]$ decreases.

Sometimes we may be given the pH value of a solution and asked to calculate the H^+ ion concentration. In that case, we need to take the antilog of Equation (16.5) as follows:

$$[H^+] = 10^{-\mathrm{pH}} \tag{16.6}$$

Be aware that the definition of pH just shown, and indeed all the calculations involving solution concentrations (expressed either as molarity or molality) discussed in previous chapters, are subject to error because we have implicitly assumed ideal behavior. In reality, ion-pair formation and other types of intermolecular interactions may affect the actual concentrations of species in solution. The situation is analogous to the relationships between ideal gas behavior and the behavior of real gases discussed in Chapter 5. Depending on temperature, volume, and amount and type of gas present, the measured gas pressure may differ from that calculated using the ideal gas equation. Similarly, the actual or "effective" concentration of a solute may not be what we think it is, knowing the amount of substance originally dissolved in solution. Just as we have the van der Waals and other equations to reconcile discrepancies between the ideal gas and nonideal gas behavior, we can account for nonideal behavior in solution.

One way is to replace the concentration term with *activity,* which is the effective concentration. Strictly speaking, then, the pH of solution should be defined as

$$\mathrm{pH} = -\log a_{H^+} \tag{16.7}$$

In dilute solutions, molarity is numerically equal to activity.

where a_{H^+} is the activity of the H^+ ion. As mentioned in Chapter 15 (see p. 516), for an ideal solution activity is numerically equal to concentration. For real solutions,

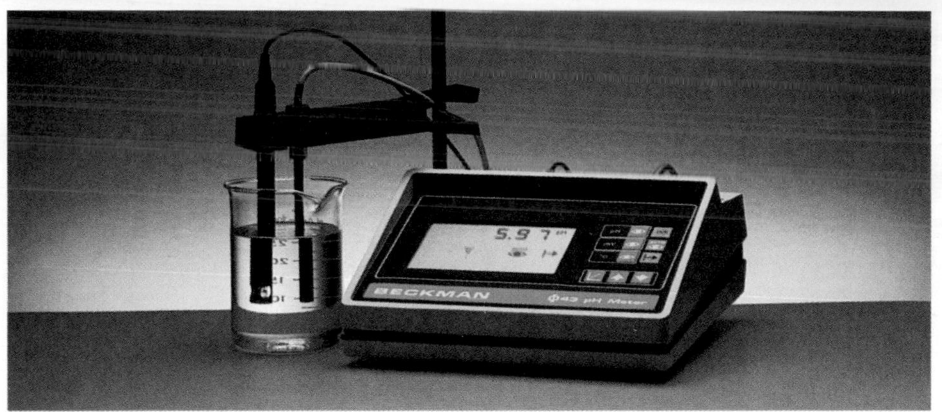

Figure 16.2
A pH meter is commonly used in the laboratory to determine the pH of a solution. Although many pH meters have scales marked with values from 1 to 14, pH values can, in fact, be less than 1 and greater than 14.

activity usually differs from concentration, sometimes appreciably. Knowing the solute concentration, there are reliable ways based on thermodynamics for estimating its activity, but the details are beyond the scope of this text. Keep in mind, therefore, that the measured pH of a solution is usually not the same as that calculated from Equation (16.5) because the concentration of the H^+ ion in molarity is not numerically equal to its activity value. Although we will continue to use concentration in our discussion, it is important to know that this approach will give us only an approximation of the chemical processes that actually take place in the solution phase.

In the laboratory, the pH of a solution is measured with a pH meter (Figure 16.2). Table 16.1 lists the pHs of a number of common fluids. As you can see, the pH of body fluids varies greatly, depending on location and function. The low pH (high acidity) of gastric juices facilitates digestion whereas a higher pH of blood is necessary for the transport of oxygen.

A pOH scale analogous to the pH scale can be devised using the negative logarithm of the hydroxide ion concentration of a solution. Thus, we define pOH as

$$\text{pOH} = -\log[\text{OH}^-] \tag{16.8}$$

If we are given the pOH value of a solution and asked to calculate the OH^- ion concentration, we can take the antilog of Equation (16.8) as follows

$$[\text{OH}^-] = 10^{-\text{pOH}} \tag{16.9}$$

Now consider again the ion-product constant for water at 25°C:

$$[\text{H}^+][\text{OH}^-] = K_w = 1.0 \times 10^{-14}$$

Taking the negative logarithm of both sides, we obtain

$$-(\log[\text{H}^+] + \log[\text{OH}^-]) = -\log(1.0 \times 10^{-14})$$
$$-\log[\text{H}^+] - \log[\text{OH}^-] = 14.00$$

From the definitions of pH and pOH we obtain

$$\text{pH} + \text{pOH} = 14.00 \tag{16.10}$$

Equation (16.10) provides us with another way to express the relationship between the H^+ ion concentration and the OH^- ion concentration.

Table 16.1
The pHs of Some Common Fluids

Sample	pH Value
Gastric juice in the stomach	1.0–2.0
Lemon juice	2.4
Vinegar	3.0
Grapefruit juice	3.2
Orange juice	3.5
Urine	4.8–7.5
Water exposed to air*	5.5
Saliva	6.4–6.9
Milk	6.5
Pure water	7.0
Blood	7.35–7.45
Tears	7.4
Milk of magnesia	10.6
Household ammonia	11.5

*Water exposed to air for a long period of time absorbs atmospheric CO_2 to form carbonic acid, H_2CO_3.

EXAMPLE 16.3

The concentration of H^+ ions in a bottle of table wine was 4.1×10^{-4} *M* right after the cork was removed. Only half of the wine was consumed. The other half, after it had been standing open to the air for a month, was found to have a hydrogen ion concentration equal to 2.3×10^{-3} *M*. Calculate the pH of the wine on these two occasions.

Strategy We are given the H^+ ion concentration and are asked to calculate the pH of the solution. What is the definition of pH?

Solution According to Equation (16.5), $pH = -\log[H^+]$. When the bottle was first opened, $[H^+] = 4.1 \times 10^{-4}$ *M*, which we substitute in Equation (16.5):

$$pH = -\log[H^+]$$
$$= -\log(4.1 \times 10^{-4}) = 3.39$$

In each case, the pH has only two significant figures. The two digits to the right of the decimal in 3.39 tell us that there are two significant figures in the original number (see Appendix 3).

On the second occasion, $[H^+] = 2.3 \times 10^{-3}$ *M*, so that

$$pH = -\log(2.3 \times 10^{-3}) = 2.64$$

Comment The increase in hydrogen ion concentration (or decrease in pH) is largely the result of the conversion of some of the alcohol (ethanol) to acetic acid, a reaction that takes place in the presence of molecular oxygen.

Similar problems: 16.17(a), (d).

Practice Exercise Nitric acid (HNO_3) is used in the production of fertilizer, dyes, drugs, and explosives. Calculate the pH of a HNO_3 solution having a hydrogen ion concentration of 0.76 *M*.

EXAMPLE 16.4

The pH of rainwater collected in a certain region of the northeastern United States on a particular day was 5.23. Calculate the H^+ ion concentration of the rainwater.

Strategy Here we are given the pH of a solution and asked to calculate $[H^+]$. Because pH is defined as $pH = -\log[H^+]$, we can solve for $[H^+]$ by taking the antilog of the pH; that is, $[H^+] = 10^{-pH}$, as shown in Equation (16.6).

Solution From Equation (16.5),

$$pH = -\log[H^+] = 5.23$$

Therefore,

$$\log[H^+] = -5.23$$

Scientific calculators have an antilog function that is sometimes labeled INV log or 10^x.

To calculate $[H^+]$, we need to take the antilog of −5.23:

$$[H^+] = 10^{-5.23} = 5.9 \times 10^{-6}\ M$$

Check Because the pH is between 5 and 6, we can expect $[H^+]$ to be between 1×10^{-5} and 1×10^{-6} *M*. Therefore, the answer is reasonable.

Similar problems: 16.16(a), (b).

Practice Exercise The pH of a certain orange juice is 3.33. Calculate the H^+ ion concentration.

EXAMPLE 16.5

In a NaOH solution $[OH^-]$ is 7.2×10^{-5} *M*. Calculate the pH of the solution.

Strategy Solving this problem takes two steps. First, we need to calculate the pOH using Equation (16.8). Next, we use Equation (16.10) to calculate the pH of the solution.

Solution We use Equation (16.8):

$$\begin{aligned} \text{pOH} &= -\log[OH^-] \\ &= -\log(7.2 \times 10^{-5}) \\ &= 4.14 \end{aligned}$$

Now we use Equation (16.10)

$$\begin{aligned} \text{pH} + \text{pOH} &= 14.00 \\ \text{pH} &= 14.00 - \text{pOH} \\ &= 14.00 - 4.14 = \boxed{9.86} \end{aligned}$$

Alternatively, we can use the ion-product constant of water, $K_w = [H^+][OH^-]$ to calculate $[H^+]$, and then we can calculate the pH from the $[H^+]$. Try it.

Check The answer shows that the solution is basic (pH > 7), which is consistent with a NaOH solution.

Similar problem: 16.17(b).

Practice Exercise The OH^- ion concentration of a blood sample is 2.5×10^{-7} *M*. What is the pH of the blood?

REVIEW OF CONCEPTS

Which is more acidic: a solution where $[H^+] = 2.5 \times 10^{-3}$ *M* or a solution with a pOH = 11.6?

16.4 Strength of Acids and Bases

In reality, no acids are known to ionize completely in water.

Strong acids *are strong electrolytes which,* for practical purposes, *are assumed to ionize completely in water* (Figure 16.3). Most of the strong acids are inorganic acids: hydrochloric acid (HCl), nitric acid (HNO_3), perchloric acid ($HClO_4$), and sulfuric acid (H_2SO_4):

$$\begin{aligned} HCl(aq) + H_2O(l) &\longrightarrow H_3O^+(aq) + Cl^-(aq) \\ HNO_3(aq) + H_2O(l) &\longrightarrow H_3O^+(aq) + NO_3^-(aq) \\ HClO_4(aq) + H_2O(l) &\longrightarrow H_3O^+(aq) + ClO_4^-(aq) \\ H_2SO_4(aq) + H_2O(l) &\longrightarrow H_3O^+(aq) + HSO_4^-(aq) \end{aligned}$$

Note that H_2SO_4 is a diprotic acid; we show only the first stage of ionization here. At equilibrium, solutions of strong acids will not contain any nonionized acid molecules.

Most acids are ***weak acids,*** which *ionize only to a limited extent in water.* At equilibrium, aqueous solutions of weak acids contain a mixture of nonionized acid molecules, H_3O^+ ions, and the conjugate base. Examples of weak acids are hydrofluoric acid (HF), acetic acid (CH_3COOH), and the ammonium ion (NH_4^+). The limited ionization of weak acids is related to the equilibrium constant for ionization, which we will study in the next section.

Zn reacts more vigorously with a strong acid like HCl (left) than with a weak acid like CH_3COOH (right) of the same concentration because there are more H^+ ions in the former solution.

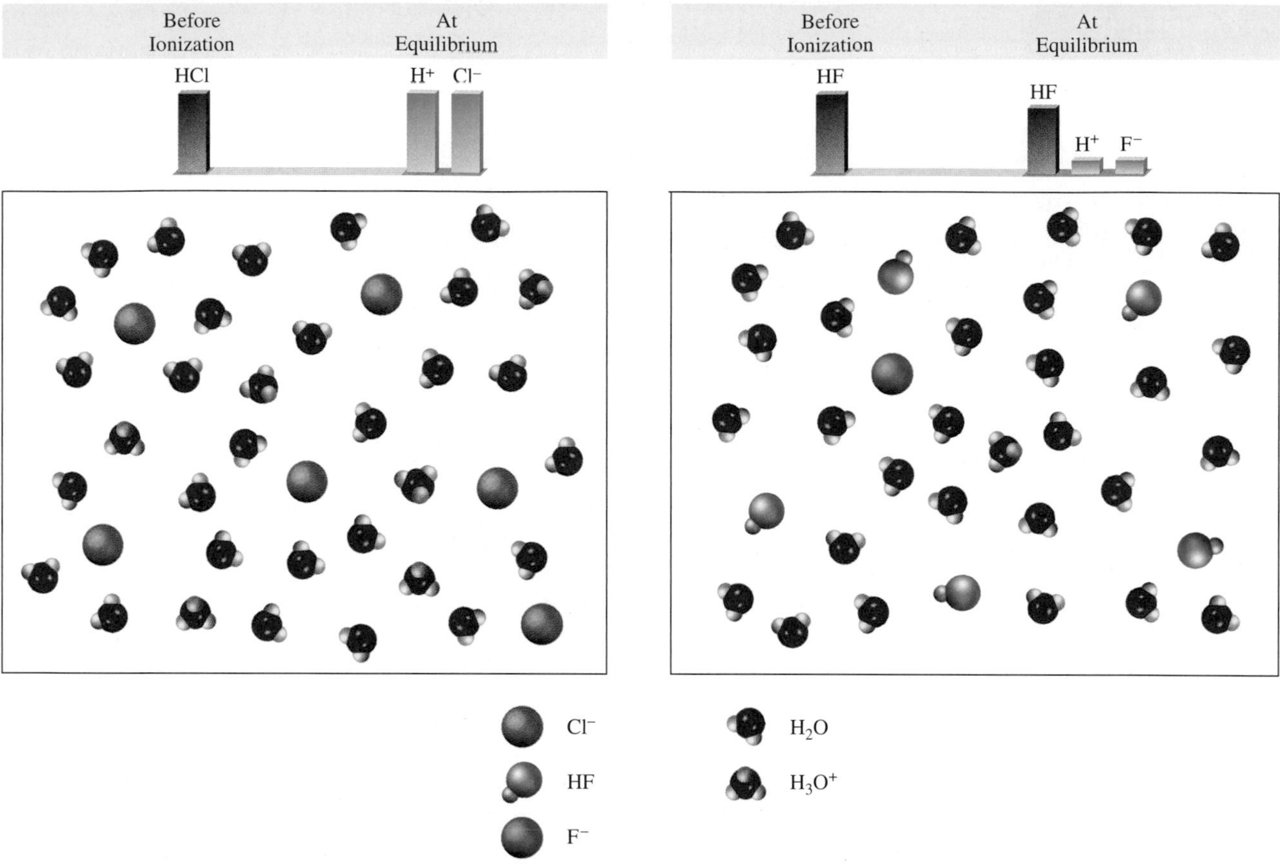

Figure 16.3
The extent of ionization of a strong acid such as HCl (left) and a weak acid such as HF (right). Initially, there were 6 HCl and 6 HF molecules present. The strong acid is assumed to be completely ionized in solution. The proton exists in solution as the hydronium ion (H_3O^+).

Like strong acids, ***strong bases*** *are all strong electrolytes that ionize completely in water.* Hydroxides of alkali metals and certain alkaline earth metals are strong bases. [All alkali metal hydroxides are soluble. Of the alkaline earth hydroxides, $Be(OH)_2$ and $Mg(OH)_2$ are insoluble; $Ca(OH)_2$ and $Sr(OH)_2$ are slightly soluble; and $Ba(OH)_2$ is soluble.] Some examples of strong bases are

$$NaOH(s) \xrightarrow{H_2O} Na^+(aq) + OH^-(aq)$$
$$KOH(s) \xrightarrow{H_2O} K^+(aq) + OH^-(aq)$$
$$Ba(OH)_2(s) \xrightarrow{H_2O} Ba^{2+}(aq) + 2OH^-(aq)$$

Strictly speaking, these metal hydroxides are not Brønsted bases because they cannot accept a proton. However, the hydroxide ion (OH^-) formed when they ionize *is* a Brønsted base because it can accept a proton:

$$H_3O^+(aq) + OH^-(aq) \longrightarrow 2H_2O(l)$$

Thus, when we call NaOH or any other metal hydroxide a base, we are actually referring to the OH^- species derived from the hydroxide.

Weak bases, like weak acids, *are weak electrolytes.* Ammonia ionizes in water as follows:

$$NH_3(aq) + H_2O(l) \rightleftharpoons NH_4^+(aq) + OH^-(aq)$$

Table 16.2 Relative Strengths of Conjugate Acid-Base Pairs

	Acid	Conjugate Base	
Acid strength increases ↑ — Strong acids	$HClO_4$ (perchloric acid)	ClO_4^- (perchlorate ion)	Base strength increases ↓
	HI (hydroiodic acid)	I^- (iodide ion)	
	HBr (hydrobromic acid)	Br^- (bromide ion)	
	HCl (hydrochloric acid)	Cl^- (chloride ion)	
	H_2SO_4 (sulfuric acid)	HSO_4^- (hydrogen sulfate ion)	
	HNO_3 (nitric acid)	NO_3^- (nitrate ion)	
	H_3O^+ (hydronium ion)	H_2O (water)	
Weak acids	HSO_4^- (hydrogen sulfate ion)	SO_4^{2-} (sulfate ion)	
	HF (hydrofluoric acid)	F^- (fluoride ion)	
	HNO_2 (nitrous acid)	NO_2^- (nitrite ion)	
	HCOOH (formic acid)	$HCOO^-$ (formate ion)	
	CH_3COOH (acetic acid)	CH_3COO^- (acetate ion)	
	NH_4^+ (ammonium ion)	NH_3 (ammonia)	
	HCN (hydrocyanic acid)	CN^- (cyanide ion)	
	H_2O (water)	OH^- (hydroxide ion)	
	NH_3 (ammonia)	NH_2^- (amide ion)	

In this reaction, NH_3 acts as a base by accepting a proton from water to form NH_4^+ and OH^- ions. It is a weak base because only a small fraction of the molecules undergo this reaction.

Table 16.2 lists some important conjugate acid-base pairs, in order of their relative strengths. Conjugate acid-base pairs have the following properties:

1. If an acid is strong, its conjugate base has no measurable strength. Thus, the Cl^- ion, which is the conjugate base of the strong acid HCl, is an extremely weak base.
2. H_3O^+ is the strongest acid that can exist in aqueous solution. Acids stronger than H_3O^+ react with water to produce H_3O^+ and their conjugate bases. Thus, HCl, which is a stronger acid than H_3O^+, reacts with water completely to form H_3O^+ and Cl^-:

$$HCl(aq) + H_2O(l) \longrightarrow H_3O^+(aq) + Cl^-(aq)$$

Acids weaker than H_3O^+ react with water to a much smaller extent, producing H_3O^+ and their conjugate bases. For example, the following equilibrium lies largely to the left:

$$HF(aq) + H_2O(l) \rightleftharpoons H_3O^+(aq) + F^-(aq)$$

3. The OH^- ion is the strongest base that can exist in aqueous solution. Bases stronger than OH^- react with water to produce OH^- and their conjugate acids. For example, the oxide ion (O^{2-}) is a stronger base than OH^-, so it reacts with water completely as follows:

$$O^{2-}(aq) + H_2O(l) \longrightarrow 2OH^-(aq)$$

For this reason the oxide ion does not exist in aqueous solutions.

EXAMPLE 16.6

Calculate the pH of (a) a 2.5×10^{-3} *M* HCl solution and (b) a 0.045 *M* $Ba(OH)_2$ solution.

Strategy Keep in mind that HCl is a strong acid and $Ba(OH)_2$ is a strong base. Thus, these species are completely ionized and no HCl or $Ba(OH)_2$ will be left in solution.

Solution

(a) The ionization of HCl is

Recall that $H^+(aq)$ is the same as $H_3O^+(aq)$.

$$HCl(aq) \longrightarrow H^+(aq) + Cl^-(aq)$$

The concentrations of all the species (HCl, H^+, and Cl^-) before and after ionization can be represented as follows:

We use the ICE method for solving equilibrium concentrations as shown in Section 15.3 (p. 523).

	$HCl(aq)$	$\longrightarrow$	$H^+(aq)$	+	$Cl^-(aq)$
Initial (*M*):	2.5×10^{-3}		0.0		0.0
Change (*M*):	-2.5×10^{-3}		$+2.5 \times 10^{-3}$		$+2.5 \times 10^{-3}$
Equilibrium (*M*):	0.0		2.5×10^{-3}		2.5×10^{-3}

A positive (+) change represents an increase and a negative (–) change indicates a decrease in concentration. Thus,

$$[H^+] = 2.5 \times 10^{-3}\ M$$
$$pH = -\log(2.5 \times 10^{-3})$$
$$= 2.60$$

(b) $Ba(OH)_2$ is a strong base; each $Ba(OH)_2$ unit produces two OH^- ions:

$$Ba(OH)_2(aq) \longrightarrow Ba^{2+}(aq) + 2OH^-(aq)$$

The changes in concentrations of all the species can be represented as follows:

	$Ba(OH)_2(aq)$	$\longrightarrow$	$Ba^{2+}(aq)$	+	$2OH^-(aq)$
Initial (*M*):	0.045		0.00		0.00
Change (*M*):	−0.045		+0.045		+2(0.045)
Equilibrium (*M*):	0.00		0.045		0.090

Thus,

$$[OH^-] = 0.090\ M$$
$$pOH = -\log 0.090 = 1.05$$

Therefore, from Equation (16.10)

$$pH = 14.00 - pOH$$
$$= 14.00 - 1.05$$
$$= 12.95$$

Check Note that in both (a) and (b) we have neglected the contribution of the auto-ionization of water to $[H^+]$ and $[OH^-]$ because 1.0×10^{-7} *M* is so small compared with 2.5×10^{-3} *M* and 0.090 *M*.

Similar problems: 16.17(a), (c).

Practice Exercise Calculate the pH of a 1.8×10^{-2} *M* $Ba(OH)_2$ solution.

EXAMPLE 16.7

Predict the direction of the following reaction in aqueous solution:

$$HF(aq) + NO_2^-(aq) \rightleftharpoons HNO_2(aq) + F^-(aq)$$

Strategy The problem is to determine whether, at equilibrium, the reaction will be shifted to the right, favoring F^- and HNO_2, or to the left, favoring NO_2^- and HF. Which of the two is a stronger acid and hence a stronger proton donor: HNO_2 or HF? Which of the two is a stronger base and hence a stronger proton acceptor: F^- or NO_2^-? Remember that the stronger the acid, the weaker its conjugate base.

Solution In Table 16.2 we see that HF is a stronger acid than HNO_2. Thus, NO_2^- is a stronger base than F^-. The net reaction will proceed from left to right because HF is a better proton donor than HNO_2 (and NO_2^- is a better proton acceptor than F^-).

Check Table 16.2 has the acids arranged in order of increasing acid strength from the bottom to the top. Thus, acids higher in the acid column indicate a stronger proton donor than those listed below them.

Similar problems: 16.35, 16.36.

Practice Exercise Predict whether the equilibrium constant for the following reaction is greater than or smaller than 1:

$$CH_3COOH(aq) + HCOO^-(aq) \rightleftharpoons CH_3COO^-(aq) + HCOOH(aq)$$

REVIEW OF CONCEPTS

(a) List in order of decreasing concentration of all the ionic and molecular species in the following acid solutions: (i) HNO_3 and (ii) HF.

(b) List in order of decreasing concentration of all the ionic and molecular species in the following base solutions: (i) NH_3 and (ii) KOH.

16.5 Weak Acids and Acid Ionization Constants

Animation:
Acid Ionization

As we have seen, there are relatively few strong acids. The vast majority of acids are weak acids. Consider a weak monoprotic acid, HA. Its ionization in water is represented by

$$HA(aq) + H_2O(l) \rightleftharpoons H_3O^+(aq) + A^-(aq)$$

or simply

$$HA(aq) \rightleftharpoons H^+(aq) + A^-(aq)$$

Ka = acid ionization constant

The equilibrium expression for this ionization is

$$K_a = \frac{[H_3O^+][A^-]}{[HA]} \quad \text{or} \quad K_a = \frac{[H^+][A^-]}{[HA]} \qquad (16.11)$$

All concentrations in this equation are equilibrium concentrations.

where K_a, the ***acid ionization constant***, is the *equilibrium constant for the ionization of an acid*. At a given temperature, the strength of the acid HA is measured quantitatively by the magnitude of K_a. The larger K_a, the stronger the acid—that is, the greater the concentration of H^+ ions at equilibrium due to its ionization. Keep in mind, however, that only weak acids have K_a values associated with them.

Table 16.3 Ionization Constants of Some Weak Acids and Their Conjugate Bases at 25°C

Name of Acid	Formula	Structure	K_a	Conjugate Base	K_b
Hydrofluoric acid	HF	H—F	7.1×10^{-4}	F^-	1.4×10^{-11}
Nitrous acid	HNO_2	O═N—O—H	4.5×10^{-4}	NO_2^-	2.2×10^{-11}
Acetylsalicylic acid (aspirin)	$C_9H_8O_4$	(benzene ring)—C(═O)—O—H; (benzene ring)—O—C(═O)—CH_3	3.0×10^{-4}	$C_9H_7O_4^-$	3.3×10^{-11}
Formic acid	HCOOH	H—C(═O)—O—H	1.7×10^{-4}	$HCOO^-$	5.9×10^{-11}
Ascorbic acid*	$C_6H_8O_6$	H—O—C═C—OH; H—C; C═O; O; CHOH; CH_2OH	8.0×10^{-5}	$C_6H_7O_6^-$	1.3×10^{-10}
Benzoic acid	C_6H_5COOH	(benzene ring)—C(═O)—O—H	6.5×10^{-5}	$C_6H_5COO^-$	1.5×10^{-10}
Acetic acid	CH_3COOH	CH_3—C(═O)—O—H	1.8×10^{-5}	CH_3COO^-	5.6×10^{-10}
Hydrocyanic acid	HCN	H—C≡N	4.9×10^{-10}	CN^-	2.0×10^{-5}
Phenol	C_6H_5OH	(benzene ring)—O—H	1.3×10^{-10}	$C_6H_5O^-$	7.7×10^{-5}

*For ascorbic acid it is the upper left hydroxyl group that is associated with this ionization constant.

The back endpaper gives an index to all the useful tables and figures in the text.

Table 16.3 lists a number of weak acids and their K_a values at 25°C in order of decreasing acid strength. Although all these acids are weak, within the group there is great variation in their strengths. For example, K_a for HF (7.1×10^{-4}) is about 1.5 million times that for HCN (4.9×10^{-10}).

Generally, we can calculate the hydrogen ion concentration or pH of an acid solution at equilibrium, given the initial concentration of the acid and its K_a value. Alternatively, if we know the pH of a weak acid solution and its initial concentration, we can determine its K_a. The basic approach for solving these problems, which deal with equilibrium concentrations, is the same one outlined in Chapter 15. However, because acid ionization represents a major category of chemical equilibrium in aqueous solution, we will develop a systematic procedure for solving this type of problem that will also help us to understand the chemistry involved.

Suppose we are asked to calculate the pH of a 0.50 *M* HF solution at 25°C. The ionization of HF is given by

$$HF(aq) \rightleftharpoons H^+(aq) + F^-(aq)$$

From Table 16.3 we write

$$K_a = \frac{[H^+][F^-]}{[HF]} = 7.1 \times 10^{-4}$$

The first step is to identify all the species present in solution that may affect its pH. Because weak acids ionize to a small extent, at equilibrium the major species present are nonionized HF and some H^+ and F^- ions. Another major species is H_2O, but its very small K_w (1.0×10^{-14}) means that water is not a significant contributor to the H^+ ion concentration. Therefore, unless otherwise stated, we will always ignore the H^+ ions produced by the autoionization of water. Note that we need not be concerned with the OH^- ions that are also present in solution. The OH^- concentration can be determined from Equation (16.4) after we have calculated $[H^+]$.

We can summarize the changes in the concentrations of HF, H^+, and F^- according to the ICE method shown on p. 523 as follows:

	$HF(aq)$ $\rightleftharpoons$	$H^+(aq)$ +	$F^-(aq)$
Initial (M):	0.50	0.00	0.00
Change (M):	$-x$	$+x$	$+x$
Equilibrium (M):	$0.50 - x$	x	x

The equilibrium concentrations of HF, H^+, and F^-, expressed in terms of the unknown x, are substituted into the ionization constant expression to give

$$K_a = \frac{(x)(x)}{0.50 - x} = 7.1 \times 10^{-4}$$

Rearranging this expression, we write

$$x^2 + 7.1 \times 10^{-4}x - 3.6 \times 10^{-4} = 0$$

This is a quadratic equation, which can be solved using the quadratic formula (see Appendix 3). Or we can try using a shortcut to solve for x. Because HF is a weak acid and weak acids ionize only to a slight extent, we reason that x must be small compared to 0.50. Therefore, we can make the approximation

$$0.50 - x \approx 0.50$$

The sign ≈ means "approximately equal to." An analogy of the approximation is a truck loaded with coal. Losing a few lumps of coal on a delivery trip will not appreciably change the overall mass of the load.

Now the ionization constant expression becomes

$$\frac{x^2}{0.50 - x} \approx \frac{x^2}{0.50} = 7.1 \times 10^{-4}$$

Rearranging, we get

$$x^2 = (0.50)(7.1 \times 10^{-4}) = 3.55 \times 10^{-4}$$
$$x = \sqrt{3.55 \times 10^{-4}} = 0.019\ M$$

Thus, we have solved for x without having to use the quadratic equation. At equilibrium, we have

$$[HF] = (0.50 - 0.019)\ M = 0.48\ M$$
$$[H^+] = 0.019\ M$$
$$[F^-] = 0.019\ M$$

and the pH of the solution is

$$\text{pH} = -\log(0.019) = 1.72$$

How good is this approximation? Because K_a values for weak acids are generally known to an accuracy of only ±5%, it is reasonable to require x to be less than 5% of 0.50, the number from which it is subtracted. In other words, the approximation is valid if the following expression is equal to or less than 5%:

$$\frac{0.019\ M}{0.50\ M} \times 100\% = 3.8\%$$

Thus, the approximation we made is acceptable.

Now consider a different situation. If the initial concentration of HF is 0.050 M, and we use the above procedure to solve for x, we would get 6.0×10^{-3} M. However, the following test shows that this answer is not a valid approximation because it is greater than 5% of 0.050 M:

$$\frac{6.0 \times 10^{-3}\ M}{0.050\ M} \times 100\% = 12\%$$

In this case, we can solve for x by using the quadratic equation.

We start by writing the ionization expression in terms of the unknown x:

$$\frac{x^2}{0.050 - x} = 7.1 \times 10^{-4}$$
$$x^2 + 7.1 \times 10^{-4}x - 3.6 \times 10^{-5} = 0$$

This expression fits the quadratic equation $ax^2 + bx + c = 0$. Using the quadratic formula, we write

$$\begin{aligned} x &= \frac{-b \pm \sqrt{b^2 - 4ac}}{2a} \\ &= \frac{-7.1 \times 10^{-4} \pm \sqrt{(7.1 \times 10^{-4})^2 - 4(1)(-3.6 \times 10^{-5})}}{2(1)} \\ &= \frac{-7.1 \times 10^{-4} \pm 0.012}{2} \\ &= 5.6 \times 10^{-3}\ M \quad \text{or} \quad -6.4 \times 10^{-3}\ M \end{aligned}$$

The second solution ($x = -6.4 \times 10^{-3}$ M) is physically impossible because the concentration of ions produced as a result of ionization cannot be negative. Choosing $x = 5.6 \times 10^{-3}$ M, we can solve for [HF], $[H^+]$, and $[F^-]$ as follows:

$$[\text{HF}] = (0.050 - 5.6 \times 10^{-3})\ M = 0.044\ M$$
$$[\text{H}^+] = 5.6 \times 10^{-3}\ M$$
$$[\text{F}^-] = 5.6 \times 10^{-3}\ M$$

The pH of the solution, then, is

$$\text{pH} = -\log(5.6 \times 10^{-3}) = 2.25$$

In summary, the main steps for solving weak acid ionization problems are

1. Identify the major species that can affect the pH of the solution. In most cases, we can ignore the ionization of water. We omit the hydroxide ion because its concentration is determined by that of the H^+ ion.

2. Express the equilibrium concentrations of these species in terms of the initial concentration of the acid and a single unknown x, which represents the change in concentration.
3. Write the acid ionization constant (K_a) in terms of the equilibrium concentrations. First solve for x by the approximate method. If the approximation is not valid, use the quadratic equation to solve for x.
4. Having solved for x, calculate the equilibrium concentrations of all species and/or the pH of the solution.

EXAMPLE 16.8

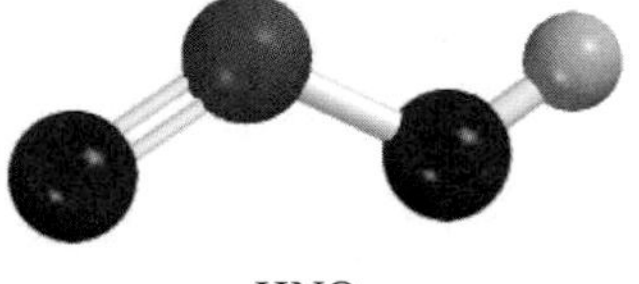

HNO_2

Calculate the pH of a 0.036 *M* nitrous acid (HNO_2) solution:

$$HNO_2(aq) \rightleftharpoons H^+(aq) + NO_2^-(aq)$$

Strategy Recall that a weak acid only partially ionizes in water. We are given the initial concentration of a weak acid and asked to calculate the pH of the solution at equilibrium. It is helpful to make a sketch to keep track of the pertinent species.

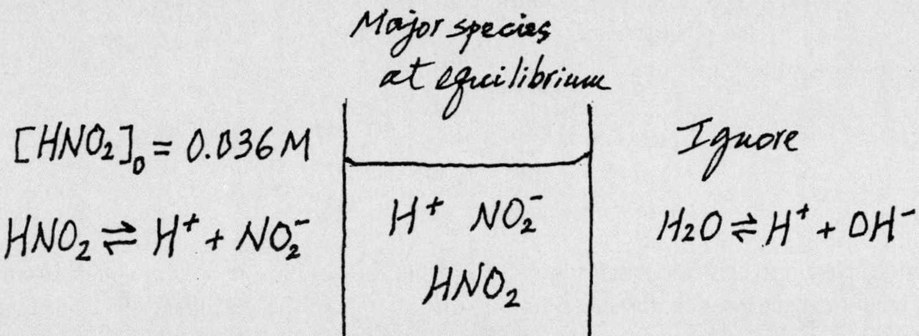

As in Example 16.6, we ignore the ionization of H_2O so the major source of H^+ ions is the acid. The concentration of OH^- ions is very small as we would expect from an acidic solution so it is present as a minor species.

Solution We follow the procedure already outlined.

Step 1: The species that can affect the pH of the solution are HNO_2, H^+, and the conjugate base NO_2^-. We ignore water's contribution to $[H^+]$.

Step 2: Letting x be the equilibrium concentration of H^+ and NO_2^- ions in mol/L, we summarize:

	$HNO_2(aq)$	$\rightleftharpoons$	$H^+(aq)$	+	$NO_2^-(aq)$
Initial (*M*):	0.036		0.00		0.00
Change (*M*):	$-x$		$+x$		$+x$
Equilibrium (*M*):	$0.036 - x$		x		x

Step 3: From Table 16.3 we write

$$K_a = \frac{[H^+][NO_2^-]}{[HNO_2]}$$

$$4.5 \times 10^{-4} = \frac{x^2}{0.036 - x}$$

(Continued)

Applying the approximation $0.036 - x \approx 0.036$, we obtain

$$4.5 \times 10^{-4} = \frac{x^2}{0.036 - x} \approx \frac{x^2}{0.036}$$
$$x^2 = 1.62 \times 10^{-5}$$
$$x = 4.0 \times 10^{-3}\ M$$

To test the approximation,

$$\frac{4.0 \times 10^{-3}\ M}{0.036\ M} \times 100\% = 11\%$$

Because this is greater than 5%, our approximation is not valid and we must solve the quadratic equation, as follows:

$$x^2 + 4.5 \times 10^{-4}x - 1.62 \times 10^{-5} = 0$$
$$x = \frac{-4.5 \times 10^{-4} \pm \sqrt{(4.5 \times 10^{-4})^2 - 4(1)(-1.62 \times 10^{-5})}}{2(1)}$$
$$= 3.8 \times 10^{-3}\ M \quad \text{or} \quad -4.3 \times 10^{-3}\ M$$

The second solution is physically impossible, because the concentration of ions produced as a result of ionization cannot be negative. Therefore, the solution is given by the positive root, $x = 3.8 \times 10^{-3}\ M$.

Step 4: At equilibrium,

$$[H^+] = 3.8 \times 10^{-3}\ M$$
$$pH = -\log(3.8 \times 10^{-3})$$
$$= 2.42$$

Check Note that the calculated pH indicates that the solution is acidic, which is what we would expect for a weak acid solution. Also, compare the calculated pH with that of a 0.036 M strong acid solution such as HCl to convince yourself of the difference between a strong acid and a weak acid.

Similar problem: 16.45.

Practice Exercise What is the pH of a 0.122 M monoprotic acid whose K_a is 5.7×10^{-4}?

One way to determine K_a of an acid is to measure the pH of the acid solution of known concentration at equilibrium. Example 16.9 shows this approach.

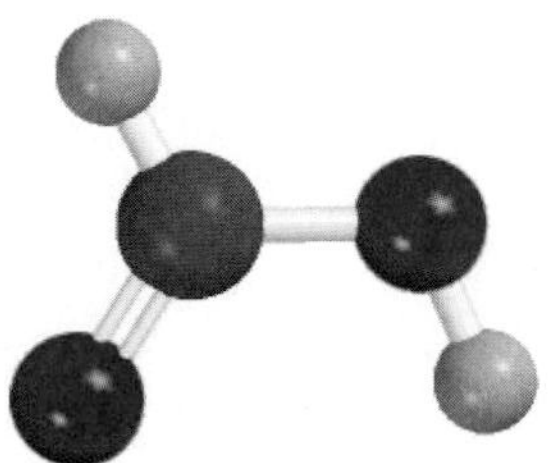

HCOOH

EXAMPLE 16.9

The pH of a 0.10 M solution of formic acid (HCOOH) is 2.39. What is the K_a of the acid?

Strategy Formic acid is a weak acid. It only partially ionizes in water. Note that the concentration of formic acid refers to the initial concentration, before ionization has started. The pH of the solution, on the other hand, refers to the equilibrium state. To calculate K_a, then, we need to know the concentrations of all three species: $[H^+]$,

(Continued)

$[HCOO^-]$, and $[HCOOH]$ at equilibrium. As usual, we ignore the ionization of water. The following sketch summarizes the situation.

Major species at equilibrium

$[HCOOH]_0 = 0.10\ M$

$HCOOH \rightleftharpoons H^+ + HCOO^-$

H^+ $HCOO^-$ $HCOOH$

pH = 2.39

$[H^+] = 10^{-2.39}$

Solution We proceed as follows.

Step 1: The major species in solution are HCOOH, H^+, and the conjugate base $HCOO^-$.

Step 2: First we need to calculate the hydrogen ion concentration from the pH value

$$\text{pH} = -\log[H^+]$$
$$2.39 = -\log[H^+]$$

Taking the antilog of both sides, we get

$$[H^+] = 10^{-2.39} = 4.1 \times 10^{-3}\ M$$

Next we summarize the changes:

	$HCOOH(aq)$	$\rightleftharpoons$	$H^+(aq)$	+	$HCOO^-(aq)$
Initial (M):	0.10		0.00		0.00
Change (M):	-4.1×10^{-3}		$+4.1 \times 10^{-3}$		$+4.1 \times 10^{-3}$
Equilibrium (M):	$(0.10 - 4.1 \times 10^{-3})$		4.1×10^{-3}		4.1×10^{-3}

Note that because the pH and hence the H^+ ion concentration is known, it follows that we also know the concentrations of HCOOH and $HCOO^-$ at equilibrium.

Step 3: The ionization constant of formic acid is given by

$$K_a = \frac{[H^+][HCOO^-]}{[HCOOH]}$$
$$= \frac{(4.1 \times 10^{-3})(4.1 \times 10^{-3})}{(0.10 - 4.1 \times 10^{-3})}$$
$$= 1.8 \times 10^{-4}$$

Check The K_a value differs slightly from the one listed in Table 16.3 because of the rounding-off procedure we used in the calculation.

Similar problem: 16.43.

Practice Exercise The pH of a 0.060 M weak monoprotic acid is 3.44. Calculate the K_a of the acid.

Percent Ionization

We have seen that the magnitude of K_a indicates the strength of an acid. Another measure of the strength of an acid is its ***percent ionization,*** which is defined as

$$\text{percent ionization} = \frac{\text{ionized acid concentration at equilibrium}}{\text{initial concentration of acid}} \times 100\% \qquad (16.12)$$

We can compare the strengths of acids in terms of percent ionization only if concentrations of the acids are the same.

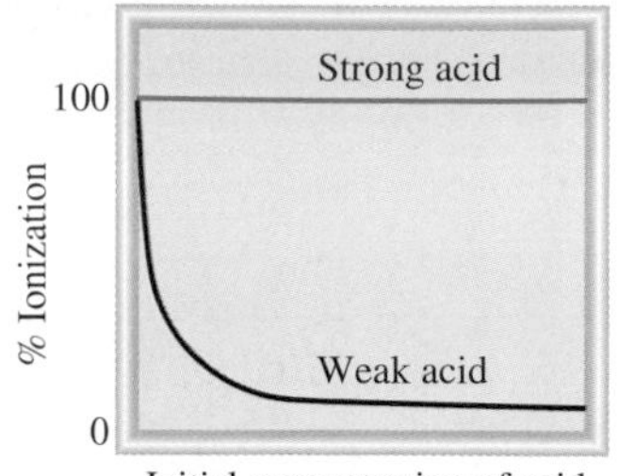

Figure 16.4
Dependence of percent ionization on initial concentration of acid. Note that at very low concentrations, all acids (weak and strong) are almost completely ionized.

The stronger the acid, the greater the percent ionization. For a monoprotic acid HA, the concentration of the acid that undergoes ionization is equal to the concentration of the H^+ ions or the concentration of the A^- ions at equilibrium. Therefore, we can write the percent ionization as

$$\text{percent ionization} = \frac{[H^+]}{[HA]_0} \times 100\%$$

where $[H^+]$ is the concentration at equilibrium and $[HA]_0$ is the initial concentration.

Referring to Example 16.8, we see that the percent ionization of a 0.036 *M* HNO_2 solution is

$$\text{percent ionization} = \frac{3.8 \times 10^{-3}\,M}{0.036\,M} \times 100\% = 11\%$$

Thus, only about one out of every 9 HNO_2 molecules has ionized. This is consistent with the fact that HNO_2 is a weak acid.

The extent to which a weak acid ionizes depends on the initial concentration of the acid. The more dilute the solution, the greater the percent ionization (Figure 16.4). In qualitative terms, when an acid is diluted, the concentration of the "particles" in the solution is reduced. According to Le Châtelier's principle (see Section 15.4), this reduction in particle concentration (the stress) is counteracted by shifting the reaction to the side with more particles; that is, the equilibrium shifts from the nonionized acid side (one particle) to the side containing the H^+ ion and the conjugate base (two particles). Consequently, the percent ionization of the acid increases.

Diprotic and Polyprotic Acids

Diprotic and polyprotic acids may yield more than one hydrogen ion per molecule. These acids ionize in a stepwise manner, that is, they lose one proton at a time. An ionization constant expression can be written for each ionization stage. Consequently, two or more equilibrium constant expressions must often be used to calculate the concentrations of species in the acid solution. For example, for H_2CO_3 we write

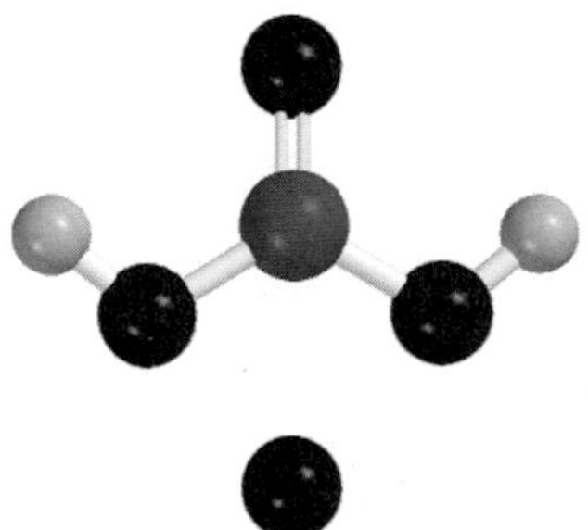

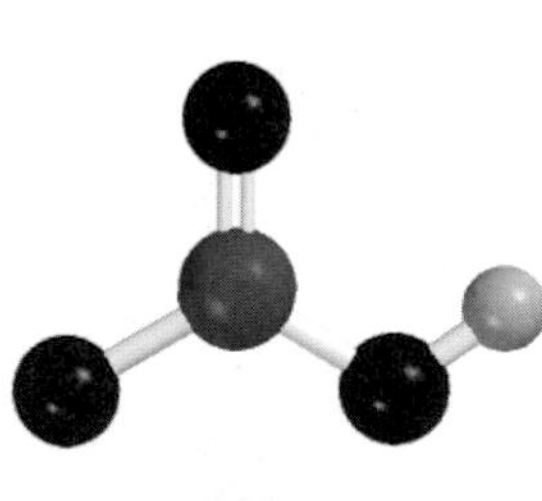

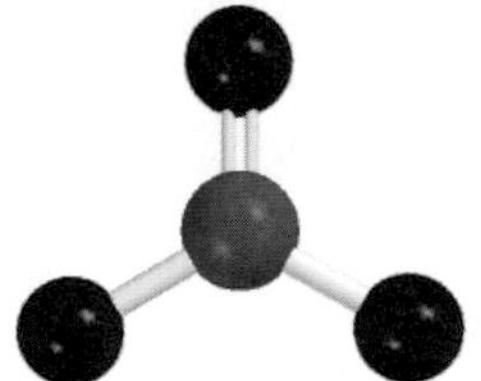

Top to bottom: H_2CO_3, HCO_3^-, and CO_3^{2-}.

$$H_2CO_3(aq) \rightleftharpoons H^+(aq) + HCO_3^-(aq) \qquad K_{a_1} = \frac{[H^+][HCO_3^-]}{[H_2CO_3]}$$

$$HCO_3^-(aq) \rightleftharpoons H^+(aq) + CO_3^{2-}(aq) \qquad K_{a_2} = \frac{[H^+][CO_3^{2-}]}{[HCO_3^-]}$$

Note that the conjugate base in the first ionization stage becomes the acid in the second ionization stage.

Table 16.4 shows the ionization constants of several diprotic acids and a polyprotic acid. For a given acid, the first ionization constant is much larger than the second ionization constant, and so on. This trend is reasonable because it is easier to remove an H^+ ion from a neutral molecule than to remove another H^+ from a negatively charged ion derived from the molecule.

Table 16.4 Ionization Constants of Some Diprotic Acids and a Polyprotic Acid and Their Conjugate Bases at 25°C

Name of Acid	Formula	Structure	K_a	Conjugate Base	K_b
Sulfuric acid	H_2SO_4	H—O—S(=O)(=O)—O—H	Very large	HSO_4^-	Very small
Hydrogen sulfate ion	HSO_4^-	H—O—S(=O)(=O)—O^-	1.3×10^{-2}	SO_4^{2-}	7.7×10^{-13}
Oxalic acid	$H_2C_2O_4$	H—O—C(=O)—C(=O)—O—H	6.5×10^{-2}	$HC_2O_4^-$	1.5×10^{-13}
Hydrogen oxalate ion	$HC_2O_4^-$	H—O—C(=O)—C(=O)—O^-	6.1×10^{-5}	$C_2O_4^{2-}$	1.6×10^{-10}
Sulfurous acid*	H_2SO_3	H—O—S(=O)—O—H	1.3×10^{-2}	HSO_3^-	7.7×10^{-13}
Hydrogen sulfite ion	HSO_3^-	H—O—S(=O)—O^-	6.3×10^{-8}	SO_3^{2-}	1.6×10^{-7}
Carbonic acid	H_2CO_3	H—O—C(=O)—O—H	4.2×10^{-7}	HCO_3^-	2.4×10^{-8}
Hydrogen carbonate ion	HCO_3^-	H—O—C(=O)—O^-	4.8×10^{-11}	CO_3^{2-}	2.1×10^{-4}
Hydrosulfuric acid	H_2S	H—S—H	9.5×10^{-8}	HS	1.1×10^{-7}
Hydrogen sulfide ion†	HS^-	H—S^-	1×10^{-19}	S^{2-}	1×10^{5}
Phosphoric acid	H_3PO_4	H—O—P(=O)(—O—H)—O—H	7.5×10^{-3}	$H_2PO_4^-$	1.3×10^{-12}
Dihydrogen phosphate ion	$H_2PO_4^-$	H—O—P(=O)(—O—H)—O^-	6.2×10^{-8}	HPO_4^{2-}	1.6×10^{-7}
Hydrogen phosphate ion	HPO_4^{2-}	H—O—P(=O)(O^-)—O^-	4.8×10^{-13}	PO_4^{3-}	2.1×10^{-2}

*H_2SO_3 has never been isolated and exists in only minute concentration in aqueous solution of SO_2. The K_a value here refers to the process $SO_2(g) + H_2O(l) \rightleftharpoons H^+(aq) + HSO_3^-(aq)$.

†The ionization constant of HS^- is very low and difficult to measure. The value listed here is only an estimate.

$H_2C_2O_4$

EXAMPLE 16.10

Oxalic acid ($H_2C_2O_4$) is a poisonous substance used chiefly as a bleaching and cleansing agent (for example, to remove bathtub rings). Calculate the concentrations of all the species present at equilibrium in a 0.10 *M* solution.

Strategy Determining the equilibrium concentrations of the species of a diprotic acid in aqueous solution is more involved than for a monoprotic acid. We follow the same procedure as that used for a monoprotic acid for each stage, as in Example 16.8. Note that the conjugate base from the first stage of ionization becomes the acid for the second stage ionization.

Solution We proceed according to the following steps:

Step 1: The major species in solution at this stage are the nonionized acid, H^+ ions, and the conjugate base, $HC_2O_4^-$.

Step 2: Letting x be the equilibrium concentration of H^+ and $HC_2O_4^-$ ions in mol/L, we summarize:

	$H_2C_2O_4(aq)$	$\rightleftharpoons$	$H^+(aq)$	+	$HC_2O_4^-(aq)$
Initial (*M*):	0.10		0.00		0.00
Change (*M*):	$-x$		$+x$		$+x$
Equilibrium (*M*):	$0.10 - x$		x		x

Step 3: Table 16.4 gives us

$$K_a = \frac{[H^+][HC_2O_4^-]}{[H_2C_2O_4]}$$

$$6.5 \times 10^{-2} = \frac{x^2}{0.10 - x}$$

Applying the approximation $0.10 - x \approx 0.10$. we obtain

$$6.5 \times 10^{-2} = \frac{x^2}{0.10 - x} \approx \frac{x^2}{0.10}$$

$$x^2 = 6.5 \times 10^{-3}$$

$$x = 8.1 \times 10^{-2}\,M$$

To test the approximation,

$$\frac{8.1 \times 10^{-2}\,M}{0.10\,M} \times 100\% = 81\%$$

Clearly the approximation is not valid. Therefore, we must solve the quadratic equation

$$x^2 + 6.5 \times 10^{-2}x - 6.5 \times 10^{-3} = 0$$

The result is $x = 0.054\,M$.

Step 4: When the equilibrium for the first stage of ionization is reached, the concentrations are

$$[H^+] = 0.054\,M$$

$$[HC_2O_4^-] = 0.054\,M$$

$$[H_2C_2O_4] = (0.10 - 0.054)\,M = 0.046\,M$$

Next we consider the second stage of ionization.

Step 1: At this stage, the major species are $HC_2O_4^-$, which acts as the acid in the second stage of ionization, H^+, and the conjugate base $C_2O_4^{2-}$.

(Continued)

Step 2: Letting y be the equilibrium concentration of H^+ and $C_2O_4^{2-}$ ions in mol/L, we summarize:

	$HC_2O_4^-(aq)$	$\rightleftharpoons$	$H^+(aq)$	$+$	$C_2O_4^{2-}(aq)$
Initial (M):	0.054		0.054		0.00
Change (M):	$-y$		$+y$		$+y$
Equilibrium (M):	$0.054 - y$		$0.054 + y$		y

Step 3: Table 16.4 gives us

$$K_a = \frac{[H^+][C_2O_4^{2-}]}{[HC_2O_4^-]}$$

$$6.1 \times 10^{-5} = \frac{(0.054 + y)(y)}{(0.054 - y)}$$

Applying the approximation $0.054 + y \approx 0.054$ and $0.054 - y \approx 0.054$, we obtain

$$\frac{(0.054)(y)}{(0.054)} = y = 6.1 \times 10^{-5}\ M$$

and we test the approximation,

$$\frac{6.1 \times 10^{-5}\ M}{0.054\ M} \times 100\% = 0.11\%$$

The approximation is valid.

Step 4: At equilibrium,

$$[H_2C_2O_4] = 0.046\ M$$
$$[HC_2O_4^-] = (0.054 - 6.1 \times 10^{-5})\ M = 0.054\ M$$
$$[H^+] = (0.054 + 6.1 \times 10^{-5})\ M = 0.054\ M$$
$$[C_2O_4^{2-}] = 6.1 \times 10^{-5}\ M$$
$$[OH^-] = 1.0 \times 10^{-14}/0.054 = 1.9 \times 10^{-13}\ M$$

Similar problem: 16.52.

Practice Exercise Calculate the concentrations of $H_2C_2O_4$, $HC_2O_4^-$, $C_2O_4^{2-}$, and H^+ ions in a 0.20 M oxalic acid solution.

Example 16.10 shows that for diprotic acids, if $K_{a_1} \gg K_{a_2}$, then we can assume that the concentration of H^+ ions is the product of only the first stage of ionization. Furthermore, the concentration of the conjugate base for the second-stage ionization is *numerically* equal to K_{a_2}.

Phosphoric acid (H_3PO_4) is a polyprotic acid with three ionizable hydrogen atoms:

$$H_3PO_4(aq) \rightleftharpoons H^+(aq) + H_2PO_4^-(aq) \qquad K_{a_1} = \frac{[H^+][H_2PO_4^-]}{[H_3PO_4]} = 7.5 \times 10^{-3}$$

$$H_2PO_4^-(aq) \rightleftharpoons H^+(aq) + HPO_4^{2-}(aq) \qquad K_{a_2} = \frac{[H^+][HPO_4^{2-}]}{[H_2PO_4^-]} = 6.2 \times 10^{-8}$$

$$HPO_4^{2-}(aq) \rightleftharpoons H^+(aq) + PO_4^{3-}(aq) \qquad K_{a_3} = \frac{[H^+][PO_4^{3-}]}{[HPO_4^{2-}]} = 4.8 \times 10^{-13}$$

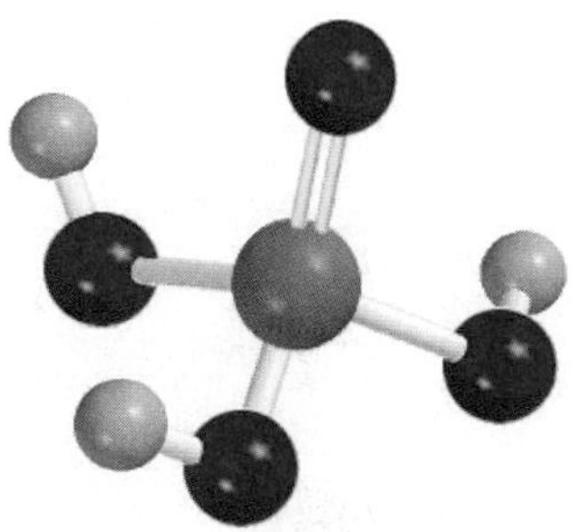
H_3PO_4

We see that phosphoric acid is a weak polyprotic acid and that its ionization constants decrease markedly for the second and third stages. Thus we can predict that, in a solution containing phosphoric acid, the concentration of the nonionized acid is the highest, and the only other species present in significant concentrations are H^+ and $H_2PO_4^-$ ions.

Review of Concepts

Which of the diagrams shown here represents a solution of sulfuric acid? Water molecules have been omitted for clarity.

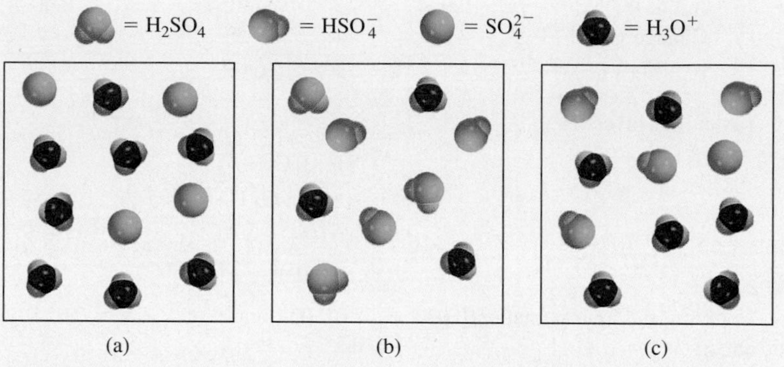

16.6 Weak Bases and Base Ionization Constants

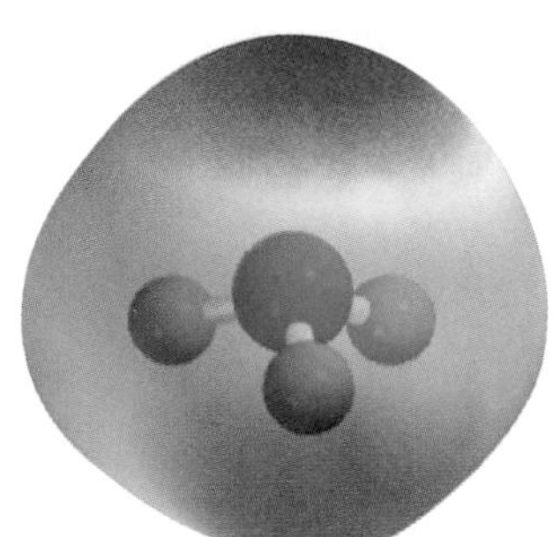

Electrostatic potential map of NH_3. The lone pair (red color) on the N atom accounts for ammonia's basicity.

The ionization of weak bases is treated in the same way as the ionization of weak acids. When ammonia dissolves in water, it undergoes the reaction

$$NH_3(aq) + H_2O(l) \rightleftharpoons NH_4^+(aq) + OH^-(aq)$$

The equilibrium constant is given by

$$K = \frac{[NH_4^+][OH^-]}{[NH_3][H_2O]}$$

The production of hydroxide ions in this *base ionization reaction* means that $[OH^-] > [H^+]$ and therefore pH > 7.

Compared with the total concentration of water, very few water molecules are consumed by this reaction, so we can treat $[H_2O]$ as a constant. Thus, we can write the ***base ionization constant (K$_b$),*** which is *the equilibrium constant for the ionization reaction,* as

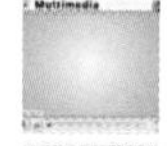

Animation: Base Ionization

$$K_b = K[H_2O] = \frac{[NH_4^+][OH^-]}{[NH_3]}$$
$$= 1.8 \times 10^{-5}$$

Table 16.5 lists a number of common weak bases and their ionization constants. Note that the basicity of all these compounds is attributable to the lone pair of electrons on the nitrogen atom. The ability of the lone pair to accept a H^+ ion makes these substances Brønsted bases.

EXAMPLE 16.11

What is the pH of a 0.40 *M* ammonia solution?

Strategy The procedure here is similar to the one used for a weak acid (see Example 16.8). From the ionization of ammonia, we see that the major species in solution at

(Continued)

Table 16.5 Ionization Constants of Some Weak Bases and Their Conjugate Acids at 25°C

Name of Base	Formula	Structure	K_b*	Conjugate Acid	K_a
Ethylamine	$C_2H_5NH_2$	$CH_3-CH_2-\ddot{N}H-H$	5.6×10^{-4}	$C_2H_5\overset{+}{N}H_3$	1.8×10^{-11}
Methylamine	CH_3NH_2	$CH_3-\ddot{N}H-H$	4.4×10^{-4}	$CH_3\overset{+}{N}H_3$	2.3×10^{-11}
Ammonia	NH_3	$H-\ddot{N}H-H$	1.8×10^{-5}	NH_4^+	5.6×10^{-10}
Pyridine	C_5H_5N	$C_5H_5N\!:$ (ring)	1.7×10^{-9}	$C_5H_5\overset{+}{N}H$	5.9×10^{-6}
Aniline	$C_6H_5NH_2$	$C_6H_5-\ddot{N}H-H$	3.8×10^{-10}	$C_6H_5\overset{+}{N}H_3$	2.6×10^{-5}
Caffeine	$C_8H_{10}N_4O_2$	(ring structure with H_3C, CH_3, $C=O$, $C-H$, $\ddot{N}$)	5.3×10^{-14}	$C_8H_{11}\overset{+}{N}_4O_2$	0.19
Urea	$(NH_2)_2CO$	$H-\ddot{N}H-C(=O)-\ddot{N}H-H$	1.5×10^{-14}	$H_2NCO\overset{+}{N}H_3$	0.67

*The nitrogen atom with the lone pair accounts for each compound's basicity. In the case of urea, K_b can be associated with either nitrogen atom.

equilibrium are NH_3, NH_4^+, and OH^-. The hydrogen ion concentration is very small as we would expect from a basic solution, so it is present as a minor species. As before, we ignore the ionization of water. We make a sketch to keep track of the pertinent species as follows:

Major species at equilibrium

$[NH_3]_0 = 0.40\ M$

$NH_3 + H_2O \rightleftharpoons NH_4^+ + OH^-$

NH_4^+ OH^- NH_3

Ignore

$H_2O \rightleftharpoons H^+ + OH^-$

(Continued)

Solution We proceed according to the following steps:

Step 1: The major species in an ammonia solution are NH_3, NII_4^+, and OH^-. We ignore the very small contribution to OH^- concentration by water.

Step 2: Letting x be the equilibrium concentration of NH_4^+ and OH^- ions in mol/L, we summarize:

	$NH_3(aq)$ + $H_2O(l)$ ⇌	$NH_4^+(aq)$ +	$OH^-(aq)$
Initial (M):	0.40	0.00	0.00
Change (M):	$-x$	$+x$	$+x$
Equilibrium (M):	$0.40 - x$	x	x

Step 3: Table 16.5 gives us K_b:

$$K_b = \frac{[NH_4^+][OH^-]}{[NH_3]}$$

$$1.8 \times 10^{-5} = \frac{x^2}{0.40 - x}$$

Applying the approximation $0.40 - x \approx 0.40$, we obtain

$$1.8 \times 10^{-5} = \frac{x^2}{0.40 - x} \approx \frac{x^2}{0.40}$$

$$x^2 = 7.2 \times 10^{-6}$$

$$x = 2.7 \times 10^{-3}\ M$$

The 5 percent rule (p. 558) also applies to bases.

To test the approximation, we write

$$\frac{2.7 \times 10^{-3}\ M}{0.40\ M} \times 100\% = 0.68\%$$

Therefore, the approximation is valid.

Step 4: At equilibrium, $[OH^-] = 2.7 \times 10^{-3}\ M$. Thus,

$$\text{pOH} = -\log(2.7 \times 10^{-3})$$
$$= 2.57$$

From Equation (16.10)

$$\text{pH} = 14.00 - 2.57$$
$$= 11.43$$

Check Note that the pH calculated is basic, which is what we would expect from a weak base solution. Compare the calculated pH with that of a 0.40 M strong base solution, such as KOH, to convince yourself of the difference between a strong base and a weak base.

Similar problem: 16.55.

Practice Exercise Calculate the pH of a 0.26 M methylamine solution (see Table 16.5).

Review of Concepts

Consider the following three solutions of equal concentration. Using data in Table 16.5, rank the three solutions from most basic to least basic: (a) aniline, (b) methylamine, (c) caffeine.

16.7 The Relationship Between Conjugate Acid-Base Ionization Constants

An important relationship between the acid ionization constant and the ionization constant of its conjugate base can be derived as follows, using acetic acid as an example:

$$CH_3COOH(aq) \rightleftharpoons H^+(aq) + CH_3COO^-(aq)$$

$$K_a = \frac{[H^+][CH_3COO^-]}{[CH_3COOH]}$$

The conjugate base, CH_3COO^-, reacts with water according to the equation

$$CH_3COO^-(aq) + H_2O(l) \rightleftharpoons CH_3COOH(aq) + OH^-(aq)$$

and we can write the base ionization constant as

$$K_b = \frac{[CH_3COOH][OH^-]}{[CH_3COO^-]}$$

The product of these two ionization constants is given by

$$\begin{aligned} K_aK_b &= \frac{[H^+][CH_3COO^-]}{[CH_3COOH]} \times \frac{[CH_3COOH][OH^-]}{[CH_3COO^-]} \\ &= [H^+][OH^-] \\ &= K_w \end{aligned}$$

This result may seem strange at first, but it can be understood by realizing that the sum of reactions (1) and (2) here is simply the autoionization of water.

$$\begin{array}{llr} (1) & CH_3COOH(aq) \rightleftharpoons H^+(aq) + CH_3COO^-(aq) & K_a \\ (2) & CH_3COO^-(aq) + H_2O(l) \rightleftharpoons CH_3COOH(aq) + OH^-(aq) & K_b \\ \hline (3) & H_2O(l) \rightleftharpoons H^+(aq) + OH^-(aq) & K_w \end{array}$$

This example illustrates one of the rules for chemical equilibria: When two reactions are added to give a third reaction, the equilibrium constant for the third reaction is the product of the equilibrium constants for the two added reactions. Thus, for any conjugate acid-base pair it is always true that

$$K_aK_b = K_w \qquad (16.13)$$

Expressing Equation (16.13) as

To find K_a, you must use the K_b of the conjugate base formed in the ionization of the acid and vice versa.

$$K_a = \frac{K_w}{K_b} \qquad K_b = \frac{K_w}{K_a}$$

enables us to draw an important conclusion: The stronger the acid (the larger K_a), the weaker its conjugate base (the smaller K_b), and vice versa (see Tables 16.3, 16.4, and 16.5).

We can use Equation (16.13) to calculate the K_b of the conjugate base (CH_3COO^-) of CH_3COOH as follows. We find the K_a value of CH_3COOH in Table 16.3 and write

$$\begin{aligned} K_b &= \frac{K_w}{K_a} \\ &= \frac{1.0 \times 10^{-14}}{1.8 \times 10^{-5}} \\ &= 5.6 \times 10^{-10} \end{aligned}$$

REVIEW OF CONCEPTS

Consider the following two acids and their ionization constants:

$$\begin{array}{ll} HCOOH & K_a = 1.7 \times 10^{-4} \\ HCN & K_a = 4.9 \times 10^{-10} \end{array}$$

Which conjugate base ($HCOO^-$ or CN^-) is stronger?

16.8 Molecular Structure and the Strength of Acids

The strength of an acid depends on a number of factors, such as the properties of the solvent, the temperature, and, of course, the molecular structure of the acid. When we compare the strengths of two acids, we can eliminate some variables by considering their properties in the same solvent and at the same temperature and concentration. Then we can focus on the structure of the acids.

Let us consider a certain acid HX. The strength of the acid is measured by its tendency to ionize:

$$HX \longrightarrow H^+ + X^-$$

Two factors influence the extent to which the acid undergoes ionization. One is the strength of the H—X bond—the stronger the bond, the more difficult it is for the HX molecule to break up and hence the *weaker* the acid. The other factor is the polarity of the H—X bond. The difference in the electronegativities between H and X results in a polar bond like

$$\overset{\delta+}{H}—\overset{\delta-}{X}$$

If the bond is highly polarized, that is, if there is a large accumulation of positive and negative charges on the H and X atoms, HX will tend to break up into H^+ and X^- ions. So a high degree of polarity characterizes a *stronger* acid. Next we will consider some examples in which either bond strength or bond polarity plays a prominent role in determining acid strength.

Hydrohalic Acids

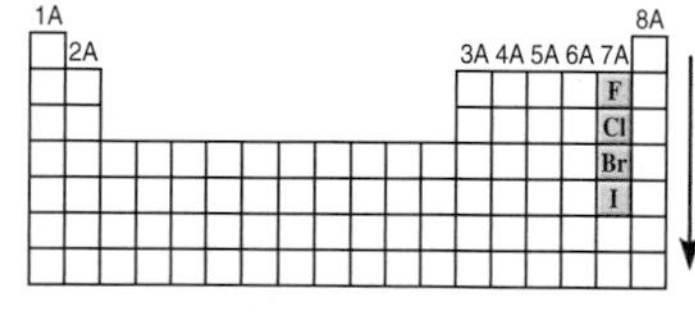

Strength of hydrohalic acids increases from HF to HI.

The halogens form a series of binary acids called the hydrohalic acids (HF, HCl, HBr, and HI). Of this series, which factor (bond strength or bond polarity) is the predominant factor in determining the strength of the binary acids? Consider first the strength of the H—X bond in each of these acids. Table 16.6 shows that HF has the highest bond

Table 16.6 Bond Enthalpies for Hydrogen Halides and Acid Strength for Hydrohalic Acids

Bond	Bond Enthalpy (kJ/mol)	Acid Strength
H—F	568.2	weak
H—Cl	431.9	strong
H—Br	366.1	strong
H—I	298.3	strong

enthalpy of the four hydrogen halides, and HI has the lowest bond enthalpy. It takes 568.2 kJ/mol to break the H—F bond and only 298.3 kJ/mol to break the H—I bond. Based on bond enthalpy, HI should be the strongest acid because it is easiest to break the bond and form the H^+ and I^- ions. Second, consider the polarity of the H—X bond. In this series of acids, the polarity of the bond decreases from HF to HI because F is the most electronegative of the halogens (see Figure 9.4). Based on bond polarity, then, HF should be the strongest acid because of the largest accumulation of positive and negative charges on the H and F atoms. Thus, we have two competing factors to consider in determining the strength of binary acids. The fact that HI is a strong acid and that HF is a weak acid indicates that bond enthalpy is the predominant factor in determining the acid strength of binary acids. In this series of binary acids, the weaker the bond, the stronger the acid so that the strength of the acids increases as follows:

$$\text{HF} \ll \text{HCl} < \text{HBr} < \text{HI}$$

Oxoacids

Now let us consider the oxoacids. Oxoacids, as we learned in Chapter 2, contain hydrogen, oxygen, and one other element Z, which occupies a central position. Figure 16.5 shows the Lewis structures of several common oxoacids. As you can see, these acids are characterized by the presence of one or more O—H bonds. The central atom Z might also have other groups attached to it:

To review the nomenclature of inorganic acids, see Section 2.8.

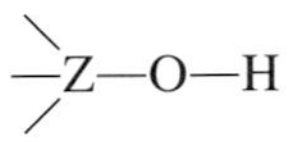

$$\rangle\!\!-\text{Z}-\text{O}-\text{H}$$

If Z is an electronegative element, or is in a high oxidation state, it will attract electrons, thus making the Z—O bond more covalent and the O—H bond more polar. Consequently, the tendency for the hydrogen to be donated as a H^+ ion increases:

As the oxidation number of an atom becomes larger, its ability to draw electrons in a bond toward itself increases.

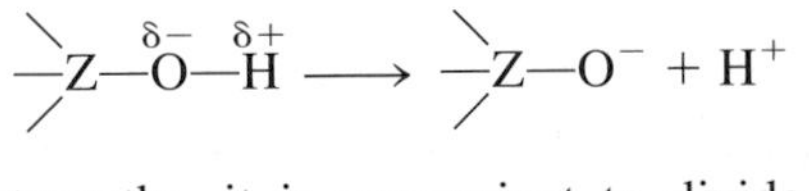

$$\rangle\!\!-\text{Z}-\overset{\delta-}{\text{O}}-\overset{\delta+}{\text{H}} \longrightarrow \rangle\!\!-\text{Z}-\text{O}^- + \text{H}^+$$

To compare their strengths, it is convenient to divide the oxoacids into two groups.

1. *Oxoacids Having Different Central Atoms That Are from the Same Group of the Periodic Table and That Have the Same Oxidation Number.* Within this group,

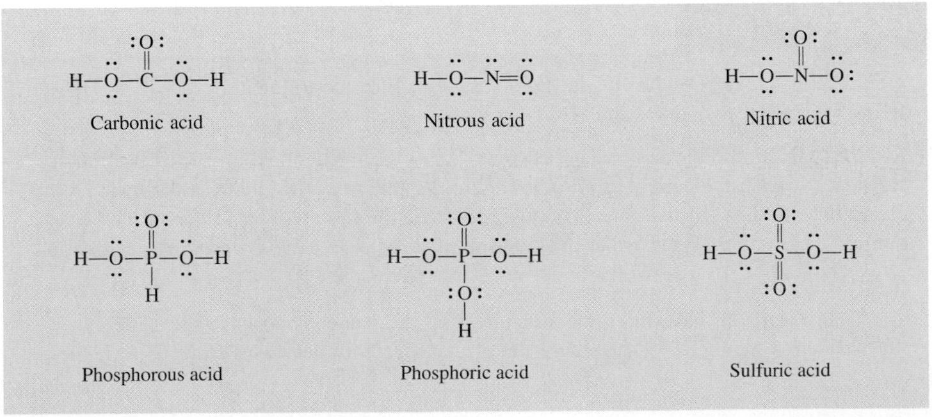

Carbonic acid | Nitrous acid | Nitric acid

Phosphorous acid | Phosphoric acid | Sulfuric acid

Figure 16.5
Lewis structures of some common oxoacids. For simplicity, the formal charges have been omitted.

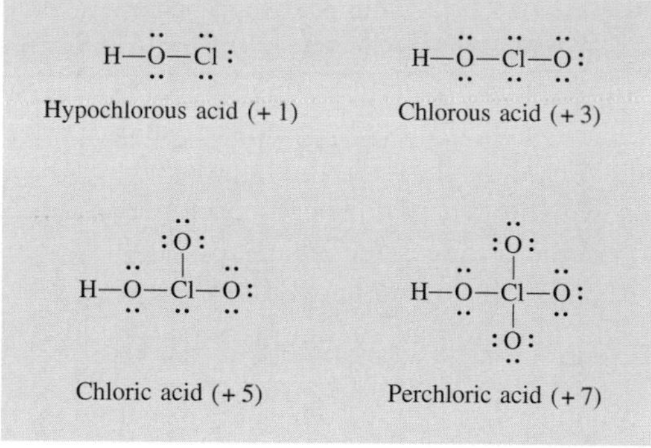

Figure 16.6
Lewis structures of the oxoacids of chlorine. The oxidation number of the Cl atom is shown in parentheses. For simplicity, the formal charges have been omitted. Note that although hypochlorous acid is written as HClO, the H atom is bonded to the O atom.

acid strength increases with increasing electronegativity of the central atom, as $HClO_3$ and $HBrO_3$ illustrate:

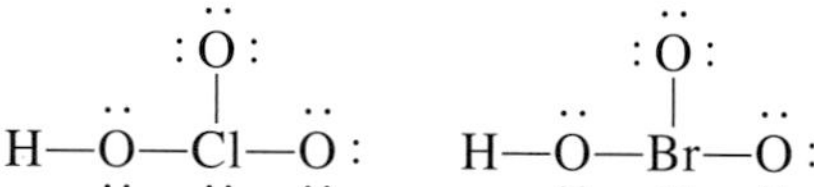

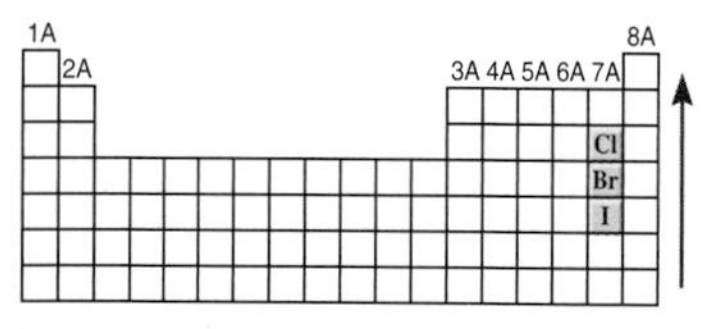

Strength of halogen-containing oxoacids having the same number of O atoms increases from bottom to top.

Cl and Br have the same oxidation number, +5. However, because Cl is more electronegative than Br, it attracts the electron pair it shares with oxygen (in the Cl—O—H group) to a greater extent than Br does. Consequently, the O—H bond is more polar in chloric acid than in bromic acid and ionizes more readily. Thus, the relative acid strengths are

$$HClO_3 > HBrO_3$$

2. *Oxoacids Having the Same Central Atom but Different Numbers of Attached Groups.* Within this group, acid strength increases as the oxidation number of the central atom increases. Consider the oxoacids of chlorine shown in Figure 16.6. In this series, the ability of chlorine to draw electrons away from the OH group (thus making the O—H bond more polar) increases with the number of electronegative O atoms attached to Cl. Thus, $HClO_4$ is the strongest acid because it has the largest number of O atoms attached to Cl, and the acid strength decreases as follows:

$$HClO_4 > HClO_3 > HClO_2 > HClO$$

EXAMPLE 16.12

Predict the relative strengths of the oxoacids in each of the following groups: (a) HClO, HBrO, and HIO; (b) HNO_3 and HNO_2.

Strategy Examine the molecular structure. In (a) the three acids have similar structure but differ only in the central atom (Cl, Br, and I). Which central atom is the most electronegative? In (b) the acids have the same central atom (N) but differ in the number of O atoms. What is the oxidation number of N in each of these two acids?

Solution

(a) These acids all have the same structure, and the halogens all have the same oxidation number (+1). Because the electronegativity decreases from Cl to I, the

(Continued)

polarity of the X—O bond (where X denotes a halogen atom) increases from HClO to HIO, and the polarity of the O—H bond decreases from HClO to HIO. Thus, the acid strength decreases as follows:

$$HClO > HBrO > HIO$$

(b) The structures of HNO_3 and HNO_2 are shown in Figure 16.5. Because the oxidation number of N is +5 in HNO_3 and +3 in HNO_2, HNO_3 is a stronger acid than HNO_2.

Practice Exercise Which of the following acids is weaker: $HBrO_3$ or $HBrO_4$?

Similar problem: 16.62.

REVIEW OF CONCEPTS

Rank the following acids from strongest to weakest: (a) H_2TeO_3, (b) H_2SO_3, (c) H_2SeO_3.

16.9 Acid-Base Properties of Salts

As defined in Section 4.3, a salt is an ionic compound formed by the reaction between an acid and a base. Salts are strong electrolytes that completely dissociate in water and in some cases react with water. The term ***salt hydrolysis*** describes *the reaction of an anion or a cation of a salt, or both, with water.* Salt hydrolysis usually affects the pH of a solution.

The word "hydrolysis" is derived from the Greek words *hydro,* meaning "water," and *lysis,* meaning "to split apart."

Salts That Produce Neutral Solutions

It is generally true that salts containing an alkali metal ion or alkaline earth metal ion (except Be^{2+}) and the conjugate base of a strong acid (for example, Cl^-, Br^-, and NO_3^-) do not undergo hydrolysis to an appreciable extent, and their solutions are assumed to be neutral. For instance, when $NaNO_3$, a salt formed by the reaction of NaOH with HNO_3, dissolves in water, it dissociates completely as

In reality, all positive ions give acid solutions in water.

$$NaNO_3(s) \xrightarrow{H_2O} Na^+(aq) + NO_3^-(aq)$$

The hydrated Na^+ ion neither donates nor accepts H^+ ions. The NO_3^- ion is the conjugate base of the strong acid HNO_3, and it has no affinity for H^+ ions. Consequently, a solution containing Na^+ and NO_3^- ions is neutral, with a pH of 7.

The mechanism by which metal ions produce acid solutions is discussed on p. 576.

Salts That Produce Basic Solutions

The dissociation of sodium acetate (CH_3COONa) in water is given by

$$CH_3COONa(s) \xrightarrow{H_2O} Na^+(aq) + CH_3COO^-(aq)$$

The hydrated Na^+ ion has no acidic or basic properties. The acetate ion CH_3COO^-, however, is the conjugate base of the weak acid CH_3COOH and therefore has an affinity for H^+ ions. The hydrolysis reaction is given by

$$CH_3COO^-(aq) + H_2O(l) \rightleftharpoons CH_3COOH(aq) + OH^-(aq)$$

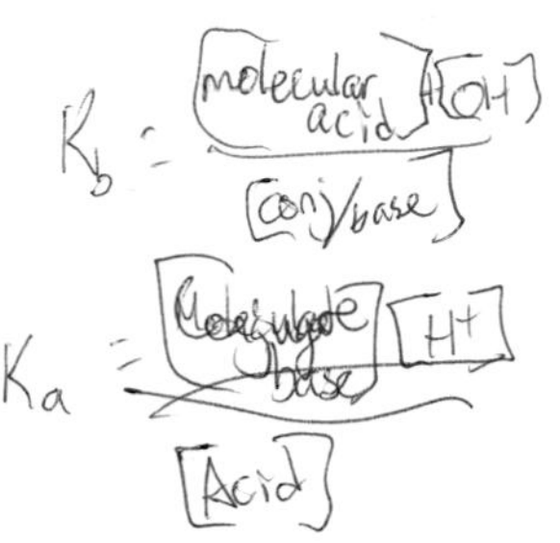

Because this reaction produces OH^- ions, the sodium acetate solution will be basic. The equilibrium constant for this hydrolysis reaction is the base ionization constant expression for CH_3COO^-, so we write

$$K_b = \frac{[CH_3COOH][OH^-]}{[CH_3COO^-]} = 5.6 \times 10^{-10}$$

Salts That Produce Acidic Solutions

When a salt derived from a strong acid and a weak base dissolves in water, the solution becomes acidic. For example, consider the process

$$NH_4Cl(s) \xrightarrow{H_2O} NH_4^+(aq) + Cl^-(aq)$$

The Cl^- ion has no affinity for H^+ ions. The ammonium ion NH_4^+ is the weak conjugate acid of the weak base NH_3 and ionizes as

$$NH_4^+(aq) + H_2O(l) \rightleftharpoons NH_3(aq) + H_3O^+(aq)$$

or simply

$$NH_4^+(aq) \rightleftharpoons NH_3(aq) + H^+(aq)$$

Because this reaction produces H^+ ions, the pH of the solution decreases. As you can see, the hydrolysis of the NH_4^+ ion is the same as the ionization of the NH_4^+ acid. The equilibrium constant (or ionization constant) for this process is given by

By coincidence, K_a of NH_4^+ has the same numerical value as K_b of CH_3COO^-.

$$K_a = \frac{[NH_3][H^+]}{[NH_4^+]} = \frac{K_w}{K_b} = \frac{1.0 \times 10^{-14}}{1.8 \times 10^{-5}} = 5.6 \times 10^{-10}$$

In solving salt hydrolysis problems, we follow the same procedure we used for weak acids and weak bases.

EXAMPLE 16.13

Calculate the pH of a 0.25 *M* solution of potassium fluoride (KF). What is the percent hydrolysis?

Strategy What is a salt? In solution, KF dissociates completely into K^+ and F^- ions. The K^+ ion does not react with water and has no effect on the pH of the solution because potassium is an alkali metal. The F^- ion is the conjugate base of the weak acid HF (Table 16.3). Therefore, we expect that it will react to a certain extent with water to produce HF and OH^-, and the solution will be basic.

Solution

Step 1: Because we started with a 0.25 *M* potassium fluoride solution, the concentrations of the ions are also equal to 0.25 *M* after dissociation:

	$KF(aq)$	$\longrightarrow$	$K^+(aq)$	+	$F^-(aq)$
Initial (*M*):	0.25		0		0
Change (*M*):	−0.25		+0.25		+0.25
Equilibrium (*M*):	0		0.25		0.25

(Continued)

Of these ions, only the fluoride ion will react with water

$$F^-(aq) + H_2O(l) \rightleftharpoons HF(aq) + OH^-(aq)$$

At equilibrium, the major species in solution are HF, F^-, and OH^-. The concentration of the H^+ ion is very small as we would expect for a basic solution, so it is treated as a minor species. We ignore the ionization of water.

Step 2: Let x be the equilibrium concentrations of HF and OH^- ions in mol/L, we summarize

	$F^-(aq)$ + $H_2O(l)$ $\rightleftharpoons$	$HF(aq)$ +	$OH^-(aq)$
Initial (M):	0.25	0	0
Change (M):	$-x$	$+x$	$+x$
Equilibrium (M):	$0.25 - x$	x	x

Step 3: From the preceding discussion and Table 16.3 we write the equilibrium constant of hydrolysis, or the base ionization constant, as

$$K_b = \frac{[HF][OH^-]}{[F^-]}$$

$$1.4 \times 10^{-11} = \frac{x^2}{0.25 - x}$$

Because K_b is very small and the initial concentration of the base is large, we can apply the approximation $0.25 - x \approx 0.25$:

$$1.4 \times 10^{-11} = \frac{x^2}{0.25 - x} \approx \frac{x^2}{0.25}$$

$$x = 1.9 \times 10^{-6}\ M$$

Step 4: At equilibrium

$$[OH^-] = 1.9 \times 10^{-6}\ M$$

$$pOH = -\log(1.9 \times 10^{-6})$$

$$= 5.72$$

$$pH = 14.00 - 5.72$$

$$= 8.28$$

Thus, the solution is basic, as we would expect. The percent hydrolysis is given by

$$\%\ \text{hydrolysis} = \frac{[F^-]_{\text{hydrolyzed}}}{[F^-]_{\text{initial}}}$$

$$= \frac{1.9 \times 10^{-6}}{0.25} \times 100\%$$

$$= 0.00076\%$$

Check The result shows that only a very small amount of the anion undergoes hydrolysis. Note that the calculation of percent hydrolysis takes the same form as the test for the approximation, which is valid in this case.

Similar problem: 16.73.

Practice Exercise Calculate the pH of a 0.24 M sodium formate solution (HCOONa).

Metal Ion Hydrolysis

Salts that contain small, highly charged metal cations (for example, Al^{3+}, Cr^{3+}, Fe^{3+}, Bi^{3+}, and Be^{2+}) and the conjugate bases of strong acids also produce an acidic solution. For example, when aluminum chloride ($AlCl_3$) dissolves in water, the Al^{3+} ions

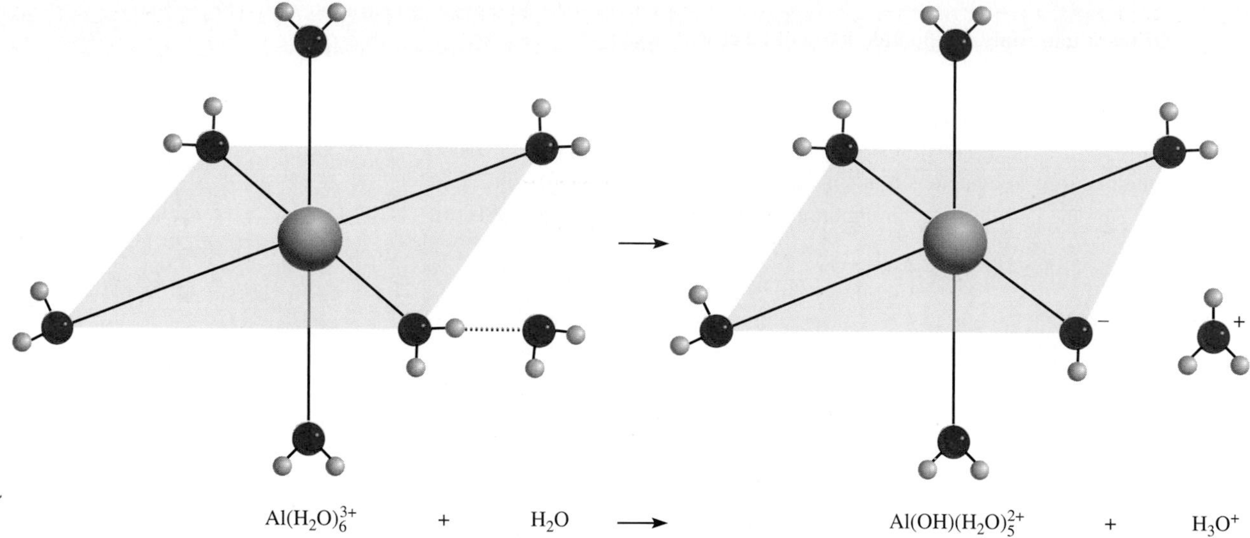

Figure 16.7

The six H_2O molecules surround the Al^{3+} ion octahedrally. The attraction of the small Al^{3+} ion for the lone pairs on the oxygen atoms is so great that the O—H bonds in a H_2O molecule attached to the metal cation are weakened, allowing the loss of a proton (H^+) to an incoming H_2O molecule. This hydrolysis of the metal cation makes the solution acidic.

take the hydrated form $Al(H_2O)_6^{3+}$ (Figure 16.7). Let us consider one bond between the metal ion and the oxygen atom of one of the six water molecules in $Al(H_2O)_6^{3+}$:

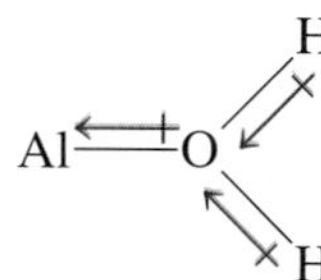

The positively charged Al^{3+} ion draws electron density toward itself, making the O—H bond more polar. Consequently, the H atoms have a greater tendency to ionize than those in water molecules not involved in hydration. The resulting ionization process can be written as

The hydrated Al^{3+} qualifies as a proton donor and thus a Brønsted acid in this reaction.

$$Al(H_2O)_6^{3+}(aq) + H_2O(l) \rightleftharpoons Al(OH)(H_2O)_5^{2+}(aq) + H_3O^+(aq)$$

or simply

$$Al(H_2O)_6^{3+}(aq) \rightleftharpoons Al(OH)(H_2O)_5^{2+}(aq) + H^+(aq)$$

The equilibrium constant for the metal cation hydrolysis is given by

Note that $Al(H_2O)_6^{3+}$ is roughly as strong an acid as CH_3COOH.

$$K_a = \frac{[Al(OH)(H_2O)_5^{2+}][H^+]}{[Al(H_2O)_6^{3+}]} = 1.3 \times 10^{-5}$$

Note that the $Al(OH)(H_2O)_5^{2+}$ species can undergo further ionization as

$$Al(OH)(H_2O)_5^{2+}(aq) \rightleftharpoons Al(OH)_2(H_2O)_4^{+}(aq) + H^+(aq)$$

and so on. However, generally it is sufficient to consider only the first step of hydrolysis.

The extent of hydrolysis is greatest for the smallest and most highly charged ions because a "compact" highly charged ion is more effective in polarizing the

Table 16.7 Acid-Base Properties of Salts

Type of Salt	Examples	Ions That Undergo Hydrolysis	pH of Solution
Cation from strong base; anion from strong acid	NaCl, KI, KNO_3, RbBr, $BaCl_2$	None	≈ 7
Cation from strong base; anion from weak acid	CH_3COONa, KNO_2	Anion	> 7
Cation from weak base; anion from strong acid	NH_4Cl, NH_4NO_3	Cation	< 7
Cation from weak base; anion from weak acid	NH_4NO_2, CH_3COONH_4, NH_4CN	Anion and cation	< 7 if $K_b < K_a$ ≈ 7 if $K_b \approx K_a$ > 7 if $K_b > K_a$
Small, highly charged cation; anion from strong acid	$AlCl_3$, $Fe(NO_3)_3$	Hydrated cation	< 7

O—H bond and facilitating ionization. This is the reason that relatively large ions of low charge such as Na^+ and K^+ do not undergo hydrolysis to any appreciable extent.

Salts in Which Both the Cation and the Anion Hydrolyze

So far we have considered salts in which only one ion undergoes hydrolysis. For salts derived from a weak acid and a weak base, both the cation and the anion hydrolyze. However, whether a solution containing such a salt is acidic, basic, or neutral depends on the relative strengths of the weak acid and the weak base. Because the mathematics associated with this type of system is rather involved, we will focus on making qualitative predictions about these solutions based on the following guidelines:

- **$K_b > K_a$.** If K_b for the anion is greater than K_a for the cation, then the solution must be basic because the anion will hydrolyze to a greater extent than the cation. At equilibrium, there will be more OH^- ions than H^+ ions.
- **$K_b < K_a$.** Conversely, if K_b for the anion is smaller than K_a for the cation, the solution will be acidic because cation hydrolysis will be more extensive than anion hydrolysis.
- **$K_b \approx K_a$.** If K_a is approximately equal to K_b, the solution will be nearly neutral.

Table 16.7 summarizes the behavior in aqueous solution of the salts discussed in this section.

EXAMPLE 16.14

Predict whether the following solutions will be acidic, basic, or nearly neutral: (a) NH_4I, (b) $NaNO_2$, (c) $FeCl_3$, (d) NH_4F.

Strategy In deciding whether a salt will undergo hydrolysis, ask yourself the following questions: Is the cation a highly charged metal ion or an ammonium ion? Is the anion the conjugate base of a weak acid? If yes to either question, then hydrolysis will occur. In cases where both the cation and the anion react with water, the pH of

(Continued)

the solution will depend on the relative magnitudes of K_a for the cation and K_b for the anion (see Table 16.7).

Solution We first break up the salt into its cation and anion components and then examine the possible reaction of each ion with water.

(a) The cation is NH_4^+, which will hydrolyze to produce NH_3 and H^+. The I^- anion is the conjugate base of the strong acid HI. Therefore, I^- will not hydrolyze and the solution is acidic.

(b) The Na^+ cation does not hydrolyze. The NO_2^- is the conjugate base of the weak acid HNO_2 and will hydrolyze to give HNO_2 and OH^-. The solution will be basic.

(c) Fe^{3+} is a small metal ion with a high charge and hydrolyzes to produce H^+ ions. The Cl^- does not hydrolyze. Consequently, the solution will be acidic.

(d) Both the NH_4^+ and F^- ions will hydrolyze. From Tables 16.5 and 16.3 we see that the K_a of NH_4^+ (5.6×10^{-10}) is greater than the K_b for F^- (1.4×10^{-11}). Therefore, the solution will be acidic.

Similar problem: 16.69.

Practice Exercise Predict whether the following solutions will be acidic, basic, or nearly neutral: (a) $LiClO_4$, (b) Na_3PO_4, (c) $Bi(NO_3)_3$, (d) NH_4CN.

Finally we note that some anions can act either as an acid or as a base. For example, the bicarbonate ion (HCO_3^-) can ionize or undergo hydrolysis as follows (see Table 16.4):

$$HCO_3^-(aq) + H_2O(l) \rightleftharpoons H_3O^+(aq) + CO_3^{2-}(aq) \qquad K_a = 4.8 \times 10^{-11}$$
$$HCO_3^-(aq) + H_2O(l) \rightleftharpoons H_2CO_3(aq) + OH^-(aq) \qquad K_b = 2.4 \times 10^{-8}$$

Because $K_b > K_a$, we predict that the hydrolysis reaction will outweigh the ionization process. Thus, a solution of sodium bicarbonate ($NaHCO_3$) will be basic.

Review of Concepts

The diagrams shown here represent solutions of three salts NaX (X = A, B, or C). (a) Which X^- has the weakest conjugate acid? (b) Arrange the three X^- anions in order of increasing base strength. The Na^+ ion and water molecules have been omitted for clarity.

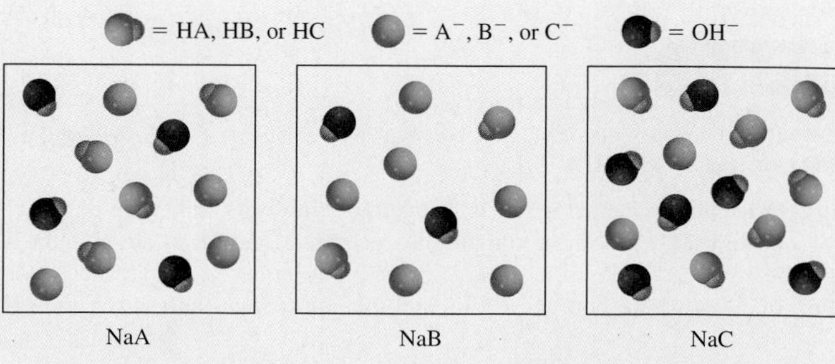

16.10 Acidic, Basic, and Amphoteric Oxides

As we saw in Chapter 8, oxides can be classified as acidic, basic, or amphoteric. Our discussion of acid-base reactions would be incomplete without examining the properties of these compounds.

Figure 16.8 shows the formulas of a number of oxides of the representative elements in their highest oxidation states. Note that all alkali metal oxides and all alkaline earth metal oxides except BeO are basic. Beryllium oxide and several metallic oxides in Groups 3A and 4A are amphoteric. Nonmetallic oxides in which the oxidation number of the representative element is high are acidic (for example, N_2O_5, SO_3, and Cl_2O_7). Those in which the oxidation number of the representative element is low (for example, CO and NO) show no measurable acidic properties. No nonmetallic oxides are known to have basic properties.

The basic metallic oxides react with water to form metal hydroxides:

$$Na_2O(s) + H_2O(l) \longrightarrow 2NaOH(aq)$$
$$BaO(s) + H_2O(l) \longrightarrow Ba(OH)_2(aq)$$

The reactions between acidic oxides and water are

$$CO_2(g) + H_2O(l) \rightleftharpoons H_2CO_3(aq)$$
$$SO_3(g) + H_2O(l) \rightleftharpoons H_2SO_4(aq)$$
$$N_2O_5(g) + H_2O(l) \rightleftharpoons 2HNO_3(aq)$$
$$P_4O_{10}(s) + 6H_2O(l) \rightleftharpoons 4H_3PO_4(aq)$$
$$Cl_2O_7(g) + H_2O(l) \rightleftharpoons 2HClO_4(aq)$$

A forest damaged by acid rain.

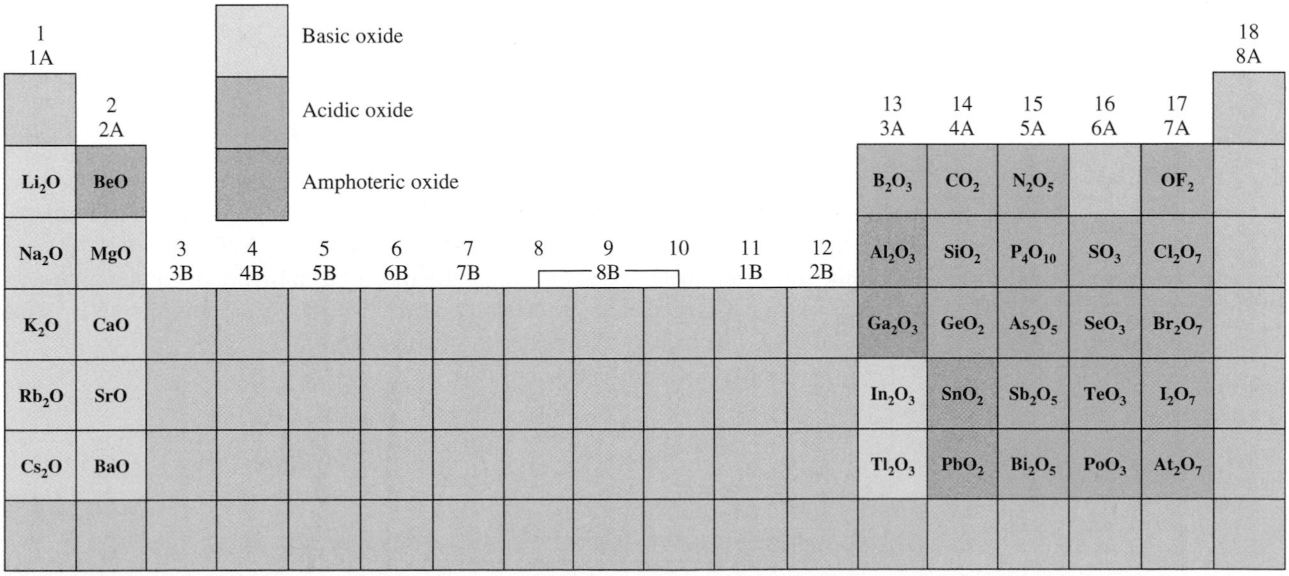

The reaction between CO_2 and H_2O explains why when pure water is exposed to air (which contains CO_2) it gradually reaches a pH of about 5.5 (Figure 16.9). The reaction between SO_3 and H_2O is largely responsible for acid rain.

Basic oxide

Acidic oxide

Amphoteric oxide

1 1A	2 2A	13 3A	14 4A	15 5A	16 6A	17 7A	18 8A
Li_2O	BeO	B_2O_3	CO_2	N_2O_5		OF_2	
Na_2O	MgO	Al_2O_3	SiO_2	P_4O_{10}	SO_3	Cl_2O_7	
K_2O	CaO	Ga_2O_3	GeO_2	As_2O_5	SeO_3	Br_2O_7	
Rb_2O	SrO	In_2O_3	SnO_2	Sb_2O_5	TeO_3	I_2O_7	
Cs_2O	BaO	Tl_2O_3	PbO_2	Bi_2O_5	PoO_3	At_2O_7	

3 3B, 4 4B, 5 5B, 6 6B, 7 7B, 8, 9 8B, 10, 11 1B, 12 2B

Figure 16.8
Oxides of the representative elements in their highest oxidation states.

Figure 16.9
(Left) A beaker of water to which a few drops of bromothymol blue indicator have been added. (Right) As dry ice is added to the water, the CO_2 reacts to form carbonic acid, which turns the solution acidic and changes the color from blue to yellow.

Reactions between acidic oxides and bases and those between basic oxides and acids resemble normal acid-base reactions in that the products are a salt and water:

$$\underset{\text{acidic oxide}}{CO_2(g)} + \underset{\text{base}}{2NaOH(aq)} \longrightarrow \underset{\text{salt}}{Na_2CO_3(aq)} + \underset{\text{water}}{H_2O(l)}$$

$$\underset{\text{basic oxide}}{BaO(s)} + \underset{\text{acid}}{2HNO_3(aq)} \longrightarrow \underset{\text{salt}}{Ba(NO_3)_2(aq)} + \underset{\text{water}}{H_2O(l)}$$

As Figure 16.8 shows, aluminum oxide (Al_2O_3) is amphoteric. Depending on the reaction conditions, it can behave either as an acidic oxide or as a basic oxide. For example, Al_2O_3 acts as a base with hydrochloric acid to produce a salt ($AlCl_3$) and water:

$$Al_2O_3(s) + 6HCl(aq) \longrightarrow 2AlCl_3(aq) + 3H_2O(l)$$

and acts as an acid with sodium hydroxide:

$$Al_2O_3(s) + 2NaOH(aq) + 3H_2O(l) \longrightarrow 2NaAl(OH)_4(aq)$$

Note that only a salt, $NaAl(OH)_4$ [containing the Na^+ and $Al(OH)_4^-$ ions], is formed in the latter reaction; no water is produced. Nevertheless, this reaction can still be classified as an acid-base reaction because Al_2O_3 neutralizes NaOH.

The higher the oxidation number of the metal, the more covalent the compound; the lower the oxidation number, the more ionic the compound.

Some transition metal oxides in which the metal has a high oxidation number act as acidic oxides. Consider manganese(VII) oxide (Mn_2O_7) and chromium(VI) oxide (CrO_3), both of which react with water to produce acids:

$$Mn_2O_7(l) + H_2O(l) \longrightarrow \underset{\text{permanganic acid}}{2HMnO_4(aq)}$$

$$CrO_3(s) + H_2O(l) \longrightarrow \underset{\text{chromic acid}}{H_2CrO_4(aq)}$$

REVIEW OF CONCEPTS

Arrange the following oxides in order of increasing basicity: K_2O, Al_2O_3, BaO.

16.11 Lewis Acids and Bases

Figure 16.10
A Lewis acid-base reaction involving BF_3 and NH_3.

Acid-base properties so far have been discussed in terms of the Brønsted theory. To behave as a Brønsted base, for example, a substance must be able to accept protons. By this definition both the hydroxide ion and ammonia are bases:

$$H^+ + {}^-:\ddot{\underset{..}{O}}—H \longrightarrow H—\ddot{\underset{..}{O}}—H$$

$$H^+ + \begin{array}{c} H \\ | \\ :N—H \\ | \\ H \end{array} \longrightarrow \left[\begin{array}{c} H \\ | \\ H—N—H \\ | \\ H \end{array} \right]^+$$

In each case, the atom to which the proton becomes attached possesses at least one unshared pair of electrons. This characteristic property of the OH^- ion, of NH_3, and of other Brønsted bases suggests a more general definition of acids and bases.

The American chemist G. N. Lewis formulated such a definition. According to Lewis's definition, a base is *a substance that can donate a pair of electrons,* and an acid is *a substance that can accept a pair of electrons.* For example, in the protonation of ammonia, NH_3 acts as a ***Lewis base*** because it donates a pair of electrons to the proton H^+, which acts as a ***Lewis acid*** by accepting the pair of electrons. A Lewis acid-base reaction, therefore, is one that involves the donation of a pair of electrons from one species to another. Such a reaction does not produce a salt and water.

> All Brønsted bases are Lewis bases.

> Lewis acids are either deficient in electrons (cations) or the central atom has a vacant valence orbital.

The significance of the Lewis concept is that it is much more general than other definitions; it includes as acid-base reactions many reactions that do not involve Brønsted acids. Consider, for example, the reaction between boron trifluoride (BF_3) and ammonia (Figure 16.10):

$$\underset{\text{acid}}{\begin{array}{c} F \\ | \\ F—B \\ | \\ F \end{array}} + \underset{\text{base}}{\begin{array}{c} H \\ | \\ :N—H \\ | \\ H \end{array}} \longrightarrow \begin{array}{c} F\ \ H \\ |\ \ \ | \\ F—B—N—H \\ |\ \ \ | \\ F\ \ H \end{array}$$

In Section 10.4 we saw that the B atom in BF_3 is sp^2-hybridized. The vacant, unhybridized $2p_z$ orbital accepts the pair of electrons from NH_3. So BF_3 functions as an acid according to the Lewis definition, even though it does not contain an ionizable proton. Note that a coordinate covalent bond is formed between the B and N atoms (see p. 306).

> A coordinate covalent bond is always formed in a Lewis acid-base reaction.

Another Lewis acid containing boron is boric acid (H_3BO_3). Boric acid (a weak acid used in eyewash) is an oxoacid with the following structure:

$$\begin{array}{c} H \\ | \\ :O: \\ | \\ H—\ddot{\underset{..}{O}}—B—\ddot{\underset{..}{O}}—H \end{array}$$

H_3BO_3

Note that boric acid does not ionize in water to produce an H^+ ion. Instead, its reaction with water is

$$B(OH)_3(aq) + H_2O(l) \rightleftharpoons B(OH)_4^-(aq) + H^+(aq)$$

In this Lewis acid-base reaction, boric acid accepts a pair of electrons from the hydroxide ion that is derived from the H_2O molecule.

The hydration of carbon dioxide to produce carbonic acid

$$CO_2(g) + H_2O(l) \rightleftharpoons H_2CO_3(aq)$$

can be understood in the Lewis framework as follows: The first step involves donation of a lone pair on the oxygen atom in H_2O to the carbon atom in CO_2. An orbital is vacated on the C atom to accommodate the lone pair by removal of the electron pair in the C—O pi bond. These shifts of electrons are indicated by the curved arrows.

Therefore, H_2O is a Lewis base and CO_2 is a Lewis acid. Next, a proton is transferred onto the O atom bearing a negative charge to form H_2CO_3.

EXAMPLE 16.15

Identify the Lewis acid and Lewis base in each of the following reactions:

(a) $C_2H_5OC_2H_5 + AlCl_3 \rightleftharpoons (C_2H_5)_2OAlCl_3$

(b) $Hg^{2+}(aq) + 4CN^-(aq) \rightleftharpoons Hg(CN)_4^{2-}(aq)$

Strategy In Lewis acid-base reactions, the acid is usually a cation or an electron-deficient molecule, whereas the base is an anion or a molecule containing an atom with lone pairs. (a) Draw the molecular structure for $C_2H_5OC_2H_5$. What is the hybridization state of Al in $AlCl_3$? (b) Which ion is likely to be an electron acceptor? An electron donor?

Solution

(a) The Al is sp^2-hybridized in $AlCl_3$ with an empty $2p_z$ orbital. It is electron-deficient, sharing only six electrons. Therefore, the Al atom has a tendency to gain two electrons to complete its octet. This property makes $AlCl_3$ a Lewis acid. On the other hand, the lone pairs on the oxygen atom in $C_2H_5OC_2H_5$ make the compound a Lewis base:

What are the formal charges on Al and O in the product?

(b) Here the Hg^{2+} ion accepts four pairs of electrons from the CN^- ions. Therefore Hg^{2+} is the Lewis acid and CN^- is the Lewis base.

Similar problem: 16.80.

Practice Exercise Identify the Lewis acid and Lewis base in the reaction

$$Co^{3+}(aq) + 6NH_3(aq) \rightleftharpoons Co(NH_3)_6^{3+}(aq)$$

Review of Concepts

Which of the following cannot behave as a Lewis base? (a) NH_3, (b) OF_2, (c) CH_4, (d) OH^-, (e) Fe^{3+}.

Key Equations

$K_w = [H^+][OH^-] = 1.0 \times 10^{-14}$	(16.4)	Ion-product constant of water.
$pH = -\log [H^+]$	(16.5)	Definition of pH of a solution.
$[H^+] = 10^{-pH}$	(16.6)	Calculating H^+ ion concentration from pH.
$pOH = -\log [OH^-]$	(16.8)	Definition of pOH of a solution.
$[OH^-] = 10^{-pOH}$	(16.9)	Calculating OH^- ion concentration from pOH.
$pH + pOH = 14.00$	(16.10)	Another form of Equaton (16.4).

$$\text{percent ionization} = \frac{\text{ionized acid concentration at equilibrium}}{\text{initial concentration of acid}} \times 100\% \qquad (16.12)$$

$K_aK_b = K_w$	(16.13)	Relationship between the acid and base ionization constants of a conjugate acid-base pair.

Summary of Facts and Concepts

1. Brønsted acids donate protons, and Brønsted bases accept protons. These are the definitions that normally underlie the use of the terms "acid" and "base."
2. The acidity of an aqueous solution is expressed as its pH, which is defined as the negative logarithm of the hydrogen ion concentration (in mol/L).
3. At 25°C, an acidic solution has pH < 7, a basic solution has pH > 7, and a neutral solution has pH $= 7$.
4. In aqueous solution, the following are classified as strong acids: $HClO_4$, HI, HBr, HCl, H_2SO_4 (first stage of ionization), and HNO_3. Strong bases in aqueous solution include hydroxides of alkali metals and of alkaline earth metals (except beryllium).
5. The acid ionization constant K_a increases with acid strength. K_b similarly expresses the strengths of bases.
6. Percent ionization is another measure of the strength of acids. The more dilute a solution of a weak acid, the greater the percent ionization of the acid.
7. The product of the ionization constant of an acid and the ionization constant of its conjugate base is equal to the ion-product constant of water.
8. The relative strengths of acids can be explained qualitatively in terms of their molecular structures.
9. Most salts are strong electrolytes that dissociate completely into ions in solution. The reaction of these ions with water, called salt hydrolysis, can produce acidic or basic solutions. In salt hydrolysis, the conjugate bases of weak acids yield basic solutions, and the conjugate acids of weak bases yield acidic solutions.
10. Small, highly charged metal ions, such as Al^{3+} and Fe^{3+}, hydrolyze to yield acidic solutions.
11. Most oxides can be classified as acidic, basic, or amphoteric. Metal hydroxides are either basic or amphoteric.
12. Lewis acids accept pairs of electrons and Lewis bases donate pairs of electrons. The term "Lewis acid" is generally reserved for substances that can accept electron pairs but do not contain ionizable hydrogen atoms.

Key Words

Questions and Problems

Brønsted Acids and Bases

Review Questions

16.1 Define Brønsted acids and bases. How do the Brønsted definitions differ from Arrhenius's definitions of acids and bases?

16.2 For a species to act as a Brønsted base, an atom in the species must possess a lone pair of electrons. Explain why this is so.

Problems

16.3 Classify each of these species as a Brønsted acid or base, or both: (a) H_2O, (b) OH^-, (c) H_3O^+, (d) NH_3, (e) NH_4^+, (f) NH_2^-, (g) NO_3^-, (h) CO_3^{2-}, (i) HBr, (j) HCN.

16.4 What are the names and formulas of the conjugate bases of these acids: (a) HNO_2, (b) H_2SO_4, (c) H_2S, (d) HCN, (e) HCOOH (formic acid)?

16.5 Identify the acid-base conjugate pairs in each of these reactions:

(a) $CH_3COO^- + HCN \rightleftharpoons CH_3COOH + CN^-$

(b) $HCO_3^- + HCO_3^- \rightleftharpoons H_2CO_3 + CO_3^{2-}$

(c) $H_2PO_4^- + NH_3 \rightleftharpoons HPO_4^{2-} + NH_4^+$

(d) $HClO + CH_3NH_2 \rightleftharpoons CH_3NH_3^+ + ClO^-$

(e) $CO_3^{2-} + H_2O \rightleftharpoons HCO_3^- + OH^-$

(f) $CH_3COO^- + H_2O \rightleftharpoons CH_3COOH + OH^-$

16.6 Give the conjugate acid of each of these bases: (a) HS^-, (b) HCO_3^-, (c) CO_3^{2-}, (d) $H_2PO_4^-$, (e) HPO_4^{2-}, (f) PO_4^{3-}, (g) HSO_4^-, (h) SO_4^{2-}, (i) NO_2^-, (j) SO_3^{2-}.

16.7 Give the conjugate base of each of these acids: (a) $CH_2ClCOOH$, (b) HIO_4, (c) H_3PO_4, (d) $H_2PO_4^-$, (e) HPO_4^{2-}, (f) H_2SO_4, (g) HSO_4^-, (h) HCOOH, (i) HSO_3^-, (j) NH_4^+, (k) H_2S, (l) HS^-, (m) HClO.

16.8 Oxalic acid ($H_2C_2O_4$) has the following structure:

```
O=C—OH
  |
O=O—OH
```

An oxalic acid solution contains these species in varying concentrations: $H_2C_2O_4$, $HC_2O_4^-$, $C_2O_4^{2-}$, and H^+. (a) Draw Lewis structures of $HC_2O_4^-$ and $C_2O_4^{2-}$. (b) Which of the four species listed here can act only as acids, which can act only as bases, and which can act as both acids and bases?

pH and pOH Calculations

Review Questions

16.9 What is the ion-product constant for water?

16.10 Write an equation relating $[H^+]$ and $[OH^-]$ in solution at 25°C.

16.11 The ion-product constant for water is 1.0×10^{-14} at 25°C and 3.8×10^{-14} at 40°C. Is the process

$$H_2O(l) \rightleftharpoons H^+(aq) + OH^-(aq)$$

endothermic or exothermic?

16.12 Define pH. Why do chemists normally choose to discuss the acidity of a solution in terms of pH rather than hydrogen ion concentration, $[H^+]$?

16.13 The pH of a solution is 6.7. From this statement alone, can you conclude that the solution is acidic? If not, what additional information would you need? Can the pH of a solution be zero or negative? If so, give examples to illustrate these values.

16.14 Define pOH. Write an equation relating pH and pOH.

Problems

16.15 Calculate the hydrogen ion concentration for solutions with these pH values: (a) 2.42, (b) 11.21, (c) 6.96, (d) 15.00.

16.16 Calculate the hydrogen ion concentration in moles per liter for each of these solutions: (a) a solution whose pH is 5.20, (b) a solution whose pH is 16.00; (c) a solution whose hydroxide concentration is 3.7×10^{-9} *M*.

16.17 Calculate the pH of each of these solutions: (a) 0.0010 *M* HCl, (b) 0.76 *M* KOH, (c) 2.8×10^{-4} *M* $Ba(OH)_2$, (d) 5.2×10^{-4} *M* HNO_3.

16.18 Calculate the pH of water at 40°C, given that K_w is 3.8×10^{-14} at this temperature.

16.19 Complete this table for a solution:

pH	$[H^+]$	Solution is
< 7		
	$< 1.0 \times 10^{-7} M$	
		Neutral

16.20 Fill in the word "acidic," "basic," or "neutral" for these solutions:

(a) pOH > 7; solution is ______.

(b) pOH = 7; solution is ______.

(c) pOH < 7; solution is ______.

16.21 The pOH of a solution is 9.40. Calculate the hydrogen ion concentration of the solution.

16.22 Calculate the number of moles of KOH in 5.50 mL of a 0.360 *M* KOH solution. What is the pOH of the solution?

16.23 A solution is made by dissolving 18.4 g of HCl in 662 mL of water. Calculate the pH of the solution. (Assume that the volume of the solution is also 662 mL.)

16.24 How much NaOH (in grams) is needed to prepare 546 mL of solution with a pH of 10.00?

Strengths of Acids and Bases

Review Questions

16.25 Explain what is meant by the strength of an acid.

16.26 Without referring to the text, write the formulas of four strong acids and four weak acids.

16.27 What are the strongest acid and strongest base that can exist in water?

16.28 H_2SO_4 is a strong acid, but HSO_4^- is a weak acid. Account for the difference in strength of these two related species.

Problems

16.29 Which of the following diagrams best represents a strong acid, such as HCl, dissolved in water? Which represents a weak acid? Which represents a very weak acid? (The hydrated proton is shown as a hydronium ion. Water molecules are omitted for clarity.)

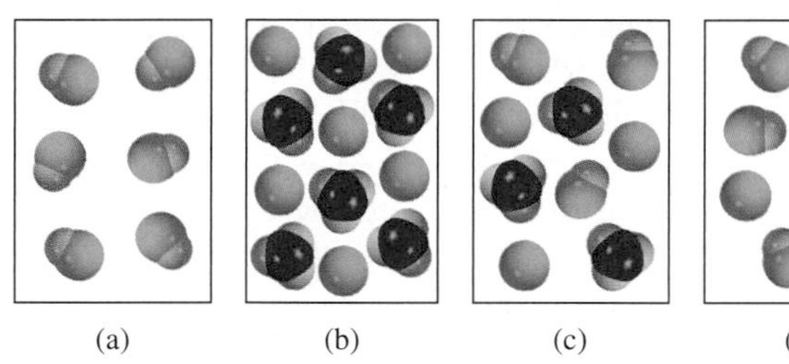

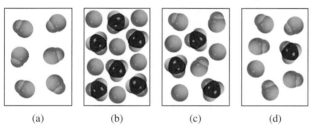

16.30 (1) Which of the following diagrams represents a solution of a weak diprotic acid? (2) Which diagrams represent chemically implausible situations? (The hydrated proton is shown as a hydronium ion. Water molecules are omitted for clarity.)

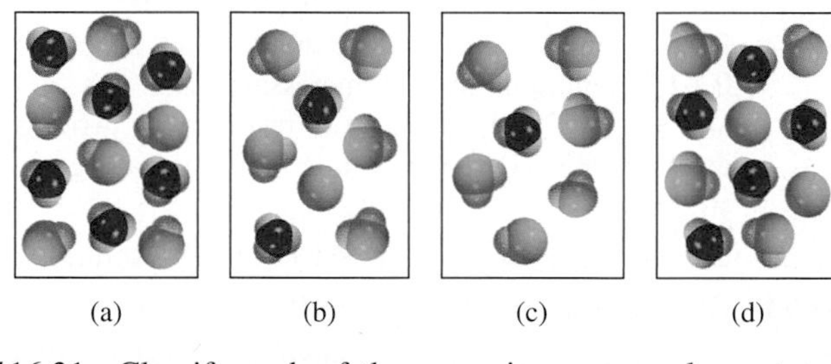

16.31 Classify each of these species as a weak or strong acid: (a) HNO_3, (b) HF, (c) H_2SO_4, (d) HSO_4^-, (e) H_2CO_3, (f) HCO_3^-, (g) HCl, (h) HCN, (i) HNO_2.

16.32 Classify each of these species as a weak or strong base: (a) LiOH, (b) CN^-, (c) H_2O, (d) ClO_4^-, (e) NH_2^-.

16.33 Which of these statements is/are true regarding a 0.10 *M* solution of a weak acid HA?

(a) The pH is 1.00.

(b) $[H^+] \gg [A^-]$

(c) $[H^+] = [A^-]$

(d) The pH is less than 1.

16.34 Which of these statements is/are true regarding a 1.0 *M* solution of a strong acid HA?

(a) $[A^-] > [H^+]$

(b) The pH is 0.00.

(c) $[H^+] = 1.0\ M$

(d) $[HA] = 1.0\ M$

16.35 Predict the direction that predominates in this reaction:

$$F^-(aq) + H_2O(l) \rightleftharpoons HF(aq) + OH^-(aq)$$

16.36 Predict whether this reaction will proceed from left to right to any measurable extent:

$$CH_3COOH(aq) + Cl^-(aq) \longrightarrow$$

Weak Acid Ionization Constants

Review Questions

16.37 What does the ionization constant tell us about the strength of an acid?

16.38 List the factors on which the K_a of a weak acid depends.

16.39 Why do we normally not quote K_a values for strong acids such as HCl and HNO_3? Why is it necessary to specify temperature when giving K_a values?

16.40 Which of the following solutions has the highest pH? (a) 0.40 *M* HCOOH, (b) 0.40 *M* $HClO_4$, (c) 0.40 *M* CH_3COOH.

Problems

16.41 Calculate the concentrations of all the species (HCN, H^+, CN^-, and OH^-) in a 0.15 *M* HCN solution.

16.42 A 0.0560-g quantity of acetic acid is dissolved in enough water to make 50.0 mL of solution. Calculate the concentrations of H^+, CH_3COO^-, and CH_3COOH at equilibrium. (K_a for acetic acid $= 1.8 \times 10^{-5}$.)

16.43 The pH of an HF solution is 6.20. Calculate the ratio [conjugate base]/[acid] for HF at this pH.

16.44 What is the original molarity of a solution of formic acid (HCOOH) whose pH is 3.26 at equilibrium?

16.45 Calculate the pH of a 0.060 *M* HF solution.

16.46 Calculate the percent ionization of hydrofluoric acid at these concentrations: (a) 0.60 *M*, (b) 0.080 *M*, (c) 0.0046 *M*, (d) 0.00028 *M*. Comment on the trends.

16.47 A 0.040 *M* solution of a monoprotic acid is 14 percent ionized. Calculate the ionization constant of the acid.

16.48 (a) Calculate the percent ionization of a 0.20 *M* solution of the monoprotic acetylsalicylic acid (aspirin). ($K_a = 3.0 \times 10^{-4}$.) (b) The pH of gastric juice in the stomach of a certain individual is 1.00. After a few aspirin tablets have been swallowed, the concentration of acetylsalicylic acid in the stomach is 0.20 *M*. Calculate the percent ionization of the acid under these conditions.

Diprotic and Polyprotic Acids

Review Questions

16.49 Malonic acid [$CH_2(COOH)_2$] is a diprotic acid. Explain what that means.

16.50 Write all the species (except water) that are present in a phosphoric acid solution. Indicate which species can act as a Brønsted acid, which as a Brønsted base, and which as both a Brønsted acid and a Brønsted base.

Problems

16.51 What are the concentrations of HSO_4^-, SO_4^{2-}, and H^+ in a 0.20 *M* $KHSO_4$ solution? (*Hint:* H_2SO_4 is a strong acid: K_a for $HSO_4^- = 1.3 \times 10^{-2}$.)

16.52 Calculate the concentrations of H^+, HCO_3^-, and CO_3^{2-} in a 0.025 *M* H_2CO_3 solution.

Weak Base Ionization Constants; K_a–K_b Relationship

Review Questions

16.53 Use NH_3 to illustrate what we mean by the strength of a base.

16.54 Write the equation relating K_a for a weak acid and K_b for its conjugate base. Use NH_3 and its conjugate acid NH_4^+ to derive the relationship between K_a and K_b.

Problems

16.55 Calculate the pH for each of these solutions: (a) 0.10 *M* NH_3, (b) 0.050 *M* pyridine.

16.56 The pH of a 0.30 *M* solution of a weak base is 10.66. What is the K_b of the base?

16.57 What is the original molarity of a solution of ammonia whose pH is 11.22?

16.58 In a 0.080 *M* NH_3 solution, what percent of the NH_3 is present as NH_4^+?

Molecular Structure and the Strength of Acids

Review Questions

16.59 List four factors that affect the strength of an acid.

16.60 How does the strength of an oxoacid depend on the electronegativity and oxidation number of the central atom?

Problems

16.61 Predict the acid strengths of the following compounds: H_2O, H_2S, and H_2Se.

16.62 Compare the strengths of the following pairs of acids: (a) H_2SO_4 and H_2SeO_4, (b) H_3PO_4 and H_3AsO_4.

16.63 Which of the following is the stronger acid: $CH_2ClCOOH$ or $CHCl_2COOH$? Explain your choice.

16.64 Consider the following compounds:

C_6H_5—O—H CH_3—O—H

phenol   methanol

Experimentally, phenol is found to be a stronger acid than methanol. Explain this difference in terms of the structures of the conjugate bases. (*Hint:* A more stable conjugate base favors ionization. Only one of the conjugate bases can be stabilized by resonance.)

Acid-Base Properties of Salt Solutions

Review Questions

16.65 Define salt hydrolysis. Categorize salts according to how they affect the pH of a solution.

16.66 Explain why small, highly charged metal ions are able to undergo hydrolysis.

16.67 Al^{3+} is not a Brønsted acid, but $Al(H_2O)_6^{3+}$ is. Explain.

16.68 Specify which of these salts will undergo hydrolysis: KF, $NaNO_3$, NH_4NO_2, $MgSO_4$, KCN, C_6H_5COONa, RbI, Na_2CO_3, $CaCl_2$, HCOOK.

16.69 Predict the pH (>7, <7, or ≈ 7) of the aqueous solutions containing the following salts: (a) KBr, (b) $Al(NO_3)_3$, (c) $BaCl_2$, (d) $Bi(NO_3)_3$.

16.70 Which ion of the alkaline earth metals is most likely to undergo hydrolysis?

Problems

16.71 A certain salt, MX (containing the M^+ and X^- ions), is dissolved in water, and the pH of the resulting solution is 7.0. Can you say anything about the strengths of the acid and the base from which the salt is derived?

16.72 In a certain experiment a student finds that the pHs of 0.10 *M* solutions of three potassium salts KX, KY, and KZ are 7.0, 9.0, and 11.0, respectively. Arrange the acids HX, HY, and HZ in order of increasing acid strength.

16.73 Calculate the pH of a 0.36 *M* CH_3COONa solution.

16.74 Calculate the pH of a 0.42 *M* NH_4Cl solution.

16.75 Predict whether a solution containing the salt K_2HPO_4 will be acidic, neutral, or basic. (*Hint:* You need to consider both the ionization and hydrolysis of HPO_4^{2-}.)

16.76 Predict the pH (>7, <7, ≈ 7) of a $NaHCO_3$ solution.

Lewis Acids and Bases

Review Questions

16.77 What are the Lewis definitions of an acid and a base? In what way are they more general than the Brønsted definitions?

16.78 In terms of orbitals and electron arrangements, what must be present for a molecule or an ion to act as a Lewis acid (use H^+ and BF_3 as examples)? What must be present for a molecule or ion to act as a Lewis base (use OH^- and NH_3 as examples)?

Problems

16.79 Classify each of these following species as a Lewis acid or a Lewis base: (a) CO_2, (b) H_2O, (c) I^-, (d) SO_2, (e) NH_3, (f) OH^-, (g) H^+, (h) BCl_3.

16.80 Describe this reaction according to the Lewis theory of acids and bases:

$$AlCl_3(s) + Cl^-(aq) \longrightarrow AlCl_4^-(aq)$$

16.81 Which would be considered a stronger Lewis acid: (a) BF_3 or BCl_3, (b) Fe^{2+} or Fe^{3+}? Explain.

16.82 All Brønsted acids are Lewis acids, but the reverse is not true. Give two examples of Lewis acids that are not Brønsted acids.

Additional Problems

16.83 Classify these following oxides as acidic, basic, amphoteric, or neutral: (a) CO_2, (b) K_2O, (c) CaO, (d) N_2O_5, (e) CO, (f) NO, (g) SnO_2, (h) SO_3, (i) Al_2O_3, (j) BaO.

16.84 A typical reaction between an antacid and the hydrochloric acid in gastric juice is

$$NaHCO_3(aq) + HCl(aq) \longrightarrow NaCl(aq) + H_2O(l) + CO_2(g)$$

Calculate the volume (in liters) of CO_2 generated from 0.350 g of $NaHCO_3$ and excess gastric juice at 1.00 atm and 37.0°C.

16.85 To which of the following would the addition of an equal volume of 0.60 *M* NaOH lead to a solution having a lower pH? (a) water, (b) 0.30 *M* HCl, (c) 0.70 *M* KOH, (d) 0.40 *M* $NaNO_3$.

16.86 The pH of a 0.0642 *M* solution of a monoprotic acid is 3.86. Is this a strong acid?

16.87 Like water, ammonia undergoes autoionization in liquid ammonia:

$$NH_3 + NH_3 \rightleftharpoons NH_4^+ + NH_2^-$$

(a) Identify the Brønsted acids and Brønsted bases in this reaction. (b) What species correspond to H^+ and OH^-, and what is the condition for a neutral solution?

16.88 HA and HB are both weak acids although HB is the stronger of the two. Will it take more volume of a 0.10 *M* NaOH solution to neutralize 50.0 mL of 0.10 *M* HB than 50.0 mL of 0.10 *M* HA?

16.89 A 1.87-g sample of Mg reacts with 80.0 mL of a HCl solution whose pH is −0.544. What is the pH of the solution after all the Mg has reacted? Assume volume of solution is constant.

16.90 The three common chromium oxides are CrO, Cr_2O_3, and CrO_3. If Cr_2O_3 is amphoteric, what can you say about the acid-base properties of CrO and CrO_3?

16.91 Most of the hydrides of Group 1A and Group 2A metals are ionic (the exceptions are BeH_2 and MgH_2, which are covalent compounds). (a) Describe the reaction between the hydride ion (H^-) and water in terms of a Brønsted acid-base reaction. (b) The same reaction can also be classified as a redox reaction. Identify the oxidizing and reducing agents.

16.92 Use the data in Table 16.3 to calculate the equilibrium constant for the following reaction:

$$CH_3COOH(aq) + NO_2^-(aq) \rightleftharpoons CH_3COO^-(aq) + HNO_2(aq)$$

16.93 Calculate the pH of a 0.20 *M* ammonium acetate (CH_3COONH_4) solution.

16.94 Novocaine, used as a local anesthetic by dentists, is a weak base ($K_b = 8.91 \times 10^{-6}$). What is the ratio of the concentration of the base to that of its acid in the blood plasma (pH = 7.40) of a patient?

16.95 In the vapor phase, acetic acid molecules associate to a certain extent to form dimers:

$$2CH_3COOH(g) \rightleftharpoons (CH_3COOH)_2(g)$$

At 51°C, the pressure of a certain acetic acid vapor system is 0.0342 atm in a 360-mL flask. The vapor is condensed and neutralized with 13.8 mL of 0.0568 *M* NaOH. (a) Calculate the degree of dissociation (α) of the dimer under these conditions:

$$(CH_3COOH)_2 \rightleftharpoons 2CH_3COOH$$

(*Hint:* See Problem 15.65 for general procedure.) (b) Calculate the equilibrium constant K_P for the reaction in (a).

16.96 Calculate the concentrations of all the species in a 0.100 *M* Na_2CO_3 solution.

16.97 Henry's law constant for CO_2 at 38°C is 2.28×10^{-3} mol/L · atm. Calculate the pH of a solution of CO_2 at 38°C in equilibrium with the gas at a partial pressure of 3.20 atm.

16.98 Hydrocyanic acid (HCN) is a weak acid and a deadly poisonous compound that, in the gaseous form (hydrogen cyanide), is used in gas chambers. Why is it dangerous to treat sodium cyanide with acids (such as HCl) without proper ventilation?

16.99 A solution of formic acid (HCOOH) has a pH of 2.53. How many grams of formic acid are there in 100.0 mL of the solution?

16.100 Calculate the pH of a 1-L solution containing 0.150 mole of CH_3COOH and 0.100 mole of HCl.

16.101 You are given two beakers containing separately an aqueous solution of strong acid (HA) and an aqueous solution of weak acid (HB) of the same concentration.

Describe how you would compare the strengths of these two acids by (a) measuring the pH, (b) measuring electrical conductance, (c) studying the rate of hydrogen gas evolution when these solutions are reacted with an active metal such as Mg or Zn.

16.102 Use Le Châtelier's principle to predict the effect of the following changes on the extent of hydrolysis of sodium nitrite ($NaNO_2$) solution: (a) HCl is added; (b) NaOH is added; (c) NaCl is added; (d) the solution is diluted.

16.103 The disagreeable odor of fish is mainly due to organic compounds (RNH_2) containing an amino group, $—NH_2$, in which R is the rest of the molecule. Amines are bases just like ammonia. Explain why putting some lemon juice on fish can greatly reduce the odor.

16.104 A 0.400 *M* formic acid (HCOOH) solution freezes at −0.758°C. Calculate the K_a of the acid at that temperature. (*Hint:* Assume that molarity is equal to molality. Carry your calculations to three significant figures and round off to two for K_a.)

16.105 Both the amide ion (NH_2^-) and the nitride ion (N^{3-}) are stronger bases than the hydroxide ion and hence do not exist in aqueous solutions. (a) Write the equations showing the reactions of these ions with water, and identify the Brønsted acid and base in each case. (b) Which of the two is the stronger base?

16.106 The atmospheric sulfur dioxide (SO_2) concentration over a certain region is 0.12 ppm by volume. Calculate the pH of the rainwater as a result of this pollutant. Assume that the dissolution of SO_2 does not affect its pressure and that the pH of rainwater is solely due to this compound.

16.107 Explain the action of smelling salt, which is ammonium carbonate [$(NH_4)_2CO_3$]. (*Hint:* The thin film of aqueous solution that lines the nasal passage is slightly basic.)

16.108 Which of the following is the stronger base: NF_3 or NH_3? (*Hint:* F is more electronegative than H.)

16.109 Which of the following is a stronger base: NH_3 or PH_3? (*Hint:* The N—H bond is stronger than the P—H bond.)

16.110 How many milliliters of a strong monoprotic acid solution at pH = 4.12 must be added to 528 mL of the same acid solution at pH = 5.76 to change its pH to 5.34? Assume that the volumes are additive.

16.111 When chlorine reacts with water, the resulting solution is weakly acidic and reacts with $AgNO_3$ to give a white precipitate. Write balanced equations to represent these reactions. Explain why manufacturers of household bleaches add bases such as NaOH to their products to increase their effectiveness.

16.112 Calculate the concentrations of all species in a 0.100 *M* H_3PO_4 solution.

16.113 A solution of methylamine (CH_3NH_2) has a pH of 10.64. How many grams of methylamine are there in 100.0 mL of the solution?

16.114 The diagrams here show three weak acids HA (A = X, Y, or Z) in solution. (a) Arrange the acids in order of increasing K_a. (b) Arrange the conjugate bases in increasing order of K_b. (c) Calculate the percent ionization of each acid. (d) Which of the 0.1 *M* sodium salt solutions (NaX, NaY, or NaZ) has the lowest pH? (The hydrated proton is shown as a hydronium ion. Water molecules are omitted for clarity.)

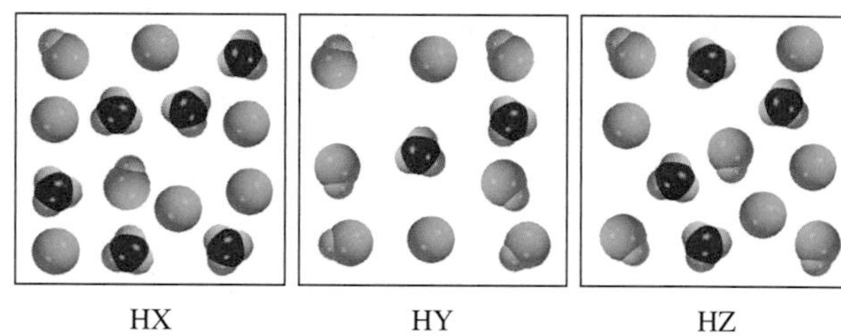

16.115 Calculate the pH and percent ionization of a 0.80 *M* HNO_2 solution.

16.116 Consider the two weak acids HX (molar mass = 180 g/mol) and HY (molar mass = 78.0 g/mol). If a solution of 16.9 g/L of HX has the same pH as one containing 9.05 g/L of HY, which is the stronger acid?

Special Problems

16.117 At 28°C and 0.982 atm, a gaseous compound HA has a density of 1.16 g/L. A quantity of 2.03 g of this compound is dissolved in water and diluted to exactly one liter. If the pH of the solution is 5.22 (due to the ionization of HA) at 25°C, calculate K_a of the acid.

16.118 About half of the hydrochloric acid produced annually in the United States (3.0 billion pounds) is used for metal pickling. This process involves the removal of metal oxide layers from metal surfaces to prepare them for coating. (a) Write the overall and net ionic equations for the reaction between iron(III) oxide, which represents the rust layer over iron, and HCl. Identify the Brønsted acid and base. (b) Hydrochloric acid is also used to remove scale (which is mostly

$CaCO_3$) from water pipes. Hydrochloric acid reacts with calcium carbonate in two stages; the first stage forms the bicarbonate ion, which then reacts further to form carbon dioxide. Write equations for these two stages and for the overall reaction. (c) Hydrochloric acid is used to recover oil from the ground. It dissolves rocks (often $CaCO_3$) so that the oil can flow more easily. In one process, a 15 percent (by mass) HCl solution is injected into an oil well to dissolve the rocks. If the density of the acid solution is 1.073 g/mL, what is the pH of the solution?

16.119 Hemoglobin (Hb) is a blood protein that is responsible for transporting oxygen. It can exist in the protonated form of HbH^+. The binding of oxygen can be represented by the simplified equation

$$HbH^+ + O_2 \rightleftharpoons HbO_2 + H^+$$

(a) What form of hemoglobin is favored in the lungs where oxygen concentration is highest? (b) In body tissues, where carbon dioxide is released as a result of metabolism, the medium is more acidic because of the formation of carbonic acid. What form of hemoglobin is favored under this condition? (c) When a person hyperventilates, the concentration of CO_2 in his or her blood decreases. How does this action affect the above equilibrium? Frequently a person who is hyperventilating is advised to breathe into a paper bag. Why does this action help the individual?

16.120 A 1.294-g sample of a metal carbonate (MCO_3) is reacted with 500 mL of a 0.100 *M* HCl solution. The excess HCl acid is then neutralized by 32.80 mL of 0.588 *M* NaOH. Identify M.

16.121 Prove the statement that when the concentration of a weak acid HA decreases by a factor of 10, its percent ionization increases by a factor of $\sqrt{10}$. State any assumptions.

16.122 Calculate the pH of a solution that is 1.00 *M* HCN and 1.00 *M* HF. Compare the concentration (in molarity) of the CN^- ion in this solution with that in a 1.00 *M* HCN solution. Comment on the difference.

16.123 Teeth enamel consists largely of hydroxyapatite [$Ca_5(PO_4)_3OH$]. When it dissolves in water (a process called *demineralization*), it dissociates as follows:

$$Ca_5(PO_4)_3OH \longrightarrow 5Ca^{2+} + 3PO_4^{3-} + OH^-$$

The reverse process, called *remineralization,* is the body's natural defense against tooth decay. Acids produced from food remove the OH^- ions and thereby weaken the enamel layer. Most toothpastes contain a flouride compound such as NaF or SnF_2. What is the function of these compounds in preventing tooth decay?

16.124 Use the van't Hoff equation (see Problem 15.100) and the data in Appendix 2 to calculate the pH of water at its normal boiling point.

Answers to Practice Exercises

16.1 (1) H_2O (acid) and OH^- (base); (2) HCN (acid) and CN^- (base). **16.2** 7.7×10^{-15} *M*. **16.3** 0.12. **16.4** 4.7×10^{-4} *M*. **16.5** 7.40. **16.6** 12.56. **16.7** Smaller than 1. **16.8** 2.09. **16.9** 2.2×10^{-6}. **16.10** $[H_2C_2O_4] = 0.11$ *M*, $[HC_2O_4^-] = 0.086$ *M*. $[C_2O_4^{2-}] = 6.1 \times 10^{-5}$ *M*, $[H^+] = 0.086$ *M*. **16.11** 12.03. **16.12** $HBrO_3$. **16.13** 8.58. **16.14** (a) pH $\approx$ 7, (b) pH $>$ 7, (c) pH $<$ 7, (d) pH $>$ 7. **16.15** Lewis acid: Co^{3+}; Lewis base: NH_3.

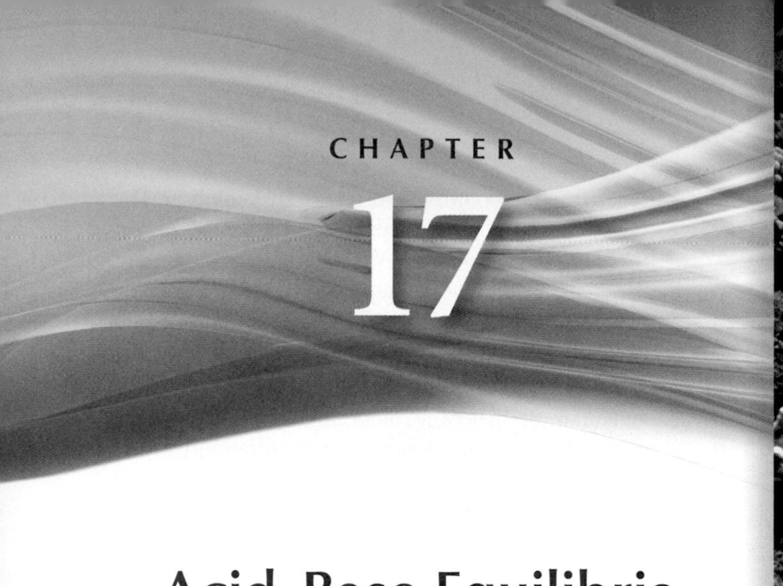

CHAPTER

17

Acid-Base Equilibria and Solubility Equilibria

The precipitation of calcium carbonate ($CaCO_3$) results in the formation of coral exoskeletons. Increasing acidification of the oceans enhances the dissolution of these structures.

Chapter Outline

Student Interactive Activities

Animations
Buffer Solutions (17.2)
Acid-Base Titrations (17.3)

Electronic Homework
Example Practice Problems
End of Chapter Problems

Essential Concepts

Buffer Solutions A buffer solution contains a weak acid and a salt derived from the acid. To maintain a relatively constant pH, the acid and base components of the buffer solution react with added acid or base. Buffer solutions play an important role in many chemical and biological processes.

Acid-Base Titrations The characteristics of an acid-base titration depend on the strength of the acid and base involved. Different indicators are used to determine the end point of a titration.

Solubility Equilibria Another application of the equilibrium concept is the solubility equilibria of sparingly soluble salts, which are expressed as the solubility product. The solubility of such a substance can be affected by the presence of a common cation or anion, or the pH. Complex-ion formation, an example of the Lewis acid-base type reaction, increases the solubility of an insoluble salt.

17.1 Homogeneous Versus Heterogeneous Solution Equilibria

In Chapter 16 we saw that weak acids and weak bases never ionize completely in water. Thus, at equilibrium a weak acid solution, for example, contains nonionized acid as well as H^+ ions and the conjugate base. Nevertheless, all of these species are dissolved, so that the system is an example of homogeneous equilibrium (see Chapter 15).

Another important type of equilibrium, which we will study in the second half of the chapter, involves the dissolution and precipitation of slightly soluble substances. These processes are examples of heterogeneous equilibria; that is, they pertain to reactions in which the components are in more than one phase. But first we will conclude our discussion of acid-base equilibria by considering buffer solutions and taking a closer look at acid-base titrations.

17.2 Buffer Solutions

A **buffer solution** is *a solution of (1) a weak acid or a weak base and (2) its salt; both components must be present. The solution has the ability to resist changes in pH upon the addition of small amounts of either acid or base.* Buffers are very important to chemical and biological systems. The pH in the human body varies greatly from one fluid to another; for example, the pH of blood is about 7.4, whereas the gastric juice in our stomachs has a pH of about 1.5. These pH values, which are crucial for the proper functioning of enzymes and the balance of osmotic pressure, are maintained by buffers in most cases.

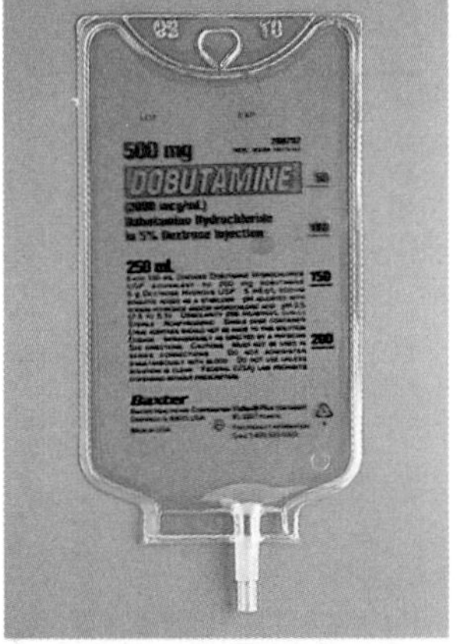

500 mg
DOBUTAMINE
Baxter
EACH 100 mL CONTAINS DOBUTAMINE HYDROCHLORIDE USP EQUIVALENT TO 200 mg DOBUTAMINE 5 g DEXTROSE HYDROUS USP 5 mEq/L SODIUM BISULFITE ADDED AS A STABILIZER pH ADJUSTED WITH SODIUM HYDROXIDE AND/OR HYDROCHLORIC ACID pH 3.5 (2.5 TO 5.5) OSMOLARITY 266 mOsmol/L (CALC) STERILE NONPYROGENIC SINGLE DOSE CONTAINER DRUG ADDITIVES SHOULD NOT BE MADE TO THIS SOLUTION DOSAGE INTRAVENOUSLY AS DIRECTED BY A PHYSICIAN SEE DIRECTIONS CAUTIONS MUST NOT BE USED IN

Fluids for intravenous injection must include buffer systems to maintain the proper blood pH.

A buffer solution must contain a relatively large concentration of acid to react with any OH^- ions that may be added to it and must contain a similar concentration of base to react with any added H^+ ions. Furthermore, the acid and the base components of the buffer must not consume each other in a neutralization reaction. These requirements are satisfied by an acid-base conjugate pair (a weak acid and its conjugate base or a weak base and its conjugate acid).

A simple buffer solution can be prepared by adding comparable amounts of acetic acid (CH_3COOH) and sodium acetate (CH_3COONa) to water. The equilibrium concentrations of both the acid and the conjugate base (from CH_3COONa) are assumed to be the same as the starting concentrations. This is so because (1) CH_3COOH is a weak acid and the extent of hydrolysis of the CH_3COO^- ion is very small and (2) the presence of CH_3COO^- ions suppresses the ionization of CH_3COOH, and the presence of CH_3COOH suppresses the hydrolysis of the CH_3COO^- ions.

Animation:
Buffer solutions

A solution containing these two substances has the ability to neutralize either added acid or added base. Sodium acetate, a strong electrolyte, dissociates completely in water:

$$CH_3COONa(s) \xrightarrow{H_2O} CH_3COO^-(aq) + Na^+(aq)$$

If an acid is added, the H^+ ions will be consumed by the conjugate base in the buffer, CH_3COO^-, according to the equation

$$CH_3COO^-(aq) + H^+(aq) \longrightarrow CH_3COOH(aq)$$

If a base is added to the buffer system, the OH^- ions will be neutralized by the acid in the buffer:

$$CH_3COOH(aq) + OH^-(aq) \longrightarrow CH_3COO^-(aq) + H_2O(l)$$

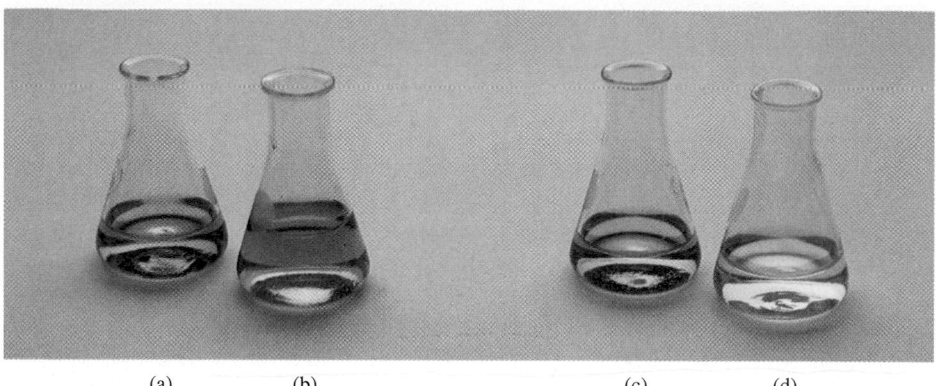

(a) (b) (c) (d)

Figure 17.1

The acid-base indicator bromophenol blue (added to all solutions shown) is used to illustrate buffer action. The indicator's color is blue-purple above pH 4.6 and yellow below pH 3.0. (a) A buffer solution made up of 50 mL of 0.1 M CH_3COOH and 50 mL of 0.1 M CH_3COONa. The solution has a pH of 4.7 and turns the indicator blue-purple. (b) After the addition of 40 mL of 0.1 M HCl solution to the solution in (a), the color remains blue-purple. (c) A 100-mL CH_3COOH solution whose pH is 4.7. (d) After the addition of 6 drops (about 0.3 mL) of 0.1 M HCl solution, the color turns yellow. Without buffer action, the pH of the solution decreases rapidly to less than 3.0 upon the addition of 0.1 M HCl.

The *buffering capacity,* that is, the effectiveness of the buffer solution, depends on the amounts of acid and conjugate base from which the buffer is made. The larger the amount, the greater the buffering capacity.

In general, a buffer system can be represented as salt/acid or conjugate base/acid. Thus, the sodium acetate–acetic acid buffer system can be written as CH_3COONa/CH_3COOH or CH_3COO^-/CH_3COOH. Figure 17.1 shows this buffer system in action.

EXAMPLE 17.1

Which of the following solutions can be classified as buffer systems? (a) KH_2PO_4/H_3PO_4, (b) $NaClO_4/HClO_4$, (c) C_5H_5N/C_5H_5NHCl (C_5H_5N is pyridine; its K_b is given in Table 16.5). Explain your answer.

Strategy What constitutes a buffer system? Which of the preceding solutions contains a weak acid and its salt (containing the weak conjugate base)? Which of the preceding solutions contains a weak base and its salt (containing the weak conjugate acid)? Why is the conjugate base of a strong acid not able to neutralize an added acid?

Solution The criteria for a buffer system is that we must have a weak acid and its salt (containing the weak conjugate base) or a weak base and its salt (containing the weak conjugate acid).

(a) H_3PO_4 is a weak acid, and its conjugate base, $H_2PO_4^-$, is a weak base (see Table 16.4). Therefore, this is a buffer system.

(b) Because $HClO_4$ is a strong acid, its conjugate base, ClO_4^-, is an extremely weak base. This means that the ClO_4^- ion will not combine with a H^+ ion in solution to form $HClO_4$. Thus, the system cannot act as a buffer system.

(c) As Table 16.5 shows, C_5H_5N is a weak base and its conjugate acid, $C_5H_5\overset{+}{N}H$ (the cation of the salt C_5H_5NHCl), is a weak acid. Therefore, this is a buffer system.

Similar problems: 17.5, 17.6.

Practice Exercise Which of the following are buffer systems? (a) KF/HF, (b) KBr/HBr, (c) $Na_2CO_3/NaHCO_3$.

EXAMPLE 17.2

(a) Calculate the pH of a buffer system containing 1.0 *M* CH_3COOH and 1.0 *M* CH_3COONa. (b) What is the pH of the buffer system after the addition of 0.10 mole of gaseous HCl to 1.0 L of the solution? Assume that the volume of the solution does not change when the HCl is added.

Strategy (a) The pH of the buffer solution before the addition of HCl can be calculated from the ionization of CH_3COOH. Note that because both the acid and the sodium salt of the acid are present, the initial concentrations of CH_3COOH and CH_3COO^- (from CH_3COONa) are both 1.0 *M*. The K_a of CH_3COOH is 1.8×10^{-5} (see Table 16.3). (b) It is helpful to make a sketch of the changes that occur in this case.

HCl

Buffer soln

$[CH_3COOH] = 1.0\,M$

$[CH_3COO^-] = 1.0\,M$

$H^+ \; Cl^-$

Buffer action in (b)

$CH_3COO^- + H^+ \rightarrow CH_3COOH$

Solution

(a) We summarize the concentrations of the species at equilibrium as follows:

	$CH_3COOH(aq)$	$\rightleftharpoons$ $H^+(aq)$	+ $CH_3COO^-(aq)$
Initial (*M*):	1.0	0	1.0
Change (*M*):	$-x$	$+x$	$+x$
Equilibrium (*M*):	$1.0 - x$	x	$1.0 + x$

$$K_a = \frac{[H^+][CH_3COO^-]}{[CH_3COOH]}$$

$$1.8 \times 10^{-5} = \frac{(x)(1.0 + x)}{(1.0 - x)}$$

Assuming $1.0 + x \approx 1.0$ and $1.0 - x \approx 1.0$, we obtain

$$1.8 \times 10^{-5} = \frac{(x)(1.0 + x)}{(1.0 - x)} \approx \frac{x(1.0)}{1.0}$$

or

$$x = [H^+] = 1.8 \times 10^{-5}\,M$$

Thus,

$$pH = -\log(1.8 \times 10^{-5}) = 4.74$$

Recall that the presence of CH_3COOH suppresses the hydrolysis of CH_3COO^- and the presence of CH_3COO^- suppresses the ionization of CH_3COOH.

When the concentrations of the acid and the conjugate base are the same, the pH of the buffer is equal to the pK_a of the acid.

(b) When HCl is added to the solution, the initial changes are

	$HCl(aq)$	$\longrightarrow$ $H^+(aq)$	+ $Cl^-(aq)$
Initial (mol):	0.10	0	0
Change (mol):	−0.10	+0.10	+0.10
Final (mol):	0	0.10	0.10

The Cl^- ion is a spectator ion in solution because it is the conjugate base of a strong acid.

The H^+ ions provided by the strong acid HCl react completely with the conjugate base of the buffer, which is CH_3COO^-. At this point it is more convenient to work with moles rather than molarity. The reason is that in some

(Continued)

cases the volume of the solution may change when a substance is added. A change in volume will change the molarity, but not the number of moles. The neutralization reaction is summarized next:

	$CH_3COO^-(aq)$	$+\ H^+(aq)$	$\longrightarrow CH_3COOH(aq)$
Initial (mol):	1.0	0.10	1.0
Change (mol):	−0.10	−0.10	+0.10
Final (mol):	0.90	0	1.1

Finally, to calculate the pH of the buffer after neutralization of the acid, we convert back to molarity by dividing moles by 1.0 L of solution.

	$CH_3COOH(aq)$	$\rightleftharpoons H^+(aq)$	$+\ CH_3COO^-(aq)$
Initial (*M*):	1.1	0	0.90
Change (*M*):	$-x$	$+x$	$+x$
Equilibrium (*M*):	$1.1 - x$	x	$0.90 + x$

$$K_a = \frac{[H^+][CH_3COO^-]}{[CH_3COOH]}$$

$$1.8 \times 10^{-5} = \frac{(x)(0.90 + x)}{1.1 - x}$$

Assuming $0.90 + x \approx 0.90$ and $1.1 - x \approx 1.1$, we obtain

$$1.8 \times 10^{-5} = \frac{(x)(0.90 + x)}{1.1 - x} \approx \frac{x(0.90)}{1.1}$$

or

$$x = [H^+] = 2.2 \times 10^{-5}\ M$$

Similar problem: 17.14.

Thus,

$$pH = -\log(2.2 \times 10^{-5}) = 4.66$$

Practice Exercise Calculate the pH of the 0.30 *M* NH_3/0.36 *M* NH_4Cl buffer system. What is the pH after the addition of 20.0 mL of 0.050 *M* NaOH to 80.0 mL of the buffer solution?

In the buffer solution examined in Example 17.2, there is a decrease in pH (the solution becomes more acidic) as a result of added HCl. We can also compare the changes in H^+ ion concentration as follows

$$\text{Before addition of HCl: } [H^+] = 1.8 \times 10^{-5}\ M$$
$$\text{After addition of HCl: } [H^+] = 2.2 \times 10^{-5}\ M$$

Thus, the H^+ ion concentration increases by a factor of

$$\frac{2.2 \times 10^{-5}\ M}{1.8 \times 10^{-5}\ M} = 1.2$$

To appreciate the effectiveness of the CH_3COONa/CH_3COOH buffer, let us find out what would happen if 0.10 mol HCl were added to 1 L of water, and compare the increase in H^+ ion concentration.

$$\text{Before addition of HCl: } [H^+] = 1.0 \times 10^{-7}\ M$$
$$\text{After addition of HCl: } [H^+] = 0.10\ M$$

As a result of the addition of HCl, the H^+ ion concentration increases by a factor of

$$\frac{0.10\ M}{1.0 \times 10^{-7}\ M} = 1.0 \times 10^6$$

amounting to a millionfold increase! This comparison shows that a properly chosen buffer solution can maintain a fairly constant H^+ ion concentration, or pH (Figure 17.2).

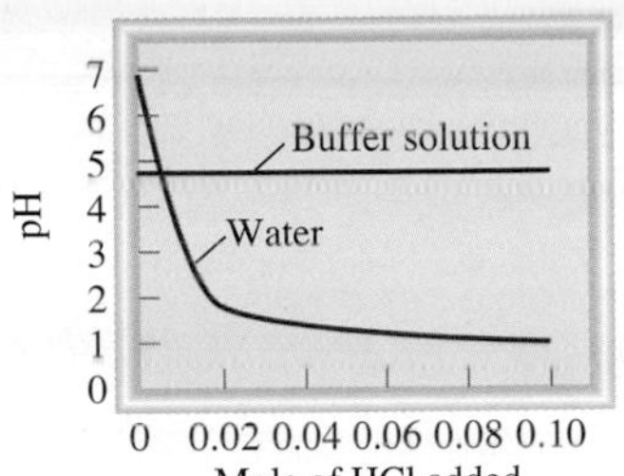

Figure 17.2
A comparison of the change in pH when 0.10 mol HCl is added to pure water and to an acetate buffer solution, as described in Example 17.2.

Preparing a Buffer Solution with a Specific pH

Now suppose we want to prepare a buffer solution with a specific pH. How do we go about it? Referring to the acetic acid–sodium acetate buffer system, we can write the equilibrium constant as

$$K_a = \frac{[CH_3COO^-][H^+]}{[CH_3COOH]}$$

Note that this expression holds whether we have only acetic acid or a mixture of acetic acid and sodium acetate in solution. Rearranging the equation gives

$$[H^+] = \frac{K_a[CH_3COOH]}{[CH_3COO^-]}$$

Taking the negative logarithm of both sides, we obtain

$$-\log [H^+] = -\log K_a - \log \frac{[CH_3COOH]}{[CH_3COO^-]}$$

or

$$-\log [H^+] = -\log K_a + \log \frac{[CH_3COO^-]}{[CH_3COOH]}$$

So

$$\mathrm{pH} = \mathrm{p}K_a + \log \frac{[CH_3COO^-]}{[CH_3COOH]} \quad (17.1)$$

in which

$$\mathrm{p}K_a = -\log K_a \quad (17.2)$$

$\mathrm{p}K_a$ is related to K_a as pH is related to $[H^+]$. Remember that the stronger the acid (that is, the larger the K_a), the smaller the $\mathrm{p}K_a$.

Equation (17.1) is called the *Henderson-Hasselbalch equation*. In a more general form, it can be expressed as

$$\mathrm{pH} = \mathrm{p}K_a + \log \frac{[\text{conjugate base}]}{[\text{acid}]} \quad (17.3)$$

Keep in mind that $\mathrm{p}K_a$ is a constant, but the ratio of the two concentration terms in Equation (17.3) depends on a particular solution.

If the molar concentrations of the acid and its conjugate base are approximately equal, that is, [acid] ≈ [conjugate base], then

$$\log \frac{[\text{conjugate base}]}{[\text{acid}]} \approx 0$$

or

$$\mathrm{pH} \approx \mathrm{p}K_a$$

Thus, to prepare a buffer solution, we choose a weak acid whose pK_a is close to the desired pH. This choice not only gives the correct pH value of the buffer system, but also ensures that we have *comparable* amounts of the acid and its conjugate base present; both are prerequisites for the buffer system to function effectively.

EXAMPLE 17.3

Describe how you would prepare a "phosphate buffer" with a pH of about 7.45.

Strategy For a buffer to function effectively, the concentrations of the acid component must be roughly equal to the conjugate base component. According to Equation (17.3), when the desired pH is close to the pK_a of the acid, that is, when pH $\approx pK_a$,

$$\log \frac{[\text{conjugate base}]}{[\text{acid}]} \approx 0$$

or

$$\frac{[\text{conjugate base}]}{[\text{acid}]} \approx 1$$

Solution Because phosphoric acid is a triprotic acid, we write the three stages of ionization as follows. The K_a values are obtained from Table 16.4 and the pK_a values are found by applying Equation (17.2).

$$H_3PO_4(aq) \rightleftharpoons H^+(aq) + H_2PO_4^-(aq) \qquad K_{a_1} = 7.5 \times 10^{-3};\ pK_{a_1} = 2.12$$

$$H_2PO_4^-(aq) \rightleftharpoons H^+(aq) + HPO_4^{2-}(aq) \qquad K_{a_2} = 6.2 \times 10^{-8};\ pK_{a_2} = 7.21$$

$$HPO_4^{2-}(aq) \rightleftharpoons H^+(aq) + PO_4^{3-}(aq) \qquad K_{a_3} = 4.8 \times 10^{-13};\ pK_{a_3} = 12.32$$

The most suitable of the three buffer systems is $HPO_4^{2-}/H_2PO_4^-$, because the pK_a of the acid $H_2PO_4^-$ is closest to the desired pH. From the Henderson-Hasselbach equation we write

$$\text{pH} = pK_a + \log \frac{[\text{conjugate base}]}{[\text{acid}]}$$

$$7.45 = 7.21 + \log \frac{[HPO_4^{2-}]}{[H_2PO_4^-]}$$

$$\log \frac{[HPO_4^{2-}]}{[H_2PO_4^-]} = 0.24$$

Taking the antilog, we obtain

$$\frac{[HPO_4^{2-}]}{[H_2PO_4^-]} = 10^{0.24} = 1.7$$

Thus, one way to prepare a phosphate buffer with a pH of 7.45 is to dissolve disodium hydrogen phosphate (Na_2HPO_4) and sodium dihydrogen phosphate (NaH_2PO_4) in a mole ratio of 1.7:1.0 in water. For example, we could dissolve 1.7 moles of Na_2HPO_4 and 1.0 mole of NaH_2PO_4 in enough water to make up a 1-L solution.

Similar problems: 17.15, 17.16.

Practice Exercise How would you prepare a liter of "carbonate buffer" at a pH of 10.10? You are provided with carbonic acid (H_2CO_3), sodium hydrogen carbonate ($NaHCO_3$), and sodium carbonate (Na_2CO_3). See Table 16.4 for K_a values.

REVIEW OF CONCEPTS

The following diagrams represent solutions containing a weak acid HA and/or its sodium salt NaA. Which solutions can act as a buffer? Which solution has the greatest buffer capacity? The Na^+ ions and water molecules are omitted for clarity.

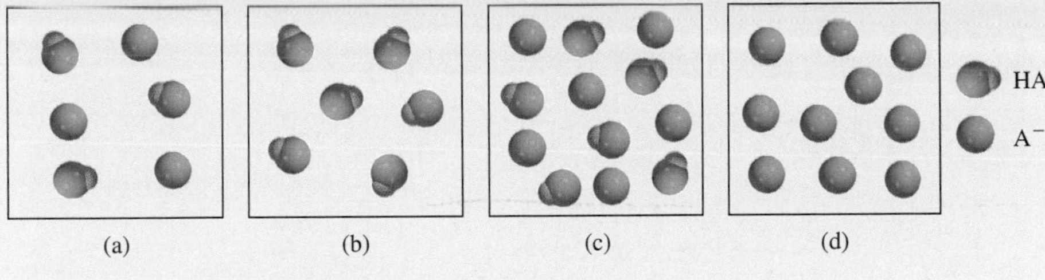

17.3 A Closer Look at Acid-Base Titrations

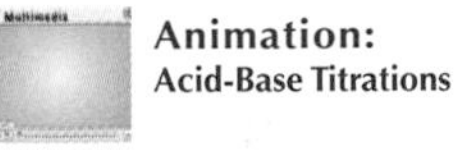

Animation:
Acid-Base Titrations

Having discussed buffer solutions, we can now look in more detail at the quantitative aspects of acid-base titrations (see Section 4.6). We will consider three types of reactions: (1) titrations involving a strong acid and a strong base, (2) titrations involving a weak acid and a strong base, and (3) titrations involving a strong acid and a weak base. Titrations involving a weak acid and a weak base are complicated by the hydrolysis of both the cation and the anion of the salt formed. These titrations will not be dealt with here. Figure 17.3 shows the arrangement for monitoring the pH during the course of a titration.

Strong Acid–Strong Base Titrations

The reaction between a strong acid (say, HCl) and a strong base (say, NaOH) can be represented by

$$NaOH(aq) + HCl(aq) \longrightarrow NaCl(aq) + H_2O(l)$$

or in terms of the net ionic equation

$$H^+(aq) + OH^-(aq) \longrightarrow H_2O(l)$$

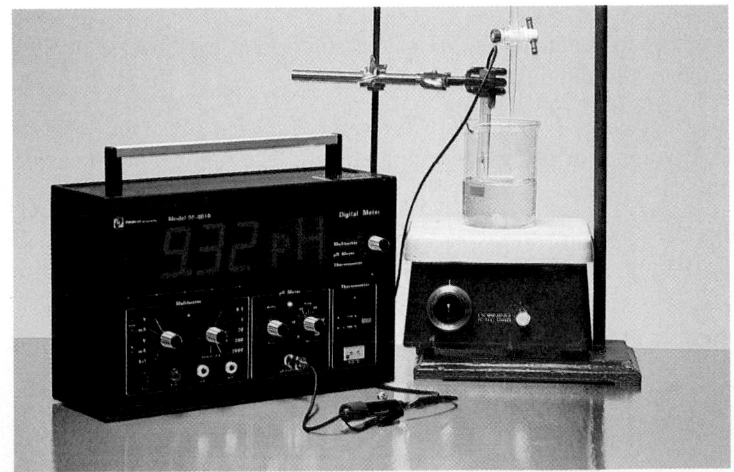

Figure 17.3
A pH meter is used to monitor an acid-base titration.

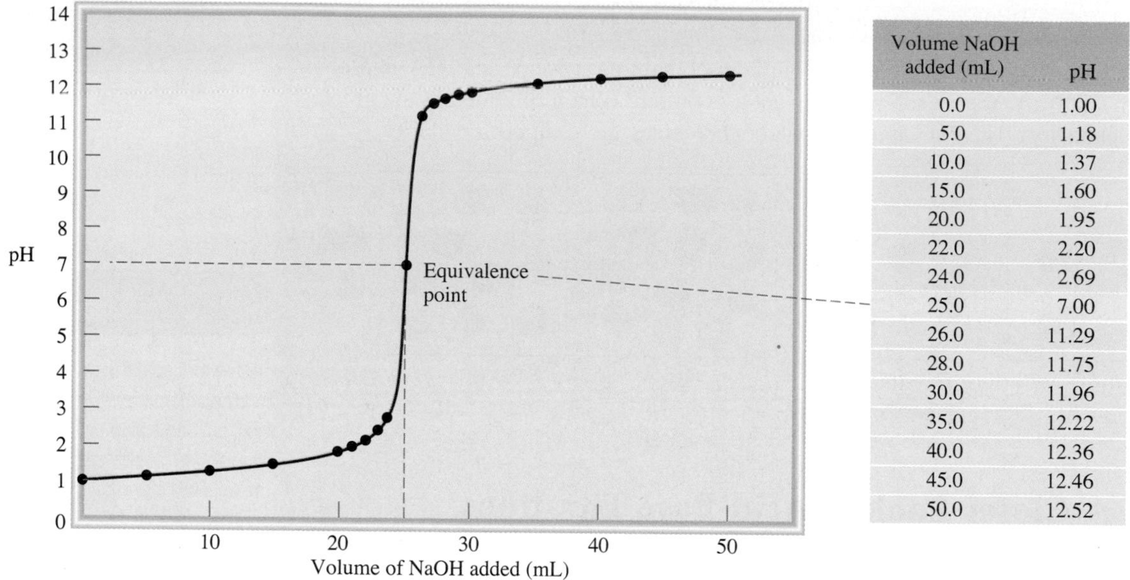

Volume NaOH added (mL)	pH
0.0	1.00
5.0	1.18
10.0	1.37
15.0	1.60
20.0	1.95
22.0	2.20
24.0	2.69
25.0	7.00
26.0	11.29
28.0	11.75
30.0	11.96
35.0	12.22
40.0	12.36
45.0	12.46
50.0	12.52

Figure 17.4
pH profile of a strong acid–strong base titration. A 0.100 M NaOH solution is added from a buret to 25.0 mL of a 0.100 M HCl solution in an Erlenmeyer flask (see Figure 4.18). This curve is sometimes referred to as a titration curve.

Consider the addition of a 0.100 *M* NaOH solution (from a buret) to an Erlenmeyer flask containing 25.0 mL of 0.100 *M* HCl. For convenience, we will use only three significant figures for volume and concentration and two significant figures for pH. Figure 17.4 shows the pH profile of the titration (also known as the titration curve). Before the addition of NaOH, the pH of the acid is given by −log (0.100), or 1.00. When NaOH is added, the pH of the solution increases slowly at first. Near the equivalence point the pH begins to rise steeply, and at the equivalence point (that is, the point at which equimolar amounts of acid and base have reacted) the curve rises almost vertically. In a strong acid–strong base titration both the hydrogen ion and hydroxide ion concentrations are very small at the equivalence point (approximately 1×10^{-7} *M*); consequently, the addition of a single drop of the base can cause a large increase in $[OH^-]$ and in the pH of the solution. Beyond the equivalence point, the pH again increases slowly with the addition of NaOH.

It is possible to calculate the pH of the solution at every stage of titration. Here are three sample calculations:

1. *After the addition of 10.0 mL of 0.100 M NaOH to 25.0 mL of 0.100 M HCl.* The total volume of the solution is 35.0 mL. The number of moles of NaOH in 10.0 mL is

A faster way to calculate the number of moles of NaOH is to write

$$10.0 \text{ mL} \times \frac{0.100 \text{ mol}}{1000 \text{ mL}} = 1.0 \times 10^{-3} \text{ mol}$$

$$10.0 \text{ mL} \times \frac{0.100 \text{ mol NaOH}}{1 \text{ L NaOH}} \times \frac{1 \text{ L}}{1000 \text{ mL}} = 1.00 \times 10^{-3} \text{ mol}$$

The number of moles of HCl originally present in 25.0 mL of solution is

$$25.0 \text{ mL} \times \frac{0.100 \text{ mol HCl}}{1 \text{ L HCl}} \times \frac{1 \text{ L}}{1000 \text{ mL}} = 2.50 \times 10^{-3} \text{ mol}$$

Thus, the amount of HCl left after partial neutralization is $(2.50 \times 10^{-3}) - (1.00 \times 10^{-3})$, or 1.50×10^{-3} mol. Next, the concentration of H^+ ions in 35.0 mL of solution is found as follows:

Keep in mind that 1 mol NaOH ≏ 1 mol HCl.

$$\frac{1.50 \times 10^{-3}\ \text{mol HCl}}{35.0\ \text{mL}} \times \frac{1000\ \text{mL}}{1\ \text{L}} = 0.0429\ \text{mol HCl/L}$$
$$= 0.0429\ M\ \text{HCl}$$

Thus, $[H^+] = 0.0429\ M$, and the pH of the solution is

$$\text{pH} = -\log 0.0429 = 1.37$$

2. *After the addition of 25.0 mL of 0.100 M NaOH to 25.0 mL of 0.100 M HCl.* This is a simple calculation, because it involves a complete neutralization reaction and the salt (NaCl) does not undergo hydrolysis. At the equivalence point, $[H^+] = [OH^-] = 1.00 \times 10^{-7}\ M$ and the pH of the solution is 7.00.

Neither Na^+ nor Cl^- undergoes hydrolysis.

3. *After the addition of 35.0 mL of 0.100 M NaOH to 25.0 mL of 0.100 M HCl.* The total volume of the solution is now 60.0 mL. The number of moles of NaOH added is

$$35.0\ \text{mL} \times \frac{0.100\ \text{mol NaOH}}{1\ \text{L NaOH}} \times \frac{1\ \text{L}}{1000\ \text{mL}} = 3.50 \times 10^{-3}\ \text{mol}$$

The number of moles of HCl in 25.0 mL solution is 2.50×10^{-3} mol. After complete neutralization of HCl, the number of moles of NaOH left is $(3.50 \times 10^{-3}) - (2.50 \times 10^{-3})$, or 1.00×10^{-3} mol. The concentration of NaOH in 60.0 mL of solution is

$$\frac{1.00 \times 10^{-3}\ \text{mol NaOH}}{60.0\ \text{mL}} \times \frac{1000\ \text{mL}}{1\ \text{L}} = 0.0167\ \text{mol NaOH/L}$$
$$= 0.0167\ M\ \text{NaOH}$$

Thus, $[OH^-] = 0.0167\ M$ and $\text{pOH} = -\log 0.0167 = 1.78$. Hence, the pH of the solution is

$$\begin{aligned}\text{pH} &= 14.00 - \text{pOH}\\ &= 14.00 - 1.78\\ &= 12.22\end{aligned}$$

Weak Acid–Strong Base Titrations

Consider the neutralization reaction between acetic acid (a weak acid) and sodium hydroxide (a strong base):

$$CH_3COOH(aq) + NaOH(aq) \longrightarrow CH_3COONa(aq) + H_2O(l)$$

This equation can be simplified to

$$CH_3COOH(aq) + OH^-(aq) \longrightarrow CH_3COO^-(aq) + H_2O(l)$$

The acetate ion undergoes hydrolysis as follows:

$$CH_3COO^-(aq) + H_2O(l) \rightleftharpoons CH_3COOH(aq) + OH^-(aq)$$

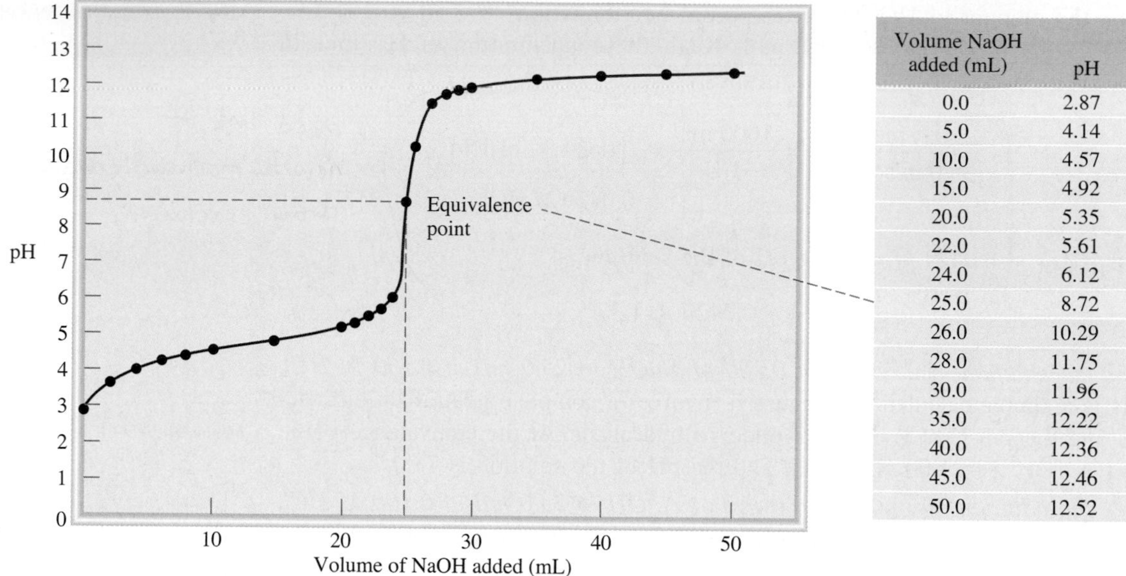

Volume NaOH added (mL)	pH
0.0	2.87
5.0	4.14
10.0	4.57
15.0	4.92
20.0	5.35
22.0	5.61
24.0	6.12
25.0	8.72
26.0	10.29
28.0	11.75
30.0	11.96
35.0	12.22
40.0	12.36
45.0	12.46
50.0	12.52

Figure 17.5
pH profile of a weak acid–strong base titration. A 0.100 M NaOH solution is added from a buret to 25.0 mL of a 0.100 M CH_3COOH solution in an Erlenmeyer flask. Due to the hydrolysis of the salt formed, the pH at the equivalence point is greater than 7.

Therefore, at the equivalence point, when we only have sodium acetate present, the pH will be *greater than* 7 as a result of the excess OH^- ions formed (Figure 17.5). Note that this situation is analogous to the hydrolysis of sodium acetate (CH_3COONa) (see p. 573).

EXAMPLE 17.4

Calculate the pH in the titration of 25.0 mL of 0.100 *M* acetic acid by sodium hydroxide after the addition to the acid solution of (a) 10.0 mL of 0.100 *M* NaOH, (b) 25.0 mL of 0.100 *M* NaOH, (c) 35.0 mL of 0.100 *M* NaOH.

Strategy The reaction between CH_3COOH and NaOH is

$$CH_3COOH(aq) + NaOH(aq) \longrightarrow CH_3COONa(aq) + H_2O(l)$$

We see that 1 mol $CH_3COOH \simeq$ 1 mol NaOH. Therefore, at every stage of the titration we can calculate the number of moles of base reacting with the acid, and the pH of the solution is determined by the excess acid or base left over. At the equivalence point, however, the neutralization is complete and the pH of the solution will depend on the extent of the hydrolysis of the salt formed, which is CH_3COONa.

Solution

(a) The number of moles of NaOH in 10.0 mL is

$$10.0 \text{ mL} \times \frac{0.100 \text{ mol NaOH}}{1 \text{ L NaOH soln}} \times \frac{1 \text{ L}}{1000 \text{ mL}} = 1.00 \times 10^{-3} \text{ mol}$$

The number of moles of CH_3COOH originally present in 25.0 mL of solution is

$$25.0 \text{ mL} \times \frac{0.100 \text{ mol } CH_3COOH}{1 \text{ L } CH_3COOH \text{ soln}} \times \frac{1 \text{ L}}{1000 \text{ mL}} = 2.50 \times 10^{-3} \text{ mol}$$

(Continued)

We work with moles at this point because when two solutions are mixed, the solution volume increases. As the volume increases, molarity will change but the number of moles will remain the same. The changes in number of moles are summarized next:

	$CH_3COOH(aq)$ +	$NaOH(aq)$ ⟶	$CH_3COONa(aq) + H_2O(l)$
Initial (mol):	2.50×10^{-3}	1.00×10^{-3}	0
Change (mol):	-1.00×10^{-3}	-1.00×10^{-3}	$+1.00 \times 10^{-3}$
Final (mol):	1.50×10^{-3}	0	1.00×10^{-3}

At this stage we have a buffer system made up of CH_3COOH and CH_3COO^- (from the salt, CH_3COONa). To calculate the pH of the solution, we write

$$K_a = \frac{[H^+][CH_3COO^-]}{[CH_3COOH]}$$

$$[H^+] = \frac{[CH_3COOH]K_a}{[CH_3COO^-]}$$

$$= \frac{(1.50 \times 10^{-3})(1.8 \times 10^{-5})}{1.00 \times 10^{-3}} = 2.7 \times 10^{-5}\ M$$

Because the volume of the solution is the same for CH_3COOH and CH_3COO^-, the ratio of the number of moles present is equal to the ratio of their molar concentrations.

Therefore, $\quad pH = -\log(2.7 \times 10^{-5}) = 4.57$

(b) These quantities (that is, 25.0 mL of 0.100 *M* NaOH reacting with 25.0 mL of 0.100 *M* CH_3COOH) correspond to the equivalence point. The number of moles of NaOH in 25.0 mL of the solution is

$$25.0\ \text{mL} \times \frac{0.100\ \text{mol NaOH}}{1\ \text{L NaOH soln}} \times \frac{1\ \text{L}}{1000\ \text{mL}} = 2.50 \times 10^{-3}\ \text{mol}$$

The changes in number of moles are summarized next:

	$CH_3COOH(aq)$ +	$NaOH(aq)$ ⟶	$CH_3COONa(aq) + H_2O(l)$
Initial (mol):	2.50×10^{-3}	2.50×10^{-3}	0
Change (mol):	-2.50×10^{-3}	-2.50×10^{-3}	$+2.50 \times 10^{-3}$
Final (mol):	0	0	2.50×10^{-3}

At the equivalence point, the concentrations of both the acid and the base are zero. The total volume is (25.0 + 25.0) mL or 50.0 mL, so the concentration of the salt is

$$[CH_3COONa] = \frac{2.50 \times 10^{-3}\ \text{mol}}{50.0\ \text{mL}} \times \frac{1000\ \text{mL}}{1\ \text{L}}$$

$$= 0.0500\ \text{mol/L} = 0.0500\ M$$

The next step is to calculate the pH of the solution that results from the hydrolysis of the CH_3COO^- ions. Following the procedure described in Example 16.13 and looking up the base ionization constant (K_b) for CH_3COO^- in Table 16.3, we write

$$K_b = 5.6 \times 10^{-10} = \frac{[CH_3COOH][OH^-]}{[CH_3COO^-]} = \frac{x^2}{0.0500 - x}$$

$$x = [OH^-] = 5.3 \times 10^{-6}\ M,\ pH = 8.72$$

(c) After the addition of 35.0 mL of NaOH, the solution is well past the equivalence point. The number of moles of NaOH originally present is

$$35.0\ \text{mL} \times \frac{0.100\ \text{mol NaOH}}{1\ \text{L NaOH soln}} \times \frac{1\ \text{L}}{1000\ \text{mL}} = 3.50 \times 10^{-3}\ \text{mol}$$

(Continued)

The changes in number of moles are summarized next:

	$CH_3COOH(aq)$ +	$NaOH(aq)$ ⟶	$CH_3COONa(aq)$ + $H_2O(l)$
Initial (mol):	2.50×10^{-3}	3.50×10^{-3}	0
Change (mol):	-2.50×10^{-3}	-2.50×10^{-3}	$+2.50 \times 10^{-3}$
Final (mol):	0	1.00×10^{-3}	2.50×10^{-3}

At this stage we have two species in solution that are responsible for making the solution basic: OH^- and CH_3COO^- (from CH_3COONa). However, because OH^- is a much stronger base than CH_3COO^-, we can safely neglect the hydrolysis of the CH_3COO^- ions and calculate the pH of the solution using only the concentration of the OH^- ions. The total volume of the combined solutions is (25.0 + 35.0) mL or 60.0 mL, so we calculate OH^- concentration as follows:

$$[OH^-] = \frac{1.00 \times 10^{-3}\text{ mol}}{60.0\text{ mL}} \times \frac{1000\text{ mL}}{1\text{ L}}$$

$$= 0.0167\text{ mol/L} = 0.0167\ M$$

$$pOH = -\log [OH^-] = -\log 0.0167 = 1.78$$

$$pH = 14.00 - 1.78 = 12.22$$

Similar problem: 17.21(b).

Practice Exercise Exactly 100 mL of 0.10 *M* nitrous acid (HNO_2) are titrated with a 0.10 *M* NaOH solution. Calculate the pH for (a) the initial solution, (b) the point at which 80 mL of the base has been added, (c) the equivalence point, (d) the point at which 105 mL of the base has been added.

Strong Acid–Weak Base Titrations

Consider the titration of HCl, a strong acid, with NH_3, a weak base:

$$HCl(aq) + NH_3(aq) \longrightarrow NH_4Cl(aq)$$

or simply

$$H^+(aq) + NH_3(aq) \longrightarrow NH_4^+(aq)$$

The pH at the equivalence point is *less than* 7 due to the hydrolysis of the NH_4^+ ion:

$$NH_4^+(aq) + H_2O(l) \rightleftharpoons NH_3(aq) + H_3O^+(aq)$$

or simply

$$NH_4^+(aq) \rightleftharpoons NH_3(aq) + H^+(aq)$$

Because of the volatility of an aqueous ammonia solution, it is more convenient to add hydrochloric acid from a buret to the ammonia solution. Figure 17.6 shows the titration curve for this experiment.

Review of Concepts

For which of the following titrations will the pH at the equivalence point not be neutral? (a) HCOOH with KOH, (b) HI with KOH, (c) NaOH with HF, (d) NaOH with HNO_3.

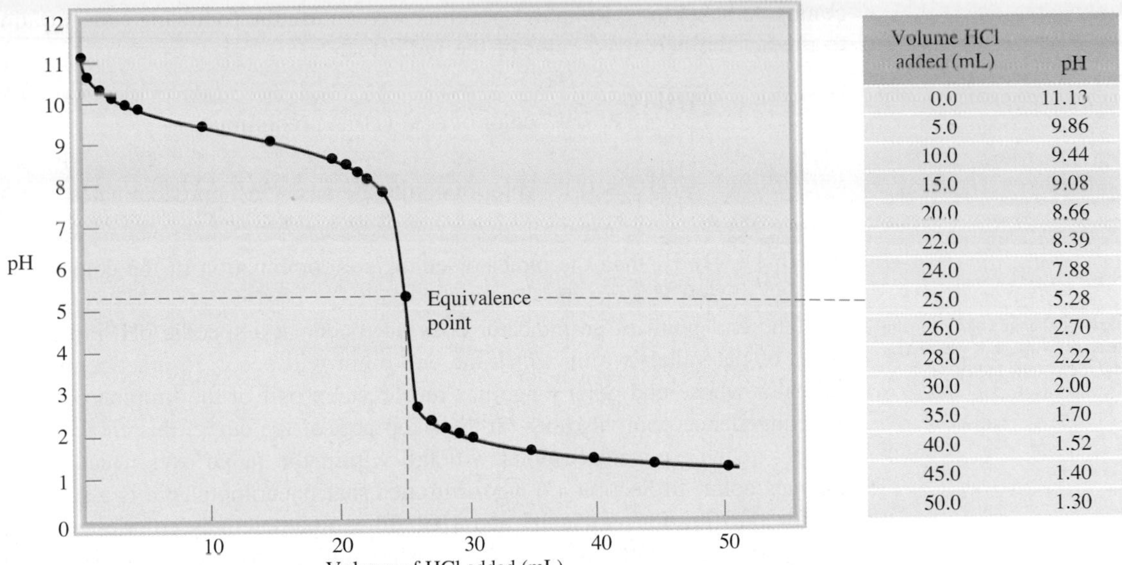

Volume HCl added (mL)	pH
0.0	11.13
5.0	9.86
10.0	9.44
15.0	9.08
20.0	8.66
22.0	8.39
24.0	7.88
25.0	5.28
26.0	2.70
28.0	2.22
30.0	2.00
35.0	1.70
40.0	1.52
45.0	1.40
50.0	1.30

Figure 17.6
pH profiles of a strong acid–weak base titration. A 0.100 M HCl solution is added from a buret to 25.0 mL of a 0.100 M NH_3 solution in an Erlenmeyer flask. As a result of salt hydrolysis, the pH at the equivalence point is lower than 7.

17.4 Acid-Base Indicators

The equivalence point, as we have seen, is the point at which the number of moles of OH^- ions added to a solution is equal to the number of moles of H^+ ions originally present. To determine the equivalence point in a titration, then, we must know exactly how much volume of a base to add from a buret to an acid in a flask. One way to achieve this goal is to add a few drops of an acid-base indicator to the acid solution at the start of the titration. You will recall from Chapter 4 that an indicator has distinctly different colors in its nonionized and ionized forms. These two forms are related to the pH of the solution in which the indicator is dissolved. The ***end point*** of a titration *occurs when the indicator changes color.* However, not all indicators change color at the same pH, so the choice of indicator for a particular titration depends on the nature of the acid and base used in the titration (that is, whether they are strong or weak). By choosing the proper indicator for a titration, we can use the end point to determine the equivalence point, as we will see next.

An indicator is usually a weak organic acid or organic base.

Let us consider a weak monoprotic acid that we will call HIn. To be an effective indicator, HIn and its conjugate base, In^-, must have distinctly different colors. In solution, the acid ionizes to a small extent:

$$\mathrm{HIn}(aq) \rightleftharpoons \mathrm{H}^+(aq) + \mathrm{In}^-(aq)$$

If the indicator is in a sufficiently acidic medium, the equilibrium, according to Le Châtelier's principle, shifts to the left and the predominant color of the indicator is that of the nonionized form (HIn). On the other hand, in a basic medium the equilibrium shifts to the right and the color of the solution will be due mainly to that of the conjugate base (In^-). Roughly speaking, we can use the following concentration ratios to predict the perceived color of the indicator:

$$\frac{[HIn]}{[In^-]} \geq 10 \quad \text{color of acid (HIn) predominates}$$

$$\frac{[HIn]}{[In^-]} \leq 0.1 \quad \text{color of conjugate base } (In^-) \text{ predominates}$$

If $[HIn] \approx [In^-]$, then the indicator color is a combination of the colors of HIn and In^-.

Typical indicators change color over the pH range given pH = $pK_a \pm 1$, where K_a is the acid ionization of the indicator.

The end point of an indicator does not occur at a specific pH; rather, there is a range of pH values within which the end point will occur. In practice, we choose an indicator whose end point range lies on the steep part of the titration curve. Because the equivalence point also lies on the steep part of the curve, this choice ensures that the pH at the equivalence point will fall within the range over which the indicator changes color. In Section 4.6 we mentioned that phenolphthalein is a suitable indicator for the titration of NaOH and HCl. Phenolphthalein is colorless in acidic and neutral solutions, but reddish pink in basic solutions. Measurements show that at pH < 8.3 the indicator is colorless but that it begins to turn reddish pink when the pH exceeds 8.3. As shown in Figure 17.4, the steepness of the pH curve near the equivalence point means that the addition of a very small quantity of NaOH (say, 0.05 mL, which is about the volume of a drop from the buret) brings about a large rise in the pH of the solution. What is important, however, is the fact that the steep portion of the pH profile includes the range over which phenolphthalein changes from colorless to reddish pink. Whenever such a correspondence occurs, the indicator can be used to locate the equivalence point of the titration (Figure 17.7).

Many acid-base indicators are plant pigments. For example, by boiling chopped red cabbage in water we can extract pigments that exhibit many different colors at various pHs (Figure 17.8). Table 17.1 lists a number of indicators commonly used in acid-base titrations. The choice of a particular indicator depends on the strength of the acid and base to be titrated.

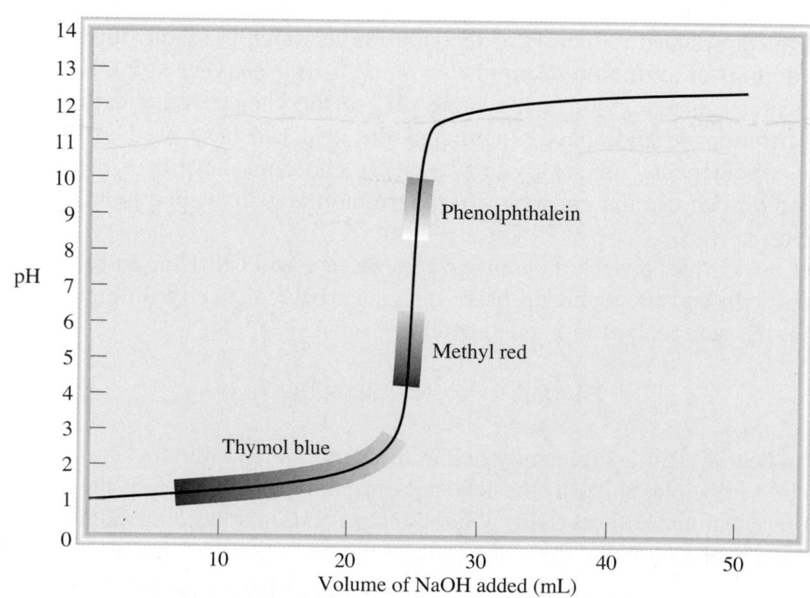

Figure 17.7
The titration curve of a strong acid with a strong base. Because the regions over which the indicators methyl red and phenolphthalein change color along the steep portion of the curve, they can be used to monitor the equivalence point of the titration. Thymol blue, on the other hand, cannot be used for the same purpose because the color change does not match the steep portion of the titration curve (see Table 17.1).

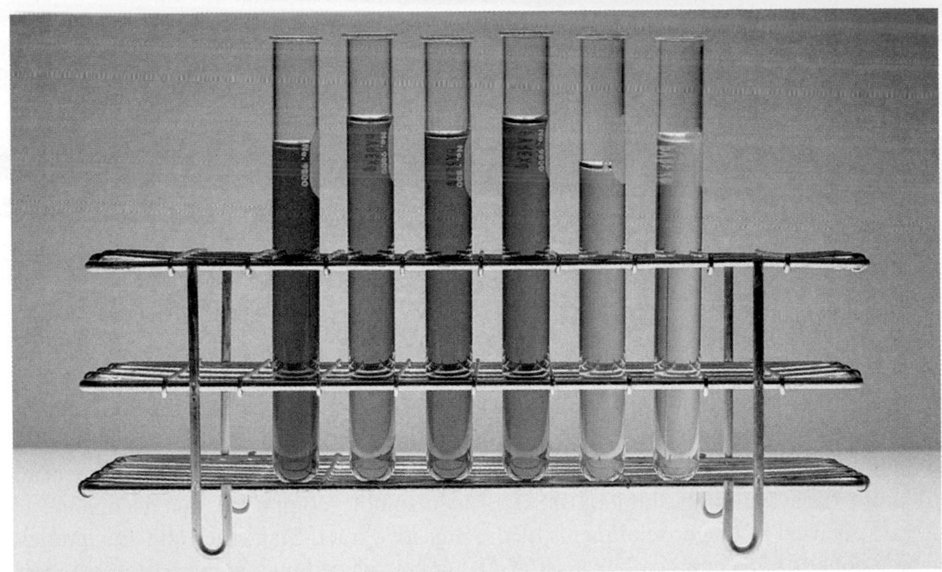

Figure 17.8
Solutions containing extracts of red cabbage (obtained by boiling the cabbage in water) produce different colors when treated with an acid and a base. The pH of the solutions increases from left to right.

EXAMPLE 17.5

Which indicator or indicators listed in Table 17.1 would you use for the acid-base titrations shown in (a) Figure 17.4, (b) Figure 17.5, and (c) Figure 17.6.

Strategy The choice of an indicator for a particular titration is based on the fact that its pH range for color change must overlap the steep portion of the titration curve. Otherwise we cannot use the color change to locate the equivalence point.

Solution

(a) Near the equivalence point, the pH of the solution changes abruptly from 4 to 10. Therefore, all the indicators except thymol blue, bromophenol blue, and methyl orange are suitable for use in the titration.

(b) Here the steep portion covers the pH range between 7 and 10; therefore, the suitable indicators are cresol red and phenolphthalein.

(Continued)

Table 17.1 Some Common Acid-Base Indicators

	Color		
Indicator	**In Acid**	**In Base**	**pH Range***
Thymol blue	Red	Yellow	1.2–2.8
Bromophenol blue	Yellow	Bluish purple	3.0–4.6
Methyl orange	Orange	Yellow	3.1–4.4
Methyl red	Red	Yellow	4.2–6.3
Chlorophenol blue	Yellow	Red	4.8–6.4
Bromothymol blue	Yellow	Blue	6.0–7.6
Cresol red	Yellow	Red	7.2–8.8
Phenolphthalein	Colorless	Reddish pink	8.3–10.0

*The pH range is defined as the range over which the indicator changes from the acid color to the base color.

(c) Here the steep portion of the pH curve covers the pH range between 3 and 7; therefore, the suitable indicators are bromophenol blue, methyl orange, methyl red, and chlorophenol blue.

Similar problem: 17.29.

Practice Exercise Referring to Table 17.1, specify which indicator or indicators you would use for the following titrations: (a) HBr versus CH_3NH_2, (b) HNO_3 versus NaOH, (c) HNO_2 versus KOH.

17.5 Solubility Equilibria

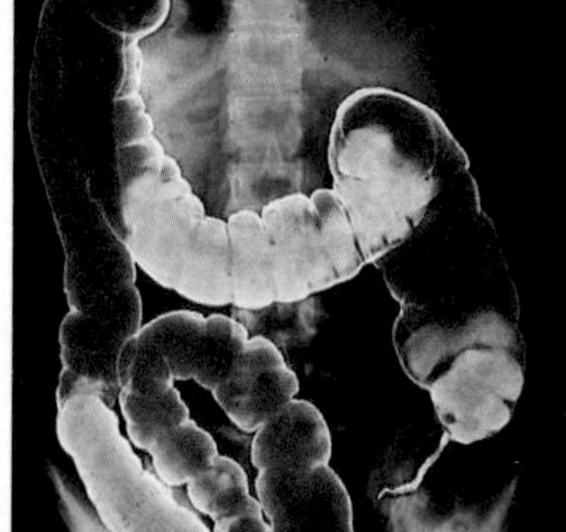

$BaSO_4$ imaging of the human large intestine.

Precipitation reactions are important in industry, medicine, and everyday life. For example, the preparation of many essential industrial chemicals such as sodium carbonate (Na_2CO_3) makes use of precipitation reactions. The dissolving of tooth enamel, which is mainly made of hydroxyapatite [$Ca_5(PO_4)_3OH$], in an acidic medium leads to tooth decay. Barium sulfate ($BaSO_4$), an insoluble compound that is opaque to X rays, is used to diagnose ailments of the digestive tract. Stalactites and stalagmites, which consist of calcium carbonate ($CaCO_3$), are produced by a precipitation reaction, and so are many foods, such as fudge.

The general rules for predicting the solubility of ionic compounds in water were introduced in Section 4.2. Although useful, these solubility rules do not enable us to make quantitative predictions about how much of a given ionic compound will dissolve in water. To develop a quantitative approach, we start with what we already know about chemical equilibrium.

Downward-growing stalactites and upward-growing stalagmites.

Solubility Product

Consider a saturated solution of silver chloride that is in contact with solid silver chloride. The solubility equilibrium can be represented as

$$AgCl(s) \rightleftharpoons Ag^+(aq) + Cl^-(aq)$$

Because salts such as AgCl are treated as strong electrolytes, all the AgCl that dissolves in water is assumed to dissociate completely into Ag^+ and Cl^- ions. We know from Chapter 15 that for heterogeneous reactions the concentration of the solid is a constant. Thus, we can write the equilibrium constant for the dissociation of AgCl as

$$K_{sp} = [Ag^+][Cl^-]$$

in which K_{sp} is called the solubility product constant or simply the solubility product. In general, the ***solubility product*** of a compound is *the product of the molar concentrations of the constituent ions, each raised to the power of its stoichiometric coefficient in the equilibrium equation.*

Because each AgCl unit contains only one Ag^+ ion and one Cl^- ion, its solubility product expression is particularly simple to write. The following cases are more complex.

We ignore both ion pair formation and salt hydrolysis (see pp. 457 and 557).

- MgF_2

$$MgF_2(s) \rightleftharpoons Mg^{2+}(aq) + 2F^-(aq) \qquad K_{sp} = [Mg^{2+}][F^-]^2$$

- Ag_2CO_3

$$Ag_2CO_3(s) \rightleftharpoons 2Ag^+(aq) + CO_3^{2-}(aq) \qquad K_{sp} = [Ag^+]^2[CO_3^{2-}]$$

- $Ca_3(PO_4)_2$

$$Ca_3(PO_4)_2(s) \rightleftharpoons 3Ca^{2+}(aq) + 2PO_4^{3-}(aq) \qquad K_{sp} = [Ca^{2+}]^3[PO_4^{3-}]^2$$

Table 17.2 Solubility Products of Some Slightly Soluble Ionic Compounds at 25°C

Compound	K_{sp}	Compound	K_{sp}
Aluminum hydroxide [$Al(OH)_3$]	1.8×10^{-33}	Lead(II) chromate ($PbCrO_4$)	2.0×10^{-14}
Barium carbonate ($BaCO_3$)	8.1×10^{-9}	Lead(II) fluoride (PbF_2)	4.1×10^{-8}
Barium fluoride (BaF_2)	1.7×10^{-6}	Lead(II) iodide (PbI_2)	1.4×10^{-8}
Barium sulfate ($BaSO_4$)	1.1×10^{-10}	Lead(II) sulfide (PbS)	3.4×10^{-28}
Bismuth sulfide (Bi_2S_3)	1.6×10^{-72}	Magnesium carbonate ($MgCO_3$)	4.0×10^{-5}
Cadmium sulfide (CdS)	8.0×10^{-28}	Magnesium hydroxide [$Mg(OH)_2$]	1.2×10^{-11}
Calcium carbonate ($CaCO_3$)	8.7×10^{-9}	Manganese(II) sulfide (MnS)	3.0×10^{-14}
Calcium fluoride (CaF_2)	4.0×10^{-11}	Mercury(I) chloride (Hg_2Cl_2)	3.5×10^{-18}
Calcium hydroxide [$Ca(OH)_2$]	8.0×10^{-6}	Mercury(II) sulfide (HgS)	4.0×10^{-54}
Calcium phosphate [$Ca_3(PO_4)_2$]	1.2×10^{-26}	Nickel(II) sulfide (NiS)	1.4×10^{-24}
Chromium(III) hydroxide [$Cr(OH)_3$]	3.0×10^{-29}	Silver bromide (AgBr)	7.7×10^{-13}
Cobalt(II) sulfide (CoS)	4.0×10^{-21}	Silver carbonate (Ag_2CO_3)	8.1×10^{-12}
Copper(I) bromide (CuBr)	4.2×10^{-8}	Silver chloride (AgCl)	1.6×10^{-10}
Copper(I) iodide (CuI)	5.1×10^{-12}	Silver iodide (AgI)	8.3×10^{-17}
Copper(II) hydroxide [$Cu(OH)_2$]	2.2×10^{-20}	Silver sulfate (Ag_2SO_4)	1.4×10^{-5}
Copper(II) sulfide (CuS)	6.0×10^{-37}	Silver sulfide (Ag_2S)	6.0×10^{-51}
Iron(II) hydroxide [$Fe(OH)_2$]	1.6×10^{-14}	Strontium carbonate ($SrCO_3$)	1.6×10^{-9}
Iron(III) hydroxide [$Fe(OH)_3$]	1.1×10^{-36}	Strontium sulfate ($SrSO_4$)	3.8×10^{-7}
Iron(II) sulfide (FeS)	6.0×10^{-19}	Tin(II) sulfide (SnS)	1.0×10^{-26}
Lead(II) carbonate ($PbCO_3$)	3.3×10^{-14}	Zinc hydroxide [$Zn(OH)_2$]	1.8×10^{-14}
Lead(II) chloride ($PbCl_2$)	2.4×10^{-4}	Zinc sulfide (ZnS)	3.0×10^{-23}

Table 17.2 lists the K_{sp} values for a number of salts of low solubility. Soluble salts such as NaCl and KNO_3, which have very large K_{sp} values, are not listed in the table.

For the dissolution of an ionic solid in aqueous solution, any one of the following conditions may exist: (1) The solution is unsaturated, (2) the solution is saturated, or (3) the solution is supersaturated. Following the procedure in Section 15.3, we use Q, called the *ion product,* to represent the product of the molar concentrations of the ions raised to the power of their stoichiometric coefficients. Thus, for an aqueous solution containing Ag^+ and Cl^- ions at 25°C,

$$Q = [Ag^+]_0[Cl^-]_0$$

The subscript 0 reminds us that these are initial concentrations and do not necessarily correspond to those at equilibrium. The possible relationships between Q and K_{sp} are

Depending on how a solution is made up, $[Ag^+]$ may or may not be equal to $[Cl^-]$.

$$Q < K_{sp}$$
$$[Ag^+]_0[Cl^-]_0 < 1.6 \times 10^{-10}$$
Unsaturated solution

$$Q = K_{sp}$$
$$[Ag^+][Cl^-] = 1.6 \times 10^{-10}$$
Saturated solution

$$Q > K_{sp}$$
$$[Ag^+]_0[Cl^-]_0 > 1.6 \times 10^{-10}$$
Supersaturated solution; AgCl will precipitate out until the product of the ionic concentrations is equal to 1.6×10^{-10}

Review of Concepts

The following diagrams represent solutions of AgCl, which may also contain ions such as Na^+ and NO_3^- (not shown) that do not affect the solubility of AgCl. If (a) represents a saturated solution of AgCl, classify the other solutions as unsaturated, saturated, or supersaturated.

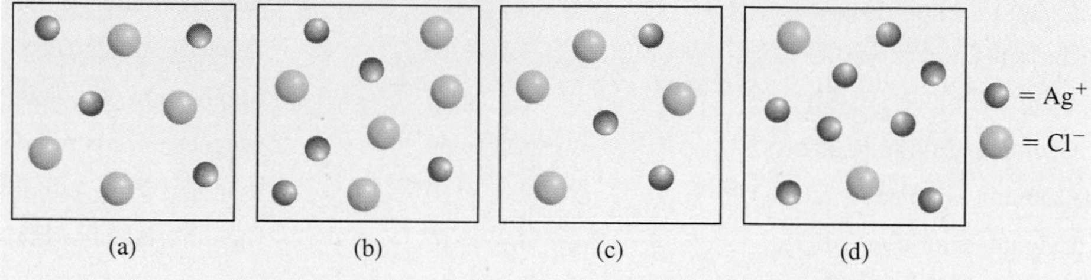

Molar Solubility and Solubility

The value of K_{sp} indicates the solubility of an ionic compound—the smaller the value, the less soluble the compound in water. However, in using K_{sp} values to compare solubilities, you should choose compounds that have similar formulas, such as AgCl and ZnS, or CaF_2 and $Fe(OH)_2$. There are two other quantities that express a substance's solubility: ***molar solubility,*** which is *the number of moles of solute in 1 L of a saturated solution (moles per liter),* and ***solubility,*** which is *the number of grams of solute in 1 L of a saturated solution (grams per liter).* Note that all these expressions refer to the concentration of saturated solutions at some given temperature (usually 25°C). Figure 17.9 shows the relationships among solubility, molar solubility, and K_{sp}.

Both molar solubility and solubility are convenient to use in the laboratory. We can use them to determine K_{sp} by following the steps outlined in Figure 17.9(a).

EXAMPLE 17.6

The solubility of calcium sulfate ($CaSO_4$) is found to be 0.67 g/L. Calculate the value of K_{sp} for calcium sulfate.

(Continued)

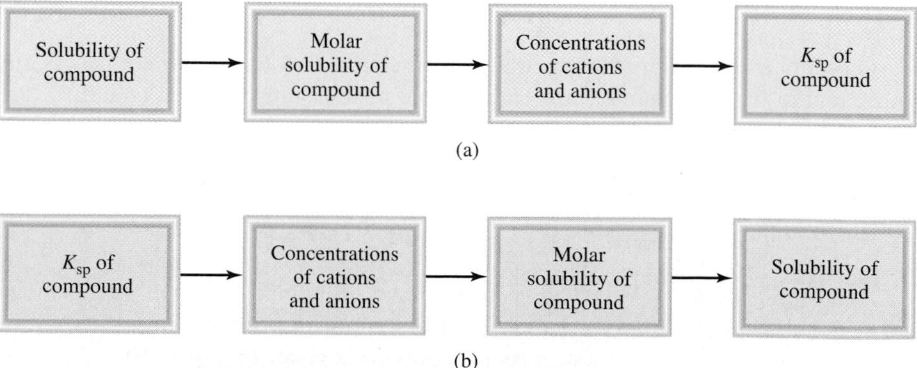

Figure 17.9
Sequence of steps (a) for calculating K_{sp} from solubility data and (b) for calculating solubility from K_{sp} data.

Strategy We are given the solubility of $CaSO_4$ and asked to calculate its K_{sp}. The sequence of conversion steps, according to Figure 17.9(a), is

$$\begin{matrix}\text{solubility of} \\ CaSO_4 \text{ in g/L}\end{matrix} \longrightarrow \begin{matrix}\text{molar solubility} \\ \text{of } CaSO_4\end{matrix} \longrightarrow \begin{matrix}[Ca^{2+}] \text{ and} \\ [SO_4^{2-}]\end{matrix} \longrightarrow \begin{matrix}K_{sp} \text{ of} \\ CaSO_4\end{matrix}$$

Solution Consider the dissociation of $CaSO_4$ in water. Let s be the molar solubility (in mol/L) of $CaSO_4$.

	$CaSO_4(s) \rightleftharpoons$	$Ca^{2+}(aq)$ +	$SO_4^{2-}(aq)$
Initial (M):		0	0
Change (M):	$-s$	$+s$	$+s$
Equilibrium (M):		s	s

The solubility product for $CaSO_4$ is

$$K_{sp} = [Ca^{2+}][SO_4^{2-}] = s^2$$

First we calculate the number of moles of $CaSO_4$ dissolved in 1 L of solution

$$\frac{0.67 \text{ g } CaSO_4}{1 \text{ L soln}} \times \frac{1 \text{ mol } CaSO_4}{136.2 \text{ g } CaSO_4} = 4.9 \times 10^{-3} \text{ mol/L} = s$$

From the solubility equilibrium we see that for every mole of $CaSO_4$ that dissolves, 1 mole of Ca^{2+} and 1 mole of SO_4^{2-} are produced. Thus, at equilibrium

$$[Ca^{2+}] = 4.9 \times 10^{-3} M \text{ and } [SO_4^{2-}] = 4.9 \times 10^{-3} M$$

Now we can calculate K_{sp}:

$$\begin{aligned} K_{sp} &= [Ca^{2+}][SO_4^{2-}] \\ &= (4.9 \times 10^{-3})(4.9 \times 10^{-3}) \\ &= 2.4 \times 10^{-5} \end{aligned}$$

Practice Exercise The solubility of lead chromate ($PbCrO_4$) is 4.5×10^{-5} g/L. Calculate the solubility product of this compound.

Calcium sulfate is used as a drying agent and in the manufacture of paints, ceramics, and paper. A hydrated form of calcium sulfate, called plaster of Paris, is used to make casts for broken bones.

Similar problem: 17.41.

Sometimes we are given the value of K_{sp} for a compound and asked to calculate the compound's molar solubility. For example, the K_{sp} of silver bromide (AgBr) is 7.7×10^{-13}. We can calculate its molar solubility by the same procedure as outlined for acid ionization constants. First, we identify the species present at equilibrium. Here we have Ag^+ and Br^- ions. Let s be the molar solubility (in mol/L) of AgBr. Because one unit of AgBr yields one Ag^+ and one Br^- ion, at equilibrium both $[Ag^+]$ and $[Br^-]$ are equal to s. We summarize the changes in concentrations as follows:

	$AgBr(s) \rightleftharpoons$	$Ag^+(aq)$ +	$Br^-(aq)$
Initial (M):		0.00	0.00
Change (M):	$-s$	$+s$	$+s$
Equilibrium (M):		s	s

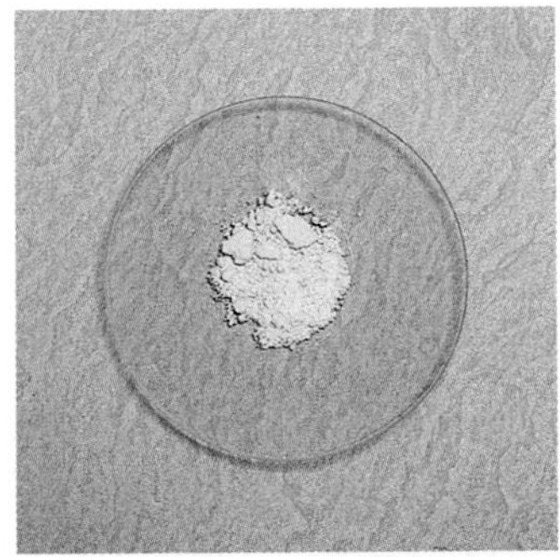

Silver bromide is used in photographic emulsions.

From Table 17.2 we write

$$\begin{aligned} K_{sp} &= [Ag^+][Br^-] \\ 7.7 \times 10^{-13} &= (s)(s) \\ s &= \sqrt{7.7 \times 10^{-13}} = 8.8 \times 10^{-7} M \end{aligned}$$

Therefore, at equilibrium

$$[Ag^+] = 8.8 \times 10^{-7} M$$
$$[Br^-] = 8.8 \times 10^{-7} M$$

Thus, the molar solubility of AgBr also is 8.8×10^{-7} *M*. Knowing the molar solubility will enable us to calculate the solubility in g/L, as shown in Example 17.7.

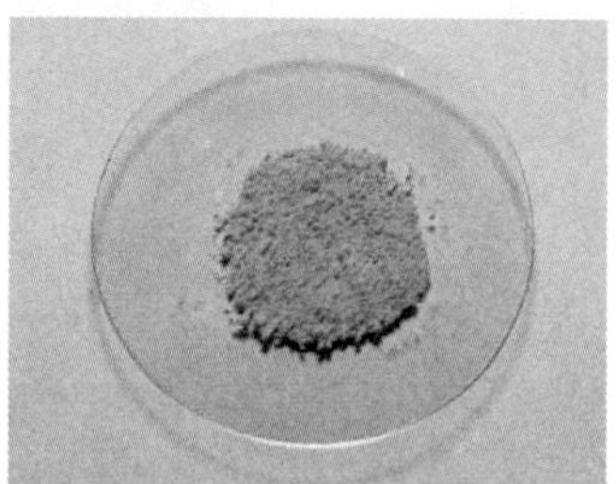

Copper(II) hydroxide is used as a pesticide and to treat seeds.

EXAMPLE 17.7

Using the data in Table 17.2, calculate the solubility of copper(II) hydroxide, $Cu(OH)_2$, in g/L.

Strategy We are given the K_{sp} of $Cu(OH)_2$ and asked to calculate its solubility in g/L. The sequence of conversion steps, according to Figure 17.9(b), is

$$\begin{matrix} K_{sp} \text{ of} \\ Cu(OH)_2 \end{matrix} \longrightarrow \begin{matrix} [Cu^{2+}] \text{ and} \\ [OH^-] \end{matrix} \longrightarrow \begin{matrix} \text{molar solubility} \\ \text{of } Cu(OH)_2 \end{matrix} \longrightarrow \begin{matrix} \text{solubility of} \\ Cu(OH)_2 \text{ in g/L} \end{matrix}$$

Solution Consider the dissociation of $Cu(OH)_2$ in water:

	$Cu(OH)_2(s)$ $\rightleftharpoons$	$Cu^{2+}(aq)$ +	$2OH^-(aq)$
Initial (*M*):		0	0
Change (*M*):	$-s$	$+s$	$+2s$
Equilibrium (*M*):		s	$2s$

Note that the molar concentration of OH^- is twice that of Cu^{2+}. The solubility product of $Cu(OH)_2$ is

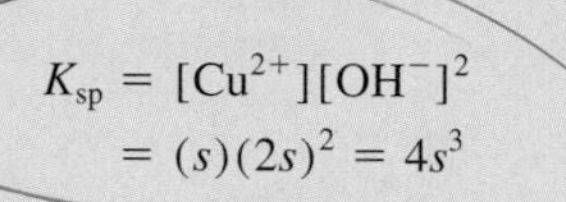

$$K_{sp} = [Cu^{2+}][OH^-]^2$$
$$= (s)(2s)^2 = 4s^3$$

From the K_{sp} value in Table 17.2, we solve for the molar solubility of $Cu(OH)_2$ as follows:

$$2.2 \times 10^{-20} = 4s^3$$
$$s^3 = \frac{2.2 \times 10^{-20}}{4} = 5.5 \times 10^{-21}$$

Hence, $s = 1.8 \times 10^{-7} M$

Finally, from the molar mass of $Cu(OH)_2$ and its molar solubility, we calculate the solubility in g/L:

$$\text{solubility of } Cu(OH)_2 = \frac{1.8 \times 10^{-7} \text{ mol } Cu(OH)_2}{1 \text{ L soln}} \times \frac{97.57 \text{ g } Cu(OH)_2}{1 \text{ mol } Cu(OH)_2}$$
$$= 1.8 \times 10^{-5} \text{ g/L}$$

Similar problem: 17.42.

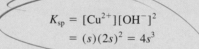

Practice Exercise Calculate the solubility of silver chloride (AgCl) in g/L.

As Examples 17.6 and 17.7 show, solubility and solubility product are related. If we know one, we can calculate the other, but each quantity provides different information. Table 17.3 shows the relationship between molar solubility and solubility product for a number of ionic compounds.

When carrying out solubility and/or solubility product calculations, keep in mind these important points:

- The solubility is the quantity of a substance that dissolves in a certain quantity of water to produce a saturated solution. In solubility equilibria calculations, it is

Table 17.3 Relationship Between K_{sp} and Molar Solubility (s)

Compound	K_{sp} Expression	Cation	Anion	Relation Between K_{sp} and s
$AgCl$	$[Ag^+][Cl^-]$	s	s	$K_{sp} = s^2; s = (K_{sp})^{\frac{1}{2}}$
$BaSO_4$	$[Ba^{2+}][SO_4^{2-}]$	s	s	$K_{sp} = s^2; s = (K_{sp})^{\frac{1}{2}}$
Ag_2CO_3	$[Ag^+]^2[CO_3^{2-}]$	$2s$	s	$K_{sp} = 4s^3; s = \left(\frac{K_{sp}}{4}\right)^{\frac{1}{3}}$
PbF_2	$[Pb^{2+}][F^-]^2$	s	$2s$	$K_{sp} = 4s^3; s = \left(\frac{K_{sp}}{4}\right)^{\frac{1}{3}}$
$Al(OH)_3$	$[Al^{3+}][OH^-]^3$	s	$3s$	$K_{sp} = 27s^4; s = \left(\frac{K_{sp}}{27}\right)^{\frac{1}{4}}$
$Ca_3(PO_4)_2$	$[Ca^{2+}]^3[PO_4^{3-}]^2$	$3s$	$2s$	$K_{sp} = 108s^5; s = \left(\frac{K_{sp}}{108}\right)^{\frac{1}{5}}$

usually expressed as *grams* of solute per liter of solution. Molar solubility is the number of *moles* of solute per liter of solution.

- The solubility product is an equilibrium constant.
- Molar solubility, solubility, and solubility product all refer to a *saturated solution.*

Predicting Precipitation Reactions

From a knowledge of the solubility rules (see Section 4.2) and the solubility products listed in Table 17.2, we can predict whether a precipitate will form when we mix two solutions or add a soluble compound to a solution. This ability often has practical value. In industrial and laboratory preparations, we can adjust the concentrations of ions until the ion product exceeds K_{sp} in order to obtain a given compound (in the form of a precipitate). The ability to predict precipitation reactions is also useful in medicine. For example, kidney stones, which can be extremely painful, consist largely of calcium oxalate, CaC_2O_4 ($K_{sp} = 2.3 \times 10^{-9}$). The normal physiological concentration of calcium ions in blood plasma is about 5 mM (1 mM = 1×10^{-3} M). Oxalate ions ($C_2O_4^{2-}$), derived from oxalic acid present in many vegetables such as rhubarb and spinach, react with the calcium ions to form insoluble calcium oxalate, which can gradually build up in the kidneys. Proper adjustment of a patient's diet can help to reduce precipitate formation.

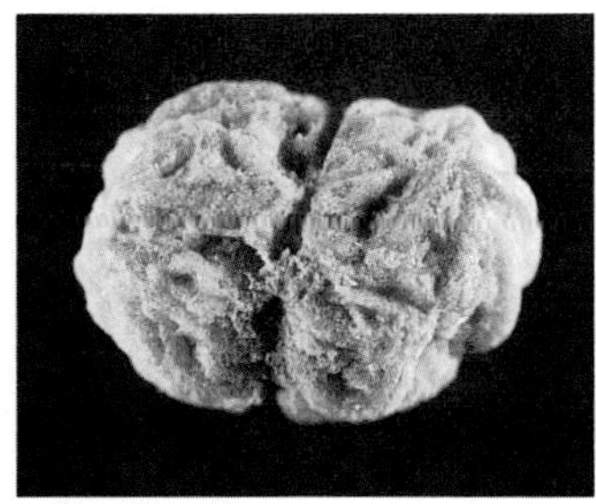

A kidney stone.

EXAMPLE 17.8

Exactly 200 mL of 0.0040 M $BaCl_2$ are added to exactly 600 mL of 0.0080 M K_2SO_4. Will a precipitate form?

Strategy Under what condition will an ionic compound precipitate from solution? The ions in solution are Ba^{2+}, Cl^-, K^+, and SO_4^{2-}. According to the solubility rules listed in Table 4.2 (p. 101), the only precipitate that can form is $BaSO_4$. From the information given, we can calculate $[Ba^{2+}]$ and $[SO_4^{2-}]$ because we know the number of moles of the ions in the original solutions and the volume of the combined solution. Next we

(Continued)

calculate the ion product Q ($Q = [Ba^{2+}]_0[SO_4^{2-}]_0$) and compare the value of Q with K_{sp} of $BaSO_4$ to see if a precipitate will form, that is, if the solution is supersaturated. It is helpful to make a sketch of the situation.

200 mL 0.0040 M $BaCl_2$ → ; 600 mL 0.0080 M K_2SO_4 → ; Compare Q with K_{sp} ; $[Ba^{2+}]_0 = ?$; $[SO_4^{2-}]_0 = ?$; Total volume = 200 + 600 = 800 mL

Solution The number of moles of Ba^{2+} present in the original 200 mL of solution is

$$200 \text{ mL} \times \frac{0.0040 \text{ mol } Ba^{2+}}{1 \text{ L soln}} \times \frac{1 \text{ L}}{1000 \text{ mL}} = 8.0 \times 10^{-4} \text{ mol } Ba^{2+}$$

We assume that the volumes are additive.

The total volume after combining the two solutions is 800 mL. The concentration of Ba^{2+} in the 800 mL volume is

$$[Ba^{2+}] = \frac{8.0 \times 10^{-4} \text{ mol}}{800 \text{ mL}} \times \frac{1000 \text{ mL}}{1 \text{ L soln}}$$
$$= 1.0 \times 10^{-3}\ M$$

The number of moles of SO_4^{2-} in the original 600 mL solution is

$$600 \text{ mL} \times \frac{0.0080 \text{ mol } SO_4^{2-}}{1 \text{ L soln}} \times \frac{1 \text{ L}}{1000 \text{ mL}} = 4.8 \times 10^{-3} \text{ mol } SO_4^{2-}$$

The concentration of SO_4^{2-} in the 800 mL of the combined solution is

$$[SO_4^{2-}] = \frac{4.8 \times 10^{-3} \text{ mol}}{800 \text{ mL}} \times \frac{1000 \text{ mL}}{1 \text{ L soln}}$$
$$= 6.0 \times 10^{-3}\ M$$

Now we must compare Q with K_{sp}. From Table 17.2,

$$BaSO_4(s) \rightleftharpoons Ba^{2+}(aq) + SO_4^{2-}(aq) \qquad K_{sp} = 1.1 \times 10^{-10}$$

As for Q,

$$Q = [Ba^{2+}]_0[SO_4^{2-}]_0 = (1.0 \times 10^{-3})(6.0 \times 10^{-3})$$
$$= 6.0 \times 10^{-6}$$

Therefore,

$$Q > K_{sp}$$

The solution is supersaturated because the value of Q indicates that the concentrations of the ions are too large. Thus, some of the $BaSO_4$ will precipitate out of solution until

Similar problem: 17.45.

$$[Ba^{2+}][SO_4^{2-}] = 1.1 \times 10^{-10}$$

Practice Exercise If 2.00 mL of 0.200 M NaOH are added to 1.00 L of 0.100 M $CaCl_2$, will precipitation occur?

17.6 The Common Ion Effect and Solubility

As we have noted, the solubility product is an equilibrium constant; precipitation of an ionic compound from solution occurs whenever the ion product exceeds K_{sp} for that substance. In a saturated solution of AgCl, for example, the ion product $[Ag^+][Cl^-]$ is, of course, equal to K_{sp}. Furthermore, simple stoichiometry tells us that $[Ag^+] = [Cl^-]$. But this equality does not hold in all situations.

Suppose we study a solution containing two dissolved substances that share a common ion, say, AgCl and $AgNO_3$. In addition to the dissociation of AgCl, the following process also contributes to the total concentration of the common silver ions in solution:

$$AgNO_3(s) \xrightarrow{H_2O} Ag^+(aq) + NO_3^-(aq)$$

If $AgNO_3$ is added to a saturated AgCl solution, the increase in $[Ag^+]$ will make the ion product greater than the solubility product:

$$Q = [Ag^+]_0[Cl^-]_0 > K_{sp}$$

To reestablish equilibrium, some AgCl will precipitate out of the solution, as Le Châtelier's principle would predict, until the ion product is once again equal to K_{sp}. The effect of adding a common ion, then, is a *decrease* in the solubility of the salt (AgCl) in solution. Note that in this case $[Ag^+]$ is no longer equal to $[Cl^-]$ at equilibrium; rather, $[Ag^+] > [Cl^-]$.

At a given temperature, only the solubility of a compound is altered (decreased) by the common ion effect. Its solubility product, which is an equilibrium constant, remains the same whether or not other substances are present in the solution.

EXAMPLE 17.9

Calculate the solubility of silver chloride (in g/L) in a 8.3×10^{-3} *M* silver nitrate solution.

Strategy This is a common ion problem. The common ion here is Ag^+, which is supplied by both AgCl and $AgNO_3$. Remember that the presence of the common ion will affect only the solubility of AgCl (in g/L), but not the K_{sp} value because it is an equilibrium constant.

Solution

Step 1: The relevant species in solution are Ag^+ ions (from both AgCl and $AgNO_3$) and Cl^- ions. The NO_3^- ions are spectator ions.

Step 2: Because $AgNO_3$ is a soluble strong electrolyte, it dissociates completely:

$$\begin{array}{cccc} AgNO_3(s) \xrightarrow{H_2O} & Ag^+(aq) & + & NO_3^-(aq) \\ & 8.3 \times 10^{-3}\,M & & 8.3 \times 10^{-3}\,M \end{array}$$

Let *s* be the molar solubility of AgCl in $AgNO_3$ solution. We summarize the changes in concentration as follows:

	$AgCl(s)$	$\rightleftharpoons$	$Ag^+(aq)$	+	$Cl^-(aq)$
Initial (*M*):			8.3×10^{-3}		0.00
Change (*M*):	$-s$		$+s$		$+s$
Equilibrium (*M*):			$(8.3 \times 10^{-3} + s)$		s

Step 3:

$$K_{sp} = [Ag^+][Cl^-]$$

$$1.6 \times 10^{-10} = (8.3 \times 10^{-3} + s)(s)$$

(Continued)

Because AgCl is quite insoluble and the presence of Ag^+ ions from $AgNO_3$ further lowers the solubility of AgCl, s must be very small compared with 8.3×10^{-3}. Therefore, applying the approximation $8.3 \times 10^{-3} + s \approx 8.3 \times 10^{-3}$, we obtain

$$1.6 \times 10^{-10} = (8.3 \times 10^{-3})s$$
$$s = 1.9 \times 10^{-8}\ M$$

Step 4: At equilibrium,

$$[Ag^+] = (8.3 \times 10^{-3} + 1.9 \times 10^{-8})\ M \approx 8.3 \times 10^{-3}\ M$$
$$[Cl^-] = 1.9 \times 10^{-8}\ M$$

and so our approximation was justified in step 3. Because all the Cl^- ions must come from AgCl, the amount of AgCl dissolved in $AgNO_3$ solution also is 1.9×10^{-8} *M*. Then, knowing the molar mass of AgCl (143.4 g), we can calculate the solubility of AgCl as follows:

$$\text{solubility of AgCl in AgNO}_3\text{ solution} = \frac{1.9 \times 10^{-8}\text{ mol AgCl}}{1\text{ L soln}} \times \frac{143.4\text{ g AgCl}}{1\text{ mol AgCl}}$$
$$= 2.7 \times 10^{-6}\text{ g/L}$$

Check The solubility of AgCl in pure water is 1.9×10^{-3} g/L (see the Practice Exercise in Example 17.7). Therefore, the lower solubility (2.7×10^{-6} g/L) in the presence of $AgNO_3$ is reasonable. You should also be able to predict the lower solubility using Le Châtelier's principle. Adding Ag^+ ions shifts the solubility equilibrium to the left, thus decreasing the solubility of AgCl.

Similar problem: 17.50.

Practice Exercise Calculate the solubility in g/L of AgBr in (a) pure water and in (b) 0.0010 *M* NaBr.

REVIEW OF CONCEPTS

For each pair of solutions, determine in which one $PbCl_2$ would be more soluble? (a) NaCl or NaBr, (b) $Pb(NO_3)_2$ or $Ca(NO_3)_2$.

17.7 Complex Ion Equilibria and Solubility

Lewis acids and bases are discussed in Section 16.11.

Lewis acid-base reactions in which a metal cation (electron-pair acceptor) combines with a Lewis base (electron-pair donor) result in the formation of complex ions:

$$\underset{\text{acid}}{Ag^+(aq)} + \underset{\text{base}}{2NH_3(aq)} \rightleftharpoons Ag(NH_3)_2^+(aq)$$

Thus, we can define a ***complex ion*** as *an ion containing a central metal cation bonded to one or more molecules or ions.* Complex ions are crucial to many chemical and biological processes. Here we will consider the effect of complex ion formation on solubility. In Chapter 20 we will discuss the chemistry of complex ions in more detail.

According to our definition, $Co(H_2O)_6^{2+}$ itself is a complex ion. When we write $Co(H_2O)_6^{2+}$, we mean the hydrated Co^{2+} ion.

Transition metals have a particular tendency to form complex ions. For example, a solution of cobalt(II) chloride is pink because of the presence of the $Co(H_2O)_6^{2+}$ ions (Figure 17.10). When HCl is added, the solution turns blue as a result of the formation of the complex ion $CoCl_4^{2-}$:

$$Co^{2+}(aq) + 4Cl^-(aq) \rightleftharpoons CoCl_4^{2-}(aq)$$

Figure 17.10
(Left) An aqueous cobalt(II) chloride solution. The pink color is due to the presence of $Co(H_2O)_6^{2+}$ *ions. (Right) After the addition of HCl solution, the solution turns blue because of the formation of the complex* $CoCl_4^{2-}$ *ions.*

Copper(II) sulfate ($CuSO_4$) dissolves in water to produce a blue solution. The hydrated copper(II) ions are responsible for this color; many other sulfates (Na_2SO_4, for example) are colorless. Adding a *few drops* of concentrated ammonia solution to a $CuSO_4$ solution causes a light-blue precipitate, copper(II) hydroxide, to form (Figure 17.11):

$$Cu^{2+}(aq) + 2OH^-(aq) \longrightarrow Cu(OH)_2(s)$$

in which the OH^- ions are supplied by the ammonia solution. If an *excess* of NH_3 is then added, the blue precipitate redissolves to produce a beautiful dark-blue solution, this time as a result of the formation of the complex ion $Cu(NH_3)_4^{2+}$, also shown in Figure 17.11:

$$Cu(OH)_2(s) + 4NH_3(aq) \rightleftharpoons Cu(NH_3)_4^{2+}(aq) + 2OH^-(aq)$$

Thus, the formation of the complex ion $Cu(NH_3)_4^{2+}$ increases the solubility of $Cu(OH)_2$.

A measure of the tendency of a metal ion to form a particular complex ion is given by the ***formation constant*** $\boldsymbol{K_f}$ (also called the *stability constant*), which is *the equilibrium constant for complex ion formation.* The larger K_f is, the more stable the complex ion is. Table 17.4 lists the formation constants of a number of complex ions.

The formation of the $Cu(NH_3)_4^{2+}$ ion can be expressed as

$$Cu^{2+}(aq) + 4NH_3(aq) \rightleftharpoons Cu(NH_3)_4^{2+}(aq)$$

Figure 17.11
Left: An aqueous solution of copper(II) sulfate. Center: After the addition of a few drops of a concentrated aqueous ammonia solution, a light-blue precipitate of $Cu(OH)_2$ *is formed. Right: When more concentrated aqueous ammonia solution is added, the* $Cu(OH)_2$ *precipitate dissolves to form the dark-blue complex ion* $Cu(NH_3)_4^{2+}$.

Table 17.4 Formation Constants of Selected Complex Ions in Water at 25°C

Complex Ion	Equilibrium Expression	Formation Constant (K_f)
$Ag(NH_3)_2^+$	$Ag^+ + 2NH_3 \rightleftharpoons Ag(NH_3)_2^+$	1.5×10^7
$Ag(CN)_2^-$	$Ag^+ + 2CN^- \rightleftharpoons Ag(CN)_2^-$	1.0×10^{21}
$Cu(CN)_4^{2-}$	$Cu^{2+} + 4CN^- \rightleftharpoons Cu(CN)_4^{2-}$	1.0×10^{25}
$Cu(NH_3)_4^{2+}$	$Cu^{2+} + 4NH_3 \rightleftharpoons Cu(NH_3)_4^{2+}$	5.0×10^{13}
$Cd(CN)_4^{2-}$	$Cd^{2+} + 4CN^- \rightleftharpoons Cd(CN)_4^{2-}$	7.1×10^{16}
CdI_4^{2-}	$Cd^{2+} + 4I^- \rightleftharpoons CdI_4^{2-}$	2.0×10^6
$HgCl_4^{2-}$	$Hg^{2+} + 4Cl^- \rightleftharpoons HgCl_4^{2-}$	1.7×10^{16}
HgI_4^{2-}	$Hg^{2+} + 4I^- \rightleftharpoons HgI_4^{2-}$	2.0×10^{30}
$Hg(CN)_4^{2-}$	$Hg^{2+} + 4CN^- \rightleftharpoons Hg(CN)_4^{2-}$	2.5×10^{41}
$Co(NH_3)_6^{3+}$	$Co^{3+} + 6NH_3 \rightleftharpoons Co(NH_3)_6^{3+}$	5.0×10^{31}
$Zn(NH_3)_4^{2+}$	$Zn^{2+} + 4NH_3 \rightleftharpoons Zn(NH_3)_4^{2+}$	2.9×10^9

for which the formation constant is

$$K_f = \frac{[Cu(NH_3)_4^{2+}]}{[Cu^{2+}][NH_3]^4} = 5.0 \times 10^{13}$$

The very large value of K_f in this case indicates the great stability of the complex ion in solution and accounts for the very low concentration of copper(II) ions at equilibrium.

EXAMPLE 17.10

A 0.22-mole quantity of $CuSO_4$ is added to a liter of 1.35 M NH_3 solution. What is the concentration of Cu^{2+} ions at equilibrium?

Strategy The addition of $CuSO_4$ to the NH_3 solution results in complex ion formation

$$Cu^{2+}(aq) + 4NH_3(aq) \rightleftharpoons Cu(NH_3)_4^{2+}(aq)$$

From Table 17.4 we see that the formation constant (K_f) for this reaction is very large; therefore, the reaction lies mostly to the right. At equilibrium, the concentration of Cu^{2+} will be very small. As a good approximation, we can assume that essentially all the dissolved Cu^{2+} ions end up as $Cu(NH_3)_4^{2+}$ ions. How many moles of NH_3 will react with 0.22 mole of Cu^{2+}? How many moles of $Cu(NH_3)_4^{2+}$ will be produced? A very small amount of Cu^{2+} will be present at equilibrium. Set up the K_f expression for the preceding equilibrium to solve for $[Cu^{2+}]$.

Solution The amount of NH_3 consumed in forming the complex ion is 4×0.22 mol, or 0.88 mol. (Note that 0.22 mol Cu^{2+} is initially present in solution and four NH_3 molecules are needed to form a complex ion with one Cu^{2+} ion.) The concentration of NH_3 at equilibrium is therefore $(1.35 - 0.88)$ mol/L soln or 0.47 M, and that of $Cu(NH_3)_4^{2+}$ is 0.22 mol/L soln or 0.22 M, the same as the initial concentration of Cu^{2+}. [There is a 1:1 mole ratio between Cu^{2+} and $Cu(NH_3)_4^{2+}$.] Because $Cu(NH_3)_4^{2+}$

(Continued)

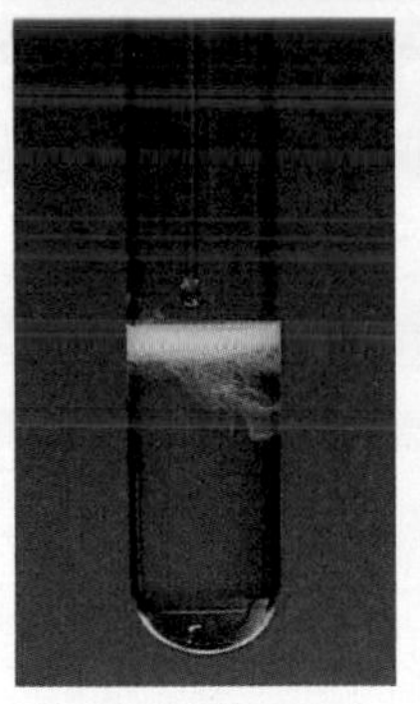
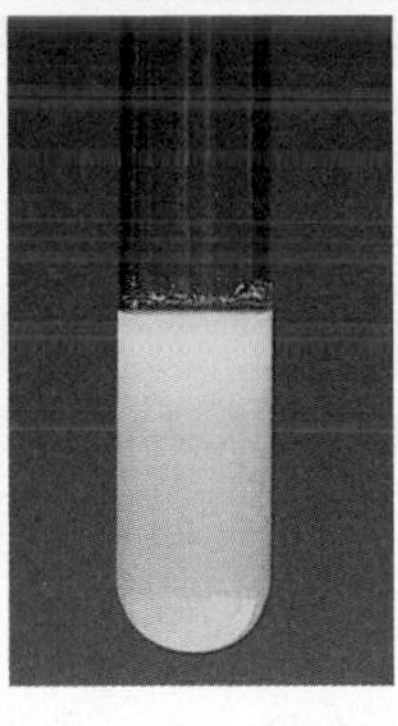

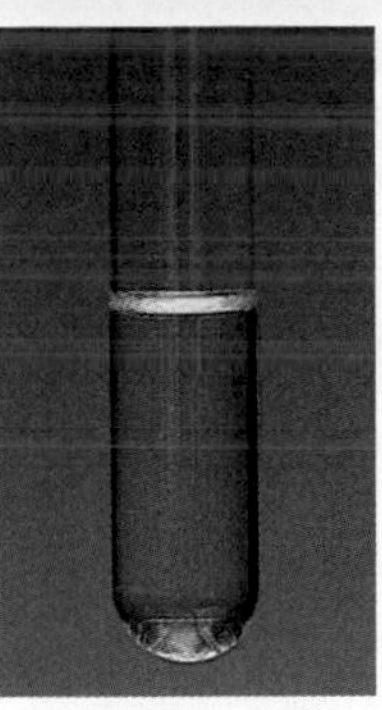

Figure 17.12 *(Left to right) Formation of $Al(OH)_3$ precipitate when NaOH solution is added to an $Al(NO_3)_3$ solution. With further addition of NaOH solution, the precipitate dissolves due to the formation of the complex ion $Al(OH)_4^-$.*

does dissociate to a slight extent, we call the concentration of Cu^{2+} ions at equilibrium x and write

$$K_f = \frac{[Cu(NH_3)_4^{2+}]}{[Cu^{2+}][NH_3]^4}$$

$$5.0 \times 10^{13} = \frac{0.22}{x(0.47)^4}$$

Solving for x and keeping in mind that the volume of the solution is 1 L, we obtain

$$x = [Cu^{2+}] = 9.0 \times 10^{-14}\ M$$

The very small value of $[Cu^{2+}]$ justifies our approximation.

Similar problem: 17.57.

Practice Exercise If 2.50 g of $CuSO_4$ are dissolved in 9.0×10^2 mL of 0.30 M NH_3, what are the concentrations of Cu^{2+}, $Cu(NH_3)_4^{2+}$, and NH_3 at equilibrium?

Finally, we note that there is a class of hydroxides, called *amphoteric hydroxides,* which can react with both acids and bases. Examples are $Al(OH)_3$, $Pb(OH)_2$, $Cr(OH)_3$, $Zn(OH)_2$, and $Cd(OH)_2$. For example, aluminum hydroxide reacts with acids and bases as

All amphoteric hydroxides are insoluble compounds.

$$Al(OH)_3(s) + 3H^+(aq) \longrightarrow Al^{3+}(aq) + 3H_2O(l)$$
$$Al(OH)_3(s) + OH^-(aq) \rightleftharpoons Al(OH)_4^-(aq)$$

The increase in solubility of $Al(OH)_3$ in a basic medium is the result of the formation of the complex ion $[Al(OH)_4^-]$ in which $Al(OH)_3$ acts as the Lewis acid and OH^- acts as the Lewis base (Figure 17.12). Other amphoteric hydroxides behave in a similar manner.

Review of Concepts

Which compound, when added to water, will increase the solubility of CdS? (a) $LiNO_3$, (b) Na_2SO_4, (c) KCN, (d) $NaClO_3$.

17.8 Application of the Solubility Product Principle to Qualitative Analysis

In Section 4.6, we discussed the principle of gravimetric analysis, by which we measure the amount of an ion in an unknown sample. Here we will briefly discuss ***qualitative analysis,*** *the determination of the types of ions present in a solution.* We will focus on the cations.

Table 17.5 Separation of Cations into Groups According to Their Precipitation Reactions with Various Reagents

Group	Cation	Precipitating Reagents	Insoluble Compound	K_{sp}
1	Ag^+	HCl	AgCl	1.6×10^{-10}
	Hg_2^{2+}	↓	Hg_2Cl_2	3.5×10^{-18}
	Pb^{2+}		$PbCl_2$	2.4×10^{-4}
2	Bi^{3+}	H_2S	Bi_2S_3	1.6×10^{-72}
	Cd^{2+}	in acidic	CdS	8.0×10^{-28}
	Cu^{2+}	solutions	CuS	6.0×10^{-37}
	Hg^{2+}	↓	HgS	4.0×10^{-54}
	Sn^{2+}		SnS	1.0×10^{-26}
3	Al^{3+}	H_2S	$Al(OH)_3$	1.8×10^{-33}
	Co^{2+}	in basic	CoS	4.0×10^{-21}
	Cr^{3+}	solutions	$Cr(OH)_3$	3.0×10^{-29}
	Fe^{2+}	↓	FeS	6.0×10^{-19}
	Mn^{2+}		MnS	3.0×10^{-14}
	Ni^{2+}		NiS	1.4×10^{-24}
	Zn^{2+}		ZnS	3.0×10^{-23}
4	Ba^{2+}	Na_2CO_3	$BaCO_3$	8.1×10^{-9}
	Ca^{2+}	↓	$CaCO_3$	8.7×10^{-9}
	Sr^{2+}		$SrCO_3$	1.6×10^{-9}
5	K^+	No precipitating	None	
	Na^+	reagent	None	
	NH_4^+		None	

Do not confuse the groups in Table 17.5, which are based on solubility products, with those in the periodic table, which are based on the electron configurations of the elements.

Twenty common cations can be analyzed readily in aqueous solution. These cations can be divided into five groups according to the solubility products of their insoluble salts (Table 17.5). Because an unknown solution may contain any one or up to all 20 ions, analysis must be carried out systematically from group 1 through group 5. Let us consider the general procedure for separating these ions by adding precipitating reagents to an unknown solution.

- *Group 1 cations.* When dilute HCl is added to the unknown solution, only the Ag^+, Hg_2^{2+}, and Pb^{2+} ions precipitate as insoluble chlorides. The other ions, whose chlorides are soluble, remain in solution.
- *Group 2 cations.* After the chloride precipitates have been removed by filtration, hydrogen sulfide is reacted with the unknown acidic solution. Under this condition, the concentration of the S^{2-} ion in solution is negligible. Therefore, the precipitation of metal sulfides is best represented as

$$M^{2+}(aq) + H_2S(aq) \rightleftharpoons MS(s) + 2H^+(aq)$$

Adding acid to the solution shifts this equilibrium to the left so that only the least soluble metal sulfides, that is, those with the smallest K_{sp} values, will precipitate out of solution. These are Bi_2S_3, CdS, CuS, HgS, and SnS.

Figure 17.13
Left to right: Flame colors of lithium, sodium, potassium, and copper.

- *Group 3 cations.* At this stage, sodium hydroxide is added to the solution to make it basic. In a basic solution, the preceding equilibrium shifts to the right. Therefore, the more soluble sulfides (CoS, FeS, MnS, NiS, and ZnS) now precipitate out of solution. Note that the Al^{3+} and Cr^{3+} ions actually precipitate as the hydroxides $Al(OH)_3$ and $Cr(OH)_3$, rather than as the sulfides, because the hydroxides are less soluble. The solution is then filtered to remove the insoluble sulfides and hydroxides.
- *Group 4 cations.* After all the group 1, 2, and 3 cations have been removed from solution, sodium carbonate is added to the basic solution to precipitate Ba^{2+}, Ca^{2+}, and Sr^{2+} ions as $BaCO_3$, $CaCO_3$, and $SrCO_3$. These precipitates too are removed from solution by filtration.
- *Group 5 cations.* At this stage, the only cations possibly remaining in solution are Na^+, K^+, and NH_4^+. The presence of NH_4^+ can be determined by adding sodium hydroxide:

$$NaOH(aq) + NH_4^+(aq) \longrightarrow Na^+(aq) + H_2O(l) + NH_3(g)$$

The ammonia gas is detected either by noting its characteristic odor or by observing a piece of wet red litmus paper turning blue when placed above (not in contact with) the solution. To confirm the presence of Na^+ and K^+ ions, we usually use a flame test, as follows: A piece of platinum wire (chosen because platinum is inert) is moistened with the solution and is then held over a Bunsen burner flame. Each type of metal ion gives a characteristic color when heated in this manner. For example, the color emitted by Na^+ ions is yellow, that of K^+ ions is violet, and that of Cu^{2+} ions is green (Figure 17.13).

Because NaOH is added in group 3 and Na_2CO_3 is added in group 4, the flame test for Na^+ ions is carried out using the original solution.

Figure 17.14 summarizes this scheme for separating metal ions.

Two points regarding qualitative analysis must be mentioned. First, the separation of the cations into groups is made as selective as possible; that is, the anions that are added as reagents must be such that they will precipitate the fewest types of cations. For example, all the cations in group 1 also form insoluble sulfides. Thus, if H_2S were reacted with the solution at the start, as many as seven different sulfides might precipitate out of solution (group 1 *and* group 2 sulfides), an undesirable outcome. Second, the removal of cations at each step must be carried out as completely as possible.

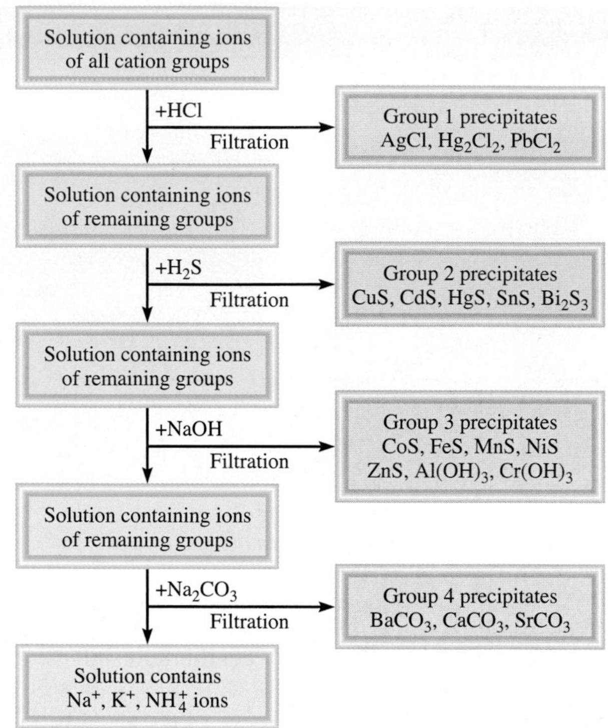

Figure 17.14
A flow chart for the separation of cations in qualitative analysis.

For example, if we do not add enough HCl to the unknown solution to remove all the group 1 cations, they will precipitate with the group 2 cations as insoluble sulfides; this too would interfere with further chemical analysis and lead us to draw erroneous conclusions.

REVIEW OF CONCEPTS

A white solid could either be $Pb(NO_3)_2$ or $Mg(NO_3)_2$. What one reagent could be added to an aqueous solution of the solid to determine its identity?

Key Equations

$$pK_a = -\log K_a \quad (17.2)$$ Definition of pK_a.

$$pH = pK_a + \log \frac{[\text{conjugate base}]}{[\text{acid}]} \quad (17.3)$$ Henderson-Hasselbalch equation.

Summary of Facts and Concepts

1. Equilibria involving weak acids or weak bases in aqueous solution are homogeneous. Solubility equilibria are examples of heterogeneous equilibria.
2. A buffer solution is a combination of a weak acid and its weak conjugate base; the solution reacts with small amounts of added acid or base in such a way that the pH of the solution remains nearly constant. Buffer systems play a vital role in maintaining the pH of body fluids.
3. The pH at the equivalence point of an acid-base titration depends on the hydrolysis of the salt formed in the

neutralization reaction. For strong acid–strong base titrations, the pH at the equivalence point is 7; for weak acid–strong base titrations, the pH at the equivalence point is greater than 7; for strong acid–weak base titrations, the pH at the equivalence point is less than 7. Acid-base indicators are weak organic acids or bases that change color at the end point in an acid-base neutralization reaction.

4. The solubility product K_{sp} expresses the equilibrium between a solid and its ions in solution. Solubility can be found from K_{sp} and vice versa. The presence of a common ion decreases the solubility of a salt.
5. Complex ions are formed in solution by the combination of a metal cation with a Lewis base. The formation constant K_f measures the tendency toward the formation of a specific complex ion. Complex ion formation can increase the solubility of an insoluble substance.
6. Qualitative analysis is the identification of cations and anions in solution. It is based largely on the principles of solubility equilibria.

Key Words

Buffer solution, p. 591
Complex ion, p. 614
End point, p. 603
Formation constant (K_f), p. 615
Molar solubility, p. 608
Qualitative analysis, p. 617
Solubility, p. 608
Solubility product (K_{sp}), p. 606

Questions and Problems

Buffer Solutions

Review Questions

17.1 Define buffer solution.

17.2 Define pK_a for a weak acid and explain the relationship between the value of the pK_a and the strength of the acid. Do the same for pK_b and a weak base.

17.3 Which of the following has the greatest buffer capacity? (a) 0.40 *M* CH_3COONa/0.20 *M* CH_3COOH, (b) 0.40 *M* CH_3COONa/0.60 *M* CH_3COOH, (c) 0.30 *M* CH_3COONa/0.60 *M* CH_3COOH.

17.4 The pK_bs for the bases X^-, Y^-, and Z^- are 2.72, 8.66, and 4.57, respectively. Arrange the following acids in order of increasing strength: HX, HY, HZ.

Problems

17.5 Specify which of these systems can be classified as a buffer system: (a) KCl/HCl, (b) NH_3/NH_4NO_3, (c) Na_2HPO_4/NaH_2PO_4.

17.6 Specify which of these systems can be classified as a buffer system: (a) KNO_2/HNO_2, (b) $KHSO_4/H_2SO_4$, (c) HCOOK/HCOOH.

17.7 The pH of a bicarbonate–carbonic acid buffer is 8.00. Calculate the ratio of the concentration of carbonic acid to that of the bicarbonate ion.

17.8 Calculate the pH of these two buffer solutions: (a) 2.0 *M* CH_3COONa/2.0 *M* CH_3COOH, (b) 0.20 *M* CH_3COONa/0.20 *M* CH_3COOH. Which is the more effective buffer? Why?

17.9 Calculate the pH of the buffer system 0.15 *M* NH_3/0.35 *M* NH_4Cl.

17.10 What is the pH of the buffer 0.10 *M* Na_2HPO_4/0.15 *M* KH_2PO_4?

17.11 The pH of a sodium acetate–acetic acid buffer is 4.50. Calculate the ratio $[CH_3COO^-]/[CH_3COOH]$.

17.12 The pH of blood plasma is 7.40. Assuming the principal buffer system is HCO_3^-/H_2CO_3, calculate the ratio $[HCO_3^-]/[H_2CO_3]$. Is this buffer more effective against an added acid or an added base?

17.13 Calculate the pH of 1.00 L of the buffer 0.80 *M* CH_3NH_2/1.00 *M* CH_3NH_3Cl before and after the addition of (a) 0.070 mol NaOH and (b) 0.11 mol HCl. (See Table 16.5 for K_a value.)

17.14 Calculate the pH of 1.00 L of the buffer 1.00 *M* CH_3COONa/1.00 *M* CH_3COOH before and after the addition of (a) 0.080 mol NaOH and (b) 0.12 mol HCl. (Assume that there is no change in volume.)

17.15 A diprotic acid, H_2A, has the following ionization constants: $K_{a_1} = 1.1 \times 10^{-3}$ and $K_{a_2} = 2.5 \times 10^{-6}$. To make up a buffer solution of pH 5.80, which combination would you choose: NaHA/H_2A or Na_2A/NaHA?

17.16 A student wishes to prepare a buffer solution at pH = 8.60. Which of these weak acids should she choose and why: HA ($K_a = 2.7 \times 10^{-3}$), HB ($K_a = 4.4 \times 10^{-6}$), or HC ($K_a = 2.6 \times 10^{-9}$)?

17.17 The diagrams shown here contain one or more of the compounds: H_2A, NaHA, and Na_2A, where H_2A is a weak diprotic acid. (1) Which of the solutions can act as buffer solutions? (2) Which solution is the most effective buffer solution? Water molecules and Na^+ ions have been omitted for clarity.

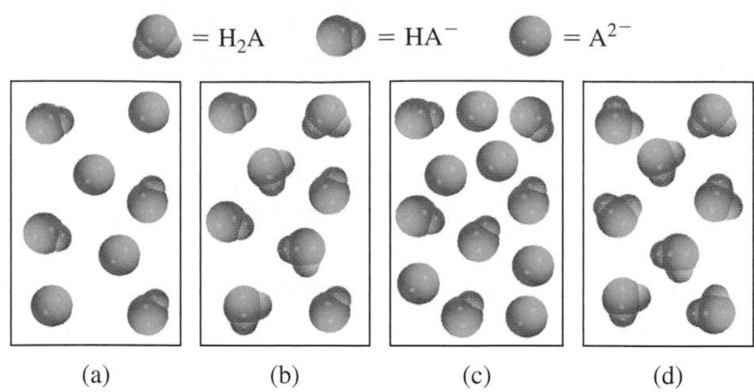

17.18 The diagrams shown here represent solutions containing a weak acid HA (pK_a = 5.00) and its sodium salt NaA. (1) Calculate the pH of the solutions. (2) What is the pH after the addition of 0.1 mol H^+ ions to solution (a)? (3) What is the pH after the addition of 0.1 mol OH^- ions to solution (d)? Treat each sphere as 0.1 mol.

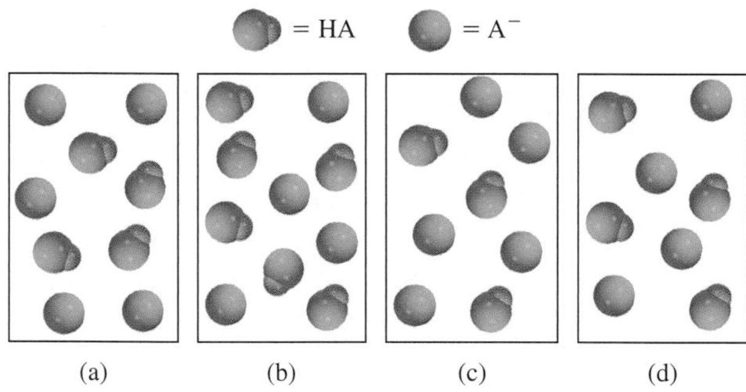

Acid-Base Titrations

Problems

17.19 A 0.2688-g sample of a monoprotic acid neutralizes 16.4 mL of 0.08133 *M* KOH solution. Calculate the molar mass of the acid.

17.20 A 5.00-g quantity of a diprotic acid is dissolved in water and made up to exactly 250 mL. Calculate the molar mass of the acid if 25.0 mL of this solution required 11.1 mL of 1.00 *M* KOH for neutralization. Assume that both protons of the acid are titrated.

17.21 Calculate the pH at the equivalence point for these titrations: (a) 0.10 *M* HCl versus 0.10 *M* NH_3, (b) 0.10 *M* CH_3COOH versus 0.10 *M* NaOH.

17.22 A sample of 0.1276 g of an unknown monoprotic acid was dissolved in 25.0 mL of water and titrated with 0.0633 *M* NaOH solution. The volume of base required to reach the equivalence point was 18.4 mL. (a) Calculate the molar mass of the acid. (b) After 10.0 mL of base had been added to the titration, the pH was determined to be 5.87. What is the K_a of the unknown acid?

17.23 The diagrams shown here represent solutions at different stages in the titration of a weak acid HA with NaOH. Identify the solution that corresponds to (1) the initial stage before the addition of NaOH, (2) halfway to the equivalence point, (3) the equivalence point, (4) beyond the equivalence point. Is the pH greater than, less than, or equal to 7 at the equivalence point? Water molecules and Na^+ ions have been omitted for clarity.

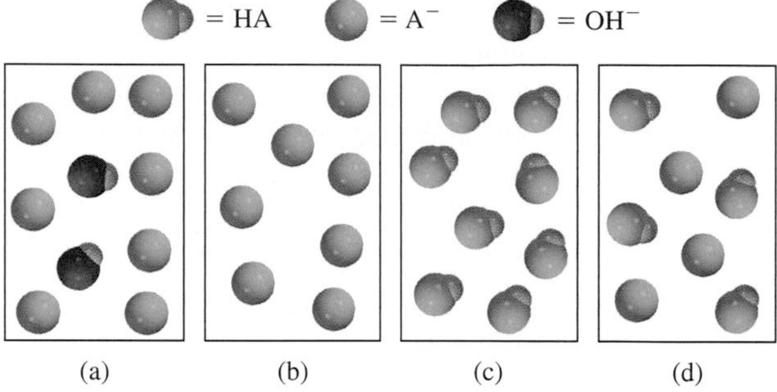

17.24 The diagrams shown here represent solutions at various stages in the titration of a weak base B (such as NH_3) with HCl. Identify the solution that corresponds to (1) the initial stage before the addition of HCl, (2) halfway to the equivalence point, (3) the equivalence point, (4) beyond the equivalence point. Is the pH greater than, less than, or equal to 7 at the equivalence point? Water molecules and Cl^- ions have been omitted for clarity.

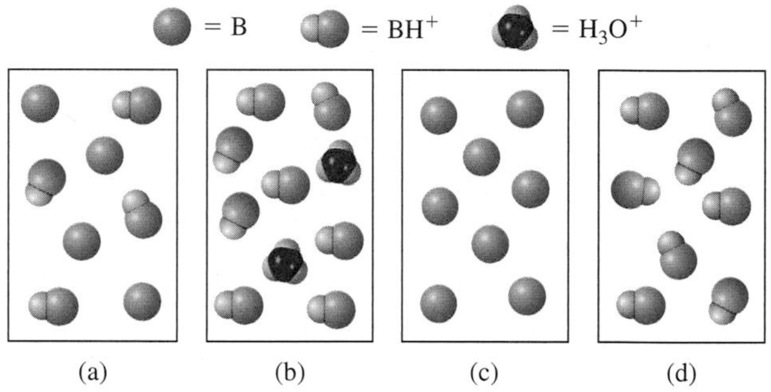

Acid-Base Indicators

Review Questions

17.25 Explain how an acid-base indicator works in a titration.

17.26 What are the criteria for choosing an indicator for a particular acid-base titration?

Problems

17.27 The amount of indicator used in an acid-base titration must be small. Why?

17.28 A student carried out an acid-base titration by adding NaOH solution from a buret to an Erlenmeyer flask containing HCl solution and using phenolphthalein as indicator. At the equivalence point, he observed a faint reddish-pink color. However, after a few minutes, the solution gradually turned colorless. What do you suppose happened?

17.29 Referring to Table 17.1, specify which indicator or indicators you would use for the following titrations: (a) HCOOH versus NaOH, (b) HCl versus KOH, (c) HNO_3 versus NH_3.

17.30 The ionization constant K_a of an indicator HIn is 1.0×10^{-6}. The color of the nonionized form is red and that of the ionized form is yellow. What is the color of this indicator in a solution whose pH is 4.00? (*Hint:* The color of an indicator can be estimated by considering the ratio $[HIn]/[In^-]$. If the ratio is equal to or greater than 10, the color will be that of the nonionized form. If the ratio is equal to or smaller than 0.1, the color will be that of the ionized form.)

Solubility and Solubility Product

Review Questions

17.31 Define solubility, molar solubility, and solubility product. Explain the difference between solubility and the solubility product of a slightly soluble substance such as $BaSO_4$.

17.32 Why do we usually not quote the K_{sp} values for soluble ionic compounds?

17.33 Write balanced equations and solubility product expressions for the solubility equilibria of these compounds: (a) CuBr, (b) ZnC_2O_4, (c) Ag_2CrO_4, (d) Hg_2Cl_2, (e) $AuCl_3$, (f) $Mn_3(PO_4)_2$.

17.34 Write the solubility product expression for the ionic compound A_xB_y.

17.35 How can we predict whether a precipitate will form when two solutions are mixed?

17.36 Silver chloride has a larger K_{sp} than silver carbonate (see Table 17.2). Does this mean that the former also has a larger molar solubility than the latter?

Problems

17.37 Calculate the concentration of ions in these saturated solutions:

(a) $[I^-]$ in AgI solution with $[Ag^+] = 9.1 \times 10^{-9}$ *M*

(b) $[Al^{3+}]$ in $Al(OH)_3$ with $[OH^-] = 2.9 \times 10^{-9}$ *M*

17.38 From the solubility data given, calculate the solubility products for these compounds:

(a) SrF_2, 7.3×10^{-2} g/L

(b) Ag_3PO_4, 6.7×10^{-3} g/L

17.39 The molar solubility of $MnCO_3$ is 4.2×10^{-6} *M*. What is K_{sp} for this compound?

17.40 Using data from Table 17.2, calculate the molar solubility of calcium phosphate, which is a component of bones.

17.41 The solubility of an ionic compound M_2X_3 (molar mass = 288 g) is 3.6×10^{-17} g/L. What is K_{sp} for the compound?

17.42 Using data from Table 17.2, calculate the solubility of CaF_2 in g/L.

17.43 What is the pH of a saturated zinc hydroxide solution?

17.44 The pH of a saturated solution of a metal hydroxide MOH is 9.68. Calculate the K_{sp} for the compound.

17.45 A sample of 20.0 mL of 0.10 *M* $Ba(NO_3)_2$ is added to 50.0 mL of 0.10 *M* Na_2CO_3. Will $BaCO_3$ precipitate?

17.46 A volume of 75 mL of 0.060 *M* NaF is mixed with 25 mL of 0.15 *M* $Sr(NO_3)_2$. Calculate the concentrations in the final solution of NO_3^-, Na^+, Sr^{2+}, and F^-. (K_{sp} for $SrF_2 = 2.0 \times 10^{-10}$.)

The Common Ion Effect

Review Questions

17.47 How does a common ion affect solubility? Use Le Châtelier's principle to explain the decrease in solubility of $CaCO_3$ in a Na_2CO_3 solution.

17.48 The molar solubility of AgCl in 6.5×10^{-3} *M* $AgNO_3$ is 2.5×10^{-8} *M*. In deriving K_{sp} from these data, which of these assumptions are reasonable?

(a) K_{sp} is the same as solubility.

(b) K_{sp} of AgCl is the same in 6.5×10^{-3} *M* $AgNO_3$ as in pure water.

(c) Solubility of AgCl is independent of the concentration of $AgNO_3$.

(d) $[Ag^+]$ in solution does not change significantly on the addition of AgCl to 6.5×10^{-3} *M* $AgNO_3$.

(e) $[Ag^+]$ in solution after the addition of AgCl to 6.5×10^{-3} *M* $AgNO_3$ is the same as it would be in pure water.

Problems

17.49 How many grams of $CaCO_3$ will dissolve in 3.0×10^2 mL of 0.050 *M* $Ca(NO_3)_2$?

17.50 The solubility product of $PbBr_2$ is 8.9×10^{-6}. Determine the molar solubility (a) in pure water, (b) in 0.20 *M* KBr solution, (c) in 0.20 *M* $Pb(NO_3)_2$ solution.

17.51 Calculate the molar solubility of AgCl in a solution made by dissolving 10.0 g of $CaCl_2$ in enough water to make up a 1.00 L solution.

17.52 Calculate the molar solubility of $BaSO_4$ (a) in water and (b) in a solution containing 1.0 *M* SO_4^{2-} ions.

Complex Ions

Review Questions

17.53 Explain the formation of complexes in Table 17.4 in terms of Lewis acid-base theory.

17.54 Give an example to illustrate the general effect of complex ion formation on solubility.

Problems

17.55 Write the formation constant expressions for these complex ions: (a) $Zn(OH)_4^{2-}$, (b) $Co(NH_3)_6^{3+}$, (c) HgI_4^{2-}.

17.56 Explain, with balanced ionic equations, why (a) CuI_2 dissolves in ammonia solution, (b) AgBr dissolves in NaCN solution, (c) Hg_2Cl_2 dissolves in KCl solution.

17.57 If 2.50 g of $CuSO_4$ are dissolved in 9.0×10^2 mL of 0.30 M NH_3, what are the concentrations of Cu^{2+}, $Cu(NH_3)_4^{2+}$, and NH_3 at equilibrium?

17.58 Calculate the concentrations of Cd^{2+}, $Cd(CN)_4^{2-}$, and CN^- at equilibrium when 0.50 g of $Cd(NO_3)_2$ dissolves in 5.0×10^2 mL of 0.50 M NaCN.

17.59 If NaOH is added to 0.010 M Al^{3+}, which will be the predominant species at equilibrium: $Al(OH)_3$ or $Al(OH)_4^-$? The pH of the solution is 14.00. [K_f for $Al(OH)_4^- = 2.0 \times 10^{33}$.]

17.60 Calculate the molar solubility of AgI in a 1.0 M NH_3 solution. (*Hint:* You need to consider two different types of equilibria.)

Qualitative Analysis

Review Questions

17.61 Outline the general principle of qualitative analysis.

17.62 Give two examples of metal ions in each group (1 through 5) in the qualitative analysis scheme.

Problems

17.63 In a group 1 analysis, a student obtained a precipitate containing both AgCl and $PbCl_2$. Suggest one reagent that would enable her to separate AgCl(s) from $PbCl_2$(s).

17.64 In a group 1 analysis, a student adds hydrochloric acid to the unknown solution to make $[Cl^-] = 0.15$ M. Some $PbCl_2$ precipitates. Calculate the concentration of Pb^{2+} remaining in solution.

17.65 Both KCl and NH_4Cl are white solids. Suggest one reagent that would enable you to distinguish between these two compounds.

17.66 Describe a simple test that would enable you to distinguish between $AgNO_3$(s) and $Cu(NO_3)_2$(s).

Additional Problems

17.67 A quantity of 0.560 g of KOH is added to 25.0 mL of 1.00 M HCl. Excess Na_2CO_3 is then added to the solution. What mass (in grams) of CO_2 is formed?

17.68 A volume of 25.0 mL of 0.100 M HCl is titrated against a 0.100 M NH_3 solution added to it from a buret. Calculate the pH values of the solution (a) after 10.0 mL of NH_3 solution have been added, (b) after 25.0 mL of NH_3 solution have been added, (c) after 35.0 mL of NH_3 solution have been added.

17.69 The buffer range is defined by the equation pH = $pK_a \pm 1$. Calculate the range of the ratio [conjugate base]/[acid] that corresponds to this equation.

17.70 The pK_a of the indicator methyl orange is 3.46. Over what pH range does this indicator change from 90% HIn to 90% In^-?

17.71 Sketch the titration curve of a weak acid versus a strong base such as that shown in Figure 17.5. On your graph indicate the volume of base used at the equivalence point and also at the half-equivalence point, that is, the point at which half of the base has been added. Show how you can measure the pH of the solution at the half-equivalence point. Using Equation (17.3), explain how you can determine the pK_a of the acid by this procedure.

17.72 A 200-mL volume of NaOH solution was added to 400 mL of a 2.00 M HNO_2 solution. The pH of the mixed solution was 1.50 units greater than that of the original acid solution. Calculate the molarity of the NaOH solution.

17.73 The pK_a of butyric acid (HBut) is 4.7. Calculate K_b for the butyrate ion (But^-).

17.74 A solution is made by mixing exactly 500 mL of 0.167 M NaOH with exactly 500 mL 0.100 M CH_3COOH. Calculate the equilibrium concentrations of H^+, CH_3COOH, CH_3COO^-, OH^-, and Na^+.

17.75 $Cd(OH)_2$ is an insoluble compound. It dissolves in a NaOH solution. Write a balanced ionic equation for this reaction. What type of reaction is this?

17.76 Calculate the pH of the 0.20 M NH_3/0.20 M NH_4Cl buffer. What is the pH of the buffer after the addition of 10.0 mL of 0.10 M HCl to 65.0 mL of the buffer?

17.77 For which of these reactions is the equilibrium constant called a solubility product?
(a) $Zn(OH)_2(s) + 2OH^-(aq) \rightleftharpoons Zn(OH)_4^{2-}(aq)$
(b) $3Ca^{2+}(aq) + 2PO_4^{3-}(aq) \rightleftharpoons Ca_3(PO_4)_2(s)$
(c) $CaCO_3(s) + 2H^+(aq) \rightleftharpoons Ca^{2+}(aq) + H_2O(l) + CO_2(g)$
(d) $PbI_2(s) \rightleftharpoons Pb^{2+}(aq) + 2I^-(aq)$

17.78 A student mixes 50.0 mL of 1.00 M $Ba(OH)_2$ with 86.4 mL of 0.494 M H_2SO_4. Calculate the mass of $BaSO_4$ formed and the pH of the mixed solution.

17.79 A 2.0-L kettle contains 116 g of calcium carbonate as boiler scale. How many times would the kettle have to be completely filled with distilled water to remove all of the deposit at 25°C?

17.80 Equal volumes of 0.12 M $AgNO_3$ and 0.14 M $ZnCl_2$ solution are mixed. Calculate the equilibrium concentrations of Ag^+, Cl^-, Zn^{2+}, and NO_3^-.

17.81 Calculate the solubility (in grams per liter) of Ag_2CO_3.

17.82 Find the approximate pH range suitable for the separation of Fe^{3+} and Zn^{2+} by precipitation of $Fe(OH)_3$ from a solution that is initially 0.010 M in Fe^{3+} and Zn^{2+}.

17.83 Which of these ionic compounds will be more soluble in acid solution than in water: (a) $BaSO_4$, (b) $PbCl_2$, (c) $Fe(OH)_3$, (d) $CaCO_3$? Explain. (*Hint:* For each salt, determine any possible reaction between the anion and H^+ ions.)

17.84 Which of these substances will be more soluble in acid solution than in pure water: (a) CuI, (b) Ag_2SO_4, (c) $Zn(OH)_2$, (d) BaC_2O_4, (e) $Ca_3(PO_4)_2$? Explain. (*Hint:* For each salt, determine any possible reaction between the anion and H^+ ions.)

17.85 What is the pH of a saturated solution of aluminum hydroxide?

17.86 The molar solubility of $Pb(IO_3)_2$ in a 0.10 *M* $NaIO_3$ solution is 2.4×10^{-11} mol/L. What is K_{sp} for $Pb(IO_3)_2$?

17.87 The solubility product of $Mg(OH)_2$ is 1.2×10^{-11}. What minimum OH^- concentration must be attained (for example, by adding NaOH) to make the Mg^{2+} concentration in a solution of $Mg(NO_3)_2$ less than 1.0×10^{-10} *M*?

17.88 Calculate whether a precipitate will form if 2.00 mL of 0.60 *M* NH_3 are added to 1.0 L of 1.0×10^{-3} *M* $FeSO_4$.

17.89 Both Ag^+ and Zn^{2+} form complex ions with NH_3. Write balanced equations for the reactions. However, $Zn(OH)_2$ is soluble in 6 *M* NaOH, and AgOH is not. Explain.

17.90 When a KI solution was added to a solution of mercury(II) chloride, a precipitate [mercury(II) iodide] was formed. A student plotted the mass of the precipitate formed versus the volume of the KI solution added and obtained the graph shown here. Explain the appearance of the graph.

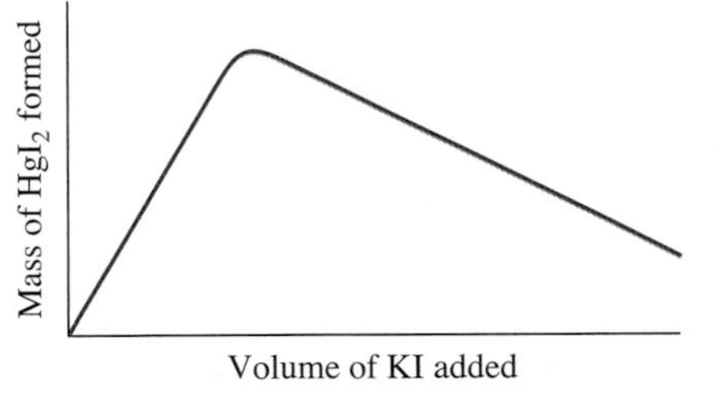

17.91 Barium is a toxic substance that can cause serious deterioration of the heart's function. In a barium enema procedure, a patient drinks an aqueous suspension of 20 g $BaSO_4$. If this substance were to equilibrate with the 5.0 L of the blood in the patient's body, how many grams of $BaSO_4$ will dissolve in the blood? For a good estimate, we may assume that the temperature is 25°C. Why is $Ba(NO_3)_2$ not chosen for this procedure?

17.92 The pK_a of phenolphthalein is 9.10. Over what pH range does this indicator change from 95% HIn to 95% In^-?

17.93 Look up the K_{sp} values for $BaSO_4$ and $SrSO_4$ in Table 17.2. Calculate $[Ba^{2+}]$, $[Sr^{2+}]$, and $[SO_4^{2-}]$ in a solution that is saturated with both compounds.

17.94 Solid NaI is slowly added to a solution that is 0.010 *M* in Cu^+ and 0.010 *M* in Ag^+. (a) Which compound will begin to precipitate first? (b) Calculate $[Ag^+]$ when CuI just begins to precipitate. (c) What percentage of Ag^+ remains in solution at this point?

17.95 Radiochemical techniques are useful in estimating the solubility product of many compounds. In one experiment, 50.0 mL of a 0.010 *M* $AgNO_3$ solution containing a silver isotope with a radioactivity of 74,025 counts per min per mL were mixed with 100 mL of a 0.030 *M* $NaIO_3$ solution. The mixed solution was diluted to 500 mL and filtered to remove all of the $AgIO_3$ precipitate. The remaining solution was found to have a radioactivity of 44.4 counts per min per mL. What is the K_{sp} of $AgIO_3$?

17.96 The molar mass of a certain metal carbonate, MCO_3, can be determined by adding an excess of HCl acid to react with the carbonate and then "back-titrating" the remaining acid with NaOH. (a) Write an equation for these reactions. (b) In a certain experiment, 20.00 mL of 0.0800 *M* HCl were added to a 0.1022-g sample of MCO_3. The excess HCl required 5.64 mL of 0.1000 *M* NaOH for neutralization. Calculate the molar mass of the carbonate and identify M.

17.97 Acid-base reactions usually go to completion. Confirm this statement by calculating the equilibrium constant for each of the following cases: (a) a strong acid reacting with a strong base, (b) a strong acid reacting with a weak base (NH_3), (c) a weak acid (CH_3COOH) reacting with a strong base, (d) a weak acid (CH_3COOH) reacting with a weak base (NH_3). (*Hint:* Strong acids exist as H^+ ions and strong bases exist as OH^- ions in solution. You need to look up the K_a, K_b, and K_w values.)

17.98 Calculate *x*, the number of molecules of water in oxalic acid hydrate, $H_2C_2O_4 \cdot xH_2O$, from the following data: 5.00 g of the compound is made up to exactly 250 mL solution and 25.0 mL of this solution requires 15.9 mL of 0.500 *M* NaOH solution for neutralization.

17.99 Describe how you would prepare 1 L of the buffer 0.20 *M* CH_3COONa/0.20 *M* CH_3COOH by (a) mixing a solution of CH_3COOH with a solution of CH_3COONa, (b) reacting a solution of CH_3COOH with a solution of NaOH, and (c) reacting a solution of CH_3COONa with a solution of HCl.

17.100 What reagents would you employ to separate these pairs of ions in solution: (a) Na^+ and Ba^{2+}, (b) K^+ and Pb^{2+}, (c) Zn^{2+} and Hg^{2+}?

17.101 $CaSO_4$ ($K_{sp} = 2.4 \times 10^{-5}$) has a larger K_{sp} value than that of Ag_2SO_4 ($K_{sp} = 1.4 \times 10^{-5}$). Does it follow that $CaSO_4$ also has greater solubility (g/L)?

17.102 How many milliliters of 1.0 *M* NaOH must be added to 200 mL of 0.10 *M* NaH_2PO_4 to make a buffer solution with a pH of 7.50?

17.103 The maximum allowable concentration of Pb^{2+} ions in drinking water is 0.05 ppm (that is, 0.05 g of Pb^{2+} in 1 million g of water). Is this guideline exceeded if an underground water supply is at equilibrium with the mineral anglesite, $PbSO_4$ ($K_{sp} = 1.6 \times 10^{-8}$)?

17.104 Which of these solutions has the highest $[H^+]$: (a) 0.10 *M* HF, (b) 0.10 *M* HF in 0.10 *M* NaF, or (c) 0.10 *M* HF in 0.10 *M* SbF_5? (*Hint:* SbF_5 reacts with F^- to form the complex ion SbF_6^-.)

17.105 Distribution curves show how the fractions of nonionized acid and its conjugate base vary as a function of pH of the medium. Plot distribution curves for CH_3COOH and its conjugate base CH_3COO^- in solution. Your graph should show fraction as the *y* axis and pH as the *x* axis. What are the fractions and pH at the point these two curves intersect?

17.106 Water containing Ca^{2+} and Mg^{2+} ions is called *hard water* and is unsuitable for some household and industrial use because these ions react with soap to form insoluble salts, or curds. One way to remove the Ca^{2+} ions from hard water is by adding washing soda ($Na_2CO_3 \cdot 10H_2O$). (a) The molar solubility of $CaCO_3$ is 9.3×10^{-5} *M*. What is its molar solubility in a 0.050 *M* Na_2CO_3 solution? (b) Why are Mg^{2+} ions not removed by this procedure? (c) The Mg^{2+} ions are removed as $Mg(OH)_2$ by adding slaked lime $[Ca(OH)_2]$ to the water to produce a saturated solution. Calculate the pH of a saturated $Ca(OH)_2$ solution. (d) What is the concentration of Mg^{2+} ions at this pH? (e) In general, which ion (Ca^{2+} or Mg^{2+}) would you remove first? Why?

17.107 (a) Referring to Figure 17.6, describe how you would determine the pK_b of the base. (b) Derive an analogous Henderson-Hasselbalch equation relating pOH to pK_b of a weak base B and its conjugate acid HB^+. Sketch a titration curve showing the variation of the pOH of the base solution versus the volume of a strong acid added from a buret. Describe how you would determine the pK_b from this curve.

17.108 A 25.0-mL of 0.20 *M* HF solution is titrated with a 0.20 *M* NaOH solution. Calculate the volume of NaOH solution added when the pH of the solution is (a) 2.85, (b) 3.15, (c) 11.89. Ignore salt hydrolysis.

Special Problems

17.109 One of the most commonly used antibiotics is penicillin G (benzylpenicillinic acid), which has the following structure:

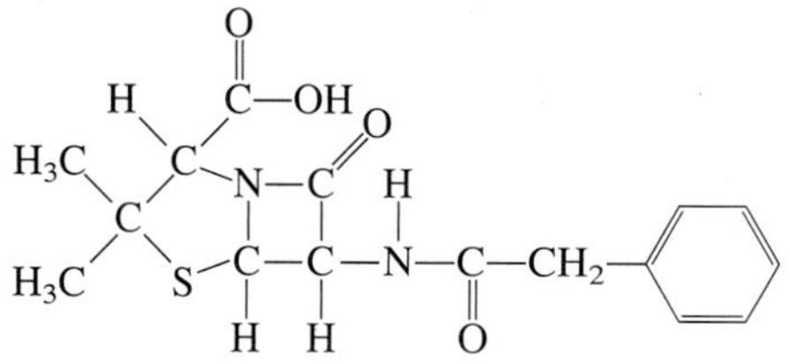

It is a weak monoprotic acid:

$$HP \rightleftharpoons H^+ + P^- \qquad K_a = 1.64 \times 10^{-3}$$

in which HP denotes the parent acid and P^- the conjugate base. Penicillin G is produced by growing molds in fermentation tanks at 25°C and a pH range of 4.5 to 5.0. The crude form of this antibiotic is obtained by extracting the fermentation broth with an organic solvent in which the acid is soluble. (a) Identify the acidic hydrogen atom. (b) In one stage of purification, the organic extract of the crude penicillin G is treated with a buffer solution at pH = 6.50. What is the ratio of the conjugate base of penicillin G to the acid at this pH? Would you expect the conjugate base to be more soluble in water than the acid? (c) Penicillin G is not suitable for oral administration, but the sodium salt (NaP) is because it is soluble. Calculate the pH of a 0.12 *M* NaP solution formed when a tablet containing the salt is dissolved in a glass of water.

17.110 Amino acids are the building blocks of proteins. These compounds contain at least one amino group and one carboxyl group. Consider glycine, whose structure is shown in Figure 11.18. Depending on the pH of the solution, glycine can exist in one of three possible forms:

Fully protonated: $\overset{+}{N}H_3—CH_2—COOH$
Dipolar ion: $\overset{+}{N}H_3—CH_2—COO^-$
Fully ionized: $NH_2—CH_2—COO^-$

Predict the predominant form of glycine at pH 1.0, 7.0, and 12.0. The pK_a of the carboxyl group is 2.3 and that of the ammonium group is 9.6. [*Hint:* Use Equation (17.3).]

17.111 One way to distinguish a buffer solution with an acid solution is by dilution. (a) Consider a buffer solution made of 0.500 *M* CH_3COOH and 0.500 *M* CH_3COONa. Calculate its pH and the pH after it has been diluted 10-fold. (b) Compare the result in (a) with the pHs of a 0.500 M CH_3COOH solution before and after it has been diluted 10-fold.

17.112 A sample of 0.96 L of HCl at 372 mmHg and 22°C is bubbled into 0.034 L of 0.57 M NH_3. What is the pH of the resulting solution? Assume the volume of solution remains constant and that the HCl is totally dissolved in the solution.

17.113 Histidine is one of the 20 amino acids found in proteins. Shown here is a fully protonated histidine molecule where the numbers denote the pK_a values of the acidic groups.

9.17 (on $H_3\overset{+}{N}$), 1.82 (on —OH)

$H_3\overset{+}{N}$—CH—C(=O)—OH

CH$_2$ (attached to CH)

6.00 (on $\overset{+}{H}N$)

imidazole ring: $\overset{+}{H}N$ … NH

(a) Show stepwise ionization of histidine in solution. (*Hint:* The H^+ ion will first come off from the strongest acid group followed by the next strongest acid group and so on.) (b) A dipolar ion is one in which the species has an equal number of positive and negative charges. Identify the dipolar ion in (a). (c) The pH at which the dipolar ion predominates is called the isoelectric point, denoted by pI. The isoelectric point is the average of the pK_a values leading to and following the formation of the dipolar ion. Calculate the pI of histidine. (d) The histidine group plays an important role in buffering blood. Which conjugate acid-base pair shown in (a) is responsible for this action?

17.114 A 1.0-L saturated silver carbonate solution at 5°C is treated with enough hydrochloric acid to decompose the compound. The carbon dioxide generated is collected in a 19-mL vial and exerts a pressure of 114 mmHg at 25°C. What is the K_{sp} of Ag_2CO_3 at 5°C?

17.115 The titration curve shown here represents the titration of a weak diprotic acid (H_2A) versus NaOH. (a) Label the major species present at the marked points. (b) Estimate the pK_{a_1} and pK_{a_2} values of the acid.

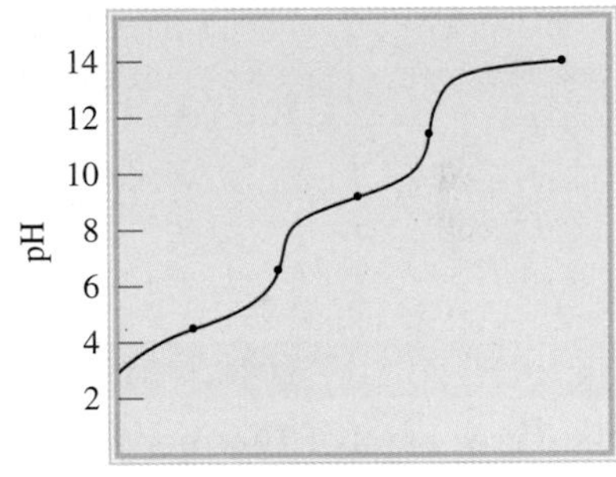

Volume of NaOH added

Answers to Practice Exercises

17.1 (a) and (c). **17.2** 9.17, 9.20. **17.3** Weigh out Na_2CO_3 and $NaHCO_3$ in a mole ratio of 0.60 to 1.0. Dissolve in enough water to make up a 1-L solution. **17.4** (a) 2.19, (b) 3.95, (c) 8.02, (d) 11.39. **17.5** (a) Bromophenol blue, methyl orange, methyl red, and chlorophenol blue; (b) all except thymol blue, bromophenol blue, and methyl orange; (c) cresol red and phenolphthalein. **17.6** 2.0×10^{-14}. **17.7** 1.9×10^{-3} g/L. **17.8** No. **17.9** (a) 1.7×10^{-4} g/L, (b) 1.4×10^{-7} g/L. **17.10** $[Cu^{2+}] = 1.2 \times 10^{-13}$ M, $[Cu(NH_3)_4^{2+}] = 0.017$ M, $[NH_3] = 0.23$ M.

Thermodynamics

The operation of steam engines and other heat engines that convert thermal energy to work is based on the laws of thermodynamics.

Chapter Outline

Student Interactive Activities

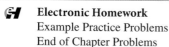

Electronic Homework
Example Practice Problems
End of Chapter Problems

Essential Concepts

Laws of Thermodynamics The laws of thermodynamics have been successfully applied to the study of chemical and physical processes. The first law of thermodynamics is based on the law of conservation of energy. The second law of thermodynamics deals with natural or spontaneous processes. The function that predicts the spontaneity of a reaction is entropy. The second law states that for a spontaneous process, the change in the entropy of the universe must be positive. The third law enables us to determine absolute entropy values.

Gibbs Free Energy Gibbs free energy helps us to determine the spontaneity of a reaction by focusing only on the system. The change in Gibbs free energy for a process is made up of two terms: a change in enthalpy and a change in entropy times temperature. At constant temperature and pressure, a decrease in Gibbs free energy signals a spontaneous reaction. The change in the standard Gibbs free energy can be related to the equilibrium constant of a reaction.

Thermodynamics in Living Systems Many reactions of biological importance are nonspontaneous. By coupling such reactions to those that have a negative Gibbs free energy change with the aid of enzymes, the net reaction can be made to proceed to yield the desired products.

18.1 The Three Laws of Thermodynamics

In Chapter 6 we encountered the first of three laws of thermodynamics, which says that energy can be converted from one form to another, but it cannot be created or destroyed. One measure of these changes is the amount of heat given off or absorbed by a system during a constant-pressure process, which chemists define as a change in enthalpy (ΔH).

The second law of thermodynamics explains why chemical processes tend to favor one direction. The third law is an extension of the second law and will be examined briefly in Section 18.4.

18.2 Spontaneous Processes

One of the main objectives in studying thermodynamics, as far as chemists are concerned, is to be able to predict whether or not a reaction will occur when reactants are brought together under a specific set of conditions (for example, at a certain temperature, pressure, and concentration). This knowledge is important whether one is synthesizing compounds in a research laboratory, manufacturing chemicals on an industrial scale, or trying to understand the intricate biological processes in a cell. A reaction that *does* occur under the given set of conditions is called a *spontaneous reaction.* If a reaction does not occur under specified conditions, it is said to be nonspontaneous. We observe spontaneous physical and chemical processes every day, including many of the following examples:

A spontaneous reaction does not necessarily mean an instantaneous reaction.

- A waterfall runs downhill, but never up, spontaneously.
- A lump of sugar spontaneously dissolves in a cup of coffee, but dissolved sugar does not spontaneously reappear in its original form.
- Water freezes spontaneously below 0°C, and ice melts spontaneously above 0°C (at 1 atm).
- Heat flows from a hotter object to a colder one, but the reverse never happens spontaneously.
- The expansion of a gas into an evacuated bulb is a spontaneous process [Figure 18.1(a)]. The reverse process, that is, the gathering of all the molecules into one bulb, is not spontaneous [Figure 18.1(b)].

A spontaneous and a nonspontaneous process.

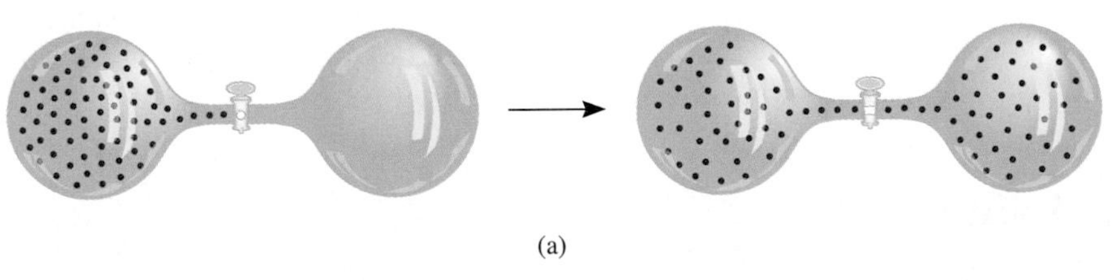

(a)

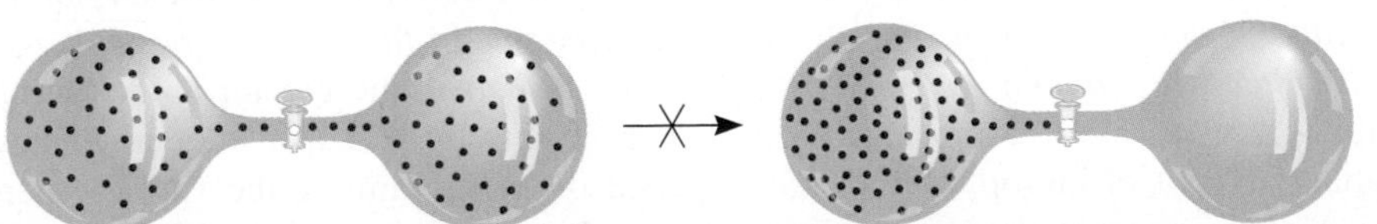

(b)

Figure 18.1
(a) A spontaneous process.
(b) A nonspontaneous process.

- A piece of sodium metal reacts violently with water to form sodium hydroxide and hydrogen gas. However, hydrogen gas does not react with sodium hydroxide to form water and sodium.
- Iron exposed to water and oxygen forms rust, but rust does not spontaneously change back to iron.

These examples show that processes that occur spontaneously in one direction cannot, under the same conditions, also take place spontaneously in the opposite direction.

If we assume that spontaneous processes occur so as to decrease the energy of a system, we can explain why a ball rolls downhill and why springs in a clock unwind. Similarly, a large number of exothermic reactions are spontaneous. An example is the combustion of methane:

Because of the activation energy barrier, an input of energy is needed to get this reaction started.

$$CH_4(g) + 2O_2(g) \longrightarrow CO_2(g) + 2H_2O(l) \quad \Delta H° = -890.4 \text{ kJ/mol}$$

Another example is the acid-base neutralization reaction:

$$H^+(aq) + OH^-(aq) \longrightarrow H_2O(l) \quad \Delta H° = -56.2 \text{ kJ/mol}$$

But consider a solid-to-liquid phase transition such as

$$H_2O(s) \longrightarrow H_2O(l) \quad \Delta H° = 6.01 \text{ kJ/mol}$$

In this case, the assumption that spontaneous processes always decrease a system's energy fails. Experience tells us that ice melts spontaneously above 0°C even though the process is endothermic. Another example that contradicts our assumption is the dissolution of ammonium nitrate in water:

$$NH_4NO_3(s) \xrightarrow{H_2O} NH_4^+(aq) + NO_3^-(aq) \quad \Delta H° = 25 \text{ kJ/mol}$$

When heated, HgO decomposes to give Hg and O_2.

This process is spontaneous, and yet it is also endothermic. The decomposition of mercury(II) oxide is an endothermic reaction that is nonspontaneous at room temperature, but it becomes spontaneous when the temperature is raised:

$$2HgO(s) \longrightarrow 2Hg(l) + O_2(g) \quad \Delta H° = 90.7 \text{ kJ/mol}$$

From a study of the examples mentioned and many more cases, we come to the following conclusion: Exothermicity favors the spontaneity of a reaction but does not guarantee it. Just as it is possible for an endothermic reaction to be spontaneous, it is possible for an exothermic reaction to be nonspontaneous. In other words, we cannot decide whether or not a chemical reaction will occur spontaneously solely on the basis of energy changes in the system. To make this kind of prediction we need another thermodynamic quantity, which turns out to be *entropy.*

18.3 Entropy

In order to predict the spontaneity of a process, we need to introduce a new thermodynamic quantity called entropy. ***Entropy (S)*** is often described as *a measure of how spread out or dispersed the energy of a system is among the different possible ways that system can contain energy.* The greater the dispersal, the greater is the entropy. Most processes are accompanied by a change in entropy. A cup of hot water has a certain amount of entropy due to the dispersal of energy among the various energy states of the water molecules (for example, energy states associated with the translational, rotational, and vibrational motions of the water molecules). If left standing on

Translational motion is motion through space of the whole molecule.

a table, the water loses heat to the cooler surroundings. Consequently, there is an increase in entropy because of the dispersal of energy over a great many energy states of the air molecules.

As another example, consider the situation depicted in Figure 18.1. Before the valve is opened, the system possesses a certain amount of entropy. Upon opening the valve, the gas molecules now have access to the combined volume of both bulbs. A larger volume for movement results in a narrowing of the gap between translational energy levels of the molecules. Consequently, the entropy of the system increases because closely spaced energy levels leads to a greater dispersal among the energy levels.

Microstates and Entropy

Before we introduce the second law of thermodynamics, which relates entropy change (increase) to spontaneous processes, it is useful to first provide a proper definition of entropy. To do so let us consider a simple system of four molecules distributed between two equal compartments, as shown in Figure 18.2. There is only one way to arrange all the molecules in the left compartment, four ways to have three molecules in the left compartment and one in the right compartment, and six ways to have two molecules in each of the two compartments. The eleven possible ways of distributing the molecules are called microscopic states or microstates and each set of similar microstates is called a distribution.† As you can see, distribution III is the most probable because there are six microstates or six ways to achieve it and distribution I is the least probable because it has one microstate and therefore there is only one way to achieve it. Based on this analysis, we conclude that the probability of occurrence of a particular distribution (state) depends on the number of ways (microstates) in which

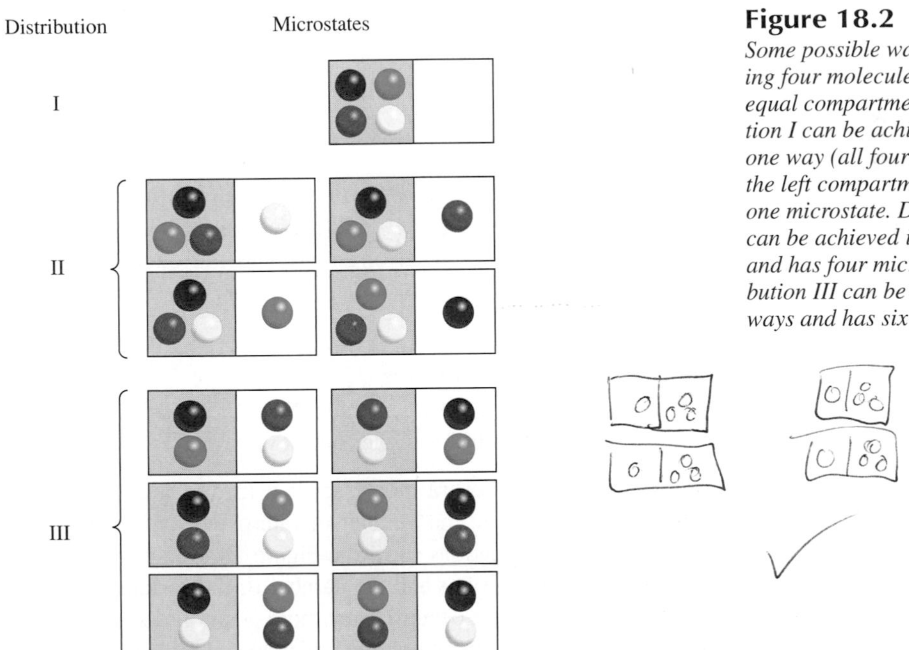

Figure 18.2
Some possible ways of distributing four molecules between two equal compartments. Distribution I can be achieved in only one way (all four molecules in the left compartment) and has one microstate. Distribution II can be achieved in four ways and has four microstates. Distribution III can be achieved in six ways and has six microstates.

†Actually there are still other possible ways to distribute the four molecules between the two compartments. We can have all four molecules in the right compartment (one way) and three molecules in the right compartment and one molecule in the left compartment (four ways). However, the distributions shown in Figure 18.2 are sufficient for our discussion.

the distribution can be achieved. As the number of molecules approaches macroscopic scale, it is not difficult to see that they will be evenly distributed between the two compartments because this distribution has many, many more microstates than all other distributions.

Engraved on Ludwig Boltzmann's tombstone in Vienna is his famous equation. The "log" stands for "$\log_e$," which is the natural logarithm or ln.

In 1868 Boltzmann showed that the entropy of a system is related to the natural log of the number of microstates (W):

$$S = k \ln W \tag{18.1}$$

where k is called the Boltzmann constant (1.38×10^{-23} J/K). Thus, the larger the W, the greater is the entropy of the system. Like enthalpy, entropy is a state function (see Section 6.3). Consider a certain process in a system. The entropy change for the process, ΔS, is

$$\Delta S = S_f - S_i \tag{18.2}$$

where S_i and S_f are the entropies of the system in the initial and final states, respectively. From Equation (18.1) we can write

$$\begin{aligned}\Delta S &= k \ln W_f - k \ln W_i \\ &= k \ln \frac{W_f}{W_i}\end{aligned} \tag{18.3}$$

where W_i and W_f are the corresponding numbers of microstates in the initial and final state. Thus, if $W_f > W_i$, $\Delta S > 0$ and the entropy of the system increases.

REVIEW OF CONCEPTS

Referring to the footnote on p. 631, draw the missing distributions in Figure 18.2.

Changes in Entropy

Earlier we described the increase in entropy of a system as a result of the increase in the dispersal of energy. There is a connection between the qualitative description of entropy in terms of dispersal of energy and the quantitative definition of entropy in terms of microstates given by Equation (18.1). We conclude that

- A system with fewer microstates (smaller W) among which to spread its energy (small dispersal) has a lower entropy.
- A system with more microstates (larger W) among which to spread its energy (large dispersal) has a higher entropy.

Next, we will study several processes that lead to a change in entropy of a system in terms of the change in the number of microstates of the system.

Consider the situations shown in Figure 18.3. In a solid the atoms or molecules are confined to fixed positions and the number of microstates is small. Upon melting, these atoms or molecules can occupy many more positions as they move away from the lattice points. Consequently, the number of microstates increases because there are now many more ways to arrange the particles. Therefore, we predict this "order $\longrightarrow$ disorder" phase transition to result in an increase in entropy because the number of microstates has increased. Similarly, we predict the vaporization process will also lead to an increase in the entropy of the system. The increase will be considerably greater than that for melting, however, because molecules in the gas phase occupy much more space, and

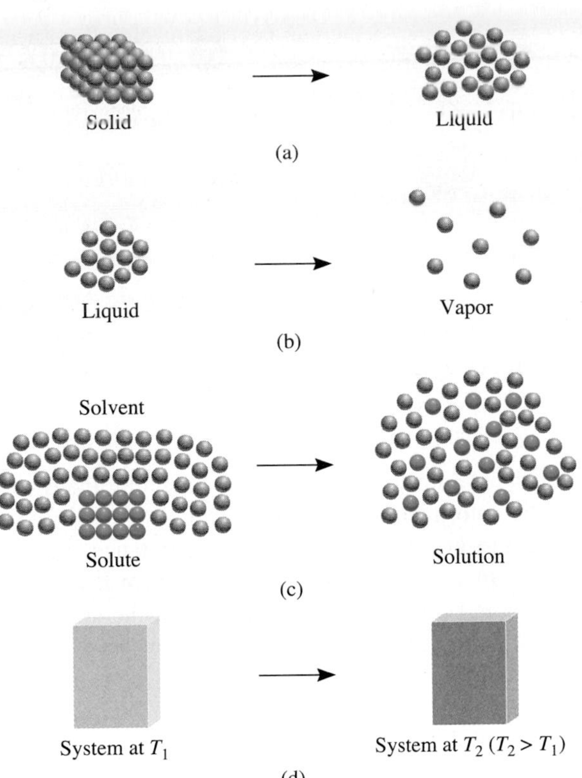

Figure 18.3
Processes that lead to an increase in entropy of the system: (a) melting: $S_{liquid} > S_{solid}$; (b) vaporization: $S_{vapor} > S_{liquid}$; (c) dissolving: $S_{soln} > S_{solute} + S_{solvent}$; (d) heating: $S_{T_2} > S_{T_1}$.

therefore there are far more microstates than in the liquid phase. The solution process usually leads to an increase in entropy. When a sugar crystal dissolves in water, the highly ordered structure of the solid and part of the ordered structure of water break down. Thus, the solution has a greater number of microstates than the pure solute and pure solvent combined. When an ionic solid such as NaCl dissolves in water, there are two contributions to entropy increase: the solution process (mixing of solute with solvent) and the dissociation of the compound into ions:

$$\text{NaCl}(s) \xrightarrow{\text{H}_2\text{O}} \text{Na}^+(aq) + \text{Cl}^-(aq)$$

More particles lead to a greater number of microstates. However, we must also consider hydration, which causes water molecules to become more ordered around the ions. This process decreases entropy because it reduces the number of microstates of the solvent molecules. For small, highly charged ions such as Al^{3+} and Fe^{3+}, the decrease in entropy due to hydration can outweigh the increase in entropy due to mixing and dissociation so that the entropy change for the overall process can actually be negative. Heating also increases the entropy of a system. In addition to translational motion, molecules can also execute rotational motions and vibrational motions (Figure 18.4). As the temperature is increased, the energies associated with all types of molecular motion increase. This increase in energy is distributed or dispersed among the quantized energy levels. Consequently, more microstates become available at a higher temperature; therefore, the entropy of a system always increases with increasing temperature.

Standard Entropy

Equation (18.1) provides a useful molecular interpretation of entropy, but is normally not used to calculate the entropy of a system because it is difficult to determine the

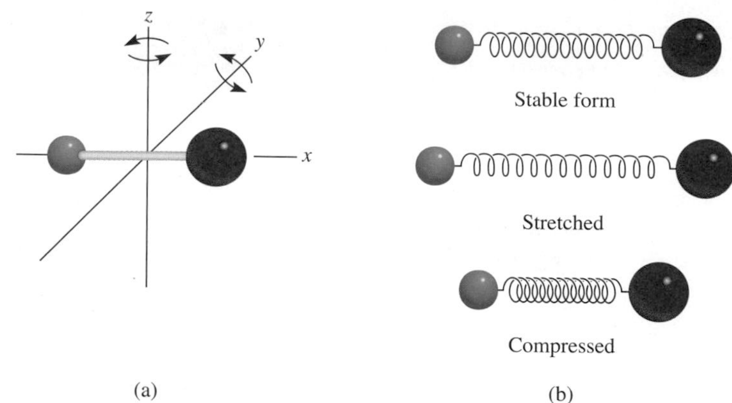

Figure 18.4
(a) A diatomic molecule can rotate about the y- and z-axes (the x-axis is along the bond). (b) Vibrational motion of a diatomic molecule. Chemical bonds can be stretched and compressed like a spring.

Table 18.1

Standard Entropy Values ($S°$) for Some Substances at 25°C

Substance	$S°$ (J/K · mol)
$H_2O(l)$	69.9
$H_2O(g)$	188.7
$Br_2(l)$	152.3
$Br_2(g)$	245.3
$I_2(s)$	116.7
$I_2(g)$	260.6
C (diamond)	2.4
C (graphite)	5.69
CH_4 (methane)	186.2
C_2H_6 (ethane)	229.5
He(g)	126.1
Ne(g)	146.2

number of microstates for a macroscopic system containing many molecules. Instead, entropy is obtained by calorimetric methods. In fact, as we will see shortly, it is possible to determine the absolute value of entropy of a substance, called absolute entropy, something we cannot do for energy or enthalpy. *Standard entropy* is the absolute entropy of a substance at 1 atm and 25°C. (Recall that the standard state refers only to 1 atm. The reason for specifying 25°C is that many processes are carried out at room temperature.) Table 18.1 lists standard entropies of a few elements and compounds; Appendix 2 provides a more extensive listing. The units of entropy are J/K or J/K · mol for 1 mole of the substance. We use joules rather than kilojoules because entropy values are typically quite small. Entropies of elements and compounds are all positive (that is, $S° > 0$). By contrast, the standard enthalpy of formation ($\Delta H_f°$) for elements in their stable form is arbitrarily set equal to zero, and for compounds it may be positive or negative.

Referring to Table 18.1, we see that the standard entropy of water vapor is greater than that of liquid water. Similarly, bromine vapor has a higher standard entropy than liquid bromine, and iodine vapor has a greater standard entropy than solid iodine. For different substances in the same phase, molecular complexity determines which ones have higher entropies. Both diamond and graphite are solids, but diamond has a more ordered structure and hence a smaller number of microstates (see Figure 12.24). Therefore, diamond has a smaller standard entropy than graphite. Consider the natural gases methane and ethane. Ethane has a more complex structure and hence more ways to execute molecular motions, which also increase its microstates. Therefore, ethane has a greater standard entropy than methane. Both helium and neon are monatomic gases, which cannot execute rotational or vibrational motions, but neon has a greater standard entropy than helium because its molar mass is greater. Heavier atoms have more closely spaced energy levels so there is a greater distribution of the atoms' energy among the energy levels. Consequently, there are more microstates associated with these atoms.

EXAMPLE 18.1

Predict whether the entropy change is greater or less than zero for each of the following processes: (a) freezing ethanol, (b) evaporating a beaker of liquid bromine at room temperature, (c) dissolving glucose in water, (d) cooling nitrogen gas from 80°C to 20°C.

(Continued)

Strategy To determine the entropy change in each case, we examine whether the number of microstates of the system increases or decreases. The sign of ΔS will be positive if there is an increase in the number of microstates and negative if the number of microstates decreases.

Solution

(a) Upon freezing, the ethanol molecules are held rigidly in position. This phase transition reduces the number of microstates and therefore the entropy decreases; that is, $\Delta S < 0$.

(b) Evaporating bromine increases the number of microstates because the Br_2 molecules can occupy many more positions in nearly empty space. Therefore $\Delta S > 0$.

(c) Glucose is a nonelectrolyte. The solution process leads to a greater dispersal of matter due to the mixing of glucose and water molecules so we expect $\Delta S > 0$.

(d) The cooling process decreases various molecular motions. This leads to a decrease in microstates and so $\Delta S < 0$.

Bromine is a fuming liquid at room temperature.

Similar problem: 18.5.

Practice Exercise How does the entropy of a system change for each of the following processes? (a) condensing water vapor, (b) forming sucrose crystals from a supersaturated solution, (c) heating hydrogen gas from 60°C to 80°C, and (d) subliming dry ice.

REVIEW OF CONCEPTS

For which of the following physical changes is ΔS positive? (a) condensing ether vapor, (b) melting iron, (c) subliming solid iodine, (d) freezing benzene.

18.4 The Second Law of Thermodynamics

The connection between entropy and the spontaneity of a reaction is expressed by the ***second law of thermodynamics:*** *The entropy of the universe increases in a spontaneous process and remains unchanged in an equilibrium process.* Because the universe is made up of the system and the surroundings, the entropy change in the universe (ΔS_{univ}) for any process is the *sum* of the entropy changes in the system (ΔS_{sys}) and in the surroundings (ΔS_{surr}). Mathematically, we can express the second law of thermodynamics as follows:

Just talking about entropy increases its value in the universe.

For a spontaneous process: $$\Delta S_{univ} = \Delta S_{sys} + \Delta S_{surr} > 0 \quad (18.4)$$

For an equilibrium process: $$\Delta S_{univ} = \Delta S_{sys} + \Delta S_{surr} = 0 \quad (18.5)$$

For a spontaneous process, the second law says that ΔS_{univ} must be greater than zero, but it does not place a restriction on either ΔS_{sys} or ΔS_{surr}. Thus, it is possible for either ΔS_{sys} or ΔS_{surr} to be negative, as long as the sum of these two quantities is greater than zero. For an equilibrium process, ΔS_{univ} is zero. In this case, ΔS_{sys} and ΔS_{surr} must be equal in magnitude, but opposite in sign. What if for some hypothetical process we find that ΔS_{univ} is negative? What this means is that the process is not spontaneous in the direction described. Rather, it is spontaneous in the *opposite* direction.

Entropy Changes in the System

To calculate ΔS_{univ}, we need to know both ΔS_{sys} and ΔS_{surr}. Let us focus first on ΔS_{sys}. Suppose that the system is represented by the following reaction:

$$aA + bB \longrightarrow cC + dD$$

As is the case for the enthalpy of a reaction [see Equation (6.17)], the ***standard entropy of reaction*** ΔS°_{rxn} is given by *the difference in standard entropies between products and reactants:*

$$\Delta S^\circ_{rxn} = [cS^\circ(C) + dS^\circ(D)] - [aS^\circ(A) + bS^\circ(B)] \quad (18.6)$$

or, in general, using Σ to represent summation and m and n for the stoichiometric coefficients in the reaction,

$$\Delta S^\circ_{rxn} = \Sigma nS^\circ(\text{products}) - \Sigma mS^\circ(\text{reactants}) \quad (18.7)$$

The standard entropy values of a large number of compounds have been measured in J/K · mol. To calculate ΔS°_{rxn} (which is ΔS_{sys}), we look up their values in Appendix 2 and proceed according to Example 18.2.

EXAMPLE 18.2

From the standard entropy values in Appendix 2, calculate the standard entropy changes for the following reactions at 25°C.

(a) $CaCO_3(s) \longrightarrow CaO(s) + CO_2(g)$

(b) $N_2(g) + 3H_2(g) \longrightarrow 2NH_3(g)$

(c) $H_2(g) + Cl_2(g) \longrightarrow 2HCl(g)$

Strategy To calculate the standard entropy of a reaction, we look up the standard entropies of reactants and products in Appendix 2 and apply Equation (18.7). As in the calculation of enthalpy of reaction [see Equation (6.18)], the stoichiometric coefficients have no units, so ΔS°_{rxn} is expressed in units of J/K · mol.

Solution

(a)
$$\begin{aligned}\Delta S^\circ_{rxn} &= [S^\circ(CaO) + S^\circ(CO_2)] - [S^\circ(CaCO_3)] \\ &= [(39.8\ \text{J/K}\cdot\text{mol}) + (213.6\ \text{J/K}\cdot\text{mol})] - (92.9\ \text{J/K}\cdot\text{mol}) \\ &= 160.5\ \text{J/K}\cdot\text{mol}\end{aligned}$$

Thus, when 1 mole of $CaCO_3$ decomposes to form 1 mole of CaO and 1 mole of gaseous CO_2, there is an increase in entropy equal to 160.5 J/K · mol.

(b)
$$\begin{aligned}\Delta S^\circ_{rxn} &= [2S^\circ(NH_3)] - [S^\circ(N_2) + 3S^\circ(H_2)] \\ &= (2)(193\ \text{J/K}\cdot\text{mol}) - [(192\ \text{J/K}\cdot\text{mol}) + (3)(131\ \text{J/K}\cdot\text{mol})] \\ &= -199\ \text{J/K}\cdot\text{mol}\end{aligned}$$

This result shows that when 1 mole of gaseous nitrogen reacts with 3 moles of gaseous hydrogen to form 2 moles of gaseous ammonia, there is a decrease in entropy equal to −199 J/K · mol.

(c)
$$\begin{aligned}\Delta S^\circ_{rxn} &= [2S^\circ(HCl)] - [S^\circ(H_2) + S^\circ(Cl_2)] \\ &= (2)(187\ \text{J/K}\cdot\text{mol}) - [(131\ \text{J/K}\cdot\text{mol}) + (223\ \text{J/K}\cdot\text{mol})] \\ &= 20\ \text{J/K}\cdot\text{mol}\end{aligned}$$

Thus, the formation of 2 moles of gaseous HCl from 1 mole of gaseous H_2 and 1 mole of gaseous Cl_2 results in a small increase in entropy equal to 20 J/K · mol.

(Continued)

Comment The ΔS°_{rxn} values all apply to the system.

Similar problems: 18.11 and 18.12.

Practice Exercise Calculate the standard entropy change for the following reactions at 25°C:

(a) $2CO(g) + O_2(g) \longrightarrow 2CO_2(g)$
(b) $3O_2(g) \longrightarrow 2O_3(g)$
(c) $2NaHCO_3(s) \longrightarrow Na_2CO_3(s) + H_2O(l) + CO_2(g)$

The results of Example 18.2 are consistent with those observed for many other reactions. Taken together, they support the following general rules:

We omit the subscript rxn for simplicity.

- If a reaction produces more gas molecules than it consumes [Example 18.2(a)], ΔS° is positive.
- If the total number of gas molecules diminishes [Example 18.2(b)], ΔS° is negative.
- If there is no net change in the total number of gas molecules [Example 18.2(c)], then ΔS° may be positive or negative, but will be relatively small numerically.

These conclusions make sense, given that gases invariably have greater entropy than liquids and solids. For reactions involving only liquids and solids, predicting the sign of ΔS° is more difficult, but in many such cases an increase in the total number of molecules and/or ions is accompanied by an increase in entropy.

EXAMPLE 18.3

Predict whether the entropy change of the system in each of the following reactions is positive or negative.

(a) $2H_2(g) + O_2(g) \longrightarrow 2H_2O(l)$
(b) $NH_4Cl(s) \longrightarrow NH_3(g) + HCl(g)$
(c) $H_2(g) + Br_2(g) \longrightarrow 2HBr(g)$

Strategy We are asked to predict, not calculate, the sign of entropy change in the reactions. The factors that lead to an increase in entropy are (1) a transition from a condensed phase to the vapor phase and (2) a reaction that produces more product molecules than reactant molecules in the same phase. It is also important to compare the relative complexity of the product and reactant molecules. In general, the more complex the molecular structure, the greater the entropy of the compound.

Solution

(a) Two reactant molecules combine to form one product molecule. Even though H_2O is a more complex molecule than either H_2 and O_2, the fact that there is a net decrease of one molecule and gases are converted to liquid ensures that the number of microstates will be diminished and hence ΔS° is negative.

(b) A solid is converted to two gaseous products. Therefore, ΔS° is positive.

(c) The same number of molecules is involved in the reactants as in the product. Furthermore, all molecules are diatomic and therefore of similar complexity. As a result, we cannot predict the sign of ΔS°, but we know that the change must be quite small in magnitude.

Similar problems: 18.13 and 18.14.

Practice Exercise Discuss qualitatively the sign of the entropy change expected for each of the following processes:

(a) $I_2(s) \longrightarrow 2I(g)$
(b) $2Zn(s) + O_2(g) \longrightarrow 2ZnO(s)$
(c) $N_2(g) + O_2(g) \longrightarrow 2NO(g)$

Review of Concepts

Consider the gas-phase reaction of A_2 (blue) and B_2 (orange) to form AB_3.

(a) Write a balanced equation for the reaction.

(b) What is the sign of ΔS for the reaction?

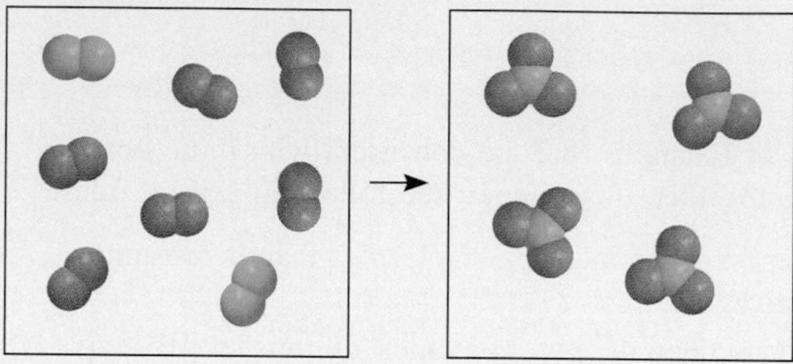

Entropy Changes in the Surroundings

Next we see how ΔS_{surr} is calculated. When an exothermic process takes place in the system, the heat transferred to the surroundings enhances motion of the molecules in the surroundings. Consequently, there is an increase in the number of microstates and the entropy of the surroundings increases. Conversely, an endothermic process in the system absorbs heat from the surroundings and so decreases the entropy of the surroundings because molecular motion decreases (Figure 18.5). For constant-pressure processes, the heat change is equal to the enthalpy change of the system, ΔH_{sys}. Therefore, the change in entropy of the surroundings, ΔS_{surr}, is proportional to ΔH_{sys}:

$$\Delta S_{surr} \propto -\Delta H_{sys}$$

The minus sign is used because if the process is exothermic, ΔH_{sys} is negative and ΔS_{surr} is a positive quantity, indicating an increase in entropy. On the other hand, for an endothermic process, ΔH_{sys} is positive and the negative sign ensures that the entropy of the surroundings decreases.

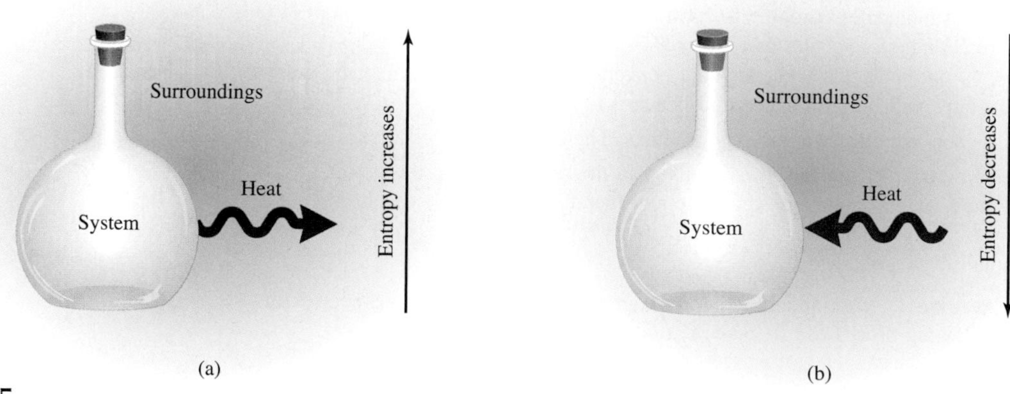

Figure 18.5

(a) An exothermic process transfers heat from the system to the surroundings and results in an increase in the entropy of the surroundings. (b) An endothermic process absorbs heat from the surroundings and thereby decreases the entropy of the surroundings.

The change in entropy for a given amount of heat absorbed also depends on the temperature. If the temperature of the surroundings is high, the molecules are already quite energetic. Therefore, the absorption of heat from an exothermic process in the system will have relatively little impact on molecular motion and the resulting increase in entropy of the surroundings will be small. However, if the temperature of the surroundings is low, then the addition of the same amount of heat will cause a more drastic increase in molecular motion and hence a larger increase in entropy. By analogy, someone coughing in a crowded restaurant will not disturb too many people, but someone coughing in a library definitely will. From the inverse relationship between ΔS_{surr} and temperature T (in kelvins)—that is, the higher the temperature, the smaller the ΔS_{surr} and vice versa—we can rewrite the preceding relationship as

$$\Delta S_{surr} = \frac{-\Delta H_{sys}}{T} \qquad (18.8)$$

This equation, which can be derived from the laws of thermodynamics, assumes that both the system and the surroundings are at temperature *T*.

Let us now apply the procedure for calculating ΔS_{sys} and ΔS_{surr} to the synthesis of ammonia and ask whether the reaction is spontaneous at 25°C:

$$N_2(g) + 3H_2(g) \longrightarrow 2NH_3(g) \qquad \Delta H^\circ_{rxn} = -92.6 \text{ kJ/mol}$$

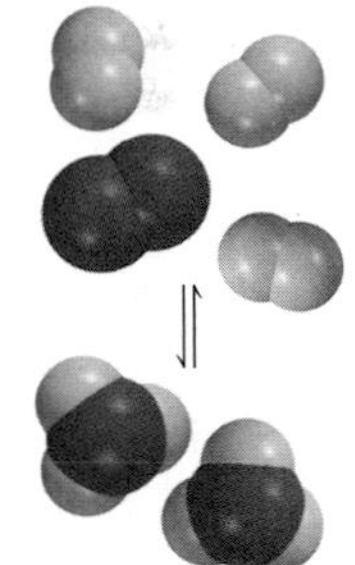

The synthesis of NH_3 from H_2 and N_2.

From Example 18.2(b) we have $\Delta S_{sys} = -199$ J/K · mol, and substituting ΔH_{sys} (−92.6 kJ/mol) in Equation (18.8), we obtain

$$\Delta S_{surr} = \frac{-(-92.6 \times 1000)\text{ J/mol}}{298\text{ K}} = 311 \text{ J/K} \cdot \text{mol}$$

The change in entropy of the universe is

$$\begin{aligned}\Delta S_{univ} &= \Delta S_{sys} + \Delta S_{surr}\\ &= -199 \text{ J/K} \cdot \text{mol} + 311 \text{ J/K} \cdot \text{mol}\\ &= 112 \text{ J/K} \cdot \text{mol}\end{aligned}$$

Because ΔS_{univ} is positive, we predict that the reaction is spontaneous at 25°C. It is important to keep in mind that just because a reaction is spontaneous does not mean that it will occur at an observable rate. The synthesis of ammonia is, in fact, extremely slow at room temperature. Thermodynamics can tell us whether a reaction will occur spontaneously under specific conditions, but it does not say how fast it will occur. Reaction rates are the subject of chemical kinetics (see Chapter 14).

The Third Law of Thermodynamics and Absolute Entropy

Finally, it is appropriate to consider the *third law of thermodynamics* briefly in connection with the determination of entropy values. So far we have related entropy to microstates—the greater the number of microstates a system possesses, the larger is the entropy of the system. Consider a perfect crystalline substance at absolute zero (0 K). Under these conditions, molecular motions are kept at a minimum and the number of microstates (W) is one (there is only one way to arrange the atoms or molecules to form a perfect crystal). From Equation (18.1) we write

$$\begin{aligned}S &= k \ln W\\ &= k \ln 1 = 0\end{aligned}$$

According to the ***third law of thermodynamics,*** *the entropy of a perfect crystalline substance is zero at the absolute zero of temperature.* As the temperature increases, the freedom of motion increases and hence also the number of microstates. Thus, the entropy of any substance at a temperature above 0 K is greater than zero. Note also that if the crystal is impure or if it has defects, then its entropy is greater than zero even at 0 K because it would not be perfectly ordered and the number of microstates would be greater than one.

The important point about the third law of thermodynamics is that it enables us to determine the *absolute* entropies of substances. Starting with the knowledge that the entropy of a pure crystalline substance is zero at absolute zero, we can measure the increase in entropy of the substance when it is heated from 0 K to, say, 298 K. The change in entropy, ΔS, is given by

$$\begin{aligned}\Delta S &= S_f - S_i \\ &= S_f\end{aligned}$$

The entropy increase can be calculated from the temperature change and heat capacity of the substance, plus any phase changes.

because S_i is zero. The entropy of the substance at 298 K, then, is given by ΔS or S_f, which is called the absolute entropy because this is the *true* value and not a value derived using some arbitrary reference as in the case of standard enthalpy of formation. Thus, the entropy values quoted so far and those listed in Appendix 2 are all absolute entropies. Because measurements are carried out at 1 atm, we usually refer to absolute entropies as standard entropies. In contrast, we cannot have the absolute energy or enthalpy of a substance because the zero of energy or enthalpy is undefined. Figure 18.6 shows the change (increase) in entropy of a substance with temperature. At absolute zero, it has a zero entropy value (assuming that it is a perfect crystalline substance). As it is heated, its entropy increases gradually because of greater molecular motion. At the melting point, there is a sizable increase in entropy as the liquid state is formed. Further heating increases the entropy of the liquid again due to

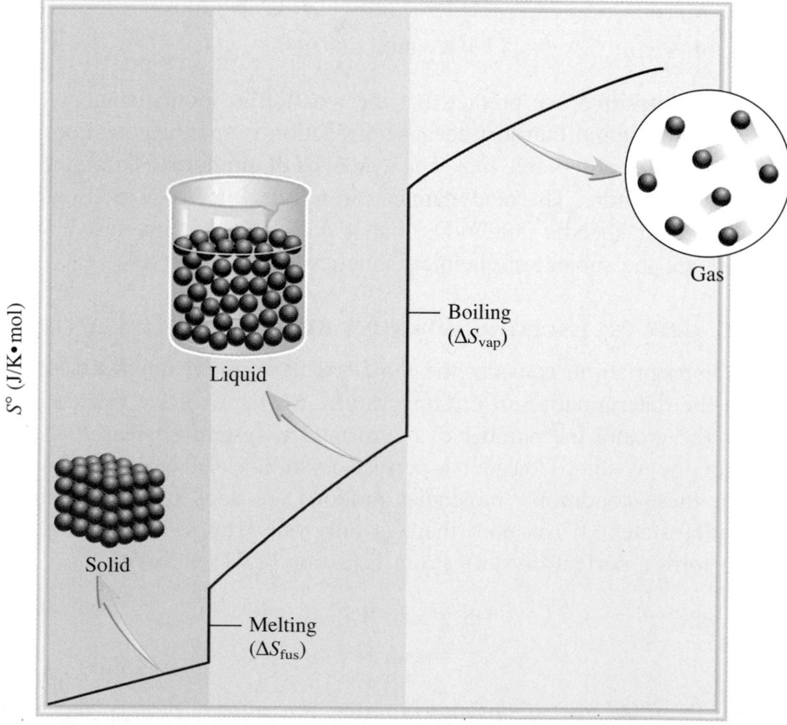

Figure 18.6
Entropy increase of a substance as the temperature rises from absolute zero.

enhanced molecular motion. At the boiling point there is a large increase in entropy as a result of the liquid-to-vapor transition. Beyond that temperature, the entropy of the gas continues to rise with increasing temperature.

18.5 Gibbs Free Energy

The second law of thermodynamics tells us that a spontaneous reaction increases the entropy of the universe; that is, $\Delta S_{univ} > 0$. In order to determine the sign of ΔS_{univ} for a reaction, however, we would need to calculate both ΔS_{sys} and ΔS_{surr}. In general, we are usually concerned only with what happens in a particular system. Therefore, we need another thermodynamic function to help us determine whether a reaction will occur spontaneously if we consider only the system itself.

From Equation (18.4), we know that for a spontaneous process, we have

$$\Delta S_{univ} = \Delta S_{sys} + \Delta S_{surr} > 0$$

Substituting $-\Delta H_{sys}/T$ for ΔS_{surr}, we write

$$\Delta S_{univ} = \Delta S_{sys} - \frac{\Delta H_{sys}}{T} > 0$$

Multiplying both sides of the equation by T gives

$$T\Delta S_{univ} = -\Delta H_{sys} + T\Delta S_{sys} > 0$$

Now we have a criterion for a spontaneous reaction that is expressed only in terms of the properties of the system (ΔH_{sys} and ΔS_{sys}) and we can ignore the surroundings. For convenience, we can change the preceding equation by multiplying it throughout by -1 and replacing the $>$ sign with $<$:

$$-T\Delta S_{univ} = \Delta H_{sys} - T\Delta S_{sys} < 0$$

The change in unequal sign when we multiply the equation by -1 follows from the fact that $1 > 0$ and $-1 < 0$.

This equation says that for a process carried out at constant pressure and temperature T, if the changes in enthalpy and entropy of the system are such that $\Delta H_{sys} - T\Delta S_{sys}$ is less than zero, the process must be spontaneous.

In order to express the spontaneity of a reaction more directly, we introduce another thermodynamic function called ***Gibbs free energy (G),*** or simply ***free energy*** (after the American physicist Josiah Willard Gibbs):

$$G = H - TS \qquad (18.9)$$

A commemorative stamp honoring Gibbs.

All quantities in Equation (18.9) pertain to the system, and T is the temperature of the system. You can see that G has units of energy (both H and TS are in energy units). Like H and S, G is a state function.

The change in free energy (ΔG) of a system for a *constant-temperature* process is

$$\Delta G = \Delta H - T\Delta S \qquad (18.10)$$

We omit the subscript sys for simplicity.

In this context free energy is *the energy available to do work.* Thus, if a particular reaction is accompanied by a release of usable energy (that is, if ΔG is negative), this fact alone guarantees that it is spontaneous, and there is no need to worry about what happens to the rest of the universe.

The word "free" in the term "free energy" does not mean without cost.

Note that we have merely organized the expression for the entropy change of the universe and equating the free-energy change of the system (ΔG) with $-T\Delta S_{univ}$, so that we can focus on changes in the system. We can now summarize the conditions

Table 18.2

Conventions for Standard States

State of Matter	Standard State
Gas	1 atm pressure
Liquid	Pure liquid
Solid	Pure solid
Elements*	$\Delta G_f^\circ = 0$
Solution	1 molar concentration

*The most stable allotropic form at 25°C and 1 atm.

for spontaneity and equilibrium at constant temperature and pressure in terms of ΔG as follows:

$\Delta G < 0$ The reaction is spontaneous in the forward direction.
$\Delta G > 0$ The reaction is nonspontaneous. The reaction is spontaneous in the opposite direction.
$\Delta G = 0$ The system is at equilibrium. There is no net change.

Standard Free-Energy Changes

The ***standard free-energy of reaction*** **(ΔG°_{rxn})** is the *free-energy change for a reaction when it occurs under standard-state conditions, when reactants in their standard states are converted to products in their standard states.* Table 18.2 summarizes the conventions used by chemists to define the standard states of pure substances as well as solutions. To calculate (ΔG°_{rxn}) we start with the equation

$$aA + bB \longrightarrow cC + dD$$

The standard free-energy change for this reaction is given by

$$\Delta G^\circ_{rxn} = [c\Delta G_f^\circ(C) + d\Delta G_f^\circ(D)] - [a\Delta G_f^\circ(A) + b\Delta G_f^\circ(B)] \quad (18.11)$$

or, in general,

$$\Delta G^\circ_{rxn} = \Sigma n\Delta G_f^\circ(\text{products}) - \Sigma m\Delta G_f^\circ(\text{reactants}) \quad (18.12)$$

where m and n are stoichiometric coefficients. The term ΔG_f° is the ***standard free energy of formation*** of a compound, that is, *the free-energy change that occurs when 1 mole of the compound is synthesized from its elements in their standard states.* For the combustion of graphite:

$$C(\text{graphite}) + O_2(g) \longrightarrow CO_2(g)$$

the standard free-energy change [from Equation (18.12)] is

$$\Delta G^\circ_{rxn} = \Delta G_f^\circ(CO_2) - [\Delta G_f^\circ(C, \text{graphite}) + \Delta G_f^\circ(O_2)]$$

As in the case of the standard enthalpy of formation (p. 197), we define the standard free energy of formation of any element in its stable allotropic form at 1 atm and 25°C as zero. Thus,

$$\Delta G_f^\circ(C, \text{graphite}) = 0 \qquad \text{and} \qquad \Delta G_f^\circ(O_2) = 0$$

Therefore, the standard free-energy change for the reaction in this case is equal to the standard free energy of formation of CO_2:

$$\Delta G^\circ_{rxn} = \Delta G_f^\circ(CO_2)$$

Appendix 2 lists the values of ΔG_f° for a number of compounds.

EXAMPLE 18.4

Calculate the standard free-energy changes for the following reactions at 25°C.

(a) $CH_4(g) + 2O_2(g) \longrightarrow CO_2(g) + 2H_2O(l)$

(b) $2MgO(s) \longrightarrow 2Mg(s) + O_2(g)$

Strategy To calculate the standard free-energy change of a reaction, we look up the standard free energies of formation of reactants and products in Appendix 2 and apply

(Continued)

Equation (18.12). Note that all the stoichiometric coefficients have no units so ΔG°_{rxn} is expressed in units of kJ/mol, and ΔG°_f for O_2 is zero because it is the stable allotropic element at 1 atm and 25°C.

Solution

(a) According to Equation (18.12), we write

$$\Delta G^\circ_{rxn} = [\Delta G^\circ_f(CO_2) + 2\Delta G^\circ_f(H_2O)] - [\Delta G^\circ_f(CH_4) + 2\Delta G^\circ_f(O_2)]$$

We insert the appropriate values from Appendix 2:

$$\Delta G^\circ_{rxn} = [(-394.4\text{ kJ/mol}) + (2)(-237.2\text{ kJ/mol})] - [(-50.8\text{ kJ/mol}) + (2)(0\text{ kJ/mol})]$$

$$= -818.0\text{ kJ/mol}$$

(b) The equation is

$$\Delta G^\circ_{rxn} = [2\Delta G^\circ_f(Mg) + \Delta G^\circ_f(O_2)] - [2\Delta G^\circ_f(MgO)]$$

From data in Appendix 2 we write

$$\Delta G^\circ_{rxn} = [(2)(0\text{ kJ/mol}) + (0\text{ kJ/mol})] - [(2)(-569.6\text{ kJ/mol})]$$

$$= 1139\text{ kJ/mol}$$

Similar problems: 18.17 and 18.18.

Practice Exercise Calculate the standard free-energy changes for the following reactions at 25°C:

(a) $H_2(g) + Br_2(l) \longrightarrow 2HBr(g)$

(b) $2C_2H_6(g) + 7O_2(g) \longrightarrow 4CO_2(g) + 6H_2O(l)$

Applications of Equation (18.10)

In order to predict the sign of ΔG, according to Equation (18.10) we need to know both ΔH and ΔS. A negative ΔH (an exothermic reaction) and a positive ΔS (a reaction that results in an increase in the microstates of the system) tend to make ΔG negative, although temperature may also influence the *direction* of a spontaneous reaction. The four possible outcomes of this relationship are:

- If both ΔH and ΔS are positive, then ΔG will be negative only when the $T\Delta S$ term is greater in magnitude than ΔH. This condition is met when T is large.
- If ΔH is positive and ΔS is negative, ΔG will always be positive, regardless of temperature.
- If ΔH is negative and ΔS is positive, then ΔG will always be negative regardless of temperature.
- If ΔH is negative and ΔS is negative, then ΔG will be negative only when $T\Delta S$ is smaller in magnitude than ΔH. This condition is met when T is small.

The temperatures that will cause ΔG to be negative for the first and last cases depend on the actual values of ΔH and ΔS of the system. Table 18.3 summarizes the effects of the possibilities just described.

Review of Concepts

(a) Under what circumstances will an endothermic reaction proceed spontaneously?

(b) Explain why, in many reactions in which both the reactant and product species are in the solution phase, ΔH often gives a good hint about the spontaneity of a reaction at 298 K.

Table 18.3 Factors Affecting the Sign of ΔG in the Relationship $\Delta G = \Delta H - T\Delta S$

ΔH	ΔS	ΔG	Example
+	+	Reaction proceeds spontaneously at high temperatures. At low temperatures, reaction is spontaneous in the reverse direction.	$2HgO(s) \longrightarrow 2Hg(l) + O_2(g)$
+	−	ΔG is always positive. Reaction is spontaneous in the reverse direction at all temperatures.	$3O_2(g) \longrightarrow 2O_3(g)$
−	+	ΔG is always negative. Reaction proceeds spontaneously at all temperatures.	$2H_2O_2(l) \longrightarrow 2H_2O(l) + O_2(g)$
−	−	Reaction proceeds spontaneously at low temperatures. At high temperatures, the reverse reaction becomes spontaneous.	$NH_3(g) + HCl(g) \longrightarrow NH_4Cl(s)$

Before we apply the change in free energy to predict reaction spontaneity, it is useful to distinguish between ΔG and $\Delta G°$. Suppose we carry out a reaction in solution with all the reactants in their standard states (that is, all at 1 *M* concentration). As soon as the reaction starts, the standard-state condition no longer exists for the reactants or the products because their concentrations are different from 1 *M*. Under nonstandard state conditions, we must use the sign of ΔG rather than that of $\Delta G°$ to predict the direction of the reaction. The sign of $\Delta G°$, on the other hand, tells us whether the products or the reactants are favored when the reacting system reaches equilibrium. Thus, a negative value of $\Delta G°$ indicates that the reaction favors product formation whereas a positive value of $\Delta G°$ indicates that there will be more reactants than products at equilibrium.

In Section 18.6 we will see an equation relating $\Delta G°$ to the equilibrium constant *K*.

We will now consider two specific applications of Equation (18.10).

Temperature and Chemical Reactions

Calcium oxide (CaO), also called quicklime, is an extremely valuable inorganic substance used in steelmaking, production of calcium metal, the paper industry, water treatment, and pollution control. It is prepared by decomposing limestone ($CaCO_3$) in a kiln at a high temperature (Figure 18.7):

Le Châtelier's principle predicts that the forward, endothermic reaction is favored by heating.

$$CaCO_3(s) \rightleftharpoons CaO(s) + CO_2(g)$$

The reaction is reversible, and CaO readily combines with CO_2 to form $CaCO_3$. The pressure of CO_2 in equilibrium with $CaCO_3$ and CaO increases with temperature. In the industrial preparation of quicklime, the system is never maintained at equilibrium; rather, CO_2 is constantly removed from the kiln to shift the equilibrium from left to right, promoting the formation of calcium oxide.

The important information for the practical chemist is the temperature at which the decomposition of $CaCO_3$ becomes appreciable (that is, the temperature at which the reaction begins to favor products). We can make a reliable estimate of that temperature as follows. First we calculate $\Delta H°$ and $\Delta S°$ for the reaction at 25°C, using the data in Appendix 2. To determine $\Delta H°$ we apply Equation (6.17):

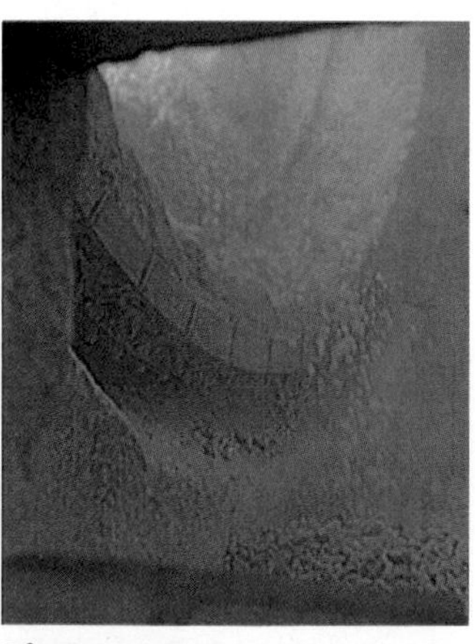

Figure 18.7
The production of CaO from $CaCO_3$ in a rotary kiln.

$$\begin{aligned}\Delta H° &= [\Delta H_f°(CaO) + \Delta H_f°(CO_2)] - [\Delta H_f°(CaCO_3)]\\ &= [(-635.6\ \text{kJ/mol}) + (-393.5\ \text{kJ/mol})] - (-1206.9\ \text{kJ/mol})\\ &= 177.8\ \text{kJ/mol}\end{aligned}$$

Next we apply Equation (18.6) to find $\Delta S°$

$$\begin{aligned}\Delta S° &= [S°(CaO) + S°(CO_2)] - S°(CaCO_3)\\ &= [(39.8\ J/K\cdot mol) + (213.6\ J/K\cdot mol)] - (92.9\ J/K\cdot mol)\\ &= 160.5\ J/K\cdot mol\end{aligned}$$

From Equation (18.10)

$$\Delta G° = \Delta H° - T\Delta S°$$

we obtain

$$\begin{aligned}\Delta G° &= 177.8\ kJ/mol - (298\ K)(160.5\ J/K\cdot mol)\left(\frac{1\ kJ}{1000\ J}\right)\\ &= 130.0\ kJ/mol\end{aligned}$$

Because $\Delta G°$ is a large positive quantity, we conclude that the reaction is not favored for product formation at 25°C (or 298 K). Indeed, the pressure of CO_2 is so low at room temperature that it cannot be measured. In order to make $\Delta G°$ negative, we first have to find the temperature at which $\Delta G°$ is zero; that is,

$$0 = \Delta H° - T\Delta S°$$

or

$$\begin{aligned}T &= \frac{\Delta H°}{\Delta S°}\\ &= \frac{(177.8\ kJ/mol)(1000\ J/1\ kJ)}{160.5\ J/K\cdot mol}\\ &= 1108\ K\ \text{or}\ 835°C\end{aligned}$$

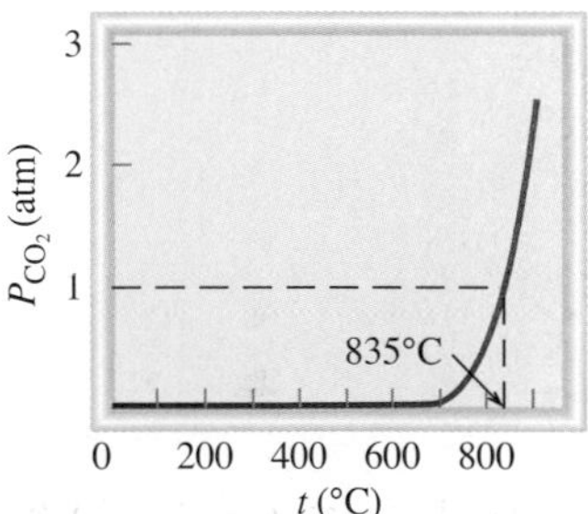

Figure 18.8
Equilibrium pressure of CO_2 from the decomposition of $CaCO_3$, as a function of temperature. This curve is calculated by assuming that $\Delta H°$ and $\Delta S°$ of the reaction do not change with temperature.

At a temperature higher than 835°C, $\Delta G°$ becomes negative, indicating that the reaction now favors the formation of CaO and CO_2. For example, at 840°C, or 1113 K,

$$\begin{aligned}\Delta G° &= \Delta H° - T\Delta S°\\ &= 177.8\ kJ/mol - (1113\ K)\ (160.5\ J/K\cdot mol)\left(\frac{1\ kJ}{1000\ J}\right)\\ &= -0.8\ kJ/mol\end{aligned}$$

Two points are worth making about such a calculation. First, we used the $\Delta H°$ and $\Delta S°$ values at 25°C to calculate changes that occur at a much higher temperature. Because both $\Delta H°$ and $\Delta S°$ change with temperature, this approach will not give us an accurate value of $\Delta G°$, but it is good enough for "ball park" estimates. Second, we should not be misled into thinking that nothing happens below 835°C and that at 835°C $CaCO_3$ suddenly begins to decompose. Far from it. The fact that $\Delta G°$ is a positive value at some temperature below 835°C does not mean that no CO_2 is produced, but rather that the pressure of the CO_2 gas formed at that temperature will be below 1 atm (its standard-state value; see Table 18.2). As Figure 18.8 shows, the pressure of CO_2 at first increases very slowly with temperature; it becomes easily measurable above 700°C. The significance of 835°C is that this is the temperature at which the equilibrium pressure of CO_2 reaches 1 atm. Above 835°C, the equilibrium pressure of CO_2 exceeds 1 atm.

The equilibrium constant of this reaction is $K_P = P_{CO_2}$.

Phase Transitions

At the temperature at which a phase transition occurs (that is, at the melting point or boiling point) the system is at equilibrium ($\Delta G = 0$), so Equation (18.10) becomes

$$\Delta G = \Delta H - T\Delta S$$
$$0 = \Delta H - T\Delta S$$

or

$$\Delta S = \frac{\Delta H}{T}$$

Let us first consider the ice-water equilibrium. For the ice $\longrightarrow$ water transition, ΔH is the molar heat of fusion (see Table 12.7), and T is the melting point. The entropy change is therefore

$$\Delta S_{\text{ice} \longrightarrow \text{water}} = \frac{6010 \text{ J/mol}}{273 \text{ K}} = 22.0 \text{ J/K} \cdot \text{mol}$$

Thus, when 1 mole of ice melts at 0°C, there is an increase in entropy of 22.0 J/K · mol. The increase in entropy is consistent with the increase in microstates from solid to liquid. Conversely, for the water $\longrightarrow$ ice transition, the decrease in entropy is given by

The melting of ice is an endothermic process (ΔH is positive), and the freezing of water is exothermic (ΔH is negative).

$$\Delta S_{\text{water} \longrightarrow \text{ice}} = \frac{-6010 \text{ J/mol}}{273 \text{ K}} = -22.0 \text{ J/K} \cdot \text{mol}$$

In the laboratory we normally carry out unidirectional changes, that is, either ice to water or water to ice transition. We can calculate entropy change in each case using the equation $\Delta S = \Delta H/T$ as long as the temperature remains at 0°C. The same procedure can be applied to the water $\longrightarrow$ steam transition. In this case ΔH is the heat of vaporization and T is the boiling point of water.

Liquid and solid benzene in equilibrium at 5.5°C.

EXAMPLE 18.5

The molar heats of fusion and vaporization of benzene are 10.9 kJ/mol and 31.0 kJ/mol, respectively. Calculate the entropy changes for the solid $\longrightarrow$ liquid and liquid $\longrightarrow$ vapor transitions for benzene. At 1 atm pressure, benzene melts at 5.5°C and boils at 80.1°C.

Strategy At the melting point, liquid and solid benzene are at equilibrium, so $\Delta G = 0$. From Equation (18.10) we have $\Delta G = 0 = \Delta H - T\Delta S$ or $\Delta S = \Delta H/T$. To calculate the entropy change for the solid benzene $\longrightarrow$ liquid benzene transition, we write $\Delta S_{\text{fus}} = \Delta H_{\text{fus}}/T_{\text{f}}$. Here ΔH_{fus} is positive for an endothermic process, so ΔS_{fus} is also positive, as expected for a solid to liquid transition. The same procedure applies to the liquid benzene $\longrightarrow$ vapor benzene transition. What temperature unit should be used?

Solution The entropy change for melting 1 mole of benzene at 5.5°C is

$$\Delta S_{\text{fus}} = \frac{\Delta H_{\text{fus}}}{\Delta T_{\text{f}}} = \frac{(10.9 \text{ kJ/mol})(1000 \text{ J/1 kJ})}{(5.5 + 273) \text{ K}} = 39.1 \text{ J/K} \cdot \text{mol}$$

(Continued)

Similarly, the entropy change for boiling 1 mole of benzene at 80.1°C is

$$\begin{aligned}\Delta S_{vap} &= \frac{\Delta H_{vap}}{T_{bp}} \\ &= \frac{(31.0 \text{ kJ/mol})(1000 \text{ J/1 kJ})}{(80.1 + 273) \text{ K}} \\ &= 87.8 \text{ J/K} \cdot \text{mol}\end{aligned}$$

Check Because vaporization creates more microstates than the melting process, $\Delta S_{vap} > \Delta S_{fus}$.

Similar problem: 18.60.

Practice Exercise The molar heats of fusion and vaporization of argon are 1.3 kJ/mol and 6.3 kJ/mol, and argon's melting point and boiling point are −190°C and −186°C, respectively. Calculate the entropy changes for fusion and vaporization.

REVIEW OF CONCEPTS

Consider the sublimation of iodine (I_2) at 45°C in a closed flask shown here. If the enthalpy of sublimation is 62.4 kJ/mol, what is the ΔS for sublimation?

18.6 Free Energy and Chemical Equilibrium

As mentioned earlier, during the course of a chemical reaction not all the reactants and products will be at their standard states. Under this condition, the relationship between ΔG and $\Delta G°$, which can be derived from thermodynamics, is

$$\Delta G = \Delta G° + RT \ln Q \qquad (18.13)$$

where R is the gas constant (8.314 J/K · mol), T is the absolute temperature of the reaction, and Q is the reaction quotient (see p. 522). We see that ΔG depends on two quantities: $\Delta G°$ and $RT \ln Q$. For a given reaction at temperature T the value of $\Delta G°$ is fixed but that of $RT \ln Q$ is not, because Q varies according to the composition of the reaction mixture. Let us consider two special cases:

Case 1: A large negative value of $\Delta G°$ will tend to make ΔG also negative. Thus, the net reaction will proceed from left to right until a significant amount of product has been formed. At that point, the $RT \ln Q$ term will become positive enough to match the negative $\Delta G°$ term.

Case 2: A large positive $\Delta G°$ term will tend to make ΔG also positive. Thus, the net reaction will proceed from right to left until a significant amount of reactant has been formed. At that point, the $RT \ln Q$ term will become negative enough to match the positive $\Delta G°$ term.

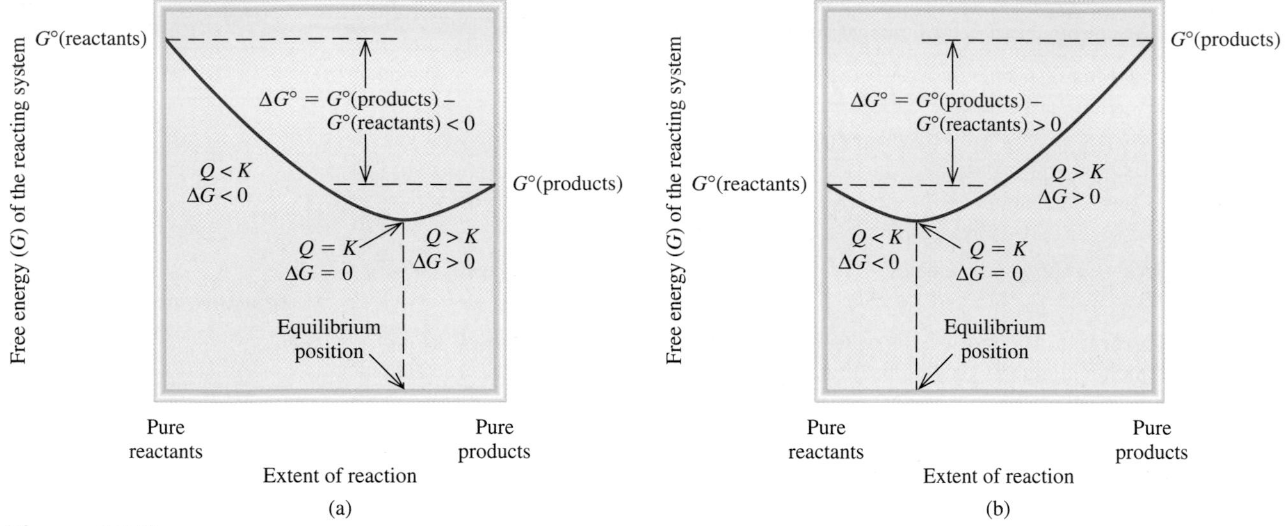

Figure 18.9

(a) $\Delta G^\circ < 0$. At equilibrium, there is a significant conversion of reactants to products. (b) $\Delta G^\circ > 0$. At equilibrium, reactants are favored over products. In both cases, the net reaction toward equilibrium is from left to right (reactants to products) if $Q < K$ and right to left (products to reactants) if $Q > K$.

Sooner or later a reversible reaction will reach equilibrium.

At equilibrium, by definition, $\Delta G = 0$ and $Q = K$, where K is the equilibrium constant. Thus,

$$0 = \Delta G^\circ + RT \ln K$$

or

$$\Delta G^\circ = -RT \ln K \quad (18.14)$$

In this equation, K_P is used for gases and K_c for reactions in solution. Note that the larger the K is, the more negative ΔG° is. For chemists, Equation (18.14) is one of the most important equations in thermodynamics because it enables us to find the equilibrium constant of a reaction if we know the change in standard free energy and vice versa.

It is significant that Equation (18.14) relates the equilibrium constant to the *standard* free-energy change ΔG° rather than to the *actual* free-energy change ΔG. The actual free-energy change of the system varies as the reaction progresses and becomes zero at equilibrium. On the other hand, ΔG° is a constant for a particular reaction at a given temperature. Figure 18.9 shows plots of the free energy of a reacting system versus the extent of the reaction for two types of reactions. As you can see, if $\Delta G^\circ < 0$, the products are favored over reactants at equilibrium. Conversely, if $\Delta G^\circ > 0$, there will be more reactants than products at equilibrium. Table 18.4 summarizes the three possible

Table 18.4 Relation Between ΔG° and K as Predicted by the Equation $\Delta G^\circ = -RT \ln K$

K	ln K	ΔG°	Comments
> 1	Positive	Negative	Products are favored over reactants at equilibrium.
$= 1$	0	0	Products and reactants are equally favored at equilibrium.
< 1	Negative	Positive	Reactants are favored over products at equilibrium.

relations between $\Delta G°$ and K, as predicted by Equation (18.14). Remember this important distinction: It is the sign of ΔG and not that of $\Delta G°$ that determines the direction of reaction spontaneity. The sign of $\Delta G°$ tells us only the relative amounts of products and reactants when equilibrium is reached, not the direction of the net reaction.

For reactions having very large or very small equilibrium constants, it is generally very difficult, if not impossible, to measure the K values by monitoring the concentrations of all the reacting species. Consider, for example, the formation of nitric oxide from molecular nitrogen and molecular oxygen:

$$N_2(g) + O_2(g) \rightleftharpoons 2NO(g)$$

At 25°C, the equilibrium constant K_P is

$$K_P = \frac{P_{NO}^2}{P_{N_2}P_{O_2}} = 4.0 \times 10^{-31}$$

The very small value of K_P means that the concentration of NO at equilibrium will be exceedingly low. In such a case the equilibrium constant is more conveniently obtained from $\Delta G°$. (As we have seen, $\Delta G°$ can be calculated from $\Delta H°$ and $\Delta S°$.) On the other hand, the equilibrium constant for the formation of hydrogen iodide from molecular hydrogen and molecular iodine is near unity at room temperature:

$$H_2(g) + I_2(g) \rightleftharpoons 2HI(g)$$

For this reaction it is easier to measure K_P and then calculate $\Delta G°$ using Equation (18.14) than to measure $\Delta H°$ and $\Delta S°$ and use Equation (18.10).

EXAMPLE 18.6

Using data listed in Appendix 2, calculate the equilibrium constant (K_P) for the following reaction at 25°C:

$$2H_2O(l) \rightleftharpoons 2H_2(g) + O_2(g)$$

Strategy According to Equation (18.14), the equilibrium constant for the reaction is related to the standard free energy-change; that is, $\Delta G° = -RT \ln K$. Therefore, we first need to calculate $\Delta G°$ by following the procedure in Example 18.4. Then we can calculate K_P. What temperature unit should be used?

Solution According to Equation (18.12),

$$\begin{aligned}\Delta G°_{rxn} &= [2\Delta G°_f(H_2) + \Delta G°_f(O_2)] - [2\Delta G°_f(H_2O)] \\ &= [(2)(0\ \text{kJ/mol}) + (0\ \text{kJ/mol})] - [(2)(-237.2\ \text{kJ/mol})] \\ &= 474.4\ \text{kJ/mol}\end{aligned}$$

Using Equation (18.14)

$$\Delta G°_{rxn} = -RT \ln K_P$$

$$474.4\ \text{kJ/mol} \times \frac{1000\ \text{J}}{1\ \text{kJ}} = -(8.314\ \text{J/K} \cdot \text{mol})(298\ \text{K}) \ln K_P$$

$$\ln K_P = -191.5$$

$$K_P = e^{-191.5} = 7 \times 10^{-84}$$

To calculate K_P, enter −191.5 on your calculator and then press the key labeled "e" or "inv(erse) ln x."

(Continued)

Comment This extremely small equilibrium constant is consistent with the fact that water does not decompose into hydrogen and oxygen gases at 25°C. Thus, a large positive $\Delta G°$ favors reactants over products at equilibrium.

Similar problems: 18.23 and 18.26.

Practice Exercise Calculate the equilibrium constant (K_P) for the reaction at 25°C

$$2O_3(g) \longrightarrow 3O_2(g)$$

EXAMPLE 18.7

In Chapter 17 we discussed the solubility product of slightly soluble substances. Using the solubility product of silver chloride at 25°C (1.6×10^{-10}), calculate $\Delta G°$ for the process

$$AgCl(s) \rightleftharpoons Ag^+(aq) + Cl^-(aq)$$

Strategy According to Equation (18.14), the equilibrium constant for the reaction is related to standard free-energy change; that is, $\Delta G° = -RT \ln K$. Because this is a heterogeneous equilibrium, the solubility product (K_{sp}) is the equilibrium constant. We calculate the standard free energy change from the K_{sp} value of AgCl. What temperature unit should be used?

Solution The solubility equilibrium for AgCl is

$$AgCl(s) \rightleftharpoons Ag^+(aq) + Cl^-(aq)$$

$$K_{sp} = [Ag^+][Cl^-] = 1.6 \times 10^{-10}$$

Using Equation (18.14) we obtain

$$\begin{aligned}\Delta G° &= -(8.314 \text{ J/K} \cdot \text{mol})(298 \text{ K}) \ln (1.6 \times 10^{-10}) \\ &= 5.6 \times 10^4 \text{ J/mol} \\ &= 56 \text{ kJ/mol}\end{aligned}$$

Check The large, positive $\Delta G°$ indicates that AgCl is slightly soluble and that the equilibrium lies mostly to the left.

Similar problem: 18.25.

Practice Exercise Calculate $\Delta G°$ for the following process at 25°C:

$$BaF_2(s) \rightleftharpoons Ba^{2+}(aq) + 2F^-(aq)$$

The K_{sp} of BaF_2 is 1.7×10^{-6}.

EXAMPLE 18.8

The equilibrium constant (K_P) for the reaction

$$N_2O_4(g) \rightleftharpoons 2NO_2(g)$$

is 0.113 at 298 K, which corresponds to a standard free-energy change of 5.40 kJ/mol. In a certain experiment, the initial pressures are $P_{NO_2} = 0.122$ atm and $P_{N_2O_4} = 0.453$ atm. Calculate ΔG for the reaction at these pressures and predict the direction of the net reaction.

Strategy From the information given we see that neither the reactant nor the product is at its standard state of 1 atm. To determine the direction of the net reaction, we need to calculate the free-energy change under nonstandard-state conditions (ΔG) using Equation (18.13) and the given $\Delta G°$ value. Note that the partial pressures are expressed as dimensionless quantities in the reaction quotient Q_P.

(Continued)

Solution Equation (18.13) can be written as

$$\begin{aligned}\Delta G &= \Delta G^\circ + RT \ln Q_P \\ &= \Delta G^\circ + RT \ln \frac{P_{NO_2}^2}{P_{N_2O_4}} \\ &= 5.40 \times 10^3 \text{ J/mol} + (8.314 \text{ J/K} \cdot \text{mol})(298 \text{ K}) \times \ln \frac{(0.122)^2}{0.453} \\ &= 5.40 \times 10^3 \text{ J/mol} - 8.46 \times 10^3 \text{ J/mol} \\ &= -3.06 \times 10^3 \text{ J/mol} = -3.06 \text{ kJ/mol}\end{aligned}$$

Because $\Delta G < 0$, the net reaction proceeds from left to right to reach equilibrium.

Check Note that although $\Delta G^\circ > 0$, the reaction can be made to favor product formation initially by having a small concentration (pressure) of the product compared to that of the reactant. Confirm the prediction by showing that $Q_P < K_P$.

Similar problems: 18.27 and 18.28.

Practice Exercise The ΔG° for the reaction

$$H_2(g) + I_2(g) \rightleftharpoons 2HI(g)$$

is 2.60 kJ/mol at 25°C. In one experiment, the initial pressures are $P_{H_2} = 4.26$ atm, $P_{I_2} = 0.024$ atm, and $P_{HI} = 0.23$ atm. Calculate ΔG for the reaction and predict the direction of the net reaction.

REVIEW OF CONCEPTS

A reaction is determined to have a positive ΔH° and a negative ΔS°. Is the value of K for this reaction greater than 1, equal to 1, or less than 1?

18.7 Thermodynamics in Living Systems

Many biochemical reactions have a positive ΔG° value, yet they are essential to the maintenance of life. In living systems these reactions are coupled to an energetically favorable process, one that has a negative ΔG° value. The principle of *coupled reactions* is based on a simple concept: we can use a thermodynamically favorable reaction to drive an unfavorable one. Consider an industrial process. Suppose we wish to extract zinc from the ore sphalerite (ZnS). The following reaction will not work because it has a large positive ΔG° value:

$$ZnS(s) \longrightarrow Zn(s) + S(s) \qquad \Delta G^\circ = 198.3 \text{ kJ/mol}$$

On the other hand, the combustion of sulfur to form sulfur dioxide is favored because of its large negative ΔG° value:

$$S(s) + O_2(g) \longrightarrow SO_2(g) \qquad \Delta G^\circ = -300.1 \text{ kJ/mol}$$

By coupling the two processes we can bring about the separation of zinc from zinc sulfide. In practice, this means heating ZnS in air so that the tendency of S to form SO_2 will promote the decomposition of ZnS:

$$\begin{array}{lr} ZnS(s) \longrightarrow Zn(s) + S(s) & \Delta G^\circ = 198.3 \text{ kJ/mol} \\ S(s) + O_2(g) \longrightarrow SO_2(g) & \Delta G^\circ = -300.1 \text{ kJ/mol} \\ \hline ZnS(s) + O_2(g) \longrightarrow Zn(s) + SO_2(g) & \Delta G^\circ = -101.8 \text{ kJ/mol} \end{array}$$

A mechanical analog for coupled reactions. We can make the smaller weight move upward (a nonspontaneous process) by coupling it with the falling of a larger weight.

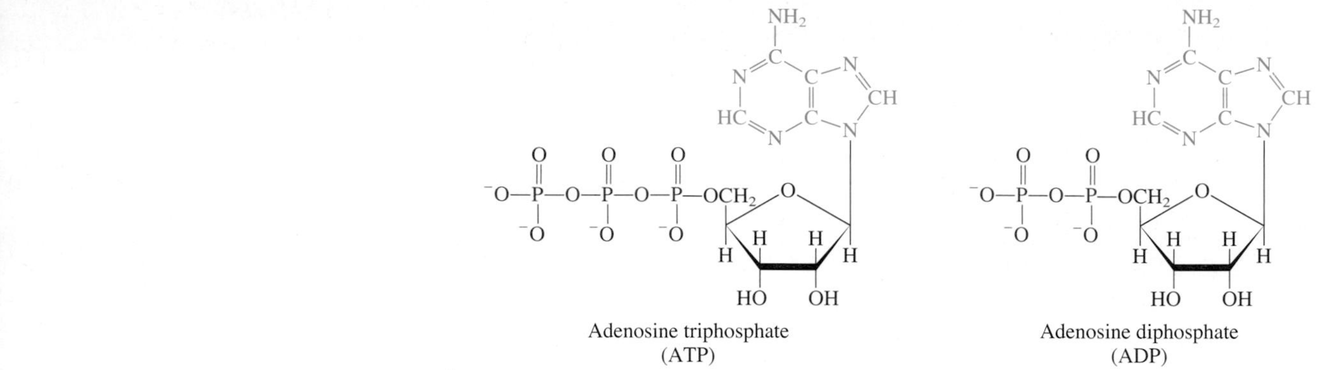

Adenosine triphosphate (ATP)

Adenosine diphosphate (ADP)

Figure 18.10
Structure of ATP and ADP in ionized forms. The adenine group is in blue, the ribose group in black, and the phosphate group in red. Note that ADP has one fewer phosphate group than ATP.

Coupled reactions play a crucial role in our survival. In biological systems, enzymes facilitate a wide variety of nonspontaneous reactions. For example, in the human body, food molecules, represented by glucose ($C_6H_{12}O_6$), are converted to carbon dioxide and water during metabolism with a substantial release of free energy:

$$C_6H_{12}O_6(s) + 6O_2(g) \longrightarrow 6CO_2(g) + 6H_2O(l) \quad \Delta G^\circ = -2880 \text{ kJ/mol}$$

In a living cell, this reaction does not take place in a single step (as burning glucose in a flame would); rather, the glucose molecule is broken down with the aid of enzymes in a series of steps. Much of the free energy released along the way is used to synthesize adenosine triphosphate (ATP) from adenosine diphosphate (ADP) and phosphoric acid (Figure 18.10):

$$\text{ADP} + H_3PO_4 \longrightarrow \text{ATP} + H_2O \quad \Delta G^\circ = +31 \text{ kJ/mol}$$

The function of ATP is to store free energy until it is needed by cells. Under appropriate conditions, ATP undergoes hydrolysis to give ADP and phosphoric acid, with a release of 31 kJ of free energy, which can be used to drive energetically unfavorable reactions, such as protein synthesis.

Proteins are polymers made of amino acids. The stepwise synthesis of a protein molecule involves the joining of individual amino acids. Consider the formation of the dipeptide (a two-amino-acid unit) alanylglycine from alanine and glycine. This reaction represents the first step in the synthesis of a protein molecule:

$$\text{Alanine} + \text{Glycine} \longrightarrow \text{Alanylglycine} \quad \Delta G^\circ = +29 \text{ kJ/mol}$$

As you can see, this reaction does not favor the formation of product, and so only a little of the dipeptide would be formed at equilibrium. However, with the aid of an enzyme, the reaction is coupled to the hydrolysis of ATP as follows:

$$\text{ATP} + H_2O + \text{Alanine} + \text{Glycine} \longrightarrow \text{ADP} + H_3PO_4 + \text{Alanylglycine}$$

The overall free-energy change is given by $\Delta G^\circ = -31 \text{ kJ/mol} + 29 \text{ kJ/mol} = -2 \text{ kJ/mol}$, which means that the coupled reaction now favors the formation of product, and an appreciable amount of alanylglycine will be formed under this condition. Figure 18.11 shows the ATP-ADP interconversions that act as energy storage (from metabolism) and free-energy release (from ATP hydrolysis) to drive essential reactions.

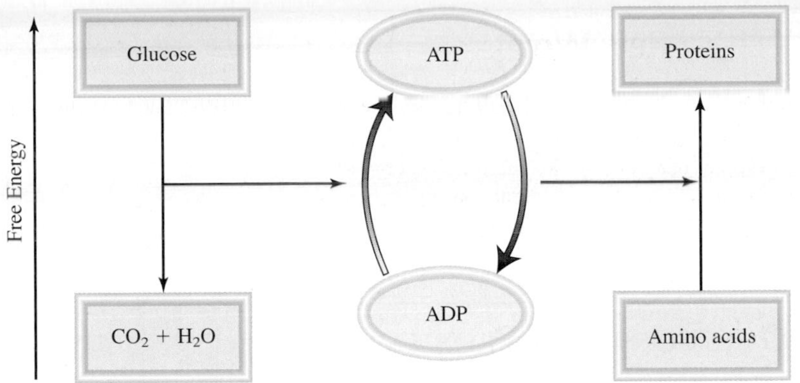

Figure 18.11 *Schematic representation of ATP synthesis and coupled reactions in living systems. The conversion of glucose to carbon dioxide and water during metabolism releases free energy. The released free energy is used to convert ADP into ATP. The ATP molecules are then used as an energy source to drive unfavorable reactions, such as protein synthesis from amino acids.*

Key Equations

$S = k \ln W$	(18.1)	Relating entropy to number of microstates.
$\Delta S_{univ} = \Delta S_{sys} + \Delta S_{surr} > 0$	(18.4)	The second law of thermodynamics (spontaneous process).
$\Delta S_{univ} = \Delta S_{sys} + \Delta S_{surr} = 0$	(18.5)	The second law of thermodynamics (equilibrium process).
$\Delta S^\circ_{rxn} = \Sigma n S^\circ(\text{products}) - \Sigma m S^\circ(\text{reactants})$	(18.7)	Standard entropy change of a reaction.
$G = H - TS$	(18.9)	Definition of Gibbs free energy.
$\Delta G = \Delta H - T\Delta S$	(18.10)	Free-energy change at constant temperature.
$\Delta G^\circ_{rxn} = \Sigma n \Delta G^\circ_f(\text{products}) - \Sigma m \Delta G^\circ_f(\text{reactants})$	(18.12)	Standard free-energy change of a reaction.
$\Delta G = \Delta G^\circ + RT \ln Q$	(18.13)	Relationship between free-energy change and standard free-energy change and reaction quotient.
$\Delta G^\circ = -RT \ln K$	(18.14)	Relationship between standard free-energy change and the equilibrium constant.

Summary of Facts and Concepts

1. Entropy is usually described in terms of dispersal of energy of a system. Any spontaneous process must lead to a net increase in entropy in the universe (second law of thermodynamics).
2. The standard entropy of a chemical reaction can be calculated from the absolute entropies of reactants and products.
3. The third law of thermodynamics states that the entropy of a perfect crystalline substance is zero at 0 K. This law enables us to measure the absolute entropies of substances.
4. Under conditions of constant temperature and pressure, the free-energy change ΔG is less than zero for a spontaneous process and greater than zero for a nonspontaneous process. For an equilibrium process, $\Delta G = 0$.
5. For a chemical or physical process at constant temperature and pressure, $\Delta G = \Delta H - T\Delta S$. This equation can be used to predict the spontaneity of a process.
6. The standard free-energy change for a reaction, ΔG°, can be calculated from the standard free energies of formation of reactants and products.
7. The equilibrium constant of a reaction and the standard free-energy change of the reaction are related by the equation $\Delta G^\circ = -RT \ln K$.
8. Many biological reactions are nonspontaneous. They are driven by the hydrolysis of ATP, for which ΔG° is negative.

Key Words

Entropy (S), p. 630
Free energy (G), p. 641
Gibbs free energy (G), p. 641
Second law of thermodynamics, p. 635
Standard entropy of reaction (ΔS°_{rxn}), p. 636
Standard free energy of formation (ΔG°_f), p. 642
Standard free energy of reaction (ΔG°_{rxn}), p. 642
Third law of thermodynamics, p. 639

Questions and Problems

Spontaneous Processes and Entropy

Review Questions

18.1 Explain what is meant by a spontaneous process. Give two examples each of spontaneous and nonspontaneous processes.

18.2 Which of the following processes are spontaneous and which are nonspontaneous? (a) dissolving table salt (NaCl) in hot soup; (b) climbing Mt. Everest; (c) spreading fragrance in a room by removing the cap from a perfume bottle; (d) separating helium and neon from a mixture of the gases.

18.3 Which of the following processes are spontaneous and which are nonspontaneous at a given temperature?
(a) $NaNO_3(s) \xrightarrow{H_2O} NaNO_3(aq)$ saturated soln
(b) $NaNO_3(s) \xrightarrow{H_2O} NaNO_3(aq)$ unsaturated soln
(c) $NaNO_3(s) \xrightarrow{H_2O} NaNO_3(aq)$ supersaturated soln

18.4 Define entropy. What are the units of entropy?

18.5 How does the entropy of a system change for each of the following processes?
(a) A solid melts.
(b) A liquid freezes.
(c) A liquid boils.
(d) A vapor is converted to a solid.
(e) A vapor condenses to a liquid.
(f) A solid sublimes.
(g) Urea dissolves in water.

Problems

18.6 Consider the situation shown in Figure 18.1(a). After the valve is opened, the probability of finding one molecule in either bulb is $\frac{1}{2}$ (because the bulbs have the same volume) and that of finding the molecule in the total volume is one. The probability of finding two molecules in the same bulb is the product of the individual probabilities; that is, $\frac{1}{2} \times \frac{1}{2}$ or $\frac{1}{4}$. Extend the calculation to finding 100 molecules in the same bulb. Based on your results, explain why it is highly improbable for the process shown in Figure 18.1(b) to occur spontaneously when the number of molecules becomes very large, say, 6×10^{23}.

The Second Law of Thermodynamics

Review Questions

18.7 State the second law of thermodynamics in words and express it mathematically.

18.8 State the third law of thermodynamics and explain its usefulness in calculating entropy values.

Problems

18.9 For each pair of substances listed here, choose the one having the larger standard entropy value at 25°C. The same molar amount is used in the comparison. Explain the basis for your choice. (a) $Li(s)$ or $Li(l)$; (b) $C_2H_5OH(l)$ or $CH_3OCH_3(l)$ (*Hint:* Which molecule can hydrogen-bond?); (c) $Ar(g)$ or $Xe(g)$; (d) $CO(g)$ or $CO_2(g)$; (e) $O_2(g)$ or $O_3(g)$; (f) $NO_2(g)$ or $N_2O_4(g)$.

18.10 Arrange the following substances (1 mole each) in order of increasing entropy at 25°C: (a) $Ne(g)$, (b) $SO_2(g)$, (c) $Na(s)$, (d) $NaCl(s)$, (e) $H_2(g)$. Give the reasons for your arrangement.

18.11 Using the data in Appendix 2, calculate the standard entropy changes for the following reactions at 25°C:
(a) $S(s) + O_2(g) \longrightarrow SO_2(g)$
(b) $MgCO_3(s) \longrightarrow MgO(s) + CO_2(g)$

18.12 Using the data in Appendix 2, calculate the standard entropy changes for the following reactions at 25°C:
(a) $H_2(g) + CuO(s) \longrightarrow Cu(s) + H_2O(g)$
(b) $2Al(s) + 3ZnO(s) \longrightarrow Al_2O_3(s) + 3Zn(s)$
(c) $CH_4(g) + 2O_2(g) \longrightarrow CO_2(g) + 2H_2O(l)$

18.13 Without consulting Appendix 2, predict whether the entropy change is positive or negative for each of the following reactions. Give reasons for your predictions.
(a) $2KClO_4(s) \longrightarrow 2KClO_3(s) + O_2(g)$
(b) $H_2O(g) \longrightarrow H_2O(l)$
(c) $2Na(s) + 2H_2O(l) \longrightarrow 2NaOH(aq) + H_2(g)$
(d) $N_2(g) \longrightarrow 2N(g)$

18.14 State whether the sign of the entropy change expected for each of the following processes will be positive or negative, and explain your predictions.
(a) $PCl_3(l) + Cl_2(g) \longrightarrow PCl_5(s)$
(b) $2HgO(s) \longrightarrow 2Hg(l) + O_2(g)$
(c) $H_2(g) \longrightarrow 2H(g)$
(d) $U(s) + 3F_2(g) \longrightarrow UF_6(s)$

Gibbs Free Energy

Review Questions

18.15 Define free energy. What are its units?

18.16 Why is it more convenient to predict the direction of a reaction in terms of ΔG_{sys} instead of ΔS_{univ}? Under what conditions can ΔG_{sys} be used to predict the spontaneity of a reaction?

Problems

18.17 Calculate $\Delta G°$ for the following reactions at 25°C:
(a) $N_2(g) + O_2(g) \longrightarrow 2NO(g)$
(b) $H_2O(l) \longrightarrow H_2O(g)$
(c) $2C_2H_2(g) + 5O_2(g) \longrightarrow 4CO_2(g) + 2H_2O(l)$
(*Hint:* Look up the standard free energies of formation of the reactants and products in Appendix 2.)

18.18 Calculate $\Delta G°$ for the following reactions at 25°C:
(a) $2Mg(s) + O_2(g) \longrightarrow 2MgO(s)$
(b) $2SO_2(g) + O_2(g) \longrightarrow 2SO_3(g)$
(c) $2C_2H_6(g) + 7O_2(g) \longrightarrow 4CO_2(g) + 6H_2O(l)$
See Appendix 2 for thermodynamic data.

18.19 From the values of ΔH and ΔS, predict which of the following reactions would be spontaneous at 25°C: Reaction A: $\Delta H = 10.5$ kJ/mol, $\Delta S = 30$ J/K · mol; reaction B: $\Delta H = 1.8$ kJ/mol, $\Delta S = -113$ J/K · mol. If either of the reactions is nonspontaneous at 25°C, at what temperature might it become spontaneous?

18.20 Find the temperatures at which reactions with the following ΔH and ΔS values would become spontaneous: (a) $\Delta H = -126$ kJ/mol, $\Delta S = 84$ J/K · mol; (b) $\Delta H = -11.7$ kJ/mol, $\Delta S = -105$ J/K · mol.

Free Energy and Chemical Equilibrium

Review Questions

18.21 Explain the difference between ΔG and $\Delta G°$.

18.22 Explain why Equation (18.14) is of great importance in chemistry.

Problems

18.23 Calculate K_P for the following reaction at 25°C:

$$H_2(g) + I_2(g) \rightleftharpoons 2HI(g) \qquad \Delta G° = 2.60 \text{ kJ/mol}$$

18.24 For the autoionization of water at 25°C,

$$H_2O(l) \rightleftharpoons H^+(aq) + OH^-(aq)$$

K_w is 1.0×10^{-14}. What is $\Delta G°$ for the process?

18.25 Consider the following reaction at 25°C:

$$Fe(OH)_2(s) \rightleftharpoons Fe^{2+}(aq) + 2OH^-(aq)$$

Calculate $\Delta G°$ for the reaction. K_{sp} for $Fe(OH)_2$ is 1.6×10^{-14}.

18.26 Calculate $\Delta G°$ and K_P for the following equilibrium reaction at 25°C.

$$2H_2O(g) \rightleftharpoons 2H_2(g) + O_2(g)$$

18.27 (a) Calculate $\Delta G°$ and K_P for the following equilibrium reaction at 25°C. The $\Delta G_f°$ values are 0 for $Cl_2(g)$, −286 kJ/mol for $PCl_3(g)$, and −325 kJ/mol for $PCl_5(g)$.

$$PCl_5(g) \rightleftharpoons PCl_3(g) + Cl_2(g)$$

(b) Calculate ΔG for the reaction if the partial pressures of the initial mixture are $P_{PCl_5} = 0.0029$ atm, $P_{PCl_3} = 0.27$ atm, and $P_{Cl_2} = 0.40$ atm.

18.28 The equilibrium constant (K_P) for the reaction

$$H_2(g) + CO_2(g) \rightleftharpoons H_2O(g) + CO(g)$$

is 4.40 at 2000 K. (a) Calculate $\Delta G°$ for the reaction. (b) Calculate ΔG for the reaction when the partial pressures are $P_{H_2} = 0.25$ atm, $P_{CO_2} = 0.78$ atm, $P_{H_2O} = 0.66$ atm, and $P_{CO} = 1.20$ atm.

18.29 Consider the decomposition of calcium carbonate:

$$CaCO_3(s) \rightleftharpoons CaO(s) + CO_2(g)$$

Calculate the pressure in atm of CO_2 in an equilibrium process (a) at 25°C and (b) at 800°C. Assume that $\Delta H° = 177.8$ kJ/mol and $\Delta S° = 160.5$ J/K · mol for the temperature range.

18.30 The equilibrium constant K_P for the reaction

$$CO(g) + Cl_2(g) \rightleftharpoons COCl_2(g)$$

is 5.62×10^{35} at 25°C. Calculate $\Delta G_f°$ for $COCl_2$ at 25°C.

18.31 At 25°C, $\Delta G°$ for the process

$$H_2O(l) \rightleftharpoons H_2O(g)$$

is 8.6 kJ/mol. Calculate the vapor pressure of water at this temperature.

18.32 Calculate $\Delta G°$ for the process

$$C(\text{diamond}) \longrightarrow C(\text{graphite})$$

Is the formation of graphite from diamond favored at 25°C? If so, why is it that diamonds do not become graphite on standing?

Thermodynamics in Living Systems

Review Questions

18.33 What is a coupled reaction? What is its importance in biological reactions?

18.34 What is the role of ATP in biological reactions?

Problems

18.35 Referring to the metabolic process involving glucose on p. 652, calculate the maximum number of moles of ATP that can be synthesized from ADP from the breakdown of one mole of glucose.

18.36 In the metabolism of glucose, the first step is the conversion of glucose to glucose 6-phosphate:

$$\text{glucose} + H_3PO_4 \longrightarrow \text{glucose 6-phosphate} + H_2O \qquad \Delta G° = 13.4 \text{ kJ/mol}$$

Because $\Delta G°$ is positive, this reaction does not favor the formation of products. Show how this reaction can be made to proceed by coupling it with the hydrolysis of ATP. Write an equation for the coupled reaction and estimate the equilibrium constant for the coupled process.

Additional Problems

18.37 Explain the following nursery rhyme in terms of the second law of thermodynamics.

Humpty Dumpty sat on a wall;
Humpty Dumpty had a great fall.
All the King's horses and all the King's men
Couldn't put Humpty together again.

18.38 Calculate ΔG for the reaction

$$H_2O(l) \rightleftharpoons H^+(aq) + OH^-(aq)$$

at 25°C for the following conditions:

(a) $[H^+] = 1.0 \times 10^{-7}\ M$, $[OH^-] = 1.0 \times 10^{-7}\ M$
(b) $[H^+] = 1.0 \times 10^{-3}\ M$, $[OH^-] = 1.0 \times 10^{-4}\ M$
(c) $[H^+] = 1.0 \times 10^{-12}\ M$, $[OH^-] = 2.0 \times 10^{-8}\ M$
(d) $[H^+] = 3.5\ M$, $[OH^-] = 4.8 \times 10^{-4}\ M$

18.39 Which of the following thermodynamic functions are associated only with the first law of thermodynamics: S, U, G, and H?

18.40 A student placed 1 g of each of three compounds A, B, and C in a container and found that after 1 week no change had occurred. Offer some possible explanations for the fact that no reactions took place. Assume that A, B, and C are totally miscible liquids.

18.41 Give a detailed example of each of the following, with an explanation: (a) a thermodynamically spontaneous process; (b) a process that would violate the first law of thermodynamics; (c) a process that would violate the second law of thermodynamics; (d) an irreversible process; (e) an equilibrium process.

18.42 Predict the signs of ΔH, ΔS, and ΔG of the system for the following processes at 1 atm: (a) ammonia melts at −60°C, (b) ammonia melts at −77.7°C, (c) ammonia melts at −100°C. (The normal melting point of ammonia is −77.7°C.)

18.43 Consider the following facts: Water freezes spontaneously at −5°C and 1 atm, and ice has a more ordered structure than liquid water. Explain how a spontaneous process can lead to a decrease in entropy.

18.44 Ammonium nitrate (NH_4NO_3) dissolves spontaneously and endothermically in water. What can you deduce about the sign of ΔS for the solution process?

18.45 Calculate the equilibrium pressure of CO_2 due to the decomposition of magnesium carbonate ($MgCO_3$) at 25°C.

18.46 (a) Trouton's rule states that the ratio of the molar heat of vaporization of a liquid (ΔH_{vap}) to its boiling point in kelvins is approximately 90 J/K · mol. Use the following data to show that this is the case and explain why Trouton's rule holds true:

	t_{bp}(°C)	ΔH_{vap}(kJ/mol)
Benzene	80.1	31.0
Hexane	68.7	30.8
Mercury	357	59.0
Toluene	110.6	35.2

(b) Use the values in Table 12.5 to calculate the same ratio for ethanol and water. Explain why Trouton's rule does not apply to these two substances as well as it does to other liquids.

18.47 Referring to Problem 18.46, explain why the ratio is considerably smaller than 90 J/K · mol for liquid HF.

18.48 Carbon monoxide (CO) and nitric oxide (NO) are polluting gases contained in automobile exhaust. Under suitable conditions, these gases can be made to react to form nitrogen (N_2) and the less harmful carbon dioxide (CO_2). (a) Write an equation for this reaction. (b) Identify the oxidizing and reducing agents. (c) Calculate the K_P for the reaction at 25°C. (d) Under normal atmospheric conditions, the partial pressures are $P_{N_2} = 0.80$ atm, $P_{CO_2} = 3.0 \times 10^{-4}$ atm, $P_{CO} = 5.0 \times 10^{-5}$ atm, and $P_{NO} = 5.0 \times 10^{-7}$ atm. Calculate Q_P and predict the direction toward which the reaction will proceed. (e) Will raising the temperature favor the formation of N_2 and CO_2?

18.49 For reactions carried out under standard-state conditions, Equation (18.10) takes the form $\Delta G° = \Delta H° - T\Delta S°$. (a) Assuming $\Delta H°$ and $\Delta S°$ are independent of temperature, derive the equation

$$\ln\frac{K_2}{K_1} = \frac{\Delta H°}{R}\left(\frac{T_2 - T_1}{T_1 T_2}\right)$$

where K_1 and K_2 are the equilibrium constants at T_1 and T_2, respectively. (b) Given that at 25°C K_c is 4.63×10^{-3} for the reaction

$$N_2O_4(g) \rightleftharpoons 2NO_2(g) \qquad \Delta H° = 58.0 \text{ kJ/mol}$$

calculate the equilibrium constant at 65°C.

18.50 The K_{sp} of AgCl is given in Table 17.2. What is its value at 60°C? [*Hint:* You need the result of Problem 18.49(a) and the data in Appendix 2 to calculate $\Delta H°$.]

18.51 Under what conditions does a substance have a standard entropy of zero? Can a substance ever have a negative standard entropy?

18.52 Water gas, a mixture of H_2 and CO, is a fuel made by reacting steam with red-hot coke (a by-product of coal distillation):

$$H_2O(g) + C(s) \rightleftharpoons CO(g) + H_2(g)$$

From the data in Appendix 2, estimate the temperature at which the reaction begins to favor the formation of products.

18.53 Consider the following Brønstead acid-base reaction at 25°C:

$$HF(aq) + Cl^-(aq) \rightleftharpoons HCl(aq) + F^-(aq)$$

(a) Predict whether K will be greater or smaller than unity. (b) Does $\Delta S°$ or $\Delta H°$ make a greater contribution to $\Delta G°$? (c) Is $\Delta H°$ likely to be positive or negative?

18.54 Crystallization of sodium acetate from a supersaturated solution occurs spontaneously (see p. 437). What can you deduce about the signs of ΔS and ΔH?

18.55 Consider the thermal decomposition of $CaCO_3$:

$$CaCO_3(s) \rightleftharpoons CaO(s) + CO_2(g)$$

The equilibrium vapor pressures of CO_2 are 22.6 mmHg at 700°C and 1829 mmHg at 950°C. Calculate the standard enthalpy of the reaction. [*Hint:* See Problem 18.49(a).]

18.56 A certain reaction is spontaneous at 72°C. If the enthalpy change for the reaction is 19 kJ/mol, what is the *minimum* value of ΔS (in J/K · mol) for the reaction?

18.57 Predict whether the entropy change is positive or negative for each of these reactions:
(a) $Zn(s) + 2HCl(aq) \longrightarrow ZnCl_2(aq) + H_2(g)$
(b) $O(g) + O(g) \longrightarrow O_2(g)$
(c) $NH_4NO_3(s) \longrightarrow N_2O(g) + 2H_2O(g)$
(d) $2H_2O_2(l) \longrightarrow 2H_2O(l) + O_2(g)$

18.58 The reaction $NH_3(g) + HCl(g) \longrightarrow NH_4Cl(s)$ proceeds spontaneously at 25°C even though there is a decrease in the number of microstates of the system (gases are converted to a solid). Explain.

18.59 Use the following data to determine the normal boiling point, in kelvins, of mercury. What assumptions must you make in order to do the calculation?

Hg(l): $\Delta H_f° = 0$ (by definition)
$S° = 77.4$ J/K · mol

Hg(g): $\Delta H_f° = 60.78$ kJ/mol
$S° = 174.7$ J/K · mol

18.60 The molar heat of vaporization of ethanol is 39.3 kJ/mol and the boiling point of ethanol is 78.3°C. Calculate ΔS for the vaporization of 0.50 mol ethanol.

18.61 A certain reaction is known to have a $\Delta G°$ value of −122 kJ/mol. Will the reaction necessarily occur if the reactants are mixed together?

18.62 In the Mond process for the purification of nickel, carbon monoxide is reacted with heated nickel to produce $Ni(CO)_4$, which is a gas and can therefore be separated from solid impurities:

$$Ni(s) + 4CO(g) \rightleftharpoons Ni(CO)_4(g)$$

Given that the standard free energies of formation of $CO(g)$ and $Ni(CO)_4(g)$ are −137.3 kJ/mol and −587.4 kJ/mol, respectively, calculate the equilibrium constant of the reaction at 80°C. Assume that $\Delta G_f°$ is temperature independent.

18.63 Calculate $\Delta G°$ and K_P for the following processes at 25°C:
(a) $H_2(g) + Br_2(l) \rightleftharpoons 2HBr(g)$
(b) $\frac{1}{2}H_2(g) + \frac{1}{2}Br_2(l) \rightleftharpoons HBr(g)$

Account for the differences in $\Delta G°$ and K_P obtained for (a) and (b).

18.64 Calculate the pressure of O_2 (in atm) over a sample of NiO at 25°C if $\Delta G° = 212$ kJ/mol for the reaction

$$NiO(s) \rightleftharpoons Ni(s) + \tfrac{1}{2}O_2(g)$$

18.65 Comment on the statement: "Just talking about entropy increases its value in the universe."

18.66 For a reaction with a negative $\Delta G°$ value, which of the following statements is false? (a) The equilibrium constant K is greater than one, (b) the reaction is spontaneous when all the reactants and products are in their standard states, and (c) the reaction is always exothermic.

18.67 Consider the reaction

$$N_2(g) + O_2(g) \rightleftharpoons 2NO(g)$$

Given that $\Delta G°$ for the reaction at 25°C is 173.4 kJ/mol, (a) calculate the standard free energy of formation of NO, and (b) calculate K_P of the reaction. (c) One of the starting substances in smog formation is NO. Assuming that the temperature in a running automobile engine is 1100°C, estimate K_P for the above reaction. (d) As farmers know, lightning helps to produce a better crop. Why?

18.68 Heating copper(II) oxide at 400°C does not produce any appreciable amount of Cu:

$$CuO(s) \rightleftharpoons Cu(s) + \tfrac{1}{2}O_2(g) \quad \Delta G° = 127.2 \text{ kJ/mol}$$

However, if this reaction is coupled to the conversion of graphite to carbon monoxide, it becomes spontaneous. Write an equation for the coupled process and calculate the equilibrium constant for the coupled reaction.

18.69 The internal combustion engine of a 1200-kg car is designed to run on octane (C_8H_{18}), whose enthalpy of combustion is 5510 kJ/mol. If the car is moving up a slope, calculate the maximum height (in meters) to which the car can be driven on 1.0 gallon of the fuel. Assume that the engine cylinder temperature is 2200°C and the exit temperature is 760°C, and neglect all forms of friction. The mass of 1 gallon of fuel is 3.1 kg. [*Hint:* The efficiency of the internal combustion engine, defined as work performed by the engine divided by the energy input, is given by $(T_2 - T_1)/T_2$, where T_2 and T_1 are the engine's operating temperature and exit temperature (in kelvins). The work done in moving the car over a vertical distance is mgh, where m is the mass of the car in kg, g the acceleration due to gravity (9.81 m/s^2), and h the height in meters.]

18.70 A carbon monoxide (CO) crystal is found to have entropy greater than zero at absolute zero of temperature. Give two possible explanations for this observation.

18.71 (a) Over the years there have been numerous claims about "perpetual motion machines," machines that will produce useful work with no input of energy. Explain why the first law of thermodynamics prohibits the possibility of such a machine existing. (b) Another kind of machine, sometimes called a "perpetual motion of the second kind," operates as follows. Suppose an ocean liner sails by scooping up water from the ocean and then extracting heat from the water, converting the heat to electric power to run the ship, and dumping the water back into the ocean. This process does not violate the first law of thermodynamics, for no energy is created—energy from the ocean is just converted to electrical energy. Show that the second law of thermodynamics prohibits the existence of such a machine.

18.72 The activity series in Section 4.4 shows that reaction (a) is spontaneous while reaction (b) is nonspontaneous at 25°C:

(a) $Fe(s) + 2H^+ \longrightarrow Fe^{2+}(aq) + H_2(g)$

(b) $Cu(s) + 2H^+ \longrightarrow Cu^{2+}(aq) + H_2(g)$

Use the data in Appendix 2 to calculate the equilibrium constant for these reactions and hence confirm that the activity series is correct.

18.73 The rate constant for the elementary reaction

$$2NO(g) + O_2(g) \longrightarrow 2NO_2(g)$$

is $7.1 \times 10^9/M^2 \cdot$ s at 25°C. What is the rate constant for the reverse reaction at the same temperature?

18.74 The following reaction was described as the cause of sulfur deposits formed at volcanic sites:

$$2H_2S(g) + SO_2(g) \rightleftharpoons 3S(s) + 2H_2O(g)$$

It may also be used to remove SO_2 from powerplant stack gases. (a) Identify the type of redox reaction it is. (b) Calculate the equilibrium constant (K_P) at 25°C and comment on whether this method is feasible for removing SO_2. (c) Would this procedure become more or less effective at a higher temperature?

18.75 Describe two ways that you could measure $\Delta G°$ of a reaction.

18.76 The following reaction represents the removal of ozone in the stratosphere:

$$2O_3(g) \rightleftharpoons 3O_2(g)$$

Calculate the equilibrium constant (K_P) for the reaction. In view of the magnitude of the equilibrium constant, explain why this reaction is not considered a major cause of ozone depletion in the absence of man-made pollutants such as the nitrogen oxides and CFCs? Assume the temperature of the stratosphere to be −30°C and $\Delta G_f°$ to be temperature independent.

18.77 A 74.6-g ice cube floats in the Arctic Sea. The temperature and pressure of the system and surroundings are at 1 atm and 0°C. Calculate ΔS_{sys}, ΔS_{surr}, and ΔS_{univ} for the melting of the ice cube. What can you conclude about the nature of the process from the value of ΔS_{univ}? (The molar heat of fusion of water is 6.01 kJ/mol.)

18.78 Comment on the feasibility of extracting copper from its ore chalcocite (Cu_2S) by heating:

$$Cu_2S(s) \longrightarrow 2Cu(s) + S(s)$$

Calculate the $\Delta G°$ for the overall reaction if the above process is coupled to the conversion of sulfur to sulfur dioxide, given that $\Delta G_f°(Cu_2S) = -86.1$ kJ/mol.

18.79 Active transport is the process in which a substance is transferred from a region of lower concentration to one of higher concentration. This is a nonspontaneous process and must be coupled to a spontaneous process, such as the hydrolysis of ATP. The concentrations of K^+ ions in the blood plasma and in nerve cells are 15 mM and 400 mM, respectively (1 mM = 1×10^{-3} M). Use Equation (18.13) to calculate ΔG for the process at the physiological temperature of 37°C:

$$K^+(15\ mM) \longrightarrow K^+(400\ mM)$$

In this calculation, the $\Delta G°$ term can be set to zero. What is the justification for this step?

18.80 Large quantities of hydrogen are needed for the synthesis of ammonia. One preparation of hydrogen involves the reaction between carbon monoxide and steam at 300°C in the presence of a copper-zinc catalyst:

$$CO(g) + H_2O(g) \rightleftharpoons CO_2(g) + H_2(g)$$

Calculate the equilibrium constant (K_P) for the reaction and the temperature at which the reaction favors the formation of CO and H_2O. Will a larger K_P be attained at the same temperature if a more efficient catalyst is used?

18.81 Consider two carboxylic acids (acids that contain the —COOH group): CH_3COOH (acetic acid, $K_a = 1.8 \times 10^{-5}$) and $CH_2ClCOOH$ (chloroacetic acid, $K_a = 1.4 \times 10^{-3}$). (a) Calculate $\Delta G°$ for the ionization of these acids at 25°C (b) From the equation $\Delta G° = \Delta H° - T\Delta S°$, we see that the contributions to the $\Delta G°$ term are an enthalpy term ($\Delta H°$) and a temperature times entropy term ($T\Delta S°$). These contributions are listed below for the two acids:

	$\Delta H°$(kJ/mol)	$T\Delta S°$(kJ/mol)
CH_3COOH	−0.57	−27.6
$CH_2ClCOOH$	−4.7	−21.1

Which is the dominant term in determining the value of $\Delta G°$ (and hence K_a of the acid)? (c) What processes contribute to $\Delta H°$? (Consider the ionization of the acids as a Brønsted acid-base reaction.) (d) Explain why the $T\Delta S°$ term is more negative for CH_3COOH.

18.82 One of the steps in the extraction of iron from its ore (FeO) is the reduction of iron(II) oxide by carbon monoxide at 900°C:

$$FeO(s) + CO(g) \rightleftharpoons Fe(s) + CO_2(g)$$

If CO is allowed to react with an excess of FeO, calculate the mole fractions of CO and CO_2 at equilibrium. State any assumptions.

18.83 Derive the following equation

$$\Delta G = RT \ln (Q/K)$$

where Q is the reaction quotient and describe how you would use it to predict the spontaneity of a reaction.

18.84 The sublimation of carbon dioxide at −78°C is given by

$$CO_2(s) \longrightarrow CO_2(g) \quad \Delta H_{sub} = 25.2 \text{ kJ/mol}$$

Calculate ΔS_{sub} when 84.8 g of CO_2 sublimes at this temperature.

18.85 Entropy has sometimes been described as "time's arrow" because it is the property that determines the forward direction of time. Explain.

18.86 Referring to Figure 18.1, we see that the probability of finding all 100 molecules in the same flask is 8×10^{-31} (see Problem 18.6). Assuming that the age of the universe is 13 billion years, calculate the time in seconds during which this event can be observed.

18.87 A student looked up the $\Delta G_f°$, $\Delta H_f°$, and $S°$ values for CO_2 in Appendix 2. Plugging these values into Equation (18.10), he found that $\Delta G_f° \neq \Delta H_f° - TS°$ at 298 K. What is wrong with his approach?

18.88 Consider the following reaction at 298 K:

$$2H_2(g) + O_2(g) \longrightarrow 2H_2O(l)$$
$$\Delta H° = -571.6 \text{ kJ/mol}$$

Calculate ΔS_{sys}, ΔS_{surr}, and ΔS_{univ} for the reaction.

18.89 As an approximation, we can assume that proteins exist either in the native (or physiologically functioning) state and the denatured state

$$\text{native} \rightleftharpoons \text{denatured}$$

The standard molar enthalpy and entropy of the denaturation of a certain protein are 512 kJ/mol and 1.60 kJ/K · mol, respectively. Comment on the signs and magnitudes of these quantities, and calculate the temperature at which the process favors the denatured state.

18.90 Which of the following are not state functions: S, H, q, w, T?

18.91 Which of the following is not accompanied by an increase in the entropy of the system? (a) mixing of two gases at the same temperature and pressure, (b) mixing of ethanol and water, (c) discharging a battery, (d) expansion of a gas followed by compression to its original temperature, pressure, and volume.

18.92 Hydrogenation reactions (for example, the process of converting C=C bonds to C—C bonds in food industry) are facilitated by the use of a transition metal catalyst, such as Ni or Pt. The initial step is the adsorption, or binding, of hydrogen gas onto the metal surface. Predict the signs of ΔH, ΔS, and ΔG when hydrogen gas is adsorbed onto the surface of Ni metal.

Special Problems

18.93 The standard free energies of formation ($\Delta G_f°$) of the three isomers of pentane (see p. 365) in the gas phase are: n-pentane: −8.37 kJ/mol; 2-methylbutane: −14.8 kJ/mol; 2,2-dimethylpropane: −15.2 kJ/mol. (a) Determine the mole percent of these molecules in an equilibrium mixture at 25°C. (b) How does the stability of these molecules depend on the extent of branching?

18.94 Carry out the following experiments: Quickly stretch a rubber band (at least 0.5 cm wide) and press it against your lips. You will feel a warming effect. Next, carry out the reverse procedure; that is, first stretch a rubber band and hold it in position for a few seconds. Now quickly release the tension and press the rubber band against your lips. You will feel a cooling effect. (a) Apply Equation (18.10) to these processes to determine the signs of ΔG, ΔH, and hence ΔS in each case. (b) From the signs of ΔS, what can you conclude about the structure of rubber molecules?

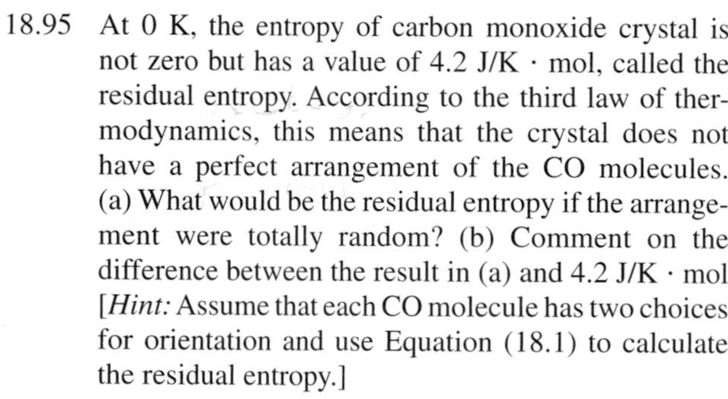

18.95 At 0 K, the entropy of carbon monoxide crystal is not zero but has a value of 4.2 J/K · mol, called the residual entropy. According to the third law of thermodynamics, this means that the crystal does not have a perfect arrangement of the CO molecules. (a) What would be the residual entropy if the arrangement were totally random? (b) Comment on the difference between the result in (a) and 4.2 J/K · mol [*Hint:* Assume that each CO molecule has two choices for orientation and use Equation (18.1) to calculate the residual entropy.]

18.96 Comment on the correctness of the analogy sometimes used to relate a student's dormitory room becoming untidy to an increase in entropy.

18.97 The standard enthalpy of formation and the standard entropy of gaseous benzene are 82.93 kJ/mol and 269.2 J/K · mol, respectively. Calculate $\Delta H°$, $\Delta S°$, and $\Delta G°$ for the process at 25°C. Comment on your answers.

$$C_6H_6(l) \longrightarrow C_6H_6(g)$$

18.98 The following diagram shows the variation of the equilibrium constant with temperature for the reaction

$$I_2(g) \rightleftharpoons 2I(g)$$

Calculate $\Delta G°$, $\Delta H°$, and $\Delta S°$ for the reaction at 872 K. (*Hint:* See Problem 18.49.)

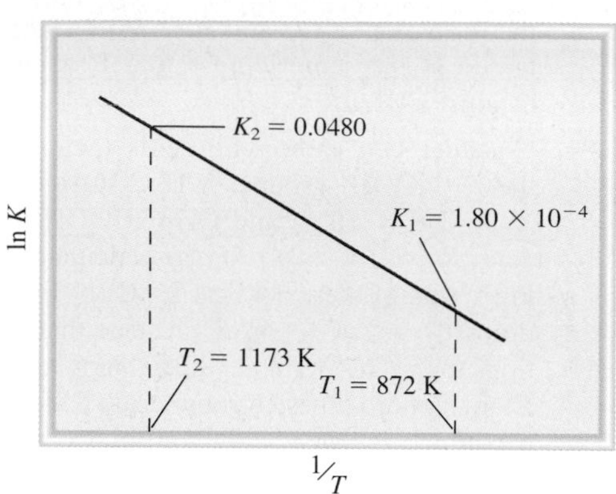

18.99 The boiling point of benzene is 80.1°C. Estimate (a) its molar heat of vaporization and (b) its vapor pressure at 74°C. (*Hint:* See Problems 18.46 and 18.49.)

18.100 Consider the gas-phase reaction between A_2 (green) and B_2 (red) to form AB at 298 K:

$$A_2(g) + B_2(g) \rightleftharpoons 2AB(g) \quad \Delta G° = -3.4 \text{ kJ/mol}$$

(1) Which of the following reaction mixtures is at equilibrium?

(2) Which of the following reaction mixtures has a negative ΔG value?

(3) Which of the following reaction mixtures has a positive ΔG value?

The partial pressures of the gases in each frame are equal to the number of A_2, B_2, and AB molecules times 0.10 atm. Round your answers to two significant figures.

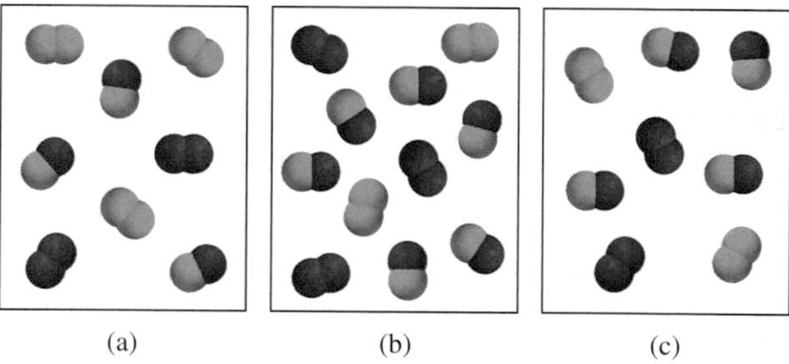

(a) (b) (c)

Answers to Practice Exercises

18.1 (a) Entropy decreases, (b) entropy decreases, (c) entropy increases, (d) entropy increases. **18.2** (a) −173.6 J/K · mol, (b) −139.8 J/K · mol, (c) 215.3 J/K · mol. **18.3** (a) $\Delta S > 0$, (b) $\Delta S < 0$, (c) $\Delta S \approx 0$. **18.4** (a) −106.4 kJ/mol, (b) −2935.0 kJ/mol. **18.5** $\Delta S_{fus} = 16$ J/K · mol; $\Delta S_{vap} = 72$ J/K · mol. **18.6** 2×10^{57}. **18.7** 33 kJ/mol. **18.8** $\Delta G = 0.97$ kJ/mol; direction is from right to left.

Electric cars with rechargeable batteries represent a viable alternative to standard automobiles with internal combustion engines.

CHAPTER

19

Redox Reactions and Electrochemistry

Chapter Outline

Student Interactive Activities

Animations
Galvanic Cells (19.2)

Electrocnic Homework
Example Practice Problems
End of Chapter Problems

Essential Concepts

Redox Reactions and Electrochemical Cells Equations representing redox reactions can be balanced using the half-reaction method. These reactions involve the transfer of electrons from a reducing agent to an oxidizing agent. Using separate compartments, such a reaction can be used to generate electrons in an arrangement called a galvanic cell.

Thermodynamics of Galvanic Cells The voltage measured in a galvanic cell can be broken down into the electrode potentials of the anode (where oxidation takes place) and cathode (where reduction takes place). This voltage can be related to the Gibbs free energy change and the equilibrium constant of the redox process. The Nernst equation relates the cell voltage to the cell voltage under standard-state conditions and the concentrations of reacting species.

Batteries Batteries are electrochemical cells that can supply direct electric current at a constant voltage. There are many different types of batteries used in automobiles, flashlights, and pacemakers. Fuel cells are a special type of electrochemical cell that generates electricity by the oxidation of hydrogen or hydrocarbons.

Corrosion Corrosion is a spontaneous redox reaction that results in the formation of rust from iron, silver sulfide from silver, and patina (copper carbonate) from copper. Corrosion causes enormous damage to buildings, structures, ships, and cars. Many methods have been devised to prevent or minimize the effect of corrosion.

Electrolysis Electrolysis is the process in which electrical energy is used to cause a nonspontaneous redox reaction to occur. The quantitative relationship between the current supplied and the products formed is provided by Faraday. Electrolysis is the major method for producing active metals and nonmetals and many essential industrial chemicals.

19.1 Redox Reactions

***Electrochemistry** is the branch of chemistry that deals with the interconversion of electrical energy and chemical energy.* Electrochemical processes are redox (oxidation-reduction) reactions in which the energy released by a spontaneous reaction is converted to electricity or in which electrical energy is used to cause a nonspontaneous reaction to occur. Although redox reactions were discussed in Chapter 4, it is helpful to review some of the basic concepts that will come up again in this chapter.

In redox reactions, electrons are transferred from one substance to another. The reaction between magnesium metal and hydrochloric acid is an example of a redox reaction:

Rules for assigning oxidation numbers are presented in Section 4.4.

$$\overset{0}{Mg}(s) + 2\overset{+1}{H}Cl(aq) \longrightarrow \overset{+2}{Mg}Cl_2(aq) + \overset{0}{H_2}(g)$$

Recall that the numbers above the elements are the oxidation numbers of the elements. The loss of electrons by an element during oxidation is marked by an increase in the element's oxidation number. In reduction, there is a decrease in oxidation number resulting from a gain of electrons by an element. In the preceding reaction Mg metal is oxidized and H^+ ions are reduced; the Cl^- ions are spectator ions.

Balancing Redox Equations

Equations for redox reactions like the preceding one are relatively easy to balance. However, in the laboratory we often encounter more complex redox reactions involving oxoanions such as chromate (CrO_4^{2-}), dichromate ($Cr_2O_7^{2-}$), permanganate (MnO_4^-), nitrate (NO_3^-), and sulfate (SO_4^{2-}). In principle, we can balance any redox equation using the procedure outlined in Section 3.7, but there are some special techniques for handling redox reactions, techniques that also give us insight into electron transfer processes. Here we will discuss one such procedure, called the *half-reaction method.* In this approach, the overall reaction is divided into two half-reactions, one for oxidation and one for reduction. The equations for the two half-reactions are balanced separately and then added together to give the overall balanced equation.

Suppose we are asked to balance the equation showing the oxidation of Fe^{2+} ions to Fe^{3+} ions by dichromate ions ($Cr_2O_7^{2-}$) in an acidic medium. As a result, the $Cr_2O_7^{2-}$ ions are reduced to Cr^{3+} ions. The following steps will help us balance the equation.

Step 1: *Write the unbalanced equation for the reaction in ionic form.*

$$Fe^{2+} + Cr_2O_7^{2-} \longrightarrow Fe^{3+} + Cr^{3+}$$

Step 2: *Separate the equation into two half-reactions.*

$$\text{Oxidation:} \quad \overset{+2}{Fe^{2+}} \longrightarrow \overset{+3}{Fe^{3+}}$$

$$\text{Reduction:} \quad \overset{+6}{Cr_2}O_7^{2-} \longrightarrow \overset{+3}{Cr^{3+}}$$

Step 3: *Balance each half-reaction for number and type of atoms and charges. For reactions in an acidic medium, add H_2O to balance the O atoms and H^+ to balance the H atoms.*

In an oxidation half-reaction, electrons appear as a product; in a reduction half-reaction, electrons appear as a reactant.

Oxidation half-reaction: The atoms are already balanced. To balance the charge, we add an electron to the right-hand side of the arrow:

$$Fe^{2+} \longrightarrow Fe^{3+} + e^-$$

Reduction half-reaction: Because the reaction takes place in an acidic medium, we add seven H_2O molecules to the right-hand side of the arrow to balance the O atoms:

$$Cr_2O_7^{2-} \longrightarrow 2Cr^{3+} + 7H_2O$$

To balance the H atoms, we add 14 H^+ ions on the left-hand side:

$$14H^+ + Cr_2O_7^{2-} \longrightarrow 2Cr^{3+} + 7H_2O$$

There are now 12 positive charges on the left-hand side and only six positive charges on the right-hand side. Therefore, we add six electrons on the left

$$14H^+ + Cr_2O_7^{2-} + 6e^- \longrightarrow 2Cr^{3+} + 7H_2O$$

Step 4: *Add the two half-reactions together and balance the final equation by inspection. The electrons on both sides must cancel. If the oxidation and reduction half-reactions contain different numbers of electrons, we need to multiply one or both half-reactions to equalize the number of electrons.*

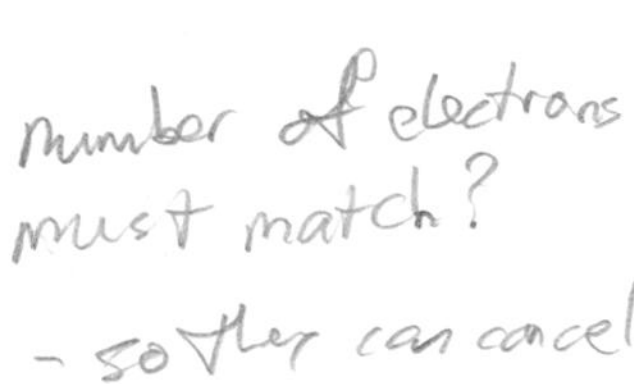
number of electrons
must match?
- so they can cancel

Here we have only one electron for the oxidation half-reaction and six electrons for the reduction half-reaction, so we need to multiply the oxidation half-reaction by 6 and write

$$\begin{array}{rcl} 6(Fe^{2+} & \longrightarrow & Fe^{3+} + e^-) \\ 14H^+ + Cr_2O_7^{2-} + 6e^- & \longrightarrow & 2Cr^{3+} + 7H_2O \\ \hline 6Fe^{2+} + 14H^+ + Cr_2O_7^{2-} + \cancel{6e^-} & \longrightarrow & 6Fe^{3+} + 2Cr^{3+} + 7H_2O + \cancel{6e^-} \end{array}$$

The electrons on both sides cancel, and we are left with the balanced net ionic equation:

$$6Fe^{2+} + 14H^+ + Cr_2O_7^{2-} \longrightarrow 6Fe^{3+} + 2Cr^{3+} + 7H_2O$$

This reaction can be carried out by dissolving potassium dichromate and iron(II) sulfate in a dilute sulfuric acid solution.

Step 5: *Verify that the equation contains the same type and numbers of atoms and the same charges on both sides of the equation.*

A final check shows that the resulting equation is "atomically" and "electrically" balanced.

For reactions in a basic medium, we proceed through step 4 as if the reaction were carried out in a acidic medium. Then, for every H^+ ion we add an equal number of OH^- ions to *both* sides of the equation. Where H^+ and OH^- ions appear on the same side of the equation, we combine the ions to give H_2O. Example 19.1 illustrates this procedure.

EXAMPLE 19.1

Write a balanced ionic equation to represent the oxidation of iodide ion (I^-) by permanganate ion (MnO_4^-) in basic solution to yield molecular iodine (I_2) and manganese(IV) oxide (MnO_2).

Strategy We follow the preceding procedure for balancing redox equations. Note that the reaction takes place in a basic medium.

Solution

Step 1: The unbalanced equation is

$$MnO_4^- + I^- \longrightarrow MnO_2 + I_2$$

(Continued)

Step 2: The two half-reactions are

$$\text{Oxidation:} \qquad \overset{-1}{\text{I}} \longrightarrow \overset{0}{\text{I}_2}$$

$$\text{Reduction:} \qquad \overset{+7}{\text{MnO}_4} \longrightarrow \overset{+4}{\text{MnO}_2}$$

Step 3: We balance each half-reaction for number and type of atoms and charges.
Oxidation half-reaction: We first balance the I atoms:

$$2\text{I}^- \longrightarrow \text{I}_2$$

To balance charges, we add two electrons to the right-hand side of the equation:

$$2\text{I}^- \longrightarrow \text{I}_2 + 2e^-$$

Reduction half-reaction: To balance the O atoms, we add two H_2O molecules on the right:

$$\text{MnO}_4^- \longrightarrow \text{MnO}_2 + 2\text{H}_2\text{O}$$

To balance the H atoms, we add four H^+ ions on the left:

$$\text{MnO}_4^- + 4\text{H}^+ \longrightarrow \text{MnO}_2 + 2\text{H}_2\text{O}$$

There are three net positive charges on the left, so we add three electrons to the same side to balance the charges:

$$\text{MnO}_4^- + 4\text{H}^+ + 3e^- \longrightarrow \text{MnO}_2 + 2\text{H}_2\text{O}$$

Step 4: We now add the oxidation and reduction half reactions to give the overall reaction. To equalize the number of electrons, we need to multiply the oxidation half-reaction by 3 and the reduction half-reaction by 2 as follows:

$$\begin{array}{r} 3(2\text{I}^- \longrightarrow \text{I}_2 + 2e^-) \\ 2(\text{MnO}_4^- + 4\text{H}^+ + 3e^- \longrightarrow \text{MnO}_2 + 2\text{H}_2\text{O}) \\ \hline 6\text{I}^- + 2\text{MnO}_4^- + 8\text{H}^+ + \cancel{6e^-} \longrightarrow 3\text{I}_2 + 2\text{MnO}_2 + 4\text{H}_2\text{O} + \cancel{6e^-} \end{array}$$

The electrons on both sides cancel, and we are left with the balanced net ionic equation:

$$6\text{I}^- + 2\text{MnO}_4^- + 8\text{H}^+ \longrightarrow 3\text{I}_2 + 2\text{MnO}_2 + 4\text{H}_2\text{O}$$

This is the balanced equation in an acidic medium. However, because the reaction is carried out in a basic medium, for every H^+ ion we need to add an equal number of OH^- ions to both sides of the equation:

$$6\text{I}^- + 2\text{MnO}_4^- + 8\text{H}^+ + 8\text{OH}^- \longrightarrow 3\text{I}_2 + 2\text{MnO}_2 + 4\text{H}_2\text{O} + 8\text{OH}^-$$

Finally, combining the H^+ and OH^- ions to form water, we obtain

$$6\text{I}^- + 2\text{MnO}_4^- + 4\text{H}_2\text{O} \longrightarrow 3\text{I}_2 + 2\text{MnO}_2 + 8\text{OH}^-$$

To carry out this reaction, mix KI and $KMnO_4$ in a basic medium.

Step 5: A final check shows that the equation is balanced in terms of both atoms and charges.

Similar problems: 19.1, 19.2.

Practice Exercise Balance the following equation for the reaction in an acidic medium by the half-reaction method:

$$\text{Fe}^{2+} + \text{MnO}_4^- \longrightarrow \text{Fe}^{3+} + \text{Mn}^{2+}$$

Review of Concepts

For the following reaction in acidic solution, what is the coefficient for NO_2 when the equation is balanced?

$$\text{Sn} + \text{NO}_3^- \longrightarrow \text{SnO}_2 + \text{NO}_2$$

19.2 Galvanic Cells

In Section 4.4 we saw that when a piece of zinc metal is placed in a $CuSO_4$ solution, Zn is oxidized to Zn^{2+} ions while Cu^{2+} ions are reduced to metallic copper (see Figure 4.13):

$$Zn(s) + Cu^{2+}(aq) \longrightarrow Zn^{2+}(aq) + Cu(s)$$

The electrons are transferred directly from the reducing agent (Zn) to the oxidizing agent (Cu^{2+}) in solution. However, if we physically separate the oxidizing agent from the reducing agent, the transfer of electrons can take place via an external conducting medium (a metal wire). As the reaction progresses, it sets up a constant flow of electrons and hence generates electricity (that is, it produces electrical work such as driving an electric motor).

Animation: Galvanic Cells

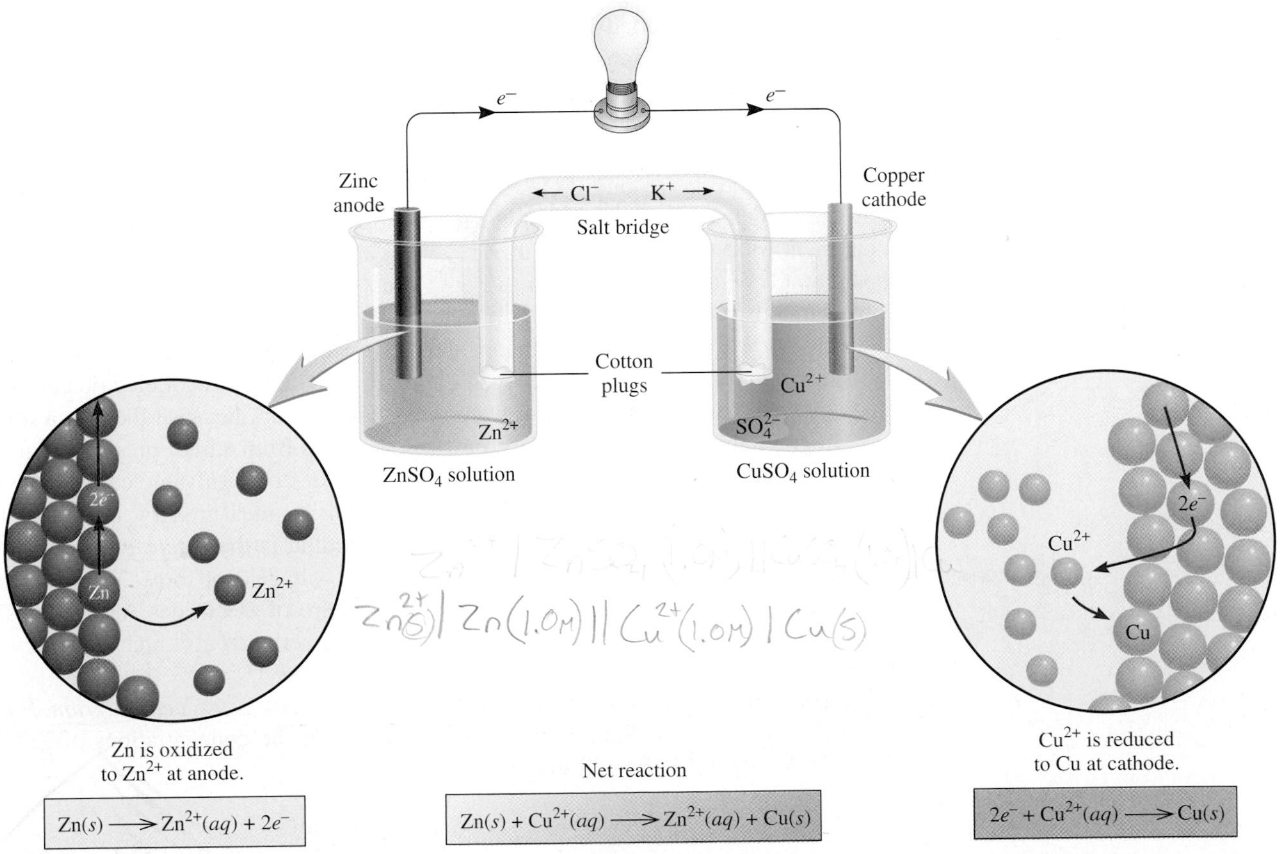

The experimental apparatus for generating electricity through the use of a spontaneous reaction is called a ***galvanic cell*** or *voltaic cell,* after the Italian scientists Luigi Galvani and Alessandro Volta, who constructed early versions of the device. Figure 19.1 shows the essential components of a galvanic cell. A zinc bar is immersed in a $ZnSO_4$ solution, and a copper bar is immersed in a $CuSO_4$ solution. The cell

e^- e^-

Zinc anode

$\leftarrow Cl^-$ $K^+ \rightarrow$

Salt bridge

Copper cathode

Cotton plugs

Zn^{2+} Cu^{2+} SO_4^{2-}

$ZnSO_4$ solution

$CuSO_4$ solution

Zn is oxidized to Zn^{2+} at anode.

$Zn(s) \longrightarrow Zn^{2+}(aq) + 2e^-$

Net reaction

$Zn(s) + Cu^{2+}(aq) \longrightarrow Zn^{2+}(aq) + Cu(s)$

Cu^{2+} is reduced to Cu at cathode.

$2e^- + Cu^{2+}(aq) \longrightarrow Cu(s)$

Figure 19.1
A galvanic cell. The salt bridge (an inverted U tube) containing a KCl solution provides an electrically conducting medium between two solutions. The openings of the U tube are loosely plugged with cotton balls to prevent the KCl solution from flowing into the containers while allowing the anions and cations to move across. The lightbulb is lit as electrons flow externally from the Zn electrode (anode) to the Cu electrode (cathode).

operates on the principle that the oxidation of Zn to Zn^{2+} and the reduction of Cu^{2+} to Cu can be made to take place simultaneously in separate locations with the transfer of electrons between them occurring through an external wire. The zinc and copper bars are called *electrodes.* This particular arrangement of electrodes (Zn and Cu) and solutions ($ZnSO_4$ and $CuSO_4$) is called the Daniell cell. By definition, the ***anode*** in a galvanic cell is *the electrode at which oxidation occurs* and the ***cathode*** is *the electrode at which reduction occurs.*

Alphabetically anode precedes cathode and oxidation precedes reduction. Therefore, anode is where oxidation occurs and cathode is where reduction takes place.

For the Daniell cell, the ***half-cell reactions,*** that is, *the oxidation and reduction reactions at the electrodes,* are

Half-cell reactions are similar to the half-reactions discussed earlier.

$$\text{Zn electrode (anode):} \quad Zn(s) \longrightarrow Zn^{2+}(aq) + 2e^-$$
$$\text{Cu electrode (cathode):} \quad Cu^{2+}(aq) + 2e^- \longrightarrow Cu(s)$$

Note that unless the two solutions are separated from each other, the Cu^{2+} ions will react directly with the zinc bar:

$$Cu^{2+}(aq) + Zn(s) \longrightarrow Cu(s) + Zn^{2+}(aq)$$

and no useful electrical work will be obtained.

Salt bridge - allows anions and cations to flow through the electric current instead of reacting directly. Allows for no build up of positive or negative energy

To complete the electrical circuit, the solutions must be connected by a conducting medium through which the cations and anions can move from one electrode compartment to the other. This requirement is satisfied by a *salt bridge,* which, in its simplest form, is an inverted U tube containing an inert electrolyte solution, such as KCl or NH_4NO_3, whose ions will not react with other ions in solution or with the electrodes (see Figure 19.1). During the course of the overall redox reaction, electrons flow externally from the anode (Zn electrode) through the wire and lightbulb to the cathode (Cu electrode). In the solution, the cations (Zn^{2+}, Cu^{2+}, and K^+) move toward the cathode, while the anions (SO_4^{2-} and Cl^-) move toward the anode. Without the salt bridge connecting the two solutions, the buildup of positive charge in the anode compartment (due to the formation of Zn^{2+} ions) and negative charge in the cathode compartment (created when some of the Cu^{2+} ions are reduced to Cu) would quickly prevent the cell from operating.

An electric current flows from the anode to the cathode because there is a difference in electrical potential energy between the electrodes. This flow of electric current is analogous to that of water down a waterfall, which occurs because there is a difference in gravitational potential energy, or the flow of gas from a high-pressure region to a low-pressure region. The voltage across the electrodes of a galvanic cell is called the ***cell voltage,*** or *cell potential.* Experimentally, this is measured by a voltmeter (Figure 19.2). Another common term for the cell potential is the ***electromotive force*** or ***emf (E),*** which, despite the name, is a measure of voltage, not force. We will see that the voltage of a cell depends not only on the nature of electrodes and the ions, but also on the concentrations of the ions and the temperature at which the cell is operated.

The conventional notation for representing galvanic cells is the *cell diagram.* For the Daniell cell shown in Figure 19.1, if we assume that the concentrations of Zn^{2+} and Cu^{2+} ions are 1 *M*, the cell diagram is

$$Zn(s)|Zn^{2+}(1\ M)||Cu^{2+}(1\ M)|Cu(s)$$

The single vertical line represents a phase boundary. For example, the zinc electrode is a solid and the Zn^{2+} ions (from $ZnSO_4$) are in solution. Thus, we draw a line between Zn and Zn^{2+} to show the phase boundary. The double vertical lines denote the salt bridge. By convention, the anode is written first, to the left of the double lines

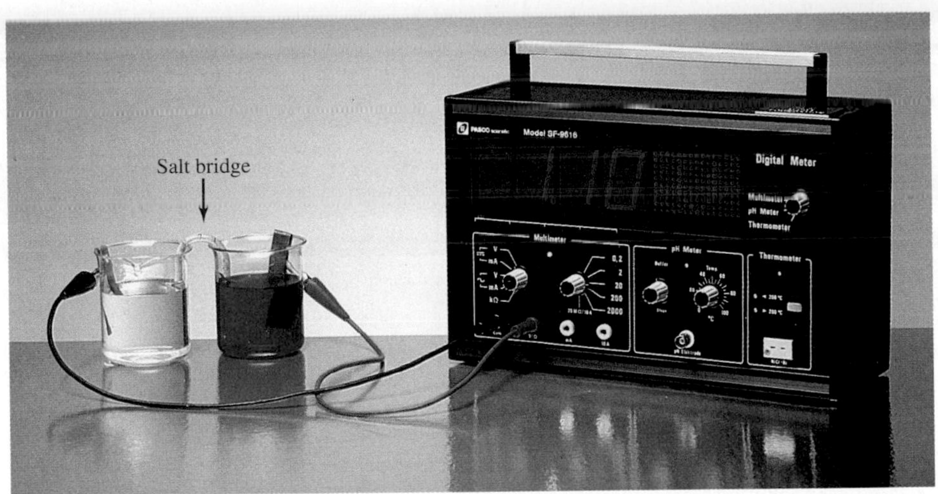

Figure 19.2
Practical setup of the galvanic cell described in Figure 19.1. Note the U tube (salt bridge) connecting the two beakers. When the concentrations of $ZnSO_4$ and $CuSO_4$ are 1 molar (1 M) at 25°C, the cell voltage is 1.10 V. No current flows between the electrodes during a voltage measurement.

and the other components appear in the order in which we would encounter them in moving from the anode to the cathode.

Review of Concepts

Write the cell diagram for the following redox reaction.

$$3Fe^{2+}(aq) + 2Al(s) \longrightarrow 3Fe(s) + 2Al^{3+}(aq)$$

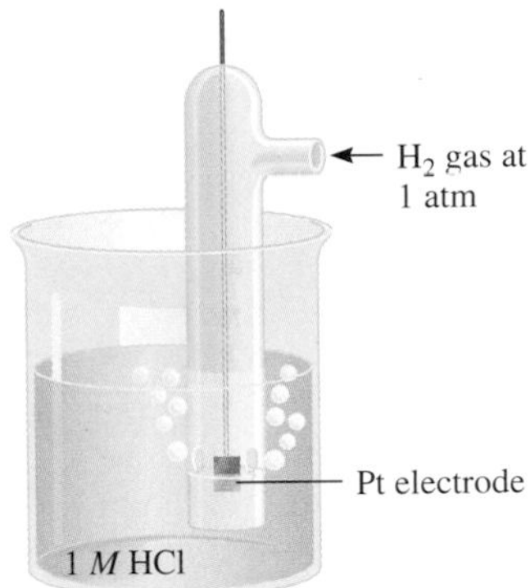

Figure 19.3
A hydrogen electrode operating under standard-state conditions. Hydrogen gas at 1 atm is bubbled through a 1 M HCl solution. The platinum electrode is part of the hydrogen electrode.

19.3 Standard Reduction Potentials

When the concentrations of the Cu^{2+} and Zn^{2+} ions are both 1.0 *M*, we find that the voltage or emf of the Daniell cell is 1.10 V at 25°C (see Figure 19.2). This voltage must be related directly to the redox reactions, but how? Just as the overall cell reaction can be thought of as the sum of two half-cell reactions, the measured emf of the cell can be treated as the sum of the electrical potentials at the Zn and Cu electrodes. Knowing one of these electrode potentials, we could obtain the other by subtraction (from 1.10 V). It is impossible to measure the potential of just a single electrode, but if we arbitrarily set the potential value of a particular electrode at zero, we can use it to determine the relative potentials of other electrodes. The hydrogen electrode, shown in Figure 19.3, serves as the reference for this purpose. Hydrogen gas is bubbled into a hydrochloric acid solution at 25°C. The platinum electrode has two functions. First, it provides a surface on which the dissociation of hydrogen molecules can take place:

The choice of an arbitrary reference for measuring electrode potential is analogous to choosing the surface of the ocean as the reference for altitude, calling it zero meters, and then referring to any terrestrial altitude as being a certain number of meters above or below sea level.

$$H_2 \longrightarrow 2H^+ + 2e^-$$

Second, it serves as an electrical conductor to the external circuit.

Under standard-state conditions (when the pressure of H_2 is 1 atm and the concentration of the HCl solution is 1 *M*; see Table 18.2), the potential for the reduction of H^+ at 25°C is taken to be *exactly* zero:

Standard states are defined in Table 18.2 (p. 642).

$$2H^+(1\ M) + 2e^- \longrightarrow H_2(1\ \text{atm}) \qquad E° = 0\ \text{V}$$

The superscript "°" denotes standard-state conditions, and $E°$ is the ***standard reduction potential,*** or *the voltage associated with a reduction reaction at an electrode*

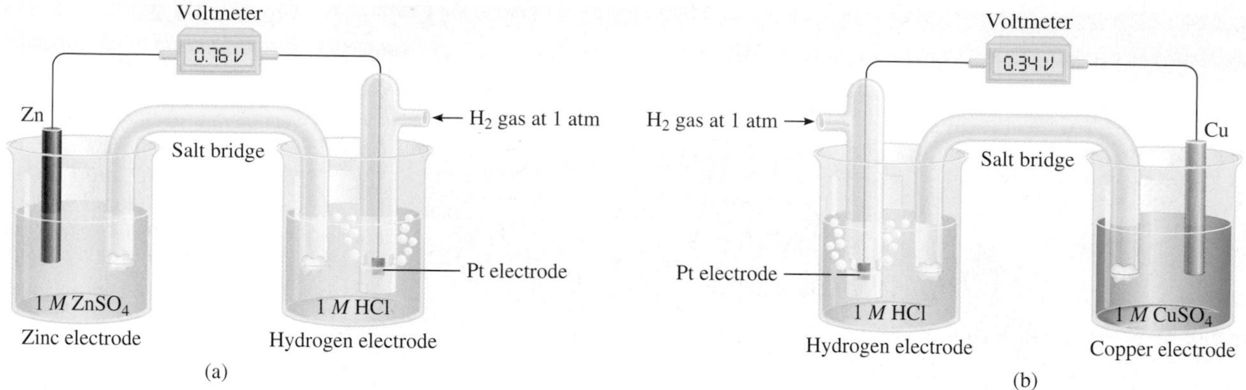

Figure 19.4
(a) A cell consisting of a zinc electrode and a hydrogen electrode. (b) A cell consisting of a copper electrode and a hydrogen electrode. Both cells are operating under standard-state conditions. Note that in (a) the SHE acts as the cathode, but in (b) it is the anode.

when all solutes are 1 M and all gases are at 1 atm. Thus, the standard reduction potential of the hydrogen electrode is defined as zero. The hydrogen electrode is called the *standard hydrogen electrode (SHE).*

We can use the SHE to measure the potentials of other kinds of electrodes. For example, Figure 19.4(a) shows a galvanic cell with a zinc electrode and a SHE. In this case the zinc electrode is the anode and the SHE is the cathode. We deduce this fact from the decrease in mass of the zinc electrode during the operation of the cell, which is consistent with the loss of zinc to the solution caused by the oxidation reaction:

$$\mathrm{Zn}(s) \longrightarrow \mathrm{Zn}^{2+}(aq) + 2e^-$$

The cell diagram is

$$\mathrm{Zn}(s)|\mathrm{Zn}^{2+}(1\,M)||\mathrm{H}^+(1\,M)|\mathrm{H}_2(1\text{ atm})|\mathrm{Pt}(s)$$

As mentioned earlier, the Pt electrode provides the surface on which the reduction takes place. When all the reactants are in their standard states (that is, H_2 at 1 atm, H^+ and Zn^{2+} ions at 1 M), the emf of the cell is 0.76 V at 25°C. We can write the half-cell reactions as follows:

$$\begin{array}{rrl} \text{Anode (oxidation):} & \mathrm{Zn}(s) & \longrightarrow \mathrm{Zn}^{2+}(1\,M) + 2e^- \\ \text{Cathode (reduction):} & 2\mathrm{H}^+(1\,M) + 2e^- & \longrightarrow \mathrm{H}_2(1\text{ atm}) \\ \hline \text{Overall:} & \mathrm{Zn}(s) + 2\mathrm{H}^+(1\,M) & \longrightarrow \mathrm{Zn}^{2+}(1\,M) + \mathrm{H}_2(1\text{ atm}) \end{array}$$

By convention, the ***standard emf*** of the cell, E°_{cell}, which is composed of a contribution from the anode and a contribution from the cathode, is given by

$$E^\circ_{\text{cell}} = E^\circ_{\text{cathode}} - E^\circ_{\text{anode}} \tag{19.1}$$

where *both* E°_{cathode} and E°_{anode} are the standard reduction potentials of the electrodes. For the Zn-SHE cell, we write

$$E^\circ_{\text{cell}} = E^\circ_{\mathrm{H^+/H_2}} - E^\circ_{\mathrm{Zn^{2+}/Zn}}$$
$$0.76\text{ V} = 0 - E^\circ_{\mathrm{Zn^{2+}/Zn}}$$

where the subscript H^+/H_2 means $2H^+ + 2e^- \longrightarrow H_2$ and the subscript Zn^{2+}/Zn means $Zn^{2+} + 2e^- \longrightarrow Zn$. Thus, the standard reduction potential of zinc, $E^\circ_{\mathrm{Zn^{2+}/Zn}}$, is −0.76 V.

The standard electrode potential of copper can be obtained in a similar fashion, by using a cell with a copper electrode and a SHE [Figure 19.4(b)]. In this case, the copper electrode is the cathode because its mass increases during the operation of the cell, as is consistent with the reduction reaction:

$$Cu^{2+}(aq) + 2e^- \longrightarrow Cu(s)$$

The cell diagram is

$$Pt(s)|H_2(1\ atm)|H^+(1\ M)||Cu^{2+}(1\ M)|Cu(s)$$

and the half-cell reactions are

$$\begin{array}{rrl} \text{Anode (oxidation):} & H_2(1\ atm) & \longrightarrow 2H^+(1\ M) + 2e^- \\ \text{Cathode (reduction):} & Cu^{2+}(1\ M) + 2e^- & \longrightarrow Cu(s) \\ \hline \text{Overall:} & H_2(1\ atm) + Cu^{2+}(1\ M) & \longrightarrow 2H^+(1\ M) + Cu(s) \end{array}$$

Under standard-state conditions and at 25°C, the emf of the cell is 0.34 V, so we write

$$\begin{aligned} E^\circ_{cell} &= E^\circ_{cathode} - E^\circ_{anode} \\ 0.34\ V &= E^\circ_{Cu^{2+}/Cu} - E^\circ_{H^+/H_2} \\ &= E^\circ_{Cu^{2+}/Cu} - 0 \end{aligned}$$

In this case, the standard reduction potential of copper, $E^\circ_{Cu^{2+}/Cu}$, is 0.34 V, where the subscript means $Cu^{2+} + 2e^- \longrightarrow Cu$.

For the Daniell cell shown in Figure 19.1, we can now write

$$\begin{array}{rrl} \text{Anode (oxidation):} & Zn(s) & \longrightarrow Zn^{2+}(1\ M) + 2e^- \\ \text{Cathode (reduction):} & Cu^{2+}(1\ M) + 2e^- & \longrightarrow Cu(s) \\ \hline \text{Overall:} & Zn(s) + Cu^{2+}(1\ M) & \longrightarrow Zn^{2+}(1\ M) + Cu(s) \end{array}$$

The emf of the cell is

$$\begin{aligned} E^\circ_{cell} &= E^\circ_{cathode} - E^\circ_{anode} \\ &= E^\circ_{Cu^{2+}/Cu} - E^\circ_{Zn^{2+}/Zn} \\ &= 0.34\ V - (-0.76\ V) \\ &= 1.10\ V \end{aligned}$$

As in the case of ΔG° (p. 648), we can use the sign of E° to predict the extent of a redox reaction. A positive E° means the redox reaction will favor the formation of products at equilibrium. Conversely, a negative E° means that more reactants than products will be formed at equilibrium. We will examine the relationships among E°_{cell}, ΔG°, and K later in this chapter.

Table 19.1 lists standard reduction potentials for a number of half-cell reactions. By definition, the SHE has an E° value of 0.00 V. Below the SHE the negative standard reduction potentials increase, and above it the positive standard reduction potentials increase. It is important to know the following points about the table in calculations:

The activity series in Figure 4.14 is based on data given in Table 19.1.

1. The E° values apply to the half-cell reactions as read in the forward (left to right) direction.
2. The more positive E° is, the greater the tendency for the substance to be reduced. For example, the half-cell reaction

$$F_2(1\ atm) + 2e^- \longrightarrow 2F^-(1\ M) \qquad E^\circ = 2.87\ V$$

Table 19.1 Standard Reduction Potentials at 25°C*

Half-Reaction	$E°$(V)
$F_2(g) + 2e^- \longrightarrow 2F^-(aq)$	+2.87
$O_3(g) + 2H^+(aq) + 2e^- \longrightarrow O_2(g) + H_2O$	+2.07
$Co^{3+}(aq) + e^- \longrightarrow Co^{2+}(aq)$	+1.82
$H_2O_2(aq) + 2H^+(aq) + 2e^- \longrightarrow 2H_2O$	+1.77
$PbO_2(s) + 4H^+(aq) + SO_4^{2-}(aq) + 2e^- \longrightarrow PbSO_4(s) + 2H_2O$	+1.70
$Ce^{4+}(aq) + e^- \longrightarrow Ce^{3+}(aq)$	+1.61
$MnO_4^-(aq) + 8H^+(aq) + 5e^- \longrightarrow Mn^{2+}(aq) + 4H_2O$	+1.51
$Au^{3+}(aq) + 3e^- \longrightarrow Au(s)$	+1.50
$Cl_2(g) + 2e^- \longrightarrow 2Cl^-(aq)$	+1.36
$Cr_2O_7^{2-}(aq) + 14H^+(aq) + 6e^- \longrightarrow 2Cr^{3+}(aq) + 7H_2O$	+1.33
$MnO_2(s) + 4H^+(aq) + 2e^- \longrightarrow Mn^{2+}(aq) + 2H_2O$	+1.23
$O_2(g) + 4H^+(aq) + 4e^- \longrightarrow 2H_2O$	+1.23
$Br_2(l) + 2e^- \longrightarrow 2Br^-(aq)$	+1.07
$NO_3^-(aq) + 4H^+(aq) + 3e^- \longrightarrow NO(g) + 2H_2O$	+0.96
$2Hg^{2+}(aq) + 2e^- \longrightarrow Hg_2^{2+}(aq)$	+0.92
$Hg_2^{2+}(aq) + 2e^- \longrightarrow 2Hg(l)$	+0.85
$Ag^+(aq) + e^- \longrightarrow Ag(s)$	+0.80
$Fe^{3+}(aq) + e^- \longrightarrow Fe^{2+}(aq)$	+0.77
$O_2(g) + 2H^+(aq) + 2e^- \longrightarrow H_2O_2(aq)$	+0.68
$MnO_4^-(aq) + 2H_2O + 3e^- \longrightarrow MnO_2(s) + 4OH^-(aq)$	+0.59
$I_2(s) + 2e^- \longrightarrow 2I^-(aq)$	+0.53
$O_2(g) + 2H_2O + 4e^- \longrightarrow 4OH^-(aq)$	+0.40
$Cu^{2+}(aq) + 2e^- \longrightarrow Cu(s)$	+0.34
$AgCl(s) + e^- \longrightarrow Ag(s) + Cl^-(aq)$	+0.22
$SO_4^{2-}(aq) + 4H^+(aq) + 2e^- \longrightarrow SO_2(g) + 2H_2O$	+0.20
$Cu^{2+}(aq) + e^- \longrightarrow Cu^+(aq)$	+0.15
$Sn^{4+}(aq) + 2e^- \longrightarrow Sn^{2+}(aq)$	+0.13
$2H^+(aq) + 2e^- \longrightarrow H_2(g)$	0.00
$Pb^{2+}(aq) + 2e^- \longrightarrow Pb(s)$	−0.13
$Sn^{2+}(aq) + 2e^- \longrightarrow Sn(s)$	−0.14
$Ni^{2+}(aq) + 2e^- \longrightarrow Ni(s)$	−0.25
$Co^{2+}(aq) + 2e^- \longrightarrow Co(s)$	−0.28
$PbSO_4(s) + 2e^- \longrightarrow Pb(s) + SO_4^{2-}(aq)$	−0.31
$Cd^{2+}(aq) + 2e^- \longrightarrow Cd(s)$	−0.40
$Fe^{2+}(aq) + 2e^- \longrightarrow Fe(s)$	−0.44
$Cr^{3+}(aq) + 3e^- \longrightarrow Cr(s)$	−0.74
$Zn^{2+}(aq) + 2e^- \longrightarrow Zn(s)$	−0.76
$2H_2O + 2e^- \longrightarrow H_2(g) + 2OH^-(aq)$	−0.83
$Mn^{2+}(aq) + 2e^- \longrightarrow Mn(s)$	−1.18
$Al^{3+}(aq) + 3e^- \longrightarrow Al(s)$	−1.66
$Be^{2+}(aq) + 2e^- \longrightarrow Be(s)$	−1.85
$Mg^{2+}(aq) + 2e^- \longrightarrow Mg(s)$	−2.37
$Na^+(aq) + e^- \longrightarrow Na(s)$	−2.71
$Ca^{2+}(aq) + 2e^- \longrightarrow Ca(s)$	−2.87
$Sr^{2+}(aq) + 2e^- \longrightarrow Sr(s)$	−2.89
$Ba^{2+}(aq) + 2e^- \longrightarrow Ba(s)$	−2.90
$K^+(aq) + e^- \longrightarrow K(s)$	−2.93
$Li^+(aq) + e^- \longrightarrow Li(s)$	−3.05

Increasing strength as oxidizing agent (↑)

Increasing strength as reducing agent (↓)

*For all half-reactions the concentration is 1 *M* for dissolved species and the pressure is 1 atm for gases. These are the standard-state values

has the highest positive $E°$ value among all the half-cell reactions. Thus, F_2 is the *strongest* oxidizing agent because it has the greatest tendency to be reduced. At the other extreme is the reaction

$$Li^+(1\,M) + e^- \longrightarrow Li(s) \qquad E° = -3.05\text{ V}$$

which has the most negative $E°$ value. Thus, Li^+ is the *weakest* oxidizing agent because it is the most difficult species to reduce. Conversely, we say that F^- is the weakest reducing agent and Li metal is the strongest reducing agent. Under standard-state conditions, the oxidizing agents (the species on the left-hand side of the half-reactions in Table 19.1) increase in strength from bottom to top and the reducing agents (the species on the right-hand side of the half-reactions) increase in strength from top to bottom.

3. The half-cell reactions are reversible. Depending on the conditions, any electrode can act either as an anode or as a cathode. Earlier we saw that the SHE is the cathode (H^+ is reduced to H_2) when coupled with zinc in a cell and that it becomes the anode (H_2 is oxidized to H^+) when used in a cell with copper.
4. Under standard-state conditions, any species on the left of a given half-cell reaction will react spontaneously with a species that appears on the right of any half-cell reaction located *below* it in Table 19.1. This principle is sometimes called the *diagonal rule.* In the case of the Daniell cell,

$$Cu^{2+}(1\,M) + 2e^- \longrightarrow Cu(s) \qquad E° = 0.34\text{ V}$$
$$Zn^{2+}(1\,M) + 2e^- \longrightarrow Zn(s) \qquad E° = -0.76\text{ V}$$

The diagonal red line shows that Cu^{2+} is the oxidizing agent and Zn is the reducing agent.

We see that the substance on the left of the first half-cell reaction is Cu^{2+} and the substance on the right in the second half-cell reaction is Zn. Therefore, as we saw earlier, Zn spontaneously reduces Cu^{2+} to form Zn^{2+} and Cu.

5. Changing the stoichiometric coefficients of a half-cell reaction *does not* affect the value of $E°$ because electrode potentials are intensive properties. This means that the value of $E°$ is unaffected by the size of the electrodes or the amount of solutions present. For example,

$$I_2(s) + 2e^- \longrightarrow 2I^-(1\,M) \qquad E° = 0.53\text{ V}$$

but $E°$ does not change if we multiply the half-reaction by 2:

$$2I_2(s) + 4e^- \longrightarrow 4I^-(1\,M) \qquad E° = 0.53\text{ V}$$

6. Like ΔH, ΔG, and ΔS, the sign of $E°$ changes but its magnitude remains the same when we reverse a reaction.

As Examples 19.2 and 19.3 show, Table 19.1 enables us to predict the outcome of redox reactions under standard-state conditions, whether they take place in a galvanic cell, where the reducing agent and oxidizing agent are physically separated from each other, or in a beaker, where the reactants are all mixed together.

EXAMPLE 19.2

Predict what will happen if molecular bromine (Br_2) is added to a solution containing NaCl and NaI at 25°C. Assume all species are in their standard states.

Strategy To predict what redox reaction(s) will take place, we need to compare the standard reduction potentials of Cl_2, Br_2, and I_2 and apply the diagonal rule.

(Continued)

Solution From Table 19.1, we write the standard reduction potentials as follows:

$$Cl_2(1\ \text{atm}) + 2e^- \longrightarrow 2Cl^-(1\ M) \qquad E° = 1.36\ \text{V}$$
$$Br_2(l) + 2e^- \longrightarrow 2Br^-(1\ M) \qquad E° = 1.07\ \text{V}$$
$$I_2(s) + 2e^- \longrightarrow 2I^-(1\ M) \qquad E° = 0.53\ \text{V}$$

Applying the diagonal rule we see that Br_2 will oxidize I^- but will not oxidize Cl^-. Therefore, the only redox reaction that will occur appreciably under standard-state conditions is

$$\text{Oxidation:} \qquad 2I^-(1\ M) \longrightarrow I_2(s) + 2e^-$$
$$\text{Reduction:} \qquad Br_2(l) + 2e^- \longrightarrow 2Br^-(1\ M)$$
$$\text{Overall:} \qquad 2I^-(1\ M) + Br_2(l) \longrightarrow I_2(s) + 2Br^-(1\ M)$$

Check We can confirm our conclusion by calculating $E°_{cell}$. Try it. Note that the Na^+ ions are inert and do not enter into the redox reaction.

Similar problems: 19.14, 19.17.

Practice Exercise Can Sn reduce $Zn^{2+}(aq)$ under standard-state conditions?

EXAMPLE 19.3

A galvanic cell consists of a Mg electrode in a 1.0 *M* $Mg(NO_3)_2$ solution and a Ag electrode in a 1.0 *M* $AgNO_3$ solution. Calculate the standard emf of this cell at 25°C.

Strategy At first it may not be clear how to assign the electrodes in the galvanic cell. From Table 19.1 we write the standard reduction potentials of Ag and Mg and apply the diagonal rule to determine which is the anode and which is the cathode.

Solution The standard reduction potentials are

$$Ag^+(1.0\ M) + e^- \longrightarrow Ag(s) \qquad E° = 0.80\ \text{V}$$
$$Mg^{2+}(1.0\ M) + 2e^- \longrightarrow Mg(s) \qquad E° = -2.37\ \text{V}$$

Applying the diagonal rule, we see that Ag^+ will oxidize Mg:

$$\text{Anode (oxidation):} \qquad Mg(s) \longrightarrow Mg^{2+}(1.0\ M) + 2e^-$$
$$\text{Cathode (reduction):} \qquad 2Ag^+(1.0\ M) + 2e^- \longrightarrow 2Ag(s)$$
$$\text{Overall:} \qquad Mg(s) + 2Ag^+(1.0\ M) \longrightarrow Mg^{2+}(1.0\ M) + 2Ag(s)$$

Note that in order to balance the overall equation we multiplied the reduction of Ag^+ by 2. We can do so because, as an intensive property, $E°$ is not affected by this procedure. We find the emf of the cell by using Equation (19.1) and Table 19.1:

$$\begin{aligned} E°_{cell} &= E°_{cathode} - E°_{anode} \\ &= E°_{Ag^+/Ag} - E°_{Mg^{2+}/Mg} \\ &= 0.80\ \text{V} - (-2.37\ \text{V}) \\ &= 3.17\ \text{V} \end{aligned}$$

Check The positive value of $E°$ shows that the forward reaction is favored.

Similar problems: 19.11, 19.12.

Practice Exercise What is the standard emf of a galvanic cell made of a Cd electrode in a 1.0 *M* $Cd(NO_3)_2$ solution and a Cr electrode in a 1.0 *M* $Cr(NO_3)_3$ solution at 25°C?

REVIEW OF CONCEPTS

Which of the following metals will react with (that is, be oxidized by) HNO_3, but not with HCl: Cu, Zn, Ag?

Final material

19.4 Thermodynamics of Redox Reactions

Our next step is to see how E°_{cell} is related to thermodynamic quantities such as ΔG° and K. In a galvanic cell, chemical energy is converted to electrical energy to do electrical work. Electrical energy in this case is the product of the emf of the cell and the total electrical charge (in coulombs) that passes through the cell:

$$\text{electrical energy} = \text{volts} \times \text{coulombs}$$
$$= \text{joules}$$

1 J = 1 V × 1 C

The total charge is determined by the number of electrons that pass through the cell, so we have

$$\text{total charge} = \text{number of } e^- \times \text{charge of one } e^-$$

F = the charge of one mole of electrons

In general, it is more convenient to express the total charge in molar quantities. The charge of one mole of electrons is called the ***Faraday constant (F)***, after the English chemist and physicist Michael Faraday, where

$$1\,F = 6.022 \times 10^{23}\ e^-/\text{mol } e^- \times 1.602 \times 10^{-19}\ \text{C}/e^-$$
$$F = 9.647 \times 10^4\ \text{C/mol } e^-$$

In most calculations, we round the Faraday constant to 96,500 C/mol e⁻.

Therefore, the total charge can now be expressed as nF, where n is the number of moles of electrons exchanged between the reducing agent and the oxidizing agent in the overall redox equation.

The measured emf (E_{cell}) is the *maximum* voltage the cell can achieve. It is given by the electrical work done (w_{ele}) divided by the total charge; that is,

$$E_{cell} = \frac{-w_{ele}}{\text{total charge}} = \frac{-w_{ele}}{nF}$$

or

$$w_{ele} = -nFE_{cell}$$

The negative sign indicates that the electrical work is done by the system (galvanic cell) on the surroundings. In Chapter 18 we defined free energy as the energy available to do work. Specifically, the change in free energy (ΔG) represents the maximum amount of useful work that can be obtained in a reaction:

The sign convention for electrical work is the same as that for *P-V* work, discussed in Section 6.3.

$$\Delta G = w_{max} = w_{ele}$$

Therefore, we can write

$$\Delta G = -nFE_{cell} \tag{19.2}$$

Both n and F are positive quantities and ΔG is negative for a spontaneous process, so E_{cell} must be positive. For reactions in which reactants and products are in their standard states, Equation (19.2) becomes

$$\Delta G^\circ = -nFE^\circ_{cell} \tag{19.3}$$

Now we can relate E°_{cell} to the equilibrium constant (K) of a redox reaction. In Section 18.5 we saw that the standard free-energy change ΔG° for a reaction is related to its equilibrium constant as follows [see Equation (18.14)]:

$$\Delta G^\circ = -RT \ln K$$

Therefore, if we combine Equations (18.14) and (19.3) we obtain

$$-nFE^\circ_{cell} = -RT \ln K$$

Table 19.2 Relationships Among $\Delta G°$, K, and $E°_{cell}$

$\Delta G°$	K	$E°_{cell}$	Reaction Under Standard-State Conditions
Negative	>1	Positive	Favors formation of products.
0	=1	0	Reactants and products are equally favored.
Positive	<1	Negative	Favors formation of reactants.

Solving for $E°_{cell}$

$$E°_{cell} = \frac{RT}{nF}\ln K \qquad (19.4)$$

When T = 298 K, Equation (19.4) can be simplified by substituting for R and F:

In calculations involving F, we sometimes omit the symbol e^-.

$$E°_{cell} = \frac{(8.314 \text{ J/K} \cdot \text{mol})(298 \text{ K})}{n(96{,}500 \text{ J/V} \cdot \text{mol})}\ln K$$

$$E°_{cell} = \frac{0.0257 \text{ V}}{n}\ln K \qquad (19.5)$$

Alternatively, Equation (19.5) can be written using the base-10 logarithm of K:

$$E°_{cell} = \frac{0.0592 \text{ V}}{n}\log K \qquad (19.6)$$

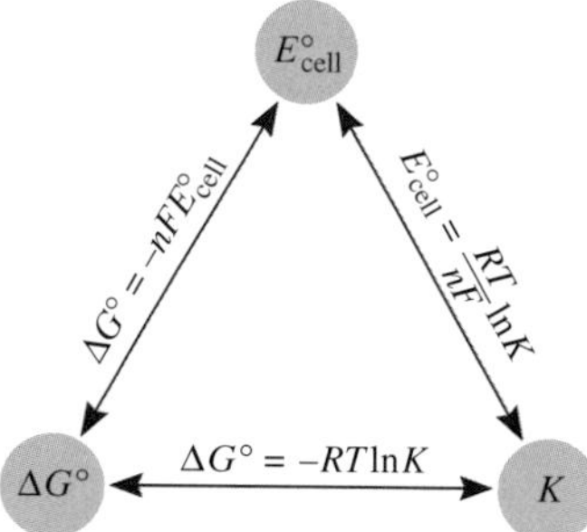

Figure 19.5
Relationships among $E°_{cell}$, K, and $\Delta G°$.

Thus, if any one of the three quantities $\Delta G°$, K, or $E°_{cell}$ is known, the other two can be calculated using Equation (18.14), Equation (19.3), or Equation (19.4) (Figure 19.5). We summarize the relationships among $\Delta G°$, K, and $E°_{cell}$ and characterize the spontaneity of a redox reaction in Table 19.2. For simplicity, we sometimes omit the subscript "cell" in $E°$ and E.

EXAMPLE 19.4

Calculate the equilibrium constant for the following reaction at 25°C:

$$Sn(s) + 2Cu^{2+}(aq) \rightleftharpoons Sn^{2+}(aq) + 2Cu^{+}(aq)$$

Strategy The relationship between the equilibrium constant K and the standard emf is given by Equation (19.5): $E°_{cell} = (0.0257 \text{ V}/n)\ln K$. Thus, if we can determine the standard emf, we can calculate the equilibrium constant. We can determine the $E°_{cell}$ of a hypothetical galvanic cell made up of two couples (Sn^{2+}/Sn and Cu^{2+}/Cu^{+}) from the standard reduction potentials in Table 19.1.

Solution The half-cell reactions are

$$\text{Anode (oxidation):} \quad Sn(s) \longrightarrow Sn^{2+}(aq) + 2e^-$$
$$\text{Cathode (reduction):} \quad 2Cu^{2+}(aq) + 2e^- \longrightarrow 2Cu^{+}(aq)$$

$$\begin{aligned} E°_{cell} &= E°_{cathode} - E°_{anode} \\ &= E°_{Cu^{2+}/Cu^{+}} - E°_{Sn^{2+}/Sn} \\ &= 0.15 \text{ V} - (-0.14 \text{ V}) \\ &= 0.29 \text{ V} \end{aligned}$$

(Continued)

Equation (19.5) can be written

$$\ln K = \frac{nE^\circ}{0.0257\ \text{V}}$$

In the overall reaction we find $n = 2$. Therefore,

$$\ln K = \frac{(2)(0.29\ \text{V})}{0.0257\ \text{V}} = 22.6$$

$$K = e^{22.6} = 7 \times 10^9$$

Similar problems: 19.21, 19.22.

Practice Exercise Calculate the equilibrium constant for the following reaction at 25°C:

$$Fe^{2+}(aq) + 2Ag(s) \rightleftharpoons Fe(s) + 2Ag^+(aq)$$

EXAMPLE 19.5

Calculate the standard free-energy change for the following reaction at 25°C:

$$2Au(s) + 3Ca^{2+}(1.0\ M) \longrightarrow 2Au^{3+}(1.0\ M) + 3Ca(s)$$

Strategy The relationship between the standard free energy change and the standard emf of the cell is given by Equation (19.3): $\Delta G^\circ = -nFE^\circ_{cell}$. Thus, if we can determine E°_{cell}, we can calculate ΔG°. We can determine the E°_{cell} of a hypothetical galvanic cell made up of two couples (Au^{3+}/Au and Ca^{2+}/Ca) from the standard reduction potentials in Table 19.1.

Solution The half-cell reactions are

$$\text{Anode (oxidation):}\quad 2Au(s) \longrightarrow 2Au^{3+}(1.0\ M) + 6e^-$$
$$\text{Cathode (reduction):}\quad 3Ca^{2+}(1.0\ M) + 6e^- \longrightarrow 3Ca(s)$$

$$\begin{aligned} E^\circ_{cell} &= E^\circ_{cathode} - E^\circ_{anode} \\ &= E^\circ_{Ca^{2+}/Ca} - E^\circ_{Au^{3+}/Au} \\ &= -2.87\ \text{V} - 1.50\ \text{V} \\ &= -4.37\ \text{V} \end{aligned}$$

Now we use Equation (19.3):

$$\Delta G^\circ = -nFE^\circ$$

The overall reaction shows that $n = 6$, so

$$\begin{aligned} \Delta G^\circ &= -(6)(96{,}500\ \text{J/V} \cdot \text{mol})(-4.37\text{V}) \\ &= 2.53 \times 10^6\ \text{J/mol} \\ &= 2.53 \times 10^3\ \text{kJ/mol} \end{aligned}$$

Check The large positive value of ΔG° tells us that the reaction favors the reactants at equilibrium. The result is consistent with the fact that E° for the galvanic cell is negative.

Similar problem: 19.24.

Practice Exercise Calculate ΔG° for the following reaction at 25°C:

$$2Al^{3+}(aq) + 3Mg(s) \rightleftharpoons 2Al(s) + 3Mg^{2+}(aq)$$

Review of Concepts

What is the sign of E° for a redox reaction where K is less than one?

19.5 The Effect of Concentration on Cell Emf

So far we have focused on redox reactions in which reactants and products are in their standard states, but standard-state conditions are often difficult, and sometimes impossible, to maintain. However, there is a mathematical relationship between the emf of a galvanic cell and the concentration of reactants and products in a redox reaction under nonstandard-state conditions. This equation is derived next.

The Nernst Equation

Consider a redox reaction of the type

$$a\text{A} + b\text{B} \longrightarrow c\text{C} + d\text{D}$$

From Equation (18.13)

$$\Delta G = \Delta G^\circ + RT \ln Q$$

Because $\Delta G = -nFE$ and $\Delta G^\circ = -nFE^\circ$, the equation can be expressed as

$$-nFE = -nFE^\circ + RT \ln Q$$

Dividing the equation through by $-nF$, we get

$$E = E^\circ - \frac{RT}{nF} \ln Q \qquad (19.7)$$

where Q is the reaction quotient (see Section 15.3). Equation (19.7) is known as the ***Nernst equation*** (after the German chemist Walther Nernst). At 298 K, Equation (19.7) can be rewritten as

Note that the Nernst equation is used to calculate the cell voltage under nonstandard state conditions.

$$E = E^\circ - \frac{0.0257\ \text{V}}{n} \ln Q \qquad (19.8)$$

or, expressing Equation (19.8) using the base-10 logarithm of Q:

$$E = E^\circ - \frac{0.0592\ \text{V}}{n} \log Q \qquad (19.9)$$

During the operation of a galvanic cell, electrons flow from the anode to the cathode, resulting in product formation and a decrease in reactant concentration. Thus, Q increases, which means that E decreases. Eventually, the cell reaches equilibrium. At equilibrium, there is no net transfer of electrons, so $E = 0$ and $Q = K$, where K is the equilibrium constant.

The Nernst equation enables us to calculate E as a function of reactant and product concentrations in a redox reaction. For example, for the Daniell cell in Figure 19.1,

$$\text{Zn}(s) + \text{Cu}^{2+}(aq) \longrightarrow \text{Zn}^{2+}(aq) + \text{Cu}(s)$$

The Nernst equation for this cell at 25°C can be written as

Remember that concentrations of pure solids (and pure liquids) do not appear in the expression for Q (see p. 519).

$$E = 1.10\ \text{V} - \frac{0.0257\ \text{V}}{2} \ln \frac{[\text{Zn}^{2+}]}{[\text{Cu}^{2+}]}$$

If the ratio $[Zn^{2+}]/[Cu^{2+}]$ is less than 1, ln $([Zn^{2+}]/[Cu^{2+}])$ is a negative number, so that the second term on the right-hand side of the preceding equation is positive. Under this condition E is greater than the standard emf $E°$. If the ratio is greater than 1, E is smaller than $E°$.

EXAMPLE 19.6

Predict whether the following reaction would proceed spontaneously as written at 298 K:

$$Co(s) + Fe^{2+}(aq) \longrightarrow Co^{2+}(aq) + Fe(s)$$

given that $[Co^{2+}] = 0.25\ M$ and $[Fe^{2+}] = 0.94\ M$.

Strategy Because the reaction is not run under standard-state conditions (concentrations are not 1 *M*), we need Nernst's equation [Equation (19.8)] to calculate the emf (*E*) of a hypothetical galvanic cell and determine the spontaneity of the reaction. The standard emf (*E*°) can be calculated using the standard reduction potentials in Table 19.1. Remember that solids do not appear in the reaction quotient (*Q*) term in the Nernst equation. Note that 2 moles of electrons are transferred per mole of reaction, that is, $n = 2$.

Solution The half-cell reactions are

$$\begin{aligned} &\text{Anode (oxidation):} & Co(s) &\longrightarrow Co^{2+}(aq) + 2e^- \\ &\text{Cathode (reduction):} & Fe^{2+}(aq) + 2e^- &\longrightarrow Fe(s) \end{aligned}$$

$$\begin{aligned} E°_{cell} &= E°_{cathode} - E°_{anode} \\ &= E°_{Fe^{2+}/Fe} - E°_{Co^{2+}/Co} \\ &= -0.44\ V - (-0.28\ V) \\ &= -0.16\ V \end{aligned}$$

From Equation (19.8) we write

$$\begin{aligned} E &= E° - \frac{0.0257\ V}{n} \ln Q \\ &= E° - \frac{0.0257\ V}{n} \ln \frac{[Co^{2+}]}{[Fe^{2+}]} \\ &= -0.16\ V - \frac{0.0257\ V}{2} \ln \frac{0.25}{0.94} \\ &= -0.16\ V + 0.017\ V \\ &= -0.14\ V \end{aligned}$$

Because E is negative, the reaction is not spontaneous in the direction written.

Similar problems: 19.29, 19.30.

Practice Exercise Will the following reaction occur spontaneously at 25°C, given that $[Fe^{2+}] = 0.60\ M$ and $[Cd^{2+}] = 0.010\ M$?

$$Cd(s) + Fe^{2+}(aq) \longrightarrow Cd^{2+}(aq) + Fe(s)$$

Now suppose we want to determine at what ratio of $[Co^{2+}]$ to $[Fe^{2+}]$ the reaction in Example 19.6 would become spontaneous. We can use Equation (19.8) as follows:

$$E = E° - \frac{0.0257\ V}{n} \ln Q$$

When $E = 0$, $Q = K$.

We first set E equal to zero, which corresponds to the equilibrium situation.

$$0 = -0.16\text{ V} - \frac{0.0257\text{ V}}{2}\ln\frac{[\text{Co}^{2+}]}{[\text{Fe}^{2+}]}$$

$$\ln\frac{[\text{Co}^{2+}]}{[\text{Fe}^{2+}]} = -12.5$$

$$\frac{[\text{Co}^{2+}]}{[\text{Fe}^{2+}]} = e^{-12.5} = K$$

or

$$K = 4 \times 10^{-6}$$

Thus, for the reaction to be spontaneous, the ratio $[Co^{2+}]/[Fe^{2+}]$ must be smaller than 4×10^{-6} so that E would become positive.

As Example 19.7 shows, if gases are involved in the cell reaction, their concentrations should be expressed in atm.

EXAMPLE 19.7

Consider the galvanic cell shown in Figure 19.4(a). In a certain experiment, the emf (E) of the cell is found to be 0.54 V at 25°C. Suppose that $[Zn^{2+}] = 2.5\ M$ and $P_{H_2} = 1.0$ atm. Calculate the molar concentration of H^+.

Strategy The equation that relates standard emf and nonstandard emf is the Nernst equation. The overall cell reaction is

$$\text{Zn}(s) + 2\text{H}^+(?\ M) \longrightarrow \text{Zn}^{2+}(2.5\ M) + \text{H}_2(1.0\text{ atm})$$

Given the emf of the cell (E), we apply the Nernst equation to solve for $[H^+]$. Note that 2 moles of electrons are transferred per mole of reaction; that is, $n = 2$.

Solution As we saw earlier (p. 668), the standard emf ($E°$) for the cell is 0.76 V. From Equation (19.8) we write

The concentrations in Q are divided by their standard-state value of 1 M and pressure is divided by 1 atm.

$$E = E° - \frac{0.0257\text{ V}}{n}\ln Q$$

$$= E° - \frac{0.0257\text{ V}}{n}\ln\frac{[\text{Zn}^{2+}]P_{\text{H}_2}}{[\text{H}^+]^2}$$

$$0.54\text{ V} = 0.76\text{ V} - \frac{0.0257\text{ V}}{2}\ln\frac{(2.5)(1.0)}{[\text{H}^+]^2}$$

$$-0.22\text{ V} = \frac{0.0257\text{ V}}{2}\ln\frac{2.5}{[\text{H}^+]^2}$$

$$17.1 = \ln\frac{2.5}{[\text{H}^+]^2}$$

$$e^{17.1} = \frac{2.5}{[\text{H}^+]^2}$$

Similar problem: 19.32.

$$[\text{H}^+] = \sqrt{\frac{2.5}{3 \times 10^7}} = 3 \times 10^{-4}\ M$$

Practice Exercise What is the emf of a galvanic cell consisting of a Cd^{2+}/Cd half-cell and a $Pt/H^+/H_2$ half-cell if $[Cd^{2+}] = 0.20\ M$, $[H^+] = 0.16\ M$, and $P_{H_2} = 0.80$ atm?

Review of Concepts

Consider the following cell diagram:

$$Mg(s)|MgSO_4(0.40\ M)||NiSO_4(0.60\ M)|Ni(s)$$

Calculate the cell voltage at 25°C. How does the cell voltage change when (a) $[Mg^{2+}]$ is decreased by a factor of 4 and (b) $[Ni^{2+}]$ is decreased by a factor of 3?

Example 19.7 shows that a galvanic cell whose cell reaction involves H^+ ions can be used to measure $[H^+]$ or pH. The pH meter described in Section 16.3 is based on this principle. However, the hydrogen electrode (see Figure 19.3) is normally not employed in laboratory work because it is awkward to use. Instead, it is replaced by a *glass electrode,* shown in Figure 19.6. The electrode consists of a very thin glass membrane that is permeable to H^+ ions. A silver wire coated with silver chloride is immersed in a dilute hydrochloric acid solution. When the electrode is placed in a solution whose pH is different from that of the inner solution, the potential difference that develops between the two sides of the membrane can be monitored using a reference electrode. The emf of the cell made up of the glass electrode and the reference electrode is measured with a voltmeter that is calibrated in pH units.

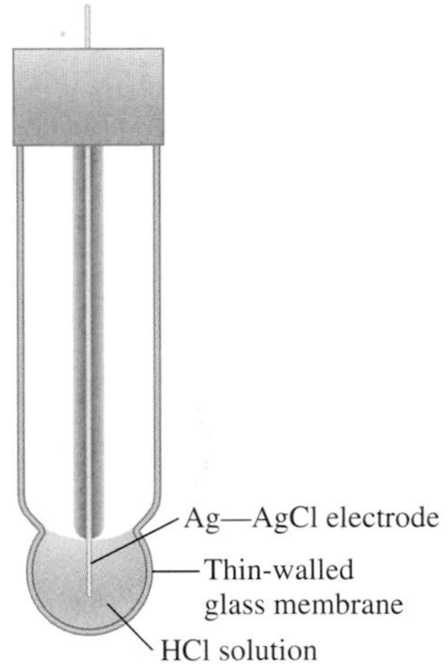

Figure 19.6
A glass electrode that is used in conjunction with a reference electrode in a pH meter.

Concentration Cells

Because electrode potential depends on ion concentrations, it is possible to construct a galvanic cell from two half-cells composed of the *same* material but differing in ion concentrations. Such a cell is called a *concentration cell.*

Consider a situation in which zinc electrodes are put into two aqueous solutions of zinc sulfate at 0.10 *M* and 1.0 *M* concentrations. The two solutions are connected by a salt bridge, and the electrodes are joined by a piece of wire in an arrangement like that shown in Figure 19.1. According to Le Châtelier's principle, the tendency for the reduction

$$Zn^{2+}(aq) + 2e^- \longrightarrow Zn(s)$$

increases with increasing concentration of Zn^{2+} ions. Therefore, reduction should occur in the more concentrated compartment and oxidation should take place on the more dilute side. The cell diagram is

$$Zn(s)|Zn^{2+}(0.10\ M)||Zn^{2+}(1.0\ M)|Zn(s)$$

and the half-reactions are

$$\begin{array}{rrcl} \text{Oxidation:} & Zn(s) & \longrightarrow & Zn^{2+}(0.10\ M) + 2e^- \\ \text{Reduction:} & Zn^{2+}(1.0\ M) + 2e^- & \longrightarrow & Zn(s) \\ \hline \text{Overall:} & Zn^{2+}(1.0\ M) & \longrightarrow & Zn^{2+}(0.10\ M) \end{array}$$

The emf of the cell is

$$E = E° - \frac{0.0257\ V}{2} \ln \frac{[Zn^{2+}]_{dil}}{[Zn^{2+}]_{conc}}$$

where the subscripts "dil" and "conc" refer to the 0.10 *M* and 1.0 *M* concentrations, respectively. The $E°$ for this cell is zero (the same electrode and the same type of ions are involved), so

$$E = 0 - \frac{0.0257\text{ V}}{2}\ln\frac{0.10}{1.0}$$
$$= 0.0296\text{ V}$$

The emf of concentration cells is usually small and decreases continually during the operation of the cell as the concentrations in the two compartments approach each other. When the concentrations of the ions in the two compartments are the same, E becomes zero, and no further change occurs.

A biological cell can be compared to a concentration cell for the purpose of calculating its *membrane potential.* Membrane potential is the electrical potential that exists across the membrane of various kinds of cells, including muscle cells and nerve cells. It is responsible for the propagation of nerve impulses and heartbeat. A membrane potential is established whenever there are unequal concentrations of the same type of ion in the interior and exterior of a cell. For example, the concentrations of K^+ ions in the interior and exterior of a nerve cell are 400 m*M* and 15 m*M*, respectively. Treating the situation as a concentration cell and applying the Nernst equation for just one kind of ions, we can write

1 m*M* = 1×10^{-3} *M*.

$$E = E° - \frac{0.0257\text{ V}}{1}\ln\frac{[K^+]_{ex}}{[K^+]_{in}}$$
$$= -(0.0257\text{ V})\ln\frac{15}{400}$$
$$= 0.084\text{ V or } 84\text{ mV}$$

where "ex" and "in" denote exterior and interior. Note that we have set $E° = 0$ because the same type of ion is involved. Thus, an electrical potential of 84 mV exists across the membrane due to the unequal concentrations of K^+ ions.

19.6 Batteries

A ***battery*** is *a galvanic cell, or a series of combined galvanic cells, that can be used as a source of direct electric current at a constant voltage.* Although the operation of a battery is similar in principle to that of the galvanic cells described in Section 19.2, a battery has the advantage of being completely self-contained and requiring no auxiliary components such as salt bridges. Here we will discuss several types of batteries that are in widespread use.

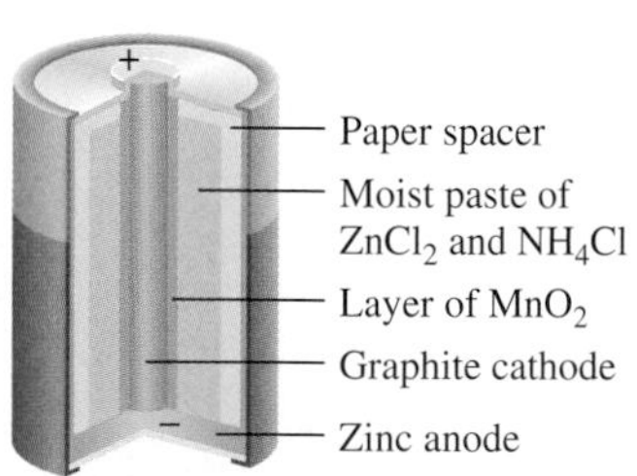

Figure 19.7
Interior section of a dry cell of the kind used in flashlights and transistor radios. Actually, the cell is not completely dry, as it contains a moist electrolyte paste.

The Dry Cell Battery

The most common dry cell, that is, a cell without a fluid component, is the *Leclanché cell* used in flashlights and transistor radios. The anode of the cell consists of a zinc can or container that is in contact with manganese dioxide (MnO_2) and an electrolyte. The electrolyte consists of ammonium chloride and zinc chloride in water, to which starch is added to thicken the solution to a pastelike consistency so that it is less likely to leak (Figure 19.7). A carbon rod serves as the cathode, which is immersed in the electrolyte in the center of the cell. The cell reactions are

$$\begin{aligned}
\text{Anode:} \quad & Zn(s) \longrightarrow Zn^{2+}(aq) + 2e^- \\
\text{Cathode:} \quad & 2NH_4^+(aq) + 2MnO_2(s) + 2e^- \longrightarrow Mn_2O_3(s) + 2NH_3(aq) + H_2O(l) \\
\hline
\text{Overall:} \quad & Zn(s) + 2NH_4^+(aq) + 2MnO_2(s) \longrightarrow Zn^{2+}(aq) + 2NH_3(aq) + H_2O(l) + Mn_2O_3(s)
\end{aligned}$$

Actually, this equation is an oversimplification of a complex process. The voltage produced by a dry cell is about 1.5 V.

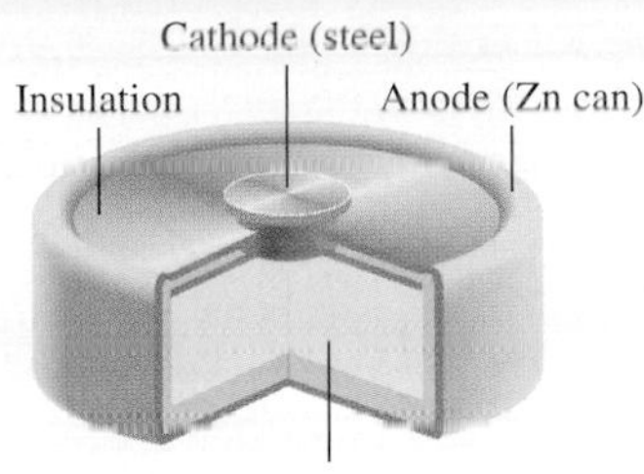

Figure 19.8
Interior section of a mercury battery.

The Mercury Battery

The mercury battery is used extensively in medicine and electronic industries and is more expensive than the common dry cell. Contained in a stainless steel cylinder, the mercury battery consists of a zinc anode (amalgamated with mercury) in contact with a strongly alkaline electrolyte containing zinc oxide and mercury(II) oxide (Figure 19.8). The cell reactions are

$$\begin{aligned}
\text{Anode:} \quad & Zn(Hg) + 2OH^-(aq) \longrightarrow ZnO(s) + H_2O(l) + 2e^- \\
\text{Cathode:} \quad & HgO(s) + H_2O(l) + 2e^- \longrightarrow Hg(l) + 2OH^-(aq) \\
\hline
\text{Overall:} \quad & Zn(Hg) + HgO(s) \longrightarrow ZnO(s) + Hg(l)
\end{aligned}$$

Because there is no change in electrolyte composition during operation—the overall cell reaction involves only solid substances—the mercury battery provides a more constant voltage (1.35 V) than the Leclanché cell. It also has a considerably higher capacity and longer life. These qualities make the mercury battery ideal for use in pacemakers, hearing aids, electric watches, and light meters.

The Lead Storage Battery

The lead storage battery commonly used in automobiles consists of six identical cells joined together in series. Each cell has a lead anode and a cathode made of lead dioxide (PbO_2) packed on a metal plate (Figure 19.9). Both the cathode and the anode

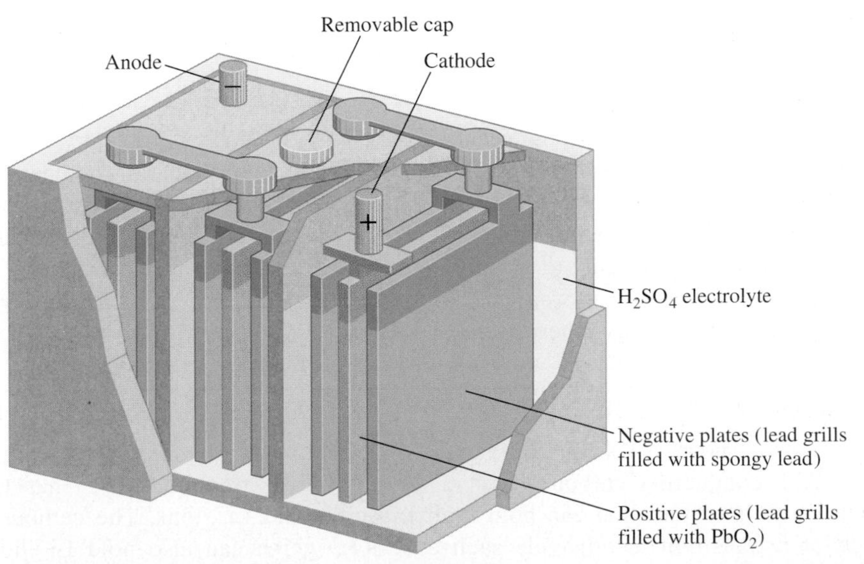

Figure 19.9
Interior section of a lead storage battery. Under normal operating conditions, the concentration of the sulfuric acid solution is about 38 percent by mass.

are immersed in an aqueous solution of sulfuric acid, which acts as the electrolyte. The cell reactions are

$$\begin{array}{lrcl} \text{Anode:} & Pb(s) + SO_4^{2-}(aq) & \longrightarrow & PbSO_4(s) + 2e^- \\ \text{Cathode:} & PbO_2(s) + 4H^+(aq) + SO_4^{2-}(aq) + 2e^- & \longrightarrow & PbSO_4(s) + 2H_2O(l) \\ \hline \text{Overall:} & Pb(s) + PbO_2(s) + 4H^+(aq) + 2SO_4^{2-}(aq) & \longrightarrow & 2PbSO_4(s) + 2H_2O(l) \end{array}$$

Under normal operating conditions, each cell produces 2 V; a total of 12 V from the six cells is used to power the ignition circuit of the automobile and its other electrical systems. The lead storage battery can deliver large amounts of current for a short time, such as the time it takes to start up the engine.

Unlike the Leclanché cell and the mercury battery, the lead storage battery is rechargeable. Recharging the battery means reversing the normal electrochemical reaction by applying an external voltage at the cathode and the anode. (This kind of process is called *electrolysis,* see p. 687.) The reactions that replenish the original materials are

$$\begin{array}{lrcl} & PbSO_4(s) + 2e^- & \longrightarrow & Pb(s) + SO_4^{2-}(aq) \\ & PbSO_4(s) + 2H_2O(l) & \longrightarrow & PbO_2(s) + 4H^+(aq) + SO_4^{2-}(aq) + 2e^- \\ \hline \text{Overall:} & 2PbSO_4(s) + 2H_2O(l) & \longrightarrow & Pb(s) + PbO_2(s) + 4H^+(aq) + 2SO_4^{2-}(aq) \end{array}$$

The overall reaction is exactly the opposite of the normal cell reaction.

Two aspects of the operation of a lead storage battery are worth noting. First, because the electrochemical reaction consumes sulfuric acid, the degree to which the battery has been discharged can be checked by measuring the density of the electrolyte with a *hydrometer,* as is usually done at gas stations. The density of the fluid in a "healthy," fully charged battery should be equal to or greater than 1.2 g/mL. Second, people living in cold climates sometimes have trouble starting their cars because the battery has "gone dead." Thermodynamic calculations show that the emf of many galvanic cells decreases with decreasing temperature. However, for a lead storage battery, the temperature coefficient is about 1.5×10^{-4} V/°C; that is, there is a decrease in voltage of 1.5×10^{-4} V for every degree drop in temperature. Thus, even allowing for a 40°C change in temperature, the decrease in voltage amounts to only 6×10^{-3} V, which is about

$$\frac{6 \times 10^{-3}\text{ V}}{12\text{ V}} \times 100\% = 0.05\%$$

of the operating voltage, an insignificant change. The real cause of a battery's apparent breakdown is an increase in the viscosity of the electrolyte as the temperature decreases. For the battery to function properly, the electrolyte must be fully conducting. However, the ions move much more slowly in a viscous medium, so the resistance of the fluid increases, leading to a decrease in the power output of the battery. If an apparently "dead battery" is warmed to near room temperature on a frigid day, it recovers its ability to deliver normal power.

The Lithium-Ion Battery

Figure 19.10 shows a schematic diagram of a lithium-ion battery. The anode is made of a conducting carbonaceous material, usually graphite, which has tiny spaces in its structure that can hold both Li atoms and Li^+ ions. The cathode is made of a transition metal oxide such as CoO_2, which can also hold Li^+ ions.

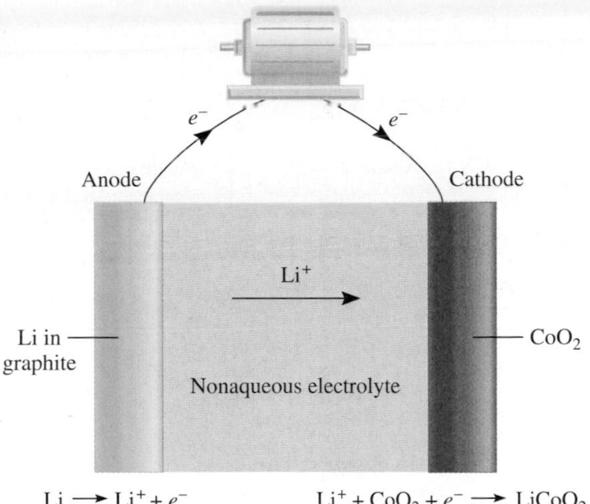

Figure 19.10
A lithium-ion battery. Lithium atoms are embedded in the graphite, which serves as the anode and CoO_2 is the cathode. During operation, Li^+ ions migrate through the nonaqueous electrolyte from the anode to the cathode while electrons flow externally from the anode to the cathode to complete the circuit.

Because of the high reactivity of the metal, nonaqueous electrolyte (organic solvent plus dissolved salt) must be used. During the discharge of the battery, the half-cell reactions are

$$\begin{array}{lrcl} \text{Anode (oxidation):} & Li(s) & \longrightarrow & Li^+ + e^- \\ \text{Cathode (reduction):} & Li^+ + CoO_2 + e^- & \longrightarrow & LiCoO_2(s) \\ \hline \text{Overall:} & Li(s) + CoO_2 & \longrightarrow & LiCoO_2(s) \qquad E_{\text{cell}} = 3.4\text{ V} \end{array}$$

The advantage of the battery is that lithium has the most negative standard reduction potential (see Table 19.1) and hence the greatest reducing strength. Furthermore, lithium is the lightest metal so that only 6.941 g of Li (its molar mass) are needed to produce 1 mole of electrons. A lithium-ion battery can be recharged literally hundreds of times without deterioration. These desirable characteristics make it suitable for use in cellular telephones, digital cameras, and laptop computers.

Fuel Cells

Fossil fuels are a major source of energy, but conversion of fossil fuel into electrical energy is a highly inefficient process. Consider the combustion of methane:

$$CH_4(g) + 2O_2(g) \longrightarrow CO_2(g) + 2H_2O(l) + \text{energy}$$

To generate electricity, heat produced by the reaction is first used to convert water to steam, which then drives a turbine that drives a generator. An appreciable fraction of the energy released in the form of heat is lost to the surroundings at each step; even the most efficient power plant converts only about 40 percent of the original chemical energy into electricity. Because combustion reactions are redox reactions, it is more desirable to carry them out directly by electrochemical means, thereby greatly increasing the efficiency of power production. This objective can be accomplished by a device known as a ***fuel cell,*** *a galvanic cell that requires a continuous supply of reactants to keep functioning.*

In its simplest form, a hydrogen-oxygen fuel cell consists of an electrolyte solution, such as potassium hydroxide solution, and two inert electrodes. Hydrogen and oxygen gases are bubbled through the anode and cathode compartments

A car powered by hydrogen fuel cells manufactured by General Motors.

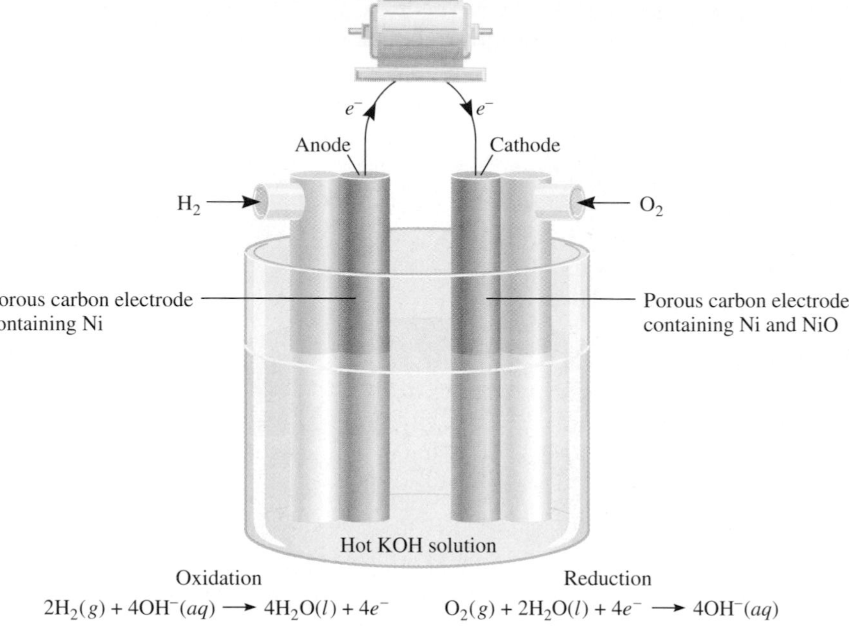

Figure 19.11
A hydrogen-oxygen fuel cell. The Ni and NiO embedded in the porous carbon electrodes are electrocatalysts.

(Figure 19.11), where the following reactions take place:

$$\begin{array}{lrcl} \text{Anode:} & 2H_2(g) + 4OH^-(aq) & \longrightarrow & 4H_2O(l) + 4e^- \\ \text{Cathode:} & O_2(g) + 2H_2O(l) + 4e^- & \longrightarrow & 4OH^-(aq) \\ \hline \text{Overall:} & 2H_2(g) + O_2(g) & \longrightarrow & 2H_2O(l) \end{array}$$

The standard emf of the cell can be calculated as follows, with data from Table 19.1:

$$\begin{aligned} E^\circ_{\text{cell}} &= E^\circ_{\text{cathode}} - E^\circ_{\text{anode}} \\ &= 0.40\text{ V} - (-0.83\text{ V}) \\ &= 1.23\text{ V} \end{aligned}$$

Thus, the cell reaction is spontaneous under standard-state conditions. Note that the reaction is the same as the hydrogen combustion reaction, but the oxidation and reduction are carried out separately at the anode and the cathode. Like platinum in the standard hydrogen electrode, the electrodes have a twofold function. They serve as electrical conductors, and they provide the necessary surfaces for the initial decomposition of the molecules into atomic species, prior to electron transfer. They are *electrocatalysts.* Metals such as platinum, nickel, and rhodium are good electrocatalysts.

In addition to the H_2-O_2 system, a number of other fuel cells have been developed. Among these is the propane-oxygen fuel cell. The half-cell reactions are

$$\begin{array}{lrcl} \text{Anode:} & C_3H_8(g) + 6H_2O(l) & \longrightarrow & 3CO_2(g) + 20H^+(aq) + 20e^- \\ \text{Cathode:} & 5O_2(g) + 20H^+(aq) + 20e^- & \longrightarrow & 10H_2O(l) \\ \hline \text{Overall:} & C_3H_8(g) + 5O_2(g) & \longrightarrow & 3CO_2(g) + 4H_2O(l) \end{array}$$

The overall reaction is identical to the burning of propane in oxygen.

Unlike batteries, fuel cells do not store chemical energy. Reactants must be constantly resupplied, and products must be constantly removed from a fuel cell. In this respect, a fuel cell resembles an engine more than it does a battery.

Properly designed fuel cells may be as much as 70 percent efficient, about twice as efficient as an internal combustion engine. In addition, fuel-cell generators are free of the noise, vibration, heat transfer, thermal pollution, and other problems normally associated with conventional power plants. Nevertheless, fuel cells are not yet in widespread use. A major problem lies in the lack of cheap electrocatalysts able to function efficiently for long periods of time without contamination. The most successful application of fuel cells to date has been in space vehicles (Figure 19.12).

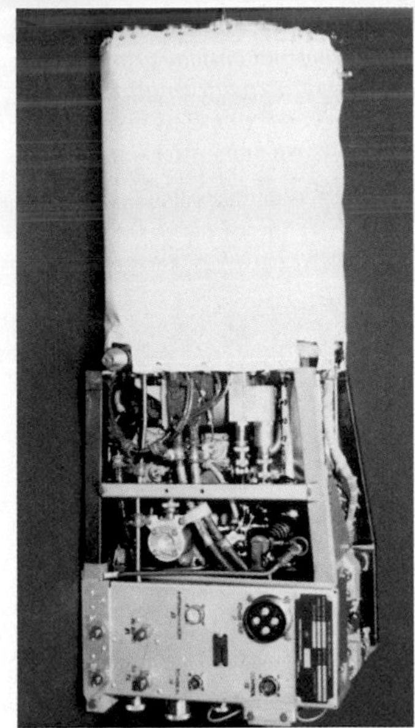

Figure 19.12
A hydrogen-oxygen fuel cell used in the space program. The pure water produced by the cell is consumed by the astronauts.

19.7 Corrosion

Corrosion is the term usually applied to *the deterioration of metals by an electrochemical process.* We see many examples of corrosion around us. Rust on iron, tarnish on silver, and the green patina formed on copper and brass are a few of them. Corrosion causes enormous damage to buildings, bridges, ships, and cars. The cost of metallic corrosion to the U.S. economy has been estimated to be well over 300 billion dollars a year! This section discusses some of the fundamental processes that occur in corrosion and methods used to protect metals against it.

By far the most familiar example of corrosion is the formation of rust on iron. Oxygen gas and water must be present for iron to rust. Although the reactions involved are quite complex and not completely understood, the main steps are believed to be as follows. A region of the metal's surface serves as the anode, where oxidation occurs:

$$Fe(s) \longrightarrow Fe^{2+}(aq) + 2e^-$$

The electrons given up by iron reduce atmospheric oxygen to water at the cathode, which is another region of the same metal's surface:

$$O_2(g) + 4H^+(aq) + 4e^- \longrightarrow 2H_2O(l)$$

The overall redox reaction is

$$2Fe(s) + O_2(g) + 4H^+(aq) \longrightarrow 2Fe^{2+}(aq) + 2H_2O(l)$$

With data from Table 19.1, we find the standard emf for this process:

$$\begin{aligned} E^\circ_{cell} &= E^\circ_{cathode} - E^\circ_{anode} \\ &= 1.23\ V - (-0.44\ V) \\ &= 1.67\ V \end{aligned}$$

The positive standard emf means that the process will occur spontaneously.

Note that this reaction occurs in an acidic medium; the H^+ ions are supplied in part by the reaction of atmospheric carbon dioxide with water to form H_2CO_3.

The Fe^{2+} ions formed at the anode are further oxidized by oxygen:

$$4Fe^{2+}(aq) + O_2(g) + (4 + 2x)H_2O(l) \longrightarrow 2Fe_2O_3 \cdot xH_2O(s) + 8H^+(aq)$$

This hydrated form of iron(III) oxide is known as rust. The amount of water associated with the iron oxide varies, so we represent the formula as $Fe_2O_3 \cdot xH_2O$.

Figure 19.13 shows the mechanism of rust formation. The electrical circuit is completed by the migration of electrons and ions; this is why rusting occurs so rapidly in salt water. In cold climates, salts (NaCl or $CaCl_2$) spread on roadways to melt ice and snow are a major cause of rust formation on automobiles.

Metallic corrosion is not limited to iron. Consider aluminum, a metal used to make many useful things, including airplanes and beverage cans. Aluminum has a

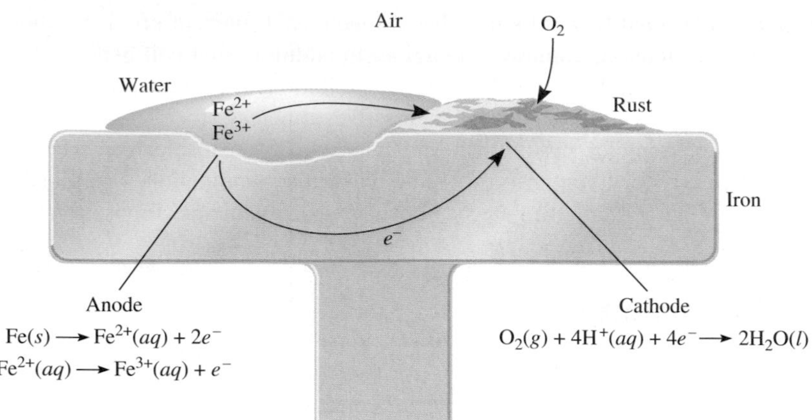

Figure 19.13
The electrochemical process involved in rust formation. The H^+ ions are supplied by H_2CO_3, which forms when CO_2 dissolves in water.

much greater tendency to oxidize than iron does; in Table 19.1 we see that Al has a more negative standard reduction potential than Fe. Based on this fact alone, we might expect to see airplanes slowly corrode away in rainstorms, and soda cans transformed into piles of corroded aluminum. These processes do not occur because the layer of insoluble aluminum oxide (Al_2O_3) that forms on its surface when the metal is exposed to air serves to protect the aluminum underneath from further corrosion. The rust that forms on the surface of iron, however, is too porous to protect the underlying metal.

Coinage metals such as copper and silver also corrode, but much more slowly.

$$Cu(s) \longrightarrow Cu^{2+}(aq) + 2e^-$$
$$Ag(s) \longrightarrow Ag^+(aq) + e^-$$

In normal atmospheric exposure, copper forms a layer of copper carbonate ($CuCO_3$), a green substance also called patina, that protects the metal underneath from further corrosion. Likewise, silverware that comes into contact with foodstuffs develops a layer of silver sulfide (Ag_2S).

A number of methods have been devised to protect metals from corrosion. Most of these methods are aimed at preventing rust formation. The most obvious approach is to coat the metal surface with paint. However, if the paint is scratched, pitted, or dented to expose even the smallest area of bare metal, rust will form under the paint layer. The surface of iron metal can be made inactive by a process called *passivation*. A thin oxide layer is formed when the metal is treated with a strong oxidizing agent such as concentrated nitric acid. A solution of sodium chromate is often added to cooling systems and radiators to prevent rust formation.

The tendency for iron to oxidize is greatly reduced when it is alloyed with certain other metals. For example, in stainless steel, an alloy of iron and chromium, a layer of chromium oxide forms that protects the iron from corrosion.

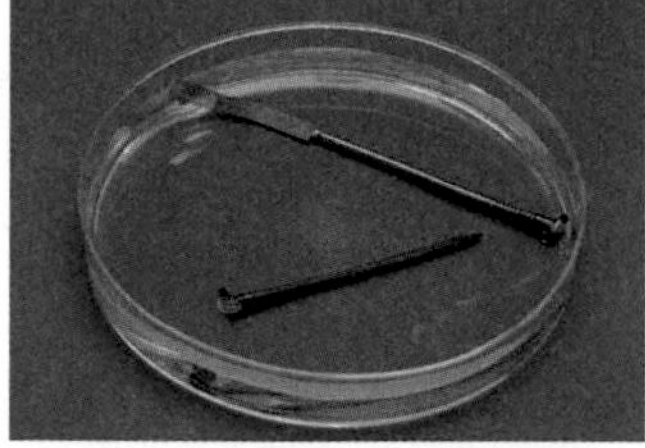

Figure 19.14
An iron nail that is cathodically protected by a piece of zinc strip does not rust in water, while an iron nail without such protection rusts readily.

An iron container can be covered with a layer of another metal such as tin or zinc. A "tin" can is made by applying a thin layer of tin over iron. Rust formation is prevented as long as the tin layer remains intact. However, once the surface has been scratched, rusting occurs rapidly. If we look up the standard reduction potentials, according to the diagonal rule, we find that iron acts as the anode and tin as the cathode in the corrosion process:

$$Sn^{2+}(aq) + 2e^- \longrightarrow Sn(s) \qquad E° = -0.14 \text{ V}$$
$$Fe^{2+}(aq) + 2e^- \longrightarrow Fe(s) \qquad E° = -0.44 \text{ V}$$

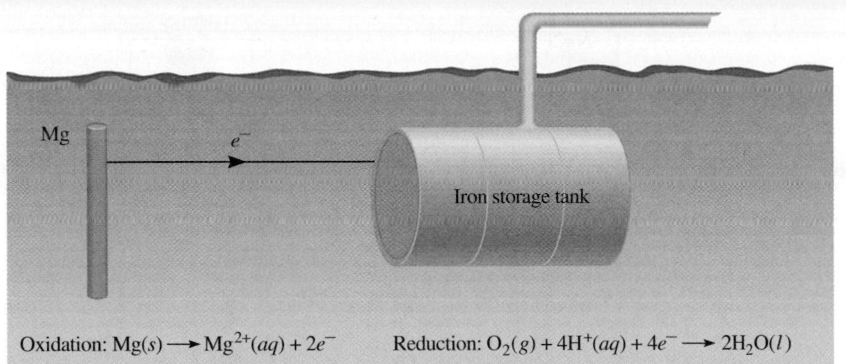

Figure 19.15
Cathodic protection of an iron storage tank (cathode) by magnesium, a more electropositive metal (anode). Because only the magnesium is depleted in the electrochemical process, it is sometimes called the sacrificial anode.

The protective process is different for zinc-plated, or *galvanized,* iron. Zinc is more easily oxidized than iron (see Table 19.1):

$$Zn^{2+}(aq) + 2e^- \longrightarrow Zn(s) \qquad E^\circ = -0.76\text{ V}$$

So even if a scratch exposes the iron, the zinc is still attacked. In this case, the zinc metal serves as the anode and the iron is the cathode.

Cathodic protection is a process in which the metal that is to be protected from corrosion is made the cathode in what amounts to a galvanic cell. Figure 19.14 shows how an iron nail can be protected from rusting by connecting the nail to a piece of zinc. Without such protection, an iron nail quickly rusts in water. Rusting of underground iron pipes and iron storage tanks can be prevented or greatly reduced by connecting them to metals such as zinc and magnesium, which oxidize more readily than iron (Figure 19.15).

19.8 Electrolysis

In contrast to spontaneous redox reactions, which result in the conversion of chemical energy into electrical energy, ***electrolysis*** is the process in which *electrical energy is used to cause a nonspontaneous chemical reaction to occur.* An ***electrolytic cell*** is *an apparatus for carrying out electrolysis.* The same principles underlie electrolysis and the processes that take place in galvanic cells. Here we will discuss three examples of electrolysis based on those principles. Then we will look at the quantitative aspects of electrolysis.

Electrolysis of Molten Sodium Chloride

In its molten state, sodium chloride, an ionic compound, can be electrolyzed to form sodium metal and chlorine. Figure 19.16(a) is a diagram of a *Downs cell,* which is used for large-scale electrolysis of NaCl. In molten NaCl, the cations and anions are the Na^+ and Cl^- ions, respectively. Figure 19.16(b) is a simplified diagram showing the reactions that occur at the electrodes. The electrolytic cell contains a pair of electrodes connected to the battery. The battery serves as an "electron pump," driving electrons to the cathode, where reduction occurs, and withdrawing electrons from the anode, where oxidation occurs. The reactions at the electrodes are

$$\begin{array}{rrcl} \text{Anode (oxidation):} & 2Cl^-(l) & \longrightarrow & Cl_2(g) + 2e^- \\ \text{Cathode (reduction):} & 2Na^+(l) + 2e^- & \longrightarrow & 2Na(l) \\ \hline \text{Overall:} & 2Na^+(l) + 2Cl^-(l) & \longrightarrow & 2Na(l) + Cl_2(g) \end{array}$$

This process is a major source of pure sodium metal and chlorine gas.

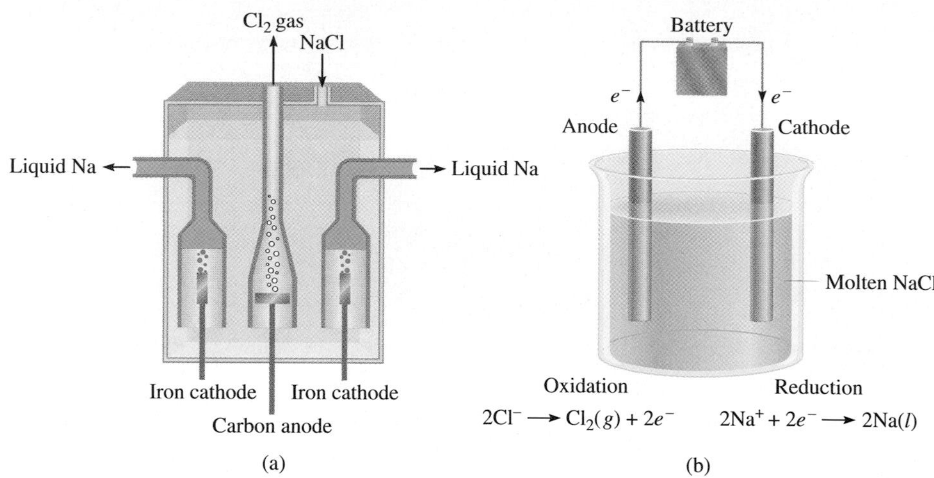

Figure 19.16 *(a) A practical arrangement called a Downs cell for the electrolysis of molten NaCl (m.p. = 801°C). The sodium metal formed at the cathodes is in the liquid state. Because liquid sodium metal is less dense than molten NaCl, the sodium floats to the surface, as shown, and is collected. Chlorine gas forms at the anode and is collected at the top. (b) A simplified diagram showing the electrode reactions during the electrolysis of molten NaCl. The battery is needed to drive the nonspontaneous reactions.*

Theoretical estimates show that the $E°$ value for the overall process is about −4 V, which means that this is a nonspontaneous process. Therefore, a *minimum* of 4 V must be supplied by the battery to carry out the reaction. In practice, a higher voltage is necessary because of inefficiencies in the electrolytic process and because of overvoltage, to be discussed shortly.

Figure 19.17 *Apparatus for small-scale electrolysis of water. The volume of hydrogen gas generated (left column) is twice that of oxygen gas (right column).*

Electrolysis of Water

Water in a beaker under atmospheric conditions (1 atm and 25°C) will not spontaneously decompose to form hydrogen and oxygen gas because the standard free-energy change for the reaction is a large positive quantity:

$$2H_2O(l) \longrightarrow 2H_2(g) + O_2(g) \qquad \Delta G° = 474.4 \text{ kJ/mol}$$

However, this reaction can be induced in a cell like the one shown in Figure 19.17. This electrolytic cell consists of a pair of electrodes made of a nonreactive metal, such as platinum, immersed in water. When the electrodes are connected to the battery, nothing happens because there are not enough ions in pure water to carry much of an electric current. (Remember that at 25°C, pure water has only 1×10^{-7} *M* H^+ ions and 1×10^{-7} *M* OH^- ions.) On the other hand, the reaction occurs readily in a 0.1 *M* H_2SO_4 solution because there are a sufficient number of ions to conduct electricity. Immediately, gas bubbles begin to appear at both electrodes.

Figure 19.18 shows the electrode reactions. The process at the anode is

$$2H_2O(l) \longrightarrow O_2(g) + 4H^+(aq) + 4e^-$$

while at the cathode we have

$$H^+(aq) + e^- \longrightarrow \tfrac{1}{2}H_2(g)$$

The overall reaction is given by

$$\begin{array}{rl} \text{Anode (oxidation):} & 2H_2O(l) \longrightarrow O_2(g) + 4H^+(aq) + 4e^- \\ \text{Cathode (reduction):} & 4[H^+(aq) + e^- \longrightarrow \tfrac{1}{2}H_2(g)] \\ \hline \text{Overall:} & 2H_2O(l) \longrightarrow 2H_2(g) + O_2(g) \end{array}$$

What is the minimum voltage needed for this electrolytic process?

Note that no net H_2SO_4 is consumed.

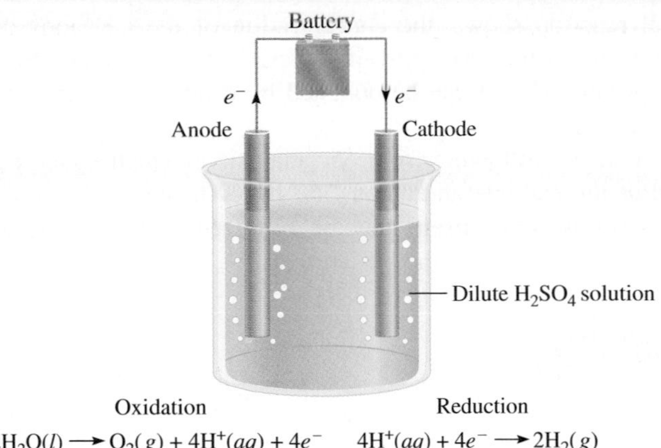

Figure 19.18
A diagram showing the electrode reactions during the electrolysis of water.

Electrolysis of an Aqueous Sodium Chloride Solution

This is the most complicated of the three examples of electrolysis considered here because aqueous sodium chloride solution contains several species that could be oxidized and reduced. The oxidation reactions that might occur at the anode are

$$(1)\quad 2Cl^-(aq) \longrightarrow Cl_2(g) + 2e^-$$
$$(2)\quad 2H_2O(l) \longrightarrow O_2(g) + 4H^+(aq) + 4e^-$$

Referring to Table 19.1, we find

$$Cl_2(g) + 2e^- \longrightarrow 2Cl^-(aq) \qquad E° = 1.36\text{ V}$$
$$O_2(g) + 4H^+(aq) + 4e^- \longrightarrow 2H_2O(l) \qquad E° = 1.23\text{ V}$$

The standard reduction potentials of (1) and (2) are not very different, but the values do suggest that H_2O should be preferentially oxidized at the anode. However, by experiment we find that the gas liberated at the anode is Cl_2, not O_2! In studying electrolytic processes, we sometimes find that the voltage required for a reaction is considerably higher than the electrode potential indicates. The ***overvoltage*** is *the difference between the electrode potential and the actual voltage required to cause electrolysis.* The overvoltage for O_2 formation is quite high. Therefore, under normal operating conditions Cl_2 gas is actually formed at the anode instead of O_2.

Because Cl_2 is more easily reduced than O_2, it follows that it would be more difficult to oxidize Cl^- than H_2O at the anode.

The reductions that might occur at the cathode are

$$(3)\quad 2H^+(aq) + 2e^- \longrightarrow H_2(g) \qquad E° = 0.00\text{ V}$$
$$(4)\quad 2H_2O(l) + 2e^- \longrightarrow H_2(g) + 2OH^-(aq) \qquad E° = -0.83\text{ V}$$
$$(5)\quad Na^+(aq) + e^- \longrightarrow Na(s) \qquad E° = -2.71\text{ V}$$

Reaction (5) is ruled out because it has a very negative standard reduction potential. Reaction (3) is preferred over (4) under standard-state conditions. At a pH of 7 (as is the case for a NaCl solution), however, they are equally probable. We generally use (4) to describe the cathode reaction because the concentration of H^+ ions is too low (about 1×10^{-7} M) to make (3) a reasonable choice.

Thus, the half-cell reactions in the electrolysis of aqueous sodium chloride are

$$\begin{array}{rl} \text{Anode (oxidation):} & 2Cl^-(aq) \longrightarrow Cl_2(g) + 2e^- \\ \text{Cathode (reduction):} & 2H_2O(l) + 2e^- \longrightarrow H_2(g) + 2OH^-(aq) \\ \hline \text{Overall:} & 2H_2O(l) + 2Cl^-(aq) \longrightarrow H_2(g) + Cl_2(g) + 2OH^-(aq) \end{array}$$

As the overall reaction shows, the concentration of the Cl^- ions decreases during electrolysis and that of the OH^- ions increases. Therefore, in addition to H_2 and Cl_2, the useful by-product NaOH can be obtained by evaporating the aqueous solution at the end of the electrolysis.

Keep in mind the following from our analysis of electrolysis: cations are likely to be reduced at the cathode and anions are likely to be oxidized at the anode, and in aqueous solutions water itself may be oxidized and/or reduced. The outcome depends on the nature of other species present.

EXAMPLE 19.8

An aqueous Na_2SO_4 solution is electrolyzed, using the apparatus shown in Figure 19.17. If the products formed at the anode and cathode are oxygen gas and hydrogen gas, respectively, describe the electrolysis in terms of the reactions at the electrodes.

The SO_4^{2-} ion is the conjugate base of the weak acid HSO_4^- ($K_a = 1.3 \times 10^{-2}$). However, the extent to which SO_4^{2-} hydrolyzes is negligible. Also, the SO_4^{2-} ion is not oxidized at the anode.

Strategy Before we look at the electrode reactions, we should consider the following facts: (1) Because Na_2SO_4 does not hydrolyze, the pH of the solution is close to 7. (2) The Na^+ ions are not reduced at the cathode and the SO_4^{2-} ions are not oxidized at the anode. These conclusions are drawn from the electrolysis of water in the presence of sulfuric acid and in aqueous sodium chloride solution, as discussed earlier. Therefore, both the oxidation and reduction reactions involve only water molecules.

Solution The electrode reactions are

$$\text{Anode:} \quad 2H_2O(l) \longrightarrow O_2(g) + 4H^+(aq) + 4e^-$$

$$\text{Cathode:} \quad 2H_2O(l) + 2e^- \longrightarrow H_2(g) + 2OH^-(aq)$$

The overall reaction, obtained by doubling the cathode reaction coefficients and adding the result to the anode reaction, is

$$6H_2O(l) \longrightarrow 2H_2(g) + O_2(g) + 4H^+(aq) + 4OH^-(aq)$$

If the H^+ and OH^- ions are allowed to mix, then

$$4H^+(aq) + 4OH^-(aq) \longrightarrow 4H_2O(l)$$

and the overall reaction becomes

Similar problem: 19.44.

$$2H_2O(l) \longrightarrow 2H_2(g) + O_2(g)$$

Practice Exercise An aqueous solution of $Mg(NO_3)_2$ is electrolyzed. What are the gaseous products at the anode and cathode?

Electrolysis has many important applications in industry, mainly in the extraction and purification of metals. We will discuss some of these applications in Section 19.9.

REVIEW OF CONCEPTS

Complete the following electrolytic cell by labeling the electrodes and showing the half-cell reactions. Explain why the signs of the anode and cathode are opposite to those in a galvanic cell.

Quantitative Aspects of Electrolysis

The quantitative treatment of electrolysis was developed primarily by Faraday. He observed that the mass of product formed (or reactant consumed) at an electrode is proportional to both the amount of electricity transferred at the electrode and the molar mass of the substance in question. For example, in the electrolysis of molten NaCl, the cathode reaction tells us that one Na atom is produced when one Na^+ ion accepts an electron from the electrode. To reduce 1 mole of Na^+ ions, we must supply Avogadro's number (6.02×10^{23}) of electrons to the cathode. On the other hand, the stoichiometry of the anode reaction shows that oxidation of two Cl^- ions yields one chlorine molecule. Therefore, the formation of 1 mole of Cl_2 results in the transfer of 2 moles of electrons from the Cl^- ions to the anode. Similarly, it takes 2 moles of electrons to reduce 1 mole of Mg^{2+} ions and 3 moles of electrons to reduce 1 mole of Al^{3+} ions:

$$Mg^{2+} + 2e^- \longrightarrow Mg$$
$$Al^{3+} + 3e^- \longrightarrow Al$$

In an electrolysis experiment, we generally measure the current (in amperes, A) that passes through an electrolytic cell in a given period of time. The relationship between charge (in coulombs, C) and current is

$$1\ C = 1\ A \times 1\ s$$

that is, a coulomb is the quantity of electrical charge passing any point in the circuit in 1 second when the current is 1 ampere.

Figure 19.19 shows the steps involved in calculating the quantities of substances produced in electrolysis. Let us illustrate the approach by considering molten $CaCl_2$ in an electrolytic cell. Suppose a current of 0.452 A is passed through the cell for 1.50 h. How much product will be formed at the anode and at the cathode? In solving electrolysis problems of this type, the first step is to determine which species will be oxidized at the anode and which species will be reduced at the cathode. Here the choice is straightforward because we only have Ca^{2+} and Cl^- ions in molten $CaCl_2$. Thus, we write the half- and overall cell reactions as

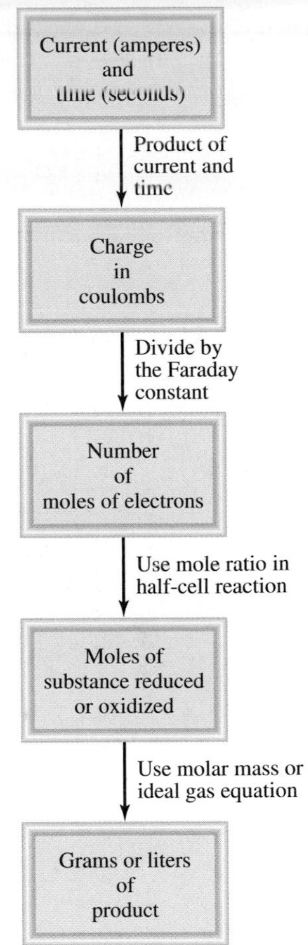

Figure 19.19
Steps involved in calculating amounts of substances reduced or oxidized in electrolysis.

$$\begin{aligned} \text{Anode (oxidation):}\quad & 2Cl^-(l) \longrightarrow Cl_2(g) + 2e^- \\ \text{Cathode (reduction):}\quad & Ca^{2+}(l) + 2e^- \longrightarrow Ca(l) \\ \hline \text{Overall:}\quad & Ca^{2+}(l) + 2Cl^-(l) \longrightarrow Ca(l) + Cl_2(g) \end{aligned}$$

The quantities of calcium metal and chlorine gas formed depend on the number of electrons that pass through the electrolytic cell, which in turn depends on current × time, or charge:

$$?\ C = 0.452\ \cancel{A} \times 1.50\ \cancel{h} \times \frac{3600\ \cancel{s}}{1\ \cancel{h}} \times \frac{1\ C}{1\ \cancel{A} \cdot \cancel{s}} = 2.44 \times 10^3\ C$$

Because 1 mol e^- = 96,500 C and 2 mol e^- are required to reduce 1 mole of Ca^{2+} ions, the mass of Ca metal formed at the cathode is calculated as follows:

$$?\ g\ Ca = 2.44 \times 10^3\ \cancel{C} \times \frac{1\ \cancel{mol\ e^-}}{96{,}500\ \cancel{C}} \times \frac{1\ \cancel{mol\ Ca}}{2\ \cancel{mol\ e^-}} \times \frac{40.08\ g\ Ca}{1\ \cancel{mol\ Ca}} = 0.507\ g\ Ca$$

The anode reaction indicates that 1 mole of chlorine is produced per 2 mol e^- of electricity. Hence the mass of chlorine gas formed is

$$?\ g\ Cl_2 = 2.44 \times 10^3\ \cancel{C} \times \frac{1\ \cancel{mol\ e^-}}{96{,}500\ \cancel{C}} \times \frac{1\ \cancel{mol\ Cl_2}}{2\ \cancel{mol\ e^-}} \times \frac{70.90\ g\ Cl_2}{1\ \cancel{mol\ Cl_2}} = 0.896\ g\ Cl_2$$

EXAMPLE 19.9

A current of 1.72 A is passed through an electrolytic cell containing a dilute sulfuric acid solution for 6.42 h. Write the half-cell reactions and calculate the volume of gases generated at STP.

Strategy Earlier (see p. 688) we saw that the half-cell reactions for the process are

$$\begin{array}{lrcl} \text{Anode (oxidation):} & 2H_2O(l) & \longrightarrow & O_2(g) + 4H^+(aq) + 4e^- \\ \text{Cathode (reduction):} & 4[H^+(aq) + e^- & \longrightarrow & \frac{1}{2}H_2(g)] \\ \hline \text{Overall:} & 2H_2O(l) & \longrightarrow & 2H_2(g) + O_2(g) \end{array}$$

According to Figure 19.19, we carry out the following conversion steps to calculate the quantity of O_2 in moles:

$$\text{current} \times \text{time} \longrightarrow \text{coulombs} \longrightarrow \text{moles of } e^- \longrightarrow \text{moles of } O_2$$

Then, using the ideal gas equation we can calculate the volume of O_2 in liters at STP. A similar procedure can be used for H_2.

Solution First we calculate the number of coulombs of electricity that pass through the cell:

$$?\ \text{C} = 1.72\ \cancel{\text{A}} \times 6.42\ \cancel{\text{h}} \times \frac{3600\ \cancel{\text{s}}}{1\ \cancel{\text{h}}} \times \frac{1\ \text{C}}{1\ \cancel{\text{A}} \cdot \cancel{\text{s}}} = 3.98 \times 10^4\ \text{C}$$

Next, we convert number of coulombs to number of moles of electrons

$$3.98 \times 10^4\ \cancel{\text{C}} \times \frac{1\ \text{mol}\ e^-}{96{,}500\ \cancel{\text{C}}} = 0.412\ \text{mol}\ e^-$$

From the oxidation half-reaction we see that 1 mol $O_2 \simeq 4$ mol e^-. Therefore, the number of moles of O_2 generated is

$$0.412\ \cancel{\text{mol}\ e^-} \times \frac{1\ \text{mol}\ O_2}{4\ \cancel{\text{mol}\ e^-}} = 0.103\ \text{mol}\ O_2$$

The volume of 0.103 mol O_2 at STP is given by

$$V = \frac{nRT}{P}$$

$$= \frac{(0.103\ \text{mol})(0.0821\ \text{L} \cdot \text{atm/K} \cdot \text{mol})(273\ \text{K})}{1\ \text{atm}} = 2.31\ \text{L}$$

The procedure for hydrogen is similar. To simplify, we combine the first two steps to calculate the number of moles of H_2 generated:

$$3.98 \times 10^4\ \cancel{\text{C}} \times \frac{1\ \cancel{\text{mol}\ e^-}}{96{,}500\ \cancel{\text{C}}} \times \frac{1\ \text{mol}\ H_2}{2\ \cancel{\text{mol}\ e^-}} = 0.206\ \text{mol}\ H_2$$

The volume of 0.206 mol H_2 at STP is given by

$$V = \frac{nRT}{P}$$

$$= \frac{(0.206\ \text{mol})(0.0821\ \text{L} \cdot \text{atm/K} \cdot \text{mol})(273\ \text{K})}{1\ \text{atm}}$$

$$= 4.62\ \text{L}$$

Check Note that the volume of H_2 is twice that of O_2 (see Figure 19.17), which is what we would expect based on Avogadro's law (at the same temperature and pressure, volume is directly proportional to the number of moles of gases).

Similar problem: 19.49.

Practice Exercise A constant current is passed through an electrolytic cell containing molten $MgCl_2$ for 18 h. If 4.8×10^5 g of Cl_2 are obtained, what is the current in amperes?

Review of Concepts

In the electrolysis of molten $CaCl_2$, a current of 1.24 A is passed through the cell for 2.0 h. What is the mass of Ca produced at the cathode?

19.9 Electrometallurgy

Electrolysis methods are useful for obtaining a pure metal from its ores or for refining (purifying) the metal. Collectively, these processes are called *electrometallurgy.* In Section 19.8 we saw how an active metal, sodium, can be obtained by electrolytically reducing its cation in the molten NaCl salt (p. 687). Here we will consider two other examples.

Production of Aluminum Metal

Aluminum is usually prepared from bauxite ore ($Al_2O_3 \cdot 2H_2O$). The ore is first treated to remove various impurities and then heated to obtain the anhydrous Al_2O_3. The oxide is dissolved in molten cryolite (Na_3AlF_6) in a Hall electrolytic cell (Figure 19.20). The cell contains a series of carbon anodes; the cathode is also made of carbon and constitutes the lining inside the cell. The solution is electrolyzed to produce aluminum and oxygen gas:

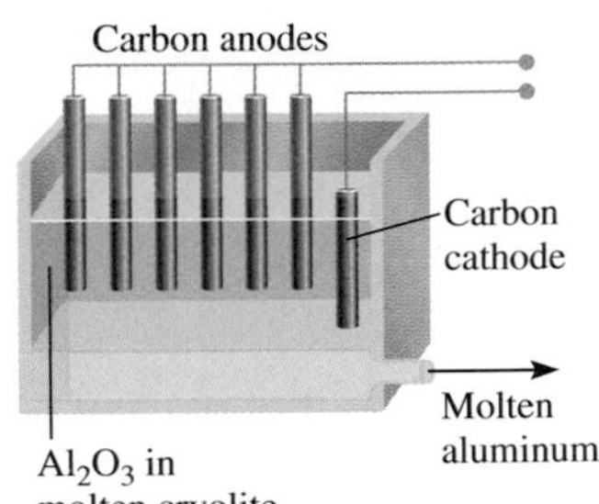

Figure 19.20 *Electrolytic production of aluminum based on the Hall process.*

$$\begin{array}{lrcl} \text{Anode:} & 3[2O^{2-} & \longrightarrow & O_2(g) + 4e^-] \\ \text{Cathode:} & 4[Al^{3+} + 3e^- & \longrightarrow & Al(l)] \\ \hline \text{Overall:} & 2Al_2O_3 & \longrightarrow & 4Al(l) + 3O_2(g) \end{array}$$

Oxygen gas reacts with the carbon anodes at 1000°C (the melting point of cryolite) to form carbon monoxide, which escapes as a gas. The liquid aluminum metal (m.p. 660°C) sinks to the bottom of the vessel, from which it can be drained.

Purification of Copper Metal

The copper metal obtained from its ores usually contains a number of impurities such as zinc, iron, silver, and gold. The more electropositive metals are removed by an electrolysis process in which the impure copper acts as the anode and *pure* copper acts as the cathode in a sulfuric acid solution containing Cu^{2+} ions (Figure 19.21). The half-reactions are

$$\begin{array}{lrcl} \text{Anode:} & Cu(s) & \longrightarrow & Cu^{2+}(aq) + 2e^- \\ \text{Cathode:} & Cu^{2+}(aq) + 2e^- & \longrightarrow & Cu(s) \end{array}$$

Reactive metals in the copper anode, such as iron and zinc, are also oxidized at the anode and enter the solution as Fe^{2+} and Zn^{2+} ions. They are not reduced at the cathode, however. The less electropositive metals, such as gold and silver, are not oxidized at the anode. Eventually, as the copper anode oxidizes, these metals fall to the bottom of the cell. Thus, the net result of this electrolysis process is the transfer of copper from the anode to the cathode. Copper prepared this way has a purity greater than 99.5 percent. It is interesting to note that the metal impurities (mostly silver and gold) from the copper anode are valuable by-products, the sale of which often pays for the electricity used to drive the electrolysis.

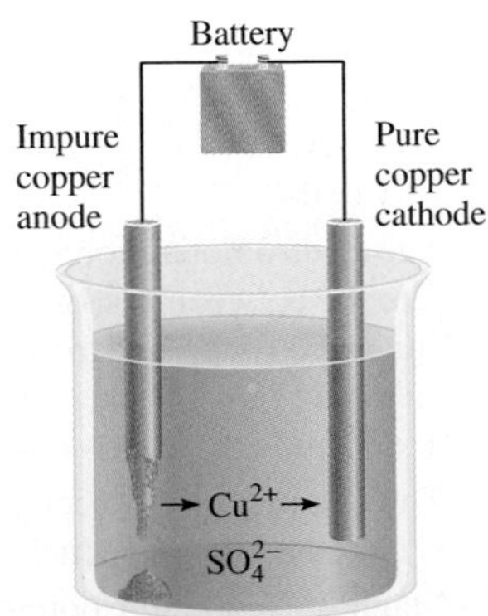

Figure 19.21 *Electrolytic purification of copper.*

Key Equations

$E^\circ_{cell} = E^\circ_{cathode} - E^\circ_{anode}$	(19.1)	Calculating the standard emf of a galvanic cell.
$\Delta G = -nFE_{cell}$	(19.2)	Relating free-energy change to the emf of the cell.
$\Delta G^\circ = -nFE^\circ_{cell}$	(19.3)	Relating the standard free-energy change to the standard emf of the cell.
$E^\circ_{cell} = \frac{RT}{nF} \ln K$	(19.4)	Relating the standard emf of the cell to the equilibrium constant.
$E^\circ_{cell} = \frac{0.0257 \text{ V}}{n} \ln K$	(19.5)	Relating the standard emf of the cell to the equilibrium constant at 298 K.
$E^\circ_{cell} = \frac{0.0592 \text{ V}}{n} \log K$	(19.6)	Relating the standard emf of the cell to the equilibrium constant at 298 K.
$E = E^\circ - \frac{RT}{nF} \ln Q$	(19.7)	The Nernst equation. For calculating the emf of a cell under nonstandard-state conditions.
$E = E^\circ - \frac{0.0257 \text{ V}}{n} \ln Q$	(19.8)	The Nernst equation. For calculating the emf of a cell under nonstandard-state conditions at 298 K.
$E = E^\circ - \frac{0.0592 \text{ V}}{n} \log Q$	(19.9)	The Nernst equation. For calculating the emf of a cell under non-standard-state conditions at 298 K.

Summary of Facts and Concepts

1. Redox reactions involve the transfer of electrons. Equations representing redox processes can be balanced using the half-reaction method.
2. All electrochemical reactions involve the transfer of electrons and therefore are redox reactions.
3. In a galvanic cell, electricity is produced by a spontaneous chemical reaction. Oxidation and reduction take place separately at the anode and cathode, respectively, and the electrons flow through an external circuit.
4. The two parts of a galvanic cell are the half-cells, and the reactions at the electrodes are the half-cell reactions. A salt bridge allows ions to flow between the half-cells.
5. The electromotive force (emf) of a cell is the voltage difference between the two electrodes. In the external circuit, electrons flow from the anode to the cathode in a galvanic cell. In solution, the anions move toward the anode and the cations move toward the cathode.
6. The quantity of electricity carried by 1 mole of electrons is called a faraday, which is equal to 96,500 C.
7. Standard reduction potentials show the relative likelihood of half-cell reduction reactions and can be used to predict the products, direction, and spontaneity of redox reactions between various substances.
8. The decrease in free energy of the system in a spontaneous redox reaction is equal to the electrical work done by the system on the surroundings, or $\Delta G = -nFE$.
9. The equilibrium constant for a redox reaction can be found from the standard electromotive force of a cell.
10. The Nernst equation gives the relationship between the cell emf and the concentrations of the reactants and products under nonstandard-state conditions.
11. Batteries, which consist of one or more galvanic cells, are used widely as self-contained power sources. Some of the better-known batteries are the dry cell, such as the Leclanché cell, the mercury battery, the lithium-ion battery, and the lead storage battery used in automobiles. Fuel cells produce electrical energy from a continuous supply of reactants.
12. The corrosion of metals, such as the rusting of iron, is an electrochemical phenomenon.
13. Electric current from an external source is used to drive a nonspontaneous chemical reaction in an electrolytic cell. The amount of product formed or reactant consumed depends on the quantity of electricity transferred at the electrode.
14. Electrolysis plays an important role in obtaining pure metals from their ores and in purifying metals.

Key Words

Questions and Problems

Balancing Redox Equations

Problems

19.1 Balance the following redox equations by the half-reaction method:
(a) $H_2O_2 + Fe^{2+} \longrightarrow Fe^{3+} + H_2O$ (in acidic solution)
(b) $Cu + HNO_3 \longrightarrow Cu^{2+} + NO + H_2O$ (in acidic solution)
(c) $CN^- + MnO_4^- \longrightarrow CNO^- + MnO_2$ (in basic solution)
(d) $Br_2 \longrightarrow BrO_3^- + Br^-$ (in basic solution)
(e) $S_2O_3^{2-} + I_2 \longrightarrow I^- + S_4O_6^{2-}$ (in acidic solution)

19.2 Balance the following redox equations by the half-reaction method:
(a) $Mn^{2+} + H_2O_2 \longrightarrow MnO_2 + H_2O$ (in basic solution)
(b) $Bi(OH)_3 + SnO_2^{2-} \longrightarrow SnO_3^{2-} + Bi$ (in basic solution)
(c) $Cr_2O_7^{2-} + C_2O_4^{2-} \longrightarrow Cr^{3+} + CO_2$ (in acidic solution)
(d) $ClO_3^- + Cl^- \longrightarrow Cl_2 + ClO_2$ (in acidic solution)

Galvanic Cells and Standard Emfs

Review Questions

19.3 Define the following terms: anode, cathode, cell voltage, electromotive force, standard reduction potential.

19.4 Describe the basic features of a galvanic cell. Why are the two components of the cell separated from each other?

19.5 What is the function of a salt bridge? What kind of electrolyte should be used in a salt bridge?

19.6 What is a cell diagram? Write the cell diagram for a galvanic cell consisting of an Al electrode placed in a 1 M $Al(NO_3)_3$ solution and a Ag electrode placed in a 1 M $AgNO_3$ solution.

19.7 What is the difference between the half-reactions discussed in redox processes in Chapter 4 and the half-cell reactions discussed in Section 19.2?

19.8 After operating a Daniell cell (see Figure 19.1) for a few minutes, a student notices that the cell emf begins to drop. Why?

19.9 Use the information in Table 2.1, and calculate the Faraday constant.

19.10 Discuss the spontaneity of an electrochemical reaction in terms of its standard emf ($E°_{cell}$).

Problems

19.11 Calculate the standard emf of a cell that uses the Mg/Mg^{2+} and Cu/Cu^{2+} half-cell reactions at 25°C. Write the equation for the cell reaction that occurs under standard-state conditions.

19.12 Calculate the standard emf of a cell that uses Ag/Ag^+ and Al/Al^{3+} half-cell reactions. Write the cell reaction that occurs under standard-state conditions.

19.13 Predict whether Fe^{3+} can oxidize I^- to I_2 under standard-state conditions.

19.14 Which of the following reagents can oxidize H_2O to $O_2(g)$ under standard-state conditions? $H^+(aq)$, $Cl^-(aq)$, $Cl_2(g)$, $Cu^{2+}(aq)$, $Pb^{2+}(aq)$, $MnO_4^-(aq)$ (in acid).

19.15 Consider the following half-reactions:

$$MnO_4^-(aq) + 8H^+(aq) + 5e^- \longrightarrow Mn^{2+}(aq) + 4H_2O(l)$$

$$NO_3^-(aq) + 4H^+(aq) + 3e^- \longrightarrow NO(g) + 2H_2O(l)$$

Predict whether NO_3^- ions will oxidize Mn^{2+} to MnO_4^- under standard-state conditions.

19.16 Predict whether the following reactions would occur spontaneously in aqueous solution at 25°C. Assume that the initial concentrations of dissolved species are all 1.0 M.
(a) $Ca(s) + Cd^{2+}(aq) \longrightarrow Ca^{2+}(aq) + Cd(s)$
(b) $2Br^-(aq) + Sn^{2+}(aq) \longrightarrow Br_2(l) + Sn(s)$
(c) $2Ag(s) + Ni^{2+}(aq) \longrightarrow 2Ag^+(aq) + Ni(s)$
(d) $Cu^+(aq) + Fe^{3+}(aq) \longrightarrow Cu^{2+}(aq) + Fe^{2+}(aq)$

19.17 Which species in each pair is a better oxidizing agent under standard-state conditions? (a) Br_2 or Au^{3+}, (b) H_2 or Ag^+, (c) Cd^{2+} or Cr^{3+}, (d) O_2 in acidic media or O_2 in basic media.

19.18 Which species in each pair is a better reducing agent under standard-state conditions? (a) Na or Li, (b) H_2 or I_2, (c) Fe^{2+} or Ag, (d) Br^- or Co^{2+}.

Spontaneity of Redox Reactions

Review Questions

19.19 Write the equations relating $\Delta G°$ and K to the standard emf of a cell. Define all the terms.

19.20 The $E°$ value of one cell reaction is positive and that of another cell reaction is negative. Which cell reaction will proceed toward the formation of more products at equilibrium?

Problems

19.21 What is the equilibrium constant for the following reaction at 25°C?

$$Mg(s) + Zn^{2+}(aq) \rightleftharpoons Mg^{2+}(aq) + Zn(s)$$

19.22 The equilibrium constant for the reaction

$$Sr(s) + Mg^{2+}(aq) \rightleftharpoons Sr^{2+}(aq) + Mg(s)$$

is 2.69×10^{12} at 25°C. Calculate $E°$ for a cell made up of Sr/Sr^{2+} and Mg/Mg^{2+} half-cells.

19.23 Use the standard reduction potentials to find the equilibrium constant for each of the following reactions at 25°C:

(a) $Br_2(l) + 2I^-(aq) \rightleftharpoons 2Br^-(aq) + I_2(s)$
(b) $2Ce^{4+}(aq) + 2Cl^-(aq) \rightleftharpoons Cl_2(g) + 2Ce^{3+}(aq)$
(c) $5Fe^{2+}(aq) + MnO_4^-(aq) + 8H^+(aq) \rightleftharpoons Mn^{2+}(aq) + 4H_2O(l) + 5Fe^{3+}(aq)$

19.24 Calculate $\Delta G°$ and K_c for the following reactions at 25°C:

(a) $Mg(s) + Pb^{2+}(aq) \rightleftharpoons Mg^{2+}(aq) + Pb(s)$
(b) $Br_2(l) + 2I^-(aq) \rightleftharpoons 2Br^-(aq) + I_2(s)$
(c) $O_2(g) + 4H^+(aq) + 4Fe^{2+}(aq) \rightleftharpoons 2H_2O(l) + 4Fe^{3+}(aq)$
(d) $2Al(s) + 3I_2(s) \rightleftharpoons 2Al^{3+}(aq) + 6I^-(aq)$

19.25 Under standard-state conditions, what spontaneous reaction will occur in aqueous solution among the ions Ce^{4+}, Ce^{3+}, Fe^{3+}, and Fe^{2+}? Calculate $\Delta G°$ and K_c for the reaction.

19.26 Given that $E° = 0.52$ V for the reduction $Cu^+(aq) + e^- \longrightarrow Cu(s)$, calculate $E°$, $\Delta G°$, and K for the following reaction at 25°C:

$$2Cu^+(aq) \longrightarrow Cu^{2+}(aq) + Cu(s)$$

The Effect of Concentration on Cell Emf

Review Questions

19.27 Write the Nernst equation and explain all the terms.

19.28 Write the Nernst equation for the following processes at some temperature T:

(a) $Mg(s) + Sn^{2+}(aq) \longrightarrow Mg^{2+}(aq) + Sn(s)$
(b) $2Cr(s) + 3Pb^{2+}(aq) \longrightarrow 2Cr^{3+}(aq) + 3Pb(s)$

Problems

19.29 What is the potential of a cell made up of Zn/Zn^{2+} and Cu/Cu^{2+} half-cells at 25°C if $[Zn^{2+}] = 0.25\ M$ and $[Cu^{2+}] = 0.15\ M$?

19.30 Calculate $E°$, E, and ΔG for the following cell reactions.

(a) $Mg(s) + Sn^{2+}(aq) \longrightarrow Mg^{2+}(aq) + Sn(s)$
$[Mg^{2+}] = 0.045\ M$, $[Sn^{2+}] = 0.035\ M$
(b) $3Zn(s) + 2Cr^{3+}(aq) \longrightarrow 3Zn^{2+}(aq) + 2Cr(s)$
$[Cr^{3+}] = 0.010\ M$, $[Zn^{2+}] = 0.0085\ M$

19.31 Calculate the standard potential of the cell consisting of the Zn/Zn^{2+} half-cell and the SHE. What will the emf of the cell be if $[Zn^{2+}] = 0.45\ M$, $P_{H_2} = 2.0$ atm, and $[H^+] = 1.8\ M$?

19.32 What is the emf of a cell consisting of a Pb^{2+}/Pb half-cell and a $Pt/H^+/H_2$ half-cell if $[Pb^{2+}] = 0.10\ M$, $[H^+] = 0.050\ M$, and $P_{H_2} = 1.0$ atm?

19.33 Referring to the arrangement in Figure 19.1, calculate the $[Cu^{2+}]/[Zn^{2+}]$ ratio at which the following reaction is spontaneous at 25°C:

$$Cu(s) + Zn^{2+}(aq) \longrightarrow Cu^{2+}(aq) + Zn(s)$$

19.34 Calculate the emf of the following concentration cell:

$$Mg(s)|Mg^{2+}(0.24\ M)\|Mg^{2+}(0.53M)|Mg(s)$$

Batteries and Fuel Cells

Review Questions

19.35 Explain the differences between a primary galvanic cell—one that is not rechargeable—and a storage cell (for example, the lead storage battery), which is rechargeable.

19.36 Discuss the advantages and disadvantages of fuel cells over conventional power plants in producing electricity.

Problems

19.37 The hydrogen-oxygen fuel cell is described in Section 19.6. (a) What volume of $H_2(g)$, stored at 25°C at a pressure of 155 atm, would be needed to run an electric motor drawing a current of 8.5 A for 3.0 h? (b) What volume (liters) of air at 25°C and 1.00 atm will have to pass into the cell per minute to run the motor? Assume that air is 20 percent O_2 by volume and that all the O_2 is consumed in the cell. The other components of air do not affect the fuel-cell reactions. Assume ideal gas behavior.

19.38 Calculate the standard emf of the propane fuel cell discussed on p. 684 at 25°C, given that $\Delta G_f°$ for propane is -23.5 kJ/mol.

Corrosion

Review Questions

19.39 Steel hardware, including nuts and bolts, is often coated with a thin plating of cadmium. Explain the function of the cadmium layer.

19.40 "Galvanized iron" is steel sheet that has been coated with zinc; "tin" cans are made of steel sheet coated with tin. Discuss the functions of these coatings and the electrochemistry of the corrosion reactions that occur if an electrolyte contacts the scratched surface of a galvanized iron sheet or a tin can.

19.41 Tarnished silver contains Ag_2S. The tarnish can be removed by placing silverware in an aluminum pan containing an inert electrolyte solution, such as NaCl. Explain the electrochemical principle for this procedure. [The standard reduction potential for the half-cell reaction $Ag_2S(s) + 2e^- \longrightarrow 2Ag(s) + S^{2-}(aq)$ is −0.71 V.]

19.42 How does the tendency of iron to rust depend on the pH of solution?

Electrolysis

Review Questions

19.43 What is the difference between a galvanic cell (such as a Daniell cell) and an electrolytic cell?

19.44 Describe the electrolysis of an aqueous solution of KNO_3.

Problems

19.45 The half-reaction at an electrode is

$$Mg^{2+}(\text{molten}) + 2e^- \longrightarrow Mg(s)$$

Calculate the number of grams of magnesium that can be produced by supplying 1.00 *F* to the electrode.

19.46 Consider the electrolysis of molten barium chloride, $BaCl_2$. (a) Write the half-reactions. (b) How many grams of barium metal can be produced by supplying 0.50 A for 30 min?

19.47 Considering only the cost of electricity, would it be cheaper to produce a ton of sodium or a ton of aluminum by electrolysis?

19.48 If the cost of electricity to produce magnesium by the electrolysis of molten magnesium chloride is \$155 per ton of metal, what is the cost (in dollars) of the electricity necessary to produce (a) 10.0 tons of aluminum, (b) 30.0 tons of sodium, (c) 50.0 tons of calcium?

19.49 One of the half-reactions for the electrolysis of water is

$$2H_2O(l) \longrightarrow O_2(g) + 4H^+(aq) + 4e^-$$

If 0.076 L of O_2 is collected at 25°C and 755 mmHg, how many faradays of electricity had to pass through the solution?

19.50 How many faradays of electricity are required to produce (a) 0.84 L of O_2 at exactly 1 atm and 25°C from aqueous H_2SO_4 solution; (b) 1.50 L of Cl_2 at 750 mmHg and 20°C from molten NaCl; (c) 6.0 g of Sn from molten $SnCl_2$?

19.51 Calculate the amounts of Cu and Br_2 produced in 1.0 h at inert electrodes in a solution of $CuBr_2$ by a current of 4.50 A.

19.52 In the electrolysis of an aqueous $AgNO_3$ solution, 0.67 g of Ag is deposited after a certain period of time. (a) Write the half-reaction for the reduction of Ag^+. (b) What is the probable oxidation half-reaction? (c) Calculate the quantity of electricity used, in coulombs.

19.53 A steady current was passed through molten $CoSO_4$ until 2.35 g of metallic cobalt was produced. Calculate the number of coulombs of electricity used.

19.54 A constant electric current flows for 3.75 h through two electrolytic cells connected in series. One contains a solution of $AgNO_3$ and the second a solution of $CuCl_2$. During this time 2.00 g of silver are deposited in the first cell. (a) How many grams of copper are deposited in the second cell? (b) What is the current flowing, in amperes?

19.55 What is the hourly production rate of chlorine gas (in kg) from an electrolytic cell using aqueous NaCl electrolyte and carrying a current of 1.500×10^3 A? The anode efficiency for the oxidation of Cl^- is 93.0 percent.

19.56 Chromium plating is applied by electrolysis to objects suspended in a dichromate solution, according to the following (unbalanced) half-reaction:

$$Cr_2O_7^{2-}(aq) + e^- + H^+(aq) \longrightarrow Cr(s) + H_2O(l)$$

How long (in hours) would it take to apply a chromium plating 1.0×10^{-2} mm thick to a car bumper with a surface area of 0.25 m^2 in an electrolytic cell carrying a current of 25.0 A? (The density of chromium is 7.19 g/cm^3.)

19.57 The passage of a current of 0.750 A for 25.0 min deposited 0.369 g of copper from a $CuSO_4$ solution. From this information, calculate the molar mass of copper.

19.58 A quantity of 0.300 g of copper was deposited from a $CuSO_4$ solution by passing a current of 3.00 A through the solution for 304 s. Calculate the value of the faraday constant.

19.59 In a certain electrolysis experiment, 1.44 g of Ag were deposited in one cell (containing an aqueous $AgNO_3$ solution), while 0.120 g of an unknown metal X was deposited in another cell (containing an aqueous XCl_3 solution) in series with the $AgNO_3$ cell. Calculate the molar mass of X.

19.60 One of the half-reactions for the electrolysis of water is

$$2H^+(aq) + 2e^- \longrightarrow H_2(g)$$

If 0.845 L of H_2 is collected at 25°C and 782 mmHg, how many faradays of electricity had to pass through the solution?

Additional Problems

19.61 For each of the following redox reactions, (i) write the half-reactions; (ii) write a balanced equation for the whole reaction, (iii) determine in which direction the reaction will proceed spontaneously under standard-state conditions:

(a) $H_2(g) + Ni^{2+}(aq) \longrightarrow H^+(aq) + Ni(s)$

(b) $MnO_4^-(aq) + Cl^-(aq) \longrightarrow Mn^{2+}(aq) + Cl_2(g)$ (in acid solution)

(c) $Cr(s) + Zn^{2+}(aq) \longrightarrow Cr^{3+}(aq) + Zn(s)$

19.62 The oxidation of 25.0 mL of a solution containing Fe^{2+} requires 26.0 mL of 0.0250 *M* $K_2Cr_2O_7$ in acidic solution. Balance the following equation and calculate the molar concentration of Fe^{2+}:

$$Cr_2O_7^{2-} + Fe^{2+} + H^+ \longrightarrow Cr^{3+} + Fe^{3+}$$

19.63 The SO_2 present in air is mainly responsible for the phenomenon of acid rain. The concentration of SO_2 can be determined by titrating against a standard permanganate solution as follows:

$$5SO_2 + 2MnO_4^- + 2H_2O \longrightarrow 5SO_4^{2-} + 2Mn^{2+} + 4H^+$$

Calculate the number of grams of SO_2 in a sample of air if 7.37 mL of 0.00800 *M* $KMnO_4$ solution are required for the titration.

19.64 A sample of iron ore weighing 0.2792 g was dissolved in an excess of a dilute acid solution. All the iron was first converted to Fe(II) ions. The solution then required 23.30 mL of 0.0194 *M* $KMnO_4$ for oxidation to Fe(III) ions. Calculate the percent by mass of iron in the ore.

19.65 The concentration of a hydrogen peroxide solution can be conveniently determined by titration against a standardized potassium permanganate solution in an acidic medium according to the following unbalanced equation:

$$MnO_4^- + H_2O_2 \longrightarrow O_2 + Mn^{2+}$$

(a) Balance the above equation. (b) If 36.44 mL of a 0.01652 *M* $KMnO_4$ solution are required to completely oxidize 25.00 mL of a H_2O_2 solution, calculate the molarity of the H_2O_2 solution.

19.66 Oxalic acid ($H_2C_2O_4$) is present in many plants and vegetables. (a) Balance the following equation in acid solution:

$$MnO_4^- + C_2O_4^{2-} \longrightarrow Mn^{2+} + CO_2$$

(b) If a 1.00-g sample of $H_2C_2O_4$ requires 24.0 mL of 0.0100 *M* $KMnO_4$ solution to reach the equivalence point, what is the percent by mass of $H_2C_2O_4$ in the sample?

19.67 Complete the following table. State whether the cell reaction is spontaneous, nonspontaneous, or at equilibrium.

E	Δ*G*	Cell Reaction
> 0		
	>0	
= 0		

19.68 Calcium oxalate (CaC_2O_4) is insoluble in water. This property has been used to determine the amount of Ca^{2+} ions in blood. The calcium oxalate isolated from blood is dissolved in acid and titrated against a standardized $KMnO_4$ solution, as described in Problem 19.66. In one test it is found that the calcium oxalate isolated from a 10.0-mL sample of blood requires 24.2 mL of 9.56×10^{-4} *M* $KMnO_4$ for titration. Calculate the number of milligrams of calcium per milliliter of blood.

19.69 From the following information, calculate the solubility product of AgBr:

$$Ag^+(aq) + e^- \longrightarrow Ag(s) \qquad E° = 0.80\ V$$
$$AgBr(s) + e^- \longrightarrow Ag(s) + Br^-(aq) \qquad E° = 0.07\ V$$

19.70 Consider a galvanic cell composed of the SHE and a half-cell using the reaction $Ag^+(aq) + e^- \longrightarrow Ag(s)$. (a) Calculate the standard cell potential. (b) What is the spontaneous cell reaction under standard-state conditions? (c) Calculate the cell potential when $[H^+]$ in the hydrogen electrode is changed to (i) 1.0×10^{-2} *M* and (ii) 1.0×10^{-5} *M*, all other reagents being held at standard-state conditions. (d) Based on this cell arrangement, suggest a design for a pH meter.

19.71 A galvanic cell consists of a silver electrode in contact with 346 mL of 0.100 *M* $AgNO_3$ solution and a magnesium electrode in contact with 288 mL of 0.100 *M* $Mg(NO_3)_2$ solution. (a) Calculate *E* for the cell at 25°C. (b) A current is drawn from the cell until 1.20 g of silver have been deposited at the silver electrode. Calculate *E* for the cell at this stage of operation.

19.72 Explain why chlorine gas can be prepared by electrolyzing an aqueous solution of NaCl but fluorine gas cannot be prepared by electrolyzing an aqueous solution of NaF.

19.73 Calculate the emf of the following concentration cell at 25°C:

$$Cu(s)|Cu^{2+}(0.080\ M)||Cu^{2+}(1.2\ M)|Cu(s)$$

19.74 The cathode reaction in the Leclanché cell is given by

$$2MnO_2(s) + Zn^{2+}(aq) + 2e^- \longrightarrow ZnMn_2O_4(s)$$

If a Leclanché cell produces a current of 0.0050 A, calculate how many hours this current supply will last if there are initially 4.0 g of MnO_2 present in the cell. Assume that there is an excess of Zn^{2+} ions.

19.75 Suppose you are asked to verify experimentally the electrode reactions shown in Example 19.8. In addition to the apparatus and the solution, you are also given two pieces of litmus paper, one blue and the other red. Describe what steps you would take in this experiment.

19.76 For a number of years it was not clear whether mercury(I) ions existed in solution as Hg^+ or as Hg_2^{2+}. To distinguish between these two possibilities, we could set up the following system:

$$Hg(l)|\text{soln A}\|\text{soln B}|Hg(l)$$

where soln A contained 0.263 g mercury(I) nitrate per liter and soln B contained 2.63 g mercury(I) nitrate per liter. If the measured emf of such a cell is 0.0289 V at 18°C, what can you deduce about the nature of the mercury(I) ions?

19.77 An aqueous KI solution to which a few drops of phenolphthalein have been added is electrolyzed using an apparatus like the one shown here:

Describe what you would observe at the anode and the cathode. (*Hint:* Molecular iodine is only slightly soluble in water, but in the presence of I^- ions, it forms the brown color of I_3^- ions.)

19.78 A piece of magnesium metal weighing 1.56 g is placed in 100.0 mL of 0.100 *M* $AgNO_3$ at 25°C. Calculate $[Mg^{2+}]$ and $[Ag^+]$ in solution at equilibrium. What is the mass of the magnesium left? The volume remains constant.

19.79 Describe an experiment that would enable you to determine which is the cathode and which is the anode in a galvanic cell using copper and zinc electrodes.

19.80 An acidified solution was electrolyzed using copper electrodes. A constant current of 1.18 A caused the anode to lose 0.584 g after 1.52×10^3 s. (a) What is the gas produced at the cathode and what is its volume at STP? (b) Given that the charge of an electron is 1.6022×10^{-19} C, calculate Avogadro's number. Assume that copper is oxidized to Cu^{2+} ions.

19.81 In a certain electrolysis experiment involving Al^{3+} ions, 60.2 g of Al is recovered when a current of 0.352 A is used. How many minutes did the electrolysis last?

19.82 Consider the oxidation of ammonia:

$$4NH_3(g) + 3O_2(g) \longrightarrow 2N_2(g) + 6H_2O(l)$$

(a) Calculate the $\Delta G°$ for the reaction. (b) If this reaction were used in a fuel cell, what would the standard cell potential be?

19.83 A galvanic cell is constructed by immersing a piece of copper wire in 25.0 mL of a 0.20 *M* $CuSO_4$ solution and a zinc strip in 25.0 mL of a 0.20 *M* $ZnSO_4$ solution. (a) Calculate the emf of the cell at 25°C and predict what would happen if a small amount of concentrated NH_3 solution were added to (i) the $CuSO_4$ solution and (ii) the $ZnSO_4$ solution. Assume that the volume in each compartment remains constant at 25.0 mL. (b) In a separate experiment, 25.0 mL of 3.00 *M* NH_3 are added to the $CuSO_4$ solution. If the emf of the cell is 0.68 V, calculate the formation constant (K_f) of $Cu(NH_3)_4^{2+}$.

19.84 In an electrolysis experiment, a student passes the same quantity of electricity through two electrolytic cells, one containing a silver salt and the other a gold salt. Over a certain period of time, she finds that 2.64 g of Ag and 1.61 g of Au are deposited at the cathodes. What is the oxidation state of gold in the gold salt?

19.85 People living in cold-climate countries where there is plenty of snow are advised not to heat their garages in the winter. What is the electrochemical basis for this recommendation?

19.86 Given that

$$2Hg^{2+}(aq) + 2e^- \longrightarrow Hg_2^{2+}(aq) \qquad E° = 0.92\text{ V}$$
$$Hg_2^{2+}(aq) + 2e^- \longrightarrow 2Hg(l) \qquad E° = 0.85\text{ V}$$

calculate $\Delta G°$ and K for the following process at 25°C:

$$Hg_2^{2+}(aq) \longrightarrow Hg^{2+}(aq) + Hg(l)$$

(The preceding reaction is an example of a *disproportionation reaction* in which an element in one oxidation state is both oxidized and reduced.)

19.87 Fluorine (F_2) is obtained by the electrolysis of liquid hydrogen fluoride (HF) containing potassium fluoride (KF). (a) Write the half-cell reactions and the overall reaction for the process. (b) What is the purpose of KF? (c) Calculate the volume of F_2 (in liters) collected at 24.0°C and 1.2 atm after electrolyzing the solution for 15 h at a current of 502 A.

19.88 A 300-mL solution of NaCl was electrolyzed for 6.00 min. If the pH of the final solution was 12.24, calculate the average current used.

19.89 Industrially, copper is purified by electrolysis. The impure copper acts as the anode, and the cathode is made of pure copper. The electrodes are immersed in a $CuSO_4$ solution. During electrolysis, copper at the anode enters the solution as Cu^{2+} while Cu^{2+} ions are reduced at the cathode. (a) Write half-cell reactions and the overall reaction for the electrolytic process. (b) Suppose the anode was contaminated with Zn and Ag. Explain what happens to these impurities during electrolysis. (c) How many hours will it take to obtain 1.00 kg of Cu at a current of 18.9 A?

19.90 An aqueous solution of a platinum salt is electrolyzed at a current of 2.50 A for 2.00 h. As a result, 9.09 g of metallic Pt are formed at the cathode. Calculate the charge on the Pt ions in this solution.

19.91 Consider a galvanic cell consisting of a magnesium electrode in contact with 1.0 *M* $Mg(NO_3)_2$ and a cadmium electrode in contact with 1.0 *M* $Cd(NO_3)_2$. Calculate $E°$ for the cell, and draw a diagram showing the cathode, anode, and direction of electron flow.

19.92 A current of 6.00 A passes through an electrolytic cell containing dilute sulfuric acid for 3.40 h. If the volume of O_2 gas generated at the anode is 4.26 L (at STP), calculate the charge (in coulombs) on an electron.

19.93 Gold will not dissolve in either concentrated nitric acid or concentrated hydrochloric acid. However, the metal does dissolve in a mixture of the acids (one part HNO_3 and three parts HCl by volume), called *aqua regia.* (a) Write a balanced equation for this reaction. (*Hint:* Among the products are $HAuCl_4$ and NO_2.) (b) What is the function of HCl?

19.94 Explain why most useful galvanic cells give voltages of no more than 1.5 to 2.5 V. What are the prospects for developing practical galvanic cells with voltages of 5 V or more?

19.95 A silver rod and a SHE are dipped into a saturated aqueous solution of silver oxalate, $Ag_2C_2O_4$, at 25°C. The measured potential difference between the rod and the SHE is 0.589 V, the rod being positive. Calculate the solubility product constant for silver oxalate.

19.96 Zinc is an amphoteric metal; that is, it reacts with both acids and bases. The standard reduction potential is −1.36 V for the reaction

$$Zn(OH)_4^{2-}(aq) + 2e^- \longrightarrow Zn(s) + 4OH^-(aq)$$

Calculate the formation constant (K_f) for the reaction

$$Zn^{2+}(aq) + 4OH^-(aq) \rightleftharpoons Zn(OH)_4^{2-}(aq)$$

19.97 Use the data in Table 19.1 to determine whether or not hydrogen peroxide will undergo disproportionation in an acid medium: $2H_2O_2 \longrightarrow 2H_2O + O_2$.

19.98 The magnitudes (but *not* the signs) of the standard reduction potentials of two metals X and Y are

$$Y^{2+} + 2e^- \longrightarrow Y \quad |E°| = 0.34\ V$$

$$X^{2+} + 2e^- \longrightarrow X \quad |E°| = 0.25\ V$$

where the || notation denotes that only the magnitude (but not the sign) of the $E°$ value is shown. When the half-cells of X and Y are connected, electrons flow from X to Y. When X is connected to a SHE, electrons flow from X to SHE. (a) Are the $E°$ values of the half-reactions positive or negative? (b) What is the standard emf of a cell made up of X and Y?

19.99 A galvanic cell is constructed as follows. One half-cell consists of a platinum wire immersed in a solution containing 1.0 *M* Sn^{2+} and 1.0 *M* Sn^{4+}; the other half-cell has a thallium rod immersed in a solution of 1.0 *M* Tl^+. (a) Write the half-cell reactions and the overall reaction. (b) What is the equilibrium constant at 25°C? (c) What is the cell voltage if the Tl^+ concentration is increased tenfold? ($E°_{Tl^+/Tl} = -0.34$ V.)

19.100 Given the standard reduction potential for Au^{3+} in Table 19.1 and

$$Au^+(aq) + e^- \longrightarrow Au(s) \quad E° = 1.69\ V$$

answer the following questions. (a) Why does gold not tarnish in air? (b) Will the following disproportionation occur spontaneously?

$$3Au^+(aq) \longrightarrow Au^{3+}(aq) + 2Au(s)$$

(c) Predict the reaction between gold and fluorine gas.

19.101 Calculate $E°$ for the reactions of mercury with (a) 1 *M* HCl and (b) 1 *M* HNO_3. Which acid will oxidize Hg to Hg_2^{2+} under standard-state conditions? Can you identify which test tube below contains HNO_3 and Hg and which contains HCl and Hg?

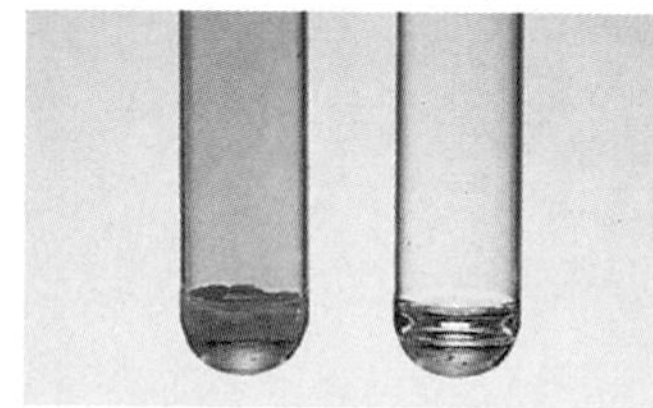

Based on your answer, explain why ingestion of a very small quantity of mercury is not considered too harmful.

19.102 When 25.0 mL of a solution containing both Fe^{2+} and Fe^{3+} ions is titrated with 23.0 mL of 0.0200 *M* $KMnO_4$ (in dilute sulfuric acid), all of the Fe^{2+} ions are oxidized to Fe^{3+} ions. Next, the solution is treated with Zn metal to convert all of the Fe^{3+} ions to Fe^{2+} ions. Finally, 40.0 mL of the same $KMnO_4$ solution are added to the solution in order to oxidize the Fe^{2+} ions to Fe^{3+}. Calculate the molar concentrations of Fe^{2+} and Fe^{3+} in the original solution.

19.103 Consider the Daniell cell in Figure 19.1. When viewed externally, the anode appears negative and the cathode positive (electrons are flowing from the anode to the cathode). Yet in solution anions are moving toward the anode, which means that it must appear positive to the anions. Because the anode cannot simultaneously be negative and positive, give an explanation for this apparently contradictory situation.

19.104 Lead storage batteries are rated by ampere hours, that is, the number of amperes they can deliver in an hour. (a) Show that 1 A · h = 3600 C. (b) The lead anodes of a certain lead storage battery have a total mass of 406 g. Calculate the maximum theoretical capacity of the battery in ampere hours. Explain why in practice we can never extract this much energy from the battery. (*Hint:* Assume all of the lead will be used up in the electrochemical reaction and refer to the electrode reactions on p. 682.) (c) Calculate E°_{cell} and ΔG° for the battery.

19.105 The concentration of sulfuric acid in the lead storage battery of an automobile over a period of time has decreased from 38.0 percent by mass (density = 1.29 g/mL) to 26.0 percent by mass (1.19 g/mL). Assume the volume of the acid remains constant at 724 mL. (a) Calculate the total charge in coulombs supplied by the battery. (b) How long (in hours) will it take to recharge the battery back to the original sulfuric acid concentration using a current of 22.4 A.

19.106 Consider a Daniell cell operating under nonstandard-state conditions. Suppose that the cell's reaction is multiplied by 2. What effect does this have on each of the following quantities in the Nernst equation? (a) E, (b) E°, (c) Q, (d) ln Q, and (e) n?

19.107 A spoon was silver-plated electrolytically in a $AgNO_3$ solution. (a) Sketch a diagram for the process. (b) If 0.884 g of Ag was deposited on the spoon at a constant current of 18.5 mA, how long (in minutes) did the electrolysis take?

19.108 Comment on whether F_2 will become a stronger oxidizing agent with increasing H^+ concentration.

19.109 In recent years there has been much interest in electric cars. List some advantages and disadvantages of electric cars compared to automobiles with internal combustion engines.

19.110 Calculate the pressure of H_2 (in atm) required to maintain equilibrium with respect to the following reaction at 25°C:

$$Pb(s) + 2H^+(aq) \rightleftharpoons Pb^{2+}(aq) + H_2(g)$$

Given that $[Pb^{2+}] = 0.035\ M$ and the solution is buffered at pH 1.60.

19.111 Because all alkali metals react with water, it is not possible to measure the standard reduction potentials of these metals directly as in the case of, say, zinc. An indirect method is to consider the following hypothetical reaction

$$Li^+(aq) + \tfrac{1}{2}H_2(g) \longrightarrow Li(s) + H^+(aq)$$

Use the appropriate equation presented in this chapter and the thermodynamic data in Appendix 2, calculate E° for $Li^+(aq) + e^- \longrightarrow Li(s)$ at 298 K. Compare your result with that listed in Table 19.1. (See back endpaper for the Faraday constant.)

19.112 A galvanic cell using Mg/Mg^{2+} and Cu/Cu^{2+} half-cells operates under standard-state conditions at 25°C and each compartment has a volume of 218 mL. The cell delivers 0.22 A for 31.6 h. (a) How many grams of Cu are deposited? (b) What is the $[Cu^{2+}]$ remaining?

19.113 Shown here is a galvanic cell connected to an electrolytic cell. Label the electrodes (anodes and cathodes) and show the movement of electrons along the wires and cations and anions in solutions. For simplicity, the salt bridge is not shown for the galvanic cell.

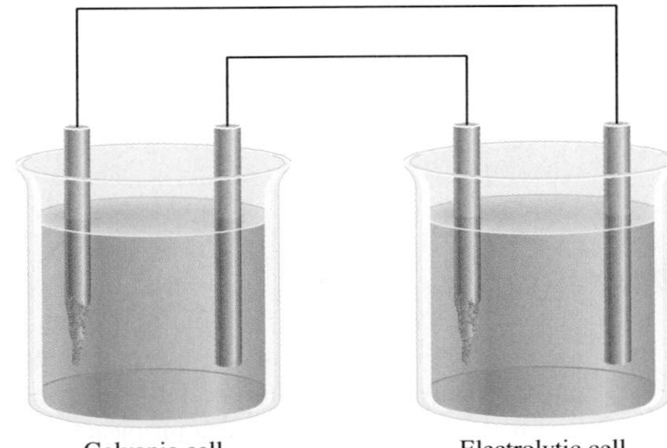

19.114 Given the following standard reduction potentials, calculate the ion-product, K_w, for water at 25°C:

$$2H^+(aq) + 2e^- \longrightarrow H_2(g) \qquad E^\circ = 0.00\ V$$

$$2H_2O(l) + 2e^- \longrightarrow H_2(g) + 2OH^-(aq) \qquad E^\circ = -0.83\ V$$

Special Problems

19.115 Fluorine is a highly reactive gas that attacks water to form HF and other products. Follow the procedure in Problem 19.111, to show how you can determine indirectly the standard reduction for fluorine as shown in Table 19.1

19.116 A piece of magnesium ribbon and a copper wire are partially immersed in a 0.1 M HCl solution in a beaker. The metals are joined externally by another piece of metal wire. Bubbles are seen to evolve at both the Mg and Cu surfaces. (a) Write equations representing the reactions occurring at the metals. (b) What visual evidence would you seek to show that Cu is not oxidized to Cu^{2+}? (c) At some stage, NaOH solution is added to the beaker to neutralize the HCl acid. Upon further addition of NaOH, a white precipitate forms. What is it?

19.117 The zinc-air battery shows much promise for electric cars because it is lightweight and rechargeable:

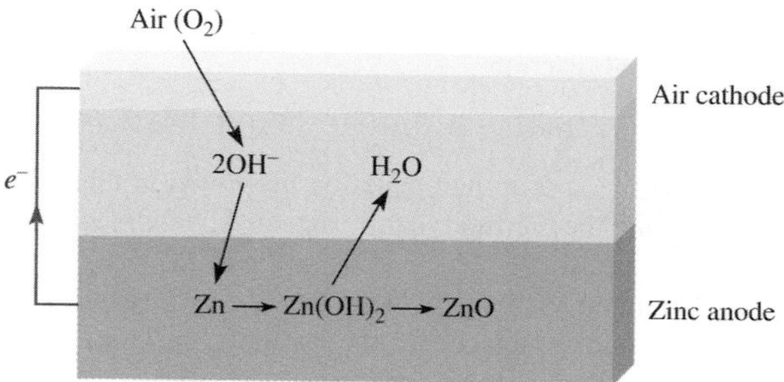

The net transformation is $Zn(s) + \frac{1}{2}O_2(g) \rightarrow ZnO(s)$. (a) Write the half-reactions at the zinc-air electrodes and calculate the standard emf of the battery at 25°C. (b) Calculate the emf under actual operating conditions when the partial pressure of oxygen is 0.21 atm. (c) What is the energy density (measured as the energy in kilojoules that can be obtained from 1 kg of the metal) of the zinc electrode? (d) If a current of 2.1×10^5 A is to be drawn from a zinc-air battery system, what volume of air (in liters) would need to be supplied to the battery every second? Assume that the temperature is 25°C and the partial pressure of oxygen is 0.21 atm.

19.118 Calculate the equilibrium constant for the following reaction at 298 K:

$$Zn(s) + Cu^{2+}(aq) \rightleftharpoons Zn^{2+}(aq) + Cu(s)$$

19.119 A construction company is installing an iron culvert (a long cylindrical tube) that is 40.0 m long with a radius of 0.900 m. To prevent corrosion, the culvert must be galvanized. This process is carried out by first passing an iron sheet of appropriate dimensions through an electrolytic cell containing Zn^{2+} ions, using graphite as the anode and the iron sheet as the cathode. If the voltage is 3.26 V, what is the cost of electricity for depositing a layer 0.200 mm thick if the efficiency of the process is 95 percent? The electricity rate is \$0.12 per kilowatt hour (kWh), where 1 W = 1 J/s and the density of Zn is 7.14 g/cm³.

19.120 A 9.00×10^2-mL 0.200 *M* MgI_2 was electrolyzed. As a result, hydrogen gas was generated at the cathode and iodine was formed at the anode. The volume of hydrogen collected at 26°C and 779 mmHg was 1.22×10^3 mL. (a) Calculate the charge in coulombs consumed in the process. (b) How long (in min) did the electrolysis last if a current of 7.55 A was used? (c) A white precipitate was formed in the process. What was it and what was its mass in grams? Assume the volume of the solution was constant.

19.121 To remove the tarnish (Ag_2S) on a silver spoon, a student carried out the following steps. First, she placed the spoon in a large pan filled with water so the spoon was totally immersed. Next, she added a few tablespoonfuls of baking soda (sodium bicarbonate), which readily dissolved. Finally, she placed some aluminum foil at the bottom of the pan in contact with the spoon and then heated the solution to about 80°C. After a few minutes, the spoon was removed and rinsed with cold water. The tarnish was gone and the spoon regained its original shiny appearance. (a) Describe with equations the electrochemical basis for the procedure. (b) Adding NaCl instead of $NaHCO_3$ would also work because both compounds are strong electrolytes. What is the added advantage of using $NaHCO_3$? (*Hint:* Consider the pH of the solution.) (c) What is the purpose of heating the solution? (d) Some commercial tarnish removers containing a fluid (or paste) that is a dilute HCl solution. Rubbing the spoon with the fluid will also remove the tarnish. Name two disadvantages of using this procedure compared to the one described here.

19.122 As mentioned on p. 679, a concentration cell ceases to operate when the concentrations of the two cell compartments are equal. At this stage, is it possible to generate an emf from the cell by adjusting another parameter without changing the concentrations? Explain.

19.123 The nitrite ion (NO_2^-) in soil is oxidized to nitrate ion (NO_3^-) by the bacteria *Nitrobacter agilis* in the presence of oxygen. The half-reduction reactions are

$$NO_3^- + 2H^+ + 2e^- \longrightarrow NO_2^- + H_2O \quad E° = 0.42\ V$$
$$O_2 + 4H^+ + 4e^- \longrightarrow 2H_2O \quad E° = 1.23\ V$$

Calculate the yield of ATP synthesis per mole of nitrite oxidized. (*Hint:* See Section 18.7.)

Answers to Practice Exercises

19.1 $5Fe^{2+} + MnO_4^- + 8H^+ \longrightarrow 5Fe^{3+} + Mn^{2+} + 4H_2O$. **19.2** No. **19.3** 0.34 V. **19.4** 1×10^{-42}. **19.5** $\Delta G° = -4.1 \times 10^2$ kJ/mol. **19.6** Yes, $E = +0.01$ V. **19.7** 0.38 V. **19.8** Anode, O_2; cathode, H_2. **19.9** 2.0×10^4 A.

The presence of trace amounts of transition metal ions such as Cr^{3+}, Fe^{2+}, and Fe^{3+} in normally colorless minerals leads to the various brilliant colors of precious gemstones.

CHAPTER

20

The Chemistry of Coordination Compounds

Chapter Outline

Student Interactive Activities

Animations
Absorption of Color (20.4)

Electronic Homework
Example Practice Problems
End of Chapter Problems

Essential Concepts

Coordination Compounds A coordination compound contains one or more complex ions in which a small number of molecules or ions surround a central metal atom or ion, usually of the transition metal family. Common geometries of coordination compounds are linear, tetrahedral, square planar, and octahedral.

Bonding in Coordination Compounds Crystal field theory explains the bonding in a complex ion in terms of electrostatic forces. The approach of the ligands toward the metal causes a splitting in energy in the five *d* orbitals. The extent of the splitting, called crystal-field splitting, depends on the nature of the ligands. Crystal field theory successfully accounts for the color and magnetic properties of many complex ions.

Coordination Compounds in Living Systems Coordination compounds play many important roles in animals and plants. They are also used as therapeutic drugs.

20.1 Properties of the Transition Metals

Transition metals typically have incompletely filled *d* subshells or readily give rise to ions with incompletely filled *d* subshells (Figure 20.1). (The Group 2B metals—Zn, Cd, and Hg—do not have this characteristic electron configuration and so, although they are sometimes called transition metals, they really do not belong in this category.) This attribute is responsible for several notable properties, including distinctive coloring, formation of paramagnetic compounds, catalytic activity, and especially a great tendency to form complex ions. In this chapter, we focus on the first-row elements from scandium to copper, the most common transition metals (Figure 20.2). Table 20.1 lists some of their properties.

As we read across any period from left to right, atomic numbers increase, electrons are added to the outer shell, and the nuclear charge increases by the addition of protons. In the third-period elements—sodium to argon—the outer electrons weakly shield one another from the extra nuclear charge. Consequently, atomic radii decrease rapidly from sodium to argon, and the electronegativities and ionization energies increase steadily (see Figures 8.5, 8.9, and 9.5).

For the transition metals, the trends are different. Looking at Table 20.1 we see that the nuclear charge, of course, increases from scandium to copper, but electrons are being added to the inner 3*d* subshell. These 3*d* electrons shield the 4*s* electrons from the increasing nuclear charge somewhat more effectively than outer-shell electrons can shield one another, so the atomic radii decrease less rapidly. For the same reason, electronegativities and ionization energies increase only slightly from scandium across to copper compared with the increases from sodium to argon.

Although the transition metals are less electropositive (or more electronegative) than the alkali and alkaline earth metals, their standard reduction potentials suggest that all of them except copper should react with strong acids such as hydrochloric acid to produce hydrogen gas. However, most transition metals are inert toward acids or react slowly with them because of a protective layer of oxide. A case in point is

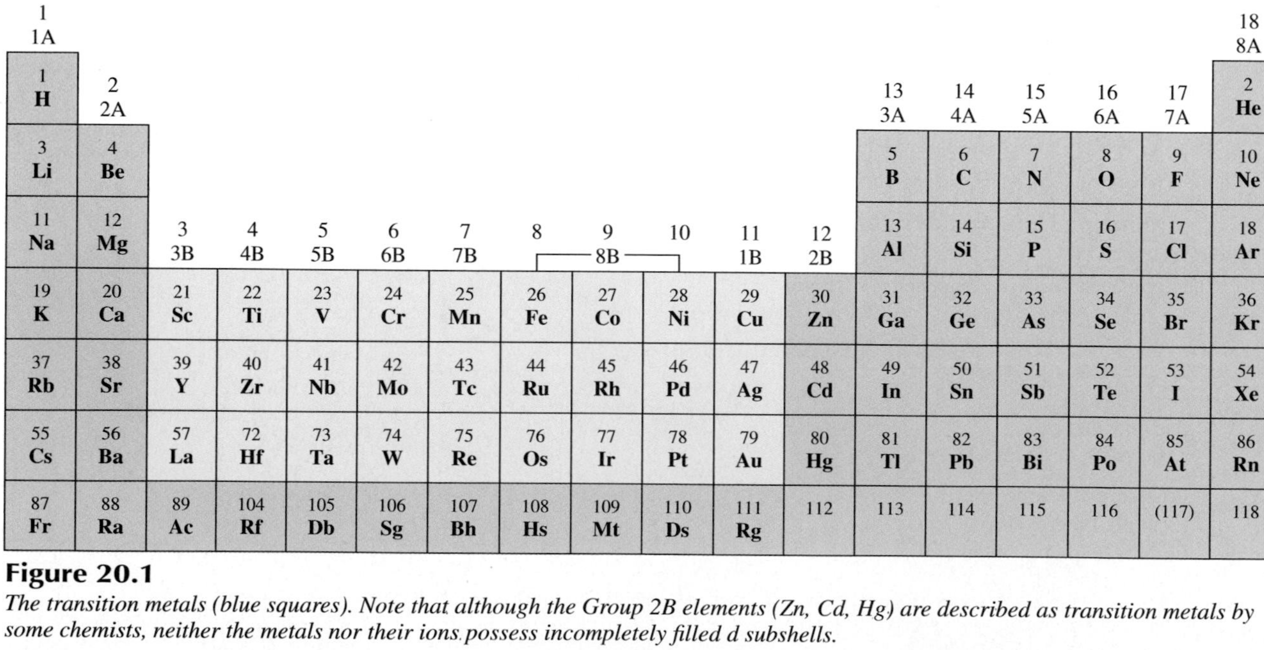

1 1A	2 2A	3 3B	4 4B	5 5B	6 6B	7 7B	8 8B	9 8B	10 8B	11 1B	12 2B	13 3A	14 4A	15 5A	16 6A	17 7A	18 8A
1 **H**																	2 **He**
3 **Li**	4 **Be**											5 **B**	6 **C**	7 **N**	8 **O**	9 **F**	10 **Ne**
11 **Na**	12 **Mg**											13 **Al**	14 **Si**	15 **P**	16 **S**	17 **Cl**	18 **Ar**
19 **K**	20 **Ca**	21 **Sc**	22 **Ti**	23 **V**	24 **Cr**	25 **Mn**	26 **Fe**	27 **Co**	28 **Ni**	29 **Cu**	30 **Zn**	31 **Ga**	32 **Ge**	33 **As**	34 **Se**	35 **Br**	36 **Kr**
37 **Rb**	38 **Sr**	39 **Y**	40 **Zr**	41 **Nb**	42 **Mo**	43 **Tc**	44 **Ru**	45 **Rh**	46 **Pd**	47 **Ag**	48 **Cd**	49 **In**	50 **Sn**	51 **Sb**	52 **Te**	53 **I**	54 **Xe**
55 **Cs**	56 **Ba**	57 **La**	72 **Hf**	73 **Ta**	74 **W**	75 **Re**	76 **Os**	77 **Ir**	78 **Pt**	79 **Au**	80 **Hg**	81 **Tl**	82 **Pb**	83 **Bi**	84 **Po**	85 **At**	86 **Rn**
87 **Fr**	88 **Ra**	89 **Ac**	104 **Rf**	105 **Db**	106 **Sg**	107 **Bh**	108 **Hs**	109 **Mt**	110 **Ds**	111 **Rg**	112	113	114	115	116	(117)	118

Figure 20.1

The transition metals (blue squares). Note that although the Group 2B elements (Zn, Cd, Hg) are described as transition metals by some chemists, neither the metals nor their ions possess incompletely filled d subshells.

Figure 20.2
The first-row transition metals.

chromium: Despite a rather negative standard reduction potential, it is quite inert chemically because of the formation on its surfaces of chromium(III) oxide, Cr_2O_3. Consequently, chromium is commonly used as a protective and noncorrosive plating on other metals. On automobile bumpers and trim, chromium plating serves a decorative as well as a functional purpose.

Electron Configurations

The electron configurations of the first-row transition metals were discussed in Section 7.9. Calcium has the electron configuration $[Ar]4s^2$. From scandium across to copper, electrons are added to the $3d$ orbitals. Thus, the outer electron configuration of scandium is $4s^23d^1$, that of titanium is $4s^23d^2$, and so on. The two exceptions are chromium and copper, whose outer electron configurations are $4s^13d^5$ and $4s^13d^{10}$, respectively. These irregularities are the result of the extra stability associated with half-filled and completely filled $3d$ subshells.

When the first-row transition metals form cations, electrons are removed first from the 4s orbitals and then from the $3d$ orbitals. (This is the opposite of the order in which orbitals are filled in neutral atoms.) For example, the outer electron configuration of Fe^{2+} is $3d^6$, not $4s^23d^4$.

Table 20.1 Electron Configurations and Other Properties of the First-Row Transition Metals

	Sc	Ti	V	Cr	Mn	Fe	Co	Ni	Cu
Electron configuration									
M	$4s^23d^1$	$4s^23d^2$	$4s^23d^3$	$4s^13d^5$	$4s^23d^5$	$4s^23d^6$	$4s^23d^7$	$4s^23d^8$	$4s^13d^{10}$
M^{2+}	—	$3d^2$	$3d^3$	$3d^4$	$3d^5$	$3d^6$	$3d^7$	$3d^8$	$3d^9$
M^{3+}	[Ar]	$3d^1$	$3d^2$	$3d^3$	$3d^4$	$3d^5$	$3d^6$	$3d^7$	$3d^8$
Electronegativity	1.3	1.5	1.6	1.6	1.5	1.8	1.9	1.9	1.9
Ionization energy (kJ/mol)									
First	631	658	650	652	717	759	760	736	745
Second	1235	1309	1413	1591	1509	1561	1645	1751	1958
Third	2389	2650	2828	2986	3250	2956	3231	3393	3578
Radius (pm)									
M	162	147	134	130	135	126	125	124	128
M^{2+}	—	90	88	85	80	77	75	69	72
M^{3+}	81	77	74	64	66	60	64	—	—
Standard reduction potential (V)*	−2.08	−1.63	−1.2	−0.74	−1.18	−0.44	−0.28	−0.25	0.34

*The half-reaction is $M^{2+}(aq) + 2e^- \longrightarrow M(s)$ (except for Sc and Cr, where the ions are Sc^{3+} and Cr^{3+}, respectively).

Oxidation States

As noted in Chapter 4, the transition metals exhibit variable oxidation states in their compounds. Figure 20.3 shows the oxidation states from scandium to copper. Note that the common oxidation states for each element include +2, +3, or both. The +3 oxidation states are more stable at the beginning of the series, whereas toward the

Sc	Ti	V	Cr	Mn	Fe	Co	Ni	Cu
				+7				
			+6	+6	+6			
		+5	+5	+5	+5			
	+4	+4	+4	+4	+4	+4		
+3	+3	+3	+3	+3	+3	+3	+3	+3
	+2	+2	+2	+2	+2	+2	+2	+2
								+1

Figure 20.3
Oxidation states of the first-row transition metals. The most stable oxidation numbers are shown in color. The zero oxidation state is encountered in some compounds, such as $Ni(CO)_4$ and $Fe(CO)_5$.

end, the +2 oxidation states are more stable. The reason is that the ionization energies increase gradually from left to right. However, the third ionization energy (when an electron is removed from the 3*d* orbital) increases more sharply than the first and second ionization energies. Because it takes more energy to remove the third electron from the metals near the end of the row than from those near the beginning, the metals near the end tend to form M^{2+} ions rather than M^{3+} ions.

The highest oxidation state is +7, for manganese ($4s^23d^5$). Transition metals usually exhibit their highest oxidation states in compounds with very electronegative elements such as oxygen and fluorine—for example, VF_5, CrO_3, and Mn_2O_7.

Recall that oxides in which the metal has a high oxidation number are covalent and acidic, whereas those in which the metal has a low oxidation number are ionic and basic (see Section 16.10).

Review of Concepts

Locate the transition metal atoms and ions in the periodic table shown here. Atoms: (1) $[Kr]5s^24d^5$. (2) $[Xe]6s^24f^{14}5d^4$. Ions: (3) $[Ar]3d^3$ (a +4 ion). (4) $[Xe]4f^{14}5d^8$ (a +3 ion). (See Table 7.3)

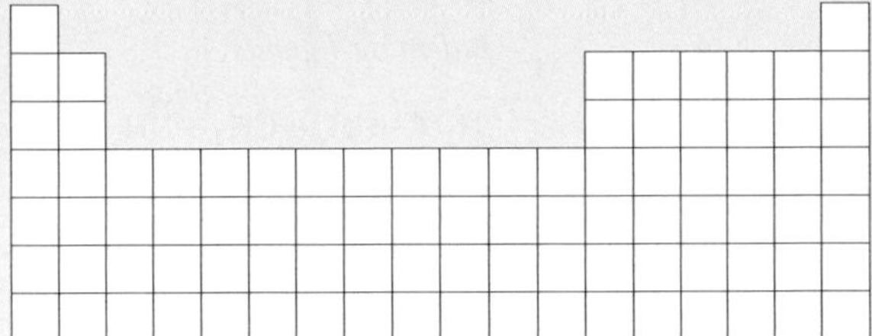

20.2 Coordination Compounds

Transition metals have a distinct tendency to form complex ions (see p. 614). A ***coordination compound*** *typically consists of a complex ion and counter ion.* [Note that some coordination compounds such as $Fe(CO)_5$ do not contain complex ions.] Our understanding of the nature of coordination compounds stems from the classic work of the Swiss chemist Alfred Werner, who prepared and characterized many coordination compounds. In 1893, at the age of 26, Werner proposed what is now commonly referred to as *Werner's coordination theory.*

A complex ion contains a central metal ion bonded to one or more ions or molecules (see Section 17.7). A complex ion can be either a cation or an anion.

Nineteenth-century chemists were puzzled by a certain class of reactions that seemed to violate valence theory. For example, the valences of the elements in cobalt(III) chloride and in ammonia seem to be completely satisfied, and yet these two substances react to form a stable compound having the formula $CoCl_3 \cdot 6NH_3$. To explain this behavior, Werner postulated that most elements exhibit two types of valence: *primary valence* and *secondary valence*. In modern terminology, primary valence corresponds to the oxidation number and secondary valence to the coordination number of the element. In $CoCl_3 \cdot 6NH_3$, according to Werner, cobalt has a primary valence of 3 and a secondary valence of 6.

Today we use the formula $[Co(NH_3)_6]Cl_3$ to indicate that the ammonia molecules and the cobalt atom form a complex ion; the chloride ions are not part of the complex but are held to it by ionic forces. Most, but not all, of the metals in coordination compounds are transition metals.

The molecules or ions that surround the metal in a complex ion are called ***ligands*** (Table 20.2). The interactions between a metal atom and the ligands can be thought of as Lewis acid-base reactions. As we saw in Section 16.11, a Lewis base

Ligands act as Lewis bases by donating electrons to metals, which act as Lewis acids.

Table 20.2 Some Common Ligands

Name	Structure
	Monodentate ligands
Ammonia	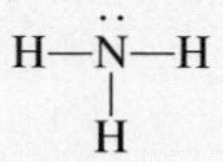 H—N̈—H with H below N
Carbon monoxide	:C≡O:
Chloride ion	$:\ddot{Cl}:^-$
Cyanide ion	$[:C\equiv N:]^-$
Thiocyanate ion	$[:\ddot{S}—C\equiv N:]^-$
Water	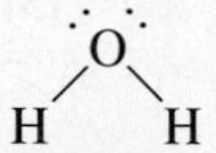 H—Ö—H
	Bidentate ligands
Ethylenediamine	$H_2\ddot{N}—CH_2—CH_2—\ddot{N}H_2$
Oxalate ion	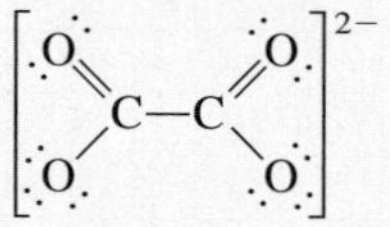 $[(O)(O)C—C(O)(O)]^{2-}$
	Polydentate ligand
Ethylenediaminetetraacetate ion (EDTA)	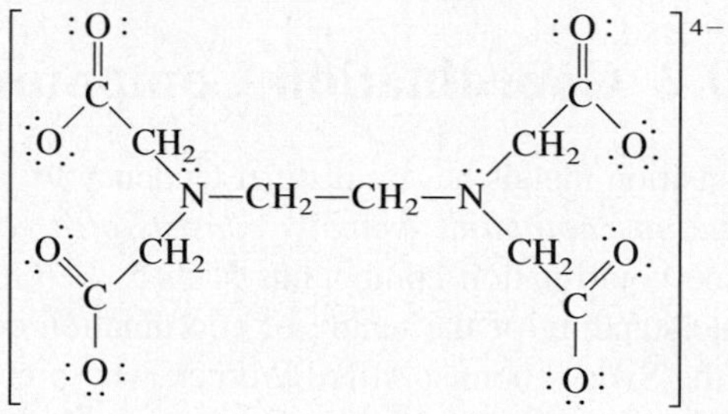 $[(^-OOCCH_2)_2\ddot{N}—CH_2—CH_2—\ddot{N}(CH_2COO^-)_2]^{4-}$

is a substance capable of donating one or more electron pairs. Every ligand has at least one unshared pair of valence electrons, as these examples show:

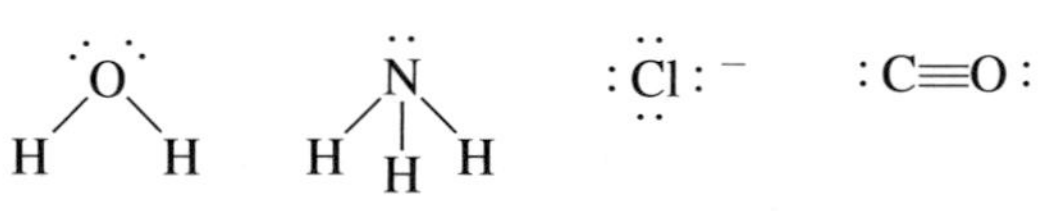

$$H_2\ddot{O} \qquad \ddot{N}H_3 \qquad :\ddot{Cl}:^- \qquad :C\equiv O:$$

Therefore, ligands play the role of Lewis bases. On the other hand, a transition metal atom (in either its neutral or positively charged state) acts as a Lewis acid, accepting (and sharing) pairs of electrons from the Lewis bases. Thus, the metal-ligand bonds are usually coordinate covalent bonds (see Section 9.9).

The atom in a ligand that is bound directly to the metal atom is known as the ***donor atom.*** For example, nitrogen is the donor atom in the $[Cu(NH_3)_4]^{2+}$ complex ion. The ***coordination number*** in coordination compounds is defined as *the number of donor atoms surrounding the central metal atom in a complex ion*. For example, the coordination number of Ag^+ in $[Ag(NH_3)_2]^+$ is 2, that of Cu^{2+} in $[Cu(NH_3)_4]^{2+}$

In a crystal lattice the coordination number of an atom (or ion) is defined as the number of atoms (or ions) surrounding the atom (or ion).

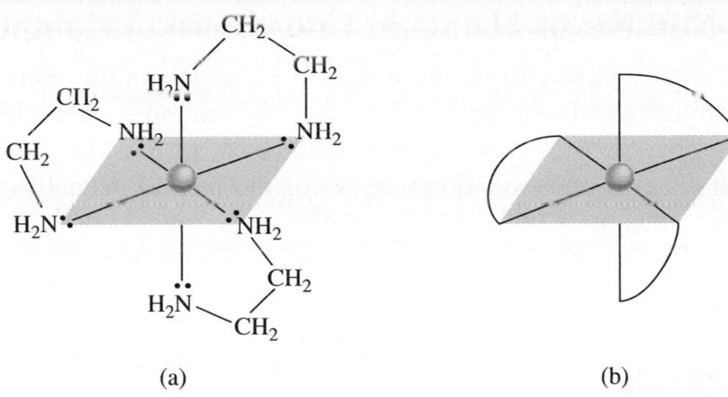

Figure 20.4
(a) Structure of a metal-ethylenediamine complex cation such as $[Co(en)_3]^{2+}$. Each ethylenediamine molecule provides two N donor atoms and is therefore a bidentate ligand. (b) Simplified structure of the same complex cation.

is 4, and that of Fe^{3+} in $[Fe(CN)_6]^{3-}$ is 6. The most common coordination numbers are 4 and 6, but coordination numbers such as 2 and 5 are also known.

Depending on the number of donor atoms present, ligands are classified as *monodentate, bidentate, or polydentate* (see Table 20.2). H_2O and NH_3 are monodentate ligands with only one donor atom each. One bidentate ligand is ethylenediamine (sometimes abbreviated "en"):

$$H_2\ddot{N}—CH_2—CH_2—\ddot{N}H_2$$

The two nitrogen atoms can coordinate with a metal atom, as shown in Figure 20.4.

Bidentate and polydentate ligands are also called ***chelating agents*** because of *their ability to hold the metal atom like a claw* (from the Greek *chele,* meaning "claw"). One example is ethylenediaminetetraacetate ion (EDTA), a polydentate ligand used to treat metal poisoning (Figure 20.5). Six donor atoms enable EDTA to form a very stable complex ion with lead, which is removed from the blood and excreted from the body. EDTA is also used to clean up spills of radioactive metals.

Review of Concepts

What is the difference between these two compounds: $CrCl_3 \cdot 6H_2O$ and $[Cr(H_2O)_6]Cl_3$?

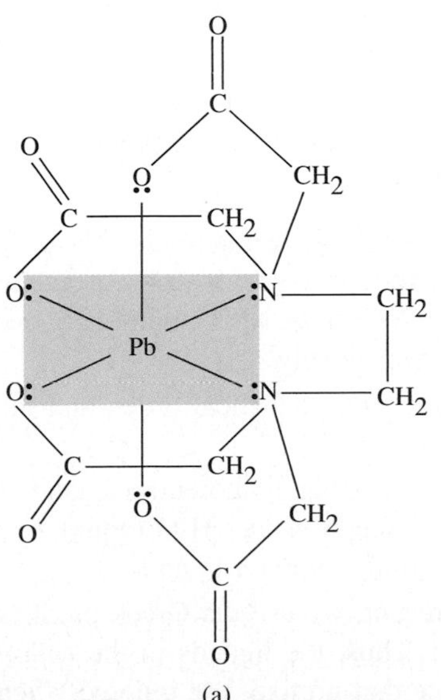

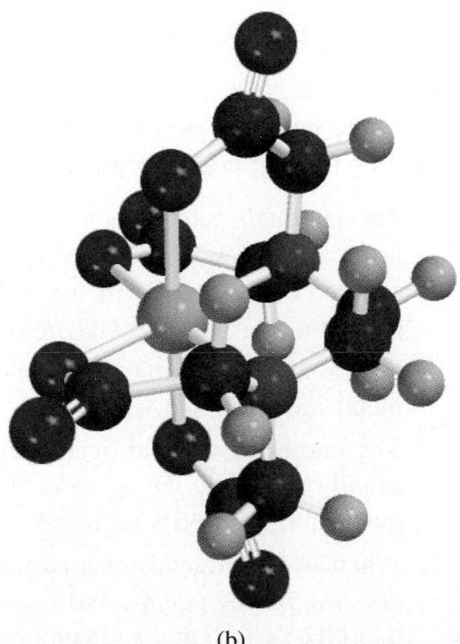

Figure 20.5
(a) EDTA complex of lead. The complex bears a net charge of −2 because each O donor atom has one negative charge and the lead ion carries two positive charges. Only the lone pairs that participate in bonding are shown. (b) Molecular model of the Pb^{2+}–EDTA complex. The green sphere is the Pb^{2+} ion. Note the octahedral geometry around the Pb^{2+} ion.

Oxidation Number of Metals in Coordination Compounds

The net charge of a complex ion is the sum of the charges on the central metal atom and its surrounding ligands. In the $[PtCl_6]^{2-}$ ion, for example, each chloride ion has an oxidation number of -1, so the oxidation number of Pt must be $+4$. If the ligands do not bear net charges, the oxidation number of the metal is equal to the charge of the complex ion. Thus, in $[Cu(NH_3)_4]^{2+}$ each NH_3 is neutral, so the oxidation number of Cu is $+2$.

EXAMPLE 20.1

Specify the oxidation number of the central metal atom in each of the following compounds: (a) $[Ru(NH_3)_5(H_2O)]Cl_2$, (b) $[Cr(NH_3)_6](NO_3)_3$, (c) $[Fe(CO)_5]$, and (d) $K_4[Fe(CN)_6]$.

Strategy The oxidation number of the metal atom is equal to its charge. First we examine the anion or the cation that electrically balances the complex ion. This step gives us the net charge of the complex ion. Next, from the nature of the ligands (charged or neutral species) we can deduce the net charge of the metal and hence its oxidation number.

Solution

(a) Both NH_3 and H_2O are neutral species. Because each chloride ion carries a -1 charge, and there are two Cl^- ions, the oxidation number of Ru must be $+2$.

(b) Each nitrate ion has a charge of -1; therefore, the cation must be $[Cr(NH_3)_6]^{3+}$. NH_3 is neutral, so the oxidation number of Cr is $+3$.

(c) Because the CO species are neutral, the oxidation number of Fe is zero.

(d) Each potassium ion has a charge of $+1$; therefore, the anion is $[Fe(CN)_6]^{4-}$. Next, we know that each cyanide group bears a charge of -1, so Fe must have an oxidation number of $+2$.

Similar problems: 20.13, 20.14.

Practice Exercise Write the oxidation numbers of the metals in the compound $K[Au(OH)_4]$.

Naming Coordination Compounds

Now that we have discussed the various types of ligands and the oxidation numbers of metals, our next step is to learn what to call these coordination compounds. The rules for naming coordination compounds are as follows:

1. The cation is named before the anion, as in other ionic compounds. The rule holds regardless of whether the complex ion bears a net positive or a negative charge. For example, in the $K_3[Fe(CN)_6]$ and $[Co(NH_3)_4Cl_2]Cl$ compound, we name the K^+ and $[Co(NH_3)_4Cl_2]^+$ cations first, respectively.
2. Within a complex ion, the ligands are named first, in alphabetical order, and the metal ion is named last.
3. The names of anionic ligands end with the letter *o*, whereas a neutral ligand is usually called by the name of the molecule. The exceptions are H_2O (aqua), CO (carbonyl), and NH_3 (ammine). Table 20.3 lists some common ligands.
4. When several ligands of a particular kind are present, we use the Greek prefixes *di-*, *tri-*, *tetra-*, *penta-*, and *hexa-* to name them. Thus, the ligands in the cation $[Co(NH_3)_4Cl_2]^+$ are "tetraamminedichloro." (Note that prefixes are ignored when

Table 20.3 Names of Common Ligands in Coordination Compounds

Ligand	Name of Ligand in Coordination Compound
Bromide, Br^-	Bromo
Chloride, Cl^-	Chloro
Cyanide, CN^-	Cyano
Hydroxide, OH^-	Hydroxo
Oxide, O^{2-}	Oxo
Carbonate, CO_3^{2-}	Carbonato
Nitrite, NO_2^-	Nitro
Oxalate, $C_2O_4^{2-}$	Oxalato
Ammonia, NH_3	Ammine
Carbon monoxide, CO	Carbonyl
Water, H_2O	Aqua
Ethylenediamine	Ethylenediamine
Ethylenediaminetetraacetate	Ethylenediaminetetraacetato

alphabetizing ligands.) If the ligand itself contains a Greek prefix, we use the prefixes *bis* (2), *tris* (3), and *tetrakis* (4) to indicate the number of ligands present. For example, the ligand ethylenediamine already contains *di*; therefore, if two such ligands are present the name is *bis(ethylenediamine)*.

5. The oxidation number of the metal is written in Roman numerals following the name of the metal. For example, the Roman numeral III is used to indicate the +3 oxidation state of chromium in $[Cr(NH_3)_4Cl_2]^+$, which is called tetraamminedichlorochromium(III) ion.
6. If the complex is an anion, its name ends in *-ate*. For example, in $K_4[Fe(CN)_6]$ the anion $[Fe(CN)_6]^{4-}$ is called hexacyanoferrate(II) ion. Note that the Roman numeral II indicates the oxidation state of iron. Table 20.4 gives the names of anions containing metal atoms.

Table 20.4 Name of Anions Containing Metal Atoms

Metal	Name of Metal in Anionic Complex
Aluminum	Aluminate
Chromium	Chromate
Cobalt	Cobaltate
Copper	Cuprate
Gold	Aurate
Iron	Ferrate
Lead	Plumbate
Manganese	Manganate
Molybdenum	Molybdate
Nickel	Nickelate
Silver	Argentate
Tin	Stannate
Tungsten	Tungstate
Zinc	Zincate

EXAMPLE 20.2

Write the systematic names of the following coordination compounds: (a) $Ni(CO)_4$, (b) $NaAuF_4$, (c) $K_3[Fe(CN)_6]$, (d) $[Cr(en)_3]Cl_3$.

Strategy We follow the preceding procedure for naming coordination compounds and refer to Tables 20.3 and 20.4 for names of ligands and anions containing metal atoms.

Solution

(a) The CO ligands are neutral species and therefore the Ni atom bears no net charge. The compound is called tetracarbonylnickel(0), or more commonly, nickel tetracarbonyl.

(b) The sodium cation has a positive charge; therefore, the complex anion has a negative charge (AuF_4^-). Each fluoride ion has a negative charge so the oxidation

(Continued)

number of gold must be +3 (to give a net negative charge). The compound is called sodium tetrafluoroaurate(III) .

(c) The complex ion is the anion and it bears three negative charges because each potassium ion bears a +1 charge. Looking at $[Fe(CN)_6]^{3-}$, we see that the oxidation number of Fe must be +3 because each cyanide ion bears a −1 charge (−6 total). The compound is potassium hexacyanoferrate(III) . This compound is commonly called potassium ferricyanide .

(d) As we noted earlier, *en* is the abbreviation for the ligand ethylenediamine. Because there are three chloride ions each with a −1 charge, the cation is $[Cr(en)_3]^{3+}$. The *en* ligands are neutral so the oxidation number of Cr must be +3. Because there are three *en* groups present and the name of the ligand already contains *di* (see rule 4), the compound is called *tris*(ethylenediamine)chromium(III) chloride .

Similar problems: 20.15, 20.16.

Practice Exercise What is the systematic name of $[Cr(H_2O)_4Cl_2]Cl$?

EXAMPLE 20.3

Write the formulas for the following compounds: (a) pentaamminechlorocobalt(III) chloride, (b) dichlorobis(ethylenediamine)platinum(IV) nitrate, (c) sodium hexanitrocobaltate(III).

Strategy We follow the preceding procedure and refer to Tables 20.3 and 20.4 for names of ligands and anions containing metal atoms.

Solution

(a) The complex cation contains five NH_3 groups, a Cl^- ion, and a Co ion having a +3 oxidation number. The net charge of the cation must be +2, $[Co(NH_3)_5Cl]^{2+}$. Two chloride anions are needed to balance the positive charges. Therefore, the formula of the compound is $[Co(NH_3)_5Cl]Cl_2$.

(b) There are two chloride ions (−1 each), two *en* groups (neutral), and a Pt ion with an oxidation number of +4. The net charge on the cation must be +2, $[Pt(en)_2Cl_2]^{2+}$. Two nitrate ions are needed to balance the +2 charge of the complex cation. Therefore, the formula of the compound is $[Pt(en)_2Cl_2](NO_3)_2$.

(c) The complex anion contains six nitro groups (−1 each) and a cobalt ion with an oxidation number of +3. The net charge on the complex anion must be −3, $[Co(NO_2)_6]^{3-}$. Three sodium cations are needed to balance the −3 charge of the complex anion. Therefore, the formula of the compound is $Na_3[Co(NO_2)_6]$.

Similar problems: 20.17, 20.18.

Practice Exercise Write the formula for the following compound: *tris*(ethylene-diamine)cobalt(III) sulfate.

REVIEW OF CONCEPTS

A student writes the name of the compound $[Co(NH_3)_5(H_2O)]Br_2$ as aquapenta-amminecobalt dibromide. Is this correct? If not, provide a proper systematic name.

20.3 Geometry of Coordination Compounds

Figure 20.6 shows four different geometric arrangements for metal atoms with monodentate ligands. In these diagrams we see that the structure and the coordination number of the metal atom relate to each other as follows:

Coordination Number	Structure
2	Linear
4	Tetrahedral or square planar
6	Octahedral

Stereoisomers are *compounds that are made up of the same types and numbers of atoms bonded together in the same sequence but with different spatial arrangements.* There are two types of stereoisomers: geometric isomers and optical isomers (enantiomers). Coordination compounds may exhibit one or both types of isomerism. Note, however, that many coordination compounds do not have stereoisomers.

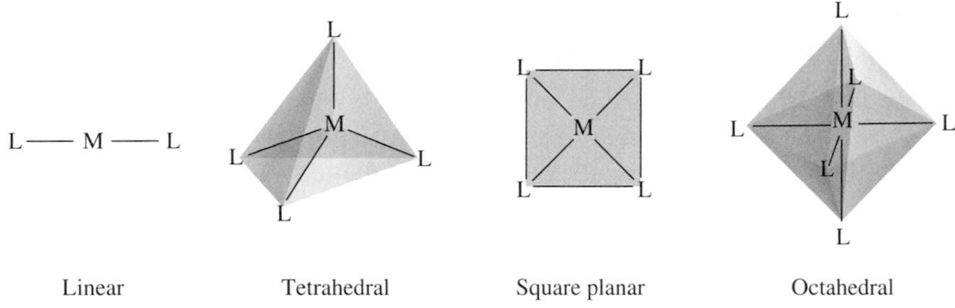

Figure 20.6
Common geometries of complex ions. In each case, M is a metal and L is a monodentate ligand.

Coordination Number = 2

The $[Ag(NH_3)_2]^+$ complex ion, formed by the reaction between Ag^+ ions and ammonia (see Table 17.4), has a coordination number of 2 and a linear geometry. Other examples are $[CuCl_2]^-$ and $[Au(CN)_2]^-$.

Coordination Number = 4

There are two types of geometries with a coordination number of 4. The $[Zn(NH_3)_4]^{2+}$ and $[CoCl_4]^{2-}$ ions have tetrahedral geometry, whereas the $[Pt(NH_3)_4]^{2+}$ ion has the square planar geometry. In Chapter 11 we discussed geometric isomers of alkenes (see p. 373). Square planar complex ions with two different monodentate ligands can also exhibit geometric isomerism. Figure 20.7 shows the *cis* and *trans* isomers of diamminedichloroplatinum(II). Note that although the types of bonds are the same in

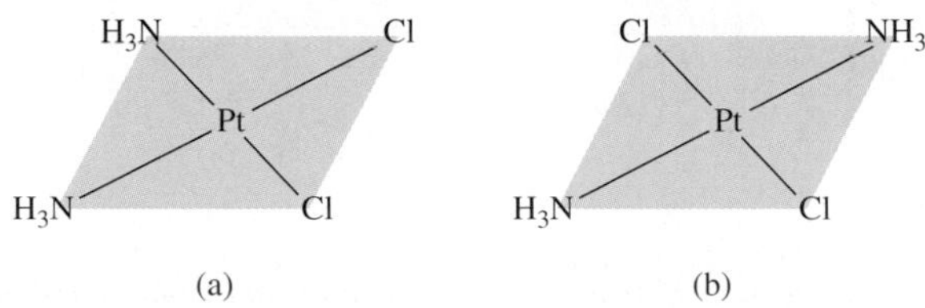

Figure 20.7
The (a) cis and (b) trans isomers of diamminedichloroplatinum(II), $Pt(NH_3)_2Cl_2$. Note that the two Cl atoms are adjacent to each other in the cis isomer and diagonally across from each other in the trans isomer.

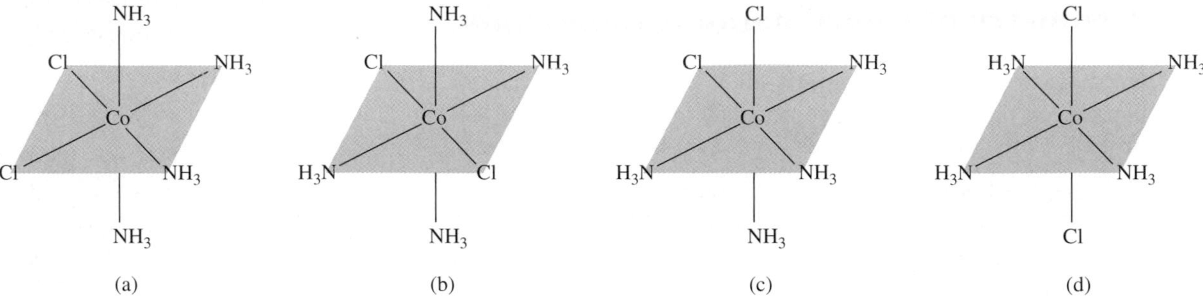

Figure 20.8
The (a) cis and (b) trans isomers of tetraamminedichlorocobalt(III) ion, $[Co(NH_3)_4Cl_2]^+$. The structure shown in (c) can be generated by rotating that in (a), and the structure shown in (d) can be generated by rotating that in (b). The ion has only two geometric isomers, (a) [or (c)] and (b) [or (d)].

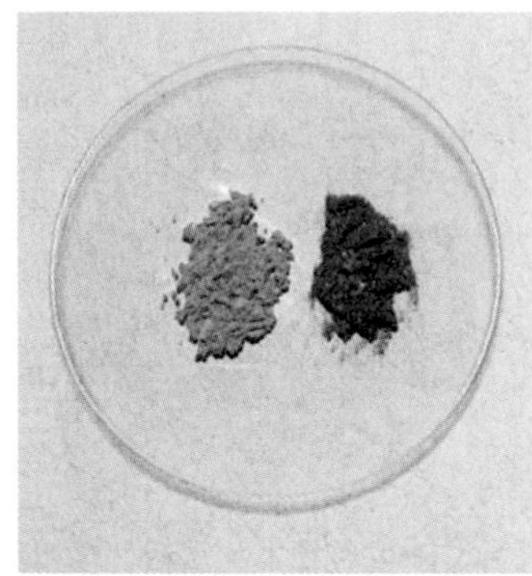

Figure 20.9
Left: cis-tetraamminedichlorocobalt(III) chloride. Right: trans-tetraamminedichlorocobalt(III) chloride.

both isomers (two Pt—N and two Pt—Cl bonds), the spatial arrangements are different. These two isomers have different properties (melting point, boiling point, color, solubility in water, and dipole moment).

Coordination Number = 6

Complex ions with a coordination number of 6 all have octahedral geometry (see Section 10.1). Geometric isomers are possible in octahedral complexes when two or more different ligands are present. An example is the tetraamminedichlorocobalt(III) ion shown in Figure 20.8. The two geometric isomers have different colors and other properties even though they have the same ligands and the same number and types of bonds (Figure 20.9).

In addition to geometric isomers, certain octahedral complex ions can also give rise to enantiomers, as discussed in Chapter 11. Figure 20.10 shows the *cis* and *trans* isomers of dichlorobis(ethylenediamine)cobalt(III) ion and their mirror images. Careful examination reveals that the *trans* isomer and its mirror image are superimposable, but the *cis* isomer and its mirror image are not. Therefore, the *cis* isomer and its mirror image are enantiomers. It is interesting to note that unlike the case in most organic compounds, there are no asymmetric carbon atoms in these compounds.

Review of Concepts

How many geometric isomers of the $[CoBr_2(en)(NH_3)_2]^+$ ion are possible?

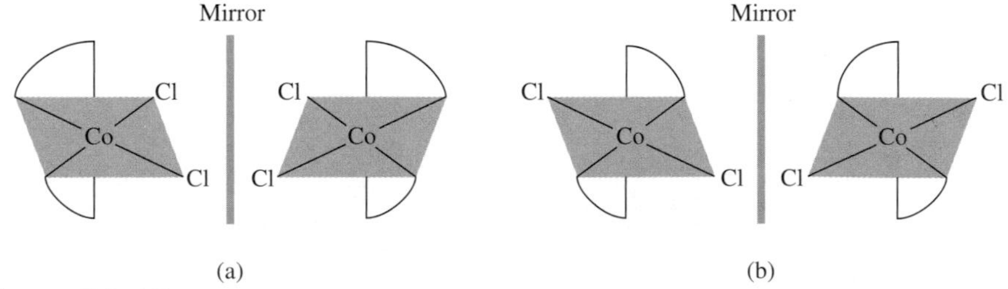

Figure 20.10
The (a) cis and (b) trans isomers of dichlorobis(ethylenediamine)cobalt(III) ion, $[Co(en)_2Cl_2]^+$, and their mirror images. If you could rotate the mirror image in (b) 90° clockwise about the vertical position and place the ion over the trans isomer, you would find that the two are superimposable. No matter how you rotated the cis isomer and its mirror image in (a), however, you could not superimpose one on the other.

20.4 Bonding in Coordination Compounds: Crystal Field Theory

A satisfactory theory of bonding in coordination compounds must account for properties such as color and magnetism, as well as stereochemistry and bond strength. No single theory as yet does all this for us. Rather, several different approaches have been applied to transition metal complexes. We will consider only one of them here—crystal field theory—because it accounts for both the color and magnetic properties of many coordination compounds.

We will begin our discussion of crystal field theory with the most straightforward case, namely, complex ions with octahedral geometry. Then we will see how it is applied to tetrahedral and square-planar complexes.

Crystal Field Splitting in Octahedral Complexes

Crystal field theory explains the bonding in complex ions purely in terms of electrostatic forces. In a complex ion, two types of electrostatic interaction come into play. One is the attraction between the positive metal ion and the negatively charged ligand or the negatively charged end of a polar ligand. This is the force that binds the ligands to the metal. The second type of interaction is electrostatic repulsion between the lone pairs on the ligands and the electrons in the *d* orbitals of the metals.

The name "crystal field" is associated with the theory used to explain the properties of solid, crystalline materials. The same theory is used to study coordination compounds.

As we saw in Chapter 7, *d* orbitals have different orientations, but in the absence of external disturbance they all have the same energy. In an octahedral complex, a central metal atom is surrounded by six lone pairs of electrons (on the six ligands), so all five *d* orbitals experience electrostatic repulsion. The magnitude of this repulsion depends on the orientation of the *d* orbital that is involved. Take the $d_{x^2-y^2}$ orbital as an example. In Figure 20.11, we see that the lobes of this orbital point toward corners of the octahedron along the *x* and *y* axes, where the lone-pair electrons are positioned. Thus, an electron residing in this orbital would experience a greater repulsion from the ligands than an electron would in, say, the d_{xy} orbital. For this reason, the energy of the $d_{x^2-y^2}$ orbital is increased relative to the d_{xy}, d_{yz}, and d_{xz} orbitals. The d_{z^2} orbital's energy is also greater, because its lobes are pointed at the ligands along the *z* axis.

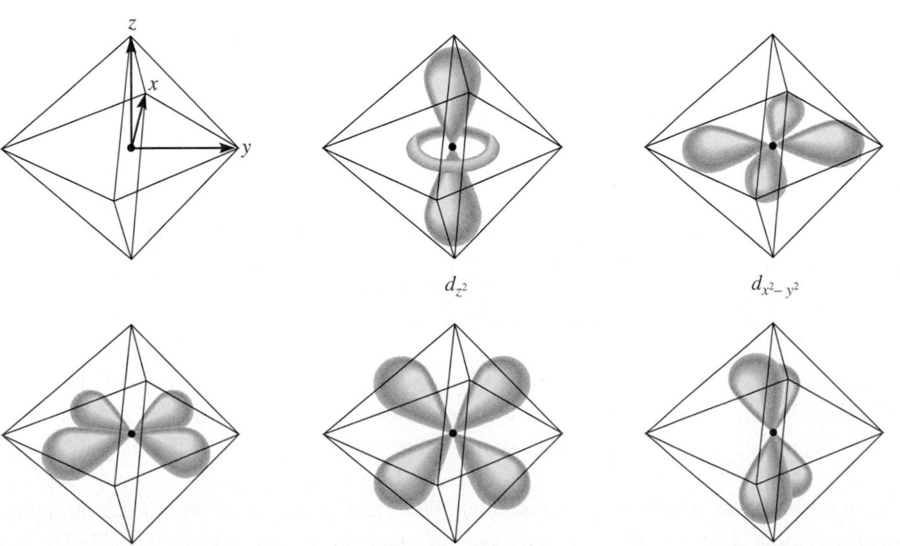

Figure 20.11
The five d orbitals in an octahedral environment. The metal atom (or ion) is at the center of the octahedron, and the six lone pairs on the donor atoms of the ligands are at the corners.

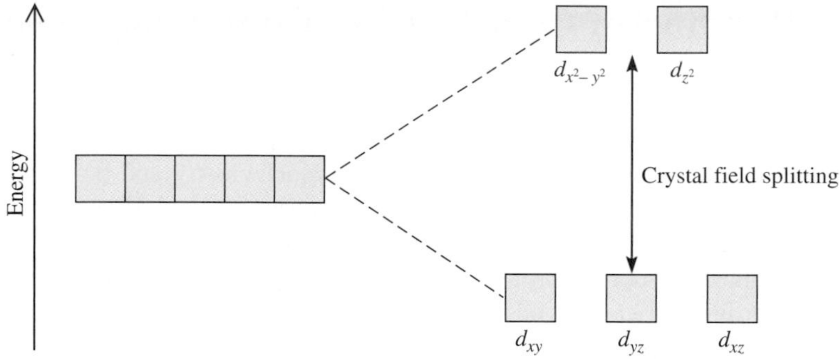

Figure 20.12
Crystal field splitting between d orbitals in an octahedral complex.

As a result of these metal-ligand interactions, the five *d* orbitals in an octahedral complex are split between two sets of energy levels: a higher level with two orbitals ($d_{x^2-y^2}$ and d_{z^2}) having the same energy and a lower level with three equal-energy orbitals (d_{xy}, d_{yz}, and d_{xz}), as shown in Figure 20.12. The ***crystal field splitting (Δ)*** is *the energy difference between two sets of d orbitals in a metal atom when ligands are present.* The magnitude of Δ depends on the metal and the nature of the ligands: it has a direct effect on the color and magnetic properties of complex ions.

Color

Animation: Absorption of Color

In Chapter 7 we learned that white light, such as sunlight, is a combination of all colors. A substance appears black if it absorbs all the visible light that strikes it. If it absorbs no visible light, it is white or colorless. An object appears green if it absorbs all light but reflects the green component. An object also looks green if it reflects all colors except red, the *complementary* color of green (Figure 20.13).

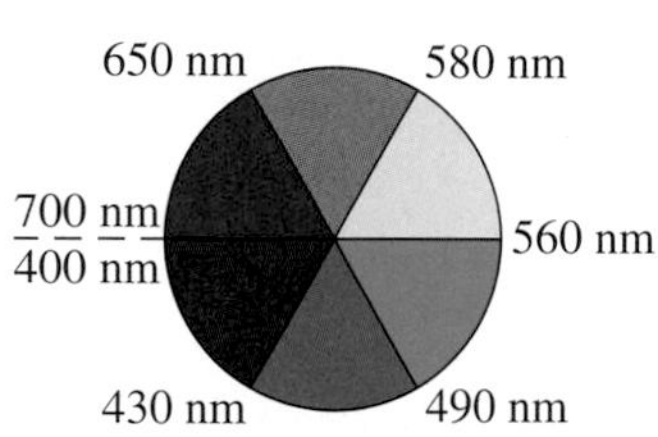

Figure 20.13
A color wheel with appropriate wavelengths. The compound that absorbs in the green region will appear red, the complementary color of green.

What has been said of reflected light also applies to transmitted light (that is, the light that passes through the medium, for example, a solution). Consider the hydrated cupric ion, $[Cu(H_2O)_6]^{2+}$, which absorbs light in the orange region of the spectrum so that a solution of $CuSO_4$ appears blue to us. Recall from Chapter 7 that when the energy of a photon is equal to the difference between the ground state and an excited state, absorption occurs as the photon strikes the atom (or ion or compound), and an electron is promoted to a higher level. This knowledge enables us to calculate the energy change involved in the electron transition. The energy of a photon, given by Equation (7.2), is

$$E = h\nu$$

where h represents Planck's constant (6.63×10^{-34} J · s) and ν is the frequency of the radiation, which is 5.00×10^{14}/s for a wavelength of 600 nm. Here $E = \Delta$, so we have

$$\begin{aligned}\Delta &= h\nu \\ &= (6.63 \times 10^{-34}\ \text{J} \cdot \text{s})(5.00 \times 10^{14}/\text{s}) \\ &= 3.32 \times 10^{-19}\ \text{J}\end{aligned}$$

(Note that this is the energy absorbed by one ion.) If the wavelength of the photon absorbed by an ion lies outside the visible region, then the transmitted light looks the same (to us) as the incident light—white—and the ion appears colorless.

A *d*-to-*d* transition must occur for a transition metal complex to show color. Therefore, ions with d^0 or d^{10} electron configurations are usually colorless.

The best way to measure crystal field splitting is to use spectroscopy to determine the wavelength at which light is absorbed. The $[Ti(H_2O)_6]^{3+}$ ion provides a straightforward example, because Ti^{3+} has only one 3*d* electron (Figure 20.14). The $[Ti(H_2O)_6]^{3+}$ ion absorbs light in the visible region of the spectrum (Figure 20.15). The wavelength

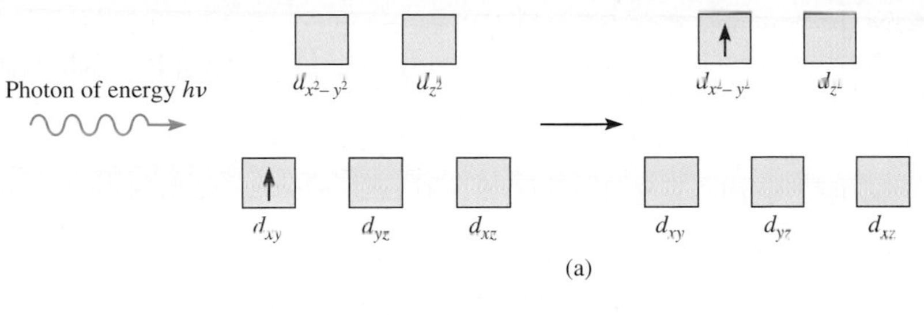

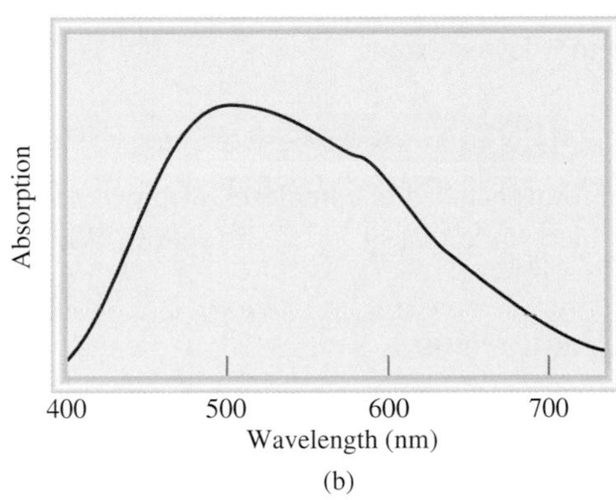

Figure 20.14
(a) The process of photon absorption and (b) a graph of the absorption spectrum of $[Ti(H_2O)_6]^{3+}$. The energy of the incoming photon is equal to the crystal field splitting. The maximum absorption peak in the visible region occurs at 498 nm.

corresponding to maximum absorption is 498 nm [Figure 20.14(b)]. This information enables us to calculate the crystal field splitting as follows. We start by writing

$$\Delta = h\nu \tag{20.1}$$

Also

$$\nu = \frac{c}{\lambda}$$

where c is the speed of light and λ is the wavelength. Therefore,

$$\Delta = \frac{hc}{\lambda} = \frac{(6.63 \times 10^{-34}\ \text{J} \cdot \text{s})(3.00 \times 10^8\ \text{m/s})}{(498\ \text{nm})(1 \times 10^{-9}\ \text{m/1 nm})}$$
$$= 3.99 \times 10^{-19}\ \text{J}$$

Equation (7.3) shows that $E = hc/\lambda$.

This is the energy required to excite *one* $[Ti(H_2O)_6]^{3+}$ ion. To express this energy difference in the more convenient units of kilojoules per mole, we write

$$\Delta = (3.99 \times 10^{-19}\ \text{J/ion})(6.02 \times 10^{23}\ \text{ions/mol})$$
$$= 240{,}000\ \text{J/mol}$$
$$= 240\ \text{kJ/mol}$$

Figure 20.15
Colors of some of the first-row transition metal ions in solution. From left to right: Ti^{3+}, Cr^{3+}, Mn^{2+}, Fe^{3+}, Co^{2+}, Ni^{2+}, Cu^{2+}. The Sc^{3+} and V^{5+} ions are colorless.

Aided by spectroscopic data for a number of complexes, all having the same metal ion but different ligands, chemists calculated the crystal splitting for each ligand and established a ***spectrochemical series,*** which is *a list of ligands arranged in increasing order of their abilities to split the d orbital energy levels:*

The order in the spectrochemical series is the same no matter which metal atom (or ion) is present.

$$I^- < Br^- < Cl^- < OH^- < F^- < H_2O < NH_3 < \text{en} < CN^- < CO$$

These ligands are arranged in the order of increasing value of Δ. CO and CN^- are called *strong-field ligands,* because they cause a large splitting of the *d* orbital energy levels. The halide ions and hydroxide ion are *weak-field ligands*, because they split the *d* orbitals to a lesser extent.

REVIEW OF CONCEPTS

The Cr^{3+} ion forms octahedral complexes with two neutral ligands X and Y. The color of CrX_6^{3+} is blue while that of CrY_6^{3+} is yellow. Which is a stronger field ligand?

Magnetic Properties

The magnitude of the crystal field splitting also determines the magnetic properties of a complex ion. The $[Ti(H_2O)_6]^{3+}$ ion, having only one *d* electron, is always paramagnetic. However, for an ion with several *d* electrons, the situation is less clearcut. Consider, for example, the octahedral complexes $[FeF_6]^{3-}$ and $[Fe(CN)_6]^{3-}$ (Figure 20.16). The electron configuration of Fe^{3+} is $[Ar]3d^5$, and there are two possible ways to distribute the five *d* electrons among the *d* orbitals. According to Hund's rule (see Section 7.8), maximum stability is reached when the electrons are placed in five separate orbitals with parallel spins. But this arrangement can be achieved only at a cost; two of the five electrons must be promoted to the higher-energy $d_{x^2-y^2}$ and d_{z^2} orbitals. No such energy investment is needed if all five electrons enter the d_{xy}, d_{yz}, and d_{xz} orbitals. According to Pauli's exclusion principle (p. 233), there will be only one unpaired electron present in this case.

Figure 20.17 shows the distribution of electrons among *d* orbitals that results in low- and high-spin complexes. The actual arrangement of the electrons is determined by the amount of stability gained by having maximum parallel spins versus the investment in energy required to promote electrons to higher *d* orbitals. Because F^- is a weak-field ligand, the five *d* electrons enter five separate *d* orbitals with parallel spins to create a high-spin complex (see Figure 20.16). On the other hand, the cyanide ion is a strong-field ligand, so it is energetically preferable for all five electrons to be in the lower orbitals and therefore a low-spin complex is formed. High-spin complexes are more paramagnetic than low-spin complexes.

The magnetic properties of a complex ion depend on the number of unpaired electrons present.

The actual number of unpaired electrons (or spins) in a complex ion can be found by magnetic measurements, and in general, experimental findings support predictions

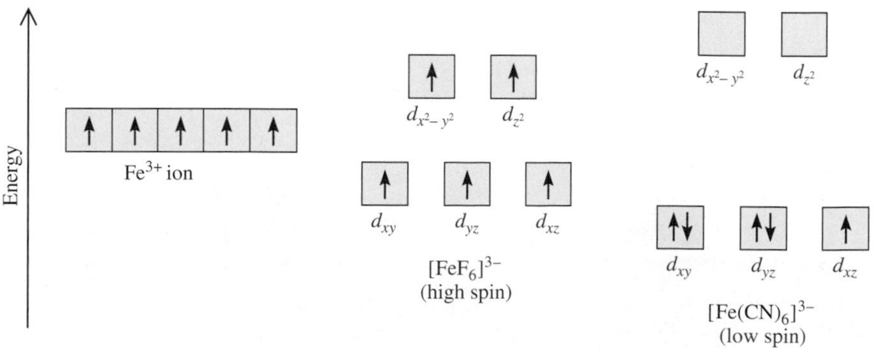

Figure 20.16
Energy-level diagrams for the Fe^{3+} ion and for the $[FeF_6]^{3-}$ and $[Fe(CN)_6]^{3-}$ complex ions.

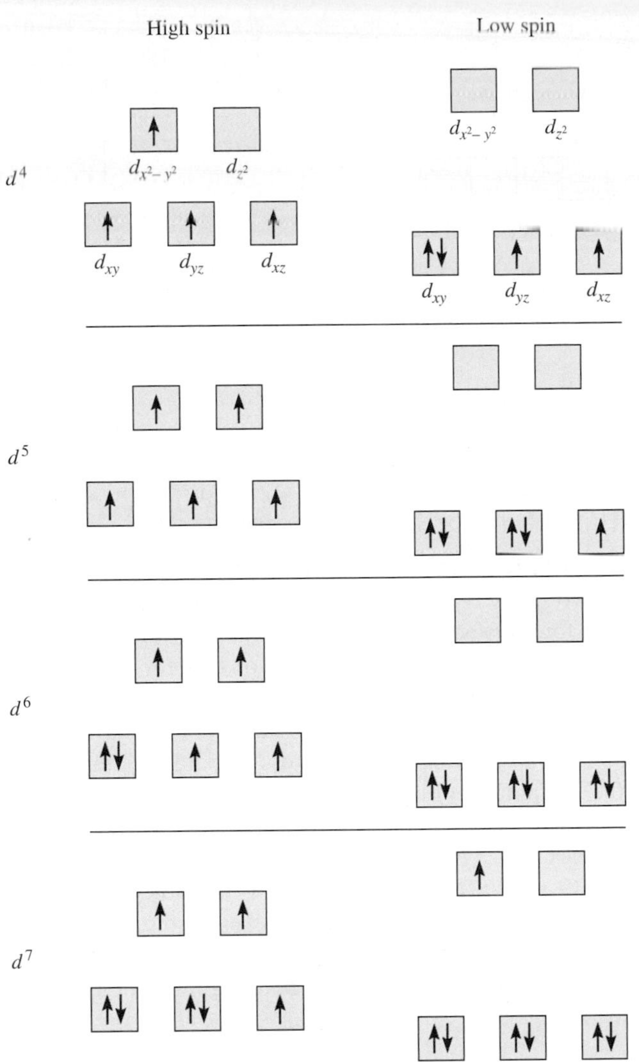

Figure 20.17
Orbital diagrams for the high-spin and low-spin octahedral complexes corresponding to the electron configurations d^4, d^5, d^6, and d^7. No such distinctions can be made for d^1, d^2, d^3, d^8, d^9, and d^{10}.

based on crystal field splitting. However, a distinction between low- and high-spin complexes can be made only if the metal ion contains more than three and fewer than eight *d* electrons, as shown in Figure 20.17.

EXAMPLE 20.4

Predict the number of unpaired spins in the $[Cr(en)_3]^{2+}$ ion.

Strategy The magnetic properties of a complex ion depend on the strength of the ligands. Strong-field ligands, which cause a high degree of splitting among the *d* orbital energy levels, result in low-spin complexes. Weak-field ligands, which cause a small degree of splitting among the *d* orbital energy levels, result in high-spin complexes.

Solution The electron configuration of Cr^{2+} is $[Ar]3d^4$. Because en is a strong-field ligand, we expect $[Cr(en)_3]^{2+}$ to be a low-spin complex. According to Figure 20.17, all four electrons will be placed in the lower-energy *d* orbitals (d_{xy}, d_{yz}, and d_{xz}) and there will be a total of two unpaired spins.

Practice Exercise How many unpaired spins are in $[Mn(H_2O)_6]^{2+}$? (H_2O is a weak-field ligand.)

Similar problem: 20.30.

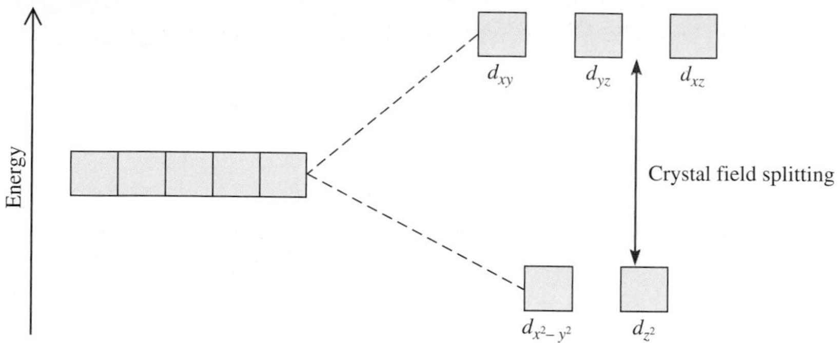

Figure 20.18

Crystal field splitting between d orbitals in a tetrahedral complex.

Tetrahedral and Square-Planar Complexes

So far we have concentrated on octahedral complexes. The splitting of the d orbital energy levels in two other types of complexes—tetrahedral and square-planar—can also be accounted for satisfactorily by the crystal field theory. In fact, the splitting pattern for a tetrahedral ion is just the reverse of that for octahedral complexes. In this case, the d_{xy}, d_{yz}, and d_{xz} orbitals are more closely directed at the ligands and therefore have more energy than the $d_{x^2-y^2}$ and d_{z^2} orbitals (Figure 20.18). Most tetrahedral complexes are high-spin complexes. Presumably, the tetrahedral arrangement reduces the magnitude of metal-ligand interactions, resulting in a smaller Δ value. This is a reasonable assumption because the number of ligands is smaller in a tetrahedral complex.

As Figure 20.19 shows, the splitting pattern for square-planar complexes is the most complicated. Clearly, the $d_{x^2-y^2}$ orbital possesses the highest energy (as in the octahedral case), and the d_{xy} orbital the next highest. However, the relative placement of the d_{z^2} and the d_{xz} and d_{yz} orbitals cannot be determined simply by inspection and must be calculated.

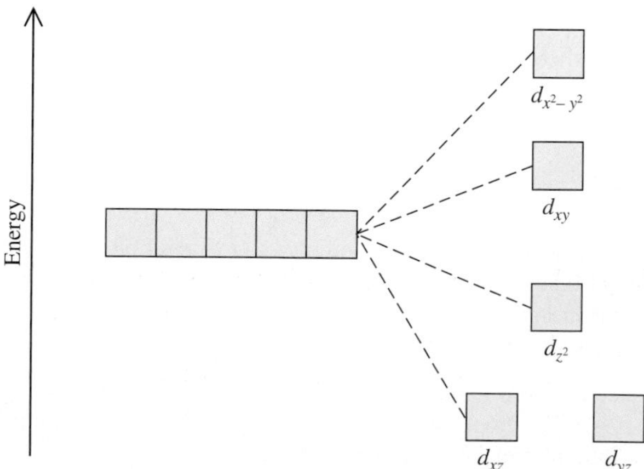

Figure 20.19

Energy-level diagram for a square-planar complex. Because there are more than two energy levels, we cannot define crystal field splitting as we can for octahedral and tetrahedral complexes.

20.5 Reactions of Coordination Compounds

Complex ions undergo ligand exchange (or substitution) reactions in solution. The rates of these reactions vary widely, depending on the nature of the metal ion and the ligands.

In studying ligand exchange reactions, it is often useful to distinguish between the stability of a complex ion and its tendency to react, which we call *kinetic lability*. Stability in this context is a thermodynamic property, which is measured in terms of the species' formation constant K_f (see p. 615). For example, we say that the complex ion tetracyanonickelate(II) is stable because it has a large formation constant ($K_f \approx 1 \times 10^{30}$)

$$Ni^{2+} + 4CN^- \rightleftharpoons [Ni(CN)_4]^{2-}$$

By using cyanide ions labeled with the radioactive isotope carbon-14, chemists have shown that $[Ni(CN)_4]^{2-}$ undergoes ligand exchange very rapidly in solution. The following equilibrium is established almost as soon as the species are mixed:

$$[Ni(CN)_4]^{2-} + 4{*}CN^- \rightleftharpoons [Ni({*}CN)_4]^{2-} + 4CN^-$$

At equilibrium, there is a distribution of $*CN^-$ ions in the complex ion.

where the asterisk denotes a ^{14}C atom. Complexes like the tetracyanonickelate(II) ion are termed ***labile complexes*** because they *undergo rapid ligand exchange reactions*. Thus, a thermodynamically stable species (that is, one that has a large formation constant) is not necessarily unreactive. (In Section 14.4 we saw that the smaller the activation energy, the larger the rate constant, and hence the greater the rate.)

A complex that is thermodynamically *unstable* in acidic solution is $[Co(NH_3)_6]^{3+}$. The equilibrium constant for the following reaction is about 1×10^{20}:

$$[Co(NH_3)_6]^{3+} + 6H^+ + 6H_2O \rightleftharpoons [Co(H_2O)_6]^{3+} + 6NH_4^+$$

When equilibrium is reached, the concentration of the $[Co(NH_3)_6]^{3+}$ ion is very low. However, this reaction requires several days to complete because of the inertness of the $[Co(NH_3)_6]^{3+}$ ion. This is an example of an ***inert complex,*** *a complex ion that undergoes very slow exchange reactions* (on the order of hours or even days). It shows that a thermodynamically unstable species is not necessarily chemically reactive. The rate of reaction is determined by the energy of activation, which is high in this case.

Most complex ions containing Co^{3+} Cr^{3+}, and Pt^{2+} are kinetically inert. Because they exchange ligands very slowly, they are easy to study in solution. As a result, our knowledge of the bonding, structure, and isomerism of coordination compounds has come largely from studies of these compounds.

20.6 Coordination Compounds in Living Systems

Coordination compounds play many roles in animals and plants. They are essential in the storage and transport of oxygen, as electron transfer agents, as catalysts, and in photosynthesis. Here we will briefly discuss the coordination compounds containing the porphyrin group and cisplatin as an anticancer drug.

Hemoglobin and Related Compounds

Hemoglobin functions as an oxygen carrier for metabolic processes. The molecule contains four folded long chains called *subunits*. Hemoglobin carries oxygen in the

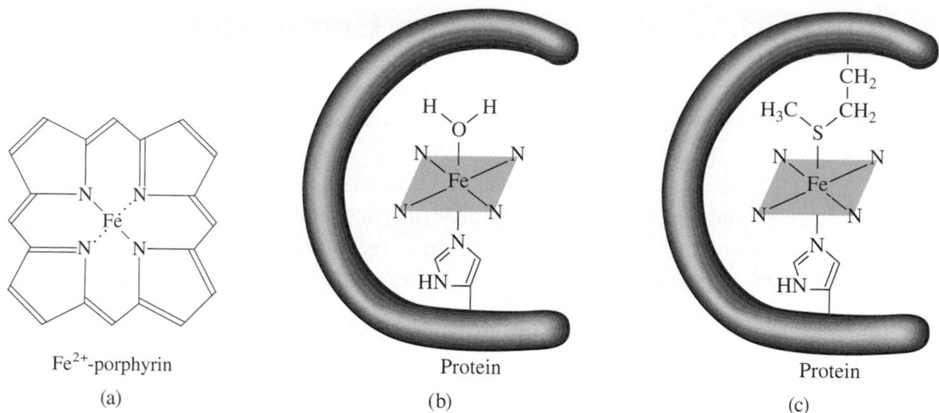

Figure 20.20
(a) Structure of Fe^{2+}-porphyrin. (b) The heme group in hemoglobin. The Fe^{2+} ion is coordinated with the nitrogen atoms of the heme group. The ligand below the porphyrin is the histidine group, which is part of the protein. The sixth ligand is a water molecule, which can be replaced by oxygen. (c) The heme group in cyctochromes. The ligands above and below the porphyrin are the methionine group and histidine group of the protein molecule.

blood from the lungs to the tissues, where it delivers the oxygen molecules to myoglobin. Myoglobin, which is made up of only one subunit, stores oxygen for metabolic processes in muscle.

The *heme* group in each subunit is a complex ion formed between a Fe^{2+} ion and a porphyrin group [Figure 20.20(a)]. The Fe^{2+} ion is coordinated to the four nitrogen atoms in the porphyrin group and also to a nitrogen donor atom in a ligand that is part of the protein molecule. The sixth ligand is a water molecule, which binds to the ion on the other side of the planar ring to complete the octahedral geometry [Figure 20.20(b)]. In this state, the molecule is called *deoxyhemoglobin* and imparts a bluish tinge to venous blood. The water ligand can be replaced readily by molecular oxygen to form the red *oxyhemoglobin* found in arterial blood.

The iron-heme complex is present in another class of proteins called the *cytochromes*. Here too, the iron forms an octahedral complex, but both the fifth and sixth ligands are part of the protein structure [Figure 20.20(c)]. Because the ligands are firmly bound to the metal ion, they cannot be displaced by oxygen or other ligands. Instead, the cytochromes act as electron carriers, which are essential to metabolic processes. In cytochromes, iron undergoes rapid reversible redox processes:

$$Fe^{3+} + e^- \rightleftharpoons Fe^{2+}$$

which are coupled to the oxidation of organic molecules such as the carbohydrates.

The chlorophyll molecule, which is necessary for plant photosynthesis, also contains the porphyrin ring, but in this case the metal ion is Mg^{2+} rather than Fe^{2+}.

Cisplatin

In the mid-1960s, scientists discovered that *cis*-diamminedichloroplatinum(II), *cis*-$[Pt(NH_3)_2Cl_2]$, also called cisplatin, is an effective drug for certain types of cancer (see Figure 20.7). The mechanism for the action of cisplatin is the chelation of DNA (deoxyribonucleic acid). Cisplatin binds to DNA by forming cross-links in which the two chloride ions on cisplatin are replaced by nitrogen donor atoms on the DNA

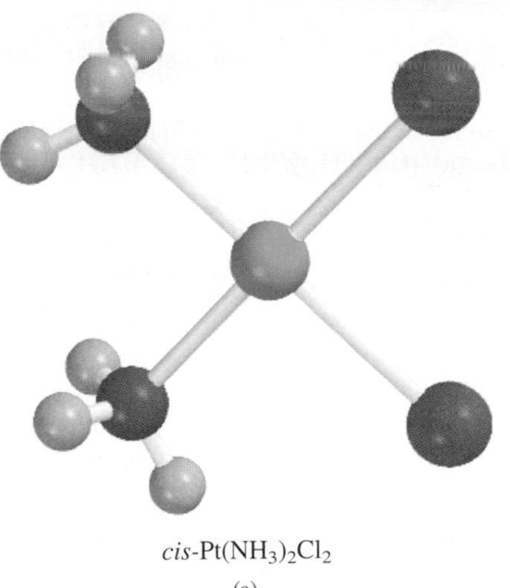
cis-Pt$(NH_3)_2Cl_2$

(a)

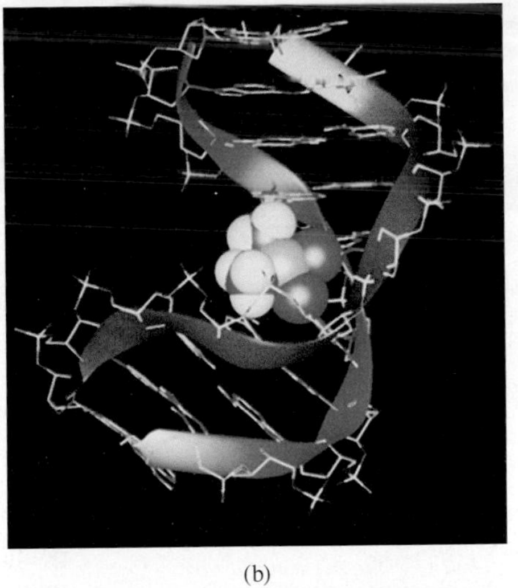
(b)

Figure 20.21
(a) cis-$Pt(NH_3)_2Cl_2$. (b) Cisplatin disrupts DNA replication and transcription by binding to the double helix. The structure of this major DNA adduct, depicted here, was elucidated by Professor Stephen Lippard's group at MIT.

molecule (Figure 20.21). This action leads to a mistake (mutation) in the DNA's replication and the eventual destruction of the cancerous cell. Interestingly, the geometric isomer, *trans*-$[Pt(NH_3)_2Cl_2]$, has no anticancer effect because it lacks the ability to bind to DNA.

Key Equation

$\Delta = h\nu$ (20.1) Calculating crystal-field splitting.

Summary of Facts and Concepts

1. Transition metals usually have incompletely filled d subshells and a pronounced tendency to form complexes. Compounds that contain complex ions are called coordination compounds.
2. The first-row transition metals (scandium to copper) are the most common of all the transition metals; their chemistry is characteristic, in many ways, of the entire group.
3. Complex ions consist of a metal ion surrounded by ligands. The donor atoms in the ligands each contribute an electron pair to the central metal ion in a complex.
4. Coordination compounds may exist as geometric isomers and/or enantiomers.
5. Crystal field theory explains bonding in complexes in terms of electrostatic interactions. According to crystal field theory, the d orbitals are split into two higher-energy and three lower-energy orbitals in an octahedral complex. The energy difference between these two sets of d orbitals is the crystal field splitting.
6. Strong-field ligands cause a large crystal field splitting, and weak-field ligands cause a small splitting. Electron spins tend to be parallel with weak-field ligands and paired with strong-field ligands, where a greater investment of energy is required to promote electrons into the high-lying d orbitals.
7. Complex ions undergo ligand exchange reactions in solution.
8. Coordination compounds occur in nature and are used as therapeutic drugs.

Key Words

Chelating agent, p. 709
Coordination compound, p. 707
Coordination number, p. 708
Crystal field splitting (Δ), p. 716
Donor atom, p. 708
Inert complex, p. 721
Labile complex, p. 721
Ligand, p. 707
Spectrochemical series, p. 718
Stereoisomers, p. 713

Questions and Problems

Properties of Transition Metals

Review Questions

20.1 What distinguishes a transition metal from a representative metal?

20.2 Why is zinc not considered a transition metal?

20.3 Explain why atomic radii decrease very gradually from scandium to copper.

20.4 Without referring to the text, write the ground-state electron configurations of the first-row transition metals. Explain any irregularities.

20.5 Write the electron configurations of these ions: V^{5+}, Cr^{3+}, Mn^{2+}, Fe^{3+}, Cu^{2+}, Sc^{3+}, Ti^{4+}.

20.6 Why do transition metals have more oxidation states than other elements? Give the highest oxidation states for scandium to copper.

20.7 As we read across the first-row transition metals from left to right, the +2 oxidation state becomes more stable in comparison with the +3 state. Why is this so?

20.8 Chromium exhibits several oxidation states in its compounds, whereas aluminum exhibits only the +3 oxidation state. Explain.

Coordination Compounds: Nomenclature; Oxidation Number

Review Questions

20.9 Define the following terms: coordination compound, ligand, donor atom, coordination number, chelating agent.

20.10 Describe the interaction between a donor atom and a metal atom in terms of a Lewis acid-base reaction.

Problems

20.11 Complete these statements for the complex ion $[Co(en)_2(H_2O)CN]^{2+}$. (a) The term "en" is the abbreviation for ______. (b) The oxidation number of Co is ______. (c) The coordination number of Co is ______. (d) ______ is a bidentate ligand.

20.12 Complete these statements for the complex ion $[Cr(C_2O_4)_2(H_2O)]^{2-}$. (a) The oxidation number of Cr is ______. (b) The coordination number of Cr is ______. (c) ______ is a bidentate ligand.

20.13 Give the oxidation numbers of the metals in these species: (a) $K_3[Fe(CN)_6]$, (b) $K_3[Cr(C_2O_4)_3]$, (c) $[Ni(CN)_4]^{2-}$.

20.14 Give the oxidation numbers of the metals in these species: (a) Na_2MoO_4, (b) $MgWO_4$, (c) $K_4[Fe(CN)_6]$.

20.15 What are the systematic names for these ions and compounds: (a) $[Co(NH_3)_4Cl_2]^+$, (b) $Cr(NH_3)_3Cl_3$, (c) $[Co(en)_2Br_2]^+$, (d) $Fe(CO)_5$?

20.16 What are the systematic names for these ions and compounds: (a) $[cis\text{-}Co(en)_2Cl_2]^+$, (b) $[Pt(NH_3)_5Cl]Cl_3$, (c) $[Co(NH_3)_6]Cl_3$, (d) $[Co(NH_3)_5Cl]Cl_2$, (e) *trans*-$Pt(NH_3)_2Cl_2$?

20.17 Write the formulas for each of these ions and compounds: (a) tetrahydroxozincate(II), (b) pentaaquachlorochromium(III) chloride, (c) tetrabromocuprate(II), (d) ethylenediaminetetraacetatoferrate(II).

20.18 Write the formulas for each of these ions and compounds: (a) bis(ethylenediamine)dichlorochromium(III), (b) pentacarbonyliron(0), (c) potassium tetracyanocuprate(II), (d) tetraammineaquachlorocobalt(III) chloride.

Structure of Coordination Compounds

Problems

20.19 Draw structures of all the geometric and optical isomers of each of these cobalt complexes:
(a) $[Co(NH_3)_6]^{3+}$
(b) $[Co(NH_3)_5Cl]^{2+}$
(c) $[Co(NH_3)_4Cl_2]^+$
(d) $[Co(en)_3]^{3+}$
(e) $[Co(C_2O_4)_3]^{3-}$

20.20 How many geometric isomers are in these species: (a) $[Co(NH_3)_2Cl_4]^-$, (b) $[Co(NH_3)_3Cl_3]$?

20.21 A student prepared a cobalt complex that has one of the following structures: $[Co(NH_3)_6]Cl_3$, $[Co(NH_3)_5Cl]Cl_2$, or $[Co(NH_3)_4Cl_2]Cl$. Explain how the student would distinguish among these possibilities by an electrical conductance experiment. At the student's disposal are three strong electrolytes: NaCl, $MgCl_2$, and $FeCl_3$, which may be used for comparison purposes.

20.22 The complex ion $[Ni(CN)_2Br_2]^{2-}$ has a square planar geometry. Draw the structures of the geometric isomers of this complex.

Bonding, Color, Magnetism

Review Questions

20.23 Briefly describe the crystal field theory. Define the following terms: crystal field splitting, high-spin complex, low-spin complex, spectrochemical series.

20.24 What is the origin of color in a compound?

20.25 Compounds containing the Sc^{3+} ion are colorless, whereas those containing the Ti^{3+} ion are colored. Explain.

20.26 What factors determine whether a given complex will be diamagnetic or paramagnetic?

Problems

20.27 For the same type of ligands, explain why the crystal field splitting for an octahedral complex is always greater than that for a tetrahedral complex.

20.28 Transition metal complexes containing CN^- ligands are often yellow, whereas those containing H_2O ligands are often green or blue. Explain.

20.29 The $[Ni(CN)_4]^{2-}$ ion, which has a square planar geometry, is diamagnetic, whereas the $[NiCl_4]^{2-}$ ion, which has a tetrahedral geometry, is paramagnetic. Show the crystal field splitting diagrams for those two complexes.

20.30 Predict the number of unpaired electrons in these complex ions: (a) $[Cr(CN)_6]^{4-}$, (b) $[Cr(H_2O)_6]^{2+}$.

20.31 The absorption maximum for the complex ion $[Co(NH_3)_6]^{3+}$ occurs at 470 nm. (a) Predict the color of the complex and (b) calculate the crystal field splitting in kilojoules per mole.

20.32 A solution made by dissolving 0.875 g of $Co(NH_3)_4Cl_3$ in 25.0 g of water freezes 0.56°C below the freezing point of pure water. Calculate the number of moles of ions produced when 1 mole of $Co(NH_3)_4Cl_3$ is dissolved in water, and suggest a structure for the complex ion present in this compound.

Reactions of Coordination Compounds

Review Questions

20.33 Define the terms (a) labile complex, (b) inert complex.

20.34 Explain why a thermodynamically stable species may be chemically reactive and a thermodynamically unstable species may be unreactive.

Problems

20.35 Oxalic acid, $H_2C_2O_4$, is sometimes used to clean rust stains from sinks and bathtubs. Explain the chemistry underlying this cleaning action.

20.36 The $[Fe(CN)_6]^{3-}$ complex is more labile than the $[Fe(CN)_6]^{4-}$ complex. Suggest an experiment that would prove that $[Fe(CN)_6]^{3-}$ is a labile complex.

20.37 Aqueous copper(II) sulfate solution is blue in color. When aqueous potassium fluoride is added, a green precipitate is formed. When aqueous potassium chloride is added instead, a bright-green solution is formed. Explain what is happening in these two cases.

20.38 When aqueous potassium cyanide is added to a solution of copper(II) sulfate, a white precipitate, soluble in an excess of potassium cyanide, is formed. No precipitate is formed when hydrogen sulfide is bubbled through the solution at this point. Explain.

20.39 A concentrated aqueous copper(II) chloride solution is bright green in color. When diluted with water, the solution becomes light blue. Explain.

20.40 In a dilute nitric acid solution, Fe^{3+} reacts with thiocyanate ion (SCN^-) to form a dark-red complex:

$$[Fe(H_2O)_6]^{3+} + SCN^- \rightleftharpoons H_2O + [Fe(H_2O)_5NCS]^{2+}$$

The equilibrium concentration of $[Fe(H_2O)_5NCS]^{2+}$ may be determined by how darkly colored the solution is (measured by a spectrometer). In one such experiment, 1.0 mL of 0.20 *M* $Fe(NO_3)_3$ was mixed with 1.0 mL of 1.0×10^{-3} *M* KSCN and 8.0 mL of dilute HNO_3. The color of the solution quantitatively indicated that the $[Fe(H_2O)_5NCS]^{2+}$ concentration was 7.3×10^{-5} *M*. Calculate the formation constant for $[Fe(H_2O)_5NCS]^{2+}$.

Additional Problems

20.41 Explain these facts: (a) Copper and iron have several oxidation states, whereas zinc exists in only one. (b) Copper and iron form colored ions, whereas zinc does not.

20.42 The formation constant for the reaction $Ag^+ + 2NH_3 \rightleftharpoons [Ag(NH_3)_2]^+$ is 1.5×10^7 and that for the reaction $Ag^+ + 2CN^- \rightleftharpoons [Ag(CN)_2]^-$ is 1.0×10^{21} at 25°C (see Table 17.4). Calculate the equilibrium constant at 25°C for the reaction

$$[Ag(NH_3)_2]^+ + 2CN^- \rightleftharpoons [Ag(CN)_2]^- + 2NH_3$$

20.43 Hemoglobin is the oxygen-carrying protein. In each hemoglobin molecule, there are four heme groups. In each heme group an Fe(II) ion is octahedrally bound to five N atoms and to either a water molecule (in deoxyhemoglobin) or an oxygen molecule (in oxyhemoglobin). Oxyhemoglobin is bright red, whereas deoxyhemoglobin is purple. Show that the difference in color can be accounted for qualitatively on the basis of high-spin and low-spin complexes. (*Hint:* O_2 is a strong-field ligand.)

20.44 Hydrated Mn^{2+} ions are practically colorless (see Figure 20.15) even though they possess five 3*d* electrons. Explain. (*Hint:* Electronic transitions in which there is a change in the number of unpaired electrons do not occur readily.)

20.45 Which of these hydrated cations are colorless: $Fe^{2+}(aq)$, $Zn^{2+}(aq)$, $Cu^+(aq)$, $Cu^{2+}(aq)$, $V^{5+}(aq)$, $Ca^{2+}(aq)$, $Co^{2+}(aq)$, $Sc^{3+}(aq)$, $Pb^{2+}(aq)$? Explain your choice.

20.46 In each of these pairs of complexes, choose the one that absorbs light at a longer wavelength: (a) $[Co(NH_3)_6]^{2+}$, $[Co(H_2O)_6]^{2+}$; (b) $[FeF_6]^{3-}$, $[Fe(CN)_6]^{3-}$; (c) $[Cu(NH_3)_4]^{2+}$, $[CuCl_4]^{2-}$.

20.47 A student in 1895 prepared three chromium coordination compounds having the same formulas of $CrCl_3(H_2O)_6$ with these properties:

Color	Cl^- Ions in Solution per Formula Unit
Violet	3
Light green	2
Dark green	1

Write modern formulas for these compounds and suggest a method for confirming the number of Cl^- ions present in solution in each case. (*Hint:* Some of the compounds may exist as hydrates, which are compounds that have a specific number of water molecules attached to them. The Cr has a coordination number of 6 in all three compounds.)

20.48 Complex ion formation has been used to extract gold, which exists in nature in the uncombined state. To separate it from other solid impurities, the ore is treated with a sodium cyanide (NaCN) solution in the presence of air to dissolve the gold by forming the soluble complex ion $[Au(CN)_2]^-$. (a) Balance the following equation:

$$Au + CN^- + O_2 + H_2O \longrightarrow [Au(CN)_2]^- + OH^-$$

(b) The gold is obtained by reducing the complex ion with zinc metal. Write a balanced ionic equation for this process. (c) What is the geometry and coordination number of the $[Au(CN)_2]^-$ ion?

20.49 Aqueous solutions of $CoCl_2$ are generally either light pink or blue. Low concentrations and low temperatures favor the pink form while high concentrations and high temperatures favor the blue form. Adding hydrochloric acid to a pink solution of $CoCl_2$ causes the solution to turn blue; the pink color is restored by the addition of $HgCl_2$. Account for these observations.

20.50 Which is a stronger oxidizing agent, Mn^{3+} or Cr^{3+}? Explain.

20.51 Suggest a method that would enable you to distinguish between *cis*-$Pt(NH_3)_2Cl_2$ and *trans*-$Pt(NH_3)_2Cl_2$.

20.52 The label of a certain brand of mayonnaise lists EDTA as a food preservative. How does EDTA prevent the spoilage of mayonnaise?

20.53 You are given two solutions containing $FeCl_2$ and $FeCl_3$ at the same concentration. One solution is light yellow and the other one is brown. Identify these solutions based only on color.

20.54 The K_f for the complex ion formation between Pb^{2+} and $EDTA^{4-}$

$$Pb^{2+} + EDTA^{4-} \rightleftharpoons Pb(EDTA)^{2-}$$

is 1.0×10^{18} at 25°C. Calculate $[Pb^{2+}]$ at equilibrium in a solution containing 1.0×10^{-3} *M* Pb^{2+} and 2.0×10^{-3} *M* $EDTA^{4-}$.

20.55 Manganese forms three low-spin complex ions with the cyanide ion with the formulas $[Mn(CN)_6]^{5-}$, $[Mn(CN)_6]^{4-}$, and $[Mn(CN)_6]^{3-}$. For each complex ion, determine the oxidation number of Mn and the number of unpaired *d* electrons present.

Special Problems

20.56 How many geometric isomers can the following square planar complex have?

```
a   b
 \ /
  Pt
 / \
d   c
```

20.57 Carbon monoxide binds to the iron atom in hemoglobin some 200 times more strongly than oxygen. This is the reason why CO is a toxic substance. The metal-to-ligand sigma bond is formed by donating a lone pair from the donor atom to an empty sp^3d^2 orbital on Fe. (a) On the basis of electronegativities, would you expect the C or O atom to form the bond to Fe? (b) Draw a diagram illustrating the overlap of the orbitals involved in bonding.

20.58 $[Pt(NH_3)_2Cl_2]$ is found to exist in two geometric isomers designated I and II, which react with oxalic acid as follows:

$$I + H_2C_2O_4 \longrightarrow [Pt(NH_3)_2C_2O_4]$$

$$II + H_2C_2O_4 \longrightarrow [Pt(NH_3)_2(HC_2O_4)_2]$$

Comment on the structures of I and II.

20.59 The compound 1,1,1-trifluoroacetylacetone (tfa) is a bidentate ligand:

```
     O      O
     ||     ||
CF3CCH2CCH3
```

It forms a tetrahedral complex with Be^{2+} and a square planar complex with Cu^{2+}. Draw structures of these complex ions and identify the type of isomers exhibited by these ions.

20.60 Commercial silver-plating operations frequently use a solution containing the complex $Ag(CN)_2^-$ ion. Because the formation constant (K_f) is quite large, this procedure ensures that the free Ag^+ concentration in solution is low for uniform electrodeposition. In one process, a chemist added 9.0 L of 5.0 *M* NaCN to 90.0 L of 0.20 *M* $AgNO_3$. Calculate the concentration of free Ag^+ ions at equilibrium. See Table 17.4 for K_f value.

20.61 Draw qualitative diagrams for the crystal-field splittings in (a) a linear complex ion ML_2, (b) a trigonal-planar complex ion ML_3, and (c) a trigonal- bipyramidal complex ion ML_5.

20.62 (a) The free Cu(I) ion is unstable in solution and has a tendency to disproportionate:

$$2Cu^+(aq) \rightleftharpoons Cu^{2+}(aq) + Cu(s)$$

Use the information in Table 19.1 (p. 670) to calculate the equilibrium constant for the reaction. (b) Based on your result in (a), explain why most Cu(I) compounds are insoluble.

20.63 Consider the following two ligand exchange reactions:

$$[Co(H_2O)_6]^{3+} + 6NH_3 \rightleftharpoons [Co(NH_3)_6]^{3+} + 6H_2O$$
$$[Co(H_2O)_6]^{3+} + 3en \rightleftharpoons [Co(en)_3]^{3+} + 6H_2O$$

(a) Which of the reactions should have a larger $\Delta S°$?

(b) Given that the Co—N bond strength is approximately the same in both complexes, which reaction will have a larger equilibrium constant? Explain your choices.

20.64 Copper is also known to exist in +3 oxidation state, which is believed to be involved in some biological electron transfer reactions. (a) Would you expect this oxidation state of copper to be stable? Explain. (b) Name the compound K_3CuF_6 and predict the geometry of the complex ion and its magnetic properties. (c) Most of the known Cu(III) compounds have square planar geometry. Are these compounds diamagnetic or paramagnetic?

Answers to Practice Exercises

20.1 K: +1; Au: +3.
20.2 Tetraaquadichlorochromium(III) chloride.
20.3 $[Co(en)_3]_2(SO_4)_3$. **20.4** 5.

CHAPTER

21

Nuclear Chemistry

The core of the world's largest superconducting solenoid magnet at the European Organization for Nuclear Research (CERN)'s Large Hadron Collider (LHC) particle accelerator.

Chapter Outline

Student Interactive Activities

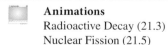

Animations
Radioactive Decay (21.3)
Nuclear Fission (21.5)

Electronic Homework
Example Practice Problems
End of Chapter Problems

Essential Concepts

Nuclear Stability To maintain nuclear stability, the ratio of neutrons-to-protons must fall within a certain range. A quantitative measure of nuclear stability is nuclear binding energy, which is the energy required to break up a nucleus into its component protons and neutrons. Nuclear binding energy can be calculated from the masses of protons and neutrons and that of the nucleus using Einstein's mass-energy equivalence relationship.

Natural Radioactivity and Nuclear Transmutation Unstable nuclei undergo spontaneous decay with the emission of radiation and particles. All nuclear decays obey first-order kinetics. The half-lives of several radioactive nuclei have been used to date objects. Stable nuclei can also be made radioactive by bombardment with elementary particles or atomic nuclei. Many new elements have been created artificially in particle accelerators where such bombardments occur.

Nuclear Fission and Nuclear Fusion Certain nuclei, when bombarded with neutrons, undergo fission to produce smaller nuclei, additional neutrons, and a large amount of energy. When enough nuclei are present to reach critical mass, a nuclear chain reaction, a self-sustaining sequence of nuclear fission reactions, takes place. Nuclear fissions find applications in the construction of atomic bombs and nuclear reactors. Nuclear fusion is the process in which nuclei of light elements are made to fuse at very high temperatures to form a heavier nucleus. Such a process releases even a greater amount of energy than nuclear fission and is used in making hydrogen, or thermonuclear, bombs.

Uses of Isotopes Isotopes, especially radioactive isotopes, are used as tracers to study the mechanisms of chemical and biological reactions and as medical diagnostic tools.

Biological Effects of Radiation The penetrating and harmful effects of radiation on biological systems have been thoroughly studied and are well understood.

21.1 The Nature of Nuclear Reactions

With the exception of hydrogen ($^1_1$H), all nuclei contain two kinds of fundamental particles, called *protons* and *neutrons*. Some nuclei are unstable; they emit particles and/or electromagnetic radiation spontaneously (see Section 2.2). The name for this phenomenon is *radioactivity*. All elements having an atomic number greater than 83 are radioactive. For example, the isotope of polonium, polonium-210 ($^{210}_{84}$Po), decays spontaneously to $^{206}_{82}$Pb by emitting an α particle.

Another type of radioactivity, known as ***nuclear transmutation,*** *results from the bombardment of nuclei by neutrons, protons, or other nuclei.* An example of a nuclear transmutation is the conversion of atmospheric $^{14}_7$N to $^{14}_6$C and $^1_1$H, which results when the nitrogen isotope captures a neutron (from the sun). In some cases, heavier elements are synthesized from lighter elements. This type of transmutation occurs naturally in outer space, but it can also be achieved artificially, as we will see in Section 21.4.

Radioactive decay and nuclear transmutation are ***nuclear reactions,*** which differ significantly from ordinary chemical reactions. Table 21.1 summarizes the differences.

Balancing Nuclear Equations

To discuss nuclear reactions in any depth, we need to understand how to write and balance the equations. Writing a nuclear equation differs somewhat from writing equations for chemical reactions. In addition to writing the symbols for various chemical elements, we must also explicitly indicate protons, neutrons, and electrons. In fact, we must show the numbers of protons and neutrons present in *every* species in such an equation.

The symbols for elementary particles are as follows:

$^1_1$p or $^1_1$H	$^1_0$n	$^{\ 0}_{-1}e$ or $^{\ 0}_{-1}\beta$	$^{\ 0}_{+1}e$ or $^{\ 0}_{+1}\beta$	$^4_2$He or $^4_2\alpha$
proton	neutron	electron	positron	α particle

In accordance with the notation used in Section 2.3, the superscript in each case denotes the mass number (the total number of neutrons and protons present) and the subscript is the atomic number (the number of protons). Thus, the "atomic number"

Table 21.1 Comparison of Chemical Reactions and Nuclear Reactions

Chemical Reactions	Nuclear Reactions
1. Atoms are rearranged by the breaking and forming of chemical bonds.	1. Elements (or isotopes of the same elements) are converted from one to another.
2. Only electrons in atomic or molecular orbitals are involved in the breaking and forming of bonds.	2. Protons, neutrons, electrons, and other elementary particles may be involved.
3. Reactions are accompanied by absorption or release of relatively small amounts of energy.	3. Reactions are accompanied by absorption or release of tremendous amounts of energy.
4. Rates of reaction are influenced by temperature, pressure, concentration, and catalysts.	4. Rates of reaction normally are not affected by temperature, pressure, and catalysts.

of a proton is 1, because there is one proton present, and the "mass number" is also 1, because there is one proton but no neutrons present. On the other hand, the "mass number" of a neutron is 1, but its "atomic number" is zero, because there are no protons present. For the electron, the "mass number" is zero (there are neither protons nor neutrons present), but the "atomic number" is −1, because the electron possesses a unit negative charge.

The symbol ${}_{-1}^{0}e$ represents an electron in or from an atomic orbital. The symbol ${}_{-1}^{0}\beta$ represents an electron that, although physically identical to any other electron, comes from a nucleus (in a decay process in which a neutron is converted to a proton and an electron) and not from an atomic orbital. The ***positron*** *has the same mass as the electron, but bears a +1 charge.* The α particle has two protons and two neutrons, so its atomic number is 2 and its mass number is 4.

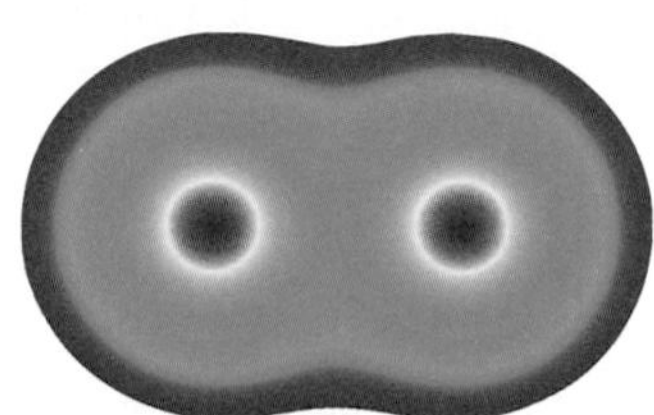

A positron is the antiparticle of the electron. In 2007 physicists prepared dipositronium (Ps_2), which contains only electrons and positrons. The diagram here shows the central nuclear positions containing positrons (red) surrounded by electrons (green). The Ps_2 species exists for less than a nanosecond before the electron and positron annihilate each other with the emission of γ rays.

In balancing any nuclear equation, we observe these rules:

- The total number of protons plus neutrons in the products and in the reactants must be the same (conservation of mass number).
- The total number of nuclear charges in the products and in the reactants must be the same (conservation of atomic number).

If we know the atomic numbers and mass numbers of all the species but one in a nuclear equation, we can identify the unknown species by applying these rules, as shown in Example 21.1, which illustrates how to balance nuclear decay equations.

EXAMPLE 21.1

Balance the following nuclear equations (that is, identify the product X):

(a) ${}_{84}^{212}Po \longrightarrow {}_{82}^{208}Pb + X$

(b) ${}_{55}^{137}Cs \longrightarrow {}_{56}^{137}Ba + X$

Keep in mind that nuclear equations are often not balanced electrically.

Strategy In balancing nuclear equations, note that the sum of atomic numbers and that of mass numbers must match on both sides of the equation.

Solution

(a) The mass number and atomic number are 212 and 84, respectively, on the left-hand side and 208 and 82, respectively, on the right-hand side. Thus, X must have a mass number of 4 and an atomic number of 2, which means that it is an α particle. The balanced equation is

$${}_{84}^{212}Po \longrightarrow {}_{82}^{208}Pb + {}_{2}^{4}\alpha$$

(b) In this case, the mass number is the same on both sides of the equation, but the atomic number of the product is 1 more than that of the reactant. Thus, X must have a mass number of 0 and an atomic number of −1, which means that it is a β particle. The only way this change can come about is to have a neutron in the Cs nucleus transformed into a proton and an electron; that is, ${}_{0}^{1}n \longrightarrow {}_{1}^{1}p + {}_{-1}^{0}\beta$ (note that this process does not alter the mass number). Thus, the balanced equation is

We use the ${}_{-1}^{0}\beta$ notation here because the electron came from the nucleus.

$${}_{55}^{137}Cs \longrightarrow {}_{56}^{137}Ba + {}_{-1}^{0}\beta$$

Check Note that the equation in (a) and (b) are balanced for nuclear particles but not for electrical charges. To balance the charges, we would need to add two electrons on the right-hand side of (a) and express barium as a cation (Ba^{+}) in (b).

Similar problems: 21.5, 21.6.

Practice Exercise Identify X in the following nuclear equation:

$${}_{33}^{78}As \longrightarrow {}_{-1}^{0}\beta + X$$

REVIEW OF CONCEPTS

Write a nuclear equation depicting the formation of a positron from a proton.

21.2 Nuclear Stability

The nucleus occupies a very small portion of the total volume of an atom, but it contains most of the atom's mass because both the protons and the neutrons reside there. In studying the stability of the atomic nucleus, it is helpful to know something about its density, because it tells us how tightly the particles are packed together. As a sample calculation, let us assume that a nucleus has a radius of 5×10^{-3} pm and a mass of 1×10^{-22} g. These figures correspond roughly to a nucleus containing 30 protons and 30 neutrons. Density is mass/volume, and we can calculate the volume from the known radius (the volume of a sphere is $\frac{4}{3}\pi r^3$, where r is the radius of the sphere). First we convert the pm units to cm. Then we calculate the density in g/cm^3:

$$r = 5 \times 10^{-3} \text{ pm} \times \frac{1 \times 10^{-12} \text{ m}}{1 \text{ pm}} \times \frac{100 \text{ cm}}{1 \text{ m}} = 5 \times 10^{-13} \text{ cm}$$

$$\text{density} = \frac{\text{mass}}{\text{volume}} = \frac{1 \times 10^{-22} \text{ g}}{\frac{4}{3}\pi r^3} = \frac{1 \times 10^{-22} \text{ g}}{\frac{4}{3}\pi(5 \times 10^{-13} \text{ cm})^3}$$
$$= 2 \times 10^{14} \text{ g/cm}^3$$

This is an exceedingly high density. The highest density known for an element is 22.6 g/cm^3, for osmium (Os). Thus, the average atomic nucleus is roughly 9×10^{12} (or 9 trillion) times more dense than the densest element known!

To dramatize the almost incomprehensibly high density, it has been suggested that it is equivalent to packing the mass of all the world's automobiles into one thimble.

The enormously high density of the nucleus prompts us to wonder what holds the particles together so tightly. From electrostatic interaction we know that like charges repel and unlike charges attract one another. We would thus expect the protons to repel one another strongly, particularly when we consider how close they must be to each other. This indeed is so. However, in addition to the repulsion, there are also short-range attractions between proton and proton, proton and neutron, and neutron and neutron. The stability of any nucleus is determined by the difference between electrostatic repulsion and the short-range attraction. If repulsion outweighs attraction, the nucleus disintegrates, emitting particles and/or radiation. If attractive forces prevail, the nucleus is stable.

The principal factor that determines whether a nucleus is stable is the *neutron-to-proton ratio (n/p)*. For stable atoms of elements having low atomic number, the n/p value is close to 1. As the atomic number increases, the neutron-to-proton ratios of the stable nuclei become greater than 1. This deviation at higher atomic numbers arises because a larger number of neutrons is needed to counteract the strong repulsion among the protons and stabilize the nucleus. The following rules are useful in predicting nuclear stability:

1. Nuclei that contain 2, 8, 20, 50, 82, or 126 protons or neutrons are generally more stable than nuclei that do not possess these numbers. For example, there are 10 stable isotopes of tin (Sn) with the atomic number 50 and only 2 stable isotopes of antimony (Sb) with the atomic number 51. The numbers 2, 8, 20, 50, 82, and 126 are called *magic numbers*. The significance of these numbers for nuclear stability is similar to the numbers of electrons associated with the very stable noble gases (that is, 2, 10, 18, 36, 54, and 86 electrons).

Table 21.2 Number of Stable Isotopes with Even and Odd Numbers of Protons and Neutrons

Protons	Neutrons	Number of Stable Isotopes
Odd	Odd	4
Odd	Even	50
Even	Odd	53
Even	Even	164

2. Nuclei with even numbers of both protons and neutrons are generally more stable than those with odd numbers of these particles (Table 21.2).
3. All isotopes of the elements with atomic numbers higher than 83 are radioactive. All isotopes of technetium (Tc, $Z = 43$) and promethium (Pm, $Z = 61$) are radioactive.

Figure 21.1 shows a plot of the number of neutrons versus the number of protons in various isotopes. The stable nuclei are located in an area of the graph known as the *belt of stability.* Most radioactive nuclei lie outside this belt. Above the stability belt, the nuclei have higher neutron-to-proton ratios than those within the belt (for the same number of protons). To lower this ratio (and hence move down toward

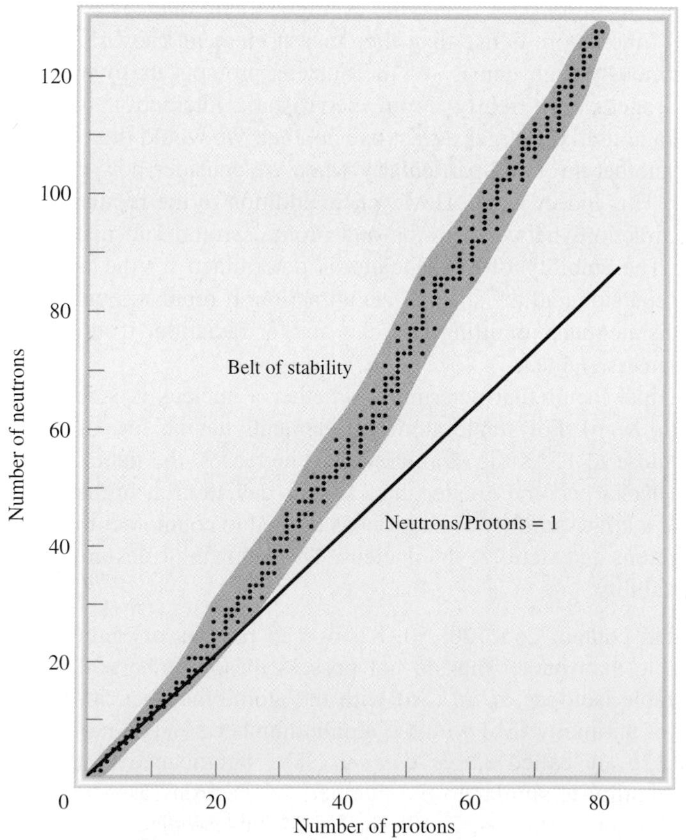

Figure 21.1
Plot of neutrons versus protons for various stable isotopes, represented by dots. The straight line represents the points at which the neutron-to-proton ratio equals 1. The shaded area represents the belt of stability.

the belt of stability), these nuclei undergo the following process, called *β-particle emission:*

$$^{1}_{0}\text{n} \longrightarrow {}^{1}_{1}\text{p} + {}^{0}_{-1}\beta$$

Beta-particle emission leads to an increase in the number of protons in the nucleus and a simultaneous decrease in the number of neutrons. Some examples are

$$^{14}_{6}\text{C} \longrightarrow {}^{14}_{7}\text{N} + {}^{0}_{-1}\beta$$
$$^{40}_{19}\text{K} \longrightarrow {}^{40}_{20}\text{Ca} + {}^{0}_{-1}\beta$$
$$^{97}_{40}\text{Zr} \longrightarrow {}^{97}_{41}\text{Nb} + {}^{0}_{-1}\beta$$

Below the stability belt, the nuclei have lower neutron-to-proton ratios than those in the belt (for the same number of protons). To increase this ratio (and hence move up toward the belt of stability), these nuclei either emit a positron

$$^{1}_{1}\text{p} \longrightarrow {}^{1}_{0}\text{n} + {}^{0}_{+1}\beta$$

or undergo electron capture. An example of positron emission is

$$^{38}_{19}\text{K} \longrightarrow {}^{38}_{18}\text{Ar} + {}^{0}_{+1}\beta$$

Electron capture is the capture of an electron—usually a 1*s* electron—by the nucleus. The captured electron combines with a proton to form a neutron so that the atomic number decreases by one while the mass number remains the same. This process has the same net effect as positron emission:

$$^{37}_{18}\text{Ar} + {}^{0}_{-1}e \longrightarrow {}^{37}_{17}\text{Cl}$$
$$^{55}_{26}\text{Fe} + {}^{0}_{-1}e \longrightarrow {}^{55}_{25}\text{Mn}$$

We use ${}^{0}_{-1}e$ rather than ${}^{0}_{-1}\beta$ here because the electron came from an atomic orbital and not from the nucleus.

REVIEW OF CONCEPTS

The following isotopes are unstable. Use Figure 21.1 to predict whether they will undergo beta decay or positron emission. (a) ^{13}B. (b) ^{188}Au. Write a nuclear equation for each case.

Nuclear Binding Energy

A quantitative measure of nuclear stability is the ***nuclear binding energy,*** which is *the energy required to break up a nucleus into its component protons and neutrons.* This quantity represents the conversion of mass to energy that occurs during an exothermic nuclear reaction.

The concept of nuclear binding energy evolved from studies of nuclear properties showing that the masses of nuclei are always less than the sum of the masses of the *nucleons,* which is a general term for the protons and neutrons in a nucleus. For example, the $^{19}_{9}$F isotope has an atomic mass of 18.9984 amu. The nucleus has 9 protons and 10 neutrons and therefore a total of 19 nucleons. Using the known masses of the $^{1}_{1}$H atom (1.007825 amu) and the neutron (1.008665 amu), we can carry out the following analysis. The mass of 9 $^{1}_{1}$H atoms (that is, the mass of 9 protons and 9 electrons) is

$$9 \times 1.007825 \text{ amu} = 9.070425 \text{ amu}$$

and the mass of 10 neutrons is

$$10 \times 1.008665 \text{ amu} = 10.08665 \text{ amu}$$

Therefore, the atomic mass of a $^{19}_{9}F$ atom calculated from the known numbers of electrons, protons, and neutrons is

$$9.070425 \text{ amu} + 10.08665 \text{ amu} = 19.15708 \text{ amu}$$

There is no change in the electron's mass because it is not a nucleon.

which is larger than 18.9984 amu (the measured mass of $^{19}_{9}F$) by 0.1587 amu.

The difference between the mass of an atom and the sum of the masses of its protons, neutrons, and electrons is called the ***mass defect.*** Relativity theory tells us that the loss in mass shows up as energy (heat) given off to the surroundings. Thus, the formation of $^{19}_{9}F$ is exothermic. Einstein's *mass-energy equivalence relationship* states that

This is the only equation listed in the Bartlett's quotations.

$$E = mc^2 \tag{21.1}$$

where E is energy, m is mass, and c is the speed of light. We can calculate the amount of energy released by writing

$$\Delta E = (\Delta m)c^2 \tag{21.2}$$

where ΔE and Δm are defined as follows:

$$\Delta E = \text{energy of product} - \text{energy of reactants}$$
$$\Delta m = \text{mass of product} - \text{mass of reactants}$$

Thus, for the change in mass we have

$$\begin{aligned}\Delta m &= 18.9984 \text{ amu} - 19.15708 \text{ amu} \\ &= -0.1587 \text{ amu}\end{aligned}$$

Because $^{19}_{9}F$ has a mass that is less than the mass calculated from the number of electrons and nucleons present, Δm is a negative quantity. Consequently, ΔE is also a negative quantity; that is, energy is released to the surroundings as a result of the formation of the fluorine-19 nucleus. So we calculate ΔE as follows:

$$\begin{aligned}\Delta E &= (-0.1587 \text{ amu})(3.00 \times 10^8 \text{ m/s})^2 \\ &= -1.43 \times 10^{16} \text{ amu m}^2/\text{s}^2\end{aligned}$$

With the conversion factors

$$\begin{aligned}1 \text{ kg} &= 6.022 \times 10^{26} \text{ amu} \\ 1 \text{ J} &= 1 \text{ kg m}^2/\text{s}^2\end{aligned}$$

When you apply Equation (21.2), remember to express the mass defect in kilograms because $1 \text{ J} = 1 \text{ kg} \cdot \text{m}^2/\text{s}^2$.

we obtain

$$\begin{aligned}\Delta E &= \left(-1.43 \times 10^{16} \frac{\text{amu} \cdot \text{m}^2}{\text{s}^2}\right) \times \left(\frac{1.00 \text{ kg}}{6.022 \times 10^{26} \text{ amu}}\right) \times \left(\frac{1 \text{ J}}{1 \text{ kg m}^2/\text{s}^2}\right) \\ &= -2.37 \times 10^{-11} \text{ J}\end{aligned}$$

The nuclear binding energy is a positive quantity.

This is the amount of energy released when one fluorine-19 nucleus is formed from 9 protons and 10 neutrons. The nuclear binding energy of the nucleus is 2.37×10^{-11} J, which is the amount of energy needed to decompose the nucleus into separate protons and neutrons. In the formation of 1 mole of fluorine nuclei, for instance, the energy released is

$$\begin{aligned}\Delta E &= (-2.37 \times 10^{-11}\ \text{J})(6.022 \times 10^{23}/\text{mol}) \\ &= -1.43 \times 10^{13}\ \text{J/mol} \\ &= -1.43 \times 10^{10}\ \text{kJ/mol}\end{aligned}$$

The nuclear binding energy, therefore, is 1.43×10^{10} kJ for 1 mole of fluorine-19 nuclei, which is a tremendously large quantity when we consider that the enthalpies of ordinary chemical reactions are of the order of only 200 kJ. The procedure we have followed can be used to calculate the nuclear binding energy of any nucleus.

As we have noted, nuclear binding energy is an indication of the stability of a nucleus. However, in comparing the stability of any two nuclei we must account for the fact that they have different numbers of nucleons. For this reason it is more meaningful to use the *nuclear binding energy per nucleon,* defined as

$$\text{nuclear binding energy per nucleon} = \frac{\text{nuclear binding energy}}{\text{number of nucleons}} \qquad (21.3)$$

For the fluorine-19 nucleus,

$$\begin{aligned}\text{nuclear binding energy per nucleon} &= \frac{2.37 \times 10^{-11}\ \text{J}}{19\ \text{nucleons}} \\ &= 1.25 \times 10^{-12}\ \text{J/nucleon}\end{aligned}$$

The nuclear binding energy per nucleon enables us to compare the stability of all nuclei on a common basis. Figure 21.2 shows the variation of nuclear binding energy per nucleon plotted against mass number. As you can see, the curve rises rather steeply. The highest binding energies per nucleon belong to elements with intermediate mass numbers—between 40 and 100—and are greatest for elements in the iron, cobalt, and nickel region (the Group 8B elements) of the periodic table. This means that the *net* attractive forces among the particles (protons and neutrons) are greatest for the nuclei of these elements.

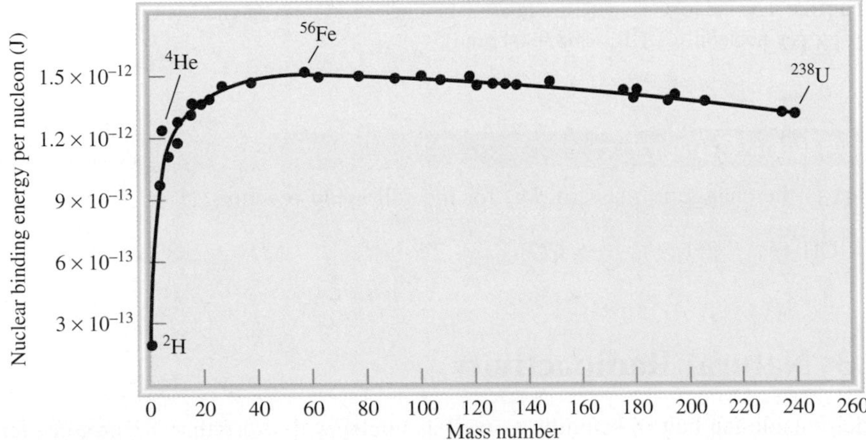

Figure 21.2
Plot of nuclear binding energy per nucleon versus mass number.

EXAMPLE 21.2

The atomic mass of $^{205}_{81}Tl$ is 204.9744 amu. Calculate the nuclear binding energy of this nucleus and the corresponding nuclear binding energy per nucleon.

Strategy To calculate the nuclear binding energy, we first determine the difference between the mass of the nucleus and the mass of all the protons and neutrons, which gives us the mass defect. Next, we apply Einstein's mass-energy relationship $[\Delta E = (\Delta m)c^2]$.

Solution There are 81 protons and 124 neutrons in the thallium nucleus. The mass of 81 ^1_1H atoms is

$$81 \times 1.007825 \text{ amu} = 81.63383 \text{ amu}$$

and the mass of 124 neutrons is

$$124 \times 1.008665 \text{ amu} = 125.07446 \text{ amu}$$

Therefore, the predicted mass for $^{205}_{81}Tl$ is $81.63383 + 125.07446 = 206.70829$ amu, and the mass defect is

$$\begin{aligned}\Delta m &= 204.9744 \text{ amu} - 206.70829 \text{ amu}\\ &= -1.7339 \text{ amu}\end{aligned}$$

The energy released is

$$\begin{aligned}\Delta E &= (\Delta m)c^2\\ &= (-1.7339 \text{ amu})(3.00 \times 10^8 \text{ m/s})^2\\ &= -1.56 \times 10^{17} \text{ amu} \cdot \text{m}^2/\text{s}^2\end{aligned}$$

Let's convert to a more familiar energy unit of joules. Recall that $1 \text{ J} = 1 \text{ kg} \cdot \text{m}^2/\text{s}^2$. Therefore, we need to convert amu to kg:

$$\begin{aligned}\Delta E &= -1.56 \times 10^{17} \frac{\cancel{\text{amu}} \cdot \text{m}^2}{\text{s}^2} \times \frac{1.00 \cancel{\text{g}}}{6.022 \times 10^{23} \cancel{\text{amu}}} \times \frac{1 \text{ kg}}{1000 \cancel{\text{g}}}\\ &= -2.59 \times 10^{-10} \frac{\text{kg} \cdot \text{m}^2}{\text{s}^2} = -2.59 \times 10^{-10} \text{ J}\end{aligned}$$

The neutron-to-proton ratio is 1.5, which places thallium-205 in the belt of stability.

Thus, the nuclear binding energy is 2.59×10^{-10} J. The nuclear binding energy per nucleon is obtained as follows:

$$\frac{2.59 \times 10^{-10} \text{ J}}{205 \text{ nucleons}} = 1.26 \times 10^{-12} \text{ J/nucleon}$$

Similar problems: 21.19, 21.20.

Practice Exercise Calculate the nuclear binding energy (in J) and the nuclear binding energy per nucleon of $^{209}_{83}Bi$ (208.9804 amu).

REVIEW OF CONCEPTS

What is the change in mass (in kg) for the following reaction?

$$CH_4(g) + 2O_2(g) \longrightarrow CO_2(g) + 2H_2O(l) \qquad \Delta H° = -890.4 \text{ kJ/mol}$$

21.3 Natural Radioactivity

Animation: Radioactive Decay

Nuclei outside the belt of stability, as well as nuclei with more than 83 protons, tend to be unstable. The spontaneous emission by unstable nuclei of particles or electromagnetic radiation, or both, is known as radioactivity. The main types of radiation

Table 21.3 The Uranium Decay Series*

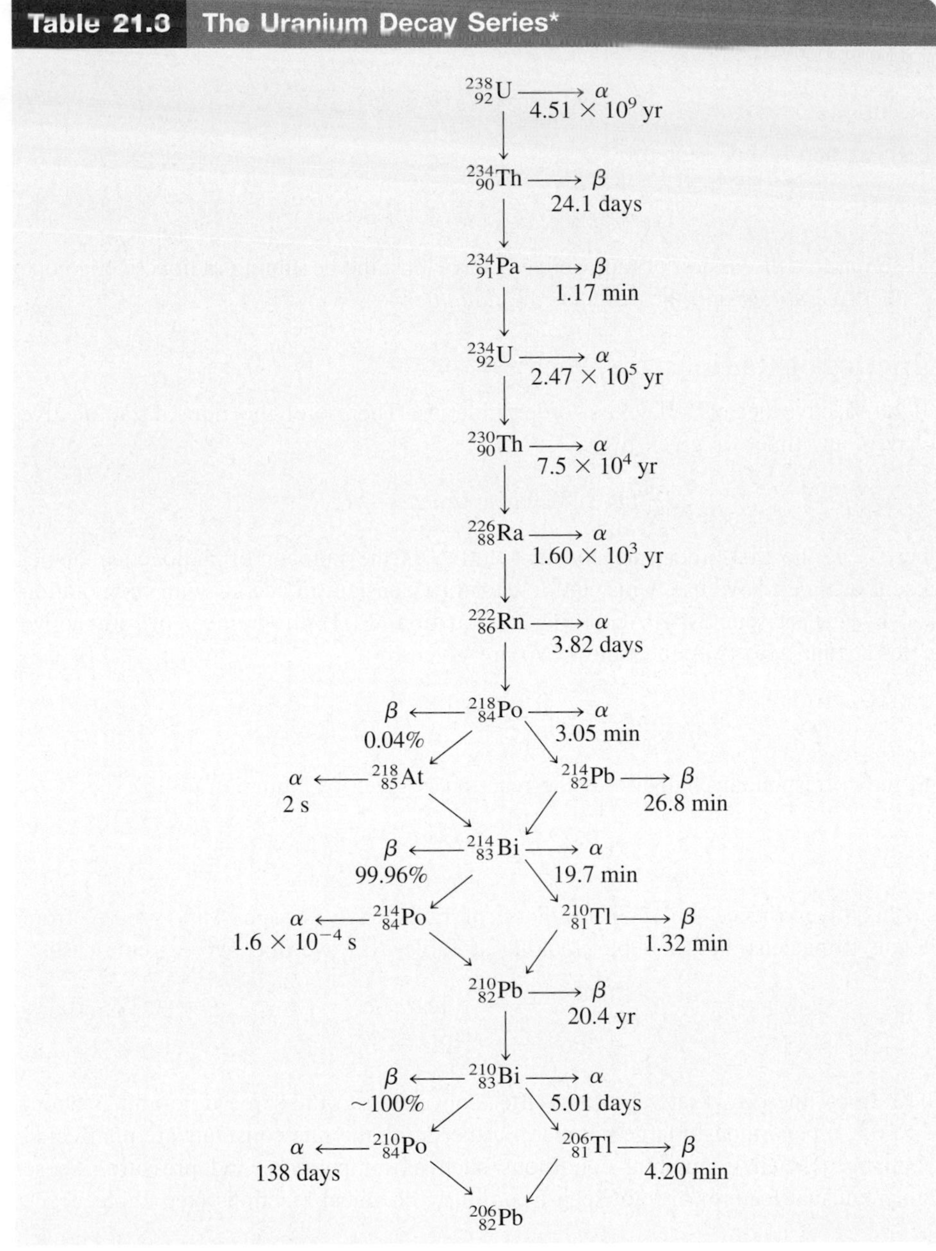

*The times denote the half-lives.

are: α particles (or doubly charged helium nuclei, He^{2+}); β particles (or electrons); γ rays, which are very-short-wavelength (0.1 nm to 10^{-4} nm) electromagnetic waves; positron emission; and electron capture.

The disintegration of a radioactive nucleus is often the beginning of a ***radioactive decay series,*** which is *a sequence of nuclear reactions that ultimately result in the formation of a stable isotope.* Table 21.3 shows the decay series of naturally occurring uranium-238, which involves 14 steps. This decay scheme, known as the *uranium decay series,* also shows the half-lives of all the products.

It is important to be able to balance the equation of a nuclear reaction for each of the steps in a radioactive decay series. For example, the first step in the uranium decay series is the decay of uranium-238 to thorium-234, with the emission of an α particle. Hence, the reaction is

$$^{238}_{92}\text{U} \longrightarrow {}^{234}_{90}\text{Th} + {}^{4}_{2}\alpha$$

The next step is represented by

$$^{234}_{90}\text{Th} \longrightarrow {}^{234}_{91}\text{Pa} + {}^{0}_{-1}\beta$$

and so on. In a discussion of radioactive decay steps, the beginning radioactive isotope is called the *parent* and the product, the *daughter.*

Kinetics of Radioactive Decay

All radioactive decays obey first-order kinetics. Therefore, the rate of radioactive decay at any time t is given by

$$\text{rate of decay at time } t = \lambda N$$

where λ is the first-order rate constant and N is the number of radioactive nuclei present at time t. (We use λ instead of k for rate constant in accord with the notation used by nuclear scientists.) According to Equation (14.3), the number of radioactive nuclei at time zero (N_0) and time t (N_t) are given by

$$\ln\frac{N_t}{N_0} = -\lambda t$$

and the corresponding half-life of the reaction is given by Equation (14.5):

$$t_{\frac{1}{2}} = \frac{0.693}{\lambda}$$

The half-lives (hence the rate constants) of radioactive isotopes vary greatly from nucleus to nucleus. For example, looking at Table 21.3, we find two extreme cases:

We do not have to wait 4.51×10^9 yr to make a half-life measurement of uranium-238. Its value can be calculated from the rate constant using Equation (14.5).

$$^{238}_{92}\text{U} \longrightarrow {}^{234}_{90}\text{Th} + {}^{4}_{2}\alpha \qquad t_{\frac{1}{2}} = 4.51 \times 10^9 \text{ yr}$$

$$^{214}_{84}\text{Po} \longrightarrow {}^{210}_{82}\text{Pb} + {}^{4}_{2}\alpha \qquad t_{\frac{1}{2}} = 1.6 \times 10^{-4} \text{ s}$$

The ratio of these two rate constants after conversion to the same time unit is about 1×10^{21}, an enormously large number. Furthermore, the rate constants are unaffected by changes in environmental conditions such as temperature and pressure. These highly unusual features are not seen in ordinary chemical reactions (see Table 21.1).

Dating Based on Radioactive Decay

The half-lives of radioactive isotopes have been used as "atomic clocks" to determine the ages of certain objects. Some examples of dating by radioactive decay measurements will be described here.

Radiocarbon Dating

The carbon-14 isotope is produced when atmospheric nitrogen is bombarded by cosmic rays:

$$^{14}_{7}\text{N} + {}^{1}_{0}\text{n} \longrightarrow {}^{14}_{6}\text{C} + {}^{1}_{1}\text{H}$$

The radioactive carbon-14 isotope decays according to the equation

$$^{14}_{6}\text{C} \longrightarrow {}^{14}_{7}\text{N} + {}^{0}_{-1}\beta \qquad t_{\frac{1}{2}} = 5730 \text{ yr}$$

The carbon-14 isotopes enter the biosphere as CO_2, which is taken up in plant photosynthesis. Plant-eating animals in turn exhale carbon-14 in CO_2. Eventually, carbon-14 participates in many aspects of the carbon cycle. The ^{14}C lost by radioactive decay is constantly replenished by the production of new isotopes in the atmosphere until a dynamic equilibrium is established whereby the ratio of ^{14}C to ^{12}C remains constant in living matter. But when an individual plant or an animal dies, the carbon-14 isotope in it is no longer replenished, so the ratio decreases as ^{14}C decays. This same change occurs when carbon atoms are trapped in coal, petroleum, or wood preserved underground, and in mummified bodies. After a number of years, there are proportionately fewer ^{14}C nuclei in a mummy than in a living person.

The age of the Shroud of Turin was shown by carbon-14 dating to be between A.D. 1260 and A.D. 1390, and therefore the shroud cannot be the burial cloth of Jesus Christ.

The decreasing ratio of ^{14}C to ^{12}C can be used to estimate the age of a specimen. Using Equation (14.3), we can write

$$\ln \frac{N_0}{N_t} = \lambda t$$

in which N_0 and N_t are the number of ^{14}C nuclei present at $t = 0$ and $t = t$, and λ is the first-order rate constant ($1.21 \times 10^{-4}\ \text{yr}^{-1}$). Because the decay rate is proportional to the amount of the radioactive isotope present, we have

$$\begin{aligned} t &= \frac{1}{\lambda} \ln \frac{N_0}{N_t} \\ &= \frac{1}{1.21 \times 10^{-4}\text{yr}^{-1}} \ln \frac{\text{decay rate of fresh sample}}{\text{decay rate of old sample}} \end{aligned}$$

Thus, by measuring the decay rates of the fresh sample and the old sample, we can calculate t, which is the age of the old sample. Radiocarbon dating is a valuable tool for estimating the age of objects (containing C atoms) dating back 1000 to 50,000 years.

Dating Using Uranium-238 Isotopes

Because some of the intermediate products in the uranium decay series have very long half-lives (see Table 21.3), this series is particularly suitable for estimating the age of rocks in the earth and of extraterrestrial objects. The half-life for the first step ($^{238}_{92}U$ to $^{234}_{90}Th$) is 4.51×10^9 yr. This is about 20,000 times the second largest value (that is, 2.47×10^5 yr), which is the half-life for $^{234}_{92}U$ to $^{230}_{90}Th$. Therefore, as a good approximation, we can assume that the half-life for the overall process (that is, from $^{238}_{92}U$ to $^{206}_{82}Pb$) is governed solely by the first step:

We can think of the first step as the rate-determining step in the overall process.

$$^{238}_{92}U \longrightarrow {}^{206}_{82}Pb + 8\,{}^{4}_{2}\alpha + 6\,{}^{0}_{-1}\beta \qquad t_{\frac{1}{2}} = 4.51 \times 10^9\ \text{yr}$$

In naturally occurring uranium minerals, we should and do find some lead-206 isotopes formed by radioactive decay. Assuming that no lead was present when the mineral was formed and that the mineral has not undergone chemical changes that would enable the lead-206 isotope to be separated from the parent uranium-238, it is possible to estimate the age of the rocks from the mass ratio of $^{206}_{82}Pb$ to $^{238}_{92}U$. The preceding equation tells us that for every mole, or 238 g, of uranium that undergoes complete decay, 1 mole, or 206 g, of lead is formed. If only half a mole of uranium-238 has undergone decay, the mass ratio $^{206}Pb/^{238}U$ becomes

$$\frac{206\ \text{g}/2}{238\ \text{g}/2} = 0.866$$

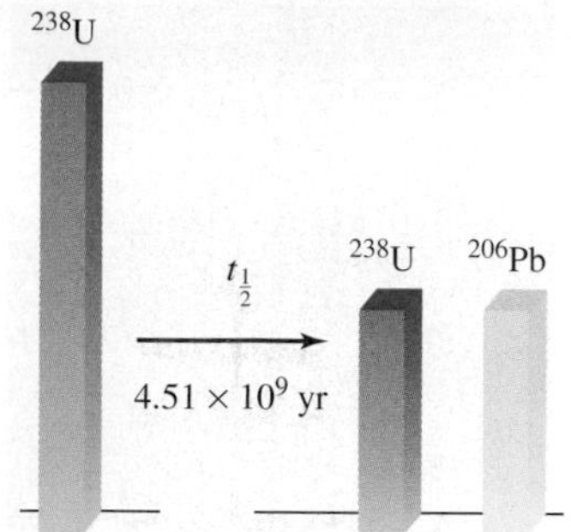

Figure 21.3
After one half-life, half of the original uranium-238 is converted to lead-206.

and the process would have taken a half-life of 4.51×10^9 yr to complete (Figure 21.3). Ratios lower than 0.866 mean that the rocks are less than 4.51×10^9 yr old, and higher ratios suggest a greater age. Interestingly, studies based on the uranium series as well as other decay series put the age of the oldest rocks and, therefore, probably the age of Earth itself at 4.5×10^9, or 4.5 billion, years.

Dating Using Potassium-40 Isotopes

This is one of the most important techniques employed in geochemistry. The radioactive potassium-40 isotope decays by several different modes, but the relevant one as far as dating is concerned is that of electron capture:

$$^{40}_{19}\text{K} + {}^{0}_{-1}e \longrightarrow {}^{40}_{18}\text{Ar} \qquad t_{\frac{1}{2}} = 1.2 \times 10^9 \text{ yr}$$

The accumulation of gaseous argon-40 is used to gauge the age of a specimen. When a potassium-40 atom in a mineral decays, argon-40 is trapped in the lattice of the mineral and can escape only if the material is melted. Melting, therefore, is the procedure for analyzing a mineral sample in the laboratory. The amount of argon-40 present can be conveniently measured with a mass spectrometer (see p. 68). Knowing the ratio of argon-40 to potassium-40 in the mineral and the half-life of decay makes it possible to establish the ages of rocks ranging from millions to billions of years old.

REVIEW OF CONCEPTS

Iron-59 (yellow spheres) decays to cobalt (blue spheres) via beta decay with a half-life of 45.1 d. (a) Write a balanced nuclear equation for the process. (b) From the following diagram, determine how many half-lives have elapsed.

21.4 Nuclear Transmutation

The scope of nuclear chemistry would be rather narrow if study were limited to natural radioactive elements. An experiment performed by Rutherford in 1919, however, suggested the possibility of producing radioactivity artificially. When he bombarded a sample of nitrogen with α particles, the following reaction took place:

$$^{14}_{7}\text{N} + {}^{4}_{2}\alpha \longrightarrow {}^{17}_{8}\text{O} + {}^{1}_{1}\text{p}$$

Note that the ^{17}O isotope is not radioactive.

An oxygen-17 isotope was produced with the emission of a proton. This reaction demonstrated for the first time the feasibility of converting one element into another, by the process of nuclear transmutation. Nuclear transmutation differs from radioactive decay in that the former is brought about by the collision of two particles.

The preceding reaction can be abbreviated as $^{14}_{7}\text{N}(\alpha,\text{p})^{17}_{8}\text{O}$. Note that in the parentheses the bombarding particle is written first, followed by the ejected particle.

EXAMPLE 21.3

Write the balanced equation for the nuclear reaction $^{56}_{26}\text{Fe}(\text{d},\alpha)^{54}_{25}\text{Mn}$, where d represents the deuterium nucleus (that is, $^{2}_{1}\text{H}$).

Strategy To write the balanced nuclear equation, remember that the first isotope $^{56}_{26}\text{Fe}$ is the reactant and the second isotope $^{54}_{25}\text{Mn}$ is the product. The first symbol in parentheses (d) is the bombarding particle and the second symbol in parentheses (α) is the particle emitted as a result of nuclear transmutation.

Solution The abbreviation tells us that when iron-56 is bombarded with a deuterium nucleus, it produces the manganese-54 nucleus plus an α particle, $^{4}_{2}\text{He}$. Thus, the equation for this reaction is

$$^{56}_{26}\text{Fe} + {}^{2}_{1}\text{H} \longrightarrow {}^{4}_{2}\alpha + {}^{54}_{25}\text{Mn}$$

Check Make sure that the sum of mass numbers and the sum of atomic numbers are the same on both sides of the equation.

Similar problems: 21.33, 21.34.

Practice Exercise Write a balanced equation for $^{106}_{46}\text{Pd}(\alpha,\text{p})^{109}_{47}\text{Ag}$.

Although light elements are generally not radioactive, they can be made so by bombarding their nuclei with appropriate particles. As we saw earlier, the radioactive carbon-14 isotope can be prepared by bombarding nitrogen-14 with neutrons. Tritium, $^{3}_{1}\text{H}$, is prepared according to the following bombardment:

$$^{6}_{3}\text{Li} + {}^{1}_{0}\text{n} \longrightarrow {}^{3}_{1}\text{H} + {}^{4}_{2}\alpha$$

Tritium decays with the emission of β particles:

$$^{3}_{1}\text{H} \longrightarrow {}^{3}_{2}\text{He} + {}^{0}_{-1}\beta \qquad t_{\frac{1}{2}} = 12.5 \text{ yr}$$

Many synthetic isotopes are prepared by using neutrons as projectiles. This approach is particularly convenient because neutrons carry no charges and therefore are not repelled by the targets—the nuclei. In contrast, when the projectiles are positively charged particles (for example, protons or α particles), they must have considerable kinetic energy to overcome the electrostatic repulsion between themselves and the target nuclei. The synthesis of phosphorus from aluminum is one example:

$$^{27}_{13}\text{Al} + {}^{4}_{2}\alpha \longrightarrow {}^{30}_{15}\text{P} + {}^{1}_{0}\text{n}$$

A *particle accelerator* uses electric and magnetic fields to increase the kinetic energy of charged species so that a reaction will occur (Figure 21.4). Alternating the polarity (that is, + and −) on specially constructed plates causes the particles to accelerate along a spiral path. When they have the energy necessary to initiate the desired nuclear reaction, they are guided out of the accelerator into a collision with a target substance.

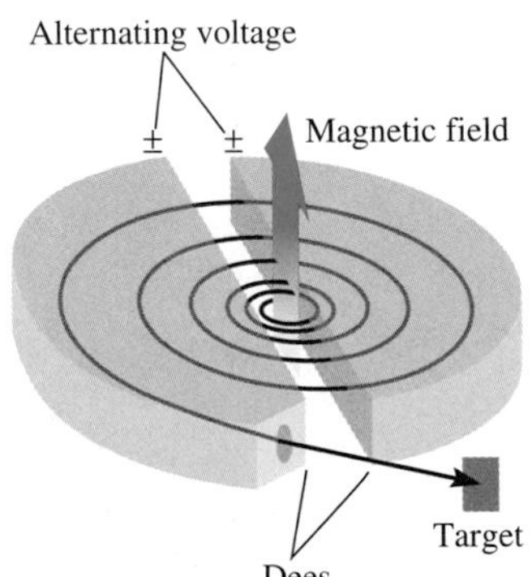

Figure 21.4
Schematic diagram of a cyclotron particle accelerator. The particle (an ion) to be accelerated starts at the center and is forced to move in a spiral path through the influence of electric and magnetic fields until it emerges at a high velocity. The magnetic fields are perpendicular to the plane of the dees (so-called because of their shape), which are hollow and serve as electrodes.

Various designs have been developed for particle accelerators, one of which accelerates particles along a linear path of about 3 km (Figure 21.5). It is now possible to accelerate particles to a speed well above 90 percent of the speed of light. (According to Einstein's theory of relativity, it is impossible for a particle to move *at* the speed of light. The only exception is the photon, which has a zero rest mass.) The extremely energetic particles produced in accelerators are employed by physicists to smash atomic nuclei to fragments. Studying the debris from such disintegrations provides valuable information about nuclear structure and binding forces.

Figure 21.5
A section of a particle accelerator.

Table 21.4 The Transuranium Elements

Atomic Number	Name	Symbol	Preparation
93	Neptunium	Np	${}^{238}_{92}\text{U} + {}^{1}_{0}\text{n} \longrightarrow {}^{239}_{93}\text{Np} + {}^{0}_{-1}\beta$
94	Plutonium	Pu	${}^{239}_{93}\text{Np} \longrightarrow {}^{239}_{94}\text{Pu} + {}^{0}_{-1}\beta$
95	Americium	Am	${}^{239}_{94}\text{Pu} + {}^{1}_{0}\text{n} \longrightarrow {}^{240}_{95}\text{Am} + {}^{0}_{-1}\beta$
96	Curium	Cm	${}^{239}_{94}\text{Pu} + {}^{4}_{2}\alpha \longrightarrow {}^{242}_{96}\text{Cm} + {}^{1}_{0}\text{n}$
97	Berkelium	Bk	${}^{241}_{95}\text{Am} + {}^{4}_{2}\alpha \longrightarrow {}^{243}_{97}\text{Bk} + 2{}^{1}_{0}\text{n}$
98	Californium	Cf	${}^{242}_{96}\text{Cm} + {}^{4}_{2}\alpha \longrightarrow {}^{245}_{98}\text{Cf} + {}^{1}_{0}\text{n}$
99	Einsteinium	Es	${}^{238}_{92}\text{U} + 15{}^{1}_{0}\text{n} \longrightarrow {}^{253}_{99}\text{Es} + 7{}^{0}_{-1}\beta$
100	Fermium	Fm	${}^{238}_{92}\text{U} + 17{}^{1}_{0}\text{n} \longrightarrow {}^{255}_{100}\text{Fm} + 8{}^{0}_{-1}\beta$
101	Mendelevium	Md	${}^{253}_{99}\text{Es} + {}^{4}_{2}\alpha \longrightarrow {}^{256}_{101}\text{Md} + {}^{1}_{0}\text{n}$
102	Nobelium	No	${}^{246}_{96}\text{Cm} + {}^{12}_{6}\text{C} \longrightarrow {}^{254}_{102}\text{No} + 4{}^{1}_{0}\text{n}$
103	Lawrencium	Lr	${}^{252}_{98}\text{Cf} + {}^{10}_{5}\text{B} \longrightarrow {}^{257}_{103}\text{Lr} + 5{}^{1}_{0}\text{n}$
104	Rutherfordium	Rf	${}^{249}_{98}\text{Cf} + {}^{12}_{6}\text{C} \longrightarrow {}^{257}_{104}\text{Rf} + 4{}^{1}_{0}\text{n}$
105	Dubnium	Db	${}^{249}_{98}\text{Cf} + {}^{15}_{7}\text{N} \longrightarrow {}^{260}_{105}\text{Db} + 4{}^{1}_{0}\text{n}$
106	Seaborgium	Sg	${}^{249}_{98}\text{Cf} + {}^{18}_{8}\text{O} \longrightarrow {}^{263}_{106}\text{Sg} + 4{}^{1}_{0}\text{n}$
107	Bohrium	Bh	${}^{209}_{83}\text{Bi} + {}^{54}_{24}\text{Cr} \longrightarrow {}^{262}_{107}\text{Bh} + {}^{1}_{0}\text{n}$
108	Hassium	Hs	${}^{208}_{82}\text{Pb} + {}^{58}_{26}\text{Fe} \longrightarrow {}^{265}_{108}\text{Hs} + {}^{1}_{0}\text{n}$
109	Meitnerium	Mt	${}^{209}_{83}\text{Bi} + {}^{58}_{26}\text{Fe} \longrightarrow {}^{266}_{109}\text{Mt} + {}^{1}_{0}\text{n}$
110	Darmstadtium	Ds	${}^{208}_{82}\text{Pb} + {}^{62}_{28}\text{Ni} \longrightarrow {}^{269}_{110}\text{Ds} + {}^{1}_{0}\text{n}$
111	Roentgenium	Rg	${}^{209}_{83}\text{Bi} + {}^{64}_{28}\text{Ni} \longrightarrow {}^{272}_{111}\text{Rg} + {}^{1}_{0}\text{n}$

The Transuranium Elements

Particle accelerators made it possible to synthesize the so-called ***transuranium elements,*** *elements with atomic numbers greater than 92.* Neptunium ($Z = 93$) was first prepared in 1940. Since then, 24 other transuranium elements have been synthesized. All isotopes of these elements are radioactive. Table 21.4 lists the transuranium elements up to $Z = 111$ and the reactions through which they are formed.

Review of Concepts

Element 118, known currently by its IUPAC systematic name ununoctium (symbol: Uo), was first created in 2006 in Dubna, Russia. The nuclear reaction used to produce this element was $^{249}_{98}Cf(^{48}_{20}Ca,X)^{294}_{118}Uuo$. Determine the product X and write the balanced equation for this nuclear reaction.

21.5 Nuclear Fission

Animation: Nuclear Fission

Nuclear fission is the process in which *a heavy nucleus (mass number > 200) divides to form smaller nuclei of intermediate mass and one or more neutrons.* Because the heavy nucleus is less stable than its products (see Figure 21.2), this process releases a large amount of energy.

The first nuclear fission reaction to be studied was that of uranium-235 bombarded with slow neutrons, whose speed is comparable to that of air molecules at room temperature. Under these conditions, uranium-235 undergoes fission, as shown in Figure 21.6. Actually, this reaction is very complex: More than 30 different elements have been found among the fission products (Figure 21.7). A representative reaction is

$$^{235}_{92}U + ^{1}_{0}n \longrightarrow ^{90}_{38}Sr + ^{143}_{54}Xe + 3^{1}_{0}n$$

Although many heavy nuclei can be made to undergo fission, only the fission of naturally occurring uranium-235 and of the artificial isotope plutonium-239 has any practical importance. Table 21.5 shows the nuclear binding energies of uranium-235 and its fission products. As the table shows, the binding energy per nucleon for uranium-235 is less than the sum of the binding energies for strontium-90 and xenon-143. Therefore, when a uranium-235 nucleus is split into two smaller nuclei, a certain amount of energy is released. Let us estimate the magnitude of this

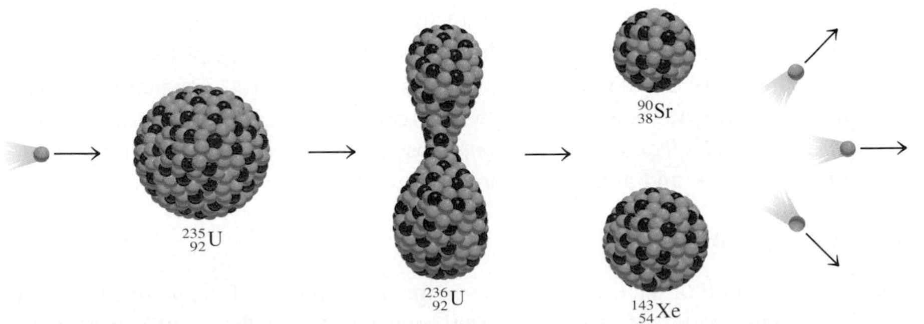

Figure 21.6
Nuclear fission of ^{235}U. When a ^{235}U nucleus captures a neutron (green sphere), it undergoes fission to yield two smaller nuclei. On the average, 2.4 neutrons are emitted for every ^{235}U nucleus that divides.

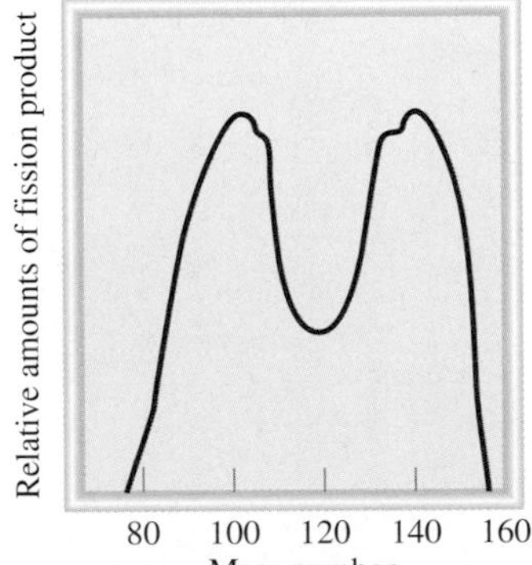

Figure 21.7
Relative yields of the products resulting from the fission of ^{235}U, as a function of mass number.

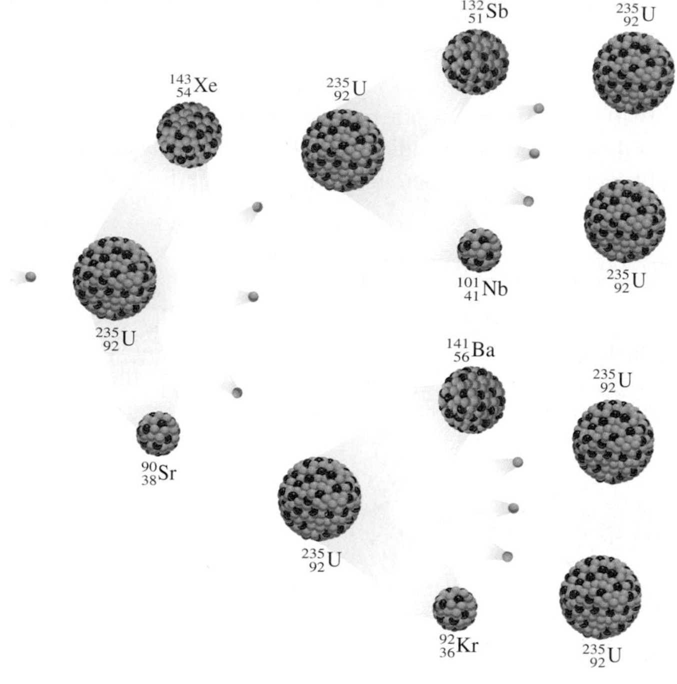

Figure 21.8
If a critical mass is present, many of the neutrons emitted during the fission process will be captured by other ^{235}U nuclei and a chain reaction will occur.

Table 21.5
Nuclear Binding Energies of ^{235}U and Its Fission Products

	Nuclear Binding Energy
^{235}U	2.82×10^{-10} J
^{90}Sr	1.23×10^{-10} J
^{143}Xe	1.92×10^{-10} J

energy. The difference between the binding energies of the reactants and products is $(1.23 \times 10^{-10} + 1.92 \times 10^{-10})$ J $-$ (2.82×10^{-10}) J, or 3.3×10^{-11} J per uranium-235 nucleus. For 1 mole of uranium-235, the energy released would be $(3.3 \times 10^{-11}) \times (6.02 \times 10^{23})$, or 2.0×10^{13} J. This is an extremely exothermic reaction, considering that the heat of combustion of 1 ton of coal is only about 5×10^{7} J.

The significant feature of uranium-235 fission is not just the enormous amount of energy released, but the fact that more neutrons are produced than are originally captured in the process. This property makes possible a ***nuclear chain reaction,*** which is *a self-sustaining sequence of nuclear fission reactions.* The neutrons generated during the initial stages of fission can induce fission in other uranium-235 nuclei, which in turn produce more neutrons, and so on. In less than a second, the reaction can become uncontrollable, liberating a tremendous amount of heat to the surroundings.

Figure 21.8 shows a nuclear chain reaction. For such a chain reaction to occur, enough uranium-235 must be present in the sample to capture the neutrons generated in the fission reaction. Otherwise, many of the neutrons will escape from the sample and the chain reaction will not occur. In this situation, the mass of the sample is said to be *subcritical.* When the amount of the fissionable material is equal to or greater than the ***critical mass,*** *the minimum mass of fissionable material required to generate a self-sustaining nuclear chain reaction,* most of the neutrons will be captured by uranium-235 nuclei, and a chain reaction will occur.

The Atomic Bomb

The first application of nuclear fission was in the development of the atomic bomb. How is such a bomb made and detonated? The crucial factor in the bomb's design is the determination of the critical mass for the bomb. A small atomic bomb is equivalent to 20,000 tons of TNT (trinitrotoluene). Because 1 ton of TNT releases about 4×10^{9} J of energy, 20,000 tons would produce 8×10^{13} J. Earlier we saw that 1 mole, or 235 g, of uranium-235 liberates 2.0×10^{13} J of energy when it undergoes fission. Thus, the mass of the isotope present in a small bomb must be at least

$$235 \text{ g} \times \frac{8 \times 10^{13} \text{ J}}{2.0 \times 10^{13} \text{ J}} \approx 1 \text{ kg}$$

For obvious reasons, an atomic bomb is never assembled with the critical mass already present. Instead, the critical mass is formed by using a conventional explosive, such as TNT, to force the fissionable sections together, as shown in Figure 21.9. Neutrons from a source at the center of the device trigger the nuclear chain reaction. Uranium-235 was the fissionable material in the bomb dropped on Hiroshima, Japan, on August 6, 1945. Plutonium-239 was used in the bomb exploded over Nagasaki three days later. The fission reactions generated were similar in these two cases, as was the extent of the destruction.

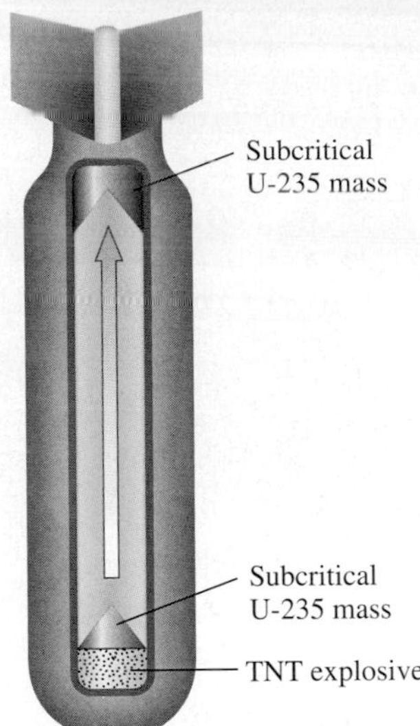

Figure 21.9
Schematic diagram of an atomic bomb. The TNT explosives are set off first. The explosion forces the sections of fissionable material together to form an amount considerably larger than the critical mass.

Nuclear Reactors

A peaceful but controversial application of nuclear fission is the generation of electricity using heat from a controlled chain reaction in a nuclear reactor. Currently, nuclear reactors provide about 20 percent of the electrical energy in the United States. This is a small but by no means negligible contribution to the nation's energy production. Several different types of nuclear reactors are in operation; we will briefly discuss the main features of three of them, along with their advantages and disadvantages.

Light Water Reactors

Most of the nuclear reactors in the United States are *light water reactors.* Figure 21.10 is a schematic diagram of such a reactor, and Figure 21.11 shows the refueling process in the core of a nuclear reactor.

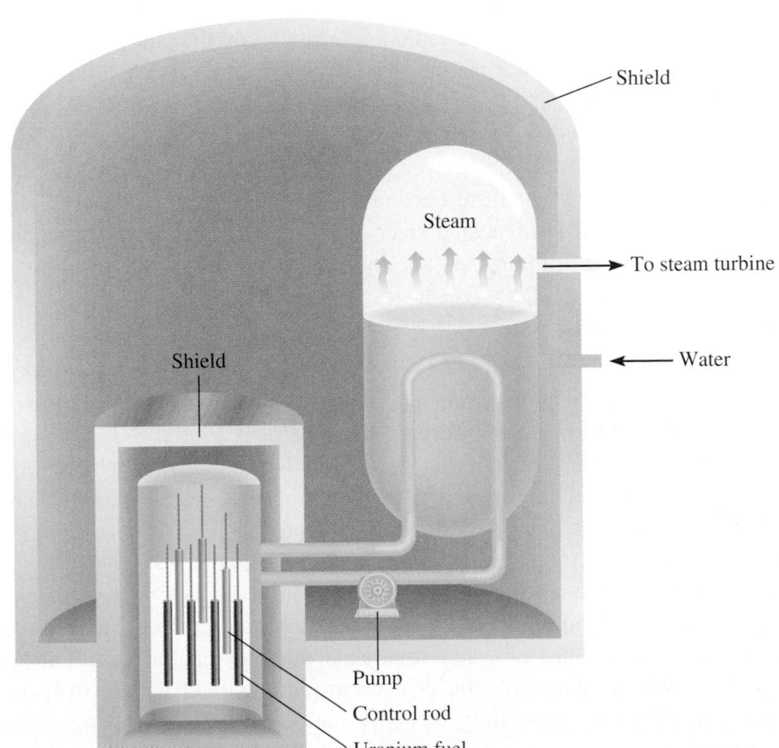

Figure 21.10
Schematic diagram of a nuclear fission reactor. The fission process is controlled by cadmium or boron rods. The heat generated by the process is used to produce steam for the generation of electricity via a heat exchange system.

Figure 21.11
Refueling the core of a nuclear reactor.

Figure 21.12
Uranium oxide, U_3O_8.

An important aspect of the fission process is the speed of the neutrons. Slow neutrons split uranium-235 nuclei more efficiently than do fast ones. Because fission reactions are highly exothermic, the neutrons produced usually move at high velocities. For greater efficiency, they must be slowed down before they can be used to induce nuclear disintegration. To accomplish this goal, scientists use ***moderators,*** which are *substances that can reduce the kinetic energy of neutrons.* A good moderator must satisfy several requirements: It should be nontoxic and inexpensive (as very large quantities of it are necessary); and it should resist conversion into a radioactive substance by neutron bombardment. Furthermore, it is advantageous for the moderator to be a fluid so that it can also be used as a coolant. No substance fulfills all these requirements, although water comes closer than many others that have been considered. Nuclear reactors that use light water (H_2O) as a moderator are called light water reactors because ${}^1_1H$ is the lightest isotope of the element hydrogen.

The nuclear fuel consists of uranium, usually in the form of its oxide, U_3O_8 (Figure 21.12). Naturally occurring uranium contains about 0.7 percent of the uranium-235 isotope, which is too low a concentration to sustain a small-scale chain reaction. For effective operation of a light water reactor, uranium-235 must be enriched to a concentration of 3 or 4 percent. In principle, the main difference between an atomic bomb and a nuclear reactor is that the chain reaction that takes place in a nuclear reactor is kept under control at all times. The factor limiting the rate of the reaction is the number of neutrons present. This can be controlled by lowering cadmium or boron control rods between the fuel elements. These rods capture neutrons according to the equations

$$ {}^{113}_{48}Cd + {}^1_0n \longrightarrow {}^{114}_{48}Cd + \gamma $$
$$ {}^{10}_5B + {}^1_0n \longrightarrow {}^7_3Li + {}^4_2\alpha $$

where γ denotes gamma rays. Without the control rods the reactor core would melt from the heat generated and release radioactive materials into the environment.

Nuclear reactors have rather elaborate cooling systems that absorb the heat given off by the nuclear reaction and transfer it outside the reactor core, where it is used to produce enough steam to drive an electric generator. In this respect, a nuclear power plant is similar to a conventional power plant that burns fossil fuel. In both cases, large quantities of cooling water are needed to condense steam for reuse. Thus, most nuclear power plants are built near a river or a lake. Unfortunately this method of cooling causes thermal pollution (see Section 13.4).

Heavy Water Reactors

Another type of nuclear reactor uses D_2O, or heavy water, as the moderator, rather than H_2O. Deuterium absorbs neutrons much less efficiently than does ordinary hydrogen. Because fewer neutrons are absorbed, the reactor is more efficient and does not require enriched uranium. The fact that deuterium is a less efficient moderator has a negative impact on the operation of the reactor, because more neutrons leak out of the reactor. However, this is not a serious disadvantage.

The main advantage of a heavy water reactor is that it eliminates the need for building expensive uranium enrichment facilities. However, D_2O must be prepared by either fractional distillation or electrolysis of ordinary water, which can be very expensive considering the amount of water used in a nuclear reactor. In countries where hydroelectric power is abundant, the cost of producing D_2O by electrolysis can be reasonably low. Canada is currently the only nation successfully using heavy water nuclear reactors. The fact that no enriched uranium is required in a heavy water reactor enables a country to enjoy the benefits of nuclear power without undertaking work that is closely associated with weapons technology.

Breeder Reactors

A ***breeder reactor*** uses uranium fuel, but unlike a conventional nuclear reactor, it *produces more fissionable materials than it uses.*

We know that when uranium-238 is bombarded with fast neutrons, the following reactions take place:

$$^{238}_{92}\text{U} + {}^{1}_{0}\text{n} \longrightarrow {}^{239}_{92}\text{U}$$

$$^{239}_{92}\text{U} \longrightarrow {}^{239}_{93}\text{Np} + {}^{0}_{-1}\beta \qquad t_{\frac{1}{2}} = 23.4 \text{ min}$$

$$^{239}_{93}\text{Np} \longrightarrow {}^{239}_{94}\text{Pu} + {}^{0}_{-1}\beta \qquad t_{\frac{1}{2}} = 2.35 \text{ days}$$

Plutonium-239 forms plutonium oxide, which can be readily separated from uranium.

In this manner the nonfissionable uranium-238 is transmuted into the fissionable isotope plutonium-239 (Figure 21.13).

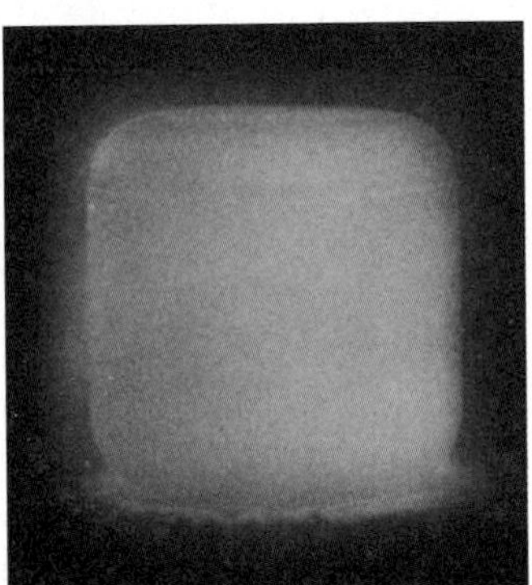

Figure 21.13
The red glow of the radioactive plutonium oxide, PuO_2.

In a typical breeder reactor, nuclear fuel containing uranium-235 or plutonium-239 is mixed with uranium-238 so that breeding takes place within the core. For every uranium-235 (or plutonium-239) nucleus undergoing fission, more than one neutron is captured by uranium-238 to generate plutonium-239. Thus, the stockpile of fissionable material can be steadily increased as the starting nuclear fuels are consumed. It takes about 7 to 10 yr to regenerate the sizable amount of material needed to refuel the original reactor and to fuel another reactor of comparable size. This interval is called the *doubling time.*

At present, the United States does not have a single operating breeder reactor, and only a few have been built in other countries, such as France and Russia. One problem is economics; breeder reactors are more expensive to build than conventional reactors. There are also more technical difficulties associated with the construction of such reactors. As a result, the future of breeder reactors, in the United States at least, is rather uncertain.

Hazards of Nuclear Energy

Many people, including environmentalists, regard nuclear fission as a highly undesirable method of energy production. Many fission products such as strontium-90 are dangerous radioactive isotopes with long half-lives. Plutonium-239, used as a nuclear fuel and produced in breeder reactors, is one of the most toxic substances known. It is an alpha emitter with a half-life of 24,400 yr.

Plutonium is chemically toxic in addition to being radioactive.

Accidents, too, present many dangers. An accident at the Three Mile Island reactor in Pennsylvania in 1979 first brought the potential hazards of nuclear plants to public attention. In this instance, very little radiation escaped the reactor, but the plant remained closed for more than a decade while repairs were made and safety issues addressed. Only a few years later, on April 26, 1986, a reactor at the Chernobyl nuclear plant in Ukraine surged out of control, resulting in a chemical explosion and fire. This accident released much radioactive material into the environment. People working near the plant died within weeks as a result of the exposure to the intense radiation. The long-term effect of the radioactive fallout from this incident has not yet been clearly assessed, although agriculture and dairy farming were affected by the fallout. The number of potential cancer deaths attributable to the radiation contamination is estimated to be between a few thousand and more than 100,000.

Molten glass is poured over nuclear waste before burial.

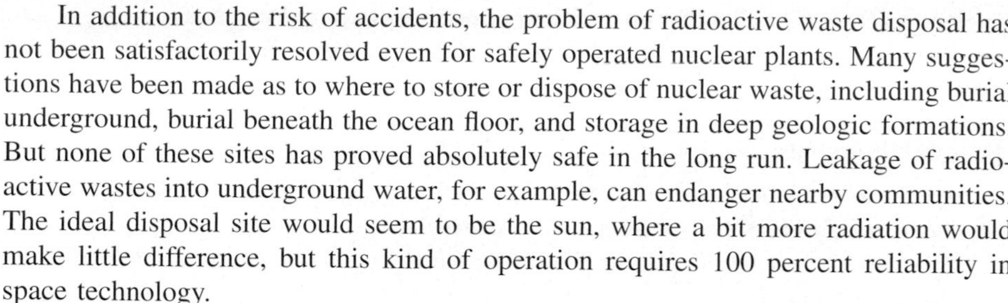

In addition to the risk of accidents, the problem of radioactive waste disposal has not been satisfactorily resolved even for safely operated nuclear plants. Many suggestions have been made as to where to store or dispose of nuclear waste, including burial underground, burial beneath the ocean floor, and storage in deep geologic formations. But none of these sites has proved absolutely safe in the long run. Leakage of radioactive wastes into underground water, for example, can endanger nearby communities. The ideal disposal site would seem to be the sun, where a bit more radiation would make little difference, but this kind of operation requires 100 percent reliability in space technology.

Because of the hazards, the future of nuclear reactors is clouded. What was once hailed as the ultimate solution to our energy needs in the twenty-first century is now being debated and questioned by both the scientific community and laypeople. It seems likely that the controversy will continue for some time.

21.6 Nuclear Fusion

In contrast to the nuclear fission process, ***nuclear fusion,*** *the combining of small nuclei into larger ones,* is largely exempt from the waste disposal problem.

Figure 21.2 showed that for the lightest elements, nuclear stability increases with increasing mass number. This behavior suggests that if two light nuclei combine or fuse together to form a larger, more stable nucleus, an appreciable amount of energy will be released in the process. This is the basis for ongoing research into the harnessing of nuclear fusion for the production of energy.

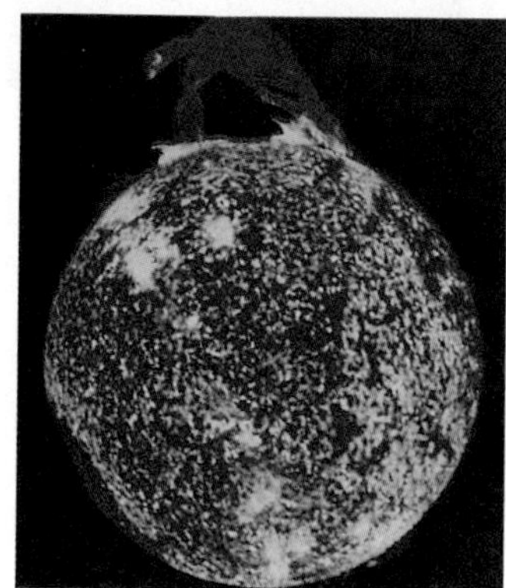

Nuclear fusion keeps the temperature in the interior of the sun at about 15 million °C.

Nuclear fusion occurs constantly in the sun. The sun is made up mostly of hydrogen and helium. In its interior, where temperatures reach about 15 million degrees Celsius, the following fusion reactions are believed to take place:

$$^{1}_{1}\text{H} + ^{2}_{1}\text{H} \longrightarrow ^{3}_{2}\text{He}$$

$$^{3}_{2}\text{He} + ^{3}_{2}\text{He} \longrightarrow ^{4}_{2}\text{He} + 2^{1}_{1}\text{H}$$

$$^{1}_{1}\text{H} + ^{1}_{1}\text{H} \longrightarrow ^{2}_{1}\text{H} + ^{0}_{+1}\beta$$

Because *fusion reactions take place only at very high temperatures,* they are often called ***thermonuclear reactions.***

Fusion Reactors

A major concern in choosing the proper nuclear fusion process for energy production is the temperature necessary to carry out the process. Some promising reactions are

Reaction	Energy Released
$^{2}_{1}\text{H} + ^{2}_{1}\text{H} \longrightarrow ^{3}_{1}\text{H} + ^{1}_{1}\text{H}$	6.3×10^{-13} J
$^{2}_{1}\text{H} + ^{3}_{1}\text{H} \longrightarrow ^{4}_{2}\text{He} + ^{1}_{0}\text{n}$	2.8×10^{-12} J
$^{6}_{3}\text{Li} + ^{2}_{1}\text{H} \longrightarrow 2^{4}_{2}\text{He}$	3.6×10^{-12} J

These reactions take place at extremely high temperatures, on the order of 100 million degrees Celsius, to overcome the repulsive forces between the nuclei. The first reaction is particularly attractive because the world's supply of deuterium is virtually inexhaustible. The total volume of water on Earth is about 1.5×10^{21} L. Because the natural abundance of deuterium is 1.5×10^{-2} percent, the total amount of deuterium present is roughly 4.5×10^{21} g, or 5.0×10^{15} tons. The cost of preparing deuterium is minimal compared with the value of the energy released by the reaction.

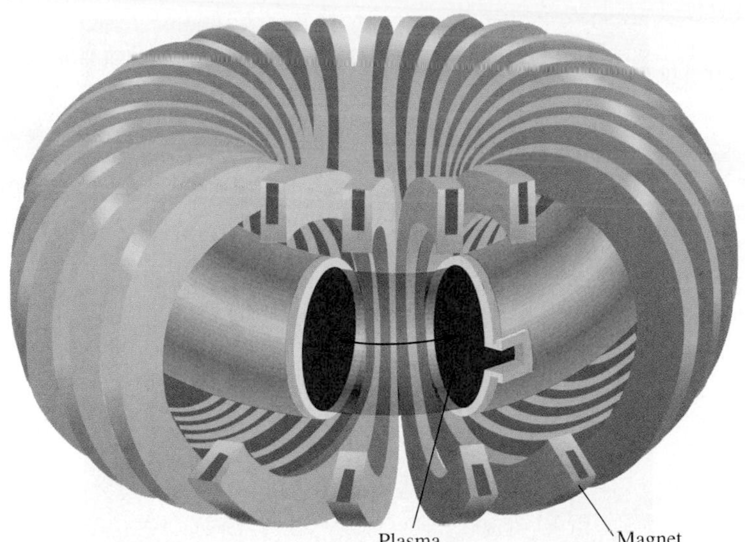

Figure 21.14
A magnetic plasma confinement design called tokamak.

In contrast to the fission process, nuclear fusion looks like a very promising energy source, at least "on paper." Although thermal pollution would be a problem, fusion has the following advantages: (1) The fuels are cheap and almost inexhaustible and (2) the process produces little radioactive waste. If a fusion machine were turned off, it would shut down completely and instantly, without any danger of a meltdown.

If nuclear fusion is so great, why isn't there even one fusion reactor producing energy? Although we command the scientific knowledge to design such a reactor, the technical difficulties have not yet been solved. The basic problem is finding a way to hold the nuclei together long enough, and at the appropriate temperature, for fusion to occur. At temperatures of about 100 million degrees Celsius, molecules cannot exist, and most or all of the atoms are stripped of their electrons. This *state of matter, a gaseous mixture of positive ions and electrons,* is called ***plasma.*** The problem of containing this plasma is a formidable one. What solid container can exist at such temperatures? None, unless the amount of plasma is small; but then the solid surface would immediately cool the sample and quench the fusion reaction. One approach to solving this problem is to use *magnetic confinement.* Because a plasma consists of charged particles moving at high speeds, a magnetic field will exert force on it. As Figure 21.14 shows, the plasma moves through a doughnut-shaped tunnel, confined by a complex magnetic field. Thus, the plasma never comes in contact with the walls of the container.

Another promising design employs high-power lasers to initiate the fusion reaction. In test runs, a number of laser beams transfer energy to a small fuel pellet, heating it and causing it to *implode,* that is, to collapse inward from all sides and compress into a small volume (Figure 21.15). Consequently, fusion occurs. Like the magnetic confinement approach, laser fusion presents a number of technical difficulties that still need to be overcome before it can be put to practical use on a large scale.

The Hydrogen Bomb

The technical problems inherent in the design of a nuclear fusion reactor do not affect the production of a hydrogen bomb, also called a thermonuclear bomb. In this case, the objective is all power and no control. Hydrogen bombs do not contain

Figure 21.15
This small-scale fusion reaction was created at the Lawrence Livermore National Laboratory using a powerful laser, Nova.

Figure 21.16
Explosion of a thermonuclear bomb.

gaseous hydrogen or gaseous deuterium; they contain solid lithium deuteride (LiD), which can be packed very tightly. The detonation of a hydrogen bomb occurs in two stages—first a fission reaction and then a fusion reaction. The required temperature for fusion is achieved with an atomic bomb. Immediately after the atomic bomb explodes, the following fusion reactions occur, releasing vast amounts of energy (Figure 21.16):

$$^{6}_{3}\text{Li} + {}^{2}_{1}\text{H} \longrightarrow 2\,^{4}_{2}\alpha$$
$$^{2}_{1}\text{H} + {}^{2}_{1}\text{H} \longrightarrow {}^{3}_{1}\text{H} + {}^{1}_{1}\text{H}$$

There is no critical mass in a fusion bomb, and the force of the explosion is limited only by the quantity of reactants present. Thermonuclear bombs are described as being "cleaner" than atomic bombs because the only radioactive isotopes they produce are tritium, which is a weak β-particle emitter ($t_{\frac{1}{2}} = 12.5$ yr), and the products of the fission starter. Their damaging effects on the environment can be aggravated, however, by incorporating in the construction some nonfissionable material such as cobalt. Upon bombardment by neutrons, cobalt-59 is converted to cobalt-60, which is a very strong γ-ray emitter with a half-life of 5.2 yr. The presence of radioactive cobalt isotopes in the debris or fallout from a thermonuclear explosion would be fatal to those who survived the initial blast.

21.7 Uses of Isotopes

Radioactive and stable isotopes alike have many applications in science and medicine. We have previously described the use of isotopes in dating artifacts (Section 21.3). In this section, we will discuss a few more examples.

Structural Determination

The formula of the thiosulfate ion is $S_2O_3^{2-}$. For some years chemists were uncertain as to whether the two sulfur atoms occupied equivalent positions in the ion. The thiosulfate ion is prepared by treatment of the sulfite ion with elemental sulfur:

$$SO_3^{2-}(aq) + S(s) \longrightarrow S_2O_3^{2-}(aq)$$

When thiosulfate is treated with dilute acid, the reaction is reversed. The sulfite ion is reformed and elemental sulfur precipitates:

$$S_2O_3^{2-}(aq) \xrightarrow{H^+} SO_3^{2-}(aq) + S(s) \qquad (21.4)$$

If this sequence is started with elemental sulfur enriched with the radioactive sulfur-35 isotope, the isotope acts as a "label" for S atoms. All the labels are found in the sulfur precipitate in Equation (21.4); none of them appears in the final sulfite ions. Clearly, then, the two atoms of sulfur in $S_2O_3^{2-}$ are not structurally equivalent, as would be the case if the structure were

$$\left[:\ddot{O}-\ddot{S}-\ddot{O}-\ddot{S}-\ddot{O}: \right]^{2-}$$

Otherwise, the radioactive isotope would be present in both the elemental sulfur precipitate and the sulfite ion. Based on spectroscopic studies, we now know that the structure of the thiosulfate ion is

$$\left[\begin{array}{c} :S: \\ \| \\ :\ddot{O}-S-\ddot{O}: \\ \| \\ :O: \end{array} \right]^{2-}$$

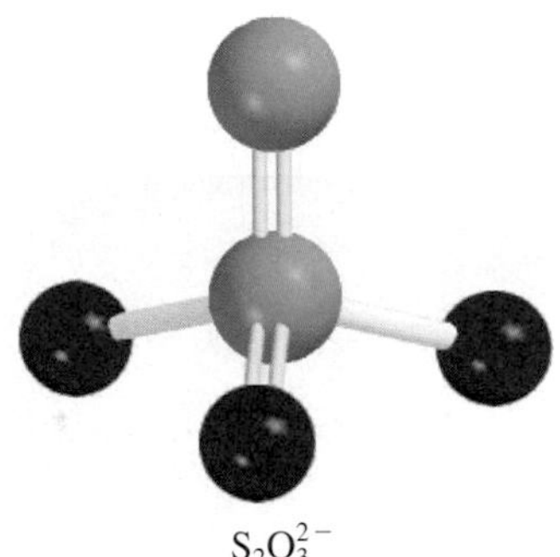

$S_2O_3^{2-}$

Study of Photosynthesis

The study of photosynthesis is also rich with isotope applications. The overall photosynthesis reaction can be represented as

$$6CO_2 + 6H_2O \longrightarrow C_6H_{12}O_6 + 6O_2$$

The radioactive ^{14}C isotope has helped to determine the path of carbon in photosynthesis. Starting with $^{14}CO_2$, it was possible to isolate the intermediate products during photosynthesis and measure the amount of radioactivity of each carbon-containing compound. In this manner, the path from CO_2 through various intermediate compounds to carbohydrate could be clearly charted. *Isotopes, especially radioactive isotopes that are used to trace the path of the atoms of an element in a chemical or biological process,* are called ***tracers.***

Isotopes in Medicine

Tracers are used also for diagnosis in medicine. Sodium-24 (a β emitter with a half-life of 14.8 h) injected into the bloodstream as a salt solution can be monitored to trace the flow of blood and detect possible constrictions or obstructions in the circulatory system. Iodine-131 (a β emitter with a half-life of 8 days) has been used to test the activity of the thyroid gland. A malfunctioning thyroid can be detected by giving the patient a drink of a solution containing a known amount of $Na^{131}I$ and measuring the radioactivity just above the thyroid to see if the iodine is absorbed at the normal rate. Of course, the amounts of radioisotope used in the human body must always be kept small; otherwise, the patient might suffer permanent damage from the high-energy radiation. Another radioactive isotope of iodine, iodine-125 (a γ-ray emitter), is used to image the thyroid gland (Figure 21.17).

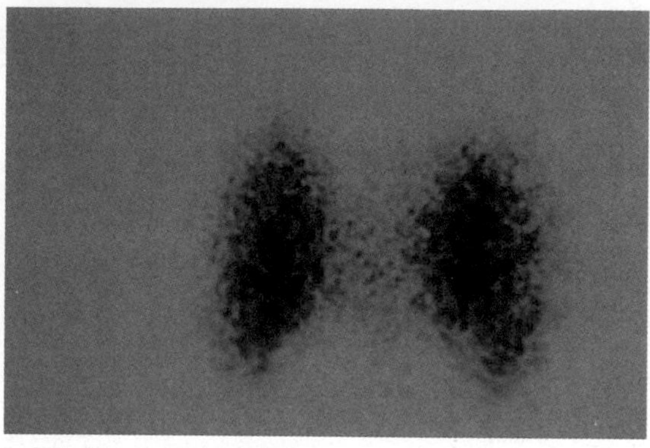
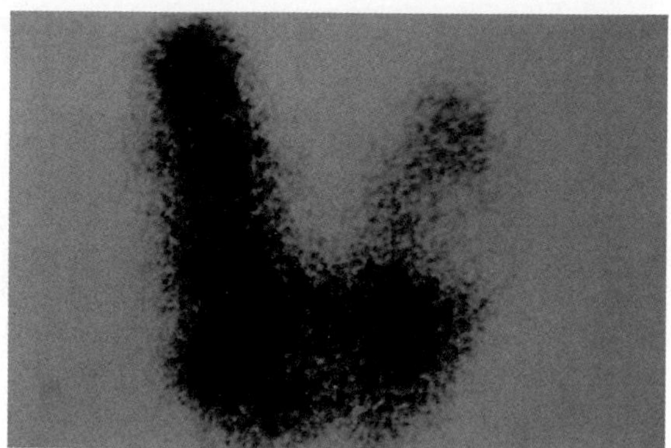

Figure 21.17
After ingesting $Na^{125}I$, the uptake of the radioactive iodine by the thyroid gland in a patient is monitored with a scanner. The photos show a normal thyroid gland (left) and an enlarged thyroid gland (right).

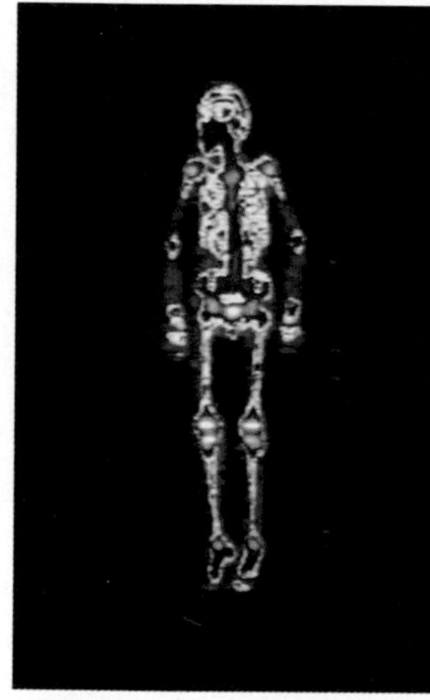

Image of a person's skeleton obtained using $^{99m}_{43}Tc$.

Technetium, the first artificially prepared element, is one of the most useful elements in nuclear medicine. Although technetium is a transition metal, all its isotopes are radioactive. In the laboratory, it is prepared by the nuclear reactions

$$^{98}_{42}\text{Mo} + {}^{1}_{0}\text{n} \longrightarrow {}^{99}_{42}\text{Mo}$$
$$^{99}_{42}\text{Mo} \longrightarrow {}^{99m}_{43}\text{Tc} + {}^{0}_{-1}\beta$$

where the superscript m denotes that the technetium-99 isotope is produced in its excited nuclear state. This isotope has a half-life of about 6 h, decaying by γ radiation to technetium-99 in its nuclear ground state. Thus, it is a valuable diagnostic tool. The patient either drinks or is injected with a solution containing ^{99m}Tc. By detecting the γ rays emitted by ^{99m}Tc, doctors can obtain images of organs such as the heart, liver, and lungs.

A major advantage of using radioactive isotopes as tracers is that they are easy to detect. Their presence even in very small amounts can be detected by photographic techniques or by devices known as counters. Figure 21.18 is a diagram of a Geiger counter, an instrument widely used in scientific work and medical laboratories to detect radiation.

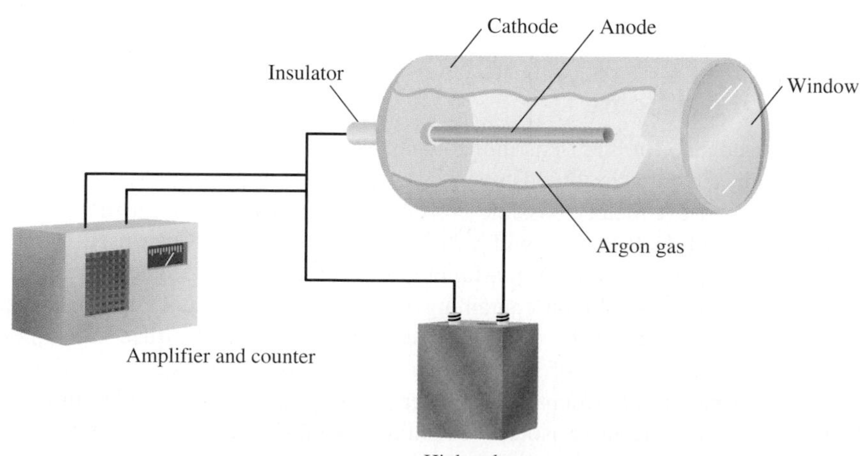

Figure 21.18
Schematic diagram of a Geiger counter. Radiation (α, β, or γ rays) entering through the window ionized the argon gas to generate a small current flow between the electrodes. This current is amplified and is used to flash a light or operate a counter with a clicking sound.

21.8 Biological Effects of Radiation

In this section we will examine briefly the effects of radiation on biological systems. But first let us define quantitative measures of radiation. The fundamental unit of radioactivity is the *curie* (Ci); 1 Ci corresponds to exactly 3.70×10^{10} nuclear disintegrations per second. This decay rate is equivalent to that of 1 g of radium. A *millicurie* (mCi) is one-thousandth of a curie. Thus, 10 mCi of a carbon-14 sample is the quantity that undergoes

$$(10 \times 10^{-3})(3.70 \times 10^{10}) = 3.70 \times 10^{8}$$

disintegrations per second.

The intensity of radiation depends on the number of disintegrations as well as on the energy and type of radiation emitted. One common unit for the absorbed dose of radiation is the *rad* (radiation *a*bsorbed *d*ose), which is the amount of radiation that results in the absorption of 1×10^{-2} J per kilogram of irradiated material. The biological effect of radiation also depends on the part of the body irradiated and the type of radiation. For this reason, the rad is often multiplied by a factor called *RBE* (*r*elative *b*iological *e*ffectiveness). The RBE is approximately 1 for beta and gamma radiation and about 10 for alpha radiation. To measure the biological damage, which depends on dose rate, total dose, and the type of tissue affected, we introduce another term called a *rem* (*r*oentgen *e*quivalent for *m*an), given by

$$\text{number of rems} = (\text{number of rads})(\text{RBE}) \qquad (21.5)$$

Of the three types of nuclear radiation, alpha particles usually have the least penetrating power. Beta particles are more penetrating than alpha particles, but less so than gamma rays. Gamma rays have very short wavelengths and high energies. Furthermore, because they carry no charge, they cannot be stopped by shielding materials as easily as alpha and beta particles. However, if alpha or beta emitters are ingested, their damaging effects are greatly aggravated because the organs will be constantly subject to damaging radiation at close range. For example, strontium-90, a beta emitter, can replace calcium in bones, where it does the greatest damage.

Table 21.6 lists the average amounts of radiation an American receives every year. It should be pointed out that for short-term exposures to radiation, a dosage

Table 21.6 Average Yearly Radiation Doses for Americans

Source	Dose (mrem/yr)*
Cosmic rays	20–50
Ground and surroundings	25
Human body†	26
Medical and dental X rays	50–75
Air travel	5
Fallout from weapons tests	5
Nuclear waste	2
Total	133–188

*1 mrem = 1 millirem = 1×10^{-3} rem.
†The radioactivity in the body comes from food and air.

of 50–200 rem will cause a decrease in white blood cell counts and other complications, while a dosage of 500 rem or greater may result in death within weeks. Current safety standards permit nuclear workers to be exposed to no more than 5 rem per year and specify a maximum of 0.5 rem of human-made radiation per year for the general public.

The chemical basis of radiation damage is that of ionizing radiation. Radiation of either particles or gamma rays can remove electrons from atoms and molecules in its path, leading to the formation of ions and radicals. Radicals are usually short-lived and highly reactive. For example, when water is irradiated with gamma rays, the following reactions take place:

$$H_2O \xrightarrow{\text{radiation}} H_2O^+ + e^-$$

$$H_2O^+ + H_2O \longrightarrow H_3O^+ + \underset{\text{hydroxyl radical}}{\cdot OH}$$

The electron (in the hydrated form) can subsequently react with water or with a hydrogen ion to form atomic hydrogen, and with oxygen to produce the superoxide ion, O_2^- (a radical):

$$e^- + O_2 \longrightarrow \cdot O_2^-$$

In the tissues, the superoxide ions and other free radicals attack cell membranes and a host of organic compounds, such as enzymes and DNA molecules. Organic compounds can themselves be directly ionized and destroyed by high-energy radiation.

It has long been known that exposure to high-energy radiation can induce cancer in humans and other animals. Cancer is characterized by uncontrolled cellular growth. On the other hand, it is also well established that cancer cells can be destroyed by proper radiation treatment. In radiation therapy, a compromise is sought. The radiation to which the patient is exposed must be sufficient to destroy cancer cells without killing too many normal cells and, it is hoped, without inducing another form of cancer.

Radiation damage to living systems is generally classified as *somatic* or *genetic*. Somatic injuries are those that affect the organism during its own lifetime. Sunburn, skin rash, cancer, and cataracts are examples of somatic damage. Genetic damage means inheritable changes or gene mutations. For example, a person whose chromosomes have been damaged or altered by radiation may have deformed offspring.

Chromosomes are parts of the cell structure that contain the genetic material (DNA).

Key Equations

$E = mc^2$ (21.1) Einstein's mass-energy equivalence relationship.

$\Delta E = (\Delta m)c^2$ (21.2) Relation between mass defect and energy released.

$$\text{nuclear binding energy per nucleon} = \frac{\text{nuclear binding energy}}{\text{number of nucleons}} \quad (21.3)$$

Summary of Facts and Concepts

1. Nuclear chemistry is the study of changes in atomic nuclei. Such changes are termed nuclear reactions. Radioactive decay and nuclear transmutation are nuclear reactions.
2. For stable nuclei of low atomic number, the neutron-to-proton ratio is close to 1. For heavier stable nuclei, the ratio becomes greater than 1. All nuclei with 84 or more

protons are unstable and radioactive. Nuclei with even atomic numbers are more stable than those with odd atomic numbers. A quantitative measure of nuclear stability is the nuclear binding energy, which can be calculated from a knowledge of the mass defect of the nucleus.

3. Radioactive nuclei emit α particles, β particles, positrons, or γ rays. The equation for a nuclear reaction includes the particles emitted, and both the mass numbers and the atomic numbers must balance. Uranium-238 is the parent of a natural radioactive decay series. A number of radioactive isotopes, such as ^{238}U and ^{14}C, can be used to date objects. Artificially radioactive elements are created by the bombardment of other elements by accelerated neutrons, protons, or α particles.
4. Nuclear fission is the splitting of a large nucleus into smaller nuclei plus neutrons. When these neutrons are captured efficiently by other nuclei, an uncontrollable chain reaction can occur. Nuclear reactors use the heat from a controlled nuclear fission reaction to produce power. The three important types of reactors are light water reactors, heavy water reactors, and breeder reactors.
5. Nuclear fusion, the type of reaction that occurs in the sun, is the combination of two light nuclei to form one heavy nucleus. Fusion takes place only at very high temperatures—so high that controlled large-scale nuclear fusion has so far not been achieved.
6. Radioactive isotopes are easy to detect and thus make excellent tracers in chemical reactions and in medical practice. High-energy radiation damages living systems by causing ionization and the formation of reactive radicals.

Key Words

Questions and Problems

Nuclear Reactions

Review Questions

21.1 How do nuclear reactions differ from ordinary chemical reactions?

21.2 What are the steps in balancing nuclear equations?

21.3 What is the difference between $^{0}_{-1}e$ and $^{0}_{-1}\beta$?

21.4 What is the difference between an electron and a positron?

Problems

21.5 Complete these nuclear equations and identify X in each case:
(a) $^{26}_{12}Mg + ^{1}_{1}p \longrightarrow ^{4}_{2}\alpha + X$
(b) $^{59}_{27}Co + ^{2}_{1}H \longrightarrow ^{60}_{27}Co + X$
(c) $^{235}_{92}U + ^{1}_{0}n \longrightarrow ^{94}_{36}Kr + ^{139}_{56}Ba + 3X$
(d) $^{53}_{24}Cr + ^{4}_{2}\alpha \longrightarrow ^{1}_{0}n + X$
(e) $^{20}_{8}O \longrightarrow ^{20}_{9}F + X$

21.6 Complete these nuclear equations and identify X in each case:
(a) $^{135}_{53}I \longrightarrow ^{135}_{54}Xe + X$
(b) $^{40}_{19}K \longrightarrow ^{0}_{-1}\beta + X$
(c) $^{59}_{27}Co + ^{1}_{0}n \longrightarrow ^{56}_{25}Mn + X$
(d) $^{235}_{92}U + ^{1}_{0}n \longrightarrow ^{99}_{40}Zr + ^{135}_{52}Te + 2X$

Nuclear Stability

Review Questions

21.7 State the general rules for predicting nuclear stability.

21.8 What is the belt of stability?

21.9 Why is it impossible for the isotope $^{2}_{2}He$ to exist?

21.10 Define nuclear binding energy, mass defect, and nucleon.

21.11 How does Einstein's equation, $E = mc^2$, enable us to calculate nuclear binding energy?

21.12 Why is it preferable to use nuclear binding energy per nucleon for a comparison of the stabilities of different nuclei?

Problems

21.13 The radius of a uranium-235 nucleus is about 7.0×10^{-3} pm. Calculate the density of the nucleus in g/cm^3. (Assume the atomic mass is 235 amu.)

21.14 For each pair of isotopes listed, predict which one is less stable: (a) $^{6}_{3}Li$ or $^{9}_{3}Li$, (b) $^{23}_{11}Na$ or $^{25}_{11}Na$, (c) $^{48}_{20}Ca$ or $^{48}_{21}Sc$.

21.15 For each pair of elements listed, predict which one has more stable isotopes: (a) Co or Ni, (b) F or Se, (c) Ag or Cd.

21.16 In each pair of isotopes shown, indicate which one you would expect to be radioactive: (a) $^{20}_{10}Ne$ and $^{17}_{10}Ne$, (b) $^{40}_{20}Ca$ and $^{45}_{20}Ca$, (c) $^{95}_{42}Mo$ and $^{92}_{43}Tc$, (d) $^{195}_{80}Hg$ and $^{196}_{80}Hg$, (e) $^{209}_{83}Bi$ and $^{242}_{96}Cm$.

21.17 Given that

$$H(g) + H(g) \longrightarrow H_2(g) \qquad \Delta H° = -436.4\ \text{kJ}$$

calculate the change in mass (in kg) per mole of H_2 formed.

21.18 Estimates show that the total energy output of the sun is 5×10^{26} J/s. What is the corresponding mass loss in kg/s of the sun?

21.19 Calculate the nuclear binding energy (in J) and the nuclear binding energy per nucleon of the following isotopes: (a) $^{7}_{3}Li$ (7.01600 amu) and (b) $^{35}_{17}Cl$ (34.95952 amu).

21.20 Calculate the nuclear binding energy (in J) and the nuclear binding energy per nucleon of the following isotopes: (a) $^{4}_{2}He$ (4.0026 amu) and (b) $^{184}_{74}W$ (183.9510 amu).

Natural Radioactivity

Review Questions

21.21 Discuss factors that lead to nuclear decay.

21.22 Outline the principle for dating materials using radioactive isotopes.

Problems

21.23 Fill in the blanks in these radioactive decay series:

(a) $^{232}Th \xrightarrow{\alpha}$ ____ $\xrightarrow{\beta}$ ____ $\xrightarrow{\beta} {}^{228}Th$

(b) $^{235}U \xrightarrow{\alpha}$ ____ $\xrightarrow{\beta}$ ____ $\xrightarrow{\alpha} {}^{227}Ac$

(c) ____ $\xrightarrow{\alpha} {}^{233}Pa \xrightarrow{\beta}$ ____ $\xrightarrow{\alpha}$ ____

21.24 A radioactive substance undergoes decay as:

Time (days)	Mass (g)
0	500
1	389
2	303
3	236
4	184
5	143
6	112

Calculate the first-order decay constant and the half-life of the reaction.

21.25 The radioactive decay of Tl-206 to Pb-206 has a half-life of 4.20 min. Starting with 5.00×10^{22} atoms of Tl-206, calculate the number of such atoms left after 42.0 min.

21.26 A freshly isolated sample of ^{90}Y was found to have an activity of 9.8×10^{5} disintegrations per minute at 1:00 P.M. on December 3, 2009. At 2:15 P.M. on December 17, 2009, its activity was redetermined and found to be 2.6×10^{4} disintegrations per minute. Calculate the half-life of ^{90}Y.

21.27 Why do radioactive decay series obey first-order kinetics?

21.28 In the thorium decay series, thorium-232 loses a total of 6 α particles and 4 β particles in a 10-stage process. What is the final isotope produced?

21.29 Strontium-90 is one of the products of the fission of uranium-235. This strontium isotope is radioactive, with a half-life of 28.1 yr. Calculate how long (in years) it will take for 1.00 g of the isotope to be reduced to 0.200 g by decay.

21.30 Consider the decay series

$$A \longrightarrow B \longrightarrow C \longrightarrow D$$

where A, B, and C are radioactive isotopes with half-lives of 4.50 s, 15.0 days, and 1.00 s, respectively, and D is nonradioactive. Starting with 1.00 mole of A, and none of B, C, or D, calculate the number of moles of A, B, C, and D left after 30 days.

Nuclear Transmutation

Review Questions

21.31 What is the difference between radioactive decay and nuclear transmutation?

21.32 How is nuclear transmutation achieved in practice?

Problems

21.33 Write balanced nuclear equations for these reactions and identify X:

(a) $X(p,\alpha)^{12}_{6}C$, (b) $^{27}_{13}Al(d,\alpha)X$, (c) $^{55}_{25}Mn(n,\gamma)X$.

21.34 Write balanced nuclear equations for these reactions and identify X:

(a) $^{80}_{34}Se(d,p)X$, (b) $X(d,2p)^{9}_{3}Li$, (c) $^{10}_{5}B(n,\alpha)X$.

21.35 Describe how you would prepare astatine-211, starting with bismuth-209.

21.36 A long-cherished dream of alchemists was to produce gold from cheaper and more abundant elements. This dream was finally realized when $^{198}_{80}Hg$ was converted into gold by neutron bombardment. Write a balanced equation for this reaction.

Nuclear Fission

Review Questions

21.37 Define nuclear fission, nuclear chain reaction, and critical mass.

21.38 Which isotopes can undergo nuclear fission?

21.39 Explain how an atomic bomb works.

21.40 Explain the functions of a moderator and a control rod in a nuclear reactor.

21.41 Discuss the differences between a light water and a heavy water nuclear fission reactor. What are the advantages of a breeder reactor over a conventional nuclear fission reactor?

21.42 What makes water particularly suitable for use as moderator in a nuclear reactor?

Nuclear Fusion

Review Questions

21.43 Define nuclear fusion, thermonuclear reaction, and plasma.

21.44 Why do heavy elements such as uranium undergo fission while light elements such as hydrogen and lithium undergo fusion?

21.45 How does a hydrogen bomb work?

21.46 What are the advantages of a fusion reactor over a fission reactor? What are the practical difficulties in operating a large-scale fusion reactor?

Uses of Isotopes

Problems

21.47 Describe how you would use a radioactive iodine isotope to demonstrate that the following process is in dynamic equilibrium:

$$PbI_2(s) \rightleftharpoons Pb^{2+}(aq) + 2I^-(aq)$$

21.48 Consider this redox reaction:

$$IO_4^-(aq) + 2I^-(aq) + H_2O(l) \longrightarrow I_2(s) + IO_3^-(aq) + 2OH^-(aq)$$

When KIO_4 is added to a solution containing iodide ions labeled with radioactive iodine-128, all the radioactivity appears in I_2 and none in the IO_3^- ion. What can you deduce about the mechanism for the redox process?

21.49 Explain how you might use a radioactive tracer to show that ions are not completely motionless in crystals.

21.50 Each molecule of hemoglobin, the oxygen carrier in blood, contains four Fe atoms. Explain how you would use the radioactive $^{59}_{26}Fe$ ($t_{\frac{1}{2}}$ = 46 days) to show that the iron in a certain food is converted into hemoglobin.

Additional Problems

21.51 How does a Geiger counter work?

21.52 Nuclei with an even number of protons and an even number of neutrons are more stable than those with an odd number of protons and/or an odd number of neutrons. What is the significance of the even numbers of protons and neutrons in this case?

21.53 Tritium, ^3H, is radioactive and decays by electron emission. Its half-life is 12.5 yr. In ordinary water the ratio of ^1H to ^3H atoms is 1.0×10^{17} to 1. (a) Write a balanced nuclear equation for tritium decay. (b) How many disintegrations will be observed per minute in a 1.00-kg sample of water?

21.54 (a) What is the activity, in millicuries, of a 0.500-g sample of $^{237}_{93}Np$? (This isotope decays by α-particle emission and has a half-life of 2.20×10^6 yr.) (b) Write a balanced nuclear equation for the decay of $^{237}_{93}Np$.

21.55 These equations are for nuclear reactions that are known to occur in the explosion of an atomic bomb. Identify X.

(a) $^{235}_{92}U + ^1_0n \longrightarrow ^{140}_{56}Ba + 3^1_0n + X$
(b) $^{235}_{92}U + ^1_0n \longrightarrow ^{144}_{55}Cs + ^{90}_{37}Rb + 2X$
(c) $^{235}_{92}U + ^1_0n \longrightarrow ^{87}_{35}Br + 3^1_0n + X$
(d) $^{235}_{92}U + ^1_0n \longrightarrow ^{160}_{62}Sm + ^{72}_{30}Zn + 4X$

21.56 Calculate the nuclear binding energies, in J/nucleon, for these species: (a) ^{10}B (10.0129 amu), (b) ^{11}B (11.00931 amu), (c) ^{14}N (14.00307 amu), (d) ^{56}Fe (55.9349 amu).

21.57 Write complete nuclear equations for these processes: (a) tritium, ^3H, undergoes β decay; (b) ^{242}Pu undergoes α-particle emission; (c) ^{131}I undergoes β decay; (d) ^{251}Cf emits an α particle.

21.58 The nucleus of nitrogen-18 lies above the stability belt. Write an equation for a nuclear reaction by which nitrogen-18 can achieve stability.

21.59 Why is strontium-90 a particularly dangerous isotope for humans?

21.60 How are scientists able to tell the age of a fossil?

21.61 After the Chernobyl accident, people living close to the nuclear reactor site were urged to take large amounts of potassium iodide as a safety precaution. What is the chemical basis for this action?

21.62 Astatine, the last member of Group 7A, can be prepared by bombarding bismuth-209 with α particles. (a) Write an equation for the reaction. (b) Represent the equation in the abbreviated form, as discussed in Section 21.4.

21.63 To detect bombs that may be smuggled onto airplanes, the Federal Aviation Administration (FAA) will soon require all major airports in the United States to install thermal neutron analyzers. The thermal neutron analyzer will bombard baggage with low-energy neutrons, converting some of the nitrogen-14 nuclei to nitrogen-15, with simultaneous emission of γ rays. Because nitrogen content is usually high in explosives, detection of a high dosage of

γ rays will suggest that a bomb may be present. (a) Write an equation for the nuclear process. (b) Compare this technique with the conventional X-ray detection method.

21.64 Explain why achievement of nuclear fusion in the laboratory requires a temperature of about 100 million degrees Celsius, which is much higher than that in the interior of the sun (15 million degrees Celsius).

21.65 Tritium contains one proton and two neutrons. There is no proton-proton repulsion present in the nucleus. Why, then, is tritium radioactive?

21.66 The carbon-14 decay rate of a sample obtained from a young tree is 0.260 disintegration per second per gram of the sample. Another wood sample prepared from an object recovered at an archaeological excavation gives a decay rate of 0.186 disintegration per second per gram of the sample. What is the age of the object?

21.67 The usefulness of radiocarbon dating is limited to objects no older than 50,000 years. What percent of the carbon-14, originally present in the sample, remains after this period of time?

21.68 The radioactive potassium-40 isotope decays to argon-40 with a half-life of 1.2×10^9 yr. (a) Write a balanced equation for the reaction. (b) A sample of moon rock is found to contain 18 percent potassium-40 and 82 percent argon by mass. Calculate the age of the rock in years.

21.69 Both barium (Ba) and radium (Ra) are members of Group 2A and are expected to exhibit similar chemical properties. However, Ra is not found in barium ores. Instead, it is found in uranium ores. Explain.

21.70 Nuclear waste disposal is one of the major concerns of the nuclear industry. In choosing a safe and stable environment to store nuclear wastes, consideration must be given to the heat released during nuclear decay. As an example, consider the β decay of ^{90}Sr (89.907738 amu):

$$^{90}_{38}Sr \longrightarrow {}^{90}_{39}Y + {}^{0}_{-1}\beta \qquad t_{\frac{1}{2}} = 28.1 \text{ yr}$$

The ^{90}Y (89.907152 amu) further decays as follows:

$$^{90}_{39}Y \longrightarrow {}^{90}_{40}Zr + {}^{0}_{-1}\beta \qquad t_{\frac{1}{2}} = 64 \text{ h}$$

Zirconium-90 (89.904703 amu) is a stable isotope. (a) Use the mass defect to calculate the energy released (in joules) in each of the preceding two decays. (The mass of the electron is 5.4857×10^{-4} amu.) (b) Starting with 1 mole of ^{90}Sr, calculate the number of moles of ^{90}Sr that will decay in a year. (c) Calculate the amount of heat released (in kilojoules) corresponding to the number of moles of ^{90}Sr decayed to ^{90}Zr in (b).

21.71 Which of the following poses a greater health hazard: a radioactive isotope with a short half-life or a radioactive isotope with a long half-life? Explain. [Assume same type of radiation (α or β) and comparable energetics per particle emitted.]

21.72 As a result of being exposed to the radiation released during the Chernobyl nuclear accident, the dose of iodine-131 in a person's body is 7.4 mC (1 mC = 1×10^{-3} Ci). Use the relationship rate = λN to calculate the number of atoms of iodine-131 to which this radioactivity corresponds. (The half-life of ^{131}I is 8.1 days.)

21.73 Bismuth-214 is an α-emitter with a half-life of 19.7 min. A 5.26-mg sample of the isotope is placed in a sealed, evacuated flask of volume 20.0 mL at 40°C. Assuming that all the α particles generated are converted to helium gas and that the other decay product is nonradioactive, calculate the pressure (in mmHg) inside the flask after 78.8 min. Use 214 amu for the atomic mass of bismuth.

21.74 From the definition of curie, calculate Avogadro's number, given that the molar mass of ^{226}Ra is 226.03 g/mol and that it decays with a half-life of 1.6×10^3 yr.

21.75 Since 1998, elements 112, 114, and 116 have been synthesized. Element 112 was created by bombarding ^{208}Pb with ^{62}Zn; element 114 was created by bombarding ^{244}Pu with ^{48}Ca; element 116 was created by bombarding ^{248}Cm with ^{48}Ca. Write an equation for each synthesis. Predict the chemical properties of these elements. Use X for element 112, Y for element 114, and Z for element 116.

21.76 Sources of energy on Earth include fossil fuels, geothermal, gravitational, hydroelectric, nuclear fission, nuclear fusion, solar, and wind. Which of these have a "nuclear origin," either directly or indirectly?

21.77 A person received an anonymous gift of a decorative box which he placed on his desk. A few months later he became ill and died shortly afterward. After investigation, the cause of his death was linked to the box. The box was air-tight and had no toxic chemicals on it. What might have killed the man?

21.78 Identify two of the most abundant radioactive elements that exist on Earth. Explain why they are still present. (You may need to consult a handbook of chemistry.)

21.79 (a) Calculate the energy released when a ^{238}U isotope decays to ^{234}Th. The atomic masses are given by: ^{238}U: 238.0508 amu; ^{234}Th: 234.0436 amu; ^{4}He: 4.0026 amu. (b) The energy released in (a) is transformed into the kinetic energy of the recoiling ^{234}Th nucleus and the α particle. Which of the two will move away faster? Explain.

21.80 Cobalt-60 is an isotope used in diagnostic medicine and cancer treatment. It decays with γ ray emission. Calculate the wavelength of the radiation in nanometers if the energy of the γ ray is 2.4×10^{-13} J/photon.

21.81 Americium-241 is used in smoke detectors because it has a long half-life (458 yr) and its emitted α particles are energetic enough to ionize air molecules. Given the schematic diagram of a smoke detector below, explain how it works.

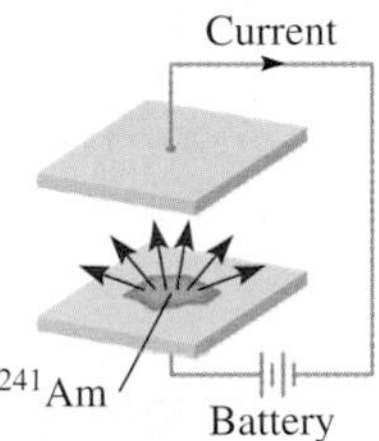

21.82 The constituents of wine contain, among others, carbon, hydrogen, and oxygen atoms. A bottle of wine was sealed about 6 yr ago. To confirm its age, which of the isotopes would you choose in a radioactive dating study? The half-lives of the isotopes are: ^{14}C: 5730 yr; ^{15}O: 124 s; ^{3}H: 12.5 yr. Assume that the activities of the isotopes were known at the time the bottle was sealed.

21.83 Name two advantages of a nuclear-powered submarine over a conventional submarine.

21.84 In 1997, a scientist at a nuclear research center in Russia placed a thin shell of copper on a sphere of highly enriched uranium-235. Suddenly, there was a huge burst of radiation, which turned the air blue. Three days later, the scientist died of radiation damage. Explain what caused the accident. (*Hint:* Copper is an effective metal for reflecting neutrons.)

21.85 A radioactive isotope of copper decays as follows:

$$^{64}\text{Cu} \longrightarrow {}^{64}\text{Zn} + {}^{0}_{-1}\beta \qquad t_{\frac{1}{2}} = 12.8 \text{ h}$$

Starting with 84.0 g of ^{64}Cu, calculate the quantity of ^{64}Zn produced after 18.4 h.

21.86 A 0.0100-g sample of a radioactive isotope with a half-life of 1.3×10^{9} yr decays at the rate of 2.9×10^{4} dpm. Calculate the molar mass of the isotope.

Special Problems

21.87 Describe, with appropriate equations, nuclear processes that lead to the formation of the noble gases He, Ne, Ar, Kr, Xe, and Rn. (*Hint:* Helium is formed from radioactive decay, neon is formed from the positron emission of ^{22}Na, the formation of Ar, Xe, and Rn are discussed in the chapter, and Kr is produced from the fission of ^{235}U.)

21.88 Write an essay on the pros and cons of nuclear power (based on nuclear fission), paying particular attention to its effect on global warming, nuclear reactor safety and weapon risks, and nuclear waste disposal.

21.89 The half-life of ^{27}Mg is 9.50 min. (a) Initially there were 4.20×10^{12} ^{27}Mg nuclei present. How many ^{27}Mg nuclei are left 30.0 min later? (b) Calculate the ^{27}Mg activities (in Ci) at $t = 0$ and $t = 30.0$ min. (c) What is the probability that any one ^{27}Mg nucleus decays during a 1-s interval? What assumption is made in this calculation?

21.90 The radioactive isotope ^{238}Pu, used in pacemakers, decays by emitting an alpha particle with a half-life of 86 yr. (a) Write an equation for the decay process. (b) The energy of the emitted alpha particle is 9.0×10^{-13} J, which is the energy per decay. Assume that all the alpha particle energy is used to run the pacemaker, calculate the power output at $t = 0$ and $t = 10$ yr. Initially 1.0 mg of ^{238}Pu was present in the pacemaker (*Hint:* After 10 yr, the activity of the isotope decreases by 8.0 percent. Power is measured in watts or J/s.).

21.91 (a) Assume nuclei are spherical in shape, show that its radius (r) is proportional to the cube root of mass number (A). (b) In general, the radius of a nucleus is given by $r = r_0 A^{\frac{1}{3}}$, where r_0, the proportionality constant, is given by 1.2×10^{-15} m. Calculate the volume of the ^{238}U nucleus.

21.92 The quantity of a radioactive material is often measured by its activity (measured in curies or millicuries) rather than by its mass. In a brain scan procedure, a 70-kg patient is injected with 20.0 mCi of ^{99m}Tc, which decays by emitting γ-ray photons with a half-life of 6.0 h. Given that the RBE of these photons is 0.98 and only two-thirds of the photons are absorbed by the body, calculate the rem dose received by the patient. Assume all of the ^{99m}Tc nuclei decay while in the body. The energy of a gamma photon is 2.29×10^{-14} J.

21.93 Modern designs of atomic bombs contain, in addition to uranium or plutonium, small amounts of tritium and deuterium to boost the power of explosion. What is the role of tritium and deuterium in these bombs?

21.94 What is the source of heat for volcanic activities on Earth?

21.95 Alpha particles produced from radioactive decays eventually pick up electrons from the surroundings to form helium atoms. Calculate the volume (mL) of He collected at STP when 1.00 g of pure ^{226}Ra is stored in a closed container for 100 yr. (*Hint:* Focusing only on half-lives that are short compared to 100 yr and ignoring minor decay schemes in Table 21.3, first show that there are 5 α particles generated per ^{226}Ra decay to ^{206}Pb.)

21.96 An electron and a positron are accelerated to nearly the speed of light before colliding in a particle accelerator. The ensuing collision produces an exotic particle having a mass many times that of a proton. Does the result violate the law of conservation of mass?

21.97 In 2006, an ex-KGB agent was murdered in London. Subseqvent investigation showed that the cause of death was poisoning with the radioactive isotope ^{210}Po, which was added to his drinks/food. (a) ^{210}Po is prepared by bombarding ^{209}Bi with neutrons. Write an equation for the reaction. (b) Who discovered the element polonium? (*Hint:* See Appendix 4.) (c) The half-life of ^{210}Po is 138 d. It decays with the emission of an α particle. Write an equation for the decay process. (d) Calculate the energy of an emitted α particle. Assume both the parent and daughter nuclei to have zero kinetic energy. The atomic masses are: ^{210}Po (209.98285 amu), ^{206}Pb (205.97444 amu), $^{4}_{2}\alpha$ (4.00150 amu). (e) Ingestion of 1 μg of ^{210}Po could prove fatal. What is the total energy released by this quantity of ^{210}Po?

Answers to Practice Exercises

21.1 $^{78}_{34}$Se. **21.2** 2.63×10^{-10} J; 1.26×10^{-12} J/nucleon. **21.3** $^{106}_{46}\text{Pd} + ^{4}_{2}\alpha \longrightarrow ^{109}_{47}\text{Ag} + ^{1}_{1}\text{p}$.

The strength of one kind of polymer called Lexan is so great that it is used to make bullet-proof windows.

CHAPTER

22

Organic Polymers—Synthetic and Natural

Chapter Outline

Student Interactive Activities

 **Electronic Homework**
End of Chapter Problems

Essential Concepts

Synthetic Organic Polymers Many organic polymers have been synthesized by various chemical processes. They mimic and sometimes surpass the properties of those that occur naturally. Nylon is the best known of all synthetic organic polymers.

Proteins Proteins are natural polymers made of amino acids. They perform a host of functions ranging from catalysis, transport and storage of vital substances, coordinated motion, and protection against diseases. The complex structures of proteins have been analyzed in terms of their primary, secondary, tertiary, and quaternary structures. The three-dimensional integrity of protein molecules is maintained by various intermolecular forces and hydrogen bonding.

Nucleic Acids Deoxyribonucleic acid (DNA) carries all the genetic information, and ribonucleic acid (RNA) controls the synthesis of proteins. The elucidation of the double helical structure of DNA is one of the major accomplishments in science in the twentieth century.

22.1 Properties of Polymers

A ***polymer*** is *a molecular compound distinguished by a high molar mass, ranging into thousands and millions of grams, and made up of many repeating units.* The physical properties of these so-called macromolecules differ greatly from those of small, ordinary molecules, and special techniques are required to study them.

Naturally occurring polymers include proteins, nucleic acids, cellulose (polysaccharides), and rubber (polyisoprene). Most synthetic polymers are organic compounds. Familiar examples are nylon, poly(hexamethylene adipamide); Dacron, poly(ethylene terephthalate); and Lucite or Plexiglas, poly(methyl methacrylate).

The development of polymer chemistry began in the 1920s with the investigation into a puzzling behavior of certain materials, including wood, gelatin, cotton, and rubber. For example, when rubber, with the known empirical formula of C_5H_8, was dissolved in an organic solvent, the solution displayed several unusual properties—high viscosity, low osmotic pressure, and negligible freezing-point depression. These observations strongly suggested the presence of solutes of very high molar mass, but chemists were not ready at that time to accept the idea that such giant molecules could exist. Instead, they postulated that materials such as rubber consist of aggregates of small molecular units, like C_5H_8 or $C_{10}H_{16}$, held together by intermolecular forces. This misconception persisted for a number of years, until the German chemist Hermann Staudinger clearly showed that these so-called aggregates are, in fact, enormously large molecules, each of which contains many thousands of atoms held together by covalent bonds.

Once the structures of these macromolecules were understood, the way was open for manufacturing polymers, which now pervade almost every aspect of our daily lives. About 90 percent of today's chemists, including biochemists, work with polymers.

22.2 Synthetic Organic Polymers

Because of their size, we might expect molecules containing thousands of carbon and hydrogen atoms to form an enormous number of structural and geometric isomers (if C═C bonds are present). However, these molecules are made up of ***monomers,*** *simple repeating units,* and this type of composition severely restricts the number of possible isomers. Synthetic polymers are created by joining monomers together, one at a time, by means of addition reactions and condensation reactions.

Addition Reactions

Addition reactions were described on p. 375.

Addition reactions involve unsaturated compounds containing double or triple bonds, particularly C═C and C≡C. Hydrogenation and reactions of hydrogen halides and halogens with alkenes and alkynes are examples of addition reactions.

In Chapter 11 we saw that polyethylene is formed by addition reaction. Polyethylene is an example of a ***homopolymer,*** which is *a polymer made up of only one type of monomer.* Other homopolymers that are synthesized by the radical mechanism are Teflon, polytetrafluoroethylene and poly(vinyl chloride) (PVC):

n is in the hundreds or thousands.

$$-\!\!(CF_2-CF_2)\!\!-_n \qquad -\!\!(CH_2-\underset{\displaystyle Cl}{\underset{|}{CH}})\!\!-_n$$

Teflon PVC

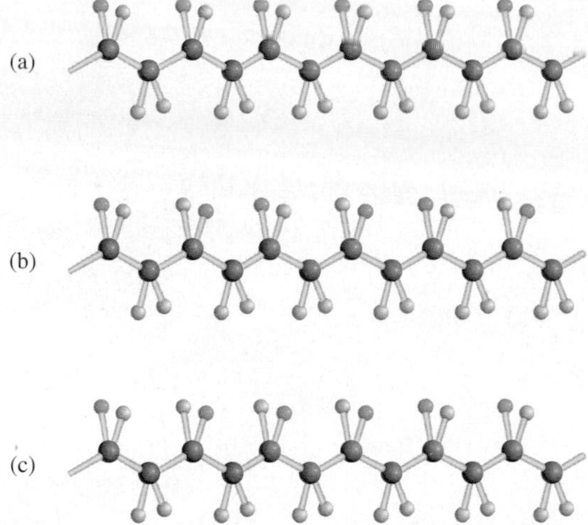

Figure 22.1
Stereoisomers of polymers. When the R group (green sphere) is CH_3, the polymer is polypropene. (a) When the R groups are all on one side of the chain, the polymer is said to be isotactic. (b) When the R groups alternate from side to side, the polymer is said to be syndiotactic. (c) When the R groups are disposed at random, the polymer is atactic.

The chemistry of polymers is more complex if the starting units are asymmetric:

$$\underset{\text{propene}}{\begin{array}{ccc}H_3C & & H\\ & C{=}C & \\ H & & H\end{array}} \qquad \underset{\text{polypropene}}{\left(\begin{array}{ccc}CH_3 & & H\\ | & & |\\ C & \text{---} & C\\ | & & |\\ H & & H\end{array}\right)_n}$$

Several geometric isomers can result from an addition reaction of propenes (Figure 22.1). If the additions occur randomly, we obtain *atactic* polypropenes, which do not pack together well. These polymers are rubbery, amorphous, and relatively weak. Two other possibilities are an *isotactic* structure, in which the R groups are all on the same side of the asymmetric carbon atoms, and a *syndiotactic* form, in which the R groups alternate on the left and right of the asymmetric carbons. Of these, the isotactic isomer has the highest melting point and greatest crystallinity and is endowed with superior mechanical properties.

Rubber is probably the best known organic polymer and the only true hydrocarbon polymer found in nature. It is formed by the radical addition of the monomer isoprene. Actually, polymerization can result in either poly-*cis*-isoprene or poly-*trans*-isoprene—or a mixture of both, depending on reaction conditions:

Stereoisomerism was discussed in Section 20.3.

$$\underset{\text{isoprene}}{n\,CH_2{=}\overset{\overset{\displaystyle CH_3}{|}}{C}{-}CH{=}CH_2} \longrightarrow \underset{\text{poly-}cis\text{-isoprene}}{\left(\begin{array}{ccc}CH_3 & & H\\ & C{=}C & \\ CH_2 & & CH_2\end{array}\right)_n} \quad \text{and/or} \quad \underset{\text{poly-}trans\text{-isoprene}}{\left(\begin{array}{ccc}CH_2 & & H\\ & C{=}C & \\ CH_3 & & CH_2\end{array}\right)_n}$$

Note that in the *cis* isomer the two CH_2 groups are on the same side of the C=C bond, whereas the same groups are across from each other in the *trans* isomer.

Figure 22.2
Latex (aqueous suspension of rubber particles) being collected from a rubber tree.

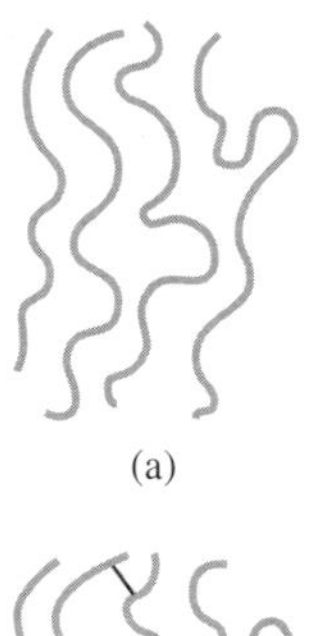

(a)

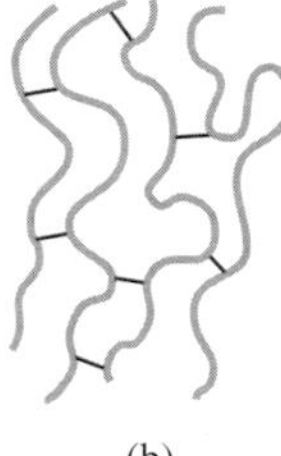

(b)

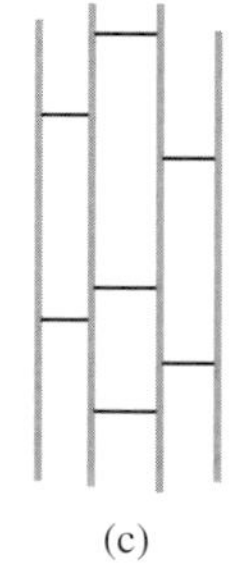

(c)

Figure 22.3
Rubber molecules ordinarily are bent and convoluted. Parts (a) and (b) represent the long chains before and after vulcanization, respectively; (c) shows the alignment of molecules when stretched. Without vulcanization these molecules would slip past one another, and rubber's elastic properties would be gone.

Natural rubber is poly-*cis*-isoprene, which is extracted from the tree *Hevea brasiliensis* (Figure 22.2).

An unusual and very useful property of rubber is its elasticity. Rubber will stretch up to 10 times its length and, if released, will return to its original size. In contrast, a piece of copper wire can be stretched only a small percentage of its length and still return to its original size. Unstretched rubber is amorphous. Stretched rubber, however, possesses a fair amount of crystallinity and order.

The elastic property of rubber is due to the flexibility of the long-chain molecules. In the bulk state, however, rubber is a tangle of polymeric chains, and if the external force is strong enough, individual chains slip past one another, thereby causing the rubber to lose most of its elasticity. In 1839, the American chemist Charles Goodyear discovered that natural rubber could be cross-linked with sulfur (using zinc oxide as the catalyst) to prevent chain slippage (Figure 22.3). His process, known as *vulcanization,* paved the way for many practical and commercial uses of rubber, such as in automobile tires and dentures.

During World War II a shortage of natural rubber in the United States prompted an intensive program to produce synthetic rubber. Most synthetic rubbers (called *elastomers*) are made from petroleum products such as ethylene, propene, and butadiene. For example, chloroprene molecules polymerize readily to form polychloroprene, commonly known as *neoprene,* which has properties that are comparable or even superior to those of natural rubber:

$$\underset{\text{chloroprene}}{H_2C{=}CCl{-}CH{=}CH_2} \qquad \underset{\text{polychloroprene}}{\left(\begin{array}{c} -CH_2 \quad\quad H \\ \;\;\diagdown\;\;\;\diagup \\ C{=}C \\ \;\;\diagup\;\;\;\diagdown \\ Cl \quad\quad CH_2- \end{array} \right)_n}$$

Another important synthetic rubber is formed by the addition of butadiene to styrene in a 3:1 ratio to give styrene-butadiene rubber (SBR). Because styrene and butadiene are different monomers, SBR is called a ***copolymer,*** which is *a polymer containing*

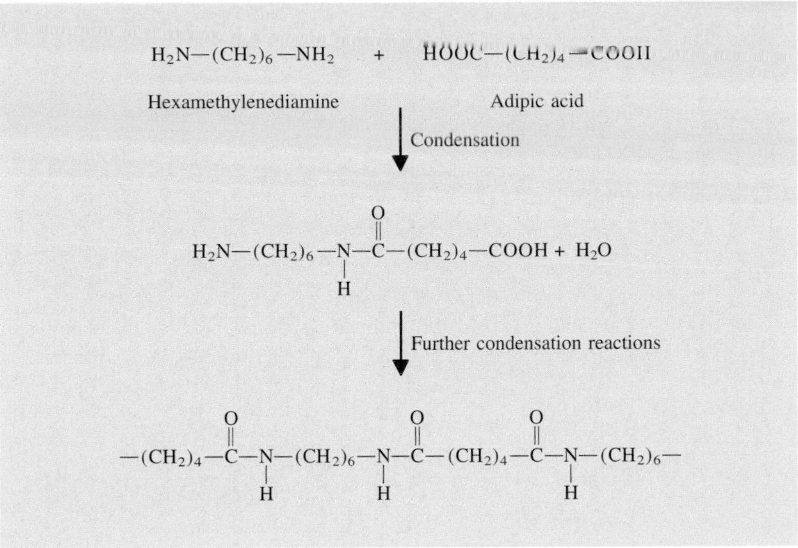

$$H_2N{-}(CH_2)_6{-}NH_2 \quad + \quad HOOC{-}(CH_2)_4{-}COOH$$

Hexamethylenediamine — Adipic acid

$$\downarrow \text{Condensation}$$

$$H_2N{-}(CH_2)_6{-}\underset{\substack{|\\H}}{N}{-}\overset{\substack{O\\\|}}{C}{-}(CH_2)_4{-}COOH + H_2O$$

$$\downarrow \text{Further condensation reactions}$$

$$-(CH_2)_4{-}\overset{\substack{O\\\|}}{C}{-}\underset{\substack{|\\H}}{N}{-}(CH_2)_6{-}\underset{\substack{|\\H}}{N}{-}\overset{\substack{O\\\|}}{C}{-}(CH_2)_4{-}\overset{\substack{O\\\|}}{C}{-}\underset{\substack{|\\H}}{N}{-}(CH_2)_6-$$

Figure 22.4
The formation of nylon by the condensation reaction between hexamethylenediamine and adipic acid.

Figure 22.5
The nylon rope trick. Adding a solution of adipoyl chloride (an adipic acid derivative in which the OH groups have been replaced by Cl groups) in cyclohexane to an aqueous solution of hexamethylenediamine causes nylon to form at the interface of the two solutions, which do not mix. It can then be drawn off.

two or more different monomers. Table 22.1 shows a number of common and familiar homopolymers and one copolymer produced by addition reactions.

Condensation Reactions

Condensation reaction was defined on p. 384.

One of the best-known polymer condensation processes is the reaction between hexamethylenediamine and adipic acid, shown in Figure 22.4. The final product, called nylon 66 (because there are six carbon atoms each in hexamethylenediamine and adipic acid), was first made by the American chemist Wallace Carothers at DuPont in 1931. The versatility of nylons is so great that the annual production of nylons and related substances now amounts to several billion pounds. Figure 22.5 shows how nylon 66 is prepared in the laboratory.

Condensation reactions are also used in the manufacture of Dacron (polyester)

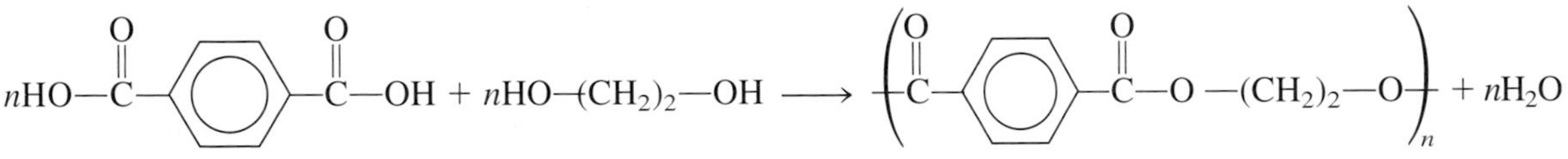

$$n\text{HO}{-}\overset{\substack{O\\\|}}{C}{-}C_6H_4{-}\overset{\substack{O\\\|}}{C}{-}\text{OH} + n\text{HO}{-}(CH_2)_2{-}\text{OH} \longrightarrow \left(\overset{\substack{O\\\|}}{C}{-}C_6H_4{-}\overset{\substack{O\\\|}}{C}{-}O{-}(CH_2)_2{-}O \right)_n + nH_2O$$

terephthalic acid — 1,2-ethylene glycol — Dacron

Polyesters are used in fibers, films, and plastic bottles.

Review of Concepts

Food wraps such as Saran were originally formulated using polyvinylidene chloride, whose general structure is shown here:

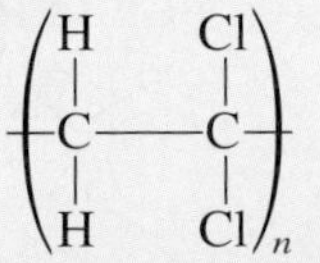

$$\left(\begin{array}{cc} H & Cl \\ | & | \\ -C- & -C- \\ | & | \\ H & Cl \end{array} \right)_n$$

Draw the structure of the monomer used to prepare this polymer. Is this a condensation or addition polymer?

Table 22.1 Some Monomers and Their Common Synthetic Polymers

Monomer		Polymer	
Formula	**Name**	**Name and Formula**	**Uses**
$H_2C{=}CH_2$	Ethylene	Polyethylene $-\!(CH_2{-}CH_2)\!-_n$	Plastic piping, bottles, electrical insulation, toys
$H_2C{=}C(H)(CH_3)$	Propene	Polypropene $\left(CH(CH_3){-}CH_2{-}CH(CH_3){-}CH_2\right)_n$	Packaging film, carpets, crates for soft-drink bottles, lab wares, toys
$H_2C{=}C(H)(Cl)$	Vinyl chloride	Poly(vinyl chloride) (PVC) $-\!(CH_2{-}CH(Cl))\!-_n$	Piping, siding, gutters, floor tile, clothing, toys
$H_2C{=}C(H)(CN)$	Acrylonitrile	Polyacrylonitrile (PAN) $\left(CH_2{-}CH(CN)\right)_n$	Carpets, knitwear
$F_2C{=}CF_2$	Tetrafluoro-ethylene	Polytetrafluoroethylene (Teflon) $-\!(CF_2{-}CF_2)\!-_n$	Coating on cooking utensils, electrical insulation, bearings
$H_2C{=}C(COOCH_3)(CH_3)$	Methyl methacrylate	Poly(methyl methacrylate) (Plexiglas) $-\!(CH_2{-}C(COOCH_3)(CH_3))\!-_n$	Optical equipment, home furnishing
$H_2C{=}C(H)(C_6H_5)$	Styrene	Polystyrene $-\!(CH_2{-}CH(C_6H_5))\!-_n$	Containers, thermal insulation (ice buckets, water coolers), toys
$H_2C{=}C(H){-}C(H){=}CH_2$	Butadiene	Polybutadiene $-\!(CH_2CH{=}CHCH_2)\!-_n$	Tire tread, coating resin
See above structures	Butadiene and styrene	Styrene-butadiene rubber (SBR) $-\!(CH(C_6H_5){-}CH_2{-}CH_2{-}CH{=}CH{-}CH_2)\!-_n$	Synthetic rubber

Bubble gums contain synthetic styrene-butadiene rubber.

22.3 Proteins

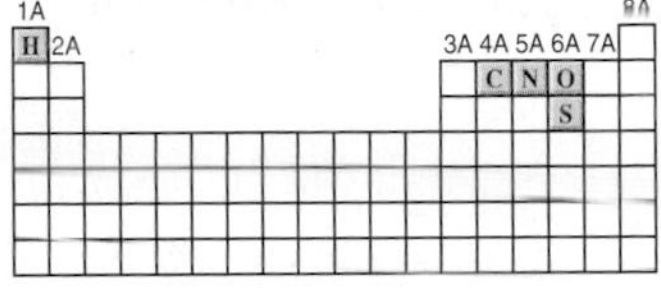

Elements in proteins.

Proteins are *polymers of amino acids;* they play a key role in nearly all biological processes. Enzymes, the catalysts of biochemical reactions, are mostly proteins. Proteins also facilitate a wide range of other functions, such as transport and storage of vital substances, coordinated motion, mechanical support, and protection against diseases. The human body contains an estimated 100,000 different kinds of proteins, each of which has a specific physiological function. As we will see in this section, the chemical composition and structure of these complex natural polymers are the basis of their specificity.

Amino Acids

Proteins have high molar masses, ranging from about 5000 g to 1×10^7 g, and yet the percent composition by mass of the elements in proteins is remarkably constant: carbon, 50 to 55 percent; hydrogen, 7 percent; oxygen, 23 percent; nitrogen, 16 percent; and sulfur, 1 percent.

The basic structural units of proteins are *amino acids.* An ***amino acid*** is *a compound that contains at least one amino group (—NH_2) and at least one carboxyl group (—COOH):*

$$-N(H)(H) \qquad\qquad -C(=O)-O-H$$

amino group   carboxyl group

Twenty different amino acids are the building blocks of all the proteins in the human body. Table 22.2 shows the structures of these vital compounds, along with their three-letter abbreviations.

Amino acids in solution at neutral pH exist as *dipolar ions,* meaning that the proton on the carboxyl group has migrated to the amino group. Consider glycine, the simplest amino acid. The un-ionized form and the dipolar ion of glycine are shown here:

$$H-\underset{H}{\overset{NH_2}{C}}-COOH \qquad\qquad H-\underset{H}{\overset{NH_3^+}{C}}-COO^-$$

un-ionized form   dipolar ion

The first step in the synthesis of a protein molecule is a condensation reaction between an amino group on one amino acid and a carboxyl group on another amino acid. The molecule formed from the two amino acids is called a *dipeptide,* and the bond joining them together is a *peptide bond:*

$$^{+}H_3N-\underset{R_1}{\overset{H}{C}}-\overset{O}{\overset{\|}{C}}-O^- + {}^{+}H_3N-\underset{R_2}{\overset{H}{C}}-\overset{O}{\overset{\|}{C}}-O^- \rightleftharpoons {}^{+}H_3N-\underset{R_1}{\overset{H}{C}}-\overset{O}{\overset{\|}{C}}-\underset{H}{N}-\underset{R_2}{\overset{H}{C}}-\overset{O}{\overset{\|}{C}}-O^- + H_2O$$

peptide bond

It is interesting to compare this reaction with the one shown in Figure 22.4.

where R_1 and R_2 represent a H atom or some other group; —CO—NH— is called the *amide group.* Because the equilibrium of the reaction joining two amino acids lies to the left, the process is coupled to the hydrolysis of ATP (see p. 652).

Table 22.2 The 20 Amino Acids Essential to Living Organisms*

Name	Abbreviation	Structure
Alanine	Ala	$H_3C-CH(NH_3^+)-COO^-$
Arginine	Arg	$H_2N-C(=NH)-NH-CH_2-CH_2-CH_2-CH(NH_3^+)-COO^-$
Asparagine	Asn	$H_2N-C(=O)-CH_2-CH(NH_3^+)-COO^-$
Aspartic acid	Asp	$HOOC-CH_2-CH(NH_3^+)-COO^-$
Cysteine	Cys	$HS-CH_2-CH(NH_3^+)-COO^-$
Glutamic acid	Glu	$HOOC-CH_2-CH_2-CH(NH_3^+)-COO^-$
Glutamine	Gln	$H_2N-C(=O)-CH_2-CH_2-CH(NH_3^+)-COO^-$
Glycine	Gly	$H-CH(NH_3^+)-COO^-$
Histidine	His	$HC{=}C(-N{=}CH-NH-)-CH_2-CH(NH_3^+)-COO^-$ (imidazole ring: HC=C, N, NH, C–H)
Isoleucine	Ile	$H_3C-CH_2-CH(CH_3)-CH(NH_3^+)-COO^-$

(Continued)

*The shaded portion is the R group of the amino acid. The amino acids are shown as dipolar ions.

Table 22.2 The 20 Amino Acids Essential to Living Organisms—Cont.

Name	Abbreviation	Structure
Leucine	Leu	$(H_3C)_2CH{-}CH_2{-}CH(NH_3^+){-}COO^-$
Lysine	Lys	$H_2N{-}CH_2{-}CH_2{-}CH_2{-}CH_2{-}CH(NH_3^+){-}COO^-$
Methionine	Met	$H_3C{-}S{-}CH_2{-}CH_2{-}CH(NH_3^+){-}COO^-$
Phenylalanine	Phe	$C_6H_5{-}CH_2{-}CH(NH_3^+){-}COO^-$
Proline	Pro	$H_2\overset{+}{N}{-}CH(COO^-)$, ring: $H_2C{-}CH_2{-}CH_2$
Serine	Ser	$HO{-}CH_2{-}CH(NH_3^+){-}COO^-$
Threonine	Thr	$H_3C{-}CH(OH){-}CH(NH_3^+){-}COO^-$
Tryptophan	Trp	indole $C{-}CH_2{-}CH(NH_3^+){-}COO^-$ (indole ring: C=CH, N–H)
Tyrosine	Tyr	$HO{-}C_6H_4{-}CH_2{-}CH(NH_3^+){-}COO^-$
Valine	Val	$(H_3C)_2CH{-}CH(NH_3^+){-}COO^-$

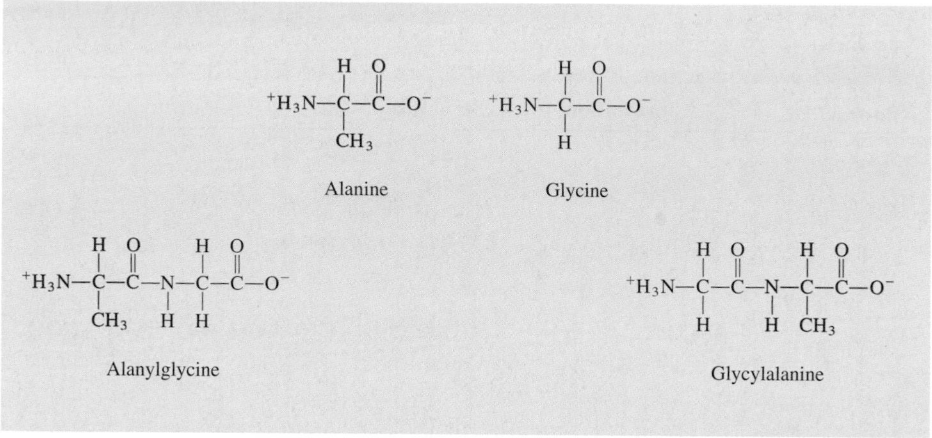

Figure 22.6
The formation of two dipeptides from two different amino acids. Alanylglycine is different from glycylalanine in that in alanylglycine the amino and methyl groups are bonded to the same carbon atom.

$$^{+}H_3N{-}\overset{H}{\underset{CH_3}{C}}{-}\overset{O}{\overset{\|}{C}}{-}O^- \qquad ^{+}H_3N{-}\overset{H}{\underset{H}{C}}{-}\overset{O}{\overset{\|}{C}}{-}O^-$$

Alanine Glycine

$$^{+}H_3N{-}\overset{H}{\underset{CH_3}{C}}{-}\overset{O}{\overset{\|}{C}}{-}\underset{H}{N}{-}\overset{H}{\underset{H}{C}}{-}\overset{O}{\overset{\|}{C}}{-}O^- \qquad ^{+}H_3N{-}\overset{H}{\underset{H}{C}}{-}\overset{O}{\overset{\|}{C}}{-}\underset{H}{N}{-}\overset{H}{\underset{CH_3}{C}}{-}\overset{O}{\overset{\|}{C}}{-}O^-$$

Alanylglycine Glycylalanine

Either end of a dipeptide can engage in a condensation reaction with another amino acid to form a *tripeptide,* a *tetrapeptide,* and so on. The final product, the protein molecule, is a *polypeptide;* it can also be thought of as a polymer of amino acids.

An amino acid unit in a polypeptide chain is called a *residue.* Typically, a polypeptide chain contains 100 or more amino acid residues. The sequence of amino acids in a polypeptide chain is written conventionally from left to right, starting with the amino-terminal residue and ending with the carboxyl-terminal residue. Let us consider a dipeptide formed from glycine and alanine. Figure 22.6 shows that alanylglycine and glycylalanine are different molecules. With 20 different amino acids to choose from, 20^2, or 400, different dipeptides can be generated. Even for a very small protein such as insulin, which contains only 50 amino acid residues, the number of chemically different structures that is possible is of the order of 20^{50} or 10^{65}! This is an incredibly large number when you consider that the total number of atoms in our galaxy is about 10^{68}. With so many possibilities for protein synthesis, it is remarkable that generation after generation of cells can produce identical proteins for specific physiological functions.

Protein Structure

The type and number of amino acids in a given protein along with the sequence or order in which these amino acids are joined together determine the protein's structure. In the 1930s Linus Pauling and his coworkers conducted a systematic investigation of protein structure. First they studied the geometry of the basic repeating group, that is, the amide group, which is represented by the following resonance structures:

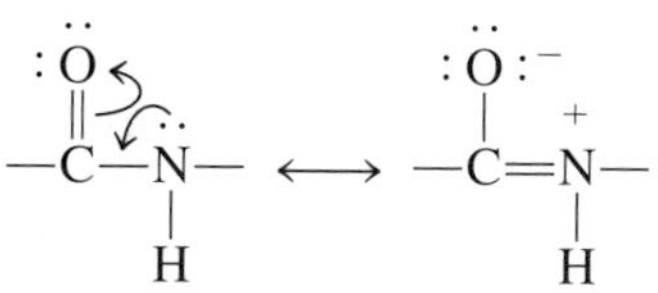

$$-\overset{:\ddot{O}}{\overset{\|}{C}}-\underset{H}{\ddot{N}}- \longleftrightarrow -\overset{:\ddot{O}:^-}{\overset{|}{C}}=\underset{H}{\overset{+}{N}}-$$

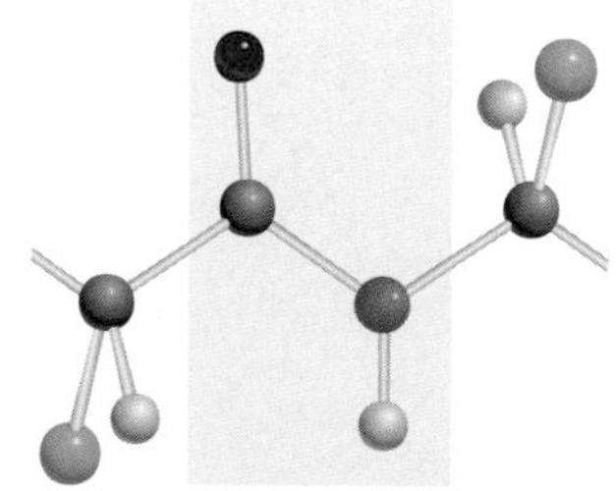

Figure 22.7
The planar amide group in protein. Rotation about the peptide bond in the amide group is hindered by its double-bond character. The black atoms represent carbon; blue, nitrogen; red, oxygen; green, R group; and gray, hydrogen.

Because it is more difficult (that is, it would take more energy) to twist a double bond than a single bond, the four atoms in the amide group become locked in the same plane (Figure 22.7). Figure 22.8 depicts the repeating amide group in a polypeptide chain.

On the basis of models and X-ray diffraction data, Pauling deduced that there are two common structures for protein molecules, called the *α helix* and the *β-pleated* sheet. The α-helical structure of a polypeptide chain is shown in Figure 22.9. The helix is stabilized by *intramolecular* hydrogen bonds between the NH and CO groups of the main chain, giving rise to an overall rodlike shape. The CO group of each amino acid

```
 H  O      H  O      H  O      H  O
 |  ||     |  ||     |  ||     |  ||
—C——C——N——C——C——N——C——C——N——C——C——N—
 |         |  |      |  |      |  |      |
 R         H  R      H  R      H  H      H
```

Figure 22.8

A polypeptide chain. Note the repeating units of the amide group. The symbol R represents part of the structure characteristic of the individual amino acids. For glycine, R is simply a H atom.

is hydrogen-bonded to the NH group of the amino acid that is four residues away in the sequence. In this manner all the main-chain CO and NH groups take part in hydrogen bonding. X-ray studies have shown that the structure of a number of proteins, including myoglobin and hemoglobin, is to a great extent α-helical in nature.

The β-pleated structure is markedly different from the α helix in that it is like a sheet rather than a rod. The polypeptide chain is almost fully extended, and each chain forms many *intermolecular* hydrogen bonds with adjacent chains. Figure 22.10 shows the two different types of β-pleated structures, called *parallel* and *antiparallel.* Silk molecules possess the β structure. Because its polypeptide chains are already in extended form, silk lacks elasticity and extensibility, but it is quite strong due to the many intermolecular hydrogen bonds.

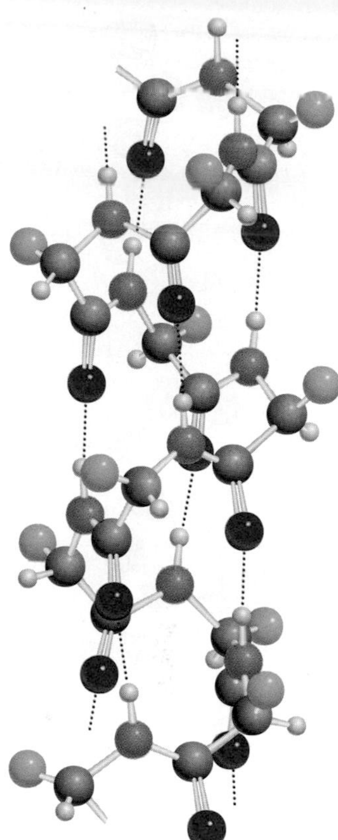

Figure 22.9

The α-helical structure of a polypeptide chain. The structure is held in position by intramolecular hydrogen bonds, shown as dotted lines. For color key, see Figure 22.7.

It is customary to divide protein structure into four levels of organization. The *primary structure* refers to the unique amino acid sequence of the polypeptide chain. The *secondary structure* includes those parts of the polypeptide chain that are stabilized by a regular pattern of hydrogen bonds between the CO and NH groups of the backbone, for example, the α helix. The term *tertiary structure* applies to the three-dimensional structure stabilized by dispersion forces, hydrogen bonding, and other intermolecular forces. It differs from secondary structure in that the amino acids taking part in these interactions may be far apart in the polypeptide chain. A protein molecule may be made up of more than one polypeptide chain. Thus, in addition to the various interactions *within* a chain that give rise to the secondary and tertiary structures, we must also consider the interaction *between* chains. The overall arrangement of the polypeptide chains is called the *quaternary structure.* For example, the hemoglobin molecule consists of four separate polypeptide chains, or *subunits.* These subunits are held together by van der Waals forces and ionic forces (Figure 22.11).

Intermolecular forces play an important role in the secondary, tertiary, and quaternary structure of proteins.

Pauling's work was a great triumph in protein chemistry. It showed for the first time how to predict a protein structure purely from a knowledge of the geometry of its fundamental building blocks—amino acids. However, there are many proteins whose structures do not correspond to the α-helical or β structure. Chemists now know that the three-dimensional structures of these biopolymers are maintained by several types of intermolecular forces in addition to hydrogen bonding (Figure 22.12). The delicate balance of the various interactions can be appreciated by considering an example: When glutamic acid, one of the amino acid residues in two of the four polypeptide chains in hemoglobin, is replaced by valine, another amino acid, the protein molecules aggregate to form insoluble polymers, causing the disease known as sickle cell anemia.

In spite of all the forces that give proteins their structural stability, most proteins have a certain amount of flexibility. Enzymes, for example, are flexible enough to change their geometry to fit substrates of various sizes and shapes. Another interesting example of protein flexibility is found in the binding of hemoglobin to oxygen. Each of the four polypeptide chains in hemoglobin contains a heme group that can bind to an oxygen molecule (see Section 20.6). In deoxyhemoglobin, the affinity of each of

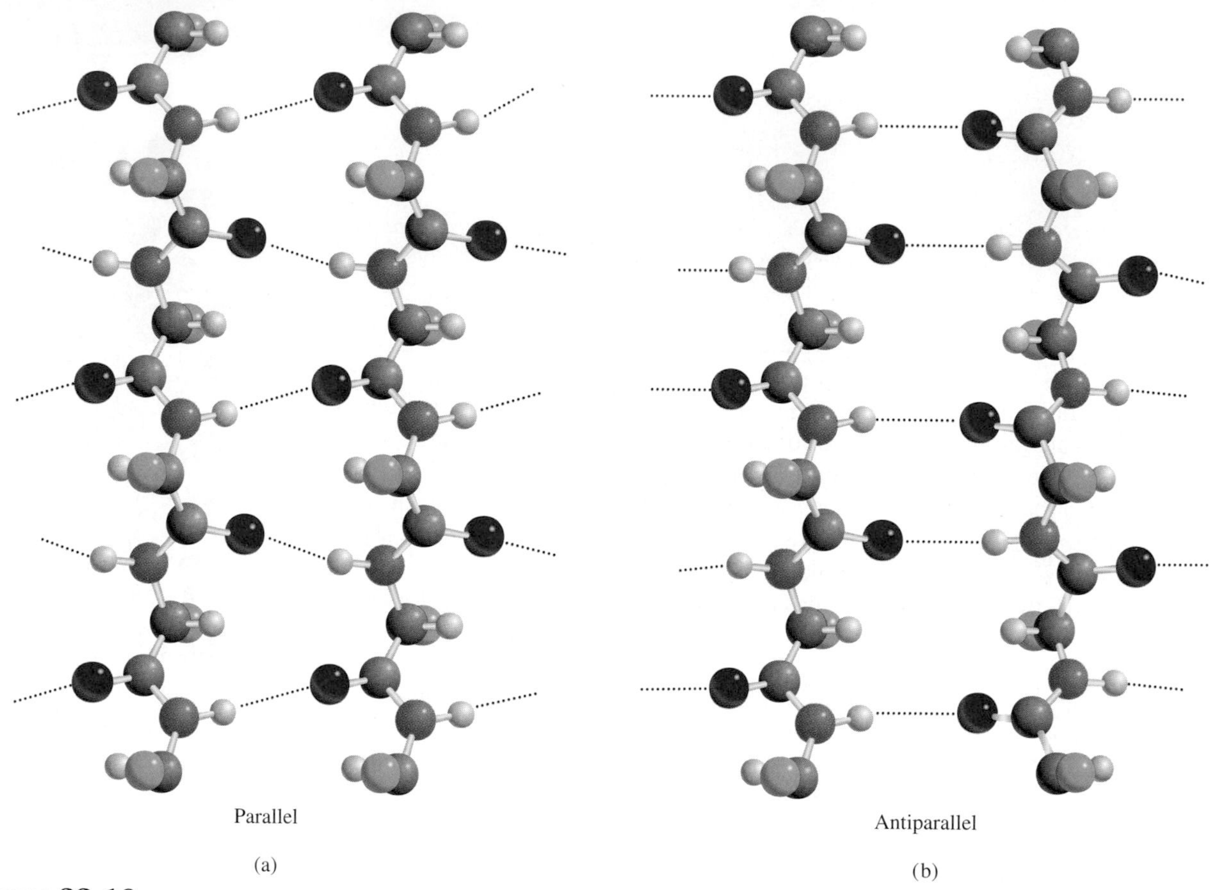

Figure 22.10
Hydrogen bonds (a) in a parallel β-pleated sheet structure, in which all the polypeptide chains are oriented in the same direction, and (b) in an antiparallel β-pleated sheet, in which adjacent polypeptide chains run in opposite directions. For color key, see Figure 22.7.

the heme groups for oxygen is about the same. However, as soon as one of the heme groups becomes oxygenated, the affinity of the other three hemes for oxygen is greatly enhanced. This phenomenon, called *cooperativity,* makes hemoglobin a particularly suitable substance for the uptake of oxygen in the lungs. By the same token, once a fully oxygenated hemoglobin molecule releases an oxygen molecule (to myoglobin in the tissues), the other three oxygen molecules will depart with increasing ease. The cooperative nature of the binding is such that information about the presence (or absence) of oxygen molecules is transmitted from one subunit to another along the polypeptide chains, a process made possible by the flexibility of the three-dimensional structure (Figure 22.13). It is believed that the Fe^{2+} ion has too large a radius to fit into the porphyrin ring of deoxyhemoglobin. When O_2 binds to Fe^{2+}, however, the ion shrinks somewhat so that it can fit into the plane of the ring. As the ion slips into the ring, it pulls the histidine residue toward the ring and thereby sets off a sequence of structural changes from one subunit to another. Although the details of the changes are not clear, biochemists believe that this is how the binding of an oxygen molecule to one heme group affects another heme group. The structural changes drastically affect the affinity of the remaining heme groups for oxygen molecules.

Hard-boiling an egg denatures the proteins in the egg white.

When proteins are heated above body temperature or when they are subjected to unusual acid or base conditions or treated with special reagents called *denaturants,* they lose some or all of their tertiary and secondary structure. Called ***denatured proteins,*** proteins in this state *no longer exhibit normal biological*

Ala
His
Val
Pro
Primary structure
Secondary structure
Tertiary structure
Quaternary structure

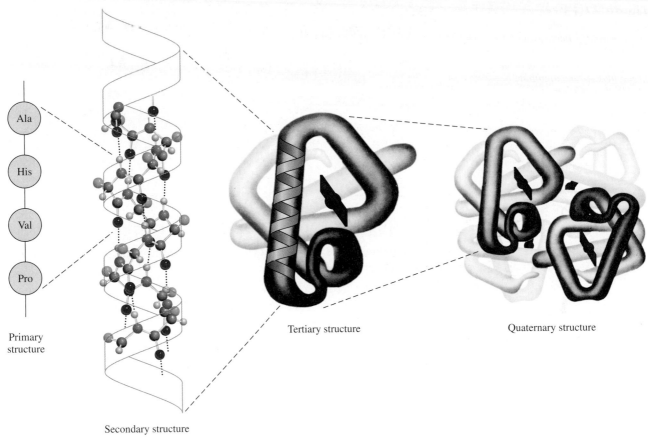

Figure 22.11
The primary, secondary, tertiary, and quaternary structure of the hemoglobin molecule.

activities. Figure 22.14 shows the variation of rate with temperature for a typical enzyme-catalyzed reaction. Initially, the rate increases with increasing temperature, as we would expect. Beyond the optimum temperature, however, the enzyme begins to denature and the rate falls rapidly. If a protein is denatured under mild conditions, its original structure can often be regenerated by removing the denaturant or by restoring the temperature to normal conditions. This process is called *reversible denaturation.*

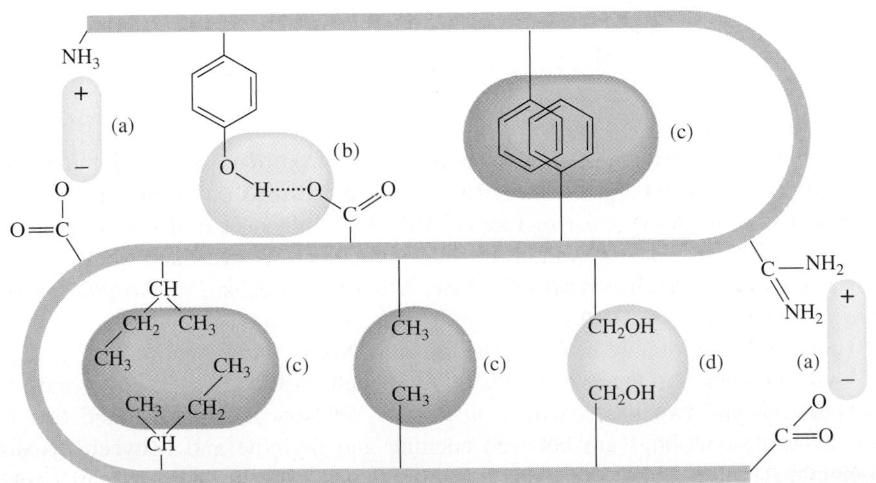

Figure 22.12
Intermolecular forces in a protein molecule: (a) ionic forces, (b) hydrogen bonding, (c) dispersion forces, and (d) dipole-dipole forces.

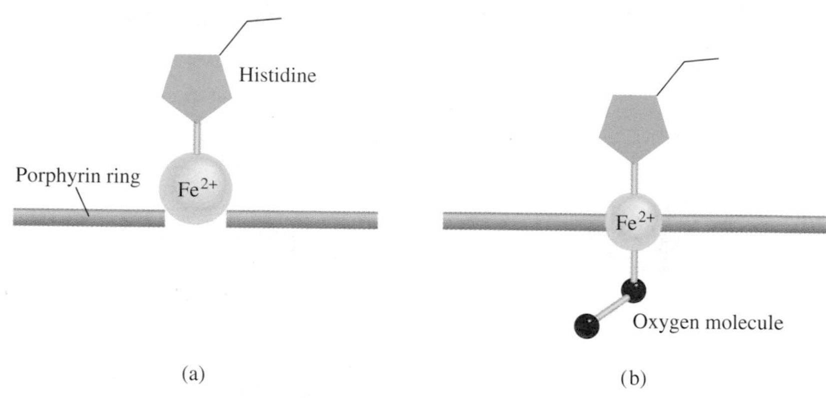

Figure 22.13
The structural changes that occur when the heme group in hemoglobin binds to an oxygen molecule. (a) The heme group in deoxyhemoglobin. (b) Oxyhemoglobin.

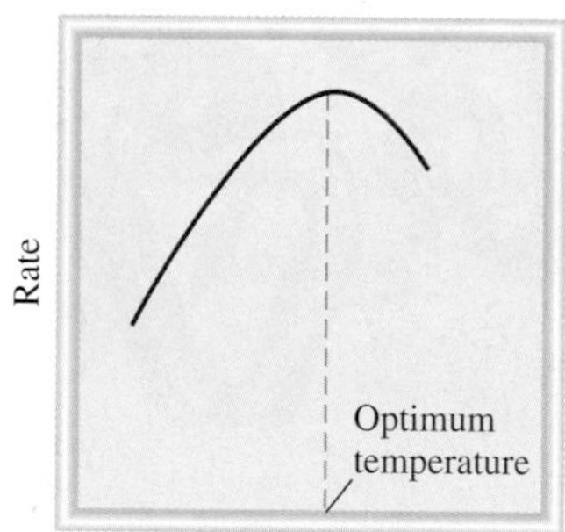

Figure 22.14
Dependence of the rate of an enzyme-catalyzed reaction on temperature. Above the optimum temperature at which an enzyme is most effective, its activity drops off as a consequence of denaturation.

REVIEW OF CONCEPTS

Draw the structures of the dipeptides that can be formed from the reaction between the amino acids cysteine and serine.

22.4 Nucleic Acids

Nucleic acids are *high-molar-mass polymers that play an essential role in protein synthesis.* ***Deoxyribonucleic acid (DNA)*** and ***ribonucleic acid (RNA)*** are *the two types of nucleic acid.* DNA molecules are among the largest molecules known; they have molar masses of up to tens of billions of grams. On the other hand, RNA molecules vary greatly in size, some having a molar mass of about 25,000 g. Compared with proteins, which are made of up to 20 different amino acids, nucleic acids are fairly simple in composition. A DNA or RNA molecule contains only four types of building blocks: purines, pyrimidines, furanose sugars, and phosphate groups (Figure 22.15). Each purine or pyrimidine is called a *base.*

In the 1940s, the American biochemist Erwin Chargaff studied DNA molecules obtained from various sources and observed certain regularities. *Chargaff's rules,* as his findings are now known, describe these patterns:

An electron micrograph of a DNA molecule. The double-helical structure is evident.

If the DNA molecules from all the cells in a human were stretched and joined end to end, the length would be about 100 times the distance to the sun!

1. The amount of adenine (a purine) is equal to that of thymine (a pyrimidine); that is, A = T, or A/T = 1.
2. The amount of cytosine (a pyrimidine) is equal to that of guanine (a purine); that is, C = G, or C/G = 1.
3. The total number of purine bases is equal to the total number of pyrimidine bases; that is, A + G = C + T.

Based on chemical analyses and information obtained from X-ray diffraction measurements, the American biologist James Watson and the British biologist Francis Crick formulated the double-helical structure for the DNA molecule in 1953. Watson and Crick determined that the DNA molecule has two helical strands. Each strand is made up of ***nucleotides,*** which *consist of a base, a deoxyribose, and a phosphate group linked together* (Figure 22.16).

The key to the double-helical structure of DNA is the formation of hydrogen bonds between bases in the two strands of a molecule. Although hydrogen bonds can form between any two bases, called *base pairs,* Watson and Crick found that the most favorable couplings are between adenine and thymine and between cytosine and guanine (Figure 22.17). Note that this scheme is consistent with Chargaff's rules,

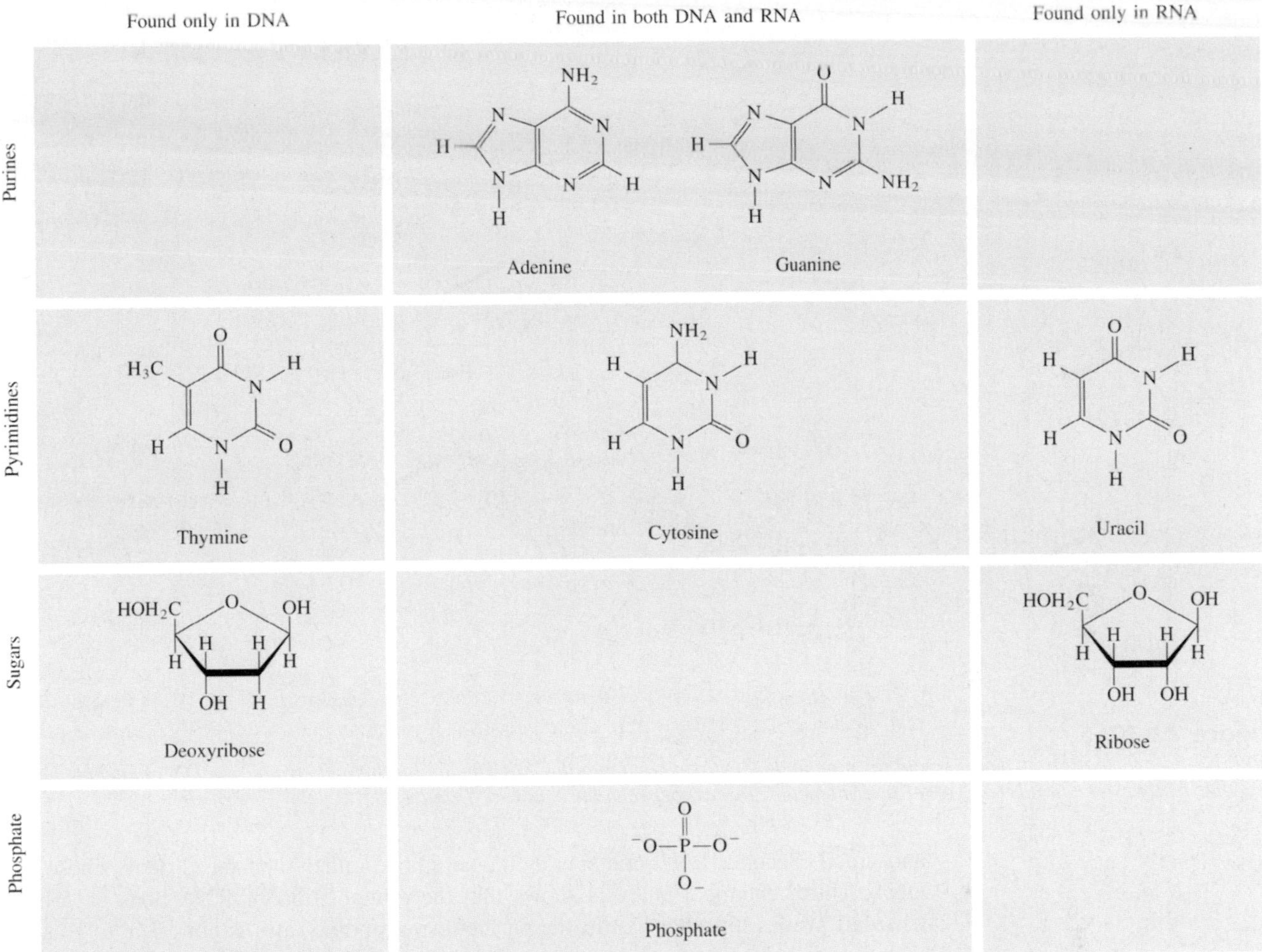

Figure 22.15
The components of the nucleic acids DNA and RNA.

because every purine base is hydrogen-bonded to a pyrimidine base, and vice versa (A + G = C + T). Other attractive forces such as dipole-dipole interactions and van der Waals forces between the base pairs also help to stabilize the double helix.

The structure of RNA differs from that of DNA in several respects. First, as shown in Figure 22.15, the four bases found in RNA molecules are adenine, cytosine, guanine,

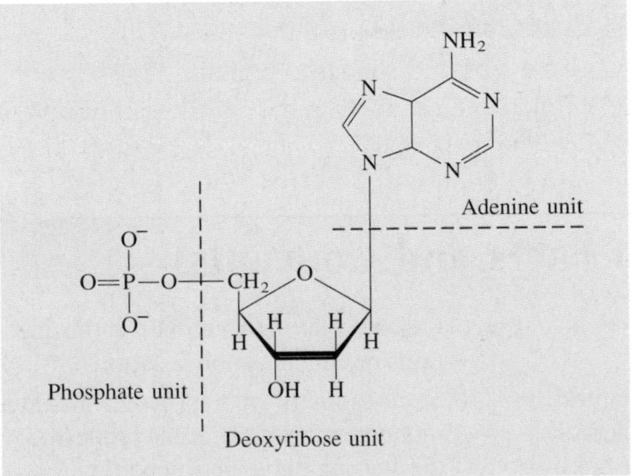

Figure 22.16
Structure of a nucleotide, one of the repeating units in DNA.

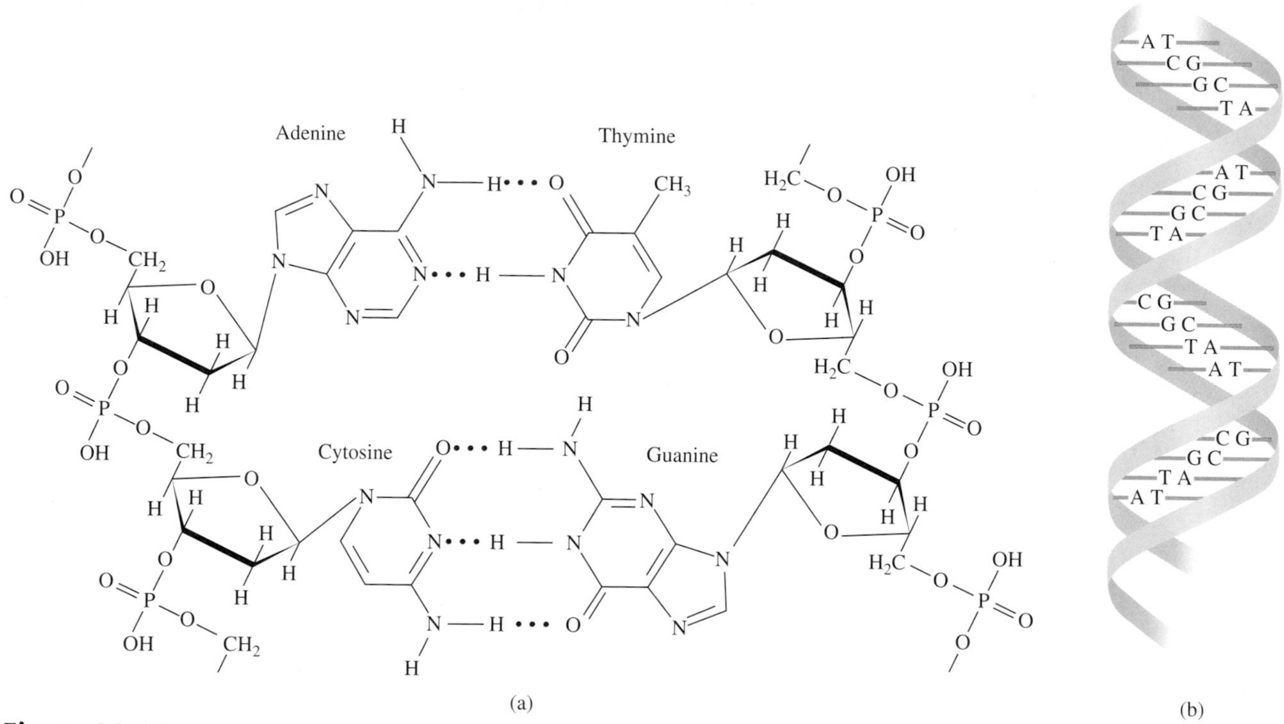

(a) (b)

Figure 22.17

(a) Base-pair formation by adenine and thymine and by cytosine and guanine. (b) The double-helical strand of a DNA molecule held together by hydrogen bonds (and other intermolecular forces) between base pairs A-T and C-G.

and uracil. Second, RNA contains the sugar ribose rather than the 2-deoxyribose of DNA. Third, chemical analysis shows that the composition of RNA does not obey Chargaff's rules. In other words, the purine-to-pyrimidine ratio is not equal to 1 as in the case of DNA. This and other evidence rule out a double-helical structure. In fact, the RNA molecule exists as a single-strand polynucleotide. There are actually three types of RNA molecules—messenger RNA (*m*RNA), ribosomal RNA (*r*RNA), and transfer RNA (*t*RNA). These RNAs have similar nucleotides but differ from one another in molar mass, overall structure, and biological functions.

In the 1980s, chemists discovered that certain RNAs can function as enzymes.

DNA and RNA molecules direct the synthesis of proteins in the cell, a subject that is beyond the scope of this book. Introductory texts in biochemistry and molecular biology explain this process.

Review of Concepts

A DNA sample was analyzed and shown to contain 27.4 percent A and 22.6 percent C. Determine the percentages of G and T in the sample.

Summary of Facts and Concepts

1. Polymers are large molecules made up of small, repeating units called monomers.
2. Proteins, nucleic acids, cellulose, and rubber are natural polymers. Nylon, Dacron, and Lucite are examples of synthetic polymers.
3. Organic polymers can be synthesized via addition reactions or condensation reactions.
4. Stereoisomers of a polymer made up of asymmetric monomers have different properties, depending on how the starting units are joined together.

5. Synthetic rubbers include polychloroprene and styrene-butadiene rubber, which is a copolymer of styrene and butadiene.
6. Structure determines the function and properties of proteins. To a great extent, hydrogen bonding and other intermolecular forces determine the structure of proteins.
7. The primary structure of a protein is its amino acid sequence. Secondary structure is the shape defined by hydrogen bonds joining the CO and NH groups of the amino acid backbone. Tertiary and quaternary structures are the three-dimensional folded arrangements of proteins that are stabilized by hydrogen bonds and other intermolecular forces.
8. Nucleic acids—DNA and RNA—are high-molar-mass polymers that carry genetic instructions for protein synthesis in cells. Nucleotides are the building blocks of DNA and RNA. DNA nucleotides each contain a purine or pyrimidine base, a deoxyribose molecule, and a phosphate group. RNA nucleotides are similar but contain different bases and ribose instead of deoxyribose.

Key Words

Questions and Problems

Synthetic Organic Polymers

Review Questions

22.1 Define the following terms: monomer, polymer, homopolymer, copolymer.

22.2 Name 10 objects that contain synthetic organic polymers.

22.3 Calculate the molar mass of a particular polyethylene sample, $\text{-(}CH_2\text{—}CH_2\text{)-}_n$, where $n = 4600$.

22.4 Describe the two major mechanisms of organic polymer synthesis.

22.5 Discuss the three isomers of polypropene.

22.6 In Chapter 13 you learned about the colligative properties of solutions. Which of the colligative properties is suitable for determining the molar mass of a polymer? Why?

Problems

22.7 Teflon is formed by a radical addition reaction involving the monomer tetrafluoroethylene. Show the mechanism for this reaction.

22.8 Vinyl chloride, $H_2C\text{=}CHCl$, undergoes copolymerization with 1,1-dichloroethylene, $H_2C\text{=}CCl_2$, to form a polymer commercially known as Saran. Draw the structure of the polymer, showing the repeating monomer units.

22.9 Kevlar is a copolymer used in bullet-proof vests. It is formed in a condensation reaction between the following two monomers:

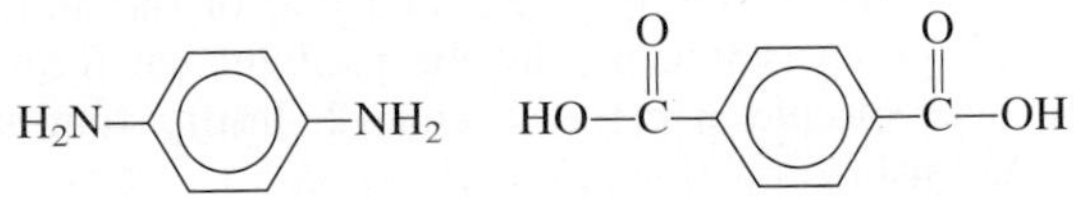

$H_2N\text{—}C_6H_4\text{—}NH_2$ $\quad$ $HO\text{—}\overset{O}{\overset{\|}{C}}\text{—}C_6H_4\text{—}\overset{O}{\overset{\|}{C}}\text{—}OH$

Sketch a portion of the polymer chain showing several monomer units. Write the overall equation for the condensation reaction.

22.10 Describe the formation of polystyrene.

22.11 Deduce plausible monomers for polymers with the following repeating units:
(a) $\text{-(}CH_2\text{—}CF_2\text{)-}_n$
(b) 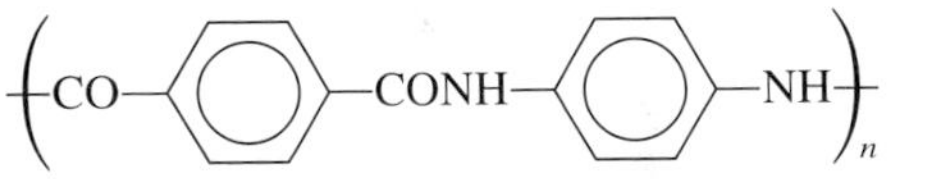 $\text{-(}CO\text{—}C_6H_4\text{—}CONH\text{—}C_6H_4\text{—}NH\text{)-}_n$

22.12 Deduce plausible monomers for polymers with the following repeating units:
(a) $\text{-(}CH_2\text{—}CH\text{=}CH\text{—}CH_2\text{)-}_n$
(b) $\text{-(}CO\text{-(}CH_2\text{)-}_6NH\text{)-}_n$

Proteins

Review Questions

22.13 Discuss the characteristics of an amide group and its importance in protein structure.

22.14 What is the α-helical structure in proteins?

22.15 Describe the β-pleated structure present in some proteins.

22.16 Discuss the main functions of proteins in living systems.

22.17 Briefly explain the phenomenon of cooperativity exhibited by the hemoglobin molecule in binding oxygen.

22.18 Why is sickle-cell anemia called a molecular disease?

Problems

22.19 Draw the structures of the dipeptides that can be formed from the reaction between the amino acids glycine and alanine.

22.20 Draw the structures of the dipeptides that can be formed from the reaction between the amino acids glycine and lysine.

22.21 The amino acid glycine can be condensed to form a polymer called polyglycine. Draw the repeating monomer unit.

22.22 The following are data obtained on the rate of product formation of an enzyme-catalyzed reaction:

Temperature (°C)	Rate of Product Formation (M/s)
10	0.0025
20	0.0048
30	0.0090
35	0.0086
45	0.0012

Comment on the dependence of rate on temperature. (No calculations are required.)

Nucleic Acids

Review Questions

22.23 Describe the structure of a nucleotide.

22.24 What is the difference between ribose and deoxyribose?

22.25 What are Chargaff's rules?

22.26 Describe the role of hydrogen bonding in maintaining the double-helical structure of DNA.

Additional Problems

22.27 Discuss the importance of hydrogen bonding in biological systems. Use proteins and nucleic acids as examples.

22.28 Proteins vary widely in structure, whereas nucleic acids have rather uniform structures. How do you account for this major difference?

22.29 If untreated, fevers of 104°F or higher may lead to brain damage. Why?

22.30 The "melting point" of a DNA molecule is the temperature at which the double-helical strand breaks apart. Suppose you are given two DNA samples. One sample contains 45 percent C-G base pairs while the other contains 64 percent C-G base pairs. The total number of bases is the same in each sample. Which of the two samples has a higher melting point? Why?

22.31 When fruits such as apples and pears are cut, the exposed parts begin to turn brown. This is the result of an oxidation reaction catalyzed by enzymes present in the fruit. Often the browning action can be prevented or slowed by adding a few drops of lemon juice to the exposed areas. What is the chemical basis for this treatment?

22.32 "Dark meat" and "white meat" are one's choices when eating a turkey. Explain what causes the meat to assume different colors. (*Hint:* The more active muscles in a turkey have a higher rate of metabolism and need more oxygen.)

22.33 Nylon can be destroyed easily by strong acids. Explain the chemical basis for the destruction. (*Hint:* The products are the starting materials of the polymerization reaction.)

22.34 Despite what you may have read in science fiction novels or seen in horror movies, it is extremely unlikely that insects can ever grow to human size. Why? (*Hint:* Insects do not have hemoglobin molecules in their blood.)

22.35 How many different tripeptides can be formed by lysine and alanine?

22.36 Chemical analysis shows that hemoglobin contains 0.34 percent Fe by mass. What is the minimum possible molar mass of hemoglobin? The actual molar mass of hemoglobin is four times this minimum value. What conclusion can you draw from these data?

22.37 The folding of a polypeptide chain depends not only on its amino acid sequence but also on the nature of the solvent. Discuss the types of interactions that might occur between water molecules and the amino acid residues of the polypeptide chain. Which groups would be exposed on the exterior of the protein in contact with water and which groups would be buried in the interior of the protein?

22.38 What kind of intermolecular forces are responsible for the aggregation of hemoglobin molecules that leads to sickle-cell anemia?

22.39 Draw structures of the nucleotides containing the following components: (a) deoxyribose and cytosine, (b) ribose and uracil.

22.40 When a nonapeptide (containing nine amino acid residues) isolated from rat brains was hydrolyzed, it gave the following smaller peptides as identifiable products: Gly-Ala-Phe, Ala-Leu-Val, Gly-Ala-Leu, Phe-Glu-His, and His-Gly-Ala. Reconstruct the amino acid sequence in the nonapeptide, giving your reasons. (Remember the convention for writing peptides.)

22.41 At neutral pH amino acids exist as dipolar ions. Using glycine as an example, and given that the pK_a of the carboxyl group is 2.3 and that of the ammonium group is 9.6, predict the predominant form of the molecule at pH 1, 7, and 12. Justify your answers using Equation (17.3).

22.42 In Lewis Carroll's tale "Through the Looking Glass," Alice wonders whether "looking-glass milk" on the other side of the mirror would be fit to drink. Based on your knowledge of chirality and enzyme action, comment on the validity of Alice's concern.

22.43 The enthalpy change in the denaturation of a certain protein is 125 kJ/mol. If the entropy change is 397 J/K · mol, calculate the minimum temperature at which the protein would denature spontaneously.

22.44 When deoxyhemoglobin crystals are exposed to oxygen, they shatter. On the other hand, deoxymyoglobin crystals are unaffected by oxygen. Explain. (Myoglobin is made up of only one of the four subunits, or polypeptide chains, in hemoglobin.)

Special Problems

22.45 Nylon was designed to be a synthetic silk. (a) The average molar mass of a batch of nylon 66 is 12,000 g/mol. How many monomer units are there in this sample? (b) Which part of nylon's structure is similar to a polypeptide's structure? (c) How many different tripeptides (made up of three amino acids) can be formed from the amino acids alanine (Ala), glycine (Gly), and serine (Ser), which account for most of the amino acids in silk?

22.46 In protein synthesis, the selection of a particular amino acid is determined by the so-called genetic code, or a sequence of three bases in DNA. Will a sequence of only two bases unambiguously determine the selection of 20 amino acids found in proteins? Explain.

22.47 Consider the fully protonated amino acid valine:

```
          +9.62
     CH3 NH3
      |   |          2.32
  H—C—C—COOH
      |   |
     CH3 H
```

where the numbers denote the pK_a values. (a) Which of the two groups ($-\overset{+}{N}H_3$ or $-COOH$) is more acidic? (b) Calculate the predominant form of valine at pH 1.0, 7.0, and 12.0. (c) Calculate the isoelectric point of valine. (*Hint:* See Problem 17.113.)

22.48 Consider the formation of a dimeric protein

$$2P \longrightarrow P_2$$

At 25°C, we have $\Delta H° = 17$ kJ/mol and $\Delta S° = 65$ J/K · mol. Is the dimerization favored at this temperature? Comment on the effect of lowering the temperature. Does your result explain why some enzymes lose their activities under cold conditions?

22.49 The following diagram shows the structure of the enzyme ribonuclease in its native form. The three-dimensional protein structure is maintained in part by the disulfide bonds (—S—S—) between the amino acid residues (each color sphere represents an S atom). Using certain denaturants, the compact structure is destroyed and the disulfide bonds are converted to sulfhydryl groups (—SH) shown on the right of the arrow. (a) Describe the bonding scheme in the disulfide bond in terms of hybridization. (b) Which amino acid in Table 22.2 contains the —SH group? (c) Predict the signs of ΔH and ΔS for the denaturation process. If denaturation is induced by a change in temperature, show why a rise in temperature would favor denaturation. (d) The sulfhydryl groups can be oxidized (that is, removing the H atoms) to form the disulfide bonds. If the formation of the disulfide bonds is totally random between any two —SH groups, what is the fraction of the regenerated protein structures that corresponds to the native form? (e) An effective remedy to deodorize a dog that has been sprayed by a skunk is to rub the affected areas with a solution of an oxidizing agent such as hydrogen peroxide. What is the chemical basis for this action? (*Hint:* An odiferous component of a skunk's secretion is 2-butene-1-thiol, $CH_3CH{=}CHCH_2SH$.)

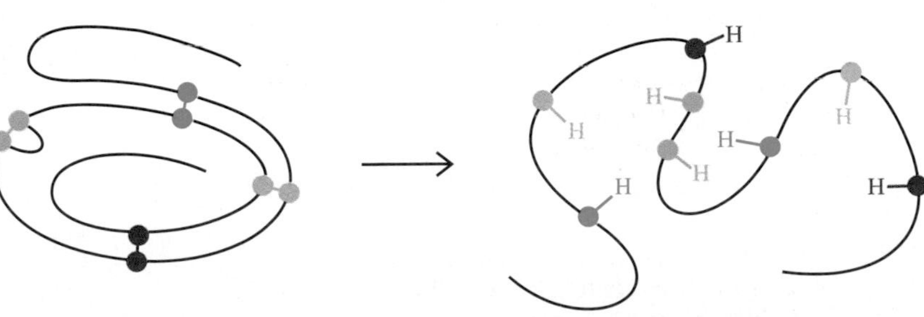

Native form

Denatured form

Appendix 1

Units for the Gas Constant

In this appendix, we will see how the gas constant R can be expressed in units J/K · mol. Our first step is to derive a relationship between atm and pascal. We start with

$$\begin{aligned}\text{pressure} &= \frac{\text{force}}{\text{area}} \\ &= \frac{\text{mass} \times \text{acceleration}}{\text{area}} \\ &= \frac{\text{volume} \times \text{density} \times \text{acceleration}}{\text{area}} \\ &= \text{length} \times \text{density} \times \text{acceleration}\end{aligned}$$

By definition, the standard atmosphere is the pressure exerted by a column of mercury exactly 76 cm high of density 13.5951 g/cm^3, in a place where acceleration due to gravity is 980.665 cm/s^2. However, to express pressure in N/m^2 it is necessary to write

$$\begin{aligned}\text{density of mercury} &= 1.35951 \times 10^4 \text{ kg/m}^3 \\ \text{acceleration due to gravity} &= 9.80665 \text{ m/s}^2\end{aligned}$$

The standard atmosphere is given by

$$\begin{aligned}1 \text{ atm} &= (0.76 \text{ m Hg})(1.35951 \times 10^4 \text{ kg/m}^3)(9.80665 \text{ m/s}^2) \\ &= 101{,}325 \text{ kg m/m}^2 \cdot \text{s}^2 \\ &= 101{,}325 \text{ N/m}^2 \\ &= 101{,}325 \text{ Pa}\end{aligned}$$

From Section 5.4 we see that the gas constant R is given by 0.082057 L · atm/K · mol. Using the conversion factors

$$\begin{aligned}1 \text{ L} &= 1 \times 10^{-3} \text{ m}^3 \\ 1 \text{ atm} &= 101{,}325 \text{ N/m}^2\end{aligned}$$

we write

$$\begin{aligned}R &= \left(0.082057 \frac{\text{L atm}}{\text{K mol}}\right)\left(\frac{1 \times 10^{-3}\text{m}^3}{1 \text{ L}}\right)\left(\frac{101{,}325 \text{ N/m}^2}{1 \text{ atm}}\right) \\ &= 8.314 \frac{\text{N m}}{\text{K mol}} \\ &= 8.314 \frac{\text{J}}{\text{K mol}}\end{aligned}$$

and

$$\begin{aligned}1 \text{ L} \cdot \text{atm} &= (1 \times 10^{-3} \text{ m}^3)(101{,}325 \text{ N/m}^2) \\ &= 101.3 \text{ N m} \\ &= 101.3 \text{ J}\end{aligned}$$

Appendix 2

Selected Thermodynamic Data at 1 atm and 25°C*

Inorganic Substances

Substance	ΔH_f° (kJ/mol)	ΔG_f° (kJ/mol)	S° (J/K · mol)
$Ag^+(aq)$	105.9	77.1	73.9
$AgCl(s)$	−127.0	−109.7	96.1
$Al(s)$	0	0	28.3
$Al^{3+}(aq)$	−524.7	−481.2	−313.38
$Al_2O_3(s)$	−1669.8	−1576.4	50.99
$Br_2(l)$	0	0	152.3
$Br^-(aq)$	−120.9	−102.8	80.7
$HBr(g)$	−36.2	−53.2	198.48
C(graphite)	0	0	5.69
C(diamond)	1.90	2.87	2.4
$CO(g)$	−110.5	−137.3	197.9
$CO_2(g)$	−393.5	−394.4	213.6
$CO_2(aq)$	−412.9	−386.2	121.3
$CO_3^{2-}(aq)$	−676.3	−528.1	−53.1
$HCO_3^-(aq)$	−691.1	−587.1	94.98
$H_2CO_3(aq)$	−699.7	−623.2	187.4
$CS_2(g)$	115.3	65.1	237.8
$CS_2(l)$	87.3	63.6	151.0
$HCN(aq)$	105.4	112.1	128.9
$CN^-(aq)$	151.0	165.69	117.99
$(NH_2)_2CO(s)$	−333.19	−197.15	104.6
$(NH_2)_2CO(aq)$	−319.2	−203.84	173.85
$Ca(s)$	0	0	41.6
$Ca^{2+}(aq)$	−542.96	−553.0	−55.2
$CaO(s)$	−635.6	−604.2	39.8
$Ca(OH)_2(s)$	−986.6	−896.8	76.2
$CaF_2(s)$	−1214.6	−1161.9	68.87
$CaCl_2(s)$	−794.96	−750.19	113.8
$CaSO_4(s)$	−1432.69	−1320.3	106.69
$CaCO_3(s)$	−1206.9	−1128.8	92.9

*The ΔH_f°, ΔG_f°, and S° values for ions are based on the reference states $\Delta H_f^\circ(H^+) = 0$, $\Delta G_f^\circ(H^+) = 0$, and $S^\circ(H^+) = 0$.

(*Continued*)

Inorganic Substances—Cont.

Substance	ΔH_f° (kJ/mol)	ΔG_f° (kJ/mol)	S° (J/K · mol)
$Cl_2(g)$	0	0	223.0
$Cl^-(aq)$	−167.2	−131.2	56.5
$HCl(g)$	−92.3	−95.27	187.0
$Cu(s)$	0	0	33.3
$Cu^+(aq)$	51.88	50.2	40.6
$Cu^{2+}(aq)$	64.39	64.98	−99.6
$CuO(s)$	−155.2	−127.2	43.5
$Cu_2O(s)$	−166.69	−146.36	100.8
$CuS(s)$	−48.5	−49.0	66.5
$CuSO_4(s)$	−769.86	−661.9	113.39
$F_2(g)$	0	0	203.34
$F^-(aq)$	−329.1	−276.48	−9.6
$HF(g)$	−271.6	−270.7	173.5
$Fe(s)$	0	0	27.2
$Fe^{2+}(aq)$	−87.86	−84.9	−113.39
$Fe^{3+}(aq)$	−47.7	−10.5	−293.3
$FeO(s)$	−272.0	−255.2	60.8
$Fe_2O_3(s)$	−822.2	−741.0	90.0
$Fe(OH)_2(s)$	−568.19	−483.55	79.5
$Fe(OH)_3(s)$	−824.25	?	?
$H(g)$	218.2	203.2	114.6
$H_2(g)$	0	0	131.0
$H^+(aq)$	0	0	0
$OH^-(aq)$	−229.94	−157.30	−10.5
$H_2O(g)$	−241.8	−228.6	188.7
$H_2O(l)$	−285.8	−237.2	69.9
$H_2O_2(l)$	−187.6	−118.1	?
$I_2(s)$	0	0	116.7
$I^-(aq)$	55.9	51.67	109.37
$HI(g)$	25.9	1.30	206.3
$K(s)$	0	0	63.6
$K^+(aq)$	−251.2	−282.28	102.5
$KOH(s)$	−425.85	?	?
$KCl(s)$	−435.87	−408.3	82.68
$KClO_3(s)$	−391.20	−289.9	142.97
$KClO_4(s)$	−433.46	−304.18	151.0
$KBr(s)$	−392.17	−379.2	96.4
$KI(s)$	−327.65	−322.29	104.35
$KNO_3(s)$	−492.7	−393.1	132.9

(Continued)

Inorganic Substances—Cont.

Substance	ΔH_f° (kJ/mol)	ΔG_f° (kJ/mol)	S° (J/K · mol)
$Li(s)$	0	0	28.0
$Li^+(aq)$	−278.46	−293.8	14.2
$Li_2O(s)$	−595.8	?	?
$LiOH(s)$	−487.2	−443.9	50.2
$Mg(s)$	0	0	32.5
$Mg^{2+}(aq)$	−461.96	−456.0	−117.99
$MgO(s)$	−601.8	−569.6	26.78
$Mg(OH)_2(s)$	−924.66	−833.75	63.1
$MgCl_2(s)$	−641.8	−592.3	89.5
$MgSO_4(s)$	−1278.2	−1173.6	91.6
$MgCO_3(s)$	−1112.9	−1029.3	65.69
$N_2(g)$	0	0	191.5
$N_3^-(aq)$	245.18	?	?
$NH_3(g)$	−46.3	−16.6	193.0
$NH_4^+(aq)$	−132.80	−79.5	112.8
$NH_4Cl(s)$	−315.39	−203.89	94.56
$NH_3(aq)$	−80.3	−263.76	111.3
$N_2H_4(l)$	50.4	?	?
$NO(g)$	90.4	86.7	210.6
$NO_2(g)$	33.85	51.8	240.46
$N_2O_4(g)$	9.66	98.29	304.3
$N_2O(g)$	81.56	103.6	219.99
$HNO_3(aq)$	−207.4	−111.3	146.4
$Na(s)$	0	0	51.05
$Na^+(aq)$	−239.66	−261.87	60.25
$Na_2O(s)$	−415.89	−376.56	72.8
$NaCl(s)$	−411.0	−384.0	72.38
$NaI(s)$	−288.0	?	?
$Na_2SO_4(s)$	−1384.49	−1266.8	149.49
$NaNO_3(s)$	−466.68	−365.89	116.3
$Na_2CO_3(s)$	−1130.9	−1047.67	135.98
$NaHCO_3(s)$	−947.68	−851.86	102.09
$O(g)$	249.4	230.1	160.95
$O_2(g)$	0	0	205.0
$O_3(aq)$	−12.09	16.3	110.88
$O_3(g)$	142.2	163.4	237.6
P(white)	0	0	44.0
P(red)	−18.4	13.8	29.3
$PO_4^{3-}(aq)$	−1284.07	−1025.59	−217.57
$P_4O_{10}(s)$	−3012.48	?	?
$PH_3(g)$	9.25	18.2	210.0

(Continued)

Inorganic Substances—Cont.

Substance	ΔH_f° (kJ/mol)	ΔG_f° (kJ/mol)	S° (J/K · mol)
$HPO_4^{2-}(aq)$	−1298.7	−1094.1	−35.98
$H_2PO_4^-(aq)$	−1302.48	−1135.1	89.1
S(rhombic)	0	0	31.88
S(monoclinic)	0.30	0.10	32.55
$SO_2(g)$	−296.1	−300.4	248.5
$SO_3(g)$	−395.2	−370.4	256.2
$SO_3^{2-}(aq)$	−624.25	−497.06	43.5
$SO_4^{2-}(aq)$	−907.5	−741.99	17.15
$H_2S(g)$	−20.15	−33.0	205.64
$HSO_3^-(aq)$	−627.98	−527.3	132.38
$HSO_4^-(aq)$	−885.75	−752.87	126.86
$H_2SO_4(l)$	−811.3	−690.0	156.9
$H_2SO_4(aq)$	−909.3	−744.5	20.1
$SF_6(g)$	−1096.2	−1105.3	291.8
$Zn(s)$	0	0	41.6
$Zn^{2+}(aq)$	−152.4	−147.2	−112.1
$ZnO(s)$	−348.0	−318.2	43.9
$ZnCl_2(s)$	−415.89	−369.26	108.37
$ZnS(s)$	−202.9	−198.3	57.7
$ZnSO_4(s)$	−978.6	−871.6	124.7

Organic Substances

Substance	Formula	ΔH_f° (kJ/mol)	ΔG_f° (kJ/mol)	S° (J/K · mol)
Acetic acid(*l*)	CH_3COOH	−484.2	−389.45	159.8
Acetaldehyde(*g*)	CH_3CHO	−166.35	−139.08	264.2
Acetone(*l*)	CH_3COCH_3	−246.8	−153.55	198.7
Acetylene(*g*)	C_2H_2	226.6	209.2	200.8
Benzene(*l*)	C_6H_6	49.04	124.5	172.8
Butane(*g*)	C_4H_{10}	−124.7	−15.7	310.0
Ethanol(*l*)	C_2H_5OH	−276.98	−174.18	161.0
Ethane(*g*)	C_2H_6	−84.7	−32.89	229.5
Ethylene(*g*)	C_2H_4	52.3	68.1	219.5
Formic acid(*l*)	$HCOOH$	−409.2	−346.0	129.0
Glucose(*s*)	$C_6H_{12}O_6$	−1274.5	−910.56	212.1
Methane(*g*)	CH_4	−74.85	−50.8	186.2
Methanol(*l*)	CH_3OH	−238.7	−166.3	126.8
Propane(*g*)	C_3H_8	−103.9	−23.5	269.9
Sucrose(*s*)	$C_{12}H_{22}O_{11}$	−2221.7	−1544.3	360.2

APPENDIX 3

Mathematical Operations

Logarithms

Common Logarithms

The concept of the logarithm is an extension of the concept of exponents, which is discussed in Chapter 1. The *common,* or base-10, logarithm of any number is the power to which 10 must be raised to equal the number. These examples illustrate this relationship:

Logarithm	Exponent
$\log 1 = 0$	$10^0 = 1$
$\log 10 = 1$	$10^1 = 10$
$\log 100 = 2$	$10^2 = 100$
$\log 10^{-1} = -1$	$10^{-1} = 0.1$
$\log 10^{-2} = -2$	$10^{-2} = 0.01$

In each case, the logarithm of the number can be obtained by inspection.

Because the logarithms of numbers are exponents, they have the same properties as exponents. Thus, we have

Logarithm	Exponent
$\log AB = \log A + \log B$	$10^A \times 10^B = 10^{A+B}$
$\log \frac{A}{B} = \log A - \log B$	$\frac{10^A}{10^B} = 10^{A-B}$

Furthermore, $\log A^n = n \log A$.

Now suppose we want to find the common logarithm of 6.7×10^{-4}. On most electronic calculators, the number is entered first and then the log key is pressed. This operation gives us

$$\log 6.7 \times 10^{-4} = -3.17$$

Note that there are as many digits *after* the decimal point as there are significant figures in the original number. The original number has two significant figures and the "17" in -3.17 tells us that the log has two significant figures. The 3 in 3.17 serves only to locate the decimal point in the number 6.7×10^{-4}. Other examples are

Number	Common Logarithm
62	1.79
0.872	-0.0595
1.0×10^{-7}	-7.00

Sometimes (as in the case of pH calculations) it is necessary to obtain the number whose logarithm is known. This procedure is known as taking the antilogarithm; it is simply the reverse of taking the logarithm of a number. Suppose in a certain calculation we have pH = 1.46 and are asked to calculate $[H^+]$. From the definition of pH (pH = $-\log [H^+]$) we can write

$$[H^+] = 10^{-1.46}$$

Many calculators have a key labeled $\log^{-1}$ or INV log to obtain antilogs. Other calculators have a 10^x or y^x key (in which x corresponds to -1.46 in our example and y is 10 for base-10 logarithm). Therefore, we find that $[H^+] = 0.035\ M$.

Natural Logarithms

Logarithms taken to the base e instead of 10 are known as natural logarithms (denoted by ln or $\log_e$); e is equal to 2.7183. The relationship between common logarithms and natural logarithms is

$$\log 10 = 1 \qquad 10^1 = 10$$
$$\ln 10 = 2.303 \qquad e^{2.303} = 10$$

Thus,

$$\ln x = 2.303 \log x$$

To find the natural logarithm of 2.27, say, we first enter the number on the electronic calculator and then press the ln key to get

$$\ln 2.27 = 0.820$$

If no ln key is provided, we can proceed as

$$\begin{aligned} 2.303 \log 2.27 &= 2.303 \times 0.356 \\ &= 0.820 \end{aligned}$$

Sometimes we may be given the natural logarithm and asked to find the number it represents. For example,

$$\ln x = 59.7$$

On many calculators, we simply enter the number and press the e key:

$$e^{59.7} = 9 \times 10^{25}$$

The Quadratic Equation

A quadratic equation takes the form

$$ax^2 + bx + c = 0$$

If coefficients a, b, and c are known, then x is given by

$$x = \frac{-b \pm \sqrt{b^2 - 4ac}}{2a}$$

Suppose we have this quadratic equation:

$$2x^2 + 5x - 12 = 0$$

Solving for x, we write

$$\begin{aligned} x &= \frac{-5 \pm \sqrt{(5)^2 - 4(2)(-12)}}{2(2)} \\ &= \frac{-5 \pm \sqrt{25 + 96}}{4} \end{aligned}$$

Therefore,

$$x = \frac{-5 + 11}{4} = \frac{3}{2}$$

and

$$x = \frac{-5 - 11}{4} = -4$$

APPENDIX 4

The Elements and the Derivation of Their Names and Symbols*

Element	Symbol	Atomic No.	Atomic Mass†	Date of Discovery	Discoverer and Nationality‡	Derivation
Actinium	Ac	89	(227)	1899	A. Debierne (Fr.)	Gr. *aktis,* beam or ray
Aluminum	Al	13	26.98	1827	F. Wochler (Ge.)	Alum, the aluminum compound in which it was discovered; derived from L. *alumen,* astringent taste
Americium	Am	95	(243)	1944	A. Ghiorso (USA) R. A. James (USA) G. T. Seaborg (USA) S. G. Thompson (USA)	The Americas
Antimony	Sb	51	121.8	Ancient		L. *antimonium* (*anti,* opposite of; *monium,* isolated condition), so named because it is a tangible (metallic) substance which combines readily; symbol, L. *stibium,* mark
Argon	Ar	18	39.95	1894	Lord Raleigh (GB) Sir William Ramsay (GB)	Gr. *argos,* inactive
Arsenic	As	33	74.92	1250	Albertus Magnus (Ge.)	Gr. *aksenikon,* yellow pigment; L. *arsenicum,* orpiment; the Greeks once used arsenic trisulfide as a pigment
Astatine	At	85	(210)	1940	D. R. Corson (USA) K. R. MacKenzie (USA) E. Segre (USA)	Gr. *astatos,* unstable
Barium	Ba	56	137.3	1808	Sir Humphry Davy (GB)	barite, a heavy spar, derived from Gr. *barys,* heavy
Berkelium	Bk	97	(247)	1950	G. T. Seaborg (USA) S. G. Thompson (USA) A. Ghiorso (USA)	Berkeley, Calif.

Source: From "The Elements and Derivation of Their Names and Symbols," G. P. Dinga, *Chemistry* **41** (2), 20–22 (1968). Copyright by the American Chemical Society.

*At the time this table was drawn up, only 103 elements were known to exist.

†The atomic masses given here correspond to the 1961 values of the Commission on Atomic Weights. Masses in parentheses are those of the most stable or most common isotopes.

‡The abbreviations are (Ar.) Arabic; (Au.) Austrian; (Du.) Dutch; (Fr.) French; (Ge.) German; (GB) British; (Gr.) Greek; (H.) Hungarian; (I.) Italian; (L.) Latin; (P.) Polish; (R.) Russian; (Sp.) Spanish; (Swe.) Swedish; (USA) American.

(*Continued*)

Element	Symbol	Atomic No.	Atomic Mass	Date of Discovery	Discoverer and Nationality	Derivation
Beryllium	Be	4	9.012	1828	F. Woehler (Ge.) A. A. B. Bussy (Fr.)	Fr. L. *beryl,* sweet
Bismuth	Bi	83	209.0	1753	Claude Geoffroy (Fr.)	Ge. *bismuth,* probably a distortion of *weisse masse* (white mass) in which it was found
Boron	B	5	10.81	1808	Sir Humphry Davy (GB) J. L. Gay-Lussac (Fr.) L. J. Thenard (Fr.)	The compound borax, derived from Ar. *buraq,* white
Bromine	Br	35	79.90	1826	A. J. Balard (Fr.)	Gr. *bromos,* stench
Cadmium	Cd	48	112.4	1817	Fr. Stromeyer (Ge.)	Gr. *kadmia,* earth; L. *cadmia,* calamine (because it is found along with calamine)
Calcium	Ca	20	40.08	1808	Sir Humphry Davy (GB)	L. *calx,* lime
Californium	Cf	98	(249)	1950	G. T. Seaborg (USA) S. G. Thompson (USA) A. Ghiorso (USA) K. Street, Jr. (USA)	California
Carbon	C	6	12.01	Ancient		L. *carbo,* charcoal
Cerium	Ce	58	140.1	1803	J. J. Berzelius (Swe.) William Hisinger (Swe.) M. H. Klaproth (Ge.)	Asteroid Ceres
Cesium	Cs	55	132.9	1860	R. Bunsen (Ge.) G. R. Kirchhoff (Ge.)	L. *caesium,* blue (cesium was discovered by its spectral lines, which are blue)
Chlorine	Cl	17	35.45	1774	K. W. Scheele (Swe.)	Gr. *chloros,* light green
Chromium	Cr	24	52.00	1797	L. N. Vauquelin (Fr.)	Gr. *chroma,* color (because it is used in pigments)
Cobalt	Co	27	58.93	1735	G. Brandt (Ge.)	Ge. *Kobold,* goblin (because the ore yielded cobalt instead of the expected metal, copper, it was attributed to goblins)
Copper	Cu	29	63.55	Ancient		L. *cuprum,* copper, derived from *cyprium,* Island of Cyprus, the main source of ancient copper
Curium	Cm	96	(247)	1944	G. T. Seaborg (USA) R. A. James (USA) A. Ghiorso (USA)	Pierre and Marie Curie
Dysprosium	Dy	66	162.5	1886	Lecoq de Boisbaudran (Fr.)	Gr. *dysprositos,* hard to get at
Einsteinium	Es	99	(254)	1952	A. Ghiorso (USA)	Albert Einstein
Erbium	Er	68	167.3	1843	C. G. Mosander (Swe.)	Ytterby, Sweden, where many rare earths were discovered

(*Continued*)

Element	Symbol	Atomic No.	Atomic Mass	Date of Discovery	Discoverer and Nationality	Derivation
Europium	Eu	63	152.0	1896	E. Demarcay (Fr.)	Europe
Fermium	Fm	100	(253)	1953	A. Ghiorso (USA)	Enrico Fermi
Fluorine	F	9	19.00	1886	H. Moissan (Fr.)	Mineral fluorspar, from L. *fluere,* flow (because fluorspar was used as a flux)
Francium	Fr	87	(223)	1939	Marguerite Perey (Fr.)	France
Gadolinium	Gd	64	157.3	1880	J. C. Marignac (Fr.)	Johan Gadolin, Finnish rare earth chemist
Gallium	Ga	31	69.72	1875	Lecoq de Boisbaudran (Fr.)	L. *Gallia,* France
Germanium	Ge	32	72.59	1886	Clemens Winkler (Ge.)	L. *Germania,* Germany
Gold	Au	79	197.0	Ancient		L. *aurum,* shining dawn
Hafnium	Hf	72	178.5	1923	D. Coster (Du.) G. von Hevesey (H.)	L. *Hafnia,* Copenhagen
Helium	He	2	4.003	1868	P. Janssen (spectr) (Fr.) Sir William Ramsay (isolated) (GB)	Gr. *helios,* sun (because it was first discovered in the sun's spectrum)
Holmium	Ho	67	164.9	1879	P. T. Cleve (Swe.)	L. *Holmia,* Stockholm
Hydrogen	H	1	1.008	1766	Sir Henry Cavendish (GB)	Gr. *hydro,* water; *genes,* forming (because it produces water when burned with oxygen)
Indium	In	49	114.8	1863	F. Reich (Ge.) T. Richter (Ge.)	Indigo, because of its indigo blue lines in the spectrum
Iodine	I	53	126.9	1811	B. Courtois (Fr.)	Gr. *iodes,* violet
Iridium	Ir	77	192.2	1803	S. Tennant (GB)	L. *iris,* rainbow
Iron	Fe	26	55.85	Ancient		L. *ferrum,* iron
Krypton	Kr	36	83.80	1898	Sir William Ramsay (GB) M. W. Travers (GB)	Gr. *kryptos,* hidden
Lanthanum	La	57	138.9	1839	C. G. Mosander (Swe.)	Gr. *lanthanein,* concealed
Lawrencium	Lr	103	(257)	1961	A. Ghiorso (USA) T. Sikkeland (USA) A. E. Larsh (USA) R. M. Latimer (USA)	E. O. Lawrence (USA), inventor of the cyclotron
Lead	Pb	82	207.2	Ancient		Symbol, L. *plumbum,* lead, meaning heavy
Lithium	Li	3	6.941	1817	A. Arfvedson (Swe.)	Gr. *lithos,* rock (because it occurs in rocks)
Lutetium	Lu	71	175.0	1907	G. Urbain (Fr.) C. A. von Welsbach (Au.)	*Lutetia,* ancient name for Paris
Magnesium	Mg	12	24.31	1808	Sir Humphry Davy (GB)	*Magnesia,* a district in Thessaly; possibly derived from L. *magnesia*
Manganese	Mn	25	54.94	1774	J. G. Gahn (Swe.)	L. *magnes,* magnet

(*Continued*)

Element	Symbol	Atomic No.	Atomic Mass	Date of Discovery	Discoverer and Nationality	Derivation
Mendelevium	Md	101	(256)	1955	A. Ghiorso (USA) G. R. Choppin (USA) G. T. Seaborg (USA) B. G. Harvey (USA) S. G. Thompson (USA)	Mendeleev, Russian chemist who prepared the periodic chart and predicted properties of undiscovered elements
Mercury	Hg	80	200.6	Ancient		Symbol, L. *hydrargyrum,* liquid silver
Molybdenum	Mo	42	95.94	1778	G. W. Scheele (Swe.)	Gr. *molybdos,* lead
Neodymium	Nd	60	144.2	1885	C. A. von Welsbach (Au.)	Gr. *neos,* new; *didymos,* twin
Neon	Ne	10	20.18	1898	Sir William Ramsay (GB) M. W. Travers (GB)	Gr. *neos,* new
Neptunium	Np	93	(237)	1940	E. M. McMillan (USA) P. H. Abelson (USA)	Planet Neptune
Nickel	Ni	28	58.69	1751	A. F. Cronstedt (Swe.)	Swe. *kopparnickel,* false copper; also Ge. *nickel,* referring to the devil that prevented copper from being extracted from nickel ores
Niobium	Nb	41	92.91	1801	Charles Hatchett (GB)	Gr. *Niobe,* daughter of Tantalus (niobium was considered identical to tantalum, named after *Tantalus,* until 1884; originally called columbium, with symbol Cb)
Nitrogen	N	7	14.01	1772	Daniel Rutherford (GB)	Fr. *nitrogene,* derived from L. *nitrum,* native soda, or Gr. *nitron,* native soda, and Gr. *genes,* forming
Nobelium	No	102	(253)	1958	A. Ghiorso (USA) T. Sikkeland (USA) J. R. Walton (USA) G. T. Seaborg (USA)	Alfred Nobel
Osmium	Os	76	190.2	1803	S. Tennant (GB)	Gr. *osme,* odor
Oxygen	O	8	16.00	1774	Joseph Priestley (GB) C. W. Scheele (Swe.)	Fr. *oxygene,* generator of acid, derived from Gr. *oxys,* acid, and L. *genes,* forming (because it was once thought to be a part of all acids)
Palladium	Pd	46	106.4	1803	W. H. Wollaston (GB)	Asteroid Pallas
Phosphorus	P	15	30.97	1669	H. Brandt (Ge.)	Gr. *phosphoros,* light bearing
Platinum	Pt	78	195.1	1735 1741	A. de Ulloa (Sp.) Charles Wood (GB)	Sp. *platina,* silver

(*Continued*)

Element	Symbol	Atomic No.	Atomic Mass	Date of Discovery	Discoverer and Nationality	Derivation
Plutonium	Pu	94	(242)	1940	G. T. Seaborg (USA) E. M. McMillan (USA) J. W. Kennedy (USA) A. C. Wahl (USA)	Planet Pluto
Polonium	Po	84	(210)	1898	Marie Curie (P.)	Poland
Potassium	K	19	39.10	1807	Sir Humphry Davy (GB)	Symbol, L. *kalium,* potash
Praseodymium	Pr	59	140.9	1885	C. A. von Welsbach (Au.)	Gr. *prasios,* green; *didymos,* twin
Promethium	Pm	61	(147)	1945	J. A. Marinsky (USA) L. E. Glendenin (USA) C. D. Coryell (USA)	Gr. mythology, *Prometheus* the Greek Titan who stole fire from heaven,
Protactinium	Pa	91	(231)	1917	O. Hahn (Ge.) L. Meitner (Au.)	Gr. *protos,* first; *actinium* (because it disintegrates into actinium)
Radium	Ra	88	(226)	1898	Pierre and Marie Curie (Fr.; P.)	L. *radius,* ray
Radon	Rn	86	(222)	1900	F. E. Dorn (Ge.)	Derived from radium with suffix "on" common to inert gases (once called nitron, meaning shining, with symbol Nt)
Rhenium	Re	75	186.2	1925	W. Noddack (Ge.) I. Tacke (Ge.) Otto Berg (Ge.)	L. *Rhenus,* Rhine
Rhodium	Rh	45	102.9	1804	W. H. Wollaston (GB)	Gr. *rhodon,* rose (because some of its salts are rose-colored)
Rubidium	Rb	37	85.47	1861	R. W. Bunsen (Ge.) G. Kirchoff (Ge.)	L. *rubidius,* dark red (discovered with the spectroscope, its spectrum shows red lines)
Ruthenium	Ru	44	101.1	1844	K. K. Klaus (R.)	L. *Ruthenia,* Russia
Samarium	Sm	62	150.4	1879	Lecoq de Boisbaudran (Fr.)	Samarskite, after Samarski, a Russian engineer
Scandium	Sc	21	44.96	1879	L. F. Nilson (Swe.)	Scandinavia
Selenium	Se	34	78.96	1817	J. J. Berzelius (Swe.)	Gr. *selene,* moon (because it resembles tellurium, named for the earth)
Silicon	Si	14	28.09	1824	J. J. Berzelius (Swe.)	L. *silex, silicis,* flint
Silver	Ag	47	107.9	Ancient		Symbol, L. *argentum,* silver
Sodium	Na	11	22.99	1807	Sir Humphry Davy (GB)	L. *sodanum,* headache remedy; symbol, L. *natrium,* soda
Strontium	Sr	38	87.62	1808	Sir Humphry Davy (GB)	Strontian, Scotland, derived from mineral strontionite
Sulfur	S	16	32.07	Ancient		L. *sulphurium* (Sanskrit, *sulvere*)

(*Continued*)

Element	Symbol	Atomic No.	Atomic Mass	Date of Discovery	Discoverer and Nationality	Derivation
Tantalum	Ta	73	180.9	1802	A. G. Ekeberg (Swe.)	Gr. mythology, *Tantalus,* because of difficulty in isolating it (Tantalus, son of Zeus, was punished by being forced to stand up to his chin in water that receded whenever he tried to drink)
Technetium	Tc	43	(99)	1937	C. Perrier (I.)	Gr. *technetos,* artificial (because it was the first artificial element)
Tellurium	Te	52	127.6	1782	F. J. Müller (Au.)	L. *tellus,* earth
Terbium	Tb	65	158.9	1843	C. G. Mosander (Swe.)	Ytterby, Sweden
Thallium	Tl	81	204.4	1861	Sir William Crookes (GB)	Gr. *thallos,* a budding twig (because its spectrum shows a bright green line)
Thorium	Th	90	232.0	1828	J. J. Berzelius (Swe.)	Mineral thorite, derived from *Thor,* Norse god of war
Thulium	Tm	69	168.9	1879	P. T. Cleve (Swe.)	*Thule,* early name for Scandinavia
Tin	Sn	50	118.7	Ancient		Symbol, L. *stannum,* tin
Titanium	Ti	22	47.88	1791	W. Gregor (GB)	Gr. giants, the Titans, and L. *titans,* giant deities
Tungsten	W	74	183.9	1783	J. J. and F. de Elhuyar (Sp.)	Swe. *tung sten,* heavy stone; symbol, wolframite, a mineral
Uranium	U	92	238.0	1789 1841	M. H. Klaproth (Ge.) E. M. Peligot (Fr.)	Planet Uranus
Vanadium	V	23	50.94	1801 1830	A. M. del Rio (Sp.) N. G. Sefstrom (Swe.)	*Vanadis,* Norse goddess of love and beauty
Xenon	Xe	54	131.3	1898	Sir William Ramsay (GB) M. W. Travers (GB)	Gr. *xenos,* stranger
Ytterbium	Yb	70	173.0	1907	G. Urbain (Fr.)	Ytterby, Sweden
Yttrium	Y	39	88.91	1843	C. G. Mosander (Swe.)	Ytterby, Sweden
Zinc	Zn	30	65.39	1746	A. S. Marggraf (Ge.)	Ge. *zink,* of obscure origin
Zirconium	Zr	40	91.22	1789	M. H. Klaproth (Ge.)	Zircon, in which it was found, derived from Ar. *zargum,* gold color

GLOSSARY

The number in parentheses is the number of the section in which the term first appears.

A

absolute temperature scale. A temperature scale on which absolute zero (0 K) is the lowest temperature (also called the Kelvin temperature scale). (5.3)

absolute zero. Theoretically the lowest attainable temperature. (5.3)

accuracy. The closeness of a measurement to the true value of the quantity that is being measured. (1.6)

acid. A substance that yields hydrogen ions (H^+) when dissolved in water. (2.7)

acid ionization constant (K_a). The equilibrium constant for acid ionization. (16.5)

actinide series. Elements that have incompletely filled 5*f* subshells or readily give rise to cations that have incompletely filled 5*f* subshells. (7.9)

activated complex. The species temporarily formed by reactant molecules as a result of a collision before they form the product. (14.4)

activation energy. The minimum amount of energy required to initiate a chemical reaction. (14.4)

activity series. A summary of the results of many possible displacement reactions. (4.4)

actual yield. The amount of product actually obtained in a reaction. (3.10)

addition reaction. A reaction in which one molecule is added to another. (11.2)

adhesion. Attraction between unlike molecules. (12.3)

alcohol. An organic compound containing the hydroxyl group (—OH). (11.4)

aldehydes. Compounds with a carbonyl functional group and the general formula RCHO, in which R is an H atom, an alkyl, or an aromatic hydrocarbon group. (11.4)

aliphatic hydrocarbons. Hydrocarbons that do not contain the benzene group or the benzene ring. (11.1)

alkali metals. The Group 1A elements (Li, Na, K, Rb, Cs, and Fr). (2.4)

alkaline earth metals. The Group 2A elements (Be, Mg, Ca, Sr, Ba, and Ra). (2.4)

alkanes. Hydrocarbons having the general formula C_nH_{2n+2}, in which $n = 1, 2, \ldots$. (11.2)

alkenes. Hydrocarbons that contain one or more carbon-carbon double bonds. They have the general formula C_nH_{2n} in which $n = 2, 3, \ldots$. (11.2)

alkynes. Hydrocarbons that contain one or more carbon-carbon triple bonds. They have the general formula C_nH_{2n-2}, in which $n = 2, 3, \ldots$. (11.2)

allotropes. Two or more forms of the same element that differ significantly in chemical and physical properties. (2.6)

alpha particles. See alpha rays.

alpha (α) rays. Helium ions with a charge of +2. (2.2)

amines. Organic bases that have the functional group $—NR_2$, in which R may be H, an alkyl group, or an aromatic hydrocarbon group. (11.4)

amino acid. A special kind of carboxylic acid that contains at least one carboxyl group (—COOH) and at least one amino group ($—NH_2$). (22.3)

amorphous solid. A solid that lacks a regular three-dimensional arrangement of atoms or molecules. (12.5)

amphoteric oxide. An oxide that exhibits both acidic and basic properties. (8.6)

amplitude. The vertical distance from the middle of a wave to the peak or trough. (7.1)

anion. An ion with a net negative charge. (2.5)

anode. The electrode at which oxidation occurs. (19.2)

antibonding molecular orbital. A molecular orbital that is of higher energy and lower stability than the atomic orbitals from which it was formed. (10.6)

aqueous solution. A solution in which the solvent is water. (4.1)

aromatic hydrocarbon. A hydrocarbon that contains one or more benzene rings. (11.1)

atmospheric pressure. The pressure exerted by Earth's atmosphere. (5.2)

atom. The basic unit of an element that can enter into chemical combination. (2.2)

atomic mass. The mass of an atom in atomic mass units. (3.1)

atomic mass unit (amu). A mass exactly equal to one-twelfth the mass of one carbon-12 atom. (3.1)

atomic number (*Z*). The number of protons in the nucleus of an atom. (2.3)

atomic orbital. The wave function of an electron in an atom. (7.5)

atomic radius. One-half the distance between the nuclei in two adjacent atoms of the same element in a metal. For elements that exist as diatomic units, the atomic radius is one-half the distance between the nuclei of two atoms in a particular molecule. (8.3)

Aufbau principle. As protons are added one by one to the nucleus to build up the elements, electrons similarly are added to the atomic orbitals. (7.9)

Avogadro's law. At constant pressure and temperature, the volume of a gas is directly proportional to the number of moles of the gas present. (5.3)

Avogadro's number (N_A). 6.022×10^{23}; the number of particles in a mole. (3.2)

B

barometer. An instrument that measures atmospheric pressure. (5.2)

base. A substance that yields hydroxide ions (OH^-) when dissolved in water. (2.7)

base ionization constant (K_b). The equilibrium constant for the ionization of a base. (16.6)

battery. A galvanic cell or a series of several connected galvanic cells that can be used as a source of direct electric current at a constant voltage. (19.6)

beta particles. See beta rays.

beta (β) rays. Streams of electrons emitted during the decay of certain radioactive substances. (2.2)

bimolecular reaction. An elementary step involving two molecules that is part of a reaction mechanism. (14.5)

binary compounds. Compounds containing just two elements. (2.7)

boiling point. The temperature at which the vapor pressure of a liquid is equal to the external atmospheric pressure. (12.6)

bond enthalpy. The enthalpy change required to break a bond in a mole of gaseous molecules. (9.10)

bond length. The distance between the centers of two bonded atoms in a molecule. (9.4)

bond order. The difference between the number of electrons in bonding molecular orbitals and antibonding molecular orbitals, divided by two. (10.6)

bonding molecular orbital. A molecular orbital that is of lower energy and greater stability than the atomic orbitals from which it was formed. (10.6)

Born-Haber cycle. The cycle that relates lattice energies of ionic compounds to ionization energies, electron affinities, and other atomic and molecular properties. (9.3)

boundary surface diagram. Diagram of the region containing about 90 percent of the electron density in an atomic orbital. (7.7)

Boyle's law. The volume of a fixed amount of gas is inversely proportional to the gas pressure at constant temperature. (5.3)

breeder reactor. A nuclear reactor that produces more fissionable material than it uses. (21.5)

Brønsted acid. A substance capable of donating a proton in a reaction. (4.3)

Brønsted base. A substance capable of accepting a proton in a reaction. (4.3)

buffer solution. A solution of (a) a weak acid or base and (b) its salt; both components must be present. A buffer solution has the ability to resist changes in pH when small amounts of either acid or base are added to it. (17.2)

C

calorimetry. The measurement of heat changes. (6.5)

carboxylic acids. Acids that contain the carboxyl group (—COOH). (11.4)

catalyst. A substance that increases the rate of a chemical reaction by providing an alternate reaction pathway without being consumed during the reaction. (14.6)

cathode. The electrode at which reduction occurs. (19.2)

cation. An ion with a net positive charge. (2.5)

cell voltage. Difference in electrical potential between the anode and the cathode of a galvanic cell. (19.2)

Charles's and Gay-Lussac's law. See Charles's law.

Charles's law. The volume of a fixed amount of gas is directly proportional to the absolute temperature of the gas when the pressure is held constant. (5.3)

chelating agent. A substance that forms complex ions with metal ions in solution. (20.2)

chemical energy. Energy stored within the structural units of chemical substances. (6.1)

chemical equation. An equation that uses chemical symbols to show what happens during a chemical reaction. (3.7)

chemical equilibrium. A state in which the rates of the forward and reverse reactions are equal and no net changes can be observed. (4.1, 15.1)

chemical formula. An expression showing the chemical composition of a compound in terms of the symbols for the atoms of the elements involved. (2.6)

chemical kinetics. The area of chemistry concerned with the speeds, or rates, at which chemical reactions occur. (14.1)

chemical property. Any property of a substance that cannot be studied without converting the substance into some other substance. (1.4)

chemical reaction. Chemical change. (3.7)

chemistry. The science that studies the properties of substances and how substances react with one another. (1.3)

chiral. Compounds or ions that are not superimposable with their mirror images. (11.5)

closed system. A system that allows the exchange of energy (usually in the form of heat) but not mass with its surroundings. (6.2)

closest packing. The most efficient arrangements for packing atoms, molecules, or ions in a crystal. (12.4)

cohesion. The intermolecular attraction between like molecules. (12.3)

colligative properties. Properties of solutions that depend on the number of solute particles in solution and not on the nature of the solute. (13.6)

combination reaction. A reaction in which two or more substances combine to form a single product. (4.4)

combustion reaction. A reaction in which a substance reacts with oxygen, usually with the release of heat and light to produce a flame. (4.4)

complex ion. An ion containing a central metal cation bonded to one or more molecules or ions. (17.7)

compound. A substance composed of two or more elements chemically united in fixed proportions. (1.3)

concentration of a solution. The amount of solute present in a given quantity of solution. (4.5)

condensation. The phenomenon of going from the gaseous state to the liquid state. (12.6)

condensation reaction. The joining of two molecules and the elimination of a small molecule, usually water. (11.4)

conformations. Different spatial arrangements of a molecule that are generated by rotation about single bonds. (11.2)

conjugate acid-base pair. An acid and its conjugate base or a base and its conjugate acid. (16.1)

coordinate covalent bond. A bond in which the pair of electrons is supplied by one of the two bonded atoms. (9.9)

coordination compound. A neutral species containing a complex ion. (20.2)

coordination number. In a crystal lattice it is defined as the number of atoms (or ions) surrounding an atom (or ion) (12.4). In coordination compounds it is defined as the number of donor atoms surrounding the central metal atom in a complex. (20.2)

copolymer. A polymer containing two or more different monomers. (22.2)

corrosion. The deterioration of metals by an electrochemical process. (19.7)

Coulomb's law. The potential energy between two ions is directly proportional to the product of their charges and inversely proportional to the distance between them. (9.3)

covalent bond. A bond in which two electrons are shared by two atoms. (9.4)

covalent compounds. Compounds containing only covalent bonds. (9.4)

critical mass. The minimum mass of fissionable material required to generate a self-sustaining nuclear chain reaction. (21.5)

critical pressure (P_c). The minimum pressure necessary to bring about liquefaction at the critical temperature. (12.6)

critical temperature (T_c). The temperature above which a gas will not liquefy. (12.6)

crystal-field splitting. The energy difference between two sets of *d* orbitals of a metal atom in the presence of ligands. (20.4)

crystalline solid. A solid that possesses rigid and long-range structural order; its atoms, molecules, or ions occupy specific positions. (12.4)

crystallization. The process in which dissolved solute comes out of solution and forms crystals. (13.1)

cycloalkanes. Hydrocarbons having the general formula C_nH_{2n} in which $n = 3, 4, \ldots.$ (11.2)

D

Dalton's law of partial pressures. The total pressure of a mixture of gases is just the sum of the pressures that each gas would exert if it were present alone. (5.5)

decomposition reaction. The breakdown of a compound into two or more components. (4.4)

delocalized molecular orbital. A molecular orbital that is not confined between two adjacent bonding atoms but actually extends over three or more atoms. (11.3)

denatured protein. Protein that does not exhibit normal biological activities. (22.3)

density. The mass of a substance divided by its volume. (1.5)

deoxyribonucleic acid (DNA). A type of nucleic acid. (22.4)

deposition. The process in which vapor molecules are converted directly to the solid phase. (12.6)

diagonal relationship. Similarities between pairs of elements in different groups and periods of the periodic table. (8.6)

diamagnetic. Repelled by a magnet; a diamagnetic substance contains only paired electrons. (7.8)

diatomic molecule. A molecule that consists of two atoms. (2.5)

diffusion. The gradual mixing of molecules of one gas with the molecules of another by virtue of their kinetic properties. (5.6)

dilution. A procedure for preparing a less concentrated solution from a more concentrated solution. (4.5)

dipole moment. The product of charge and the distance between the charges in a molecule. (10.2)

dipole-dipole forces. Forces that act between polar molecules. (12.2)

diprotic acid. Each unit of the acid yields two hydrogen ions. (4.3)

dispersion forces. The attractive forces that arise as a result of temporary dipoles induced in the atoms or molecules. (12.2)

displacement reaction. A reaction in which an atom or an ion in a compound is replaced by an atom of another element. (4.4)

donor atom. The atom in a ligand that is bonded directly to the metal atom. (20.2)

double bond. A covalent bond in which two atoms share two pairs of electrons. (9.4)

dynamic equilibrium. The condition in which the rate of a forward process is exactly balanced by the rate of the reverse process. (12.6)

E

effective nuclear charge (Z_{eff}). The nuclear charge felt by an electron when both the actual charge (Z) and the repulsive effect (shielding) of the other electrons are taken into account. (8.3)

effusion. The process by which a gas under pressure escapes from one compartment of a container to another by passing through a small opening. (5.6)

electrochemistry. The branch of chemistry that deals with the interconversion of electrical energy and chemical energy. (19.1)

electrolysis. A process in which electrical energy is used to bring about a nonspontaneous chemical reaction. (19.8)

electrolyte. A substance that, when dissolved in water, results in a solution that can conduct electricity. (4.1)

electrolytic cell. An apparatus for carrying out electrolysis. (19.8)

electromagnetic radiation. The emission and transmission of energy in the form of electromagnetic waves. (7.1)

electromagnetic wave. A wave that has an electric field component and a mutually perpendicular magnetic field component. (7.1)

electromotive force (emf) (E). The voltage difference between electrodes. (19.2)

electron. A subatomic particle that has a very low mass and carries a single negative electric charge. (2.2)

electron affinity. The negative of the energy change that takes place when an electron is accepted by an atom (or an ion) in the gaseous state. (8.5)

electron configuration. The distribution of electrons among the various orbitals in an atom or molecule. (7.8)

electron density. The probability that an electron will be found at a particular region in an atomic orbital. (7.5)

electronegativity. The ability of an atom to attract electrons toward itself in a chemical bond. (9.5)

element. A substance that cannot be separated into simpler substances by chemical means. (1.3)

elementary steps. A series of simple reactions that represent the overall progress of a reaction at the molecular level. (14.5)

emission spectrum. The continuous or line spectrum of electromagnetic radiation emitted by a substance. (7.3)

empirical formula. An expression using chemical symbols to show the types of elements in a substance and the simplest ratios of the different kinds of atoms. (2.6)

enantiomers. Compounds and their nonsuperimposable mirror images. (11.5)

end point. Occurs in a titration when the indicator changes color. (17.4)

endothermic processes. Processes that absorb heat from the surroundings. (6.2)

energy. The capacity to do work or to produce change. (6.1)

enthalpy (H). A thermodynamic quantity used to describe heat changes taking place at constant pressure. (6.4)

enthalpy of reaction (ΔH). The difference between the enthalpies of the products and the enthalpies of the reactants. (6.4)

entropy (S). A measure of how spread out or dispersed the energy of a system is among the different possible ways that the system can contain energy. (18.3)

enzyme. A biological catalyst. (14.6)

equilibrium constant. A number equal to the ratio of the equilibrium concentrations of products to the equilibrium concentrations of reactants, each raised to the power of its stoichiometric coefficient. (15.1)

equilibrium vapor pressure. The vapor pressure measured for a dynamic equilibrium of condensation and evaporation. (12.6)

equivalence point. The point at which an acid is completely reacted with or neutralized by a base. (4.6)

esters. Compounds that have the general formula RCOOR′, in which R can be H or an alkyl group or an aromatic hydrocarbon group and R′ is an alkyl group or an aromatic hydrocarbon group. (11.4)

ether. An organic compound containing the R—O—R′ linkage, in which R and R′ are alkyl and/or aromatic hydrocarbon groups. (11.4)

evaporation. The escape of molecules from the surface of a liquid; also called vaporization. (12.6)

excess reagent. A reactant present in a quantity greater than necessary to react with the amount of the limiting reagent present. (3.9)

excited level (or state). A state that has higher energy than the ground state of the system. (7.3)

exothermic processes. Processes that give off heat to the surroundings. (6.2)

extensive property. A property that depends on how much matter is being considered. (1.4)

F

family. The elements in a vertical column of the periodic table. (2.4)

Faraday constant (F). Charge contained in 1 mole of electrons, equivalent to 96,485 coulombs. (19.4)

first law of thermodynamics. Energy can be converted from one form to another, but cannot be created or destroyed. (6.3)

first-order reaction. A reaction whose rate depends on reactant concentration raised to the first power. (14.3)

formal charge. The electrical charge difference between the number of valence electrons in an isolated atom and the number of electrons assigned to that atom in a Lewis structure. (9.7)

formation constant (K_f). The equilibrium constant for the complex ion formation. (17.7)

free energy (G). The energy available to do useful work. (18.5)

frequency (ν). The number of waves that pass through a particular point per unit time. (7.1)

fuel cell. A galvanic cell that requires a continuous supply of reactants to keep functioning. (19.6)

functional group. That part of a molecule characterized by a special arrangement of atoms that is largely responsible for the chemical behavior of the parent molecule. (11.1)

G

galvanic cell. An electrochemical cell that generates electricity by means of a spontaneous redox reaction. (19.2)

gamma (γ) rays. High-energy radiation. (2.2)

gas constant (R). The constant that appears in the ideal gas equation ($PV = nRT$). It is expressed as 0.08206 L · atm/K · mol, or 8.314 J/K · mol. (5.4)

geometric isomers. Compounds with the same type and number of atoms and the same chemical bonds but different spatial arrangements; such isomers cannot be interconverted without breaking a chemical bond. (11.2)

Gibbs free energy. See free energy.

Graham's law of diffusion. Under the same conditions of temperature and pressure, the rates of diffusion of gases are inversely proportional to the square roots of their molar masses. (5.6)

gravimetric analysis. An experimental procedure that involves the measurement of mass to identify an unknown component of a substance. (4.6)

ground level (or state). The lowest energy state of a system. (7.3)

group. The elements in a vertical column of the periodic table. (2.4)

H

half-cell reactions. Oxidation and reduction reactions that occur at the electrodes. (19.2)

half-life. The time required for the concentration of a reactant to decrease to half its initial concentration. (14.3)

half-reaction. A reaction that explicitly shows electrons involved in either oxidation or reduction. (4.4)

halogens. The nonmetallic elements in Group 7A (F, Cl, Br, I, and At). (2.4)

heat. Transfer of energy between two bodies that are at different temperatures. (6.2)

heat capacity (*C*). The amount of heat required to raise the temperature of a given quantity of a substance by one degree Celsius. (6.5)

Heisenberg uncertainty principle. It is impossible to know simultaneously both the momentum and the position of a particle with certainty. (7.5)

Henry's law. The solubility of a gas in a liquid is proportional to the pressure of the gas over the solution. (13.5)

Hess's law. When reactants are converted to products, the change in enthalpy is the same whether the reaction takes place in one step or in a series of steps. (6.6)

heterogeneous equilibrium. An equilibrium state in which the reacting species are not all in the same phase. (15.2)

heterogeneous mixture. The individual components of such a mixture remain physically separate and can be seen as separate components. (1.3)

homogeneous equilibrium. An equilibrium condition in which all reacting species are in the same phase. (15.2)

homogeneous mixture. The composition of the mixture is the same throughout the solution. (1.3)

homonuclear diatomic molecule. A diatomic molecule containing atoms of the same element. (10.6)

homopolymer. A polymer that is made from only one type of monomer. (22.2)

Hund's rule. The most stable arrangement of electrons in atomic subshells is the one with the greatest number of parallel spins. (7.8)

hybrid orbitals. Atomic orbitals obtained when two or more nonequivalent orbitals of the same atom combine before covalent bond formation. (10.4)

hybridization. The process of mixing the atomic orbitals in an atom (usually the central atom) to generate a set of new atomic orbitals before covalent bond formation. (10.4)

hydrates. Compounds that have a specific number of water molecules attached to them. (2.7)

hydration. A process in which an ion or a molecule is surrounded by water molecules arranged in a specific manner. (4.1)

hydrocarbons. Compounds made up of only carbon and hydrogen. (11.1)

hydrogen bond. A special type of dipole-dipole interaction between the hydrogen atom bonded to an atom of a very electronegative element (F, N, O) and another atom of one of the three electronegative elements. (12.2)

hydrogenation. The addition of hydrogen, especially to compounds with double and triple carbon-carbon bonds. (11.2)

hydronium ion. H_3O^+. (4.3)

hypothesis. A tentative explanation for a set of observations. (1.2)

I

ideal gas. A hypothetical gas whose pressure-volume-temperature behavior can be completely accounted for by the ideal gas equation. (5.4)

ideal gas equation. An equation expressing the relationships among pressure, volume, temperature, and amount of gas ($PV = nRT$, in which R is the gas constant). (5.4)

ideal solution. Any solution that obeys Raoult's law. (13.6)

indicators. Substances that have distinctly different colors in acidic and basic media. (4.6)

induced dipole. The separation of positive and negative charges in an atom (or a nonpolar molecule) caused by the proximity of an ion or a polar molecule. (12.2)

inert complex. A complex ion that undergoes very slow ligand exchange reactions. (20.5)

intensive property. A property that does not depend on how much matter is being considered. (1.4)

intermediate. A species that appears in the mechanism of the reaction (that is, in the elementary steps) but not in the overall balanced equation. (14.5)

intermolecular forces. Attractive forces that exist among molecules. (12.2)

International System of Units. A revised metric system (abbreviated SI) that is widely used in scientific research. (1.5)

intramolecular forces. Forces that hold atoms together in a molecule. (12.2)

ion. An atom or group of atoms that has a net positive or negative charge. (2.5)

ion pair. A species made up of at least one cation and at least one anion held together by electrostatic forces. (13.6)

ion-dipole forces. Forces that operate between an ion and a dipole. (12.2)

ionic bond. The electrostatic force that holds ions together in an ionic compound. (9.2)

ionic compound. Any neutral compound containing cations and anions. (2.5)

ionic equation. An equation that shows dissolved ionic compounds in terms of their free ions. (4.2)

ionic radius. The radius of a cation or an anion as measured in an ionic compound. (8.3)

ionization energy. The minimum energy required to remove an electron from an isolated atom (or an ion) in its ground state. (8.4)

ion-product constant. Product of hydrogen ion concentration and hydroxide ion concentration (both in molarity) at a particular temperature. (16.2)

isoelectronic. Ions, or atoms and ions, that possess the same number of electrons, and hence the same ground-state electron configuration, are said to be isoelectronic. (8.2)

isolated system. A system that does not allow the transfer of either mass or energy to or from its surroundings. (6.2)

isotopes. Atoms having the same atomic number but different mass numbers. (2.3)

J

Joule. Unit of energy given by newtons × meters. (5.6)

K

Kelvin temperature scale. See absolute temperature scale.

ketones. Compounds with a carbonyl functional group and the general formula RR′CO, in which R and R′ are alkyl and/or aromatic hydrocarbon groups. (11.4)

kinetic energy (KE). Energy available because of the motion of an object. (5.6)

kinetic molecular theory of gases. A theory that describes the physical behavior of gases at the molecular level. (5.6)

L

labile complex. Complexes that undergo rapid ligand exchange reactions. (20.5)

lanthanide series. Elements that have incompletely filled 4*f* subshells or readily give rise to cations that have incompletely filled 4*f* subshells. (7.9)

lattice energy. The energy required to completely separate one mole of a solid ionic compound into gaseous ions. (9.3)

lattice points. The positions occupied by atoms, molecules, or ions that define the geometry of a unit cell. (12.4)

law. A concise verbal or mathematical statement of a relationship between phenomena that is always the same under the same conditions. (1.2)

law of conservation of energy. The total quantity of energy in the universe is constant. (6.1)

law of conservation of mass. Matter can be neither created nor destroyed. (2.1)

law of definite proportions. Different samples of the same compound always contain its constituent elements in the same proportions by mass. (2.1)

law of multiple proportions. If two elements can combine to form more than one type of compound, the masses of one element that combine with a fixed mass of the other element are in ratios of small whole numbers. (2.1)

Le Châtelier's principle. If an external stress is applied to a system at equilibrium, the system will adjust itself in such a way as to partially offset the stress. (15.4)

Lewis acid. A substance that can accept a pair of electrons. (16.11)

Lewis base. A substance that can donate a pair of electrons. (16.11)

Lewis dot symbol. The symbol of an element with one or more dots that represent the number of valence electrons in an atom of the element. (9.1)

Lewis structure. A representation of covalent bonding using Lewis symbols. Shared electron pairs are shown either as lines or as pairs of dots between two atoms, and lone pairs are shown as pairs of dots on individual atoms. (9.4)

ligand. A molecule or an ion that is bonded to the metal ion in a complex ion. (20.2)

limiting reagent. The reactant used up first in a reaction. (3.9)

line spectrum. Spectrum produced when radiation is absorbed or emitted by a substance only at some wavelengths. (7.3)

liter. The volume occupied by 1 cubic decimeter. (1.5)

lone pairs. Valence electrons that are not involved in covalent bond formation. (9.4)

M

macroscopic properties. Properties that can be measured directly. (1.5)

manometer. A device used to measure the pressure of gases. (5.2)

many-electron atoms. Atoms that contain two or more electrons. (7.5)

mass. A measure of the quantity of matter contained in an object. (1.5)

mass defect. The difference between the mass of an atom and the sum of the masses of its protons, neutrons, and electrons. (21.2)

mass number. The total number of neutrons and protons present in the nucleus of an atom. (2.3)

matter. Anything that occupies space and possesses mass. (1.3)

melting point. The temperature at which solid and liquid phases coexist in equilibrium. (12.6)

metalloid. An element with properties intermediate between those of metals and nonmetals. (2.4)

metals. Elements that are good conductors of heat and electricity and have the tendency to form positive ions in ionic compounds. (2.4)

metathesis reaction. A reaction that involves the exchange of parts between two compounds. (4.2)

microscopic properties. Properties that must be measured indirectly with the aid of a microscope or other special instrument. (1.5)

miscible. Two liquids that are completely soluble in each other in all proportions are said to be miscible. (13.2)

mixture. A combination of two or more substances in which the substances retain their identities. (1.3)

moderator. A substance that can reduce the kinetic energy of neutrons. (21.5)

molality. The number of moles of solute dissolved in 1 kilogram of solvent. (13.3)

molar concentration. See molarity.

molar heat of fusion (ΔH_{fus}). The energy (in kilojoules) required to melt 1 mole of a solid. (12.6)

molar heat of sublimation (ΔH_{sub}). The energy (in kilojoules) required to sublime 1 mole of a solid. (12.6)

molar heat of vaporization (ΔH_{vap}). The energy (in kilojoules) required to vaporize 1 mole of a liquid. (12.6)

molar mass ($\mathcal{M}$). The mass (in grams or kilograms) of 1 mole of atoms, molecules, or other particles. (3.2)

molar solubility. The number of moles of solute in 1 liter of a saturated solution (mol/L). (17.5)

molarity (M). The number of moles of solute in 1 liter of solution. (4.5)

mole (mol). The amount of substance that contains as many elementary entities (atoms, molecules, or other particles) as there are atoms in exactly 12 grams (or 0.012 kilograms) of the carbon-12 isotope. (3.2)

mole fraction. Ratio of the number of moles of one component of a mixture to the total number of moles of all components in the mixture. (5.5)

mole method. An approach for determining the amount of product formed in a reaction. (3.8)

molecular equations. Equations in which the formulas of the compounds are written as though all species existed as molecules or whole units. (4.2)

molecular formula. An expression showing the exact numbers of atoms of each element in a molecule. (2.6)

molecular mass. The sum of the atomic masses (in amu) present in a given molecule. (3.3)

molecular orbital. An orbital that results from the interaction of atomic orbitals of the bonding atoms. (10.6)

molecularity of a reaction. The number of molecules reacting in an elementary step. (14.5)

molecule. An aggregate of at least two atoms in a definite arrangement held together by special forces. (2.5)

monatomic ion. An ion that contains only one atom. (2.5)

monomer. Simple repeating units in a polymer. (22.2)

monoprotic acid. Each unit of the acid yields one hydrogen ion. (4.3)

multiple bonds. Bonds formed when two atoms share two or more pairs of electrons. (9.4)

N

Nernst equation. The relation between the emf of a galvanic cell and the standard emf and the concentrations of the oxidizing and reducing agents. (19.5)

net ionic equation. An equation that includes only the ionic species that actually take part in the reaction. (4.2)

neutralization reaction. A reaction between an acid and a base. (4.3)

neutron. A subatomic particle that bears no net electric charge. Its mass is slightly greater than a proton's. (2.2)

Newton (N). The SI unit for force. (5.2)

noble gas core. The noble gas that most nearly precedes the element being considered; used in writing electron configurations. (7.9)

noble gases. Nonmetallic elements in Group 8A (He, Ne, Ar, Kr, Xe, and Rn). (2.4)

node. A point at which the amplitude of a wave is zero. (7.4)

nonelectrolyte. A substance that, when dissolved in water, gives a solution that is not electrically conducting. (4.1)

nonmetals. Elements that are usually poor conductors of heat and electricity. (2.4)

nonpolar molecule. A molecule that does not possess a dipole moment. (10.2)

nonvolatile. Does not have a measurable vapor pressure. (13.6)

nuclear binding energy. The energy required to break up a nucleus into protons and neutrons. (21.2)

nuclear chain reaction. A self-sustaining sequence of nuclear fission reactions. (21.5)

nuclear fission. The process in which a heavy nucleus (mass number > 200) divides to form small nuclei of intermediate mass and one or more neutrons. (21.5)

nuclear fusion. The combining of small nuclei into larger ones. (21.6)

nuclear reaction. A reaction involving change in an atomic nucleus. (21.1)

nuclear transmutation. The change undergone by a nucleus as a result of bombardment by neutrons or other particles. (21.1)

nucleic acid. High molar mass polymers that play an essential role in protein synthesis. (22.4)

nucleotide. The repeating unit in each strand of a DNA molecule which consists of a base-deoxyribose-phosphate group. (22.4)

nucleus. The central core of an atom. (2.2)

O

octet rule. An atom other than hydrogen tends to form bonds until it is surrounded by eight valence electrons. (9.4)

open system. A system that can exchange mass and energy (usually in the form of heat) with its surroundings. (6.2)

organic chemistry. The branch of chemistry that deals with carbon compounds. (11.1)

osmosis. The net movement of solvent molecules through a semipermeable membrane from a pure solvent or from a dilute solution to a more concentrated solution. (13.6)

osmotic pressure (π). The pressure required to stop osmosis. (13.6)

overvoltage. The additional voltage required to cause electrolysis. (19.8)

oxidation number. The number of charges an atom would have in a molecule if electrons were transferred completely in the direction indicated by the difference in electronegativity. (4.4)

oxidation reaction. The half-reaction that represents the loss of electrons in a redox process. (4.4)

oxidation state. See oxidation number.

oxidation-reduction reaction. See redox reaction.

oxidizing agent. A substance that can accept electrons from another substance or increase the oxidation numbers of another substance. (4.4)

oxoacid. An acid containing hydrogen, oxygen, and another element (the central element). (2.7)

oxoanion. An anion derived from an oxoacid. (2.7)

P

paramagnetic. Attracted to a magnet. A paramagnetic substance contains one or more unpaired electrons. (7.8)

partial pressure. The pressure of one component in a mixture of gases. (5.5)

Pascal (Pa). A pressure of one newton per square meter (1 N/m^2). (5.2)

Pauli exclusion principle. No two electrons in an atom can have the same four quantum numbers. (7.8)

percent composition. The percent by mass of each element in a compound. (3.5)

percent ionization. The ratio of ionized acid concentration at equilibrium to the initial concentration of acid. (16.5)

percent by mass. The ratio of the mass of a solute to the mass of the solution, multiplied by 100%. (13.3)

percent yield. The ratio of the actual yield of a reaction to the theoretical yield, multiplied by 100%. (3.10)

period. A horizontal row of the periodic table. (2.4)

periodic table. A tabular arrangement of the elements by similarities in properties and by increasing atomic number. (2.4)

pH. The negative logarithm of the hydrogen ion concentration in an aqueous solution. (16.3)

phase. A homogeneous part of a system that is in contact with other parts of the system but separated from them by a well-defined boundary. (12.6)

phase change. Transformation from one phase to another. (12.6)

phase diagram. A diagram showing the conditions at which a substance exists as a solid, liquid, and vapor. (12.7)

photoelectric effect. A phenomenon in which electrons are ejected from the surface of certain metals exposed to light of at least a certain minimum frequency. (7.2)

photon. A particle of light. (7.2)

physical equilibrium. An equilibrium in which only physical properties change. (15.1)

physical property. Any property of a substance that can be observed without transforming the substance into some other substance. (1.4)

pi bond (π bond). A covalent bond formed by sideways overlapping orbitals; its electron density is concentrated above and below the plane of the nuclei of the bonding atoms. (10.5)

pi molecular orbital. A molecular orbital in which the electron density is concentrated above and below the line joining the two nuclei of the bonding atoms. (10.6)

plasma. A gaseous state of matter consisting of positive ions and electrons. (21.6)

polar covalent bond. In such a bond, the electrons spend more time in the vicinity of one atom than the other. (9.5)

polar molecule. A molecule that possesses a dipole moment. (10.2)

polarimeter. The instrument for studying interaction between plane-polarized light and chiral molecules. (11.5)

polarizability. The ease with which the electron distribution in the atom (or molecule) can be distorted. (12.2)

polyatomic ion. An ion that contains more than one atom. (2.5)

polyatomic molecule. A molecule that consists of more than two atoms. (2.5)

polymer. A molecular compound distinguished by a high molar mass and made up of many repeating units. (22.1)

positron. A particle that has the same mass as the electron but bears a +1 charge. (21.1)

potential energy. Energy available by virtue of an object's position. (6.1)

precipitate. An insoluble solid that separates from a supersaturated solution. (4.2)

precipitation reaction. A reaction characterized by the formation of a precipitate. (4.2)

precision. The closeness of agreement of two or more measurements of the same quantity. (1.6)

pressure. Force applied per unit area. (5.2)

product. The substance formed as a result of a chemical reaction. (3.7)

protein. A polymer of amino acids. (22.3)

proton. A subatomic particle having a single positive electric charge. The mass of a proton is about 1840 times that of an electron. (2.2)

Q

qualitative. Consisting of general observations about the system. (1.2)

qualitative analysis. The determination of the types of ions present in a solution. (17.8)

quantitative. Comprising numbers obtained by various measurements of the system. (1.2)

quantum. The smallest quantity of energy that can be emitted or absorbed in the form of electromagnetic radiation. (7.1)

quantum numbers. Numbers that describe the distribution of electrons in atoms. (7.6)

R

racemic mixture. An equimolar mixture of two enantiomers. (11.5)

radiant energy. Energy transmitted in the form of waves. (6.1)

radiation. The emission and transmission of energy through space in the form of particles and/or waves. (2.2)

radical. A species that contains an unpaired electron. (11.2)

radioactive decay series. A sequence of nuclear reactions that ultimately result in the formation of a stable isotope. (21.3)

radioactivity. The spontaneous breakdown of a nucleus by the emission of particles and/or radiation. (2.2)

Raoult's law. The partial pressure of the solvent over a solution is given by the product of the vapor pressure of the pure solvent and the mole fraction of the solvent in the solution. (13.6)

rare earth series. See lanthanide series.

rare gases. See noble gases.

rate constant (k). Proportionality constant relating reaction rate to the concentrations of reactants. (14.2)
rate law. An expression relating the rate of a reaction to the rate constant and the concentrations of the reactants. (14.2)
rate-determining step. The slowest step in the sequence of steps leading to the formation of products. (14.5)
reactants. The starting substances in a chemical reaction. (3.7)
reaction mechanism. The sequence of elementary steps that leads to product formation. (14.5)
reaction order. The sum of the powers to which all reactant concentrations appearing in the rate law are raised. (14.2)
reaction quotient (Q_c). A number equal to the ratio of product concentrations to reactant concentrations, each raised to the power of its stoichiometric coefficient at some point other than equilibrium. (15.3)
reaction rate. The change in concentration of reactant or product with time. (14.1)
redox reaction. A reaction in which there is either a transfer of electrons or a change in the oxidation numbers of the substances taking part in the reaction. Also called oxidation-reduction reaction. (4.4)
reducing agent. A substance that can donate electrons to another substance or decrease the oxidation numbers in another substance. (4.4)
reduction reaction. The half-reaction that represents the gain of electrons in a redox process. (4.4)
representative elements. Elements in Groups 1A through 7A, all of which have at least an incompletely filled *s* or *p* subshell of the highest principal quantum number. (8.2)
resonance. The use of two or more Lewis structures to represent a particular molecule. (9.8)
resonance structure. One of two or more alternative Lewis structures for a single molecule that cannot be described fully with a single Lewis structure. (9.8)
reversible reaction. A reaction that can occur in both directions. (4.1)
ribonucleic acid (RNA). A type of nucleic acid. (22.4)
root-mean-square (rms) speed (u_{rms}). A measure of the average molecular speed at a given temperature. (5.6)

S

salt. An ionic compound made up of a cation other than H^+ and an anion other than OH^- or O^{2-}. (4.3)
salt hydrolysis. The reaction of the anion or cation, or both, of a salt with water. (16.9)
saponification. Soapmaking. (11.4)
saturated hydrocarbons. Hydrocarbons that contain only single covalent bonds. (11.2)
saturated solution. At a given temperature, the solution that results when the maximum amount of a substance has dissolved in a solvent. (13.1)
scientific method. A systematic approach to research. (1.2)
second law of thermodynamics. The entropy of the universe increases in a spontaneous process and remains unchanged in an equilibrium process. (18.4)
second-order reaction. A reaction whose rate depends on the concentration of one reactant raised to the second power or on the concentrations of two different reactants, each raised to the first power. (14.3)
semipermeable membrane. A membrane that allows solvent molecules to pass through, but blocks the movement of solute molecules. (13.6)
sigma bond (σ bond). A covalent bond formed by orbitals overlapping end-to-end; its electron density is concentrated between the nuclei of the bonding atoms. (10.5)
sigma molecular orbital. A molecular orbital in which the electron density is concentrated around a line between the two nuclei of the bonding atoms. (10.6)
significant figures. The number of meaningful digits in a measured or calculated quantity. (1.6)
single bond. Two atoms held together by one electron pair are joined by a single bond. (9.4)
solubility. The maximum amount of solute that can be dissolved in a given quantity of solvent at a specific temperature. (4.2, 17.5)
solubility product (K_{sp}). The product of the molar concentrations of constituent ions, each raised to the power of its stoichiometric coefficient in the equilibrium equation. (17.5)
solute. The substance present in the smaller amount in a solution. (4.1)
solution. A homogeneous mixture of two or more substances. (4.1)
solvation. The process in which an ion or molecule is surrounded by solvent molecules arranged in an ordered manner. (13.2)
solvent. The substance present in the larger amount in a solution. (4.1)
specific heat (s). The amount of heat energy required to raise the temperature of one gram of a substance by one degree Celsius. (6.5)
spectator ions. Ions that are not involved in the overall reaction. (4.2)
spectrochemical series. A list of ligands arranged in order of their abilities to split the *d*-orbital energies. (20.4)
standard atmospheric pressure (1 atm). The pressure that supports a column of mercury exactly 76 cm high at 0°C at sea level. (5.2)
standard emf ($E°$). The difference of the standard reduction potentials of the substance that undergoes reduction and the substance that undergoes oxidation in a redox process. (19.3)
standard enthalpy of formation (ΔH_f°). The heat change that results when 1 mole of a compound is formed from its elements in their standard states. (6.6)
standard enthalpy of reaction (ΔH_{rxn}°). The enthalpy change that occurs when a reaction is carried out under standard-state conditions. (6.6)
standard entropy of reaction (ΔS_{rxn}°). The entropy change when the reaction is carried out under standard-state conditions. (18.4)
standard free energy of formation (ΔG_f°). The free-energy change when 1 mole of a compound is synthesized from its elements in their standard states. (18.5)
standard free energy of reaction (ΔG_{rxn}°). The free energy change when the reaction is carried out under standard-state conditions. (18.5)
standard reduction potential. The voltage measured as a reduction reaction occurs at the electrode when all solutes are 1 *M* and all gases are at 1 atm. (19.3)
standard solution. A solution of accurately known concentration. (4.6)
standard state. The condition of 1 atm of pressure. (6.6)
standard temperature and pressure (STP). 0°C and 1 atm. (5.4)
state function. A property that is determined by the state of the system. (6.3)
state of a system. The values of all pertinent macroscopic variables (for example, composition, volume, pressure, and temperature) of a system. (6.3)
stereoisomers. Compounds that are made up of the same types and numbers of atoms bonded together in the same sequence but with different spatial arrangements. (20.3)
stoichiometric amounts. The exact molar amounts of reactants and products that appear in a balanced chemical equation. (3.9)
stoichiometry. The mass relationships among reactants and products in chemical reactions. (3.8)
strong acid. An acid that is a strong electrolyte. (16.4)
strong base. A base that is a strong electrolyte. (16.4)
structural formula. A representation that shows how atoms are bonded to one another in a molecule. (2.6)
structural isomers. Molecules that have the same molecular formula but different structures. (11.2)

sublimation. The process in which molecules go directly from the solid phase into the vapor phase. (12.6)

substance. A form of matter that has a definite or constant composition (the number and type of basic units present) and distinct properties. (1.3)

supersaturated solution. A solution that contains more of the solute than is present in a saturated solution. (13.1)

surface tension. The amount of energy required to stretch or increase the surface of a liquid by a unit area. (12.3)

surroundings. The rest of the universe outside a system. (6.2)

system. Any specific part of the universe that is of interest to us. (6.2)

T

termolecular reaction. An elementary step involving three molecules. (14.5)

ternary compounds. Compounds consisting of three elements. (2.7)

theoretical yield. The amount of product predicted by the balanced equation when all of the limiting reagent has reacted. (3.10)

theory. A unifying principle that explains a body of facts and/or those laws that are based on them. (1.2)

thermal energy. Energy associated with the random motion of atoms and molecules. (6.1)

thermal pollution. The heating of the environment to temperatures that are harmful to its living inhabitants. (13.4)

thermochemical equation. An equation that shows both the mass and enthalpy relations. (6.4)

thermochemistry. The study of heat changes in chemical reactions. (6.2)

thermodynamics. The scientific study of the interconversion of heat and other forms of energy. (6.3)

thermonuclear reactions. Nuclear fusion reactions that occur at very high temperatures. (21.6)

third law of thermodynamics. The entropy of a perfect crystalline substance is zero at the absolute zero of temperature. (18.4)

titration. The gradual addition of a solution of accurately known concentration to another solution of unknown concentration until the chemical reaction between the two solutions is complete. (4.6)

tracers. Isotopes, especially radioactive isotopes, that are used to trace the path of the atoms of an element in a chemical or biological process. (21.7)

transition metals. Elements that have incompletely filled *d* subshells or readily give rise to cations that have incompletely filled *d* subshells. (7.9)

transition state. See activated complex.

transuranium elements. Elements with atomic numbers greater than 92. (21.4)

triple bond. A covalent bond in which two atoms share three pairs of electrons. (9.4)

triple point. The point at which the vapor, liquid, and solid states of a substance are in equilibrium. (12.7)

triprotic acid. Each unit of the acid yields three hydrogen ions. (4.3)

U

unimolecular reaction. An elementary step involving one molecule. (14.5)

unit cell. The basic repeating unit of the arrangement of atoms, molecules, or ions in a crystalline solid. (12.4)

unsaturated hydrocarbons. Hydrocarbons that contain carbon-carbon double bonds or carbon-carbon triple bonds. (11.2)

unsaturated solution. A solution that contains less solute than it has the capacity to dissolve. (13.1)

V

valence electrons. The outer electrons of an atom, which are the ones involved in chemical bonding. (8.2)

valence shell. The outermost electron-occupied shell of an atom, which holds the electrons that are usually involved in bonding. (10.1)

valence-shell electron-pair repulsion (VSEPR) model. A model that accounts for the geometrical arrangements of shared and unshared electron pairs around a central atom in terms of the repulsive forces between electron pairs. (10.1)

van der Waals equation. An equation that describes the relationships among P, V, n, and T for a nonideal gas. (5.7)

van der Waals forces. The collective name for certain attractive forces between atoms and molecules, namely, dipole-dipole, dipole-induced dipole, and dispersion forces. (12.2)

vaporization. The escape of molecules from the surface of a liquid; also called evaporation. (12.6)

viscosity. A measure of a fluid's resistance to flow. (12.3)

volatile. Having a measurable vapor pressure. (13.6)

volume. Length cubed. (1.5)

W

wave. A vibrating disturbance by which energy is transmitted. (7.1)

wavelength (λ). The distance between identical points on successive waves. (7.1)

weak acid. An acid that is a weak electrolyte. (16.4)

weak base. A base that is a weak electrolyte. (16.4)

weight. The force that gravity exerts on an object. (1.5)

work. Directed energy change resulting from a process. (6.1)

X

X-ray diffraction. The scattering of X rays by the units of a regular crystalline solid. (12.4)

ANSWERS
to Even-Numbered Problems

Chapter 1

1.8 (a) Physical change. (b) Chemical change. (c) Physical change. (d) Chemical change. (e) Physical change. **1.10** (a) Extensive. (b) Intensive. (c) Intensive. **1.12** (a) Compound. (b) Element. (c) Compound. (d) Element. **1.18** 1.30×10^3g. **1.20** (a) (i) 386 K. (ii) 3.10×10^2 K. (iii) 6.30×10^2 K. (b) (i) −196°C. (ii) −269°C. (iii) 328°C. **1.22** (a) 7.49×10^{-1}. (b) 8.026×10^2. (c) 6.21×10^{-7}. **1.24** (a) 0.00003256. (b) 6,030,000. **1.26** (a) 1.8×10^{-2}. (b) 1.14×10^{10}. (c) -5×10^4. (d) 1.3×10^3. **1.28** (a) Three. (b) One. (c) One or two. (d) Two. **1.30** (a) 1.28. (b) 3.18×10^{-3} mg. (c) 8.14×10^7 dm. (d) 3.8 m/s. **1.32** (a) 1.10×10^8 mg. (b) 6.83×10^{-5} m^3. **1.34** Student B is the most accurate; student C is the most precise. **1.36** (a) 81 in/s. (b) 1.2×10^2 m/min. (c) 7.4 km/h. **1.38** (a) 8.35×10^{12} mi. (b) 2.96×10^3 cm. (c) 9.8×10^8 ft/s. (d) 8.6°C. (e) −459.67°F. (f) 7.12×10^{-5} m^3. (g) 7.2×10^3 L. **1.40** 6.25×10^{-4} g/cm^3. **1.42** 4.35×10^7 ton. **1.44** 2.6 g/cm^3. **1.46** 0.882 cm. **1.48** 10.5 g/cm^3. **1.50** 767 mph. **1.52** 75.6°F ± 0.2°F. **1.54** 500 mL. **1.56** 5.5×10^{10} L. **1.58** −40°. **1.60** (a) 0.5%. (b) 3.1%. **1.62** 6.0×10^{12} g Au; \$2.0 × 10^{14}. **1.64** 30 times. **1.66** 1.450×10^{-2} mm. **1.68** 2.3×10^4 kg NaF; 99%. **1.70** 4.2×10^{-19} g/L. **1.72** (a) First, data was collected. Then a hypothesis was formulated. (b) (i) Is there a similarly high Ir content at different locations on Earth? (ii) Are there simultaneous extinctions of other large species in addition to dinosaurs? (c) Yes, because the hypothesis has survived many experimental tests. (d) 5.0×10^{14} kg, 5.5×10^{11} tons; 4×10^3 m. **1.74** 1725 sets; 4.576×10^4 g. **1.76** 4.97×10^4 g; alloy must be homogeneous in composition. **1.78** 0.13 L.

Chapter 2

2.8 0.62 mi. **2.12** 145. **2.14** ^{15}N: 7 protons, 7 electrons, and 8 neutrons; ^{33}S: 16 protons, 16 electrons, and 17 neutrons; ^{63}Cu: 29 protons, 29 electrons, and 34 neutrons; ^{84}Sr: 38 protons, 38 electrons, and 46 neutrons; ^{130}Ba: 56 protons, 56 electrons, and 74 neutrons; ^{186}W: 74 protons, 74 electrons, and 112 neutrons; ^{202}Hg: 80 protons, 80 electrons, and 122 neutrons. **2.16** (a) $^{186}_{74}$W. (b) $^{201}_{80}$Hg. **2.20** A: Alkali metals (Li, K); B: alkaline earth metals (Ca, Ba); C: halogens (Cl, I); D: noble gases (Ne, Xe). **2.22** (a) The metallic properties increase. (b) The metallic properties decrease. **2.24** Na, K; N, P; F, Cl. **2.30** (a) Diatomic molecule and compound. (b) Polyatomic molecule and compound. (c) Polyatomic molecule and element. **2.32** (a) H_2 and F_2. (b) HCl and CO. (c) P_4 and S_8. (d) H_2O and $C_6H_{12}O_6$. **2.34** (protons, electrons): K^+ (19, 18); Mg^{2+} (12, 10); Fe^{3+} (26, 23); Br^- (35, 36); Mn^{2+} (25, 23); C^{4-} (6, 10); Cu^{2+} (29, 27). **2.42** (a) $AlBr_3$. (b) $NaSO_2$. (c) N_2O_5. (d) $K_2Cr_2O_7$. **2.44** C_2H_6O. **2.46** Ionic: NaBr, BaF_2, CsCl. Molecular: CH_4, CCl_4, ICl, NF_3. **2.48** (a) Potassium hypochlorite. (b) Silver carbonate. (c) Iron(II) chloride. (d) Potassium permanganate. (e) Cesium chlorate. (f) Hypoiodous acid. (g) Iron(II) oxide. (h) Iron(III) oxide. (i) Titanium(IV) chloride. (j) Sodium hydride. (k) Lithium nitride. (l) Sodium oxide. (m) Sodium peroxide. (n) Iron(III) chloride hexahydrate. **2.50** CuCN. (b) $Sr(ClO_2)_2$. (c) $HClO_4$. (d) HI. (e) $Na_2NH_4PO_4$. (f) $PbCO_3$. (g) SnF_2. (h) P_4S_{10}. (i) HgO. (j) Hg_2I_2. (k) $CoCl_2 \cdot 6H_2O$. **2.52** (c). **2.54** (a) H or H_2? (b) NaCl is an ionic compound. **2.56** (a) Molecule and compound. (b) Element and molecule. (c) Element. (d) Molecule and compound. (c) Element. (f) Element and molecule. (g) Element and molecule. (h) Molecule and compound. (i) Compound. (j) Element. (k) Element and molecule. (l) Compound. **2.58** (a) CO_2 (solid). (b) NaCl. (c) N_2O. (d) $CaCO_3$. (e) CaO. (f) $Ca(OH)_2$. (g) $NaHCO_3$. (h) $Mg(OH)_2$. **2.60** (a) Metals in Groups 1A, 2A, and aluminum and nonmetals such as nitrogen, oxygen, and the halogens. (b) The transition metals. **2.62** ^{23}Na. **2.64** (a) 28. (b) 30. (c) 28. (d) 30. **2.66** H_2, N_2, O_2, O_3, F_2, Cl_2, He, Ne, Ar, Kr, Xe, Rn. **2.68** He, Ne, and Ar are chemically inert and do not react with other elements. **2.70** All isotopes of radium are radioactive. It is a radioactive decay product of ^{238}U. **2.72** (a) NaH (sodium hydride). (b) B_2O_3 (diboron trioxide). (c) Na_2S (sodium sulfide), (d) AlF_3 (aluminum fluoride). (e) OF_2 (oxygen difluoride). (f) $SrCl_2$ (strontium chloride). **2.74** (a) Bromine. (b) Radon. (c) Selenium. (d) Rubidium. (e) Lead. **2.76** 1.91×10^{-8} g. The change in mass is too small to measure accurately. **2.78** (a) Yes. (b) Ethane: CH_3, C_2H_6. Acetylene: CH, C_2H_2. **2.80** Manganese (Mn).

Chapter 3

3.6 92.5%. **3.8** 5.1×10^{24} amu. **3.12** 5.8×10^3 light-yr. **3.14** 9.96×10^{-15} mol Co. **3.16** 3.01×10^3 g Au. **3.18** (a) 1.244×10^{-22} g/As atom. (b) 9.746×10^{-23} g/Ni atom. **3.20** 2.98×10^{22} Cu atoms. **3.22** Pb. **3.24** (a) 73.89 g. (b) 76.15 g. (c) 119.37 g. (d) 176.12 g. (e) 101.11 g. (f) 100.95 g. **3.26** 4.51×10^{21} molecules. **3.28** N: 3.37×10^{26} atoms; C: 1.69×10^{26} atoms; O: 1.69×10^{26} atoms; H: 6.74×10^{26} atoms. **3.30** 8.56×10^{22} molecules. **3.34** 7. **3.40** C: 10.06%; H: 0.8442%; Cl: 89.07%. **3.42** NH_3. **3.44** $C_2H_3NO_5$. **3.46** 39.3 g S. **3.48** 5.97 g F. **3.50** (a) CH_2O. (b) KCN. **3.52** C_6H_6. **3.54** $C_5H_8O_4NNa$. **3.60** (a) $2N_2O_5 \longrightarrow 2N_2O_4 + O_2$. (b) $2KNO_3 \longrightarrow 2KNO_2 + O_2$. (c) $NH_4NO_3 \longrightarrow N_2O + 2H_2O$. (d) $NH_4NO_2 \longrightarrow N_2 + 2H_2O$. (e) $2NaHCO_3 \longrightarrow Na_2CO_3 + H_2O + CO_2$. (f) $P_4O_{10} + 6H_2O \longrightarrow 4H_3PO_4$. (g) $2HCl + CaCO_3 \longrightarrow CaCl_2 + H_2O + CO_2$. (h) $2Al + 3H_2SO_4 \longrightarrow Al_2(SO_4)_3 + 3H_2$. (i) $CO_2 + 2KOH \longrightarrow K_2CO_3 + H_2O$. (j) $CH_4 + 2O_2 \longrightarrow CO_2 + 2H_2O$. (k) $Be_2C + 4H_2O \longrightarrow 2Be(OH)_2 + CH_4$. (l) $3Cu + 8HNO_3 \longrightarrow 3Cu(NO_3)_2 + 2NO + 4H_2O$. (m) $S + 6HNO_3 \longrightarrow H_2SO_4 + 6NO_2 + 2H_2O$. (n) $2NH_3 + 3CuO \longrightarrow 3Cu + N_2 + 3H_2O$. **3.64** (d). **3.66** 1.01 mol. **3.68** 20 mol. **3.70** (a) $2NaHCO_3 \longrightarrow Na_2CO_3 + CO_2 + H_2O$. (b) 78.3 g. **3.72** 255.9 g; 0.324 L. **3.74** 0.294 mol. **3.76** (a) $NH_4NO_3 \longrightarrow N_2O + 2H_2O$. (b) 20 g N_2O. **3.78** 18.0 g. **3.82** 6 mol NH_3; 1 mol H_2. **3.84** 0.709 g NO_2; O_3; 6.9×10^{-3} mol NO. **3.86** HCl; 23.4 g. **3.90** (a) 7.05 g. (b) 92.9%. **3.92** 20.6 g; 28.6%. **3.94** (b). **3.96** Cl_2O_7. **3.98** 18. **3.100** 65.4 amu; Zn. **3.102** 89.6%. **3.104** CH_2O; $C_6H_{12}O_6$. **3.106** (a) Mn_3O_4. (b) $3MnO_2 \longrightarrow Mn_3O_4 + O_2$. **3.108** Mg_3N_2 (magnesium nitride). **3.110** $C_3H_7NO_2S$. C: 29.74%; H: 5.823%; N: 11.56%; O: 26.41%; S: 26.47%. **3.112** 4%. **3.114** (a) The less abundant isotope was undergoing radioactive decay. (b) 16 amu: CH_4; 17 amu: NH_3; 18 amu: H_2O; 64 amu: SO_2. (c) C_3H_8. A fragment (CH_3) can break off from C_3H_8, but not from CO_2. (d) ±0.030 amu. (e) Do a mass spectrometry analysis of gold samples and compare the types and amounts of impurity elements present from different sources. **3.116** 9.6 g Fe_2O_3; 7.4 g $KClO_3$.

Chapter 4

4.8 (c). **4.10** (a) Strong electrolyte. (b) Nonelectrolyte. (c) Weak electrolyte. (d) Strong electrolyte. **4.12** (b) and (c). Ions have no mobility in the solid. **4.14** HCl does not ionize in benzene. **4.18** (b). **4.20** (a) Insoluble. (b) Soluble. (c) Soluble. (d) Insoluble. (e) Soluble. **4.22** (a) Ionic: $2Na^+(aq) + S^{2-}(aq) + Zn^{2+}(aq) + 2Cl^-(aq) \longrightarrow ZnS(s) + 2Na^+(aq) + 2Cl^-(aq)$. Net ionic: $Zn^{2+}(aq) + S^{2-}(aq) \longrightarrow ZnS(s)$. (b) Ionic: $6K^+(aq) + 2PO_4^{3-}(aq) + 3Sr^{2+}(aq) + 6NO_3^-(aq) \longrightarrow Sr_3(PO_4)_2(s) + 6K^+(aq) + 6NO_3^-(aq)$. Net ionic: $3Sr^{2+}(aq) + 2PO_4^{3-}(aq) \longrightarrow Sr_3(PO_4)_2(s)$. (c) Ionic: $Mg^{2+}(aq) + 2NO_3^-(aq) + 2Na^+(aq) + 2OH^-(aq) \longrightarrow Mg(OH)_2(s) + 2Na^+(aq) + 2NO_3^-(aq)$. Net ionic: $Mg^{2+}(aq) + 2OH^-(aq) \longrightarrow Mg(OH)_2(s)$. **4.24** (a) Add chloride ions. (b) Add hydroxide ions. (c) Add carbonate ions. (d) Add sulfate ions. **4.32** (a) Brønsted base. (b) Brønsted base. (c) Brønsted acid. (d) Brønsted base and Brønsted acid. **4.34** (a) $CH_3COOH + K^+ + OH^- \longrightarrow K^+ + CH_3COO^- + H_2O$; $CH_3COOH + OH^- \longrightarrow CH_3COO^- + H_2O$. (b) $H_2CO_3 + 2Na^+ + 2OH^- \longrightarrow 2Na^+ + CO_3^{2-} + 2H_2O$; $H_2CO_3 + 2OH^- \longrightarrow CO_3^{2-} + 2H_2O$. (c) $2H^+ + 2NO_3^- + Ba^{2+} + 2OH^- \longrightarrow Ba^{2+} + 2NO_3^- + 2H_2O$; $2H^+ + 2OH^- \longrightarrow 2H_2O$. **4.40** (a) $Fe \longrightarrow Fe^{3+} + 3e^-$; $O_2 + 4e^- \longrightarrow 2O^{2-}$. Oxidizing agent: O_2; reducing agent: Fe. (b) $2Br^- \longrightarrow Br_2 + 2e^-$; $Cl_2 + 2e^- \longrightarrow 2Cl^-$. Oxidizing agent: Cl_2; reducing agent: Br^-. (c) $Si \longrightarrow Si^{4+} + 4e^-$; $F_2 + 2e^- \longrightarrow 2F^-$. Oxidizing agent: F_2; reducing agent: Si. (d) $H_2 \longrightarrow 2H^+ + 2e^-$; $Cl_2 + 2e^- \longrightarrow 2Cl^-$. Oxidizing agent: Cl_2; reducing agent: H_2. **4.42** (a) +5. (b) +1. (c) +3. (d) +5. (e) +5. (f) +5. **4.44** All are zero. **4.46** (a) −3. (b) $-\frac{1}{2}$. (c) −1. (d) +4. (e) +3. (f) −2. (g) +3. (h) +6. **4.48** (b) and (d). **4.50** (a) No reaction. (b) No reaction. (c) $Mg + CuSO_4 \longrightarrow MgSO_4 + Cu$. (d) $Cl_2 + 2KBr \longrightarrow Br_2 + 2KCl$. **4.54** Dissolve 15.0 g $NaNO_3$ in enough water to make up 250 mL. **4.56** 10.8 g. **4.58** (a) 1.37 *M*. (b) 0.426 *M*. (c) 0.716 *M*. **4.60** (a) 6.50 g. (b) 2.45 g. (c) 2.65 g. (d) 7.36 g. (e) 3.95 g. **4.64** 0.0433 *M*. **4.66** 126 mL. **4.68** 1.09 *M*. **4.72** 35.72%. **4.74** 2.31×10^{-4} *M*. **4.78** 0.217 *M*. **4.80** (a) 6.00 mL. (b) 8.00 mL. **4.82** The Ba^{2+} ions combine with the SO_4^{2-} ions to form $BaSO_4$ precipitate. **4.84** Physical test: Only the NaCl solution would conduct electricity. Chemical test: Add $AgNO_3$ solution. Only the NaCl solution would give AgCl precipitate. **4.86** Mg, Na, Ca, Ba, K, or Li. **4.88** Oxidation number of C is +2 in CO and +4 (maximum) in CO_2. **4.90** 1.26 *M*. **4.92** 0.171 *M*. **4.94** 0.115 *M*. **4.96** 0.80 L. **4.98** 1.73 *M*. **4.100** $NaHCO_3$. NaOH is caustic and expensive. **4.102** 44.11% NaCl; 55.89% KCl. **4.104** (a) The precipitate $CaSO_4$ formed over Ca prevents the Ca from reacting with the sulfuric acid. (b) Aluminum is protected by a tenacious oxide layer (Al_2O_3). (c) These metals react more readily with water. (d) The metal should be placed below Fe and above H. **4.106** (a) 8.312×10^{-7} *M*. (b) 3.284×10^{-5} g. **4.108** 4.99 grains. **4.110** (a) $NH_4^+ + OH^- \longrightarrow NH_3 + H_2O$. (b) 97.99%. **4.112** Because the volume of the solution changes (increases or decreases) when the solid dissolves. **4.114** $FeCl_2 \cdot 4H_2O$. **4.116** (a) Reactions (2) and (4). (b) Reaction (5). (c) Reaction (3). (d) Reaction (1).

Chapter 5

5.14 0.797 atm; 80.8 kPa. **5.18** (1) b. (2) a. (3) c. (4) a. **5.20** 53 atm. **5.22** (a) 0.69 L. (b) 61 atm. **5.24** 1.3×10^2 K. **5.26** ClF_3. **5.32** 6.2 atm. **5.34** 472°C. **5.36** 1.9 atm. **5.38** 0.82 L. **5.40** 33.6 mL. **5.42** 6.1×10^{-3} atm. **5.44** 35.1 g/mol. **5.46** N_2: 2.1×10^{22} molecules; O_2: 5.7×10^{21} molecules; Ar: 3×10^{20} atoms. **5.48** 2.98 g/L. **5.50** SF_4. **5.52** 18 g. **5.54** P_2F_4. **5.58** (a) 0.89 atm. (b) 1.4 L. **5.60** 349 mmHg. **5.62** (a) Right box. (b) Left box. **5.64** 19.8 g. **5.66** N_2: 217 mmHg; H_2: 650 mmHg. **5.74** N_2: 472 m/s; O_2: 441 m/s; O_3: 360 m/s. **5.76** Average speed: 2.7 m/s; rms speed: 2.8 m/s. Squaring favors the larger values compared to just taking the average value. **5.82** Not ideal. Ideal gas pressure is 164 atm. **5.84** 1.7×10^2 L. CO_2: 0.49 atm; N_2: 0.25 atm; H_2O: 0.41 atm; O_2: 0.041 atm. **5.86** (a) $NH_4NO_2(s) \longrightarrow N_2(g) + 2H_2O(l)$. (b) 0.273 g. **5.88** No. An ideal gas would not condense. **5.90** Higher in the winter because of reduced photosynthesis. **5.92** (a) 0.86 L. (b) $NH_4HCO_3(s) \longrightarrow NH_3(g) + CO_2(g) + H_2O(l)$. Advantage: more gases (CO_2 and NH_3) generated; disadvantage: odor of ammonia! **5.94** 3.88 L. **5.96** (a) $C_3H_8(g) + 5O_2(g) \longrightarrow 3CO_2(g) + 4H_2O(g)$. (b) 11.4 L. **5.98** O_2: 0.166 atm; NO_2: 0.333 atm. **5.100** (a) $CO_2(g) + CaO(s) \longrightarrow CaCO_3(s)$; $CO_2(g) + BaO(s) \longrightarrow BaCO_3(s)$. (b) 10.5% CaO; 89.5% BaO. **5.102** 1.7×10^{12} O_2 molecules. **5.104** Ne, because it has the smallest *a* value. **5.106** 53.4%. **5.108** 1072 mmHg. **5.110** 43.8 g/mol; CO_2. **5.112** (a) No. Temperature is a statistical concept. (b) (i) Both are the same. (ii) The He atoms in V_1 collide with the walls more frequently. (c) (i) The rms speed is greater at T_2. (ii) The He atoms at T_2 collide with the walls more frequently and with greater force. (d) (i) False. (ii) True. (iii) True. **5.114** (a) u_{mp}: 421 m/s; u_{rms}: 515 m/s. (b) 1200 K. **5.116** (a) $8(4\pi r^3/3)$. (b) $4N_A(4\pi r^3/3)$. The excluded volume is 4 times the volume of the atoms. **5.118** NO. **5.120** (b).

Chapter 6

6.16 (a) 0. (b) −9.5 J. (c) −18 J. **6.18** 48 J. **6.20** -3.1×10^3 J. **6.26** 1.57×10^4 kJ. **6.28** 0; −553.8 kJ/mol. **6.32** A. **6.34** 728 kJ. **6.36** 50.7°C. **6.38** 26.3°C. **6.46** O_2. **6.48** (a) $\Delta H_f^\circ[Br_2(l)] = 0$; $\Delta H_f^\circ[Br_2(g)] > 0$. (b) $\Delta H_f^\circ[I_2(s)] = 0$; $\Delta H_f^\circ[I_2(g)] > 0$. **6.50** Measure ΔH° for the formation of Ag_2O from Ag and O_2 and of $CaCl_2$ from Ca and Cl_2. **6.52** (a) −167.2 kJ/mol. (b) −56.2 kJ/mol. **6.54** (a) −1411 kJ/mol. (b) −1124 kJ/mol. **6.56** 218.2 kJ/mol. **6.58** 4.51 kJ/g. **6.60** 2.70×10^2 kJ. **6.62** −84.6 kJ/mol. **6.64** −847.6 kJ/mol. **6.66** $\Delta H_2 - \Delta H_1$. **6.68** (a) −336.5 kJ/mol. (b) NH_3. **6.70** 43.6 kJ. **6.72** 0. **6.74** −350.7 kJ/mol. **6.76** 0.492 J/g · °C. **6.78** The first (exothermic) reaction can be used to promote the second (endothermic) reaction. **6.80** 1.09×10^4 L. **6.82** 5.60 kJ/mol. **6.84** (a). **6.86** (a) A more fully packed freezer has a greater mass and hence a larger heat capacity. (b) Tea or coffee has a greater amount of water, which has a higher specific heat than noodles. **6.88** 1.84×10^3 kJ. **6.90** 3.0×10^9. **6.92** 5.35 kJ/°C. **6.94** -5.2×10^6 kJ. **6.96** (a) 1.4×10^2 kJ. (b) 3.9×10^2 kJ. **6.98** 104 g. **6.100** 9.9×10^8 J; 304°C. **6.102** (a) $CaC_2 + 2H_2O \longrightarrow C_2H_2 + Ca(OH)_2$. (b) 1.51×10^3 kJ. **6.104** (a) $\Delta U = w$. The system does work on the surroundings and decreases its internal energy. The water cools to form snow. (b) Just the opposite of (a). (c) First set $q = (1/2)mu^2$. Next, show stopping distance, *d*, to be proportional to *q* and hence to u^2. As *u* increases to 2*u*, *d* increases to $4u^2$, which is proportional to 4*d*. **6.106** The deserts, unlike coastal regions, have very low humidity and cannot retain heat because sand and rocks have smaller specific heat, than water. **6.108** 96.21%. **6.110** (a) CH. (b) 49 kJ/mol. **6.112** −101.3 J. Yes, because in a cyclic process, the change in a state function must be zero. **6.114** -9.75×10^3 J/mol.

Chapter 7

7.8 (a) 6.58×10^{14}/s. (b) 1.22×10^8 nm. **7.10** 2.5 min. **7.12** 4.95×10^{14}/s. **7.16** (a) 4.0×10^2 nm. (b) 5.0×10^{-19} J. **7.18** 1.2×10^2 nm (UV). **7.20** (a) 3.70×10^2 nm; 3.70×10^{-7} m. (b) UV. (c) 5.38×10^{-19} J. **7.22** 8.16×10^{-19} J. **7.26** Use a prism.

7.28 Compare the emission spectra with those on Earth of known elements. **7.30** 3.027×10^{-19} J. **7.32** 6.17×10^{14}/s; 486 nm. **7.34** 5. **7.40** 1.37×10^{-6} nm. **7.42** 1.7×10^{-23} nm. **7.54** $l = 2$: $m_l = -2, -1, 0, 1, 2$. $l = 1$: $m_l = -1, 0, 1$. $l = 0$: $m_l = 0$. **7.56** (a) $n = 3$, $l = 0$, $m_l = 0$. (b) $n = 4$, $l = 1$, $m_l = -1, 0, 1$. (c) $n = 3$, $l = 2$, $m_l = -2, -1, 0, 1, 2$. In all cases, $m_s = +\frac{1}{2}$ or $-\frac{1}{2}$. **7.58** Differ in orientation only. **7.60** 6*s*, 6*p*, 6*d*, 6*f*, 6*g*, and 6*h*. **7.62** $2n^2$. **7.64** (a) 3. (b) 6. (c) 0. **7.66** There is no shielding in an H atom. **7.68** (a) $2s < 2p$. (b) $3p < 3d$. (c) $3s < 4s$. (d) $4d < 5f$. **7.80** Al: $1s^22s^22p^63s^23p^1$. B: $1s^22s^22p^1$. F: $1s^22s^22p^5$. **7.82** B(1), Ne(0), P(3), Sc(1), Mn(5), Se(2), Kr(0), Fe(4), Cd(0), I(1), Pb(2). **7.84** Ge: $[\text{Ar}]4s^23d^{10}4p^2$. Fe: $[\text{Ar}]4s^23d^6$. Zn: $[\text{Ar}]4s^23d^{10}$. Ni: $[\text{Ar}]4s^23d^8$. W: $[\text{Xe}]6s^24f^{14}5d^4$. Tl: $[\text{Xe}]6s^24f^{14}5d^{10}6p^1$. **7.86** S^+. **7.92** $[\text{Kr}]5s^24d^5$. **7.94** (a) An *e* in a 2*s* and an *e* in each 2*p* orbital. (b) 2 *e* each in a 4*p*, a 4*d*, and a 4*f* orbital. (c) 2 *e* in each of the 5 3*d* orbitals. (d) An *e* in a 2*s* orbital. (e) 2 *e* in a 4*f* orbital. **7.96** Wave properties. **7.98** (a) 1.05×10^{-25} nm. (b) 8.86 nm. **7.100** (a) 1.20×10^{18} photons. (b) 3.76×10^8 W. **7.102** 419 nm. In principle, yes; in practice, no. **7.104** 3.0×10^{19} photons. **7.106** He^+: 164 nm, 121 nm, 109 nm, 103 nm (all in the UV region). H: 657 nm, 487 nm, 434 nm, 411 nm (all in the visible region). **7.108** 1.2×10^2 photons. **7.110** Yellow light will generate more electrons; blue light will generate electrons with greater kinetic energy. **7.112** 7.39×10^{-2} nm. **7.114** 0.929 pm; 3.23×10^{20}/s. **7.116** (b) and (d). **7.118** 10.6 nm. **7.120** 3.87×10^5 m/s.

7.122 (a)

$$n = 2,\ E_2 = \frac{5}{2}h\nu$$
$$n = 1,\ E_1 = \frac{3}{2}h\nu$$
$$n = 0,\ E_0 = \frac{1}{2}h\nu$$

(*E* increasing upward)

(b) 5.74×10^{-20} J. (c) If the molecules were not vibrating, then the uncertainty in its velocity and hence its momentum must be zero. However, the uncertainty in position for a nonvibrating molecule would also be zero because the bond distance is fixed. Thus, knowing both the momentum and position of the molecule with certainty simultaneously violates the Heisenberg uncertainty principle. **7.124** 1.06 nm.

Chapter 8

8.16 (a) $1s^22s^22p^63s^23p^5$. (b) Representative element. (c) Paramagnetic. **8.18** (a) and (d); (b) and (e); (c) and (f). **8.20** (a) Group 1A. (b) Group 5A. (c) Group 8A. (d) Group 8B. **8.22** Fe. **8.28** (a) [Ne]. (b) [Ne]. (c) [Ar]. (d) [Ar]. (e) [Ar]. (f) $[\text{Ar}]3d^6$. (g) $[\text{Ar}]3d^9$. (h) $[\text{Ar}]3d^{10}$. **8.30** (a) Cr^{3+}. (b) Sc^{3+}. (c) Rh^{3+}. (d) Ir^{3+}. **8.32** Be^{2+} and He; F^- and N^{3-}; Fe^{2+} and Co^{3+}; S^{2-} and Ar. **8.38** Na > Mg > Al > P > Cl. **8.40** F. **8.42** The effective nuclear charge on the outermost electrons increases from left to right as a result of incomplete shielding by the inner electrons. **8.44** $Mg^{2+} < Na^+ < F^- < O^{2-} < N^{3-}$. **8.46** Te^{2-}. **8.48** -199.4°C. **8.52** The $3p^1$ electron of Al is effectively shielded by the inner electrons and $3s^2$ electrons. **8.54** 2080 kJ/mol is for $1s^22s^22p^6$. **8.56** 8.40×10^6 kJ/mol. **8.60** Cl. **8.62** Alkali metals have ns^1 and so can accept another electron. **8.66** Low ionization energy, reacts with water to form FrOH, and with oxygen to form oxide and superoxide. **8.68** The ns^1 electron of Group 1B metals is incompletely shielded by the inner *d* electrons and therefore they have much higher first ionization energies. **8.70** (a) $Li_2O(s) + H_2O(l) \longrightarrow 2LiOH(aq)$. (b) $CaO(s) + H_2O(l) \longrightarrow Ca(OH)_2(aq)$. (c) $CO_2(g) + H_2O(l) \longrightarrow H_2CO_3(aq)$. **8.72** BaO. Ba is more metallic. **8.74** (a) Bromine. (b) Nitrogen. (c) Rubidium. (d) Magnesium. **8.76** (a) $Mg^{2+} < Na^+ < F^- < O^{2-}$. (b) $O^{2-} < F^- < Na^+ < Mg^{2+}$. **8.78** M is potassium (K) and X is bromine (Br_2). **8.80** O^+ and N; Ar and S^{2-}; Ne and N^{3-}; Zn and As^{3+}; Cs^+ and Xe. **8.82** (a) and (d). **8.84** First: bromine; second: iodine; third: chlorine; fourth: fluorine. **8.86** Fluorine. **8.88** H^-. **8.90** Li_2O (basic), BeO (amphoteric), B_2O_3 (acidic), CO_2 (acidic), N_2O_5 (acidic). **8.92** It can form H^+ and H^- ions. **8.94** 418 kJ/mol. Vary the frequency. **8.96** About 23°C. **8.98** (a) Mg [in $Mg(OH)_2$]. (b) Liquid Na. (c) Mg (in $MgSO_4 \cdot 7H_2O$). (d) Na (in $NaHCO_3$). (e) K (in KNO_3). (f) Mg. (g) Ca (in CaO). (h) Ca. (i) Na (in NaCl); Ca (in $CaCl_2$). **8.100** Physical properties: solid, high m.p.; $2NaAt + 2H_2SO_4 \longrightarrow At_2 + SO_2 + Na_2SO_4 + 2H_2O$. **8.102** 242 nm. **8.104** (a) F_2. (b) Na. (c) B. (d) N_2. (e) Al. **8.106** (a) Use a discharge tube and compare the emission spectrum of Ar with other elements. (b) Ar is inert. (c) With the discovery of Ar, Ramsay realized there may be other unreactive gases belonging to the same periodic group. (d) Being a light gas, helium's concentration in the atmosphere is very low. It is chemically inert. (e) Radon is unreactive toward most elements and its has a short half-life. **8.108** 419 nm. **8.110** To prepare NH_3, isolate N_2 from air and obtain H_2 by electrolyzing water. Combine N_2 and H_2 to form NH_3. To prepare HNO_3, first combine N_2 and O_2 to form NO, which is further oxidized to NO_2. Next, react NO_2 with water to form HNO_2 and HNO_3. Finally, combine NH_3 and HNO_3 to form NH_4NO_3.

Chapter 9

9.16 (a) RbI, rubidium iodide. (b) Cs_2SO_4, cesium sulfate. (c) Sr_3N_2, strontium nitride. (d) Al_2S_3, aluminum sulfide.

9.18 (a) $\cdot\text{Sr}\cdot + \cdot\ddot{\text{Se}}\cdot \longrightarrow \text{Sr}^{2+}\ :\ddot{\text{Se}}:^{2-}$

(b) $\cdot\text{Ca}\cdot + 2\text{H}\cdot \longrightarrow \text{Ca}^{2+}\ 2\text{H}:^{-}$

(c) $3\text{Li}\cdot + \cdot\dot{\text{N}}\cdot \longrightarrow 3\text{Li}^{+}\ :\ddot{\text{N}}:^{3-}$

(d) $2\cdot\dot{\text{Al}}\cdot + 3\cdot\ddot{\text{S}}\cdot \longrightarrow 2\text{Al}^{3+}\ 3:\ddot{\text{S}}:^{2-}$

9.20 (a) BF_3, covalent. Boron triflouride. (b) KBr, ionic. Potassium bromide. **9.26** 2195 kJ/mol. **9.34** C—H < Br—H < F—H < Li—Cl < Na—Cl < K—F. **9.36** Cl—Cl < Br—Cl < Si—C < Cs—F. **9.38** (a) Covalent. (b) Polar covalent. (c) Ionic. (d) Polar covalent.

9.42 (a) $[:\ddot{\text{O}}—\ddot{\text{O}}:]^{2-}$

(b) $[:\text{C}\equiv\text{C}:]^{2-}$

(c) $[:\text{N}\equiv\text{O}:]^{+}$

(d) $[\text{NH}_4]^+$: N bonded to four H atoms (H—N—H, with H above and below N), bracketed, charge +

9.44 (a) Neither O atom has a complete octet; one H atom forms a double bond.

(b) H—C(H)(H)—C(=Ö:)—Ö—H

9.48 (1) The C atom has a lone pair. (2) The negative charge is on the less electronegative C atom.

9.50 $\ddot{\text{O}}=\text{Cl}(-\ddot{\text{O}}:^{-})—\ddot{\text{O}}:^{-}$ (+ on Cl) $\longleftrightarrow$ $^{-}:\ddot{\text{O}}—\text{Cl}(=\ddot{\text{O}})—\ddot{\text{O}}:^{-}$ (+ on Cl) $\longleftrightarrow$ $^{-}:\ddot{\text{O}}—\text{Cl}(-\ddot{\text{O}}:^{-})=\ddot{\text{O}}$ (+ on Cl)

9.52 $H-\overset{H}{C}=\overset{+}{N}=N^- \longleftrightarrow H-\overset{H}{\overset{-}{C}}-\overset{+}{N}\equiv N:$

9.54 $O=C=S \longleftrightarrow :\overset{-}{O}-C\equiv\overset{+}{S}: \longleftrightarrow :\overset{+}{O}\equiv C-\overset{-}{S}:$

9.60 $^{+}Cl=\overset{2-}{Be}=Cl^{+}$

9.62 $Cl-\overset{Cl}{Sb}(Cl)-Cl$ No.

9.64 $:Cl-Al(Cl)-Cl: + :Cl:^- \longrightarrow [:Cl-Al(Cl)_2-Cl:]^-$

9.68 303.0 kJ/mol. **9.70** (a) −2759 kJ/mol. (b) −2855 kJ/mol. **9.72** Covalent: SiF_4, PF_5, SF_6, ClF_3; ionic: NaF, MgF_2, AlF_3. **9.74** KF is a solid; has a high melting point; is an electrolyte. CO_2 is a gas; it is a molecular compound.

9.76 $:\overset{-}{N}=\overset{+}{N}=\overset{-}{N}: \longleftrightarrow {}^{2-}:N-\overset{+}{N}\equiv N: \longleftrightarrow :N\equiv\overset{+}{N}-N:^{2-}$

9.78 (a) $AlCl_4^-$. (b) AlF_6^{3-}. (c) $AlCl_3$. **9.80** C has incomplete octet in CF_2; C has an expanded octet in CH_5; F and H can only form a single bond; the I atoms are too large to surround the P atom. **9.82** (a) False. (b) True. (c) False. (d) False. **9.84** −67 kJ/mol. **9.86** N_2. **9.88** NH_4^+ and CH_4; CO and N_2; $B_3N_3H_6$ and C_6H_6.

9.90 $H-\overset{-}{N}(H): + H-O(H): \longrightarrow H-N(H)-H + {}^{-}:O-H$

9.92 F cannot form an expanded octet.

9.94 $H-C(H)_2-N=C=O \longleftrightarrow H-C(H)_2-\overset{+}{N}\equiv C-O:^-$

9.96 (a) Important. (b) Important. (c) Not important. (d) Not important.

9.98 $F-C(Cl)_2-Cl$ $\quad F-C(Cl)_2-F$ $\quad H-C(F)(Cl)-F$ $\quad F-C(F)(H)-C(F)(H)-F$

9.100 (a) −9.2 kJ/mol. (b) −9.2 kJ/mol. **9.102** (a) ${}^{-}:C\equiv O:^{+}$ (b) $:N\equiv O:^{+}$ (c) ${}^{-}:C\equiv N:$ (d) $:N\equiv N:$ **9.104** True.

9.106 $:\overset{-}{C}\equiv\overset{+}{N}-O: \longleftrightarrow {}^{2-}C=\overset{+}{N}=O \longleftrightarrow {}^{3-}:C-\overset{+}{N}\equiv O:^{+}$

The importance decreases from left to right. **9.108** 347 kJ/mol. **9.110** (a) −107 kJ/mol. (b) −92.6 kJ/mol.

9.112 (a) $H-C(H)(Cl)-C(H)(Cl)-H$ (ethylene dichloride) $\quad$ $H_2C=C(H)Cl$ (vinyl chloride)

C—C bond covalent; C—H and C—Cl bonds polar.

(b) $-C(H)_2-C(H)(Cl)-C(H)(H)-C(H)(Cl)-C(H)(H)-C(H)(Cl)-$ (chain with H above each C and H, Cl, H, Cl, H, Cl below)

(c) -1.2×10^6 kJ. This process generates a great deal of heat. The design would need to incorporate a means of cooling the reaction vessel.

9.114 (a) $H{-}H\ H^+ \longleftrightarrow H^+\ H{-}H \longleftrightarrow H^+\ H-H$ (b) −413 kJ/mol.

9.116 $:N\equiv\overset{+}{N}-N=\overset{+}{N}=N^- \longleftrightarrow {}^{-}N=\overset{+}{N}=N-\overset{+}{N}\equiv N: \longleftrightarrow :N\equiv\overset{+}{N}-\overset{2-}{N}-\overset{+}{N}\equiv N:$

9.118 The repulsion between lone pairs on adjacent atoms weakens the bond. There are two lone pairs on each O atom in H_2O_2. The repulsion is the greatest; it has the smallest bond enthalpy (about 142 kJ/mol). There is one lone pair on each N atom in N_2H_4; it has the intermediate bond enthalpy (about 193 kJ/mol). There are no lone pairs on the C atoms in C_2H_6; it has the greatest bond enthalpy (about 347 kJ/mol).

Chapter 10

10.8 (a) Trigonal planar. (b) Linear. (c) Tetrahedral. **10.10** (a) Tetrahedral. (b) Bent. (c) Trigonal planar. (d) Linear. (e) Square planar. (f) Tetrahedral. (g) Trigonal bipyramidal. (h) Trigonal pyramidal. (i) Tetrahedral. **10.12** $SiCl_4$, CI_4, $CdCl_4^{2-}$. **10.18** Electronegativity decreases from F to I. **10.20** Higher. **10.22** (b) = (d) < (c) < (a). **10.32** sp^3 for both Si atoms. **10.34** B: sp^2 to sp^3; N: remains sp^3. **10.36** (a) sp^3. (b) sp^3, sp^2, sp^2. (c) sp^3, sp, sp, sp^3. (d) sp^3, sp^2. (e) sp^3, sp^2. **10.38** sp. **10.40** sp^3d. **10.42** Nine pi bonds and nine sigma bonds. **10.44** IF_4^-. **10.50** Electron spins must be paired in H_2. **10.52** $Li_2^- = Li_2^+ < Li_2$. **10.54** B_2^+. **10.56** Only MO theory predicts O_2 to be paramagnetic. **10.58** $O_2^{2-} < O_2^- < O_2 < O_2^+$. **10.60** B_2 contains a pi bond; C_2 contains two pi bonds. **10.62** Linear. Dipole moment measurement. **10.64** The large size of Si results in poor sideways overlap of *p* orbitals to form pi bonds. **10.66** XeF_3^+: T-shaped; XeF_5^+: square pyramidal; SbF_6^-: octahedral. **10.68** (a) 180°. (b) 120°. (c) 109.5°. (d) About 109.5°. (e) 180°. (f) About 120°. (g) About 109.5°. (h) 109.5°. **10.70** sp^3d. **10.72** ICl_2^- and $CdBr_2$. **10.74** (a) sp^2. (b) Molecule on the right. **10.76** The pi bond in *cis*-dichloroethylene prevents rotation. **10.78** Si has 3*d* orbitals so water can add to Si (valence shell expansion). **10.80** C: sp^2; N: N atom that forms a double bond is sp^2, the others are sp^3. O is sp^2. **10.82** 43.6%.

10.84 (a) $Cl_2Al(\mu\text{-}Cl)_2AlCl_2$ (two Al atoms bridged by two Cl atoms, each Al bonded to two terminal Cl atoms)

(b) The hybridization of Al in $AlCl_3$ is sp^2. The molecule is trigonal planar. The hybridization of Al in Al_2Cl_6 is sp^3.

(c) The geometry about each Al atom is tetrahedral.

(d) The molecules are nonpolar; they do not possess a dipole moment.

10.86 The double bonded C atoms (4) are sp^2 hybridized. The rest are all sp^3 hybridized. **10.88** (a) A σ bond is formed by end-to-end overlaps. Rotation will not break this overlap. A π bond is formed by sideways overlap. The two 90° rotations will break and then reform the π bond. (b) The π bond is weaker because of the lesser extent of overlap. (c) 444 nm.

10.90 H
H
H
H

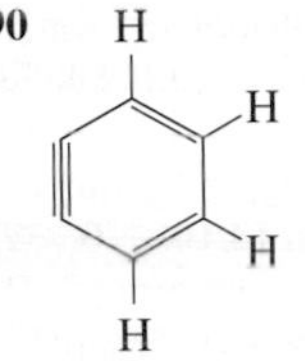

The normal bond angle for a hexagon is 60°. However, the triple bond requires an angle of 180° (linear) and therefore there is a great deal of strain in the molecule. Consequently, the molecule is very reactive (breaking the bond to relieve the strain). **10.92** The Lewis structure shows four pairs of electrons on the two oxygen atoms. From Table 10.5 we see that these eight valence electrons are placed in the σ_{2p_x}, π_{2p_y}, π_{2p_z}, $\pi^{\star}_{2p_y}$, and $\pi^{\star}_{2p_z}$ orbitals. For all the electrons to be paired, energy is needed to flip the spin in one of the antibonding molecular orbitals ($\pi^{\star}_{2p_y}$ or $\pi^{\star}_{2p_z}$). According to Hund's rule, this arrangement is less stable than the ground-state configuration. Hence, the Lewis structure shown actually corresponds to an excited state of the oxygen molecule.
10.94 (a) Although the O atoms are sp^3-hybridized, they are locked in a planar structure by the benzene rings. The molecule is symmetrical and therefore does not possess a dipole moment. (b) 20 σ bonds and 6 π bonds. **10.96** Both (a) and (c) possess a dipole moment.

Chapter 11

11.12 $CH_3CH_2CH_2CH_2CH_2Cl$; $CH_3CH_2CH_2CHClCH_3$; $CH_3CH_2CHClCH_2CH_3$.
11.14 H, CH_3 / C=C / Br, H Br, CH_3 / C=C / H, H
H, CH_2Br / C=C / H, H H, H / C / H, C—C, H / H, Br
11.16 (a) Alkene or cycloalkane. (b) Alkyne. (c) Alkane. (d) Like (a). (e) Alkyne.
11.18
CH_3 / H, H / H, H / CH_3 — A
CH_3 / H, CH_3 / H, H / H — B
CH_3, H / H, CH_3 / H, H — C
CH_3, CH_3 / H, H / H, H — D

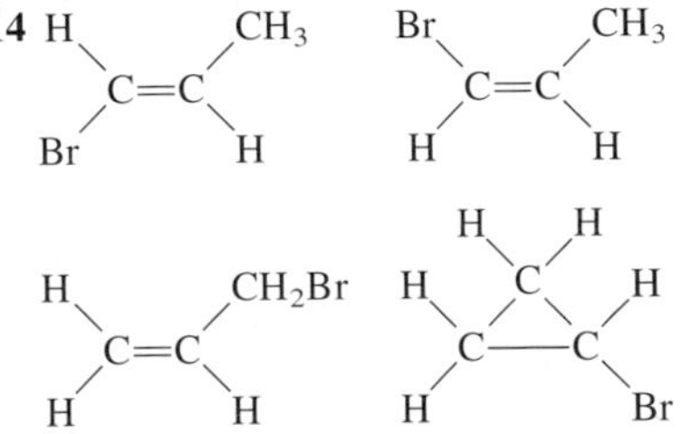

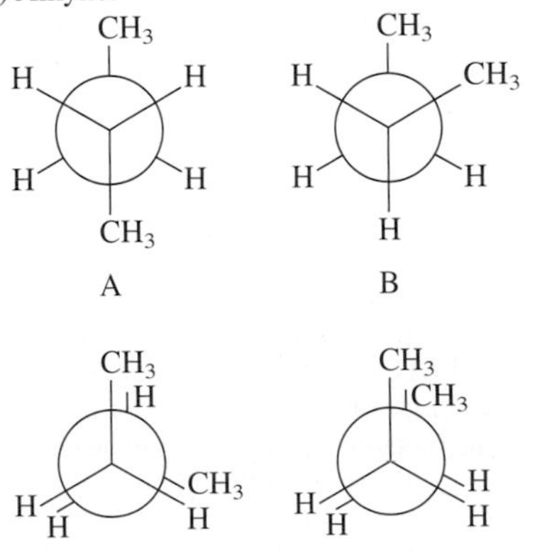

Stability decreases from A to D. **11.20** No. The bond angles are too small. **11.22** (a) is alkane and (b) is alkene. Only an alkene reacts with a hydrogen halide. **11.24** −630.8 kJ/mol. **11.26** (a) *cis*-1,2-dichlorocyclopropane. (b) *trans*-1,2-dichlorocyclopropane.
11.28 (a) 2-methylpentane. (b) 2,3,4-trimethylhexane. (c) 3-ethylhexane. (d) 3-bromo-1-pentene. (e) 2-pentyne.
11.32 (a) 1-chloro-4-methylbenzene. (b) 1-ethyl-3-nitrobenzene. (c) 1,2,4,5-tetramethylbenzene. (d) 3-phenyl-1- butene.
11.36 (a) Ether. (b) Amine. (c) Aldehyde. (d) Ketone. (e) Carboxylic acid. (f) Alcohol. (g) Amino acid (amine and carboxylic acid). **11.38** $HCOOH + CH_3OH \longrightarrow HCOOCH_3 + H_2O$. Methyl formate. **11.40** $(CH_3)_2CH—O—CH_3$.
11.42 (a) Ketone. (b) Ester. (c) Ether. **11.46** (a) The C atoms bonded to the methyl group and the amino group and the H atom. (b) The C atoms bonded to Br. **11.48** −174 kJ/mol.
11.50 $\text{+}CF_2—CF_2\text{+}_n$. **11.52** (a) Rubbing alcohol. (b) Vinegar. (c) Moth balls. (d) Organic synthesis. (e) Solvent. (f) Antifreeze. (g) Natural gas. (h) Synthetic polymer. **11.54** (a) 3. (b) 16. (c) 6.
11.56 (a) C: 15.81 mg; H: 1.33 mg; O: 3.49 mg. (b) C_6H_6O. (c) Phenol. **11.58** Empirical and molecular formula: $C_5H_{10}O$.

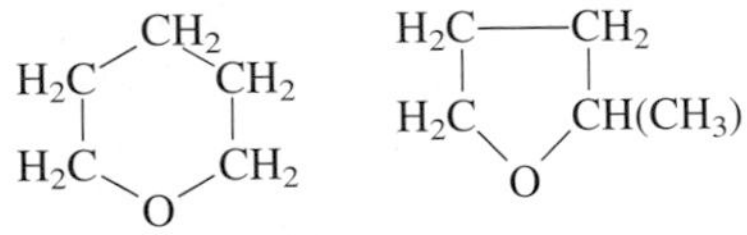

CH_2 / H_2C, CH_2 / H_2C, CH_2 / O H_2C—CH_2 / H_2C, $CH(CH_3)$ / O
$CH_2{=}CH—CH_2—O—CH_2—CH_3$

11.60 (a) CH_3, H / C=C / H, C_2H_5 (b) H, C_2H_5 / C=C / H, C_2H_5
(c) CH_3, H / C=C / H, CH—C_3H_7 / C_2H_5 (d) H—C≡C—CH—CH_3 (with phenyl on CH)

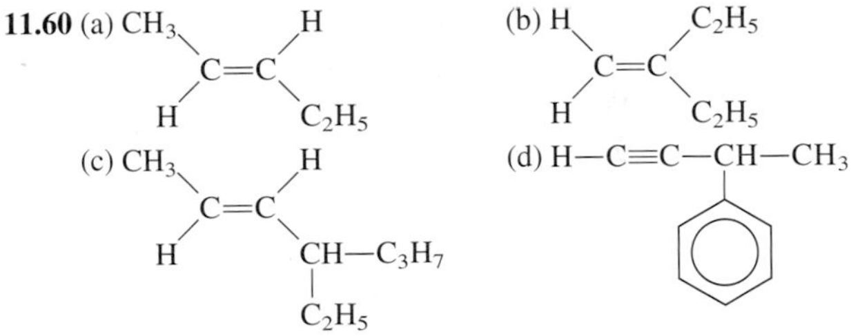

11.62 CH_3CH_2CHO. **11.64** (a) Alcohol. (b) Ether. (c) Aldehyde. (d) Carboxylic acid. (e) Amine. **11.66** The acids in lemon juice convert the amines to the ammonium salts, which have very low vapor pressures. **11.68** Methane (CH_4); ethanol (C_2H_5OH); methanol (CH_3OH); isopropanol (C_3H_7OH); ethylene glycol (CH_2OHCH_2OH); naphthalene ($C_{10}H_8$); acetic acid (CH_3COOH).
11.70 (a) One. (b) Two. (c) Five. **11.72** Br_2 dissociates into Br atoms, which react with CH_4 to form CH_3Br and HBr.
11.74 (a) Reaction between glycerol and carboxylic acid to form an ester. (b) Fat or oil (see problem for structure) + NaOH(*aq*) $\longrightarrow$ glycerol + 3 $RCOO^-Na^+$ (soap). (c) Molecules having more C=C bonds are harder to pack tightly. Consequently, they have a lower melting point. (d) Use H_2 and a catalyst. (e) 123.
11.76 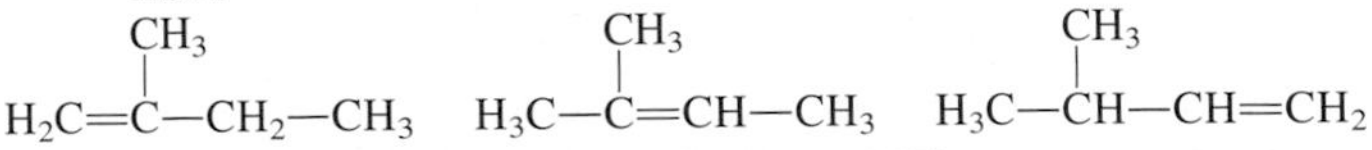
$H_2C{=}C(CH_3)—CH_2—CH_3$ $H_3C—C(CH_3){=}CH—CH_3$ $H_3C—CH(CH_3)—CH{=}CH_2$
11.78 $CH_3CH_2CH_2OH$ or $CH_3CH(OH)CH_3$.

Chapter 12

12.8 Methane; it has the lowest boiling point. **12.10** (a) Dispersion forces. (b) Dispersion and dipole-dipole forces. (c) Same as (b). (d) Ionic and dispersion forces. (e) Dispersion forces. **12.12** (e).
12.14 Only 1-butanol can form hydrogen bonds so it has a higher boiling point. **12.16** (a) Xe (greater dispersion forces). (b) CS_2 (greater dispersion forces). (c) Cl_2 (greater dispersion forces). (d) LiF (ionic compound). (e) NH_3 (hydrogen bond).
12.18 (a) Hydrogen bonding, dispersion forces. (b) Dispersion forces. (c) Dispersion forces. (d) The attractive forces of the covalent bond. **12.20** The compound on the left can form hydrogen bonds with itself (intramolecular hydrogen bonding). **12.32** Its viscosity is between that of ethanol and glycerol. **12.36** Ionic.
12.38 Covalent. **12.40** Li: metallic; Be: metallic; B: molecular

(An elemental form of boron is B_{12}.); C: covalent; N: molecular; O: (molecular); F: molecular; Ne: (atomic). **12.42** Covalent: Si, C, B. Molecular: Se_8, HBr, CO_2, P_4O_6, SiH_4. **12.44** Simple cubic: one sphere; body-centered cubic: two spheres; face-centered cubic: four spheres. **12.46** 6.20×10^{23} atom/mol. **12.48** 458 pm. **12.50** XY_3. **12.52** Each C atom in diamond is covalently bonded to four other C atoms. Graphite has delocalized electrons. **12.72** 2670 kJ. **12.74** 47.03 kJ/mol. **12.76** First step: freezing; second step: sublimation. **12.78** Additional heat is liberated when steam condenses at 100°C. **12.80** 331 mmHg. **12.84** (a) 3. (b) Rhombic. (c) From rhombic to monoclinic to liquid sulfur. **12.86** (a) Water will boil and turn into steam. (b) Water vapor will turn into ice. (c) Liquid water will vaporize. **12.88** (a) Dispersion and hydrogen bonding. (b) Dispersion. (c) Ionic, hydrogen bonding (in HF), dispersion. (d) Metallic bond. **12.90** HF molecules form hydrogen bonds; HI molecules do not. **12.92** (a) Solid. (b) Vapor. **12.94** The critical temperature of CO_2 is 31°C. The liquid CO_2 in the fire extinguisher is converted to critical fluid on a hot summer day. **12.96** Initially water boils under reduced pressure. During boiling, water cools and it freezes. Eventually it sublimes. **12.98** (a) Two. (b) Diamond. (c) Apply high pressure and high temperature. **12.100** W: Au; X: PbS; Y: I_2; Z: SiO_2. **12.102** 8.3×10^{-3} atm. **12.104** Smaller ions have high charge density and are more effective in ion-dipole interaction. **12.106** $3Hg + O_3 \longrightarrow 3HgO$. Conversion to solid HgO changes its surface tension. **12.108** 66.8%. **12.110** 1.69 g/cm^3. **12.112** When water freezes, the heat released protects the fruits. Also, the ice forms help to insulate the fruits by keeping the temperature at 0°C. **12.114** The water vapor generated from the combustion of methane condenses on the cold beaker outside. **12.116** The ice condenses the water vapor inside. Because the water is still hot, it will begin to boil at reduced pressure. (Air must first be driven off in the beginning.) **12.118** 6.019×10^{23} Fe atoms/mol. **12.120** Na.

Chapter 13

13.8 Cyclohexane cannot form hydrogen bonds with ethanol. **13.10** The longer the chain gets, the more nonpolar the molecule becomes. The —OH group can form hydrogen bonds with water, but the rest of the molecule cannot. **13.14** (a) 25.9 g. (b) 1.72×10^3 g. **13.16** (a) 2.68 *m*. (b) 7.82 *m*. **13.20** 5.0×10^2 *m*; 18.3 *M*. **13.22** (a) 2.41 *m*. (b) 2.13 *M*. (c) 0.0587 L. **13.26** 45.9 g. **13.32** Oxygen gas was driven off by boiling. **13.34** At the bottom of the mine, the carbon dioxide pressure is greater and the dissolved gas is not released from the solution. Coming up, the pressure decreases and carbon dioxide is released from the solution. **13.36** 0.28 L. **13.48** 1.3×10^3 g. **13.50** Ethanol: 30.0 mmHg; propanol: 26.3 mmHg. **13.52** 128 g. **13.54** 0.59 *m*. **13.56** 120 g/mol; $C_4H_8O_4$. **13.58** −8.6°C. **13.60** 4.3×10^2 g/mol; $C_{24}H_{20}P_4$. **13.62** 1.75×10^4 g/mol. **13.64** 343 g/mol. **13.68** Boiling point, vapor pressure, osmotic pressure. **13.70** (b) > (c) > (a). **13.72** 0.9420 *m*. **13.74** 7.6 atm. **13.76** 1.6 atm. **13.78** 3.5 atm. **13.80** (a) 104 mmHg. (b) 116 mmHg. **13.82** 2.95×10^3 g/mol. **13.84** No. Compound is assumed to be pure, monomeric, and a nonelectrolyte. **13.86** 12.3 *M*. **13.88** Soln A: 124 g/mol; Soln B: 248 g/mol. Dimerization in benzene. **13.90** (a) Boiling under reduced pressure. (b) CO_2 boils off, expands, and cools, condensing water vapor to form fog. **13.92** (a) Seawater contains many solutes. (b) Solubility of CO_2 in water decreases at reduced pressure. (c) Density of solution is close to that of water. (d) Molality is independent of temperature. (e) Methanol is volatile. **13.94** C_6H_{12}: 36%; $C_{10}H_8$: 64%. **13.96** First row: positive, positive; second row: $A \longleftrightarrow A$, $B \longleftrightarrow B < A \longleftrightarrow B$, negative; third row: $A \longleftrightarrow A$, $B \longleftrightarrow B = A \longleftrightarrow B$, no deviation. **13.98** Ethanol and water have an intermolecular attraction that results in an overall smaller volume. **13.100** 14.2%. **13.102** (a) $X_A = 0.524$, $X_B = 0.476$. (b) $P_A =$ 50 mmHg, $P_B = 20$ mmHg. (c) $P_A = 67$ mmHg, $P_B = 12$ mmHg. **13.104** A: 1.9×10^2 mmHg; B: 4.0×10^2 mmHg. **13.106** −0.737°C. **13.108** Valinomycin contains both polar and nonpolar groups. The polar groups bind the K^+ ions and the nonpolar —CH_3 groups enable the molecule to pass through the nonpolar lipid barrier of the cell. This is how the K^+ ions are transported across the cell membrane into the cell to offset the ionic balance.

Chapter 14

14.6 (a) 0.049 *M*/s. (b) 0.025 *M*/s. **14.14** (a) Rate = $k[F_2][ClO_2]$. (b) $1.2/M \cdot s$. (c) $2.4 \times 10^{-4} M/s$. **14.16** (a) Rate = $k[X]^2[Y]^3$; fifth order. (b) 0.68 *M*/s. **14.18** (a) 0.046 s^{-1}. (b) 0.13 $M^{-1}\,s^{-1}$. **14.20** Three half-lives or $2.08/k$. **14.22** (a) 0.0198 s^{-1}. (b) 151 s. **14.24** (a) 3.6 s. (b) 3.0 s, 6.6 s. **14.26** (a) 4:3:6. (b) No effect. (c) Same. **14.34** 135 kJ/mol. **14.36** 371°C. **14.38** 51.0 kJ/mol. **14.48** (a) Rate = $k[X_2][Y]$. (b) Z does not appear in the rate-determining step. (c) $X_2 + Y \longrightarrow XY + X$ (slow); $X + Z \longrightarrow XZ$ (fast). **14.56** Rate = $(k_1k_2/k_{-1})[E][S]$. **14.58** 0.0347 s^{-1}. **14.60** Area of large sphere: 22.6 cm^2; area of eight small spheres: 44.9 cm^2. The sphere with the larger surface area is the more effective catalyst. **14.62** $[H_2O]$ can be treated as a constant. **14.64** (a) Rate = $k[H^+][CH_3COCH_3]$. (b) $3.8 \times 10^{-3}/M \cdot s$. **14.66** 2.63 atm. **14.68** $M^{-2}s^{-1}$. **14.70** 56.4 min. **14.72** (b), (d), and (e). **14.74** 0.098%. **14.76** (a) Increased. (b) Decreased. (c) Decreased. (d) Increased. **14.78** 0.0896 min^{-1}. **14.80** 1.12×10^3 min. **14.82** (a) Rate = $k[X][Y]^2$. (b) $1.9 \times 10^{-2}\,M^{-2}\,s^{-1}$. **14.84** 3.1. **14.86** During the first 10 min or so the engine is relatively cold so the reaction will be slow. **14.90** At very high H_2 concentration, rate = $(k_1/k_2)[NO]^2$; at very low H_2 concentration, rate = $k_1[NO]^2[H_2]$. **14.92** $NO_2 + NO_2 \longrightarrow NO_3 + NO$ (slow); $NO_3 + CO \longrightarrow NO_2 + CO_2$ (fast). **14.94** At high pressures, all the sites on W are occupied so the rate is zero order in $[PH_3]$. **14.96** (a) $\Delta[B]/\Delta t = k_1[A] - k_2[B]$. (b) $[B] = (k_1/k_2)[A]$. **14.98** Use H_2O enriched with the ^{18}O isotope. Only mechanism (a) would give acetic acid with that isotope and only mechanism (b) would give methanol with that isotope. **14.100** (1) (b) < (c) < (a). (2) (a): −40 kJ/mol; (b): 20 kJ/mol; (c) −20 kJ/mol. **14.102** $H_2 + ICl \longrightarrow HCl + HI$ (slow); $HI + ICl \longrightarrow HCl + I_2$ (fast). **14.104** (a) 2.5×10^{-5} *M*/s. (b) 2.5×10^{-5} *M*/s. (c) 8.3×10^{-6} *M*. **14.106** (a) Drinking too much alcohol means all the active sites of the enzyme are occupied and the excess alcohol will damage the central nervous system. (b) Both ethanol and methanol will compete for the same active site of the enzyme, leading to methanol's discharge from the body. **14.108** $t_{\frac{1}{2}} = C(1/[A]_0^{n-1})$. Substituting $n = 0$, 1, and 2 for zero-, first-, and second-order reactions. **14.110** Second order. $0.42/M \cdot$ min. **14.112** (d). **14.114** During a cardiac arrest, there is a diminished rate of oxygen reaching the brain. Lowering body temperature will reduce the metabolic rate of oxygen needed for the brain, thereby reducing cell damage in the brain.

Chapter 15

15.8 (a) $K_P = P_{CO_2}P_{H_2O}$. (b) $K_P = P_{SO_2}^2 P_{O_2}$. **15.12** (a) $A + C \rightleftharpoons AC$. (b) $A + D \rightleftharpoons AD$. **15.14** 1.08×10^7. **15.16** 1.1×10^{-5}. **15.18** (a) 0.082. (b) 0.29. **15.20** 0.105; 2.05×10^{-3}. **15.22** 7.09×10^{-3}. **15.24** 3.3. **15.26** 0.0353. **15.30** $[N_2]$ and $[H_2]$ will decrease and $[NH_3]$ will increase. **15.32** 0.50 atm and 0.020 atm. **15.34** $[I] = 8.58 \times 10^{-4}$ *M*; $[I_2] = 0.0194$ *M*. **15.36** (a) 0.52. (b) $[CO_2] =$ 0.48 *M*; $[H_2] = 0.020$ *M*; $[CO] = 0.075$ *M*; $[H_2O] = 0.065$ *M*. **15.38** $[H_2] = [CO_2] = 0.05$ *M*; $[H_2O] = [CO] = 0.11$ *M*.

15.44 (a) Shift to the right. (b) No effect. (c) No effect. **15.46** (a) No effect. (b) No effect. (c) Shift to the left. (d) No effect. (e) Shift to the left. **15.48** (a) Shift to the right. (b) Shift to the left. (c) Shift to the right. (d) Shift to the left. (e) No effect. **15.50** No change. **15.52** (a) Shift to the right. (b) No effect. (c) No effect. (d) Shift to the left. (e) Shift to the right. (f) Shift to the left. (g) Shift to the right. **15.54** (a) NO: 0.24 atm; Cl_2: 0.12 atm. (b) 0.017. **15.56** (a) No. (b) Yes. **15.58** (a) 8×10^{-44}. (b) The reaction needs an input of energy to get started. **15.60** (a) 1.7. (b) $P_A = 0.69$ atm; $P_B = 0.81$ atm. **15.62** 1.5×10^5. **15.64** H_2: 0.28 atm; Cl_2: 0.049 atm; HCl: 1.67 atm. **15.66** 50 atm. **15.68** 3.84×10^{-2}. **15.70** 3.13. **15.72** N_2: 0.860 atm; H_2: 0.366 atm; NH_3: 4.40×10^{-3} atm. **15.74** (a) 1.16. (b) 53.7%. **15.76** (a) 0.49. (b) 0.23. (c) 0.037. (d) 0.037 mol. **15.78** $[H_2] = 0.070\ M$; $[I_2] = 0.182\ M$; $[HI] = 0.825\ M$. **15.80** (c). **15.82** (a) Color deepens. (b) Increases. (c) Decreases. (d) Increases. (e) Unchanged. **15.84** K is more volatile than Na. Therefore, its removal shifts the equilibrium to the right. **15.86** (a) High temperature and low pressure. (b) (i) 1.4×10^5; (ii) CH_4; H_2O: 2 atm; CO: 13 atm; H_2: 39 atm. **15.88** 1.2. **15.90** (a) React Ni with CO above 50°C. (b) The decomposition is endothermic. Heat $Ni(CO)_4$ above 200°C. **15.92** (a) $K_P = 2.6 \times 10^{-6}$; $K_c = 1.1 \times 10^{-7}$. (b) 2.2 g; 22 mg/m^3; yes. **15.94** $[NH_3] = 0.042\ M$; $[N_2] = 0.086\ M$; $[H_2] = 0.26\ M$. **15.96** 1.3 atm. **15.98** $[N_2O_4] = 0.996\ M$, $[NO_2] = 0.0678\ M$. The color will be darker. **15.100** (a) Assuming $\Delta H° > 0$ and $T_2 > T_1$, the equation predicts that $K_2 > K_1$, according to Le Châtelier's principle. (b) 43.4 kJ/mol. **15.102** 4.0. **15.104** (a) The plot curves toward higher pressure at low values of $1/V$. (b) The plot curves toward higher volume as T increases.

Chapter 16

16.4 (a) Nitrite ion: NO_2^-. (b) Hydrogen sulfate ion: HSO_4^-. (c) Hydrogen sulfide ion: HS^-. (d) Cyanide ion: CN^-. (e) Formate ion: $HCOO^-$. **16.6** (a) H_2S. (b) H_2CO_3. (c) HCO_3^-. (d) H_3PO_4. (e) $H_2PO_4^-$. (f) HPO_4^{2-}. (g) H_2SO_4. (h) HSO_4^-. (i) HNO_2. (j) HSO_3^-. **16.8** (a) $^-O-\overset{\overset{\displaystyle O}{\|}}{C}-\overset{\overset{\displaystyle O}{\|}}{C}-OH$ $\quad$ $^-O-\overset{\overset{\displaystyle O}{\|}}{C}-\overset{\overset{\displaystyle O}{\|}}{C}-O^-$
(b) Acid: H^+ and $H_2C_2O_4$; base: $C_2O_4^{2-}$; both acid and base: $HC_2O_4^-$. **16.16** (a) $6.3 \times 10^{-6}\ M$. (b) $1.0 \times 10^{-16}\ M$. (c) $2.7 \times 10^{-6}\ M$. **16.18** 6.72. **16.20** (a) Acidic. (b) Neutral. (c) Basic. **16.22** 1.98×10^{-3} mol; 0.444. **16.24** 2.2×10^{-3} g. **16.30** (1) (c). (2) (b) and (d). **16.32** (a) Strong base. (b) Weak base. (c) Weak base. (d) Weak base. (e) Strong base. **16.34** (a) False. (b) True. (c) True. (d) False. **16.36** To the left. **16.42** $[H^+] = [CH_3COO^-] = 5.8 \times 10^{-4}\ M$; $[CH_3COOH] = 0.0181\ M$. **16.44** $2.3 \times 10^{-3}\ M$. **16.46** (a) 3.5%. (b) 9.0%. (c) 33%. (d) 79%. Extent of ionization increases with dilution. **16.48** (a) 3.9%. (b) 0.30%. **16.52** $[H^+] = [HCO_3^-] = 1.0 \times 10^{-4}\ M$; $[CO_3^{2-}] = 4.8 \times 10^{-11}\ M$. **16.56** 7.1×10^{-7}. **16.58** 1.5%. **16.62** (a) $H_2SO_4 > H_2SeO_4$. (b) $H_3PO_4 > H_3AsO_4$. **16.64** Only the anion of phenol can be stabilized by resonance. **16.72** HZ < HY < HX. **16.74** 4.82. **16.76** >7. **16.80** $AlCl_3$ is a Lewis acid, Cl^- is a Lewis base. **16.82** CO_2 and BF_3. **16.84** 0.106 L. **16.86** No. **16.88** No. **16.90** CrO is ionic and basic; CrO_3 is covalent and acidic. **16.92** 4.0×10^{-2}. **16.94** 0.028. **16.96** $[Na^+] = 0.200\ M$; $[HCO_3^-] = 4.6 \times 10^{-3}\ M$; $[H_2CO_3] = 2.4 \times 10^{-8}\ M$; $[OH^-] = 4.6 \times 10^{-3}\ M$; $[H^+] = 2.2 \times 10^{-12}\ M$. **16.98** $NaCN + HCl \longrightarrow NaCl + HCN$. HCN is a very weak acid so it escapes into the vapor phase. **16.100** 1.000. **16.102** (a) Increases. (b) Decreases. (c) No effect. (d) Increases. **16.104** 1.6×10^{-4}. **16.106** 4.40. **16.108** NH_3. **16.110** 21 mL. **16.112** $[H^+] = [H_2PO_4^-] = 0.0239\ M$; $[H_3PO_4] = 0.076\ M$; $[HPO_4^{2-}] = 6.2 \times 10^{-8}\ M$; $[PO_4^{3-}] = 1.2 \times 10^{-18}\ M$. **16.114** (a) HY < HZ < HX. (b) $X^- < Z^- < Y^-$. (c) HX: 75%; HY: 25%, HZ: 50%. (d) NaX. **16.116** HX. **16.118** (a) $Fe_2O_3(s) + 6HCl(aq) \longrightarrow 2FeCl_3(aq) + 3H_2O(l)$; $Fe_2O_3(s) + 6H^+(aq) \longrightarrow 2Fe^{3+}(aq) + 3H_2O(l)$. (b) First stage: $CaCO_3(s) + HCl(aq) \longrightarrow Ca^{2+}(aq) + HCO_3^-(aq) + Cl^-(aq)$. Second stage: $HCl(aq) + HCO_3^-(aq) \longrightarrow CO_2(g) + Cl^-(aq) + H_2O(l)$. Overall: $CaCO_3(s) + 2HCl(aq) \longrightarrow CaCl_2(aq) + H_2O(l) + CO_2(g)$. (c) −0.64. **16.120** Mg. **16.122** pH = 1.57. $[CN^-] = 1.8 \times 10^{-8}\ M$. $[CN^-] = 2.2 \times 10^{-5}\ M$. $[CN^-]$ is greater in 1.00 M HCN solution as predicted by Le Châtelier's principle. **16.124** 6.02.

Chapter 17

17.6 (a) and (c). **17.8** 4.74; (a) Higher concentration. **17.10** 7.03. **17.12** 10; more effective against an added acid. **17.14** (a) 4.82. (b) 4.64. **17.16** HC. pK_a is closest to pH. **17.18** (1) (a) 5.10. (b) 4.82. (c) 5.22. (d) 5.00. (2) 4.90. (3) 5.22. **17.20** 90.1 g/mol. **17.22** (a) 110 g/mol. (b) 1.6×10^{-6}. **17.24** (1) (c). (2) (a). (3) (d). (4) (b). **17.28** The atmospheric CO_2 is converted to H_2CO_3, which neutralizes the NaOH. **17.30** Red. **17.38** (a) 7.8×10^{-10}. (b) 1.8×10^{-18}. **17.40** $2.6 \times 10^{-6}\ M$. **17.42** 1.7×10^{-2} g/L. **17.44** 2.3×10^{-9}. **17.46** $[NO_3^-] = 0.076\ M$; $[Na^+] = 0.045\ M$; $[Sr^{2+}] = 1.6 \times 10^{-2}\ M$; $[F^-] = 1.1 \times 10^{-4}\ M$. **17.50** (a) 0.013 M. (b) $2.2 \times 10^{-4}\ M$. (c) $3.3 \times 10^{-3}\ M$. **17.52** (a) $1.0 \times 10^{-5}\ M$. (b) $1.1 \times 10^{-10}\ M$. **17.56** Complex ion formations. (a) $[Cu(NH_3)_4]^{2+}$. (b) $[Ag(CN)_2]^-$. (c) $[HgCl_4]^{2-}$. For equations for the formation of the complex ions see Table 17.4. **17.58** $[Cd^{2+}] = 1.1 \times 10^{-18}\ M$; $[Cd(CN)_4^{2-}] = 4.2 \times 10^{-3}\ M$; $[CN^-] = 0.48\ M$. **17.60** $3.5 \times 10^{-5}\ M$. **17.64** 0.011 M. **17.66** Chloride ions will only precipitate Ag^+ ions, or a flame test for Cu^{2+} ions. **17.68** (a) 1.37. (b) 5.28. (c) 8.85. **17.70** 2.51–4.41. **17.72** 1.28 M. **17.74** $[H^+] = 3.0 \times 10^{-13}\ M$; $[CH_3COO^-] = 0.0500\ M$; $[CH_3COOH] = 8.4 \times 10^{-10}\ M$; $[OH^-] = 0.0335\ M$; $[Na^+] = 0.0835\ M$. **17.76** 9.25; 9.18. **17.78** 9.97 g; 13.04. **17.80** $[Ag^+] = 2.0 \times 10^{-9}\ M$; $[Cl^-] = 0.080\ M$; $[Zn^{2+}] = 0.070\ M$; $[NO_3^-] = 0.060\ M$. **17.82** pH is greater than 2.68 but less than 8.11. **17.84** All except (a). **17.86** 2.4×10^{-13}. **17.88** A precipitate of $Fe(OH)_2$ will form. **17.90** The original precipitate is HgI_2. At higher concentration of KI, the complex ion HgI_4^{2-} forms so mass of HgI_2 decreases. **17.92** 7.82–10.38. **17.94** (a) AgI. (b) $1.6 \times 10^{-7}\ M$. (c) 0.0016%. **17.96** (a) $MCO_3 + 2HCl \longrightarrow MCl_2 + H_2O + CO_2$, $HCl + NaOH \longrightarrow NaCl + H_2O$. (b) 197 g/mol; Ba. **17.98** 2. **17.100** (a) Sulfate. (b) Sulfide. (c) Iodide. **17.102** 13 mL. **17.104** (c). **17.106** (a) $1.7 \times 10^{-7}\ M$. (b) Because $MgCO_3$ is fairly soluble. (c) 12.40. (d) $1.9 \times 10^{-8}\ M$. (e) Ca^{2+} because it is present in larger amounts. **17.108** (a) 8.4 mL. (b) 12.5 mL. (c) 27.0 mL. **17.110** At pH = 1.0: $^+NH_3-CH_2-COOH$; at pH = 7.0: $^+NH_3-CH_2-COO^-$; at pH = 12.0: $NH_2-CH_2-COO^-$. **17.112** 4.75. **17.114** 6.2×10^{-12}.

Chapter 18

18.6 8×10^{-31}. The probability is practically zero for 1 mole of molecules. **18.10** (c) < (d) < (e) < (a) < (b). Solids have smaller entropies than gases. More complex structures have higher entropies. **18.12** (a) 47.5 J/K · mol. (b) −12.5 J/K · mol. (c) −242.8 J/K · mol. **18.14** (a) $\Delta S < 0$. (b) $\Delta S > 0$. (c) $\Delta S > 0$. (d) $\Delta S < 0$. **18.18** (a) −1139 kJ/mol. (b) −140.0 kJ/mol. (c) −2935 kJ/mol. **18.20** (a) At all temperatures. (b) Below 111 K. **18.24** 8.0×10^1 kJ/mol. **18.26** 4.572×10^2 kJ/mol; 7.2×10^{-81}. **18.28** (a) −24.6 kJ/mol. (b) −1.33 kJ/mol. **18.30** −341 kJ/mol. **18.32** −2.87 kJ/mol. The process has a high activation energy. **18.36** 1×10^3; glucose + ATP $\longrightarrow$ glucose 6-phosphate + ADP.

18.38 (a) 0. (b) 4.0×10^4 J/mol. (c) -3.2×10^4 J/mol. (d) 6.4×10^4 J/mol. **18.40** (a) No reaction is possible because $\Delta G > 0$. (b) The reaction has a very large activation energy and therefore a slow rate. (c) Reactants and products already at their equilibrium concentrations. **18.42** In all cases, $\Delta H > 0$ and $\Delta S > 0$. $\Delta G < 0$ for (a), $= 0$ for (b), and > 0 for (c). **18.44** $\Delta S > 0$. **18.46** (a) Most liquids have similar structure so the changes in entropy from liquid to vapor are similar. (b) ΔS_{vap} are larger for ethanol and water because of hydrogen bonding (there are fewer microstates in these liquids). **18.48** (a) $2CO + 2NO \longrightarrow 2CO_2 + N_2$. (b) Oxidizing agent: NO; reducing agent: CO. (c) 3×10^{120}. (d) 1.2×10^{14}; From left to right. (e) No. **18.50** 2.6×10^{-9}. **18.52** 976 K. **18.54** $\Delta S < 0$; $\Delta H < 0$. **18.56** 55 J/K · mol. **18.58** Increase in entropy of the surroundings offsets the decrease in entropy of the system. **18.60** 56 J/K. **18.62** 4.5×10^5. **18.64** 4.8×10^{-75} atm. **18.66** (a) True. (b) True. (c) False. **18.68** $C + CuO \rightleftharpoons CO + Cu$; 6.1. **18.70** Crystal structure has disorder or impurity. **18.72** (a) 7.6×10^{14}. (b) 4.1×10^{-12}. **18.74** (a) A reverse disproportionation reaction. (b) 8.2×10^{15}. Yes, a large K makes this an efficient process. (c) Less effective. **18.76** 1.8×10^{70}. Reaction has a large activation energy. **18.78** Heating the ore alone is not a feasible process. -214.3 kJ/mol. **18.80** $K_P = 36$; 981 K. No. **18.82** $X_{CO} = 0.45$; $X_{CO_2} = 0.55$. Use ΔG°_f values at 25°C for 900°C. **18.84** 249 J/K. **18.86** 3×10^{-13} s. **18.88** $\Delta S_{sys} = -327$ J/K · mol; $\Delta S_{surr} = 1918$ J/K · mol; $\Delta S_{univ} = 1591$ J/K · mol. **18.90** q, w. **18.92** $\Delta H < 0$; $\Delta S < 0$; $\Delta G < 0$. **18.94** (a) $\Delta H > 0$ (endothermic) and $\Delta G < 0$ (spontaneous). Therefore, $\Delta S > 0$. (b) The rubber molecules become more entangled (resulting in a greater number of microstates) upon contraction. **18.96** Analogy is inappropriate. Entropy is a measure of the dispersal of molecules among available energy levels. The entropy of the room is the same whether it is tidy or not. **18.98** $\Delta G^\circ = 62.5$ kJ/mol; $\Delta H^\circ = 157.8$ kJ/mol; $\Delta S^\circ = 109$ J/K · mol. **18.100** (1) (c). (2) (a). (3) (b).

Chapter 19

19.2 (a) $Mn^{2+} + H_2O_2 + 2OH^- \longrightarrow MnO_2 + 2H_2O$. (b) $2Bi(OH)_3 + 3SnO_2^{2-} \longrightarrow 2Bi + 3H_2O + 3SnO_3^{2-}$. (c) $Cr_2O_7^{2-} + 14H^+ + 3C_2O_4^{2-} \longrightarrow 2Cr^{3+} + 6CO_2 + 7H_2O$. (d) $2Cl^- + 2ClO_3^- + 4H^+ \longrightarrow Cl_2 + 2ClO_2 + 2H_2O$. **19.12** 2.46 V; $Al + 3Ag^+ \longrightarrow 3Ag + Al^{3+}$. **19.14** $Cl_2(g)$ and $MnO_4^-(aq)$. **19.16** Only (a) and (d) are spontaneous. **19.18** (a) Li. (b) H_2. (c) Fe^{2+}. (d) Br^-. **19.22** 0.368 V. **19.24** (a) -432 kJ/mol; 5×10^{75}. (b) -104 kJ/mol; 2×10^{18}. (c) -178 kJ/mol; 1×10^{31}. (d) -1.27×10^3 kJ/mol; 1×10^{222}. **19.26** 0.37 V; -36 kJ/mol; 2×10^6. **19.30** (a) 2.23 V; 2.23 V; -430 kJ/mol. (b) 0.02 V; 0.04 V; -23 kJ/mol. **19.32** 0.083 V. **19.34** 0.010 V. **19.38** 1.09 V. **19.46** (a) anode: $2Cl^-(aq) \longrightarrow Cl_2(g) + 2e^-$, cathode: $Ba^{2+}(aq) + 2e^- \longrightarrow Ba(s)$. (b) 0.64 g. **19.48** (a) $\$2.10 \times 10^3$. (b) $\$2.46 \times 10^3$. (c) $\$4.70 \times 10^3$. **19.50** (a) 0.14 *F*. (b) 0.123 *F*. (c) 0.10 *F*. **19.52** (a) $Ag^+ + e^- \longrightarrow Ag$. (b) $2H_2O \longrightarrow O_2 + 4H^+ + 4e^-$. (c) 6.0×10^2 C. **19.54** (a) 0.589 g Cu. (b) 0.133 A. **19.56** 2.3 h. **19.58** 9.66×10^4 C. **19.60** 0.0710 *F*. **19.62** 0.156 *M*; $Cr_2O_7^{2-} + 6Fe^{2+} + 14H^+ \longrightarrow 2Cr^{3+} + 6Fe^{3+} + 7H_2O$. **19.64** 45.1%. **19.66** (a) $2MnO_4^- + 16H^+ + 5C_2O_4^{2-} \longrightarrow 2Mn^{2+} + 10CO_2 + 8H_2O$. (b) 5.40%. **19.68** 0.231 mg Ca^{2+}/mL blood. **19.70** (a) 0.80 V. (b) $2Ag^+ + H_2 \longrightarrow 2Ag + 2H^+$. (c) (i) 0.92 V; (ii) 1.10 V. (d) The cell operates as a pH meter. **19.72** Fluorine gas reacts with water. **19.74** 2.5×10^2 h. **19.76** Hg_2^{2+}. **19.78** $[Mg^{2+}] = 0.0500$ *M*, $[Ag^+] = 7 \times 10^{-55}$ *M*, 1.44 g. **19.80** (a) 0.206 L H_2. (b) 6.09×10^{23}/mol e^-. **19.82** (a) -1357 kJ/mol. (b) 1.17 V. **19.84** +3. **19.86** 6.8 kJ/mol; 0.064. **19.88** 1.4 A. **19.90** +4. **19.92** 1.60×10^{-19} C/e^-. **19.94** A cell made of Li^+/Li and F_2/F^- gives the maximum voltage of 5.92 V. Reactive oxidizing and reducing agents are hard to handle. **19.96** 2×10^{20}. **19.98** (a) E° for X is negative; E° for Y is positive. (b) 0.59 V. **19.100** (a) The reduction potential of O_2 is insufficient to oxidize gold. (b) Yes. (c) $2Au + 3F_2 \longrightarrow 2AuF_3$. **19.102** $[Fe^{2+}] = 0.0920$ *M*, $[Fe^{3+}] = 0.0680$ *M*. **19.104** (b) 105 A · h. The concentration of H_2SO_4 keeps decreasing. (c) 2.01 V; -3.88×10^5 J/mol. **19.106** (a) Unchanged. (b) Unchanged. (c) Squared. (d) Doubled. (e) Doubled. **19.108** Stronger. **19.110** 4.4×10^2 atm. **19.112** (a) 8.2 g. (b) 0.40 *M*. **19.114** 1×10^{-14}. **19.116** (a) Anode (Mg): $Mg \longrightarrow Mg^{2+} + 2e^-$ (also: $Mg + 2HCl \longrightarrow MgCl_2 + H_2$). Cathode (Cu): $2H^+ + 2e^- \longrightarrow H_2$. (b) The solution does not turn blue. (c) $Mg(OH)_2$. **19.118** 2×10^{37}. **19.120** (a) 9.83×10^3 C. (b) 21.7 min. (c) 2.97 g $Mg(OH)_2$. **19.122** Yes. By heating one of the electrodes it is possible to generate a small emf because E° depends on temperature.

Chapter 20

20.12 (a) +2. (b) 6. (c) Oxalate. **20.14** (a) Na: +1; Mo: +6. (b) Mg: +2; W: +6. (c) K: +1; Fe: +2. **20.16** (a) *cis*-dichlorobis(ethylenediamine)cobalt(III). (b) pentaamminechloroplatinum(IV) chloride. (c) hexaamminecobalt(III) chloride. (d) pentaamminechlorocobalt(III) chloride. (e) *trans*-diamminedichloroplatinum(II). **20.18** (a) $[Cr(en)_2Cl_2]^+$. (b) $Fe(CO)_5$. (c) $K_2[Cu(CN)_4]$. (d) $[Co(NH_3)_4(H_2O)Cl]Cl_2$. **20.20** (a) 2. (b) 2. **20.22**

Br CN / Ni / Br CN	NC Br / Ni / Br CN
cis isomer	*trans* isomer

20.28 CN^- is a strong-field ligand. Absorption occurs near the high-energy end of the spectrum (blue) and the complex ion looks yellow. H_2O is a weaker-field ligand and absorption occurs in the orange or red part of the spectrum. Consequently, the complex ion appears green or blue. **20.30** (a) two unpaired spins. (b) four unpaired spins. **20.32** 2; $[Co(NH_3)_4Cl_2]Cl$. **20.36** Use $^{14}CN^-$ label (in NaCN). **20.38** First $Cu(CN)_2$ (white) is formed. It redissolves as $Cu(CN)_4^{2-}$. **20.40** 1.4×10^2. **20.42** 6.7×10^{13}. **20.44** The $Mn(H_2O)_6^{2+}$ ion is a high-spin complex, containing five unpaired spins. **20.46** These will absorb at longer wavelength: (a) $[Co(H_2O)_6]^{2+}$, (b) $[FeF_6]^{3-}$, (c) $[CuCl_4]^{2-}$. **20.48** (a) $4Au(s) + 8CN^-(aq) + O_2(g) + 2H_2O(l) \longrightarrow 4[Au(CN)_2]^-(aq) + 4OH^-(aq)$. (b) $Zn(s) + 2[Au(CN)_2]^-(aq) \longrightarrow [Zn(CN)_4]^{2-}(aq) + 2Au(s)$. (c) Linear (Au is *sp*-hybridized). **20.50** Mn^{3+} ($3d^4$) because it is less stable than Cr^{3+} ($3d^5$). **20.52** EDTA sequesters metal ions like Ca^{2+} and Mg^{2+} that are essential for bacterial growth and function. **20.54** 1.0×10^{-18} *M*. **20.56** 3. **20.58** I is the *cis* isomer and II is the *trans* isomer. **20.60** $[Ag^+] = 2.2 \times 10^{-20}$ *M*. **20.62** (a) 2.7×10^6. (b) Free Cu^+ ions are unstable in solution. The only stable Cu(I) compounds are insoluble. **20.64** (a) No. It is easily reduced to Cu^{2+}. (b) Potassium hexafluorocuprate(III). It is paramagnetic and has an octahedral geometry. (c) Diamagnetic.

Chapter 21

21.6 (a) ${}_{-1}^{0}\beta$. (b) ${}_{20}^{40}Ca$. (c) ${}_{2}^{4}\alpha$. (d) ${}_{0}^{1}n$. **21.14** (a) ${}_{3}^{9}Li$. (b) ${}_{11}^{25}Na$. (c) ${}_{21}^{48}Sc$. **21.16** (a) ${}_{10}^{17}Ne$. (b) ${}_{20}^{45}Ca$. (c) ${}_{43}^{92}Tc$. (d) ${}_{80}^{195}Hg$. (e) ${}_{96}^{242}Cm$. **21.18** 6×10^9 kg/s. **21.20** (a) 4.55×10^{-12} J; 1.14×10^{-12} J/nucleon. (b) 2.36×10^{-10} J; 1.28×10^{-12} J/nucleon. **21.24** 0.250 d^{-1}; 2.77 d. **21.26** 2.7 d. **21.28** ${}_{82}^{208}Pb$. **21.30** A: 0; B: 0.25 mole; C: 0; D: 0.75 mole. **21.34** (a) ${}_{34}^{80}Se + {}_{1}^{2}H \longrightarrow {}_{1}^{1}p + {}_{34}^{81}Se$. (b) ${}_{4}^{9}Be + {}_{1}^{2}H \longrightarrow 2{}_{1}^{1}p + {}_{3}^{9}Li$. (c) ${}_{5}^{10}B + {}_{0}^{1}n \longrightarrow {}_{2}^{4}\alpha + {}_{3}^{7}Li$.

21.36 $^{198}_{80}Hg + ^{1}_{0}n \longrightarrow ^{198}_{79}Au + ^{1}_{1}p$. **21.48** IO_3^- is only formed from IO_4^-. **21.50** Incorporate ^{59}Fe into a person's body. After a few days isolate red blood cells and monitor radioactivity from the hemoglobin molecules. **21.52** An analogous Pauli exclusion principle for nucleons. **21.54** (a) 0.343 mCi. (b) $^{237}_{93}Np \longrightarrow ^{4}_{2}\alpha + ^{233}_{91}Pa$. **21.56** (a) 1.039×10^{-12} J/nucleon. (b) 1.111×10^{-12} J/nucleon. (c) 1.199×10^{-12} J/nucleon. (d) 1.410×10^{-12} J/nucleon. **21.58** $^{18}_{7}N \longrightarrow ^{18}_{8}O + ^{0}_{-1}\beta$. **21.60** Radioactive dating. **21.62** (a) $^{209}_{83}Bi + ^{4}_{2}\alpha \longrightarrow ^{211}_{85}At + 2^{1}_{0}n$. (b) $^{209}_{83}Bi(\alpha,2n)^{211}_{85}At$. **21.64** The sun exerts a much greater gravity on the particles. **21.66** 2.77×10^3 yr. **21.68** (a) $^{40}_{19}K \longrightarrow ^{40}_{18}Ar + ^{0}_{+1}\beta$. (b) 3.0×10^9 yr. **21.70** (a) 5.59×10^{-15} J; 2.84×10^{-13} J. (b) 0.024 mol. (c) 4.26×10^6 kJ. **21.72** 2.7×10^{14} ^{131}I atoms. **21.74** 5.9×10^{23}/mol. **21.76** All except gravitational. **21.78** ^{238}U and ^{232}Th. **21.80** 8.3×10^{-4} nm. **21.82** $^{3}_{1}H$. **21.84** The reflected neutrons induced a nuclear chain reaction. **21.86** 2.1×10^2 g/mol. **21.88** Most of the hazards of a nuclear power plant are discussed in Section 21.5. However, it contributes much less to global warming than a coal-fired power plant. **21.90** (a) $^{238}_{94}Pu \longrightarrow ^{4}_{2}He + ^{234}_{92}U$. (b) $t = 0$: 0.58 mW; $t = 10$ yr: 0.53 mW. **21.92** 0.49 rem. **21.94** Heat is generated from radioactive decay of long half-life isotopes. **21.96** No. The kinetic energy generated during the collision is converted to mass.

Chapter 22

22.8 $\text{+}CH_2\text{—}CHCl\text{—}CH_2\text{—}CCl_2\text{+}$. **22.10** By an addition reaction involving styrene monomers. **22.12** (a) $CH_2\text{=}CH\text{—}CH\text{=}CH_2$. (b) $HO_2C(CH_2)_6NH_2$. **22.22** At 35°C the enzyme begins to denature. **22.28** Proteins are made of 20 amino acids. Nucleic acids are made of only four building blocks (purines, pyrimidines, sugar, phosphate group). **22.30** C-G base pairs have three hydrogen bonds; A-T base pairs have two hydrogen bonds. The sample with the higher C-G pairs has the higher mp. **22.32** Leg muscles are active, have a high metabolic rate and hence a high concentration of myoglobin. The iron content in Mb makes the meat look dark. **22.34** Insects have blood that contains no hemoglobin. It is unlikely that a human-sized insect could obtain sufficient oxygen for metabolism by diffusion. **22.36** There are four Fe atoms per hemoglobin molecule. **22.38** Mostly dispersion forces. **22.40** Gly-Ala-Phe-Glu-His-Gly-Ala-Leu-Val. **22.42** No. Enzymes only act on one of the two enantiomers of a compound. **22.44** When deoxyhemoglobin binds to oxygen, there is a structural change due to the cooperativity effect, which causes the crystal to break up. Because myoglobin is made up of only one subunit, there is no structural change as deoxymyoglobin is converted to oxymyoglobin. **22.46** A sequence of only 2 bases to define a particular amino acid has a total of 4^2 or 16 possible combinations, not enough to cover the 20 amino acids. A sequence of 3 bases has 4^3 or 64 combinations, which are more than enough. **22.48** Yes, because $\Delta G° < 0$. As temperature is lowered, $\Delta G°$ becomes less negative. The reaction becomes spontaneous in the reverse direction and eventually denaturation occurs.

Credits

About the Author

Author photos (top): © Margaret A. Chang; (bottom): © Robin Overby.

Table of Contents

p. vii(top): © T. Sasaki/Rice University; p. vii(center): © C. Powell, P. Fowler & D. Perkins/Photo Researchers; p. vii(bottom): © McGraw-Hill Higher Education Inc./Charles Winter, Photographer; p. viii(top): © National Oceanic and Atmospheric Administration/ Department of Commerce.; p. viii(center): © NOAA Satellite and Information Service, National Environmental Satellite, Data, and Information Service (NESCIS); p. viii(bottom): © U.S. Marine Corps photo by Lance Cpl. Ronald Stauffer; p. ix(top): From "Confinement of Electrons to Quantum Corrals on a Metal Surface," M.F. Crommie, C.P. Lutz and D.M. Eigler, Science 262, 218–220 (Oct. 8, 1993). Image originally created by IBM Corporation.; p. ix(center): © SPL/Photo Researchers; p. ix(bottom): From G.N. Lewis, Valence, Dover Publications, Inc., New York 1966; p. x(top): Courtesy Aneel Aggarwal, Ph.D., and Stephen C. Harrison, Department of Biochemistry and Molecular Biology, Harvard University, from A.K. Aggarwal., et al., November 1988. Recognition of a DNA operator by the repressor of phage 434: A view at high resolution. Science 242, pp. 899-907. @ 1988 American Association for the Advancement of Science.; p. x(11): © Stockdisc/PunchStock; p. x(12): © The McGraw-Hill Companies, Inc./Ken Karp, Photographer; p. x(bottom): © Richard Megna/ Fundamental Photographs; p. xi(top): © NASA; p. xi(center): © Kent Knudson/PhotoLink/ Getty Images RF; p. xi(bottom): © Kristin Brochmann/Fundamental Photographs; p. xii(top): © Frank & Joyce Burek/Getty Images; p. xii(center): © Mike McCormac/ Alamy; p. xii(bottom): © Jim West/Jim West Photography ; p. xiii(top): © PjrFoto/studio/ Alamy; p. xiii(center): © AP Photo/Keystone, Martial Trezzini; p. xiii(bottom): © The McGraw-Hill Companies, Inc./Ken Karp, Photographer

Chapter 1

Opener: © T. Sasaki/Rice University; figure 1.2: © The McGraw-Hill Companies, Inc./ Charles D. Winters/Timeframe Photography, Inc., photographer; figure 1.3a-b: © The McGraw-Hill Companies, Inc./Ken Karp, Photographer; p. 7: © The McGraw-Hill Companies, Inc./Ken Karp, Photographer; p. 10: © NASA; p. 11: © Comstock Royalty Free; p. 12: © The McGraw-Hill Companies, Inc./Ken Karp, Photographer; figure 1.9: Courtesy of Mettler; p. 20: © Leonard Lessin/ Photo Researchers; p. 21: © Charles Falco/ Photo Researchers; p. 22: © Charles D. Winter/ Photo Researchers

Chapter 2

Opener: © C. Powell, P. Fowler & D. Perkins/ Photo Researchers; figure 2.4a-c: © Richard Megna/Fundamental Photographs; p. 35: © The Image Bank/Getty Images; figure 2.12c: © E.R. Degginger/Color-Pic; p. 43: © Andrew Lambert/Photo Researchers; pp. 45, 52: © The McGraw-Hill Companies, Inc./Ken Karp, Photographer

Chapter 3

Opener: © McGraw-Hill Higher Education Inc./Charles Winter, Photographer; p. 62: © Andrew Popper; figure 3.1: © The McGraw-Hill Companies, Inc./Photo by Stephen Frisch; p. 64: © E.R. Degginger/Color-Pic; p. 65: © L.V. Bergman/The Bergman Collection; p. 66: © Nikreates/Alamy; p. 67: © American Gas Association; p. 68: © The McGraw-Hill Companies, Inc./Ken Karp, Photographer; p. 72: © Ward's Natural Science Establishment; p. 77: © The McGraw-Hill Companies, Inc./ Ken Karp, Photographer; p. 79: Courtesy of Scott MacLaren, Center for Microanalysis of Materials, University of Illinois at Urbana-Champaign; p. 82: © The McGraw-Hill Companies, Inc./Ken Karp, Photographer; p. 86: © PhotoLink/Photodisc/Getty Images

Chapter 4

Opener: © National Oceanic and Atmospheric Administration/Department of Commerce.; figures 4.1(all), 4.4, 4.6: © The McGraw-Hill Companies, Inc./Ken Karp, Photographer; figures 4.3, 4.5, 4.9: © McGraw-Hill Higher Education Inc./Charles Winter, Photographer; pp. 104, 107: © The McGraw-Hill Companies, Inc./Ken Karp, Photographer; p. 114(top): © Richard Megna/Fundamental Photographs; p. 114 (bottom): © McGraw-Hill Higher Education Inc./Charles Winter, Photographer; figure 4.11a-b: © The McGraw-Hill Companies, Inc./Ken Karp, Photographer; figure 4.12a-c: © The McGraw-Hill Companies, Inc./Photo by Stephen Frisch; figure 4.13(both): © The McGraw-Hill Companies, Inc./Ken Karp, Photographer; p. 118: © Mula & Haramaty/Phototake; pp. 119, 121, 125(top): © The McGraw-Hill Companies, Inc./Ken Karp, Photographer; figures 4.17a-c, 4.18a-b: © The McGraw-Hill Companies, Inc./ Ken Karp, Photographer; p. 135: Courtesy Dow Chemical USA

Chapter 5

Opener: © NOAA Satellite and Information Service, National Environmental Satellite, Data, and Information Service (NESCIS); p. 137: © The McGraw-Hill Companies, Inc./ Ken Karp, Photographer; figure 5.10: © The McGraw-Hill Companies, Inc./Ken Karp, Photographer; p. 152: Courtesy of General Motors Corporation; p. 162(top): © NASA; figure 5.17: © The McGraw-Hill Companies, Inc./Ken Karp, Photographer

Chapter 6

Opener: © U.S. Marine Corps photo by Lance Cpl. Ronald Stauffer; p. 177: © Jacques Jangoux/Photo Researchers; p. 178: © Edward Kinsman/Photo Researchers; figure 6.2: © UPI/Corbis; p. 179: © McGraw-Hill Higher Education Inc./Charles Winter, Photographer; p. 181: © Richard Megna/Fundamental Photographs; p. 187: © McGraw-Hill Higher Education Inc./Stephen Frisch, Photographer; p. 189: © Richard Megna/Fundamental Photographs; p. 190: © McGraw-Hill Higher Education Inc./Stephen Frisch, Photographer; p. 197(top): © The McGraw-Hill Companies, Inc./Ken Karp, Photographer; p. 197(bottom): Courtesy of the Diamond Information Center; pp. 199, 200: © E.R. Degginger/Color-Pic; p. 201: © Orgo-Thermite

Chapter 7

Opener: © Crommie, Lutz & Eigler. Image originally created by IBM Corporation. Courtesy of IBM Research Division; p. 212: © Volvox /Photolibrary; figure 7.3: © B.S.I.P./ Custom Medical Stock Photo; p. 219: © The McGraw-Hill Companies, Inc./Photo by Stephen Frisch; figure 7.6: © Joel Gordon 1979; figure 7.12(both): © Educational Development Center

Chapter 8

Opener: © SPL/Photo Researchers; p. 252: © McGraw-Hill Higher Education Inc./Charles Winter, Photographer; figure 8.11(Li): © The McGraw-Hill Companies, Inc./Ken Karp, Photographer; figure 8.11(K): © Al Fenn/Time & Life Pictures/Getty Images; figure 8.11(Rb, Cs, Be, Mg, Ca, Sr, Ba): © L.V. Bergman/The Bergman Collection; figure 8.13(Al): © The McGraw-Hill Companies, Inc./Charles D. Winters/Timeframe Photography, Inc., photographer; figure 8.13(Ga): © McGraw-Hill

Higher Education Inc./Charles Winter, Photographer; figure 8.13(B, In): © L.V. Bergman/The Bergman Collection; figure 8.14(all): © The McGraw-Hill Companies, Inc./Charles D. Winters/Timeframe Photography, Inc., photographer; figure 8.15(Sb, Se8): © The McGraw-Hill Companies, Inc./Charles D. Winters/Timeframe Photography, Inc., photographer; figure 8.15(N2): © Joe McNally; figure 8.15(P): © Al Fenn/Time & Life Pictures/Getty Images; figure 8.15(As, Bi): © L.V. Bergman/The Bergman Collection; figure 8.16(S8): © L.V. Bergman/The Bergman Collection; figure 8.16(Te): © The McGraw-Hill Companies, Inc./Ken Karp, Photographer; figure 8.17: © Joel Gordon 1979; figure 8.18(all): © The McGraw-Hill Companies, Inc./Ken Karp, Photographer; figure 8.19a-b: © Neil Bartlett; figure 8.20: Courtesy of Argonne National Laboratory; figure 8.11: © The McGraw-Hill Companies, Inc./Charles D. Winter/Timeframe Photography, Inc., photographer

Chapter 9

Opener: From G.N. Lewis, Valence, Dover Publications, Inc., New York 1966; p. 287: © The McGraw-Hill Companies, Inc./Ken Karp, Photographer; p. 288: © Wards Natural Science Establishment; p. 300: © The McGraw-Hill Companies, Inc./Ken Karp, Photographer; p. 304: © Courtesy of James O. Schreck, Professor of Chemistry, University of Northern Colorado

Chapter 10

Opener: Courtesy Aneel Aggarwal, Ph.D., and Stephen C. Harrison, Department of Biochemistry and Molecular Biology, Harvard University, from A.K. Aggarwal., et al., November 1988. Recognition of a DNA operator by the repressor of phage 434: A view at high resolution. Science 242, pp. 899-907. @1988 American Association for the Advancement of Science; figure 10.20: © Donald Clegg

Chapter 11

Opener: © Stockdisc/PunchStock; p. 365: © J.H. Robinson/Photo Researchers; p. 366: Reprinted courtesy of the American Petroleum Institute; p. 374: © Steve Gschmeissner/SPL/Photo Researchers; figure 11.11: © The McGraw-Hill Companies, Inc./Ken Karp, Photographer; p. 377: © The McGraw-Hill Companies, Inc./Ken Karp, Photographer; p. 378: © E.R. Degginger/Color-Pic; p. 379: © IBM's Almaden Research Center; pp. 383, 384, 386: © The McGraw-Hill Companies, Inc./Ken Karp, Photographer; p. 388: © BioPhoto/Photo Researchers; figure 11.23: © Joel Gordon 1979

Chapter 12

Opener, figure 12.9: © The McGraw-Hill Companies, Inc./Ken Karp, Photographer; p. 407(center): © Hermann Eisenbeiss/Photo Researchers; figure 12.11: © The McGraw-Hill Companies, Inc./Ken Karp, Photographer; p. 413: © Tony Mendoza/Stock Boston; p. 417: © Bryan Quintard/Jacqueline McBride/Lawrence Livermore National Laboratory; p. 418: © L.V. Bergman/The Bergman Collection; p. 419: © Grant Heilman/Grant Heilman Photography; figures 12.30a-d, 12.34, p. 426: © The McGraw-Hill Companies, Inc./Ken Karp, Photographer; p. 434: © Tony Ranze/AFP/Getty Images

Chapter 13

Opener: © Richard Megna/Fundamental Photographs; figure 13.1(all): © The McGraw-Hill Companies, Inc./Ken Karp, Photographer; p. 445: © The McGraw-Hill Companies, Inc./Ken Karp, Photographer; p. 450: © Hank Morgan/Photo Researchers; p. 451: © The McGraw-Hill Companies, Inc./Ken Karp, Photographer; p. 454: © John Mead/Science Photo Library/Photo Researchers; figures 13.10a-c: © David Phillips/Photo Researchers

Chapter 14

Opener: © NASA; figures 14.3, 14.6: © The McGraw-Hill Companies, Inc./Ken Karp, Photographer; p. 477: © The McGraw-Hill Companies, Inc./Ken Karp, Photographer; figure 14.18: © The McGraw-Hill Companies, Inc./Ken Karp, Photographer; figure 14.22: Courtesy of Johnson Matthey; figure 14.24: Courtesy of General Motors Corporation

Chapter 15

Opener: © Kent Knudson/PhotoLink/Getty Images RF; p. 511(both): © The McGraw-Hill Companies, Inc./Ken Karp, Photographer; p. 518: © Collection Varin-Visage Jacana/Photo Researchers; figures 15.6, 15.9a-b, 15.10: © The McGraw-Hill Companies, Inc./Ken Karp, Photographer

Chapter 16

Opener: © Kristin Brochmann/Fundamental Photographs; figure 16.2: © The McGraw-Hill Companies, Inc./Ken Karp, Photographer; p. 551: © The McGraw-Hill Companies, Inc./Photo by Stephen Frisch; p. 579: © Michael Melford; figures 16.9(both), 16.10: © The McGraw-Hill Companies, Inc./Ken Karp, Photographer

Chapter 17

Opener: © Frank & Joyce Burek/Getty Images; p. 591: © The McGraw-Hill Companies, Inc./Ken Karp, Photographer; figures 17.1, 17.3, 17.8: © The McGraw-Hill Companies, Inc./Ken Karp, Photographer; p. 606(top): © CNRI/SPL/Science Source/Photo Researchers; p. 606 (bottom): © Bjorn Bolstad/Peter Arnold; pp. 609(both), 610: © The McGraw-Hill Companies, Inc./Ken Karp, Photographer; p. 611: © Runk/Schoenberger/Grant Heilman Photography; figures 17.10, 17.11, 17.12(all), 17.13(all): © The McGraw-Hill Companies, Inc./Ken Karp, Photographer

Chapter 18

Opener: © Mike McCormac/Alamy; p. 629: © Harry Bliss. Originally published on the cover of the New Yorker Magazine; p. 630: © The McGraw-Hill Companies, Inc./Ken Karp, Photographer; p. 632: © Matthias K. Gobbert/University of Maryland, Baltimore County/Dept. of Mathematics and Statistics; p. 635: © The McGraw-Hill Companies, Inc./Ken Karp, Photographer; p. 641: © U.S. Postal Service; figure 18.7: Courtesy National Lime Association; pp. 646, 647: © The McGraw-Hill Companies, Inc./Ken Karp, Photographer

Chapter 19

Opener: © Jim West/Jim West Photography; figure 19.2: © The McGraw-Hill Companies, Inc./Ken Karp, Photographer; p. 683: © AP/Wide World Photos; figure 19.12: © NASA; figure 19.14: © The McGraw-Hill Companies, Inc./Ken Karp, Photographer; figure 19.17: © The McGraw-Hill Companies, Inc./Photo by Stephen Frisch; pp. 699, 700: © The McGraw-Hill Companies, Inc./Ken Karp, Photographer

Chapter 20

Opener: © PjrFoto/studio/Alamy; figure 20.2(Cu): © L.V. Bergman/The Bergman Collection; figure 20.2(All others): © The McGraw-Hill Companies, Inc./Ken Karp, Photographer; figures 20.9, 20.15: © The McGraw-Hill Companies, Inc./Ken Karp, Photographer; figure 20.21: © Stephen J. Lippard, Chemistry Department, Massachusetts Institute of Technology

Chapter 21

Opener: © AP Photo/Keystone, Martial Trezzini; p. 739: © Francois Lochon/Gamma Press; figure 21.5: © Fermilab Visual Media Services; figure 21.11: © Pierre Kopp/Corbis; figure 21.12: © M. Lazarus/Photo Researchers; figure 21.13: © Los Alamos National Laboratory; p. 748(bottom): © NASA; p. 748(top): © U.S. Department of Energy/Science Photo Library/Photo Researchers; figure 21.15: © Lawrence Livermore National Laboratory; figure 21.16: © Department of Defense, Still Media Records Center, US Navy Photo; p. 752(bottom): © Alexander Tsiaras/Photo Researchers; figure 21.17(both): © SIU/Visuals Unlimited

Chapter 22

Opener: © The McGraw-Hill Companies, Inc./Ken Karp, Photographer; figure 22.2: © Neil Rabinowitz/Corbis; figure 22.5: © E.R. Degginger/Color-Pic; p. 766: © Richard Hutchings/Photo Researchers; p. 774: Courtesy of Lawrence Berkeley Laboratory

Index

A

D

E

F

G

H

I

N

O

T

Fundamental Constants

Avogadro's number	6.0221415×10^{23}
Electron charge (e)	$1.60217653 \times 10^{-19}$ C
Electron mass	$9.1093826 \times 10^{-28}$ g
Faraday constant (F)	96,485.3383 C/mol e^-
Gas constant (R)	8.314 J/K · mol (0.08206 L · atm/K · mol)
Planck's constant (h)	$6.6260693 \times 10^{-34}$ J · s
Proton mass	1.672621×10^{-24} g
Neutron mass	$1.67492728 \times 10^{-24}$ g
Speed of light in vacuum	2.99792458×10^{8} m/s

Useful Conversion Factors and Relationships

1 lb = 453.6 g
1 in = 2.54 cm (exactly)
1 mi = 1.609 km
1 km = 0.6215 mi
1 pm = 1×10^{-12} m = 1×10^{-10} cm
1 gal = 3.785 L = 4 quarts
1 atm = 760 mmHg = 760 torr = 101,325 N/m^2 = 101,325 Pa
1 cal = 4.184 J (exactly)
1 L · atm = 101.325 J
1 J = 1 C × 1 V

$$?^\circ\text{C} = (^\circ\text{F} - 32^\circ\text{F}) \times \frac{5^\circ\text{C}}{9^\circ\text{F}}$$

$$?^\circ\text{F} = \frac{9^\circ\text{F}}{5^\circ\text{C}} \times (^\circ\text{C}) + (32^\circ\text{F})$$

$$?\text{K} = (^\circ\text{C} + 273.15^\circ\text{C}) \times \left(\frac{1\ \text{K}}{1^\circ\text{C}}\right)$$

Color Codes for Molecular Models

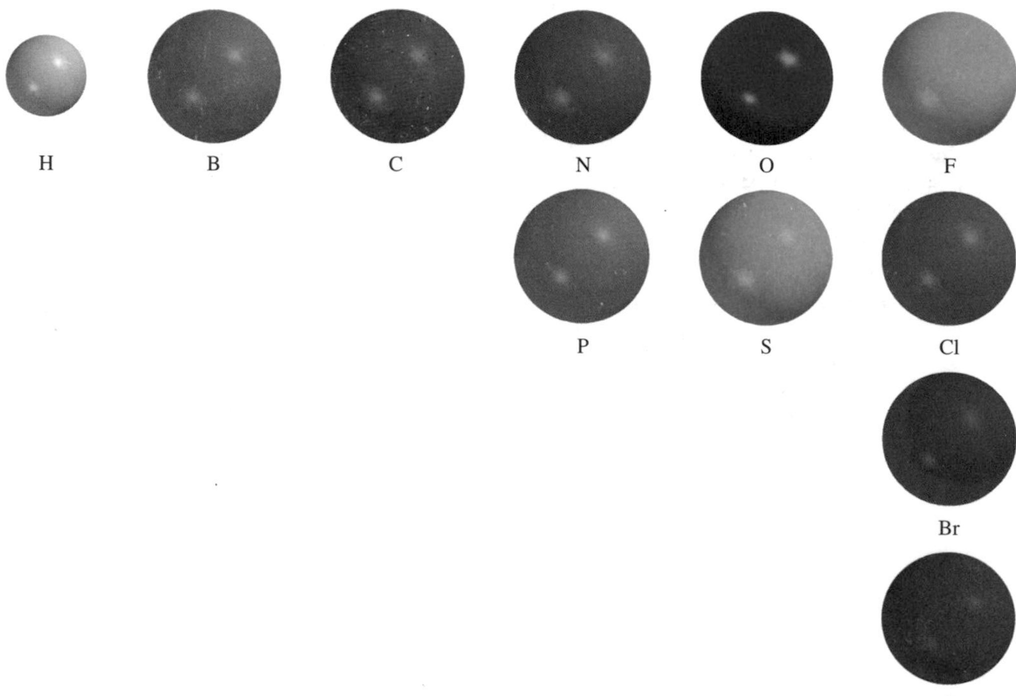